CRC HANDBOOK OF Tables for the Use of Order Statistics in Estimation

H. Leon Harter
Champaign, Illinois

N. Balakrishnan
Department of Mathematics and Statistics
McMaster University
Ontario, Canada

CRC Press
Boca Raton New York London Tokyo

Library of Congress Cataloging-in-Publication Data

Catalog record is available from the Library of Congress.

International Standard Book Number 0-8493-9452-X
Printed in the United States of America 1 2 3 4 5 6 7 8 9 0
Printed on acid-free paper

To Alice Lauretta

To Julia Kahlian

Preface

This Handbook is a revision of the first author's *Order Statistics and Their Use in Testing and Estimation*, Volume 2: *Estimates Based on Order Statistics of Samples from Various Populations* (published in 1970 through the U.S. Government Printing Office), of which the Preface reads as follows:

Following a suggestion by Dr. Paul R. Rider, whose sage advice and constant encouragement over a period of a dozen years of close daily association were most helpful, the author began in 1957 a study of the use of sample quasi-ranges in estimating the standard deviation of a normal population. While the computation of the efficient estimate based on the sample standard deviation does not seem too difficult to a statistician, it appeared that many users of statistics considered it so and therefore often resorted to subjective and quite inefficient methods of estimation. So it was evident that there was a need for an objective estimator which would be both reasonably efficient and simple to compute. Estimators based on quasi-ranges had been investigated by Mosteller and by Cadwell, and had been shown to have the desired properties, but up to that time they had been used relatively little, due mainly to the lack of suitable tables. The author therefore undertook the computation, for sample size $n \leq 100$, of the most efficient estimators based on one quasi-range and on linear combinations of two quasi-ranges. It was necessary first to compute tables of expected values, variances, and covariances of quasi-ranges. The author was able to tabulate the expected values to six decimal places on the Burroughs E101 computer, but found that machine inadequate to the task of tabulating the variances and covariances. The latter job was accomplished and the estimators and their efficiencies were calculated on the Univac Scientific (ERA 1103) computer with the programming assistance of Eugene H. Guthrie. The results were published in a 1958 Air Force technical report and in condensed form in a 1959 journal article. In addition, both versions contained results on the use of the sample range in obtaining the efficient estimator of the standard deviation of a rectangular population. The technical report also contained some results on the use of quasi-ranges in estimating the standard deviation of an exponential population, but these were omitted from the journal article because, as the author had become aware, quasi-ranges do not yield good estimators of the standard deviation of an asymmetric population. In fact, better estimators can be obtained from a single order statistic of a sample from a one-parameter exponential population or from a linear combination of two order statistics of a sample from a two-parameter exponential population.

Meanwhile, the author had begun in 1960 an investigation of exact confidence bounds for the standard deviation of rectangular, one-parameter exponential, and normal populations, based respectively on the sample range, one order statistic, and a suitably chosen quasi-range. Approximate confidence bounds, obtained by the use of partially distribution-free methods, had already been tabulated by Leone, Rutenberg, and Topp, but the author was convinced that increases in effectiveness large enough to warrant the greater computational effort could be obtained by tabulating the cumulative distribution functions of the statistics involved, interpolating inversely to find the required percentage points, and taking the reciprocals of the percentage points as coefficients of the statistics in the exact confidence bounds. The computations for the rectangular population were programmed by Eugene H. Guthrie for the Univac Scientific (ERA 1103A) computer, and the results were published in a 1961 Air Force technical report and in condensed form in a 1961 journal article. For the normal population some preliminary numerical analysis, most of which was performed by Edwin L. Godfrey, was required. Eugene H. Guthrie did most of the programming for the IBM 7090 computer. The results were published in a 1963 Air Force technical report and in condensed form in a 1964 journal article. In these publications the efficiency of an interval estimator was defined and a criterion based on maximizing the efficiency instead of the effectiveness was proposed; in the technical report the results obtained by use of the two criteria were compared. For the exponential population, the author wrote his own FORTRAN programs for the IBM 7090 and 7094 computers. The results based on the criterion of maximizing the effectiveness were published in 1963 in the *Rider Anniversary Volume* and those based on maximizing the efficiency in a 1964 journal article. By-products of these studies were new tables of the incomplete Gamma-function ratio and of percentage points of the chi-square and Beta distributions, which were published by the author in book form in 1964.

The author, with the assistance of Eugene H. Guthrie, who programmed the computations for the Univac Scientific (ERA 1103A) computer, tabulated the expected values of normal order statistics for all samples sizes n up through 100 and for selected values of n up through 400. The results were published in a 1960 Air Force technical report and in condensed form in a 1961 journal article. During the next two or three years, the author received several inquiries as to the availability of similar tables for samples from populations other than the normal. In order to satisfy a need for such tables, he undertook the computation of expected values of exponential, Weibull, and Gamma order statistics, this time writing his own FORTRAN programs for the IBM 7094 and 1620 computers. In the cases of the Weibull and Gamma order statistics, these were modifications of programs supplied by Captains Ronald J. Quayle and Robert C. Karns, respectively. The results were published in a 1964 Air Force technical report. The expected values of normal order statistics have applications to tests of hypotheses based on normal theory as well as to problems of estimation, so they could have been included in Volume 1, but for convenience they are included in this volume with those for samples from other populations.

Late in 1963 or early in 1964 the author began what has proved to be a very fruitful collaboration with Professor Albert H. Moore of the Air Force Institute of Technology on problems involving the estimation of parameters of life distributions by the use of order statistics. The ensuing investigation has followed two separate but related lines. The first of these involves the use of functional relationships between the one-parameter exponential population and other populations (Weibull, extreme-value, Pareto, and limited) to obtain estimators, based on one order statistic or on the first (or last) m order statistics of a sample of size n ($\geq m$), of one parameter of the other population, assuming that its other parameter(s) is (are) known, by a transformation of the known results for the exponential population. Henry E. Fettis gave invaluable assistance in the evaluation of integrals and other mathematical manipulations. The second and more extensive line of investigation has been concerned with iterative maximum-likelihood estimation, from doubly censored samples, of the parameters of normal, three-parameter lognormal, three-parameter Weibull, three-parameter Gamma, logistic, Type I extreme-value, and four-parameter generalized Gamma populations. Professor Moore collaborated with the author in all but the last of these. In each case, the likelihood function and its first and second partial derivatives with respect to the parameters were written down, the first partial derivatives were equated to zero to obtain the likelihood equations, and a FORTRAN program was written to solve the likelihood equations iteratively on the IBM 7094 computer. The information matrix, whose elements are limits, as the sample size $n \to \infty$, of the negatives of the expected values of the second partial derivatives of the likelihood function with respect to the parameters, was inverted numerically on the IBM 7094 computer, by the use of a FORTRAN program, to obtain the asymptotic variances and covariances of the maximum-likelihood estimators, which were then tabulated. For several of the populations, Monte Carlo studies were conducted for samples of small to moderate size in order to obtain information concerning the rate of convergence of the variances and covariances to their asymptotic values, and to compare the mean square errors of the maximum-likelihood estimators with those of other estimators. The results have been published in several journal articles which have appeared since 1965 and in a 1966 Air Force technical report.

All of the above results, except those included in the author's 1964 book, have been collected in this volume, which contains essentially nothing new, though some of the results, especially the longer tables, have previously appeared only in Air Force technical reports, not in the open literature. Practically all of the tables in this volume are exactly as they appeared in previously published technical reports and journal articles, and some parts of the text have been lifted from the same sources with only minor revisions. Other parts of the text have been completely rewritten to provide a more unified treatment and to avoid unnecessary repetitions. Despite the paucity of new material, the author believes that a useful purpose will be served by assembling all these results, both theory and tables, in a single hardcover volume that will be more readily and permanently accessible than the separate sources.

References to related work by other authors, especially that performed after the author's own, have been purposely held to a minimum, both in Volume 1 and in the present volume. The author plans to remedy this deficiency in Volume 3, which will be devoted to a chronological annotated bibliography, with lists of references and citations.

The author wishes to express his appreciation to all the persons whose specific contributions have been acknowledged above, and to Miss Eva Brandenburg for typing the manuscript; also to Mr. Donald S. Clemm for ruling the tables and to TSgt. James R. Marshall for preparing the overlays and figures.

Wright-Patterson Air Force Base, Ohio
May 1969

In addition to the initial distribution of 400 copies by the Aerospace Research Laboratories, 892 copies of the 1970 volume were sold during the period from 1970 to 1976. Requests for that volume are still being received, even though it has been out of print since 1976. Since that time, there have been many developments, several of them due to the second author, which merit attention. Several populations not covered in the 1970 volume have been found to be useful in modeling the lifetimes of devices and organisms; in particular, the log-Gamma population has been shown to be more tractable than the since 1976. Since that time, there have been many developments, several of them due to the second author, which merit attention. Several populations not covered in the 1970 volume have been found to be useful in modeling the lifetimes of devices and organisms; in particular, the log-Gamma population has been shown to be more tractable than the related generalized Gamma population. The widespread use of best linear unbiased estimators, together with other applications such as quantile plots, goodness-of-fit tests, and robust inference, indicates a need for tables of expected values, variances and covariances of the order statistics of samples from various populations. In view of these facts, we decided to collaborate on a revision of the 1970 volume, the proposal for which was accepted by CRC Press.

New material includes tables of variances and covariances of normal, exponential, Weibull and Gamma order statistics to complement the tables of expected values given in the 1970 volume, and tables of expected values, variances and covariances of order statistics of samples from eleven other populations, together with discussions of the methods of computation of the tables and their applications. Also added was material on maximum-likelihood estimation, from censored samples, of the parameters of generalized extreme-value and log-Gamma populations, the latter replacing more voluminous material on the generalized Gamma population. Also omitted to make room for the new material were tables of the probability integrals of the quasi-ranges of samples from a normal population and of the range of samples from a rectangular population, whose main use is in computing tables of percentage points (still included).

We thank Mr. Robert Stern (CRC Press, Boca Raton) for supporting this project and Ms. Nora Konopka, Ms. Mary Kugler, and Ms. Andrea Demby (CRC Press, Boca Raton) for assisting in the production of this volume. Our sincere thanks also goes to Mrs. Debbie Iscoe (Hamilton, Canada) for the excellent typesetting of the manuscript.

Acknowledgements Thanks are offered to the Institute of Mathematical Statistics, the American Statistical Association, the American Society for Quality Control, the Biometrika Trust, the Institute of Electrical and Electronics Engineers, Gordon and Breach Publishers, Marcel Dekker, Inc., the Japanese Union of Scientists and Engineers, and the editors of *Sankhyā* and *Scandinavian Actuarial Journal* for providing permission to reproduce some previously published tables. Tables C1.2, C14.1, C14.2, D1, D3, F3, F4, and G3 have been reproduced with permission from the Institute of Mathematical Statistics. Tables C9.1, C9.2, C10.1, C10.2, D4 C,D, E1, E2, E3, F2, F5, F6, G2, G4, and G5 have been reproduced with permission from the American Statistical Association and the American Society for Quality Control. Tables F1 and G1 have been reproduced with permission from the Biometrika Trust. Table D2 has been reproduced with permission from the IEEE. Tables C15.1 and C15.2 have been reproduced with permission from Gordon and Breach Publishers. Tables C16.1 and C16.2 have been reproduced with permission from Marcel Dekker, Inc. Tables C6.1 and C6.2 have been reproduced with permission from the editor of *Sankhyā*. Tables C13.1 and C13.2 have been reproduced with permission from the editor of *Scandinavian Actuarial Journal*. Tables C11.1 and C11.2 have been reproduced with permission from the JUSE.

Champaign, Illinois, USA
Hamilton, Ontario, Canada

October 1995

Contents

List of Tables

Introduction

Consider a sample consisting of n observations drawn at random from some specified or unspecified population. If the sample values are arranged in order from the smallest, x_1, to the largest, x_n, so that $x_1 \leq x_2 \leq \cdots \leq x_n$, then x_1 is called the *first order statistic*, x_2 the *second order statistic*, ... , and x_n the nth *order statistic* of the sample. The difference, $w = x_n - x_1$, between the nth (largest) and first (smallest) order statistics is called the sample *range*. The range of the $n - 2r$ observations remaining after the r largest and the r smallest have been discarded (censored), which is the difference, $w_r = x_{n-r} - x_{r+1}$, between the $(n - r)$th and $(r + 1)$th order statistics, is called the rth *quasi-range* of the sample. Other functions of order statistics, such as the median, the midrange, the studentized range, and the studentized maximum modulus, are in common use, but they will not be defined here, since this volume will deal only with the range, the quasi-ranges, and the individual order statistics. More specifically, we shall deal with estimates of population parameters based on the range of samples from a rectangular population, quasi-ranges of samples from a normal population, and one or more order statistics of samples from various populations.

Every statistical population is characterized by one or more parameters whose values specify the population completely, once the form of the population is known. The familiar normal population has two parameters, the mean and the standard deviation, while the one-parameter negative exponential population has, as the name implies, just one parameter, which is both the mean and the standard deviation. A mathematical expression which estimates a population parameter as a function of a sample statistic is called an *estimator*, while a numerical value obtained from an estimator by substituting a numerical value for the sample statistic is called an *estimate*.

Distinction should be made at the outset between *point estimation* and *interval estimation*. In the former the result is a single value, while in the latter it is an interval with which is associated a specified level of confidence that the population parameter lies in this interval. It is desirable that a point estimator should be *unbiased*; that is, that its expected value should be equal to the true value of the parameter. It is also desirable that an unbiased point estimator should be *efficient*; that is, that its variance should be smaller than that of any other unbiased estimator. Minimum-variance unbiased estimators are sometimes spoken of as *conventional* estimators. In certain situations, however, it may be desirable to use *substitute* estimators, which at best are unbiased but inefficient and at worst may even be biased. The *efficiency* of an unbiased substitute estimator is defined as the ratio, expressed as a percentage, of the variance of the conventional estimator (if one exists) to that of the substitute estimator; the efficiency of a biased estimator is not defined. Interval estimators (confidence intervals) may be either *exact* or *approximate*, depending upon whether the level of confidence is exactly as specified or only approximately so; the requirement that interval estimators be exact is, in a certain sense, analogous to the requirement that point estimators be unbiased. We have defined the *efficiency* of an interval estimator by a logical extension of the corresponding definition for a point estimator.

In a few cases estimators based on order statistics are the efficient estimators, but more often they are substitute estimators which sacrifice some efficiency in the interest of computational simplicity and/or robustness in the presence of outliers. In life testing, they are often used to obtain estimates of the parameters of the life distribution before all of the items placed on test have failed. Point estimators based on order statistics may be best linear unbiased estimators, based on all available observations from complete or censored samples or on one or more observations chosen in some optimal manner, or they may be maximum-likelihood estimators, based on censored samples, of a single parameter or of two or more parameters jointly. Interval estimators may be based on percentage points of order statistics or of functions of order statistics, or they may be based on maximum-likelihood estimators and their asymptotic variances and covariances.

Chapter 1 of this volume will deal with point and interval estimation, from sample quasi-ranges, of the standard deviation of a normal population, and will include discussions of the method of computation of the necessary tables and of criteria for best interval estimators. The tables themselves will be given in Appendix A. A similar discussion of point and interval estimation, from the sample range (which provides efficient estimators), of the standard deviation of a rectangular population will be given in Chapter 2 and the related tables in Appendix B.

Chapter 3 will deal with the method of computation of expected values, variances and covariances of order statistics from various populations and with their use in both testing and estimation. The following populations will be included: normal, exponential, Weibull, Gamma, lognormal, log-Weibull, log-Gamma, Cauchy, double exponential, half normal, logistic, Pareto, Student's t, half logistic, and generalized logistic. The related tables, together with a one-page table of moments of exponential, Weibull and Gamma populations, will be given in Appendix C.

Chapter 4 will deal with point estimation, from one or two order statistics, of one or two parameters of an exponential population and with interval estimation, from a single order statistic, of the parameter of a one-parameter exponential population. The criteria developed in Chapter 1 for best substitute interval estimators will be applied, and the method of computation of the necessary tables will be discussed. The tables themselves will be given in Appendix D.

Chapter 5 will deal with conditional maximum-likelihood estimation, from the m available order statistics of singly censored samples, of certain parameters of populations related to the exponential. Both point and interval estimators will be given for the scale parameter of a Weibull population with known location and shape parameters, for the location parameter of a Type I extreme-value population with known scale parameter, for the scale parameters of Type II extreme-value populations with known shape parameters, and for the shape parameters of Pareto and limited populations with known location parameters. All of these estimators will be derived by transforming the known estimators for the parameter of a one-parameter exponential population. The computation of unbiasing factors, biases, variances of unbiased estimators, and efficiencies relative to the Cramér-Rao lower bound will be discussed. The tables themselves will be given in Appendix E. A similar treatment of the estimators of the same parameters from a single order statistic, again obtained by transformation of the known estimators for the parameter of a one-parameter exponential population, will be given in Chapter 6, which will also include a comparison of the one-order-statistic and m-order-statistic estimators.

Chapter 7 will deal with joint maximum-likelihood estimation, from doubly censored samples, of the parameters of normal, three-parameter lognormal, three-parameter Weibull, three-parameter Gamma, logistic, Type I extreme value, log-Gamma, and three-parameter generalized extreme value populations. In each case the likelihood function will be written down and its first and second derivatives with respect to the parameters will be worked out, an iterative procedure will be given for solution on an electronic computer of the likelihood equations obtained by equating the first derivatives to zero, and the information matrix, whose elements are the negatives of the limits, as the sample size tends to infinity, of the expected values of the second partial derivatives, will be inverted numerically to obtain the matrix of asymptotic variances and covariances. A discussion will be given of Monte Carlo studies conducted for several of these populations in order to check the applicability of the asymptotic theory to samples of small or moderate size and to compare the maximum-likelihood estimators with other estimators. Most of the chapter will be devoted to point estimation, but the final section will contain some remarks on interval estimation of the parameters of the above-mentioned populations. Tables of asymptotic variances and covariances of the maximum-likelihood estimators will be given in Appendix F, and tables summarizing the results of the Monte Carlo studies in Appendix G.

Each chapter will be divided into two or more self-contained sections. Equations will be numbered consecutively within sections. A list of references will also be given at the end of each section.

1

Quasi-Ranges of Samples from a Normal Population

1.1 Point Estimation of the Population Standard Deviation σ [1]

1.1.1 Introduction

Both engineers and operations analysts often make use of statistical methods, but many of them seem to have an aversion to computing standard deviations. Attempts have been made to estimate the standard deviation σ of a population, assumed to be normal, by plotting the data on normal probability paper, drawing the best-fitting straight line "by eye", and estimating σ as the difference between the 84 percent point and the 50 percent point. This is a highly subjective procedure, and moreover one which is highly sensitive to departures from normality, so it is not surprising that it often gives quite inaccurate results. It is well known that, for small samples, the population standard deviation can be estimated quite efficiently from the sample range. However, the efficiency of the estimator based on the range decreases rather rapidly as the sample size increases, being less than 35 percent for samples of 100. There appears to be a need for substitute estimators which are reasonably efficient for moderate sample sizes, yet much simpler to compute than the efficient estimator based on the sample standard deviation. This chapter will be concerned with estimators based on sample quasi-ranges, which satisfy these requirements quite well. Such estimators were proposed by Mosteller (1946), but for a dozen years thereafter they were used relatively little, due mainly to the lack of suitable tables.

The rth quasi-range, w_r, of a sample of size n is defined as the range of $(n-2r)$ sample values, omitting the r largest and the r smallest. Symbolically,

$$w_r = x_{n-r} - x_{r+1} \ , \tag{1.1.1}$$

where $x_1 \leq x_2 \leq \cdots \leq x_n$ are the ordered sample values. Cadwell (1953) has shown that the range, w_0, is the most efficient statistic of this type for sample sizes up through $n = 17$, beyond which point w_1 is optimum up through $n = 31$, where w_2 becomes better. Cadwell (1953) has also proposed the use of linear combinations of quasi-ranges.

This section will deal with the most efficient unbiased estimators of the standard deviation of a normal population, based on one sample quasi-range and on linear combinations of two sample quasi-ranges.

[1] Earlier versions of this material were published by Harter (1958, 1959).

1.1.2 Method of Computation of Auxiliary Tables

Expected Values of Quasi-Ranges

In order to determine the factor by which the rth quasi-range, w_r, must be multiplied in order to obtain an unbiased estimator of the population standard deviation, it is necessary to know the expected value of the rth quasi-range for samples of n from a standard normal population, which is given by the equation [see Cadwell (1953), p. 606]

$$E(w_r) = 2(r+1)\binom{n}{r+1}\int_{-\infty}^{\infty} x\left[\frac{1}{2}-\Phi(x)\right]^r\left[\frac{1}{2}+\Phi(x)\right]^{n-r-1}\phi(x)\,dx\ , \quad (1.1.2)$$

where $\phi(x) = (1/\sqrt{2\pi})\exp(-x^2/2)$ and $\Phi(x) = \int_0^x \phi(x)\,dx$. Expected values of the range (to five decimal places) have been tabulated for $n = 2(1)1000$ by Tippett (1925). Cadwell (1953) has tabulated $E(w_1)$ to four decimal places for $n = 10(1)30$. The first author has computed tables of $E(w_r)$, accurate to within a unit in the sixth decimal place, for $r = 0(1)8$ and $n = (2r+2)(1)100$, using the Burroughs E101 computer. The trapezoidal rule was employed for the numerical integration. The results are given in Table A1.

Variances of Quasi-Ranges

In order to determine the variance of unbiased estimators based on quasi-ranges (and hence their efficiency), it is necessary to know the variance of the rth quasi-range for samples of n from a standard normal populations, which is given by the equation

$$\operatorname{var} w_r = E(w_r^2) - [E(w_r)]^2\ , \quad (1.1.3)$$

where [see Cadwell (1953), p. 604 for the probability density function]

$$E(w_r^2) = \frac{n!}{(n-2r-2)!(r!)^2}\int_{-\infty}^{\infty}\left\{\int_0^{\infty} F(x, w_r)\,dw_r\right\}\left[\frac{1}{2}+\Phi(x)\right]^r\phi(x)\,dx\ , \quad (1.1.4)$$

in which

$$F(x, w_r) = w_r^2\left[\frac{1}{2}-\Phi(x+w_r)\right]^r[\Phi(x+w_r)-\Phi(x)]^{n-2r-2}\phi(x+w_r)\ . \quad (1.1.5)$$

Tippett (1925) and others have computed approximate values of the variance of the range for a few values of n. Cadwell (1953) has tabulated $\operatorname{var} w_1$ to four decimal places for $n = 10(1)30$. The first author, with the assistance of Eugene H. Guthrie, has computed tables of $\operatorname{var} w_r$, accurate to within a unit in the fifth decimal place, for $r = 0(1)8$ and $n = (2r+2)(1)100$, using the Univac Scientific (ERA 1103) computer. A seven-point integration formula was used for a few cases where the trapezoidal rule did not give sufficient accuracy. The results are given in Table A2. Table A3 gives values of the standard deviation of the rth quasi-range, accurate to within 2 units in the fifth decimal place, for the same values of r and n.

Covariances of Quasi-Ranges

In order to determine the variance of unbiased estimators based on linear combinations of two quasi-ranges (and hence their efficiency), it is necessary to know not only the variances of the two quasi-ranges but also their covariance. The covariance of the rth and r'th quasi-ranges for samples of n from a standard normal population is given by the equation

$$\operatorname{cov}(w_r, w_{r'}) = E(w_r w_{r'}) - E(w_r)E(w_{r'}) \tag{1.1.6}$$

in which

$$E(w_r w_{r'}) = 2[E(x_{r+1}x_{r'+1}) - E(x_{r+1}x_{n-r'})] \ , \tag{1.1.7}$$

where the expected value of the product of the kth and k'th order statistics [see Wilks (1948), p. 20 for their joint probability density function] is given by the equation

$$E(x_k x_{k'}) = \frac{n!}{(k-1)!(k'-k-1)!(n-k)!} \cdot \int_{-\infty}^{\infty} \left\{ \int_{x_k}^{\infty} G(x_k, x_{k'})\, dx_{k'} \right\} H(x_k)\, dx_k \ , \tag{1.1.8}$$

in which

$$G(x_k, x_{k'}) = x_{k'}[\Phi(x_{k'}) - \Phi(x_k)]^{k'-k-1} \left[\frac{1}{2} - \Phi(x_{k'})\right]^{n-k'} \phi(x_{k'}) \ , \tag{1.1.9}$$

and

$$H(x_k) = x_k \left[\frac{1}{2} + \Phi(x_k)\right]^{k-1} \phi(x_k) \ . \tag{1.1.10}$$

Godwin (1949) has tabulated (to five decimal places) the covariances of all order statistics for samples of $n = 2(1)10$. Teichroew (1956) has computed more extensive and more precise tables, accurate to ten decimal places, for $n = 2(1)20$. The first author, again with the assistance of Eugene H. Guthrie, undertook the task of computing $\operatorname{cov}(w_r, w_{r'})$ for $0 \leq r < r' \leq 8$ and $n = (2r'+2)(1)100$, accurate to within a unit in the fifth decimal place, using the Univac Scientific (ERA 1103) computer, with a seven-point integration formula. Finding that the complete tabulation required too much machine time, he decided to limit the computations to those values required to determine the most efficient estimators of population standard deviation based on two adjacent quasi-ranges and on any two quasi-ranges ($r < r' \leq 8$) for $n = 4(1)100$, together with the numerical values of the efficiency of these estimators.

1.1.3 Unbiased Estimators of σ

The Estimators Themselves

The efficient unbiased estimator of population standard deviation σ is the one based on the sample standard deviation s, and given by the equation

$$\hat{\sigma} = s/c_2 \ , \tag{1.1.11}$$

where

$$s = \left[\sum (x - \bar{x})^2/(n-1)\right]^{1/2} \tag{1.1.12}$$

and

$$c_2 = [2/(n-1)]^{1/2}\Gamma\left(\frac{n}{2}\right)\Big/\Gamma\left(\frac{n-1}{2}\right) . \tag{1.1.13}$$

The unbiased estimator of σ based on one sample quasi-range is given by the equation

$$\tilde{\sigma}_r = w_r/E(w_r) , \tag{1.1.14}$$

while the unbiased estimator of σ based on a linear combination of two sample quasi-ranges is given by the equation

$$\tilde{\sigma}_{r,r'} = \frac{w_r + \lambda_{r,r'} w_{r'}}{E(w_r) + \lambda_{r,r'} E(w_{r'})} \tag{1.1.15}$$

where $\lambda_{r,r'}$ is a weighting factor. In Equations (1.1.14) and (1.1.15), the expected values are understood to be those for samples drawn from $N(0,1)$, the standard normal population.

Variances of the Estimators

The variance of the efficient estimator $\hat{\sigma}$ of population standard deviation σ is given by the equation

$$\operatorname{var}\hat{\sigma} = \frac{\operatorname{var} s}{c_2^2} = \frac{E(s^2) - [E(s)]^2}{c_2^2} = \frac{\sigma^2 - (c_2\sigma)^2}{c_2^2} = \frac{1 - c_2^2}{c_2^2}\,\sigma^2 . \tag{1.1.16}$$

The variance of the estimator $\tilde{\sigma}_r$ based on one quasi-range is given by the equation

$$\operatorname{var}\tilde{\sigma}_r = \frac{\operatorname{var} w_r}{[E(w_r)]^2} , \tag{1.1.17}$$

while the variance of the estimator $\tilde{\sigma}_{r,r'}$ based on a linear combination of two quasi-ranges is given by the equation

$$\operatorname{var}\tilde{\sigma}_{r,r'} = \frac{\operatorname{var} w_r + 2\lambda_{r,r'}\operatorname{cov}(w_r, w_{r'}) + \lambda_{r,r'}^2 \operatorname{var} w_{r'}}{[E(w_r) + \lambda_{r,r'}E(w_{r'})]^2} . \tag{1.1.18}$$

Efficiencies of the Estimators

The efficiency of the efficient estimator $\hat{\sigma}$ is by definition 1 (100 percent). The efficiency of a substitute estimator is defined as the ratio of the variance of the efficient estimator to the variance of the substitute estimator. Thus the efficiency of $\tilde{\sigma}_r$ is given by

$$\operatorname{Eff}\tilde{\sigma}_r = \frac{\operatorname{var}\hat{\sigma}}{\operatorname{var}\tilde{\sigma}_r} \tag{1.1.19}$$

while the efficiency of $\tilde{\sigma}_{r,r'}$ is given by

$$\text{Eff}\,\tilde{\sigma}_{r,r'} = \frac{\text{var}\,\hat{\sigma}}{\text{var}\,\tilde{\sigma}_{r,r'}} . \tag{1.1.20}$$

By varying the weighting factor $\lambda_{r,r'}$, one may obtain a one-parameter family of unbiased estimators $\tilde{\sigma}_{r,r'}$. However, there is just one value of $\lambda_{r,r'}$ which minimizes $V_{r,r'} = \text{var}\,\tilde{\sigma}_{r,r'}$, and hence maximizes $\text{Eff}\,\tilde{\sigma}_{r,r'}$. This value of $\lambda_{r,r'}$ which maximizes the efficiency of the estimator may be obtained by setting $dV_{r,r'}/d\lambda_{r,r'} = 0$ and solving for $\lambda_{r,r'}$, and is given by

$$\lambda_{r,r'} = \frac{E(w_{r'})\,\text{var}\,w_r - E(w_r)\,\text{cov}\,(w_r\,, w_{r'})}{E(w_r)\,\text{var}\,w_{r'} - E(w_{r'})\,\text{cov}\,(w_r\,, w_{r'})} . \tag{1.1.21}$$

Table A4 gives the efficiency of estimators based on w_r for $r = 0(1)8$ and $n = (2r + 2)(1)100$, accurate to within 0.01 percent. Table A5 gives the most efficient estimator based on one sample quasi-range, together with its efficiency, for $n = 2(1)100$, also the most efficient estimators based on linear combinations of two adjacent quasi-ranges and of any two quasi-ranges $(r < r' \leq 8)$, together with their efficiencies, for $n = 4(1)100$, and the efficient estimator based on the sample standard deviation. For the estimators based on one sample quasi-range, the numerical coefficients $1/E(w_r)$ are accurate to within a unit in the sixth decimal place, and the efficiencies are accurate to within 0.01 percent. For the estimators based on a linear combination of two quasi-ranges, the numerical coefficients $1/E(w_r + \lambda_{r,r'}w_{r'})$ are accurate to within a unit in the fourth decimal place, the values of $\lambda_{r,r'}$ are accurate to within a unit in the third decimal place, and the efficiencies are accurate to within 0.01 percent. The efficiency of the estimators based on quasi-ranges is shown graphically by Figure 1.1. It will be noted that for $n < 56$ the estimator based on the best linear combination of any two quasi-ranges $(r < r' \leq 8)$ always involves the range $(r = 0)$, with $1 \leq r' \leq 5$, while the best such estimator for $56 \leq n \leq 100$ is $\tilde{\sigma}_{1,8}$, with the efficiency dropping to 82.71 percent for $n = 100$. It seems likely that slightly better estimators for n near 100 could be obtained by dropping the restriction $r' \leq 8$, but it is doubtful whether the increase in efficiency would exceed 1 percent, which would hardly justify the additional computation of expected values, variances, and covariances required to obtain such estimators. One may wonder why the best estimator based on a linear combination of two quasi-ranges is not the one based on a linear combination of the two quasi-ranges which do best individually. A little reflection, however, will convince one that these two quasi-ranges, which are certainly adjacent ones, are too highly correlated to do best together.

1.1.4 Comparison with Grubbs-Weaver Estimators

Grubbs and Weaver (1947) have proposed estimators of the population standard deviation based on a weighted average of the ranges of random subgroups of the complete sample. Since the optimum size of such subgroups is 8, the sample is divided into subgroups which are as nearly as possible of size 8. If the sample size is an integral multiple of 8, all subgroups are of size 8; otherwise, some subgroups will not be of size 8, since no observations are discarded. The Grubbs-Weaver estimator is always more efficient than the estimator $\tilde{\sigma}_r$ based on one sample quasi-range, except for $n < 12$, when it is identical with $\tilde{\sigma}_r$, both using the range of the complete sample, and always less efficient than the estimator based on the best linear combination of any two quasi-ranges, at least for samples up through size 100. The Grubbs-Weaver estimator is less efficient than the estimator based on the best linear combination of two adjacent quasi-ranges for $n \leq 34$,

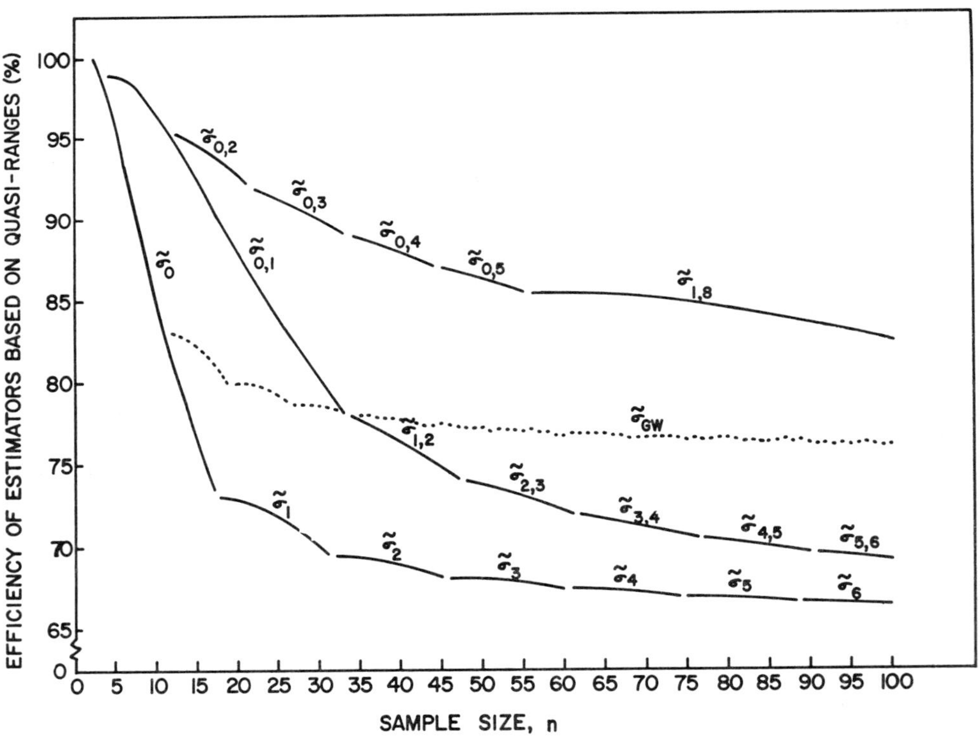

FIGURE 1.1
Efficiency of Estimators of Standard Deviation for Normal Population

but more efficient for $n \geq 35$. It can be shown that the asymptotic efficiency (as sample size approaches infinity) of the Grubbs-Weaver estimator is 75.38 percent. No similar figure for the efficiency of estimators based on quasi-ranges is available. The efficiency of the Grubbs-Weaver estimator $\tilde{\sigma}_{GW}$ for sample sizes up through 100 is shown in Figure 1.1 along with the efficiencies of the estimators based on quasi-ranges. The irregularities in the efficiency of $\tilde{\sigma}_{GW}$ are due partly to the inherent nature of the estimators and partly to the fact that the number of decimal places carried by Grubbs and Weaver is sufficient to yield values of the efficiency accurate only to within about 0.1 percent.

1.1.5 Numerical Example

As an example of the use of estimators based on sample quasi-ranges, consider the following data, given by Morse and Kimball (1950, p. 134) and assumed to come from a normal population, which represent the deviation (in one dimension) from the aiming point of the mean point of impact of salvos of two projectiles:

−237	−23	Quasi-ranges:
−133	−13	$w_0 = 270 - (-237) = 507$
−93	−10	$w_1 = 209 - (-133) = 342$
−77	57	$w_2 = 173 - (-93) = 266$
−75	65	Sample standard deviation: $s = 127.2$
−70	142	Estimates of population standard deviation:
−66	154	$\tilde{\sigma}_1 = .355214\, w_1 = 121.5$
−65	173	$\tilde{\sigma}_{0,1} = .12670(w_0 + 1.4769\, w_1) = 128.2$
−34	209	$\tilde{\sigma}_{0,2} = .14192(w_0 + 1.4640\, w_2) = 127.2$
−28	270	$\hat{\sigma} = s/.986934 = 128.9$

Morse and Kimball (1950) plotted the data on normal probability paper, fitted a straight line "by eye", and estimated the standard deviation as the difference between the 84 percent point and the 50 percent point, the result being 161, a value nearly 25 percent greater than the efficient estimate. Much better results could have been obtained by using an estimator based on a single quasi-range, and still better ones by using an estimator based on a linear combination of two quasi-ranges. In addition to giving much better results, arranging the data in order and making the simple calculations shown above is easier than plotting on normal probability paper (though one may want to do the latter for other reasons). Moreover, it is really not necessary to arrange all the data in order; it would suffice in this example to pick out the three largest and the three smallest values.

1.1.6 References

1. Cadwell, J.H. (1953). The distribution of quasi-range in samples from a normal population. *Annals of Mathematical Statistics* **24**, 603-613.
2. Godwin, H.J. (1949). Some low moments of order statistics. *Annals of Mathematical Statistics* **20**, 279-285.
3. Grubbs, F.E. and Weaver, C.L. (1947). The best unbiased estimate of population standard deviation based on group ranges. *Journal of the American Statistical Association* **42**, 224-241.
4. Harter, H. Leon (1958). The Use of Sample Quasi-Ranges in Estimating Population Standard Deviation. Wright Air Development Center Technical Report 58-200.

Wright-Patterson Air Force Base. AD 151200.

5. Harter, H. Leon (1959). The use of sample quasi-ranges in estimating population standard deviation. *Annals of Mathematical Statistics* **30**, 980-999.

6. Morse, P.M. and Kimball, G.E. (1950). *Methods of Operations Research.* Technology Press, Cambridge, Mass.

7. Mosteller, F. (1946). On some useful inefficient statistics. *Annals of Mathematical Statistics* **17**, 377-408.

8. Teichroew, D. (editor) (1956). Tables of expected values of order statistics and products of order statistics for samples of size twenty and less from the normal distribution. *Annals of Mathematical Statistics* **27**, 410-426.

9. Tippett, L.H.C. (1925). On the extreme individuals and the range of samples taken from a normal population. *Biometrika* **17**, 364-387.

10. Wilks, S.S. (1948). Order statistics. *Bulletin of the American Mathematical Society* **54**, 6-50.

1.2 Interval Estimation of the Population Standard Deviation σ [2]

1.2.1 Introduction

Since the appearance of a justification by E.S. Pearson and Haines (1935), it has been common practice to use the range of small samples to estimate the population standard deviation σ. For the normal population, the efficiency of estimators based on the sample range decreases quite rapidly as the sample size n increases. Mosteller (1946) advanced the suggestion, which was latent in a paper by Karl Pearson (1920), of using quasi-ranges of larger samples to estimate σ. Harter (1958, 1959) has tabulated the most efficient unbiased estimators of σ based on one quasi-range for samples of size $n = 2(1)100$ from normal, rectangular, and exponential populations and on linear combinations of two quasi-ranges for samples of size $n = 4(1)100$ from the normal population. The tables for the normal and rectangular populations are included as Tables A5 and B1 of the present volume, and the related theory is given in Chapter 1, Section 1.1 and Chapter 2, Section 2.1. Chu, Leone, and Topp (1957) have proposed a procedure, using sample quasi-ranges, of setting confidence bounds for the population standard deviation. Leone, Rutenberg, and Topp (1961) have obtained approximate confidence bounds for the standard deviation of normal, exponential, and rectangular populations by first applying distribution-free methods and then imposing the distribution. Though the basic procedure is sound, the use of distribution-free methods yields approximate confidence intervals which are much longer than the exact ones given in the present volume.

For the normal population, which is the one under consideration here, the range of small samples yields rather good point and interval estimators of σ, but for larger sample sizes, better estimators of both kinds can be obtained from quasi-ranges, with the optimum order of the quasi-range increasing with sample size. In order to determine the coefficients of the rth quasi-range w_r in the exact confidence bounds for σ, it has been necessary to tabulate the probability integral of the rth quasi-range, then perform inverse interpolation to determine the percentage points of the rth quasi-range, and finally take the reciprocals of the percentage points. By this method substitute confidence bounds have been determined which compare rather favorably with the conventional confidence bounds based on the sample standard deviation.

Various criteria have been proposed for use in deciding which of a number of substitute confidence bounds or confidence intervals is best. Among these are effectiveness, effectivity, and efficiency, which are defined and discussed in some detail in subsection 1.2.2. The mathematical formulation is given in subsection 1.2.3 for the probability integral of the (standardized) rth quasi-range $W_r = w_r/\sigma$ and for the effectiveness and the efficiency of interval estimators based on w_r. Subsection 1.2.4 describes the method of computation of tables of these functions and tables of percentage points of W_r and coefficients of w_r in exact lower confidence bounds for σ. A comparison of these tables with other tables of substitute confidence intervals for σ is given in subsection 1.2.5.

The following tables were computed, and all but the first are included in Appendix A: (1) an eight-decimal-place table of the probability integral of the (standardized) rth quasi-range $W_r = w_r/\sigma$ for $r = 0(1)8$ and $n = (2r+2)(1)20(2)40(10)100$, with W_r at intervals of 0.05; (2) a six-decimal-place table (Table A6) of the percentage points of the rth quasi-range W_r corresponding to cumulative probabilities $P =$ .0001, .0005, .001, .005, .01, .025, .05, .1(.1).9, .95, .975, .99, .995, .999, .9995, .9999 for the above values of r

[2] Earlier versions of this material were published by Harter (1963, 1964b).

and n; (3) a table (Table A7), to seven significant figures or six decimal places, whichever is less accurate, of the coefficients of the rth quasi-range w_r in the exact lower confidence bounds for σ for the above values of P, r, and n; (4) a table (Table A8) of the upper and lower $(1-P)$ confidence bounds of the central $(1-2P)$ confidence interval for σ, based on the rth quasi-range, where r is optimized in some sense. Table A8 is divided into four parts according to whether r is chosen so as to maximize (a) the effectiveness F_u of the substitute upper confidence bound, (b) the effectiveness F_i of the substitute central confidence interval, (c) the efficiency E_u of the substitute upper confidence bound, or (d) the efficiency E_i of the substitute central confidence interval. Parts (b) and (d) give results only for those pairs of values of n and $(1-P)$ for which the optimum values of r differ from those in parts (a) and (c), respectively. Values of the effectiveness (F_u and F_i) and the efficiency (E_u and E_i) are given to the nearest tenth of a percent.

1.2.2 Criteria for Best Substitute Interval Estimators

General Discussion

A random subset S of the possible values of a parameter θ is by definition a $(1-\alpha)$ confidence set for θ if $P_\theta(\theta \in S) \geq 1-\alpha$ for all θ. If the equality holds, the confidence set is said to be exact; otherwise it is called approximate. If the parameter θ is that of a continuous distribution, it is usually possible to find exact confidence sets, though it is sometimes easier to find approximate ones. If, on the other hand, the distribution is discrete, it is often impossible to find exact confidence sets with specified confidence $1-\alpha$. Whenever exact confidence sets can be found, they are certainly to be preferred to approximate ones. If a confidence set consists of all the values in an interval, it is called a confidence interval. Confidence intervals may be either one-sided, consisting of all values below an upper bound $\bar{\theta}$ (or above a lower bound $\underline{\theta}$), or two-sided, consisting of all values between a lower bound $\underline{\theta}$ and an upper bound $\bar{\theta}$. If the confidence associated with the upper bound $\bar{\theta}$ is the same as that associated with the lower bound $\underline{\theta}$, the two-sided confidence interval is called an equal-tailed or central confidence interval.

Confidence sets (confidence $1-\alpha$) are equivalent to a family of tests (significance level α) of the hypothesis $H(\theta')$ against alternatives $\bar{H}(\theta')$ for varying θ'. In the two-sided case, $H(\theta')$ is the hypothesis $\theta = \theta'$ and $\bar{H}(\theta')$ is the hypothesis $\theta \neq \theta'$. In the one-sided case the hypotheses are $H(\theta') : \theta \leq \theta'$ and $\bar{H}(\theta') : \theta > \theta'$ or $H(\theta') : \theta \geq \theta'$ and $\bar{H}(\theta') : \theta < \theta'$. The $(1-\alpha)$-confidence sets are said to be unbiased if the corresponding tests are unbiased, that is if $P_\theta(\theta' \in S) \leq 1-\alpha$ for all θ' such that $\theta \in \bar{H}(\theta')$ and for all θ, so that the probability of covering the false values specified by the alternative hypothesis does not exceed the confidence level. An unbiased confidence set S for θ is said to be uniformly most accurate (Neyman-shortest) if it minimizes $P_\theta(\theta' \in S)$ for all θ' such that $\theta \in \bar{H}(\theta')$ and for all θ. A confidence set satisfying this condition is optimum in the sense that among all unbiased $(1-\alpha)$-confidence sets it uniformly minimizes the probability of covering false values specified by $\bar{H}(\theta')$.

Various criteria have been proposed for deciding which of a number of procedures for determining central confidence intervals is best; most of these can be extended to procedures for determining upper or lower confidence bounds. Perhaps the most widely used criterion is that of minimizing the expected length of the confidence interval. Pratt (1961) has shown that minimizing the expected length of the confidence interval is equivalent to minimizing the integral, over all possible false values, of the probability of covering false values. For one-sided confidence intervals, the expected length is minimized by minimizing (maximizing) the expected value of the upper (lower) confidence bound. The relation proved by Pratt still holds if all possible false values are considered. But if only false values specified by $\bar{H}(\theta')$ are considered, minimizing the integral of the probability of including such false values is equivalent to minimizing the expected excess (shortage) of the upper (lower) confidence bound as compared with the true value [see Madansky (1962)]. It turns out that criteria based on minimizing expected length, expected excess (shortage), or the probability of including false values are not entirely satisfactory, even though they have been widely used. Lehmann (1959) says, " Short intervals are desirable when they cover the true parameter value but not necessarily otherwise." Faced with a situation in which a confidence interval fails to cover the true value, one might prefer an interval made longer by extending it in the direction of the true value and thus including additional false values nearer the true value. Madansky (1962) has given an example in which minimizing the expected length of a confidence interval is not a good criterion.

Possible alternative procedures are to minimize the mean (absolute) deviation or the mean squared deviation of the upper (lower) confidence bound from the true value in the one-sided case, or in the case of a central confidence interval, to minimize the sum of the mean (absolute) deviations or the sum of the mean squared deviations of the upper and lower confidence bounds from the true value. In the sequel, the criterion based upon minimizing the expected length and these two alternative criteria will be discussed with particular reference to the problem of choosing, not the best possible interval estimator (the conventional estimator), but the best substitute interval estimator of a specified class.

Effectiveness, Effectivity, and Efficiency

The effectiveness of a substitute central confidence interval is defined as the ratio, expressed as a percentage, of the expected length of the conventional confidence interval to that of the substitute interval. A logical extension, in the case of a parameter which cannot be negative, results in the definition of the effectiveness of a substitute upper confidence bound as the ratio, expressed as a percentage, of the expected value of the conventional upper confidence bound to that of the substitute bound. The effectivity of a substitute upper confidence bound is defined as the ratio, expressed as a percentage, of the mean (absolute) deviation of the conventional upper confidence bound from the true parameter value to the mean (absolute) deviation of the substitute upper confidence bound from the true parameter value. Similarly, the effectivity of a substitute central confidence interval is defined as the ratio, expressed as a percentage, of the sum of the mean (absolute) deviations of the conventional upper and lower confidence bounds from the true parameter value to the sum of the mean (absolute) deviations of the substitute upper and lower confidence bounds from the true parameter value. The efficiency of a substitute upper confidence bound is defined as the ratio, expressed as a percentage, of the mean squared deviation of the conventional upper confidence bound from the true parameter value to the mean squared deviation of the substitute upper confidence bound from the true parameter value. Similarly, the efficiency of a substitute central confidence interval is defined as the ratio, expressed as a percentage, of the sum of the mean squared deviations of the conventional upper and lower confidence bounds from the true parameter value to the sum of the mean squared deviations of the substitute upper and lower confidence bounds from the true parameter value.

The above definitions will now be given in symbolic form. Let $(\underline{\theta}, \bar{\theta})$ be the conventional confidence interval, $\underline{\theta}$ and $\bar{\theta}$ being the conventional lower and upper confidence bounds, respectively, and let $(\underline{\theta}', \bar{\theta}')$ be a substitute confidence interval, $\underline{\theta}'$ and $\bar{\theta}'$ being substitute lower and upper confidence bounds, respectively, for the parameter θ. Let E_1, E_2, and E_3 represent effectiveness, effectivity, and efficiency, respectively, all expressed as percentages, with a second subscript, u for upper confidence bounds and i for confidence intervals. Then the equations which give the symbolic definitions are as follows:

$$E_{1u} = 100\ E(\bar{\theta})/E(\bar{\theta}') , \tag{1.2.1}$$

$$E_{2u} = 100\ E|\bar{\theta} - \theta|/E|\bar{\theta}' - \theta| , \tag{1.2.2}$$

$$E_{3u} = 100\ E[(\bar{\theta} - \theta)^2]/E[(\bar{\theta}' - \theta)^2] , \tag{1.2.3}$$

$$E_{1i} = 100\ E(\bar{\theta} - \underline{\theta})/E(\bar{\theta}' - \underline{\theta}') , \tag{1.2.4}$$

$$E_{2i} = 100[E|\bar{\theta} - \theta| + E|\underline{\theta} - \theta|]/[E|\bar{\theta}' - \theta| + E|\underline{\theta}' - \theta|] , \tag{1.2.5}$$

$$E_{3i} = 100\{E[(\bar{\theta} - \theta)^2] + E[(\underline{\theta} - \theta)^2]\}/\{E[(\bar{\theta}' - \theta)^2] + E[(\underline{\theta}' - \theta)^2]\} . \tag{1.2.6}$$

The definition of effectiveness was given by Leone, Rutenberg, and Topp (1961), while the definitions of effectivity and efficiency are due to Harter (1964b), though the latter is a logical extension of the corresponding definition for point estimators. The criterion of maximizing the effectiveness is the same as that of minimizing the expected length of the confidence interval. We have already seen that this criterion is not entirely satisfactory, so we pass on to the others.

Since the numerators of the expressions for effectivity are constant, the criterion of maximizing the effectivity is equivalent to the previously mentioned criterion of minimizing the denominators of these expressions. Use of this procedure overcomes the objections raised to minimizing the expected length of the confidence interval, the probability of including false values, or the expected excess. Its superiority can be seen from the fact that it penalizes estimators for the occurrence of cases in which the upper (lower) confidence bound falls short of (exceeds) the true parameter value, whereas minimizing the expected length of the confidence interval or the probability of including false values actually rewards such occurrences and minimizing the expected excess (shortage) or the probability of including false values greater (less) than the true value ignores them. Essentially, this criterion minimizes the sum of the expected excess and the expected shortage for the single confidence bound in the case of a one-sided interval or the sum of their sums for the two confidence bounds in the case of a two-sided interval, thus requiring the bound or bounds to be as near as possible to the true value regardless of whether they fall above it or below it. Despite these desirable properties, this criterion is nevertheless not the final answer, since it depends upon the mathematical manipulation of expressions involving absolute values, which are notoriously intractable.

The criterion of maximizing the efficiency is equivalent to the previously mentioned criterion of minimizing the expression in the denominator, since the numerator is constant. This criterion has all the advantages mentioned above for the criterion of maximizing the effectivity, and in addition is mathematically much more tractable, since the absolute values have been replaced by squares. We shall see later that is has certain other nice properties from a practical viewpoint, and these confirm the indication of theoretical considerations that it is the best criterion so far advanced.

1.2.3 Mathematical Formulation

Probability Integral of rth Quasi-Range $W_r = w_r/\sigma$

The probability density function of the (standardized) rth quasi-range for samples of size n from $N(\mu, 1)$ is given [see Cadwell (1953)] by

$$f(W_r, n) = \frac{n!}{(n-2r-2)!(r!)^2} \int_{-\infty}^{\infty} \left[\frac{1}{2} + \Phi(x)\right]^r \left[\frac{1}{2} - \Phi(x+W_r)\right]^r \cdot [\Phi(x+W_r) - \Phi(x)]^{n-2r-2} \phi(x)\phi(x+W_r)dx \ , \quad (1.2.7)$$

where $\Phi(x) = \int_0^x \phi(x)dx$ and $\phi(x) = (2\pi)^{-1/2}e^{-x^2/2}$. Integrating with respect to W_r between the limits 0 and W_r, we obtain the probability integral of the rth quasi-range in the form

$$P(W_r, n) = \frac{n!}{(n-2r-2)!(r!)^2} \int_0^{W_r} \int_{-\infty}^{\infty} \left[\frac{1}{2} + \Phi(x)\right]^r \left[\frac{1}{2} - \Phi(x+W_r)\right]^r \cdot [\Phi(x+W_r) - \Phi(x)]^{n-2r-2} \phi(x)\phi(x+W_r)dx\, dW_r \ . \quad (1.2.8)$$

If we change the order of integration, this becomes

$$P(W_r, n) = \frac{n!}{(n-2r-2)!(r!)^2} \int_{-\infty}^{\infty} \left[\frac{1}{2} + \Phi(x)\right]^r \phi(x) \int_0^{W_r} \left[\frac{1}{2} - \Phi(x+W_r)\right]^r \cdot [\Phi(x+W_r) - \Phi(x)]^{n-2r-2} \phi(x+W_r)dW_r\, dx \ . \quad (1.2.9)$$

Integration by parts r times then yields

$$P(W_r, n) = \int_{-\infty}^{\infty} \sum_{k=0}^{r} \frac{n(n-1)\cdots(n-2r+k)}{r!(r-k)!} \left[\frac{1}{2} - \Phi(x+W_r)\right]^{r-k} \cdot [\Phi(x+W_r) - \Phi(x)]^{n-2r+k-1} \left[\frac{1}{2} + \Phi(x)\right]^r \phi(x)dx \ , \quad (1.2.10)$$

from which the probability integral may be computed by numerical integration. The percentage points of W_r, whose reciprocals are the coefficients of w_r in the exact confidence bounds for σ, may then be obtained by inverse interpolation.

Effectiveness of Interval Estimators of σ Based on w_r

If we consider the standardized population ($\sigma = 1$), which we may do without loss of generality, the conventional upper and lower confidence bounds (confidence $1-P$) for σ are given by $B_u = [(n-1)s^2/\chi^2_{P,n-1}]^{1/2}$ and $B_\ell = [(n-1)s^2/\chi^2_{1-P,n-1}]^{1/2}$, respectively, where $s^2 = \sum(x-\bar{x})^2/(n-1)$ is the sample variance and $\chi^2_{P,n-1}$ and $\chi^2_{1-P,n-1}$ are the percentage points of the chi-square distribution with $(n-1)$ degrees of freedom, corresponding to cumulative probabilities P and $1-P$, respectively. Similarly, the substitute upper and lower confidence bounds (confidence $1-P$), based on the rth quasi-range w_r, are given by $B_{ur} = w_r/w_{r,P}$ and $B_{\ell r} = w_r/w_{r,1-P}$, respectively, where $w_{r,P}$ and $w_{r,1-P}$ are the percentage points of the distribution of w_r corresponding to cumulative probabilities P and $1-P$, respectively.

Since $E(s) = [2/(n-1)]^{1/2}\Gamma(n/2)/\Gamma[(n-1)/2]$, the expected values of the conventional upper and lower confidence bounds are given by

$$E(B_u) = \{2^{1/2}\Gamma(n/2)/\Gamma[(n-1)/2]\}\{1/[\chi^2_{P,n-1}]^{1/2}\} \tag{1.2.11}$$

and

$$E(B_\ell) = \{2^{1/2}\Gamma(n/2)/\Gamma[(n-1)/2]\}\{1/[\chi^2_{1-P,n-1}]^{1/2}\} \tag{1.2.12}$$

respectively, and hence the expected length of the conventional central confidence interval is given by

$$E(B_u - B_\ell) = \{2^{1/2}\Gamma(n/2)/\Gamma[(n-1)/2]\}\{1/[\chi^2_{P,n-1}]^{1/2} - 1/[\chi^2_{1-P,n-1}]^{1/2}\}. \tag{1.2.13}$$

The expected values of the substitute upper and lower confidence bounds are given by

$$E(B_{ur}) = (1/w_{r,P})E(w_r) \tag{1.2.14}$$

and

$$E(B_{\ell r}) = (1/w_{r,1-P})E(w_r) \tag{1.2.15}$$

respectively, and hence the expected length of the substitute central confidence interval is given by

$$E(B_{ur} - B_{\ell r}) = [(1/w_{r,P}) - (1/w_{r,1-P})]E(w_r)\ . \tag{1.2.16}$$

Then it follows from the definition of effectiveness that the effectiveness $F_u = E_{1u}$ of the substitute upper confidence bound is given by

$$F_u = 100\ E(B_u)/E(B_{ur})\ , \tag{1.2.17}$$

where $E(B_u)$ and $E(B_{ur})$ are given by Equations (1.2.11) and (1.2.14), respectively, and that the effectiveness $F_i = E_{1i}$ of the substitute central confidence interval is given by

$$F_i = 100\ E(B_u - B_\ell)/E(B_{ur} - B_{\ell r})\ , \tag{1.2.18}$$

where $E(B_u - B_\ell)$ and $E(B_{ur} - B_{\ell r})$ are given by Equations (1.2.13) and (1.2.16), respectively.

Efficiency of Interval Estimators of σ Based on w_r

If we again assume, without loss of generality, that $\sigma = 1$, the mean squared deviation of the conventional upper confidence bound from the true parameter value may be written in the form

$$\begin{aligned} E(D^2_{c,u}) &= E[(B_u - 1)^2] = E\{([(n-1)s^2/\chi^2_{P,n-1}]^{1/2} - 1)^2\} \\ &= \{\chi^2_{P,n-1} - 2[(n-1)\chi^2_{P,n-1}]^{1/2}E(s) + (n-1)E(s^2)\}/\chi^2_{P,n-1}\ . \end{aligned} \tag{1.2.19}$$

Since $E(s) = [2/(n-1)]^{1/2}\Gamma[(n-1)/2]$ and $E(s^2) = 1$, Equation (1.2.19) may be written in the form

$$E(D^2_{c,u}) = \left\{\chi^2_{P,n-1} - \frac{2^{3/2}\Gamma(n/2)[\chi^2_{P,n-1}]^{1/2}}{\Gamma[(n-1)/2]} + n - 1\right\}\Big/\chi^2_{P,n-1}\ . \tag{1.2.20}$$

Similarly, the mean squared deviation of the conventional lower confidence bound from the true parameter value is given by

$$E(D^2_{c,\ell}) = \left\{\chi^2_{1-P,n-1} - \frac{2^{3/2}\Gamma(n/2)[\chi^2_{1-P,n-1}]^{1/2}}{\Gamma[(n-1)/2]} + n - 1\right\} \Big/ \chi^2_{1-P,n-1} \,. \tag{1.2.21}$$

The mean squared deviation of the substitute upper confidence bound, based on the rth quasi-range, from the true parameter value may be written in the form

$$\begin{aligned} E(D^2_{s,u}) &= E[(B_{ur} - 1)^2] = E[(w_r/w_{r,P} - 1)^2] \\ &= [w^2_{r,P} - 2w_{r,P}E(w_r) + E(w_r^2)]/w^2_{r,P} \,. \end{aligned} \tag{1.2.22}$$

Since $E(w_r^2) = E^2(w_r) + \text{var}\, w_r$, Equation (1.2.22) may be written in the form

$$E(D^2_{s,u}) = [w^2_{r,P} - 2w_{r,P}E(w_r) + E^2(w_r) + \text{var}\, w_r]/w^2_{r,P} \,. \tag{1.2.23}$$

Similarly, the mean squared deviation of the substitute lower confidence bound, based on the rth quasi-range, from the true parameter value may be written in the form

$$E(D^2_{s,\ell}) = [w^2_{r,1-P} - 2w_{r,1-P}E(w_r) + E^2(w_r) + \text{var}\, w_r]/w^2_{r,1-P} \,. \tag{1.2.24}$$

The efficiency $E_u = E_{3u}$ of the substitute upper confidence bound is given by

$$E_u = 100\; E(D^2_{c,u})/E(D^2_{s,u}) \tag{1.2.25}$$

and the efficiency $E_i = E_{3i}$ of the substitute central confidence interval is given by

$$E_i = 100[E(D^2_{c,u}) + E(D^2_{c,\ell})]/[E(D^2_{s,u}) + E(D^2_{s,\ell})] \,. \tag{1.2.26}$$

1.2.4 Method of Computation of the Tables

Probability Integral of rth Quasi-Range $W_r = w_r/\sigma$

Values of $P(W_r, n)$ were computed from Equation (1.2.10) for W_r at intervals of 0.01, starting at $W = 0.01$ and continuing until $P(W_{r,n}) > 0.999999995$, for $r = 0(1)8$ and $n = (2r+2)(1)20(2)40(10)100$. Double precision arithmetic was used on the IBM 7090 computer, and the trapezoidal rule was employed for the numerical integration. Input values of $\phi(x)$ and $\Phi(x)$ were taken from tables computed by the WPA Mathematical Tables Project (1942). To obtain values of the percentage points, the computation of which is described in the next paragraph, it was necessary to use these values of $P(W_r, n)$. Because of the presence of both truncation errors and rounding errors, these values are not accurate to the full extent of double precision (approximately 16 significant decimal digits), but they are accurate to at least 10 decimal places, since the numerical integration was performed by starting with interval $h(x) = 0.32$ and halving the interval until successive values differed by no more that 10^{-10}.

Percentage Points of rth Quasi-Range $W_r = w_r/\sigma$

The percentage points of the rth quasi-range corresponding to cumulative probabilities $P =$ 0.0001, 0.0005, 0.001, 0.005, 0.01, 0.025, 0.05, 0.1(0.1)0.9, 0.95, 0.975, 0.99, 0.995, 0.999, 0.9995, 0.9999 for the above values of r and n were computed from the values of the probability integral (computed as described in the preceding paragraph) by inverse interpolation on the IBM 7090, by use of Aitken's method with double precision. First linear (2-point) interpolation was used, then quadratic (3-point), and the number of points was increased in steps of one until successive interpolated values agreed to 24 significant binary bits (corresponding to a little more than 7 significant decimal digits) or until 20 points had been used in the interpolation. In case the tolerance had not been met by the time 20-point interpolation had been performed, the probability integral was subtabulated at subintervals of 0.001 in the interval (of width 0.01) containing the required percentage point and in the two adjacent intervals, with error no greater than 10^{-10} in the subtabulated values. Then the inverse interpolation was repeated, and if the tolerance still could not be met, alternate subtabulation (at a subinterval 1/10 as large as before) and inverse interpolation were continued until the tolerance had been met. Since direct interpolation is more accurate than inverse interpolation in this case, the calculated percentage points were checked by direct interpolation. This was done by checking whether the probabilities found by direct interpolation for values 5×10^{-7} more and 5×10^{-7} less than each calculated percentage point bracket the required probability. In the few cases for which the check failed, the calculated percentage points were adjusted so that the bracketing was achieved. The resulting percentage points were rounded to six decimal places and the rounded values were punched on IBM cards. The results are shown in Table A6, which was reproduced from a listing of the cards on the IBM 407 tabulator.

Coefficients of w_r in Exact Confidence Bounds for σ

The coefficients of the rth quasi-range w_r in the exact lower confidence bounds (confidence P) for σ are the reciprocals of the percentage points of the (standardized) rth quasi-range W_r for the same values of P, r, and n. These coefficients were calculated, double-precision arithmetic being used on the IBM 7090, by taking the reciprocals of the unrounded and unadjusted values of the corresponding percentage points. Values of the coefficients were to be rounded to seven significant digits or six decimal places, whichever is less accurate, but before this was done 5 was added and subtracted in the position of the leftmost digit to be discarded, reciprocals were taken, probabilities were found by direct interpolation, and bracketing was checked as above. After adjustment if necessary to achieve bracketing and after rounding, the rounded values were punched on IBM cards and listed on the IBM 407 tabulator, as reproduced in Table A7.

Effectiveness of Interval Estimators of σ Based on w_r

The expected values of the conventional upper confidence bounds and the expected lengths of the conventional central confidence intervals were computed from Equations (1.2.11) and (1.2.13), respectively, for $n =$ 2(1)20(2)40(10)100 and $P =$ 0.0001, .0005, .001, .005, .01, .025, .05, .1(.1).5, use being made of a table of percentage points of the chi-square distribution computed by Harter (1964a). The expected values of the substitute upper confidence bounds and the expected lengths of the substitute central confidence intervals were computed from Equations (1.2.14) and (1.2.16), respectively, for the same values of n and P with $r = 0(1)\min([2n-2], 8)$, with the aid of tables of expected values of w_r (Table A1) and coefficients of w_r in exact lower confidence bounds for σ (Table A7). The effectivenesses of the substitute upper confidence bounds and the substitute central confidence intervals were then computed from Equations (1.2.17) and (1.2.18), respectively, all computations being performed on the IBM 7090 computer, and the results were rounded to the nearest tenth of one percent. These results, together with the corresponding values of n, r, $1-P$, and $1-2P$ and the coefficients of w_r in the exact upper and lower confidence bounds for σ, were punched on IBM cards. These cards were sorted manually. For each pair of values of n and $1-P$, the card for that value of r which maximizes the effectiveness of the upper confidence bound (confidence $1-P$) was placed in deck A. In most cases (all but 59 out of 385), the same value of r also maximizes the effectiveness of the central confidence interval (confidence $1-2P$). For each of the 59 cases in which this is not true, an asterisk was punched after the effectiveness of the central confidence interval in the card in deck A, and the card for the value of r which does maximize the effectiveness of the central confidence interval was placed in deck B. All cards not belonging either to deck A or to deck B were discarded. Table A8A was reproduced from a listing of deck A and Table A8B from a listing of deck B on the IBM 407 tabulator.

Efficiency of Interval Estimators of σ Based on w_r

The mean squared deviations of the conventional upper and lower confidence bounds from the true parameter value were computed from Equations (1.2.20) and (1.2.21), respectively, for the values of n and P listed in the preceding paragraph, use again being made of the table of percentage points of chi-square. The mean squared deviations of the substitute upper and lower confidence bounds from the true parameter value were computed from Equations (1.2.23) and (1.2.24), respectively, for the values of n, r, and P listed in the preceding paragraph, with the aid of tables of the expected value and the variance of w_r (Tables A1 and A2), percentage points of w_r (Table A6), and coefficients of w_r in exact lower confidence bounds for σ (Table A7). The efficiencies of the substitute upper confidence bounds and the substitute central confidence intervals were then computed from Equations (1.2.25) and (1.2.26), respectively, all computations being performed on the IBM 7090 computer, and the results were rounded to the nearest tenth of one percent. These results, together with the corresponding values of n, r, $1-P$, and $1-2P$ and the coefficients of w_r in the exact upper and lower confidence bounds for σ, were punched on IBM cards. These cards were sorted manually. For each pair of values of n and $1-P$, the card for that value of r which maximizes the efficiency of the upper confidence bound (confidence $1-P$) was placed in deck C. In most cases (all but 22 out of 385), the same value of r also maximizes the efficiency of the central confidence interval (confidence $1-2P$). For each of the 22 cases in which this is not true, an asterisk was punched after the efficiency of the central confidence interval in the card in deck C, and the card for the value of r which does maximize the efficiency of the central confidence interval was placed in deck D. All cards not belonging either to deck C or to deck D were discarded. Table A8C was reproduced from a listing of deck C and Table A8D from a listing of deck D on the IBM 407 tabulator.

In addition to the theoretical advantages mentioned earlier, an examination of Table A8 shows that the criterion of maximizing the efficiency has the following practical advantages over that of maximizing the effectiveness for the problem under consideration: (a) there are fewer cases (22 vs. 59) in which the optimum value of r for the central confidence interval differs from that for the upper confidence bound; (b) the optimum values of r agree more closely with those for point estimation (see Table A5); and (c) for a given value of n, the optimum value of r varies less with changes in P. The criterion of maximizing the effectivity has not been tried because of computational difficulties.

1.2.5 Comparison with Other Tables

The effectiveness of the shortest exact central confidence intervals based on one quasi-range never falls below 81 percent for any case included in Tables A8A and A8B, and for small samples it is much higher (100 percent for samples of size 2). This compares with an effectiveness which never reaches 49 percent and is considerably lower for small samples in the case of the approximate central confidence intervals of Leone, Rutenberg, and Topp (1961). This marked difference in effectiveness is due to the superiority of the exact method based on the probability integral and percentage points of the particular distribution under study over the approximation based on using distribution-free methods and afterward imposing the particular distribution. The effectiveness of the exact central confidence intervals could have been made slightly higher by optimizing the upper and lower confidence bounds separately as Leone, Rutenberg, and Topp did for their approximate confidence intervals. It was decided that the small increase in effectiveness that would have resulted was not worth the additional complication of having the upper and lower confidence bounds based on different quasi-ranges. In comparing results, it should be noted that Leone, Rutenberg, and Topp call the range the 1st than the 0th quasi-range, so that their quasi-range of order $r+1$ corresponds to the rth quasi-range in the present volume.

The values given in Table A8A for the effectiveness of the lowest exact upper confidence bounds never fall below 92 percent. No comparable values are available in other tables.

The efficiency of the exact central confidence intervals based on one quasi-range never falls below 65 percent and that of the exact upper confidence bounds based on one quasi-range never falls below 64 percent for any case included in Tables A8C and A8D. Again no directly comparable values are available from other tables, though it is interesting to note that the efficiency of interval estimators is approximately the same as that of the corresponding point estimators. Since the efficiency is the ratio of two mean squared deviations, the effectivity is the ratio of two mean (absolute) deviations, and the effectiveness is the ratio of two lengths, one would expect the efficiency to be roughly equal to the square of the effectivity and, in the case of central confidence intervals, also roughly equal to the square of the effectiveness. The values in Table A8 confirm this latter approximation. In the case of upper confidence bounds, however, it does not work out this way; since the effectiveness of an upper confidence bound is defined as that of a confidence interval extending from a fixed lower bound (0) to the upper bound, and since the upper confidence bound approaches the true value of the parameter as $n \to \infty$, the effectiveness of a substitute upper confidence bound approaches 100 percent as $n \to \infty$.

Chapter 2 and Appendix B of this volume will give 100 percent effective (or efficient) upper confidence bounds and central confidence intervals, based on the sample range, for the standard deviation of a rectangular population. The results for rectangular and normal populations demonstrate the feasibility of interval estimation of the standard deviation of symmetric populations from sample quasi-ranges. The situation is different for asymmetric populations; for the one-parameter negative exponential population, for example (see Chapter 4 and Appendix D), better point and interval estimators can be obtained from a single order statistic than from a quasi-range.

1.2.6 References

1. Cadwell, J.H. (1953). The distribution of quasi-ranges in samples from a normal population. *Annals of Mathematical Statistics* **24**, 603-613.
2. Chu, J.T., Leone, F.C., and Topp, C.W. (1957). Some uses of quasi-ranges, II. (abstract). *Annals of Mathematical Statistics* **28**, 530-531.
3. Harter, H. Leon (1958). The Use of Sample Quasi-Ranges in Estimating Population Standard Deviation. Wright Air Development Center Technical Report 58-200. Wright-Patterson Air Force Base. AD 151200.
4. Harter, H. Leon (1959). The use of sample quasi-ranges in estimating population standard deviation. *Annals of Mathematical Statistics* **30**, 980-999.
5. Harter, H. Leon (1963). The Use of Sample Ranges and Quasi-Ranges in Setting Exact Confidence Bounds for the Population Standard Deviation. Aeronautical Research Laboratories Technical Report ARL 31, Part II. Wright-Patterson Air Force Base. AD 412352.
6. Harter, H. Leon (1964a). *New Tables of the Incomplete Gamma-Function Ratio and of Percentage Points of the Chi-Square and Beta Distributions.* U.S. Government Printing Office, Washington.
7. Harter, H. Leon (1964b). Criteria for best substitute interval estimators, with an application to the normal distribution. *Journal of the American Statistical Association* **59**, 1133-1140.
8. Lehmann, E.L. (1959). *Testing Statistical Hypotheses.* John Wiley & Sons, Inc., New York.
9. Leone, F.C., Rutenberg, Y.H., and Topp, C.W. (1961). The use of sample quasi-ranges in setting confidence intervals for the population standard deviation. *Journal of the American Statistical Association* **56**, 260-272.
10. Madansky, Albert (1962). More on length of confidence intervals. *Journal of the American Statistical Association* **57**, 586-589.
11. Mosteller, Frederick (1946). On some useful "inefficient" statistics. *Annals of Mathematical Statistics* **17**, 377-408.
12. Pearson, E.S. and Haines, Joan (1935). The use of range in place of standard deviation in small samples. *Journal of the Royal Statistical Society, Supplement* **2**, 83-98.
13. Pearson, Karl (1920). On the probable errors of frequency constants, Part III. *Biometrika* **13**, 113-132.
14. Pratt, John W. (1961). Length of confidence intervals. *Journal of the American Statistical Association* **56**, 549-567.
15. WPA Mathematical Tables Project (1942). *Tables of Probability Functions*, Volume II. WPA New Reading Materials Project for the National Bureau of Standards, New York.

2

The Range of Samples From a Rectangular Population

2.1 Point Estimation of the Population Standard Deviation σ [1]

2.1.1 Unbiased Estimator When Population is Known to be Rectangular

For the standard rectangular population (mean zero and variance one), the probability density function is

$$f(x) = 1/2\sqrt{3}\ , \qquad (-\sqrt{3}, \sqrt{3})\ . \tag{2.1.1}$$

It can be shown that the probability density function of the rth quasi-range, w_r, for samples of n is

$$g(w_r) = \frac{1}{(2\sqrt{3})^n B(n-2r-1, 2r+2)}\ w_r^{n-2r-2}(2\sqrt{3} - w_r)^{2r+1}\ . \tag{2.1.2}$$

Then the expected value of w_r is given by

$$E(w_r) = \int_0^{2\sqrt{3}} w_r g(w_r) dw_r = \frac{n-2r-1}{n+1}\ (2\sqrt{3})\ , \tag{2.1.3}$$

and the expected value of w_r^2 is given by

$$E(w_r^2) = \int_0^{2\sqrt{3}} w_r^2 g(w_r) dw_r = \frac{12(n-2r)(n-2r-1)}{(n+1)(n+2)}\ . \tag{2.1.4}$$

From Equations (2.1.3) and (2.1.4), one can obtain the variance of w_r, which is given by

$$\text{var}\ w_r = E(w_r^2) - [E(w_r)]^2 = \frac{12(2r+2)(n-2r-1)}{(n+1)^2(n+2)}\ . \tag{2.1.5}$$

An unbiased estimator of the standard deviation of a rectangular population is given by

$$\tilde{\sigma}_r = w_r / E(w_r)\ , \tag{2.1.6}$$

where $E(w_r)$ is understood to be the expected value for $R(0,1)$, the standard rectangular population whose probability density function is given by Equation (2.1.1). The variance of $\tilde{\sigma}_r$ is

$$\text{var}\ \tilde{\sigma}_r = \frac{\text{var}\ w_r}{[E(w_r)]^2} = \frac{2r+2}{(n+2)(n-2r-1)}\ . \tag{2.1.7}$$

[1] Earlier versions of this material were published by Harter (1958, 1959).

From Equation (2.1.7), it is evident that the range is more efficient than any of the quasi-ranges for estimating σ, since increasing r both increases the numerator and decreases the denominator of var $\tilde{\sigma}_r$. As a matter of fact, it can be shown that the range is both an efficient and a sufficient statistic for estimating the standard deviation of a rectangular population. Table B1 gives the unbiased estimator $\tilde{\sigma}_0$ for $n = 2(1)100$. The numerical coefficients $1/E(w_0)$ are accurate to within a unit in the sixth decimal place. Since the efficiency is always 100 percent, it is not given in the table.

2.1.2 Bias When Estimators Which Assume Normality Are Used

Section 2.1.1 covers the case when the population being sampled is known to be rectangular. Suppose, however, that the population is of this type, but the investigator who is interested in estimating the standard deviation is not aware of this fact, and proceeds to use one of the estimators which assume normality. In this case, the estimator is no longer unbiased. The bias of an estimator, based on one sample quasi-range, which assumes normality, when the population being sampled is actually of some other type is given by

$$B_o = \frac{E_o(w_r) - E_n(w_r)}{E_n(w_r)} \,. \tag{2.1.8}$$

The bias of an estimator, based on a linear combination of two sample quasi-ranges, which assumes normality, when the population being sampled is actually of some other type is given by

$$B_o = \frac{[E_o(w_r) + \lambda_{r,r'} E_o(w_{r'})] - [E_n(w_r) + \lambda_{r,r'} E_n(w_{r'})]}{E_n(w_r) + \lambda_{r,r'} E_n(w_{r'})} \,. \tag{2.1.9}$$

In Equations (2.1.8) and (2.1.9), E_n represents the expected value for the normal population, while E_o represents the expected value for the other population, both populations having variance one. Table B1 gives the bias B_r for a rectangular population when the estimators of Table A5, which assume normality, are used. The values of the bias are accurate to within 0.01 percent.

2.1.3 References

1. Harter, H. Leon (1958). The Use of Sample Quasi-Ranges in Estimating Population Standard Deviation. Wright Air Development Center Technical Report 58-200. Wright-Patterson Air Force Base. AD 151200.
2. Harter, H. Leon (1959). The use of sample quasi-ranges in estimating population standard deviation. *Annals of Mathematical Statistics* **30**, 980-999.

2.2 Interval Estimation of the Population Standard Deviation σ [2]

2.2.1 Introduction

A general discussion of point and interval estimation of the population standard deviation σ, based on the sample range and quasi-ranges, has been given in Chapter 1 and will not be repeated here. For the rectangular population, which is the one under consideration in this chapter, the efficient point estimator of the population standard deviation is the one based on the sample range. Similarly, interval estimators based on the sample range are more effective (efficient) than those based on any of the sample quasi-ranges. In order to determine the coefficients of the sample range w in the exact lower confidence bounds for σ, it has been necessary to tabulate the probability integral of the range, then perform inverse interpolation to determine the percentage points of the range, and finally take the reciprocals of the percentage points. A discussion is given of the method of computation. The following tables were computed and, except for the first, are included in Appendix B: (1) An eight-decimal-place table of the probability integral of the range for $W =$ 0.01(0.01)3.46(0.001)3.464 and $n =$ 2(1)20(2)40(10)100; (2) a six-decimal-place table (Table B2) of the percentage points of the range corresponding to cumulative probabilities $P =$ 0.0001, 0.0005, 0.001, 0.005, 0.01, 0.025, 0.05, 0.1(0.1)0.9, 0.95, 0.975, 0.99, 0.995, 0.999, 0.9995, 0.9999 for the same values of n; and (3) a table (Table B3), to seven significant figures or six decimal places, whichever is less accurate, of the coefficients of the sample range w in the exact lower confidence bounds for σ for the above values of P and n. The exact confidence intervals between bounds corresponding to P and $(1-P)$ for $P =$ 0.995, 0.990, 0.975, 0.950, 0.900 are much shorter than the approximate confidence intervals obtained by Leone, Rutenberg, and Topp (1961), especially for n small and P small ($1 - P$ large).

2.2.2 Method of Computation of the Tables

Probability Integral of the Range $W = w/\sigma$

The probability integral of the range W for samples of size n from $R(\mu, 1)$, the rectangular population with mean μ and variance 1, may be written [see Kenney and Keeping (1951), p. 192] in the form

$$P(W,n) = \left(\frac{W}{2\sqrt{3}}\right)^{n-1}\left[1+(n-1)\left(1-\frac{W}{2\sqrt{3}}\right)\right] . \tag{2.2.1}$$

Values of $P(W,n)$ were computed from Equation (2.2.1) for $W =$ 0.01(0.01)3.46(0.001) 3.464 and $n =$ 2(1)20(2)40(10)100. Double-precision, floating-point arithmetic was used on the Univac Scientific (ERA 1103A) computer.

[2] Earlier versions of this material were published by Harter (1961a,b).

Percentage Points of the Range $W = w/\sigma$

The percentage points of the range corresponding to cumulative probabilities $P =$ 0.0001, 0.0005, 0.001, 0.005, 0.01, 0.025, 0.05, 0.1(0.1)0.9, 0.95, 0.975, 0.99, 0.995, 0.999, 0.9995, 0.9999 for the above values of n were computed from the table of the probability integral by inverse interpolation on the ERA 1103A, using the following iterative procedure:

1. In the table of the probability integral of the range (computed as described in the preceding paragraph) for the desired value of n, find the two successive probabilities, y_0 and y_1, between which the required cumulative probability P lies. Call the two corresponding arguments (ranges) x_0 and x_1.
2. Perform linear inverse interpolation to find an approximation x to the required W, using the relation
$$x = \frac{(x_1 - x_0)(P - y_0)}{y_1 - y_0} + x_0 \ .$$
3. Take, as the tolerance for W, $T = x \cdot 2^{-27}$.
4. Compute $P(x, n)$ by substituting x for W in Equation (2.2.1).
5. (a) If $P(x, n) > P$, replace x_1 by x and y_1 by $P(x, n)$, then repeat step (2).
   (b) If $P(x, n) < P$, replace x_0 by x and y_0 by $P(x, n)$, then repeat step (2).
6. If the absolute value of the difference between the latest x and the next preceding x is less than T, set W equal to the latest x. Otherwise, using the latest x, repeat steps 4 and 5 until this condition is satisfied, then set W equal to the last x.

The resulting percentage points were rounded to six decimal places, and the rounded values were punched on IBM cards. The results are shown in Table B2, which was reproduced from a listing of the cards on the IBM 407 tabulator.

Coefficients of w in Exact Confidence Bounds for σ

The coefficients of the sample range w in the lower confidence bounds (confidence P) for the population standard deviation σ were found by taking the reciprocals of the unrounded percentage points of the range for the same values of P and n. The results were rounded to seven significant figures or six decimal places, whichever is less accurate, and the rounded values were punched on IBM cards. The results are shown in Table B3, which was reproduced from a listing of the cards on the IBM 407 tabulator.

Check on the Tabular Values

The percentage points in Table B2 were checked by verifying that $P(W - 5 \times 10^{-7}, n) < P$ and that $P(W + 5 \times 10^{-7}, n) > P$, using Equation (2.2.1). In a similar manner, the coefficients of w in lower confidence bounds for σ (Table B3) were checked by adding and subtracting 5 in the position immediately to the right of that occupied by the least significant digit, taking reciprocals, and then verifying that the corresponding probabilities, as computed from Equation (2.2.1), bracket the required probability P.

2.2.3 Effectiveness of Confidence Intervals

Leone, Rutenberg, and Topp (1961) have defined the effectiveness of confidence intervals for σ as the ratio of the expected length $E(\ell)$ of the (exact) intervals based on the efficient estimator of σ to the expected length $E(\ell_0)$ of the intervals under consideration. In the case of the rectangular population, the efficient estimator of σ is the one based on the sample range, and hence $E(\ell)$ is the expected length of the exact confidence intervals for which the coefficients of w are given in Table B3. The effectiveness of the approximate confidence intervals proposed by Leone, Rutenberg, and Topp is the ratio of $E(\ell)$ to the expected length $E(\ell_0)$ of the intervals between their approximate lower confidence bounds corresponding to cumulative probabilities P and $1 - P$. Numerical values of the effectiveness (%) of the approximate confidence intervals $1 - P =$ 0.995, 0.990, 0.975, 0.950, 0.900 and $n = 10(10)100$ are as follows:

$1-P$	$n = 10$	$n = 20$	$n = 30$	$n = 40$	$n = 50$	$n = 60$	$n = 70$	$n = 80$	$n = 90$	$n = 100$
0.995	12.94	42.78	49.89	53.04	54.82	55.96	56.76	57.34	57.79	58.14
0.990	21.61	45.78	51.71	54.37	55.88	56.85	57.52	58.02	58.40	58.71
0.975	31.79	49.19	53.63	55.65	56.80	57.54	58.06	58.44	58.74	58.97
0.950	38.14	50.90	54.25	55.78	56.66	57.23	57.63	57.93	58.15	58.33
0.900	42.35	50.82	53.10	54.16	54.77	55.16	55.44	55.64	55.80	55.93

2.2.4 Example

Let w be the range of a sample of size $n = 30$ from a rectangular population. Find the lower 97.5 percent confidence bound, the upper 97.5 percent confidence bound, and the two-sided 95 percent confidence interval (with equal tails) for the population standard deviation σ.

By referring to Table B3, one finds that the exact lower 97.5 percent confidence bound is $0.291055\,w$ and that the exact upper 97.5 percent confidence bound (lower 2.5 percent confidence bound) is $0.348713\,w$. Obviously, then, the two-sided 95 percent confidence interval (with equal tails) is the interval $(0.291055\,w, 0.348713\,w)$.

It is interesting to compare these results with those obtained by Leone, Rutenberg, and Topp, and to look at the exact confidence associated with their approximate confidence bounds. By referring to their Table III, one finds that their approximate lower 97.5 percent confidence bound is $0.288917\,w$ and that their approximate upper 97.5 percent confidence bound (lower 2.5 percent confidence bound) is $0.396424\,w$. It can be seen from Table B3 that the actual confidence associated with their lower bound is greater than 0.9995 (approximately 0.9997) and that the actual confidence associated with their upper bound is greater than 0.9990 (approximately 0.9991). Thus, their two-sided approximate 95 percent confidence interval $(0.288917\,w, 0.396424\,w)$ has unequal tails, and the actual confidence associated with it is approximately $0.9997 - (1 - 0.9991) = 0.9988$.

2.2.5 References

1. Harter, H. Leon (1961a). The Use of Sample Quasi-Ranges in Setting Exact Confidence Bounds for the Population Standard Deviation. Aeronautical Research Laboratory Technical Report ARL 31, Part I. Wright-Patterson Air Force Base. AD 260325.
2. Harter, H. Leon (1961b). The use of sample ranges in setting exact confidence bounds for the standard deviation of a rectangular population. *Journal of the American Statistical Association* **56**, 601-609.
3. Kenney, J.F. and Keeping, E.S. (1951). *Mathematics of Statistics*, Part II, Second

Edition. D. Van Nostrand Company, Inc., New York.

4. Leone, F.C., Rutenberg, Y.H., and Topp, C.W. (1961). The use of sample quasi-ranges in setting confidence intervals for the population standard deviation. *Journal of the American Statistical Association* **56**, 260-272.

3

Expected Values, Variances and Covariances of Order Statistics of Samples from Various Populations

3.1 Normal Population[1]

3.1.1 History

The problem of order statistics has received a great deal of attention from statisticians dating at least as far back as the early years of this century. Karl Pearson (1902) published a solution of a generalization of a problem proposed by Galton (1902). The generalized problem is that of finding the average difference between the pth and the $(p+1)$st individuals in a sample of size n when the sample is arranged in order of magnitude. The result is

$$\frac{n!}{(n-p)!p!}\int_{-\infty}^{\infty}\alpha^{n-p}(1-\alpha)^{p}\,dx\ , \tag{3.1.1}$$

where $\alpha = \int_{-\infty}^{x}\phi(x)dx$ and $\phi(x)$ is the probability density function of the variable x. Pearson stated a theorem, which he attributed to W.F. Sheppard, that the average differences between successive individuals are the successive terms in the binomial expansion of

$$\int_{-\infty}^{\infty}\{\alpha+(1-\alpha)\}^{n}dx\ . \tag{3.1.2}$$

In a footnote, Pearson remarked, "Clearly a knowledge of the average difference in character of two adjacent individuals involves also a knowledge of the average difference in character between any two individuals". For a symmetric population, such knowledge also involves a knowledge of the expected values of all the order statistics, since for odd sample sizes $n = 2k+1$, where k is an integer, $E(x_{k+1}) = \mu$ (the population mean), while for even sample sizes $n = 2k$, $\frac{1}{2}[E(x_k) + E(x_{k+1})] = \mu$.

Irwin (1925) gave expressions for the mean difference between the pth and qth individuals in order of magnitude and for the moments of the frequency distribution of differences between consecutive individuals. Tippett (1925) published a seven-decimal-place table of the probability integral of the largest individual in samples of size n from a normal population for $n = 3$, 5, 10 and $x = -2.6(0.2)5.8$; $n = 20$, 30, 50 and $x = -0.1(0.1)6.0$; $n = 100(100)1000$ and $x = 1.0(0.1)6.5$. The same paper included a five-decimal-place table of the mean range of samples of size $n = 2(1)1000$ from a normal population. Since

[1] Earlier versions of this material were published by Harter (1960, 1961).

the expected values of the largest and smallest individuals are numerically equal, but opposite in sign, and since their difference is the expected value of the range, the expectation of their absolute value is equal to half the expected value of the range, and hence can be obtained from Tippett's table. The expected values of normal order statistics other than the first and last were not computed until somewhat later.

Karl Pearson and Margaret V. Pearson (1931) obtained an expansion in Taylor series for $E(x_i)$, accurate to 5 or 6 decimal places for $|E(x_i)|$ not too large (say < 1). Fisher and Yates (1938, Table XX) published a two-decimal-place table of the expected values of all normal order statistics for samples of size $n = 2(1)50$. Their values are correct except for four errors of a unit in the last place, due to rounding. Hastings, Mosteller, Tukey and Winsor (1947) published a five-decimal-place table of the means and standard deviations of all order statistics for samples of size $n = 2(1)10$ from a normal population, also from a uniform population and from a selected long-tailed population. Their values for the means of normal order statistics are correct except for $n = 10$, where there are errors as large as 7 units in the last place. Their values for the standard deviations of normal order statistics are accurate to within a unit in the last place except for $n = 9$ and $n = 10$, where there are errors as large as 5 and 14 units, respectively, in the last place.

Wilks (1948) published an expository paper, summarizing work on order statistics up to that time and listing 90 references.

Godwin (1949a) published a table of the expected values of rank differences in normal samples, to 10 decimal places for $n = 2$; 9 decimal places for $n = 3$, 4; 8 decimal places for $n = 5$; 7 decimal places for $n = 6$, 7; 6 decimal places for $n = 8$; and 5 decimals places for $n = 9$, 10. Godwin (1949b) also published a seven-decimal-place table of the means and standard deviations of all normal order statistics for samples of size $n = 2(1)10$. His values for the means are accurate to within a unit in the last place, and his values for the standard deviations are accurate to within 2 units in the last place. Cadwell (1953) published a table of moments (mean, variance, β_1 and β_2) and selected percentage points of the first quasi-range for samples of size $n = 10(1)30$. His values of the means are correct except for one error of a unit in the last place, due to rounding. E.S. Pearson and Hartley (1954, Table 28) published a table of expected values of normal order statistics, to 3 decimal places for $n = 2(1)20$ and to 2 decimal places for $n = 21(1)26(2)50$; values for $n = 2(1)10$ were compiled from Godwin's table, those for $n = 11(1)20$ were freshly computed by Jean H. Thompson, while those for $n > 20$ were taken from the table by Fisher and Yates. These values are correct except for three errors of a unit in the last place, due to rounding. Harter (1959) published a 6-decimal-place table (accurate to within a unit in the last place) of the expected values of the range and of the first 8 quasi-ranges for samples of size $n = 2(1)100$ taken from a normal population. This table is included as Table A1 of the present volume and the related theory is given in Chapter 1, Section 1.1. By dividing these values by two, the expectations of the absolute values of the nine largest and the nine smallest normal deviates can be obtained.

Federer (1951) used a somewhat different approach than did most of the aforementioned authors, who depended largely on numerical integration for the determination of tabular values. Federer made use of the recurrence formula

$$E(x_{m,i+1}) = \frac{1}{i}\left\{mE(x_{m-1,i}) - (m-i)E(x_{m,i})\right\}, \tag{3.1.3}$$

where $x_{m,i}$ is the ith largest deviate from a sample of size m. Starting from Tippett's table of expected values of the range, Federer computed 3-decimal-place values of the three largest normal deviates for samples of size $n = 41(1)200$ and 2-decimal-place values of the fourth largest deviate for $n = 41(1)200$ and of the fifth largest normal deviate for $n = 41(1)100$. Because of loss of accuracy with repeated application of the recurrence formula, some of Federer's values are in error by from 1 to 3 units in the last place, and it is evident that the form of the recurrence formula given by Equation (3.1.3) is of little value in computation. E.S. Pearson has pointed out that, if written in the form

$$E(x_{m-1,i}) = \frac{1}{m}\left\{iE(x_{m,i+1}) + (m-i)E(x_{m,i})\right\}, \tag{3.1.4}$$

the recurrence formula can be used for working *downwards* with no serious accumulation of rounding errors. Similar recurrence formulas for the variance and covariance of order statistics have been obtained by Govindarajulu (1959); see Arnold & Balakrishnan (1989) for a review of all such recurrence relations for order statistics.

Teichroew (1956) published ten-decimal-place tables of the expected values of order statistics and products of order statistics from samples of size twenty and less from a normal population. Sarhan & Greenberg (1956) used his results to calculate ten-decimal-place tables of variances and covariances of normal order statistics for the same sample sizes. Yamauti (1972) published a ten-decimal-place table of the expected values and standard deviations of normal order statistics from samples of size $n = 2(1)50$ and an eight-decimal-place table of expected values of products of normal order statistics from samples of size $n = 2(1)30$. Tietjen, Kahaner & Beckman (1977) published a ten-decimal place table of variances and covariances of normal order statistics for samples of size $n = 2(1)50$. Parrish (1992a) published a 25-decimal-place table of expected values of normal order statistics from samples of size $n = 2(1)50$, and stated that a table of expected values and standard deviations covering sample sizes 2(1)50(10)200(25)500 is available. Parrish (1992b) published a table of variances and covariances of normal order statistics to 25 decimal places (d.p.) for $n = 2(1)20$ and a table of product moments of normal order statistics to 25 d.p. for $n = 20$, 20 d.p. for $n = 30$, 15 d.p. for $n = 40$, and 10 d.p. for $n = 50$. In addition to the published values, he constructed tables to 20 d.p. for $n = 21(1)29$, 15 d.p. for $n = 31(1)39$, and 10 d.p. for $n = 41(1)49$. He stated that the numbers of decimal places reported correspond generally to the indications of precision from several different numerical checks that were applied in an attempt to verify the tabular values. As stated by Parrish, his results support the results of Harter (5 d.p. tables of expected values) and Yamauti (10 d.p. tables of expected values and standard deviations and 8 d.p. tables of expected values of products), but contrast with those of Tietjen et al., which in some cases differ in the fourth decimal place for $n = 50$.

Royston (1982) has presented an algorithm for computing the expected values of normal order statistics which will yield the exact values for sample sizes 1000 and less and approximate values for sample sizes larger than 1000. Koeniger (1983) has made some further remarks on Royston's algorithm. Davis & Stephens (1978) have presented an algorithm for approximating the variance-covariance matrix of normal order statistics which seems to be particularly effective for sample sizes up to 50. Shea & Scallon (1988) have made some further remarks on the algorithm presented by Davis & Stephens. Balakrishnan (1984) has presented an algorithm for approximating the sum of squares of the expected values of normal order statistics. Miyakawa, Tamiya & Kotani (1985) have discussed an algorithm for the evaluation of moments of order statistics through an orthogonal inverse expansion.

3.1.2 Method of Computation of Tables of Expected Values

The expected value of the kth largest observation in a sample of size n from a standard normal population (mean zero and variance one) is given by the equation

$$E(x_{k|n}) = \frac{n!}{(n-k)!(k-1)!} \int_{-\infty}^{\infty} x \left[\frac{1}{2} - \Phi(x)\right]^{k-1} \left[\frac{1}{2} + \Phi(x)\right]^{n-k} \phi(x)dx \ , \quad (3.1.5)$$

where $\phi(x) = (2\pi)^{-1/2}e^{-x^2/2}$ and $\Phi(x) = \int_0^x \phi(x)dx$. The expected value of the kth smallest observation is given by the same expression preceded by a minus sign, so that for a given value of n it is necessary only to compute the expected values for $k = 1(1)[n/2]$. This was done by numerical integration on the Univac Scientific (ERA 1103A) computer, for $n = 2(1)100$ and for values of n, none of whose prime factors exceeds seven, up through $n = 400$. Values of $\log_{10} n!$ for $n = 1(1)400$ from a table by Pearson & Hartley (1954, Table 51) and values of $\phi(x) = \phi(-x)$ for $x = 0(0.05)7.60$, $2\Phi(x) = -2\Phi(-x)$ for $x = 0(0.05)5.95$, and $1 - 2\Phi(x) = 1 + 2\Phi(-x)$ for $x = 6.00(0.05)7.60$ from tables by the National Bureau of Standards (1953, Tables I and II) were read into the computer. For each pair of values of n and k, the product $I(n,k,x)$ of the multiplicative constant and the integrand in Equation (3.1.5) was determined for $x = -7.60(0.05)7.60$ by computing $e^{\log_e I(n,k,x)}$, where

$$\log_e I(n,k,x) = \log_e n! - \log_e(n-k)! - \log_e(k-1)! + \log_e x + (k-1)\log_e \left[\frac{1}{2} - \Phi(x)\right]$$
$$+(n-k)\log_e \left[\frac{1}{2} + \Phi(x)\right] + \log_e \phi(x) \ . \quad (3.1.6)$$

Fixed-point binary arithmetic was used, and the numbers were scaled so as to retain as much accuracy as possible. Since $I(n,k,x)$ is zero (to the number of places carried in the computer) for all values of n and k when $|x| > 7.60$, the resulting value of $E(x_{k|n})$, obtained by using either the trapezoidal rule or one which assigns special weights to the first seven and the last seven values of the integrand (all zero in this case), is found by summing $I(n,k,x)$ for $x = -7.60(h)7.60$ and multiplying by the interval, h. Results were computed and printed out (to seven decimal places) for $h = 0.05$ and $h = 0.10$, and agreement is sufficiently close to guarantee that the values of $E(x_{k|n})$ for $h = 0.05$ are accurate to within a unit in the fifth decimal place. Accordingly, the values for $h = 0.05$ were rounded to five decimal places, and the five-decimal-place values were punched on cards and listed on the IBM 407 tabulator. The results are reproduced as Table C1.1A.

3.1.3 Blom's Approximation

In 1954 Blom became interested in the problem of plotting points on normal probability paper and, after reading a paper published by Chernoff and Lieberman (1954), in the related problem of estimating parameters by means of linear functions of order statistics, Blom (1958) proposed approximating the ith normal order statistic (ith smallest normal deviate) for a sample of size n by means of the relation

$$E(x_i) = \Phi^{-1}\left(\frac{i-\alpha}{n-2\alpha+1}\right) \ , \quad (3.1.7)$$

where $\Phi(x) = \int_{-\infty}^x \phi(x)dx$, with $\phi(x) = (2\pi)^{-1/2}e^{-x^2/2}$. Note that $\Phi(x)$ is defined differently here than in subsection 3.1.2. It should be mentioned that there has been an argument of long standing between advocates of the approximations corresponding to

$\alpha = 0$ and $\alpha = 1/2$. However, neither is optimal. Blom tabulated the value of α required to yield the correct value of $E(x_i)$ for $i = 1(1)[n/2]$ when $n = 2(2)10(5)20$. The values of α increase as n increases, the lowest value being 0.330 for $n = 2$, $i = 1$. For a given n, α is least for $i = 1$, rises quickly to a peak for a relatively small value of i, and then drops off slowly; as an example, for $n = 20$, $\alpha = 0.374$ for $i = 1$, the peak value of α is 0.391 for $i = 3$, and α drops to 0.386 for $i = 8, 9, 10$. Blom conjectured that α always lies in the interval (0.33, 0.50). He suggested the use of $\alpha = 3/8$ as a compromise value.

If one solves Equation (3.1.7) for the value of α required to yield the correct value of $E(x_i)$ for given i and n, one obtains

$$\alpha_{i,n} = \frac{i - (n+1)\Phi[E(x_i)]}{1 - 2\Phi[E(x_i)]} \,. \tag{3.1.8}$$

Values of $\alpha_{i,n}$ for $i = 1(1)[n/2]$ when $n = 25, 50, 100, 200, 400$ have been computed on the Burroughs E101-3 computer, and the results, rounded to three decimal places, are shown in Table C1.1B. For brevity, results have been given only for values of i which are multiples of 5 for i between 25 and 100 and multiples of 10 for i between 100 and 200. A glance at the values in Table C1.1B is sufficient to show that the compromise value of 3/8 for α proposed by Blom is too low except for small values of n. It is possible, however, to do a fairly good job of estimating $E(x_i)$ by choosing one or two compromise values of α for each n. One can choose a single compromise value, α_n, for each n, to be used for all values of i, and simultaneously insure that the error in $E(x_i)$ does not exceed four units in the third decimal place. If one uses $\alpha_{1,n}$ to estimate $E(x_1)$ and $\alpha_{2,n}$ to estimate $E(x_i)$ for $i \neq 1$, the error in $E(x_i)$ will not exceed one unit in the third decimal place. Values of α_n, $\alpha_{1,n}$ and $\alpha_{2,n}$ are given in Table C1.1C for $n = 2(1)10(5)25, 50, 100, 200, 400$, along with regression equations to be used for intermediate values of n. Values of α found by substituting tabular values of n in these regression equations do not differ from the corresponding tabular values of α by more than two units in the third decimal place, and this error in α does not increase the error in $E(x_i)$ by more than one unit in the third decimal place. There is reason to believe that results for intermediate values of n will be equally good, but use of these equations for $n > 400$ is emphatically discouraged. Thus, if one wishes to interpolate for intermediate values of n, the maximum errors are five units in the third decimal place for the approximation based on a single compromise value of α and two units in the third decimal place for the approximation based on two compromise values of α. These errors compare with a maximum error of between one and two units in the third decimal place for linear interpolation between successive values of n for a given i (k) in Table C1.1A. Comparison of the maximum errors might lead to the conclusion that interpolation in Table C1.1A is always more accurate than interpolation using Blom's approximation. This would be erroneous, since the maximum error for the former occurs for large values of i (near $n/2$), while the maximum error for the latter occurs for small values of i. Interpolation using Blom's approximation for large values of i, especially when the desired value of n lies about midway between widely separated successive tabular values of n (for example, when $n = 232$), and interpolation in Table C1.1A otherwise will limit the error to no more than a unit in the third decimal place. If more accurate values are required, they should be computed in the same way that Table C1.1A was computed, as should values for $n > 400$.

In addition to Blom's approximation, there are numerous other methods of approximation for the moments of order statistics available in the literature. Arnold & Balakrishnan (1989) have provided a detailed review of all these methods of approximation. It should be mentioned here that Öztürk & Aly (1991) have recently proposed some simple approximations for the moments of order statistics.

3.1.4 Applications of Tables of Expected Values

Pearson and Hartley (1954, p. 56) have given two examples of applications of tables of expected values of normal order statistics. The first of these is concerned with estimating the weight of the five heaviest of 30 lambs at age $2\frac{1}{2}$ months, given the mean and standard deviation of the population, which is assumed to be normal. The second deals with the use of order statistics in estimating the population standard deviation. Pearson & Hartley and also Fisher & Yates (1938, Table XX, footnote) mention the use of expected values of normal order statistics in the analysis of variance of ranked data. The potential use of expected values of normal order statistics for transformation to standard normal scores preliminary to the analysis of variance was the principal motivation for the present study. In cases where only the rank of the observations is known, there is no reasonable alternative to transformation to standard normal scores, but the usefulness of this method is not restricted to such cases. When the data are known to have come from a population which does not satisfy the assumptions underlying the analysis of variance, of which normality is one, or when the data themselves give a strong indication to that effect, the experimenter seeks a transformation which will minimize or eliminate departures from the assumptions. One transformation which should be considered is the transformation to standard normal scores, and a preliminary investigation has shown that this transformation has some very desirable properties; in some cases it reduces both non-additivity and non-homogeneity of variance to lower levels than does any transformation of the form $(x+c)^p$. It has, of course, the obvious disadvantage of not being reversible, except when the raw data are ranks. As noted in subsection 3.1.3, expected values of normal order statistics may also be used to find plotting positions $F[E(x_i)]$ on normal probability paper.

Blom (1956, 1958) derived nearly best linear unbiased estimators (NBLUEs) of the parameters of location-scale distributions with cdf of the form $F[(x-\mu)/\sigma]$, including the normal distribution. The NBLUEs, which depend only on the expected values of the order statistics, often have variances only slightly greater than those of the best linear unbiased estimators (BLUEs) [see subsection 3.1.6] which depend not only on the means, but also on the variances and covariances of the order statistics. Closely related estimators (called alternative estimators) were discussed by Gupta (1952) and investigated numerically by Sarhan & Greenberg (1956), who tabulated the variances and relative efficiencies (with respect to the BLUEs) of alternative estimators of the mean and standard deviation of a normal population from censored samples of size 10 and gave a numerical example for a doubly censored sample of size 10. Prescott (1970) used the method of Gupta (1952) to obtain linear unbiased estimators of the standard deviation of a normal population from complete samples up through size $n = 50$ and doubly censored samples up through $n = 25$, for which he tabulated the coefficients for $n = 21(1)25$ by use of Harter's (1961) table of expected values of normal order statistics [our Table C1.1A].

3.1.5 Method of Computation of Tables of Variances and Covariances

Let $x_1, x_2, \ldots, x_n$ be a sample from $N(0,1)$ arranged in order of size so that $x_1 \geq x_2 \geq \cdots \geq x_n$. The means, variances, and covariances of these order statistics may be obtained from the following expressions:

$$E(x_i;n) = \frac{n!}{(i-1)!(n-i)!}\int_{-\infty}^{\infty} x\, f(x)[F(x)]^{i-1}[1-F(x)]^{n-i}dx\ , \tag{3.1.9}$$

$$E(x_i^2;n) = \frac{n!}{(i-1)!(n-i)!}\int_{-\infty}^{\infty} x^2 f(x)[F(x)]^{i-1}[1-F(x)]^{n-i}dx\ , \tag{3.1.10}$$

and

$$E(x_i x_j; n) = C \int_{-\infty}^{\infty} \int_{-\infty}^{y} xy\, f(x)\, f(y)[F(x)]^{i-1}[F(y) - F(x)]^{j-i-1}[1 - F(y)]^{n-j}\, dx\, dy\ , \tag{3.1.11}$$

where $f(x) = e^{-x^2/2}/\sqrt{2\pi}$, $F(x) = \int_{-\infty}^{x} f(t)dt$, and $C = n!/[(i-1)!(j-i-1)!(n-j)!]$. (Note that Equation (3.1.9) is equivalent to Equation (3.1.5) of subsection 3.1.2.) Once the above expected values of order statistics and of their squares and products have been computed, the variances and covariances follow:

$$\text{var}(x_i; n) = E(x_i^2; n) - [E(x_i; n)]^2\ , \tag{3.1.12}$$

$$\text{cov}(x_i, x_j; n) = E(x_i x_j; n) - E(x_i; n)E(x_j; n)\ . \tag{3.1.13}$$

As already mentioned in subsection 3.1.1, tables of variances and covariances (and/or expected values of squares and products) for various sample sizes, to various numbers of decimal places, have been computed by the following authors: Teichroew (1956), Sarhan & Greenberg (1956), Yamauti (1972), Tietjen et al. (1977), and Parrish (1992b). While all the computations were based on Equations (3.1.9) - (3.1.13) or equivalent expressions, the specific methods of computation differed considerably from one author to another; the reader is referred to the original sources for details. Our Table C1.2 for $n = 1(1)20$ was obtained by rounding the 10 d.p. table of Sarhan & Greenberg (1956) to 5 d.p.

3.1.6 Applications of Tables of Variances and Covariances

One of the most important applications of tables of variances and covariances of order statistics is in best linear unbiased estimation of the parameters of populations whose cumulative distribution functions have the form $F[(x-\mu)/\sigma]$, where μ and σ are location and scale parameters, respectively. Using the generalized least-squares theorem, Lloyd (1952) found expressions for the best linear unbiased estimators (BLUEs) μ^* and σ^* and for their variances and covariance. In matrix form, he wrote the expectations and the variance-covariance matrix

$$E[\boldsymbol{x}_{(o)}] = (\mathbf{1}\vdots\boldsymbol{\alpha})\begin{pmatrix}\mu\\ \sigma\end{pmatrix}, \qquad V(\boldsymbol{x}_{(o)}) = \boldsymbol{V}\,\sigma^2\ , \tag{3.1.14}$$

where $\mathbf{1}$ denotes a column of ones, $\boldsymbol{\alpha}$ denotes the vector of expected values, $\boldsymbol{V}$ denotes the variance-covariance matrix of the standardized order statistics of a sample from the parent population, and $\boldsymbol{x}_{(o)}$ denotes the vector of ordered observations. By solving the normal equations

$$(\mathbf{1}'\boldsymbol{\Omega}\mathbf{1})\mu^* + (\mathbf{1}'\boldsymbol{\Omega}\boldsymbol{\alpha})\sigma^* = \mathbf{1}'\boldsymbol{\Omega}\boldsymbol{x}_{(o)}$$

and (3.1.15)

$$(\mathbf{1}'\boldsymbol{\Omega}\boldsymbol{\alpha})\mu^* + (\boldsymbol{\alpha}'\boldsymbol{\Omega}\boldsymbol{\alpha})\sigma^* = \boldsymbol{\alpha}'\boldsymbol{\Omega}\boldsymbol{x}_{(o)}$$

where $\boldsymbol{\Omega} = \boldsymbol{V}^{-1}$, he obtained the estimates

$$\mu^* = -\boldsymbol{\alpha}'\boldsymbol{\Gamma}\boldsymbol{x}_{(o)}, \qquad \sigma^* = \mathbf{1}'\boldsymbol{\Gamma}\boldsymbol{x}_{(o)} \tag{3.1.16}$$

where

$$\boldsymbol{\Gamma} = \boldsymbol{\Omega}(\mathbf{1}\boldsymbol{\alpha}' - \boldsymbol{\alpha}\mathbf{1}')\boldsymbol{\Omega}/\Delta \tag{3.1.17}$$

and

$$\Delta = (\mathbf{1}'\boldsymbol{\Omega}\mathbf{1})(\boldsymbol{\alpha}'\boldsymbol{\Omega}\boldsymbol{\alpha}) - (\mathbf{1}'\boldsymbol{\Omega}\boldsymbol{\alpha})^2 \ . \tag{3.1.18}$$

By inverting the matrix of coefficients of the normal equations (3.1.15), he found the variances and covariance of μ^* and σ^* to be

$$V(\mu^*) = (\boldsymbol{\alpha}'\boldsymbol{\Omega}\boldsymbol{\alpha})\sigma^2/\Delta, \ V(\sigma^*) = (\mathbf{1}'\boldsymbol{\Omega}\mathbf{1})\sigma^2/\Delta, \ \mathrm{Cov}(\mu^*, \sigma^*) = -(\mathbf{1}'\boldsymbol{\Omega}\boldsymbol{\alpha})\sigma^2/\Delta \ . \tag{3.1.19}$$

The parent population enters into this method only through the means, variances and covariances of its order statistics. Lloyd gave simplified expressions for symmetrical distributions and for one-parameter distributions. For details, one may refer to Lloyd (1952) and also the books by David (1981) and Balakrishnan & Cohen (1991).

Sarhan & Greenberg (1962) tabulated the coefficients a_{1i} and a_{2i} in the BLUEs

$$\mu^* = \sum_{i=r_1+1}^{n-r_2} a_{1i}x_i \ , \qquad \sigma^* = \sum_{i=r_1+1}^{n-r_2} a_{2i}x_i \tag{3.1.20}$$

for doubly censored samples of size $n = 2(1)20$ from a normal population, as well as the percentage efficiencies of μ^* and σ^* for censored samples relative to uncensored samples. Hoksch (1967) extended these results to conditional estimation of σ (given μ) and Simpson (1967) extended them to simultaneous estimation of μ and σ, both using selected order statistics. Nelson & Schmee (1979) presented tables for confidence limits for the parameters, percentiles, and reliabilities of normal life distributions using BLUEs from singly failure censored data (first r order statistics of a sample of size n). The tables cover $n = 2(1)10$, $r = 2(1)n$, and various confidence levels. More detailed tables that have been prepared are described by Balakrishnan & Cohen (1991).

Other applications include the W-test for normality given by Shapiro & Wilk (1965) and the best linear invariant estimators defined by Mann (1965). The coefficients used in calculating the numerator of the W-test statistic are the (normalized) coefficients used for calculating σ^*. Best linear invariant estimators, defined as point estimators which have smallest expected loss (mean square error) among all linear estimators with risk independent of a location parameter, can be easily obtained from the best linear unbiased estimators and their covariance matrices. For complete or symmetrically censored samples from symmetric populations (including the normal), the best linear invariant estimators are the same as the best linear unbiased estimators.

3.1.7 References

1. Arnold, Barry C. and Balakrishnan, N. (1989). *Relations, Bounds and Approximations for Order Statistics*, Lecture Notes in Statistics No. 53. Springer-Verlag, New York.

2. Balakrishnan, N. (1984). Algorithm AS 200. Approximating the sum of squares of normal scores. *Journal of the Royal Statistical Society, Series C* **33**, 242-245.

3. Balakrishnan, N. and Cohen, A. Clifford, Jr. (1991). *Order Statistics and Inference: Estimation Methods.* Academic Press, San Diego.

4. Blom, Gunnar (1956). On linear estimates with nearly minimum variance. *Arkiv för Matematik* **3**, 365-369.

5. Blom, Gunnar (1958). *Statistical Estimates and Transformed Beta-Variables.* John Wiley & Sons, Inc., New York.

6. Cadwell, J.H. (1953). The distribution of quasi-ranges in samples from a normal population. *Annals of Mathematical Statistics* **24**, 603-613.

7. Chernoff, Herman and Lieberman, Gerald J. (1954). Use of normal probability paper. *Journal of the American Statistical Association* **49**, 778-785.

8. David, Herbert A. (1981). *Order Statistics* (Second edition). John Wiley & Sons, Inc., New York.

9. Davis, C.S. and Stephens, Michael A. (1978). Algorithm AS 128. Approximating the covariance matrix of normal order statistics. *Journal of the Royal Statistical Society, Series C* **27**, 206-212.

10. Federer, Walter T. (1951). Evaluation of Variance Components from a Group of Experiments with Multiple Classifications. *Iowa Agricultural Experiment Station Research Bulletin 380.*

11. Fisher, Ronald A. and Yates, Frank (1938). *Statistical Tables for Biological, Agricultural and Medical Research.* Hafner Publishing Company, Inc., New York. (Sixth edition, 1963).

12. Galton, Francis (1902). The most suitable proportion between the values of first and second prizes. *Biometrika* **1**, 385-390.

13. Godwin, H.J. (1949a). On the estimation of dispersion by linear systematic statistics. *Biometrika* **36**, 92-100.

14. Godwin, H.J. (1949b). Some low moments of order statistics. *Annals of Mathematical Statistics* **20**, 279-285.

15. Govindarajulu, Z. (1959). On moments of order statistics from normal populations (abstract). *Annals of Mathematical Statistics* **30**, 617.

16. Gupta, A.K. (1952). Estimation of the mean and standard deviation of a normal population from a censored sample. *Biometrika* **39**, 260-273.

17. Harter, H. Leon (1959). The use of sample quasi-ranges in estimating population standard deviation. *Annals of Mathematical Statistics* **30**, 980-999.

18. Harter, H. Leon (1960). Expected Values of Normal Order Statistics. *Aeronautical Research Laboratories Technical Report 60-292.* Wright-Patterson Air Force Base. AD 245261.

19. Harter, H. Leon (1961). Expected values of normal order statistics. *Biometrika* **48**, 151-165.

20. Hastings, Cecil, Jr., Mosteller, Frederick, Tukey, John W., and Winsor, Charles P. (1947). Low moments for small samples: a comparative study of order statistics. *Annals of Mathematical Statistics* **18**, 413-426.

21. Hoksch, Orville J. (1967). Conditional Estimation of the Standard Deviation of the Normal and Logistic Distributions from Doubly Censored Samples Using Selected Order Statistics. *Unpublished M.S. Thesis* (GRE/Math/67-5), Air Force Institute of Technology, Wright-Patterson AFB, Ohio.

22. Irwin, J.O. (1925). The further theory of Francis Galton's individual difference problem. *Biometrika* **17**, 100-128.

23. Koeniger, W. (1983). A remark on AS 177. Expected normal order statistics (exact and approximate). *Journal of the Royal Statistical Society, Series C* **32**, 223-224.

24. Lloyd, E.H. (1962). Generalized least-squares theorem. Chapter 3 in *Contributions to Order Statistics* (Eds., Ahmed E. Sarhan and Bernard G. Greenberg), 20-33. John Wiley & Sons, Inc., New York.

25. Mann, Nancy R. (1965). Point and Interval Estimates for Reliability Parameters when Failure Times Have the Two-Parameter Weibull Distribution. *Ph.D. dissertation*, University of California at Los Angeles. University Microfilms, Ann Arbor, Michigan.

26. Mathematical Tables Project, National Bureau of Standards (1953). *Tables of Normal Probability Functions.* Applied Mathematics Series 23. U.S. Government Printing Office, Washington.

27. Miyakawa, K., Tamiya, T., and Kotani, K. (1985). Some comments on the numerical evaluation of moments of order statistics from the normal distribution: Applications of the orthogonal inverse expansion. *TRU Mathematics, Science University of Tokyo* **21/1**, 147-159.

28. Nelson, Wayne and Schmee, Josef (1979). Inference for (log)normal life distributions from small singly censored samples and BLUEs. *Technometrics* **21**, 43-54.

29. Öztürk, A. and Aly, E.E.A.A. (1991). Simple approximation for the moments of normal order statistics, In *The Frontiers of Statistical Computation, Simulation, and Modeling*, Vol. 1 (Eds., Edward J. Dudewicz, A. Öztürk, and E.C. van der Meulen), 151-170. American Sciences Press, Columbus, Ohio.

30. Parrish, Rudolph S. (1992a). Computing expected values of normal order statistics. *Communications in Statistics - Simulation and Computation* **21**, 57-70.

31. Parrish, Rudolph S. (1992b). Computing variances and covariances of normal order statistics. *Communications in Statistics - Simulation and Computation* **21**, 71-101.

32. Pearson, E.S. and Hartley, H.O. (1954). *Biometrika Tables for Statisticians*, Volume I. Cambridge University Press, Cambridge. (Third edition, 1966).

33. Pearson, Karl (1902). Note on Francis Galton's problem. *Biometrika* **1**, 390-399.

34. Pearson, Karl and Pearson, Margaret V. (1931). On the mean character and variance of a ranked individual, and on the mean and variance of the intervals between ranked individuals. *Biometrika* **23**, 364-397.

35. Prescott, P. (1970). Estimation of the standard deviation of a normal population from doubly censored samples using normal scores. *Biometrika* **57**, 409-419.

36. Royston, J.P. (1982). Algorithm AS 177. Expected normal order statistics (exact and approximate). *Journal of the Royal Statistical Society, Series C* **31**, 161-165.

37. Sarhan, Ahmed E. and Greenberg, Bernard G. (1956). Estimation of location and scale parameters by order statistics from singly and doubly censored samples. Part I. The normal distribution up to samples of size 10. *Annals of Mathematical Statistics* **27**, 427-451.

38. Sarhan, Ahmed E. and Greenberg, Bernard G. (1962). The best linear estimates for the parameters of the normal distribution. Chapter 10C in *Contributions to Order Statistics* (Eds., Ahmed E. Sarhan and Bernard G. Greenberg), 206-269. John Wiley & Sons, Inc., New York.

39. Shapiro, Samuel S. and Wilk, Martin B. (1965). An analysis of variance test for normality. *Biometrika* **52**, 591-611.

40. Shea, B.L. and Scallon, A.J. (1988). Approximating the covariance matrix of normal order statistics. *Journal of the Royal Statistical Society, Series C* **37**, 151-155.

41. Simpson, Jackson S. (1967). Simultaneous Linear Estimation of the Mean and Standard Deviation of the Normal and Logistic Distributions by the Use of Selected Order Statistics from Doubly Censored Samples. *Unpublished M.S. The-*

sis (GRE/Math/67-10), Air Force Institute of Technology, Wright-Patterson AFB, Ohio. AD 675544.

42. Teichroew, D. (1956). Tables of expected values of order statistics and products of order statistics for samples of size twenty and less from the normal distribution. *Annals of Mathematical Statistics* **27**, 410-426.

43. Tietjen, G.L., Kahaner, D.K., and Beckman, R.J. (1977). Variances and covariances of the normal order statistics for sample sizes 2 to 50. *Selected Tables in Mathematical Statistics* **5**, 1-73. American Mathematical Society, Providence, Rhode Island.

44. Tippett, L.H.C. (1925). On the extreme individual and the range of samples taken from a normal population. *Biometrika* **17**, 364-387.

45. Wilks, S.S. (1948). Order statistics. *Bulletin of the American Mathematical Society* **54**, 6-50.

46. Yamauti, Ziro (1972). *Statistical Tables and Formulas with Computer Applications*, 33-49. Japanese Standards Association, Tokyo.

3.2 Exponential, Weibull and Gamma Populations[2]

3.2.1 Introduction

After the first author (1960, 1961) published the table of means of normal order statistics discussed in Section 3.1 and included as Table C1.1 of this volume, he received several inquiries as to the availability of similar tables for order statistics arising from populations other than the normal. Upon investigation he found that, while mathematical expressions were known for means of order statistics of samples from many other populations, few of them had been tabulated. In some cases, the absence of published tables is due to the fact that the mathematical expressions are so simple that they do not warrant any tabulation. This is certainly true in the case of the uniform (0,1) population, for which the mean of the ith order statistic of a sample of size n is simply $i/(n+1)$. A similar explicit expression (although not as simple) is also available for the mean of an order statistic from the exponential population. In other cases, such as the Weibull and Gamma populations, the lack of tables is due to the formidability of the task of computing them. For this reason, the first author believed that it should be bridged and succeeded in doing so for the exponential, Weibull, and Gamma populations. Later, tables of variances and covariances of order statistics were computed for the exponential, Weibull and Gamma populations by a number of authors.

3.2.2 Mathematical Formulations and Methods of Computation

Exponential Order Statistics

The probability density function of the one-parameter exponential population with scale parameter θ is given by

$$f(x;\theta) = \frac{1}{\theta}\, e^{-x/\theta}, \qquad x \geq 0,\ \theta > 0\ . \tag{3.2.1}$$

From the works of Sukhatme (1937) and Epstein and Sobel (1953), it is known that the expected value of the ith order statistic of a sample of size n from this population is

$$E(x_{i,n};\theta) = \theta \sum_{j=1}^{i} 1/(n-j+1)\ . \tag{3.2.2}$$

This expression has been tabulated, without loss of generality, for $\theta = 1$; if $\theta \neq 1$, the means can be computed by multiplying the tabular values, as given in Table C2.1, by θ. The values in Table C2.1 were computed from Eq. (3.2.2) with $\theta = 1$ using an IBM 7094 computer with program written in FORTRAN. Double-precision arithmetic was used to avoid any possible loss of accuracy and the results were then rounded to five decimal places.

It is also well known that the variance of $x_{i,n}$ from the exponential population with density as in (3.2.1) is

$$\text{var}(x_{i,n};\theta) = \theta^2 \sum_{j=1}^{i} 1/(n-j+1)^2 \tag{3.2.3}$$

and that, for $i < j$, the covariance of $x_{i,n}$ and $x_{j,n}$ is the same as the variance of $x_{i,n}$ (due

[2]An earlier version of this material was published by Harter (1964b).

to the additive Markov chain property of exponential order statistics). The expression in (3.2.3) has been tabulated to five decimal places for $n = 1(1)23$ and, without loss of generality, for $\theta = 1$; if $\theta \neq 1$, the variances and covariances can be computed by multiplying the tabular values, as given in Table C2.2, by θ^2.

Earlier tables of the means (and their reciprocals) and of the variances and covariances of the order statistics in samples of size $n = 1(1)10$ were given in fractions and decimals (8 decimal places) by Greenberg and Sarhan (1962a).

Weibull Order Statistics

The probability density function of the two-parameter Weibull population with scale parameter θ and shape parameter K is given by

$$f(x;\theta,K) = \frac{K}{\theta}\left(\frac{x}{\theta}\right)^{K-1} e^{-(x/\theta)^K}, \qquad x \geq 0,\ \theta > 0,\ K > 0\ . \tag{3.2.4}$$

Lieblein (1955), Leone, Rutenberg, and Topp (1960), and Quayle (1963) derived mathematical expressions, equivalent but differing slightly in form, for the mean of the ith order statistic of a sample of size n from a Weibull population with density function as in (3.2.4). [Leone, Rutenberg, and Topp also computed the means of the middle order statistics of samples of size $n = 5(2)11$ for $K = 1.5(0.5)5$, but unfortunately their results, which are given to 8 decimal places, are in error even in the fourth decimal place.] The present computations are based upon Quayle's expression which, with a slight change in notation, is

$$E(x_{i,n};\theta,K) = \theta n\binom{n-1}{i-1}\Gamma\left(1+\frac{1}{K}\right)\sum_{j=0}^{i-1}(-1)^{i-1+j}\binom{i-1}{j}\Big/(n-j)^{1+(1/K)}; \tag{3.2.5}$$

also see Balakrishnan and Cohen (1991). The expression in (3.2.5) was evaluated on the IBM 1620 computer for $n = 40$, $i = 1(1)40$, $\theta = 1$, and $K = 0.5(0.5)4(1)8$ using a FORTRAN program. In order to prevent loss of accuracy because of the heavy cancellation that occurs while summing the finite series in (3.2.5), which seems to be the computational pitfall that Leone, Rutenberg, and Topp failed to avoid, 28 decimal digits were carried in the computations. The results were rounded to 12 decimal places and then used to compute $E(x_{i,n};\theta,K)$ for $n = 39(-1)1$, $i = 1(1)n$, $\theta = 1$, and the above-mentioned choices of K by means of the recurrence relation

$$E(x_{i,n}) = \{i\,E(x_{i+1,n+1}) + (n-i+1)E(x_{i,n+1})\}/(n+1)\ ; \tag{3.2.6}$$

see Arnold and Balakrishnan (1989). Realize that the arguments θ and K of the expectations have been omitted since they remain fixed. While this downward recurrence relation is remarkably stable, in contrast with the instability of the corresponding upward recurrence relation, 11 decimal digits were carried in the computations to prevent the accumulation of round-off errors. The values were then rounded to five decimal places (a maximum of seven significant digits) and were checked by means of the identity

$$\sum_{i=1}^{n} E(x_{i,n}) = n\,\Gamma\left(1+\frac{1}{K}\right) \tag{3.2.7}$$

within the allowable rounding error. The values of the means of order statistics so calculated and checked are presented in Table C3.1. Values for $K = 1$ have been omitted since they are the same as those for samples from the exponential population, presented in Table C2.1.

Lieblein (1955) derived the following expressions for single and product moments of order statistics:

$$E(x_{i,n}^m) = \frac{n!}{(i-1)!(n-i)!}\,\Gamma\left(1+\frac{m}{K}\right)\sum_{j=0}^{i-1}(-1)^j\binom{i-1}{j}\Big/(n-i+1+j)^{1+(m/K)},$$

$$K > 0,\ m = 0, 1, 2, \ldots,\ 1 \le i \le n\ ; \qquad (3.2.8)$$

$$E(x_{i,n}x_{j,n}) = K^2\,\frac{n!}{(i-1)!(j-i-1)!(n-j)!}\sum_{k=0}^{i-1}\sum_{\ell=0}^{j-i-1}\binom{i-1}{k}\binom{j-i-1}{\ell}$$

$$\times\ \psi(j-i+k-\ell, n-j+\ell+1)\ ,$$

$$K > 0,\ 1 \le i < j \le n\ , \qquad (3.2.9)$$

where

$$\psi(t,u) = \frac{1}{K^2}\,\Gamma\left(2+\frac{2}{K}\right)B_p\left(1+\frac{1}{K},\ 1+\frac{1}{K}\right)\Big/(tu)^{1+(1/K)}, \qquad (3.2.10)$$

$p = t/(t+u)$, $K, t, u > 0$ $(t \ge u)$, and $B_p(a,b)$ is the incomplete beta function defined by

$$B_p(a,b) = \int_0^p z^{a-1}(1-z)^{b-1}dz, \qquad 0 < p < 1\ . \qquad (3.2.11)$$

For $t < u$, $\psi(t,u)$ can be computed from the identity

$$\psi(t,u) + \psi(u,t) = \frac{1}{K^2}\left\{\Gamma\left(1+\frac{1}{K}\right)\right\}^2\Big/(tu)^{1+(1/K)}\ . \qquad (3.2.12)$$

After computing the single moments and the product moments from Eqs. (3.2.8) and (3.2.9), respectively, one may compute the variances and covariances of order statistics as

$$\mathrm{var}(x_{i,n}) = E(x_{i,n}^2) - \{E(x_{i,n})\}^2 \qquad (3.2.13)$$

and

$$\mathrm{cov}(x_{i,n}, x_{j,n}) = E(x_{i,n}x_{j,n}) - E(x_{i,n})E(x_{j,n})\ . \qquad (3.2.14)$$

In their masters theses, Clark (1964) computed, and Musson (1965) spot checked, a table of means, variances and covariances of Weibull order statistics for sample sizes up to 15 for shape parameter $K = 0.5(0.25)2.0(0.5)4.0$. Both used the expressions in (3.2.8) and (3.2.9) for the computation, with two different numerical integration techniques to evaluate the incomplete beta function appearing in (3.2.10). Both carried their computations to ten decimal places. Musson concluded that Clark's values are reasonably accurate for $n \le 10$, with some loss of accuracy for $n > 10$, especially for large values of K. Table C3.2 is a five decimal-place table of variances and covariances of samples of size $n = 1(1)10$ from Weibull populations with shape parameters $K = 0.5, 1.5(0.5)4.0$, which has been constructed from Clark's table. As in the case of the table of means (Table C3.1), values for $K = 1$ have been omitted since they are the same as those for samples from the exponential population, presented in Table C2.2. The tabular values are accurate to within a unit in the last place, except possibly for occasional errors of 2 or 3 units in the last place when $n = 9$ and 10.

Weibull (1967) presented tables of means, variances and covariances of all order statistics for $n = 5(5)20$ and $K^{-1} = 0.1(0.1)1.0$. Govindarajulu and Joshi (1968) similarly tabulated means, variances and covariances of order statistics for $n = 1(1)10$ and $K = 1, 2, 2.5, 3(1)10$. Balakrishnan and Chan (1993a) have recently tabulated these quantities for n up to 20 and $K = \frac{1}{5}, \frac{1}{4}, \frac{1}{3}, \frac{1}{2}, 1.5(0.5)3, 4(2)10$, while Adatia (1994) prepared similar tables for sample sizes up to 25.

Patel (1975) and Patel and Read (1975) derived bounds on moments of linear functions of order statistics from the Weibull population. By observing from Eq. (3.2.8) that negative moments of order $m > -K$ exist for the Weibull population, Khan, Khan, and Parvez (1984) determined these negative moments for some selected choices of n and K.

Gamma Order Statistics

The probability density function of the two-parameter Gamma population with scale parameter θ and shape parameter α is given by

$$f(x; \theta, \alpha) = \frac{1}{\theta\,\Gamma(\alpha)} \left(\frac{x}{\theta}\right)^{\alpha-1} e^{-x/\theta}, \qquad x \geq 0,\ \theta > 0,\ \alpha > 0\ . \tag{3.2.15}$$

Expressions for the means, variances and covariances of Gamma order statistics cannot, in general, be written in closed form. Gupta (1960, 1962), Karns (1963), and Breiter and Krishnaiah (1967, 1968) developed computational formulas (involving numerical integration) for the moments of Gamma order statistics. Gupta used his formulas (valid only for integral values of α) to tabulate the first four raw moments (and from them the central moments) of the ith order statistic in a sample of size n for $i = 1(1)n$, $n = 1(1)10$ as well as $i = 1$, $n = 11(1)15$ when $\alpha = 1(1)5$. Although the results given by Gupta in his 1960 paper are to six digits, some of them are in error by as much as six units in the fifth digit, because of his use of a recurrence relation with alternating signs, which can result in a severe loss of accuracy due to heavy cancellation. In his 1962 paper the results have been rounded to three decimal places (a maximum of four significant digits), perhaps indicating that he had some doubt as to the accuracy in the fourth and fifth decimal places of the tables in the earlier paper. To compute the means of order statistics, Harter (1964b) used Karns' expression which is

$$E(x_{i,n}; \theta, \alpha) = \frac{n\theta}{\Gamma(\alpha)} \binom{n-1}{i-1} \int_0^\infty \left\{\frac{\Gamma(\alpha; x)}{\Gamma(\alpha)}\right\}^{i-1} \left\{1 - \frac{\Gamma(\alpha; x)}{\Gamma(\alpha)}\right\}^{n-i} e^{-x} x^\alpha \, dx\ , \tag{3.2.16}$$

where $\Gamma(\alpha; x)$ is the incomplete Gamma function defined by

$$\Gamma(\alpha; x) = \int_0^x e^{-t}\, t^{\alpha-1}\, dt\ , \qquad \alpha > 0,\ 0 < x < \infty\ , \tag{3.2.17}$$

and $\Gamma(\alpha; x)/\Gamma(\alpha)$ is the incomplete Gamma-function ratio, which has been tabulated extensively by Pearson (1922) and Harter (1964a). The expression in (3.2.16) was evaluated by numerical integration on the IBM 7094 computer for $i = 1(1)n$, $n = 1(1)40$, $\theta = 1$, and $\alpha = 0.5(0.5)4$ using a FORTRAN program. Eight decimal-place values of the incomplete Gamma-function ratio $I(u, p) = \Gamma(\alpha; x)/\Gamma(\alpha)$, where $p = \alpha - 1$ and $u = x/\sqrt{\alpha}$

from Harter (1964a, Table 1) were used as input to the program. It was found that the tabular interval of 0.1 in u was too large to give good results for the numerical integration, so that it was necessary to subtabulate using Aitken's method of interpolation, but results were still unsatisfactory for small values of u. Hence for $u \leq 1.5$, it was necessary to use a table by Pearson (1922, Table III) of the logarithms of the auxiliary function $I'(u,p) = I(u,p)/u^{p+1}$ which is easy to interpolate. The interpolated values of $I(u,p)$ were then found by multiplying the antilogarithms of the interpolated values of $\log I'(u,p)$ by u^{p+1}. In evaluating the integral in (3.2.16) by numerical integration, it was necessary to use a smaller interval for $u < 1.5$ ($x < 1.5\sqrt{\alpha}$) than for $u > 1.5$ ($x > 1.5\sqrt{\alpha}$). Hence, the integral from 0 to ∞ in (3.2.16) was expressed as the sum of the integrals from 0 to $1.5\sqrt{\alpha}$ and from $1.5\sqrt{\alpha}$ to ∞, and each of these two integrals was evaluated by use of a seven-point formula with the intervals taken small enough to insure that the sum of the two truncation errors would not produce an error in excess of 5×10^{-6} in $E(x_{i,n};\theta,\alpha)$. The computations were carried out in double precision in order to avoid any possible loss of accuracy, but the results were rounded to five decimal places (a maximum of six significant digits) and were also checked by means of the identity

$$\sum_{i=1}^{n} E(x_{i,n}) = n\alpha \tag{3.2.18}$$

within the allowable rounding error. A further check was also made by use of the recurrence relation in (3.2.6); the values computed by the use of (3.2.6) for $\alpha = 0.5(0.5)4$, $i = 1(1)n$ and $n = 1(1)39$ all agreed with the corresponding values produced by the numerical integration to within a unit in the fifth decimal place. The recurrence relation in (3.2.6) itself was not used as the basic computing formula for $n < 40$, as was done in the case of the Weibull distribution, since it was impossible to obtain (from available tables of the incomplete Gamma-function ratio) values for $n = 40$ which were sufficiently accurate to guarantee that round-off errors will not accumulate above the allowable limits. These values of means of order statistics are included in Table C4.1.

When Table C4.1 was first published by Harter (1964b), the author believed that all tabular values were accurate to within a unit in the fifth decimal place. Since that time, however, tables of the first four moments of Gamma order statistics for sample sizes up through $n = 16$ have been published by Breiter and Krishnaiah (1967). The values for the cases that are common to the two tables show good agreement, except for the means of the smallest order statistics when $\alpha = 0.5$, for which there are discrepancies as large as two units in the fifth decimal place. Moreover, the discrepancy increases with n, so that one would expect it to be even larger for sample sizes greater than 16. Harter rechecked the tabular values for $n < 16$, when $\alpha = 0.5$, by numerical integration in Eq. (3.2.16), this time making use of a double-precision computer subroutine, based on a series expansion derived by Harter (1967), to calculate the required values of the incomplete Gamma-function ratio instead of interpolating in the table of Harter (1964a). He found that the values given by Breiter and Krishnaiah (1967) are correct; hence, he carried out the computations up through $n = 40$ and, as a result, corrections as large as five units in the fifth decimal place were found to be necessary in the means of the smallest order statistics when $\alpha = 0.5$. The corrected values are given in Table C4.1. Values for $\alpha = 1$ have been omitted since they are the same as those for samples from the exponential population presented in Table C2.1 and values for $\alpha = 5$ and 6 have also been added.

In their masters theses, Bruce (1964) and Hill (1965) independently computed, by different methods, the means, variances and covariances of Gamma order statistics for samples of size $n = 1(1)20$ when $\alpha = 1(1)3$ and for $n = 1(1)15$ when $\alpha = 4(1)6$. Their results agreed to 8 decimal places. Table C4.2, based on their printout[3] gives the variances and covariances (rounded to five decimal places) of Gamma order statistics for $n = 1(1)10$ and $\alpha = 2(1)6$. Once again, values for $\alpha = 1$ have been omitted since they are the same as those for the exponential order statistics given in Table C2.2.

Breiter and Krishnaiah (1967, 1968) tabulated (to five significant digits) the first four raw moments, from which one can readily obtain the variances (but not to five decimal places), of Gamma order statistics for $\alpha = 0.5(1.0)5.5, 6(1)10$ and $n = 1(1)16$ in their 1967 report, and for $\alpha = 0.5(1.0)10.5$ and $n = 1(1)9$ in their 1968 article. Krishnaiah and Sen (1984) reprinted the 1968 tables. Prescott (1974) published a table of variances and covariances (to four decimal places) for $n = 1(1)10$ and $\alpha = 2(1)5$. Walter and Stitt (1988) presented tables of means of order statistics (to four decimal places) for $n = 1(1)25$, $i = 1, n$ and $\alpha = 1(1)10(5)20$, for $n = 1(1)5$, $i = 1(1)n$ and $\alpha = 1, 5(5)20$, and for $n = 10(5)25$, $i = 1, 5(5)n$ and $\alpha = 1, 5(5)20$. Balasooriya and Hapuarachchi (1992) recently presented (to five decimal places) more elaborate tables of means (for $n = 1(1)10(5)40$ and $\alpha = 5(1)8$) and variances and covariances (for $n = 15(5)25$ and $\alpha = 2(1)5$) of Gamma order statistics. By observing that the negative moments of order $m > -\alpha$ exist for the Gamma population, Joshi (1979) determined these negative moments for some selected choices of n and α.

Moments of Exponential, Weibull, and Gamma Populations

The mean μ, the variance σ^2, the skewness α_3, and the kurtosis α_4 of the distribution of a random variable x having a probability density function $f(x)$ are given by

$$\mu = E(x) , \tag{3.2.19}$$

$$\sigma^2 = E(x^2) - \mu^2 , \tag{3.2.20}$$

$$\alpha_3 = \{E(x^3) - 3\mu\, E(x^2) + 2\mu^3\}/\sigma^3 , \tag{3.2.21}$$

and

$$\alpha_4 = \{E(x^4) - 4\mu\, E(x^3) + 6\mu^2\, E(x^2) - 3\mu^4\}/\sigma^4 , \tag{3.2.22}$$

where

$$E(x^r) = \int_a^b x^r\, f(x)\, dx , \tag{3.2.23}$$

a and b being the lower and upper limits of the range of x.

For the one-parameter exponential population with probability density function as in (3.2.1), we find that

$$E(x^r) = \frac{1}{\theta}\int_0^\infty x^r\, e^{-x/\theta}\, dx = \theta^r\, \Gamma(r+1) = \theta^r\, r! , \tag{3.2.24}$$

whence

$$\mu = \theta,\ \sigma^2 = \theta^2,\ \alpha_3 = 2 \text{ and } \alpha_4 = 9 . \tag{3.2.25}$$

Thus, the single parameter θ is the mean as well as the standard deviation.

[3]We thank Albert H. Moore, faculty advisor to Bruce and Hill, for providing a computer printout of their results.

For the two-parameter Weibull population with probability density function as in (3.2.4), we find that

$$E(x^r) = \frac{K}{\theta} \int_0^\infty x^r \left(\frac{x}{\theta}\right)^{K-1} e^{-(x/\theta)^K} dx = \theta^r \, \Gamma\left(1 + \frac{r}{K}\right) , \tag{3.2.26}$$

whence

$$\mu = \theta \, \Gamma\left(1 + \frac{1}{K}\right) , \tag{3.2.27}$$

$$\sigma^2 = \theta^2 \left\{ \Gamma\left(1 + \frac{2}{K}\right) - \Gamma^2\left(1 + \frac{1}{K}\right) \right\} , \tag{3.2.28}$$

$$\alpha_3 = \frac{\Gamma(1 + \frac{3}{K}) - 3\,\Gamma(1 + \frac{2}{K})\Gamma(1 + \frac{1}{K}) + 2\,\Gamma^3(1 + \frac{1}{K})}{\left\{\Gamma(1 + \frac{2}{K}) - \Gamma^2(1 + \frac{1}{K})\right\}^{3/2}} , \tag{3.2.29}$$

and

$$\alpha_4 = \frac{\Gamma(1 + \frac{4}{K}) - 4\,\Gamma(1 + \frac{3}{K})\Gamma(1 + \frac{1}{K}) + 6\,\Gamma(1 + \frac{2}{K})\Gamma^2(1 + \frac{1}{K}) - 3\,\Gamma^4(1 + \frac{1}{K})}{\left\{\Gamma(1 + \frac{2}{K}) - \Gamma^2(1 + \frac{1}{K})\right\}^2} . \tag{3.2.30}$$

For the two-parameter Gamma population with probability density function as in (3.2.15), we find that

$$\begin{aligned} E(x^r) &= \frac{1}{\theta \, \Gamma(\alpha)} \int_0^\infty x^r \left(\frac{x}{\theta}\right)^{\alpha - 1} e^{-(x/\theta)} dx \\ &= \theta^r \, \Gamma(\alpha + r)/\Gamma(\alpha) = \theta^r \, \alpha(\alpha + 1) \cdots (\alpha + r - 1) , \end{aligned} \tag{3.2.31}$$

whence

$$\mu = \theta\alpha , \tag{3.2.32}$$

$$\sigma^2 = \theta^2\alpha , \tag{3.2.33}$$

$$\alpha_3 = 2/\sqrt{\alpha} , \tag{3.2.34}$$

and

$$\alpha_4 = 3(\alpha + 2)/\alpha . \tag{3.2.35}$$

For $\theta = 1$, the mean, variance, skewness, and kurtosis of the Weibull population with shape parameter $K = 0.5(0.5)4(1)8$ and of the Gamma population with shape parameter $\alpha = 0.5(0.5)4(1)6$ were computed, accurate to eight decimal places, from Eqs. (3.2.27)-(3.2.30) and (3.2.32)-(3.2.35), respectively. All computations were performed on the IBM 1620 computer using FORTRAN programs. In the case of the Weibull population, Stirling's approximation to the Gamma function was used, with a sufficient number of terms being taken into account to insure the required accuracy. The results for exponential, Weibull, and Gamma populations are presented in Table C5. For more elaborate treatment of these three populations, one may refer to Johnson, Kotz, and Balakrishnan (1994).

3.2.3 Uses of Tables of Means

Before proceeding to a discussion of the specific uses of Tables C2.1, C3.1, C4.1 and C5 described in this section, it is appropriate to make some general remarks about the usefulness of the exponential population and of the Weibull and Gamma populations (both of which are generalizations of the exponential). All three of these populations have been widely used in life-testing and reliability studies; for example, see Mann, Schafer, and Singpurwalla (1974), Bain (1978), Kalbfleisch and Prentice (1980), Lawless (1982), Nelson (1982), Cohen and Whitten (1988), Cohen (1991), Bain and Engelhardt (1991), Balakrishnan and Cohen (1991), and Balakrishnan (1995). The exponential population, being the simplest, has probably been used most, sometimes in situations where it is not even appropriate. This population has the property of constant hazard rate; that is, the conditional probability that a device whose life is exponentially distributed will fail in the time interval $(t, t + \Delta t)$, given that it has not failed until time t, is independent of the length of time t that it has already been in service. Obviously, there are many devices which do not have constant hazard rate, but a surprisingly large number of electronic and other devices do, at least over a substantial portion of the life cycle between the period of high infant mortality (decreasing hazard rate) and the period when wearout failures have begun to occur (increasing hazard rate). The distribution of the time between failures of complex equipment which has been in service for a substantial period of time, during which components that have failed have been replaced by new ones, so that the ages of the components are thoroughly mixed, also closely resembles the exponential distribution, even though the lives of the individual components may not. Nevertheless, there are many cases in which life is not exponentially distributed; in a substantial number of these cases, the life distributions can be adequately approximated by Weibull or Gamma distributions with suitable shape parameters. These two distributions have decreasing, constant, or increasing hazard rates according as their shape parameters are less than, equal to, or greater than 1. The role of various populations in life-testing, reliability, and renewal theory has been discussed in early papers by Davis (1952), Epstein and Sobel (1953, 1954), Epstein (1954, 1960a,b), Birnbaum and Saunders (1958), Drenick (1960), and Gupta and Groll (1961), and in a number of papers thereafter that are cited in the volumes that are listed above.

The Weibull population, introduced by Weibull (1939) in connection with a statistical theory of the strength of materials, has proved to be quite a useful and versatile one, with its characteristics depending upon the value of the shape parameter K. It includes as special cases the exponential population ($K = 1$) and the Rayleigh population ($K = 2$). It is positively skewed for $K < 3.6$, almost normal (slightly platykurtic) for $K = 3.6$, and negatively skewed for $K > 3.6$, as can be seen easily from Table C5. As $K \to \infty$, it approaches the "spike" population, in which all values are concentrated at a single point.

The Gamma population is also quite versatile, with its characteristics depending upon the value of the shape parameter α. When $\alpha = 1$, the Gamma variable is exponentially distributed. Whatever the value of α, the Gamma variable is equivalent to one-half the chi-square variable with degrees of freedom 2α, thus making the Gamma population especially useful for values of α which are integral multiples of one-half, corresponding to integral numbers of degrees of freedom for chi-square. The Gamma distribution is positively skewed for finite values of α, but approaches normality as $\alpha \to \infty$, as can be seen easily from Eqs. (3.2.34) and (3.2.35) and Table C5.

One of the principal uses of means of order statistics is as plotting positions on graphs constructed to test graphically the hypothesis that a particular set of data have come from a specified population, and if the hypothesis is accepted, to estimate the parameters of the population. For some populations (especially normal, exponential, and extreme value), special probability graph papers have been produced with the scale linear in one direction and based upon percentiles of the particular population in the other direction, so that data which have come from the population in question, when ordered and plotted against their cumulative probabilities, tend to fall on a straight line. The construction and use of Weibull and Gamma probability papers, which involve the added complication of a shape parameter, has been discussed by King (1971) [second edition, 1981]. These and other special papers may be purchased from TEAM (Technical and Engineering Aids for Management), Box 25, Tamworth, NH 03886. If the appropriate paper is unavailable, an alternative method is to plot the data on ordinary linear-by-linear graph paper, using the expected values of order statistics of samples of the given size from the population in question as plotting positions; Fowlkes (1987) has presented a collection of such quantile-quantile plots for a variety of populations. In the absence of knowledge of these exact plotting positions, an approximation to these quantities may be used. The question of plotting positions, in general and for specific populations, has been discussed by Chernoff and Lieberman (1954, 1956), Blom (1958), Kimball (1960), Harter (1960, 1961), Barnett (1975, 1976), and more recently by Harter (1984) and Harter and Wiegand (1985) [see also subsection 3.1.3].

Let us now consider an example of the use of the means of Weibull order statistics given in Table C3.1. Suppose that 20 components are subjected to a life-test and that the times to failure in hours are as follows:

> 154, 419, 590, 603, 770, 845, 848, 891, 899, 953, 954, 982, 1044, 1059, 1126, 1127, 1294, 1678, 1831, 1847

Suppose it is required to test graphically the hypothesis that the life of components of this type follows a Weibull distribution, and if the hypothesis is accepted, to estimate the parameters of the distribution. In Fig. 3.1, the lives of the components listed above have been plotted as the ordinates of three sets of points whose abscissas are the expected values of order statistics for samples of size 20 from Weibull populations with scale parameter 1, and shape parameters $K = 1, 2$, and 3, respectively. Successive points for each value of K have been connected by straight lines, and the first point for each value of K has been connected with the origin by a straight line. If a parabola were fitted to the points in each group, it is obvious at a glance that the parabola for $K = 1$ would be concave downward and the parabola for $K = 3$ would be concave upward. The points for $K = 2$ exhibit the irregularity characteristic of such groups of points for samples no larger than 20, but it does not appear that the best-fitting parabola would differ materially from a straight line. Therefore it is reasonable to believe that the data have arisen from a Weibull population with shape parameter K approximately equal to 2. If the expected value of a Weibull order statistic for $K = 2$ is denoted by x and the correspondingly ordered sample value by y, then the least squares regression line is found to be $y = 172.6 + 928.7\,x$. The y-intercept, 172.6 hours, is an estimate of the location parameter (the lower limit of the population), and the slope, 928.7 hours, is an estimate of the scale parameter. Since one of the sample values is 154 hours, the true value of the location parameter must be less than or equal to 154 hours. If it is assumed to be zero, so that the regression line is constrained to pass through the origin, the regression line is found to be $y = 1123.4\,x$, yielding an estimate of 1123.4 hours for the scale parameter. Both regression lines are shown in Figure 3.1. Actually, the data used in the problem are fictitious; they were obtained by random sampling from a two-parameter Weibull population with scale parameter 1000, and shape parameter 2, by use of a Monte Carlo technique. More precisely, the Monte Carlo technique was used to obtain a random sample from an exponential population with scale parameter 100; then the square roots of the sample values were taken so as to obtain

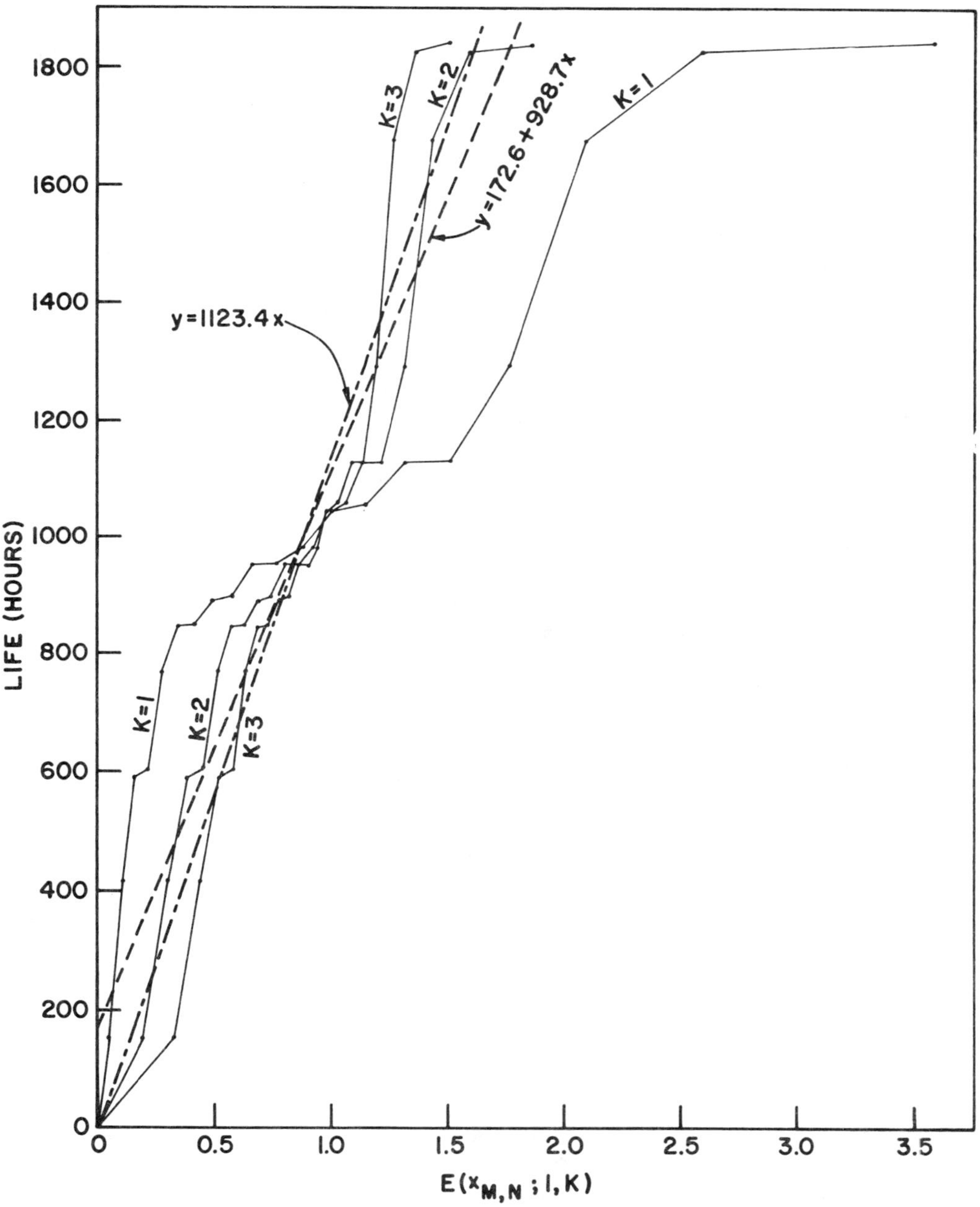

FIGURE 3.1
Graphical test of hypothesis that life test data have come from Weibull population and estimation of parameters, using expected values of order statistics

a sample from a two-parameter Weibull population with scale parameter 10 and shape parameter 2; and finally these results were multiplied by 100. This method is based upon the fact that if x has a Weibull distribution with shape parameter K, x^K is exponentially distributed. This fact can also be used to test whether or not a sample of size n, where $40 < n \leq 120$, may reasonably be supposed to have come from a Weibull population. Table C3.1 extends only through $n = 40$ and hence cannot be used for this purpose, but the data can be transformed by raising them to the Kth power, and Table C2.1 can be employed to test whether the transformed data may reasonably be supposed to have come from an exponential population. Several values of K can be tried, and the choice of K is not restricted to the values included in Table C3.1. If the result, for some value $K = K_0$, is a set of points which fall approximately along a straight line, it is reasonable to suppose that the original data came from a Weibull population with shape parameter K_0. If the shape parameter is estimated to be K_0, the location and scale parameters of the distribution of x^{K_0} (assumed to be exponential) can then be estimated, and the estimates transformed back to obtain estimates of the location and scale parameters of the population (assumed to be Weibull with shape parameter K_0) from which the original data came.

Next consider an example of the use of the means of Gamma order statistics given in Table C4.1. Hald (1952, p. 472) reports the logarithms of the permeabilities (measured in seconds) for 81 sheets of building material manufactured on 9 days by 3 machines, 3 sheets per machine per day. Before performing a two-way analysis of variance to test the effects of days, machines, and their interaction, it is desirable to test the homogeneity of the within-cell variances, each of which is computed from 3 observations and hence has associated with it 2 degrees of freedom. Instead of the usual Bartlett's test, one may perform a graphical test. If the logarithms of the permeabilities are assumed to be normal (permeabilities lognormal), the variances of subgroups of size 3 are distributed as χ^2 with 2 degrees of freedom, which is proportional to a Gamma variable with $\alpha = 1$. Therefore the ordered sums of squares of deviations (which are proportional to the within-cell variances) have been plotted as the ordinates and the means of order statistics of a sample of size 27 from a Gamma population with scale parameter 1 and shape parameter 1 (which are proportional to the order statistics of a χ^2 population with 2 degrees of freedom) as abscissas of the points in Figure 3.2. The points in Figure 3.2 appear to depart from a straight line more than one would expect if the fluctuations were purely random. The pattern is very much the same as that obtained by Wilk, Gnanadesikan, and Huyett (1962, Figure 12) who, lacking a table of means of Gamma order statistics, obtained the abscissas for their points by interpolating in their table of percentage points of the Gamma distribution with shape parameter 1 to find the values corresponding to cumulative probabilities $(i - \frac{1}{2})/n$, where $i = 1(1)n$ and $n = 27$ in this example. The present method has two advantages over their method. One is that the means are exact plotting positions, while their plotting positions are only approximations. A second advantage is that a computer is not required, since the plotting positions may be taken directly from Table C4.1 [or Table C2.1 when $\alpha = 1$, as in the above example], whereas the non-linear interpolation in the table of percentage points which must be performed in applying their method necessarily requires a computer. On the other hand, if a computer is available, then one could easily adapt their program to use means as plotting positions not only for the Gamma population, but also for the exponential and Weibull populations and even for the normal population, in the latter case using the table of Harter (1960, 1961) of means of normal order statistics, which is included in this volume as Table C1.1. A disadvantage of this method is that it can be used only for values of n and the shape parameter for which tables of means of order statistics exist. In the case of the Gamma

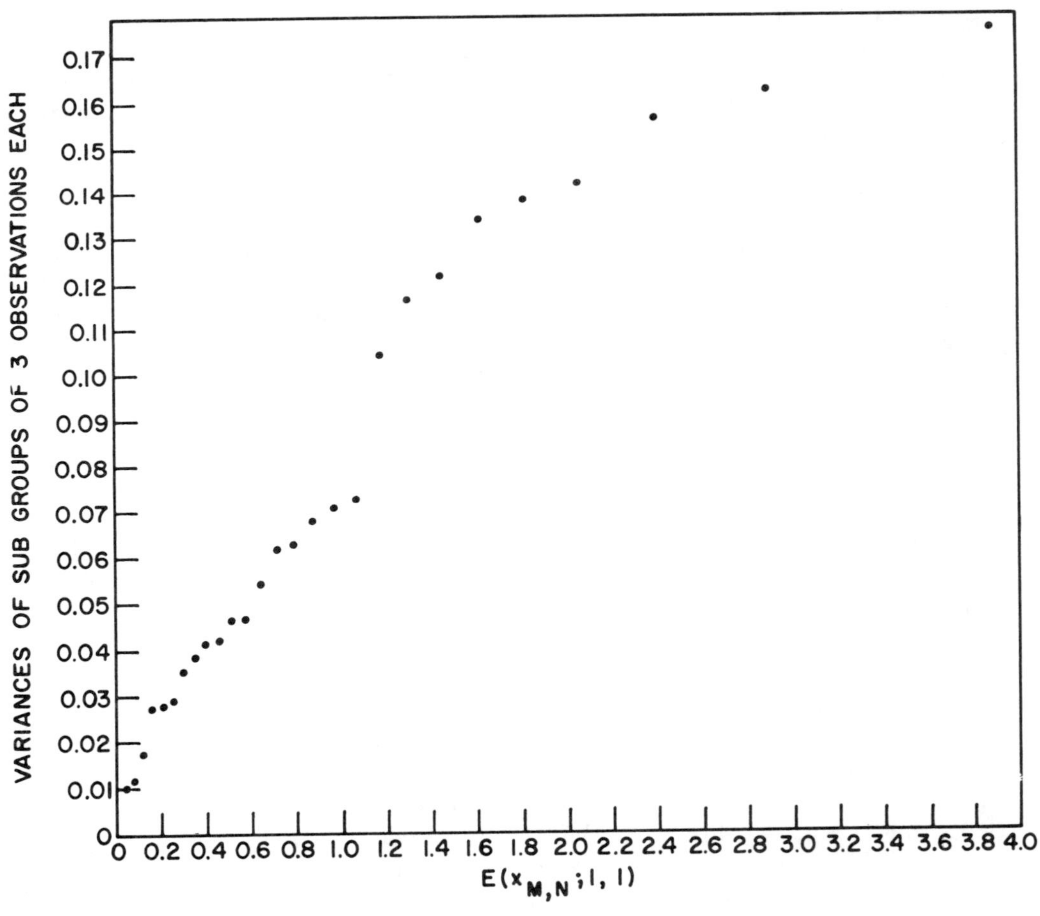

FIGURE 3.2
Graphical test of homogeneity of variance, using expected values of Gamma order statistics

population, the restriction to $\alpha = 0.5(0.5)4(1)6$ is not a serious one for the application to testing homogeneity of variance, since the tabular values of α correspond to 1(1)8(2)12 degrees of freedom for χ^2 [2(1)9(2)13 observations within each cell], which covers most cases of practical interest. Tables for $n = 1(1)40$ are adequate for experiments which do not contain more than 40 cells (combinations of level of the factors).

Another use of tables of means of order statistics is in nearly (or almost) best linear unbiased (or invariant) estimation of location and scale parameters. If tables of variances and covariances of order statistics are also available, best linear unbiased (or invariant) estimation is possible (see subsection 3.2.4) and is usually preferred, though the increase in efficiency is often small.

Epstein (1956) derived almost best linear unbiased estimators (ABLUEs), from censored samples, of parameters of exponential distributions. Blom (1958) derived nearly best linear unbiased estimators (NBLUEs), from complete or censored samples, of the parameters of one- and two-parameter exponential distributions. He determined the variances of the estimators μ^* and σ^* in the two-parameter case for sample size $n = 5$ and their relative efficiencies (97.2% for μ^* and 96.5% for σ^*).

Weibull (1959) derived the NBLUEs μ^* and β^* of the location and scale parameters of a Weibull distribution with shape parameter α [$1/\alpha = K$ in our notation] and cumulative distribution function

$$F(x) = 1 - e^{-\left(\frac{x-\mu}{\beta}\right)^{1/\alpha}}$$

and tabulated their coefficients (two or three decimal places) for $n = 2(1)15$, $\alpha = 0.1(0.1)0.6(0.2)1.0$. Using Harter's (1964b) table of means of Weibull order statistics (Table C3.1), Stump (1968) tabulated coefficients for simultaneous NBLUEs of the location and scale parameters for complete and censored samples of size ten to forty. In the same way, Davis (1969) tabulated coefficients (to five decimal places) for conditional NBLUEs $\bar{\mu}|\beta$ and $\bar{\beta}|\mu$ for the Weibull distribution with shape parameter K and cumulative distribution function

$$F(x) = 1 - e^{-\left(\frac{x-\mu}{\beta}\right)^{K}}$$

for complete and right-censored samples of size $n = 1(1)40$ when $K = 0.5(0.5)2(1)4$. Elkins (1970) presented similar results for nearly best conditional linear invariant estimators. If the shape parameter K is unknown, it must be estimated first; for one method, see Murthy (1968).

Morgan (1968) developed the necessary theory for nearly best linear unbiased estimators of the location and scale parameters of a Gamma population. He used Harter's (1964b) table of means of Gamma order statistics (Table C4.1) to tabulate the coefficients (to five decimal places) of the order statistics in simultaneous NBLUEs of the location and scale parameters of a Gamma population with known shape parameter α based on m order statistics, for $\alpha = 0.5(0.5)4$, $n = 15(1)40$, and some selected values of $m < n$. He also tabulated the efficiencies (to three decimal places). In the same manner, Davis (1970) [Gray (1970)] developed results for conditional nearly best linear unbiased [invariant] estimators.

Balakrishnan and Cohen (1991) have discussed in detail these methods of estimation based on order statistics.

3.2.4 Uses of Tables of Variances and Covariances

Tables of variances and covariances of order statistics of samples from any specified location-scale parameter population are used, along with tables of means, to determine the coefficients of the sample order statistics in best linear unbiased estimators (BLUEs) or best linear invariant estimators (BLIEs) of the population parameters.

Sarhan and Greenberg (1957) [see also Greenberg and Sarhan (1962b)] derived expressions for the best linear unbiased estimator of θ of the one-parameter exponential population with density function

$$f(x;\theta) = \frac{1}{\theta}\, e^{-x/\theta}, \qquad x \geq 0\ , \tag{3.2.36}$$

and of the location and scale parameters, μ and θ, of the two-parameter exponential population with density function

$$f(x;\mu,\theta) = \frac{1}{\theta}\, e^{-(x-\mu)/\theta}, \qquad x \geq \mu\ , \tag{3.2.37}$$

from complete and (singly or doubly) censored samples. They also presented expressions for the variances as well as the efficiencies of the estimators from censored samples relative to those from a complete sample. Sarhan and Greenberg (1957) tabulated the exact coefficients of the order statistics in the best linear unbiased estimators and the exact variances of these estimators and their percentage efficiencies (to two decimal places) for sample sizes up to 10. Harter (1961b) tabulated coefficients and variances (to six significant figures) and relative efficiencies (to five significant figures) for the best linear unbiased estimators based on k order statistics from samples of size $n = 1(1)100$ for $k = 1, 2$ in the case of the one-parameter exponential density function in (3.2.36), and for $k = 2$ in the case of the two-parameter exponential density function in (3.2.37). Kulldorff (1963) gave similar tables for $k = 3(1)5$. (For further information, see Chapter 4, Section 4.1 and Tables D1-D3.)

For the Weibull population with probability density function

$$f(x;\mu,\theta) = \frac{K}{\theta}\left(\frac{x-\mu}{\theta}\right)^{K-1} e^{-\left(\frac{x-\mu}{\theta}\right)^{K}}, \qquad x \geq \mu\ , \tag{3.2.38}$$

Govindarajulu and Joshi (1962, 1968) tabulated the coefficients of the order statistics in the best linear unbiased estimators of the location parameter μ and of the scale parameter θ, based on a sample of size $n = 2(1)12$ when the shape parameter $K = 1.5(0.5)3(1)8$. Their estimators are based on all available order statistics from complete samples and from some types of censored samples. The tabular values are given in their 1962 report to five decimal places for $n = 2(1)5$, four decimal places for $n = 6(1)9$, and three decimal places for $n = 10(1)12$; however, these have gross errors since their tables of means, variances and covariances of Weibull order statistics (in that version) had serious errors. More accurate tables (to three decimal places) are given in their 1968 paper for complete samples of size $n = 2(1)10$ for $K = 3, 5, 8, 10$. Balakrishnan and Chan (1993b) extended these tables for larger sample sizes and more choices of K.

If the shape parameter K as well as the location parameter μ is known, then the scale parameter θ of the Weibull density function in (3.2.38) can be estimated from a single order statistic. Quayle (1963) tabulated the coefficients (to six decimal places) in the best one-order-statistic estimators of θ for $K = 0.5(0.5)4(1)8$ and $n = 1(1)20(2)40$ after computing the means and variances of the order statistics.

Clark (1964) used his tables of means, variances and covariances of Weibull order statistics mentioned earlier, and reproduced in part in Table C3.2, to tabulate the coefficients (to five decimal places) in m-order-statistic estimators of the scale parameter θ of a Weibull population with known location parameter μ and shape parameter K, for $K = 0.5(0.25)2.0(0.5)4.0$, $m = 2(1)n$ and $n = 2(1)15$. Clark's m-order-statistic estimators are the best linear unbiased estimators based on the first m order statistics of a sample of size n (all available order statistics in a right-censored sample). He then compared their efficiency with that of the best one-order-statistic estimators based on $x_{r,n}$ where $r \leq m$. Musson (1965) tabulated (to five decimal places) the coefficients in simultaneous best linear unbiased estimators of the location parameter μ and the scale parameter θ, based on two order statistics $x_{r,n}$ and $x_{s,n}$, $1 \leq r < s \leq m$, and on the first m order statistics of a sample of size n from a Weibull population with known shape parameter K, for $K = 0.5(0.25)2.0(0.5)4.0$, $m = 2(1)n$ and $n = 2(1)15$.

Carlton (1965) tabulated the coefficients (to five decimal places) in the best linear unbiased estimators of the scale parameter θ of a Weibull population with known shape parameter K and location parameter μ, based on ℓ of the first m order statistics of a sample of size n, for $K = 0.5(0.25)2.0(0.5)4.0$, $m = 2(1)n$ and $n = 2(1)10$, where ℓ is the smallest integer such that the ℓ-order-statistic estimator has efficiency at least $E_{\text{min}} =$ 0.80, 0.85, 0.90, 0.95, 0.98, 0.99 relative to the m-order-statistic estimator. Salisberry (1966) tabulated the coefficients (to five decimal places) in the simultaneous best linear unbiased estimators of μ and θ of a Weibull population with known shape parameter K, based on ℓ of the first m order statistics of a sample of size n, for the same choices of K, m and n and the same definition of ℓ as that used by Carlton. Herman (1968) tabulated the coefficients (to six decimal places) in the conditional best linear unbiased estimators of $\mu|\theta$ and $\theta|\mu$ for the same values of K, m and n as well as for $n = 11(1)13$, and ℓ defined the same as above. He showed that if the same ℓ order statistics are used for both conditional estimators, the coefficients of the conditional estimators can be used to obtain those of simultaneous estimators, so that the latter need not be tabulated. Niles (1970) gave theory and tables for best simultaneous linear invariant estimators of μ and θ for selected values of K from samples of size up to 40.

Karns (1963) used his computational formulas for the first two moments of Gamma order statistics to tabulate best linear unbiased estimators (which are also maximum likelihood estimators) of the scale parameter θ of the two-parameter Gamma population with known shape parameter α, based on one order statistic of a sample of size $n = 1(1)50$ for $\alpha = 0.5(0.5)4.0$. He tabulated the coefficients to six decimal places, the variances of the estimators to five decimal places, and their percentage efficiencies to two decimal places. Bruce (1964) used his unpublished tables of means, variances and covariances of Gamma order statistics discussed earlier, and reproduced in part in Table C4.2, to tabulate the best linear unbiased estimator of θ when the shape parameter α is known, based on the first m order statistics of a sample of size n for $\alpha = 1(1)3$, $n = 1(1)20$, and $\alpha = 4(1)6$, $n = 1(1)15$ with $m = 1(1)n$. He tabulated the coefficients as well as the variances of the m-order-statistic estimators to five decimal places. He also tabulated the percentage efficiencies (to two decimal places) for the one-order-statistic estimators relative to (i) the m-order-statistic estimators and (ii) the maximum likelihood estimators for the complete sample. Hill (1965) used his unpublished tables of means, variances and covariances of order statistics to tabulate the best linear unbiased estimator of θ when the shape parameter α is known, based on ℓ of the first m order statistics of a sample of size n for $\alpha = 1(1)6$, $n = 1(1)15$, $m = 1(1)n$, and ℓ equal to the smallest number of order statistics required to attain a desired efficiency (0.99, 0.98, 0.95, 0.90, 0.85 or 0.80) relative to the m-order-statistic estimators. He tabulated the coefficients to five decimal places and the relative efficiencies to three decimal places.

Musson (1965) tabulated (to five decimal places) the coefficients in simultaneous best linear unbiased estimators of the location parameter μ and the scale parameter θ based on two order statistics, $x_{r,n}$ and $x_{s,n}$ for $1 \leq r < s \leq m \leq n$, and on the first m order statistics of a sample of size n from a Gamma population with known shape parameter α, for $\alpha = 1(1)6$, $m = 2(1)n$ and $n = 2(1)15$. Salisberry (1966) tabulated (to five decimal places) the coefficients in simultaneous best linear unbiased estimators of μ and θ based on ℓ of the first m order statistics of a sample of size n for $\alpha = 1(1)6$, $m = 3(1)n$, $n = 3(1)15$, and ℓ as defined by Hill. He also tabulated the actual relative efficiencies to three decimal places.

Tables of variances of order statistics, along with tables of means, are also useful in interval estimation of parameters of various populations including the exponential, Weibull and Gamma, from one order statistic of a random sample. Harter (1963, 1964c) found expressions for exact confidence bounds, based on the mth order statistic of a sample of size n, for the parameter of a one-parameter exponential population, and tabulated the coefficients (to six decimal places) for $n = 1(1)20(2)40$ and the values of m which maximize the effectiveness (efficiency) of the upper confidence bound (confidence $1 - P$) and of the central confidence interval (confidence $1 - 2P$), for $P = 0.0001$, 0.0005, 0.001, 0.005, 0.01, 0.025, 0.1(0.1)0.5. (For details, see Chapter 4, Section 4.2, and Table D4). Quayle (1963) and Karns (1963) provided analogous results for the most efficient upper confidence bounds and central confidence intervals based on the mth order statistic of a sample of size n for the scale parameter of a Weibull population with known shape parameter $K = 0.5(0.5)4(1)8$, and for the scale parameter of a Gamma population with known shape parameter $\alpha = 0.5(0.5)4.0$, respectively. For the two-parameter exponential population, Kaminsky (1968) found exact confidence intervals for the scale parameter θ based on the best linear unbiased estimator of θ from optimally chosen order statistics.

3.2.5 References

1. Adatia, A. (1994). Extended tables of covariances of order statistics from the Weibull distribution. *Unpublished Report*, University of Regina, Regina, Saskatchewan.

2. Arnold, Barry C. and Balakrishnan, N. (1989). *Relations, Bounds and Approximations for Order Statistics.* Lecture Notes in Statistics No. 53, Springer-Verlag, New York.

3. Bain, Lee J. (1978). *Statistical Analysis of Reliability and Life-Testing Models: Theory and Methods.* Marcel Dekker, New York.

4. Bain, Lee J. and Engelhardt, Max (1991). *Statistical Analysis of Reliability and Life-Testing Models*, Second edition. Marcel Dekker, New York.

5. Balakrishnan, N. (Ed.) (1995). *Recent Advances in Life-Testing and Reliability.* CRC Press, Boca Raton, FL.

6. Balakrishnan, N. and Chan, P.S. (1993a). Extended tables of means, variances and covariances of Weibull order statistics. *Unpublished Report*, McMaster University, Hamilton, Ontario.

7. Balakrishnan, N. and Chan, P.S. (1993b). Extended tables of the best linear unbiased estimators of the location and scale parameters of the Weibull distribution based on complete and Type II censored data. *Unpublished Report*, McMaster University, Hamilton, Ontario.

8. Balakrishnan, N. and Cohen, A. Clifford (1991). *Order Statistics and Inference: Estimation Methods.* Academic Press, San Diego, CA.

9. Balasooriya, U. and Hapuarachchi, K.P. (1992). Extended tables for the moments of gamma distribution order statistics. *IEEE Transactions on Reliability* **41**, 256-264.

10. Barnett, Vic (1975). Probability plotting methods and order statistics. *Applied Statistics* **24**, 95-108.

11. Barnett, Vic (1976). Convenient probability plotting positions for the normal dis-

tribution. *Applied Statistics* **25**, 47-50.

12. Birnbaum, Z.W. and Saunders, S.C. (1958). A statistical model for life-length of materials. *Journal of the American Statistical Association* **53**, 151-160.

13. Blom, Gunnar (1958). *Statistical Estimates and Transformed Beta-Variables.* John Wiley & Sons, New York.

14. Breiter, M.C. and Krishnaiah, P.R. (1967). Tables for the Moments of Gamma Order Statistics. *ARL Technical Report 67-0166*, Aerospace Research Laboratories, Wright-Patterson Air Force Base, AD 639729.

15. Breiter, M.C. and Krishnaiah, P.R. (1968). Tables for the moments of gamma order statistics. *Sankhyā, Series B* **30**, 59-72.

16. Bruce, Richard G. (1964). Estimation of the Scale Parameter of the Gamma Distribution by Use of m Order Statistics. *Unpublished M.S. Thesis* (GRE/Math/64-2), Air Force Institute of Technology, Wright-Patterson Air Force Base, AD 610774.

17. Carlton, Howard F. (1965). Estimation of the Scale Parameter of the Weibull Probability Density Function by Use of ℓ Order Statistics. *Unpublished M.S. Thesis* (GRE/Math/65-2), Air Force Institute of Technology, Wright-Patterson Air Force Base.

18. Chernoff, Herman and Lieberman, Gerald J. (1954). Use of normal probability paper. *Journal of the American Statistical Association* **49**, 778-785.

19. Chernoff, Herman and Lieberman, Gerald J. (1956). The use of generalized probability paper for continuous distributions. *Annals of Mathematical Statistics* **27**, 806-818.

20. Clark, Lester John, Jr. (1964). Estimation of the Scale Parameter of the Weibull Probability Function by Use of One-order and m-order Statistics. *Unpublished M.S. Thesis* (GRE/Math/64-3), Air Force Institute of Technology, Wright-Patterson Air Force Base, AD 630809.

21. Cohen, A. Clifford (1991). *Truncated and Censored Samples: Theory and Applications.* Marcel Dekker, New York.

22. Cohen, A. Clifford and Whitten, Betty J. (1988). *Parameter Estimation in Reliability and Life Span Models.* Marcel Dekker, New York.

23. Davis, D.J. (1952). An analysis of some failure data. *Journal of the American Statistical Association* **47**, 113-150.

24. Davis, Robert I. (1969). Nearly Best Linear Conditional Unbiased Estimation of the Location and Scaie Parameters of the Weibull Probability Distribution by the Use of Order Statistics. *Unpublished M.S. Thesis* (GRE/Math/69-1), Air Force Institute of Technology, Wright-Patterson Air Force Base, AD 768388.

25. Davis, Robert L. (1970). Nearly Best Linear Conditional Unbiased Estimation of the Location and Scale Parameters of the Gamma Probability Distribution by the Use of Order Statistics. *Unpublished M.S. Thesis* (GRE/Math/70-2), Air Force Institute of Technology, Wright-Patterson Air Force Base, AD 776770.

26. Drenick, R.F. (1960). Mathematical aspects of the reliability problem. *Journal of the Society for Industrial and Applied Mathematics* **8**, 125-149.

27. Elkins, Robert W. (1970). Nearly Best Linear Invariant Conditional Estimation of the Location and Scale Parameters of the Weibull Probability Distribution by the Use of Order Statistics. *Unpublished M.S. Thesis* (GRE/Math/70-4), Air Force Institute of Technology, Wright-Patterson Air Force Base, AD 776764.

28. Epstein, Benjamin (1954). Truncated life tests in the exponential case. *Annals of Mathematical Statistics* **25**, 555-564.

29. Epstein, Benjamin (1956). Simple estimators of the parameters of exponential distributions when samples are censored. *Annals of the Institute of Statistical Mathematics* **8**, 15-26.

30. Epstein, Benjamin (1960a). Tests for the validity of the assumption that the underlying distribution of life is exponential: Part I. *Technometrics* **2**, 83-101.

31. Epstein, Benjamin (1960b). Tests for the validity of the assumption that the underlying distribution of life is exponential: Part II. *Technometrics* **2**, 167-183.

32. Epstein, Benjamin and Sobel, Milton (1953). Life testing. *Journal of the American Statistical Association* **48**, 486-502.

33. Epstein, Benjamin and Sobel, Milton (1954). Some theorems relevant to life testing from an exponential distribution. *Annals of Mathematical Statistics* **25**, 373-381.

34. Fowlkes, Edward B. (1987). *A Folio of Distributions: A Collection of Theoretical Quantile-Quantile Plots.* Marcel Dekker, New York.

35. Govindarajulu, Zakkula and Joshi, Madhukar (1962). Best linear unbiased estimation of location and scale parameters of Weibull distribution using ordered observations. *Report No. 102*, Statistical Laboratory, Case Institute of Technology, Cleveland, AD 409685. Revised version (1968): *Reports of Statistical Application Research, Union of Japanese Scientists and Engineers* **15**(2), 1-14.

36. Gray, Robert M. (1970). Nearly Best Linear Invariant Conditional Estimation of the Scale and Location Parameters of the Gamma Probability Distribution. *Unpublished M.S. Thesis* (GRE/Math/70-5), Air Force Institute of Technology, Wright-Patterson Air Force Base, AD 776771.

37. Greenberg, Bernard G. and Sarhan, Ahmed E. (1962a). Exponential distribution: Moments of order statistics. In *Contributions to Order Statistics* (Eds., Ahmed E. Sarhan and Bernard G. Greenberg), 343-351, John Wiley & Sons, New York.

38. Greenberg, Bernard G. and Sarhan, Ahmed E. (1962b). Exponential distribution: Best linear unbiased estimates. In *Contributions to Order Statistics* (Eds., Ahmed E. Sarhan and Bernard G. Greenberg), 352-360, John Wiley & Sons, New York.

39. Gupta, Shanti S. (1960). Order statistics from the gamma distribution. *Technometrics* **2**, 243-262.

40. Gupta, Shanti S. (1962). Gamma distribution. In *Contributions to Order Statistics* (Eds., Ahmed E. Sarhan and Bernard G. Greenberg), 431-450, John Wiley & Sons, New York.

41. Gupta, Shanti S. and Groll, Phyllis A. (1961). Gamma distribution in acceptance sampling based on life tests. *Journal of the American Statistical Association* **56**, 942-970.

42. Hald, A. (1952). *Statistical Theory with Engineering Applications.* John Wiley & Sons, New York.

43. Harter, H. Leon (1960). Expected Values of Normal Order Statistics. *ARL Technical Report 60-292*, Aeronautical Research Laboratories, Wright-Patterson Air Force Base, AD 245261.

44. Harter, H. Leon (1961a). Expected values of normal order statistics. *Biometrika* **48**, 151-165.

45. Harter, H. Leon (1961b). Estimating the parameters of negative exponential populations from one or two order statistics. *Annals of Mathematical Statistics* **32**, 1078-1090; correction: **35** (1964), 925.

46. Harter, H. Leon (1963). Exact confidence bounds, based on one order statistic, for the parameter of a one-parameter negative exponential population. In *Rider Anniversary Volume* (Ed., H. Leon Harter), 77-100, Aerospace Research Laboratories, Wright-Patterson Air Force Base.

47. Harter, H. Leon (1964a). *New Tables of the Incomplete Gamma-Function Ratio and of Percentage Points of the Chi-Square and Beta Distributions.* U.S. Government Printing Office, Washington, DC.

48. Harter, H. Leon (1964b). Expected Values of Exponential, Weibull and Gamma Order Statistics. *ARL Technical Report 64-31*, Aerospace Research Laboratories, Wright-Patterson Air Force Base, AD 436763.

49. Harter, H. Leon (1964c). Exact confidence bounds, based on one order statistic, for the parameter of an exponential population. *Technometrics* **6**, 301-317.

50. Harter, H. Leon (1967). Series expansions for the incomplete Gamma function and its derivatives. In *Blanch Anniversary Volume* (Ed., Bertram Mond), 91-97, Aerospace Research Laboratories, Wright-Patterson Air Force Base.

51. Harter, H. Leon (1984). Another look at plotting positions. *Communications in Statistics - Theory and Methods* **13**, 1613-1633.

52. Harter, H. Leon and Wiegand, Rudolf P. (1985). A Monte Carlo study of plotting positions. *Communications in Statistics - Simulation and Computation* **14**, 317-343.

53. Herman, William J. (1965). Least-Squares Conditional Estimation of the Location Parameter of Weibull Populations. *Unpublished M.S. Thesis* (GRE/Math/65-5), Air Force Institute of Technology, Wright-Patterson Air Force Base, AD 630810.

54. Hill, Thomas D. (1965). Estimation of the Scale Parameter of the Gamma Distribution by Use of ℓ Order Statistics. *Unpublished M.S. Thesis* (GRE/Math/65-5), Air Force Institute of Technology, Wright-Patterson Air Force Base, AD 630810.

55. Johnson, Norman L., Kotz, Samuel, and Balakrishnan, N. (1994). *Continuous Univariate Distributions - Volume 1*, Second edition. John Wiley & Sons, New York.

56. Joshi, Prakash C. (1979). On the moments of gamma order statistics. *Naval Research Logistics Quarterly* **26**, 675-679.

57. Kalbfleisch, John D. and Prentice, Ross L. (1980). *The Statistical Analysis of Failure Time Data.* John Wiley & Sons, New York.

58. Kaminsky, Kenneth S. (1968). Exact and Approximate Confidence Intervals for the Scale Parameter of the Exponential Distribution using Optimally Chosen Order Statistics. *Doctoral Dissertation*, Rutgers - The State University. University Microfilms, Ann Arbor, MI.

59. Karns, Robert Clay (1963). Scale Parameter Estimation of the Gamma Probability Function Based on One Order Statistic. *Unpublished M.S. Thesis* (GRE/Math/63-8), Air Force Institute of Technology, Wright-Patterson Air Force Base, AD 425223.

60. Khan, A.H., Khan, R.U., and Parvez, S. (1984). Inverse moments of order statistics from Weibull distribution. *Scandinavian Actuarial Journal*, 91-94.

61. Kimball, Bradford F. (1960). On the choice of plotting positions on probability paper. *Journal of the American Statistical Association* **55**, 546-560.

62. King, James, R. (1971). *Probability Charts for Decision Making.* Industrial Press, New York. Second edition (1981), TEAM Publications, Tamworth, NH.

63. Krishnaiah, P.R. and Sen, Pranab K. (1984). Tables for order statistics. In *Handbook of Statistics - Volume 4: Nonparametric Methods* (Eds., P.R. Krishnaiah and Pranab K. Sen), 837-935, North-Holland, Amsterdam.

64. Kulldorff, Gunnar (1963). Estimation of one or two parameters of the exponential distribution on the basis of suitably chosen order statistics. *Annals of Mathematical Statistics* **34**, 1419-1431.

65. Lawless, Jerald F. (1982). *Statistical Models and Methods for Lifetime Data.* John Wiley & Sons, New York.

66. Leone, F.C., Rutenberg, Y.H., and Topp, C.W. (1960). Order Statistics and Estimators for the Weibull Population. *AFOSR TN 60-489*, Air Force Office of Scientific Research, Washington, DC, AD 237042.

67. Lieblein, Julius (1955). On moments of order statistics from the Weibull distribution. *Annals of Mathematical Statistics* **26**, 330-333.

68. Mann, Nancy R., Schafer, R.E., and Singpurwalla, Nozer D. (1974). *Methods for Statistical Analysis of Reliability and Life Data.* John Wiley & Sons, New York.

69. Morgan, Guy A. (1968). Simultaneous Estimation of the Scale and Location Parameters of the Gamma Probability Distribution by Use of Order Statistics. *Unpublished M.S. Thesis* (GRE/Math/68-8), Air Force Institute of Technology, Wright-Patterson Air Force Base, AD 685114.

70. Murthy, V.K. (1968). A New Method of Estimating the Weibull Shape Parameter. *ARL Technical Report 68-0076*, Aerospace Research Laboratories, Wright-Patterson Air Force Base, AD 674063.

71. Musson, Thomas A. (1965). Linear Estimation of the Location and Scale Parameters of the Weibull and Gamma Probability Distributions by the Use of Order Statistics. *Unpublished M.S. Thesis* (GRE/Math/65-9), Air Force Institute of Technology, Wright-Patterson Air Force Base, AD 630811.

72. Nelson, Wayne (1982). *Applied Life Data Analysis*, John Wiley & Sons, New York.

73. Niles, John R. (1970). Best Linear Invariant Estimation of the Location and Scale Parameters of the Weibull Distribution. *Unpublished M.S. Thesis* (GRE/Math/70-6), Air Force Institute of Technology, Wright-Patterson Air Force Base, AD 776765.

74. Patel, Jagdish K. (1975). Bounds on moments of linear functions of order statistics from Weibull and other restricted families. *Journal of the American Statistical Association* **70**, 670-674.

75. Patel, Jagdish K. and Read, Campbell B. (1975). Bounds on conditional moments of Weibull and other monotone failure rate families. *Journal of the American Statistical Association* **70**, 238-244.

76. Pearson, Karl (1922). *Tables of the Incomplete* Γ*-Function.* H.M. Stationery Office, London.

77. Prescott, P. (1974). Variances and covariances of order statistics from the gamma distribution. *Biometrika* **61**, 607-613.

78. Quayle, Ronald John (1963). Estimation of the Scale Parameter of the Weibull Probability Density Function by Use of One Order Statistic. *Unpublished M.S. Thesis* (GRE/Math/63-10), Air Force Institute of Technology, Wright-Patterson Air Force Base, AD 425228.

79. Salisberry, Rex C. (1966). Simultaneous Linear Estimation of the Location and Scale Parameters of the Gamma and Weibull Distributions by the Use of Selected

Order Statistics. *Unpublished M.S. Thesis* (GRE/Math/66-8), Air Force Institute of Technology, Wright-Patterson Air Force Base.

80. Sarhan, Ahmed E. and Greenberg, Bernard G. (1957). Tables for best linear estimates by order statistics of the parameters of single exponential distributions from singly and doubly censored samples. *Journal of the American Statistical Association* **52**, 58-87.

81. Stump, Francis B. (1968). Nearly Best Linear Unbiased Estimation of the Location and Scale Parameters of the Weibull Probability Distribution by the Use of Order Statistics. *Unpublished M.S. Thesis* (GRE/Math/68-13), Air Force Institute of Technology, Wright-Patterson Air Force Base, AD 685116.

82. Sukhatme, P.V. (1937). Tests of significance for samples of the χ^2 population with 2 degrees of freedom. *Annals of Eugenics* **8**, 52-56.

83. Walter, S.D. and Stitt, L.W. (1988). Extended tables for moments of gamma distribution order statistics. *Communications in Statistics - Simulation and Computation* **17**, 471-487.

84. Weibull, Waloddi (1939). A statistical theory of the strength of materials. *Ingeniörs Vetenskaps Akademien, Handlingar*, Number 151.

85. Weibull, Waloddi (1959). Statistical Evaluation of Data from Fatigue and Creep-Rupture Tests, Part I. Fundamental Concepts and General Methods. *WADC Technical Report 59-400, Part I*, Wright Air Development Center, Wright-Patterson Air Force Base.

86. Weibull, Waloddi (1967). Moments About Smallest Sample Value. *Air Force Materials Laboratory Technical Report 67-375*, Wright-Patterson Air Force Base.

87. Wilk, M.B., Gnanadesikan, R., and Huyett, Miss M.J. (1962). Probability plots for the gamma distribution. *Technometrics* **4**, 1-15.

3.3 Other Populations

3.3.1 Lognormal Population

The probability density function of the standard lognormal random variable x is

$$f(x) = \frac{1}{\sqrt{2\pi}\,x}\, e^{-(\ln x)^2/2}, \qquad 0 < x < \infty \ , \tag{3.3.1}$$

and the cumulative distribution function is

$$F(x) = \int_0^x f(t)\, dt \ . \tag{3.3.2}$$

Then, the kth moment about the origin of the ith order statistic from a sample of size n from the standard lognormal density in (3.3.1) is given by

$$E(x_{i,n}^k) = \frac{n!}{(i-1)!(n-i)!} \int_0^\infty x^k \{F(x)\}^{i-1}\{1-F(x)\}^{n-i} f(x)\, dx \ . \tag{3.3.3}$$

A similar integral expression may be given for the product moments of order statistics as well.

Let $\Phi(\cdot)$ and $\phi(\cdot)$ denote the standard normal cumulative distribution function and probability density function, respectively. Then, by observing that $F(x) = \Phi(\ln x)$ for $x > 0$, Eq. (3.3.3) may be rewritten upon transformation as

$$E(x_{i,n}^k) = \frac{n!}{(i-1)!(n-i)!} \int_{-\infty}^\infty e^{kx} \{\Phi(x)\}^{i-1}\{1-\Phi(x)\}^{n-i} \phi(x)\, dx \ . \tag{3.3.4}$$

Similarly, the product moments of order statistics may be written as

$$E(x_{i,n}x_{j,n}) = \frac{n!}{(i-1)!(j-i-1)!(n-j)!} \int_{-\infty}^\infty \int_{-\infty}^y e^{x+y} \{\Phi(x)\}^{i-1}\{\Phi(y)-\Phi(x)\}^{j-i-1} \times\{1-\Phi(y)\}^{n-j}\phi(x)\phi(y)\, dx\, dy \ , \tag{3.3.5}$$

for $1 \le i < j \le n$.

Gupta, McDonald, and Galarneau (1974) computed the means, $E(x_{i,n})$, and the second raw moments, $E(x_{i,n}^2)$, for $1 \le i \le n$ and $n \le 20$, from Eq. (3.3.4) by numerical integration based on Lagrangian interpolation. From these values, they computed the variances of order statistics as

$$\text{var}(x_{i,n}) = E(x_{i,n}^2) - \{E(x_{i,n})\}^2 \ . \tag{3.3.6}$$

Similarly, they computed the product moments from Eq. (3.3.5) by numerical double integration for $1 \le i < j \le n$ and $n \le 20$. They tabulated the means and variances to eight decimal places, and the product moments to five decimal places for $n \le 15$ and to three decimal places for $16 \le n \le 20$. The means of order statistics, rounded to five decimal places, are presented in Table C6.1. The variances, rounded to five decimal places, and the covariances (to five decimal places), obtained as

$$\text{cov}(x_{i,n}, x_{j,n}) = E(x_{i,n}x_{j,n}) - E(x_{i,n})E(x_{j,n}) \ , \tag{3.3.7}$$

are presented in Table C6.2 for $n \le 10$. The notations used in these tables are consistent with those in tables for other populations.

3.3.2 Log-Weibull Population

If the random variable x has the two-parameter Weibull distribution (with θ and K as the scale and shape parameters, respectively), then the random variable $y = \ln x$ has the log-Weibull distribution, which is also known as the first asymptotic distribution of smallest values, with cumulative distribution function

$$F(y;\mu,\sigma) = 1 - e^{-e^{(y-\mu)/\sigma}}, \qquad -\infty < y < \infty\ , \tag{3.3.8}$$

where $\mu = \ln\theta$ is the location parameter (mode) and $\sigma = 1/K$ is the scale parameter. The corresponding probability density function is

$$f(y;\mu,\sigma) = \frac{1}{\sigma}\, e^{(y-\mu)/\sigma - e^{(y-\mu)/\sigma}}, \qquad -\infty < y < \infty\ . \tag{3.3.9}$$

Thus, the logarithmic transformation converts a scale-and-shape parameter Weibull distribution into a more convenient location-and-scale parameter log-Weibull distribution.

The kth moment about the origin of the ith order statistic from a sample of size n from the log-Weibull distribution in (3.3.8) is

$$E(y_{i,n}^k) = \frac{n!}{(i-1)!(n-i)!}\int_{-\infty}^{\infty} y^k\{F(y)\}^{i-1}\{1-F(y)\}^{n-i} f(y)\,dy\ . \tag{3.3.10}$$

Similarly, the product moments of order statistics are given by

$$\begin{aligned} E(y_{i,n}y_{j,n}) = \frac{n!}{(i-1)!(j-i-1)!(n-j)!}\int_{-\infty}^{\infty}\int_{-\infty}^{y} xy\{F(x)\}^{i-1}\{F(y)-F(x)\}^{j-i-1} \\ \times\{1-F(y)\}^{n-j} f(x)f(y)\,dx\,dy \end{aligned} \tag{3.3.11}$$

for $1 \le i < j \le n$. The variances and covariances of order statistics may then be computed as

$$\text{var}(y_{i,n}) = E(y_{i,n}^2) - \{E(y_{i,n})\}^2 \tag{3.3.12}$$

and

$$\text{cov}(y_{i,n}, y_{j,n}) = E(y_{i,n}y_{j,n}) - E(y_{i,n})E(y_{j,n})\ . \tag{3.3.13}$$

From (3.3.10) and (3.3.11), exact explicit expressions can be derived for the single and the product moments of order statistics. It is, of course, sufficient to tabulate these quantities for the standard log-Weibull distribution (case $\mu = 0$ and $\sigma = 1$) since the means for any other choice of μ and σ can be obtained by multiplying by σ and adding μ and the variances and covariances by multiplying by σ^2.

The integrals involved in the means of log-Weibull order statistics have been computed (using numerical integration) and tabulated for sample sizes up to 100. In an unpublished report [Anonymous (1951)] prepared in the Computation Laboratory of the National Bureau of Standards [see also Lieblein and Salzer (1957)], a seven decimal place table has been presented for the mean of the ith largest value in a sample of size n [for $i = 1(1)\min(n, 26)$, $n = 1(1)10(5)60(10)100$] from the first asymptotic distribution of largest values, which is simply the negative of the mean of the ith smallest value (ith order statistic) in a sample of size n from the first asymptotic distribution of smallest values (log-Weibull distribution). Lieblein (1953) derived explicit expressions (in terms of tabulated functions) for the variances and covariances of order statistics from the first extreme value distribution. Lieblein (1954) [also see Lieblein and Zelen (1956)] tabulated (to seven or eight decimal places) the means, variances and covariances of order statistics for $n = 1(1)6$. White (1964) tabulated the means (to seven decimal places) and the variances and covariances (to eight decimal places) of all order statistics for sample sizes $n = 1(1)20$ from the log-Weibull distribution. Mann (1965) independently tabulated the means (to seven decimal places) for $n = 1(1)25$ and the variances and covariances (also to seven decimal places) for $n = 1(1)20$ [later extended through $n = 25$ by Mann (1968a)]. Means and variances computed by White and Mann agree to within a unit in the seventh decimal place, as do their covariances for $n \leq 14$. There are, however, serious disagreements in their covariances for larger values of n, the maximum discrepancies (units in the seventh decimal place) being 2, 8, 34, 148, 711 and 2067 for $n = 15$, 16, 17, 18, 19 and 20, respectively. We believe that Mann's values, which have smooth differences, are much more accurate than White's, which do not have smooth differences. White (1967) tabulated the means and variances (both to seven decimal places) for $n = 1(1)50(5)100$ by using multiple-precision arithmetic. For $n = 21(1)25$, his results agree with Mann's to within 10^{-7}. Balakrishnan and Chan (1991a) extended the tables of means, variances and covariances for n up to 30. Abridged forms of these tables have been published in a paper by Balakrishnan and Chan (1992a). Means of order statistics ($i = 1(1)n$, $n = 1(1)20$), rounded to five decimal places, are presented in Table C7.1. Variances and covariances of order statistics ($i = 1(1)n$, $j = i(1)n$, $n = 1(1)10$), rounded once again to five decimal places, are presented in Table C7.2.

Mahmoud and Ragab (1975) and Provasi (1987) discussed moments of order statistics from the log-Weibull distribution further; the last reference also discusses some approximations to the means, variances and covariances of order statistics.

3.3.3 Log-Gamma Population

The probability density function of a standardized log-Gamma random variable x is given by

$$f(x) = \frac{1}{\Gamma(\kappa)}\, e^{\kappa x - e^x}, \qquad -\infty < x < \infty,\ \kappa > 0\ , \tag{3.3.14}$$

and the corresponding cumulative distribution function is

$$F(x) = I_{e^x}(\kappa), \qquad -\infty < x < \infty,\ \kappa > 0, \tag{3.3.15}$$

where $I_t(\kappa)$ is the incomplete Gamma function defined by

$$I_t(\kappa) = \int_0^t \frac{1}{\Gamma(\kappa)}\, e^{-z}\, z^{\kappa-1} dz, \qquad 0 < t < \infty,\ \kappa > 0\ . \tag{3.3.16}$$

The single moments of order statistics from a log-Gamma distribution can then be computed as

$$E(x_{i,n}^k) = \frac{n!}{(i-1)!(n-i)!} \int_{-\infty}^{\infty} x^k \{F(x)\}^{i-1}\{1-F(x)\}^{n-i} f(x)\, dx\ . \tag{3.3.17}$$

Similarly, the product moments of order statistics can be computed as

$$E(x_{i,n}x_{j,n}) = \frac{n!}{(i-1)!(j-i-1)!(n-j)!} \int_{-\infty}^{\infty}\int_{-\infty}^{y} xy\{F(x)\}^{i-1}\{F(y)-F(x)\}^{j-i-1} \times\{1-F(y)\}^{n-j} f(x)f(y)\, dx\, dy\ . \tag{3.3.18}$$

Chan (1993) computed the first two single moments from (3.3.17) and the product moments from (3.3.18) for sample sizes $n = 1(1)25$ and shape parameter $\kappa = 0.5(0.5)2.0(1.0)$

4.0(2.0)12.0; see also Balakrishnan and Chan (1995). The computations of the single moments were performed in double precision by a FORTRAN 77 program using the 96-point Gaussian integration method [see Abramowitz and Stegun (1965)]. The double integrations involved in the computations of the product moments were performed using the IMSL subroutine DTWODQ in the VAX 8640 system. The means, variances and covariances of order statistics were then tabulated (to five decimal places) for all the choices of n and κ mentioned above. Chan (1993) and Balakrishnan and Chan (1995) have also derived exact explicit expressions for the single and product moments of order statistics from Eqs. (3.3.17) and (3.3.18), respectively, for integral values of κ. The means of order statistics for $n \le 10$ are presented in Table C8.1, and the variances and covariances of order statistics for $n \le 10$ are presented in Table C8.2.

3.3.4 Cauchy Population

The probability density function of a standard Cauchy random variable is

$$f(x) = \frac{1}{\pi(1+x^2)}\ , \qquad -\infty < x < \infty\ , \tag{3.3.19}$$

and the corresponding cumulative distribution function is

$$F(x) = \frac{1}{2} + \frac{1}{\pi}\ \tan^{-1} x\ , \qquad -\infty < x < \infty\ . \tag{3.3.20}$$

Then, the single moments of order statistics can be written as

$$E(x_{i,n}^k) = \frac{n!}{(i-1)!(n-i)!\pi^n} \int_{-\pi/2}^{\pi/2} (\tan x)^k \left(\frac{\pi}{2}+x\right)^{i-1} \left(\frac{\pi}{2}-x\right)^{n-i} dx\ . \tag{3.3.21}$$

Similarly, the product moments of order statistics can be written (for $i < j$) as

$$E(x_{i,n}x_{j,n}) = \frac{n!}{(i-1)!(j-i-1)!(n-j)!\pi^n} \int_{-\pi/2}^{\pi/2} \int_{-\pi/2}^{y} \tan x \tan y \left(\frac{\pi}{2}+x\right)^{i-1} \times (y-x)^{j-i-1} \left(\frac{\pi}{2}-y\right)^{n-j} dx\, dy\ . \tag{3.3.22}$$

From Eq. (3.3.21), it may be noted that the means of order statistics do not exist when $i = 1$ or n, and that the second raw moments of order statistics do not exist when $i = 1, 2, n-1$ or n. Thus, the means of the first and last order statistics do not exist, while the variances of the first two and last two order statistics do not exist. By making use of the standard Cauchy density function in (3.3.19), Barnett (1966) established the simple recurrence relation

$$E(x_{i,n}^2) = \frac{n}{\pi}\left\{E(x_{i,n-1}) - E(x_{i-1,n-1})\right\} - 1 \tag{3.3.23}$$

so that

$$\mathrm{var}(x_{i,n}) = \frac{n}{\pi}\left\{E(x_{i,n-1}) - E(x_{i-1,n-1})\right\} - 1 - \left\{E(x_{i,n})\right\}^2\ . \tag{3.3.24}$$

Barnett (1966) computed the means of standard Cauchy order statistics from Eq. (3.3.21) by numerical integration using a composite Simpson procedure. Because of the symmetry of the Cauchy density function in (3.3.19), we have $E(x_{i,n}) = -E(x_{n-i+1,n})$ and hence only one half of them need to be tabulated. Barnett tabulated (to four decimal places) the values of $-E(x_{2,n}), \ldots, -E(x_{[\frac{n+1}{2}],n})$ for $n = 3(1)20$, and then obtained the variances of order statistics by using Eq. (3.3.24). Evaluation of the covariances was more difficult, especially when $i = 3$ and $j = 4$. Except for this anomalous case, Barnett (1966) used a two-dimensional extension of the composite Simpson procedure to evaluate numerically the double integrals in (3.3.22) to find the product moments $E(x_{i,n}x_{j,n})$ and then computed the covariances of order statistics as

$$\mathrm{cov}(x_{i,n}, x_{j,n}) = E(x_{i,n}x_{j,n}) - E(x_{i,n})E(x_{j,n}) . \tag{3.3.25}$$

For the case when $i = 3$ and $j = 4$, Barnett (1966) transformed the integral and wrote

$$E(x_{3,n}x_{4,n}) = \frac{n!}{(n-4)!2\pi^n} \int_{-\infty}^{\infty} \frac{y}{1+y^2} \left(\frac{\pi}{2} - \tan^{-1} y\right)^{n-4} \times \int_{-\pi/2}^{\tan^{-1} y} \tan x \left(\frac{\pi}{2} + x\right)^2 dx\, dy . \tag{3.3.26}$$

He then evaluated the inner integral by Gaussian quadrature, and used a composite Simpson procedure for the outer integral, and then used Eq. (3.3.25) to find $\mathrm{cov}(x_{3,n}, x_{4,n})$, In this way, he tabulated the covariances of order statistics, $\mathrm{cov}(x_{i,n}, x_{j,n})$, to four decimal places, for $n = 5(1)16(2)20$, $i = 3(1)\left[\frac{n+1}{2}\right]$ and $j = (i+1)(1)(n-i+1)$. His table of means has been reproduced in Table C9.1 for sample sizes up to ten, and the table of variances and covariances has been reproduced in Table C9.2 once again for samples sizes up to ten.

Vaughan (1993) derived explicit expressions (as infinite series involving Bernoulli numbers) for the single and the product moments of order statistics. Joshi (1969) and Hall (1978) discussed some bounds and approximations for the single moments of order statistics.

3.3.5 Double Exponential Population

The probability density function of the standard double exponential population is

$$f(x) = \frac{1}{2}\, e^{-|x|}, \qquad -\infty < x < \infty , \tag{3.3.27}$$

and the cumulative distribution function is

$$\begin{aligned} F(x) &= \tfrac{1}{2}\, e^{x}, \qquad -\infty < x \le 0 \\ &= 1 - \tfrac{1}{2}\, e^{-x}, \; 0 \le x < \infty . \end{aligned} \tag{3.3.28}$$

Govindarajulu (1963) established some recurrence relations which express the single and the product moments of order statistics from a symmetric distribution (like the double exponential) in terms of the single and the product moments of order statistics from its folded distribution (like the exponential). Then, by making use of these relationships and the explicit expressions of means, variances and covariances of exponential order statistics presented in subsection 3.2.2, Govindarajulu (1966) derived the following closed-form expressions for the means, second raw moments, and product moments of order statistics from the standard double exponential density in (3.3.27):

$$E(x_{i,n}) = 2^{-n} \left\{ \sum_{m=n-i+1}^{n} \binom{n}{m} S_1(m, n-i+1) - \sum_{m=i}^{n} \binom{n}{m} S_1(m, i) \right\}, \tag{3.3.29}$$

$$E(x_{i,n}^2) = 2^{-n}\left[\sum_{m=n-i+1}^{n}\binom{n}{m}\{S_2(m,n-i+1)+S_1^2(m,n-i+1)\}\right.$$
$$\left.+\sum_{m=i}^{n}\binom{n}{m}\{S_2(m,i)+S_1^2(m,i)\}\right], \tag{3.3.30}$$

and

$$E(x_{i,n}x_{j,n}) = 2^{-n}\left[\sum_{m=n-i+1}^{n}\binom{n}{m}\{S_2(m,n-i+1)+S_1(m,n-i+1)S_1(m,n-j+1)\}\right.$$
$$-\sum_{m=i}^{j-1}\binom{n}{m}S_1(m,i)S_1(n-j+1,n-m)$$
$$\left.+\sum_{m=j}^{n}\binom{n}{m}\{S_2(m,i)+S_1(m,i)S_1(m,j)\}\right]. \tag{3.3.31}$$

In the above equations, $S_r(j,k) = \sum_{m=j}^{k} m^{-r}$ (for $r = 1,2$) and $S_r(j_2, j_1)$ for $j_2 \geq j_1$ is interpreted as $S_r(j_1, j_2)$. After computing the second single moments and the product moments from (3.3.30) and (3.3.31), one can readily calculate variances and covariances of order statistics. Govindarajulu (1966) tabulated the means (to six decimal places) for $i = 1 + \left[\frac{n+1}{2}\right], \ldots, n$ and $n = 2(1)20$. Values of means of the other order statistics (for $i \leq [\frac{n+1}{2}]$) can be obtained simply by the symmetry relation $E(x_{i,n}) = -E(x_{n-i+1,n})$. The values of means for $n = 30(10)100$ are available in a technical report by Govindarajulu (1964) and the values for sample sizes between the tabulated ones can be obtained by use of the recurrence formula

$$E(x_{i,n-1}) = \frac{1}{n}\{i\,E(x_{i+1,n}) + (n-i)\,E(x_{i,n})\} \tag{3.3.32}$$

without accumulating serious round-off errors. Govindarajulu (1966) also tabulated variances and covariances of order statistics for $1 \leq i \leq j \leq n$, $2 \leq i+j \leq n+1$ and $n = 2(1)20$. The other values of variances and covariances can be obtained from the doubly symmetric form of the variance-covariance matrix, i.e.

$$\text{cov}(x_{i,n}, x_{j,n}) = \text{cov}(x_{j,n}, x_{i,n}) = \text{cov}(x_{n-j+1,n}, x_{n-i+1,n})\,,$$
$$1 \leq i \leq j \leq n\,. \tag{3.3.33}$$

Govindarajulu's (1966) tabular values agree, to within a unit in the last decimal place, with the values given by Sarhan (1954) for sample sizes up to five. Table C10.1 gives means of order statistics for $i = 1 + [\frac{n+1}{2}], \ldots, n$ and $n = 1(1)20$ to five decimal places (by rounding Govindarajulu's values). Table C10.2 gives variances and covariances of order statistics for $i = 1(1)[\frac{n+1}{2}]$, $j = i(1)(n-i+1)$ and $n = 1(1)15$ to five decimal places (by rounding Govindarajulu's values).

3.3.6 Half Normal Population

The probability density function of the standard half normal population, also known as the chi (1 degree of freedom) population, is

$$f(x) = \sqrt{\frac{2}{\pi}}\, e^{-x^2/2}, \qquad x \geq 0\,, \tag{3.3.34}$$

and the cumulative distribution function is

$$F(x) = \int_0^x f(t)\,dt = 2\Phi(x) - 1\ , \qquad x \ge 0\ , \tag{3.3.35}$$

where $\Phi(\cdot)$ denotes the standard normal cumulative distribution function. For more details of this distribution, one may refer to Johnson, Kotz, and Balakrishnan (1994).

The single moments of order statistics are then given by

$$E(x_{i,n}^k) = \frac{n!}{(i-1)!(n-i)!}\int_0^\infty x^k\{F(x)\}^{i-1}\{1-F(x)\}^{n-i}f(x)\,dx\ , \tag{3.3.36}$$

and the product moments of order statistics are given by

$$E(x_{i,n}x_{j,n}) = \frac{n!}{(i-1)!(j-i-1)!(n-j)!}\int_0^\infty\int_0^y xy\{F(x)\}^{i-1}\{F(y)-F(x)\}^{j-i-1} \times\{1-F(y)\}^{n-j}f(x)\,f(y)\,dx\,dy \tag{3.3.37}$$

for $1 \le i < j \le n$. Govindarajulu (1962) used Eqs. (3.3.36) and (3.3.37), along with some basic recurrence relations, to evaluate the means of order statistics for sample sizes up to four and the second single moments and the product moments for sample sizes up to 5. He also computed the values of variances and covariances of order statistics for sample sizes up to four and tabulated them to eight decimal places.

Govindarajulu and Eisenstat (1965) tabulated the means of all half normal order statistics to six decimal places for $n = 1(1)20(10)50$, and to five decimal places for $n = 60(10)100$. Table C11.1 gives the means of order statistics to five decimal places for $n = 1(1)20$ (by rounding Govindarajulu and Eisenstat's values). They also tabulated the variances and covariances of all order statistics to five decimal places for $i = 1(1)n$, $j = i(1)n$ and $n = 1(1)20$. Table C11.2 presents these values of variances and covariances of half normal order statistics for sample sizes up to 10.

3.3.7 Logistic Population

The probability density function of a standard logistic population is

$$f(x) = e^{-x}/(1+e^{-x})^2, \qquad -\infty < x < \infty\ , \tag{3.3.38}$$

and the cumulative distribution function is

$$F(x) = 1/(1+e^{-x}), \qquad -\infty < x < \infty\ . \tag{3.3.39}$$

In this case, the first two single moments of order statistics can be easily shown to be [see Balakrishnan (1992)]

$$E(x_{i,n}) = \psi(i) - \psi(n-i+1), \qquad 1 \le i \le n \tag{3.3.40}$$

and

$$E(x_{i,n}^2) = \psi'(i) + \psi'(n-i+1) + \{\psi(i) - \psi(n-i+1)\}^2,\ 1 \le i \le n\ , \tag{3.3.41}$$

where $\psi(z) = \frac{d}{dz}\ln\Gamma(z) = \frac{\Gamma'(z)}{\Gamma(z)}$ is the psi or digamma function, and $\psi'(z) = \frac{d}{dz}\psi(z) = \frac{d^2}{dz^2}\ln\Gamma(z)$ is the trigamma function. From Eqs. (3.3.40) and (3.3.41), variances of order statistics are readily obtained as

$$\mathrm{var}(x_{i,n}) = E(x_{i,n}^2) - \{E(x_{i,n})\}^2 = \psi'(i) + \psi'(n-i+1)\ . \tag{3.3.42}$$

Similarly, the product moments of order statistics are given by

$$E(x_{i,n}x_{j,n}) = \{E(x_{j,n})\}^2 + \sum_{r=1}^{j-1}\sum_{s=1}^{r-1}(-1)^{r+i}\binom{r-1}{i-1}\binom{n}{r}\binom{j-i-r+s}{s}$$
$$\times B(s, n-r+1)E(x_{j+s-r,n+s-r})$$
$$+\binom{n}{r}\sum_{r=0}^{j-i-1}(-1)^r\binom{n-i}{r}\frac{1}{i+r}[-\psi'(n-j+1)$$
$$+\{\psi(n-j+1)-\psi(n-i-r+1)\}\{\psi(j-i-r)-\psi(n-j+1)\}] \,. \tag{3.3.43}$$

The covariances of order statistics may then be computed as

$$\operatorname{cov}(x_{i,n}, x_{j,n}) = E(x_{i,n}x_{j,n}) - E(x_{i,n})E(x_{j,n}) \,. \tag{3.3.44}$$

Due to the symmetry of the standard logistic density function in (3.3.38), we also have $E(x_{i,n}) = -E(x_{n-i+1,n})$, $\operatorname{var}(x_{i,n}) = \operatorname{var}(x_{n-i+1,n})$, and $\operatorname{cov}(x_{i,n}, x_{j,n}) = \operatorname{cov}(x_{n-j+1,n}, x_{n-i+1,n})$.

Balakrishnan and Cohen (1991) and Balakrishnan (1992) have presented several recurrence relations satisfied by the single and the product moments of logistic order statistics. These recurrence relations, established originally by Shah (1966, 1970), enable one to compute the single and the product moments of all order statistics in a very simple recursive manner without accumulating serious rounding errors. Balakrishnan and Malik (1995) employed this recursive procedure to tabulate (to ten decimal places) the means, variances and covariances of logistic order statistics for sample sizes up to fifty. Table C12.1 of means of order statistics for sample sizes up to 20 and $i = 1 + [\frac{n+1}{2}], \ldots, n$, and Table C12.2 of variances and covariances of order statistics for sample sizes up to 10 and $i = 1, \ldots, [\frac{n+1}{2}]$ and $j = i, \ldots, n-i+1$ were obtained by rounding the results of Balakrishnan and Malik to five decimal places.

Less extensive tables were presented earlier by several authors. Kjelsberg (1962) used the results of Plackett (1958), who obtained the first four cumulants of order statistics from a standard logistic distribution, to tabulate exact values of the means and variances of logistic order statistics for sample sizes up to 5. Birnbaum and Dudman (1963) tabulated the means and standard deviations of logistic order statistics to five decimal places for $n = 1(1)10$ and to four decimal places for $n = 15$, 20, 50 (all order statistics) and 100 (selected order statistics). Gupta and Shah (1965) presented exact expressions for the first four raw moments of order statistics for sample sizes $n = 1(1)10$. Shah (1966) tabulated the variances and covariances to eight decimal places for $n = 1(1)10$, and Gupta, Qureishi, and Shah (1965) extended these tables for $n = 11(1)25$. Gupta, Qureishi, and Shah (1967) gave an abridged table for $n = 10(5)25$.

3.3.8 Pareto Population

The probability density function of the Pareto population with scale parameter a and shape parameter ν is

$$f(x) = \nu a^\nu x^{-\nu-1}, \qquad x \geq a,\ a > 0,\ \nu > 0 \,, \tag{3.3.45}$$

and cumulative distribution function is

$$F(x) = 1 - a^\nu x^{-\nu}, \qquad x \geq a,\ a > 0,\ \nu > 0 \,. \tag{3.3.46}$$

Malik (1966) derived the following expressions for the means, variances and covariances of order statistics from the Pareto density function in (3.3.45):

$$E(x_{i,n}) = \frac{\Gamma(n+1)\Gamma(n-i+1-\frac{1}{\nu})a}{\Gamma(n-i+1)\Gamma(n+1-\frac{1}{\nu})}, \qquad \nu > 1, \tag{3.3.47}$$

$$\text{var}(x_{i,n}) = \frac{\Gamma(n+1)\Gamma(n-i+1-\frac{2}{\nu})a^2}{\Gamma(n-i+1)\Gamma(n+1-\frac{2}{\nu})} - \left\{\frac{\Gamma(n+1)\Gamma(n-i+1-\frac{1}{\nu})a}{\Gamma(n-i+1)\Gamma(n+1-\frac{1}{\nu})}\right\}^2, \; \nu > 2, \tag{3.3.48}$$

and

$$\begin{aligned}\text{cov}(x_{i,n}, x_{j,n}) = & \frac{\Gamma(n+1)\Gamma(n-j+1-\frac{1}{\nu})\Gamma(n-i+1-\frac{2}{\nu})a^2}{\Gamma(n-j+1)\Gamma(n-i+1-\frac{1}{\nu})\Gamma(n+1-\frac{2}{\nu})} \\ & - \frac{\{\Gamma(n+1)\}^2\Gamma(n-i+1-\frac{1}{\nu})\Gamma(n-j+1-\frac{1}{\nu})a^2}{\Gamma(n-i+1)\Gamma(n-j+1)\{\Gamma(n+1-\frac{1}{\nu})\}^2}, \; \nu > 2.\end{aligned} \tag{3.3.49}$$

Reference may also be made to Huang (1975).

Malik (1966) tabulated the means, variances and covariances of the Pareto order statistics (to four decimal places) for $\nu = 2.5(0.5)5.0$ for sample sizes up to 8 (taking the scale parameter $a = 1$). Table C13.1 of means and Table C13.2 of variances and covariances of Pareto order statistics have been reproduced from Malik's tables.

3.3.9 Student's t Population

Tiku and Kumra (1985) consider the family of symmetric distributions with probability density function

$$f(x) = C\left(1+\frac{x^2}{k}\right)^{-p}, \qquad -\infty < x < \infty, \tag{3.3.50}$$

where $k = 2p-3$, $p \geq 2$, and $C = \frac{\Gamma(p)}{\sqrt{2}}\left\{\sqrt{\pi(p-\frac{3}{2})}\;\Gamma(p-\frac{1}{2})\right\}^{-1}$. This family has mean 0, standard deviation 1, and kurtosis $\alpha_4 = 3\left(p-\frac{3}{2}\right)/\left(p-\frac{5}{2}\right)$ which assumes values between 3 and ∞. As $p \to \infty$, the density in (3.3.50) reduces to the standard normal density function. Since the distribution of $\sqrt{\frac{2p-1}{k}}\,x$ is Student's t with $\nu = 2p-1$ degrees of freedom, the probability integral is

$$F(x) = C\int_{-\infty}^{x}\left(1+\frac{z^2}{k}\right)^{-p} dz = \Pr\left(t_\nu < \sqrt{\frac{2p-1}{k}}\,x\right). \tag{3.3.51}$$

Then the means of order statistics are given by

$$E(x_{i,n}) = \frac{C\,n!}{(i-1)!(n-i)!}\int_{-\infty}^{\infty} x\left(1+\frac{x^2}{k}\right)^{-p}\{F(x)\}^{i-1}\{1-F(x)\}^{n-i}dx, \tag{3.3.52}$$

which under the transformation $x = \sqrt{k}\tan(\pi u/2)$ reduces to

$$E(x_{i,n}) = \frac{C\,n!k\pi}{4(i-1)!(n-i)!}\int_{-1}^{1}\sin(\pi u)\cos^q(\pi u/2)\{F(u)\}^{i-1}\{1-F(u)\}^{n-i}du. \tag{3.3.53}$$

Similarly, the second single moments of order statistics are given by

$$E(x_{i,n}^2) = \frac{C\,n!k\sqrt{k}\pi}{2(i-1)!(n-i)!}\int_{-1}^{1}\sin^2(\pi u)\cos^q(\pi\,u/2)\{F(u)\}^{i-1}\{1-F(u)\}^{n-i}du\ , \tag{3.3.54}$$

and the product moments of order statistics are given by

$$E(x_{i,n}x_{j,n}) = \frac{C^2\,n!(k\pi)^2}{16(i-1)!(j-i-1)!(n-j)!}\int_{-1}^{1}\int_{-1}^{v}\sin(\pi u)\sin(\pi v)\cos^q(\pi\,u/2)$$
$$\times\cos^q(\pi\,v/2)\{F(u)\}^{i-1}\{F(v)-F(u)\}^{j-i-1}\{1-F(v)\}^{n-j}du\,dv, \tag{3.3.55}$$

where $q = 2p - 4$.

Tiku and Kumra (1985) evaluated the single integrals in (3.3.53) and (3.3.54) numerically by Simpson's method, and evaluated the double integrals in (3.3.55) numerically by 32-point Gaussian quadrature. They tabulated the means and variances of order statistics for $i = 1(1)[\frac{n+1}{2}]$, $n = 1(1)20$, $p = 2.0(0.5)10.0$ [3(1)19 degrees of freedom] and the covariances of order statistics for $i = 1(1)[\frac{n+1}{2}]$, $j = (i+1)(1)(n-i+1)$, $n = 1(1)20$, $p = 2.0(0.5)10.0$ [3(1)19 degrees of freedom]. Because of the symmetry of Student's t distribution, missing values in these tables can be found from the symmetry relations

$$E(x_{i,n}) = -E(x_{n-i+1,n})\ , \tag{3.3.56}$$

$$\text{var}(x_{i,n}) = \text{var}(x_{n-i+1,n})\ , \tag{3.3.57}$$

and

$$\text{cov}(x_{i,n}, x_{j,n}) = \text{cov}(x_{n-j+1,n}, x_{n-i+1,n})\ . \tag{3.3.58}$$

In Table C14.1, means of Student's t order statistics are presented to five decimal places (by rounding Tiku and Kumra's values) for $i = 1(1)[\frac{n+1}{2}]$, $n = 1(1)20$ and degrees of freedom $\nu = 3(1)19$. In Table C14.2, variances and covariances of Student's t order statistics are presented to five decimal places (by rounding Tiku and Kumra's values) for $i = 1(1)[\frac{n+1}{2}]$, $j = i(1)(n-i+1)$, $n = 1(1)10$ and degrees of freedom $\nu = 3(1)19$.

3.3.10 Half Logistic Population

The standard half logistic population has probability density function

$$f(x) = 2\,e^{-x}/(1+e^{-x})^2, \qquad 0 \le x < \infty\ , \tag{3.3.59}$$

and cumulative distribution function

$$F(x) = (1-e^{-x})/(1+e^{-x}), \qquad 0 \le x < \infty\ . \tag{3.3.60}$$

The density function in (3.3.59) is simply the density function of the absolute value of a standard logistic random variable with density function (3.3.38).

Balakrishnan (1985) established several recurrence relations for the single moments and the product moments of order statistics from the standard half logistic distribution in (3.3.60). He used these recurrence relations, starting with $E(x) = \ln 4$ and $E(x^2) = \pi^2/3$ (the first two single moments for a sample of size 1), and the fact that $E(x_{1,2}x_{2,2}) = E(x_1x_2) = (\ln 4)^2$ (the product moment for a sample of size 2), to compute the means, variances and covariances of order statistics (to five decimal places) for all sample sizes up to 15. Table C15.1 of means of order statistics and Table C15.2 of variances and covariances of order statistics of the standard half logistic distribution in (3.3.60) have been reproduced from Balakrishnan's (1985) tables.

3.3.11 Generalized Logistic Population

The standard Type I generalized logistic population has probability density function

$$f(x) = b\, e^{-x}/(1+e^{-x})^{b+1}, \qquad -\infty < x < \infty,\ b > 0\ , \tag{3.3.61}$$

and cumulative distribution function

$$F(x) = 1/(1+e^{-x})^{b}, \qquad -\infty < x < \infty,\ b > 0\ . \tag{3.3.62}$$

The case $b = 1$ in (3.3.61) corresponds to the standard logistic density function in (3.3.38).
Then the single moments of order statistics are given by

$$E(x_{i,n}^k) = \frac{n!}{(i-1)!(n-i)!}\int_{-\infty}^{\infty} x^k \{F(x)\}^{i-1}\{1-F(x)\}^{n-i} f(x)\,dx\ , \tag{3.3.63}$$

and the product moments of order statistics are given by

$$\begin{aligned}E(x_{i,n}x_{j,n}) = {} & \frac{n!}{(i-1)!(j-i-1)!(n-j)!}\int_{-\infty}^{\infty}\int_{-\infty}^{y} xy\{F(x)\}^{i-1}\{F(y)-F(x)\}^{j-i-1} \\ & \times\{1-F(y)\}^{n-j} f(x) f(y)\,dx\,dy\ .\end{aligned} \tag{3.3.64}$$

From (3.3.63), Balakrishnan and Leung (1988a) derived the following explicit expressions for the first two single moments of order statistics:

$$E(x_{i,n}) = \frac{n!}{(i-1)!(n-i)!}\sum_{r=0}^{n-i}(-1)^r\binom{n-i}{r}\frac{1}{i+r}\,\{\psi(b(i+r)) - \psi(1)\}\ , \tag{3.3.65}$$

and

$$\begin{aligned}E(x_{i,n}^2) = {} & \frac{n!}{(i-1)!(n-i)!}\sum_{r=0}^{n-i}(-1)^r\binom{n-i}{r}\frac{1}{i+r}\,\{[\psi'(b(i+r)) + \psi'(1)] \\ & +[\psi(b(i+r)) - \psi(1)]^2\}\ ,\end{aligned} \tag{3.3.66}$$

where, as in the corresponding Eqs. (3.3.40) and (3.3.41) for the logistic population case, $\psi(\cdot)$ and $\psi'(\cdot)$ are the digamma and trigamma functions, respectively. Then the variances of order statistics are given by

$$\mathrm{var}(x_{i,n}) = E(x_{i,n}^2) - \{E(x_{i,n})\}^2\ . \tag{3.3.67}$$

Balakrishnan and Leung (1988a) similarly derived an explicit expression for the product moments of order statistics as

$$\begin{aligned}& E(x_{i,n}x_{j,n}) \\ & = \frac{\Gamma(n+1)}{\Gamma(i)}\left[\frac{\Gamma(i)}{b^2\Gamma(n+1)}\,[\{\psi(i)-\psi(n+1)\}\{\psi(j)-\psi(n+1)\} + \{\psi'(j)-\psi'(n+1)\}]\right. \\ & \quad + \frac{\Gamma(i)}{b\,\Gamma(j)}\sum_{s=1}^{\infty}\frac{\Gamma(j+\frac{s}{b})}{s\,\Gamma(n+1+\frac{s}{b})}\left[\psi(i) - \psi(j) + \psi\left(j+\frac{s}{b}\right) - \psi\left(n+1+\frac{s}{b}\right)\right]\end{aligned}$$

$$+\frac{1}{b}\sum_{r=1}^{\infty}\frac{\Gamma(i+\frac{r}{b})}{r\,\Gamma(n+1+\frac{r}{b})}\left[\psi\left(j+\frac{r}{b}\right)-\psi\left(n+1+\frac{r}{b}\right)\right]$$

$$\left.+\sum_{r=1}^{\infty}\sum_{s=1}^{\infty}\frac{\Gamma(i+\frac{r}{b})}{rs\,\Gamma(j+\frac{r}{b})}\,\frac{\Gamma(j+\frac{r+s}{b})}{\Gamma(n+1+\frac{r+s}{b})}\right] . \tag{3.3.68}$$

Then the covariances of order statistics are given by

$$\mathrm{cov}(x_{i,n}, x_{j,n}) = E(x_{i,n}x_{j,n}) - E(x_{i,n})E(x_{j,n}) . \tag{3.3.69}$$

Balakrishnan and Leung (1988b) used Eqs. (3.3.65)-(3.3.69) to tabulate (to five decimal places) the means, variances and covariances of order statistics from the Type I generalized logistic distribution in (3.3.62) for $n = 1(1)10, 15$ and $b = 1.0(0.5)5.0(1.0)8.0$. Table C16.1 of means of order statistics and Table C16.2 of variances and covariances of order statistics from the Type I generalized logistic distribution for sample sizes up to 10 have been reproduced from the tables of Balakrishnan and Leung (1988b). Values for the case $b = 1$ (standard logistic distribution) have been omitted in these tables since they are the same as those presented in Tables C12.1 and C12.2.

3.3.12 Applications of Tables

As in the case of the populations discussed earlier in Sections 3.1 and 3.2, one of the primary applications of tables of means, variances and covariances of order statistics from the populations discussed in this section is in the point estimation of the location and scale parameters of the population under consideration. If tables of variances and covariances are available in addition to tables of means, they can be used to determine the coefficients of the order statistics in the best linear unbiased or best linear invariant estimators and the variances and covariance of these estimators. If tables are available only for the means, they can be used to determine the coefficients of the order statistics in the nearly best linear unbiased or nearly best linear invariant estimators. Other applications of the tables of means, variances and covariances of order statistics include interval estimation and related hypothesis tests, plotting on probability paper, and studies of the bias and efficiency of linear "robust" estimators.

Best Linear Unbiased Estimators

Highfield (1964) presented best linear unbiased estimators of the scale parameter σ of a one-parameter lognormal population based on one order statistic and on the first m order statistics of a sample of size $n \geq m$. Highfield tabulated the coefficients of the order statistics and the values of the (variance of the estimator)/σ^2 to six decimal places and the percentage efficiencies (to two decimal places) of the estimators for $n = 2(1)15$ and $m = 1(1)n$. Among the one-order-statistic estimators, the one based on the first order statistic is the most efficient. For the two-parameter lognormal population, Gupta, McDonald, and Galarneau (1974) discussed the best linear unbiased estimators of the location and scale parameters, μ and σ, respectively, based on complete as well as Type II censored samples. They did not present any tables, but presented a numerical example for sample size $n = 5$. Galarneau and McDonald (1974) presented the coefficients of the order statistics and covariance matrix of the BLUEs for sample sizes up to 15 and all possible choices of right and left censoring. Gibbons and McDonald (1975) tabulated the coefficients (to four decimal places) and the variances and covariance of the estimators (to three decimal places) for complete samples of sizes up to 15.

Lieblein and Zelen (1956) tabulated (to seven decimal places) the coefficients of the best linear unbiased estimators of μ and σ of the log-Weibull population, based on the first m order statistics of a sample of size n for $m = 2(1)n$ and $n = 2(1)6$. White (1964) extended the tables of coefficients for n up to 20, and Mann (1965) independently did the same. Coefficients tabulated by White and by Mann agree to within 10^{-7} for $n \leq 12$ and to within 2×10^{-7} for $n = 13$, except for a few misprints in White's tables. As in the case of the covariances of order statistics (see subsection 3.3.2), which are used in calculating the coefficients of order statistics in the best linear unbiased estimators, there are serious disagreements in their coefficients for $n > 13$. Once again, Mann's results having smooth differences are believed to be more accurate than White's results which do not have smooth differences. Nevertheless, estimators using White's coefficients are unbiased and their variances are only slightly larger than those of estimators using Mann's more accurate coefficients, since the variances (and covariance) of linear estimators are not very sensitive to moderate changes in the coefficients as long as the estimators remain unbiased. Both White and Mann tabulated the variances and covariance of their estimators, which agree to within 10^{-6} for $n \leq 19$, with a maximum discrepancy of 6×10^{-6} for $n = 20$. Balakrishnan and Chan (1991b) presented tables of coefficients of order statistics in the best linear unbiased estimators and the variances and covariance of these estimators for n up to 30 based on complete as well as censored samples. An abridged version of these tables is presented in the paper by Balakrishnan and Chan (1992b).

Richardson (1966) tabulated (to five decimal places) the coefficients in the best simultaneous linear unbiased estimators of the parameters of the log-Weibull distribution, based on ℓ out of the first m order statistics of a sample of size n for $n = 3(1)20$, $m = 2(1)n$ and $\ell = 2(1)\ell_s$, where ℓ_s is the smallest value of ℓ that results in an efficiency of at least 99% relative to the m-order-statistic estimator presented by White (1964) and Mann (1965). Shelnutt (1966) [also see Shelnutt, Moore, and Harter (1973)] tabulated (to five decimal places) the coefficients in the best conditional linear estimators of the scale parameter σ when the location parameter μ is known, based on ℓ of the first m order statistics of a sample of size n for $n = 1(1)20$, $m = 1(1)n$, and $\ell = 1(1)\ell_c$, where ℓ_c is the smallest value of ℓ that results in an efficiency of at least 99% relative to the m-order-statistic estimator. Elwer (1967) tabulated (to five decimal places) the coefficients of order statistics in the best conditional linear unbiased estimators of $\sigma|\mu$, based on ℓ_m of the $(m_2 - m_1 + 1)$ available order statistics, $x_{m_1,n}, x_{m_1+1,n}, \ldots, x_{m_2,n}$, of a complete, singly censored or doubly censored sample of size n for $n = 1(1)20$, $m_1 \geq 1$, $m_2 \leq n$, and $\ell_m = 1(1)\ell_{mc}$, where ℓ_{mc} is the smallest value of ℓ_m that results in an efficiency of at least 99% relative to the $(m_2 - m_1 + 1)$-order-statistic estimator. Fratzke (1967) tabulated (to five decimal places) the coefficients in the best simultaneous linear unbiased estimators of μ and σ, based on ℓ of the $m = n - r_1 - r_2$ available order statistics $x_{r_1+1,n}, x_{r_1+2,n}, \ldots, x_{n-r_2,n}$ of a complete, singly censored, or doubly censored sample of size n for $n = 2(1)18$, $r_1 \geq 0$, $r_2 \geq 0$ and $\ell = 2(1)\ell_{ms}$, where ℓ_{ms} is the smallest value of ℓ that results in an efficiency of at least 99% relative to the m-order-statistic estimator. Hassanein (1969a) tabulated (to six decimal places) the coefficients in the best simultaneous linear unbiased estimators of μ and σ, based on two or three order statistics of a sample of size $n = 3(1)20$. He used the same order statistics for both parameters and minimized $\text{var}(\tilde{\theta})$, where

$$\theta = \begin{pmatrix} \mu \\ \sigma \end{pmatrix} \quad \text{and} \quad \text{var}(\tilde{\theta}) = \text{var}(\tilde{\mu}) + \text{var}(\tilde{\sigma}) \; . \tag{3.3.70}$$

Richardson (1966), on the other hand, minimized the variance of each estimator separately using (in most cases) different order statistics. In the few cases where they used the same order statistics, Hassanein's six decimal-place values round to Richardson's five decimal-place values. Hassanein also tabulated (to six decimal places) the coefficients in the best conditional linear unbiased estimators of $\mu|\sigma$ and of $\sigma|\mu$, based on two order statistics of a complete sample of size $n = 3(1)20$. Hassanein's six decimal-place coefficients in the estimator of $\sigma|\mu$ round to Shelnutt's five decimal-place coefficients except for $n = 15$, where Hassanein failed to maximize the relative efficiency. Chan and Mead (1971) tabulated (to four decimal places) the coefficients in the best simultaneous linear unbiased estimators of μ and σ based on $k = 2(1)4$ order statistics, and the best conditional linear unbiased estimators of $\mu|\sigma$ and of $\sigma|\mu$ based on $k = 1(1)4$ order statistics, from samples of size $n = k(1)25$. Their results for the simultaneous estimators when $k = 2, 3$ differ in some cases from those of Hassanein because they minimized the determinant of the covariance matrix of the estimators (the generalized variance) rather than its trace. They tabulated the coefficients only for those cases in which they used different order statistics than Hassanein, and stated that their results agreed to six decimal places with those of Hassanein when the same order statistics were used.

Chan [(1993), Chapter 3] discussed the best linear unbiased estimator of the location and scale parameters, μ and σ, based on Type II right-censored samples from the log-Gamma population with probability density function

$$f(x;\mu,\sigma) = \frac{1}{\Gamma(\kappa)\sigma}\exp\left\{\kappa\left(\frac{x-\mu}{\sigma}\right) - e^{(x-\mu)/\sigma}\right\}, \quad -\infty < x < \infty,\ -\infty < \mu < \infty,\ \sigma > 0\ , \tag{3.3.71}$$

where the shape parameter κ is assumed to be known. He used his tables of means, variances and covariances of order statistics (see Tables C8.1 and C8.2) to compute the coefficients of order statistics in the best linear unbiased estimators of μ and σ, which he tabulated (to five decimal places) for sample sizes $n = 2(1)10$ with the largest $0(1)(n-2)$ observations censored, and $\kappa = 0.5(0.5)2.0(1.0)4.0(2.0)12.0$. He also tabulated (to five decimal places) the variances and covariance of these best linear unbiased estimators of μ and σ. Reference may also be made to the paper by Balakrishnan and Chan (1995) in which an abridged form of these tables is presented. Chan (1993) also tabulated the k optimally selected order statistics, coefficients, variances and covariance and efficiencies (all to five decimal places) for the best linear unbiased estimators of μ and σ for $n = 10(5)25$, $k = 2(1)5$, and the same values of κ as above.

Barnett (1966) used his tables of means, variances and covariances of Cauchy order statistics (see Tables C9.1 and C9.2) to compute the coefficients for the best linear unbiased estimator of the location parameter μ of the one-parameter Cauchy population with probability density function

$$f(x;\mu) = \frac{1}{\pi\{1+(x-\mu)^2\}}, \qquad -\infty < x < \infty\ . \tag{3.3.72}$$

He tabulated the coefficients (to four decimal places) for samples of size $n = 5(1)16(2)20$. Since the two smallest and the two largest order statistics have infinite variances, they are not used in the estimators. Stark (1967) tabulated the simultaneous and conditional best linear unbiased estimators, based on selected order statistics, of the location and scale parameters (μ and σ) of the two-parameter Cauchy population with probability density function

$$f(x;\mu,\sigma) = \frac{\sigma}{\pi\{\sigma^2+(x-\mu)^2\}}, \qquad -\infty < x < \infty\ . \tag{3.3.73}$$

Weiss and Howlader (1988) extended Barnett's work by tabulating the best linear unbiased estimators of μ and σ for sample sizes up to 20. Haq (1971) and Raghunandanan and Srinivasan (1972) have discussed some simplified linear estimators for complete as well as symmetrically Type II censored samples. Chan (1970), Chan, Chan, and Mead (1973), Cane (1974) and Kariya, Sinha, and Subramanyam (1981) have all discussed the best linear unbiased estimation of the Cauchy parameters based on k optimally selected order statistics.

For the double exponential population with probability density function

$$f(x;\mu,\sigma) = \frac{1}{2\sigma}\, e^{-\left|\frac{x-\mu}{\sigma}\right|}, \qquad -\infty < x < \infty\,, \tag{3.3.74}$$

Sarhan (1954) used his tables of means, variances and covariances of order statistics to tabulate (to seven decimal places) the coefficients of order statistics in the best linear unbiased estimators of μ and σ for complete samples of size $n = 2(1)5$. Sarhan (1955) [see also Sarhan and Greenberg (1962, pp. 391-395)] extended these results to singly and doubly censored samples of size $n = 2(1)5$, for which he tabulated (to five decimal places) the coefficients as well as the variances of the estimators and the percentage efficiencies (to two decimal places) of the estimators for censored samples relative to those for complete samples. Govindarajulu (1966) used his tables of means, variances and covariances of order statistics (see Tables C10.1 and C10.2) to tabulate (to four decimal places) the coefficients of the order statistics in the best linear unbiased estimators of μ and σ and the variances of these estimators for complete as well as symmetrically Type II censored samples of size $n = 2(1)20$. Balakrishnan, Chandramouleeswaran, and Ambagaspitiya (1995) have tabulated the coefficients of order statistics in the best linear unbiased estimators of μ and σ and the variances and covariance of these estimators (all to four decimal places) for Type II right-censored samples of size $n = 2(1)20$. Balakrishnan and Chandramouleeswaran (1995) have applied these results to reliability estimation and the determination of tolerance limits for the double exponential distribution.

For the half normal (or chi with one degree of freedom) population with probability density function

$$f(x;\mu,\sigma) = \sqrt{\frac{2}{\pi\,\sigma^2}}\; e^{-(x-\mu)^2/(2\sigma^2)}\,, \qquad x \ge \mu\,, \tag{3.3.75}$$

Govindarajulu and Eisenstat (1965) used their tables of means, variances and covariances of order statistics (see Tables C11.1 and C11.2) to tabulate (to four decimal places) the coefficients of order statistics in the best linear unbiased estimators of μ and σ and the variances and covariance of these estimators for complete, singly censored and doubly censored samples of size $n = 2(1)5$ and for complete and singly censored samples of size $n = 6(1)20$.

For the logistic population with cumulative distribution function

$$F(x;\mu,\sigma) = 1/\{1 + e^{-(x-\mu)/\sigma}\}\,, \qquad -\infty < x < \infty\,, \tag{3.3.76}$$

Kjelsberg (1962) tabulated exact (and six decimal-place) coefficients of order statistics in the best linear unbiased estimators of μ and σ and the variances of these estimators (exact and six decimal-place) for samples of size $n = 1(1)5$. Gupta, Qureishi, and Shah (1965) tabulated (to four decimal places) the coefficients of order statistics and the variances and covariance of the best linear unbiased estimators, and the percentage efficiencies to two decimal places for complete, singly censored and doubly censored samples of size $n = 2(1)25$. Gupta, Qureishi, and Shah (1967) presented an abridged form of these tables for $n = 2$, 5, 10, 15, 20 and 25. Balakrishnan (1990) prepared similar tables for complete, singly censored and doubly censored samples of size $n = 2(1)25(5)40$. Beyer (1966) tabulated (to five decimal places) the coefficients of order statistics in the conditional best linear unbiased estimators of $\sigma|\mu$ from the first m order statistics and from ℓ of the first m order statistics of singly censored samples of size n for $n = 2(1)20$, $1 \leq m \leq n$ and $1 \leq \ell \leq k \leq m$, where k is the smallest value of ℓ such that the ℓ-order-statistic estimator achieves a 99% efficiency relative to the m-order-statistic estimator. He also tabulated (to five decimal places) the variances of the estimators and the efficiencies of the ℓ-order-statistic estimators relative to the m-order-statistic estimators. Hoksch (1967) presented similar results for doubly censored samples of size $n = 2(1)18$. Richardson (1966) tabulated similar results for the simultaneous best linear unbiased estimators of μ and σ based on singly censored samples of size $n = 3(1)20$ and Simpson (1967) extended these to the case of doubly censored samples of size $n = 3(1)18$. The best linear unbiased estimation of the parameters μ and σ based on k optimally selected order statistics has been discussed by Kulldorff (1964), Chan (1969), Hassanein (1969b, 1974), Chan, Chan, and Mead (1971), and Chan and Cheng (1972, 1974).

For the Pareto population with cumulative distribution function

$$F(x;\mu,\sigma) = 1 - \left(1 + \frac{x-\mu}{\sigma}\right)^{-\gamma}, \qquad x \geq \mu,\ \sigma > 0,\ \gamma > 0\ , \tag{3.3.77}$$

Kulldorff and Vännman (1973) used the means, variances and covariances of order statistics (see Tables C13.1 and C13.2) to derive the best linear unbiased estimators for the location parameter μ and/or the scale parameter σ when the shape parameter γ is known, based on complete samples. Kaminsky (1973) [also see Kaminsky and Nelson (1975)] presented the best linear unbiased estimators of μ and/or σ based on the first k order statistics. Vännman (1976) presented similar results based on optimally selected order statistics.

For the Student's t population with probability density function [see Eq. (3.3.50)]

$$f(x;\mu,\sigma) = C\left\{1 + \frac{(x-\mu)^2}{k\,\sigma^2}\right\}^{-p}, \qquad -\infty < x < \infty\ , \tag{3.3.78}$$

Tiku and Kumra (1985) presented the computational formulas for the best linear unbiased estimators of μ and σ which make use of their tables of means, variances and covariances of order statistics (see Tables C14.1 and C14.2). They gave a numerical example for $p = 5$ and $n = 15$, but presented no tables of coefficients of order statistics and variances of the best linear unbiased estimators.

For the half logistic population with probability density function

$$f(x;\mu,\sigma) = \frac{2}{\sigma}\,\frac{e^{-(x-\mu)/\sigma}}{\{1 + e^{-(x-\mu)/\sigma}\}^2}, \qquad x \geq \mu\ , \tag{3.3.79}$$

Balakrishnan and Puthenpura (1986) used the tables of means, variances and covariances of half logistic order statistics (see Tables C15.1 and C15.2) to tabulate the coefficients of order statistics in the best linear unbiased estimators of μ and σ and the variances and covariance of these estimators for complete samples of size $n = 2(1)15$. Balakrishnan and Wong (1994) extended these results to the cases of singly and doubly Type II censored samples for $n = 2(1)15$. Balakrishnan and Chan (1992c) considered the scaled half logistic distribution [case $\mu = 0$ in (3.3.79)] and tabulated the coefficients of order statistics in the best linear unbiased estimator of σ and the variance of this estimator for complete as well as censored samples. They also discussed the best linear unbiased estimator of σ based on k optimally selected order statistics.

For the Type I generalized logistic population with probability density function

$$f(x;\mu,\sigma) = \frac{b}{\sigma}\,\frac{e^{-(x-\mu)/\sigma}}{\{1+e^{-(x-\mu)/\sigma}\}^{b+1}}\,, \qquad -\infty < x < \infty,\ b > 0\ , \tag{3.3.80}$$

Balakrishnan and Leung (1988b) used their tables of means, variances and covariances of order statistics (see Tables C16.1 and C16.2) to tabulate the coefficients of order statistics (to five decimal places) in the best linear unbiased estimators of μ and σ and the variances and covariance of these estimators for complete samples of size $n = 2(1)15$, with known shape parameter $b = 1.0(0.5)5.0(1.0)8.0$.

Best Linear Invariant Estimators

Best linear unbiased estimators have some very nice properties for complete or symmetrically Type II censored samples from a symmetric distribution but for a skewed distribution (such as the log-Weibull) or even in the case of asymmetric censoring from a symmetric distribution, they often have larger mean square errors than some other linear estimators, where

$$\text{mean square error} = \text{variance} + (\text{bias})^2. \tag{3.3.81}$$

Mann (1963, 1965) proposed the use of best linear invariant estimators which have minimum mean square errors in the class of all linear estimators, and can be obtained from the best linear unbiased estimators by a simple linear transformation.

Mann (1967a,b, 1968b) discussed the best linear invariant estimators, $\tilde{\mu}$ and $\tilde{\sigma}$, of the parameters μ and σ of the log-Weibull population based on the first m order statistics of a sample of size n. Mann (1968b) showed that

$$\tilde{\mu} = \mu^* - \sigma^*\left(\frac{B_{m,n}}{1+C_{m,n}}\right) \qquad \text{and} \qquad \tilde{\sigma} = \frac{\sigma^*}{1+C_{m,n}} \tag{3.3.82}$$

and that the expected squared errors of these estimators are

$$\sigma^2\left\{A_{m,n} - \frac{B_{m,n}^2}{1+C_{m,n}}\right\} \qquad \text{and} \qquad \sigma^2\cdot\frac{C_{m,n}}{1+C_{m,n}}\,, \tag{3.3.83}$$

respectively. In the above equations, μ^* and σ^* denote the best linear unbiased estimators of μ and σ, respectively, and $\sigma^2 A_{m,n}$ and $\sigma^2 C_{m,n}$ are their variances and $\sigma^2 B_{m,n}$ is their covariance. Mann (1967a,b) tabulated (to six decimal places) the coefficients of the order statistics in the best linear invariant estimators and the expected squares and the expected cross product of the errors of these estimators for $n = 2(1)25$ and $m = 2(1)n$. Mann (1968b) applied these results, for the case in which failure time has the log-Weibull distribution, to find the best linear invariant estimators of the reliability at a specified time and of the reliable life corresponding to a given reliability.

Nearly Best Linear Unbiased (or Invariant) Estimators

When tables of variances and/or covariances of order statistics are not available, the best linear unbiased estimators and the best linear invariant estimators cannot be calculated, but tables of means of order statistics can be used to obtain nearly best linear unbiased estimators [using a standard method suggested by Blom (1958)] or nearly best linear invariant estimators. For details, one may refer to Balakrishnan and Cohen (1991).

White (1967) showed how to use his table of means of log-Weibull order statistics and ordinary least squares to obtain the nearly best linear unbiased estimators of μ and σ based on the first m order statistics of a sample of size n. White also showed how to obtain still better estimators by using his tables of means and variances and weighted least squares. These methods are quite useful for $30 < n \leq 100$, where tables of covariances of log-Weibull order statistics are not available, and hence the best linear unbiased (invariant) estimators cannot be found. Tables for the nearly best linear unbiased estimators from samples of size up to 40 have been given by Eickmann (1968) (for simultaneous estimators) and by Kingcaid (1969) and Widenhouse (1970) (for conditional estimators). Kingcaid tabulated (to five decimal places) the coefficients of order statistics in the nearly best linear unbiased estimators for singly and symmetrically censored samples of size $n = 1(1)40$. He also tabulated the percentage efficiencies (to one decimal place) relative to the best linear unbiased estimators for a few selected sample sizes and censoring choices. He pointed out that the conditional estimators can also be used to obtain simultaneous estimators [see Herman (1968)].

Barnett (1966) used his table of means of Cauchy order statistics along with the asymptotic expressions of variances and covariances of order statistics, as suggested by Blom (1958), to tabulate the coefficients of the order statistics (to four decimal places) in the nearly best linear unbiased estimator of the location parameter μ of the one-parameter Cauchy population, based on order statistics of samples (excluding the first two and the last two) of size $n = 5(1)16(2)20$. He indicated that the efficiency for this estimator, relative to the best linear unbiased estimator, is in excess of 99% for samples of size $n = 5(1)16(2)20$. Chamberlain (1968) tabulated the coefficients of the order statistics in the simultaneous nearly best linear unbiased estimators of the location and scale parameters (μ and σ, respectively) of a two-parameter Cauchy population for right-censored and symmetrically censored samples of size $n = 15(1)40$. Jonson (1969) tabulated (to six decimal places) the coefficients of the order statistics in the conditional nearly best linear unbiased estimators of μ and σ for such samples of size $n = 6(1)40(5)50$. He also verified Barnett's results for μ and extended them to include efficiencies for right-censored samples of size $n = 6(1)10(5)50$. He also tabulated the efficiencies of conditional BLUEs and simultaneous BLUEs, relative to the BLUEs of Stark (1967) for $n = 6(1)10(5)20$.

For the logistic population, Rodgers (1968) tabulated the coefficients of the order statistics in the simultaneous nearly best linear unbiased estimators of μ and σ based on doubly censored samples of size $n = 15(1)40$. Hunter (1969) tabulated the coefficients of the order statistics (to five decimal places) in the conditional nearly best linear unbiased estimators of $\mu|\sigma$ and $\sigma|\mu$ based on singly and doubly censored samples of size $n = 2(1)40$. Dunigan (1970) gave analogous results for the conditional nearly best linear invariant estimators of $\mu|\sigma$ and $\sigma|\mu$.

Other Applications

Mann (1965, 1968a,b) presented results on interval estimation and related hypothesis tests for μ and σ, reliability, and reliable life based on best linear unbiased or best linear invariant estimators for the log-Weibull distribution based on right-censored or progressively censored samples, or from two or three order statistics. These results can also be adapted, as pointed out earlier, for the related Weibull distribution. Chan [(1993), Chapter 6] presented results for the interval estimation of μ and σ of the log-Gamma population, based on equivariant point estimators including the best linear unbiased estimators.

Another application of tables of means of order statistics is in probability plotting, where the values of $F[E(x_{i,n})]$ may be used as plotting positions on an appropriate probability paper. King (1971, 1981) presented an excellent overview of the construction and use of probability papers for various populations including the normal, Weibull, Gamma, lognormal, Type I largest and Type I smallest extreme value. King also discussed plotting of hazard functions. Gross and Clark (1975) considered the plotting of survival functions, especially for logistic and log-logistic distributions, which they found useful in the analysis of medical data, and Balakrishnan (1985) suggested that plotting of half logistic survival functions may also be useful.

Tiku and Kumra (1985) applied their tables of means, variances and covariances of Student's t order statistics to goodness-of-fit tests based on spacings and to examination of the robustness of various estimators for the parameters of a normal population. Balakrishnan and Leung (1988b) applied their tables of means, variances and covariances of Type I generalized logistic order statistics to examination of the bias and variance of various robust estimators, and compared the efficiencies of these estimators. Balakrishnan and Hamada (1995) recently used the tables of means, variances and covariances of half normal order statistics in the analysis of unreplicated factorial designs.

3.3.13 References

1. Abramowitz, M. and Stegun, I.A. (Eds.) (1965). *Handbook of Mathematical Functions with Formulas, Graphs, and Mathematical Tables.* Dover Publications, New York.
2. Anonymous (1951). Tables of the First Moment of Ranked Extremes. *NBS Report 1167*, National Bureau of Standards, Washington, DC.
3. Balakrishnan, N. (1985). Order statistics from the half logistic distribution. *Journal of Statistical Computation and Simulation* **20**, 287-309.
4. Balakrishnan, N. (1990). Best linear unbiased estimates of the location and scale parameters of logistic distribution for complete and censored samples of sizes 2(1)25(5) 40. *Report*, McMaster University, Hamilton, Ontario, Canada.
5. Balakrishnan, N. (Ed.) (1992). *Handbook of the Logistic Distribution.* Marcel Dekker, New York.
6. Balakrishnan, N. and Chan, Ping-Shing (1991a). Order statistics from extreme value distribution: Tables of means, variances and covariances for samples sizes up to 30. *Report*, McMaster University, Hamilton, Ontario, Canada.
7. Balakrishnan, N. and Chan, Ping-Shing (1991b). Order statistics from extreme value distribution: Best linear unbiased estimates for samples sizes up to 30. *Report*, McMaster University, Hamilton, Ontario, Canada.
8. Balakrishnan, N. and Chan, Ping-Shing (1992a). Order statistics from extreme value distribution, I: Tables of means, variances and covariances. *Communications in Statistics - Simulation and Computation* **21**(4), 1199-1217.
9. Balakrishnan, N. and Chan, Ping-Shing (1992b). Order statistics from extreme value distribution, II: Best linear unbiased estimates and some other uses. *Communications in Statistics - Simulation and Computation* **21**(4), 1219-1246.
10. Balakrishnan, N. and Chan, Ping-Shing (1992c). Estimation for the scaled half logis-

tic distribution under Type II censoring. *Computational Statistics & Data Analysis* **13**, 123-141.

11. Balakrishnan, N. and Chan, Ping-Shing (1995). Log-Gamma order statistics and linear estimation of parameters. In *Handbook of Statistics - Vol. 16: Order Statistics and Their Applications* (Eds., C.R. Rao and N. Balakrishnan), North-Holland, Amsterdam (to appear).

12. Balakrishnan, N. and Chandramouleeswaran, M.P. (1995). Reliability estimation and tolerance limits for Laplace distribution based on censored samples. *Microelectronics and Reliability* (to appear).

13. Balakrishnan, N., Chandramouleeswaran, M.P., and Ambagaspitiya, R.S. (1995). BLUE's of location and scale parameters of Laplace distribution based on Type II censored samples and associated inference. *Microelectronics and Reliability* (to appear).

14. Balakrishnan, N. and Cohen, A. Clifford (1991). *Order Statistics and Inference: Estimation Methods.* Academic Press, San Diego, CA.

15. Balakrishnan, N. and Hamada, Michael (1995). Analyzing unreplicated factorial experiments: A review and some new proposals (with discussions). *Statistica Sinica* (to appear).

16. Balakrishnan, N. and Leung, M.Y. (1988a). Order statistics from the Type I generalized logistic distribution. *Communications in Statistics - Simulation and Computation* **17**(1), 25-50.

17. Balakrishnan, N. and Leung, M.Y. (1988b). Means, variances and covariances of order statistics, BLUE's for the Type I generalized logistic distribution, and some applications. *Communications in Statistics - Simulation and Computation* **17**(1), 51-84.

18. Balakrishnan, N. and Malik, Henrick J. (1995). Means, variances and covariances of logistic order statistics for samples sizes up to fifty. *Selected Tables in Mathematical Statistics* (to appear).

19. Balakrishnan, N. and Puthenpura, S. (1986). Best linear unbiased estimators of location and scale parameters of the half logistic distribution. *Journal of Statistical Computation and Simulation* **25**, 193-204.

20. Balakrishnan, N. and Wong, K.H.T. (1994). Best linear unbiased estimation of location and scale parameters of the half logistic distribution based on Type II censored samples. *American Journal of Mathematical and Management Sciences* **14**, 53-101.

21. Barnett, Vic D. (1966). Order statistics estimators of the location of the Cauchy distribution. *Journal of the American Statistical Association* **61**, 1205-1218.

22. Beyer, Jon N. (1966). Conditional Estimation of the Scale Parameter of the Logistic Distribution by the Use of Selected Order Statistics. *Unpublished M.S. Thesis* (GRE/Math/66-1), Air Force Institute of Technology, Wright-Patterson Air Force Base, Ohio.

23. Birnbaum, A. and Dudman, J. (1963). Logistic order statistics. *Annals of Mathematical Statistics* **34**, 658-663.

24. Blom, Gunnar (1958). *Statistical Estimates and Transformed Beta-Variables.* John Wiley & Sons, New York.

25. Cane, G.J. (1974). Linear estimation of parameters of the Cauchy distribution based on sample quantiles. *Journal of the American Statistical Association* **69**, 243-245.

26. Chamberlain, Robert B. (1968). Nearly Best Linear Unbiased Estimators of the Location and Scale Parameters of the Cauchy Distribution by the Use of Censored Order Statistics. *Unpublished M.S. Thesis* (GRE/Math/68-2), Air Force Institute of Technology, Wright-Patterson Air Force Base, Ohio, AD 685112.

27. Chan, Lai K. (1969). Linear quantile estimates of the location and scale parameters of the logistic distribution. *Statistische Hefte* **10**, 277-282.

28. Chan, Lai K. (1970). Linear estimation of location and scale parameters of Cauchy distribution based on sample quantiles. *Journal of the American Statistical Association* **65**, 851-859.

29. Chan, Lai K., Chan, N.N., and Mead, Ernest R. (1971). Best linear unbiased estimates of the parameters of the logistic distribution based on selected order statistics. *Journal of the American Statistical Association* **66**, 889-892.

30. Chan, Lai K., Chan, N.N., and Mead, Ernest R. (1973). Linear estimation of the parameters of the Cauchy distribution using selected order statistics. *Utilitas Mathematica* **3**, 311-318.

31. Chan, Lai K. and Cheng, Smiley W. (1972). Optimum spacing for the asymptotically best linear estimate of the location parameter of the logistic distribution when samples are complete or censored. *Statistische Hefte* **13**, 41-57.

32. Chan, Lai K. and Cheng, Smiley W. (1974). An algorithm for determining the asymptotically best linear estimate of the mean from multiply censored logistic data. *Journal of the American Statistical Association* **69**, 1027-1030.

33. Chan, Lai K. and Mead, Ernest R. (1971). Linear estimation of the parameters of the extreme-value distribution based on suitably chosen order statistics. *IEEE Transactions on Reliability* **R-20**, 74-83.

34. Chan, Ping-Shing (1993). A Statistical Study of Log-Gamma Distribution. *Unpublished Ph.D. Thesis*, McMaster University, Hamilton, Ontario, Canada.

35. Dunigan, John M. (1970). Nearly Best Linear Invariant Conditional Estimation of the Mean and Standard Deviation of the Logistic Distribution. *Unpublished M.S. Thesis* (GRE/Math/70-3), Air Force Institute of Technology, Wright-Patterson Air Force Base, Ohio, AD 776760.

36. Eickmann, Kenneth E. (1968). Unbiased Nearly Best Linear Estimates of the Scale and Location Parameters of the Extreme Value Distribution by the Use of Order Statistics. *Unpublished M.S. Thesis* (GRE/Math/68-5), Air Force Institute of Technology, Wright-Patterson Air Force Base, Ohio, AD 693661.

37. Elwer, Gale E. (1967). Conditional Linear Estimation of the Scale Parameter of the Extreme-Value Distribution by Use of Selected Order Statistics from Doubly Censored Samples. *Unpublished M.S. Thesis* (GRE/Math/67-1), Air Force Institute of Technology, Wright-Patterson Air Force Base, Ohio.

38. Fratzke, William F. (1967). Simultaneous Linear Estimation of the Location and Scale Parameters of the Extreme-Value Distribution Using Selected Order Statistics from Doubly Censored Samples. *Unpublished M.S. Thesis* (GRE/Math/67-2), Air Force Institute of Technology, Wright-Patterson Air Force Base, Ohio, AD 849869.

39. Galarneau, D.I. and McDonald, Gary C. (1974). Best Linear Unbiased Estimators of the Parameters of a Restricted Lognormal Distribution. *General Motors Research Report No. GMR-1656*, General Motors Research Laboratories, Warren, Michigan.

40. Gibbons, Diane I. and McDonald, Gary C. (1975). Small-sample estimation for the lognormal distribution with unit shape parameter. *IEEE Transactions on Reliability* **R-24**, 290-295.

41. Govindarajulu, Zakkula (1962). Exact lower moments of order statistics in samples from the chi-distribution (1 d.f.). *Annals of Mathematical Statistics* **33**, 1292-1305.

42. Govindarajulu, Zakkula (1963). On moments of order statistics and quasi-ranges from normal populations. *Annals of Mathematical Statistics* **34**, 633-651.

43. Govindarajulu, Zakkula (1964). Best Estimates of Location and Scale Parameters of a Double Exponential Population from Ordered Observations. *Case Technical Report No. 202*, Case Institute of Technology, Cleveland, Ohio.

44. Govindarajulu, Zakkula (1966). Best linear estimates under symmetric censoring of the parameters of a double exponential population. *Journal of the American Statistical Association* **61**, 248-258.

45. Govindarajulu, Zakkula and Eisenstat, Stan (1965). Best estimates of location and scale parameters of a chi (1 d.f.) distribution, using ordered observations. *Reports of Statistical Application Research, Japanese Union of Scientists and Engineers* **12**, 149-164.

46. Gross, A.J. and Clark, V.A. (1975). *Survival Distributions: Reliability Applications in the Biomedical Sciences.* John Wiley & Sons, New York.

47. Gupta, Shanti S., McDonald, Gary C., and Galarneau, D.I. (1974). Moments, product moments and percentage points of the order statistics from the lognormal distribution for samples of size twenty and less. *Sankhyā, Series B* **36**, 230-260.

48. Gupta, Shanti S., Qureishi, A.S., and Shah, Bhupendra K. (1965). Best Linear Unbiased Estimators of the Parameters of the Logistic Distribution Using Order Statistics. *Mimeograph Series No. 52*, Department of Statistics, Purdue University, West Lafayette, Indiana.

49. Gupta, Shanti S., Qureishi, A.S., and Shah, Bhupendra K. (1967). Best linear unbiased estimators of the parameters of the logistic distribution using order statistics. *Technometrics* **9**, 43-56.

50. Gupta, Shanti S. and Shah, Bhupendra K. (1965). Exact moments and percentage points of the order statistics and the distribution of the range from the logistic distribution. *Annals of Mathematical Statistics* **36**, 907-920.

51. Hall, Peter (1978). Some asymptotic expansions of moments of order statistics. *Stochastic Processes and their Applications* **7**, 265-275.

52. Haq, M. (1971). Estimation of the scalar parameter in a Cauchy population. *Journal of Sciences - Physical Section* **1**, 67-68.

53. Hassanein, Khatab M. (1969a). Estimation of the parameters of the extreme value distribution by use of two or three order statistics. *Biometrika* **56**, 429-436.

54. Hassanein, Khatab M. (1969b). Estimation of the parameters of the logistic distribution by sample quantiles. *Biometrika* **56**, 684-687.

55. Hassanein, Khatab M. (1974). Linear estimation of the parameters of the logistic distribution by selected order statistics for very large samples. *Statistischė Hefte* **15**, 65-70.

56. Herman, William J. (1968). Least-squares Conditional Estimation of the Location Parameter of Weibull Populations. *Unpublished M.S. Thesis* (GRE/Math/68-1), Air Force Institute of Technology, Wright-Patterson Air Force Base, Ohio, AD 684696.

57. Highfield, Eugene R. (1964). Estimation of the Scale Parameter of the Lognormal Distribution by M Order Statistics. *Unpublished M.S. Thesis* (GRE/Math/64-5), Air Force Institute of Technology, Wright-Patterson Air Force Base, Ohio, AD 610770.

58. Hoksch, Orville J. (1967). Conditional Estimation of the Standard Deviation of the Normal and Logistic Distributions from Doubly Censored Samples using Selected Order Statistics. *Unpublished M.S. Thesis* (GRE/Math/67-5), Air Force Institute of Technology, Wright-Patterson Air Force Base, Ohio.

59. Huang, J.S. (1975). A note on order statistics from Pareto distribution. *Scandinavian Actuarial Journal*, 187-190.

60. Hunter, Stephen A. (1969). Nearly Best Conditional Linear Unbiased Estimation of the Mean and Standard Deviation of the Logistic Distribution. *Unpublished M.S. Thesis* (GRE/Math/69-3), Air Force Institute of Technology, Wright-Patterson Air Force Base, Ohio, AD 768718.

61. Johnson, Norman L., Kotz, Samuel, and Balakrishnan, N. (1994). *Continuous Univariate Distributions - Volume 1*, Second edition. John Wiley & Sons, New York.

62. Jonson, Edward C. (1969). Conditional Nearly Best Linear Estimation of the Location and Scale Parameters of the Cauchy Distribution by the Use of Censored Order Statistics. *Unpublished M.S. Thesis* (GRE/Math/69-4), Air Force Institute of Technology, Wright-Patterson Air Force Base, Ohio, AD 776766.

63. Joshi, Prakash C. (1969). Bounds and approximations for the moments of order statistics. *Journal of the American Statistical Association* **64**, 1617-1624.

64. Kaminsky, Kenneth S. (1973). Best Linear Unbiased Estimation and Prediction in Censored Samples from Pareto Populations. *Report*, Bucknell University, Lewisburg, Pennsylvania.

65. Kaminsky, Kenneth S. and Nelson, Paul I. (1975). Best linear unbiased prediction of order statistics in location and scale families. *Journal of the American Statistical Association* **70**, 145-150.

66. Kariya, T., Sinha, B.K., and Subramanyam, K. (1981). Nearly Efficient Estimators Based on Order Statistics. *Technical Report No. 81-05*, University of Pittsburgh, Pittsburgh, Pennsylvania.

67. King, James R. (1971). *Probability Charts for Decision Making*. Industrial Press, New York. Second edition (1981), TEAM Publications, Tamworth, New Hampshire.

68. Kingcaid, Milton D. (1969). Unbiased Nearly Best Linear Conditional Estimates of the Scale and Location Parameters of the Extreme Value Distribution by the Use of Order Statistics. *Unpublished M.S. Thesis* (GRE/Math/69-5), Air Force Institute of Technology, Wright-Patterson Air Force Base, Ohio, AD 767874.

69. Kjelsberg, Marcus Olaf (1962). Estimation of the Parameters of the Logistic Distribution under Truncation and Censoring. *Ph.D. Thesis*, University of Minnesota, Minneapolis, Minnesota. University Microfilms, Ann Arbor, Michigan.

70. Kulldorff, Gunnar (1964). Optimum Spacing of Sample Quantiles from a Logistic Distribution and Best Linear Unbiased Estimates of its Parameters. *Technical Report*, University of Lund, Sweden.

71. Kulldorff, Gunnar and Vännman, Kerstin (1973). Estimation of the location and scale parameters of a Pareto distribution by linear functions of order statistics. *Journal of the American Statistical Association* **68**, 218-227.

72. Lieblein, Julius (1953). On the exact evaluation of the variances and covariances of order statistics in samples from the extreme-value distribution. *Annals of Mathematical Statistics* **24**, 282-287.

73. Lieblein, Julius (1954). A New Method of Analyzing Extreme-Value Data. *Technical Note 3053*, National Advisory Committee for Aeronautics, Washington, DC.

74. Lieblein, Julius and Salzer, Herbert E. (1957). Table of the first moment of ranked extremes. *Journal of Research of the National Bureau of Standards* **59**, 203-206.

75. Lieblein, Julius and Zelen, Marvin (1956). Statistical investigation of the fatigue life of deep-groove ball bearings. *Journal of Research of the National Bureau of Standards* **57**, 273-316.

76. Mahmoud, M.W. and Ragab, A. (1975). On order statistics in samples drawn from the extreme value distributions. *Mathematische Operationsforschung und Statistik, Series Statistics* **6**, 809-816.

77. Malik, Henrick John (1966). Exact moments of order statistics from the Pareto distribution. *Skandinavisk Aktuarietidskrift* **49**, 144-157.

78. Mann, Nancy R. (1963). Optimum Estimates of Parameters of Continuous Distributions. *Research Report No. 63-41*, Rocketdyne Division, North American Aviation, Inc., Canoga Park, California.

79. Mann, Nancy R. (1965). Point and Interval Estimates for Reliability Parameters when Failure Times Have the Two-Parameter Weibull Distribution. *Doctoral dissertation*, University of California, Los Angeles. University Microfilms, Inc., Ann Arbor, Michigan.

80. Mann, Nancy R. (1967a). Results on Location and Scale Parameter Estimation with Application to the Extreme-Value Distribution. *ARL Technical Report 67-0023*, Aerospace Research Laboratories, Wright-Patterson Air Force Base, Ohio, AD 653575.

81. Mann, Nancy R. (1967b). Tables for obtaining the best linear invariant estimates of parameters of the Weibull distribution. *Technometrics* **9**, 629-643. Preprinted in 1966 as *Research Report No. 66-9*, Rocketdyne Division, North American Aviation, Inc., Canoga Park, California.

82. Mann, Nancy R. (1968a). Results on Statistical Estimation and Hypothesis Testing with Application to the Weibull and Extreme-Value Distribution. *ARL Technical Report 68-0068*, Aerospace Research Laboratories, Wright-Patterson Air Force Base, Ohio, AD 672979.

83. Mann, Nancy R. (1968b). Point and interval estimation procedures for the two-parameter Weibull and extreme-value distributions. *Technometrics* **10**, 231-256.

84. Plackett, R.L. (1958). Linear estimation from censored data. *Annals of Mathematical Statistics* **29**, 131-142.

85. Provasi, C. (1987). Exact and approximate means and covariances of order statistics of the standardized extreme value distribution (I type). *Rivista di Statistica Applicata* **20**, 287-295 (in Italian).

86. Raghunandanan, K. and Srinivasan, R. (1972). Simplified estimation of parameters in a Cauchy distribution. *Statistica Neerlandica* **26**, 25-27.

87. Richardson, Everett B. (1966). Simultaneous Linear Estimation of the Location and Scale Parameters of the Extreme-Value and Logistic Distributions by the Use of Selected Order Statistics. *Unpublished M.S. Thesis* (GRE/Math/66-7), Air Force

Institute of Technology, Wright-Patterson Air Force Base, Ohio, AD 425228.

88. Rodgers, Joe W. (1968). Nearly Best Linear Unbiased Estimation of the Mean and Standard Deviation of the Logistic Distribution. *Unpublished M.S. Thesis* (GRE/Math/68-12), Air Force Institute of Technology, Wright-Patterson Air Force Base, Ohio, AD 685115.

89. Sarhan, Ahmed E. (1954). Estimation of the mean and standard deviation by order statistics. *Annals of Mathematical Statistics* **25**, 317-328.

90. Sarhan, Ahmed E. (1955). Estimation of the mean and standard deviation by order statistics, Parts II and III. *Annals of Mathematical Statistics* **26**, 505-511 and 576-592.

91. Sarhan, Ahmed E. and Greenberg, Bernard G. (Eds.) (1962). *Contributions to Order Statistics.* John Wiley & Sons, New York.

92. Shah, Bhupendra K. (1966). On the bivariate moments of order statistics from a logistic distribution. *Annals of Mathematical Statistics* **37**, 1002-1010.

93. Shah, Bhupendra K. (1970). Note on moments of logistic order statistics. *Annals of Mathematical Statistics* **41**, 2151-2152.

94. Shelnutt, J. William III (1966). Conditional Linear Estimation of the Scale Parameter of the Extreme-Value Distribution by the Use of Selected Order Statistics. *Unpublished M.S. Thesis* (GRE/Math/66-9), Air Force Institute of Technology, Wright-Patterson Air Force Base, Ohio.

95. Shelnutt, J. William III, Moore, Albert H., and Harter, H. Leon (1973). Linear estimation of the scale parameter of the first asymptotic distribution of extreme values. *IEEE Transactions on Reliability* **R-22**, 259-264.

96. Simpson, Jackson S. (1967). Simultaneous Linear Estimation of the Mean and Standard Deviation of the Normal and Logistic Distributions by the Use of Selected Order Statistics from Doubly Censored Samples. *Unpublished M.S. Thesis* (GRE/Math/67-10), Air Force Institute of Technology, Wright-Patterson Air Force Base, Ohio, AD 675544.

97. Stark, Terry M. (1967). Simultaneous and Conditional Estimation of the Location and Scale Parameters of the Cauchy Distribution by the Use of Selected Order Statistics. *Unpublished M.S. Thesis* (GRE/Math/67-11), Air Force Institute of Technology, Wright-Patterson Air Force Base, Ohio.

98. Tiku, Moti L. and Kumra, S. (1985). Expected values and variances and covariances of order statistics for a family of symmetric distributions (Student's *t*). *Selected Tables in Mathematical Statistics* **8**, 141-270.

99. Vännman, Kerstin (1976). Estimators based on order statistics from a Pareto distribution. *Journal of the American Statistical Association* **71**, 704-708.

100. Vaughan, David C. (1995). The exact values of the expected values, variances and covariances of the order statistics from the Cauchy distribution. *Journal of Statistical Computation and Simulation* (to appear).

101. Weiss, Gunter and Howlader, Hatem A. (1988). Linear scale estimation. *Journal of Statistical Computation and Simulation* **29**, 117-126.

102. White, John S. (1964). Least squares unbiased censored linear estimation for the log Weibull (extreme value) distribution. *Journal of the Industrial Mathematics Society* **1**, 21-60.

103. White, John S. (1967). The moments of log-Weibull order statistics. *Research*

Publication GMR 717, Research Laboratories, General Motors Corporation, Warren, Michigan. Abridged version (1969), *Technometrics* **11**, 373-386.

104. Widenhouse, William C. (1970). Conditional Nearly Best Linear Estimation of the Location and Scale Parameters of the First Extreme-Value Distribution. *Unpublished M.S. Thesis* (GRE/Math/70-7), Air Force Institute of Technology, Wright-Patterson Air Force Base, Ohio, AD 776769.

4

One or Two Order Statistics From an Exponential Population

4.1 Point Estimation of One or Two Parameters [1]

4.1.1 Introduction

Since the publication of papers by Mosteller (1946) and by Wilks (1948), a great deal of attention has been given to the use of order statistics in various statistical procedures, including the estimation of the parameters of various populations. Among the first to use this method for exponential populations was Halperin (1952). Since that time, Epstein and Sobel (1953), Sarhan (1954, 1955), Sarhan and Greenberg (1957, 1958), and Sarhan, Greenberg, and Ogawa (1960) have discussed various aspects of the use of order statistics in the estimation of the parameters of exponential populations. Balakrishnan and Cohen (1991) have reviewed the various developments on this topic. Most of these authors have considered best linear unbiased estimators based on all order statistics, or, in the case of truncated or censored samples, on all available order statistics. The paper by Sarhan et al. (1960) includes simplified estimators based on two order statistics, with tables for samples of any size up through $n = 20$. This section deals with the computation of more accurate tables (Tables D1-D3) for samples of any size up through $n = 100$, not only for estimators based on two order statistics, but also, in the case of the one-parameter exponential population, for estimators based on one order statistic from a complete or a censored sample. The book by Balakrishnan and Cohen (1991) presents tables necessary for the asymptotically best linear unbiased estimation for the one-parameter exponential distribution based on one up to twenty five optimally selected quantiles. The recently prepared handbook on the exponential distribution by Balakrishnan and Basu (1995) synthesizes all the pertinent details.

Previously, the first author (1958, 1959) studied estimators of the standard deviation of normal, rectangular, and one-parameter exponential populations based on sample ranges and quasi-ranges. The results for the normal population are included in Chapter 1, Section 1.1 and Tables A4 and A5 of the present volume; those for the rectangular population in Chapter 2, Section 2.1 and Table B1. The estimators are quite satisfactory for these symmetric populations, but not for the one-parameter exponential, for which it was found that estimators based on a single order statistic are more efficient. That discovery led to the further research reported in this section.

[1] Earlier versions of portions of this material were published by Harter (1961) and by Moore and Harter (1965).

4.1.2 Estimators of σ for the One-Parameter Exponential Population

Estimators Based on One Order Statistic of a Complete Sample

For the one-parameter exponential population with parameter σ, which is both the mean and the standard deviation of the population, the probability density function is $f_1(x) = (1/\sigma)\exp(-x/\sigma)$, $0 \le x < \infty$. The sample mean $\bar{x}$, which has variance σ^2/n, is the best linear unbiased estimator, and also the maximum likelihood estimator, of the parameter σ. The expected value and the variance of the kth order statistic, x_k, of a sample of size n from this population are given [see Section 3.2] by

$$E(x_k) = \sigma \sum_{i=1}^{k} a_i \tag{4.1.1}$$

and

$$\text{var}(x_k) = \sigma^2 \sum_{i=1}^{k} a_i^2 , \tag{4.1.2}$$

where $a_i = 1/(n-i+1)$. An unbiased estimator of the parameter σ, based on the order statistic x_k, is given by $\tilde{\sigma} = c_k x_k$, where

$$c_k = 1 \Big/ \sum_{i=1}^{k} a_i . \tag{4.1.3}$$

The variance of this estimator is given by

$$V_k = \sigma^2 \sum_{i=1}^{k} a_i^2 \Big/ \left(\sum_{i=1}^{k} a_i \right)^2 , \tag{4.1.4}$$

and its efficiency (relative to the best linear unbiased estimator $\bar{x}$) is $E_k = \text{var}(\bar{x})/V_k$, where, as mentioned above, $\text{var}(\bar{x}) = \sigma^2/n$. Thus the relative efficiency E_k is given by

$$E_k = \left(\sum_{i=1}^{k} a_i \right)^2 \Big/ n \sum_{i=1}^{k} a_i^2 . \tag{4.1.5}$$

The best estimator of σ, based on one order statistic x_k, is the one for that value of k which minimizes V_k (maximizes E_k). There is no analytical method for determining the value of k which yields the best estimator of σ; hence, for each value of n, V_k was computed for $k = 1(1)n$. When the best value of k for a given n had been found, the corresponding c_k and E_k were also computed. The computations were performed on the IBM 1620 computer. Table D1 gives, for $n = 1(1)100$, the value of k for the best estimator of σ, the coefficient c_k (to 6 significant figures), the coefficient, V_k/σ^2, of σ^2 in the variance V_k of the estimator (to 7 significant figures or 6 decimal places, whichever is less accurate), and the relative efficiency E_k (to 5 significant figures). The tabular values of c_k, V_k/σ^2, and E_k are accurate to within a unit in the last place given. It should be pointed out that though there is no analytical method for determining the optimal value of k for a fixed value of n, there exists an elegant analytical way of determining asymptotically the optimal quantile for the best estimator of σ; see, for example, Balakrishnan and Cohen (1991) for the relevant details and tables.

Estimators Based on Two Order Statistics of a Complete Sample

Unbiased linear estimators of the parameter σ, based on two order statistics x_l and x_m, are given by $\tilde{\sigma}_{lm} = c_l x_l + c_m x_m$, where $c_l E(x_l) + c_m E(x_m) = \sigma$, with $E(x_l)$ and $E(x_m)$ given by Equation (4.1.1), if k takes the values l and m. The variances of such estimators are given by

$$V_{lm} = c_l^2 \operatorname{var}(x_l) + c_m^2 \operatorname{var}(x_m) + 2c_l c_m \operatorname{cov}(x_l, x_m) \,, \tag{4.1.6}$$

where $\operatorname{var}(x_l)$ and $\operatorname{var}(x_m)$ are given by Equation (4.1.2), if k takes the values l and m, and $\operatorname{cov}(x_l, x_m)$ is given [see Section 3.2] by

$$\operatorname{cov}(x_l, x_m) = \sigma^2 \sum_{i=1}^{l} a_i^2 = \operatorname{var}(x_l), \qquad (l < m) \,. \tag{4.1.7}$$

It can be shown that, for given values of l and m, the values of c_l and c_m which yield the unbiased estimator $\tilde{\sigma}_{lm}$ with minimum variance are

$$c_l = 1 \Big/ \left(\sum_{i=1}^{l} a_i + \lambda \sum_{i=1}^{m} a_i \right) \tag{4.1.8}$$

and

$$c_m = \lambda c_l \,, \tag{4.1.9}$$

where

$$\lambda = \sum_{i=l+1}^{m} a_i \sum_{i=1}^{l} a_i^2 \Big/ \left(\sum_{i=1}^{l} a_i \sum_{i=1}^{m} a_i^2 - \sum_{i=1}^{m} a_i \sum_{i=1}^{l} a_i^2 \right) \,. \tag{4.1.10}$$

By substituting from Equations (4.1.2) and (4.1.7) - (4.1.9) into Equation (4.1.6), one finds that the minimum variance of $\tilde{\sigma}_{lm}$, for given l and m, is

$$V_{lm} = \sigma^2 \left[(1 + 2\lambda) \sum_{i=1}^{l} a_i^2 + \lambda^2 \sum_{i=1}^{m} a_i^2 \right] \Big/ \left(\sum_{i=1}^{l} a_i + \lambda \sum_{i=1}^{m} a_i \right)^2 \,. \tag{4.1.11}$$

The efficiency of this estimator (relative to the best linear unbiased estimator $\bar{x}$) is $E_{lm} = \operatorname{var}(\bar{x})/V_{lm}$, where, as before, $\operatorname{var}(\bar{x}) = \sigma^2/n$. Thus the relative efficiency E_{lm} is given by

$$E_{lm} = \left(\sum_{i=1}^{l} a_i + \lambda \sum_{i=1}^{m} a_i \right)^2 \Big/ n \left[(1 + 2\lambda) \sum_{i=1}^{l} a_i^2 + \lambda^2 \sum_{i=1}^{m} a_i^2 \right] \,. \tag{4.1.12}$$

The best estimator of σ, based on two order statistics x_l and x_m, is the one for those values of l and m which minimize V_{lm} (maximize E_{lm}). Once again, there is no analytical method for determining these values of l and m; hence, for each value of n, V_{lm} was computed for $l = 1(1)(n-1)$ and $m = (l+1)(1)n$. When the best values of l and m for a given n had been found, the corresponding c_l, c_m, and E_{lm} were also computed. The computations were performed on the IBM 1620 computer. Table D1 gives, for $n = 2(1)100$, the values of l and m for the best estimator of σ, the coefficients c_l and c_m (to 6 significant figures), the coefficient, V_{lm}/σ^2, of σ^2 in the variance V_{lm} of the estimator (to 7 decimal places), and the relative efficiency E_{lm} (to 5 significant figures). The tabular values of c_l, c_m, V_{lm}/σ^2, and E_{lm} are accurate to within a unit in the last place given. It needs to be mentioned yet again that though there is no analytical method for determining the optimal values of l

and m for a fixed value of n, there exists an analytical way of determining asymptotically the optimal quantiles for the best estimator of σ; see Balakrishnan and Cohen (1991) for details.

Estimators Based on One Order Statistic of a Censored Sample

The minimum-variance unbiased one-order-statistic estimator of the parameter σ from a complete sample of size n from a one-parameter exponential population has been given in this subsection as $\tilde{\sigma}_r = c_r x_r$, where x_r is the rth order statistic of the sample, r is chosen so as to minimize the variance of $\tilde{\sigma}_r$, and c_r is given in Table D1. For a censored sample in which one knows only the first m of the n ordered sample values, as in a life test of n items which is terminated at the time of the mth failure, the most efficient estimator can easily be seen to be $\tilde{\sigma}_k = c_k x_k$, where $k = \min(m, r)$. The coefficients $c_k = c(k, n)$ have been computed for $n = 1(1)40$ and $m = 1(1)n$, and are given in Table D2, together with the efficiency of the best one-order-statistic estimator with respect to the best m-order-statistic estimator given by Epstein and Sobel (1953).

4.1.3 Estimators of Parameters of the Two-Parameter Exponential Population

For the two-parameter exponential population with parameters α and σ, the probability density function is $f_2(x) = (1/\sigma)\exp[-(x-\alpha)/\sigma]$, $\alpha \le x < \infty$. The mean μ of this population is given by $\mu = \alpha + \sigma$. For a sample of size n from this population, the expected value of the kth order statistic, x_k, exceeds by α the value given in Equation (4.1.1) for the one-parameter exponential population, and thus is given by

$$E(x_k) = \alpha + \sigma \sum_{i=1}^{k} a_i \, . \tag{4.1.13}$$

The variance of x_k and the covariance of x_l and x_m are the same as for the one-parameter exponential population, and hence are given by Equations (4.1.2) and (4.1.7), respectively.

Unbiased linear estimators of the parameters α and σ and the mean μ may be obtained from any two order statistics x_l and x_m. These estimators are of the form

$$\tilde{\alpha} = c_{\alpha l} x_l + c_{\alpha m} x_m \, , \tag{4.1.14}$$

$$\tilde{\sigma} = c_{\sigma l} x_l + c_{\sigma m} x_m \, , \tag{4.1.15}$$

and

$$\tilde{\mu} = c_{\mu l} x_l + c_{\mu m} x_m \, . \tag{4.1.16}$$

It has been shown [see Sarhan, Greenberg, and Ogawa (1960)] that, for given l and m, coefficients of the best linear estimators based on two order statistics x_l and x_m are given by

$$c_{\alpha l} = 1 + c_\alpha, \qquad c_{\alpha m} = -c_\alpha \, , \tag{4.1.17}$$

$$c_{\sigma l} = -c_\sigma, \qquad c_{\sigma m} = c_\sigma \, , \tag{4.1.18}$$

and

$$c_{\mu l} = 1 + c_\alpha - c_\sigma, \qquad c_{\mu m} = c_\sigma - c_\alpha \ , \tag{4.1.19}$$

where

$$c_\alpha = \sum_{i=1}^{l} a_i \Big/ \sum_{i=l+1}^{m} a_i \tag{4.1.20}$$

and

$$c_\sigma = 1 \Big/ \sum_{i=l+1}^{m} a_i \ . \tag{4.1.21}$$

The variance of the estimator $\tilde{\sigma}$ is given by

$$V_{\tilde{\sigma}} = \sigma^2 \sum_{i=l+1}^{m} a_i^2 \Big/ \left(\sum_{i=l+1}^{m} a_i \right)^2 , \tag{4.1.22}$$

and the variances of the estimators $\tilde{\alpha}$ and $\tilde{\mu}$ are given by

$$\begin{aligned} V_{\tilde{\alpha}} &= \sigma^2 \sum_{i=1}^{l} a_i^2 + \left(\sum_{i=1}^{l} a_i \right)^2 V_{\tilde{\sigma}} \\ &= \sigma^2 \left[\sum_{i=1}^{l} a_i^2 + \left(\sum_{i=1}^{l} a_i \right)^2 \sum_{i=l+1}^{m} a_i^2 \Big/ \left(\sum_{i=l+1}^{m} a_i \right)^2 \right] \end{aligned} \tag{4.1.23}$$

and

$$\begin{aligned} V_{\tilde{\mu}} &= \sigma^2 \sum_{i=1}^{l} a_i^2 + \left(\sum_{i=1}^{l} a_i - 1 \right)^2 V_{\tilde{\sigma}} \\ &= \sigma^2 \left[\sum_{i=1}^{l} a_i^2 + \left(\sum_{i=1}^{l} a_i - 1 \right)^2 \sum_{i=l+1}^{m} a_i^2 \Big/ \left(\sum_{i=l+1}^{m} a_i \right)^2 \right] . \end{aligned} \tag{4.1.24}$$

The best linear unbiased estimators of α, σ, and μ based on all order statistics [see Sarhan and Greenberg (1956) or Balakrishnan and Cohen (1991)] are

$$\hat{\alpha} = \left[(n^2 - 1)x_1 - \sum_{i=2}^{n} x_i \right] \Big/ n(n-1) \ , \tag{4.1.25}$$

$$\hat{\sigma} = \left[\sum_{i=2}^{n} x_i - (n-1)x_1 \right] \Big/ (n-1) \ , \tag{4.1.26}$$

and

$$\hat{\mu} = \sum_{i=1}^{n} x_i / n = \bar{x} \ . \tag{4.1.27}$$

These are also the maximum-likelihood estimators. Their variances are

$$V_{\hat{\alpha}} = \sigma^2 / n(n-1) \ , \tag{4.1.28}$$

$$V_{\tilde{\sigma}} = \sigma^2/(n-1) \ , \tag{4.1.29}$$

and

$$V_{\hat{\mu}} = \sigma^2/n \ . \tag{4.1.30}$$

The efficiencies of the estimators $\tilde{\alpha}$, $\tilde{\sigma}$, and $\tilde{\mu}$ (relative to the best linear unbiased estimators $\hat{\alpha}$, $\hat{\sigma}$, and $\hat{\mu}$ based on all order statistics) are given by

$$E_{\tilde{\alpha}} = V_{\hat{\alpha}}/V_{\tilde{\alpha}} = \left(\sum_{i=l+1}^{m} a_i\right)^2 \Big/ n(n-1)\left[\sum_{i=1}^{l} a_i^2 \left(\sum_{i=l+1}^{m} a_i\right)^2 + \left(\sum_{i=1}^{l} a_i\right)^2 \sum_{i=l+1}^{m} a_i^2\right] , \tag{4.1.31}$$

$$E_{\tilde{\sigma}} = V_{\hat{\sigma}}/V_{\tilde{\sigma}} = \left(\sum_{i=l+1}^{m} a_i\right)^2 \Big/ (n-1) \sum_{i=l+1}^{m} a_i^2 , \tag{4.1.32}$$

and

$$E_{\tilde{\mu}} = V_{\hat{\mu}}/V_{\tilde{\mu}} = \left(\sum_{i=l+1}^{m} a_i\right)^2 \Big/ n\left[\sum_{i=1}^{l} a_i^2 \left(\sum_{i=l+1}^{m} a_i\right)^2 + \left(\sum_{i=1}^{l} a_i - 1\right)^2 \sum_{i=l+1}^{m} a_i^2\right] . \tag{4.1.33}$$

The best estimators of α, σ, and μ, based on two order statistics x_l and x_m, are those for the values of l and m which minimize $V_{\tilde{\alpha}}$, $V_{\tilde{\sigma}}$, and $V_{\tilde{\mu}}$, (maximize $E_{\tilde{\alpha}}$, $E_{\tilde{\sigma}}$, and $E_{\tilde{\mu}}$). It can be shown that, for a fixed value of m, the variances of the estimators are smallest when $l = 1$. It can be seen from Equations (4.1.23) and (4.1.24) that, for a fixed value of l, the value of m which minimizes $V_{\tilde{\sigma}}$ also minimizes $V_{\tilde{\alpha}}$ and $V_{\tilde{\mu}}$. There is no analytical method of determining the best value of m; hence, for each value of n, $V_{\tilde{\sigma}}$ was computed for $l = 1$ and $m = 2(1)n$. When the best value of m for a given n had been found, the corresponding c_α, c_σ, $V_{\tilde{\alpha}}$, $V_{\tilde{\mu}}$, $E_{\tilde{\alpha}}$, $E_{\tilde{\sigma}}$, and $E_{\tilde{\mu}}$ were also computed. The computations were performed on the IBM 1620 computer. Table D3 gives, for $n = 2(1)100$, the value of m for the best estimators of α, σ, and μ, the factors c_α and c_σ (to 6 significant figures or 6 decimal places, whichever is less accurate), the coefficient, $V_{\tilde{\alpha}}/\sigma^2$, of σ^2 in the variance $V_{\tilde{\alpha}}$ (to 7 significant figures or 9 decimal places, whichever is less accurate), the coefficients, $V_{\tilde{\sigma}}/\sigma^2$ and $V_{\tilde{\mu}}/\sigma^2$, of σ^2 in the variances $V_{\tilde{\sigma}}$ and $V_{\tilde{\mu}}$ (to 7 significant figures or 7 decimal places, whichever is less accurate), and the relative efficiencies $E_{\tilde{\alpha}}$, $E_{\tilde{\sigma}}$, and $E_{\tilde{\mu}}$ (to 5 significant figures). The tabular values of c_α, c_σ, $V_{\tilde{\alpha}}/\sigma^2$, $V_{\tilde{\sigma}}/\sigma^2$, $V_{\tilde{\mu}}/\sigma^2$, $E_{\tilde{\alpha}}$, $E_{\tilde{\sigma}}$, and $E_{\tilde{\mu}}$ are accurate to within a unit in the last place given.

4.1.4 Remarks

(i) The variance of the best estimator of σ for the two-parameter exponential population based on two order statistics x_l and x_m from a sample of size n is the same as the variance of the best estimator of σ for the one-parameter exponential population based on one order statistic x_k from a sample of size $n-1$, with $k = m-1$. This can be seen by a comparison of Equations (4.1.4) and (4.1.22). The relative efficiencies of these estimators are also equal, since in each case the variance of the best linear unbiased estimator based on all order statistics is $\sigma^2/(n-1)$.

(ii) For the one-parameter exponential population, the values $k = 80$, $l = 64$, and $m = 93$, with relative efficiencies $E_k = 65.093$ percent and $E_{lm} = 82.460$ percent, for $n = 100$

compare favorably with the results obtained by Sarhan, Greenberg, and Ogawa (1960), whose corresponding asymptotic values are $0.7968n$, $0.6386n$, $0.9266n$, 64.76 percent and 82.03 percent; see Balakrishnan and Cohen (1991) for such results on the estimation based on up to 25 optimally selected quantiles.

(iii) For the two-parameter exponential population, it can be seen from Equations (4.1.20) and (4.1.21) that, since $l = 1$ for the best estimators, $c_\alpha = a_1 c_\sigma = c_\sigma / n$. For convenience, however, separate columns for c_α and c_σ are given in Table D3

4.1.5 References

1. Balakrishnan, N. and Basu, Asit P. (Eds.) (1995). *The Exponential Distribution: Theory, Methods and Applications.* Gordon and Breach, Langhorne, Pennsylvania.
2. Balakrishnan, N. and Cohen, A. Clifford (1991). *Order Statistics and Inference: Estimation Methods.* Academic Press, San Diego.
3. Epstein, Benjamin and Sobel, Milton (1953). Life testing. *Journal of the American Statistical Association* **48**, 486-502.
4. Halperin, Max (1952). Maximum likelihood estimation in truncated samples. *Annals of Mathematical Statistics* **23**, 226-238.
5. Harter, H. Leon (1958). The Use of Sample Quasi-Ranges in Estimating Population Standard Deviation. *Wright Air Development Center Technical Report 58-200.* Wright-Patterson Air Force Base, Ohio, AD 151200.
6. Harter H. Leon (1959). The use of sample quasi-ranges in estimating population standard deviation. *Annals of Mathematical Statistics* **30**, 980-999.
7. Harter H. Leon (1961). Estimating the parameters of negative exponential populations from one or two order statistics. *Annals of Mathematical Statistics* **32**, 1078-1090.
8. Moore, Albert H. and Harter, H. Leon (1965). One-order statistic estimation of the scale parameters of Weibull populations. *IEEE Transactions on Reliability* **R-14**, 100-106.
9. Mosteller, Frederick (1946). On some useful "inefficient" statistics. *Annals of Mathematical Statistics* **17**, 377-408.
10. Sarhan, Ahmed E. (1954). Estimation of the mean and standard deviation by order statistics. *Annals of Mathematical Statistics* **25**, 317-328.
11. Sarhan, Ahmed E. (1955). Estimation of the mean and standard deviation by order statistics. Part III. *Annals of Mathematical Statistics* **26**, 576-592.
12. Sarhan, Ahmed E. and Greenberg, Bernard G. (1957). Tables for best linear estimates by order statistics of the parameters of single exponential distributions from singly and doubly censored samples. *Journal of the American Statistical Association* **52**, 58-87.
13. Sarhan, Ahmed E. and Greenberg, Bernard G. (1958). Estimation problems in the exponential distribution using order statistics. *Proceedings of the Statistical Techniques in Missile Evaluation Symposium*, 123-175, Virginia Polytechnic Institute.
14. Sarhan, Ahmed E., Greenberg, Bernard G., and Ogawa, Junjiro (1960). Simplified Estimates for the Exponential Distribution. *Interim Technical Report No. 1*, Grant No. DA-ORD-7, University of North Carolina.
15. Wilks, S.S. (1948). Order statistics. *Bulletin of the American Mathematical Society* **54**, 6-50.

4.2 Interval Estimation of the Parameter σ [2]

4.2.1 Introduction

During the period 1957-1969, several papers were published on point and interval estimation of parameters of various populations by the use of order statistics or differences of order statistics (ranges and quasi-ranges). Harter (1959) gave theory and extensive tables for point estimation of the population standard deviation for a rectangular population, based on the sample range, and for a normal population, based on one quasi-range and on linear combinations of two quasi-ranges. The results are included in Chapter 1, Section 1.1; Chapter 2, Section 2.1; and Tables A4, A5, and B1 of the present volume. The estimator based on the sample range is the efficient estimator for a rectangular population, while in the case of a normal population the efficiencies of the estimators based on one quasi-range and on the best linear combination of two quasi-ranges are greater than or equal to the asymptotic values, approximately 65 percent and 80 percent, respectively. For asymmetric populations, however, the efficiency of estimators based on ranges and quasi-ranges is not so high. For example, in the case of a one-parameter negative exponential population, better estimates of the parameter, which is both mean and standard deviation, can be obtained from a single order statistic than from a quasi-range. Harter (1961b) gave theory and tables for point estimation of the parameter for a one-parameter negative exponential population from one or two order statistics and of the parameters for a two-parameter negative exponential population from two order statistics. The theory is included in Section 4.1 of this chapter and the tables as Tables D1 and D3 of this volume.

Interval estimation of population parameters from sample quasi-ranges has been considered in papers by Chu (1957) and Chu, Leone, and Topp (1957). The method, outlined in the latter paper, of obtaining confidence bounds by first applying distribution-free methods and then imposing the distribution, was applied by Leone, Rutenberg, and Topp (1961), who gave tables of confidence intervals for the standard deviation of normal, one-parameter negative exponential, and rectangular populations. Harter (1961a) pointed out that the exact confidence intervals, in which the coefficients of the quasi-ranges are the reciprocals of percentage points of those quasi-ranges for the standardized population of the type under consideration, have expected length much shorter than the approximate confidence intervals given by Leone, Rutenberg, and Topp. He illustrated this point by giving tables of exact confidence intervals, based on the sample range, for the standard deviation of a rectangular population. These results are given in Chapter 2, Section 2.2 and Table B3 of the present volume. A similar tabulation of exact confidence intervals for the standard deviation of a normal population, each based on one sample quasi-range, included as Table A8 of this volume, was first published by Harter (1964c). In the latter paper, the criterion of minimizing the expected length (maximizing the effectiveness) of the substitute confidence interval has been replaced by the criterion of maximizing the efficiency of the substitute confidence interval, as defined and discussed in detail in that paper. Briefly, the superiority of the new criterion lies in the fact that it penalizes the occurrence of cases in which the upper confidence bound falls short of (or the lower confidence bound exceeds) the true parameter value, whereas the old one actually rewards such occurrences. Efficiency of interval estimators is defined by a logical extension of the definition of efficiency of point estimators. The efficiency of a substitute upper confidence bound is defined as the ratio, expressed as a percentage, of the mean squared deviation of the conventional upper confidence bound from the true parameter value to the mean

[2] Earlier versions of this material were published by Harter (1963, 1964b).

squared deviation of the substitute upper confidence bound from the true parameter value. Similarly, the efficiency of a substitute central confidence interval is defined as the ratio, expressed as a percentage, of the sum of the mean squared deviations of the conventional upper and lower confidence bounds from the true parameter value to the sum of the mean squared deviations of the substitute upper and lower confidence bounds from the true parameter value. For details concerning these criteria, the reader is referred to Chapter 1, Section 1.2 of this volume.

Just as a single order statistic will yield a more efficient point estimator of the parameter of a one-parameter negative exponential population than will a quasi-range, it will also provide more efficient upper confidence bounds and central confidence intervals. Appendix D contains a table (Table D4) of upper and lower confidence bounds (confidence $1-P$) and central confidence intervals (confidence $1-2P$) for the parameter of a one-parameter negative exponential population, based on the mth order statistic of a sample of size n, for $P = .0001, .0005, .001, .005, .01, .025, .05, .1(.1).5$ and $n = 1(1)20(2)40$ for values of m which maximize the efficiency of the upper confidence bound and/or that of the central confidence interval.

4.2.2 Mathematical Formulation

Cumulative Distribution of Transformed Variable $y = e^{-x_m}$

The probability density function $\phi(x_m)$ of the mth order statistic of a sample of size n from a population having probability density function $f(x)$ and cumulative distribution function $F(x)$ is given by

$$\phi(x_m) = [\Gamma(n+1)/\Gamma(m)\Gamma(n-m+1)][F(x_m)]^{m-1}[1-F(x_m)]^{n-m}f(x_m)\ . \tag{4.2.1}$$

If we make the substitution $y = 1 - F(x_m)$, we obtain as the probability density function of the transformed variable

$$\phi(y) = [\Gamma(n+1)/\Gamma(m)\Gamma(n-m+1)](1-y)^{m-1}y^{n-m}\ . \tag{4.2.2}$$

Integrating between the limits 0 and y, we obtain the cumulative distribution function

$$\Phi(y) = [\Gamma(n+1)/\Gamma(m)\Gamma(n-m+1)]\int_0^y (1-y)^{m-1}y^{n-m}\,dy\ , \tag{4.2.3}$$

which may be written in terms of the incomplete Beta-function ratio as

$$\Phi(y) = I_y(n-m+1, m)\ . \tag{4.2.4}$$

Note that Equation (4.2.4) holds in general for any population; in the case at hand, that of the one-parameter negative exponential population, $F(x_m) = 1 - e^{-x_m}$, so $y = e^{-x_m}$.

Conventional Interval Estimators of σ

The conventional point estimator, $\hat{\sigma}_{n,n}$, of the parameter σ from a sample of size n is simply the sample mean $\bar{x}$. Since we are working with the standardized population ($\sigma = 1$), which we may do without loss of generality, the upper and lower conventional confidence bounds (confidence $1-P$) are given by $B_u = \bar{x}/\bar{x}_P$ and $B_l = \bar{x}/\bar{x}_{1-P}$, respectively, where $\bar{x}_P$ and $\bar{x}_{1-P}$ are the percentage points of the distribution of $\bar{x}$ corresponding to cumulative probabilities P and $1-P$, respectively.

The conventional upper confidence bound (confidence $1-P$), based on the sample mean, has expected value given by

$$E_n(B_u) = 2n/\chi^2_{P,2n} \tag{4.2.5}$$

and the conventional central confidence interval (confidence $1-2P$), also based on the sample mean, has expected length given by

$$E_n(B_u - B_l) = 2n\left(1/\chi^2_{P,2n} - 1/\chi^2_{1-P,2n}\right) . \tag{4.2.6}$$

The first subscript on chi-square represents the cumulative probability and the second is the number of degrees of freedom, n being the sample size.

If only the first m ordered sample values are known, the point estimator and the confidence bounds obtained from this censored sample will, for the sake of brevity, be called the censored point estimator and the censored confidence bounds, respectively. The censored point estimator $\hat{\sigma}_{m,n}$, which is given by

$$\hat{\sigma}_{m,n} = [x_1 + x_2 + \cdots + x_m + (n-m)x_m]/m , \tag{4.2.7}$$

has a chi-square over degrees of freedom distribution with $2m$ degrees of freedom [see Epstein and Sobel (1953) or Arnold, Balakrishnan and Nagaraja (1992)]. Since this is exactly the same as the distribution of the mean $\bar{x}$ for a sample of size m, the expected value of the censored upper confidence bound and the expected length of the censored central confidence interval may be obtained by replacing n by m in Equations (4.2.5) and (4.2.6), which yields

$$E_m(B_u) = 2m/\chi^2_{P,2m} \tag{4.2.8}$$

and

$$E_m(B_u - B_l) = 2m(1/\chi^2_{P,2m} - 1/\chi^2_{1-P,2m}) . \tag{4.2.9}$$

The mean squared deviation of the conventional upper confidence bound from the true parameter value may be written in the form

$$E(D^2_{n,u}) = E[(B_u - 1)^2] = E[(\bar{x}/\bar{x}_P - 1)^2] = 1 - 2E(\bar{x})/\bar{x}_P + E(\bar{x}^2)/\bar{x}^2_P . \tag{4.2.10}$$

Since $\bar{x}$ has a chi-square over degrees of freedom distribution with $2n$ degrees of freedom, $E(\bar{x}) = E[\chi^2_{2n}/2n] = 1$ and $E(\bar{x}^2) = E\{[\chi^2_{2n}/2n]^2\} = (2n+2)/2n = (n+1)/n$, so that Equation (4.2.10) may be written in the form

$$E(D^2_{n,u}) = 1 - 4n/\chi^2_{P,2n} + 4n(n+1)/[\chi^2_{P,2n}]^2 . \tag{4.2.11}$$

Similarly, the mean squared deviation of the conventional lower confidence bound from the true parameter value is given by

$$E(D^2_{n,l}) = 1 - 4n/\chi^2_{1-P,2n} + 4n(n+1)/[\chi^2_{1-P,2n}]^2 . \tag{4.2.12}$$

The mean squared deviations of the censored upper and lower confidence bounds may be obtained by replacing n by m in Equations (4.2.11) and (4.2.12) for the conventional confidence bounds, which yields

$$E(D^2_{m,u}) = 1 - 4m/\chi^2_{P,2m} + 4m(m+1)/[\chi^2_{P,2m}]^2 \tag{4.2.13}$$

and

$$E(D^2_{m,l}) = 1 - 4m/\chi^2_{1-P,2m} + 4m(m+1)/[\chi^2_{1-P,2m}]^2 \ . \tag{4.2.14}$$

Effectiveness of Interval Estimators Based on x_m

The substitute upper and lower confidence bounds (confidence $1 - P$), based on the mth order statistic x_m, are given by $B_{um} = x_m/x_{m,P}$ and $B_{lm} = x_m/x_{m,1-P}$, respectively, where $x_{m,P}$ and $x_{m,1-P}$ are the percentage points of the distribution of x_m corresponding to cumulative probabilities P and $1 - P$, respectively.

The expected value of the substitute upper confidence bound is given by

$$E(B_{um}) = E(x_m)/x_{m,P} \tag{4.2.15}$$

and the expected length of the substitute central confidence interval is given by

$$E(B_{um} - B_{lm}) = E(x_m)(1/x_{m,P} - 1/x_{m,1-P}) \ . \tag{4.2.16}$$

The effectiveness of the substitute upper confidence bound is then

$$F_u = 100\, E_n(B_u)/E(B_{um}) \ , \tag{4.2.17}$$

while that of the substitute central confidence interval is

$$F_i = 100\, E_n(B_u - B_l)/E(B_{um} - B_{lm}) \ , \tag{4.2.18}$$

where $E_n(B_u)$ and $E_n(B_u - B_l)$ are given by Equations (4.2.5) and (4.2.6) and $E(B_{um})$ and $E(B_{um} - B_{lm})$ by Equations (4.2.15) and (4.2.16), respectively. The relative effectiveness of the substitute upper confidence bound (as compared with the censored bound) is given by

$$S_u = 100\, E_m(B_u)/E(B_{um}) \ , \tag{4.2.19}$$

and that of the substitute central confidence interval (as compared with the censored interval) is given by

$$S_i = 100\, E_m(B_u - B_l)/E(B_{um} - B_{lm}) \ , \tag{4.2.20}$$

where $E_m(B_u)$ and $E_m(B_u - B_l)$ are given by Equations (4.2.8) and (4.2.9), respectively.

Efficiency of Interval Estimators Based on x_m

The mean squared deviation of the substitute upper confidence bound, based on the mth order statistic, from the true parameter value may be written in the form

$$E(D^2_{s,u}) = E(B_{um} - 1)^2 = E[(x_m/x_{m,P} - 1)^2] = 1 - 2E(x_m)/x_{m,P} + E(x^2_m)/x^2_{m,P} \ . \tag{4.2.21}$$

Similarly, the mean squared deviation of the substitute lower confidence bound, based on the mth order statistic, from the true parameter value is given by

$$E(D^2_{s,l}) = 1 - 2E(x_m)/x_{m,1-P} + E(x^2_m)/x^2_{m,1-P} \ . \tag{4.2.22}$$

In connection with Equations (4.2.21) and (4.2.22), recall that the expected value and the variance of x_m (when $\sigma = 1$) are given [see Section 3.2] by

$$E(x_m) = \sum_{j=1}^{m} [1/(n-j+1)] \tag{4.2.23}$$

and

$$V(x_m) = \sum_{j=1}^{m} [1/(n-j+1)^2], \tag{4.2.24}$$

and that the expected value of x_m^2 is simply

$$E(x_m^2) = V(x_m) + [E(x_m)]^2 \; . \tag{4.2.25}$$

The efficiency of the substitute upper confidence bound (as compared with the conventional bound) is given by

$$E_u = 100\, E(D_{n,u}^2)/E(D_{s,u}^2) \tag{4.2.26}$$

and that of the substitute central confidence interval (as compared with the conventional interval) is given by

$$E_i = 100[E(D_{n,u}^2) + E(D_{n,l}^2)]/[E(D_{s,u}^2) + E(D_{s,l}^2)]. \tag{4.2.27}$$

The relative efficiency of the substitute upper confidence bound (as compared with the censored bound) is given by

$$R_u = 100\, E(D_{m,u}^2)/E(D_{s,u}^2) \tag{4.2.28}$$

and that of the substitute central confidence interval (as compared with the censored interval) is given by

$$R_i = 100[E(D_{m,u}^2) + E(D_{m,l}^2)]/[E(D_{s,u}^2) + E(D_{s,l}^2)]. \tag{4.2.29}$$

4.2.3 Method of Computation of Tables

Coefficients of x_m in Exact Confidence Bounds for σ

The computation of the tables required determination of the coefficients, $1/x_{m,P}$ and $1/x_{m,1-P}$ respectively, of the mth order statistic x_m in the substitute upper and lower confidence bounds B_{um} and B_{lm}. According to Equation (4.2.4), the cumulative distribution function of $y = e^{-x_m}$ is the incomplete Beta-function ratio $I_y(n-m+1, m)$. Hence the percentage points y_P of y are the percentage points of the incomplete Beta-function ratio (the Beta distribution), and from the relation between y and x_m we see that $1/x_{m,P} = -1/(ln\ y_P)$ and $1/x_{m,1-P} = -1/(ln\ y_{1-P})$. Hence the coefficients of x_m in the substitute upper and lower confidence bounds are simply the negative reciprocals of the natural logarithms of the percentage points of the incomplete Beta-function ratio. Since no sufficiently extensive and accurate table of the percentage points of the incomplete Beta-function ratio was available, it was necessary to compute such a table [see Harter (1964a)].

Effectiveness of Interval Estimators Based on x_m

The first step was the determination of the expected values of the conventional and censored upper confidence bounds and the expected lengths of the conventional and censored central confidence intervals, which was accomplished by the use of Equations (4.2.5), (4.2.8), (4.2.6), and (4.2.9), respectively. Use was made of the previously discussed new table of percentage points of the chi-square distribution [see Harter (1964a)].

The second step was the determination of the expected value of the substitute upper confidence bound and the expected length of the substitute central confidence interval, which was accomplished by the use of Equations (4.2.15) and (4.2.16), respectively.

The third and final step was the calculation of the effectivenesses and relative effectivenesses of the substitute upper confidence bounds and the substitute central confidence intervals, which required the use of Equations (4.2.17)-(4.2.20).

All of the above computations were performed on the IBM 7094 computer. One FORTRAN program was sufficient for all of the computations except those involved in obtaining the auxiliary table of percentage points of the Beta distribution, which was computed in advance and used as input to the main program, along with the table of percentage points of the chi-square distribution. In order to avoid loss of accuracy, it was necessary to use double precision for the internal arithmetic operations, even though both the input and the output were in single precision.

For each sample size n, values of the coefficients of the mth order statistic (for several values of m) in the substitute upper and lower confidence bounds were rounded to 7 significant figures or 6 decimal places, whichever is less accurate. The effectivenesses and relative effectivenesses of the substitute upper confidence bounds and the substitute central confidence intervals were rounded to the nearest tenth of one percent The results were used to determine, for each pair of values of n and $1 - P$, the value of m which maximizes the effectiveness of the upper confidence bound (confidence $1 - P$). In most cases (all but 70 out of 330), the same value of m also maximizes the effectiveness of the central confidence interval (confidence $1 - 2P$). For each of the 70 cases in which this is not true, the value of m which does maximize the effectiveness of the central confidence interval was determined. Table D4A, with suitable title and column headings, was reproduced from the former 260 cases, and Table D4B, with suitable title and the same column headings, was produced from the latter 70 cases.

Efficiency of Interval Estimators Based on x_m

The first step was the determination of the mean squared deviations of the conventional upper and lower confidence bounds and the censored upper and lower confidence bounds from the true parameter value, which was accomplished by the use of Equations (4.2.11)-(4.2.14). Use was made of the previously mentioned new table of percentage points of the chi-square distribution [see Harter (1964a)].

The second step was the determination of the mean squared deviations of the substitute upper and lower confidence bounds from the true parameter value, which was accomplished by the use of Equations (4.2.21)-(4.2.25).

The third and final step was the calculation of the efficiencies and relative efficiencies of the substitute upper confidence bounds and the substitute central confidence intervals, which required the use of Equations (4.2.26)-(4.2.29).

All of the above computations were performed on the IBM 7094 computer. One FOR-

TRAN program was sufficient for all of the computations except those involved in obtaining the auxiliary tables of percentage points of the chi-square and Beta distributions, which were used as input to the main program. In order to avoid loss of accuracy, it was necessary to use double precision for the internal arithmetic operations, even though both the input and the output were in single precision.

For each sample size n, values of the coefficients of the mth order statistic (for several values of m) in the substitute upper and lower confidence bounds were rounded to 7 significant figures or 6 decimal places, whichever is less accurate. The efficiencies and relative efficiencies of the substitute upper confidence bounds and the substitute central confidence intervals were rounded to the nearest tenth of one percent. The results were used to determine, for each pair of values of n and $1-P$, the value of m which maximizes the efficiency of the upper confidence bound (confidence $1-P$). In most cases (all but 14 out of 330), the same value of m also maximizes the efficiency of the central confidence interval (confidence $1-2P$). For each of the 14 cases in which this is not true, the value of m which does maximize the efficiency of the central confidence interval was determined. Table D4C, with suitable title and column headings, was reproduced from the former 316 cases, and Table D4D, with suitable title and the same column headings, was reproduced from the latter 14 cases.

4.2.4 Comparison with Other Tables

The effectiveness and the efficiency of the best exact substitute central confidence intervals are in the neighborhood of 80 percent and 60 percent, respectively, except for very small samples, in which case they are even higher. These compare with an effectiveness in the neighborhood of 55 percent, corresponding to an efficiency of approximately $(.55)^2 =$ 30 percent, or even less, for the approximate confidence intervals of Leone, Rutenberg, and Topp (1961). There is a twofold reason for the difference, which is due partly to the superiority of the exact method over the approximation based on distribution-free methods and partly to the use of a single order statistic instead of a quasi-range. The effectiveness and the efficiency of the exact confidence intervals could have been made slightly higher by optimizing the upper and lower confidence bounds separately as Leone, Rutenberg, and Topp did. It was decided that the small increase in effectiveness or efficiency that would have resulted was not worth the additional complication of having the upper and lower confidence bounds based on different order statistics.

The effectiveness of the best exact substitute upper confidence bounds is even higher than that of the best exact substitute central confidence intervals. The efficiency of the best exact substitute upper confidence bounds is, like that of the best exact substitute central confidence intervals, in the neighborhood of 60 percent, except for very small samples, for which it is even higher. In these cases, no other tables are available for purposes of comparison.

4.2.5 Possible Uses of the Tables

One advantage often claimed for substitute estimates is that they are easier to compute than the conventional estimates. In this case there is little advantage here, since the conventional estimates are based on the sample mean, which is itself relatively easy to compute. The principal uses of the present tables will probably be in obtaining interval estimates for the mean time-to-failure of (say) electronic components on the basis of incomplete results of a life test, without waiting for all of the components in the sample placed on test to fail. In this context, the relative efficiencies are quite meaningful, since the censored interval estimators are the best that are available at the time of occurrence of the mth failure.

4.2.6 Numerical Example

Suppose that 20 components are subjected to a life test and that the numbers of hours to failure are as follows:

2.37	58.93	80.90	108.92	167.59
17.56	71.48	90.87	112.26	282.49
34.84	71.84	91.22	126.87	335.33
36.38	79.31	96.35	127.05	341.19

Assuming that the life of components of this type has a one-parameter negative exponential distribution, one may wish to set a 99 percent upper confidence bound on the parameter σ. This may be done in a number of ways, of which three will be discussed here. The conventional bound is $\hat{\sigma}_{20,20}/(\chi^2_{.01,40}/40)$, where the conventional point estimator $\hat{\sigma}$ is simply the mean life $\bar{x}$. Numerically, the conventional bound is $116.69/(22.1643/40) = 210.59$ hours. Suppose, however, that the experimenter decides to terminate the test at the time of the seventeenth failure. Then the best bound that he/she can obtain is the censored bound $\hat{\sigma}_{17,20}/(\chi^2_{.01,34}/34)$, where the censored point estimator $\hat{\sigma}_{17,20}$ can be computed from Equation (4.2.7). Numerically, the censored bound is $110.44/(17.7891/34) = 211.09$ hours. The most efficient (effective) substitute bound $B_{u,17}$, where B_{um}/x_m can be read from Table D4C(D4A), is almost as good and is easier to compute. Numerically, the substitute bound is $1.155046(167.59) = 193.57$ hours.

The "data" for this example were actually obtained by drawing a random sample of size 20 from a one-parameter exponential population with parameter $\sigma = 100$ hours and then ordering the observations. Since the substitute bound is closer to the true value of the parameter than is either the censored bound or the conventional bound, one might be led to the erroneous conclusion that substitute bounds are better than censored or conventional ones. This may be true in some individual cases, as in the present one, because of sampling fluctuations, but it is not true on the average. The true situation can be seen by comparing expected values, which are $100/(22.1643/40) = 180.47$ hours for the conventional bound, $100/(17.7891/34) = 191.13$ hours for the censored bound, and $1.155046(176.441) = 203.80$ hours for the substitute bound, or better still by comparing efficiencies. From Table D4C, one finds that the efficiency of the substitute bound (compared with the conventional bound) is 58.3 percent and that its relative efficiency (compared with the censored bound) is 75.3 percent.

4.2.7 References

1. Arnold, Barry C., Balakrishnan, N., and Nagaraja, H.N. (1992). *A First Course in Order Statistics.* John Wiley & Sons, New York.

2. Chu, J.T. (1957). Some uses of quasi-ranges. *Annals of Mathematical Statistics* **28**, 173-180.

3. Chu, J.T., Leone, F.C., and Topp, C.W. (1957). Some uses of quasi-ranges II (abstract). *Annals of Mathematical Statistics* **28**, 530-531.

4. Epstein, Benjamin and Sobel, Milton (1953). Life testing. *Journal of the American Statistical Association* **48**, 486-502.

5. Harter, H. Leon (1959). The use of sample quasi-ranges in estimating population standard deviation. *Annals of Mathematical Statistics* **30**, 980-999.

6. Harter, H. Leon (1961a). The use of sample ranges in setting exact confidence bounds for the standard deviation of a rectangular population, *Journal of the American Statistical Association* **56**, 601-609.

7. Harter, H. Leon (1961b). Estimating the parameters of negative exponential populations from one or two order statistics. *Annals of Mathematical Statistics* **32**, 1078-1090.

8. Harter, H. Leon (1963). Exact confidence bounds, based on one order statistic, for the parameter of a one-parameter negative exponential population. *Rider Anniversary Volume* (Ed., H. Leon Harter), 77-100. Aeronautical Research Laboratories, Wright-Patterson Air Force Base.

9. Harter, H. Leon (1964a). *New Tables of the Incomplete Gamma-Function Ratio and of Percentage Points of the Chi-Square and Beta Distributions.* U.S. Government Printing Office, Washington.

10. Harter, H. Leon (1964b). Exact confidence bounds, based on one order statistic, for the parameter of an exponential population. *Technometrics* **6**, 301-317.

11. Harter, H. Leon (1964c). Criteria for best substitute interval estimators, with an application to the normal distribution. *Journal of the American Statistical Association* **59**, 1133-1140.

12. Leone, F.C., Rutenberg, Y.H., and Topp, C.W. (1961). The use of sample quasi-ranges in setting confidence intervals for the population standard deviation. *Journal of the American Statistical Association* **56**, 260-272.

5

Singly Censored Samples From Populations Related to the Exponential

5.1 Weibull Population with Known Location and Shape Parameters [1]

5.1.1 Introduction

Epstein and Sobel (1953) have pointed out the advantage of the use of ordered data from truncated tests to estimate the parameters of parent populations, and have worked out details for the exponential distribution. In particular, they have derived an estimator $\hat{\sigma}$ (which is maximum likelihood, unbiased, and minimum variance), based on the first m out of n ordered observations, of the parameter σ of an exponential population and have shown that $2m\hat{\sigma}/\sigma$ has a chi-square distribution with $2m$ degrees of freedom (independent of n). They have also given without derivation the maximum likelihood estimator $\hat{\theta}^K$, based on the first m out of n ordered observations, of θ^K, where θ is the scale parameter of a Weibull population with known shape parameter K. N.R. Mann (1963, p. 39) has stated without proof that $2m\hat{\theta}^K/\theta^K$ has a chi-square distribution with $2m$ degrees of freedom. The missing derivation and proof are supplied in the present section. Expressions are given for upper and lower confidence bounds, $\bar{\theta}$ and $\underline{\theta}$, and for the efficiencies, as defined by Harter (1964b,c) [see Chapter 1, Section 1.2 and Chapter 4, Section 4.2 of the present volume], of $\bar{\theta}$ and the central confidence interval $(\underline{\theta}, \bar{\theta})$. Brief discussions of the method of computation of a table of unbiasing factors for maximum-likelihood estimators and variances of unbiased estimators and of its use are given, as well as a numerical example which illustrates the computation of both point and interval estimates and the efficiencies of both point and interval estimators.

5.1.2 Point Estimation of the Scale Parameter θ

Maximum-Likelihood Estimator for the Scale Parameter

The probability density function of the random variable Y having a Weibull distribution with location parameter 0, scale parameter θ, and shape parameter K is given by

$$f(y) = \begin{cases} (K/\theta)(y/\theta)^{K-1}\exp[-(y/\theta)^K], & y > 0 \\ 0 \ \text{elsewhere} . \end{cases} \tag{5.1.1}$$

For a detailed discussion on the Weibull distribution, interested readers may refer to Johnson, Kotz, and Balakrishnan (1994).

[1] An earlier version of this material was published by Harter and Moore (1965).

Now if we define the random variable X by $X = Y^K$ and make the change of variable $x = y^K$ in Equation (5.1.1), we find the probability density function of the random variable X to be

$$g(x) = \begin{cases} \exp(-x/\theta^K)/\theta^K, & x > 0 \\ 0 \text{ elsewhere} , \end{cases} \tag{5.1.2}$$

which is the familiar exponential density function with parameter $\sigma = \theta^K$. Therefore if Y has a Weibull distribution with scale parameter θ and shape parameter K and if $X = Y^K$, then X is exponentially distributed with parameter $\sigma = \theta^K$. Hence a maximum likelihood m-order-statistic estimator for θ can be obtained from the "best" m-order-statistic estimator for $\sigma = \theta^K$ derived by Epstein and Sobel (1955), which is given by

$$\hat{\sigma}_{mn} = [x_{1n} + x_{2n} + \cdots + x_{mn} + (n-m)x_{mn}]/m , \tag{5.1.3}$$

where x_{in} $(i = 1, 2, \ldots, m)$ are the first m order statistics of a sample of size n from an exponential population. Now, taking the Kth root of both sides of Equation (5.1.3), we obtain

$$\hat{\sigma}_{mn}^{1/K} = \{[x_{1n} + x_{2n} + \cdots + x_{mn} + (n-m)x_{mn}]/m\}^{1/K} . \tag{5.1.4}$$

Since $x_{in} = y_{in}^K$ we can write

$$\hat{\theta}_{mn} = \hat{\sigma}_{mn}^{1/K} = \{[y_{1n}^K + y_{2n}^K + \cdots + y_{mn}^K + (n-m)y_{mn}^K]/m\}^{1/K} . \tag{5.1.5}$$

Now, since $\hat{\sigma}_{mn}$ is a maximum likelihood estimator of σ, $\hat{\theta}_{mn} = \hat{\sigma}_{mn}^{1/K}$ is a maximum likelihood estimator of $\theta = \sigma^{1/K}$.

The probability density function of the random variable $X_1 = \hat{\sigma}_{mn}$ is given by Epstein and Sobel (1953) as

$$f_m(x_1) = \begin{cases} [1/\Gamma(m)](m/\sigma)^m x_1^{m-1} \exp(-mx_1/\sigma), & x_1 > 0 \\ 0 \text{ elsewhere} . \end{cases} \tag{5.1.6}$$

Now we let $\hat{\theta}_{mn} = \hat{\sigma}_{mn}^{1/K}$ or $Y_1 = X_1^{1/K}$ and we find the probability density function of $Y_1 = \hat{\theta}_{mn}$ to be

$$g_m(y_1) = \begin{cases} [K/\Gamma(m)](m/\sigma)^m y_1^{Km-1} \exp(-my_1^K/\sigma), & y_1 > 0 \\ 0 \text{ elsewhere} . \end{cases} \tag{5.1.7}$$

Now, making the substitution $\sigma = \theta^K$ in Equation (5.1.7), we obtain

$$g_m(y_1) = \begin{cases} [K/\Gamma(m)](m/\theta^K)^m y_1^{Km-1} \exp(-my_1^K/\theta^K), & y_1 > 0 \\ 0 \text{ elsewhere} . \end{cases} \tag{5.1.8}$$

Hereafter, for simplicity, we shall denote $\hat{\theta}_{mn}$ by $\hat{\theta}$.

Unbiased Estimator for the Scale Parameter

The expected value of $\hat{\theta}$ is found by using Equation (5.1.8) to be

$$E(\hat{\theta}) = \{\theta\Gamma(m + 1/K)\}/\{m^{1/K}\Gamma(m)\} . \tag{5.1.9}$$

Hence an unbiased estimator of θ is given by

$$\tilde{\theta} = [m^{1/K}\Gamma(m)/\Gamma(m+1/K)]\hat{\theta} . \tag{5.1.10}$$

The variance of the unbiased estimator $\tilde{\theta}$ is found to be

$$\mathrm{Var}\tilde{\theta} = \theta^2\{[\Gamma(m)\Gamma(m+2/K)/\Gamma^2(m+1/K)] - 1\} . \tag{5.1.11}$$

Values of the unbiasing factor $\tilde{\theta}/\hat{\theta}$ for the maximum-likelihood estimator and of the ratio $\mathrm{Var}(\tilde{\theta})/\theta^2$ of the variance of the unbiased estimator to θ^2, expressions for which can be obtained by dividing both sides of Equation (5.1.10) by $\hat{\theta}$ and both sides of Equation (5.1.11) by θ^2, were computed for $m = 1(1)100$ and $K = 0.5(0.5)4.0(1.0)8.0$. The computations were performed on the IBM 1620 computer with FORTRAN programming, use being made of Stirling's approximation to the Gamma function. Twelve decimal digits were carried in the computations, but the values of the unbiasing factor were rounded to 6 decimal places (6 or 7 significant digits) and those of the variance to 8 decimal places (5 to 9 significant digits). The results are shown in Table E1.

Use of Table

In life-testing situations, one may wish to terminate a test without waiting for all n of the items placed on test to fail. If the life distribution is Weibull with known shape parameter K, where K is one of the values included in Table E1, and if the test is terminated at the time of the mth failure ($m \leq 100$), one can compute a maximum-likelihood estimate $\hat{\theta}$ of the scale parameter θ from Equation (5.1.5) and then multiply $\hat{\theta}$ by the unbiasing factor $\tilde{\theta}/\hat{\theta}$ given in Table E1 to obtain an unbiased estimate $\tilde{\theta}$. The ratio, $\mathrm{Var}(\tilde{\theta})/\theta^2$, of the variance of the unbiased estimator to θ^2 is also given in the table. The efficiency E_p of the unbiased point estimator based on the first m order statistics as compared with the one based on all n order statistics ($m < n \leq 100$) can be found by taking the ratio of two entries in the $\mathrm{Var}(\tilde{\theta})/\theta^2$ columns of Table E1. It can be seen that the percentage efficiency is approximately $100m/n$.

5.1.3 Interval Estimation of the Scale Parameter θ

Confidence Bounds for the Scale Parameter

From Equation (5.1.8) it can be easily seen that $2m\hat{\theta}^K/\theta^K$ has a chi-square distribution with $2m$ degrees of freedom:

$$2m\hat{\theta}^K/\theta^K = \chi^2_{2m} . \tag{5.1.12}$$

Solving for θ, we obtain

$$\theta = (2m/\chi^2_{2m})^{1/K}\hat{\theta} . \tag{5.1.13}$$

Then an upper confidence bound with confidence level $1-P$ (lower confidence bound with confidence level P) on θ is given by

$$\bar{\theta}_{1-P} = \underline{\theta}_P = (2m/\chi^2_{2m,P})^{1/K}\hat{\theta} , \tag{5.1.14}$$

where the first subscript on χ^2 is the number of degrees of freedom and the second one is the cumulative probability. The interval between lower and upper confidence bounds, each with confidence level $1-P$, will be called a central confidence interval with confidence level $1-2P$. Equations (5.1.12)-(5.1.14) remain valid when $m = n$, in which case Equation (5.1.14) is an expression for the conventional confidence bound based on all n observations.

Efficiency of Confidence Bounds and Intervals

Harter (1964b,c) [see Chapter 1, Section 1.2 and Chapter 4, Section 4.2, of the present volume] has defined the efficiency of a substitute upper confidence bound as the ratio, expressed as a percentage, of the mean squared deviation of the conventional upper confidence bound from the true parameter value to the mean squared deviation of the substitute upper confidence bound from the true parameter value. This definition may be expressed symbolically in the form

$$E_u = 100\, E[(\bar{\theta} - \theta)^2]/E[(\bar{\theta}' - \theta)^2] \,, \tag{5.1.15}$$

where E_u is the efficiency (in percent) of the substitute upper confidence bound, $E[(\bar{\theta}-\theta)^2]$ is the mean squared deviation of the conventional upper confidence bound $\bar{\theta}$ from the true value θ of the parameter, and $E[(\bar{\theta}' - \theta)^2]$ is the mean squared deviation of the substitute upper confidence bound $\bar{\theta}'$ from the true value θ of the parameter. Further, the efficiency of a substitute central confidence interval is defined as the ratio, expressed as a percentage, of the sum of the mean squared deviations of the conventional upper and lower confidence bounds from the true parameter value to the sum of the mean squared deviations of the substitute upper and lower confidence bounds from the true parameter value. This definition in symbolic form is given by

$$E_i = 100\{E[(\bar{\theta} - \theta)^2] + E[(\underline{\theta} - \theta)^2]\}/\{E[(\bar{\theta}' - \theta)^2] + E[(\underline{\theta}' - \theta)^2]\} \,, \tag{5.1.16}$$

where E_i is the efficiency (in percent) of the substitute central confidence interval, $E[(\bar{\theta} - \theta)^2]$ and $E[(\bar{\theta}' - \theta)^2]$ are defined as above, and $E[(\underline{\theta} - \theta)^2]$ and $E[(\underline{\theta}' - \theta)^2]$ are respectively the mean squared deviations of the conventional and substitute lower confidence bounds $\underline{\theta}$ and $\underline{\theta}'$ from the true value θ of the parameter.

Quayle (1963) has shown that the mean squared deviation $E[(\bar{\theta}-\theta)^2]$ of the conventional upper confidence bound with confidence level $1-P$, based on all n observations, from the true value θ of the scale parameter of a Weibull population with known shape parameter K is given by

$$\begin{aligned} E[(\bar{\theta} - \theta)^2] = 1 &- 2^{1+1/K}[\Gamma(n + 1/K)/\Gamma(n)][1/\chi^2_{2n,P}]^{1/K} \\ &+ 2^{2/K}[\Gamma(n + 2/K)/\Gamma(n)][1/\chi^2_{2n,P}]^{2/K} \end{aligned} \tag{5.1.17}$$

and that the mean squared deviation $E[(\underline{\theta} - \theta)^2]$ of the corresponding conventional lower confidence bound is found by replacing $\bar{\theta}$ by $\underline{\theta}$ and P by $1-P$ in Equation (5.1.17). Since $2n\hat{\theta}^K_{nn}/\theta^K$ is distributed as χ^2 with $2n$ degrees of freedom and $2m\hat{\theta}^k_{mn}/\theta^K$ as χ^2 with $2m$ degrees of freedom, the mean squared deviations of the substitute confidence bounds based on the first m order statistics are found as follows: $E[(\theta' - \theta)^2]$ by replacing $\bar{\theta}$ by $\bar{\theta}'$ and n by m in Equation (5.1.17); $E[(\underline{\theta}' - \theta)^2]$ by replacing $\bar{\theta}$ by $\underline{\theta}'$, n by m, and P by $1-P$ in Equation (5.1.17). Substitution of the results in Equations (5.1.15) and (5.1.16) then yields the efficiencies of the substitute upper confidence bound and the substitute central confidence interval, respectively, as compared with the conventional bound and interval based on all n observations.

5.1.4 Numerical Example

As an illustration of the use of the above results, consider a simulated life test on forty components. Suppose the observed failure times in hours are as follows:

5	33	55	65	82	102	114	142
10	34	58	65	85	103	116	143
17	36	58	66	90	106	117	151
32	54	61	67	92	107	124	158
32	55	64	68	92	114	139	195

Suppose the experimenter knows that these times are from a Weibull population with shape parameter $K = 2.0$ and wishes to obtain a point estimate and set 80 percent upper and lower confidence bounds on the scale parameter θ. The conventional confidence bounds are those based on all 40 observations, but the experimenter might not want to wait for all the components to fail and might therefore terminate the test at the time of the mth failure ($m < 40$). We can simulate this occurrence by censoring upper portions of the above ordered data. The values of the maximum likelihood estimate $\hat{\theta}$ were calculated from Equation (5.1.5) for $m = 8(8)40$, and $\tilde{\theta}$ was obtained by multiplying by the unbiasing factor $\tilde{\theta}/\hat{\theta}$ given in Table E1. Then the lower and upper 80 percent confidence bounds, $\underline{\theta}_{.80}$ and $\bar{\theta}_{.80}$, were calculated from Equation (5.1.14) with the aid of a table of percentage points of the chi-square distribution given by Harter (1964a). The intervals between paired values of $\underline{\theta}_{.80}$ and $\bar{\theta}_{.80}$ are central 60 percent confidence intervals for θ. The efficiencies, E_u and E_i, of upper confidence bounds and central confidence intervals, with confidence levels 80 percent and 60 percent, respectively, based on the first m out of n ordered observations, were calculated by substituting, in Equations (5.1.15) and (5.1.16), values of $E[(\bar{\theta} - \theta)^2]$ obtained from Equation (5.1.17) and of $E[(\bar{\theta}' - \theta)^2]$, $E[(\underline{\theta} - \theta)^2]$, and $E[(\underline{\theta}' - \theta)^2]$ obtained from Equation (5.1.17) modified as indicated in subsection 5.1.3. The efficiency E_p of the unbiased point estimator $\tilde{\theta}$ was computed as indicated in subsection 5.1.2. The results are as follows:

m	$\hat{\theta}$	θ	$\underline{\theta}_{.80}$	$\bar{\theta}_{.80}$	$E_u(\%)$	$E_i(\%)$	$E_p(\%)$
8	77.0	78.2	68.1	92.2	16.2	18.2	19.8
16	91.9	92.6	83.5	103.3	34.4	38.9	39.8
24	95.2	95.7	88.2	104.8	57.3	59.3	59.9
32	93.7	94.1	87.6	101.7	78.5	79.6	79.9
40	93.3	93.6	87.8	100.3	100.0	100.0	100.0

Note that $E_u \leq E_i \leq E_p \leq 100\,m/n$ and that $E_u \to E_i \to E_p \to 100\,m/n \to 100$ percent as $m \to n$.

5.1.5 References

1. Epstein, Benjamin and Sobel, Milton (1953). Life testing. *Journal of the American Statistical Association* **48**, 486-502.
2. Harter, H. Leon (1964a). *New Tables of the Incomplete Gamma-Function Ratio and of Percentage Points of the Chi-Square and Beta Distributions.* U.S. Government Printing Office, Washington.
3. Harter, H. Leon (1964b). Exact confidence bounds, based on one order statistic, for the parameter of an exponential population. *Technometrics* **6**, 301-317.
4. Harter, H. Leon (1964c). Criteria for best substitute interval estimators with an application to the normal distribution. *Journal of the American Statistical Association*

59, 1133-1140.

5. Harter, H. Leon and Moore, Albert H. (1965). Point and interval estimators, based on m order statistics, for the scale parameter of a Weibull population with known shape parameter. *Technometrics* **7**, 405-422.

6. Johnson, Norman L., Kotz, Samuel, and Balakrishnan, N. (1994). *Continuous Univariate Distributions - Volume 1*, Second edition. John Wiley & Sons, New York.

7. Mann, Nancy R. (1963). Optimum Estimates of Parameters of Continuous Distributions, *Rocketdyne Research Report No. 63-41*, Rocketdyne Division, North American Aviation, Inc., Canoga Park, California.

8. Quayle, Ronald John (1963). Estimation of the Scale Parameter of the Weibull Probability Density Function by Use of One Order Statistic. *Unpublished M.S. thesis*, Air Force Institute of Technology, Wright-Patterson Air Force Base. AD 425228.

5.2 Type I Extreme-Value Population with Known Scale Parameter [2]

5.2.1 Introduction

Various methods of estimating the parameters of the Type I extreme-value distribution have been proposed; see, for example, Johnson, Kotz, and Balakrishnan (1995) for an extensive review. Gumbel (1935, 1941) made an extensive study of the theory and applications of this distribution, and estimated its parameters by the method of moments. Lieblein (1953, 1954) found closed expressions for the variances and covariances of the sample order statistics, tabulated them for samples of size $n = 2(1)6$, and used them to obtain best linear unbiased estimators of the parameters. White (1964) extended Lieblein's table up through $n = 20$, Mann (1965a) independently extended it up through $n = 25$, and Balakrishnan and Chan (1992a) extended it up through $n = 30$. Mann (1963, 1965a,b) proposed the use of best linear invariant estimators, which are simple linear functions of the best linear unbiased estimators. Johns and Lieberman (1966) tabulated approximate weights for obtaining best linear invariant estimates from the first m order statistics of samples of size n for $n = 10, 15, 20, 30, 50$ and 100 and four values of m for each n. Mann (1966, 1967) tabulated exact weights for $n = 2(1)25$, $m = 2(1)n$. Balakrishnan and Chan (1992b) tabulated the best linear unbiased estimators for n up to 30 and various choices of m.

Kimball (1946, 1949) obtained maximum-likelihood estimators of the parameters of Type I extreme-value distributions for complete samples, but this method has not been widely used, probably because only iterative solution of the likelihood equations is possible when both parameters are unknown. When the scale parameter is known, however, an explicit expression for the maximum-likelihood estimator of the location parameter and the exact distribution of the estimator can be obtained; not only for complete but also for censored samples. The estimator based on the first m order statistics of a sample of size $n \geq m$ and its exact distribution are worked out in subsection 5.2.2. Expressions are obtained in subsection 5.2.2 for the bias $E = E[(\hat{u}_{mn}|b) - u]$ and the variance $V = V(\hat{u}_{mn}|b)$ of the maximum-likelihood estimator of the location parameter u when the scale parameter b is known. Numerical values of E/b ($6DP$) and V/b^2 ($6DP$) are shown in Table E2 for $m = 1(1)100$. The use of the table is discussed in subsection 5.2.2. Exact confidence bounds for the location parameter are given in subsection 5.2.3. A numerical example is given in subsection 5.2.4.

5.2.2 Point Estimation of the Location Parameter u

Maximum-Likelihood Estimator for the Location Parameter

If the random variable T has the two-parameter Weibull distribution with cumulative distribution function $F(t;\theta,K) = 1 - \exp[-(t/\theta)^K]$, where θ is the scale parameter and K is the shape parameter, then the random variable $X = \ln T$ has the Type I extreme-value distribution of smallest values with cumulative distribution function $F(x;u,b) = 1 - \exp\{-\exp[(x-u)/b]\}$, where $u = \ln\theta$ is the location parameter (mode) and $b = 1/K$ is the scale parameter. Harter and Moore (1965) have shown [see subsection 5.1.2] that the maximum-likelihood estimator of θ, based on the first m order statistics t_{in} ($i = 1, 2, \ldots, m$) of a sample of size $n \geq m$ from a Weibull distribution with known shape

[2] An earlier version of this material was published by Harter and Moore (1967).

parameter K, is given by

$$\hat{\theta}_{mn}|K = \{[t_{1n}^K + t_{2n}^K + \cdots + t_{mn}^K + (n-m)t_{mn}^K]/m\}^{1/K} \; . \tag{5.2.1}$$

It follows that the maximum-likelihood estimator of u, based on the first m order statistics x_{in} $(i = 1, 2, \ldots, m)$ of a sample of size $n \geq m$ from a Type I extreme-value distribution of smallest values with known scale parameter b, is given by

$$\begin{aligned} \hat{u}_{mn}|b &= \ln \hat{\theta}_{mn}|K = \ln(\{[t_{1n}^K + t_{2n}^K + \cdots + t_{mn}^K + (n-m)t_{mn}^K]/m\}^{1/K}) \\ &= b \; \ln\{[\exp(x_{1n}/b) + \exp(x_{2n}/b) + \\ &\quad \cdots + \exp(x_{mn}/b) + (n-m)\exp(x_{mn}/b)]/m\} \; . \end{aligned} \tag{5.2.2}$$

Harter and Moore (1965) have also shown that the probability density function of the estimator $T_1 = \hat{\theta}_{mn}|K$ is independent of n and is given by

$$g_m(t_1) = [K/\Gamma(m)](m/\theta^K)^m \; t_1^{Km-1} \exp(-mt_1^K/\theta^K), \; t_1 \geq 0 \; . \tag{5.2.3}$$

It follows that the probability density function of the estimator $X_1 = \hat{u}_{mn}|b$ is given by

$$\begin{aligned} h_m(x_1) &= [K/\Gamma(m)](m/\theta^K)^m \exp(Kmx_1)\exp[-m\exp(Kx_1)/\theta^K] \\ &= [m^m/b\Gamma(m)]\exp\{m(x_1-u)/b - m\exp[(x_1-u)/b]\} \; . \end{aligned} \tag{5.2.4}$$

Without loss of generality, we may set $u = 0$, corresponding to $\theta = 1$. Then we have

$$h_m(x_1) = [m^m/b\Gamma(m)]\exp\{m[x_1/b - \exp(x_1/b)]\} \; . \tag{5.2.5}$$

Bias and Variance of Maximum-Likelihood Estimator

When the true value of the parameter u is zero, the expected values of x_1 and x_1^2, where $x_1 = \hat{u}_{mn}|b$ is the maximum-likelihood estimator, from the first m order statistics of a sample of size n, of the location parameter u when the scale parameter b is known, are given by

$$E(x_1) = \int_{-\infty}^{\infty} x_1 \; h_m(x_1)\, dx_1 \; , \tag{5.2.6}$$

$$E(x_1^2) = \int_{-\infty}^{\infty} x_1^2 \; h_m(x_1)\, dx_1 \; , \tag{5.2.7}$$

where $h_m(x_1)$ is the probability density function of x_1. Substituting in Equations (5.2.6) and (5.2.7) the value of $h_m(x_1)$ given by Equation (5.2.5), we obtain

$$E(x_1) = [1/b\Gamma(m)]m^m \int_{-\infty}^{\infty} x_1 \; e^{m(x_1/b - e^{x_1/b})}\, dx_1 \; , \tag{5.2.8}$$

$$E(x_1^2) = [1/b\Gamma(m)]m^m \int_{-\infty}^{\infty} x_1^2 \; e^{m(x_1/b - e^{x_1/b})}\, dx_1 \; . \tag{5.2.9}$$

If we make the substitution $me^{x_1/b} = v$, these become

$$E(x_1) = [b/\Gamma(m)]\left[\int_0^{\infty} e^{-v} v^{m-1} \ln v \, dv - \ln m \int_0^{\infty} e^{-v} v^{m-1}\, dv\right] \; , \tag{5.2.10}$$

$$E(x_1^2) = [b^2/\Gamma(m)]\left[\int_0^\infty e^{-v}v^{m-1}(\ln v)^2\,dv - 2\ln m\int_0^\infty e^{-v}v^{m-1}\ln v\,dv + (\ln m)^2\int_0^\infty e^{-v}v^{m-1}\,dv\right], \tag{5.2.11}$$

which can be expressed in terms of digamma and trigamma functions as

$$E(x_1) = b[\psi(m) - \ln m], \tag{5.2.12}$$

$$E(x_1^2) = b^2\{\psi'(m) + [\psi(m) - \ln m]^2\}. \tag{5.2.13}$$

The bias $E = E[(\hat{u}_{mn}|b) - u]$ is equal to the expected value of x_1 when $u = 0$, which is given by Equation (5.2.12), and the variance $V(\hat{u}_{mn}|b) = V(x_1) = E(x_1^2) - [E(x_1)]^2$ is independent of the true value of u, so we have

$$E = E[(\hat{u}_{mn}|b) - u] = b[\psi(m) - \ln m], \tag{5.2.14}$$

$$V = V(\hat{u}_{mn}|b) = b^2\psi'(m). \tag{5.2.15}$$

Six-decimal-place values of E/b and V/b^2 are given in Table E2 for $m = 1(1)100$. The bias and the variance can be read directly from the table when $b = 1$; otherwise, they can be obtained by multiplying the tabular values by b and b^2, respectively.

Asymptotic formulas for E/b and V/b^2, obtained from the well-known asymptotic formulas for the digamma and trigamma functions [see Davis (1964)], are as follows:

$$E/b = -\frac{1}{2m} - \sum_{k=1}^{\infty}\frac{B_{2k}}{2km^{2k}}, \tag{5.2.16}$$

$$V/b^2 = \frac{1}{m} + \frac{1}{2m^2} + \sum_{k=1}^{\infty}\frac{B_{2k}}{m^{2k+1}}, \tag{5.2.17}$$

where the B's are the Bernoulli numbers. Substituting the first seven Bernoulli numbers of even order in Equations (5.2.16) and (5.2.17) we find

$$E/b = -\frac{1}{2m} - \frac{1}{12m^2} + \frac{1}{120m^4} - \frac{1}{252m^6} + \frac{1}{240m^8} - \frac{1}{132m^{10}} + \frac{691}{32760m^{12}} - \frac{1}{12m^{14}} + \cdots, \tag{5.2.18}$$

$$V/b^2 = \frac{1}{m} + \frac{1}{2m^2} + \frac{1}{6m^3} - \frac{1}{30m^5} + \frac{1}{42m^7} - \frac{1}{30m^9} + \frac{5}{66m^{11}} - \frac{691}{2730m^{13}} + \frac{7}{6m^{15}} - \cdots \tag{5.2.19}$$

Even for m as small as 10, the absolute value of the truncation error is less than 10^{-6} if one uses only the first two terms of Equation (5.2.18) and the first three terms of Equation (5.2.19), since in either case it is smaller than the absolute value of the first term neglected.

If $\hat{u}_{mn}|b$ is the maximum-likelihood estimator of u, given b, then $\tilde{u}_{mn}|b = \hat{u}_{mn}|b - E[(\hat{u}_{mn}|b) - u]$ is an unbiased estimator whose variance is the same as that of $\hat{u}_{mn}|b$.

Use of Table

Table E2 may be used to find the bias and the variance of the maximum-likelihood estimator $\hat{u}_{mn}|b$. An unbiased estimate $\tilde{u}_{mn}|b$ may be obtained by subtracting the bias from a maximum-likelihood estimate. The relative efficiency R of unbiased estimators $(\tilde{u}_{rn}|b)$ based on the first r order statistics as compared with those $(\tilde{u}_{nn}|b)$ based on the complete sample of size n may be found by taking the ratio of the variance when $m = n$ to that when $m = r$. In addition to its use in the analysis of naturally occurring Type I extreme-value data, Table E2 may also be used in the analysis of Type I extreme-value data obtained by applying a logarithmic transformation to naturally occurring Weibull data, a transformation which Mann (1965b), who considers the case in which K is not known, has advocated. Such a transformation may be applied directly to Weibull data with location parameter zero, or to Weibull data with known location parameter ($\neq 0$) after subtracting the value of the location parameter.

5.2.3 Interval Estimation of the Location Parameter u

Harter and Moore (1965) have shown [see subsection 5.1.3] that $2m(\hat{\theta}^K_{mn}|K)/\theta^K$, where $\hat{\theta}_{mn}|K$ is the maximum-likelihood estimator, based on the first m order statistics of a sample of size n, of the scale parameter θ of a Weibull distribution with known shape parameter K, has a chi-square distribution with $2m$ degrees of freedom:

$$2m(\hat{\theta}^K_{mn}|K)/\theta^K = \chi^2_{2m} \; . \tag{5.2.20}$$

It is well known that if the random variable T has the two-parameter Weibull distribution with scale parameter θ and shape parameter K, the random variable $X = \ln T$ has the Type I extreme-value distribution of smallest values with location parameter (mode) $u = \ln\theta$ and scale parameter $b = 1/K$. Hence, in Equation (5.2.20), we may set $\theta = \exp u$, $K = 1/b$, and $\hat{\theta}_{mn}|K = \exp \hat{u}_{mn}|b$, which yields

$$2m \exp[(\hat{u}_{mn}|b)/b]/\exp(u/b) = \chi^2_{2m} \; . \tag{5.2.21}$$

Solving for u, we obtain

$$u = b\ln(2m/\chi^2_{2m}) + \hat{u}_{mn}|b \; . \tag{5.2.22}$$

Then an upper confidence bound with confidence level $1 - P$ (lower confidence bound with confidence level P) on u is given by

$$\bar{u}_{1-P} = \underline{u}_P = b\ln(2m/\chi^2_{2m,P}) + \hat{u}_{mn}|b \; , \tag{5.2.23}$$

where the first subscript on χ^2 is the number of degrees of freedom and the second one is the cumulative probability. The interval $(\underline{u}_{1-P}, \bar{u}_{1-P})$ is a central confidence interval with confidence level $1-2P$. Numerical confidence bounds may be obtained with the aid of a table of percentage points of the chi-square distribution [see Harter (1964)].

5.2.4 Numerical Example

Consider the following ordered sample of size 40 from a Type I extreme-value distribution with scale parameter $b = 0.5$:

1.609	3.497	4.007	4.174	4.407	4.625	4.736	4.956
2.303	3.526	4.060	4.174	4.443	4.635	4.754	4.963
2.833	3.584	4.060	4.190	4.500	4.663	4.762	5.017
3.466	3.989	4.111	4.205	4.522	4.673	4.820	5.063
3.466	4.007	4.159	4.220	4.522	4.736	4.934	5.273

These data were obtained by taking the natural logarithms of ordered failure times (in hours) in a simulated life test of 40 components whose time-to-failure was known to follow a Weibull distribution with location parameter zero and shape parameter $K = 2.0$ [see subsection 5.1.4 for an analysis of the original data]. Suppose the life test had been terminated at the time of the mth failure, where $m = 8(8)40$. For each value of m, the following computations have been performed: (1) The maximum-likelihood estimate $\hat{u}_{mn}|(b = 0.5)$ has been obtained by the use of Equation (5.2.2); (2) the bias E of the maximum-likelihood estimator has been found by multiplying by 0.5 the value of E/b read from Table E2; (3) the unbiased estimate $\tilde{u}_{mn}|b$ has been obtained by subtracting the bias E from $\hat{u}_{mn}|b$; (4) the lower and upper 80 percent confidence bounds $\underline{u}_{.80}$ and $\bar{u}_{.80}$ for the location parameter u have been found by the use of Equation (5.2.23) with the aid of a table of percentage points of the chi-square distribution [Harter (1964)]; and (5) the relative efficiency R of $\tilde{u}_{mn}|b$ as compared with $\tilde{u}_{nn}|b$ has been found by taking the ratio of the appropriate values read from the variance column of Table E2. The results are summarized below:

m	$\hat{u}_{mn}\|b$	E (bias)	$\tilde{u}_{mn}\|b$	$\underline{u}_{.80}$	$\bar{u}_{.80}$	$R(\%)$
8	4.344	−0.032	4.376	4.221	4.524	19.0
16	4.521	−0.016	4.357	4.429	4.641	39.3
24	4.556	−0.010	4.566	4.479	4.652	59.5
32	4.541	−0.008	4.549	4.474	4.624	79.7
40	4.537	−0.006	4.543	4.476	4.610	100.0

In comparing these results with those of the analysis of the original data, the reader will note that (except for discrepancies due to rounding), $\hat{u}_{mn}|b \equiv \ln\hat{\theta}_{mn}|K$, $\underline{u}_{.80} \equiv \ln\underline{\theta}_{.80}$, and $\bar{u}_{.80} \equiv \ln\bar{\theta}_{.80}$, while $\tilde{u}_{mn}|b \doteq \ln\tilde{\theta}_{mn}|K$, where $\tilde{\theta}_{mn}|K$ is an unbiased estimate obtained from the biased estimate $\hat{\theta}_{mn}|K$. The corresponding relative efficiencies are also approximately equal, but not identical.

5.2.5 References

1. Balakrishnan, N. and Chan, Ping-Shing (1992a). Order statistics from extreme value distribution, I: Tables of means, variances and covariances. *Communications in Statistics - Simulation and Computation* **21**, 1199-1217.
2. Balakrishnan, N. and Chan, Ping-Shing (1992b). Order statistics from extreme value distribution, II: Best linear unbiased estimates and some other uses. *Communications in Statistics - Simulation and Computation* **21**, 1219-1246.
3. Davis, Philip J. (1964). Gamma function and related functions. In *Handbook of Mathematical Functions with Formulas, Graphs, and Mathematical Tables* (Eds., Milton Abramowitz and Irene Stegun), 253-293, U.S. Government Printing Office, Washington.
4. Gumbel, E.J. (1935). Les valeurs extrêmes des distributions statistiques. *Annales de l'Institut Henri Poincaré* **5**, 115-158.
5. Gumbel, E.J. (1941). The return period of flood flows. *Annals of Mathematical Statistics* **12**, 163-190.

6. Harter, H. Leon (1964). *New Tables of the Incomplete Gamma-Function Ratio and of Percentage Points of the Chi-Square and Beta Distributions.* U.S. Government Printing Office, Washington.

7. Harter, H. Leon and Moore, Albert H. (1965). Point and interval estimators, based on m order statistics, for the scale parameter of a Weibull population with known shape parameter. *Technometrics* **7**, 405-422.

8. Harter, H. Leon and Moore, Albert H. (1967). A note on estimation from a Type I extreme-value distribution. *Technometrics* **9**, 325-331.

9. Johns, M.V., Jr. and Lieberman, G.J. (1966). An exact asymptotically efficient confidence bound for reliability in the case of the Weibull distribution. *Technometrics* **8**, 135-175.

10. Johnson, Norman L., Kotz, Samuel, and Balakrishnan, N. (1995). *Continuous Univariate Distributions - Volume 2*, Second edition. John Wiley & Sons, New York.

11. Kimball, Bradford F. (1946). Sufficient statistical estimation functions for the parameters of the distribution of maximum values. *Annals of Mathematical Statistics* **7**, 299-309.

12. Kimball, Bradford F. (1949). An approximation to the sampling variance of an estimated maximum value of given frequency based on fit of doubly exponential distribution of maximum values. *Annals of Mathematical Statistics* **20**, 110-113.

13. Lieblein, Julius (1953). On the exact evaluation of the variances and covariances of order statistics in samples from the extreme-value distribution. *Annals of Mathematical Statistics* **24**, 282-287.

14. Lieblein, Julius (1954). A New Method of Analyzing Extreme-Value Data. *Technical Note 3053*, National Advisory Committee for Aeronautics.

15. Mann, Nancy R. (1963). Optimum Estimates of Parameters of Continuous Distributions. *Rocketdyne Research Report 63-41*, Rocketdyne Division, North American Aviation, Inc., Canoga Park, California.

16. Mann, Nancy R. (1965a). Point and Interval Estimates for Reliability Parameters when Failure Times Have the Two-Parameter Weibull Distribution. *Doctoral dissertation*, University of California at Los Angeles, University Microfilms, Ann Arbor, Michigan.

17. Mann, Nancy R. (1965b). Cramér-Rao Efficiencies of Best Linear Invariant Estimates of Parameters of the Extreme-Value Distribution under Type II Censoring from Above. *Rocketdyne Research Report 65-36*, Rocketdyne Division, North American Aviation, Inc., Canoga Park, California.

18. Mann, Nancy R. (1966). Tables for Obtaining the Best Linear Invariant Estimates of Parameters of the Weibull Distribution. *Rocketdyne Research Report 66-9*, Rocketdyne Division, North American Aviation, Inc., Canoga Park, California.

19. Mann, Nancy R. (1967). Results on Location and Scale Parameter Estimation with Application to the Extreme-Value Distribution. *Aerospace Research Laboratories Technical Report ARL 67-0023*, Wright-Patterson Air Force Base, AD 653575.

20. White, John S. (1964). Least squares unbiased censored linear estimation for the log Weibull (extreme value) distribution. *Journal of the Industrial Mathematics Society* **14**, 21-60.

5.3 Type II Extreme-Value Populations with Known Shape Parameters [3]

5.3.1 Introduction

Fisher and Tippett (1928) showed that the asymptotic distribution of the largest (or smallest) value in a sample of size n, as $n \to \infty$, for any population satisfying certain regularity conditions, must fall into one of three types. The populations possessing asymptotic distributions of the largest value were classified by von Mises (1936), who also gave sufficient conditions for the validity of the three asymptotic distributions. Later Gnedenko (1943) gave necessary and sufficient conditions. Gumbel (1958) made an extensive study of the theory of three asymptotic distributions and of their applications. Galambos (1987) and Arnold, Balakrishnan, and Nagaraja (1992) have provided elaborate details on the asymptotic theory of extreme order statistics. Goldstein (1963) tabled random numbers from the three asymptotic distributions of largest values. This section deals with the Type II extreme-value distributions, which have received less attention than the other types; see, Johnson, Kotz, and Balakrishnan (1995) for more information.

Fréchet (1927) derived the Type II asymptotic distribution of largest values. Gumbel (1958, 1965) presented several methods of estimating the parameters of the Type II extreme-value distributions. In subsection 5.3.2, conditional maximum-likelihood estimators, $\hat{v}_n|K$ and $\hat{v}_1|K$, for the scale parameters of the Type II asymptotic distributions of largest values (for samples censored from below) and of smallest values (for samples censored from above), both with known shape parameter K, are derived, together with their distributions. Unbiased estimators, $\tilde{v}_n|K$ and $\tilde{v}_1|K$, and their variances are derived in subsection 5.3.2. The Cramér-Rao lower bound for the variance of an unbiased estimator is derived in subsection 5.3.2. Numerical values of the unbiasing factors, the variances of the unbiased estimators, and their efficiency relative to the Cramér-Rao lower bound are shown in Table E3. In subsection 5.3.3 a simple technique for obtaining exact upper and lower confidence bounds for the scale parameters is presented. A numerical example which illustrates the computation of both point and interval estimates is given in subsection 5.3.4. Some remarks on applications are made in subsection 5.3.5.

5.3.2 Point Estimation of the Scale Parameters v_n and v_1

Maximum-Likelihood Estimators of the Scale Parameters

Epstein and Sobel (1953) derived an estimator, $\hat{\theta}$, for the scale parameter θ of the exponential distribution for samples censored from above and showed that it is maximum likelihood, sufficient, unbiased, and minimum variance. Their "best" m-order-statistic estimator is given by

$$\hat{\theta}_{mn} = [t_{1n} + t_{2n} + \cdots + t_{mn} + (n-m)t_{mn}]/m, \tag{5.3.1}$$

where t_{in} $(i = 1, 2, \ldots, m)$ is the ith order statistic of a sample of size n from a one-parameter negative exponential distribution. The probability density function of the random variable $S = \hat{\theta}_{mn}$ is given by Epstein and Sobel (1953) as

$$f_m(s) = \begin{cases} [1/\Gamma(m)](m/\theta)^m s^{m-1} \exp(-ms/\theta), & s > 0\ , \\ 0 \ \text{elsewhere}\ . \end{cases} \tag{5.3.2}$$

[3] An earlier version of this material was published by Harter and Moore (1968).

a. Type II Asymptotic Distribution of Largest Values

If the random variable T has an exponential distribution with scale parameter θ and location parameter zero, then the random variable $Y = T^{-1/K}$ has the Type II asymptotic distribution of largest values with scale parameter $v_n = \theta^{-1/K}$, shape parameter K, and location parameter zero, given by

$$F_1(y; v_n, K) = \exp[-(y/v_n)^{-K}], \qquad y > 0,\ v_n > 0, \text{ and } K > 0 . \tag{5.3.3}$$

By use of Equation (5.3.1), we obtain the following m-order-statistic estimator of v_n given K:

$$\begin{aligned} \hat{v}_n|K = \hat{\theta}_{mn}^{-1/K} &= \{[t_{1n} + t_{2n} + \cdots + t_{mn} + (n-m)t_{mn}]/m\}^{-1/K} \\ &= \{m/[y_{nn}^{-K} + y_{n-1,n}^{-K} + \cdots + y_{n-m+1,n}^{-K} + (n-m)y_{n-m+1,n}^{-K}]\}^{1/K} , \end{aligned} \tag{5.3.4}$$

where t_{in} and $y_{n-i+1,n}$ $(i = 1, 2, \ldots, m)$ are the ith smallest and ith largest observations, respectively, in samples of size n from the exponential distribution and the Type II asymptotic distribution of largest values. Then $\hat{v}_n|K$ is a maximum-likelihood estimator of the scale parameter, v_n of the Type II asymptotic distribution of largest values, since $v_n = \theta^{-1/K}$ and $\hat{\theta}_{mn}$ is a maximum-likelihood estimator of the scale parameter θ of the exponential distribution. Now, we let $\hat{v}_n|K = \hat{\theta}_{mn}^{-1/K}$ or $Z = S^{-1/K}$ and $v_n = \theta^{-1/K}$ in Equation (5.3.2), and we obtain the following expression for the probability density function of $Z = \hat{v}_n|K$:

$$g_m(z) = \begin{cases} K[1/\Gamma(m)][mv_n^K]^m z^{-mK-1} \exp[-m(v_n/z)^K], & z > 0 , \\ 0 \text{ elsewhere.} \end{cases} \tag{5.3.5}$$

b. Type II Asymptotic Distribution of Smallest Values

If the random variable T has an exponential distribution with scale parameter θ and location parameter zero, then the random variable $X = -T^{-1/K}$ has the Type II asymptotic distribution of smallest values with scale parameter $v_1 = -\theta^{-1/K}$, shape parameter K, and location parameter zero, given by

$$F_2(x; v_1, K) = 1 - \exp[-(x/v_1)^{-K}], \qquad x < 0,\ v_1 < 0, \text{ and } K > 0 . \tag{5.3.6}$$

We can use the symmetry with the distribution of largest values to obtain a maximum-likelihood estimator of the scale parameter v_1. We have $v_1 = -v_n$ and hence $\hat{v}_1|K = -\hat{v}_n|K$; but $y_{n-i+1,n} = -x_{in}$ $(i = 1, 2, \ldots, m)$, and therefore we find the maximum-likelihood estimator of v_1 given K to be

$$\hat{v}_1|K = -\{m/[(-x_{1n})^{-K} + (-x_{2n})^{-K} + \cdots + (-x_{mn})^{-K} + (n-m)(-x_{mn})^{-K}]\}^{1/K} , \tag{5.3.7}$$

where x_{in} $(i = 1, 2, \ldots, m)$ is the ith order statistic of a sample of size n from the Type II asymptotic distribution of smallest values. Now we let $u = -z$ and $v_1 = -v_n$ in Equation (5.3.5), and we find the probability density function of $U = \hat{v}_1|K$ to be

$$h_m(u) = \begin{cases} K[1/\Gamma(m)][m(-v_1)^K]^m [-u]^{-mK-1} \exp[-m(v_1/u)^K], & u < 0 , \\ 0 \text{ elsewhere .} \end{cases} \tag{5.3.8}$$

If we have a Type II asymptotic distribution of largest or smallest values with known shape parameter and known location parameter different from zero, then we can transform the original sample to a sample from the corresponding distribution with location parameter zero and proceed with the estimation as outlined in (a) or (b) above.

Unbiased Estimators and Their Variances

The expected value and the variance of the estimator $\hat{v}_n|K$ are found by using Equation (5.3.5) and the transformation of variables $z = (w/m)^{-1/K} v_n$ to be

$$E(\hat{v}_n|K) = v_n m^{1/K}\Gamma(m - 1/K)/\Gamma(m), \qquad (m - 1/K) > 0 , \tag{5.3.9}$$

and

$$\text{var}(\hat{v}_n|K) = v_n^2 m^{2/K}\{\Gamma(m)\Gamma(m-2/K) - [\Gamma(m-1/K)]^2\}/[\Gamma(m)]^2 , \quad (m - 2/K) > 0 . \tag{5.3.10}$$

An unbiased estimator of the scale parameter is given by

$$\tilde{v}_n|K = [\Gamma(m)/m^{1/K}\Gamma(m - 1/K)]\hat{v}_n|K, \qquad (m - 1/K) > 0 \tag{5.3.11}$$

and its variance is given by

$$\text{var}(\tilde{v}_n|K) = v_n^2\{\Gamma(m)\Gamma(m-2/K) - [\Gamma(m-1/K)]^2\}/[\Gamma(m-1/K)]^2 , \quad (m - 2/K) > 0 . \tag{5.3.12}$$

The expected value and the variance of the maximum-likelihood estimator of the scale parameter v_1 of the Type II asymptotic distribution of smallest values and the corresponding unbiased estimator and its variance are found by replacing v_n by v_1, $\hat{v}_n$ by $\hat{v}_1$, and $\tilde{v}_n$ by $\tilde{v}_1$ in Equations (5.3.9), (5.3.10), (5.3.11), and (5.3.12), respectively.

Cramér-Rao Lower Bound

The Cramér-Rao lower bound for the variance of an unbiased estimator of a parameter ϕ is $-1/E(\partial^2 \ln L/\partial\phi^2)$, where L is the likelihood function. Let $y_{nn}, y_{n-1,n}, \ldots, y_{n-m+1,n}$ be the m largest observations in a sample of size n from a Type II asymptotic distribution of largest values with scale parameter v_n and shape parameter K. The natural logarithm of the likelihood function $L = L(y_{nn}, y_{n-1,n}, \ldots, y_{n-m+1,n}; v_n, K)$ of the censored sample is given by

$$\ln L = \ln n! - \ln(n-m)! + \sum_{i=1}^{m} \ln(K/y_{n-i+1,n}) + K\sum_{i=1}^{m} \ln(v_n/y_{n-i+1,n}) - \sum_{i=1}^{m}(v_n/y_{n-i+1,n})^K - (n-m)(v_n/y_{n-m+1,n})^K \tag{5.3.13}$$

and its first two derivatives with respect to the parameter v_n by

$$\partial \ln L/\partial v_n = mK[1/v_n - v_n^{K-1}(\hat{v}_n|K)^{-K}] , \tag{5.3.14}$$

$$\partial^2 \ln L/\partial v_n^2 = -mK[1/v_n^2 + (K-1)v_n^{K-2}(\hat{v}_n|K)^{-K}] , \tag{5.3.15}$$

where $\hat{v}_n|K$ is given by Equation (5.3.4). The expected value of the second partial derivative is

$$E(\partial^2 \ln L/\partial v_n^2) = -mK\{1/v_n^2 + (K-1)v_n^{K-2}E[(\hat{v}_n|K)^{-K}]\} = -mK^2/v_n^2 , \tag{5.3.16}$$

since $E[(\hat{v}_n|K)^{-K}] = E(\hat{\theta}) = \theta = v_n^K$, $\hat{\theta}$ being an unbiased estimator of θ. Hence the Cramér-Rao lower bound for the variance of an unbiased estimator, based on the last m order statistics of a sample of size $n \geq m$, of the scale parameter v_n of a Type II asymptotic distribution of largest values is v_n^2/mK^2. Similarly, the Cramér-Rao lower bound for the variance of an unbiased estimator, based on the first m order statistics of a sample of size $n \geq m$, of the scale parameter v_1 of a Type II asymptotic distribution of smallest values is v_1^2/mK^2.

Table E3 gives numerical values of the unbiasing factor $\tilde{v}|K/\hat{v}|K$, the variance of the unbiased estimator $\tilde{v}|K$, and the efficiency (%) of the unbiased estimator relative to the Cramér-Rao lower bound, all for $m = 1(1)100$, $K = 0.5(0.5)4.0(1.0)8.0$, where v may be either v_n or v_1.

5.3.3 Interval Estimation of the Scale Parameters v_n and v_1

Epstein and Sobel (1953) have shown that $2m\hat{\theta}_{mn}/\theta$ has a chi-square distribution with $2m$ degrees of freedom:

$$2m\,\hat{\theta}_{mn}/\theta = \chi^2_{2m}\,. \tag{5.3.17}$$

Making the transformation of random variables, discussed in subsection 5.3.2, from the exponential distribution to the Type II asymptotic distribution of largest values, we set $\theta = v_n^{-K}$ and $\hat{\theta}_{mn} = (\hat{v}_n|K)^{-K}$ in Equation (5.3.17) and obtain

$$2m(v_n/\hat{v}_n|K)^K = \chi^2_{2m} \tag{5.3.18}$$

or

$$v_n = \hat{v}_n|K/(\chi^2_{2m}/2m)^{1/K}\,. \tag{5.3.19}$$

Then an upper confidence bound with confidence level $1-P$ (lower confidence bound with confidence level P) on v_n is given by

$$\bar{v}_{n,1-P} = \underline{v}_{n,P} = \hat{v}_n|K/(\chi^2_{2m,P}/2m)^{1/K}\,, \tag{5.3.20}$$

where the first subscript on χ^2 is the number of degrees of freedom and the second is the cumulative probability.

For the Type II asymptotic distribution of smallest values, we make use of the symmetry with the distribution of largest values and let $v_1 = -v_n$ and $\hat{v}_1|K = -\hat{v}_n|K$ in Equation (5.3.19), which yields

$$v_1 = \hat{v}_1|K/(\chi^2_{2m}/2m)^{1/K}\,. \tag{5.3.21}$$

Then an upper confidence bound with confidence level P (lower confidence bound with confidence level $1-P$) on v_1 is given by

$$\bar{v}_{1,P} = \underline{v}_{1,1-P} = \hat{v}_1|K/(\chi^2_{2m,P}/2m)^{1/K}\,. \tag{5.3.22}$$

The intervals $(\underline{v}_{n,1-P}, \bar{v}_{n,1-P})$ and $(\underline{v}_{1,1-P}, \bar{v}_{1,1-P})$ are central confidence intervals with confidence levels $1-2P$. Numerical confidence bounds may be obtained with the aid of a table of percentage points of the chi-square distribution [see, for example, Harter (1964)].

5.3.4 Numerical Example

The following ordered pseudo-random sample of size 40 was drawn from a Type II asymptotic distribution of largest values with location parameter $\epsilon = 0$ [omitted from Equations

(5.3.3)-(5.3.5)], scale parameter (characteristic largest value) $v_n = 1$, and shape parameter $K = 2$:

0.4409	0.6592	0.7695	0.8960	1.0203	1.2639	1.8244	2.6122
0.4669	0.6730	0.7815	0.9031	1.1049	1.3515	1.8397	4.7915
0.5468	0.6872	0.8199	0.9209	1.1320	1.5679	1.9473	5.1470
0.6099	0.7037	0.8648	0.9566	1.1377	1.7585	2.2080	5.4795
0.6332	0.7452	0.8913	0.9676	1.1406	1.8116	2.3687	6.1992

Assuming that the value of K is known and that ϵ is known to be zero, the maximum-likelihood estimator of v_n, based on the last m order statistics (m largest sample values) of a sample of size n, is given by Equation (5.3.4) and exact confidence bounds can be obtained from Equation (5.3.20). An unbiased estimator is given by Equation (5.3.11). These calculations for the above sample of size $n = 40$ have been performed for $m = 8(8)40$, with the following results:

m	$\hat{v}_n \vert K$	$\tilde{v}_n \vert K$	$\underline{v}_{n,.80}$	$\bar{v}_{n,.80}$
8	0.9239	0.8798	0.7713	1.0449
16	0.8349	0.8152	0.7401	0.9153
24	0.8657	0.8521	0.7865	0.9350
32	0.8831	0.8727	0.8137	0.9449
40	0.8965	0.8881	0.8339	0.9530

where $\hat{v}_n|K$ is the maximum-likelihood point estimate, $\tilde{v}_n|K$ is the corresponding unbiased estimate, and $\underline{v}_{n,.80}$ and $\bar{v}_{n,.80}$ are the lower and upper exact confidence bounds, based on the m largest values in the sample. The rows of the above table are not independent, since they are all based on the same sample (with different amounts of censoring).

5.3.5 Remarks on Applications

The Type II asymptotic distribution of largest values, which is restricted to positive values, has been used by Thom (1954) in a study of maximum wind speed, by Jenkinson (1955), Bernier (1957, 1959), and Gumbel, Benham, and Thomson (1959) in studies of floods on rivers, and by Jenkinson (1955) in a study of maximum rainfalls. This distribution is likely to find application also in life testing problems. Castillo (1988) and Johnson, Kotz, and Balakrishnan (1995) have listed more applications of this distribution. Unfortunately, the method of estimating the scale parameter (characteristic largest value) proposed in this section cannot be used when a life test is terminated before all of the items placed on test have failed, since this method permits only single censoring from below, not from above. Substitute point and interval estimators, based on one order statistic, have, however, been worked out by Moore and Harter (1967) [see Chapter 6, Section 6.3 of the present volume], and can be used in cases of censoring from above.

Applications of the Type II asymptotic distribution of smallest values, which is restricted to negative values, are not so immediately apparent, but one possible application is to extremes of drought (river stages below low water mark).

5.3.6 References

1. Arnold, Barry C., Balakrishnan, N., and Nagaraja, H.N. (1992). *A First Course in Order Statistics.* John Wiley & Sons, New York.
2. Bernier, J. (1957). Sur l'application des diverses lois limites des valeurs extrêmes au problème des débits de crue. *Revue de Statistique Appliquée* **5**, 91-101.

3. Bernier, J. (1959). Comparaison des lois de Gumbel et de Fréchet sur l'estimation des débits maxima. *La Houille Blanche* **14**, 47-56.

4. Castillo, Enrique (1988). *Extreme Value Theory in Engineering.* Academic Press, San Diego.

5. Epstein, Benjamin and Sobel, Milton (1953). Life testing. *Journal of the American Statistical Association* **48**, 486-502.

6. Fisher, R.A. and Tippett, L.H.C. (1928). Limiting forms of the frequency distribution of the largest or smallest member of a sample. *Proceedings of the Cambridge Philosophical Society* **24**, 180-190.

7. Fréchet, M. (1927). Sur la loi de probabilité de l'écart maximum. *Annales de la Societé Polonaise de Mathématique* (Cracow) **6**, 93-96.

8. Galambos, Janos (1987). *Asymptotic Theory of Extreme Order Statistics*, Second edition. Krieger, Malabar, Florida.

9. Gnendenko, B.V. (1943). Sur la distribution limite du terme maximum d'une série aléatoire. *Annals of Mathematics* (2) **44**, 423-453.

10. Goldstein, Neil (1963). Random numbers from the extreme value distributions. *Publications de l'Institut de Statistique de l'Université de Paris* **12**, 137-158.

11. Gumbel, Emil J. (1958). *Statistics of Extremes.* Columbia University Press, New York.

12. Gumbel, Emil J. (1965). A quick estimation of the parameters in Fréchet's distribution. *Review of the International Statistical Institute* **33**, 349-367.

13. Gumbel, Emil J., Benham, A.D., and Thomson, D.H. (1959). Communications on the statistical theory of floods and droughts. *Journal of the Institute of Water Engineers* (London) **13**, 71-102.

14. Harter, H. Leon (1964). *New Tables of the Incomplete Gamma-Function Ratio and of Percentage Points of the Chi-Square and Beta Distributions.* U.S. Government Printing Office, Washington.

15. Harter, H. Leon and Moore, Albert H. (1968). Conditional maximum-likelihood estimators, from singly censored samples, of the scale parameters of Type II extreme-value populations. *Technometrics* **10**, 349-359.

16. Jenkinson, A.F. (1955). The frequency distribution of the annual maximum (or minimum) values of meteorological elements. *Quarterly Journal of the Royal Meteorological Society* **87**, 158-171.

17. Johnson, Norman L., Kotz, Samuel, and Balakrishnan, N. (1995). *Continuous Univariate Distributions - Volume 2,* Second edition. John Wiley & Sons, New York.

18. von Mises, R. (1936). La distribution de la plus grande de n valeurs. *Revue Mathématique de l'Union Interbalkanique* (Athens) **1**(2), 1-20.

19. Moore, Albert H. and Harter, H. Leon (1967). One-order-statistic conditional estimators of shape parameters of limited and Pareto distribution and scale parameters of Type II asymptotic distributions of smallest and largest values. *IEEE Transactions on Reliability* **R-16**, 100-103.

20. Thom, H.C.S. (1954). Frequency of maximum wind speeds. *Proceedings of the American Society of Civil Engineers* **80**, separate, No. 539.

5.4 Pareto and Limited Populations with Known Location Parameters [4]

5.4.1 Introduction

The Pareto distribution was used by Pareto (1897) as a law to describe the distribution of income. Mandelbrot (1960) discussed various models, including various versions of the Pareto law, of an important class of economic quantities including "income". Steindl (1965) showed that numerous economic mechanisms lead, in the limit and with appropriate boundary conditions, to the Pareto distribution. Arnold (1983) and Johnson, Kotz, and Balakrishnan (1994) have reviewed the developments on the Pareto distribution. Malik (1966) presented a method of estimating the parameters of the Pareto distribution from complete samples. Muniruzzaman (1951) considered conditional estimation of and testing hypotheses concerning one of the parameters of the Pareto distribution by the method of maximum likelihood from complete samples. When the location parameter is known, however, explicit expressions for the maximum-likelihood estimators of the shape parameters of the Pareto distribution and of a related limited distribution defined by Gumbel (1958, p. 157), $\hat{K}_{mn}|\epsilon$ and $\hat{K}_{mn}|\omega$, as well as the exact distribution of the estimators, can be obtained, not only for complete but also for censored samples. The estimators based on the first m order statistics of a sample of size $n \geq m$ and their exact distribution are worked out in subsection 5.4.2. Expressions are obtained in subsection 5.4.2 for the expected values of the maximum-likelihood estimators, which are used to obtain unbiased estimators, $\tilde{K}_{mn}|\epsilon$ and $\tilde{K}_{mn}|\omega$, and for the variances of the unbiased estimators. The Cramér-Rao lower bounds for the variances of the unbiased estimators are derived in subsection 5.4.2. Exact confidence bounds for the shape parameters are given in subsection 5.4.3. In subsection 5.4.4 numerical examples, which illustrate the computation of both point and interval estimates, are given.

5.4.2 Point Estimation of the Shape Parameter K

Maximum-Likelihood Estimators of the Shape Parameters

Epstein and Sobel (1953) derived an estimator, $\hat{\theta}$, for the scale parameter of the exponential distribution for samples censored from above and showed that it is maximum likelihood, sufficient, unbiased and minimum variance. Their "best" m-order-statistic estimator is given by

$$\hat{\theta}_{mn} = [t_{1n} + t_{2n} + \cdots + t_{mn} + (n-m)t_{mn}]/m \tag{5.4.1}$$

where t_{in} $(i = 1, \ldots, m)$ is the ith order statistic of a sample of size n from an exponential distribution.

The probability density function of the random variable $S = \hat{\theta}_{mn}$ is given by Epstein and Sobel as

$$f_m(s) = \begin{cases} [1/\Gamma(m)](m/\theta)^m s^{m-1} \exp(-ms/\theta), & s > 0 \\ 0 \ \text{elsewhere.} \end{cases} \tag{5.4.2}$$

a. Pareto distribution

If T has an exponential distribution with scale parameter θ and location parameter zero, then $Y = \exp T + \epsilon$ has the Pareto distribution, as shown by Gumbel (1958, p.

[4] An earlier version of this material was published by Moore and Harter (1969).

157), with location parameter ϵ, shape parameter $K = 1/\theta$, and cumulative distribution function

$$F_1(y; \epsilon, K) = 1 - (y - \epsilon)^{-K}, \qquad (y \geq 1 + \epsilon,\ K > 0) . \tag{5.4.3}$$

Therefore an m-order-statistic estimator of K given ϵ is obtained by using Equation (5.4.1), as follows

$$\begin{aligned} \hat{K}|\epsilon = 1/\hat{\theta}_{mn} &= m/[t_{1n} + \cdots + t_{mn} + (n - m)t_{mn}] \\ &= m/[\ln(y_{1n} - \epsilon) + \cdots + \ln(y_{mn} - \epsilon) + (n - m)\ln(y_{mn} - \epsilon)], \end{aligned} \tag{5.4.4}$$

where t_{in}, y_{in} $(i = 1, 2, \ldots, m)$ are the ith order statistics from samples of size n from the exponential and Pareto distributions, respectively. Now $\hat{K}|\epsilon$ is a maximum-likelihood estimator of the shape parameter, K, of the Pareto distribution, since $K = 1/\theta$ and $\hat{\theta}$ is a maximum-likelihood estimator of the scale parameter, θ, of the exponential distribution.

Now let $\hat{K}_{mn}|\epsilon = 1/\hat{\theta}_{mn}$ or $U = 1/S$ and $K = 1/\theta$ in Equation (5.4.2) and we find the probability density of $U = \hat{K}_{mn}|\epsilon$ to be the inverted Gamma density function $g_m(u)$ as given by

$$g_m(u) = \begin{cases} [1/\Gamma(m)](mK)^m u^{-(m+1)} \exp(-mK/u), & u > 0 \\ 0 \ \text{elsewhere} . \end{cases} \tag{5.4.5}$$

The Pareto distribution (strong Pareto law) is given in a slightly different form by Mandelbrot (1960) as follows:

$$F_3(y; y_0, K) = 1 - (y/y_0)^{-\alpha}, \qquad y > y_0 > 0 . \tag{5.4.6}$$

Mandelbrot has assumed that $\epsilon = 0$ and we have assumed that $y_0 = 1$. The estimator $\hat{\alpha}|y_0$ has the same properties as the estimator $\hat{K}|\epsilon$ except that it is given by the following formula:

$$\hat{\alpha}|y_0 = m/[\ln(y_{1n}/y_0) + \cdots + \ln(y_{mn}/y_0) + (n - m)\ln(y_{mn}/y_0)] . \tag{5.4.7}$$

b. Limited Distribution

If T has an exponential distribution with scale parameter θ and location parameter zero then $X = \omega - \exp(-T)$ has the limited distribution, as shown by Gumbel (1958, p. 157), with location parameter ω, shape parameter $K = 1/\theta$ and cumulative distribution function

$$F_2(x; \omega, K) = 1 - (\omega - x)^K, \qquad (\omega - 1 \leq x \leq \omega, K > 0) . \tag{5.4.8}$$

Therefore an m-order statistic estimator of K given ω is obtained by using Equation (5.4.1), as follows:

$$\begin{aligned} \hat{K}_{mn}|\omega = 1/\hat{\theta}_{mn} &= m/[t_{1n} + \cdots + t_{mn} + (n - m)t_{mn}] \\ &= -m/[\ln(\omega - x_{1n}) + \cdots + \ln(\omega - x_{mn}) + (n - m)\ln(\omega - x_{mn})] , \end{aligned} \tag{5.4.9}$$

where t_{in}, x_{in} $(i = 1, \ldots, m)$ are the ith order statistics from samples of size n from the exponential and limited distributions, respectively. Now $\hat{K}_{mn}|\omega$ is a maximum-likelihood estimator of the shape parameter, K, of the limited distribution, since $K = 1/\theta$ and $\hat{\theta}_{mn}$ is a maximum-likelihood estimator for θ.

Now let $\hat{K}_{mn} = 1/\hat{\theta}_{mn}$ or $U = 1/S$ and $K = 1/\theta$ in Equation (5.4.2) and we find the probability density of $U = \hat{K}_{mn}|\omega$ to be given by Equation (5.4.5) above.

Unbiased Estimators and Their Variances

The expected values and variances of $\hat{K}_{mn}|\epsilon$ and $\hat{K}_{mn}|\omega$ are found, by use of Equation (5.4.5) and the transformation $Z = mK/U$ to be

$$E(\hat{K}_{mn}) = [m/(m-1)]K, \qquad m > 1 , \tag{5.4.10}$$

and

$$\text{var}(\hat{K}_{mn}) = [m^2/(m-1)^2(m-2)]K^2, \qquad m > 2 . \tag{5.4.11}$$

Hence unbiased estimators of K $(m > 1)$ are given by

$$\tilde{K}_{mn}|\epsilon = [(m-1)/m](\hat{K}_{mn}|\epsilon) \tag{5.4.12}$$

and

$$\tilde{K}_{mn}|\omega = [(m-1)/m](\hat{K}_{mn}|\omega) . \tag{5.4.13}$$

The variance of the unbiased estimators given by Equations (5.4.12) and (5.4.13) is found to be

$$\text{var}(\tilde{K}_{mn}) = K^2/(m-2), \qquad m > 2 . \tag{5.4.14}$$

Cramér-Rao Lower Bounds

The Cramér-Rao lower bound for the variances of unbiased estimators of the parameter K is $-1/E[\partial^2 \ln L/\partial K^2]$, where L is the likelihood function. The likelihood functions for the Pareto and limited distributions are as follows:

$$L_1 = [n!/(n-m)!]K^m (y_{1n}-\epsilon)^{-K-1} \cdots (y_{mn}-\epsilon)^{-K-1}(y_{mn}-\epsilon)^{-K(n-m)}, \tag{5.4.15}$$

$$L_2 = [n!/(n-m)!]K^m (\omega - x_{1n})^{K-1} \cdots (\omega - x_{mn})^{K-1}(\omega - x_{mn})^{K(n-m)}. \tag{5.4.16}$$

The first two derivatives of L_1 and L_2 with respect to the parameter K, after substitution of $\hat{K}|\epsilon$ or $\hat{K}|\omega$ for its equivalent expression, are as follows

$$\partial \ln L_1/\partial K = m/K - m/\hat{K}|\epsilon , \tag{5.4.17}$$

$$\partial \ln L_2/\partial K = m/K - m/\hat{K}|\omega , \tag{5.4.18}$$

$$\partial^2 \ln L_1/\partial K^2 = -m/K^2 , \tag{5.4.19}$$

$$\partial^2 \ln L_2/\partial K^2 = -m/K^2 . \tag{5.4.20}$$

Hence the expected values of $\partial^2 \ln L/\partial K^2$ are both $-m/K^2$, the Cramér-Rao lower bound is given by K^2/m and the relative efficiency of the unbiased estimators given by Equations (5.4.12) and (5.4.13) is $(m-2)/m$, $m > 2$. It should be noted that the variances of the estimators and also the Cramér-Rao lower bounds are the same for any $n \geq m$.

5.4.3 Interval Estimation of the Shape Parameters K

Epstein and Sobel (1953) have shown that $2m\,\hat{\theta}_{mn}/\theta$ has a chi-square distribution with $2m$ degrees of freedom

$$2m\hat{\theta}_{mn}/\theta = \chi^2_{2m} \ . \tag{5.4.21}$$

Making the transformation of random variables, discussed in subsection 5.4.2, from the Pareto distribution to the exponential distribution and setting $K = 1/\theta$ and $\hat{K}_{mn}|\epsilon = 1/\hat{\theta}_{mn}$ in Equation (5.4.21) we obtain

$$2mK/(\hat{K}_{mn}|\epsilon) = \chi^2_{2m} \ . \tag{5.4.22}$$

Solving for K, we obtain

$$K = (\hat{K}_{mn}|\epsilon)\chi^2_{2m}/2m \ . \tag{5.4.23}$$

Then an upper confidence bound with confidence P (lower confidence bound with confidence $1 - P$) on the shape parameter K of a Pareto distribution is given by

$$\underline{K}_{1-P} = \overline{K}_P = (\hat{K}_{mn}|\epsilon)\chi^2_{2m,P}/2m \ . \tag{5.4.24}$$

The interval $(\underline{K}_{1-P}, \overline{K}_{1-P})$ is a central confidence interval with confidence level $1 - 2P$. Numerical confidence bounds may be obtained with the aid of a table of percentage points of the chi-square distribution [see Harter (1964)]. Confidence bounds for the shape parameter K of a limited distribution may be found exactly the same way as above since the distribution of the estimator and the functional relationships are identical if we replace $\hat{K}_{mn}|\epsilon$ by $\hat{K}_{mn}|\omega$.

5.4.4 Numerical Examples

Example 1. The following ordered pseudo-random sample of size 40 was drawn from a Pareto distribution with parameters $\epsilon = 0$ and $K = 1$:

1.0518	1.2526	1.3995	1.6715	2.3107	3.3146	4.7971	10.9258
1.0534	1.3604	1.4715	1.9564	2.3611	3.7712	5.1568	11.1492
1.1158	1.3662	1.4767	2.0622	2.4993	3.9670	5.1841	14.7554
1.1382	1.3786	1.4889	2.1212	2.6975	4.1188	7.1695	15.9696
1.1859	1.3932	1.5929	2.2638	2.8309	4.4773	8.0258	17.8400

If we assume that the value of ϵ is known, the maximum-likelihood estimator of K, based on the first m order statistics of a sample of size n, is given by Equation (5.4.4), the unbiased estimator is given by Equation (5.4.12), and exact confidence bounds can be obtained from Equation (5.4.24). These calculations for the above sample of size $n = 40$ have been performed for $m = 8(8)40$, with the following results:

m	$\hat{K}_{mn}\|\epsilon$	$\tilde{K}_{mn}\|\epsilon$	$\underline{K}_{.80}$	$\overline{K}_{.80}$
8	0.7054	0.6172	0.4917	0.9023
16	0.9508	0.8914	0.7472	1.1429
24	0.8907	0.8536	0.7352	1.0390
32	0.9076	0.8792	0.7706	1.0392
40	0.9761	0.9517	0.8444	1.1031

where $\hat{K}_{mn}|\epsilon$ is the maximum-likelihood point estimate, $\tilde{K}_{mn}|\epsilon$ is the corresponding unbiased estimate, and $\underline{K}_{.80}$ and $\overline{K}_{.80}$ are the lower and upper exact 80 percent confidence bounds, based on the first m order statistics of the above sample.

Example 2. The following ordered pseudo-random sample of size 40 was drawn from a limited distribution with parameters $\omega = 1$ and $K = 1$:

0.0006	0.1547	0.2854	0.3935	0.4963	0.6211	0.7703	0.8634
0.0117	0.1645	0.2865	0.3964	0.5535	0.6371	0.8145	0.8821
0.0962	0.1994	0.3212	0.4372	0.5600	0.6807	0.8214	0.9219
0.1029	0.2281	0.3532	0.4526	0.5970	0.6835	0.8480	0.9648
0.1186	0.2390	0.3923	0.4888	0.5995	0.7682	0.8632	0.9970

If we assume that the value of ω is known, the maximum-likelihood estimator of K, based on the first m order statistics of a sample of size n, is given by Equation (5.4.9), the unbiased estimator is given by Equation (5.4.13), and exact confidence bounds can be obtained from Equation (5.4.24), with $\hat{K}_{mn}|\epsilon$ replaced by $\hat{K}_{mn}|\omega$, with the aid of a table of percentage points of the chi-square distribution [see, for example, Harter (1964)]. These calculations for the above sample of size $n = 40$ have been performed for $m = 8(8)40$, with the following results:

m	$\hat{K}_{mn}\|\omega$	$\tilde{K}_{mn}\|\omega$	$\underline{K}_{.80}$	$\overline{K}_{.80}$
8	0.9956	0.8712	0.6940	1.2735
16	1.0033	0.9406	0.7885	1.2060
24	0.9975	0.9559	0.8233	1.1636
32	0.9754	0.9449	0.8281	1.1168
40	0.9814	0.9569	0.8490	1.1090

where $\hat{K}_{mn}|\omega$ is the maximum-likelihood point estimate, $\tilde{K}_{mn}|\omega$ is the corresponding unbiased estimate, and $\underline{K}_{.80}$ and $\overline{K}_{.80}$ are the lower and upper exact 80 percent confidence bounds, based on the first m order statistics of the above sample.

The random variable X having the limited distribution is restricted to the interval $\omega - 1 \leq x \leq \omega$. If one takes $\omega = 1$, as in the above example, values of x are restricted to the interval $[0, 1]$. If one takes $K = 1$, as in the above example, the distribution is uniform, but the estimation procedure works equally well for more general true values of the parameter K.

5.4.5 References

1. Arnold, Barry C. (1983). *Pareto Distributions.* International Co-operative Publishing House, Fairland, Maryland.
2. Epstein, Benjamin and Sobel, Milton (1953). Life testing. *Journal of the American Statistical Association* **48**, 486-502.
3. Gumbel, Emil J. (1958). *Statistics of Extremes.* Columbia University Press, New York.
4. Harter, H. Leon (1964). *New Tables of the Incomplete Gamma-Function Ratio and of Percentage Points of the Chi-Square and Beta Distributions.* U.S. Government Printing Office, Washington.
5. Johnson, Norman L., Kotz, Samuel, and Balakrishnan, N. (1994). *Continuous Univariate Distributions - Volume 1*, Second edition. John Wiley & Sons, New York.
6. Malik, Henrick J. (1966). Estimation of the parameters of the Pareto distribution (abstract). *Annals of Mathematical Statistics* **37**, 765.
7. Mandelbrot, Benoit (1960). The Pareto-Levy law and the distribution of income. *International Economic Review* **1**, 79-106.

8. Moore, Albert H. and Harter, H. Leon (1969). Conditional maximum-likelihood estimation, from singly censored samples, of the shape parameters of Pareto and limited distributions. *IEEE Transactions on Reliability* **R-18**, 76-78.

9. Muniruzzaman, A.N.M. (1951). On some distributions in connection with Pareto's law. *Proceedings of the First Pakistan Statistical Conference*, 90-93. Punjab University Press, Lahore.

10. Steindl, Josef (1965). *Random Processes and the Growth of Firms: A Study of the Pareto Law.* Hafner Publishing Co., New York.

11. Pareto, Vilfredo (1897). *Cours d'Économie Politique*, Volume 2. F. Rouge, Lausanne and F. Pichon, Paris.

6

Single Order Statistics From Populations Related to the Exponential

6.1 Weibull Population with Known Location and Shape Parameters [1]

6.1.1 Introduction

In Chapter 5, Section 5.1, a discussion has been given of point and interval estimators, derived by Harter and Moore (1965), of the scale parameter of a Weibull population with known location and shape parameters, based on the first m of n ordered observations. However, in many practical applications, less efficient estimators, based on one order statistic, may be chosen because of their inherent simplicity. In this section, use is made of the relationship between exponential and Weibull populations to obtain point and interval estimators, based on one order statistic. This is accomplished by applying a suitable transformation to the corresponding point and interval estimators for the parameter of a one-parameter exponential population as discussed in Chapter 4 and tabulated in Tables D2 and D4, respectively. The resulting point estimator, which is consistent but biased, is compared with an unbiased estimator obtained by Quayle (1963). Results are reported of a Monte Carlo study of the ratio of the mean square errors of estimators based on m order statistics and on one order statistic. The use of Table D2 to compute a consistent estimate for the scale parameter of a Weibull population with known location and shape parameters and the use of Table D4 to compute exact confidence bounds for the Weibull scale parameter are discussed. A numerical example is given of the computation of both point and interval estimates.

6.1.2 Point Estimation of the Scale Parameter θ

A Biased Estimator Based on One Order Statistic

Harter (1961) tabulated the minimum-variance unbiased one-order-statistic estimator of the parameter of a one-parameter exponential population from a complete sample of size $n = 1(1)100$. An unbiased estimator of the parameter σ was given by Epstein and Sobel (1953) [see Chapter 4, subsection 4.1.2 of the present volume] as

$$\tilde{\sigma}_k = c_k x_k, \qquad (k = 1, 2, \ldots, n) \tag{6.1.1}$$

[1] Earlier versions of portions of this material were published by Moore and Harter (1965, 1966).

with

$$c_k = 1 \Big/ \sum_{i=1}^{k} a_i \tag{6.1.2}$$

and

$$a_i = 1/(n - i + 1) \ . \tag{6.1.3}$$

For each value of n, the coefficient c_k and the estimator variance and efficiency were tabulated for $k = r$, where r is that value of k which minimizes the estimator variance. The resulting table is included as Table D1 of the present volume. For a censored sample in which one knows only the first m of the ordered sample values, as in a life test which is terminated at the time of the mth failure, the most efficient estimator is given by Equation (6.1.1), where $k = \min(m, r)$. The coefficients $c_k = c(k, n)$ and the efficiency relative to the m-order-statistic estimator are given in Table D2 for $n = 1(1)40$ and $m = 1(1)n$.

The Weibull density function is given by

$$f(y) = (K/\theta)(y/\theta)^{K-1} \exp[-(y/\theta)^K], \qquad (y \geq 0;\ \theta, K > 0) \tag{6.1.4}$$

where the shape parameter K is assumed to be known and the location parameter to be zero. (If the location parameter is known, but not equal to zero, one can transform the data by subtracting the known value of the location parameter, so the latter assumption involves no loss of generality.) Making the change of variable $x = y^K$, we obtain

$$g(x) = \exp(-x/\theta^K)/\theta^K \ , \tag{6.1.5}$$

which is the familiar exponential density function with parameter $\sigma = \theta^K$, for which an unbiased estimator is given by

$$\tilde{\sigma}_k = c_k x_k = c_k y_k^K \ , \tag{6.1.6}$$

where x_k and y_k are the kth order statistics of samples of size n from an exponential population and from a Weibull population with shape parameter K, respectively. Taking the Kth root of the extreme members of Equation (6.1.6), we get

$$\tilde{\sigma}_k^{1/K} = c_k^{1/K} y_k = \tilde{\theta}_k \ . \tag{6.1.7}$$

Now $\tilde{\theta}_k$ is a consistent estimator for θ, the parameter of the Weibull population, since $\sigma = \theta^K$ and $\tilde{\sigma}_k$ is a consistent estimator of the parameter σ of the exponential population.

Comparison With an Unbiased Estimator

Quayle (1963) obtained the unbiased estimator, based on one order statistic of a sample of size n for the scale parameter θ of a Weibull population with known shape parameter K. The estimator is given by

$$\breve{\theta}_k = \breve{c}_k y_k \tag{6.1.8}$$

where y_k $(k = 1, 2, \ldots, n)$ is the kth order statistic of the sample and

$$\breve{c}_k = \breve{c}_{k,n,K}$$
$$= 1\Big/\left[n\binom{n-1}{k-1}\Gamma(1+1/K)\sum_{j=0}^{k-1}\binom{k-1}{j}(-1)^{k+j-1}/(n-j)^{1+1/K}\right]. \tag{6.1.9}$$

Quayle has tabulated $\breve{c}_r$, where r is the value of k chosen so as to minimize the variance of $\breve{\theta}_k$, for $n = 1(1)20(2)40$, $K = 0.5(0.5)4.0(1.0)8.0$. The estimator given earlier in this subsection is

$$\tilde{\theta}_k = \tilde{c}_k y_k \tag{6.1.10}$$

where

$$\tilde{c}_k = \tilde{c}_{k,n,K} = 1\Big/\left\{\sum_{j=0}^{k-1}[1/(n-j)]\right\}^{1/K}. \tag{6.1.11}$$

Equations (6.1.9) and (6.1.11) illustrate the advantage of the estimator found here. Since the K appears only outside the summation sign in Equation (6.1.11), we need only calculate the sum once for each pair of values of n and k. Both estimators are based on the same order statistic and therefore $\tilde{\theta}_k$ is a biased estimator except for the case in which the Weibull density reduces to the exponential density, that is when $K = 1$, for which the two estimators are identical.

Monte Carlo Study of Ratios of Mean Square Errors

Moore and Harter (1966) conjectured that the ratio of the mean square errors of the maximum-likelihood m-order-statistic estimator and of the one-order-statistic estimator for the scale parameter of a Weibull population with known shape parameter is closely approximated by the relative efficiency of the one-order-statistic estimator of the parameter of a one-parameter exponential population as compared with the m-order-statistic estimator, which had been tabulated by Moore and Harter (1965, 1966) [see Table D2 of the present volume.] (It should be noted that one may speak of relative efficiencies in the case of the exponential population, since the estimators are unbiased, but only of ratios of mean square errors in the case of the Weibull population, for which the estimators are biased.) In order to check the validity of the conjecture, a Monte Carlo study of the ratio of mean square errors was performed. One thousand random samples each of size n [$n = 1(1)40$] from a one-parameter exponential population with parameter $\sigma = 1$ were generated in the IBM 7094 computer. These were transformed into samples from a two-parameter Weibull population with scale parameter $\theta = 1$ and shape parameter $K = 2$. From each sample, the one-order-statistic estimate, based on the kth order statistic [$k = \min(m, r)$, where r is chosen so as to minimize the variance of the one-order-statistic estimator of the exponential parameter], and the m-order-statistic estimate, based on the first m order statistics [$m = 1(1)n$], of the exponential parameter σ or the Weibull scale parameter θ were computed. For each population and for each combination of m and n, the ratio of the mean square error of the m-order-statistic estimates to that of the one-order-statistic estimates was calculated. Except for fluctuations due to random sampling, the ratios of the mean square errors in the case of the exponential parameter should agree with the tabulated relative efficiencies, and it was found that the agreement was quite good. Moreover, it was found that the ratios of mean square errors in the case of the Weibull scale parameter agreed with the tabulated relative efficiencies almost as well as did those for the exponential parameter, thus confirming the conjecture.

6.1.3 Interval Estimation of the Scale Parameter θ

Harter (1964) [see Chapter 4, Section 4.2 of the present volume] has obtained exact upper and lower confidence bounds and central confidence intervals, based on the mth order statistic, x_m, of a sample of size n, for the parameter of one-parameter exponential population for a wide range of confidence levels. He tabulated [see Table D4 of the present volume] the coefficients B_{um}/x_m and B_{lm}/x_m for $n = 1(1)20(2)40$ and m optimal for a wide range of confidence levels. If we introduce the notation $D_{lm} = B_{lm}/x_m$ and $D_{um} = B_{um}/x_m$, then the exact confidence interval based on one order statistic is given by

$$D_{lm}x_m < \sigma < D_{um}x_m \ . \tag{6.1.12}$$

If the random variable Y has a Weibull distribution with scale parameter θ and shape parameter K, then we have seen in subsection 6.1.2 that the random variable $X = Y^K$ has an exponential distribution with parameter $\sigma = \theta^K$. Hence, if we let $X = Y^K$, Inequality (6.1.12) becomes

$$D_{lm}y_m^K < \sigma < D_{um}y_m^K \ . \tag{6.1.13}$$

Taking the Kth root of each member of Inequality (6.1.13), we obtain

$$D_{lm}^{1/K} y_m < \sigma^{1/K} < D_{um}^{1/K} y_m \ . \tag{6.1.14}$$

But $\theta = \sigma^{1/K}$, and therefore

$$D_{lm}^{1/K} y_m < \theta < D_{um}^{1/K} y_m \tag{6.1.15}$$

gives an exact central confidence interval for the scale parameter θ of a Weibull population with known shape parameter K, the left and right members being respectively exact lower and upper confidence bounds, with confidence levels the same as for the tabulated values D_{lm} and D_{um}.

6.1.4 Use of Tables

In life-testing situations, one may wish to terminate a test without waiting for all items to fail or one may wish to estimate the unknown parameter of the population as the test progresses. If the life distribution is Weibull, and if m out of n items placed on test have failed, one may estimate the scale parameter θ by using Equation (6.1.7) with $k = \min(m, r)$, where r is the value of k chosen so as to minimize the variance of the estimator $\tilde{\sigma}_k$, given by Equation (6.1.1), of the exponential parameter σ from a complete sample of size n. For given values of n and m [$n = 1(1)40$, $m = 1(1)n$], the values of k and $c_k = c(k, n)$ can be read from Table D2, as can the efficiency of the one-order-statistic estimator given by Equation (6.1.1) for the exponential parameter relative to the m-order-statistic estimator. To approximate the efficiency of the one-order-statistic estimator relative to the estimator based on the complete sample of size n, one would multiply the tabled efficiency by m/n. The latter estimator could, of course, be used only if testing were continued until all n items had failed. The Monte Carlo study described in subsection 6.1.2 has confirmed the conjecture that the relative efficiencies of the estimators of the exponential parameter are good approximations to the corresponding ratios of mean square errors of the estimators of the Weibull scale parameter.

Table D4 gives coefficients of optimal order statistics in exact upper and lower confidence bounds, based on one order statistic, for the parameter of a one-parameter exponential population. These coefficients may also be used to obtain exact confidence bounds, based on one order statistic, for the scale parameter of a two-parameter Weibull population with known shape parameter.

In the case of both point and interval estimators, the single order statistic which is optimal for estimating the exponential parameter is not necessarily optimal for estimating the Weibull scale parameter, but in practice its departure from optimality is negligible, as can be seen by a comparison of the estimators obtained from Tables D2 and D4 with the optimal estimators for the Weibull scale parameter given by Quayle (1963).

6.1.5 Numerical Example

Consider the following data (failure times in hours) resulting from a simulated life test on forty components:

5	33	55	65	82	102	114	142
10	34	58	65	85	103	116	143
17	36	58	66	90	106	117	151
32	54	61	67	92	107	124	158
32	55	64	68	92	114	139	195

Suppose that the experimenter knows that the above data have come from a two-parameter Weibull population with shape parameter $K = 2.0$, and that he wishes to find a point estimate and 80 percent lower and upper confidence bounds for the scale parameter θ. Harter and Moore (1965) [see Chapter 5, subsection 5.1.4 of the present volume] have previously done this for estimates based on the first m order statistics [$m = 8(8)40$]. From Tables D2 (or D1) and D4 of the present volume, one finds that, for the parameter of a one-parameter exponential population, the optimal order statistic, from a sample of size forty, for obtaining a point estimator and the 80 percent lower and upper confidence bounds is the thirty-second, with coefficients 0.640744, 0.553447, and 0.768717, respectively. Substituting these coefficients and the value of the 32nd order statistic of the above sample in Equation (6.1.7) and Inequality (6.1.15), one finds that the point estimate of the scale parameter of the Weibull population from which the above sample came is $[0.640744]^{1/2}(116) = 92.9$, the 80 percent lower confidence bound is $[0.553447]^{1/2}(116) = 86.3$, and the 80 percent upper confidence bound is $[0.768717]^{1/2}(116) = 101.7$. These may be compared with the results 93.7, 87.6, and 101.7 obtained from the first 32 order statistics, 93.3, 87.8, and 100.3 obtained from all 40 observations, and the true population parameter 100.

6.1.6 References

1. Epstein, Benjamin and Sobel, Milton (1953). Life testing. *Journal of the American Statistical Association* **48**, 486-502.
2. Harter, H. Leon (1961). Estimating the parameters of negative exponential populations from one or two order statistics. *Annals of Mathematical Statistics* **32**, 1078-1090.
3. Harter, H. Leon (1964). Exact confidence bounds, based on one order statistic, for the parameter of an exponential population. *Technometrics* **6**, 301-317.

4. Harter, H. Leon and Moore, Albert H. (1965). Point and interval estimators, based on m order statistics, for the scale parameter of a Weibull population with known shape parameter. *Technometrics* **7**, 405-422.
5. Moore, Albert H. and Harter, H. Leon (1965). One-order-statistic estimation of the scale parameters of Weibull populations. *IEEE Transactions on Reliability* **R-14**, 100-106.
6. Moore, Albert H. and Harter, H. Leon (1966). Point and interval estimation, from one order statistic, of the location parameter of an extreme-value distribution with known scale parameter and of the scale parameter of a Weibull distribution with known shape parameter. *IEEE Transactions on Reliability* **R-15**, 120-126.
7. Quayle, Ronald J. (1963). Estimation of the Scale Parameter of the Weibull Probability Density Function by Use of One Order Statistic. *Unpublished M.S. thesis*, Air Force Institute of Technology, Wright-Patterson Air Force Base. AD425228.

6.2 Type I Extreme-Value Population With Known Scale Parameter [2]

6.2.1 Introduction

In Chapter 5, Section 5.2, a discussion has been given of point and interval estimators, derived by Harter and Moore (1967), of the location parameter of a Type I extreme-value population with known scale parameter, based on the first m order statistics of a sample of size n from the distribution of smallest values. However, in many practical applications, less efficient estimators, based on one order statistic, may be chosen because of their inherent simplicity. In this section, use is made of the relationship between exponential and Type I extreme-value populations to obtain point and interval estimators, based on one order statistic. This is accomplished by applying a suitable transformation to the corresponding point and interval estimators for the parameter of a one-parameter exponential population as discussed in Chapter 4 and tabulated in Tables D2 and D4, respectively. The resulting point estimator is consistent but biased. Results are reported of a Monte Carlo study of the ratio of the mean square errors of estimates based on m order statistics and on one order statistic. The use of Table D2 to compute a consistent estimate for the location parameter of a Type I extreme-value population with known scale parameter and the use of Table D4 to compute exact confidence bounds for the extreme-value location parameter are discussed. A numerical example is given of the computation of both point and interval estimates.

6.2.2 Point Estimation of the Location Parameter u

A Biased Estimator Based on One Order Statistic

If the random variable X has the one-parameter exponential distribution with parameter σ and cumulative distribution function

$$F(x;\sigma) = 1 - \exp(-x/\sigma), \qquad (x \geq 0;\ \sigma > 0) \tag{6.2.1}$$

then the random variable $Y = b \ln X$ has the Type I extreme-value distribution of smallest values with location parameter $u = b \ln \sigma$, scale parameter b, and cumulative distribution function

$$G(y;u,b) = 1 - \exp\{-\exp[(y-u)/b]\}, \qquad (b > 0)\ . \tag{6.2.2}$$

A one-order-statistic estimator for the parameter of the exponential population was given by Epstein and Sobel (1953) [see Chapter 4, subsection 4.1.2 of the present volume] as

$$\tilde{\sigma}_k = c_k x_k\ , \tag{6.2.3}$$

where

$$c_k = 1 \Big/ \sum_{i=1}^{k} [1/(n-i+1)] \tag{6.2.4}$$

and x_k is the kth order statistic of a sample of size n from the exponential population. Values of c_k have been tabulated [see Tables D1 and D2] by Harter (1961) and by Moore

[2] An earlier version of this material was published by Moore and Harter (1966).

and Harter (1965, 1966). Therefore, an estimator of the parameter u (given b) is obtained as follows:

$$\tilde{u}_k = b \ln \tilde{\sigma}_k = b \ln(c_k x_k) = b \ln c_k + b \ln x_k \ , \tag{6.2.5}$$

and since $y = b \ln x$ we obtain

$$\tilde{u}_k = b \ln c_k + y_k \ , \tag{6.2.6}$$

where c_k is given by Equation (6.2.4) and y_k is the kth order statistic of a sample of size n from the extreme-value population. Now $\tilde{u}_k$ is a consistent estimator of the location parameter u of the extreme-value population, since $u = b \ln \sigma$ and $\tilde{\sigma}_k$ is a consistent estimator for the parameter σ of the exponential population.

Monte Carlo Study of Ratios of Mean Square Errors

Moore and Harter (1966) made the same conjecture concerning the ratio of mean square errors of the maximum-likelihood m-order-statistic estimator and of the one-order-statistic estimator in the case of the location parameter of a Type I extreme-value distribution of smallest values as in the case of the scale parameter of a Weibull population with known shape parameter, which was discussed in subsection 6.1.2. The 1000 Monte Carlo samples considered there were further transformed into samples from a Type I extreme-value population with location parameter $b = 1/2$ and scale parameter $\theta = 1$. From each sample, the one-order-statistic estimate, based on the kth order statistic [$k = \min(m, r)$, where r is chosen so as to minimize the variance of the one-order-statistic estimator of the exponential parameter], and the m-order-statistic estimate, based on the first m order statistics [$m = 1(1)n$], of the extreme-value location parameter u were computed. For each combination of m and n, the ratio of the mean square error of the m-order-statistic estimates to that of the one-order-statistic estimates was calculated. As in the case of the Weibull scale parameter, it was found that the ratios of mean square errors for the estimates agreed with the tabulated relative efficiencies for estimators of the exponential parameter almost as well as did those for estimates of the exponential parameter itself, thus confirming the conjecture.

6.2.3 Interval Estimation of the Location Parameter u

Harter (1964) has obtained [see Chapter 4, Section 4.2 of the present volume] exact upper and lower confidence bounds and central confidence intervals, based on the mth order statistic, x_m, of a sample of size n, for the parameter of a one-parameter exponential population. He tabulated [see Table D4 of the present volume] the coefficients B_{um}/x_m and B_{lm}/x_m for $n = 1(1)20(2)40$ and m optimal for a wide range of confidence levels. If we use the notation $D_{lm} = B_{lm}/x_m$ and $D_{um} = B_{um}/x_m$ introduced in subsection 6.1.3, then the exact confidence interval based on one order statistic is given by

$$D_{lm} x_m < \sigma < D_{um} x_m \ . \tag{6.2.7}$$

If we let $y = b \ln x$, so that $x = \exp(y/b)$, we obtain

$$D_{lm} \exp(y_m/b) < \sigma < D_{um} \exp(y_m/b) \ , \tag{6.2.8}$$

where y_m is the mth order statistic of a sample of size n from the Type I extreme-value distribution of smallest values. Now if we take the natural logarithm of each member of Inequality (6.2.8) and multiply by b, we obtain

$$b \ln D_{lm} + y_m < b \ln \sigma < b \ln D_{um} + y_m \; . \tag{6.2.9}$$

But $u = b \ln \sigma$ and therefore by substitution we find that

$$b \ln D_{lm} + y_m < u < b \ln D_{um} + y_m \; , \tag{6.2.10}$$

which gives an exact confidence interval for u, the left and right members being respectively exact lower and upper confidence bounds with the same levels of confidence as for the tabulated values D_{lm} and D_{um}. Hence we have a simple method of computing exact central confidence intervals or upper and lower confidence bounds, based on one order statistic, for the location parameter of the Type I extreme-value distribution of smallest values, with known scale parameter b.

6.2.4 Use of Tables

Tables D2 and D4, respectively, can be used in computing point and interval estimates, based on one order statistic, for the location parameter of a Type I extreme-value population with known scale parameter in much the same way as we have outlined in subsection 6.1.4 for the scale parameter of a Weibull population with known shape parameter. Suppose only the first m order statistics of a sample of size n are known, as is the case at the time of the mth failure in a life test of n items, known or assumed to have come from a Type I extreme-value distribution of smallest values with known scale parameter. One can find a point estimate of the location parameter u by using Equation (6.2.6) with $k = \min(m, r)$, where r is the value of k chosen so as to minimize the variance of the estimator $\tilde{\sigma}_k$ of the exponential parameter from a complete sample of size n. Again the coefficient $c_k = c(k, n)$ can be read from Table D2, as can the efficiency of the one-order-statistic estimator for the exponential parameter relative to the m-order-statistic estimator, which can be used to approximate the corresponding ratio of mean square errors for the estimators of the extreme-value location parameter. Table D4 can be used, in conjunction with Inequality (6.2.10), to calculate interval estimates of the extreme-value location parameter.

The optimal order statistics for obtaining point and interval estimators for the extreme-value location parameter, unlike those for Weibull scale parameter, have not been determined, but there is reason to believe that the departure from optimality is negligible if one uses the values from Tables D2 and D4 based on the exponential population.

6.2.5 Numerical Example

Consider the following data (obtained by taking natural logarithms of the data in the example of subsection 6.1.5) from a Type I extreme-value population with scale parameter $b = 0.5$:

1.609	3.497	4.007	4.174	4.407	4.625	4.736	4.956
2.303	3.526	4.060	4.174	4.443	4.635	4.754	4.963
2.833	3.584	4.060	4.190	4.500	4.663	4.762	5.017
3.466	3.989	4.111	4.205	4.522	4.673	4.820	5.063
3.466	4.007	4.159	4.220	4.522	4.736	4.934	5.273

Using the same tabular values as in the example of subsection 6.1.5, one finds by substitution in Equation (6.2.6) and Inequality (6.2.10) that the point estimate of the location parameter of the extreme-value population is $0.5 \ln 0.640744 + 4.754 = 4.531$, the 80 percent lower confidence bound is $0.5 \ln 0.553447 + 4.754 = 4.458$, and the 80 percent upper confidence bound is $0.5 \ln 0.768717 + 4.754 = 4.623$. These may be compared with the following results of Harter and Moore (1967) [see Chapter 5, subsection 5.2.4 of the present volume]: 4.541, 4.474, and 4.624 (based on the first 32 order statistics); 4.537, 4.476, and 4.610 (based on all 40 observations); and with the true population parameter 4.605 ($= \ln 100$). The confidence bounds are exactly (except for rounding errors) and the point estimates approximately equal to the natural logarithms of the corresponding values obtained in the example of subsection 6.1.5.

6.2.6 References

1. Epstein, Benjamin and Sobel, Milton (1953). Life testing. *Journal of the American Statistical Association* **48**, 486-502.
2. Harter, H. Leon (1961). Estimating the parameters of negative exponential populations from one or two order statistics. *Annals of Mathematical Statistics* **32**, 1078-1090.
3. Harter, H. Leon (1964). Exact confidence bounds, based on one order statistic, for the parameter of an exponential population. *Technometrics* **6**, 301-317.
4. Harter, H. Leon and Moore, Albert H. (1967). A note on estimation from a Type I extreme-value distribution. *Technometrics* **9**, 325-331.
5. Moore, Albert H. and Harter, H. Leon (1965). One-order-statistic estimation of the scale parameters of Weibull populations. *IEEE Transactions on Reliability* **R-14**, 100-106.
6. Moore, Albert H. and Harter, H. Leon (1966). Point and interval estimation, from one order statistic, of the location parameter of an extreme-value distribution with known scale parameter and of the scale parameter of a Weibull distribution with known shape parameter. *IEEE Transactions on Reliability* **R-15**, 120-126.

6.3 Type II Extreme-Value Populations With Known Shape Parameters [3]

6.3.1 Introduction

In Chapter 5, Section 5.3, a discussion has been given of point and interval estimators, derived by Harter and Moore (1968), of the scale parameters of Type II extreme-value populations with known shape parameter. These estimators are based on the first (last) m order statistics of samples of size n from Type II asymptotic distributions of smallest (largest) values. However, less efficient estimators, based on one order statistic, may be chosen because of their inherent simplicity, or, in the case of the distribution of largest values, because the data required to compute the estimates become available sooner. In this section, use is made of the relationship between exponential and Type II extreme-value populations to obtain such estimators. This is accomplished by applying a suitable transformation to the corresponding estimators for the exponential parameter as discussed in Chapter 4 and tabulated in Tables D2 and D4. The resulting point estimators are consistent but biased. A discussion is given of the relative merits of point estimators based on one and on m order statistics. A numerical example is given of the computation of both point and interval estimates.

6.3.2 Point Estimation of the Scale Parameters v_1 and v_n

Biased Estimators Based on One Order Statistic

If the random variable X has the one-parameter exponential distribution with parameter σ and cumulative distribution function

$$F(x;\sigma) = 1 - \exp(-x/\sigma), \qquad (x \geq 0,\ \sigma > 0) \tag{6.3.1}$$

then the random variable $Y = -X^{-1/K}$ has the Type II asymptotic distribution of smallest values with location parameter zero, scale parameter $v_1 = -\sigma^{-1/K}$, shape parameter K, and cumulative distribution function

$$G(y; v_1, K) = 1 - \exp[-(y/v_1)^{-K})], \qquad (y, v_1 < 0;\ K > 0)\ . \tag{6.3.2}$$

A one-order-statistic estimator for the parameter of the exponential population was given by Epstein and Sobel (1953), with coefficients tabulated by Harter (1961) [see Chapter 4, subsection 4.1.2 of the present volume] as

$$\tilde{\sigma}_k = c_k x_k\ , \tag{6.3.3}$$

where

$$c_k = 1\Big/\sum_{i=1}^{k}[1/(n-i+1)] \tag{6.3.4}$$

and x_k is the kth order statistic of a sample of size n from the exponential population. Therefore, a one-order-statistic estimator for the parameter v_1 (given K) is obtained as follows:

$$\tilde{v}_1|K = -\tilde{\sigma}_k^{-1/K} = -c_k^{-1/K} x^{-1/K} = c_k^{-1/K} y_k\ , \tag{6.3.5}$$

[3] An earlier version of this material was published by Moore and Harter (1967).

where c_k is given by Equation (6.3.4) above and tabulated in Table D2 and y_k is the kth order statistic of a sample of size n from the Type II asymptotic distribution of smallest values. Now $\tilde{v}_1|K$ is a consistent estimator of the scale parameter v_1 of the Type II asymptotic distribution of smallest values, since $v_1 = -\sigma^{-1/K}$ and $\tilde{\sigma}_k$ is a consistent estimator for the parameter σ of the exponential population.

If we have a Type II asymptotic distribution of smallest values with known location parameter not equal to zero and known shape parameter, then we can transform the original sample data to data with location parameter zero and proceed with the estimation as outlined above.

If we have a sample from a Type II asymptotic distribution of largest values, we can use the symmetry with the distribution of smallest values to obtain an estimator of the scale parameter v_n, given the shape parameter K, as follows:

$$\tilde{v}_n|K = -\tilde{v}_1|K = -c_k^{-1/K} y_k \ . \tag{6.3.6}$$

But $y_k = -z_{n-k+1}$ and therefore we obtain

$$\tilde{v}_n|K = c_k^{-1/K} z_{n-k+1} \ , \tag{6.3.7}$$

where c_k is given by Equation (6.3.4) and tabulated in Table D2 and z_{n-k+1} is the $(n-k+1)$th order statistic (kth largest value) of a sample of size n from the Type II asymptotic distribution of largest values with location parameter zero, scale parameter $v_n = \sigma^{-1/K}$, shape parameter K, and cumulative distribution function

$$H(z; v_n, K) = \exp[-(z/v_n)^{-K}], \qquad (z, v_n, K > 0) \ . \tag{6.3.8}$$

Relative Merits of Estimators Based on One and m Order Statistics

Moore and Harter (1965, 1966) have tabulated [see Table D2 of the present volume] the efficiency of the one-order-statistic estimator of the exponential parameter relative to the m-order-statistic estimator. It was verified by Monte Carlo methods [see subsections 6.1.2 and 6.2.2 of this chapter] that the ratios of the mean square errors of the m-order-statistic estimates and the one-order-statistic estimates, from 1000 samples, for both the scale parameter of a two-parameter Weibull population with known shape parameter and the location parameter of a Type I asymptotic distribution of smallest values with known scale parameter were in close agreement with these efficiencies. A comparison of the mean square errors of estimates computed from the m-order-statistic estimators [see Chapter 5, subsection 5.3.2 of the present volume] derived by Harter and Moore (1968) and the one-order-statistic estimators derived in this subsection should show the same agreement.

6.3.3 Interval Estimation of Scale Parameters v_1 and v_n

We have seen in subsection 6.3.2 that if the random variable X has the one-parameter negative exponential distribution with parameter σ, then the random variable $Y = -X^{-1/K}$ has the Type II asymptotic distribution of smallest values with location parameter zero, scale parameter $v_1 = -\sigma^{-1/K}$, and shape parameter K. In Chapter 4, Section 4.2 we have given exact upper and lower confidence bounds and central confidence intervals, obtained by Harter (1964), for the parameter σ of a one-parameter exponential population, based on the mth order statistic x_m. An exact confidence interval is given by

$$D_{lm} x_m < \sigma < D_{um} x_m \ , \tag{6.3.9}$$

where the coefficients $D_{lm} = B_{lm}/x_m$ and $D_{um} = B_{um}/x_m$ have been tabulated [see Table D4] for a wide range of confidence levels, with $n = 1(1)20(2)40$, and m optimal. If we substitute $(-y)^{-K}$ for x in Inequality (6.3.9), we obtain

$$D_{lm}(-y_m)^{-K} < \sigma < D_{um}(-y_m)^{-K} \ , \tag{6.3.10}$$

where y_m is the mth order statistic of a sample of size n from the Type II asymptotic distribution of smallest values. Now if we take the Kth root of each member of Inequality (6.3.10), multiply each member by minus one, and invert each member, we find

$$D_{lm}^{-1/K} y_m < -\sigma^{-1/K} < D_{um}^{-1/K} y_m \ . \tag{6.3.11}$$

But $v_1 = -\sigma^{-1/K}$, and therefore we obtain

$$D_{lm}^{-1/K} y_m < v_1 < D_{um}^{-1/K} y_m \ , \tag{6.3.12}$$

an exact confidence interval for v_1, the left and right members being respectively exact lower and upper confidence bounds with the same level of confidence as for the tabulated values of D_{lm} and D_{um}.

If we have a sample of size n from a Type II asymptotic distribution of largest values, we can again use the symmetry with the distribution of smallest values to obtain the confidence interval for the scale parameter v_n as follows:

$$D_{lm}^{-1/K} z_{n-m+1} > v_n > D_{um}^{-1/K} z_{n-m+1} \ , \tag{6.3.13}$$

where z_{n-m+1} is the $(n-m+1)$th order statistic (mth largest value) from the sample of size n, and the left and right members are respectively exact upper and lower confidence bounds with the same levels of confidence as for the tabulated values D_{lm} and D_{um}.

6.3.4 Numerical Example

The following ordered pseudo-random sample of size 40 was drawn from a Type II asymptotic distribution of largest values with location parameter $\epsilon = 0$ [omitted from Equation (6.3.8)], scale parameter (characteristic value) $v_n = 1$, and shape parameter $K = 2$:

0.4409	0.6592	0.7695	0.8960	1.0203	1.2639	1.8244	2.6122
0.4669	0.6730	0.7815	0.9031	1.1049	1.3515	1.8397	4.7915
0.5468	0.6872	0.8199	0.9209	1.1320	1.5679	1.9473	5.1470
0.6099	0.7037	0.8648	0.9566	1.1377	1.7585	2.2080	5.4795
0.6332	0.7452	0.8913	0.9676	1.1406	1.8116	2.3687	6.1992

If we assume that the value of K is known and that ϵ is known to be zero, a consistent point estimator of v_n is given by Equation (6.3.7). For $n = 40$, the estimator based on the 9th order statistic (the 32nd largest value in the sample) is optimal or nearly so in the class of one-order-statistic estimators. For the above sample, the estimate of v_n based on the 9th order statistic is $\tilde{v}_n|K = (0.640774)^{-1/2}(0.7037) = 0.8792$, which may be compared with $\hat{v}_n|K = 0.8831$, the maximum-likelihood m-order-statistic estimate based on the 32 highest sample values, again under the assumption that ϵ is known to be zero. Exact confidence bounds, based on one order statistic, for v_n can be found from (6.3.13). The exact upper and lower 80 percent confidence bounds for v_n based on the 9th order statistic of the above sample are $(0.553447)^{-1/2}(0.7037) = 0.9460$ and

$(0.768717)^{-1/2}(0.7037) = 0.8027$, while the corresponding bounds based on the 32 highest sample values, as given by Moore and Harter (1968), are 0.9449 and 0.8137 [see Chapter 5, subsection 5.3.4 of the present volume, where point and interval estimates based on the last $m = 8(8)40$ order statistics of the above sample are worked out.]

6.3.5 Remarks on Applications

The random variable Z having a Type II asymptotic distribution of largest values is subject to the restriction $z \geq \epsilon$, where $\epsilon \geq 0$. This distribution is likely to find application in life testing problems. Estimation based on one order statistic has the advantage that only a small proportion (approximately 20 percent plus one) of the items placed on test need to be tested to failure in order to obtain nearly optimal point and interval estimates of the parameter v_n when ϵ and K are known. By contrast, the m-order-statistic estimates discussed in Chapter 5, Section 5.3 cannot be computed until all of the sample items have failed, since they are based on the last (highest) m order statistics.

Since the Type II asymptotic distribution of smallest values is restricted to negative values, it is difficult to conceive of applications in life testing or reliability. One possible application has been suggested in Chapter 5, subsection 5.3.5.

6.3.6 References

1. Epstein, Benjamin and Sobel, Milton (1953). Life testing. *Journal of the American Statistical Association* **48**, 486-502.
2. Harter, H. Leon (1961). Estimating the parameters of negative exponential populations from one or two order statistics. *Annals of Mathematical Statistics* **32**, 1078-1090.
3. Harter, H. Leon (1964). Exact confidence bounds, based on one order statistic, for the parameter of an exponential population. *Technometrics* **6**, 301-317.
4. Harter, H. Leon and Moore, Albert H. (1968). Conditional maximum-likelihood estimators, from singly censored samples, of the scale parameters of Type II extreme-value distributions. *Technometrics* **10**, 349-359.
5. Moore, Albert H. and Harter, H. Leon (1965). One-order-statistic estimation of the scale parameters of Weibull populations. *IEEE Transactions on Reliability* **R-14**, 100-106.
6. Moore, Albert H. and Harter, H. Leon (1966). Point and interval estimation, from one order statistic, of the location parameter of an extreme-value distribution with known scale parameter and of the scale parameter of a Weibull distribution with known shape parameter. *IEEE Transactions on Reliability* **R-15**, 120-126.
7. Moore, Albert H. and Harter, H. Leon (1967). One-order-statistic conditional estimators of shape parameters of limited and Pareto distributions and scale parameters of Type II asymptotic distributions of smallest and largest values. *IEEE Transactions on Reliability* **R-16**, 100-103.

6.4 Limited and Pareto Populations with Known Location Parameters [4]

6.4.1 Introduction

In Chapter 5, Section 5.4, a discussion has been given of point and interval estimators, derived by Moore and Harter (1969), of the shape parameters of limited and Pareto populations with known location parameters, based on the first m order statistics of a sample of size n. However, less efficient estimators, based on one order statistic, may be chosen because of their inherent simplicity. In this section, use is made of the relationship between the exponential and the limited and Pareto populations to obtain such estimators. This is accomplished by applying suitable transformations to the corresponding estimators for the exponential parameter as discussed in Chapter 4 and tabulated in Tables D2 and D4. The resulting point estimators are consistent but biased. A discussion is given of the relative merits of point estimators based on one and on m order statistics. Numerical examples are given of the computation of both point and interval estimates.

6.4.2 Point Estimation of the Shape Parameters K

Biased Estimator of Shape Parameter of Limited Population

If the random variable X has one-parameter exponential distribution with parameter σ and cumulative distribution function

$$F(x;\sigma) = 1 - \exp(-x/\sigma), \qquad (x \geq 0,\ \sigma > 0) \tag{6.4.1}$$

then the random variable $Y = \omega - \exp(-X)$ has the limited distribution [see Gumbel (1958, p. 157)] with location parameter ω, shape parameter $K = 1/\sigma$, and cumulative distribution function

$$G(y;\omega,K) = 1 - (\omega - y)^K, \qquad (\omega - 1 \leq y \leq \omega;\ K > 0)\ . \tag{6.4.2}$$

A one-order-statistic estimator for the parameter of the exponential population was given by Epstein and Sobel (1953) [see Chapter 4, subsection 4.1.2 of the present volume] as

$$\tilde{\sigma}_k = c_k x_k\ , \tag{6.4.3}$$

where

$$c_k = 1 \Big/ \sum_{i=1}^{k} [1/(n-i+1)] \tag{6.4.4}$$

and x_k is the kth order statistic of a sample of size n from the exponential population. Therefore a one-order-statistic estimator for the parameter K (given ω) is obtained as follows:

$$\tilde{K}|\omega = 1/\tilde{\sigma}_k = 1/c_k x_k = -1/c_k \ln(\omega - y_k)\ , \tag{6.4.5}$$

where c_k is given by Equation (6.4.4) and tabulated in Table D2 and y_k is the kth order statistic of a sample of size n from the limited distribution. Now $\tilde{K}|\omega$ is a consistent estimator of the shape parameter K of the limited distribution, since $K = 1/\sigma$ and $\tilde{\sigma}_k$ is a consistent estimator for the exponential parameter σ.

[4] An earlier version of this material was published by Moore and Harter (1967).

Biased Estimator of Shape Parameter of Pareto Population

If the random variable X has the one-parameter exponential distribution with parameter σ and cumulative distribution function given by Equation (6.4.1), then the random variable $Z = \epsilon + \exp X$ has the Pareto distribution [see Gumbel (1958, p. 157)] with location parameter ϵ, shape parameter $K = 1/\sigma$, and cumulative distribution function

$$H(z;\epsilon,K) = 1 - (z-\epsilon)^{-K}, \qquad (z \geq 1+\epsilon,\ K > 0)\ . \tag{6.4.6}$$

Therefore a one-order-statistic estimator of K (given ϵ) is obtained as follows:

$$\tilde{K}|\epsilon = 1/\tilde{\sigma}_k = 1/c_k x_k = 1/c_k \ln(z_k - \epsilon)\ , \tag{6.4.7}$$

where c_k is given by Equation (6.4.4) and tabulated in Table D2 and z_k is the kth order statistic of a sample of size n from the Pareto distribution. Now $\tilde{K}|\epsilon$ is a consistent estimator of the shape parameter K of the Pareto distribution, since $K = 1/\sigma$ and $\tilde{\sigma}_k$ is a consistent estimator of the exponential parameter σ.

Relative Merits of Estimators Based on One and m Order Statistics

Moore and Harter (1965, 1966) have tabulated [see Table D2 of the present volume] the efficiency of the one-order-statistic estimator of the exponential parameter relative to the m-order-statistic estimator. It was verified by Monte Carlo methods [see subsections 6.1.2 and 6.2.2] that the ratios of mean square errors of the m-order-statistic estimates and the one-order-statistic estimates, from 1000 samples, for both the scale parameter of a two-parameter Weibull population with known shape parameter and the location parameter of a Type I asymptotic distribution of smallest values with known scale parameter were in close agreement with these efficiencies. A comparison of the mean square errors of the estimates computed from the m-order-statistic estimators [see Chapter 5, subsection 5.3.2 of the present volume] derived by Harter and Moore (1968) and the one-order-statistic estimators derived in this section should show the same agreement. As in the case of other one-order-statistic estimators obtained by transforming the optimal estimators for the exponential parameter, there is no guarantee that the estimators obtained here are optimal in the class of one-order-statistic estimators, but in practice the departure from optimality is negligible.

6.4.3 Interval Estimation of Shape Parameters K

Exact Confidence Bounds for Shape Parameter of Limited Population

We have seen in subsection 6.4.2 that if the random variable X has the one-parameter exponential distribution with parameter σ, then the random variable $Y = \omega - \exp(-X)$ has the limited distribution with location parameter ω and shape parameter $K = 1/\sigma$. In Chapter 4, Section 4.2 we have given exact upper and lower confidence bounds and central confidence intervals, obtained by Harter (1964), for the parameter σ of a one-parameter exponential population, based on the mth order statistic x_m. An exact confidence interval is given by

$$D_{lm} x_m < \sigma < D_{um} x_m\ , \tag{6.4.8}$$

where the coefficients $D_{lm} = B_{lm}/x_m$ and $D_{um} = B_{um}/x_m$ have been tabulated [see Table D4] for a wide range of confidence levels, with $n = 1(1)20(2)40$ and m optimal. If we set $x = -\ln(\omega - y)$ in Inequality (6.4.8), we obtain

$$-D_{lm}\ln(\omega - y_m) < \sigma < -D_{um}(\omega - y_m) , \tag{6.4.9}$$

where y_m is the mth order statistic of a sample of size n from the limited population. By inverting the members of the inequality, we obtain

$$-1/D_{lm}\ln(\omega - y_m) > 1/\sigma > -1/D_{um}\ln(\omega - y_m) . \tag{6.4.10}$$

But $K = 1/\sigma$ and therefore

$$-1/D_{lm}\ln(\omega - y_m) > K > -1/D_{um}\ln(\omega - y_m) \tag{6.4.11}$$

is an exact confidence interval for K, the left and right members of Inequality (6.4.11) being respectively exact upper and lower confidence bounds for K with the same levels of confidence as for the tabulated values D_{lm} and D_{um}.

Exact Confidence Bounds for Shape Parameter of Pareto Population

We have seen in subsection 6.4.2 that if the random variable X has the one-parameter exponential distribution with parameter σ, then the random variable $Z = \epsilon + \exp X$ has the Pareto distribution with location parameter ϵ and shape parameter $K = 1/\sigma$. If, in Inequality (6.4.8), we set $x = \ln(z - \epsilon)$, we obtain

$$D_{lm}\ln(z_m - \epsilon) < \sigma < D_{um}\ln(z_m - \epsilon) , \tag{6.4.12}$$

where z_m is the mth order statistic of a sample of size n from the Pareto population. By inverting the members of the inequality, we obtain

$$1/D_{lm}\ln(z_m - \epsilon) > 1/\sigma > 1/D_{um}\ln(z_m - \epsilon) . \tag{6.4.13}$$

But $K = 1/\sigma$ and therefore

$$1/D_{lm}\ln(z_m - \epsilon) > K > 1/D_{um}\ln(z_m - \epsilon) \tag{6.4.14}$$

is an exact confidence interval for K, the left and right members being respectively exact upper and lower confidence bounds with the same levels of confidence as for the tabulated values D_{lm} and D_{um}.

6.4.4 Numerical Examples

Example 1. The following ordered pseudo-random sample of size 40 was drawn from a limited population with parameters $\omega = 1$ and $K = 1$:

0.0006	0.1547	0.2854	0.3935	0.4963	0.6211	0.7703	0.8634
0.0117	0.1645	0.2865	0.3964	0.5535	0.6371	0.8145	0.8821
0.0962	0.1994	0.3212	0.4372	0.5600	0.6807	0.8214	0.9219
0.1029	0.2281	0.3532	0.4526	0.5970	0.6835	0.8480	0.9648
0.1186	0.2390	0.3923	0.4888	0.5995	0.7682	0.8632	0.9970

If we assume that the value of ω is known, a consistent point estimator of K is given by Equation (6.4.5), where $c_k = c(k,n)$ has been tabulated [see Tables D1 and D2 of the present volume] by Harter (1961) for complete samples of size $n = 1(1)100$ and by Moore and Harter (1965, 1966) for censored samples (first m order statistics) of size $n = 1(1)40$, $m = 1(1)n$, and k such that x_k is optimal among the available order statistics for estimating the exponential parameter σ by use of Equation (6.4.3). The same value of k is optimal or nearly so for the limited and other related populations. For a complete sample of size $n = 40$, one finds $k = 32$ and $c_k = 0.640744$. The resulting one-order-statistic estimate of K, based on the 32nd order statistic, is $\tilde{K}|\omega = -1/0.640744 \ln(1-0.8145) = 0.9264$, which may be compared with $\hat{K}|\omega = 0.9754$, the maximum-likelihood m-order-statistic estimate, given by Moore and Harter (1969) [see Chapter 5, subsection 5.4.4 of the present volume], based on the first 32 order statistics, again under the assumption that ω is known. Exact confidence bounds, based on one order statistic, for K can be found from (6.4.11), with the aid of coefficients read from Table D4, which were tabulated by Harter (1964). The exact upper and lower 80 percent confidence bounds for K, based on the 32nd order statistic of the above sample, are $-1/0.553447 \ln(1-0.8145) = 1.0725$ and $-1/0.768717 \ln(1-0.8145) = 0.7721$, while the corresponding bounds, given by Moore and Harter (1969) [see Chapter 5, subsection 5.4.4 of the present volume], based on the first 32 order statistics, are 1.1168 and 0.8281.

The random variable Y having the limited distribution is restricted to the interval $\omega - 1 \leq y \leq \omega$. If one takes $\omega = 1$, as in the above example, values of y are restricted to the interval $[0,1]$. This special case is likely to have applications in reliability problems. If one takes $K = 1$, as in the above example, the distribution is uniform, but the estimation procedure works equally well for more general true values of the parameter K.

Example 2. The following pseudo-random sample of size 40 was drawn from a Pareto distribution with parameters $\epsilon = 0$ and $K = 1$:

1.0518	1.2526	1.3995	1.6715	2.3107	3.3146	4.7971	10.9258
1.0534	1.3604	1.4715	1.9564	2.3611	3.7712	5.1568	11.1492
1.1158	1.3662	1.4767	2.0622	2.4993	3.9670	5.1841	14.7554
1.1382	1.3786	1.4889	2.1212	2.6975	4.1188	7.1695	15.9696
1.1859	1.3932	1.5929	2.2638	2.8309	4.4773	8.0258	17.8400

If we assume that the value of ϵ is known, a consistent point estimator of K is given by Equation (6.4.7). The resulting one-order-statistic estimate of K, based on the 32nd order statistic, is $\tilde{K}|\epsilon = 1/0.640744 \ln 5.1568 = 0.9515$, which may be compared with $\hat{K}|\epsilon = 0.9076$, the maximum-likelihood m-order-statistic estimate based on the first 32 order statistics, again under the assumption that ϵ is known. Exact confidence bounds, based on one order statistic, for K can be found from Inequality (6.4.14). The upper and lower 80 percent confidence bounds for K based on the 32nd order statistic of the above sample are $1/0.553447 \ln 5.1568 = 1.1015$ and $1/0.768717 \ln 5.1568 = 0.7931$, while the corresponding bounds based on the first 32 order statistics were found by Moore and Harter (1969) [see Chapter 5, subsection 5.4.4 of the present volume] to be 1.1168 and 0.8281.

The random variable Z having the Pareto distribution is subject to the restriction $z \geq 1 + \varepsilon$, with the added restriction $\varepsilon \geq 0$, though mathematically unnecessary, often imposed in practice. This distribution is likely to find application in life testing with positive guaranteed life.

6.4.5 References

1. Epstein, Benjamin and Sobel, Milton (1953). Life testing. *Journal of the American Statistical Association* **48**, 486-502.
2. Gumbel, E.J. (1958). *Statistics of Extremes.* Columbia University Press, New York.
3. Harter, H. Leon (1961). Estimating the parameters of negative exponential populations from one or two order statistics. *Annals of Mathematical Statistics* **32**, 1078-1090.
4. Harter, H. Leon (1964). Exact confidence bounds, based on one order statistic, for the parameter of an exponential population. *Technometrics* **6**, 301-317.
5. Moore, Albert H. and Harter, H. Leon (1965). One-order-statistic estimation of the scale parameters of Weibull populations. *IEEE Transactions on Reliability* **R-14**, 100-106.
6. Moore, Albert H. and Harter, H. Leon (1966). Point and interval estimation, from one order statistic, of the location parameter of an extreme-value distribution with known scale parameter and of the scale parameter of a Weibull distribution with known shape parameter. *IEEE Transactions on Reliability* **R-15**, 120-126.
7. Moore, Albert H. and Harter, H. Leon (1967). One-order-statistic conditional estimators of shape parameters of limited and Pareto distributions and scale parameters of Type II asymptotic distributions of smallest and largest values. *IEEE Transactions on Reliability* **R-16**, 100-103.
8. Moore, Albert H. and Harter, H. Leon (1969). Conditional maximum-likelihood estimation, from singly censored samples, of the shape parameters of Pareto and limited distributions. *IEEE Transactions on Reliability* **R-18**, 76-78.

7

Doubly Censored Samples From Various Populations

7.1 Maximum-Likelihood Estimation of Two Parameters of Normal Population [1]

7.1.1 Introduction

The estimation of the parameters of a censored sample from a normal population has been considered by many authors, who have used several different methods, including the method of least squares and the method of maximum likelihood.

Lloyd (1952) applied the theory of least squares estimation to an ordered sample from distributions depending on location and scale parameters only. Gupta (1952) derived best linear estimators (from samples of size $n \leq 10$) for the mean and variance using singly censored samples from normal populations and for larger values of n derived an alternative linear estimator. Sarhan and Greenberg (1956, 1958a, 1958b) estimated the mean and standard deviation of normal populations from singly and doubly censored samples ($n \leq 20$) by the method of least squares. These tables were extended by Balakrishnan (1990) for sample sizes up to 40. Saw (1959) developed simplified unbiased estimators of the mean and variance given a singly censored sample from a normal population ($n \leq 20$). Dixon (1957, 1960) derived simplified estimators of the mean and standard deviation for complete and censored normal samples which are almost as efficient as the best linear estimators ($n \leq 20$). Walsh (1956) obtained distribution-free estimators for the population mean and variance for a rather general class of continuous statistical populations using doubly censored samples. Balakrishnan and Cohen (1991) and Johnson, Kotz, and Balakrishnan (1994) have reviewed all these developments on the linear estimation of the normal parameters based on singly and doubly censored samples.

Cohen (1950) used the method of maximum likelihood to estimate the parameters of normal populations from singly and doubly truncated samples. The term "truncated samples" was used by Cohen in a sense somewhat broader than its present usage and included what are now called "censored samples." Cohen was primarily concerned, however, with Type I censoring (at a specified time) rather than Type II censoring (when a specified number of failures have occurred). Gupta (1952) found maximum likelihood equations for estimators of the parameters of a normal population from a sample censored from above (Type II censoring), and determined their asymptotic variances and covariances. Halperin (1952) proved under mild regularity conditions that the maximum-likelihood estimator of a single parameter from singly censored samples is consistent, asymptotically normally distributed, and of minimum variance for large samples and indicated his

[1] An earlier version of this material was published by Harter and Moore (1966).

results could be generalized to several parameters and more general censoring. Breakwell (1953) also obtained maximum-likelihood estimators for singly censored samples, asymptotic distributions of the estimators, and their asymptotic biases. Plackett (1958) showed that maximum-likelihood estimators are asymptotically linear and that the best linear unbiased estimators are asymptotically normal and efficient. Plackett computed a "linearized maximum-likelihood" estimator and compared it with the best linear unbiased estimator for the standard deviation of a normal population from censored samples ($n \leq 10$). In three later papers Cohen (1955, 1959, 1961) extended the results given in his 1950 paper. These results have been summarized in the books by Cohen and Whitten (1988), Cohen (1991), and Balakrishnan and Cohen (1991). The present section in part duplicates the work of Cohen and Gupta but extends the results for Type II censoring to include maximum-likelihood estimation of the parameters of a normal population from a doubly censored sample, together with a completely computerized iterative procedure, and mathematical expressions and tables for asymptotic variances and covariances. The mathematical formulation for maximum-likelihood estimation is given in subsection 7.1.2; the asymptotic variances and covariances of the estimators are given in subsection 7.1.3. A discussion of the iterative procedure for maximum-likelihood estimation is given in subsection 7.1.4. A Monte Carlo study of the maximum-likelihood estimators, together with a comparison with the best linear unbiased estimators, is given, for small samples, in subsection 7.1.5.

7.1.2 Mathematical Formulation

Consider a random sample of size n from a normal population with mean μ and standard deviation σ and let $X_{r_1+1}, \ldots, X_{n-r_2}$ be the ordered observations remaining when the r_1 smallest observations and the r_2 largest observations have been censored. The joint probability density function of these order statistics is given by:

$$f(x_{r_1+1}, \ldots, x_{n-r_2}; \mu, \sigma) = \frac{n!}{r_1! r_2!} \{\sigma\sqrt{2\pi}\}^{-m} \exp \sum_{i=r_1+1}^{n-r_2} \{-(x_i - \mu)^2 / 2\sigma^2\} \cdot \left[F\left(\frac{x_{r_1+1} - \mu}{\sigma}\right)\right]^{r_1} \left[1 - F\left(\frac{x_{n-r_2} - \mu}{\sigma}\right)\right]^{r_2}, \tag{7.1.1}$$

where $m = n - (r_1 + r_2)$ and $F(z_i) = \int_{-\infty}^{z_i} f(t)\, dt$, with $z_i = (x_i - \mu)/\sigma$ and $f(z_i) = (2\pi)^{-1/2} \exp(-z_i^2/2)$.

The natural logarithm of the likelihood function is given by

$$L = \ln \frac{n!}{r_1! r_2!} - \frac{m}{2} \ln 2\pi - m \ln \sigma - \sum_{i=r_1+1}^{n-r_2} \frac{(x_i - \mu)^2}{2\sigma^2} + r_1 \ln F\left(\frac{x_{r_1+1} - \mu}{\sigma}\right) + r_2 \ln\left[1 - F\left(\frac{x_{n-r_2} - \mu}{\sigma}\right)\right] . \tag{7.1.2}$$

The likelihood equations are:

$$\frac{\partial L}{\partial \mu} = \frac{1}{\sigma^2} \sum_{i=r_1+1}^{n-r_2} (x_i - \mu) - \frac{r_1}{\sigma} \frac{f[(x_{r_1+1} - \mu)/\sigma]}{F[(x_{r_1+1} - \mu)/\sigma]} + \frac{r_2}{\sigma} \frac{f[(x_{n-r_2} - \mu)/\sigma]}{1 - F[(x_{n-r_2} - \mu)/\sigma]} = 0 , \tag{7.1.3}$$

$$\frac{\partial L}{\partial \sigma} = -\frac{m}{\sigma} + \frac{1}{\sigma^3} \sum_{i=r_1+1}^{n-r_2} (x_i - \mu)^2 - r_1 \frac{[(x_{r_1+1} - \mu)/\sigma^2] f[(x_{r_1+1} - \mu)/\sigma]}{F[(x_{r_1+1} - \mu)/\sigma]}$$

$$+ r_2 \frac{[(x_{n-r_2} - \mu)/\sigma^2] f[(x_{n-r_2} - \mu)/\sigma]}{1 - F[(x_{n-r_2} - \mu)/\sigma]} = 0 \,. \tag{7.1.4}$$

If $m = n$, i.e., if $r_1 = r_2 = 0$, these equations have explicit solutions

$$\hat{\mu} = \sum_{i=1}^{n} x_i/n, \qquad \hat{\sigma} = \left\{ \sum_{i=1}^{n} (x_i - \hat{\mu})^2/n \right\}^{1/2} .$$

The details of the iterative procedure for determining the maximum-likelihood estimates are given in subsection 7.1.4.

7.1.3 Asymptotic Variances and Covariances

Gupta (1952) has given theoretical expressions and a table for the asymptotic variances and covariances of the maximum-likelihood estimators of the parameters of a normal population from singly censored (from above) samples. His results will be extended in this subsection to the case of doubly censored samples.

The natural logarithm of the likelihood function of a sample of size n, from a normal population with mean μ and standard deviation σ, the lowest r_1 and the highest r_2 sample values having been censored, is given by

$$L = \ln \frac{n!}{r_1! r_2!} - \frac{m}{2} \ln 2\pi - m \ln \sigma - \sum_{i=r_1+1}^{n-r_2} \frac{(x_i - \mu)^2}{2\sigma^2}$$

$$+ r_1 \ln F(z_1) + r_2 \ln[1 - F(z_2)] \,, \tag{7.1.5}$$

where

$$z_1 = (x_{r_1+1} - \mu)/\sigma, \qquad z_2 = (x_{n-r_2} - \mu)/\sigma \,,$$

$$F(z_i) = \int_{-\infty}^{z_i} f(t)\, dt, \qquad f(z_i) = \frac{1}{\sqrt{2\pi}} e^{-z_i^2/2} \,,$$

and $m = n - r_1 - r_2$. In this notation, the first partial derivatives of L are given by

$$\frac{\partial L}{\partial \mu} = \frac{1}{\sigma^2} \sum_{i=r_1+1}^{n-r_2} (x_i - \mu) - \frac{r_1}{\sigma} \frac{f(z_1)}{F(z_1)} + \frac{r_2}{\sigma} \frac{f(z_2)}{1 - F(z_2)} \,, \tag{7.1.6}$$

$$\frac{\partial L}{\partial \sigma} = -\frac{m}{\sigma} + \frac{1}{\sigma^3} \sum_{i=r_1+1}^{n-r_2} (x_i - \mu)^2 - \frac{r_1}{\sigma} \frac{z_1 f(z_1)}{F(z_1)} + \frac{r_2}{\sigma} \frac{z_2 f(z_2)}{1 - F(z_2)} \,. \tag{7.1.7}$$

The second partial derivatives of L are given by

$$\frac{\partial^2 L}{\partial \mu^2} = -\frac{m}{\sigma^2} - \frac{r_1}{\sigma^2} \frac{f(z_1)}{F(z_1)} \left[z_1 + \frac{f(z_1)}{F(z_1)} \right] + \frac{r_2}{\sigma^2} \frac{f(z_2)}{1 - F(z_2)} \left[z_2 - \frac{f(z_2)}{1 - F(z_2)} \right] , \tag{7.1.8}$$

$$\frac{\partial^2 L}{\partial \mu \partial \sigma} = -\frac{2}{\sigma^3} \sum_{i=r_1+1}^{n-r_2} (x_i - \mu) - \frac{r_1}{\sigma^2} \frac{f(z_1)}{F(z_1)} \left[z_1^2 + z_1 \frac{f(z_1)}{F(z_1)} - 1 \right]$$

$$+\frac{r_2}{\sigma^2}\frac{f(z_2)}{1-F(z_2)}\left[z_2^2-z_2\frac{f(z_2)}{1-F(z_2)}-1\right], \tag{7.1.9}$$

$$\frac{\partial^2 L}{\partial\sigma^2}=\frac{m}{\sigma^2}-\frac{3}{\sigma^4}\sum_{i=r_1+1}^{n-r_2}(x_i-\mu)^2-\frac{r_1}{\sigma^2}\frac{z_1 f(z_1)}{1-F(z_1)}\left[z_1^2+z_1\frac{f(z_1)}{F(z_1)}-2\right]$$

$$+\frac{r_2}{\sigma^2}\frac{z_2 f(z_2)}{1-F(z_2)}\left[z_2^2-z_2\frac{f(z_2)}{1-F(z_2)}-2\right]. \tag{7.1.10}$$

In what follows here and in Sections 7.5 and 7.6, mathematical rigor would require that the symbol "$\rightarrow$" be replaced by "converges in probability to", that the expectations be made conditional on z_1 and z_2, and that the limits as $n \rightarrow \infty$ be replaced by limits in probability.

Now let $q_1 = r_1/n$, $q_2 = r_2/n$, and $p = 1 - q_1 - q_2 = m/n$. As $n \rightarrow \infty$ (q_1 and q_2 fixed), $z_1 \rightarrow \hat{z}_1$ where $\int_{-\infty}^{\hat{z}_1} f(t)\,dt = q_1$, $z_2 \rightarrow \hat{z}_2$ where

$$\int_{\hat{z}_2}^{\infty} f(t)\,dt = q_2, \qquad E\left(\sum_{i=r_1+1}^{n-r_2}\frac{x_i-\mu}{\sigma}\right)\rightarrow n\int_{\hat{z}_1}^{\hat{z}_2} t f(t)\,dt = -n[f(\hat{z}_2)-f(\hat{z}_1)],$$

and

$$E\left(\sum_{i=r_1+1}^{n-r_2}\left(\frac{x_i-\mu}{\sigma}\right)^2\right)\rightarrow n\int_{\hat{z}_1}^{\hat{z}_2} t^2 f(t)\,dt = n[p-\{\hat{z}_2 f(\hat{z}_2)-\hat{z}_1 f(\hat{z}_1)\}].$$

The elements of the information matrix (multiplied by $\frac{\sigma^2}{n}$) may be written as

$$\lim_{n\rightarrow\infty} -\frac{\sigma^2}{n}E\left[\frac{\partial^2 L}{\partial\mu^2}\right]=p+f(\hat{z}_1)\left[\hat{z}_1+\frac{f(\hat{z}_1)}{q_1}\right]-f(\hat{z}_2)\left[\hat{z}_2-\frac{f(\hat{z}_2)}{q_2}\right]=v^{11}, \tag{7.1.11}$$

$$\lim_{n\rightarrow\infty} -\frac{\sigma^2}{n}E\left[\frac{\partial^2 L}{\partial\mu\partial\sigma}\right]=f(\hat{z}_1)-f(\hat{z}_2)+\hat{z}_1 f(\hat{z}_1)\left[\hat{z}_1+\frac{f(\hat{z}_1)}{q_1}\right]$$

$$-\hat{z}_2 f(\hat{z}_2)\left[\hat{z}_2-\frac{f(\hat{z}_2)}{q_2}\right]=v^{12}, \tag{7.1.12}$$

$$\lim_{n\rightarrow\infty} -\frac{\sigma^2}{n}E\left[\frac{\partial^2 L}{\partial\sigma^2}\right]=2p+\hat{z}_1 f(\hat{z}_1)-\hat{z}_2 f(\hat{z}_2)+\hat{z}_1^2 f(\hat{z}_1)\left[\hat{z}_1+\frac{f(\hat{z}_1)}{q_1}\right]$$

$$-\hat{z}_2^2 f(\hat{z}_2)\left[\hat{z}_2-\frac{f(\hat{z}_2)}{q_2}\right]=v^{22}. \tag{7.1.13}$$

The asymptotic variance-covariance matrix for the estimators $\hat{\mu}$ and $\hat{\sigma}$ is then $\frac{\sigma^2}{n}[v_{ij}]$, where $[v_{ij}] = [v^{ij}]^{-1}$. If one drops the terms involving $\hat{z}_2$ from Equations (7.1.11)-(7.1.13) the results agree with those given by Gupta for the case of single censoring.

The computation of the elements v^{ij} of the information matrix (multiplied by σ^2/n), as given by Equations (7.1.11)-(7.1.13), and the inversion of this matrix to obtain the coefficients of σ^2/n in the variance-covariance matrix were performed on the IBM 1620 computer. The resulting coefficients of σ^2/n in Var($\hat{\mu}$), Cov($\hat{\mu}, \hat{\sigma}$), Var($\hat{\sigma}$), Var($\hat{\mu}|\sigma$) and Var($\hat{\sigma}|\mu$) are given in Table F1 for all combinations of q_1 and q_2 which are integral multiples of 0.1 and which are such that $q_1 + q_2 < 1$ and $q_1 \leq q_2$. Only half of the table is given since interchanging the values for q_1 and q_2 would produce no change in the tabular values except that the Cov($\hat{\mu}, \hat{\sigma}$) would change sign. Values are given to six decimal places.

The results for single censoring from above (first ten lines of Table F1), when rounded to five decimal places, agree with those of Gupta, except for slight discrepancies in the case $q_1 = 0.0$, $q_2 = 0.9$.

7.1.4 Iterative Estimation Procedure

The likelihood equations [Equations (7.1.3) and (7.1.4)] have explicit solutions only in the case of complete samples ($m = n$). For censored samples, however, iterative procedures have been developed for finding the joint maximum-likelihood estimators. These involve estimating the parameters, one at a time, in the cyclic order μ, σ, omitting a parameter if it is assumed to be known. One starts by choosing initial estimate(s) for the unknown parameter(s). At each step, the rule of false position (iterative linear interpolation) is used to determine the value of the parameter then being estimated which satisfies the appropriate likelihood equation, in which the latest estimate (or known value) of the other parameter has been substituted. Iteration continues until the results of successive steps agree to within some assigned tolerance. Experience has shown that the rate of convergence is quite rapid if the initial estimates are reasonable and the amount of censoring is not excessive.

7.1.5 Monte Carlo Study for Small Samples

There is no known analytic method of determining the variances and covariances of the joint distribution of the maximum-likelihood estimators $\hat{\mu}$ and $\hat{\sigma}$ from small samples. Furthermore, these estimators, while asymptotically unbiased, are known to be biased for small samples [except $\hat{\mu}$ when censoring is absent or symmetric], though analytic expressions for the bias are known only in the case of estimation from the complete sample ($m = n$). In order to obtain information about the small-sample properties of these estimators, a Monte Carlo study was performed on the IBM 7094 computer. For $n = 10$ and for $n = 20$, one thousand random samples of n standard normal deviates were generated, and the n deviates in each sample were arranged in order from smallest to largest. The iterative procedure described in subsection 7.1.4 was used to compute the estimates $\hat{\mu}$ and $\hat{\sigma}$, also $\hat{\mu}|\sigma$ and $\hat{\sigma}|\mu$, from the m order statistics remaining in each sample after proportions q_1 and q_2 had been censored from below and from above, respectively, where q_1 and q_2 were taken at intervals of 0.1, subject to the restrictions $q_1 \le q_2$ and $m \ge 2$. The means, variances, and covariances of the estimates from 1000 samples of size $n = 10$ are given in Table G1A, and similar results for $n = 20$ are given in Table G1B. There is no loss of generality associated with the restriction $q_1 \le q_2$, since interchanging q_1 and q_2 would produce no change in the expected tabular values except that of reversing the signs of the mean of $\hat{\mu}$ and the covariance of $\hat{\mu}$ and $\hat{\sigma}$. The rows of Table G1A (and likewise Table G1B) are not statistically independent, since they are based on the same samples (with different proportions censored).

The following tentative conclusions may be drawn from Tables G1A and G1B: (1) When $q_1 < q_2$, the estimates $\hat{\mu}$ and $\hat{\mu}|\sigma$ are negatively biased. [By symmetry, these estimates are positively biased when $q_1 > q_2$ and unbiased when $q_1 = q_2$.] (2) The estimates $\hat{\sigma}$ and $\hat{\sigma}|\mu$ are negatively biased regardless of the relative magnitude of q_1 and q_2. (3) The bias in estimating either parameter is much smaller when the other parameter is known than it is when both parameters are being estimated simultaneously. (4) The bias of $\hat{\sigma}$ (μ unknown) is approximately equal to $-1/m$.

It would be desirable to compare the variances of $\hat{\mu}$ and $\hat{\sigma}$ from samples of sizes 10 and 20 with the values which one would obtain by substituting $n = 10$ and $n = 20$

in the asymptotic values given in Table F1, as well as with the variances of the best linear unbiased estimators μ^* and σ^*. Direct comparison of variances of estimators is appropriate, however, only when all the estimators are unbiased. In order to compensate for the bias in the maximum-likelihood estimators, the mean square errors of $\hat{\mu}$, $\hat{\sigma}$, $\hat{\mu}|\sigma$, and $\hat{\sigma}|\mu$ were computed. These were compared with the variances of the best linear unbiased estimators given by Sarhan and Greenberg (1962, Table 10C.2) and with the variances of the maximum-likelihood estimators given by the asymptotic formula, which were obtained by dividing by n the values given in Table F1. The results are shown in Tables G1C and G1D, from which the following tentative conclusions may be drawn: (1) The precision of the MLE $\hat{\mu}$, when proper allowance is made for bias, closely approximates that predicted by the asymptotic formula for the variance of $\hat{\mu}$, even for m as small as 2, except in cases of strongly asymmetric censoring. (2) The precision of the MLE $\hat{\sigma}$, when proper allowance is made for bias, closely approximates that predicted by the asymptotic formula for the variance of $\hat{\sigma}$, except when m is quite small and/or censoring is strongly asymmetric. (3) Maximum likelihood estimators tend to have somewhat smaller mean square errors than best linear unbiased estimators. The difference is greatest for estimators of μ in cases of strongly asymmetric censoring and for estimators of σ when m is small and/or censoring is strongly asymmetric.

Approximate corrections for the bias of the maximum-likelihood estimators $\hat{\mu}$, $\hat{\sigma}$, $\hat{\mu}|\sigma$, and $\hat{\sigma}|\mu$ for $n = 10$ and $n = 20$ can be made by use of the means found in the Monte Carlo study and recorded in Tables G1C and G1D.

Isida and Tagami (1959) also conducted a Monte Carlo study of bias and precision in maximum-likelihood estimation of the parameters of normal populations, and reached conclusions quite similar to ours. The main differences between the two studies are that (1) their study was restricted to the case of single censoring, while ours considered both single and double censoring and (2) they studied estimates $\hat{\mu}$, $\hat{\sigma}$, and $\hat{\mu}|\sigma$, but not $\hat{\sigma}|\mu$, while we dealt with all four of these.

7.1.6 References

1. Balakrishnan, N. (1990). Best Linear Unbiased Estimates of the Mean and Standard Deviation of Normal Distribution for Complete and Censored Samples of Sizes 21(1)30(5)40. *McMaster University Report*, Hamilton, Canada.
2. Balakrishnan, N. and Cohen, A. Clifford, Jr. (1991). *Order Statistics and Inference: Estimation Methods*. Academic Press, San Diego.
3. Breakwell, J.V. (1953). On estimating both mean and standard deviation of a normal population from the lowest r out of n observations (abstract). *Annals of Mathematical Statistics* **24**, 683.
4. Cohen, A. Clifford, Jr. (1950). Estimating the mean and variance of normal populations from singly and doubly truncated samples. *Annals of Mathematical Statistics* **21**, 557-569.
5. Cohen, A. Clifford, Jr. (1955). Truncated and censored samples from normal populations. *Transactions of the American Society for Quality Control* **9**, 27-36.
6. Cohen, A. Clifford, Jr. (1959). Simplified estimators for the normal distribution when samples are singly censored or truncated. *Technometrics* **1**, 217-237.
7. Cohen, A. Clifford, Jr. (1961). Tables for maximum-likelihood estimates: singly truncated and singly censored samples. *Technometrics* **3**, 535-541.
8. Cohen, A. Clifford, Jr. (1991). *Truncated and Censored Samples: Theory and*

Applications. Marcel Dekker, New York.

9. Cohen, A. Clifford, Jr. and Whitten, Betty J. (1988). *Parameter Estimation in Reliability and Life Span Models.* Marcel Dekker, New York.

10. Dixon, Wilfrid J. (1957). Estimates of the mean and standard deviation of a normal population. *Annals of Mathematical Statistics* **28**, 806-809.

11. Dixon, Wilfrid J. (1960). Simplified estimation from censored normal samples. *Annals of Mathematical Statistics* **31**, 385-391.

12. Gupta, A.K. (1952). Estimation of the mean and standard deviation of a normal population from a censored sample. *Biometrika* **39**, 260-273.

13. Halperin, M. (1952). Maximum-likelihood estimation in truncated samples. *Annals of Mathematical Statistics* **23**, 226-238.

14. Harter, H. Leon and Moore, Albert H. (1966). Iterative maximum-likelihood estimation of the parameters of normal populations from singly and doubly censored samples. *Biometrika* **53**, 205-213.

15. Isida, M. and Tagami, S. (1959). The bias and precision in the maximum likelihood estimation of the parameters of normal populations from singly truncated samples. *Reports of Statistical Application Research, Union of Japanese Scientists and Engineers* **6**, 105-110.

16. Johnson, Norman L., Kotz, Samuel, and Balakrishnan, N. (1994). *Continuous Univariate Distributions - Volume 1*, Second edition. John Wiley & Sons, New York.

17. Lloyd, E.H. (1952). Least squares estimation of location and scale parameters using order statistics. *Biometrika* **39**, 88-95.

18. Plackett, R.L. (1958). Linear estimation from censored data. *Annals of Mathematical Statistics* **29**, 131-142.

19. Sarhan, Ahmed E. and Greenberg, Bernard G. (1956). Estimation of location and scale parameters by order statistics from singly and doubly censored samples. Part I. The normal distribution up to samples of size 10. *Annals of Mathematical Statistics* **27**, 427-457.

20. Sarhan, Ahmed E. and Greenberg, Bernard G. (1958a). Estimation of location and scale parameters by order statistics from singly and doubly censored samples. Part II. Tables for the normal distribution for samples of size $11 \leq N \leq 15$. *Annals of Mathematical Statistics* **29**, 79-105.

21. Sarhan, Ahmed E. and Greenberg, Bernard G. (1958b). Estimation of location and scale parameters by order statistics from singly and doubly censored samples. Part III. *Technical Report No. 4*, Office of Ordnance Research Project 1597.

22. Sarhan, Ahmed E. and Greenberg, Bernard G. (Eds.) (1962). *Contributions to Order Statistics.* John Wiley & Sons, New York.

23. Saw, J.G. (1959). Estimation of the normal population parameters given a singly censored sample. *Biometrika* **46**, 150-159.

24. Walsh, J.E. (1956). Estimating population mean, variance, and percentage points from truncated data. *Skandinavisk Aktuarietidskrift* **39**, 47-58.

7.2 Local-Maximum-Likelihood Estimation of Three Parameters of Lognormal Population [2]

7.2.1 Introduction

In their book on the lognormal distribution, Aitchison and Brown (1957, pp. 37-65) have devoted two entire chapters to estimation problems, one each for the two-parameter and three-parameter distributions. Similarly, five chapters of the handbook on the Lognormal Distributions edited by Crow and Shimizu (1988) have been devoted to the problem of estimation of parameters. These two books and also Chapter 14 of the volume by Johnson, Kotz, and Balakrishnan (1994) present a comprehensive summary of efforts to estimate the parameters of a lognormal population by the method of maximum likelihood, the method of moments, the method of quantiles, the graphical method, and mixed methods. The problem of estimating the parameters of a two-parameter lognormal population with known lower bound is equivalent to that of estimating the parameters of a normal population, which has been considered by a number of authors. Harter and Moore (1966a) have summarized and extended the contributions of others to the solution of that problem, with particular emphasis on the method of maximum likelihood, and have proposed an iterative procedure for obtaining maximum-likelihood estimates of the parameters from complete, singly censored, and doubly censored samples. [See Section 7.1.]

Maximum-likelihood estimation of the parameters of a three-parameter lognormal population, for complete samples, has been investigated by Wilson and Worcester (1945), Cohen (1951), Aitchison and Brown (1957, pp. 55-56), and Hill (1963). The latter has explored some unusual features of the likelihood function of the three-parameter lognormal population which had apparently gone unnoticed by earlier investigators. In particular, he has shown that there exist paths along which the likelihood function of any ordered sample $x_1, \ldots, x_n$ tends to ∞ as (τ, μ, σ) approaches $(x_1, -\infty, +\infty)$, where τ is the threshold parameter and μ and σ are the mean and the standard deviation of the parent normal population. This global maximum of the likelihood function leads to the ridiculous maximum-likelihood estimates $\hat{\tau} = x_1$, $\hat{\mu} = -\infty$, and $\hat{\sigma} = +\infty$ regardless of the sample. On the other hand, solution of the likelihood equations leads, in most cases, to local-maximum-likelihood estimates which, while not true maximum-likelihood estimates according to the usual definition, are reasonable estimates and appear to possess most of the desirable properties usually associated with maximum-likelihood estimates. Exceptions may occur in the case of small samples, for which the likelihood function may have no clearly defined local maximum.

Apparently nothing had been published prior to the paper by Harter and Moore (1966b) on the problem of estimation for truncated or censored samples from a three-parameter lognormal population, though Aitchison and Brown (1957, pp. 88-91) had discussed the two-parameter case. The present section will be devoted to local-maximum-likelihood estimation, for the three-parameter case, from singly and doubly censored as well as complete samples. An iterative estimation procedure for use on a high-speed computer will be given. This procedure will, in most cases, converge to a point where the likelihood function has a local maximum. If the sample (after censoring, if any) is small, convergence may be slow or the iterative procedure may take off along the path to infinity mentioned above. Even in the latter case, reasonable estimates can be obtained by a modification of the iterative procedure similar to that used by Harter and Moore (1965) for the three-parameter Gamma and Weibull populations. [See Section 7.3.] Further discussions on

[2] An earlier version of this material was published by Harter and Moore (1966b).

the estimation of parameters for the three-parameter lognormal model are presented by Crow and Shimizu (1988), Cohen and Whitten (1988), Cohen (1991), and Balakrishnan and Cohen (1991).

7.2.2 The Likelihood Equations

Consider a random sample of size n from a three-parameter lognormal population with parameters μ, σ, and τ (the location or threshold parameter). Let $X_{r+1}, \ldots, X_m$ be the ordered observations remaining after the $n-m$ largest and the r smallest observations have been censored. The joint probability density function of these order statistics is given by

$$f(x_{r+1}, \ldots, x_m; \mu, \sigma, \tau)$$
$$= \frac{n!}{(n-m)!r!} \prod_{i=r+1}^{m} \frac{1}{\sigma\sqrt{2\pi}(x_i-\tau)} \exp\left\{-\sum_{i=r+1}^{m} \frac{[\ln(x_i-\tau)-\mu]^2}{2\sigma^2}\right\}$$
$$\cdot \left\{1 - F\left[\frac{\ln(x_m-\tau)-\mu}{\sigma}\right]\right\}^{n-m} \left\{F\left[\frac{\ln(x_{r+1}-\tau)-\mu}{\sigma}\right]\right\}^{r} . \qquad (7.2.1)$$

The natural logarithm of the likelihood function is given by

$$L = \ln[n!/(n-m)!r!] - \frac{1}{2}(m-r)\ln 2\pi - (m-r)\ln\sigma - \sum_{i=r+1}^{m} \ln(x_i-\tau)$$
$$- \frac{1}{2}\sum_{i=r+1}^{m} z_i^2 + (n-m)\ln[1-F(z_m)] + r\ln F(z_{r+1}) , \qquad (7.2.2)$$

where $z_i = [\ln(x_i-\tau)-\mu]/\sigma$, $F(z_i) = \int_{-\infty}^{z_i} f(t)\,dt$, and $f(z_i) = (2\pi)^{-1/2}\exp(-z_i^2/2)$. The likelihood equations are

$$\frac{\partial L}{\partial \mu} = \frac{1}{\sigma}\left\{\sum_{i=r+1}^{m} z_i + (n-m)\frac{f(z_m)}{1-F(z_m)} - r\frac{f(z_{r+1})}{F(z_{r+1})}\right\} = 0 , \qquad (7.2.3)$$

$$\frac{\partial L}{\partial \sigma} = \frac{1}{\sigma}\left\{-(m-r) + \sum_{i=r+1}^{m} z_i^2 + (n-m)\frac{z_m f(z_m)}{1-F(z_m)} - r\frac{z_{r+1}f(z_{r+1})}{F(z_{r+1})}\right\} = 0, \quad (7.2.4)$$

$$\frac{\partial L}{\partial \tau} = \sum_{i=r+1}^{m} (x_i-\tau)^{-1} + \frac{1}{\sigma}\left\{\sum_{i=r+1}^{m} \frac{z_i}{x_i-r} + (n-m)\frac{f(z_m)}{(x_m-\tau)[1-F(z_m)]}\right.$$
$$\left. - r\frac{f(z_{r+1})}{(x_{r+1}-\tau)F(z_{r+1})}\right\} = 0 . \qquad (7.2.5)$$

If $m = n$ and $r = 0$, Equations (7.2.3) and (7.2.4) can be solved for μ and σ as explicit functions of τ, yielding

$$\hat{\mu} = \sum_{i=1}^{n}[\ln(x_i-\tau)]/n ,$$

$$\hat{\sigma} = \sqrt{\sum_{i=1}^{n}[\ln(x_i-\tau)-\mu]^2/n} .$$

Equation (7.2.5) cannot be solved explicitly even if $m = n$ and $r = 0$. If censoring occurs, none of the likelihood equations has an explicit solution, and it is necessary to resort to iterative solutions. The details of an iterative procedure for solving the likelihood equations are given in subsection 7.2.4.

7.2.3 Asymptotic Variances and Covariances

The second partial derivatives of L with respect to the parameters are

$$\frac{\partial^2 L}{\partial \mu^2} = \frac{1}{\sigma^2}\left\{-(m-r) + (n-m)\,\frac{f(z_m)}{1-F(z_m)}\left[z_m - \frac{f(z_m)}{1-F(z_m)}\right] - r\,\frac{f(z_{r+1})}{F(z_{r+1})}\left[z_{r+1} + \frac{f(z_{r+1})}{F(z_{r+1})}\right]\right\}, \tag{7.2.6}$$

$$\frac{\partial^2 L}{\partial \sigma^2} = \frac{1}{\sigma^2}\left\{(m-r) - 3\sum_{i=r+1}^{m} z_i^2 + (n-m)\,\frac{z_m f(z_m)}{1-F(z_m)}\left[z_m^2 - \frac{z_m f(z_m)}{1-F(z_m)} - 2\right] - r\,\frac{z_{r+1} f(z_{r+1})}{F(z_{r+1})}\left[z_{r+1}^2 + \frac{z_{r+1} f(z_{r+1})}{F(z_{r+1})} - 2\right]\right\}, \tag{7.2.7}$$

$$\frac{\partial^2 L}{\partial \tau^2} = \sum_{i=r+1}^{m} (x_i-\tau)^{-2} + \frac{1}{\sigma^2}\left\{\sum_{i=r+1}^{m} \frac{\sigma z_i - 1}{(x_i-\tau)^2} + (n-m)\,\frac{f(z_m)}{(x_m-\tau)^2[1-F(z_m)]} \cdot\left[z_m - \frac{f(z_m)}{1-F(z_m)} + \sigma\right] - r\,\frac{f(z_{r+1})}{(x_{r+1}-\tau)^2 F(z_{r+1})}\left[z_{r+1} + \frac{f(z_{r+1})}{F(z_{r+1})} + \sigma\right]\right\}, \tag{7.2.8}$$

$$\frac{\partial^2 L}{\partial \mu \partial \sigma} = \frac{1}{\sigma^2}\left\{-2\sum_{i=r+1}^{m} z_i + (n-m)\,\frac{f(z_m)}{1-F(z_m)}\left[z_m^2 - \frac{z_m f(z_m)}{1-F(z_m)} - 1\right] - r\,\frac{f(z_{r+1})}{F(z_{r+1})}\left[z_{r+1}^2 + \frac{z_{r+1} f(z_{r+1})}{F(z_{r+1})} - 1\right]\right\}, \tag{7.2.9}$$

$$\frac{\partial^2 L}{\partial \mu \partial \tau} = \frac{1}{\sigma^2}\left\{-\sum_{i=r+1}^{m} (x_i-\tau)^{-1} + (n-m)\,\frac{f(z_m)}{(x_m-\tau)[1-F(z_m)]}\left[z_m - \frac{f(z_m)}{1-F(z_m)}\right] - r\,\frac{f(z_{r+1})}{(x_{r+1}-\tau)F(z_{r+1})}\left[z_{r+1} + \frac{f(z_{r+1})}{F(z_{r+1})}\right]\right\}, \tag{7.2.10}$$

$$\frac{\partial^2 L}{\partial \sigma \partial \tau} = \frac{1}{\sigma^2}\left\{-2\sum_{i=r+1}^{m} \frac{z_i}{(x_i-\tau)} + (n-m)\,\frac{f(z_m)}{(x_m-\tau)[1-F(z_m)]}\left[z_m^2 - \frac{z_m f(z_m)}{1-F(z_m)} - 1\right] - r\,\frac{f(z_{r+1})}{(x_{r+1}-\tau)F(z_{r+1})}\left[z_{r+1}^2 + \frac{z_{r+1} f(z_{r+1})}{F(z_{r+1})} - 1\right]\right\}. \tag{7.2.11}$$

In what follows here and in Section 7.4, mathematical rigor would require that the symbol "$\rightarrow$" be replaced by "converges in probability to", that the expectations be made conditional on z_{r+1} and z_m, and that the limits as $n \rightarrow \infty$ be replaced by limits in probability. This is actually done explicitly in Section 7.3.

Let $q_1 = r/n$, $q_2 = (n-m)/n$, and $p = 1 - q_1 - q_2 = (m-r)/n$. As $n \to \infty$ (q_1 and q_2 fixed),

$$z_{r+1} \to \hat{z}_{r+1} \text{ where } \int_{-\infty}^{\hat{z}_{r+1}} f(t)\,dt = q_1, \quad z_m \to \hat{z}_m \text{ where } \int_{\hat{z}_m}^{\infty} f(t)\,dt = q_2 ,$$

$$E \sum_{i=r+1}^{m} z_i \to \int_{\hat{z}_{r+1}}^{\hat{z}_m} t f(t)\,dt = -[f(\hat{z}_m) - f(\hat{z}_{r+1})] ,$$

$$E \sum_{i=r+1}^{m} z_i^2 \to \int_{\hat{z}_{r+1}}^{\hat{z}_m} t^2 f(t)\,dt = p - [\hat{z}_m f(\hat{z}_m) - \hat{z}_{r+1} f(\hat{z}_{r+1})] ,$$

$$E \sum_{i=r+1}^{m} \{(x_i - \tau)^{-1}\} \to \int_{\hat{z}_{r+1}}^{\hat{z}_m} e^{-(\mu+\sigma t)}\, f(t)\,dt = e^{-\mu+\sigma^2/2}[F(\hat{z}_m + \sigma) - F(\hat{z}_{r+1} + \sigma)] ,$$

$$E \sum_{i=r+1}^{m} \{(x_i - \tau)^{-2}\} \to \int_{\hat{z}_{r+1}}^{\hat{z}_m} e^{-2(\mu+\sigma t)}\, f(t)\,dt = e^{2(\sigma^2-\mu)}[F(\hat{z}_m + 2\sigma) - F(\hat{z}_{r+1} + 2\sigma)] ,$$

$$\begin{aligned} E \sum_{i=r+1}^{m} \{z_i/(x_i - \tau)\} \to \int_{\hat{z}_{r+1}}^{\hat{z}_m} e^{-(\mu+\sigma t)}\, t f(t)\,dt = -\, e^{-\mu+\sigma^2/2}\{&\sigma[F(\hat{z}_m + \sigma) - F(\hat{z}_{r+1} + \sigma)] \\ &+ f(\hat{z}_m + \sigma) - f(\hat{z}_{r+1} + \sigma)\} , \end{aligned}$$

$$\begin{aligned} E \sum_{i=r+1}^{m} \{(\sigma z_i - 1)/(x_i - \tau)^2\} &\to \int_{\hat{z}_{r+1}}^{\hat{z}_m} e^{-2(\mu+\sigma t)}\, (\sigma t - 1) f(t)\,dt \\ &= -e^{2(\sigma^2-\mu)}\{(2\sigma^2 + 1)[F(\hat{z}_m + 2\sigma) - F(\hat{z}_{r+1} + 2\sigma)] \\ &\quad + \sigma[f(\hat{z}_m + 2\sigma) - f(\hat{z}_{r+1} + 2\sigma)]\} . \end{aligned}$$

The elements of the information matrix (multiplied by σ^2/n) may be written as

$$\begin{aligned} \lim_{n\to\infty} (-\sigma^2/n) E(\partial^2 L/\partial \mu^2) &= p - f(\hat{z}_m)[\hat{z}_m - f(\hat{z}_m)/q_2] \\ &\quad + f(\hat{z}_{r+1})[\hat{z}_{r+1} + f(\hat{z}_{r+1})/q_1] = v^{11} , \end{aligned} \tag{7.2.12}$$

$$\begin{aligned} \lim_{n\to\infty} (-\sigma^2/n) E(\partial^2 L/\partial \sigma^2) &= 2p - \hat{z}_m f(\hat{z}_m) - \hat{z}_m^2 f(z_m)[\hat{z}_m - f(\hat{z}_m)/q_2] + \hat{z}_{r+1} f(\hat{z}_{r+1}) \\ &\quad + \hat{z}_{r+1}^2 f(\hat{z}_{r+1})[\hat{z}_{r+1} + f(\hat{z}_{r+1})/q_1] = v^{22} , \end{aligned} \tag{7.2.13}$$

$$\begin{aligned} \lim_{n\to\infty} (-\sigma^2/n) E(\partial^2 L/\partial \tau^2) &= e^{2(\sigma^2-\mu)}\{(\sigma^2 + 1)[F(\hat{z}_m + 2\sigma) - F(\hat{z}_{r+1} + 2\sigma)] \\ &\quad + \sigma[f(\hat{z}_m + 2\sigma) - f(\hat{z}_{r+1} + 2\sigma)]\} \\ &\quad - e^{-2(\mu+\hat{z}_m\sigma)} f(\hat{z}_m)[\hat{z}_m - f(\hat{z}_m)/q_2 + \sigma] \\ &\quad + e^{-2(\mu+\hat{z}_{r+1}\sigma)} f(\hat{z}_{r+1})[\hat{z}_{r+1} + f(\hat{z}_{r+1})/q_1 + \sigma] = v^{33} , \end{aligned} \tag{7.2.14}$$

$$\lim_{n\to\infty}(-\sigma^2/n)E(\partial^2 L/\partial\mu\partial\sigma) = -f(\hat{z}_m) - \hat{z}_m f(\hat{z}_m)[\hat{z}_m - f(\hat{z}_m)/q_2] + f(\hat{z}_{r+1}) + \hat{z}_{r+1} f(\hat{z}_{r+1})[\hat{z}_{r+1} + f(\hat{z}_{r+1})/q_1] = v^{12} , \tag{7.2.15}$$

$$\lim_{n\to\infty}(-\sigma^2/n)E(\partial^2 L/\partial\mu\partial\tau) = e^{-\mu+\sigma^2/2}[F(\hat{z}_m+\sigma) - F(\hat{z}_{r+1}+\sigma)] - e^{-(\mu+\hat{z}_m\sigma)} f(\hat{z}_m)[\hat{z}_m - f(\hat{z}_m)/q_2] + e^{-(\mu+\hat{z}_{r+1}\sigma)} f(\hat{z}_{r+1})[\hat{z}_{r+1} + f(\hat{z}_{r+1})/q_1] = v^{13} , \tag{7.2.16}$$

$$\lim_{n\to\infty}(-\sigma^2/n)E(\partial^2 L/\partial\sigma\partial\tau) = -\,2e^{-\mu+\sigma^2/2}\{\sigma[F(\hat{z}_m+\sigma) - F(\hat{z}_{r+1}+\sigma)] + f(\hat{z}_m+\sigma) - f(\hat{z}_{r+1}+\sigma)\} - e^{-(\mu+\hat{z}_m\sigma)} f(\hat{z}_m)[\hat{z}_m^2 - \hat{z}_m f(\hat{z}_m)/q_2 - 1] + e^{-(\mu+\hat{z}_{r+1}\sigma)} f(\hat{z}_{r+1})[\hat{z}_{r+1} + \hat{z}_{r+1} f(\hat{z}_{r+1})/q_1 - 1] = v^{23} . \tag{7.2.17}$$

The asymptotic variance-covariance matrix for the estimators $\hat{\mu}$, $\hat{\sigma}$, and $\hat{\tau}$ is then $(\sigma^2/n)[v_{ij}]$, where $[v_{ij}] = [v^{ij}]^{-1}$. The reader will note that v^{11}, v^{22}, and v^{12} are independent of both μ and σ, while v^{33}, v^{13}, and v^{23} are dependent upon both. The dependence upon μ is only through the factor $e^{-\mu}$ in v^{13} and v^{23} and the factor $e^{-2\mu}$ in v^{33}, while the dependence upon σ is more complicated. It follows that, among the elements of the inverse matrix $[v_{ij}] = [v^{ij}]^{-1}$, v_{11}, v_{22}, and v_{12} are independent of μ, v_{13} and v_{23} depend upon μ only through the factor e^{μ}, and v_{33} depends upon μ only through the factor $e^{2\mu}$.

The computation of the elements v^{ij} of the information matrix (multiplied by σ^2/n) and the inversion of this matrix and its submatrices and multiplication by σ^2 to obtain the coefficients of $1/n$ in the asymptotic variances and covariances were all performed on the IBM 1620 computer. The resulting coefficients of $1/n$ in $\mathrm{Var}(\hat{\mu})$, $\mathrm{Var}(\hat{\sigma})$, $\mathrm{Var}(\hat{\tau})$, $\mathrm{Cov}(\hat{\mu},\hat{\sigma})$, $\mathrm{Cov}(\hat{\mu},\hat{\tau})$, and $\mathrm{Cov}(\hat{\sigma},\hat{\tau})$ are given in Table F2A for $\mu = 2$, $\sigma = 1$ and in Table F2B for $\mu = 4$, $\sigma = 2$ with censoring proportions $q_1 = 0.00, 0.01, 0.02, 0.05, 0.1$, and $q_2 = 0.0(0.1)0.9$ (excluding $q_1 = 0.1$, $q_2 = 0.9$) together with the coefficients of $1/n$ in the variances and covariances when one parameter is known and in the variances when two parameters are known. For $\sigma = 1$ or $\sigma = 2$ and other values of μ, the coefficients can be obtained from the tabular values in Table F2A or Table F2B by multiplying by the proper exponential function of μ, but for other values of σ the computations must be made afresh. The coefficients tabulated are those of the local-maximum-likelihood estimators resulting from the iterative estimation procedure which is discussed in subsection 7.2.4, not those of the global maxima. This fact is verified by the results of a Monte Carlo study, which is reported in subsection 7.2.5, comparing variances and covariances of estimates obtained from 500 random samples of each of various sizes and various degrees of censoring with those given by the asymptotic formulas.

7.2.4 Iterative Estimation Procedure

The procedure for iterative estimation on a high-speed computer involves estimating the three parameters, one at a time, in the cyclic order μ, σ, τ, omitting any assumed to be known. Assuming that the first m order statistics of a sample of size n ($m \leq n$) are known, except for the first r_0 ($0 \leq r_0 \leq m - p$, where $p = 1, 2$, or 3 is the number of parameters to be estimated), one starts by setting $r = r_0$. One then chooses initial estimates for the unknown parameters.

At each step, the rule of false position (iterative linear interpolation) is used to determine the value (if any) of the parameter then being estimated which satisfies the appropriate likelihood equation, in which the latest estimates (or known values) of the other two parameters have been substituted. For $\hat{\tau} < x_{r_0+1}$, one can always find estimates $\hat{\mu}$ (finite) and $\hat{\sigma}$ (finite and positive) in this way. In estimating τ, however, one may find that no value of τ in the permissible interval $\tau \leq x_{r_0+1}$ satisfies Equation (7.2.5). In such cases, the likelihood function is monotone increasing, so that $\hat{\tau} = x_{r_0+1}$. As $\hat{\tau} \to x_{r_0+1}$, $\hat{\mu} \to -\infty$ and $\hat{\sigma} \to +\infty$, so that the estimation is proceeding along a path to the global maximum [see Hill (1963)]. When this occurs, which is not unusual when the available sample (after censoring, if any) is small, it is still possible to obtain reasonable estimates by a slight modification of the procedure. The modification entails censoring x_{r_0+1}, the smallest observation not previously censored, and any other equal to it, thus increasing r from r_0 to $r_0 + r_1$, where $r_1 \geq 1$ is the number of observations censored at this point. Subsequently, x_{r_0+1} plays no role in the estimation procedure except as an upper bound on $\hat{\tau}$. Now the likelihood function is bounded and finite estimates $\hat{\mu}$ and $\hat{\sigma}$ are obtained. Iteration continues until the results of successive steps agree to within assigned tolerances (say 10^{-4}) or for a specified number of steps (say 550).

Use of the modified procedure calls for distributions of estimators which are conditional on the necessity for censoring the smallest previously uncensored observation(s). Lacking these, one may assert that the asymptotic variances of the estimators for the "sometimes censor" procedure are bounded below by those for the "never censor" procedure and above by those for the "always censor" procedure.

7.2.5 Monte Carlo Study for Samples of Moderate Size

In order to check the validity of the asymptotic variances and covariances determined in subsection 7.2.3 and their applicability to samples of moderate size, a Monte Carlo study was carried out on the IBM 7094 computer. Five hundred pseudo-random samples each of sizes 50, 100, and 200 from a lognormal population with parameters $\mu = 4$, $\sigma = 2$, and $\tau = 10$ were generated in the computer. The iterative procedure described in subsection 7.2.4 was used to estimate all three parameters, also every pair of parameters and every single parameter, the known values being substituted for the parameters not being estimated. In the case of samples of size 100, this was done not only for the complete samples but also for the samples with proportions $q_1 = 0.01$ censored from below and $q_2 = 0.5$ from above, both singly and in combination. The unmodified procedure converged to within the assigned tolerances of 10^{-4} in the specified 550 steps or fewer without exception. More severe censoring or the use of smaller samples can lead to a slower convergence of the iterative procedure as well as to the need to resort to the modified procedure.

The means, variances, and covariances of the estimates, based on 500 samples, are given in Table G2. A comparison of the means in Table G2 with the population parameters $\mu = 4$, $\sigma = 2$, and $\tau = 10$ and a comparison of the variances and covariances in Table G2 with the asymptotic variances and covariances obtained by dividing the coefficients in Table F2B by the sample size leads one to the following tentative conclusions:

1. The estimator $\hat{\mu}$ from complete samples has a negative bias which is approximately proportional to n^{-1}, the reciprocal of the sample size. The bias of $\hat{\mu}$ appears to be

unaffected by knowledge of σ, but it is small or non-existent if τ is known. The bias remains negative for moderate censoring.

2. The estimator $\hat{\sigma}$ has a positive bias when τ is unknown and a negative bias when τ is known. The magnitude of the bias is roughly inversely proportional to the number of observations remaining after censoring, and appears to be unaffected by knowledge of μ.
3. The estimator $\hat{\tau}$ has a positive bias which is closely proportional to n^{-1}. The magnitude of the bias appears to be unaffected by knowledge of μ and/or σ and by censoring from above, but it appears to be slightly increased by censoring even a single observation from below.
4. The variance of $\hat{\tau}$ for samples of moderate size is much larger than the value given by the asymptotic formula (see Table F2B). The excess over the asymptotic value, however, appears to be proportional to n^{-2}, whereas the asymptotic value itself is proportional to n^{-1}. Thus, for sufficiently large samples, the excess would become negligible in comparison with the asymptotic value. The excess appears to be decreased little if any by knowledge of μ and only slightly by knowledge of σ. The excess appears to be affected very little by censoring, but for censoring of even a single observation from below the asymptotic value increases markedly, so that the excess becomes smaller by comparison.
5. When τ is known, the variances of $\hat{\mu}$ and $\hat{\sigma}$ are in close agreement with the values given by the asymptotic formulas. When τ is unknown, the variances of $\hat{\mu}$ and $\hat{\sigma}$ and their covariances with $\hat{\tau}$ are somewhat larger than the values given by the asymptotic formulas. The excess over the asymptotic value is greater for $\hat{\sigma}$ than for $\hat{\mu}$, as one would expect from the fact that $\hat{\sigma}$ is more strongly correlated with $\hat{\tau}$ than is $\hat{\mu}$.

7.2.6 References

1. Aitchison, J. and Brown, J.A.C. (1957). *The Lognormal Distribution.* Cambridge University Press, Cambridge, England.
2. Balakrishnan, N. and Cohen, A. Clifford, Jr. (1991). *Order Statistics and Inference: Estimation Methods.* Academic Press, San Diego.
3. Cohen, A. Clifford, Jr. (1951). Estimating parameters of logarithmic-normal distributions by maximum likelihood. *Journal of the American Statistical Association* **46**, 206-212.
4. Cohen, A. Clifford, Jr. (1991). *Truncated and Censored Samples: Theory and Applications.* Marcel Dekker, New York.
5. Cohen, A. Clifford, Jr. and Whitten, Betty J. (1988). *Parameter Estimation in Reliability and Life Span Models.* Marcel Dekker, New York.
6. Crow, Edwin L. and Shimizu, Kunio (Eds.) (1988). *Lognormal Distributions: Theory and Applications.* Marcel Dekker, New York.
7. Harter, H. Leon and Moore, Albert H. (1965). Maximum-likelihood estimation of the parameters of Gamma and Weibull populations from complete and from censored samples. *Technometrics* **7**, 639-643.
8. Harter, H. Leon and Moore, Albert H. (1966a). Iterative maximum-likelihood estimation of the parameters of normal populations from singly and doubly censored samples. *Biometrika* **53**, 205-213.

9. Harter, H. Leon and Moore, Albert H. (1966b). Local-maximum-likelihood estimation of the parameters of three-parameter lognormal populations from complete and censored samples. *Journal of the American Statistical Association* **61**, 842-851; corrections: **61** (1966), 1247; **62** (1967), 1519-1520; and **63** (1968), 1549.

10. Hill, Bruce M. (1963). The three-parameter lognormal distribution and Bayesian analysis of a point-source epidemic. *Journal of the American Statistical Association* **58**, 72-84.

11. Johnson, Norman L., Kotz, Samuel, and Balakrishnan, N. (1994). *Continuous Univariate Distributions - Volume 1*, Second edition. John Wiley & Sons, New York.

12. Wilson, Edwin B. and Worcester, Jane (1945). The normal logarithmic transform. *Review of Economic Statistics* **27**, 17-22.

7.3 Maximum-Likelihood Estimation of Three Parameters of Weibull and Gamma Populations [3]

7.3.1 Introduction

A large number of authors have considered estimation of the parameters of Gamma (Pearson Type III) and Weibull populations by the method of moments, the method of maximum likelihood, and other methods. While the method of maximum likelihood was employed by Gauss (1809) in particular cases, its general use was first proposed by Fisher (1912). Fisher (1922) showed that for estimating the parameters of a Pearson Type III population, except when it closely approximates normality, the method of moments is inefficient, and examined in detail the properties of maximum-likelihood estimators. Further justification for the use of the method of maximum likelihood was given by Hotelling (1930) and by Doob (1934). Halperin (1952) proved under mild regularity conditions that the maximum-likelihood estimator of a single parameter from singly censored samples is consistent, asymptotically normally distributed and of minimum variance for large samples, outlined the proof of an extension of the theorem to the case of several parameters and indicated that his results would hold for more general censoring. Plackett (1958) showed that maximum-likelihood estimators are asymptotically linear and that the best linear unbiased estimators are asymptotically normal and efficient.

The method of maximum likelihood has been applied to the problem of estimating the parameters of the Gamma (Pearson Type III) population by a number of authors, including Fisher (1922), Masuyama and Kuroiwa (1951), Des Raj (1953), Chapman (1956), Greenwood and Durand (1960), Wilk, Gnanadesikan and Huyett (1962), Mickey, Mundle, Walker and Glinski (1963), and Harter and Moore (1965, 1967). These and other related developments have been reviewed in the book by Bowman and Shenton (1988) and in Chapter 17 of the volume by Johnson, Kotz and Balakrishnan (1994). The method of maximum likelihood has also been used to estimate the parameters of the Weibull population by Kao (1956, 1958), Leone, Rutenberg and Topp (1960), Dubey (1963), Lehman (1963), Ravenis (1964), Cohen (1965), and Harter and Moore (1965, 1967). Reviews of these results may be found in the book by Balakrishnan and Cohen (1991) and in Chapter 21 of the volume by Johnson, Kotz and Balakrishnan (1994). Only a few of these authors have given the information matrix. Dubey and Ravenis gave the information matrix for a complete (uncensored) sample from the three-parameter Weibull population, using a slightly different parametrization than the one we shall employ in this section. Cohen gave the information matrix for complete, singly censored, and progressively censored samples from a two-parameter Weibull population, and Harter and Moore (1967) gave it for doubly censored samples from three-parameter Weibull and Gamma populations; see also Balakrishnan and Cohen (1991), Cohen (1991), and Cohen and Whitten (1988). Three well-known populations are special cases of these. The negative exponential population is a special case (with shape parameter equal to 1) of both the Weibull population and the Gamma population. The chi-square population with ν degrees of freedom is a Gamma population with shape parameter $\alpha = \nu/2$. The Rayleigh population is a Weibull population with shape parameter $K = 2$.

In this section, maximum-likelihood estimators are given for doubly censored samples from three-parameter Weibull and Gamma populations, together with a completely computerized procedure for iterative solution of the likelihood equations and a workable modification for use when the usual procedure breaks down, which occurs when

[3] Earlier versions of this material were published by Harter and Moore (1965, 1967).

the location parameter is unknown or known to be equal to the first order statistic of the sample and the estimate of the shape parameter is less than or equal to one. The mathematical formulation is given for the Gamma population and for the Weibull population in subsection 7.3.2. The iterative procedure for solving the likelihood equations is given in subsection 7.3.3, and numerical examples in subsection 7.3.4. The elements of the maximum-likelihood information matrices for doubly censored samples from three-parameter Weibull and Gamma populations are given in subsection 7.3.5. Numerical inversion of the information matrices to obtain the asymptotic variance-covariance matrices is also discussed in subsection 7.3.5, and the results for some typical values of the shape parameters and proportions censored are given in Tables F3 and F4. Results of a Monte Carlo study for the Weibull population are reported in subsection 7.3.6 and tabulated in Table G3. Some remarks on regular and non-regular estimation are given in subsection 7.3.7.

7.3.2 Mathematical Formulation

Gamma Population

The probability density function of the random variable X having a Gamma distribution with location parameter $c \geq 0$, scale parameter θ, and shape parameter α is given by

$$g(x;c,\theta,\alpha) = [1/\Gamma(\alpha)\theta][(x-c)/\theta]^{\alpha-1}\exp[-(x-c)/\theta],\ \theta,\alpha > 0,\ x \geq c \geq 0\ . \tag{7.3.1}$$

The natural logarithm of the likelihood function of the $(m-r)$ order statistics $X_{r+1}, X_{r+2}, \ldots, X_m$ of a sample of size n is given by

$$\begin{aligned} L_{r+1,m} &= \ln n! - \ln(n-m)! - \ln r! - n\ln\Gamma(\alpha) - (m-r)\alpha\ln\theta \\ &+ (\alpha-1)\sum_{i=r+1}^{m}\ln(x_i-c) - \sum_{i=r+1}^{m}(x_i-c)/\theta \\ &+ (n-m)\ln\{\Gamma(\alpha) - \Gamma[\alpha;(x_m-c)/\theta]\} + r\ln\Gamma[\alpha;(x_{r+1}-c)/\theta], \end{aligned} \tag{7.3.2}$$

where $\Gamma(\alpha;z) = \int_0^z t^{\alpha-1}\exp(-t)dt$ is the incomplete Gamma function. The likelihood equations are obtained by equating to zero the partial derivatives of $L = L_{r+1,m}$ with respect to each of the three parameters, which are given by

$$\begin{aligned} \partial L/\partial\theta &= -(m-r)\alpha/\theta + \sum_{i=r+1}^{m}(x_i-c)/\theta^2 \\ &+ (n-m)(x_m-c)^{\alpha}\exp[-(x_m-c)/\theta]/\theta^{\alpha+1}\{\Gamma(\alpha) - \Gamma[\alpha;(x_m-c)/\theta]\} \\ &- r(x_{r+1}-c)^{\alpha}\exp[-(x_{r+1}-c)/\theta]/\theta^{\alpha+1}\Gamma[\alpha;(x_{r+1}-c)/\theta], \end{aligned} \tag{7.3.3}$$

$$\begin{aligned} \partial L/\partial\alpha &= -(m-r)\ln\theta + \sum_{i=r+1}^{m}\ln(x_i-c) - n\Gamma'(\alpha)/\Gamma(\alpha) \\ &+ (n-m)\{\Gamma'(\alpha) - \Gamma'[\alpha;(x_m-c)/\theta]\}/\{\Gamma(\alpha) - \Gamma[\alpha;(x_m-c)/\theta]\} \\ &+ r\Gamma'[\alpha;(x_{r+1}-c)/\theta]/\Gamma[\alpha;(x_{r+1}-c)/\theta], \end{aligned} \tag{7.3.4}$$

$$\partial L/\partial c = (1-\alpha)\sum_{i=r+1}^{m}(x_i-c)^{-1} + (m-r)/\theta$$

$$+ (n-m)(x_m - c)^{\alpha-1} \exp[-(x_m - c)/\theta]/\theta^\alpha\{\Gamma(\alpha) - \Gamma[\alpha;(x_m - c)/\theta]\}$$
$$- r(x_{r+1} - c)^{\alpha-1} \exp[-(x_{r+1} - c)/\theta]/\theta^\alpha \Gamma[\alpha;(x_{r+1} - c)/\theta], \qquad (7.3.5)$$

where the primes in Equation (7.3.4) indicate differentiation with respect to α.

Weibull Population

The probability density function of the random variable X having a Weibull distribution with location parameter $c \geq 0$, scale parameter θ, and shape parameter K is given by

$$g(x;c,\theta,K) = [K(x-c)^{K-1}/\theta^K]\exp\{-[(x-c)/\theta]^K\}, \ \theta, K > 0, \ x \geq c \geq 0 \ . \qquad (7.3.6)$$

The natural logarithm of the likelihood function of the $(m-r)$ order statistics $X_{r+1}, X_{r+2}, \ldots, X_m$ of a sample of size n is given by

$$L_{r+1,m} = \ln n! - \ln(n-m)! - \ln r! + (m-r)(\ln K - K \ln \theta)$$
$$+ (K-1) \sum_{i=r+1}^{m} \ln(x_i - c) - \sum_{i=r+1}^{m} [(x_i - c)/\theta]^K$$
$$- (n-m)[(x_m - c)/\theta]^K + r \ln\{1 - \exp[-(x_{r+1} - c)^K/\theta^K]\} \ . \qquad (7.3.7)$$

The likelihood equations are obtained by equating to zero the partial derivatives of $L = L_{r+1,m}$ with respect to each of the three parameters, which are given by

$$\partial L/\partial\theta = K(m-r)/\theta + K \sum_{i=r+1}^{m} (x_i - c)^K/\theta^{K+1} + K(n-m)(x_m - c)^K/\theta^{K+1}$$
$$- Kr(x_{r+1} - c)^K \exp[-(x_{r+1} - c)^K/\theta^K]/\theta^{K+1}\{1 - \exp[-(x_{r+1} - c)^K/\theta^K]\},$$
$$(7.3.8)$$

$$\partial L/\partial K = (m-r)(1/K - \ln\theta) + \sum_{i=r+1}^{m} \ln(x_i - c) - \sum_{i=r+1}^{m} [(x_i - c)/\theta]^K \ln[(x_i - c)/\theta]$$
$$- (n-m)[(x_m - c)/\theta]^K \ln[(x_m - c)/\theta] + r(x_{r+1} - c)^K \ln[(x_{r+1} - c)/\theta]$$
$$\times \exp\{-[(x_{r+1} - c)/\theta]^K\}/\theta^K\{1 - \exp[-(x_{r+1} - c)^K/\theta^K]\}, \qquad (7.3.9)$$

$$\partial L/\partial c = (1-K) \sum_{i=r+1}^{m} (x_i - c)^{-1} + K\theta^{-K} \sum_{i=r+1}^{m} (x_i - c)^{K-1} + (n-m)K\theta^{-K}(x_m - c)^{K-1}$$
$$- Kr(x_{r+1} - c)^{K-1} \exp[-(x_{r+1} - c)^K/\theta^K]/\theta^K\{1 - \exp[-(x_{r+1} - c)^K/\theta^K]\}.$$
$$(7.3.10)$$

7.3.3 Iterative Estimation Procedure

Iterative procedures have been developed for finding the joint maximum-likelihood estimators of the parameters of Gamma and Weibull populations from complete or censored samples. These involve estimating the three parameters, one at a time, in the cyclic order θ, α [or K], and c, omitting any assumed to be known. Assuming that the first m order statistics of a sample of size n ($m \leq n$) are known, one starts by setting $r = 0$ (no censoring from below). One then chooses initial estimates for the unknown parameters.

At each step, the rule of false position (iterative linear interpolation) is used to determine the value (if any) of the parameter then being estimated which satisfies the appropriate likelihood equation, in which the latest estimates (or known values) of the other two parameters have been substituted. Positive values $\hat{\theta}$ and $\hat{\alpha}$ [or $\hat{K}$] can always be found in this way. In estimating c, however, one may find that no value of c in the permissible interval $0 \leq c \leq x_1$ satisfies the likelihood equation obtained by equating to zero the right-hand side of Equation (7.3.5) [or Equation (7.3.10)]. In such cases, the likelihood function in that interval is either monotone decreasing, so that $\hat{c} = 0$, or monotone increasing, so that $\hat{c} = x_1$. The latter situation occurs when $\hat{\alpha} \leq 1$ [or $\hat{K} \leq 1$], since then the right-hand side of Equation (7.3.5) [or Equation (7.3.10)], for $r = 0$, contains only positive terms. Once that has occurred, it is impossible to continue estimation with $r = 0$, since some of the terms in the likelihood equations become infinite, so it is necessary to censor the smallest observation x_1 and any others equal to it (r observations in all). Subsequently, x_1 plays no role in the estimation procedure except as an upper bound on $\hat{c}$. Iteration continues until the results of successive steps agree to within some assigned tolerance.

7.3.4 Numerical Examples

As illustrations, consider the simulated life tests, each on forty components, summarized in Tables 7.1 and 7.2. We shall suppose that the "data" in Table 7.1 represent observed failure times (in hours). Actually, they were obtained by random sampling from Gamma and Weibull populations, each with shape parameters 2 and 3. For each set of data, the appropriate iterative estimation procedure was carried out for $m = 10(10)40$ in the following cases: (1) all parameters unknown; (2) any two parameters unknown; and (3) any one parameter unknown. The resulting estimates are shown in Table 7.2.

The iterative estimation procedures for the parameters of Gamma and Weibull populations were programmed in FORTRAN and run on the IBM 7094 computer. Most cases required only a few seconds of machine time, and even the most difficult case took only about 2 minutes.

7.3.5 Asymptotic Variances and Covariances

Weibull Maximum-Likelihood Information Matrix

The natural logarithm of the likelihood function for a sample of size n from a three-parameter Weibull population, the lowest r and the highest $n - m$ sample values having been censored, is given by Equation (7.3.7), which can be rewritten in the form

$$L_{r+1,m} = \ln n! - \ln(n-m)! - \ln r! + (m-r)(\ln K - \ln\theta) + (K-1)\sum_{i=r+1}^{m} \ln z_i - \sum_{i=r+1}^{m} z_i^K - (n-m)z_m^K + r\ln F(z_{r+1}), \quad (7.3.11)$$

where $z_i = (x_i - c)/\theta$, $F(z_i) = \int_0^{z_i} f(t)\,dt = 1 - \exp(-z_i^K)$, and $f(z_i) = K z_i^{K-1}\exp(-z_i^K)$.

In the case of complete samples, the standard procedure for finding the elements of the information matrix is to take the limits, as $n \to \infty$, of the negatives of the expected values

TABLE 7.1 - *Simulated life test data - hours to failure of ordered random samples of 40 items*

G2 - Gamma population ($\theta = 100$, $\alpha = 2$, $c = 30$)

47	56	58	64	77	79	89	128	131	142
144	149	163	166	175	176	184	184	188	190
191	204	216	227	241	250	256	256	261	273
282	283	286	297	299	338	352	353	357	495

G3 - Gamma population ($\theta = 50$, $\alpha = 3$, $c = 20$)

25	54	56	67	69	79	91	102	108	109
113	126	132	134	139	143	153	156	156	166
174	178	181	182	194	198	202	202	217	231
236	246	246	251	263	272	276	343	352	392

W2 - Weibull population ($\theta = 100$, $K = 2$, $c = 10$)

15	20	27	42	42	43	44	46	64	65
65	68	68	71	74	75	75	76	77	78
92	95	100	102	102	112	113	116	117	124
124	126	127	134	149	152	153	161	168	205

W3 - Weibull population ($\theta = 100$, $K = 3$, $c = 20$)

40.9	52.2	53.2	59.4	60.0	66.8	77.3	78.0	79.7	81.1
81.4	85.4	86.0	86.3	87.4	88.5	89.9	92.4	93.0	93.2
108.7	109.3	111.6	113.1	114.2	117.7	121.6	121.9	127.6	128.0
129.7	130.8	134.1	137.5	139.2	140.3	153.0	153.8	183.3	185.1

of the second partial derivatives of the likelihood function with respect to the parameters. In the case of censored samples, these limits do not exist in the usual sense, but they can be replaced by limits in probability [see Halperin (1952)]. Let $q_1 = r/n$, $q_2 = (n-m)/n$, and $p = 1 - q_1 - q_2 = (m-r)/n$. As $n \to \infty$ (q_1 and q_2 fixed), z_{r+1} converges in probability to $\hat{z}_{r+1}$ where $F(\hat{z}_{r+1}) = \int_0^{\hat{z}_{r+1}} f(t)\,dt = q_1$ and z_m converges in probability to $\hat{z}_m$ where $1 - F(\hat{z}_m) = \int_{\hat{z}_m}^{\infty} f(t)\,dt = q_2$. If one denotes by $E[\ldots]$ a conditional expectation given z_{r+1} and z_m, the elements of the information matrix (multiplied by $1/n$) may be written as

$$\lim \text{pr } E\left[-\frac{1}{n}\,(\partial^2 L/\partial\theta^2)\right] = \theta^{-2}\{-Kp + K(K+1)[\Gamma(2;\hat{z}_m^K) - \Gamma(2;\hat{z}_{r+1}^K] \\ + K(K+1)q_2\hat{z}_m^K \\ + \hat{z}_{r+1}f(\hat{z}_{r+1})[K\hat{z}_{r+1}^K - (K+1)q_1]/q_1\} = v^{11}\ , \tag{7.3.12}$$

$$\lim \text{pr } E\left[-\frac{1}{n}\,(\partial^2 L/\partial K^2)\right] = p/K^2 + [\Gamma''(2;\hat{z}_m^K) - \Gamma''(2;\hat{z}_{r+1}^K)]/K^2 + q_2\hat{z}_m^K(\ln\hat{z}_m)^2 \\ + \hat{z}_{r+1}f(\hat{z}_{r+1})(\ln\hat{z}_{r+1})^2[\hat{z}_{r+1}^K - q_1]/Kq_1 = v^{22}\ , \tag{7.3.13}$$

TABLE 7.2 - *Estimates of parameters from first m order statistics of samples in Table 7.1*

	m	$\hat\theta$	$\hat\theta\vert\alpha$	$\hat\theta\vert c$	$\hat\theta\vert\alpha, c$	$\hat\alpha$	$\hat\alpha\vert\theta$	$\hat\alpha\vert c$	$\hat\alpha\vert\theta, c$	$\hat c$	$\hat c\vert\theta$	$\hat c\vert\alpha$	$\hat c\vert\theta, \alpha$
G2	10	548.5	119.7	215.8	112.4	0.790	2.173	1.363	2.056	47.00	23.74	23.83	30.88
	20	68.0	98.6	103.2	99.1	3.203	1.931	1.940	1.980	0.00	32.90	30.73	30.17
	30	98.6	95.6	92.9	96.7	1.930	1.899	2.072	1.964	33.37	34.04	31.82	30.04
	40	53.9	86.3	72.8	88.8	3.851	1.776	2.441	1.900	0.00	37.90	35.00	29.53
G3	10	66.5	62.2	125.7	50.6	2.813	3.538	1.665	2.879	4.61	0.00	2.21	11.84
	20	57.9	62.0	92.5	53.8	3.227	3.611	1.952	3.020	0.00	0.00	2.77	13.08
	30	53.0	59.4	75.3	53.6	3.432	3.599	2.206	3.039	0.00	0.00	4.92	13.23
	40	45.6	55.7	60.8	51.8	3.844	3.549	2.555	3.007	0.00	0.00	8.19	13.09
	m	$\hat\theta$	$\hat\theta\vert K$	$\hat\theta\vert c$	$\hat\theta\vert K, c$	$\hat K$	$\hat K\vert\theta$	$\hat K\vert c$	$\hat K\vert\theta, c$	$\hat c$	$\hat c\vert\theta$	$\hat c\vert K$	$\hat c\vert\theta, K$
W2	10	143.2	115.2	136.6	101.3	1.257	1.586	1.372	1.704	12.05	11.72	2.87	6.76
	20	90.8	87.1	83.8	84.8	2.787	2.484	2.091	1.779	0.00	0.00	8.30	4.14
	30	98.7	100.9	96.3	96.1	1.880	1.911	1.780	1.770	7.66	6.87	5.45	5.80
	40	101.8	96.2	92.8	93.3	2.199	2.156	1.945	1.988	2.48	3.75	6.79	5.23
W3	10	95.6	91.1	93.4	92.7	1.612	1.515	2.954	2.696	37.40	38.07	21.04	16.43
	20	100.6	76.0	81.7	84.8	5.226	5.199	3.747	2.699	0.00	0.54	27.25	11.67
	30	82.3	99.5	95.8	95.7	2.143	2.844	2.719	2.686	33.12	16.88	16.31	15.94
	40	83.5	101.6	95.5	97.0	2.330	2.844	2.715	2.775	30.80	16.69	14.86	16.18

$$\begin{aligned}\text{lim pr } E\left[-\frac{1}{n}\,(\partial^2 L/\partial c^2)\right] &= \theta^{-2}\{(K-1)[\Gamma(1-2/K;\hat z_m^K) - \Gamma(1-2/K;\hat z_{r+1}^K)] \\ &\quad + K(K-1)[\Gamma(2-2/K;\hat z_m^K) - \Gamma(2-2/K;\hat z_{r+1}^K)] \\ &\quad + K(K-1)q_2\hat z_m^{K-2} \\ &\quad + f(\hat z_{r+1})[K\hat z_{r+1}^K - (K-1)q_1]/q_1\hat z_{r+1}\} = v^{33}\ ,\end{aligned} \tag{7.3.14}$$

$$\begin{aligned}\text{lim pr } E\left[-\frac{1}{n}\,(\partial^2 L/\partial\theta\partial K)\right] &= \theta^{-1}\{p - [\Gamma'(2;\hat z_m^K) - \Gamma'(2;\hat z_{r+1}^K)] \\ &\quad - [\Gamma(2;\hat z_m^K) - \Gamma(2;\hat z_{r+1}^K)] - q_2\hat z_m^K(K\ln\hat z_m + 1) \\ &\quad - \hat z_{r+1}f(\hat z_{r+1})[K\hat z_{r+1}^K \ln\hat z_{r+1} \\ &\quad - (K\ln\hat z_{r+1} + 1)q_1]/Kq_1\} = v^{12}\ ,\end{aligned} \tag{7.3.15}$$

$$\begin{aligned}\text{lim pr } E\left[-\frac{1}{n}\,(\partial^2 L/\partial\theta\partial c)\right] &= \theta^{-2}\{K^2[\Gamma(2-1/K;\hat z_m^K) - \Gamma(2-1/K;\hat z_{r+1}^K)] + K^2 q_2\hat z_m^{K-1} \\ &\quad + Kf(\hat z_{r+1}[\hat z_{r+1}^K - q_1]/q_1\} = v^{13}\ ,\end{aligned} \tag{7.3.16}$$

$$\begin{aligned}\text{lim pr } E\left[-\frac{1}{n}\,(\partial^2 L/\partial K\partial c)\right] &= \theta^{-1}\{[\Gamma(1-1/K;\hat z_m^K) - \Gamma(1-1/K;\hat z_{r+1}^K)] \\ &\quad - [\Gamma(2-1/K;\hat z_m^K) - \Gamma(2-1/K;\hat z_{r+1}^K)] \\ &\quad - [\Gamma'(2-1/K;\hat z_m^K) - \Gamma'(2-1/K;\hat z_{r+1}^K)] \\ &\quad - q_2\hat z_m^{K-1}(K\ln\hat z_m + 1) - f(\hat z_{r+1})[K\hat z_{r+1}^K \ln\hat z_{r+1} \\ &\quad - (K\ln\hat z_{r+1} + 1)q_1]/Kq_1\} = v^{23}\ ,\end{aligned} \tag{7.3.17}$$

where the definition $\Gamma(s;b) = \int_0^b t^{s-1}e^{-t}\,dt$ of the incomplete Gamma function for $s > 0$ is extended to cases in which $s \le 0$, the terms involving q_1 (q_2) drop out when $q_1(q_2) = 0$, and the primes indicate differentiation $[\Gamma'(a;b) = \partial\Gamma(s;b)/\partial s|_{s=a}]$.

Gamma Maximum-Likelihood Information Matrix

The natural logarithm of the likelihood function for a sample of size n from a three-parameter Gamma population, the lowest r and the highest $n - m$ sample values having been censored, is given by Equation (7.3.2), which can be rewritten in the form

$$\begin{aligned} L_{r+1,m} &= \ln n! - \ln(n-m)! - \ln r! - (m-r)[\ln\Gamma(\alpha) + \ln\theta] \\ &\quad + (\alpha - 1)\sum_{i=r+1}^{m} \ln z_i - \sum_{i=r+1}^{m} z_i + (n-m)\ln[1 - F(z_m)] + r\ln F(z_{r+1}) \,, \end{aligned} \tag{7.3.18}$$

where $z_i = (x_i - c)/\theta$, $F(z_i) = \int_0^{z_i} f(t)\,dt = \Gamma(\alpha; z_i)/\Gamma(\alpha)$, and $f(z_i) = z_i^{\alpha-1}\exp(-z_i)/\Gamma(\alpha)$.

As in the case of the Weibull population, the elements of the information matrix can be found by taking the expected values of the limits in probability of the negatives of the second partial derivatives of the likelihood function with respect to the parameters. Let $q_1 = r/n$, $q_2 = (n-m)/n$, and $p = 1 - q_1 - q_2 = (m-r)/n$. As $n \to \infty$ (q_1 and q_2 fixed), z_{r+1} converges in probability to $\hat{z}_{r+1}$, where $F(\hat{z}_{r+1}) = \int_0^{\hat{z}_{r+1}} f(t)\,dt = q_1$ and z_m converges in probability to $\hat{z}_m$, where $1 - F(\hat{z}_m) = \int_{\hat{z}_m}^{\infty} f(t)\,dt = q_2$. The elements of the information matrix (multiplied by $1/n$) may be written as

$$\begin{aligned} \lim \text{pr}\, E\left[-\frac{1}{n}\,(\partial^2 L/\partial\theta^2)\right] &= \theta^{-2}\{-\alpha p + 2[\Gamma(\alpha+1;\hat{z}_m) - \Gamma(\alpha+1;\hat{z}_{r+1})]/\Gamma(\alpha) \\ &\quad - \hat{z}_m f(\hat{z}_m)[q_2(\hat{z}_m - \alpha - 1) - \hat{z}_m f(\hat{z}_m)]/q_2 \\ &\quad + \hat{z}_{r+1} f(\hat{z}_{r+1})[q_1(\hat{z}_{r+1} - \alpha - 1) + \hat{z}_{r+1} f(\hat{z}_{r+1})]/q_1\} = v^{11} \,, \end{aligned} \tag{7.3.19}$$

$$\begin{aligned} \lim \text{pr}\, E\left[-\frac{1}{n}\,(\partial^2 L/\partial\alpha^2)\right] &= \{\Gamma(\alpha)\Gamma''(\alpha) - [\Gamma'(\alpha)]^2\}/[\Gamma(\alpha)]^2 \\ &\quad - \{[\Gamma(\alpha) - \Gamma(\alpha;\hat{z}_m)][\Gamma''(\alpha) - \Gamma''(\alpha;\hat{z}_m)] \\ &\quad - [\Gamma'(\alpha) - \Gamma'(\alpha;\hat{z}_m)]^2\}/q_2[\Gamma(\alpha)]^2 \\ &\quad - \{\Gamma(\alpha;\hat{z}_{r+1})\Gamma''(\alpha;\hat{z}_{r+1}) \\ &\quad - [\Gamma'(\alpha;\hat{z}_{r+1})]^2\}/q_1[\Gamma(\alpha)]^2 = v^{22} \,, \end{aligned} \tag{7.3.20}$$

$$\begin{aligned} \lim \text{pr}\, E\left[-\frac{1}{n}\,(\partial^2 L/\partial c^2)\right] &= \theta^{-2}\{(\alpha-1)[\Gamma(\alpha-2;\hat{z}_m) - \Gamma(\alpha-2;\hat{z}_{r+1})]/\Gamma(\alpha) \\ &\quad - f(\hat{z}_m)[q_2(\hat{z}_m - \alpha + 1) - \hat{z}_m f(\hat{z}_m)]/q_2\hat{z}_m \\ &\quad + f(\hat{z}_{r+1})[q_1(\hat{z}_{r+1} - \alpha + 1) + \hat{z}_{r+1} f(\hat{z}_{r+1})]/q_1\hat{z}_{r+1}\} = v^{33} \,, \end{aligned} \tag{7.3.21}$$

$$\lim \text{ pr } E\left[-\frac{1}{n}\,(\partial^2 L/\partial\theta\partial\alpha)\right] = \theta^{-1}\{p - \hat{z}_m f(\hat{z}_m)\{q_2 \ln \hat{z}_m - [\Gamma'(\alpha) - \Gamma'(\alpha;\hat{z}_m)]/\Gamma(\alpha)\}/q_2$$
$$+\ \hat{z}_{r+1} f(\hat{z}_{r+1})[q_1 \ln \hat{z}_{r+1} - \Gamma'(\alpha;\hat{z}_{r+1})/\Gamma(\alpha)]/q_1\} = v^{12}\ , \tag{7.3.22}$$

$$\lim \text{ pr } E\left[-\frac{1}{n}\,(\partial^2 L/\partial\theta\partial c)\right] = \theta^{-2}\{p - f(\hat{z}_m)[q_2(\hat{z}_m - \alpha) - \hat{z}_m f(\hat{z}_m)]/q_2$$
$$+\ f(\hat{z}_{r+1})[q_1(\hat{z}_{r+1} - \alpha) + \hat{z}_{r+1} f(\hat{z}_{r+1})]/q_1\} = v^{13}\ , \tag{7.3.23}$$

$$\lim \text{ pr } E\left[-\frac{1}{n}\,(\partial^2 L/\partial\alpha\partial c)\right] = \theta^{-1}\{[\Gamma(\alpha-1;\hat{z}_m) - \Gamma(\alpha-1;\hat{z}_{r+1})]/\Gamma(\alpha)$$
$$-\ f(\hat{z}_m)\{q_2 \ln \hat{z}_m - [\Gamma'(\alpha) - \Gamma'(\alpha;\hat{z}_m)]/\Gamma(\alpha)\}/q_2$$
$$+\ f(\hat{z}_{r+1})[q_1 \ln \hat{z}_{r+1} - \Gamma'(\alpha;\hat{z}_{r+1})/\Gamma(\alpha)]/q_1\} = v^{23}\ , \tag{7.3.24}$$

where the notation is the same as in Equations (7.3.12)-(7.3.17).

Computation of Asymptotic Variances and Covariances

For all cases in which estimation is regular, the asymptotic variance-covariance matrix for the estimators $\hat{\theta}$, $\hat{K}$ [or $\hat{\alpha}$], and $\hat{c}$ is given by $\frac{1}{n}[v_{ij}]$, where $[v_{ij}] = [v^{ij}]^{-1}$ and the v^{ij} are given by Equations (7.3.12)-(7.3.17) for the Weibull population and by Equations (7.3.19)-(7.3.24) for the Gamma population. The computation of the elements v^{ij} of the information matrix (multiplied by $1/n$) and the inversion of this matrix and its submatrices to obtain the coefficients of $1/n$ in the variance-covariance matrices for simultaneous estimation of all three parameters, or of one or two parameters when the other(s) are known, were performed on the IBM 7094 computer. Computation is quite straightforward when the shape parameter is greater than 2 and/or the location parameter is known; otherwise, one encounters quantities of the form $\Gamma(s;b) - \Gamma(s;a)$, with $s \leq 0$. These become infinite when $a = 0$ and take the indeterminate form $\infty - \infty$ when $a > 0$. In the latter case, one may use the alternate form $\int_a^b t^{s-1}e^{-t}dt$, which is finite and can be evaluated by numerical integration. Since $a = \hat{z}_{r+1}^K$ for the Weibull population and $a = \hat{z}_{r+1}$ for the Gamma population, $a = 0$ if and only if $\hat{z}_{r+1} = 0$, which is true if and only if $q_1 = 0$. Hence the asymptotic variances and covariances of the estimators have not been found when $q_1 = 0$, the location parameter is one of those being estimated, and the shape parameter K (Weibull) or α (Gamma) is less than or equal to 2. With this exception, the coefficients of $(1/n)$ times a power of the scale parameter θ in the asymptotic variances and covariances were computed for $q_1 = 0.000(0.005)0.025$, $q_2 = 0.00(0.25)0.75$, and shape parameters 1, 2, and 3, accurate to within a unit in the fifth decimal place. To attain this accuracy, it was necessary to perform the computations in double precision, since several significant digits may be lost in inverting the information matrix. The results, rounded to three decimal places, are given for Weibull populations with $K = 1(1)3$ in Table F3 and for Gamma populations with $\alpha = 1(1)3$ in Table F4. Interpolation in these tables will not always give accurate results, but a computer program can readily be used to obtain additional values.

It will be observed that the above procedure breaks down if estimation is non-regular [see subsection 7.3.7]. No systematic attempt has been made to determine the asymptotic variances and covariances of non-regular estimators, though this may be possible in some cases. It is known, for example [see Dubey (1965) and Blischke et al. (1965)], that when the shape parameter is less than 2, the first order statistic is a hyper-efficient estimator of the location parameter; in particular, for the special case of the exponential population (shape parameter 1), this estimator has variance proportional to n^{-2} rather than to n^{-1} as in the case of regular estimators.

7.3.6 Monte Carlo Study for Samples of Moderate Size

In order to check the rate of convergence of the variances and covariances to their asymptotic values, a limited Monte Carlo study was conducted in which 500 random samples each of sizes $n = 50$ and $n = 100$ were drawn from a Weibull population with location parameter $c = 20$, scale parameter $\theta = 10$, and shape parameter $K = 3$ and the iterative procedure outlined in subsection 7.3.3 was used to estimate all three parameters, also every subset of one or two parameters with the other(s) known. The means, variances and covariances of the estimates from the complete samples were then computed, and the results are shown in Table G3, where the asymptotic values of the variances and covariances, found by multiplying coefficients read from Table F3 by the proper powers of θ and dividing by n, are shown in juxtaposition with the sample values for purposes of comparison. The results shown in Table G3 lead to the tentative conclusion that when all three parameters are unknown, the variances and the absolute values of the covariances exceed their asymptotic values, with the excess closely proportional to n^{-2}, a phenomenon previously observed by Harter and Moore (1966) [see subsection 7.2.5] in the case of local-maximum-likelihood estimators of the parameters of a three-parameter lognormal population; when at least one of the parameters is known, the sample variances and covariances agree quite well with their asymptotic values, even for sample size as small as 50.

7.3.7 Remarks on Regular and Non-Regular Estimation

Mickey, Mundle, Walker and Glinski (1963) claimed that maximum-likelihood estimation of the location parameter of a Gamma population [and presumably also that of a Weibull population] from a complete sample is regular in the sense defined by Cramér (1946; p. 479) if and only if the value of the shape parameter is greater than two. Harter and Moore (1967) claimed to have verified that the regularity conditions [see Kendall and Stuart (1961, pp. 43-44)], where the parameter θ is interpreted as a vector, and the additional assumptions [see Halperin (1952)] necessary in the case of censored samples are satisfied if and only if at least one of the following conditions holds: (1) the shape parameter is greater than two; (2) the location parameter is known; (3) a proportion $q_1 > 0$ of the sample is censored from below. However, Blischke, Brady and Mundle (1970) pointed out that Cramér's regularity conditions [and even the somewhat weaker ones proposed by Huber (1964)] are not satisfied for the location parameter even when the shape parameter is greater than two. Blischke (1971) has shown, however, that new regularity conditions due to Le Cam (1970) are satisfied for the Gamma population with shape parameter greater than two when the location parameter is the only unknown parameter. It appears that the same holds also for the Weibull population, but it is not clear whether it can be extended to the case of simultaneous estimation of the location parameter and one or more other parameters. The case in which both the location and the shape parameters are unknown, so that one does not know whether the latter is greater than two, is especially nebulous. Failure of the regularity conditions to be satisfied can affect both the iterative estimation procedure and the computation of asymptotic variances and covariances. The latter is impossible for complete samples (or for samples censored only from above) from Weibull and Gamma populations with unknown location parameters and with shape parameters less than or equal to two; it becomes possible, however, as soon as some proportion $q_1 > 0$ of the sample is censored from below, as outlined in subsection 7.3.3 [see also Harter (1971)]. Therefore estimation is regular in all the cases for which asymptotic variances and covariances have been found by the methods of subsection 7.3.5 and tabulated in Tables F3 and F4, and the estimators are therefore asymptotically of minimum variance, unbiased and trivariate normal. Thus one can always insure that estimation will be regular if one is willing to pay the price of censoring one or more observations from below. Doing this unnecessarily, however, results in loss of information, and often yields estimators with substantially larger variances. It is for this reason that we have proposed the modified procedure in which the first order statistic of the sample is censored only if the location parameter is being estimated and the known or estimated value of the shape parameter is less than or equal to one, which makes such action mandatory.

7.3.8 References

1. Balakrishnan, N. and Cohen, A. Clifford, Jr. (1991). *Order Statistics and Inference: Estimation Methods.* Academic Press, San Diego.
2. Blischke, W.R. (1971). Further Results on Estimation of the Parameters of the Pearson Type III Distribution. *Aerospace Research Laboratories Technical Report ARL 71-0063,* Wright-Patterson Air Force Base. AD 728677.
3. Blischke, W.R., Brady, E.J., and Mundle, P.B. (1970). Further Results on Estimation of the Parameters of the Pearson Type III Distribution in the Regular and Nonregular Cases. *Aerospace Research Laboratories Technical Report ARL 70-0017,* Wright-Patterson Air Force Base. AD 705207.
4. Blischke, W.R., Glinski, A.M., Johns, M.V., Jr., Mundle, P.B., and Truelove, A.J. (1965). On Non-Regular Estimation, Minimum Variance Bounds and the Pearson Type III Distribution. *Aerospace Research Laboratories Technical Report ARL 65-177.* Wright-Patterson Air Force Base. AD 622274.
5. Bowman, K.O. and Shenton, L.R. (1988). *Properties of Estimators for the Gamma Distribution.* Marcel Dekker, New York.
6. Chapman, Douglas G. (1956). Estimating the parameters of a truncated Gamma distribution. *Annals of Mathematical Statistics* **27**, 498-506.
7. Cohen, A. Clifford, Jr. (1965). Maximum likelihood estimation in the Weibull distribution based on complete and on censored samples. *Technometrics* **7**, 579-588.
8. Cohen, A. Clifford, Jr. (1991). *Truncated and Censored Samples: Theory and Applications.* Marcel Dekker, New York.
9. Cohen, A. Clifford, Jr. and Whitten, Betty J. (1988). *Parameter Estimation in Reliability and Life Span Models.* Marcel Dekker, New York.
10. Cramér, Harald (1946). *Mathematical Methods of Statistics.* Princeton University Press, Princeton.

11. Des Raj (1953). Estimating the parameters of Type III populations from truncated samples. *Journal of the American Statistical Association* **48**, 336-349.

12. Doob, J.L. (1934). Probability and statistics. *Transactions of the American Mathematical Society* **36**, 759-775.

13. Dubey, Satya D. (1963). On some statistical inferences for Weibull laws (abstract). *Journal of the American Statistical Association* **58**, 549.

14. Dubey, Satya D. (1965). Hyper-efficient estimator of the location parameter of the Weibull law (abstract). *Annals of Mathematical Statistics* **36**, 734.

15. Fisher, R.A. (1912). On an absolute criterion for fitting frequency curves. *Messenger of Mathematics* **41**, 155-160.

16. Fisher, R.A. (1922). On the mathematical foundations of theoretical statistics. *Philosophical Transactions of the Royal Society of London, Series A* **222**, 309-368.

17. Gauss, Carl Friedrich (1809). *Theoria Motus Corporum Coelestium in Sectionibus Conicis Ambientium.* Perthes & Besser, Hamburg. Reprinted in *Werke*, Band VII, Königlichen Gesellschaft der Wissenschaften, Göttingen, 1906.

18. Greenwood, J. Arthur and Durand, David (1960). Aids for fitting the Gamma distribution by maximum likelihood. *Technometrics* **2**, 55-65.

19. Halperin, Max (1952). Maximum likelihood estimation in truncated samples. *Annals of Mathematical Statistics* **23**, 226-238.

20. Harter, H. Leon (1971). Some optimization problems in parameter estimation. In *Optimizing Methods in Statistics* (Ed., Jagdish S. Rustagi), pp. 33-62, Academic Press, New York.

21. Harter, H. Leon and Moore, Albert H. (1965). Maximum-likelihood estimation of the parameters of Gamma and Weibull populations from complete and from censored samples. *Technometrics* **7**, 639-643.

22. Harter, H. Leon and Moore, Albert H. (1966). Local-maximum-likelihood estimation of the parameters of three-parameter lognormal populations from complete and censored samples. *Journal of the American Statistical Association* **61**, 842-851; corrections: **61** (1966), 1247; **62** (1967), 1519-1520; and **63** (1968), 1549.

23. Harter, H. Leon and Moore, Albert H. (1967). Asymptotic variances and covariances of maximum-likelihood estimators, from censored samples, of the parameters of Weibull and Gamma populations. *Annals of Mathematical Statistics* **38**, 557-570.

24. Hotelling, Harold (1930). The consistency and ultimate distribution of optimum statistics. *Transactions of the American Mathematical Society* **32**, 847-859.

25. Huber, Peter J. (1964). Robust estimation of a location parameter. *Annals of Mathematical Statistics* **35**, 73-101.

26. Johnson, Norman L., Kotz, Samuel, and Balakrishnan, N. (1994). *Continuous Univariate Distributions - Volume 1*, Second edition. John Wiley & Sons, New York.

27. Kao, John H.K. (1956). A new life-quality measure for electron tubes. *IRE Transactions* **PGRQC-7**, 1-11.

28. Kao, John H.K. (1958). Computer methods for estimating Weibull parameters in reliability studies. *IRE Transactions* **PGRQC-13**, 15-22.

29. Kendall, Maurice G. and Stuart, Alan (1961). *The Advanced Theory of Statistics*, Volume 2. Hafner Publishing Company, New York.

30. Le Cam, L. (1970). On the assumptions used to prove asymptotic normality of maximum likelihood estimates. *Annals of Mathematical Statistics* **41**, 802-828.

31. Lehman, Eugene H., Jr. (1963). Shapes, moments and estimators of the Weibull distribution. *IEEE Transactions on Reliability* **R-12**(3), 32-38.

32. Leone, F.C., Rutenberg, Y.H., and Topp, C.W. (1960). Order Statistics and Estimators for the Weibull Distribution. *Case Statistical Laboratory Publication No. 1026* (AFOSR Report No. TN 60-489), Case Institute of Technology, Cleveland. AD 237042.

33. Masuyama, M. and Kuroiwa, Y. (1951). Table for the likelihood solutions of Gamma distributions and its medical applications. *Reports of Statistical Application Research, Union of Japanese Scientists and Engineers* **1**(1), 18-23.

34. Mickey, M.R., Mundle, P.B., Walker, D.N., and Glinski, A.M. (1963). Test Criteria for Pearson Type III Distributions. *Aeronautical Research Laboratories Technical Report ARL 63-100*, Wright-Patterson Air Force Base. AD 415831.

35. Plackett, R.L. (1958). Linear estimation from censored data. *Annals of Mathematical Statistics* **29**, 131-142.

36. Ravenis, Joseph V.J., II (1964). Estimating Weibull-distribution parameters. *Electro-Technology* (March), 46-54.

37. Wilk, Martin B., Gnanadesikan, Ramanathan, and Huyett, Marilyn J. (1962). Estimation of parameters of the gamma distribution using order statistics. *Biometrika* **49**, 525-545.

7.4 Maximum-Likelihood Estimation of Two Parameters of Logistic Population [4]

7.4.1 Introduction

The logistic population and the logistic function, which is equal to the cumulative distribution function of the logistic population when certain relations hold between the two sets of parameters, have been applied in studies of population growth by Pearl and Reed (1920), of physicochemical phenomena by Reed and Berkson (1929), of bio-assay by Wilson and Worcester (1943) and by Berkson (1944, 1951, 1957), of mental ability by Birnbaum (1958), of life test data by Plackett (1959), and of biochemical data by Gupta, Qureishi, and Shah (1965). The handbook of the Logistic Distribution edited by Balakrishnan (1992) discusses these and also various other applications of the logistic distribution.

A great variety of methods have been used to estimate the parameters of the logistic population or, equivalently, those of the logistic function. Pearl and Reed (1920) and Schultz (1930) used a Taylor series expansion to obtain a least squares solution by successive approximations, and Berkson (1944) used a modification of this method. Wilson and Worcester (1943) and Berkson (1957) used the method of maximum likelihood for estimation from complete samples. Plackett (1958) wrote down the likelihood equations for doubly censored samples and used a Taylor series expansion to obtain linearized maximum-likelihood estimators for samples which have been singly censored from above. Berkson and Hodges (1961) found a minimax estimator for the logistic function. Further work on best [or nearly best] linear unbiased estimators has been performed by Kjelsberg (1962) and by Gupta, Qureishi, and Shah (1965). The latter authors have extended, up through sample size $n = 25$, a table of variances and covariances of logistic order statistics computed by Shah (1965) up through $n = 10$ and have used the results to tabulate coefficients, variances, covariances, and relative efficiencies of best linear unbiased estimators from doubly censored samples of size $n = 2(1)25$. These tables, in turn, have been extended by Balakrishnan (1990) for samples sizes $n = 21(1)25(5)40$. (See also subsection 3.3.7 of this volume.) All these results concerning different methods of estimation of logistic parameters have been reviewed in the book by Balakrishnan (1992) and in Chapter 23 of the volume by Johnson, Kotz, and Balakrishnan (1995).

In this section an iterative procedure is given for solving the likelihood equations, which are expressed in somewhat different form than that given by Plackett (1958), for doubly censored samples. The elements of the information matrix are worked out, and this matrix is inverted numerically to obtain the asymptotic variances and covariances of the estimators, which are tabulated in Table F5 for censoring proportions $q_1 = 0.0(0.1)0.4$ from below and $q_2 = q_1(0.1)(0.9 - q_1)$ from above. Results of a Monte Carlo study to obtain information about the small-sample properties of these estimators and to compare them with the best linear unbiased estimators are reported in this section and tabulated in Table G4.

7.4.2 The Likelihood Function and Its Derivatives

The cumulative distribution function of the random variable X having the logistic distribution with location parameter (mean) μ and scale parameter (standard deviation) σ is

[4] An earlier version of this material was published by Harter and Moore (1967).

given by

$$F(x;\mu,\sigma) = \{1 + \exp[-\pi(x-\mu)/\sqrt{3}\sigma]\}^{-1} \tag{7.4.1}$$

The natural logarithm of the likelihood function of a sample of size n from such a distribution, the smallest r_1 and the largest r_2 sample values having been censored, is given by

$$\begin{aligned} L_{r_1,r_2} &= \ln n! - \ln r_1! - \ln r_2! - \sum_{i=r_1+1}^{n-r_2} z_i + 2\sum_{i=r_1+1}^{n-r_2} \ln F(z_i) + (n-r_1-r_2)\ln(\pi/\sqrt{3}\sigma) \\ &\quad + r_1 \ln F(z_{r_1+1}) + r_2 \ln[1 - F(z_{n-r_2})] \ , \end{aligned} \tag{7.4.2}$$

where $z_i = \pi(x_i - \mu)/\sqrt{3}\sigma$ and $F(z_i) = [1+\exp(-z_i)]^{-1}$, so that $f(z_i) = \exp(-z_i)[1 - \exp(-z_i)]^{-2}$ and $f'(z_i) = f(z_i)[2f(z_i)/F(z_i) - 1]$. The first partial derivatives of $L = L_{r_1,r_2}$ with respect to the parameters are given by

$$\begin{aligned} \partial L/\partial\mu &= (\pi/\sqrt{3}\sigma)\{(n-r_1-r_2) - 2\Sigma f(z_i)/F(z_i) - r_1 f(z_{r_1+1})/F(z_{r_1+1}) \\ &\quad + r_2 f(z_{n-r_2})/[1-F(z_{n-r_2})]\} \ , \end{aligned} \tag{7.4.3}$$

$$\begin{aligned} \partial L/\partial\sigma &= (1/\sigma)\{\Sigma z_i - 2\Sigma z_i f(z_i)/F(z_i) - (n-r_1-r_2) - r_1 z_{r_1+1} f(z_{r_1+1})/F(z_{r_1+1}) \\ &\quad + r_2 z_{n-r_2} f(z_{n-r_2})/[1-F(z_{n-r_2})]\} \ , \end{aligned} \tag{7.4.4}$$

where the summations are on i running from r_1+1 through $n-r_2$. The second partial derivatives of $L = L_{r_1,r_2}$ with respect to the parameters are given by

$$\begin{aligned} \partial^2 L/\partial\mu^2 &= (\pi^2/3\sigma^2)\{2\Sigma f'(z_i)/F(z_i) - 2\Sigma f^2(z_i)/F^2(z_i) + r_1 f'(z_{r_1+1})/F(z_{r_1+1}) \\ &\quad - r_1 f^2(z_{r_1+1})/F^2(z_{r_1+1}) - r_2 f'(z_{n-r_2})/[1-F(z_{n-r_2})] \\ &\quad - r_2 f^2(z_{n-r_2})/[1-F(z_{n-r_2})]^2\} \ , \end{aligned} \tag{7.4.5}$$

$$\begin{aligned} \partial^2 L/\partial\sigma^2 &= (1/\sigma^2)\{-2\Sigma z_i + 2\Sigma z_i^2 f'(z_i)/F(z_i) + 4\Sigma z_i f(z_i)/F(z_i) - 2\Sigma z_i^2 f^2(z_i)/F^2(z_i) \\ &\quad + (n-r_1-r_2) + r_1 z_{r_1+1}^2 f'(z_{r_1+1})/F(z_{r_1+1}) + 2r_1 z_{r_1+1} f(z_{r_1+1})/F(z_{r_1+1}) \\ &\quad - r_1 z_{r_1+1}^2 f^2(z_{r_1+1})/F^2(r_{r_1+1}) - r_2 z_{n-r_2}^2 f'(z_{n-r_2})/[1-F(z_{n-r_2})] \\ &\quad - 2r_2 z_{n-r_2} f(z_{n-r_2})/[1-F(z_{n-r_2})] - r_2 z_{n-r_2}^2 f^2(z_{n-r_2})/[1-F(z_{n-r_2})]^2\} \ , \end{aligned} \tag{7.4.6}$$

$$\begin{aligned} \partial^2 L/\partial\mu\partial\sigma &= (\pi/\sqrt{3}\sigma^2)\{-(n-r_1-r_2) + 2\Sigma z_i f'(z_i)/F(z_i) + 2\Sigma f(z_i)/F(z_i) \\ &\quad - 2\Sigma z_i f^2(z_i)/F^2(z_i) + r_1 z_{r_1+1} f'(z_{r_1+1})/F(z_{r_1+1}) + r_1 f(z_{r_1+1})/F(z_{r_1+1}) \\ &\quad - r_1 z_{r_1+1} f^2(z_{r_1+1})/F^2(z_{r_1+1}) - r_2 z_{n-r_2} f'(z_{n-r_2})/[1-F(z_{n-r_2})] \\ &\quad - r_2 f(z_{n-r_2})/[1-F(z_{n-r_2})] - r_2 z_{n-r_2} f^2(z_{n-r_2})/[1-F(z_{n-r_2})]^2\} \ . \end{aligned} \tag{7.4.7}$$

7.4.3 Asymptotic Variances and Covariances

Let $q_1 = r_1/n$, $q_2 = r_2/n$, and $p = 1 - q_1 - q_2 = (n - r_1 - r_2)/n$. As $n \to \infty$ (q_1 and q_2 fixed), $z_{r_1+1} \to \hat{z}_1$, where $F(\hat{z}_1) = \int_{-\infty}^{\hat{z}_1} f(t)\,dt = q_1$, and $z_{n-r_2} \to \hat{z}_2$, where

$1 - F(\hat{z}_2) = \int_{\hat{z}_2}^{\infty} f(t)\,dt = q_2$. Expressing the limits, as $n \to \infty$, of the expected values of the sums in Equations (7.4.5)-(7.4.7) as integrals between the limits $\hat{z}_1$ and $\hat{z}_2$ and performing the integrations, one obtains the following expressions for the elements of the information matrix (multiplied by σ^2/n):

$$\lim_{n\to\infty}\left(-\frac{\sigma^2}{n}\,\frac{\partial^2 L}{\partial \mu^2}\right) = \frac{\pi^2}{3}\left\{\left[\frac{2}{3}\,(1+e^t)^{-3} - (1+e^t)^{-2}\right]_{\hat{z}_1}^{\hat{z}_2} - f'(\hat{z}_1) + f^2(\hat{z}_1)/q_1 \right.$$
$$\left. + f'(\hat{z}_2) + f^2(\hat{z}_2)/q_2\right\} = v^{11}, \tag{7.4.8}$$

$$\begin{aligned}\lim_{n\to\infty}\left(-\frac{\sigma^2}{n}\,\frac{\partial^2 L}{\partial \sigma^2}\right) = &-p - \frac{1}{3}\,[2t^2e^{-2t}(1+e^{-t})^{-3} - (t^2-4t)e^{-t}(1+e^{-t})^{-2} \\ &- 4(1+e^{-t})^{-1}]_{\hat{z}_1}^{\hat{z}_2} + \frac{2}{3}\,(R_1 - R_2) \\ &- \frac{1}{3}\,[t^2(1+e^t)^{-1} + 2t\ln(1+e^{-t})]_0^{\hat{z}_2} \\ &+ \frac{1}{3}\,[t^2(1+e^{-t})^{-1} - 2t\ln(1+e^t)]_{\hat{z}_1}^0 - \hat{z}_1^2 f'(\hat{z}_1) - 2\hat{z}_1 f(\hat{z}_1) \\ &+ \hat{z}_1^2 f^2(\hat{z}_1)/q_1 + \hat{z}_2^2 f'(\hat{z}_2) + 2\hat{z}_2 f(\hat{z}_2) + \hat{z}_2^2 f^2(\hat{z}_2)/q_2 = v^{22}\ ,\end{aligned} \tag{7.4.9}$$

$$\begin{aligned}\lim_{n\to\infty}\left(-\frac{\sigma^2}{n}\,\frac{\partial^2 L}{\partial \mu\partial\sigma}\right) = \frac{\pi}{\sqrt{3}}\Big\{&p - \frac{1}{3}\,[2te^{-2t}(1+e^{-t})^{-3} - te^{-t}(1+e^{-t})^{-2} \\ &+ 2e^{-t}(1+e^{-t})^{-2} + 3(1+e^{-t})^{-1}]_{\hat{z}_1}^{\hat{z}_2} \\ &- \frac{1}{3}\,[t(1+e^t)^{-1} + \ln(1+e^{-t})]_0^{\hat{z}_2} \\ &+ \frac{1}{3}\,[t(1+e^{-t})^{-1} - \ln(1+e^t)]_{\hat{z}_1}^0 - \hat{z}_1 f'(\hat{z}_1) - f(\hat{z}_1) \\ &+ \hat{z}_1 f^2(\hat{z}_1)/q_1 + \hat{z}_2 f'(\hat{z}_2) + f(\hat{z}_2) \\ &+ \hat{z}_2 f^2(\hat{z}_2)/q_2\Big\} = v^{12}\ ,\end{aligned} \tag{7.4.10}$$

where R_1 and R_2 are definite integrals which, when the integrands are expanded in infinite series, integrated term by term, and evaluated, reduce to

$$R_1 = \begin{cases} [e^t - e^{2t}/2^2 + e^{3t}/3^2 - e^{4t}/4^2 + \ldots]_{\hat{z}_1}^0, & \hat{z}_1 < 0, \\ 0, & \hat{z}_1 = 0, \\ -[t^2/2 - e^{-t} + e^{-2t}/2^2 - e^{-3t}/3^2 + e^{-4t}/4^2 - \ldots]_0^{\hat{z}_1}, & \hat{z}_1 > 0, \end{cases}$$

and

$$R_2 = \begin{cases} [e^{-t} - e^{-2t}/2^2 + e^{-3t}/3^2 - e^{-4t}/4^2 + \ldots]_{\hat{z}_2}^0, & \hat{z}_2 > 0, \\ 0, & \hat{z}_2 = 0, \\ -[t^2/2 - e^t + e^{2t}/2^2 - e^{3t}/3^2 + e^{4t}/4^2 - \ldots]_{\hat{z}_2}^0, & \hat{z}_2 < 0. \end{cases}$$

The coefficients of σ^2/n in the asymptotic variances and covariances of $\hat{\mu}$ and $\hat{\sigma}$, which are the elements v_{ij} $(i,j = 1,2)$ of the matrix $[v_{ij}] = [v^{ij}]^{-1}$, have been computed for $q_1 = 0.0(0.1)0.4$ and $q_2 = q_1(0.1)(0.9 - q_1)$ and are given in Table F5, along with the corresponding results for $\hat{\mu}|\sigma$ and $\hat{\sigma}|\mu$, obtained by inverting submatrices of $[v^{ij}]$ consisting of a single diagonal element. Interchanging q_1 and q_2 leaves the variances and the absolute values of the covariances unchanged, but changes the signs of the covariances.

7.4.4 Iterative Estimation Procedure

The likelihood equations are obtained by equating to zero the first partial derivatives of the likelihood function, which are given by Equations (7.4.3) and (7.4.4). The likelihood equations cannot be solved explicitly, but maximum-likelihood estimates can be obtained by solving them iteratively on a computer. The iterative estimation procedure involves estimating the parameters, one at a time, in the cyclic order μ, σ, omitting a parameter if it is assumed to be known. One starts by choosing initial estimate(s) for the unknown parameter(s). At each step, the rule of false position (iterative linear interpolation) is used to determine the value of the parameter then being estimated which satisfies the appropriate likelihood equation, in which the latest estimate (or known value) of the other parameter has been substituted. Iteration continues until the results of successive steps agree to within some assigned tolerance. A FORTRAN IV program was developed for use on the IBM 7094 computer to estimate, from complete, singly censored, or doubly censored samples, the parameters of a logistic distribution. If one of the parameters is assumed to be known, a single step suffices to estimate the other. Even when both are unknown, experience has shown that the rate of convergence is quite rapid if the initial estimates are reasonable and the amount of censoring is not excessive.

7.4.5 Monte Carlo Study for Small Samples

While the maximum-likelihood estimators $\hat{\mu}$ and $\hat{\sigma}$ are asymptotically unbiased and have known asymptotic variances and covariances, they are biased for small samples (except $\hat{\mu}$ when censoring is symmetric), and there is no known analytic method of determining their biases, variances, and covariances for small samples. In order to obtain information about the small-sample properties of these estimators, and incidentally to compare them with the best linear unbiased estimators, a Monte Carlo study was performed on the IBM 7094 computer. For $n = 10$ and for $n = 20$, one thousand pseudo-random samples of size n from a standard logistic distribution ($\mu = 0$, $\sigma = 1$) were generated, and each sample was ordered from smallest to largest. The iterative procedure described in subsection 7.4.4 was used to compute the estimates $\hat{\mu}$ and $\hat{\sigma}$, also $\hat{\mu}|\sigma$ and $\hat{\sigma}|\mu$, from the m order statistics remaining in each sample after proportions q_1 and q_2 had been censored from below and from above, respectively, where q_1 and q_2 were taken at intervals of 0.1, subject to the restriction $m \geq 2$. The means, variances, and covariances of the estimates from the 1000 samples of size $n = 10$ are given in Table G4A, and similar results for $n = 20$ are given in Table G4B. The rows of Tables G4A and G4B are not statistically independent, since they are based on the same samples (with different proportions censored). The results indicate that $\hat{\sigma}$ and $\hat{\sigma}|\mu$ are negatively biased, while $\hat{\mu}$ and $\hat{\mu}|\sigma$ are negatively biased for $q_1 < q_2$, unbiased for $q_1 = q_2$, and positively biased for $q_1 > q_2$. The absolute value of the bias tends to increase as the amount of censoring increases. The bias of $\hat{\sigma}$ (μ unknown) is approximately equal to $-1/m$, where m is the number of observations remaining after censoring. The bias in estimating either parameter is much smaller when the other parameter is known than when both parameters are estimated simultaneously.

In Tables G4C and G4D the mean square errors of the maximum-likelihood estimates from the 1000 samples of size $n = 10$ and the 1000 samples of size $n = 20$ are compared with the asymptotic variances of the maximum-likelihood estimators and with the variances of the best linear unbiased estimators, which were obtained by rounding four-decimal-place values given by Gupta, Qureishi, and Shah (1965). For censoring symmetric or nearly so, the precision of the two estimators is approximately the same, but for strongly asymmetric censoring the mean square error of the maximum-likelihood estimates is significantly smaller than the variance of the best linear unbiased estimator. The mean square errors of $\hat{\mu}$, $\hat{\mu}|\sigma$, and $\hat{\sigma}$ are only slightly larger than the variances as given by the asymptotic formulas, except in cases of strongly asymmetric censoring. The mean square error of $\hat{\sigma}|\mu$ is often less, and in a few instances [censoring severe but symmetric or almost so] substantially less, than the variance as given by the asymptotic formula.

A rough estimate of the sampling fluctuation can be made by comparing the results when q_1 and q_2 are interchanged, which should theoretically be the same, except for changes in the signs of $M(\hat{\mu})$, $M(\hat{\mu}|\sigma)$, and $C(\hat{\mu}, \hat{\sigma})$ in Tables G4A and G4B.

The reader may find it interesting to compare the results of this Monte Carlo study with those of similar studies for the normal distribution by Harter and Moore (1966) [see subsection 7.1.5 and Table G1] and for the first asymptotic distribution of smallest values by Harter and Moore (1968) [see subsection 7.5.5 and Table G5]. The close agreement between the results for normal and logistic populations is not surprising in view of the fact that it has been shown by Hailey (1952) that, with the proper choice of parameters, the supremum of the absolute difference between the cumulative distribution functions of these two populations is less than one-hundredth; see also Johnson, Kotz, and Balakrishnan (1995).

7.4.6 Numerical Examples

Sarhan and Greenberg (1962) have given data resulting from an experiment in which students were learning to measure strontium-90 concentrations in samples of milk. The test substance was supposed to contain 9.22 micromicrocuries per liter. The measurements, each involving readings and calculations, were made, but, since the measurement error was known to be relatively larger at the extremes, especially the upper one, a decision was made to censor the two smallest and the three largest observations, leaving the following censored sample: 8.2, 8.4, 9.1, 9.8, 9.9. Sarhan and Greenberg assumed a normal distribution and proceeded to calculate the best linear unbiased estimates of the mean μ and the standard deviation σ from the doubly censored sample. Gupta, Qureishi, and Shah (1965) assumed a logistic rather than a normal distribution, and again calculated the best linear unbiased estimates of μ and σ. The same data have been used to calculate maximum-likelihood estimates of μ and σ, first assuming a logistic distribution and employing the iterative procedure described in subsection 7.4.4, then assuming a normal distribution and employing an iterative procedure described by Harter and Moore (1966) [see subsection 7.1.4]. The results, where μ^* and σ^* are the best linear unbiased estimates previously calculated by the other authors and $\hat{\mu}$ and $\hat{\sigma}$ are the maximum-likelihood estimates, are as follows:

Population Assumed	μ^*	$\hat{\mu}$	σ^*	$\hat{\sigma}$
Logistic	9.3031	9.2718	1.8765	1.5678
Normal	9.2900	9.2606	1.6900	1.3754

Results of the Monte Carlo studies reported in subsection 7.4.5 for the logistic population and in subsection 7.1.5 for the normal population indicate that the maximum-likelihood estimates, though negatively biased, would be expected to have slightly smaller mean square errors than the best linear unbiased estimates for the case considered in this example.

7.4.7 References

1. Balakrishnan, N. (1990). Best Linear Unbiased Estimates of the Location and Scale Parameters of Logistic Distribution for Complete and Censored Samples of Sizes 2(1)25(5)40. *McMaster University Report*, Hamilton, Canada.
2. Balakrishnan, N. (Ed.) (1992). *Handbook of the Logistic Distribution.* Marcel Dekker, New York.
3. Berkson, Joseph (1944). Application of the logistic function to bio-assay. *Journal of the American Statistical Association* **39**, 357-365.
4. Berkson, Joseph (1951). Why I prefer logits to probits. *Biometrics* **7**, 327-339.
5. Berkson, Joseph (1957). Tables for the maximum likelihood estimate of the logistic function. *Biometrics* **13**, 28-34.
6. Berkson, Joseph and Hodges, J.L., Jr. (1961). A minimax estimator for the logistic function. *Proceedings of the Fourth Berkeley Symposium on Mathematical Statistics and Probability* **4**, 77-86, University of California Press, Berkeley.
7. Birnbaum, Allan (1958). Statistical theory of tests of a mental ability (abstract). *Annals of Mathematical Statistics* **29**, 1285.
8. Gupta, S.S., Qureishi, A.S., and Shah, B.K. (1965). Best Linear Unbiased Estimators of the Parameters of the Logistic Distribution Using Order Statistics. *Mimeograph Series No. 52*, Department of Statistics, Purdue University; abridged version: *Technometrics* **9** (1967), 43-56.
9. Hailey, D.C. (1952). Estimation of the Dosage Mortality Relationship when the Cost is Subject to Error. *Technical Report Number 15*, Applied Mathematics Laboratory, Stanford University, Stanford.
10. Harter, H. Leon and Moore, Albert H. (1966). Iterative maximum-likelihood estimation of the parameters of normal populations from singly and doubly censored samples. *Biometrika* **53**, 205-213.
11. Harter, H. Leon and Moore, Albert H. (1967). Maximum-likelihood estimation, from censored samples, of the parameters of a logistic distribution. *Journal of the American Statistical Association* **62**, 675-684.
12. Harter, H. Leon and Moore, Albert H. (1968). Maximum-likelihood estimation, from doubly censored samples, of the parameters of the first asymptotic distribution of extreme values. *Journal of the American Statistical Association* **63**, 889-901.
13. Johnson, Norman L., Kotz, Samuel, and Balakrishnan, N. (1995). *Continuous Univariate Distributions - Volume 2*, Second edition. John Wiley & Sons, New York.
14. Kjelsberg, Marcus O. (1962). Estimation of the Parameters of the Logistic Distribution under Truncation and Censoring. *Ph.D. thesis*, University of Minnesota, Minneapolis. University Microfilms, Ann Arbor, Michigan.
15. Pearl, Raymond and Reed, L.J. (1920). On the rate of growth of the population of the United States since 1790 and its mathematical representation. *Proceedings of the National Academy of Sciences* **6**, 275-288.
16. Plackett, R.L. (1958). Linear estimation from censored data. *Annals of Mathematical Statistics* **29**, 131-142.
17. Plackett, R.L. (1959). The analysis of life test data. *Technometrics* **1**, 9-19.
18. Reed, L.J. and Berkson, Joseph (1929). The application of the logistic function to experimental data. *Journal of Physical Chemistry* **33**, 760-779.

19. Sarhan, Ahmed E. and Greenberg, Bernard G. (Eds.) (1962). *Contributions to Order Statistics.* John Wiley & Sons, New York.

20. Schultz, Henry (1930). The standard error of a forecast from a curve. *Journal of the American Statistical Association* **25**, 139-185.

21. Shah, B.K. (1965). On the Bivariate Moments of Order Statistics from a Logistic Distribution and Applications. *Mimeograph Series No. 48*, Department of Statistics, Purdue University; abridged version: *Annals of Mathematical Statistics* **37** (1966), 1002-1010.

22. Wilson, E.B. and Worcester, Jane (1943). The determination of L.D. 50 and its sampling error in bio-assay. *Proceedings of the National Academy of Sciences* **29**, 79-85, 114-120, 257-262.

7.5 Maximum-Likelihood Estimation of Two Parameters of Type I Extreme-Value Population [5]

7.5.1 Introduction

The extreme values in samples from a normal population were studied by von Bortkiewicz (1922a,b), by Neyman (1923), and by von Mises (1923), who introduced the notion of the characteristic largest value. Dodd (1923) found the asymptotic values of the greatest of n variates in samples from six general types of population. Tippett (1925) obtained mathematical expressions and computed a table for the probability integral of the largest of n observations from a normal population. Fréchet (1927) obtained an asymptotic distribution of the largest value, and showed that largest values in samples from different populations may have the same asymptotic distribution. Fisher and Tippett (1928) showed that the asymptotic distribution of the largest (or smallest) value in a sample of size n, as $n \to \infty$, for any population satisfying certain regularity conditions, must fall into one of three types.

Gumbel (1935) made an extensive study of the theory of extreme values of statistical distributions. The populations possessing asymptotic distributions of the largest value were classified by von Mises (1936), who also gave sufficient conditions for the validity of the three asymptotic distributions. Gumbel (1937a,b, 1941) discussed applications of extreme-value theory to such problems as extremes of human age at death, intervals between radioactive emissions, and the return period of floods. In the latter paper he estimated the parameters of an extreme-value distribution by the method of moments. Gnedenko (1943) gave necessary and sufficient conditions for the validity of the three asymptotic distributions of extreme values. The asymptotic theory of extreme order statistics has been reviewed in the books by de Haan (1970), Galambos (1978, 1987), David (1981), and Arnold, Balakrishnan and Nagaraja (1992). Castillo (1988) has illustrated various applications of the extreme value theory in engineering problems.

Kimball (1946, 1949) obtained maximum-likelihood estimators of the parameters of a Type I asymptotic distribution of largest values from complete samples. Epstein (1948a,b) showed how the theory of extreme (smallest) values can be applied in fracture problems, while Epstein and Brooks (1948) considered its implications with regard to the dielectric strength of paper capacitors. Gumbel and Freudenthal (1952) developed the statistical theory of fatigue failures, using the third asymptotic distribution of smallest values, which is identical with the Weibull distribution. Lieblein (1953, 1954) found closed expressions for the variances and covariances of the order statistics of samples from a Type I extreme-value distribution, tabulated them for samples of size $n = 2(1)6$, and used them to obtain best linear unbiased estimators of the parameters from complete samples. Lieblein (1955) extended these results to the case of censored samples. These results were further extended by Balakrishnan and Chan (1992a,b) for selected sample sizes up to 30; for more complete tables, see Balakrishnan and Chan (1992c,d).

Kimball (1956) studied the bias of the maximum-likelihood estimators of the parameters of a Type I asymptotic distribution of largest values from complete samples, obtaining an explicit expression for the bias of a modification of the maximum-likelihood estimator of the scale parameter, and compared this estimator with the best linear unbiased estimator for a sample of size $n = 6$. Aziz (1956) and Eldredge (1957), using graphical methods, applied the Type I asymptotic distribution of largest values to studies of corrosion pitting. The latter author asserted that pitting usually follows the Type I distribution, but

[5] An earlier version of this material was published by Harter and Moore (1968).

mentioned cases in which Types II and III would be applicable. Gumbel (1958) gave a comprehensive summary and bibliography of work, much of it his own, on extreme-value theory up to that time.

Epstein (1958) considered stochastic models for length of life, including extreme-value models. Epstein (1960) presented some of the basic elements of extreme-value theory, and gave examples of the application of Types I, II, and III asymptotic distributions of smallest and largest values to a great variety of physical problems. The book on order statistics edited by Sarhan and Greenberg (1962) contains a chapter by Gumbel on the theory of extreme values and sections by Lieblein on the three asymptotic distributions and on order statistics of samples from extreme-value distributions and by Gumbel on an application.

Mann (1963) proposed the use of best linear invariant estimators, which are simple linear functions of the best linear unbiased estimators, for the parameters of a Type I asymptotic distribution of smallest values. Hassanein (1964) proposed the use of nearly best linear unbiased estimators, for which he tabulated coefficients of the order statistics from singly and doubly censored samples of size $n = 1(1)10(5)25$. White (1964) extended Lieblein's table of coefficients for best linear unbiased estimators up through $n = 20$ and Mann (1965a) independently extended it up through $n = 25$, both for single censoring from above. Mann (1965b) tabulated factors for calculating Cramér-Rao efficiencies of best linear invariant and best linear unbiased estimators of the parameters from samples of size $n = 2(1)25$ with single censoring from above. [See subsection 3.3.12 of this volume.]

Posner (1965) applied extreme-value theory to error-free communication, estimating the parameters of a Type I asymptotic distribution of largest values from complete samples by the method of maximum likelihood, which he justified on the basis of its asymptotic properties. Gumbel and Mustafi (1966) criticized the use of the method of maximum likelihood, pointing out that asymptotic theory is not necessarily valid for Posner's sample size ($n = 30$) and demonstrating that the modified method of moments gives a better fit to Posner's data, no matter whether one uses the method of least squares or the Kolmogorov-Smirnov statistic as a criterion. They did not, however, offer any proof that the modified method of moments is in general superior to the method of maximum likelihood. Downton (1966) compared the efficiencies of various linear estimators of the parameters of Type I asymptotic distributions of largest and smallest values from complete samples.

Johns and Lieberman (1966) tabulated approximate weights for obtaining best linear invariant estimates from the first m order statistics of samples of size n for $n = 10$, 15, 20, 30, 50, and 100 and four values of m for each n. Weights for these asymptotically optimum linear estimates for other cases of single censoring from above can be calculated from Equations (23) and (24) of the Johns-Lieberman paper; similar expressions for the case of double censoring are easily derived; see Balakrishnan and Cohen (1991) for details. Mann (1967a,b) tabulated exact weights for $n = 2(1)25$, $m = 2(1)n$. All these developments on the estimation of parameters of the extreme-value distributions have been reviewed in Chapter 22 of the volume by Johnson, Kotz, and Balakrishnan (1995).

Harter and Moore (1967a) [see Section 5.2 of this volume] found maximum-likelihood estimators, for single censoring from above, of the location parameter of a Type I asymptotic distribution of smallest values with known scale parameter. In the present section, an iterative procedure is given for maximum-likelihood estimation, from singly or doubly censored samples, of both parameters of Type I asymptotic distributions of smallest and largest values. The elements of the information matrix are worked out, and this matrix is inverted numerically to obtain the asymptotic variances and covariances of the estimators, which are tabulated in Table F6 for censoring proportions $q_1 = 0.0(0.1)0.9$ from below and $q_2 = 0.0(0.1)(0.9 - q_1)$ from above. Results are reported of a Monte Carlo study to obtain information about the small-sample properties of these estimators and to compare them with the best linear unbiased and best linear invariant estimators. These results are tabulated in Table G5.

It should be pointed out that if the random variable T has the two-parameter Weibull distribution with scale parameter θ and shape parameter K, then the random variable $X = \ln T$ has the Type I asymptotic distribution of smallest values with location parameter (mode) $u = \ln\theta$ and scale parameter $b = 1/K$. Hence, by suitable transformations, the procedure outlined in this section could be used to estimate the parameters of a Weibull population. However, Harter and Moore (1965) have given a similar iterative procedure for direct estimation of the Weibull parameters [see Section 7.3].

7.5.2 The Likelihood Function and Its Derivatives

The cumulative distribution function of the random variable X having the first (or Type I) asymptotic distribution of smallest (largest) values with location parameter u and scale parameter b is given by

$$F(x;u,b) = \begin{cases} 1 - \exp\{-\exp[(x-u)/b]\}, & \text{(smallest values)} \\ \exp\{-\exp[-(x-u)/b]\}, & \text{(largest values)} \end{cases} \tag{7.5.1}$$

where $b > 0$. The natural logarithm L_{r_1,r_2} of the likelihood function of a random sample of size n from such a distribution, the lowest r_1 and the highest r_2 sample values having been censored, is given by

$$\begin{aligned} L_{r_1,r_2} &= \ln n! - \ln r_1! - \ln r_2! \pm \sum_{i=r_1+1}^{n-r_2} z_i - \sum_{i=r_1+1}^{n-r_2} \exp(\pm z_i) \\ &\quad - (n-r_1-r_2)\ln b + r_1 \ln F(z_{r_1+1}) + r_2 \ln[1 - F(z_{n-r_2})], \end{aligned} \tag{7.5.2}$$

where $z_i = (x_i - u)/b$ and $F(z_i) = \frac{1}{2} \pm \frac{1}{2} \mp \exp[-\exp(\pm z_i)]$, so that $f(z_i) = \exp[\pm z_i - \exp(\pm z_i)]$, the upper (lower) signs being those for the distribution of smallest (largest) values. The first partial derivatives of $L = L_{r_1,r_2}$ with respect to the parameters are given by

$$\begin{aligned} \partial L/\partial u &= \mp(n-r_1-r_2)/b \pm \Sigma \exp(\pm z_i)/b - r_1 f(z_{r_1+1})/bF(z_{r_1+1}) \\ &\quad + r_2 f(z_{n-r_2})/b[1 - F(z_{n-r_2})] \ , \end{aligned} \tag{7.5.3}$$

$$\begin{aligned} \partial L/\partial b &= \mp\Sigma z_i/b \pm \Sigma z_i \exp(\pm z_i)/b - (n-r_1-r_2)/b \\ &\quad - r_1 z_{r_1+1} f(z_{r_1+1})/bF(z_{r_1+1}) + r_2 z_{n-r_2} f(z_{n-r_2})/b[1 - F(z_{n-r_2})] \ , \end{aligned} \tag{7.5.4}$$

where the summations are on i running from $r_1 + 1$ through $n - r_2$. The second partial derivatives of $L = L_{r_1,r_2}$ with respect to the parameters are given by

$$\begin{aligned} \partial^2 L/\partial u^2 &= -\Sigma \exp(\pm z_i)/b^2 + r_1\{[\pm 1 \mp \exp(\pm z_{r_1+1})] f(z_{r_1+1}) F(z_{r_1+1}) \\ &\quad - f^2(z_{r_1+1})\}/b^2 F^2(z_{r_1+1}) - r_2\{[\pm 1 \mp \exp(\pm z_{n-r_2})] f(z_{n-r_2})[1 - F(z_{n-r_2})] \\ &\quad + f^2(z_{n-r_2})\}/b^2[1 - F(z_{n-r_2})]^2 \ , \end{aligned} \tag{7.5.5}$$

$$\begin{aligned} \partial^2 L/\partial b^2 &= \pm 2\Sigma z_i/b^2 - \Sigma z_i^2 \exp(\pm z_i)/b^2 \mp 2\Sigma z_i \exp(\pm z_i)/b^2 + (n-r_1-r_2)/b^2 \\ &\quad + r_1\{2 z_{r_1+1} f(z_{r_1+1}) F(z_{r_1+1}) + z_{r_1+1}^2 [\pm 1 \mp \exp(\pm z_{r_1+1})] f(z_{r_1+1}) F(z_{r_1+1}) \end{aligned}$$

$$- z_{r_1+1}^2 f^2(z_{r_1+1})\}/b^2 F(z_{r_1+1}) - r_2\{2z_{n-r_2} f(z_{n-r_2})[1 - F(z_{n-r_2})]$$
$$+ z_{n-r_2}^2[\pm 1 \mp \exp(\pm z_{n-r_2})] f(z_{n-r_2})[1 - F(z_{n-r_2})]$$
$$+ z_{n-r_2}^2 f^2(z_{n-r_2})\}/b^2[1 - F(z_{n-r_2})]^2, \tag{7.5.6}$$

$$\partial^2 L/\partial u \partial b = \pm(n - r_1 - r_2)/b^2 - \Sigma z_i \exp(\pm z_i)/b^2 \mp \exp(\pm z_i)/b^2$$
$$+ r_1\{f(z_{r_1+1})F(z_{r_1+1}) + z_{r_1+1}[\pm 1 \mp \exp(\pm z_{r_1+1})] f(z_{r_1+1})F(z_{r_1+1})$$
$$- z_{r_1+1} f^2(z_{r_1+1})\}/b^2 F^2(z_{r_1+1}) - r_2\{f(z_{n-r_2})[1 - F(z_{n-r_2})]$$
$$+ z_{n-r_2}[\pm 1 \mp \exp(\pm z_{n-r_2})] f(z_{n-r_2})[1 - F(z_{n-r_2})]$$
$$+ z_{n-r_2} f^2(z_{n-r_2})\}/b^2[1 - F(z_{n-r_2})]^2 . \tag{7.5.7}$$

7.5.3 Asymptotic Variances and Covariances

Let $q_1 = r_1/n$, $q_2 = r_2/n$, and $p = 1 - q_1 - q_2 = (n - r_1 - r_2)/n$. As $n \to \infty$ (q_1 and q_2 fixed), $z_{r_1+1} \to \hat{z}_1$ where $F(\hat{z}_1) = \int_{-\infty}^{\hat{z}_1} f(t)\,dt = q_1$ and $z_{n-r_2} \to \hat{z}_2$ where $1 - F(\hat{z}_2) = \int_{\hat{z}_2}^{\infty} f(t)\,dt = q_2$. For the distribution of smallest values,

$$E(\Sigma z_i) \to n \int_{\hat{z}_1}^{\hat{z}_2} t f(t)\,dt = n[\Gamma'(1; \exp \hat{z}_2) - \Gamma'(1; \exp \hat{z}_1)], \tag{7.5.8}$$

$$E(\Sigma \exp z_i) \to n \int_{\hat{z}_1}^{\hat{z}_2} e^t f(t)\,dt = n[\Gamma(2; \exp \hat{z}_2) - \Gamma(2; \exp \hat{z}_1)], \tag{7.5.9}$$

$$E(\Sigma z_i \exp z_i) \to n \int_{\hat{z}_1}^{\hat{z}_2} t\, e^t f(t)\,dt = n[\Gamma'(2; \exp \hat{z}_2) - \Gamma'(2; \exp \hat{z}_1)], \tag{7.5.10}$$

$$E(\Sigma z_i^2 \exp z_i) \to n \int_{\hat{z}_1}^{\hat{z}_2} t^2 e^t f(t)\,dt = n[\Gamma''(2; \exp \hat{z}_2) - \Gamma''(2; \exp \hat{z}_1)], \tag{7.5.11}$$

where $\Gamma(y; w) = \int_0^w t^{y-1} e^{-t} dt$ is the incomplete Gamma function, $\Gamma'(a; w) = (d/dy)[\Gamma(y; w)]|_{y=a}$ and $\Gamma''(a; w)$ is analogously defined. The elements of the information matrix (multiplied by b^2/n) may, after simplification, which involves use of the relations $\Gamma(2; w) = 1 - (1 + w)\exp(-w)$, $\exp \hat{z}_1 = -\ln(1 - q_1)$ and $\exp \hat{z}_2 = -\ln q_2$, be written as

$$\lim_{n\to\infty} (-b^2/n) E(\partial^2 L/\partial u^2) = p + (q_1^{-1} - 1)\{\ln(1 - q_1)\}^2 = v^{11} , \tag{7.5.12}$$

$$\lim_{n\to\infty} (-b^2/n) E(\partial^2 L/\partial b^2) = -p - 2\{\Gamma'(1; -\ln q_2) - \Gamma'[1; -\ln(1 - q_1)]\}$$
$$- \Gamma''(2; -\ln q_2) - \Gamma''[2; -\ln(1 - q_1)] + 2\{\Gamma'(2; -\ln q_2)$$
$$- \Gamma'[2; -\ln(1 - q_1)]\} - q_2 \ln q_2 \ln(-\ln q_2)[2 + \ln(-\ln q_2)]$$
$$+ (1 - q_1)\ln(1 - q_1)\ln[-\ln(1 - q_1)]\{2 + \ln[-\ln(1 - q_1)]$$
$$+ \ln(1 - q_1)\ln[-\ln(1 - q_1)]/q_1\} = v^{22} , \tag{7.5.13}$$

$$\lim_{n\to\infty}(-b^2/n)E(\partial^2 L/\partial u \partial b) = \Gamma'(2; -\ln q_2) - \Gamma'[2; -\ln(1-q_1)] - q_2 \ln q_2 \ln(-\ln q_2) + (1-q_1)\ln(1-q_1)\ln[-\ln(1-q_1)] + (q_1^{-1}-1)\{\ln(1-q_1)\}^2 \ln[-\ln(1-q_1)] = v^{12} \,. \quad (7.5.14)$$

The asymptotic variance-covariance matrix for the estimators $\hat{u}$ and $\hat{b}$ is then $(b^2/n)[v_{ij}]$, where $[v_{ij}] = [v^{ij}]^{-1}$. The coefficients of b^2/n in the asymptotic variances and covariances are given in Table F6 for $q_1 = 0.0(0.1)0.9$ and $q_2 = 0.0(0.1)(0.9 - q_1)$.

It can be shown that if one considers the distribution of largest values instead of the distribution of smallest values, and at the same time interchanges the values of q_1 and q_2, the values of v^{11}, v^{22}, v_{11} and v_{22} remain the same, while v^{12} and v_{12} retain the same absolute values but reverse their signs.

7.5.4 Iterative Estimation Procedure

The likelihood equations are obtained by equating to zero the first partial derivatives of the likelihood function, which are given by Equations (7.5.3) and (7.5.4). The likelihood equations cannot be solved explicitly, but maximum-likelihood estimates can be obtained by solving them iteratively on a computer. The iterative estimation procedure involves estimating the parameters, one at a time, in the cyclic order u, b, omitting a parameter if it is assumed to be known. One starts by choosing initial estimate(s) for the unknown parameter(s). At each step, the rule of false position (iterative linear interpolation) is used to determine the value of the parameter then being estimated which satisfies the appropriate likelihood equation, in which the latest estimate (or known value) of the other parameter has been substituted. Iteration continues until the results of successive steps agree to within some assigned tolerance. A FORTRAN IV program for use on the IBM 7094 computer was used to estimate, from complete, singly censored, or doubly censored samples, the parameters of a Type I asymptotic distribution of smallest (largest) values. The same program is made to work for both by including an IF statement, based on a dummy variable which is set equal to 0(1) for the distribution of smallest (largest) values, to control the signs of terms which differ for the two cases. If one of the parameters is assumed to be known, a single iteration suffices to estimate the other. Even when both are unknown, experience has shown that the rate of convergence is quite rapid if the initial estimates are reasonable and the amount of censoring is not excessive.

7.5.5 Monte Carlo Study for Small Samples

While the maximum-likelihood estimators $\hat{u}$ and $\hat{b}$ are asymptotically unbiased and have known asymptotic variances and covariances, there is no known analytic method of determining their biases, variances, and covariances for small samples. In order to obtain information about the small-sample properties of these estimators, and incidentally to compare them with the best linear unbiased and best linear invariant estimators, a Monte Carlo study was performed on the IBM 7094 computer. For $n = 10$ and for $n = 20$, two thousand pseudo-random samples of size n from a standard Type I asymptotic distribution of smallest values ($u = 0,\ b = 1$) were generated, and each sample was ordered from smallest to largest. The iterative procedure described in subsection 7.5.4 was used to compute the estimates $\hat{u}$ and $\hat{b}$, also $\hat{u}|b$ and $\hat{b}|u$, from the m order statistics remaining in each sample after proportions q_1 and q_2 had been censored from below and from above, respectively, where q_1 and q_2 were taken at intervals of 0.1, subject to the restriction

$m \geq 2$. The means, variances and covariances of the estimates from 2000 samples of size $n = 10$ are given in Table G5A, and similar results for $n = 20$ are given in Table G5B. The rows of Table G5A (and likewise Table G5B) are not statistically independent, since they are based on the same samples (with different proportions censored). The results indicate that $\hat{u}$, $\hat{b}$, $\hat{u}|b$ and $\hat{b}|u$ are all negatively biased, except that $\hat{u}$ is positively biased when q_1 is much larger than q_2. The absolute value of the bias tends to increase as the amount of censoring increases. The bias of $\hat{b}$ (u unknown) is approximately equal to $-1/m$. The bias in estimating either parameter is much smaller when the other parameter is known than when both parameters are estimated simultaneously.

In Tables G5C and G5D, the mean square errors of the maximum-likelihood estimates from the 2000 samples of sizes $n = 10$ and $n = 20$, respectively, are compared with the asymptotic variances of the maximum-likelihood estimators. In cases of single censoring from above, they are also compared with the variances of the best linear unbiased estimators and the mean square errors of the best linear invariant estimators, which were obtained by rounding six-decimal-place values given by Mann (1965b). For little or no censoring, the precision of all three estimators is approximately the same, but for severe censoring from above the mean square errors of the best linear invariant estimator and of the maximum-likelihood estimate are significantly smaller than the variance of the best linear unbiased estimator, while differing so little from each other that the difference may well be due to sampling fluctuations. The mean square errors of $\hat{u}$, $\hat{u}|b$, and $\hat{b}$ are only slightly larger than the variances as given by the asymptotic formulas, except in cases of severe censoring. The mean square error of $\hat{b}|u$ is often less, and in a few instances [censoring severe but symmetric or almost so] substantially less, than the variance as given by the asymptotic formula.

Since the mean square errors of the maximum-likelihood estimators have been shown to differ but little from those of the best linear invariant estimators and to be considerably smaller than those of the best linear unbiased estimators in some cases, especially those involving strongly asymmetric censoring, the use of the maximum-likelihood estimators has much to recommend it to anyone who has access to a computer. Listings and/or program decks for the iterative procedure proposed in this section and those for other populations proposed elsewhere in this chapter are available from the authors on request.

The reader may find it interesting to compare the results of this Monte Carlo study with those of similar studies for the normal distribution by Harter and Moore (1966) [see subsection 7.1.5 and Table G1] and for the logistic distribution by Harter and Moore (1967b) [see subsection 7.4.5 and Table G4.]

7.5.6 Numerical Example

Consider the following sample of size 40 from a Type I asymptotic distribution of smallest values whose parameters we wish to estimate:

1.609	3.466	3.989	4.060	4.174	4.407	4.522	4.673	4.762	4.963
2.303	3.497	4.007	4.111	4.190	4.443	4.625	4.736	4.820	5.017
2.833	3.526	4.007	4.159	4.205	4.500	4.635	4.736	4.934	5.063
3.466	3.584	4.060	4.174	4.220	4.522	4.663	4.754	4.956	5.273

Suppose that we decide to censor the first 4 and the last 4 observations. If we apply the procedure outlined in subsection 7.5.4, starting from initial estimates $\hat{u}_0 = 4.5$ and $\hat{b}_0 = 0.6$, the estimates converge after four iterations to $\hat{u} = 4.5509$ and $\hat{b} = 0.5024$. The results are not dependent on the initial estimates, but good initial estimates tend to decrease the number of iterations.

Since there is evidence [see Johns and Lieberman (1964)] that asymptotic theory gives a good approximation to the distribution of estimates for samples of size 40, not too severely censored, we may use the asymptotic variances given in Table F6 to obtain approximate confidence bounds on the true values of the parameters. Since b is unknown, we approximate it by $\hat{b} = 0.5024$, and find $\hat{\sigma}_u \doteq 0.5024\sqrt{1.157024/40} = 0.0855$ and $\hat{\sigma}_b \doteq 0.5024\sqrt{0.842250/40} = 0.0729$. Hence approximate 90 per cent confidence bounds on u are $4.5509 - 1.645(0.0855) = 4.4103$ and $4.5509 + 1.645(0.0855) = 4.6915$, while approximate 90 per cent confidence bounds on b are $0.5024 - 1.645(0.0729) = 0.3825$ and $0.5024 + 1.645(0.0729) = 0.6223$. Actually, the data are the natural logarithms of a pseudo-random sample from a two-parameter Weibull distribution with scale parameter 100 and shape parameter 2. Hence the true parameter values are $u = \ln 100 = 4.6052$ and $b = 1/2 = 0.5$, which are included in the approximate 90 per cent confidence intervals.

7.5.7 References

1. Arnold, Barry C., Balakrishnan, N., and Nagaraja, H.N. (1992). *A First Course in Order Statistics.* John Wiley & Sons, New York.
2. Aziz, P.M. (1956). Application of the statistical theory of extreme values to the analysis of maximum pit depth data for aluminum. *Corrosion* **12**(10), 35-46.
3. Balakrishnan, N. and Chan, Ping-Shing (1992a). Order statistics from extreme value distribution, I: Tables of means, variances and covariances. *Communications in Statistics - Simulation and Computation* **21**, 1199-1217.
4. Balakrishnan, N. and Chan, Ping-Shing (1992b). Order statistics from extreme value distribution, II: Best linear unbiased estimates and some other uses. *Communications in Statistics - Simulation and Computation* **21**, 1219-1246.
5. Balakrishnan, N. and Chan, Ping-Shing (1992c). Extended tables of means, variances and covariances of order statistics from the extreme value distribution for sample sizes up to 30. *Report*, Department of Mathematics and Statistics, McMaster University, Hamilton, Canada.
6. Balakrishnan, N. and Chan, Ping-Shing (1992d). Extended tables of best linear unbiased estimates from complete and Type-II censored samples from the extreme value distribution for sample sizes up to 30. *Report*, Department of Mathematics and Statistics, McMaster University, Hamilton, Canada.
7. Balakrishnan, N. and Cohen, A. Clifford, Jr. (1991). *Order Statistics and Inference: Estimation Methods.* Academic Press, San Diego.
8. von Bortkiewicz, L. (1922a). Variationsbreite und mittlerer Fehler. *Sitzungsberichte der Berliner Mathematischer Gesellschaft* **21**, 3-11.
9. von Bortkiewicz, L. (1922b). Die Variationsbreite beim Gauss'schen Fehlergesetz. *Nordisk Statistisk Tidskrift* **1**, 11-38; 193-220.
10. Castillo, Enrique (1988). *Extreme Value Theory in Engineering.* Academic Press, San Diego.
11. David, Herbert A. (1981). *Order Statistics*, Second edition. John Wiley & Sons, New York.
12. Dodd, Edward Lewis (1923). The greatest and the least variate under general laws of error. *Transactions of the American Mathematical Society* **25**, 525-539.
13. Downton, Frank (1966). Linear estimates of parameters in the extreme value distribution. *Technometrics* **8**, 3-17.

14. Eldredge, G.G. (1957). Analysis of corrosion pitting by extreme-value statistics and its application to oil well tubing caliper surveys. *Corrosion* **13**(1), 67-76.

15. Epstein, Benjamin (1948a). Statistical aspects of fracture problems. *Journal of Applied Physics* **19**, 140-147.

16. Epstein, Benjamin (1948b). Application of the theory of extreme values in fracture problems. *Journal of the American Statistical Association* **43**, 403-412.

17. Epstein, Benjamin (1958). Stochastic models for length of life (abstract). *Annals of Mathematical Statistics* **29**, 1283-1284.

18. Epstein, Benjamin (1960). Elements of the theory of extreme values. *Technometrics* **2**, 27-41.

19. Epstein, Benjamin and Brooks, Hamilton (1948). The theory of extreme values and its implications in the study of the dielectric strength of paper capacitors. *Journal of Applied Physics* **19**, 544-550.

20. Fisher, R.A. and Tippett, L.H.C. (1928). Limiting forms of the frequency distribution of the largest or smallest member of a sample. *Proceedings of the Cambridge Philosophical Society* **24**, 180-190.

21. Fréchet, M. (1927). Sur la loi de probabilité de l'écart maximum. *Annales de la Societé Polonaise de Mathématique* (Cracow) **6**, 93-96.

22. Galambos, Janos (1978, 1987). *The Asymptotic Theory of Extreme Order Statistics.* First edition, John Wiley & Sons, New York; Second edition, Krieger, Malabar, Florida.

23. Gnedenko, B.V. (1943). Sur la distribution limite du terme maximum d'une série aléatoire. *Annals of Mathematics* (2)**44**, 423-453.

24. Gumbel, E.J. (1935). Les valeurs extrêmes des distributions statistiques. *Annales de l'Institut Henri Poincaré* **5**, 115-158.

25. Gumbel, E.J. (1937a). *La Durée Extrême de la Vie Humaine.* Hermann et Cie., Paris.

26. Gumbel, E.J. (1973b). Les intervalles extrêmes entre les émissions radioactives. *Journal de Physique* (7)**8**, 321-329; 446-452.

27. Gumbel, E.J. (1941). The return period of flood flows. *Annals of Mathematical Statistics* **12**, 163-190.

28. Gumbel, E.J. (1958). *Statistics of Extremes.* Columbia University Press, New York.

29. Gumbel, E.J. and Freudenthal, A.M. (1952). Statistical theory of fatigue failures (abstract). *Annals of Mathematical Statistics* **23**, 136.

30. Gumbel, E.J. and Mustafi, C.K. (1966). Comments to: Edward C. Posner, "The application of extreme value theory to error free communication". *Technometrics* **8**, 363-366.

31. de Haan, L. (1970). *On Regular Variation and Its Application to the Weak Convergence of Sample Extremes.* Mathematical Centre Tract No. 32, Mathematisch Centrum, Amsterdam.

32. Harter, H. Leon and Moore, Albert H. (1965). Maximum-likelihood estimation of the parameters of Gamma and Weibull populations from complete and from censored samples. *Technometrics* **7**, 639-643.

33. Harter, H. Leon and Moore, Albert H. (1966). Iterative maximum-likelihood estimation of the parameters of normal populations from singly and doubly censored samples. *Biometrika* **53**, 205-213.

34. Harter, H. Leon and Moore, Albert H. (1967a). A note on estimation from a type I extreme-value distribution. *Technometrics* **9**, 325-331.

35. Harter, H. Leon and Moore, Albert H. (1967b). Maximum-likelihood estimation, from censored samples, of the parameters of a logistic distribution. *Journal of the American Statistical Association* **62**, 675-684.

36. Harter, H. Leon and Moore, Albert H. (1968). Maximum-likelihood estimation, from doubly censored samples, of the parameters of the first asymptotic distribution of extreme values. *Journal of the American Statistical Association* **63**, 889-901.

37. Hassanein, Khatab M. (1964). Estimation of the Parameters of the Extreme Value Distribution by Order Statistics. *Ph.D. Thesis*, University of North Carolina, Chapel Hill. AD 622257.

38. Johns, M.V., Jr. and Lieberman, G.J. (1964). An Exact Asymptotically Efficient Confidence Bound for Reliability in the Case of the Weibull Distribution. *Technical Report No. 75*, Department of Statistics, Stanford University, Stanford.

39. Johns, M.V., Jr. and Lieberman, G.J. (1966). An exact asymptotically efficient confidence bound for reliability in the case of the Weibull distribution. *Technometrics* **8**, 135-175.

40. Johnson, Norman L., Kotz, Samuel, and Balakrishnan, N. (1995). *Continuous Univariate Distributions - Volume 2*, Second edition. John Wiley & Sons, New York.

41. Kimball, Bradford F. (1946). Sufficient statistical estimating functions for the parameters of the distribution of maximum values. *Annals of Mathematical Statistics* **17**, 299-309.

42. Kimball, Bradford F. (1949). An approximation to the sampling variance of an estimated maximum value of given frequency based on fit of doubly exponential distribution of maximum values. *Annals of Mathematical Statistics* **20**, 110-113.

43. Kimball, Bradford F. (1956). The bias in certain estimates of the parameters of the extreme-value distribution. *Annals of Mathematical Statistics* **27**, 758-767.

44. Lieblein, Julius (1953). On the exact evaluation of the variances and covariances of order statistics in samples from the extreme-value distribution. *Annals of Mathematical Statistics* **24**, 282-287.

45. Lieblein, Julius (1954). A New Method of Analyzing Extreme-Value Data. *Technical Note 3053*, National Advisory Committee for Aeronautics.

46. Lieblein, Julius (1955). Estimation from "censored" samples of extreme data (abstract). *Annals of Mathematical Statistics* **26**, 156.

47. Mann, Nancy R. (1963). Optimum Estimates of Parameters of Continuous Distributions. *Rocketdyne Research Report 63-41*, North American Aviation, Inc., Canoga Park, California.

48. Mann, Nancy R. (1965a). Point and Interval Estimates for Reliability Parameters when Failure Times Have the Two-Parameter Weibull Distribution. *Doctoral dissertation*, University of California at Los Angeles. University Microfilms, Ann Arbor, Michigan.

49. Mann, Nancy R. (1965b). Cramér-Rao Efficiencies of Best Linear Invariant Estimates of Parameters of the Extreme-Value Distribution under Type II Censoring from Above. *Rocketdyne Research Report 65-36*, North American Aviation, Inc., Canoga Park, California.

50. Mann, Nancy R. (1967a). Results on Location and Scale Parameter Estimation with Application to the Extreme-Value Distribution. *Aerospace Research Laboratories Technical Report ARL 67-0023*, Wright-Patterson Air Force Base. AD 653575.

51. Mann, Nancy R. (1967b). Tables for obtaining the best linear invariant estimates of parameters of the Weibull distribution. *Technometrics* **9**, 629-645.

52. von Mises, R. (1923). Uber die Variationsbreite einer Beobachtungsreihe. *Sitzungsberichte der Berliner Mathematischer Gesellschaft* **22**, 3-8.

53. von Mises, R. (1936). La distribution de la plus grande de n valeurs. *Revue Mathématique de l'Union Interbalkanique* (Athens) **1**(2), 1-20.

54. Neyman, Jerzy (1923). Sur les valeurs theoriques de la plus grande de n erreurs. *Prace Matematyczno-Fizyczne* (Warsaw) **33**, 1 ff.

55. Posner, Edward C. (1965). The application of extreme-value theory to error-free communication. *Technometrics* **7**, 517-528.

56. Sarhan, Ahmed E. and Greenberg, Bernard G. (Eds.) (1962). *Contributions to Order Statistics*. John Wiley & Sons, New York.

57. Tippett, L.H.C. (1925). On the extreme individuals and the range of samples taken from a normal population. *Biometrika* **17**, 364-387.

58. White, John S. (1964). Least squares unbiased censored linear estimation for the log Weibull (extreme value) distribution. *Journal of the Industrial Mathematics Society* **14**, 21-60.

7.6 Maximum-Likelihood Estimation of Two Parameters of Log-Gamma Population

7.6.1 Introduction

Historical remarks by Chan [(1993), pp. 1-2], which serve as an introduction to the log-Gamma population, may be summarized as follows: The Weibull and lognormal distributions are two of the most popular models used in the analysis of survival data and reliability. These two distributions have been studied by many authors [for references, see sections 7.3 and 7.2, respectively,] but little has been written on the choice between the two models [see, however, Dumonceaux and Antle (1973)]. Farewell and Prentice (1977) suggested that embedding these two models in a single parametric family will allow the methods of ordinary parametric inference to be used for discrimination and an assessment to be made of each competing model relative to a more comprehensive one. The generalized Gamma distribution introduced by Stacy (1962) is useful in this regard. Methods of estimating its parameters were studied by many authors [e.g., Stacy and Mihram (1965), Harter (1967), and Hager and Bain (1970)]. Hager and Bain found that the algorithms for determining the maximum likelihood estimates often fail to converge. Furthermore, it was difficult to develop suitable tests or interval estimation procedures. Large-sample methods based on normal approximations to the distributions of the maximum likelihood estimators were also observed to be very poor even for sample sizes as large as 400. However, the work of Prentice (1974) clarified this problem considerably. Instead of considering the generalized Gamma variables, he considered the logarithms of these variables and found that this logarithmic transformation makes the inference on the parameters much easier. He studied the distribution of the logarithm of a generalized Gamma random variable, and pointed out that the extreme-value distribution is included in this model. Further reparametrization of the model includes the normal distribution as the limiting case. He then studied the maximum likelihood estimation and the asymptotic maximum likelihood theory.

7.6.2 The Log-Gamma Population

The probability density function of a log-Gamma random variable W is given by

$$f(w) = \frac{1}{\Gamma(\kappa)} e^{-\kappa w - e^{w}}, \qquad -\infty < w < \infty, \ \kappa > 0. \tag{7.6.1}$$

The moment generating function of the log-Gamma random variable W is

$$M_W(t) = E(e^{tW}) = \Gamma(\kappa + t)/\Gamma(\kappa) \ , \tag{7.6.2}$$

and hence the kth raw moment of W is

$$E(W^k) = \Gamma^{(k)}(\kappa)/\Gamma(\kappa) \ , \tag{7.6.3}$$

where $\Gamma^{(k)}(\kappa)$ is the kth derivative of $\Gamma(\kappa)$ with respect to κ.

It follows that the mean of W is

$$E(W) = \Gamma'(\kappa)/\Gamma(\kappa) = \psi(\kappa) \tag{7.6.4}$$

and the variance of W is

$$\mathrm{var}(W) = \frac{\Gamma''(\kappa)}{\Gamma(\kappa)} - \left(\frac{\Gamma'(\kappa)}{\Gamma(\kappa)}\right)^2 = \psi'(\kappa) \ . \tag{7.6.5}$$

Since $\psi(\kappa)$ is an increasing function of κ ($\psi'(\kappa) > 0$ for all κ), the distribution is shifted to the right as κ increases. When κ is large, $\psi(\kappa)$ and $\psi'(\kappa)$ have the series expansions [see Abramowitz and Stegun (1965)]:

$$\psi(\kappa) = \log\kappa - \frac{1}{2\kappa} - \frac{1}{12\kappa^2} + \frac{1}{120\kappa^4} - \frac{1}{252\kappa^6} + \cdots \sim \log\kappa \tag{7.6.6}$$

and

$$\psi'(\kappa) = \frac{1}{\kappa} + \frac{1}{2\kappa^2} + \frac{1}{6\kappa^3} - \frac{1}{30\kappa^5} + \frac{1}{42\kappa^7} + \cdots \sim \frac{1}{\kappa}\,. \tag{7.6.7}$$

Prentice (1974) therefore suggested a rescaled model with the transformation

$$X = \sqrt{\kappa}(W - \ln\kappa)\,. \tag{7.6.8}$$

The rescaled variable X has density function

$$f(x) = \frac{\kappa^{\kappa-1/2}}{\Gamma(\kappa)}\, e^{\sqrt{\kappa}x - \kappa\, e^{x/\sqrt{\kappa}}}, \qquad -\infty < x < \infty,\ \kappa > 0. \tag{7.6.9}$$

When κ is 1, X has an extreme-value distribution. The rescaled variable is asymptotically standard normal as $\kappa \to \infty$.

Let T be a random variable with pdf given by

$$f(t) = \frac{\beta}{\Gamma(\kappa)}\,\frac{t^{\kappa\beta-1}}{\lambda^{\kappa\beta}}\,\exp\left[-\left(\frac{t}{\lambda}\right)^{\beta}\right], \qquad t > 0, \tag{7.6.10}$$

where β, λ and κ are all positive. Then T is said to have a generalized Gamma distribution. This model, first introduced by Stacy (1962), is a very flexible three-parameter family of distributions. It is exponential when $\kappa = \beta = 1$, Weibull when $\kappa = 1$, Gamma when $\beta = 1$, and it is asymptotically lognormal as $\kappa \to \infty$. Now consider $Y = \ln T$ and set $\nu = \ln\lambda$ and $\tau = \beta^{-1}$. Then

$$W = \frac{Y-\nu}{\tau} = \frac{\ln T - \ln\lambda}{\beta^{-1}} = \beta\ln\left(\frac{T}{\lambda}\right) = \ln\left(\frac{T}{\lambda}\right)^{\beta} \tag{7.6.11}$$

has a log-Gamma distribution with pdf as given in Equation (7.6.1). Therefore the logarithmic transformation changes the generalized Gamma distribution into a log-Gamma distribution with location parameter ν and scale parameter τ, which has pdf

$$g(y;\nu,\tau) = \frac{1}{\tau}\,\frac{1}{\Gamma(\kappa)}\,\exp\left(\kappa\left(\frac{y-\nu}{\tau}\right) - e^{(y-\nu)/\tau}\right), \quad -\infty < y < \infty\,. \tag{7.6.12}$$

For the reparametrized model, the pdf is

$$g(y) = \frac{\kappa^{\kappa-1/2}}{\sigma\Gamma(\kappa)}\,\exp\left(\sqrt{\kappa}\left(\frac{y-\mu}{\sigma}\right) - \kappa\, e^{(y-\mu)/\sigma\sqrt{\kappa}}\right),$$
$$-\infty < y < \infty,\ -\infty < \mu < \infty,\ \sigma > 0,\ \kappa > 0 \tag{7.6.13}$$

due to the relationships

$$\mu = \nu + \tau\ln\kappa,\ \sigma = \tau/\kappa\,. \tag{7.6.14}$$

In Chapter 3, subsection 3.3.3, following Chan [(1993), Chapter 2], we found expressions for the expected values, variances and covariances of the order statistics of a sample of size n from the standardized log-Gamma population with pdf given by Equation (7.6.1) and described their use in computing tables. In Chapter 3, subsection 3.3.12, following

Chan [(1993), Chapter 3], we discussed the use of the tables of expected values, variances and covariances of order statistics to compute the best linear unbiased estimates of the parameters ν and τ of the log-Gamma population with pdf given by Equation (7.6.12), where κ is assumed to be known. In this section, following Chan [(1993), Chapter 5], we consider maximum likelihood estimation of the parameters μ and σ, with κ known, of the reparametrized log-Gamma population with pdf given by Equation (7.6.13).

7.6.3 The Likelihood Function and Its Derivatives

Consider a random sample of size n from the log-Gamma population with pdf as in Equation (7.6.13), where the shape parameter κ is assumed to be known. Let $Y_{r+1:n} \leq Y_{r+2:n} < \cdots \leq Y_{n-s:n}$ be the ordered observations remaining after the r smallest and the s largest observations have been censored. The likelihood function of the Type II censored sample is then [Balakrishnan and Chan (1995a,b)]

$$L = \frac{n!}{r!s!}\,[G(Y_{r+1:n})]^r[1-G(Y_{n-s:n})]^s \prod_{i=r+1}^{n-s} g(Y_{i:n}) \tag{7.6.15}$$

or equivalently,

$$L = \frac{n!}{r!s!}\,\frac{1}{\sigma^A}\,[F(X_{r+1:n})]^r[1-F(X_{n-s:n})]^s \prod_{i=r+1}^{n-s} f(X_{i:n})\ , \tag{7.6.16}$$

where $A = n-r-s$, $X = (Y-\mu)/\sigma$ is the standardized variable with the pdf $f(\cdot)$ given in Equation (7.6.9) and $F(\cdot)$ is its corresponding cdf. The log-likelihood function is given by

$$\begin{aligned}\ln L = {} & \text{const.} - A\ln\sigma + r\ln[F(X_{r+1:n})] + s\ln[1-F(X_{n-s:n})] \\ & + \sum_{i=r+1}^{n-s} \ln f(X_{i:n})\ . \end{aligned} \tag{7.6.17}$$

The derivatives of the log-likelihood function with respect to μ and σ are

$$\begin{aligned}\frac{\partial \ln L}{\partial \mu} = {} & -\frac{1}{\sigma}\Bigg\{r\,\frac{f(X_{r+1:n})}{F(X_{r+1:n})} - s\,\frac{f(X_{n-s:n})}{1-F(x_{n-s:n})} + A\sqrt{\kappa} \\ & - \sqrt{\kappa}\sum_{i=r+1}^{n-s} e^{X_{i:n}/\sqrt{\kappa}}\Bigg\} \end{aligned} \tag{7.6.18}$$

and

$$\begin{aligned}\frac{\partial \ln L}{\partial \sigma} = {} & -\frac{1}{\sigma}\Bigg\{A + rX_{r+1:n}\,\frac{f(X_{r+1:n})}{F(X_{r+1:n})} - sX_{n-s:n}\,\frac{f(X_{n-s:n})}{1-F(X_{n-s:n})} \\ & + \sqrt{\kappa}\sum_{i=r+1}^{n-s} X_{i:n} - \sqrt{\kappa}\sum_{i=r+1}^{n-s} X_{i:n}e^{X_{i:n}/\sqrt{\kappa}}\Bigg\}\ . \end{aligned} \tag{7.6.19}$$

The MLEs $\hat{\mu}$ and $\hat{\sigma}$ can be obtained by simultaneously solving the likelihood equations $\partial \ln L/\partial\mu = 0$ and $\partial \ln L/\partial\sigma = 0$. Since these two equations cannot be solved analytically, some numerical method must be employed [see Section 7.6.6].

7.6.4 Maximum-Likelihood Information Matrix

The second partial derivatives of the log-likelihood function in Equation (7.6.17) with respect to μ and σ are found to be [Balakrishnan and Chan (1995a,b)]

$$\begin{aligned}\frac{\partial^2 \ln L}{\partial \mu^2} = \frac{1}{\sigma^2}\Bigg\{ & r\,\frac{f(X_{r+1:n})}{F(X_{r+1:n})}\left[\sqrt{\kappa} - \sqrt{\kappa}\,e^{X_{r+1:n}/\sqrt{\kappa}} - \frac{f(X_{r+1:n})}{F(X_{r+1:n})}\right] \\ & - s\,\frac{f(X_{n-s:n})}{1-F(X_{n-s:n})}\left[\sqrt{\kappa} - \sqrt{\kappa}\,e^{X_{n-s:n}/\sqrt{\kappa}} + \frac{f(X_{n-s:n})}{1-F(X_{n-s:n})}\right] \\ & - \sum_{i=r+1}^{n-s} e^{X_{i:n}/\sqrt{\kappa}}\Bigg\}, \end{aligned} \tag{7.6.20}$$

$$\begin{aligned}\frac{\partial^2 \ln L}{\partial \mu \partial \sigma} = \frac{1}{\sigma^2}\Bigg\{ & r\,\frac{f(X_{r+1:n})}{F(X_{r+1:n})} + r\,X_{r+1:n}\,\frac{f(X_{r+1:n})}{F(X_{r+1:n})}\,[\sqrt{\kappa} - \sqrt{\kappa}\,e^{X_{r+1:n}/\sqrt{\kappa}}] \\ & - s\,\frac{f(X_{n-s:n})}{1-F(X_{n-s:n})} - s\,X_{n-s:n}\,\frac{f(X_{n-s:n})}{1-F(X_{n-s:n})}\,[\sqrt{\kappa} - \sqrt{\kappa}\,e^{X_{n-s:n}/\sqrt{\kappa}}] \\ & + A\sqrt{\kappa} - \sqrt{\kappa}\sum_{i=r+1}^{n-s} e^{X_{i:n}/\sqrt{\kappa}} - \sum_{i=r+1}^{n-s} X_{i:n}\,e^{X_{i:n}/\sqrt{\kappa}}\Bigg\}, \end{aligned} \tag{7.6.21}$$

$$\begin{aligned}\frac{\partial^2 \ln L}{\partial \mu \partial \sigma} = \frac{1}{\sigma^2}\Bigg\{ & A + 2r\,X_{r+1:n}\,\frac{f(X_{r+1:n})}{F(X_{r+1:n})} + r\,X_{r+1:n}^2\,\frac{f(X_{r+1:n})}{F(X_{r+1:n})} \\ & \times \left[\sqrt{\kappa} - \sqrt{\kappa}\,e^{X_{r+1:n}/\sqrt{\kappa}} - \frac{f(X_{r+1:n})}{F(X_{r+1:n})}\right] - 2s\,X_{n-s:n}\,\frac{f(X_{n-s:n})}{1-F(X_{n-s:n})} \\ & - s\,X_{n-s:n}^2\,\frac{f(X_{n-s:n})}{1-F(X_{n-s:n})}\left[\sqrt{\kappa} - \sqrt{\kappa}\,e^{X_{n-s:n}/\sqrt{\kappa}} + \frac{f(X_{n-s:n})}{1-F(X_{n-s:n})}\right] \\ & + \sqrt{\kappa}\sum_{i=r+1}^{n-s} X_{i:n} - \sqrt{\kappa}\sum_{i=r+1}^{n-s} X_{i:n}\,e^{X_{i:n}/\sqrt{\kappa}} - \sum_{i=r+1}^{n-s} X_{i:n}^2\,e^{X_{i:n}/\sqrt{\kappa}}\Bigg\}. \end{aligned} \tag{7.6.22}$$

Let $q_1 = r/n$, $q_2 = s/n$, and $p = 1 - q_1 - q_2$. As $n \to \infty$ (with q_1 and q_2 fixed), $E(X_{r+1:n}) \to \xi_{q_1}$ where $F(\xi_{q_1}) = q_1$, $E(X_{n-s:n}) \to \xi_{q_2}$ where $F(\xi_{q_2}) = 1 - q_2$, and

$$E\left(\sum_{i=r+1}^{n-s} X_{i:n}\right) \to n\int_{\xi_{q_1}}^{\xi_{q_2}} f(t)\,dt\,, \tag{7.6.23}$$

$$E\left(\sum_{i=r+1}^{n-s} e^{X_{i:n}/\sqrt{\kappa}}\right) \to n\int_{\xi_{q_1}}^{\xi_{q_2}} e^{t/\sqrt{\kappa}} f(t)\,dt\,, \tag{7.6.24}$$

$$E\left(\sum_{i=r+1}^{n-s} X_{i:n}\,e^{X_{i:n}/\sqrt{\kappa}}\right) \to n\int_{\xi_{q_1}}^{\xi_{q_2}} t\,e^{t/\sqrt{\kappa}} f(t)\,dt\,, \tag{7.6.25}$$

$$E\left(\sum_{i=r+1}^{n-s} X_{i:n}^2\,e^{X_{i:n}/\sqrt{\kappa}}\right) \to n\int_{\xi_{q_1}}^{\xi_{q_2}} t^2\,e^{t/\sqrt{\kappa}} f(t)\,dt\,. \tag{7.6.26}$$

Hence the elements of the information matrix are [Balakrishnan and Chan (1995a,b)]

$$
\begin{aligned}
V^{11} = \lim_{n\to\infty} \left(-\frac{\sigma^2}{n} \, E \, \frac{\partial^2 \ln L}{\partial \mu^2} \right) &= -f(\xi_{q_1}) \left[\sqrt{\kappa} - \sqrt{\kappa} \, e^{\xi_{q_1}/\sqrt{\kappa}} - \frac{f(\xi_{q_1})}{q_1} \right] \\
&+ f(\xi_{q_2}) \left[\sqrt{\kappa} - \sqrt{\kappa} \, e^{\xi_{q_2}/\sqrt{\kappa}} + \frac{f(\xi_{q_2})}{q_2} \right] \\
&+ \int_{\xi_{q_1}}^{\xi_{q_2}} e^{t/\sqrt{\kappa}} f(t) \, dt \, ,
\end{aligned}
\tag{7.6.27}
$$

$$
\begin{aligned}
V^{12} = \lim_{n\to\infty} \left(-\frac{\sigma^2}{n} \, E \, \frac{\partial^2 \ln L}{\partial \mu \partial \sigma} \right) &= -p\sqrt{\kappa} - f(\xi_{q_1}) + f(\xi_{q_2}) \\
&- \xi_{q_1} f(\xi_{q_1}) \left[\sqrt{\kappa} - \sqrt{\kappa} \, e^{\xi_{q_1}/\sqrt{\kappa}} - \frac{f(\xi_{q_1})}{q_1} \right] \\
&+ \xi_{q_2} f(\xi_{q_2}) \left[\sqrt{\kappa} - \sqrt{\kappa} \, e^{\xi_{q_2}/\sqrt{\kappa}} + \frac{f(\xi_{q_2})}{q_2} \right] \\
&+ \int_{\xi_{q_1}}^{\xi_{q_2}} e^{t/\sqrt{\kappa}} f(t) \, dt + \int_{\xi_{q_1}}^{\xi_{q_2}} t \, e^{t/\sqrt{\kappa}} f(t) \, dt \, ,
\end{aligned}
\tag{7.6.28}
$$

and

$$
\begin{aligned}
V^{22} = \lim_{n\to\infty} \left(-\frac{\sigma^2}{n} \, E \, \frac{\partial^2 \ln L}{\partial \sigma^2} \right) &= -p - 2\xi_{q_1} f(\xi_{q_1}) + 2\xi_{q_2} f(\xi_{q_2}) \\
&- \xi_{q_1}^2 f(\xi_{q_1}) \left[\sqrt{\kappa} - \sqrt{\kappa} \, e^{\xi_{q_1}/\sqrt{\kappa}} - \frac{f(\xi_{q_1})}{q_1} \right] \\
&+ \xi_{q_2}^2 f(\xi_{q_2}) \left[\sqrt{\kappa} - \sqrt{\kappa} \, e^{\xi_{q_2}/\sqrt{\kappa}} + \frac{f(\xi_{q_2})}{q_2} \right] \\
&- 2\sqrt{\kappa} \int_{\xi_{q_1}}^{\xi_{q_2}} t \, f(t) \, dt + 2\sqrt{\kappa} \int_{\xi_{q_1}}^{\xi_{q_2}} t \, e^{t/\sqrt{\kappa}} f(t) \, dt \\
&+ \int_{\xi_{q_1}}^{\xi_{q_2}} t^2 \, e^{t/\sqrt{\kappa}} f(t) \, dt \, .
\end{aligned}
\tag{7.6.29}
$$

7.6.5 Asymptotic Variances and Covariances

The asymptotic variance-covariance matrix of the maximum likelihood estimators of μ and σ is given by $(\sigma^2/n)[V_{ij}]$ where $[V_{ij}] = [V^{ij}]^{-1}$ with V^{ij} being given by Equations (7.6.27) - (7.6.29). The computations of V_{ij} involve the four definite integrals given in Equations (7.6.23) - (7.6.26), which can be evaluated by using the lemma stated and proved by Chan (1993) [Chapter 4 and Appendix A], yielding [also see Balakrishnan and Chan (1995a)]

$$
\begin{aligned}
V^{11} &= -f(\xi_{q_1}) \left[\sqrt{\kappa} - \sqrt{\kappa} \, e^{\xi_{q_1}/\sqrt{\kappa}} - \frac{f(\xi_{q_1})}{q_1} \right] \\
&+ f(\xi_{q_2}) \left[\sqrt{\kappa} - \sqrt{\kappa} \, e^{\xi_{q_2}/\sqrt{\kappa}} + \frac{f(\xi_{q_2})}{q_2} \right] \\
&+ p - \frac{1}{\sqrt{\kappa}} \left[f(\xi_{q_2}) - f(\xi_{q_1}) \right] \, ,
\end{aligned}
\tag{7.6.30}
$$

$$\begin{aligned}
V^{12} = {} & \frac{p}{\sqrt{\kappa}}\,(1-\kappa\ln\kappa) \\
& - \xi_{q_1} f(\xi_{q_1})\left[\sqrt{\kappa} - \sqrt{\kappa}\, e^{\xi_{q_1}/\sqrt{\kappa}} - \frac{f(\xi_{q_1})}{q_1} - \frac{1}{\sqrt{\kappa}}\right] \\
& + \xi_{q_2} f(\xi_{q_2})\left[\sqrt{\kappa} - \sqrt{\kappa}\, e^{\xi_{q_2}/\sqrt{\kappa}} + \frac{f(\xi_{q_2})}{q_2} - \frac{1}{\sqrt{\kappa}}\right] \\
& + \sqrt{\kappa}[G_\kappa(\kappa\, e^{\xi_{q_2}/\sqrt{\kappa}}) - G_\kappa(\kappa\, e^{\xi_{q_1}/\sqrt{\kappa}})]\ , \qquad (7.6.31)
\end{aligned}$$

and

$$\begin{aligned}
V^{22} = {} & p - 2p\ln\kappa + \kappa(\ln\kappa)^2 p \\
& - \xi_{q_1}^2 f(\xi_{q_1})\left[\sqrt{\kappa} - \sqrt{\kappa}\, e^{\xi_{q_1}/\sqrt{\kappa}} - \frac{f(\xi_{q_1})}{q_1} - \frac{1}{\sqrt{\kappa}}\right] \\
& + \xi_{q_2}^2 f(\xi_{q_2})\left[\sqrt{\kappa} - \sqrt{\kappa}\, e^{\xi_{q_2}/\sqrt{\kappa}} + \frac{f(\xi_{q_2})}{q_2} - \frac{1}{\sqrt{\kappa}}\right] \\
& + (2 - 2\kappa\ln\kappa)[G_\kappa(\kappa\, e^{\xi_{q_2}/\sqrt{\kappa}}) - G_\kappa(\kappa\, e^{\xi_{q_1}/\sqrt{\kappa}})] \\
& + \kappa[H_\kappa(\kappa\, e^{\xi_{q_2}/\sqrt{\kappa}}) - H_\kappa(\kappa\, e^{\xi_{q_1}/\sqrt{\kappa}})]\ , \qquad (7.6.32)
\end{aligned}$$

where

$$G_\kappa(\lambda) = \int_{-\infty}^{\lambda} \frac{y\, e^{\kappa y - e^y}}{\Gamma(\kappa)}\, dy \qquad (7.6.33)$$

and

$$H_\kappa(\lambda) = \int_{-\infty}^{\lambda} \frac{y^2 e^{\kappa y - e^y}}{\Gamma(\kappa)}\, dy\ . \qquad (7.6.34)$$

Hence, in order to obtain the values of V^{ij}, it is necessary to evaluate only two definite integrals. Chan used these values to compute the values of V^{ij} given in Equations (7.6.30)-(7.6.32) and then the asymptotic variance-covariance matrix of the MLEs $\hat{\mu}$ and $\hat{\sigma}$ for $q_1 = 0.0(0.1)0.2$, $q_2 = 0.0(0.1)0.5$ and $\kappa = 0.5(0.5)2.0(1.0)4.0(2.0)12.0$. His table of asymptotic values of V_{11}, V_{12} and V_{22} is reproduced as our Table F7; see also Balakrishnan and Chan (1995a). Comparison of these values with Monte Carlo results discussed in the next section and tabulated in Table G6 shows that the asymptotic formulae provide close approximations to the true values even for a sample of size 40.

7.6.6 Iterative Estimation Procedure and Monte Carlo Study

As pointed out in subsection 7.6.3, the likelihood equations $\partial \ln L/\partial\mu = 0$ and $\partial\ln L/\partial\sigma = 0$ cannot be solved analytically, so Chan used the Newton-Raphson iterative method to solve numerically for $\hat{\mu}$ and $\hat{\sigma}$. In order to study the performance of these estimators, he generated 3001 pseudo-random samples of size $n = 20$, 25 and 40 from the standardized log-Gamma population with $\kappa = 0.5(0.5)2.0(1.0)4.0(2.0)12.0$ and employed the Newton-Raphson procedure to solve the likelihood equations. He determined the biases, variances and covariances of the maximum likelihood estimates $\hat{\mu}$ and $\hat{\sigma}$ for $q_1 = r/n = 0.0(0.1)0.2$ and $q_2 = s/n = 0.0(0.1)0.5$. He tabulated the values of (1) Bias$(\hat{\mu})/\sigma$, (2) Bias$(\hat{\sigma})/\sigma$, (3) Var$(\hat{\mu})/\sigma^2$, (4) Var$(\hat{\sigma})/\sigma^2$ and (5) Cov$(\hat{\mu},\hat{\sigma})/\sigma^2$ determined through this Monte Carlo process for the values of n, κ, q_1 and q_2 mentioned above; see also Balakrishnan and Chan (1995a). Their results for $n = 20$ and 40 are reproduced as our Table G6.

7.6.7 Numerical Example

Lieblein and Zelen (1956) gave the following data on the logarithms of the number of revolutions (in millions) to failure for each of a group of 23 ball bearings in a fatigue test:

> 2.884, 3.365, 3.497, 3.726, 3.741, 3.820, 3.881, 3.948, 3.950, 3.991, 4.017, 4.217, 4.229, 4.229, 4.232, 4.432, 4.534, 4.591, 4.655, 4.662, 4.851, 4.852, 5.156.

Lieblein and Zelen assumed a two-parameter Weibull distribution for the original data and hence an extreme-value (log-Weibull) distribution for the log-lifetime data given above. Chan (1993), following Lawless (1980, 1982), assumed a generalized Gamma distribution for the original data and consequently a log-Gamma distribution for the given log-lifetime data. Chan found estimates of several types for the parameters μ and σ, including the BLUEs μ^* and σ^* and the MLEs $\hat{\mu}$ and $\hat{\sigma}$. Since the shape parameter κ is not actually known as assumed, Chan studied the effect of assuming various values for κ, and found it to be small over a broad range of assumed values of κ, as illustrated in the following table presenting the values of μ^*, σ^*, $\hat{\mu}$ and $\hat{\sigma}$ (with their standard errors in parentheses) for $\kappa = 3.0$ and 8.0:

κ	μ^* (s.e.(μ^*))	σ^* (s.e.(σ^*))	$\hat{\mu}$ (s.e.($\hat{\mu}$))	$\hat{\sigma}$ (s.e.($\hat{\sigma}$))
3.0	4.306 (0.111)	0.521 (0.083)	4.300 (0.106)	0.498 (0.075)
8.0	4.246 (0.112)	0.533 (0.082)	4.242 (0.107)	0.518 (0.076)

The BLUEs μ^* and σ^* in the reparametrized model were found by first finding the BLUEs ν^* and τ^* in the original model (using the method outlined in Chapter 3, subsection 3.3.12) and then using the relations $\mu = \nu + \tau \log \kappa$ and $\sigma = \tau/\sqrt{\kappa}$. In addition to the above results for the complete sample case ($n = 23$, $r = 0$, $s = 0$), Chan obtained the following results for the censored sample case ($n = 23$, $r = 0$, $s = 2$) which assumed termination of the experiment at the time of the 21st failure:

κ	μ^* (s.e.(μ^*))	σ^* (s.e.(σ^*))	$\hat{\mu}$ (s.e.($\hat{\mu}$))	$\hat{\sigma}$ (s.e.($\hat{\sigma}$))
3.0	4.302 (0.113)	0.525 (0.091)	4.302 (0.108)	0.503 (0.083)
8.0	4.187 (0.116)	0.544 (0.091)	4.247 (0.110)	0.518 (0.084)

7.6.8 References

1. Abramowitz, M. and Stegun, I.A. (Eds.) (1965). *Handbook of Mathematical Functions with Formulas, Graphs, and Mathematical Tables.* Dover Publications, New York.
2. Balakrishnan, N. and Chan, Ping-Shing (1995a). Maximum likelihood estimation for the log-Gamma distribution under Type-II censored samples and associated inference. In *Recent Advances in Life-Testing and Reliability* (Ed., N. Balakrishnan), pp. 409-437, CRC Press, Boca Raton, Florida.
3. Balakrishnan, N. and Chan, Ping-Shing (1995b). Maximum likelihood estimation for the three-parameter log-Gamma distribution. In *Recent Advances in Life-Testing and Reliability* (Ed., N. Balakrishnan), pp. 439-453, CRC Press, Boca Raton, Florida.
4. Chan, Ping-Shing (1993). A Statistical Study of the Log-Gamma Distribution. *Ph.D. thesis*, McMaster University, Hamilton, Ontario.
5. Dumonceaux, R. and Antle, C.E. (1973). Discrimination between the lognormal and the Weibull distributions. *Technometrics* **15**, 923-926.
6. Farewell, V.T. and Prentice, R.L. (1977). A study of distributional shape in life testing. *Technometrics* **19**, 69-75.

7. Hager, H.W. and Bain, L.J. (1970). Inferential procedures for the generalized Gamma distribution. *Journal of the American Statistical Association* **65**, 1601-1609.

8. Harter, H. Leon (1967). Maximum-likelihood estimation of the parameters of a four-parameter generalized Gamma population from complete and censored samples. *Technometrics* **9**, 159-165.

9. Lawless, J.F. (1980). Inference in the generalized Gamma and log-Gamma distributions. *Technometrics* **22**, 67-82.

10. Lawless, J.F. (1982). *Statistical Models & Methods for Lifetime Data.* John Wiley & Sons, New York.

11. Lieblein, J. and Zelen, M. (1956). Statistical investigation of the fatigue life of deep groove ball bearings. *Journal of Research of the National Bureau of Standards* **57**, 273-316.

12. Prentice, R.L. (1974). A log-Gamma model and its maximum likelihood estimation. *Biometrika* **61**, 539-544.

13. Stacy, E.W. (1962). A generalization of the Gamma distribution. *Annals of Mathematical Statistics* **33**, 1187-1192.

14. Stacy, E.W. and Mihram, G.A. (1965). Parameter estimation for a generalized Gamma distribution. *Technometrics* **7**, 349-358.

7.7 Maximum-Likelihood Estimation of Three Parameters of Generalized Extreme-Value Population

7.7.1 Introduction

The cumulative distribution function of the generalized extreme-value population is

$$F(x;\xi,\theta,\gamma) = \begin{cases} e^{-[1-\gamma\{(x-\xi)/\theta\}]^{1/\gamma}}, & -\infty < x \le \xi + \theta/\gamma \text{ if } \gamma > 0, \\ & \xi + \theta/\gamma \le x < \infty \text{ if } \gamma < 0, \\ e^{-e^{-(x-\xi)/\theta}}, & -\infty < x < \infty \text{ if } \gamma = 0. \end{cases} \tag{7.7.1}$$

This distribution, referred to as the von Mises type extreme-value distribution or the von Mises-Jenkinson type distribution, includes the Type I extreme-value distribution when $\gamma = 0$, the Type II distribution when $\gamma > 0$, and the Type III distribution when $\gamma < 0$. The density function corresponding to (7.7.1) is

$$f(x;\xi,\theta,\gamma) = \begin{cases} e^{-[1-\gamma\{(x-\xi)/\theta\}]^{1/\gamma}} \cdot \frac{1}{\theta}\,[1-\gamma\{(x-\xi)/\theta\}]^{\frac{1}{\gamma}-1}, & \\ & -\infty < x \le \xi + \theta/\gamma \text{ if } \gamma > 0, \\ & \xi + \theta/\gamma \le x < \infty \text{ if } \gamma < 0, \\ e^{-e^{-(x-\xi)/\theta}} \cdot \frac{1}{\theta}\, e^{-(x-\xi)/\theta}, & -\infty < x < \infty \text{ if } \gamma = 0. \end{cases} \tag{7.7.2}$$

Jenkinson (1955) and Blackman and Graff (1978) used this distribution to analyze annual maximum or minimum values of certain meteorological events. Maritz and Munro (1967) studied order statistics from this distribution and presented tables of means of order statistics for sample sizes 5 to 10 and for some selected choices of the shape parameter γ.

The maximum likelihood estimation of the parameters ξ, θ and γ has been discussed by several authors, including Jenkinson (1969), Prescott and Walden (1980, 1983), Hosking (1985), and Macleod (1989). Hosking (1985) presented a FORTRAN subroutine MLEGEV for the determination of the maximum likelihood estimates of ξ, θ and γ by the use of Newton-Raphson method and of the variance-covariance matrix of these estimates. Macleod (1989) observed that if the initial estimate for the shape parameter γ is 0, then Hosking's algorithm will attempt to calculate 1.0/0.0; to avoid this, Macleod suggested an adjustment to be made in Hosking's algorithm. Hosking, Wallis, and Wood (1985) discussed the method of probability-weighted moments, and Wang (1990) extended this method for Type II censored samples. Chen and Balakrishnan (1995) recently pointed out that this method of estimation may sometimes yield inadmissible estimates.

7.7.2 The Likelihood Equations

Suppose from a sample of size n from the generalized extreme-value distribution in (7.7.1) the smallest r and the largest $n-m$ observations have been censored. Then, the logarithm of the likelihood function $L = L(\xi,\theta,\gamma)$, of the remaining $m-r$ order statistics $X_{r+1}, X_{r+2}, \ldots, X_m$ is given by

$$L = \ln n! - \ln r! - \ln(n-m)! - (m-r)\ln\theta - (1-\gamma)\sum_{i=r+1}^{m} y_i$$

$$- \sum_{i=r+1}^{m} e^{-y_i} - r\,e^{-y_{r+1}} + (n-m)\ln\{1-\exp(e^{-y_m})\}, \tag{7.7.3}$$

where

$$y_i = -\frac{1}{\gamma}\ln\{1-\gamma(x-\xi)/\theta\} \ . \tag{7.7.4}$$

The maximum likelihood equations are obtained by equating to zero the partial derivatives of L with respect to ξ, θ and γ. These are [see Prescott and Walden (1983)]:

$$\left.\begin{aligned} \tfrac{\partial L}{\partial \theta} &= -\tfrac{1}{\theta\gamma}\,(P^*+Q^*) = 0 \\ \tfrac{\partial L}{\partial \xi} &= -\tfrac{Q^*}{\theta} = 0 \\ \tfrac{\partial L}{\partial \gamma} &= -\tfrac{1}{\gamma}\left\{R^* - \tfrac{1}{\gamma}\,(P^*+Q^*)\right\} = 0 \end{aligned}\right\} \tag{7.7.5}$$

where

$$P^* = P + P_l + P_r \ , \ Q^* = Q + Q_l + Q_r \ , \ R^* = R + R_l + R_r$$

and

$$\left.\begin{aligned} P &= (m-r) - \Sigma e^{-y_i} \\ Q &= \Sigma e^{-y_i+\gamma y_i} - (1-\gamma)\Sigma e^{\gamma y_i} \\ R &= (m-r) - \Sigma y_i + \Sigma y_i e^{-y_i} \end{aligned}\right\} \tag{7.7.6}$$

$$\left.\begin{aligned} P_l &= -r\,e^{-y_{r+1}} \\ Q_l &= -e^{\gamma y_{r+1}}\,P_l \\ R_l &= -y_{r+1}\,P_l \end{aligned}\right\} \tag{7.7.7}$$

$$\left.\begin{aligned} P_r &= (n-m)e^{-y_m}/\{\exp(e^{-y_m})-1\} \\ Q_r &= -e^{\gamma y_m}\,P_r \\ R_r &= -y_m\,P_r \end{aligned}\right\} \tag{7.7.8}$$

The summations in (7.7.6) are over the uncensored observations. For complete samples P_l, P_r, etc. are zero so that the equations reduce to (7.7.5) and (7.7.6) and are the same as those discussed by Jenkinson (1969). Equations (7.7.7) and (7.7.8) give the adjustments necessary to allow for left, right or double censoring. Left and right censoring will be indicated by terms with subscripts l and r respectively.

The three equations in (7.7.5) must be solved iteratively after the appropriate adjustments for censoring have been added to P, Q, and R in Equation (7.7.6) to give P^*, Q^*, and R^* for substitution in (7.7.5).

A suitable iterative procedure involving the matrix of negatives of the second derivatives of the logarithm of the likelihood function, proposed by Prescott and Walden (1983), is presented in the next subsection.

7.7.3 The Iterative Method

Suppose that the maximum likelihood estimate of $\boldsymbol{\alpha} = (\theta,\xi,\gamma)'$ is $\hat{\boldsymbol{\alpha}} = (\hat\theta,\hat\xi,\hat\gamma)'$ and that $\boldsymbol{\alpha}^{(i)}$ is the ith iterative solution of Eq. (7.7.5). The elements of

$$-\text{grad } L(\boldsymbol{\alpha}) = \left(-\frac{\partial L}{\partial \theta}\ ,\ -\frac{\partial L}{\partial \xi}\ ,\ -\frac{\partial L}{\partial \gamma}\right)'$$

at the ith iteration may be expanded in a Taylor series about $\hat{\boldsymbol{\alpha}}$ to give, for $i \geq 0$, the iterative formula

$$\boldsymbol{\alpha}^{(i+1)} = \boldsymbol{\alpha}^{(i)} + V(\hat{\boldsymbol{\alpha}})^{-1}\ \text{grad } L(\boldsymbol{\alpha}^{(i)}), \tag{7.7.9}$$

where

$$V(\hat{\boldsymbol{\alpha}}) = \begin{pmatrix} -\frac{\partial^2 L}{\partial \theta^2} & -\frac{\partial^2 L}{\partial \theta \partial \xi} & -\frac{\partial^2 L}{\partial \theta \partial \gamma} \\ -\frac{\partial^2 L}{\partial \theta \partial \xi} & -\frac{\partial^2 L}{\partial \xi^2} & -\frac{\partial^2 L}{\partial \xi \partial \gamma} \\ -\frac{\partial^2 L}{\partial \theta \partial \gamma} & -\frac{\partial^2 L}{\partial \xi \partial \gamma} & -\frac{\partial^2 L}{\partial \gamma^2} \end{pmatrix}$$

is the matrix of negatives of the second derivatives of the logarithm of the likelihood function evaluated at $\hat{\boldsymbol{\alpha}}$. Since the maximum likelihood estimate $\hat{\boldsymbol{\alpha}}$ is unknown, the inverse of the matrix evaluated at $\boldsymbol{\alpha}^{(i)}$ may be used as an approximation.

Prescott and Walden (1983) have pointed out that this recursive method may produce a solution which is a local maximum or sometimes the method may simply fail to converge. However, from the results of a number of simulations, they have observed that these events occur rarely.

7.7.4 Calculation of $V(\boldsymbol{\alpha}^{(i)})$

The six second derivatives in the symmetric matrix $V(\hat{\boldsymbol{\alpha}})$, obtained by differentiating $\partial L/\partial\theta$ etc. in Eqs. (7.7.5) with respect to θ, ξ, and γ, may be expressed in the form

$$\left.\begin{aligned} -\frac{\partial^2 L}{\partial \theta^2} &= \frac{1}{\theta}\,\frac{\partial L}{\partial \theta} + \frac{1}{\theta\gamma}\left[\frac{\partial P^*}{\partial \theta} + \frac{\partial Q^*}{\partial \theta}\right], -\frac{\partial^2 L}{\partial \xi^2} = \frac{1}{\theta}\,\frac{\partial Q^*}{\partial \xi}, \\ -\frac{\partial^2 L}{\partial \theta \partial \gamma} &= \frac{1}{\gamma}\left[\frac{\partial R^*}{\partial \theta} + \frac{\partial L}{\partial \theta} + \theta\,\frac{\partial^2 L}{\partial \theta^2}\right], -\frac{\partial^2 L}{\partial \xi \partial \theta} = \frac{1}{\theta\gamma}\left[\frac{\partial P^*}{\partial \xi} + \frac{\partial Q^*}{\partial \xi}\right], \\ -\frac{\partial^2 L}{\partial \gamma^2} &= \frac{1}{\gamma}\left[\frac{\partial R^*}{\partial \gamma} + \frac{\partial L}{\partial \gamma} + \theta\,\frac{\partial^2 L}{\partial \theta \partial \gamma}\right], -\frac{\partial^2 L}{\partial \xi \partial \gamma} = \frac{1}{\gamma}\left[\frac{\partial R^*}{\partial \xi} + \theta\,\frac{\partial^2 L}{\partial \xi \partial \theta}\right], \end{aligned}\right\} \quad (7.7.10)$$

where each partial derivative of P^*, Q^*, and R^* is the sum of three partial derivatives, e.g.

$$\frac{\partial P^*}{\partial \xi} = \frac{\partial P}{\partial \xi} + \frac{\partial P_l}{\partial \xi} + \frac{\partial P_r}{\partial \xi}\,.$$

The derivatives based on the uncensored observations are given by

$$\frac{\partial P}{\partial \xi} = -\frac{1}{\theta}\,\Sigma e^{-y_i+\gamma y_i}, \quad \frac{\partial P}{\partial \theta} = \frac{1}{\theta\gamma}\,\Sigma e^{-y_i} + \frac{1}{\gamma}\,\frac{\partial P}{\partial \xi},$$

$$\frac{\partial Q}{\partial \xi} = \frac{(1-\gamma)}{\theta}\left[\Sigma e^{-y_i+2\gamma y_i} + \gamma \Sigma e^{2\gamma y_i}\right],$$

$$\frac{\partial Q}{\partial \theta} = \frac{1}{\gamma}\,\frac{\partial Q}{\partial \xi} - \frac{(1-\gamma)}{\theta\gamma}\left[\Sigma e^{-y_i+\gamma y_i} + \gamma \Sigma e^{\gamma y_i}\right],$$

$$\frac{\partial R}{\partial \theta} = -\frac{1}{\theta\gamma}[(m-r) - \Sigma e^{\gamma y_i} + \Sigma y_i e^{-y_i} - \Sigma e^{-y_i} + \Sigma e^{-y_i+\gamma y_i} - \Sigma y_i e^{-y_i+\gamma y_i}],$$

$$\frac{\partial R}{\partial \gamma} = \frac{1}{\gamma}\left[\Sigma y_i - \Sigma y_i e^{-y_i} + \Sigma y_i^2\, e^{-y_i} - \theta\,\frac{\partial R}{\partial \theta}\right],$$

$$\frac{\partial R}{\partial \xi} = \frac{1}{\theta}\left[\Sigma e^{\gamma y_i}(1 - e^{-y_i} + y_i\, e^{-y_i})\right] \quad (7.7.11)$$

where the summations are over $i = r+1, \ldots, m$.

The derivatives of P_l, Q_l, and R_l which must be added if there is left hand censoring are given by

$$\frac{\partial P_l}{\partial \xi} = -\frac{Q_l}{\theta}, \quad \frac{\partial P_l}{\partial \theta} = -\frac{(P_l + Q_l)}{\theta \gamma}, \quad \frac{\partial Q_l}{\partial \xi} = \frac{(1-\gamma) Q_l e^{\gamma y_{r+1}}}{\theta},$$

$$\frac{\partial Q_l}{\partial \theta} = \frac{Q_l}{\theta} - e^{\gamma y_{r+1}} \left(\frac{Q_l}{\theta} + \frac{\partial P_l}{\partial \theta} \right), \quad \frac{\partial R_l}{\partial \theta} = \frac{\partial P_l}{\partial \theta} (1 - y_{r+1}),$$

$$\frac{\partial R_l}{\partial \gamma} = \frac{(R_l + P_l)}{\gamma^2} (1 + \gamma y_{r+1} - e^{\gamma y_{r+1}}), \quad \frac{\partial R_l}{\partial \xi} = \frac{Q_l}{\theta} (y_{r+1} - 1) . \qquad (7.7.12)$$

For samples subjected to right hand censoring adjustments in terms of the derivatives of P_r, Q_r, and R_r are required. These are given by the following equations:

$$\frac{\partial P_r}{\partial \xi} = -\frac{Q_r}{\theta} \left[1 - e^{-y_m} / \{ 1 - \exp(-e^{-y_m}) \} \right],$$

$$\frac{\partial Q_r}{\partial \xi} = -e^{\gamma y_m} \left[\frac{\partial P_r}{\partial \xi} + \frac{\gamma Q_r}{\theta} \right],$$

$$\frac{\partial P_r}{\partial \theta} = -\frac{(P_r + Q_r)}{\theta \gamma} \left[1 - e^{-y_m} / \{ 1 - \exp(-e^{-y_m}) \} \right],$$

$$\frac{\partial Q_r}{\partial \theta} = -e^{\gamma y_m} \left[\frac{\partial P_r}{\partial \theta} + \frac{(P_r + Q_r)}{\theta} \right],$$

$$\frac{\partial R_r}{\partial \theta} = -\left[y_m \frac{\partial P_r}{\partial \theta} + \frac{(P_r + Q_r)}{\theta \gamma} \right],$$

$$\frac{\partial R_r}{\partial \gamma} = \left\{ P_r + R_r \left[1 - e^{-y_m} / \{ 1 - \exp(-e^{-y_m}) \} \right] \right\} \left\{ \frac{y_m}{\gamma} + \frac{1 - e^{\gamma y_m}}{\gamma^2} \right\},$$

and

$$\frac{\partial R_r}{\partial \xi} = -\left[y_m \frac{\partial P_r}{\partial \theta} + \frac{Q_r}{\theta} \right] \qquad (7.7.13)$$

The elements of the observed information matrix $V(\boldsymbol{\alpha})$ are obtained by adding the appropriate expressions in (7.7.12) and (7.7.13) to those in (7.7.11), depending on whether the sample is left, right or doubly censored, and then substituting in (7.7.10). The six second derivatives evaluated at $\boldsymbol{\alpha}^{(i)}$ are then used to form the 3×3 symmetric matrix $V(\boldsymbol{\alpha}^{(i)})$ for use in the iterative procedure.

The calculations involved are straightforward computationally and reasonably quick even though the matrix $V(\boldsymbol{\alpha}^{(i)})$ is recomputed at each step of the iteration. In the experience of Prescott and Walden (1983), convergence to $|\text{grad } L| < 10^{-3} I$ occurs with fewer than five iterations most of the time. Atypical sample configurations may however lead to nonconvergence or convergence to a local maximum; however, for such samples it is likely that further investigation of the data would be required before any extreme-value distribution is fitted.

7.7.5 Variances and Covariances of the Estimators

The asymptotic variance-covariance matrix of the maximum likelihood estimators $\hat{\boldsymbol{\alpha}}$ of $\boldsymbol{\alpha} = (\theta, \xi, \gamma)'$ is obtained by inverting the Fisher information matrix M whose elements are the negatives of the expected values of the second order partial derivatives of the logarithm of the likelihood function. For complete samples from the generalized extreme-value distribution explicit expressions for M were given by Prescott and Walden (1980) but for censored samples these expectations do not exist in the usual sense. Let $q_1 = r/n$, $q_2 = (n-m)/n$ and $p = 1 - q_1 - q_2 = (m-r)/n$ so that q_1 and q_2 are the proportions of the observations censored from each end of the sample.

As $n \to \infty$, for q_1 and q_2 fixed, y_{r+1} converges in probability to $\hat{y}_{r+1}$ where $G(\hat{y}_{r+1}) = \int_{-\infty}^{\hat{y}_{r+1}} g(t)\,dt = q_1$ and $g(y) = e^{-y}\exp(-e^{-y})$ for $-\infty < y < \infty$.

Also y_m converges in probability to $\hat{y}_m$ where $1 - G(\hat{y}_m) = \int_{\hat{y}_m}^{\infty} g(t)\,dt = q_2$. If $E[\cdots]$ now denotes a conditional expectation given y_{r+1} and y_m, the elements of the information matrix M, multiplied by $1/n$, may be obtained as, for example,

$$\begin{aligned}
\lim \text{pr } E[-(1/n)\partial^2 L/\partial\theta^2] = (\theta\gamma)^{-2}\{&-p\gamma + (1+\gamma)[\Gamma(2; e^{-\hat{y}_{r+1}}) - \Gamma(2; e^{-\hat{y}_m})] \\
&- 2[\Gamma(2-\gamma; e^{-\hat{y}_{r+1}}) - \Gamma(2-\gamma; e^{-\hat{y}_m})] \\
&- (\gamma-1)[\Gamma(2-2\gamma; e^{-\hat{y}_{r+1}}) - \Gamma(2-2\gamma; e^{-\hat{y}_m})] \\
&+ \gamma(1-\gamma)[\Gamma(1-2\gamma; e^{-\hat{y}_{r+1}}) - \Gamma(1-2\gamma; e^{-\hat{y}_m})] \\
&+ q_1 e^{-\hat{y}_{r+1}}(1 - e^{\gamma\hat{y}_{r+1}})[(1+\gamma) + (\gamma-1)e^{\gamma\hat{y}_{r+1}}] \\
&+ q_2(1 - e^{\gamma\hat{y}_m})e^{-\hat{y}_m}\exp(e^{-\hat{y}_m})\} \qquad (7.7.14)
\end{aligned}$$

where $\Gamma(v; z) = \int_0^z t^{v-1}\, e^{-t}\, dt$ is the incomplete Gamma function.

The expressions for the remaining elements of M may similarly be derived.

To examine how close these asymptotic formulas are to the finite-sample results, Prescott and Walden (1983) carried out a simulation study involving 2000 sets of samples of size 100 from distributions with specified values of the parameters θ, ξ and γ. Six values of γ from -0.3 to 0.3 were selected to represent the range of shapes of extreme-value distributions. For each sample of 100 random deviates the maximum likelihood estimates of the three parameters were obtained from the iterative procedure described above. The inverses of the observed information matrix and the Fisher information matrix, i.e. V^{-1} and M^{-1} were evaluated at the maximum likelihood estimates for each sample and recorded.

The variances and covariances of these 2000 sets of estimates $(\hat{\theta}, \hat{\xi}, \hat{\gamma})$ were then computed, together with the theoretical asymptotic variance-covariance matrix given by the inverse of the Fisher information matrix evaluated at the specified parameter values. The averages of $V^{-1}(\hat{\boldsymbol{\alpha}})$ and $M^{-1}(\hat{\boldsymbol{\alpha}})$ over the 2000 simulations were also recorded. This procedure resulted in four matrices for each distribution considered. The true variance-covariance matrix, represented by the matrix obtained directly from the 2000 simulations, may be compared with the matrices produced by averaging $V^{-1}(\hat{\boldsymbol{\alpha}})$ and $M^{-1}(\hat{\boldsymbol{\alpha}})$ and with the theoretical asymptotic variance-covariance matrix.

As expected, the asymptotic variance-covariance matrix tended to underestimate the "true" values. In almost all cases considered the variances and covariances obtained from the Fisher information matrix also tended to underestimate the true values but were closely distributed about the theoretical asymptotic values. On the other hand the variances and the covariances obtained from the observed information matrix were closely distributed about the simulated values. As an illustration, Table 7.3, taken from Prescott and Walden (1983), shows the estimates of the variance of $\hat{\gamma}$ from six distributions considered. Similar comparisons could be made for the other variances and covariances. To obtain an indication of the variability in these results the entire simulation was repeated; thus each entry in Table 7.3 is based on 2000 samples. A comparison of these results confirms that it is preferable to use the observed information matrix to estimate the variances and covariances of the maximum likelihood estimators. This is convenient since this matrix is determined as part of the estimation procedure.

Because of the complexity of the matrix M for censored samples these simulations were based on complete samples but it is anticipated that the same conclusions would hold for the censored sample case.

It is suggested therefore that in any particular application the matrix $V^{-1}(\hat{\boldsymbol{\alpha}})$ should be used to provide an estimate of the variance-covariance matrix of the estimators $\hat{\boldsymbol{\alpha}} = (\hat{\theta}, \hat{\xi}, \hat{\gamma})'$.

TABLE 7.3 - *Comparison of estimates of variance of* $\hat{\gamma}$ *for six distributions with* $\theta = 0.2$, $\xi = 2.5$ *and* $\gamma = -0.3$ *to 0.3. All variances should be multiplied by* 10^{-5}

γ	Simulated variance	Average of $V^{-1}(\hat{\boldsymbol{\alpha}})$	Average of $M^{-1}(\hat{\boldsymbol{\alpha}})$	Asymptotic variance
0.3	466	428	260	265
	432	435	260	
0.2	461	460	325	330
	439	466	326	
0.1	516	498	397	401
	530	510	396	
−0.1	638	651	552	558
	663	646	557	
−0.2	729	734	648	644
	786	737	641	
−0.3[a]	635	816	723	737
	653	815	724	

[a] In these simulations with $\gamma = -0.3$ a higher proportion of samples resulted in failure of the iterative procedure to converge.

7.7.6 References

1. Blackman, D. and Graff, J. (1978). The analysis of annual extreme sea levels at certain ports in southern England. *Proceedings of the Institute of Civil Engineers*, Part 2, **65**, 339-357.
2. Chen, Gemai and Balakrishnan, N. (1995). The infeasibility of probability weighted moments estimation of some generalized distributions. In *Recent Advances in Life-Testing and Reliability* (Ed., N. Balakrishnan), pp. 565-573, CRC Press, Boca Raton, Florida.

3. Hosking, J.R.M. (1985). Maximum-likelihood estimation of the parameters of the generalized extreme-value distribution. *Applied Statistics* **34**, 301-310.

4. Hosking, J.R.M., Wallis, J.R., and Wood, E.F. (1985). Estimation of the generalized extreme-value distribution by the method of probability-weighted moments. *Technometrics* **27**, 251-261.

5. Jenkinson, A.F. (1955). The frequency distribution of the annual maximum (or minimum) values of meteorological elements. *Quarterly Journal of the Royal Meteorological Society* **81**, 158-171.

6. Jenkinson, A.F. (1969). Statistics of Extremes. *Technical Note No. 98*, World Meteorological Office, Geneva.

7. Macleod, A.J. (1989). Comment on "Maximum-likelihood estimation of the parameters of the generalized extreme-value distribution." *Applied Statistics* **38**, 198-199.

8. Maritz, J.S. and Munro, A.H. (1967). On the use of the generalized extreme-value distribution in estimating extreme percentiles. *Biometrics* **23**, 79-103.

9. Prescott, P. and Walden, A.T. (1980). Maximum likelihood estimation of the parameters of the generalized extreme-value distribution. *Biometrika* **67**, 723-724.

10. Prescott, P. and Walden, A.T. (1983). Maximum likelihood estimation of the parameters of the three-parameter generalized extreme-value distribution from censored samples. *Journal of Statistical Computation and Simulation* **16**, 241-250.

11. Wang, Q.J. (1990). Estimation of the GEV distribution from censored samples by method of partial probability weighted moments. *Journal of Hydrology* **120**, 103-114.

7.8 Remarks on Interval Estimation of Parameters of Above Populations

7.8.1 Confidence Bounds for Individual Parameters

If estimation is regular, so that the maximum-likelihood estimator of a single parameter is asymptotically normally distributed, and if the sample size is large enough so that the asymptotic theory is valid, the asymptotic variances tabulated in Appendix F for various populations can be used to obtain confidence bounds for individual parameters. A numerical example for a Type I asymptotic distribution of smallest values has already been given in subsection 7.5.6. The same procedure could be used for normal, two-parameter lognormal, two-parameter Weibull, two-parameter Gamma, two-parameter log-Gamma, and logistic populations, for which estimation is regular and Monte Carlo studies for small to moderate samples have shown results in close agreement with asymptotic theory. When one of the parameters being estimated is a location parameter of the threshold type, as in the case of three-parameter lognormal, three-parameter Weibull, three-parameter Gamma, and three-parameter generalized extreme-value populations, estimation is non-regular except when certain restrictive conditions are satisfied, and even under these conditions Monte Carlo studies have shown that asymptotic theory is valid only for extremely large samples, so this procedure is not recommended for these populations.

The asymptotic variances tabulated in Appendix F depend, in most cases, on a power of the scale parameter; in some cases, they also depend, in a more complicated way, on the values of one or more other parameters. If the values of the parameters on which they depend are known, these values can be used, along with the tabulated asymptotic variances, to obtain confidence bounds on each of the other parameters which are "exact" in the sense that the only approximation involved is that implicit in the assumption that asymptotic theory is valid for samples of the size under consideration. Otherwise, the best that one can do is to substitute sample estimates for unknown population parameters and proceed to obtain approximate confidence bounds, as we have done in the example of subsection 7.5.6.

7.8.2 Confidence Ellipsoids for Sets of Parameters

Let p be the number of unknown parameters of the population under consideration. Let V $(p \times p)$ be the asymptotic variance-covariance matrix, and let the maximum-likelihood estimators of the parameters be represented by the p components of the vector Y. Then, if estimation is regular, the asymptotic joint distribution of the estimators is p-variate normal and is given by

$$f(Y) = \exp\left(-\frac{1}{2}\, Y'V^{-1}Y\right) \Big/ (2\pi)^{p/2}|V|^{1/2}\,. \tag{7.8.1}$$

If the sample size is large enough so that the asymptotic theory is valid, we can use Equation (7.8.1), together with the tabulated asymptotic variances and covariances, to obtain p-dimensional confidence ellipsoids for the set of p parameters.

Appendix A

TABLES BASED ON QUASI-RANGES OF SAMPLES FROM A NORMAL POPULATION

SOURCES OF TABLES

Tables A1-A5 WADC TR 58-200 (Harter)
Tables A6-A8 ARL 31, Part II(Harter)

Table A1

EXPECTED VALUES OF QUASI-RANGES

Expected Value of the rth Quasi-Range for Samples of n from N(0, 1)

n	r = 0	r = 1	r = 2	r = 3	r = 4	r = 5	r = 6	r = 7	r = 8
2	1.128379								
3	1.692569								
4	2.058751	0.594023							
5	2.325929	0.990038							
6	2.534413	1.283510	0.403094						
7	2.704357	1.514749	0.705414						
8	2.847201	1.704450	0.945645	0.305029					
9	2.970026	1.864595	1.143942	0.549052					
10	3.077505	2.002714	1.312118	0.751529	0.245336				
11	3.172873	2.123833	1.457679	0.923957	0.449782				
12	3.258455	2.231464	1.585676	1.073686	0.624498	0.205179			
13	3.335980	2.328154	1.699669	1.205700	0.776654	0.381047			
14	3.406763	2.415805	1.802253	1.323527	0.911132	0.534594	0.176318		
15	3.471827	2.495870	1.895378	1.429755	1.031402	0.670592	0.330597		
16	3.531983	2.569488	1.980542	1.526333	1.140019	0.792446	0.467503	0.154575	
17	3.587884	2.637564	2.058922	1.614770	1.238915	0.902667	0.590373	0.291975	
18	3.640064	2.700827	2.131456	1.696250	1.329589	1.003163	0.701674	0.415471	0.137605
19	3.688963	2.759877	2.198906	1.771724	1.413223	1.095415	0.803285	0.527486	0.261450
20	3.734950	2.815208	2.261896	1.841963	1.490766	1.180594	0.896664	0.629866	0.373915
21	3.778336	2.867236	2.320945	1.907604	1.562992	1.259644	0.982970	0.724051	0.476816
22	3.819385	2.916311	2.376488	1.969174	1.630538	1.333334	1.063136	0.811183	0.571570
23	3.858323	2.962731	2.428893	2.027118	1.693938	1.402301	1.137928	0.892185	0.659305
24	3.895348	3.006755	2.478476	2.081814	1.753638	1.467076	1.207975	0.967812	0.740931
25	3.930629	3.048602	2.525506	2.133585	1.810021	1.528108	1.273807	1.038691	0.817195

Expected Value of the rth Quasi-Range for Samples of n from N(0, 1)

n	r = 0	r = 1	r = 2	r = 3	r = 4	r = 5	r = 6	r = 7	r = 8
26	3.964316	3.088468	2.570219	2.182707	1.863411	1.585779	1.335872	1.105347	0.888717
27	3.996539	3.126520	2.612819	2.229423	1.914092	1.640416	1.394549	1.168222	0.956017
28	4.027414	3.162907	2.653484	2.273942	1.962307	1.692302	1.450167	1.227697	1.019535
29	4.057044	3.197761	2.692372	2.316449	2.008271	1.741683	1.503007	1.284097	1.079647
30	4.085522	3.231200	2.729624	2.357108	2.052170	1.788775	1.553316	1.337704	1.136678
31	4.112928	3.263326	2.765363	2.396061	2.094171	1.833766	1.601311	1.388764	1.190907
32	4.139338	3.294235	2.799700	2.433439	2.134420	1.876825	1.647180	1.437492	1.242580
33	4.164817	3.324009	2.832734	2.469355	2.173049	1.918099	1.691092	1.484078	1.291910
34	4.189425	3.352725	2.864556	2.503912	2.210174	1.957721	1.733196	1.528690	1.339088
35	4.213219	3.380451	2.895245	2.537204	2.245901	1.995809	1.773626	1.571478	1.384281
36	4.236247	3.407249	2.924875	2.569314	2.280326	2.032471	1.812500	1.612575	1.427639
37	4.258554	3.433177	2.953513	2.600317	2.313532	2.067802	1.849927	1.652101	1.469294
38	4.280183	3.458286	2.981218	2.630284	2.345600	2.101890	1.886002	1.690164	1.509367
39	4.301171	3.482623	3.008047	2.659277	2.376598	2.134813	1.920814	1.726860	1.547965
40	4.321554	3.506233	3.034049	2.687353	2.406591	2.166643	1.954443	1.762279	1.585186
41	4.341364	3.529154	3.059272	2.714565	2.435639	2.197445	1.986961	1.796500	1.621118
42	4.360631	3.551424	3.083756	2.740962	2.463796	2.227280	2.018434	1.829596	1.655842
43	4.379382	3.573076	3.107544	2.766588	2.491111	2.256203	2.048923	1.861634	1.689430
44	4.397644	3.594143	3.130670	2.791484	2.517629	2.284264	2.078483	1.892675	1.721950
45	4.415439	3.614654	3.153169	2.815688	2.543394	2.311510	2.107166	1.922775	1.753463
46	4.432790	3.634635	3.175071	2.839235	2.568444	2.337984	2.135019	1.951985	1.784025
47	4.449718	3.654111	3.196406	2.862158	2.592816	2.363725	2.162086	1.980354	1.813688
48	4.466242	3.673108	3.217201	2.884486	2.616542	2.388771	2.188406	2.007924	1.842500
49	4.482379	3.691645	3.237481	2.906249	2.639654	2.413155	2.214017	2.034738	1.870505
50	4.498147	3.709744	3.257268	2.927472	2.662181	2.436910	2.238954	2.060832	1.897744

Expected Value of the rth Quasi-Range for Samples of n from N(0, 1)

n	r = 0	r = 1	r = 2	r = 3	r = 4	r = 5	r = 6	r = 7	r = 8
51	4.513562	3.727424	3.276586	2.948181	2.684150	2.460065	2.263249	2.086242	1.924256
52	4.528637	3.744702	3.295455	2.968397	2.705587	2.482647	2.286933	2.110999	1.950074
53	4.543388	3.761597	3.313894	2.988142	2.726514	2.504683	2.310032	2.135136	1.975233
54	4.557827	3.778123	3.331921	3.007438	2.746955	2.526197	2.332574	2.158679	1.999762
55	4.571967	3.794295	3.349553	3.026301	2.766929	2.547210	2.354583	2.181655	2.023691
56	4.585818	3.810128	3.366805	3.044750	2.786457	2.567746	2.376082	2.204090	2.047045
57	4.599393	3.825635	3.383695	3.062803	2.805556	2.587823	2.397093	2.226007	2.069851
58	4.612701	3.840828	3.400234	3.080474	2.824244	2.607460	2.417635	2.247427	2.092132
59	4.625752	3.855719	3.416437	3.097778	2.842538	2.626675	2.437728	2.268371	2.113909
60	4.638556	3.870319	3.432316	3.114730	2.860452	2.645484	2.457390	2.288858	2.135204
61	4.651122	3.884639	3.447884	3.131343	2.878001	2.663904	2.476638	2.308906	2.156036
62	4.663457	3.898688	3.463151	3.147629	2.895199	2.681948	2.495488	2.328534	2.176423
63	4.675569	3.912477	3.478128	3.163599	2.912058	2.699632	2.513954	2.347756	2.196382
64	4.687467	3.926014	3.492827	3.179267	2.928591	2.716968	2.532052	2.366588	2.215931
65	4.699157	3.939308	3.507255	3.194641	2.944809	2.733968	2.549794	2.385044	2.235084
66	4.710646	3.952367	3.521423	3.209732	2.960724	2.750646	2.567194	2.403139	2.253856
67	4.721941	3.965199	3.535339	3.224550	2.976347	2.767011	2.584263	2.420886	2.272261
68	4.733047	3.977811	3.549011	3.239104	2.991685	2.783075	2.601013	2.438295	2.290312
69	4.743971	3.990210	3.562448	3.253403	3.006751	2.798849	2.617456	2.455380	2.308021
70	4.754718	4.002402	3.575656	3.267455	3.021552	2.814341	2.633601	2.472152	2.325401
71	4.765294	4.014395	3.588644	3.281267	3.036096	2.829561	2.649458	2.488620	2.342462
72	4.775704	4.026195	3.601418	3.294848	3.050393	2.844517	2.665037	2.504796	2.359216
73	4.785953	4.037806	3.613984	3.308204	3.064450	2.859219	2.680347	2.520688	2.375673
74	4.796045	4.049236	3.626349	3.321343	3.078275	2.873674	2.695396	2.536305	2.391841
75	4.805985	4.060488	3.638519	3.334271	3.091874	2.887890	2.710192	2.551658	2.407731

Expected Value of the rth Quasi-Range for Samples of n from N(0, 1)

n	r = 0	r = 1	r = 2	r = 3	r = 4	r = 5	r = 6	r = 7	r = 8
76	4.815777	4.071569	3.650499	3.346994	3.105254	2.901874	2.724744	2.566752	2.423351
77	4.825426	4.082483	3.662296	3.359519	3.118422	2.915633	2.739059	2.581598	2.438709
78	4.834935	4.093235	3.673914	3.371850	3.131384	2.929174	2.753143	2.596201	2.453815
79	4.844308	4.103829	3.685358	3.383994	3.144146	2.942503	2.767004	2.610571	2.468674
80	4.853549	4.114270	3.696633	3.395957	3.156714	2.955626	2.780649	2.624712	2.483296
81	4.862661	4.124561	3.707745	3.407742	3.169094	2.968550	2.794083	2.638633	2.497686
82	4.871648	4.134708	3.718696	3.419355	3.181289	2.981279	2.807312	2.652339	2.511851
83	4.880513	4.144713	3.729492	3.430800	3.193307	2.993819	2.820343	2.665836	2.525799
84	4.889259	4.154581	3.740137	3.442082	3.205150	3.006176	2.833180	2.679131	2.539534
85	4.897890	4.164315	3.750635	3.453206	3.216825	3.018355	2.845830	2.692229	2.553063
86	4.906407	4.173918	3.760988	3.464176	3.228335	3.030359	2.858296	2.705135	2.566393
87	4.914814	4.183393	3.771203	3.474994	3.239685	3.042194	2.870585	2.717855	2.579527
88	4.923114	4.192745	3.781280	3.485666	3.250879	3.053864	2.882700	2.730393	2.592471
89	4.931308	4.201975	3.791225	3.496196	3.261921	3.065374	2.894647	2.742754	2.605232
90	4.939401	4.211087	3.801040	3.506585	3.272815	3.076727	2.906429	2.754944	2.617812
91	4.947393	4.220084	3.810729	3.516839	3.283565	3.087928	2.918051	2.766965	2.630218
92	4.955288	4.228968	3.820294	3.526960	3.294173	3.098980	2.929517	2.778824	2.642452
93	4.963087	4.237743	3.829739	3.536952	3.304644	3.109887	2.940830	2.790523	2.654521
94	4.970794	4.246410	3.839066	3.546818	3.314980	3.120652	2.951995	2.802066	2.666428
95	4.978409	4.254972	3.848278	3.556560	3.325186	3.131279	2.963015	2.813458	2.678176
96	4.985935	4.263431	3.857378	3.566181	3.335264	3.141772	2.973894	2.824702	2.689771
97	4.993374	4.271790	3.866368	3.575685	3.345216	3.152132	2.984634	2.835802	2.701215
98	5.000728	4.280051	3.875251	3.585074	3.355047	3.162364	2.995240	2.846761	2.712512
99	5.007998	4.288217	3.884029	3.594350	3.364759	3.172471	3.005713	2.857582	2.723666
100	5.015187	4.296289	3.892705	3.603517	3.374354	3.182455	3.016059	2.868269	2.734680

Table A2

VARIANCES OF QUASI-RANGES

Variance of the rth Quasi-Range for Samples of n from $N(\mu, 1)$

n	r = 0	r = 1	r = 2	r = 3	r = 4	r = 5	r = 6	r = 7	r = 8
2	.72676								
3	.78920								
4	.77406	.24902							
5	.74664	.32315							
6	.71917	.34734	.12588						
7	.69423	.35350	.18023						
8	.67212	.35231	.20590	.07602					
9	.65259	.34796	.21829	.11578					
10	.63529	.34231	.22394	.13799	.05091				
11	.61986	.33619	.22594	.15081	.08089				
12	.60603	.33003	.22589	.15830	.09943	.03649			
13	.59354	.32402	.22466	.16256	.11125	.05980			
14	.58220	.31827	.22275	.16481	.11892	.07525	.02744		
15	.57185	.31280	.22045	.16576	.12391	.08577	.04604		
16	.56235	.30763	.21796	.16588	.12713	.09305	.05902	.02138	
17	.55361	.30275	.21537	.16544	.12913	.09814	.06828	.03655	
18	.54551	.29816	.21277	.16462	.13028	.10171	.07499	.04757	.01714
19	.53799	.29382	.21019	.16356	.13084	.10419	.07991	.05572	.02973
20	.53098	.28973	.20765	.16234	.13097	.10588	.08353	.06183	.03918
21	.52442	.28586	.20518	.16101	.13079	.10700	.08620	.06646	.04637
22	.51827	.28220	.20279	.15963	.13040	.10768	.08817	.07000	.05191
23	.51249	.27874	.20048	.15822	.12984	.10805	.08959	.07271	.05622
24	.50703	.27545	.19825	.15679	.12917	.10817	.09061	.07478	.05960
25	.50188	.27233	.19611	.15537	.12841	.10810	.09131	.07637	.06226
26	.49699	.26936	.19404	.15396	.12760	.10789	.09175	.07757	.06436
27	.49236	.26653	.19205	.15258	.12675	.10757	.09201	.07848	.06602
28	.48796	.26383	.19014	.15122	.12587	.10717	.09211	.07914	.06733
29	.48377	.26126	.18830	.14989	.12498	.10671	.09209	.07961	.06836
30	.47977	.25879	.18652	.14859	.12409	.10619	.09198	.07992	.06916
31	.47595	.25644	.18481	.14732	.12319	.10565	.09178	.08011	.06977
32	.47229	.25417	.18316	.14609	.12230	.10507	.09153	.08020	.07023
33	.46879	.25201	.18158	.14489	.12141	.10448	.09123	.08020	.07056
34	.46544	.24992	.18004	.14372	.12054	.10388	.09089	.08014	.07079
35	.46221	.24792	.17857	.14259	.11969	.10327	.09052	.08002	.07094

Variance of the rth Quasi-Range for Samples of n from $N(\mu, 1)$

n	r = 0	r = 1	r = 2	r = 3	r = 4	r = 5	r = 6	r = 7	r = 8
36	.45910	.24599	.17714	.14148	.11884	.10265	.09013	.07985	.07101
37	.45612	.24413	.17576	.14042	.11801	.10204	.08972	.07965	.07103
38	.45325	.24234	.17443	.13937	.11719	.10142	.08930	.07941	.07099
39	.45048	.24062	.17313	.13836	.11640	.10081	.08886	.07916	.07091
40	.44780	.23894	.17188	.13738	.11562	.10021	.08842	.07888	.07080
41	.44521	.23733	.17067	.13642	.11486	.09961	.08797	.07858	.07066
42	.44271	.23577	.16950	.13549	.11411	.09902	.08753	.07827	.07049
43	.44029	.23426	.16837	.13458	.11337	.09844	.08708	.07796	.07031
44	.43794	.23280	.16726	.13370	.11266	.09786	.08663	.07763	.07010
45	.43567	.23137	.16619	.13284	.11196	.09730	.08619	.07730	.06988
46	.43347	.23000	.16515	.13201	.11128	.09674	.08574	.07696	.06965
47	.43133	.22867	.16414	.13119	.11061	.09619	.08530	.07662	.06942
48	.42925	.22736	.16316	.13040	.10996	.09565	.08486	.07629	.06917
49	.42724	.22611	.16220	.12963	.10932	.09512	.08443	.07594	.06892
50	.42527	.22488	.16127	.12888	.10869	.09460	.08401	.07560	.06866
51	.42336	.22369	.16036	.12814	.10808	.09409	.08359	.07526	.06840
52	.42151	.22253	.15948	.12742	.10748	.09359	.08317	.07492	.06813
53	.41970	.22140	.15862	.12673	.10690	.09310	.08276	.07458	.06787
54	.41794	.22030	.15778	.12604	.10633	.09262	.08235	.07425	.06760
55	.41621	.21923	.15696	.12537	.10577	.09215	.08195	.07392	.06733
56	.41454	.21819	.15616	.12473	.10522	.09168	.08156	.07359	.06707
57	.41291	.21716	.15538	.12408	.10469	.09122	.08117	.07327	.06680
58	.41131	.21617	.15462	.12346	.10417	.09078	.08079	.07294	.06653
59	.40976	.21519	.15388	.12286	.10365	.09034	.08042	.07262	.06627
60	.40823	.21424	.15315	.12226	.10315	.08991	.08005	.07231	.06600
61	.40674	.21332	.15244	.12167	.10265	.08948	.07968	.07200	.06574
62	.40528	.21241	.15175	.12110	.10217	.08907	.07932	.07169	.06547
63	.40387	.21153	.15107	.12055	.10170	.08866	.07897	.07138	.06522
64	.40247	.21066	.15040	.12000	.10123	.08826	.07862	.07108	.06496
65	.40111	.20981	.14975	.11946	.10078	.08787	.07828	.07079	.06470
66	.39978	.20898	.14911	.11894	.10033	.08748	.07794	.07050	.06445
67	.39847	.20816	.14849	.11843	.09989	.08710	.07761	.07021	.06420
68	.39719	.20737	.14788	.11792	.09947	.08673	.07729	.06993	.06396
69	.39594	.20659	.14728	.11743	.09904	.08636	.07697	.06965	.06371
70	.39471	.20583	.14669	.11694	.09863	.08600	.07665	.06937	.06347

Variance of the rth Quasi-Range for Samples of n from $N(\mu, 1)$

n	r = 0	r = 1	r = 2	r = 3	r = 4	r = 5	r = 6	r = 7	r = 8
71	.39351	.20508	.14612	.11647	.09822	.08565	.07634	.06910	.06324
72	.39232	.20434	.14555	.11600	.09783	.08531	.07604	.06882	.06300
73	.39116	.20362	.14500	.11554	.09743	.08496	.07574	.06856	.06276
74	.39002	.20291	.14446	.11509	.09704	.08463	.07544	.06830	.06253
75	.38890	.20222	.14392	.11465	.09667	.08430	.07515	.06804	.06230
76	.38780	.20154	.14340	.11422	.09630	.08397	.07486	.06779	.06208
77	.38672	.20087	.14289	.11379	.09593	.08365	.07458	.06753	.06186
78	.38566	.20021	.14238	.11337	.09557	.08334	.07430	.06729	.06164
79	.38463	.19957	.14188	.11296	.09522	.08303	.07403	.06704	.06142
80	.38360	.19894	.14140	.11255	.09487	.08273	.07376	.06680	.06120
81	.38260	.19832	.14092	.11216	.09453	.08242	.07349	.06656	.06099
82	.38161	.19770	.14045	.11177	.09420	.08213	.07323	.06633	.06079
83	.38064	.19711	.13999	.11138	.09386	.08184	.07297	.06610	.06058
84	.37969	.19651	.13953	.11101	.09354	.08156	.07272	.06588	.06037
85	.37874	.19593	.13908	.11064	.09322	.08127	.07247	.06565	.06018
86	.37782	.19536	.13865	.11027	.09290	.08100	.07223	.06543	.05997
87	.37691	.19480	.13820	.10991	.09259	.08072	.07198	.06521	.05978
88	.37601	.19424	.13778	.10956	.09229	.08046	.07174	.06499	.05959
89	.37513	.19370	.13736	.10920	.09199	.08019	.07150	.06479	.05939
90	.37426	.19316	.13695	.10886	.09169	.07993	.07127	.06457	.05920
91	.37340	.19264	.13654	.10852	.09139	.07967	.07104	.06437	.05901
92	.37256	.19212	.13614	.10819	.09111	.07941	.07081	.06416	.05883
93	.37173	.19160	.13575	.10786	.09082	.07916	.07059	.06396	.05865
94	.37091	.19109	.13536	.10753	.09055	.07892	.07037	.06376	.05847
95	.37010	.19060	.13498	.10721	.09027	.07868	.07015	.06357	.05829
96	.36931	.19011	.13460	.10691	.08999	.07843	.06994	.06337	.05812
97	.36852	.18963	.13423	.10659	.08973	.07820	.06973	.06318	.05794
98	.36774	.18916	.13386	.10629	.08946	.07797	.06952	.06299	.05777
99	.36699	.18868	.13350	.10599	.08920	.07773	.06931	.06281	.05760
100	.36624	.18822	.13314	.10569	.08894	.07750	.06911	.06262	.05744

Table A3

STANDARD DEVIATIONS OF QUASI-RANGES

Standard Deviation of the rth Quasi-Range for Samples of n from $N(\mu, 1)$

n	r = 0	r = 1	r = 2	r = 3	r = 4	r = 5	r = 6	r = 7	r = 8
2	.85250								
3	.88837								
4	.87981	.49902							
5	.86408	.56847							
6	.84804	.58936	.35479						
7	.83320	.59456	.42454						
8	.81983	.59356	.45377	.27572					
9	.80783	.58989	.46722	.34027					
10	.79705	.58507	.47322	.37147	.22563				
11	.78731	.57982	.47533	.38834	.28442				
12	.77848	.57448	.47528	.39787	.31533	.19101			
13	.77041	.56923	.47398	.40318	.33355	.24453			
14	.76302	.56415	.47196	.40596	.34485	.27432	.16564		
15	.75621	.55929	.46952	.40714	.35202	.29286	.21456		
16	.74990	.55465	.46686	.40729	.35655	.30504	.24294	.14623	
17	.74405	.55023	.46408	.40674	.35934	.31327	.26131	.19118	
18	.73858	.54604	.46127	.40574	.36094	.31891	.27385	.21811	.13091
19	.73348	.54205	.45846	.40443	.36172	.32278	.28268	.23604	.17243
20	.72868	.53826	.45569	.40292	.36190	.32539	.28902	.24866	.19794
21	.72417	.53466	.45297	.40127	.36166	.32710	.29361	.25781	.21533
22	.71991	.53122	.45033	.39954	.36111	.32815	.29693	.26457	.22783
23	.71588	.52796	.44775	.39777	.36033	.32870	.29932	.26964	.23711
24	.71206	.52483	.44526	.39597	.35940	.32889	.30101	.27347	.24413
25	.70843	.52185	.44284	.39417	.35834	.32878	.30217	.27635	.24952
26	.70498	.51900	.44050	.39238	.35721	.32846	.30291	.27852	.25370
27	.70169	.51626	.43823	.39061	.35602	.32798	.30333	.28014	.25695
28	.69854	.51365	.43605	.38887	.35479	.32737	.30350	.28132	.25948
29	.69554	.51113	.43393	.38716	.35353	.32666	.30347	.28215	.26145
30	.69265	.50872	.43188	.38547	.35226	.32588	.30328	.28271	.26297
31	.68989	.50640	.42990	.38383	.35098	.32504	.30296	.28304	.26414
32	.68724	.50416	.42798	.38222	.34971	.32415	.30254	.28320	.26500
33	.68468	.50200	.42612	.38064	.34845	.32324	.30205	.28320	.26564
34	.68223	.49992	.42432	.37911	.34719	.32230	.30148	.28309	.26607
35	.67986	.49791	.42257	.37761	.34596	.32136	.30087	.28288	.26635

Standard Deviation of the rth Quasi-Range for Samples of n from $N(\mu, 1)$

n	r = 0	r = 1	r = 2	r = 3	r = 4	r = 5	r = 6	r = 7	r = 8
36	.67757	.49598	.42088	.37614	.34473	.32040	.30022	.28258	.26648
37	.67537	.49410	.41924	.37472	.34353	.31944	.29953	.28222	.26651
38	.67324	.49228	.41764	.37333	.34234	.31847	.29882	.28180	.26644
39	.67118	.49053	.41609	.37197	.34117	.31751	.29810	.28135	.26629
40	.66918	.48882	.41459	.37064	.34003	.31656	.29736	.28085	.26608
41	.66724	.48716	.41312	.36935	.33890	.31561	.29661	.28032	.26582
42	.66536	.48556	.41171	.36809	.33780	.31468	.29585	.27978	.26550
43	.66354	.48401	.41032	.36686	.33671	.31375	.29509	.27921	.26515
44	.66177	.48249	.40898	.36565	.33565	.31283	.29433	.27862	.26477
45	.66005	.48101	.40766	.36448	.33461	.31192	.29357	.27802	.26436
46	.65838	.47958	.40639	.36333	.33359	.31103	.29282	.27742	.26392
47	.65676	.47819	.40514	.36221	.33258	.31015	.29206	.27681	.26347
48	.65517	.47683	.40393	.36112	.33160	.30927	.29132	.27620	.26300
49	.65363	.47551	.40274	.36004	.33063	.30842	.29058	.27558	.26252
50	.65213	.47422	.40158	.35900	.32969	.30757	.28984	.27496	.26203
51	.65066	.47296	.40045	.35797	.32876	.30674	.28911	.27434	.26153
52	.64924	.47174	.39935	.35697	.32785	.30593	.28839	.27372	.26102
53	.64784	.47053	.39827	.35599	.32696	.30512	.28768	.27310	.26051
54	.64648	.46936	.39721	.35502	.32608	.30433	.28697	.27249	.26000
55	.64515	.46822	.39618	.35408	.32523	.30356	.28628	.27189	.25948
56	.64385	.46710	.39518	.35316	.32438	.30279	.28559	.27128	.25897
57	.64258	.46601	.39418	.35226	.32356	.30203	.28490	.27068	.25845
58	.64134	.46494	.39322	.35137	.32275	.30129	.28424	.27008	.25793
59	.64012	.46389	.39227	.35051	.32195	.30056	.28358	.26949	.25742
60	.63893	.46287	.39134	.34966	.32117	.29985	.28292	.26890	.25690
61	.63776	.46186	.39043	.34882	.32040	.29914	.28228	.26833	.25639
62	.63662	.46088	.38954	.34800	.31964	.29845	.28164	.26775	.25588
63	.63550	.45992	.38868	.34720	.31890	.29776	.28102	.26718	.25538
64	.63441	.45898	.38781	.34641	.31817	.29708	.28040	.26662	.25487
65	.63333	.45805	.38698	.34564	.31746	.29643	.27979	.26607	.25437
66	.63228	.45714	.38615	.34488	.31676	.29577	.27918	.26552	.25387
67	.63124	.45625	.38534	.34413	.31606	.29513	.27859	.26496	.25338
68	.63023	.45538	.38455	.34340	.31538	.29451	.27801	.26444	.25290
69	.62924	.45452	.38377	.34268	.31471	.29388	.27743	.26390	.25242
70	.62826	.45368	.38301	.34197	.31405	.29326	.27686	.26337	.25194

Standard Deviation of the rth Quasi-Range for Samples of n from $N(\mu, 1)$

n	r = 0	r = 1	r = 2	r = 3	r = 4	r = 5	r = 6	r = 7	r = 8
71	.62730	.45286	.38225	.34127	.31341	.29266	.27630	.26286	.25147
72	.62636	.45204	.38151	.34059	.31277	.29207	.27575	.26234	.25099
73	.62543	.45125	.38079	.33992	.31214	.29149	.27520	.26184	.25052
74	.62451	.45046	.38007	.33925	.31152	.29091	.27466	.26134	.25007
75	.62362	.44969	.37937	.33860	.31091	.29034	.27414	.26084	.24961
76	.62274	.44893	.37868	.33796	.31031	.28978	.27361	.26036	.24916
77	.62187	.44818	.37800	.33733	.30973	.28923	.27309	.25987	.24872
78	.62102	.44745	.37733	.33671	.30915	.28868	.27259	.25940	.24827
79	.62018	.44673	.37667	.33610	.30858	.28815	.27209	.25892	.24783
80	.61936	.44602	.37603	.33549	.30802	.28762	.27159	.25847	.24740
81	.61855	.44533	.37539	.33490	.30745	.28710	.27110	.25800	.24697
82	.61775	.44464	.37476	.33431	.30691	.28658	.27062	.25755	.24655
83	.61696	.44397	.37415	.33374	.30637	.28608	.27014	.25711	.24612
84	.61619	.44330	.37354	.33318	.30584	.28558	.26967	.25666	.24571
85	.61542	.44264	.37293	.33262	.30531	.28508	.26920	.25622	.24531
86	.61467	.44199	.37235	.33206	.30480	.28460	.26875	.25579	.24489
87	.61393	.44136	.37176	.33153	.30429	.28412	.26829	.25536	.24449
88	.61320	.44073	.37119	.33099	.30379	.28365	.26785	.25494	.24410
89	.61248	.44011	.37063	.33046	.30329	.28318	.26740	.25453	.24370
90	.61177	.43950	.37007	.32994	.30280	.28272	.26697	.25411	.24332
91	.61107	.43890	.36952	.32943	.30231	.28226	.26654	.25371	.24293
92	.61038	.43831	.36898	.32893	.30184	.28181	.26611	.25330	.24256
93	.60970	.43772	.36844	.32842	.30137	.28136	.26569	.25290	.24218
94	.60902	.43714	.36791	.32792	.30091	.28092	.26528	.25251	.24180
95	.60836	.43657	.36739	.32744	.30044	.28049	.26487	.25212	.24144
96	.60770	.43602	.36688	.32696	.29999	.28006	.26445	.25174	.24107
97	.60706	.43547	.36637	.32649	.29955	.27964	.26406	.25136	.24071
98	.60642	.43492	.36587	.32602	.29911	.27923	.26366	.25098	.24036
99	.60579	.43438	.36538	.32556	.29866	.27881	.26328	.25062	.24001
100	.60517	.43384	.36489	.32510	.29823	.27840	.26288	.25025	.23966

Table A4

EFFICIENCY OF POINT ESTIMATORS OF σ BASED ON QUASI-RANGES

Efficiency (Percent) of Estimate of Population Standard Deviation Based on the rth Quasi-Range for Samples of n from $N(\mu, \sigma^2)$

n	r = 0	r = 1	r = 2	r = 3	r = 4	r = 5	r = 6	r = 7	r = 8
2	100.00								
3	99.19								
4	97.52	25.24							
5	95.48	39.97							
6	93.30	49.55	13.48						
7	91.12	56.14	23.88						
8	89.00	60.84	32.05	9.03					
9	86.95	64.27	38.56	16.75					
10	84.99	66.80	43.83	23.33	6.74				
11	83.13	68.67	48.14	28.97	12.80				
12	81.36	70.07	51.69	33.82	18.21	5.36			
13	79.68	71.09	54.65	38.00	23.04	10.32			
14	78.09	71.83	57.12	41.64	27.34	14.88	4.44		
15	76.57	72.35	59.20	44.80	31.19	19.05	8.62		
16	75.13	72.69	60.95	47.57	34.63	22.86	12.54	3.78	
17	73.76	72.89	62.44	50.00	37.71	26.34	16.19	7.40	
18	72.46	72.98	63.70	52.14	40.48	29.52	19.58	10.82	3.30
19	71.21	72.98	64.76	54.03	42.97	32.42	22.73	14.06	6.47
20	70.02	72.91	65.67	55.70	45.23	35.09	25.65	17.10	9.51
21	68.88	72.77	66.43	57.19	47.26	37.53	28.36	19.96	12.41
22	67.80	72.59	67.08	58.51	49.11	39.76	30.88	22.64	15.16
23	66.75	72.37	67.62	59.68	50.78	41.82	33.21	25.16	17.77
24	65.75	72.11	68.07	60.73	52.31	43.72	35.38	27.52	20.24
25	64.79	71.82	68.45	61.66	53.69	45.46	37.40	29.73	22.57
26	63.86	71.52	68.76	62.49	54.96	47.07	39.28	31.81	24.78
27	62.97	71.20	69.00	63.24	56.11	48.56	41.03	33.76	26.87
28	62.12	70.86	69.20	63.90	57.17	49.94	42.66	35.59	28.85
29	61.29	70.51	69.35	64.49	58.13	51.21	44.19	37.31	30.72
30	60.49	70.15	69.46	65.01	59.01	52.39	45.61	38.93	32.49
31	59.72	69.78	69.53	65.48	59.82	53.48	46.94	40.45	34.16
32	58.98	69.41	69.57	65.90	60.56	54.50	48.19	41.89	35.74
33	58.26	69.03	69.58	66.26	61.24	55.44	49.36	43.24	37.24
34	57.56	68.66	69.57	66.59	61.86	56.32	50.45	44.51	38.66
35	56.89	68.28	69.53	66.87	62.43	57.13	51.47	45.71	40.01

Efficiency (Percent) of Estimate of Population Standard Deviation Based on the rth Quasi-Range for Samples of n from N (μ, σ^2)

n	r = 0	r = 1	r = 2	r = 3	r = 4	r = 5	r = 6	r = 7	r = 8
36	56.23	67.89	69.48	67.12	62.95	57.89	52.44	46.35	41.29
37	55.60	67.51	69.41	67.34	63.42	58.60	53.34	47.32	42.50
38	54.98	67.13	69.32	67.53	63.86	59.26	54.19	48.33	43.66
39	54.39	66.76	69.21	67.69	64.26	59.87	54.99	49.39	44.75
40	53.81	66.38	69.10	67.82	64.63	60.44	55.74	50.30	45.79
41	53.24	66.00	68.97	67.94	64.96	60.97	56.44	51.36	46.78
42	52.70	65.63	68.83	68.03	65.27	61.46	57.11	52.47	47.72
43	52.16	65.26	68.68	68.10	65.55	61.92	57.73	53.24	48.61
44	51.64	64.89	68.53	68.16	65.80	62.35	58.32	53.[illegible]7	49.47
45	51.14	64.53	68.37	68.20	66.02	62.75	58.87	54.[illegible]6	50.28
46	50.64	64.17	68.20	68.22	66.23	63.13	59.39	55.[illegible]1	51.05
47	50.16	63.81	68.02	68.24	66.42	63.47	59.89	55.[illegible]3	51.79
48	49.70	63.46	67.84	68.23	66.59	63.80	60.35	56.52	52.49
49	49.24	63.11	67.66	68.22	66.74	64.10	60.79	57.[illegible]8	53.16
50	48.79	62.76	67.47	68.20	66.87	64.38	61.20	57.[illegible]1	53.79
51	48.36	62.42	67.28	68.17	66.99	64.64	61.58	58.[illegible]2	54.41
52	47.93	62.08	67.09	68.12	67.09	64.88	61.95	58.59	54.99
53	47.52	61.74	66.89	68.07	67.18	65.10	62.29	59.05	55.54
54	47.11	61.41	66.69	68.02	67.26	65.31	62.62	59.48	56.07
55	46.71	61.08	66.49	67.95	67.33	65.50	62.92	59.89	56.58
56	46.33	60.76	66.28	67.87	67.38	65.67	63.21	60.28	57.06
57	45.95	60.44	66.08	67.80	67.43	65.84	63.49	60.65	57.52
58	45.57	60.12	65.88	67.71	67.46	65.98	63.74	61.01	57.96
59	45.21	59.81	65.67	67.62	67.49	66.12	63.98	61.34	58.38
60	44.85	59.50	65.46	67.53	67.51	66.25	64.20	61.66	58.79
61	44.51	59.20	65.26	67.43	67.52	66.36	64.41	61.96	59.17
62	44.16	58.89	65.05	67.33	67.52	66.46	64.61	62.25	59.54
63	43.83	58.59	64.84	67.22	67.52	66.56	64.80	62.52	59.89
64	43.50	58.30	64.63	67.11	67.50	66.64	64.97	62.78	60.23
65	43.18	58.01	64.42	67.00	67.49	66.71	65.14	63.02	60.55
66	42.86	57.72	64.22	66.88	67.46	66.78	65.29	63.25	60.86
67	42.55	57.44	64.01	66.76	67.44	66.84	65.43	63.48	61.15
68	42.25	57.15	63.80	66.64	67.40	66.89	65.56	63.68	61.43
69	41.95	56.88	63.59	66.52	67.36	66.94	65.69	63.88	61.70
70	41.65	56.60	63.38	66.39	67.32	66.98	65.80	64.07	61.96

Efficiency (Percent) of Estimate of Population Standard Deviation Based on r^{th} Quasi-Range for Samples of n from $N(\mu, \sigma^2)$

n	r = 0	r = 1	r = 2	r = 3	r = 4	r = 5	r = 6	r = 7	r = 8
71	41.37	56.33	63.18	66.27	67.27	67.01	65.91	64.25	62.20
72	41.08	56.06	62.97	66.14	67.22	67.03	66.01	64.42	62.44
73	40.81	55.80	62.77	66.00	67.16	67.05	66.10	64.58	62.66
74	40.53	55.53	62.56	65.87	67.11	67.06	66.19	64.73	62.88
75	40.26	55.27	62.36	65.74	67.04	67.07	66.26	64.88	63.08
76	40.00	55.02	62.16	65.60	66.98	67.08	66.33	65.01	63.28
77	39.74	54.77	61.96	65.47	66.91	67.07	66.40	65.14	63.46
78	39.49	54.52	61.76	65.33	66.84	67.07	66.45	65.25	63.64
79	39.24	54.27	61.56	65.19	66.76	67.06	66.51	65.37	63.81
80	38.99	54.02	61.36	65.05	66.69	67.04	66.56	65.47	63.97
81	38.75	53.78	61.16	64.91	66.61	67.03	66.60	65.58	64.12
82	38.51	53.54	60.97	64.77	66.53	67.01	66.63	65.67	64.27
83	38.27	53.30	60.77	64.63	66.45	66.98	66.67	65.75	64.41
84	38.04	53.07	60.58	64.49	66.36	66.95	66.69	65.84	64.54
85	37.81	52.84	60.38	64.35	66.27	66.92	66.72	65.91	64.67
86	37.59	52.61	60.19	64.21	66.18	66.89	66.73	65.98	64.79
87	37.37	52.38	60.00	64.06	66.09	66.85	66.75	66.05	64.90
88	37.15	52.16	59.81	63.92	66.00	66.81	66.76	66.11	65.01
89	36.94	51.94	59.62	63.78	65.91	66.77	66.77	66.16	65.12
90	36.73	51.72	59.43	63.63	65.81	66.72	66.77	66.22	65.21
91	36.52	51.50	59.25	63.49	65.72	66.68	66.77	66.26	65.31
92	36.31	51.29	59.06	63.35	65.62	66.63	66.77	66.31	65.39
93	36.11	51.08	58.88	63.20	65.53	66.58	66.76	66.35	65.47
94	35.91	50.87	58.70	63.06	65.42	66.52	66.75	66.38	65.55
95	35.72	50.66	58.51	62.92	65.33	66.46	66.74	66.41	65.62
96	35.52	50.45	58.33	62.78	65.23	66.41	66.73	66.44	65.69
97	35.33	50.25	58.15	62.63	65.12	66.35	66.71	66.46	65.76
98	35.14	50.05	57.98	62.49	65.02	66.29	66.69	66.49	65.82
99	34.96	49.85	57.80	62.35	64.92	66.23	66.67	66.50	65.87
100	34.77	49.65	57.62	62.21	64.82	66.16	66.65	66.52	65.93

Table A5

MOST EFFICIENT POINT ESTIMATORS OF σ BASED ON QUASI-RANGES

[$w_r = x_{n-r} - x_{r+1} = r$th **quasi-range of sample**, $r = 0(1)8$;
$s = [\sum(x - \bar{x})^2/(n-1)]^{1/2}$ = **sample standard deviation**]

Most Efficient Unbiased Estimates of Standard Deviation of Normal Population

Sample size, n	Based on one quasi-range Estimate	Eff(%)	Based on two adjacent quasi-ranges Estimate	Eff(%)	Based on any two quasi-ranges* Estimate	Eff(%)	Efficient Estimate
2	$.886227\,w_o$	100.00					s/.797885
3	$.590818\,w_o$	99.19					s/.886227
4	$.485731\,w_o$	97.52	$.45394\,(w_o + 0.2427\,w_1)$	98.92	$.45394\,(w_o + 0.2427\,w_1)$	98.92	s/.921318
5	$.429936\,w_o$	95.48	$.37238\,(w_o + 0.3631\,w_1)$	98.84	$.37238\,(w_o + 0.3631\,w_1)$	98.84	s/.939986
6	$.394569\,w_o$	93.30	$.31803\,(w_o + 0.4752\,w_1)$	98.66	$.31803\,(w_o + 0.4752\,w_1)$	98.66	s/.951533
7	$.369774\,w_o$	91.12	$.27922\,(w_o + 0.5790\,w_1)$	98.32	$.27922\,(w_o + 0.5790\,w_1)$	98.32	s/.959369
8	$.351222\,w_o$	89.00	$.25010\,(w_o + 0.6754\,w_1)$	97.84	$.25010\,(w_o + 0.6754\,w_1)$	97.84	s/.965030
9	$.336697\,w_o$	86.95	$.22745\,(w_o + 0.7651\,w_1)$	97.23	$.22745\,(w_o + 0.7651\,w_1)$	97.23	s/.969311
10	$.324939\,w_o$	84.99	$.20931\,(w_o + 0.8489\,w_1)$	96.54	$.20931\,(w_o + 0.8489\,w_1)$	96.54	s/.972659
11	$.315172\,w_o$	83.13	$.19444\,(w_o + 0.9276\,w_1)$	95.78	$.19444\,(w_o + 0.9276\,w_1)$	95.78	s/.975350
12	$.306894\,w_o$	81.36	$.18203\,(w_o + 1.0017\,w_1)$	94.97	$.21177\,(w_o + 0.9231\,w_2)$	95.17	s/.977559
13	$.299762\,w_o$	79.68	$.17150\,(w_o + 1.0717\,w_1)$	94.12	$.19848\,(w_o + 1.0015\,w_2)$	95.00	s/.979406
14	$.293534\,w_o$	78.09	$.16244\,(w_o + 1.1381\,w_1)$	93.26	$.18704\,(w_o + 1.0762\,w_2)$	94.77	s/.980971
15	$.288033\,w_o$	76.57	$.15457\,(w_o + 1.2011\,w_1)$	92.39	$.17708\,(w_o + 1.1477\,w_2)$	94.50	s/.982316
16	$.283127\,w_o$	75.13	$.14765\,(w_o + 1.2612\,w_1)$	91.52	$.16834\,(w_o + 1.2161\,w_2)$	94.18	s/.983484
17	$.278716\,w_o$	73.76	$.14153\,(w_o + 1.3186\,w_1)$	90.65	$.16060\,(w_o + 1.2817\,w_2)$	93.82	s/.984506
18	$.370257\,w_1$	72.98	$.13606\,(w_o + 1.3736\,w_1)$	89.78	$.15369\,(w_o + 1.3448\,w_2)$	93.43	s/.985410
19	$.362335\,w_1$	72.98	$.13114\,(w_o + 1.4263\,w_1)$	88.92	$.14750\,(w_o + 1.4055\,w_2)$	93.02	s/.986214
20	$.355214\,w_1$	72.91	$.12670\,(w_o + 1.4769\,w_1)$	88.08	$.14192\,(w_o + 1.4640\,w_2)$	92.59	s/.986934
21	$.348768\,w_1$	72.77	$.12266\,(w_o + 1.5256\,w_1)$	87.24	$.13684\,(w_o + 1.5206\,w_2)$	92.14	s/.987583
22	$.342899\,w_1$	72.59	$.11897\,(w_o + 1.5726\,w_1)$	86.42	$.14637\,(w_o + 1.5298\,w_3)$	91.78	s/.988170
23	$.337526\,w_1$	72.37	$.11558\,(w_o + 1.6179\,w_1)$	85.62	$.14129\,(w_o + 1.5881\,w_3)$	91.61	s/.988705
24	$.332584\,w_1$	72.11	$.11246\,(w_o + 1.6617\,w_1)$	84.83	$.13663\,(w_o + 1.6446\,w_3)$	91.42	s/.989193
25	$.328019\,w_1$	71.82	$.10958\,(w_o + 1.7040\,w_1)$	84.05	$.13233\,(w_o + 1.6996\,w_3)$	91.21	s/.989640

*Quasi-ranges 0 through 8 only

Most Efficient Unbiased Estimates of Standard Deviation of Normal Population

Sample size, n	Based on one quasi-range Estimate	Eff(%)	Based on two adjacent quasi-ranges Estimate	Eff(%)	Based on any two quasi-ranges* Estimate	Eff(%)	Efficient Estimate
26	.323785 w_1	71.52	.10691 (w_o +1.7451 w_1)	83.29	.12836 (w_o +1.7529 w_3)	90.98	s/.990052
27	.319844 w_1	71.20	.10442 (w_o +1.7849 w_1)	82.54	.12468 (w_o +1.8050 w_3)	90.73	s/.990433
28	.316165 w_1	70.86	.10209 (w_o +1.8235 w_1)	81.81	.12126 (w_o +1.8556 w_3)	90.48	s/.990786
29	.312719 w_1	70.51	.09992 (w_o +1.8610 w_1)	81.10	.11807 (w_o +1.9050 w_3)	90.21	s/.991113
30	.309483 w_1	70.15	.09788 (w_o +1.8975 w_1)	80.40	.11508 (w_o +1.9532 w_3)	89.93	s/.991418
31	.306436 w_1	69.78	.09596 (w_o +1.9329 w_1)	79.71	.11229 (w_o +2.0002 w_3)	89.63	s/.991703
32	.357181 w_2	69.57	.09416 (w_o +1.9675 w_1)	79.04	.10967 (w_o +2.0461 w_3)	89.35	s/.991969
33	.353016 w_2	69.58	.09245 (w_o +2.0011 w_1)	78.38	.11551 (w_o +2.0673 w_4)	89.11	s/.992219
34	.349094 w_2	69.57	.14484 (w_1 +1.2398 w_2)	78.08	.11282 (w_o +2.1150 w_4)	88.97	s/.992454
35	.345394 w_2	69.53	.14201 (w_1 +1.2646 w_2)	77.82	.11028 (w_o +2.1616 w_4)	88.82	s/.992675
36	.341895 w_2	69.48	.13934 (w_1 +1.2888 w_2)	77.55	.10788 (w_o +2.2073 w_4)	88.66	s/.992884
37	.338580 w_2	69.41	.13680 (w_1 +1.3126 w_2)	77.27	.10561 (w_o +2.2520 w_4)	88.48	s/.993080
38	.335433 w_2	69.32	.13440 (w_1 +1.3358 w_2)	76.99	.10345 (w_o +2.2962 w_4)	88.31	s/.993267
39	.332442 w_2	69.21	.13211 (w_1 +1.3586 w_2)	76.71	.10141 (w_o +2.3393 w_4)	88.12	s/.993443
40	.329593 w_2	69.10	.12995 (w_1 +1.3806 w_2)	76.42	.09947 (w_o +2.3816 w_4)	87.92	s/.993611
41	.326875 w_2	68.97	.12788 (w_1 +1.4026 w_2)	76.12	.09762 (w_o +2.4232 w_4)	87.73	s/.993770
42	.324280 w_2	68.83	.12592 (w_1 +1.4236 w_2)	75.82	.09586 (w_o +2.4640 w_4)	87.52	s/.993922
43	.321798 w_2	68.68	.12403 (w_1 +1.4448 w_2)	75.53	.09418 (w_o +2.5042 w_4)	87.32	s/.994066
44	.319420 w_2	68.53	.12223 (w_1 +1.4652 w_2)	75.23	.09258 (w_o +2.5435 w_4)	87.10	s/.994203
45	.317141 w_2	68.37	.12051 (w_1 +1.4853 w_2)	74.92	.09647 (w_o +2.5741 w_5)	86.97	s/.994335
46	.352208 w_3	68.22	.11886 (w_1 +1.5051 w_2)	74.62	.09482 (w_o +2.6150 w_5)	86.85	s/.994460
47	.349387 w_3	68.24	.11727 (w_1 +1.5246 w_2)	74.32	.09323 (w_o +2.6553 w_5)	86.73	s/.994580
48	.346682 w_3	68.23	.15270 (w_2 +1.1550 w_3)	74.15	.09171 (w_o +2.6951 w_5)	86.60	s/.994695
49	.344086 w_3	68.22	.15056 (w_2 +1.1714 w_3)	74.03	.09025 (w_o +2.7341 w_5)	86.47	s/.994806
50	.341592 w_3	68.20	.14850 (w_2 +1.1877 w_3)	73.90	.08885 (w_o +2.7726 w_5)	86.33	s/.994911

*Quasi-ranges 0 through 8 only

Most Efficient Unbiased Estimates of Standard Deviation of Normal Population

Sample size, n	Based on one quasi-range Estimate	Eff(%)	Based on two adjacent quasi-ranges Estimate	Eff(%)	Based on any two quasi-ranges* Estimate	Eff(%)	Efficient Estimate
51	$.339192\ w_3$	68.17	$.14651\ (w_2 + 1.2037\ w_3)$	73.77	$.08751\ (w_o + 2.8105\ w_5)$	86.19	s/.995013
52	$.336882\ w_3$	68.12	$.14460\ (w_2 + 1.2195\ w_3)$	73.63	$.08621\ (w_o + 2.8479\ w_5)$	86.04	s/.995110
53	$.334656\ w_3$	68.07	$.14278\ (w_2 + 1.2348\ w_3)$	73.48	$.08497\ (w_o + 2.8847\ w_5)$	85.89	s/.995204
54	$.332509\ w_3$	68.02	$.14100\ (w_2 + 1.2504\ w_3)$	73.33	$.08377\ (w_o + 2.9212\ w_5)$	85.73	s/.995294
55	$.330436\ w_3$	67.95	$.13931\ (w_2 + 1.2652\ w_3)$	73.18	$.08262\ (w_o + 2.9567\ w_5)$	85.57	s/.995381
56	$.328434\ w_3$	67.87	$.13763\ (w_2 + 1.2805\ w_3)$	73.02	$.13787\ (w_1 + 1.6820\ w_8)$	85.44	s/.995465
57	$.326498\ w_3$	67.80	$.13606\ (w_2 + 1.2949\ w_3)$	72.86	$.13592\ (w_1 + 1.7062\ w_8)$	85.46	s/.995546
58	$.324625\ w_3$	67.71	$.13453\ (w_2 + 1.3093\ w_3)$	72.70	$.13403\ (w_1 + 1.7303\ w_8)$	85.47	s/.995624
59	$.322812\ w_3$	67.62	$.13304\ (w_2 + 1.3236\ w_3)$	72.53	$.13222\ (w_1 + 1.7539\ w_8)$	85.48	s/.995699
60	$.321055\ w_3$	67.53	$.13161\ (w_2 + 1.3374\ w_3)$	72.36	$.13046\ (w_1 + 1.7774\ w_8)$	85.48	s/.995772
61	$.347463\ w_4$	67.52	$.13022\ (w_2 + 1.3514\ w_3)$	72.19	$.12875\ (w_1 + 1.8006\ w_8)$	85.47	s/.995842
62	$.345399\ w_4$	67.52	$.15710\ (w_3 + 1.1114\ w_4)$	72.06	$.12710\ (w_1 + 1.8236\ w_8)$	85.46	s/.995910
63	$.343400\ w_4$	67.52	$.15534\ (w_3 + 1.1243\ w_4)$	71.99	$.12551\ (w_1 + 1.8461\ w_8)$	85.44	s/.995976
64	$.341461\ w_4$	67.50	$.15370\ (w_3 + 1.1360\ w_4)$	71.91	$.12397\ (w_1 + 1.8684\ w_8)$	85.42	s/.996040
65	$.339581\ w_4$	67.49	$.15210\ (w_3 + 1.1478\ w_4)$	71.83	$.12247\ (w_1 + 1.8907\ w_8)$	85.40	s/.996102
66	$.337755\ w_4$	67.46	$.15052\ (w_3 + 1.1598\ w_4)$	71.75	$.12102\ (w_1 + 1.9125\ w_8)$	85.36	s/.996161
67	$.335982\ w_4$	67.44	$.14895\ (w_3 + 1.1723\ w_4)$	71.67	$.11961\ (w_1 + 1.9342\ w_8)$	85.33	s/.996219
68	$.334260\ w_4$	67.40	$.14750\ (w_3 + 1.1835\ w_4)$	71.57	$.11826\ (w_1 + 1.9554\ w_8)$	85.29	s/.996276
69	$.332585\ w_4$	67.36	$.14607\ (w_3 + 1.1949\ w_4)$	71.48	$.11693\ (w_1 + 1.9765\ w_8)$	85.24	s/.996330
70	$.330956\ w_4$	67.32	$.14466\ (w_3 + 1.2064\ w_4)$	71.39	$.11564\ (w_1 + 1.9975\ w_8)$	85.20	s/.996383
71	$.329370\ w_4$	67.27	$.14333\ (w_3 + 1.2172\ w_4)$	71.29	$.11439\ (w_1 + 2.0181\ w_8)$	85.14	s/.996435
72	$.327827\ w_4$	67.22	$.14201\ (w_3 + 1.2284\ w_4)$	71.18	$.11317\ (w_1 + 2.0387\ w_8)$	85.09	s/.996485
73	$.326323\ w_4$	67.16	$.14071\ (w_3 + 1.2395\ w_4)$	71.08	$.11198\ (w_1 + 2.0593\ w_8)$	85.04	s/.996534
74	$.324857\ w_4$	67.11	$.13943\ (w_3 + 1.2510\ w_4)$	70.98	$.11083\ (w_1 + 2.0793\ w_8)$	84.98	s/.996581
75	$.346274\ w_5$	67.07	$.13821\ (w_3 + 1.2617\ w_4)$	70.87	$.10971\ (w_1 + 2.0992\ w_8)$	84.91	s/.996627

*Quasi-ranges 0 through 8 only

Most Efficient Unbiased Estimates of Standard Deviation of Normal Population

Sample size, n	Based on one quasi-range Estimate	Eff(%)	Based on two adjacent quasi-ranges Estimate	Eff(%)	Based on any two quasi-ranges* Estimate	Eff(%)	Efficient Estimate
76	$.344605\ w_5$	67.08	$.13703\ (w_3 + 1.2723\ w_4)$	70.75	$.10862\ (w_1 + 2.1188\ w_8)$	84.85	s/.996672
77	$.342979\ w_5$	67.07	$.15847\ (w_4 + 1.0948\ w_5)$	70.71	$.10756\ (w_1 + 2.1382\ w_8)$	84.78	s/.996716
78	$.341393\ w_5$	67.07	$.15704\ (w_4 + 1.1049\ w_5)$	70.66	$.10653\ (w_1 + 2.1575\ w_8)$	84.71	s/.996759
79	$.339847\ w_5$	67.06	$.15568\ (w_4 + 1.1145\ w_5)$	70.61	$.10552\ (w_1 + 2.1766\ w_8)$	84.63	s/.996800
80	$.338338\ w_5$	67.04	$.15431\ (w_4 + 1.1245\ w_5)$	70.56	$.10453\ (w_1 + 2.1956\ w_8)$	84.56	s/.996841
81	$.336865\ w_5$	67.03	$.15305\ (w_4 + 1.1334\ w_5)$	70.50	$.10357\ (w_1 + 2.2145\ w_8)$	84.48	s/.996880
82	$.335427\ w_5$	67.01	$.15176\ (w_4 + 1.1432\ w_5)$	70.44	$.10264\ (w_1 + 2.2328\ w_8)$	84.40	s/.996918
83	$.334022\ w_5$	66.98	$.15056\ (w_4 + 1.1519\ w_5)$	70.38	$.10171\ (w_1 + 2.2515\ w_8)$	84.32	s/.996956
84	$.332649\ w_5$	66.95	$.14934\ (w_4 + 1.1612\ w_5)$	70.31	$.10082\ (w_1 + 2.2696\ w_8)$	84.23	s/.996993
85	$.331306\ w_5$	66.92	$.14809\ (w_4 + 1.1714\ w_5)$	70.25	$.09996\ (w_1 + 2.2875\ w_8)$	84.15	s/.997028
86	$.329994\ w_5$	66.89	$.14697\ (w_4 + 1.1800\ w_5)$	70.18	$.09910\ (w_1 + 2.3057\ w_8)$	84.07	s/.997063
87	$.328710\ w_5$	66.85	$.14584\ (w_4 + 1.1890\ w_5)$	70.11	$.09826\ (w_1 + 2.3234\ w_8)$	83.97	s/.997097
88	$.327454\ w_5$	66.81	$.14475\ (w_4 + 1.1977\ w_5)$	70.03	$.09746\ (w_1 + 2.3406\ w_8)$	83.88	s/.997131
89	$.345465\ w_6$	66.77	$.14363\ (w_4 + 1.2071\ w_5)$	69.96	$.09665\ (w_1 + 2.3585\ w_8)$	83.79	s/.997163
90	$.344065\ w_6$	66.77	$.14259\ (w_4 + 1.2156\ w_5)$	69.88	$.09588\ (w_1 + 2.3755\ w_8)$	83.70	s/.997195
91	$.342694\ w_6$	66.77	$.16064\ (w_5 + 1.0751\ w_6)$	69.82	$.09511\ (w_1 + 2.3930\ w_8)$	83.61	s/.997226
92	$.341353\ w_6$	66.77	$.15934\ (w_5 + 1.0844\ w_6)$	69.79	$.09437\ (w_1 + 2.4096\ w_8)$	83.51	s/.997257
93	$.340040\ w_6$	66.76	$.15823\ (w_5 + 1.0916\ w_6)$	69.75	$.09364\ (w_1 + 2.4264\ w_8)$	83.42	s/.997286
94	$.338754\ w_6$	66.75	$.15705\ (w_5 + 1.0999\ w_6)$	69.71	$.09293\ (w_1 + 2.4431\ w_8)$	83.32	s/.997315
95	$.337494\ w_6$	66.74	$.15592\ (w_5 + 1.1078\ w_6)$	69.67	$.09223\ (w_1 + 2.4595\ w_8)$	83.22	s/.997344
96	$.336259\ w_6$	66.73	$.15480\ (w_5 + 1.1158\ w_6)$	69.63	$.09155\ (w_1 + 2.4761\ w_8)$	83.12	s/.997371
97	$.335049\ w_6$	66.71	$.15366\ (w_5 + 1.1244\ w_6)$	69.59	$.09087\ (w_1 + 2.4924\ w_8)$	83.02	s/.997399
98	$.333863\ w_6$	66.69	$.15261\ (w_5 + 1.1319\ w_6)$	69.54	$.09022\ (w_1 + 2.5085\ w_8)$	82.92	s/.997426
99	$.332700\ w_6$	66.67	$.15161\ (w_5 + 1.1389\ w_6)$	69.49	$.08957\ (w_1 + 2.5245\ w_8)$	82.82	s/.997452
100	$.331559\ w_6$	66.65	$.15059\ (w_5 + 1.1466\ w_6)$	69.44	$.08895\ (w_1 + 2.5401\ w_8)$	82.71	s/.997478

*Quasi-ranges 0 through 8 only

Table A6

PERCENTAGE POINTS OF THE rth QUASI-RANGE W_r FOR SAMPLES OF SIZE n FROM $N(\mu, 1)$
[P = cumulative probability = Prob($W_r \leq$ tabular value)]

PERCENTAGE POINTS OF THE 0th QUASI-RANGE

P\n	2	3	4	5	6	7
0.0001	0.000177	0.019047	0.092394	0.205489	0.334167	0.464516
0.0005	0.000886	0.042594	0.158155	0.308222	0.463700	0.612589
0.0010	0.001772	0.060245	0.199446	0.367392	0.534736	0.691347
0.0050	0.008862	0.134847	0.342702	0.554904	0.748983	0.921826
0.0100	0.017725	0.190945	0.433676	0.665015	0.869515	1.048144
0.0250	0.044319	0.303071	0.594643	0.849672	1.065952	1.250500
0.0500	0.088681	0.431402	0.759533	1.029940	1.252886	1.440141
0.1000	0.177712	0.618352	0.979366	1.261398	1.488195	1.676051
0.2000	0.358287	0.900092	1.285672	1.573441	1.799995	1.985445
0.3000	0.544926	1.138259	1.531485	1.818447	2.042028	2.223993
0.4000	0.741614	1.362597	1.756529	2.040097	2.259641	2.437704
0.5000	0.953873	1.587788	1.978321	2.256883	2.471652	2.645453
0.6000	1.190232	1.826320	2.210281	2.482427	2.691658	2.860733
0.7000	1.465738	2.094590	2.468799	2.732888	2.935559	3.099199
0.8000	1.812388	2.423529	2.783758	3.037317	3.231739	3.388684
0.9000	2.326174	2.902380	3.240446	3.478281	3.660721	3.808098
0.9500	2.771808	3.314493	3.633160	3.857656	4.030092	4.169554
0.9750	3.169822	3.682268	3.984015	4.197026	4.360906	4.493624
0.9900	3.642773	4.120303	4.402801	4.602821	4.757047	4.882166
0.9950	3.969745	4.424235	4.694087	4.885585	5.033479	5.153613
0.9990	4.653508	5.063453	5.308804	5.483754	5.619332	5.729754
0.9995	4.922533	5.316400	5.552855	5.721773	5.852849	5.959710
0.9999	5.502128	5.864157	6.082863	6.239691	6.361709	6.461391

P\n	8	9	10	11	12	13
0.0001	0.590186	0.708709	0.819433	0.922513	1.018443	1.107819
0.0005	0.751013	0.878357	0.995219	1.102585	1.201493	1.292919
0.0010	0.834826	0.965507	1.084583	1.193404	1.293251	1.385253
0.0050	1.075282	1.212115	1.334928	1.445920	1.546898	1.639328
0.0100	1.204819	1.343386	1.467033	1.578304	1.679205	1.771332
0.0250	1.410019	1.549720	1.673518	1.784355	1.884475	1.975612
0.0500	1.600413	1.739852	1.862844	1.972583	2.071455	2.161277
0.1000	1.835449	1.973327	2.094446	2.202195	2.299057	2.386902
0.2000	2.141656	2.276121	2.393843	2.498316	2.592064	2.676969
0.3000	2.376728	2.507898	2.622556	2.724195	2.815328	2.897817
0.4000	2.586852	2.714772	2.826491	2.925467	3.014177	3.094450
0.5000	2.790841	2.915438	3.024202	3.120531	3.206854	3.284960
0.6000	3.002059	3.123122	3.228778	3.322347	3.406194	3.482065
0.7000	3.235931	3.353046	3.455258	3.545785	3.626919	3.700346
0.8000	3.519834	3.632192	3.730280	3.817183	3.895093	3.965627
0.9000	3.931349	4.037023	4.129346	4.211200	4.284635	4.351158
0.9500	4.286309	4.386509	4.474124	4.551864	4.621655	4.684920
0.9750	4.604857	4.700411	4.784033	4.858286	4.924993	4.985497
0.9900	4.987183	5.077506	5.156635	5.226963	5.290196	5.347592
0.9950	5.254550	5.341439	5.417616	5.485364	5.546312	5.601663
0.9990	5.822728	5.902906	5.973307	6.036000	6.092467	6.143802
0.9995	6.049760	6.127468	6.195739	6.256567	6.311378	6.361227
0.9999	6.545529	6.618235	6.682187	6.739225	6.790667	6.837489

PERCENTAGE POINTS OF THE 0th QUASI-RANGE

P\n	14	15	16	17	18	19
0.0001	1.191258	1.269336	1.342579	1.411460	1.476395	1.537756
0.0005	1.377729	1.456683	1.530433	1.599544	1.664502	1.725726
0.0010	1.470384	1.549474	1.623228	1.692245	1.757038	1.818046
0.0050	1.724404	1.803104	1.876237	1.944472	2.008372	2.068413
0.0100	1.855958	1.934115	2.006645	2.074243	2.137488	2.196867
0.0250	2.059129	2.136114	2.207443	2.273835	2.335885	2.394088
0.0500	2.243459	2.319117	2.389145	2.454273	2.515097	2.572117
0.1000	2.467168	2.540983	2.609247	2.672690	2.731903	2.787396
0.2000	2.754467	2.825680	2.891495	2.952631	3.009675	3.063109
0.3000	2.973079	3.042214	3.106097	3.165428	3.220781	3.272627
0.4000	3.167678	3.234938	3.297083	3.354799	3.408645	3.459083
0.5000	3.356208	3.421651	3.482119	3.538281	3.590681	3.639766
0.6000	3.551278	3.614858	3.673612	3.728188	3.779115	3.826829
0.7000	3.767343	3.828898	3.885792	3.938651	3.987986	4.034216
0.8000	4.030005	4.089172	4.143877	4.194716	4.242179	4.286668
0.9000	4.411913	4.467782	4.519464	4.567519	4.612403	4.654494
0.9500	4.742732	4.795924	4.845154	4.890951	4.933745	4.973892
0.9750	5.040817	5.091743	5.138897	5.182782	5.223806	5.262307
0.9900	5.400105	5.448476	5.493291	5.535020	5.574047	5.610690
0.9950	5.652328	5.699017	5.742289	5.782597	5.820307	5.855724
0.9990	6.190836	6.234215	6.274452	6.311958	6.347071	6.380070
0.9995	6.406915	6.449067	6.488177	6.524643	6.558790	6.590889
0.9999	6.880434	6.920082	6.956890	6.991229	7.023401	7.053657

P\n	20	22	24	26	28	30
0.0001	1.595868	1.703468	1.801126	1.890368	1.972413	2.048246
0.0005	1.783582	1.890416	1.987093	2.075236	2.156123	2.230775
0.0010	1.875647	1.981896	2.077934	2.165415	2.245638	2.319635
0.0050	2.125000	2.229148	2.323063	2.408455	2.486648	2.558691
0.0100	2.252791	2.355636	2.448296	2.532490	2.609547	2.680513
0.0250	2.448862	2.549498	2.640076	2.722319	2.797547	2.866800
0.0500	2.625753	2.724238	2.812829	2.893230	2.966750	3.034415
0.1000	2.839570	2.935330	3.021432	3.099550	3.170969	3.236693
0.2000	3.113340	3.205511	3.288366	3.363534	3.432252	3.495491
0.3000	3.321364	3.410792	3.491185	3.564123	3.630811	3.692188
0.4000	3.506497	3.593503	3.671730	3.742714	3.807625	3.867378
0.5000	3.685915	3.770611	3.846776	3.915903	3.979129	4.037343
0.6000	3.871692	3.954046	4.028125	4.095376	4.156902	4.213563
0.7000	4.077692	4.157522	4.229355	4.294588	4.354286	4.409280
0.8000	4.328517	4.405388	4.474592	4.537464	4.595024	4.648069
0.9000	4.694105	4.766907	4.832495	4.892122	4.946746	4.997113
0.9500	5.011689	5.081193	5.143852	5.200850	5.253094	5.301290
0.9750	5.298566	5.365277	5.425452	5.480220	5.530445	5.576798
0.9900	5.645215	5.708769	5.766138	5.818385	5.866325	5.910592
0.9950	5.889103	5.950573	6.006087	6.056666	6.103092	6.145978
0.9990	6.411187	6.468538	6.520381	6.567656	6.611084	6.651228
0.9995	6.621165	6.676980	6.727453	6.773495	6.815801	6.854919
0.9999	7.082207	7.134873	7.182535	7.226039	7.266039	7.303044

PERCENTAGE POINTS OF THE 0th QUASI-RANGE

P\n	32	34	36	38	40	50
0.0001	2.118671	2.184353	2.245845	2.303612	2.358048	2.590591
0.0005	2.300019	2.364536	2.424887	2.481541	2.534899	2.762552
0.0010	2.388241	2.452137	2.511889	2.567966	2.620769	2.845954
0.0050	2.625423	2.687528	2.745569	2.800012	2.851256	3.069616
0.0100	2.746228	2.807370	2.864498	2.918079	2.968503	3.183321
0.0250	2.930905	2.990535	3.046240	3.098477	3.147630	3.357005
0.0500	3.097040	3.155286	3.209692	3.260710	3.308716	3.513200
0.1000	3.297516	3.354083	3.406923	3.456472	3.503096	3.701734
0.2000	3.554019	3.608457	3.659314	3.707009	3.751894	3.943206
0.3000	3.749002	3.801853	3.851234	3.897552	3.941149	4.127044
0.4000	3.922697	3.974166	4.022264	4.067386	4.109863	4.291064
0.5000	4.091247	4.141409	4.188295	4.232287	4.273708	4.450482
0.6000	4.266041	4.314888	4.360553	4.403408	4.443766	4.616090
0.7000	4.460228	4.507662	4.552018	4.593654	4.632872	4.800425
0.8000	4.697229	4.743013	4.785838	4.826050	4.863937	5.025917
0.9000	5.043816	5.087334	5.128057	5.166312	5.202370	5.356687
0.9500	5.346000	5.387678	5.426697	5.463363	5.497935	5.646026
0.9750	5.619817	5.659933	5.697502	5.732818	5.766127	5.908917
0.9900	5.951695	5.990041	6.025967	6.059751	6.091626	6.228393
0.9950	6.185810	6.222982	6.257818	6.290585	6.321508	6.454269
0.9990	6.688538	6.723377	6.756044	6.786787	6.815814	6.940585
0.9995	6.891284	6.925248	6.957102	6.987085	7.015400	7.137164
0.9999	7.337461	7.369622	7.399796	7.428210	7.455053	7.570596

P\n	60	70	80	90	100	
0.0001	2.775319	2.927786	3.057128	3.169148	3.267737	
0.0005	2.943168	3.092146	3.218493	3.327905	3.424199	
0.0010	3.024532	3.171801	3.296691	3.404839	3.500024	
0.0050	3.242655	3.385324	3.506313	3.611097	3.703340	
0.0100	3.353527	3.493863	3.612885	3.715980	3.806747	
0.0250	3.522895	3.659695	3.775746	3.876294	3.964843	
0.0500	3.675244	3.808909	3.922333	4.020632	4.107228	
0.1000	3.859209	3.989161	4.099483	4.195135	4.279426	
0.2000	4.094984	4.220324	4.326797	4.419166	4.500610	
0.3000	4.274625	4.396573	4.500222	4.590182	4.669535	
0.4000	4.435016	4.554032	4.655239	4.743118	4.820666	
0.5000	4.591013	4.707269	4.806177	4.892098	4.967946	
0.6000	4.753185	4.866667	4.963266	5.047218	5.121358	
0.7000	4.933836	5.044345	5.138468	5.220308	5.292614	
0.8000	5.155024	5.262057	5.353283	5.432652	5.502811	
0.9000	5.479870	5.582115	5.669346	5.745304	5.812497	
0.9500	5.764388	5.862730	5.946701	6.019871	6.084638	
0.9750	6.023168	6.118178	6.199362	6.270146	6.332834	
0.9900	6.337964	6.429173	6.507175	6.575233	6.635542	
0.9950	6.560722	6.649396	6.725272	6.791507	6.850223	
0.9990	7.040804	7.124400	7.196010	7.258580	7.314094	
0.9995	7.235032	7.316708	7.386704	7.447885	7.502183	
0.9999	7.663588	7.741276	7.807913	7.866200	7.917964	

PERCENTAGE POINTS OF THE 1st QUASI-RANGE

P\n	2	3	4	5	6	7
0.0001			0.000075	0.009322	0.049902	0.119575
0.0005			0.000377	0.020912	0.086002	0.180877
0.0010			0.000754	0.029647	0.108908	0.216609
0.0050			0.003776	0.067000	0.189745	0.331749
0.0100			0.007563	0.095541	0.242125	0.400604
0.0250			0.018998	0.153712	0.336645	0.517912
0.0500			0.038307	0.222048	0.435710	0.634407
0.1000			0.077924	0.324723	0.570926	0.786431
0.2000			0.161697	0.485770	0.764467	0.994952
0.3000			0.252837	0.627106	0.923401	1.160943
0.4000			0.353523	0.764001	1.071238	1.312449
0.5000			0.467063	0.904613	1.218766	1.461597
0.6000			0.598872	1.056579	1.374677	1.617559
0.7000			0.758826	1.230666	1.550044	1.791462
0.8000			0.968348	1.447940	1.765515	2.003538
0.9000			1.292551	1.770131	2.080556	2.311526
0.9500			1.583961	2.051624	2.353169	2.576824
0.9750			1.850310	2.305250	2.597607	2.814182
0.9900			2.172496	2.609525	2.890084	3.097887
0.9950			2.398059	2.821705	3.093818	3.295463
0.9990			2.875252	3.269952	3.524257	3.713081
0.9995			3.064567	3.447885	3.695270	3.879143
0.9999			3.474609	3.833970	4.066832	4.240341

P\n	8	9	10	11	12	13
0.0001	0.206016	0.299721	0.394993	0.488709	0.579270	0.665939
0.0005	0.288331	0.398416	0.506189	0.609412	0.707223	0.799444
0.0010	0.333983	0.451419	0.564597	0.671805	0.772573	0.867003
0.0050	0.473639	0.608298	0.733627	0.849487	0.956462	1.055367
0.0100	0.553371	0.695250	0.825439	0.944605	1.053844	1.154294
0.0250	0.684853	0.835718	0.971702	1.094637	1.206316	1.308311
0.0500	0.811495	0.968434	1.108101	1.233251	1.346213	1.448880
0.1000	0.972631	1.134631	1.277080	1.403663	1.517227	1.619968
0.2000	1.188410	1.353857	1.497714	1.624557	1.737709	1.839643
0.3000	1.357193	1.523450	1.667124	1.793263	1.905438	2.006255
0.4000	1.509604	1.675563	1.818380	1.943403	2.054351	2.153909
0.5000	1.658492	1.823441	1.964944	2.088549	2.198065	2.296223
0.6000	1.813251	1.976568	2.116325	2.238194	2.346041	2.442615
0.7000	1.984958	2.145938	2.283415	2.403130	2.508965	2.603669
0.8000	2.193469	2.351067	2.485429	2.602297	2.705534	2.797862
0.9000	2.495133	2.647150	2.776578	2.889057	2.988363	3.077145
0.9500	2.754341	2.901193	3.026169	3.134756	3.230619	3.316323
0.9750	2.985993	3.128102	3.249041	3.354132	3.446921	3.529893
0.9900	3.262767	3.399183	3.515318	3.616272	3.705442	3.785207
0.9950	3.455529	3.588020	3.700859	3.798984	3.885687	3.963270
0.9990	3.863204	3.987628	4.093718	4.186068	4.267742	4.340884
0.9995	4.025443	4.146774	4.250281	4.340423	4.420175	4.491623
0.9999	4.378647	4.493515	4.591627	4.677160	4.752903	4.820814

PERCENTAGE POINTS OF THE 1st QUASI-RANGE

P\n	14	15	16	17	18	19
0.0001	0.748455	0.826821	0.901174	0.971721	1.038698	1.102346
0.0005	0.886233	0.967897	1.044801	1.117318	1.185811	1.250619
0.0010	0.955447	1.038353	1.116186	1.189397	1.258398	1.323569
0.0050	1.147038	1.232258	1.311727	1.386063	1.455803	1.521417
0.0100	1.247005	1.332903	1.412788	1.487346	1.557166	1.622752
0.0250	1.401952	1.488348	1.568424	1.642953	1.712584	1.777866
0.0500	1.542783	1.629163	1.709032	1.783221	1.852420	1.917207
0.1000	1.713608	1.799504	1.878745	1.952215	2.020637	2.084616
0.2000	1.932239	2.016957	2.094947	2.167134	2.234266	2.296963
0.3000	2.097672	2.181191	2.257992	2.329010	2.395008	2.456606
0.4000	2.244074	2.326372	2.401991	2.471876	2.536787	2.597347
0.5000	2.385042	2.466055	2.540453	2.609179	2.672992	2.732510
0.6000	2.529939	2.609548	2.682627	2.750113	2.812758	2.871176
0.7000	2.689258	2.767252	2.838828	2.904910	2.966243	3.023428
0.8000	2.881271	2.957257	3.026976	3.091336	3.151063	3.206750
0.9000	3.157333	3.230377	3.297393	3.359258	3.416673	3.470207
0.9500	3.393736	3.464260	3.528972	3.588719	3.644177	3.695894
0.9750	3.604853	3.673156	3.735843	3.793732	3.847477	3.897607
0.9900	3.857295	3.923005	3.983332	4.039059	4.090812	4.139099
0.9950	4.033409	4.097359	4.156087	4.210350	4.260756	4.307797
0.9990	4.407058	4.467434	4.522916	4.574210	4.621884	4.666397
0.9995	4.556285	4.615299	4.669542	4.719705	4.766338	4.809888
0.9999	4.882318	4.938488	4.990148	5.037948	5.082407	5.123947

P\n	20	22	24	26	28	30
0.0001	1.162901	1.275613	1.378445	1.472757	1.559688	1.640185
0.0005	1.312048	1.425868	1.529206	1.623622	1.710391	1.790542
0.0010	1.385249	1.499326	1.602689	1.696984	1.783535	1.863407
0.0050	1.583311	1.697321	1.800177	1.893697	1.979308	2.058145
0.0100	1.684539	1.798165	1.900498	1.993418	2.078389	2.156572
0.0250	1.839263	1.951946	2.053211	2.145009	2.228846	2.305905
0.0500	1.978069	2.089609	2.189695	2.280319	2.363010	2.438961
0.1000	2.144651	2.254533	2.352994	2.442051	2.523246	2.597775
0.2000	2.355738	2.463185	2.559343	2.646237	2.725403	2.798030
0.3000	2.514321	2.619763	2.714068	2.799247	2.876824	2.947977
0.4000	2.654070	2.757660	2.850272	2.933899	3.010048	3.079882
0.5000	2.788245	2.890003	2.980955	3.063071	3.137835	3.206395
0.6000	2.925872	3.025717	3.114945	3.195498	3.268837	3.336089
0.7000	3.076966	3.174687	3.262013	3.340847	3.412624	3.478447
0.8000	3.258883	3.354040	3.439078	3.515854	3.585764	3.649884
0.9000	3.520329	3.611831	3.693622	3.767486	3.834762	3.896483
0.9500	3.744323	3.832756	3.911830	3.983263	4.048346	4.108072
0.9750	3.944559	4.030319	4.107030	4.176351	4.239531	4.297527
0.9900	4.184336	4.266997	4.340972	4.407852	4.468831	4.524829
0.9950	4.351876	4.432444	4.504573	4.569805	4.629300	4.683951
0.9990	4.708129	4.784457	4.852847	4.914743	4.971232	5.023155
0.9995	4.850726	4.925438	4.992402	5.053026	5.108370	5.159252
0.9999	5.162915	5.234252	5.298237	5.356201	5.409149	5.457853

PERCENTAGE POINTS OF THE 1st QUASI-RANGE

P\n	32	34	36	38	40	50
0.0001	1.715040	1.784921	1.850389	1.911917	1.969919	2.217741
0.0005	1.864929	1.934258	1.999119	2.060006	2.117346	2.361827
0.0010	1.937475	2.006462	2.070965	2.131490	2.188463	2.431184
0.0050	2.131127	2.199004	2.262394	2.321815	2.377701	2.615370
0.0100	2.228899	2.296129	2.358885	2.417689	2.472976	2.707942
0.0250	2.377132	2.443294	2.505020	2.562830	2.617163	2.847888
0.0500	2.509124	2.574267	2.635018	2.691899	2.745344	2.972189
0.1000	2.666589	2.730453	2.789994	2.845727	2.898082	3.120213
0.2000	2.865061	2.927250	2.985215	3.039460	3.090411	3.306540
0.3000	3.013634	3.074541	3.131304	3.184424	3.234315	3.445941
0.4000	3.144317	3.204087	3.259790	3.311915	3.360873	3.568559
0.5000	3.269653	3.328331	3.383017	3.434192	3.482259	3.686194
0.6000	3.398142	3.455705	3.509353	3.559560	3.606721	3.806852
0.7000	3.539185	3.595533	3.648053	3.697209	3.743386	3.939399
0.8000	3.709058	3.763962	3.815143	3.863054	3.908068	4.099216
0.9000	3.953459	4.006338	4.055644	4.101811	4.145196	4.329550
0.9500	4.163222	4.214419	4.262170	4.306891	4.348929	4.527665
0.9750	4.351095	4.400837	4.447243	4.490714	4.531586	4.705465
0.9900	4.576570	4.624632	4.669484	4.711512	4.751039	4.919315
0.9950	4.734461	4.781390	4.825195	4.866251	4.904870	5.069369
0.9990	5.071170	5.115804	5.157487	5.196573	5.233354	5.390193
0.9995	5.206315	5.250074	5.290948	5.329281	5.365361	5.519272
0.9999	5.502924	5.544850	5.584027	5.620784	5.655392	5.803163

P\n	60	70	80	90	100	
0.0001	2.414509	2.576746	2.714222	2.833150	2.937709	
0.0005	2.555491	2.714966	2.850001	2.966762	3.069390	
0.0010	2.623279	2.781384	2.915219	3.030928	3.132619	
0.0050	2.803113	2.957485	3.088093	3.200981	3.300182	
0.0100	2.893421	3.045880	3.174850	3.286316	3.384267	
0.0250	3.029877	3.179420	3.305906	3.415224	3.511293	
0.0500	3.151037	3.297974	3.422256	3.529676	3.624086	
0.1000	3.295294	3.439130	3.560801	3.665979	3.758435	
0.2000	3.476876	3.616836	3.735257	3.837654	3.927691	
0.3000	3.612753	3.749848	3.865874	3.966226	4.054487	
0.4000	3.732300	3.865907	3.980859	4.079442	4.166167	
0.5000	3.847025	3.979279	4.091271	4.188185	4.273464	
0.6000	3.964739	4.094619	4.204635	4.299867	4.383639	
0.7000	4.094109	4.221429	4.329318	4.422740	4.504993	
0.8000	4.250179	4.374485	4.479871	4.571165	4.651574	
0.9000	4.475294	4.595404	4.697307	4.785639	4.863485	
0.9500	4.669098	4.785744	4.884772	4.970660	5.046388	
0.9750	4.843174	4.956827	5.053371	5.137146	5.211044	
0.9900	5.052725	5.162923	5.256598	5.337932	5.409714	
0.9950	5.199880	5.307747	5.399484	5.479169	5.549521	
0.9990	5.514819	5.617949	5.705747	5.782076	5.849514	
0.9995	5.641647	5.742963	5.829250	5.904291	5.970611	
0.9999	5.920808	6.018308	6.101416	6.173743	6.237704	

PERCENTAGE POINTS OF THE 2nd QUASI-RANGE

P\n	2	3	4	5	6	7
0.0001					0.000047	0.006145
0.0005					0.000237	0.013799
0.0010					0.000474	0.019577
0.0050					0.002374	0.044379
0.0100					0.004757	0.063431
0.0250					0.011963	0.102524
0.0500					0.024167	0.148885
0.1000					0.049352	0.219387
0.2000					0.103230	0.331868
0.3000					0.162763	0.432330
0.4000					0.229584	0.531047
0.5000					0.306181	0.633758
0.6000					0.396642	0.746118
0.7000					0.508454	0.876387
0.8000					0.657948	1.041017
0.9000					0.895087	1.288678
0.9500					1.113259	1.507891
0.9750					1.316074	1.707236
0.9900					1.564973	1.948257
0.9950					1.741160	2.117321
0.9990					2.118073	2.476589
0.9995					2.268892	2.619843
0.9999					2.597534	2.931644

P\n	8	9	10	11	12	13
0.0001	0.034194	0.084649	0.149916	0.223248	0.300096	0.377642
0.0005	0.059081	0.128520	0.210707	0.298059	0.386216	0.472804
0.0010	0.074937	0.154232	0.244623	0.338471	0.431695	0.522231
0.0050	0.131278	0.237756	0.349211	0.458963	0.564169	0.663776
0.0100	0.168096	0.288167	0.409447	0.526270	0.636607	0.739976
0.0250	0.235120	0.374794	0.509539	0.635706	0.752624	0.860690
0.0500	0.306130	0.461680	0.606762	0.739818	0.861418	0.972704
0.1000	0.404225	0.576228	0.731494	0.871052	0.996884	1.110945
0.2000	0.546757	0.735243	0.900085	1.045390	1.174699	1.290824
0.3000	0.665464	0.863188	1.033014	1.181048	1.311799	1.428591
0.4000	0.777071	0.980880	1.153723	1.303205	1.434534	1.551397
0.5000	0.889464	1.097485	1.272170	1.422320	1.553689	1.670241
0.6000	1.009230	1.220111	1.395761	1.545972	1.676940	1.792847
0.7000	1.145013	1.357568	1.533365	1.683033	1.813132	1.928022
0.8000	1.313191	1.526075	1.701011	1.849340	1.977919	2.091239
0.9000	1.561297	1.772166	1.944359	2.089781	2.215497	2.326080
0.9500	1.777692	1.985152	2.153992	2.296277	2.419104	2.527037
0.9750	1.972780	2.176293	2.341608	2.480758	2.600786	2.706203
0.9900	2.207247	2.405295	2.565968	2.701109	2.817624	2.919928
0.9950	2.371107	2.565029	2.722293	2.854538	2.968546	3.068647
0.9990	2.718385	2.903120	3.052941	3.178950	3.287608	3.383039
0.9995	2.856674	3.037675	3.184507	3.308032	3.414572	3.508161
0.9999	3.157588	3.330475	3.470852	3.589034	3.691032	3.780685

PERCENTAGE POINTS OF THE 2nd QUASI-RANGE

P\n	14	15	16	17	18	19
0.0001	0.454210	0.528842	0.601022	0.670505	0.737211	0.801157
0.0005	0.556595	0.637010	0.713838	0.787070	0.856809	0.923217
0.0010	0.609107	0.691945	0.770685	0.845431	0.916374	0.983741
0.0050	0.757567	0.845712	0.928545	1.006461	1.079859	1.149123
0.0100	0.836553	0.926772	1.011157	1.090231	1.164491	1.234388
0.0250	0.960647	1.053311	1.139459	1.219796	1.294943	1.365444
0.0500	1.074889	1.169088	1.256277	1.337297	1.412863	1.483588
0.1000	1.214936	1.310277	1.398147	1.479518	1.555198	1.625864
0.2000	1.395966	1.491851	1.579854	1.661075	1.736409	1.806592
0.3000	1.533915	1.629673	1.717348	1.798109	1.872898	1.942482
0.4000	1.656488	1.751827	1.838967	1.919127	1.993276	2.062201
0.5000	1.774818	1.869531	1.955983	2.035424	2.108844	2.177043
0.6000	1.896655	1.990537	2.076136	2.154723	2.227302	2.294679
0.7000	2.030752	2.123544	2.208067	2.285607	2.357175	2.423580
0.8000	2.192416	2.283703	2.366783	2.442946	2.513204	2.578364
0.9000	2.424674	2.513534	2.594339	2.668370	2.736627	2.799906
0.9500	2.623199	2.709821	2.788560	2.860677	2.927155	2.988776
0.9750	2.800088	2.884640	2.961484	3.031858	3.096725	3.156851
0.9900	3.011026	3.093060	3.167615	3.235893	3.298831	3.357173
0.9950	3.157784	3.238057	3.311016	3.377839	3.439441	3.496549
0.9990	3.468044	3.544620	3.614242	3.678028	3.736849	3.791396
0.9995	3.591543	3.666671	3.734988	3.797589	3.855327	3.908878
0.9999	3.860602	3.932644	3.998184	4.058265	4.113701	4.165136

P\n	20	22	24	26	28	30
0.0001	0.862420	0.977345	1.082992	1.180394	1.270502	1.354163
0.0005	0.986480	1.104353	1.211933	1.310573	1.401433	1.485502
0.0010	1.047767	1.166724	1.274970	1.373994	1.465046	1.549173
0.0050	1.214610	1.335516	1.444806	1.544278	1.635379	1.719282
0.0100	1.300332	1.421764	1.531230	1.630653	1.721561	1.805177
0.0250	1.431773	1.553512	1.662869	1.761928	1.852314	1.935312
0.0500	1.549992	1.671575	1.780512	1.878999	1.968725	2.051020
0.1000	1.692084	1.813043	1.921152	2.018706	2.107454	2.188755
0.2000	1.872234	1.991863	2.098527	2.194603	2.281882	2.361750
0.3000	2.007493	2.125817	2.231172	2.325972	2.412024	2.490719
0.4000	2.126546	2.243551	2.347632	2.441218	2.526123	2.603737
0.5000	2.240672	2.356292	2.459067	2.551428	2.635187	2.711730
0.6000	2.357510	2.471616	2.572984	2.664041	2.746592	2.822014
0.7000	2.485478	2.597835	2.697601	2.787188	2.868385	2.942557
0.8000	2.639081	2.749247	2.847028	2.934808	3.014353	3.087007
0.9000	2.858853	2.965770	3.060639	3.145789	3.222945	3.293414
0.9500	3.046171	3.150265	3.242622	3.325520	3.400639	3.469252
0.9750	3.212854	3.314423	3.404548	3.485451	3.558770	3.625749
0.9900	3.411517	3.510094	3.597582	3.676133	3.747338	3.812400
0.9950	3.549751	3.646268	3.731945	3.808886	3.878645	3.942397
0.9990	3.842226	3.934482	4.016421	4.090043	4.156824	4.217883
0.9995	3.958787	4.049388	4.129878	4.202215	4.267845	4.327862
0.9999	4.213090	4.300185	4.377605	4.447220	4.510412	4.568226

PERCENTAGE POINTS OF THE 2nd QUASI-RANGE

P\n	32	34	36	38	40	50
0.0001	1.432111	1.504981	1.573318	1.637594	1.698217	1.957439
0.0005	1.563610	1.636461	1.704648	1.768678	1.828984	2.086092
0.0010	1.627243	1.699988	1.768022	1.831866	1.891960	2.147861
0.0050	1.796946	1.869158	1.936574	1.999743	2.059127	2.311329
0.0100	1.882493	1.954319	2.021326	2.084073	2.143031	2.393151
0.0250	2.011953	2.083074	2.149362	2.211388	2.269631	2.516384
0.0500	2.126936	2.197328	2.262894	2.324210	2.381760	2.625345
0.1000	2.263686	2.333111	2.397736	2.458140	2.514810	2.754461
0.2000	2.435295	2.503388	2.566735	2.625917	2.681417	2.915946
0.3000	2.563150	2.630184	2.692528	2.750758	2.805355	3.035975
0.4000	2.675148	2.741223	2.802661	2.860036	2.913825	3.140982
0.5000	2.782140	2.847276	2.907833	2.964379	3.017386	3.241214
0.6000	2.891379	2.955541	3.015186	3.070877	3.123080	3.343503
0.7000	3.010764	3.073849	3.132490	3.187242	3.238563	3.455269
0.8000	3.153813	3.215600	3.273033	3.326658	3.376923	3.589196
0.9000	3.358212	3.418144	3.473858	3.525881	3.574650	3.780665
0.9500	3.532351	3.590718	3.644982	3.695658	3.743170	3.943950
0.9750	3.687354	3.744346	3.797341	3.846838	3.893251	4.089463
0.9900	3.872254	3.927639	3.979150	4.027270	4.072401	4.263291
0.9950	4.001057	4.055346	4.105845	4.153028	4.197287	4.384558
0.9990	4.274088	4.326127	4.374551	4.419811	4.462280	4.642137
0.9995	4.383119	4.434288	4.481911	4.526429	4.568207	4.745204
0.9999	4.621476	4.670806	4.716734	4.759682	4.799999	4.970944

P\n	60	70	80	90	100	
0.0001	2.163278	2.332895	2.476512	2.600649	2.709699	
0.0005	2.289587	2.456960	2.598515	2.720783	2.828137	
0.0010	2.350128	2.516367	2.656900	2.778248	2.884778	
0.0050	2.510091	2.673189	2.810934	2.929808	3.034128	
0.0100	2.590044	2.751507	2.887824	3.005440	3.108645	
0.0250	2.710349	2.869291	3.003425	3.119134	3.220654	
0.0500	2.816631	2.973301	3.105485	3.219499	3.319529	
0.1000	2.942492	3.096433	3.226293	3.338296	3.436565	
0.2000	3.099821	3.250320	3.377270	3.486768	3.582848	
0.3000	3.216730	3.364665	3.489457	3.597106	3.691575	
0.4000	3.318998	3.464695	3.587610	3.693654	3.786728	
0.5000	3.416617	3.560186	3.681324	3.785850	3.877606	
0.6000	3.516247	3.657660	3.776999	3.879992	3.970418	
0.7000	3.625124	3.764203	3.881599	3.982936	4.071929	
0.8000	3.755626	3.891942	4.007042	4.106427	4.193730	
0.9000	3.942279	4.074725	4.186615	4.283273	4.368216	
0.9500	4.101552	4.230781	4.340004	4.434400	4.517388	
0.9750	4.243572	4.370003	4.476911	4.569343	4.650632	
0.9900	4.413339	4.536519	4.640737	4.730888	4.810206	
0.9950	4.531847	4.652820	4.755212	4.843815	4.921794	
0.9990	4.783775	4.900230	4.998884	5.084315	5.159552	
0.9995	4.884662	4.999373	5.096584	5.180792	5.254970	
0.9999	5.105791	5.216811	5.310966	5.392579	5.464513	

PERCENTAGE POINTS OF THE 3rd QUASI-RANGE

P\n	2	3	4	5	6	7
0.0001						
0.0005						
0.0010						
0.0050						
0.0100						
0.0250						
0.0500						
0.1000						
0.2000						
0.3000						
0.4000						
0.5000						
0.6000						
0.7000						
0.8000						
0.9000						
0.9500						
0.9750						
0.9900						
0.9950						
0.9990						
0.9995						
0.9999						

P\n	8	9	10	11	12	13
0.0001	0.000034	0.004578	0.025999	0.065554	0.118003	0.178267
0.0005	0.000172	0.010283	0.044982	0.099734	0.166274	0.238663
0.0010	0.000345	0.014594	0.057102	0.119831	0.193304	0.271414
0.0050	0.001727	0.033133	0.100327	0.185418	0.277080	0.369552
0.0100	0.003462	0.047410	0.128703	0.225224	0.325605	0.424671
0.0250	0.008710	0.076806	0.180615	0.293990	0.406653	0.514712
0.0500	0.017610	0.111832	0.235958	0.363399	0.485839	0.600820
0.1000	0.036023	0.165431	0.312956	0.455527	0.588042	0.709922
0.2000	0.075620	0.251738	0.425882	0.584483	0.727163	0.855718
0.3000	0.119691	0.329591	0.520796	0.689051	0.837560	0.969760
0.4000	0.169535	0.406746	0.610687	0.785810	0.938285	1.072840
0.5000	0.227141	0.487663	0.701795	0.882165	1.037519	1.173668
0.6000	0.295787	0.576874	0.799472	0.983972	1.141436	1.278625
0.7000	0.381494	0.681137	0.910884	1.098615	1.257536	1.395267
0.8000	0.497450	0.814058	1.049762	1.239822	1.399484	1.537166
0.9000	0.684234	1.016147	1.256196	1.447176	1.606349	1.742898
0.9500	0.858749	1.196848	1.437521	1.627538	1.785183	1.920012
0.9750	1.022942	1.362433	1.601846	1.789986	1.945631	2.078495
0.9900	1.226642	1.563996	1.800239	1.985209	2.137887	2.268022
0.9950	1.372097	1.706145	1.939382	2.121702	2.272040	2.400095
0.9990	1.686182	2.009944	2.235366	2.411282	2.556137	2.679529
0.9995	1.812806	2.131632	2.353571	2.526737	2.669350	2.790754
0.9999	2.090257	2.397381	2.611320	2.778282	2.915820	3.032907

PERCENTAGE POINTS OF THE 3rd QUASI-RANGE

P\n	14	15	16	17	18	19
0.0001	0.242683	0.308821	0.375122	0.440609	0.504685	0.567004
0.0005	0.313211	0.387719	0.460914	0.532082	0.600851	0.667059
0.0010	0.350597	0.428845	0.505061	0.578677	0.649439	0.717274
0.0050	0.460009	0.547127	0.630357	0.709551	0.784773	0.856190
0.0100	0.520138	0.611092	0.697284	0.778783	0.855809	0.928644
0.0250	0.616841	0.712791	0.802765	0.887149	0.966386	1.040919
0.0500	0.707936	0.807529	0.900192	0.986573	1.067296	1.142930
0.1000	0.821862	0.924879	1.019989	1.108123	1.190091	1.266598
0.2000	0.972136	1.078185	1.175345	1.264842	1.347684	1.424707
0.3000	1.088490	1.195996	1.294044	1.384039	1.467109	1.544166
0.4000	1.192963	1.301255	1.399696	1.489820	1.572838	1.649719
0.5000	1.294633	1.403304	1.501829	1.591843	1.674623	1.751182
0.6000	1.400016	1.508747	1.607101	1.696801	1.779176	1.855272
0.7000	1.516690	1.625156	1.723072	1.812228	1.894000	1.969461
0.8000	1.658122	1.765893	1.862991	1.951268	2.032134	2.106687
0.9000	1.862422	1.968632	2.064127	2.150810	2.230115	2.303154
0.9500	2.037778	2.142259	2.236089	2.321181	2.398974	2.470578
0.9750	2.194390	2.297111	2.389292	2.472841	2.549190	2.619442
0.9900	2.381416	2.481843	2.571915	2.653517	2.728064	2.796640
0.9950	2.511624	2.610368	2.698908	2.779111	2.852370	2.919756
0.9990	2.786908	2.881950	2.967158	3.044335	3.114828	3.179672
0.9995	2.896434	2.989975	3.073839	3.149802	3.219191	3.283022
0.9999	3.134858	3.225116	3.306054	3.379383	3.446379	3.508023

P\n	20	22	24	26	28	30
0.0001	0.627378	0.742028	0.848642	0.947697	1.039828	1.125695
0.0005	0.730664	0.850260	0.960339	1.061830	1.155671	1.242722
0.0010	0.782214	0.903811	1.015253	1.117671	1.212133	1.299589
0.0050	0.924018	1.049834	1.164026	1.268213	1.363767	1.451844
0.0100	0.997588	1.124971	1.240117	1.344853	1.440687	1.528856
0.0250	1.111166	1.240296	1.356407	1.461603	1.557565	1.645641
0.0500	1.213986	1.344109	1.460652	1.565931	1.661751	1.749540
0.1000	1.338244	1.468956	1.585572	1.690610	1.785997	1.873234
0.2000	1.496609	1.627294	1.743433	1.847738	1.942247	2.028531
0.3000	1.615964	1.746170	1.861617	1.965121	2.058782	2.144202
0.4000	1.721254	1.850774	1.965420	2.068077	2.160883	2.245461
0.5000	1.822338	1.951005	2.064743	2.166486	2.258396	2.342107
0.6000	1.925932	2.053560	2.166249	2.266969	2.357896	2.440671
0.7000	2.039471	2.165799	2.277227	2.376743	2.466533	2.548237
0.8000	2.175798	2.300385	2.410171	2.508153	2.596512	2.676882
0.9000	2.370805	2.492643	2.599903	2.695565	2.781787	2.860186
0.9500	2.536870	2.656196	2.761189	2.854795	2.939145	3.015830
0.9750	2.684463	2.801466	2.904387	2.996131	3.078795	3.153942
0.9900	2.860099	2.974271	3.074687	3.164192	3.244839	3.318155
0.9950	2.982111	3.094292	3.192956	3.280903	3.360149	3.432198
0.9990	3.239676	3.347640	3.442614	3.527291	3.603610	3.673015
0.9995	3.342093	3.448389	3.541909	3.625300	3.700471	3.768840
0.9999	3.565082	3.667787	3.758184	3.838820	3.911533	3.977690

PERCENTAGE POINTS OF THE 3rd QUASI-RANGE

P\n	32	34	36	38	40	50
0.0001	1.205920	1.281069	1.351653	1.418117	1.480856	1.749478
0.0005	1.323747	1.399413	1.470300	1.536903	1.599661	1.867325
0.0010	1.380863	1.456661	1.527596	1.594186	1.656881	1.923855
0.0050	1.533402	1.609247	1.680051	1.746384	1.808731	2.073262
0.0100	1.610380	1.686099	1.756716	1.822817	1.884899	2.147921
0.0250	1.726921	1.802295	1.872497	1.938139	1.999733	2.260182
0.0500	1.830438	1.905371	1.975095	2.040235	2.101319	2.359244
0.1000	1.953511	2.027783	2.096825	2.161279	2.221676	2.476360
0.2000	2.107818	2.181090	2.249140	2.312617	2.372061	2.622393
0.3000	2.222631	2.295062	2.362293	2.424980	2.483661	2.730595
0.4000	2.323071	2.394711	2.461182	2.523141	2.581125	2.825005
0.5000	2.418885	2.489731	2.555448	2.616687	2.673987	2.914895
0.6000	2.516562	2.586567	2.651488	2.711973	2.768559	3.006396
0.7000	2.623121	2.692180	2.756209	2.815854	2.871647	3.106098
0.8000	2.750522	2.818418	2.881357	2.939980	2.994811	3.225186
0.9000	2.932001	2.998202	3.059561	3.116708	3.170154	3.394706
0.9500	3.086067	3.150809	3.210816	3.266701	3.318969	3.538595
0.9750	3.222770	3.286215	3.345020	3.399789	3.451016	3.666306
0.9900	3.385310	3.447216	3.504601	3.558053	3.608052	3.818248
0.9950	3.498197	3.559044	3.615452	3.667998	3.717154	3.923861
0.9990	3.736608	3.795252	3.849630	3.900299	3.947710	4.147206
0.9995	3.831493	3.889276	3.942864	3.992800	4.039532	4.236225
0.9999	4.038334	4.094281	4.146180	4.194556	4.239839	4.430560

P\n	60	70	80	90	100	
0.0001	1.962927	2.138769	2.287571	2.416107	2.528946	
0.0005	2.079111	2.253159	2.400223	2.527129	2.638465	
0.0010	2.134719	2.307836	2.454022	2.580124	2.690722	
0.0050	2.281370	2.451846	2.595607	2.719509	2.828117	
0.0100	2.354504	2.523577	2.666078	2.788853	2.896450	
0.0250	2.464320	2.631201	2.771760	2.892813	2.998876	
0.0500	2.561095	2.725973	2.864783	2.984297	3.088995	
0.1000	2.675385	2.837833	2.974540	3.092219	3.195298	
0.2000	2.817750	2.977097	3.111150	3.226527	3.327585	
0.3000	2.923159	3.080174	3.212248	3.325917	3.425483	
0.4000	3.015093	3.170060	3.300404	3.412586	3.510853	
0.5000	3.102603	3.255614	3.384312	3.495083	3.592122	
0.6000	3.191667	3.342684	3.469712	3.579053	3.674850	
0.7000	3.288705	3.437554	3.562769	3.670565	3.765018	
0.8000	3.404611	3.550882	3.673948	3.779913	3.872781	
0.9000	3.569623	3.712260	3.832306	3.935703	4.026350	
0.9500	3.709725	3.849321	3.966847	4.068106	4.156905	
0.9750	3.834117	3.971055	4.086382	4.185778	4.272957	
0.9900	3.982175	4.116009	4.228771	4.325995	4.411309	
0.9950	4.085136	4.216852	4.327867	4.423611	4.507649	
0.9990	4.303010	4.430365	4.537783	4.630484	4.711896	
0.9995	4.389904	4.515567	4.621590	4.713112	4.793505	
0.9999	4.579719	4.701785	4.804841	4.893853	4.972081	

PERCENTAGE POINTS OF THE 4th QUASI-RANGE

P\n	2	3	4	5	6	7
0.0001						
0.0005						
0.0010						
0.0050						
0.0100						
0.0250						
0.0500						
0.1000						
0.2000						
0.3000						
0.4000						
0.5000						
0.6000						
0.7000						
0.8000						
0.9000						
0.9500						
0.9750						
0.9900						
0.9950						
0.9990						
0.9995						
0.9999						

P\n	8	9	10	11	12	13
0.0001			0.000027	0.003645	0.020969	0.053495
0.0005			0.000135	0.008191	0.036308	0.081495
0.0010			0.000271	0.011627	0.046114	0.097990
0.0050			0.001356	0.026420	0.081168	0.151994
0.0100			0.002719	0.037830	0.104247	0.184889
0.0250			0.006842	0.061368	0.146599	0.241922
0.0500			0.013841	0.089495	0.191931	0.299738
0.1000			0.028340	0.132700	0.255298	0.376844
0.2000			0.059611	0.202670	0.348818	0.485422
0.3000			0.094557	0.266186	0.427923	0.573976
0.4000			0.134257	0.329485	0.503232	0.656289
0.5000			0.180363	0.396224	0.579922	0.738586
0.6000			0.235605	0.470199	0.662516	0.825868
0.7000			0.305009	0.557145	0.757162	0.924525
0.8000			0.399625	0.668695	0.875739	1.046533
0.9000			0.553608	0.839647	1.053083	1.226556
0.9500			0.699036	0.993723	1.209790	1.383864
0.9750			0.837067	1.135785	1.352458	1.526040
0.9900			1.009735	1.309700	1.525417	1.697426
0.9950			1.133881	1.432923	1.647129	1.817549
0.9990			1.404008	1.697621	1.906977	2.073068
0.9995			1.513601	1.804094	2.011059	2.175157
0.9999			1.754903	2.037362	2.238517	2.397927

PERCENTAGE POINTS OF THE 4th QUASI-RANGE

P\n	14	15	16	17	18	19
0.0001	0.097337	0.148491	0.203940	0.261592	0.320042	0.378360
0.0005	0.137385	0.199178	0.263738	0.329096	0.394034	0.457810
0.0010	0.159867	0.226738	0.295525	0.364379	0.432207	0.498379
0.0050	0.229787	0.309619	0.388877	0.466195	0.540890	0.612658
0.0100	0.270448	0.356355	0.440376	0.521453	0.599134	0.673293
0.0250	0.338609	0.432970	0.523473	0.609570	0.691176	0.768428
0.0500	0.405489	0.506530	0.602033	0.691922	0.776436	0.855935
0.1000	0.492195	0.600111	0.700638	0.794254	0.881564	0.963174
0.2000	0.610862	0.725756	0.831244	0.928422	1.018316	1.101804
0.3000	0.705506	0.824483	0.932748	1.031853	1.123079	1.207478
0.4000	0.792191	0.914002	1.024135	1.124476	1.216505	1.301404
0.5000	0.877878	1.001808	1.113276	1.214443	1.306957	1.392106
0.6000	0.967888	1.093443	1.205865	1.307558	1.400316	1.485516
0.7000	1.068759	1.195531	1.308578	1.410524	1.503292	1.588345
0.8000	1.192483	1.320044	1.433341	1.535206	1.627687	1.712322
0.9000	1.373472	1.501106	1.613978	1.715134	1.806743	1.890411
0.9500	1.530492	1.657407	1.769345	1.869463	1.959991	2.042572
0.9750	1.671738	1.797543	1.908307	2.007244	2.096613	2.178070
0.9900	1.841374	1.965410	2.074452	2.171740	2.259543	2.339517
0.9950	1.959957	2.082541	2.190225	2.286247	2.372870	2.451743
0.9990	2.211595	2.330679	2.435187	2.528311	2.612274	2.688694
0.9995	2.311967	2.429541	2.532704	2.624616	2.707477	2.782889
0.9999	2.530777	2.644919	2.745052	2.834255	2.914670	2.987855

P\n	20	22	24	26	28	30
0.0001	0.435939	0.547484	0.653096	0.752371	0.845439	0.932658
0.0005	0.519988	0.638706	0.749489	0.852527	0.948351	1.037595
0.0010	0.562550	0.684322	0.797258	0.901828	0.998750	1.088778
0.0050	0.681412	0.810072	0.927732	1.035571	1.134753	1.226331
0.0100	0.743972	0.875452	0.994980	1.104058	1.204052	1.296146
0.0250	0.841559	0.976551	1.098324	1.208820	1.309682	1.402267
0.0500	0.930813	1.068228	1.191462	1.302808	1.404119	1.496882
0.1000	1.039650	1.179185	1.303586	1.415508	1.517012	1.609720
0.2000	1.179637	1.320813	1.445928	1.558002	1.659311	1.751603
0.3000	1.285918	1.427684	1.552870	1.664712	1.765610	1.857385
0.4000	1.380128	1.522034	1.647008	1.758443	1.858825	1.950027
0.5000	1.470913	1.612664	1.737230	1.848123	1.947898	2.038464
0.6000	1.564243	1.705583	1.829552	1.939760	2.038814	2.128653
0.7000	1.666816	1.807455	1.930591	2.039920	2.138089	2.227059
0.8000	1.790293	1.929797	2.051730	2.159855	2.256852	2.344698
0.9000	1.967370	2.104809	2.224713	2.330896	2.426055	2.512171
0.9500	2.118454	2.253817	2.371776	2.476151	2.569635	2.654197
0.9750	2.252870	2.386205	2.502311	2.604993	2.696927	2.780065
0.9900	2.412915	2.543669	2.657459	2.758052	2.848091	2.929500
0.9950	2.524112	2.652994	2.765125	2.864234	2.952936	3.033130
0.9990	2.758790	2.883586	2.992132	3.088061	3.173912	3.251531
0.9995	2.852058	2.975197	3.082302	3.176959	3.261676	3.338273
0.9999	3.054982	3.174494	3.278456	3.370351	3.452610	3.526999

PERCENTAGE POINTS OF THE 4th QUASI-RANGE

P\n	32	34	36	38	40	50
0.0001	1.014469	1.091327	1.163671	1.231900	1.296386	1.573033
0.0005	1.120892	1.198834	1.271954	1.340728	1.405575	1.682430
0.0010	1.172633	1.250963	1.324344	1.393284	1.458223	1.734907
0.0050	1.311223	1.390216	1.463983	1.533102	1.598065	1.873567
0.0100	1.381345	1.460493	1.534306	1.603391	1.668260	1.942825
0.0250	1.487692	1.566882	1.640604	1.709502	1.774116	2.046904
0.0500	1.582302	1.661360	1.734862	1.803478	1.867768	2.138669
0.1000	1.694919	1.773644	1.846740	1.914901	1.978706	2.247050
0.2000	1.836247	1.914332	1.986735	2.054176	2.117246	2.381997
0.3000	1.941452	2.018928	2.090708	2.157524	2.219974	2.481827
0.4000	2.033494	2.110360	2.181533	2.247752	2.309619	2.568811
0.5000	2.121287	2.197515	2.268064	2.333675	2.394955	2.651520
0.6000	2.210759	2.286289	2.356163	2.421124	2.481781	2.735594
0.7000	2.308324	2.383045	2.452146	2.516368	2.576319	2.827064
0.8000	2.424890	2.498593	2.566727	2.630032	2.689115	2.936121
0.9000	2.590739	2.662916	2.729616	2.791572	2.849382	3.090975
0.9500	2.731322	2.802156	2.867602	2.928385	2.985094	3.222050
0.9750	2.855877	2.925495	2.989811	3.049541	3.105266	3.338103
0.9900	3.003726	3.071882	3.134845	3.193318	3.247869	3.475825
0.9950	3.106246	3.173383	3.235405	3.293005	3.346744	3.571333
0.9990	3.322304	3.387294	3.447340	3.503111	3.555150	3.772721
0.9995	3.408118	3.472261	3.531528	3.586579	3.637951	3.852772
0.9999	3.594844	3.657162	3.714754	3.768259	3.818197	4.027127

P\n	60	70	80	90	100	
0.0001	1.793145	1.974486	2.127884	2.260321	2.376520	
0.0005	1.901540	2.081512	2.233466	2.364493	2.479357	
0.0010	1.953392	2.132623	2.283833	2.414152	2.528355	
0.0050	2.090017	2.267073	2.416186	2.544551	2.656955	
0.0100	2.158079	2.333946	2.481951	2.609302	2.720786	
0.0250	2.260171	2.434146	2.580424	2.706216	2.816292	
0.0500	2.350021	2.522241	2.666945	2.791332	2.900149	
0.1000	2.455980	2.626036	2.768833	2.891533	2.998851	
0.2000	2.587709	2.754969	2.895334	3.015904	3.121339	
0.3000	2.685047	2.850180	2.988717	3.107698	3.211735	
0.4000	2.769796	2.933049	3.069978	3.187570	3.290388	
0.5000	2.850339	3.011782	3.147177	3.263444	3.365107	
0.6000	2.932175	3.091766	3.225596	3.340519	3.441010	
0.7000	3.021182	3.178746	3.310873	3.424336	3.523555	
0.8000	3.127273	3.282416	3.412514	3.524243	3.621958	
0.9000	3.277890	3.429595	3.556828	3.666117	3.761717	
0.9500	3.405375	3.554186	3.679015	3.786263	3.880098	
0.9750	3.518260	3.664528	3.787253	3.892716	3.985009	
0.9900	3.652252	3.795537	3.915796	4.019171	4.109663	
0.9950	3.745197	3.886438	4.005011	4.106961	4.196223	
0.9990	3.941268	4.078277	4.193365	4.292370	4.379095	
0.9995	4.019243	4.154603	4.268334	4.366193	4.451931	
0.9999	4.189158	4.320995	4.431830	4.527244	4.610878	

PERCENTAGE POINTS OF THE 5th QUASI-RANGE

P\n	2	3	4	5	6	7
0.0001						
0.0005						
0.0010						
0.0050						
0.0100						
0.0250						
0.0500						
0.1000						
0.2000						
0.3000						
0.4000						
0.5000						
0.6000						
0.7000						
0.8000						
0.9000						
0.9500						
0.9750						
0.9900						
0.9950						
0.9990						
0.9995						
0.9999						

P\n	8	9	10	11	12	13
0.0001					0.000022	0.003028
0.0005					0.000111	0.006805
0.0010					0.000223	0.009661
0.0050					0.001116	0.021963
0.0100					0.002238	0.031462
0.0250					0.005632	0.051083
0.0500					0.011397	0.074574
0.1000					0.023350	0.110750
0.2000					0.049177	0.169564
0.3000					0.078115	0.223185
0.4000					0.111086	0.276833
0.5000					0.149498	0.333612
0.6000					0.195689	0.396792
0.7000					0.253969	0.471360
0.8000					0.333839	0.567488
0.9000					0.464777	0.715713
0.9500					0.589421	0.850145
0.9750					0.708514	0.974723
0.9900					0.858455	1.127960
0.9950					0.966852	1.236964
0.9990					1.204205	1.472171
0.9995					1.301015	1.567139
0.9999					1.515069	1.775815

PERCENTAGE POINTS OF THE 5th QUASI-RANGE

P\n	14	15	16	17	18	19
0.0001	0.017567	0.045184	0.082846	0.127282	0.175954	0.227047
0.0005	0.030434	0.068896	0.117071	0.170967	0.227889	0.286083
0.0010	0.038668	0.082886	0.136319	0.194767	0.255554	0.317006
0.0050	0.068145	0.128785	0.196332	0.266535	0.337024	0.406478
0.0100	0.087590	0.156816	0.231334	0.307127	0.382102	0.455176
0.0250	0.123349	0.205541	0.290169	0.373850	0.455028	0.533021
0.0500	0.161733	0.255090	0.348084	0.438113	0.524175	0.605972
0.1000	0.215564	0.321405	0.423426	0.520131	0.611223	0.696866
0.2000	0.295370	0.415209	0.526978	0.630687	0.726925	0.816411
0.3000	0.363190	0.492053	0.609901	0.717857	0.817134	0.908828
0.4000	0.428005	0.563735	0.686091	0.797123	0.898548	0.991761
0.5000	0.494245	0.635631	0.761613	0.875060	0.978125	1.072460
0.6000	0.565836	0.712115	0.841153	0.956577	1.060939	1.156119
0.7000	0.648171	0.798835	0.930523	1.047595	1.152979	1.248772
0.8000	0.751743	0.906443	1.040453	1.158869	1.264999	1.361153
0.9000	0.907425	1.065875	1.201810	1.321132	1.427563	1.523636
0.9500	1.045684	1.205756	1.342258	1.461582	1.567688	1.663247
0.9750	1.172052	1.332577	1.468915	1.587760	1.693221	1.788047
0.9900	1.325812	1.485895	1.621376	1.739179	1.843521	1.937208
0.9950	1.434342	1.593606	1.728150	1.844985	1.948370	2.041127
0.9990	1.666831	1.823320	1.955182	2.069478	2.170474	2.260991
0.9995	1.760217	1.915297	2.045889	2.159030	2.258971	2.348518
0.9999	1.964747	2.116331	2.243871	2.354299	2.451798	2.539126

P\n	20	22	24	26	28	30
0.0001	0.279302	0.384116	0.486325	0.584157	0.676960	0.764631
0.0005	0.344417	0.458874	0.568166	0.671247	0.767976	0.858606
0.0010	0.378082	0.496803	0.609160	0.714468	0.812833	0.904674
0.0050	0.474171	0.602968	0.722383	0.832709	0.934679	1.029122
0.0100	0.525807	0.658975	0.781367	0.893752	0.997156	1.092600
0.0250	0.607590	0.746505	0.872716	0.987673	1.092817	1.189427
0.0500	0.683534	0.826724	0.955686	1.072429	1.178723	1.276056
0.1000	0.777403	0.924734	1.056256	1.174579	1.281821	1.379679
0.2000	0.899849	1.051066	1.184838	1.304417	1.412287	1.510364
0.3000	0.993885	1.147158	1.281998	1.402060	1.510052	1.608025
0.4000	1.077894	1.232451	1.367856	1.488068	1.595961	1.693681
0.5000	1.159354	1.314733	1.450392	1.570538	1.678178	1.775537
0.6000	1.243548	1.399403	1.535068	1.654961	1.762205	1.859087
0.7000	1.336538	1.492543	1.627955	1.747386	1.854057	1.950314
0.8000	1.449022	1.604765	1.739569	1.858225	1.964048	2.059430
0.9000	1.611181	1.765859	1.899318	2.016529	2.120890	2.214838
0.9500	1.750164	1.903419	2.035384	2.151116	2.254053	2.346644
0.9750	1.874191	2.025867	2.156293	2.270563	2.372127	2.463435
0.9900	2.022221	2.171717	2.300109	2.412499	2.512330	2.602039
0.9950	2.125246	2.273074	2.399952	2.510968	2.609546	2.698110
0.9990	2.343010	2.487013	2.610497	2.718480	2.814327	2.900414
0.9995	2.429641	2.572038	2.694120	2.800862	2.895601	2.980689
0.9999	2.618217	2.757010	2.875974	2.979978	3.072282	3.155185

PERCENTAGE POINTS OF THE 5th QUASI-RANGE

P\n	32	34	36	38	40	50
0.0001	0.847333	0.925345	0.998995	1.068614	1.134524	1.418079
0.0005	0.943548	1.023262	1.098203	1.168798	1.235436	1.520416
0.0010	0.990516	1.070898	1.146332	1.217286	1.284181	1.569533
0.0050	1.116844	1.198574	1.274959	1.346565	1.413884	1.699375
0.0100	1.181014	1.263211	1.339900	1.411686	1.479092	1.764244
0.0250	1.278604	1.361277	1.438231	1.510128	1.577531	1.861737
0.0500	1.365658	1.448547	1.525566	1.597420	1.664700	1.947686
0.1000	1.469519	1.552446	1.629363	1.701017	1.768028	2.049173
0.2000	1.600151	1.682845	1.759405	1.830617	1.897130	2.175471
0.3000	1.697561	1.779908	1.856063	1.926832	1.992880	2.268841
0.4000	1.782871	1.864815	1.940531	2.010846	2.076433	2.350143
0.5000	1.864301	1.945784	2.021023	2.090854	2.155958	2.427397
0.6000	1.947334	2.028281	2.102980	2.172274	2.236851	2.505870
0.7000	2.037913	2.118209	2.192265	2.260931	2.324898	2.591174
0.8000	2.146159	2.225601	2.298828	2.366695	2.429893	2.692777
0.9000	2.300178	2.378289	2.450246	2.516901	2.578947	2.836838
0.9500	2.430701	2.507602	2.578416	2.643995	2.705023	2.958574
0.9750	2.546294	2.622074	2.691840	2.756436	2.816540	3.066194
0.9900	2.683420	2.757828	2.826318	2.889723	2.948714	3.193704
0.9950	2.778439	2.851877	2.919469	2.982038	3.040249	3.281998
0.9990	2.978481	3.049845	3.115524	3.176321	3.232883	3.467814
0.9995	3.057849	3.128384	3.193300	3.253392	3.309301	3.541540
0.9999	3.230368	3.299099	3.362360	3.420926	3.475421	3.701857

P\n	60	70	80	90	100	
0.0001	1.644147	1.830465	1.988044	2.124035	2.243296	
0.0005	1.746150	1.931516	2.087935	2.222723	2.340808	
0.0010	1.794941	1.979754	2.135556	2.269730	2.387224	
0.0050	1.923473	2.106567	2.260587	2.393034	2.508907	
0.0100	1.987478	2.169593	2.322648	2.454189	2.569222	
0.0250	2.083444	2.263959	2.415490	2.545622	2.659363	
0.0500	2.167852	2.346848	2.496971	2.625821	2.738400	
0.1000	2.267317	2.444409	2.592805	2.720103	2.831290	
0.2000	2.390846	2.565431	2.711602	2.836926	2.946351	
0.3000	2.482019	2.654674	2.799155	2.922997	3.031109	
0.4000	2.561324	2.732253	2.875242	2.997780	3.104743	
0.5000	2.636620	2.805878	2.947433	3.068725	3.174594	
0.6000	2.713050	2.880587	3.020673	3.140695	3.245452	
0.7000	2.796085	2.961729	3.100209	3.218847	3.322396	
0.8000	2.894932	3.058298	3.194857	3.311847	3.413959	
0.9000	3.035015	3.195124	3.328955	3.443614	3.543702	
0.9500	3.153348	3.310696	3.442228	3.554928	3.653320	
0.9750	3.257944	3.412855	3.542363	3.653344	3.750252	
0.9900	3.381870	3.533906	3.661034	3.770000	3.865168	
0.9950	3.467687	3.617745	3.743240	3.850825	3.944802	
0.9990	3.648332	3.794274	3.916380	4.021103	4.112616	
0.9995	3.720027	3.864358	3.985139	4.088745	4.179297	
0.9999	3.875981	4.016856	4.134802	4.236018	4.324514	

PERCENTAGE POINTS OF THE 6th QUASI-RANGE

P\n	14	15	16	17	18	19
0.0001	0.000019	0.002589	0.015114	0.039108	0.072115	0.111393
0.0005	0.000095	0.005819	0.026195	0.059671	0.101999	0.149783
0.0010	0.000189	0.008262	0.033289	0.071816	0.118827	0.170730
0.0050	0.000948	0.018791	0.058718	0.111725	0.171400	0.234030
0.0100	0.001901	0.026926	0.075516	0.136145	0.202130	0.269917
0.0250	0.004785	0.043745	0.106456	0.178676	0.253893	0.329033
0.0500	0.009685	0.063908	0.139735	0.222030	0.304975	0.386112
0.1000	0.019850	0.095016	0.186521	0.280210	0.371609	0.459151
0.2000	0.041844	0.145732	0.256119	0.362795	0.463504	0.557920
0.3000	0.066531	0.192118	0.315473	0.430685	0.537335	0.636033
0.4000	0.094715	0.238661	0.372367	0.494195	0.605348	0.707230
0.5000	0.127625	0.288060	0.430674	0.558058	0.672920	0.777376
0.6000	0.167301	0.343190	0.493867	0.626167	0.744246	0.850890
0.7000	0.217516	0.408466	0.566756	0.703591	0.824569	0.933133
0.8000	0.286597	0.492927	0.658745	0.799937	0.923610	1.033891
0.9000	0.400468	0.623796	0.797594	0.943188	1.069424	1.181198
0.9500	0.509519	0.743093	0.921426	1.069318	1.196722	1.309021
0.9750	0.614254	0.854106	1.034993	1.183992	1.311786	1.424079
0.9900	0.746797	0.991203	1.173620	1.322987	1.450591	1.562398
0.9950	0.843045	1.089059	1.271732	1.420850	1.547974	1.659192
0.9990	1.054903	1.301039	1.482552	1.630073	1.755450	1.864891
0.9995	1.141704	1.386916	1.567455	1.714020	1.838481	1.947056
0.9999	1.334330	1.576121	1.753792	1.897800	2.019943	2.126398

P\n	20	22	24	26	28	30
0.0001	0.154762	0.247876	0.343528	0.437818	0.528902	0.615978
0.0005	0.200677	0.306051	0.410903	0.512111	0.608435	0.699516
0.0010	0.225175	0.336179	0.445142	0.549379	0.647957	0.740734
0.0050	0.297477	0.422356	0.541162	0.652487	0.756241	0.852838
0.0100	0.337578	0.468770	0.591921	0.706296	0.812227	0.910394
0.0250	0.402588	0.542423	0.671381	0.789750	0.898470	0.998603
0.0500	0.464378	0.610963	0.744335	0.865661	0.976394	1.077900
0.1000	0.542358	0.695859	0.833625	0.957811	1.070421	1.173153
0.2000	0.646312	0.806873	0.948949	1.075816	1.190084	1.293810
0.3000	0.727585	0.892319	1.036825	1.165109	1.280173	1.384300
0.4000	0.801089	0.968781	1.114925	1.244093	1.359587	1.463857
0.5000	0.873066	1.043029	1.190350	1.320084	1.435781	1.540028
0.6000	0.948098	1.119869	1.268042	1.398102	1.513819	1.617904
0.7000	1.031633	1.204846	1.353587	1.483745	1.599295	1.703057
0.8000	1.133486	1.307775	1.456760	1.586726	1.701848	1.805054
0.9000	1.281620	1.456398	1.605030	1.734232	1.848387	1.950531
0.9500	1.409578	1.583968	1.731770	1.859952	1.973016	2.074059
0.9750	1.524395	1.697936	1.844669	1.971717	2.083648	2.183589
0.9900	1.662068	1.834095	1.979227	2.104699	2.215121	2.313635
0.9950	1.758222	1.928931	2.072780	2.197040	2.306331	2.403795
0.9990	1.962170	2.129546	2.270330	2.391792	2.498529	2.593655
0.9995	2.043518	2.209404	2.348865	2.469145	2.574819	2.668983
0.9999	2.220909	2.383313	2.519745	2.637357	2.740653	2.832677

PERCENTAGE POINTS OF THE 6th QUASI-RANGE

P\n	32	34	36	38	40	50
0.0001	0.698790	0.777357	0.851841	0.922468	0.989488	1.278936
0.0005	0.785410	0.866366	0.942708	1.014781	1.082926	1.375096
0.0010	0.827915	0.909851	0.986944	1.059589	1.128171	1.421293
0.0050	0.942860	1.026917	1.105591	1.179414	1.248857	1.543533
0.0100	1.001553	1.086434	1.165700	1.239942	1.309674	1.604656
0.0250	1.091151	1.177004	1.256940	1.331626	1.401633	1.696563
0.0500	1.171378	1.257849	1.338175	1.413085	1.483192	1.777623
0.1000	1.267412	1.354352	1.434926	1.509922	1.579999	1.873362
0.2000	1.388616	1.475796	1.556395	1.631265	1.701110	1.992521
0.3000	1.479243	1.566388	1.646832	1.721465	1.791017	2.080607
0.4000	1.558761	1.645744	1.725948	1.800289	1.869514	2.157298
0.5000	1.634771	1.721503	1.801398	1.875395	1.944256	2.230152
0.6000	1.712371	1.798759	1.878269	1.951860	2.020302	2.304135
0.7000	1.797112	1.883037	1.962058	2.035147	2.103086	2.384528
0.8000	1.898484	1.983752	2.062105	2.134527	2.201810	2.480233
0.9000	2.042865	2.127034	2.204307	2.275678	2.341943	2.615828
0.9500	2.165308	2.248428	2.324690	2.395094	2.460435	2.730301
0.9750	2.273784	2.355900	2.431211	2.500714	2.565201	2.831409
0.9900	2.402488	2.483344	2.557472	2.625863	2.689304	2.951085
0.9950	2.491671	2.571618	2.644899	2.712499	2.775199	3.033874
0.9990	2.679384	2.757351	2.828799	2.894695	2.955807	3.207885
0.9995	2.753834	2.830997	2.901705	2.966915	3.027391	3.276843
0.9999	2.915591	2.990985	3.060069	3.123781	3.182868	3.426621

P\n	60	70	80	90	100	
0.0001	1.510374	1.701261	1.862703	2.001987	2.124091	
0.0005	1.606893	1.797250	1.957814	2.096096	2.217173	
0.0010	1.653074	1.843066	2.003140	2.140900	2.261455	
0.0050	1.774758	1.963491	2.122092	2.258348	2.377449	
0.0100	1.835358	2.023323	2.181101	2.316553	2.434892	
0.0250	1.926216	2.112875	2.269328	2.403515	2.520673	
0.0500	2.006119	2.191497	2.346704	2.479728	2.595814	
0.1000	2.100250	2.283982	2.437640	2.569242	2.684034	
0.2000	2.217101	2.398616	2.550246	2.680022	2.793169	
0.3000	2.303298	2.483075	2.633152	2.761544	2.873455	
0.4000	2.378234	2.556440	2.705133	2.832303	2.943129	
0.5000	2.449343	2.626015	2.773371	2.899367	3.009157	
0.6000	2.521482	2.696562	2.842541	2.967337	3.076068	
0.7000	2.599805	2.773121	2.917586	3.041069	3.148649	
0.8000	2.692968	2.864147	3.006792	3.128705	3.234912	
0.9000	2.824847	2.992946	3.132992	3.252674	3.356938	
0.9500	2.936107	3.101575	3.239417	3.357216	3.459846	
0.9750	3.034336	3.197467	3.333362	3.449504	3.550698	
0.9900	3.150575	3.310935	3.444530	3.55872C	3.658229	
0.9950	3.230975	3.389420	3.521431	3.634281	3.732633	
0.9990	3.399962	3.554402	3.683113	3.793174	3.889125	
0.9995	3.466935	3.619799	3.747215	3.856185	3.951197	
0.9999	3.612428	3.761902	3.886537	3.993164	4.086164	

PERCENTAGE POINTS OF THE 7th QUASI-RANGE

P\n	14	15	16	17	18	19
0.0001			0.000016	0.002261	0.013261	0.034472
0.0005			0.000082	0.005082	0.022991	0.052625
0.0010			0.000165	0.007216	0.029223	0.063354
0.0050			0.000824	0.016418	0.051579	0.098655
0.0100			0.001652	0.023530	0.066364	0.120288
0.0250			0.004159	0.038246	0.093628	0.158022
0.0500			0.008419	0.055906	0.122998	0.196559
0.1000			0.017261	0.083189	0.164368	0.248384
0.2000			0.036410	0.127763	0.226071	0.322155
0.3000			0.057933	0.168629	0.278838	0.382970
0.4000			0.082540	0.209724	0.329538	0.439993
0.5000			0.111322	0.253437	0.381614	0.497455
0.6000			0.146088	0.302334	0.438180	0.558866
0.7000			0.190191	0.360375	0.503581	0.628826
0.8000			0.251041	0.435697	0.586344	0.716093
0.9000			0.351767	0.552868	0.711703	0.846244
0.9500			0.448687	0.660124	0.823905	0.961194
0.9750			0.542158	0.760280	0.927109	1.065964
0.9900			0.660941	0.884390	1.053440	1.193253
0.9950			0.747516	0.973236	1.143065	1.283052
0.9990			0.938924	1.166359	1.336182	1.475476
0.9995			1.017649	1.244828	1.414142	1.552833
0.9999			1.192908	1.418133	1.585569	1.722453

P\n	20	22	24	26	28	30
0.0001	0.063848	0.138147	0.222867	0.310820	0.398315	0.483498
0.0005	0.090369	0.179299	0.275458	0.372173	0.466383	0.556746
0.0010	0.105319	0.201286	0.302733	0.403394	0.500572	0.593188
0.0050	0.152094	0.266289	0.380884	0.491097	0.595306	0.693166
0.0100	0.179483	0.302412	0.423057	0.537541	0.644823	0.744929
0.0250	0.225697	0.361077	0.490090	0.610354	0.721718	0.824756
0.0500	0.271395	0.416948	0.552583	0.677311	0.791763	0.896966
0.1000	0.331136	0.487603	0.630132	0.759394	0.876901	0.984199
0.2000	0.413754	0.582032	0.731758	0.865602	0.986096	1.095355
0.3000	0.480312	0.656032	0.810134	0.946664	1.068833	1.179129
0.4000	0.541759	0.723082	0.880377	1.018797	1.142091	1.253036
0.5000	0.602928	0.788844	0.948674	1.088533	1.212632	1.323990
0.6000	0.667616	0.857503	1.019441	1.160434	1.285108	1.396706
0.7000	0.740606	0.934059	1.097797	1.239677	1.364727	1.476395
0.8000	0.830799	1.027560	1.192829	1.335344	1.460535	1.572060
0.9000	0.963936	1.163820	1.330259	1.472988	1.597887	1.708839
0.9500	1.080479	1.281757	1.448400	1.590776	1.715046	1.825232
0.9750	1.186046	1.387748	1.554065	1.695787	1.819259	1.928590
0.9900	1.313647	1.515020	1.680433	1.821035	1.943316	2.051455
0.9950	1.403319	1.604015	1.768525	1.908164	2.029490	2.136709
0.9990	1.594730	1.793029	1.955041	2.092259	2.211300	2.316382
0.9995	1.671456	1.868505	2.029342	2.165480	2.283531	2.387706
0.9999	1.839353	2.033231	2.191240	2.324848	2.440624	2.542740

PERCENTAGE POINTS OF THE 7th QUASI-RANGE

P\n	32	34	36	38	40	50
0.0001	0.565485	0.643912	0.718700	0.789917	0.857713	1.152052
0.0005	0.642760	0.724342	0.801616	0.874802	0.944160	1.242619
0.0010	0.680929	0.763846	0.842158	0.916154	0.986149	1.286190
0.0050	0.784861	0.870784	0.951393	1.027152	1.098495	1.401634
0.0100	0.838284	0.925444	1.006978	1.083426	1.155282	1.459433
0.0250	0.920240	1.008953	1.091622	1.168894	1.241337	1.546417
0.0500	0.993986	1.083790	1.167226	1.245029	1.317826	1.623199
0.1000	1.082656	1.173440	1.257530	1.335750	1.408789	1.713945
0.2000	1.195089	1.286679	1.371246	1.449705	1.522812	1.826960
0.3000	1.279483	1.371409	1.456117	1.534578	1.607590	1.910538
0.4000	1.353728	1.445787	1.530486	1.608845	1.681686	1.983318
0.5000	1.424848	1.516910	1.601503	1.679681	1.752293	2.052465
0.6000	1.497590	1.589543	1.673939	1.751861	1.824179	2.122684
0.7000	1.577164	1.668883	1.752972	1.830541	1.902479	2.198985
0.8000	1.672514	1.763818	1.847430	1.924491	1.995903	2.289806
0.9000	1.808568	1.899062	1.981826	2.058023	2.128575	2.418445
0.9500	1.924132	2.013776	2.095690	2.171052	2.240790	2.527000
0.9750	2.026622	2.115410	2.196491	2.271049	2.340015	2.622835
0.9900	2.148324	2.235994	2.316007	2.389548	2.457548	2.736209
0.9950	2.232703	2.319543	2.398774	2.471579	2.538884	2.814594
0.9990	2.410385	2.495371	2.572872	2.644059	2.709849	2.979217
0.9995	2.480875	2.565092	2.641882	2.712409	2.777584	3.044403
0.9999	2.634035	2.716537	2.791748	2.860816	2.924636	3.185881

P\n	60	70	80	90	100	
0.0001	1.388338	1.583448	1.748495	1.890867	2.015638	
0.0005	1.479992	1.675008	1.839462	1.981034	2.104927	
0.0010	1.523872	1.718717	1.882811	2.023948	2.147386	
0.0050	1.639557	1.833616	1.996548	2.136403	2.258551	
0.0100	1.697194	1.890703	2.052954	2.192108	2.313569	
0.0250	1.783633	1.976137	2.137266	2.275295	2.395683	
0.0500	1.859666	2.051133	2.211178	2.348161	2.467565	
0.1000	1.949244	2.139330	2.298002	2.433691	2.551897	
0.2000	2.060437	2.248604	2.405450	2.539455	2.656126	
0.3000	2.142449	2.329075	2.484501	2.617220	2.732730	
0.4000	2.213732	2.398946	2.553095	2.684671	2.799157	
0.5000	2.281358	2.465177	2.618084	2.748558	2.862061	
0.6000	2.349944	2.532302	2.683923	2.813263	2.925761	
0.7000	2.424384	2.605110	2.755308	2.883404	2.994803	
0.8000	2.512888	2.691619	2.840097	2.966697	3.076783	
0.9000	2.638089	2.813918	2.959924	3.084388	3.192606	
0.9500	2.743630	2.916958	3.060854	3.183507	3.290149	
0.9750	2.836740	3.007833	3.149855	3.270908	3.376161	
0.9900	2.946830	3.115256	3.255056	3.374218	3.477835	
0.9950	3.022918	3.189492	3.327755	3.445615	3.548107	
0.9990	3.182674	3.345354	3.480401	3.595542	3.695691	
0.9995	3.245924	3.407064	3.540846	3.654918	3.754148	
0.9999	3.383202	3.541016	3.672069	3.783844	3.881101	

PERCENTAGE POINTS OF THE 8th QUASI-RANGE

P\n	14	15	16	17	18	19
0.0001					0.000015	0.002006
0.0005					0.000073	0.004511
0.0010					0.000145	0.006405
0.0050					0.000729	0.014576
0.0100					0.001461	0.020895
0.0250					0.003678	0.033974
0.0500					0.007446	0.049682
0.1000					0.015269	0.073976
0.2000					0.032223	0.113731
0.3000					0.051299	0.150249
0.4000					0.073133	0.187035
0.5000					0.098705	0.226234
0.6000					0.129639	0.270162
0.7000					0.168952	0.322411
0.8000					0.223318	0.390380
0.9000					0.313613	0.496458
0.9500					0.400827	0.593897
0.9750					0.485223	0.685155
0.9900					0.592845	0.798569
0.9950					0.671528	0.879963
0.9990					0.846142	1.057426
0.9995					0.918199	1.129722
0.9999					1.079059	1.289742

P\n	20	22	24	26	28	30
0.0001	0.011813	0.057282	0.124764	0.202477	0.283872	0.365475
0.0005	0.020484	0.081121	0.162053	0.250472	0.340210	0.428310
0.0010	0.026041	0.094570	0.181998	0.275392	0.368913	0.459906
0.0050	0.045986	0.136700	0.241049	0.346907	0.449659	0.547573
0.0100	0.059188	0.161405	0.273919	0.385562	0.492486	0.593462
0.0250	0.083556	0.203147	0.327378	0.447089	0.559713	0.664804
0.0500	0.109838	0.244490	0.378376	0.504540	0.621624	0.729873
0.1000	0.146914	0.298635	0.442984	0.575948	0.697626	0.809062
0.2000	0.202329	0.373690	0.529517	0.669706	0.796132	0.910770
0.3000	0.249824	0.434292	0.597471	0.742143	0.871429	0.987933
0.4000	0.295547	0.490343	0.659140	0.807151	0.938510	1.056320
0.5000	0.342597	0.546235	0.719711	0.870433	1.003424	1.122224
0.6000	0.393798	0.605439	0.783037	0.936078	1.070415	1.189986
0.7000	0.453114	0.672352	0.853745	1.008844	1.144313	1.264481
0.8000	0.528346	0.755190	0.940230	1.097203	1.233614	1.354192
0.9000	0.642636	0.877761	1.066505	1.225172	1.362249	1.482919
0.9500	0.745246	0.985311	1.176002	1.335338	1.472450	1.592816
0.9750	0.839868	1.082918	1.274552	1.433981	1.570781	1.690632
0.9900	0.955982	1.201115	1.393052	1.552075	1.688153	1.807140
0.9950	1.038533	1.284307	1.476008	1.634469	1.769855	1.888107
0.9990	1.216855	1.462205	1.652430	1.809089	1.942602	2.059011
0.9995	1.288998	1.533623	1.722955	1.878707	2.011348	2.126933
0.9999	1.447918	1.690102	1.877014	2.030495	2.161039	2.274695

PERCENTAGE POINTS OF THE 8th QUASI-RANGE

P\n	32	34	36	38	40	50
0.0001	0.445460	0.522900	0.597361	0.668690	0.736886	1.035055
0.0005	0.513384	0.594838	0.672487	0.746364	0.816607	1.120422
0.0010	0.547212	0.630404	0.709418	0.784372	0.855472	1.161561
0.0050	0.640133	0.727354	0.809486	0.886869	0.959867	1.270755
0.0100	0.688303	0.777245	0.860686	0.939069	1.012833	1.325514
0.0250	0.762666	0.853850	0.938968	1.018611	1.093320	1.408025
0.0500	0.830008	0.922848	1.009179	1.089709	1.165062	1.480945
0.1000	0.911445	1.005883	1.093351	1.174684	1.250593	1.567214
0.2000	1.015341	1.111278	1.199761	1.281763	1.358089	1.674762
0.3000	1.093728	1.190459	1.279438	1.361727	1.438188	1.754362
0.4000	1.162935	1.260161	1.349416	1.431826	1.508299	1.823711
0.5000	1.229422	1.326967	1.416361	1.498785	1.575188	1.889623
0.6000	1.297600	1.395328	1.484751	1.567101	1.643357	1.956575
0.7000	1.372359	1.470143	1.559483	1.641658	1.717680	2.029343
0.8000	1.462160	1.559831	1.648934	1.730789	1.806439	2.115974
0.9000	1.590645	1.687874	1.776412	1.857634	1.932609	2.238691
0.9500	1.700050	1.796684	1.884573	1.965120	2.039413	2.342244
0.9750	1.797247	1.893214	1.980420	2.060282	2.133901	2.433653
0.9900	1.912834	2.007869	2.094151	2.173115	2.245866	2.541768
0.9950	1.993063	2.087374	2.172959	2.251253	2.323365	2.616496
0.9990	2.162194	2.254820	2.338809	2.415597	2.486286	2.773376
0.9995	2.229347	2.321254	2.404572	2.480732	2.550834	2.835467
0.9999	2.375334	2.465603	2.547406	2.622160	2.690952	2.970165

P\n	60	70	80	90	100	
0.0001	1.275691	1.474719	1.643153	1.788440	1.915736	
0.0005	1.362921	1.562307	1.730441	1.875131	2.001697	
0.0010	1.404717	1.604137	1.772040	1.916387	2.042565	
0.0050	1.514998	1.714128	1.881190	2.024478	2.149530	
0.0100	1.569986	1.768790	1.935320	2.078007	2.202448	
0.0250	1.652493	1.850610	2.016219	2.157927	2.281400	
0.0500	1.725100	1.922436	2.087128	2.227906	2.350482	
0.1000	1.810673	2.006904	2.170405	2.310015	2.431489	
0.2000	1.916927	2.111547	2.273425	2.411497	2.531543	
0.3000	1.995307	2.188593	2.349187	2.486071	2.605029	
0.4000	2.063437	2.255475	2.414903	2.550721	2.668716	
0.5000	2.128070	2.318860	2.477142	2.611928	2.728992	
0.6000	2.193617	2.383082	2.540170	2.673889	2.789999	
0.7000	2.264751	2.452719	2.608477	2.741018	2.856082	
0.8000	2.349307	2.535429	2.689568	2.820688	2.934495	
0.9000	2.468884	2.652290	2.804083	2.933162	3.045174	
0.9500	2.569640	2.750681	2.900458	3.027795	3.138288	
0.9750	2.658490	2.837399	2.985377	3.111170	3.220318	
0.9900	2.763488	2.939839	3.085672	3.209634	3.317192	
0.9950	2.836020	3.010584	3.154928	3.277624	3.384087	
0.9990	2.988203	3.158987	3.300200	3.420244	3.524419	
0.9995	3.048413	3.217695	3.357671	3.476669	3.579945	
0.9999	3.179006	3.345030	3.482329	3.599074	3.700412	

Table A7

COEFFICIENTS OF w_r IN EXACT LOWER CONFIDENCE BOUNDS FOR σ WITH CONFIDENCE P (UPPER BOUNDS WITH CONFIDENCE $1 - P$), BASED ON SAMPLES OF SIZE n FROM A NORMAL POPULATION

[$w_r = x_{n-r} - x_{r+1} = r$th quasi-range of sample]

COEFFICIENTS OF w_0 IN EXACT LOWER CONFIDENCE BOUNDS FOR σ

P\n	2	3	4	5	6	7
0.0001	5641.895	52.50243	10.82318	4.866452	2.992513	2.152779
0.0005	1128.379	23.47744	6.322927	3.244414	2.156566	1.632416
0.0010	564.1894	16.59896	5.013886	2.721889	1.870081	1.446451
0.0050	112.8372	7.415786	2.917986	1.802113	1.335144	1.084804
0.0100	56.41748	5.237111	2.305867	1.503725	1.150066	0.954067
0.0250	22.56389	3.299561	1.681682	1.176925	0.938129	0.799680
0.0500	11.27640	2.318021	1.316598	0.970930	0.798157	0.694376
0.1000	5.627083	1.617201	1.021069	0.792771	0.671955	0.596641
0.2000	2.791059	1.110998	0.777803	0.635550	0.555557	0.503665
0.3000	1.835113	0.878535	0.652961	0.549920	0.489709	0.449642
0.4000	1.348410	0.733893	0.569305	0.490173	0.442548	0.410222
0.5000	1.048358	0.629807	0.505479	0.443089	0.404588	0.378007
0.6000	0.840172	0.547549	0.452431	0.402832	0.371518	0.349561
0.7000	0.682250	0.477420	0.405055	0.365913	0.340651	0.322664
0.8000	0.551758	0.412621	0.359227	0.329238	0.309431	0.295100
0.9000	0.429890	0.344545	0.308599	0.287498	0.273170	0.262598
0.9500	0.360775	0.301705	0.275242	0.259225	0.248133	0.239834
0.9750	0.315475	0.271572	0.251003	0.238264	0.229310	0.222538
0.9900	0.274516	0.242701	0.227128	0.217258	0.210214	0.204827
0.9950	0.251905	0.226028	0.213034	0.204684	0.198670	0.194039
0.9990	0.214892	0.197494	0.188366	0.182357	0.177957	0.174528
0.9995	0.203147	0.188097	0.180088	0.174771	0.170857	0.167723
0.9999	0.181748	0.170527	0.164396	0.160264	0.157190	0.154765

P\n	8	9	10	11	12	13
0.0001	1.694382	1.411017	1.220356	1.083995	0.981891	0.902675
0.0005	1.331535	1.138489	1.004804	0.906960	0.832298	0.773444
0.0010	1.197855	1.035725	0.922013	0.837939	0.773245	0.721890
0.0050	0.929989	0.825004	0.749104	0.691601	0.646455	0.610006
0.0100	0.830000	0.744388	0.681648	0.633592	0.595520	0.564547
0.0250	0.709210	0.645278	0.597544	0.560427	0.530552	0.506172
0.0500	0.624839	0.574762	0.536814	0.506950	0.482752	0.462689
0.1000	0.544826	0.506758	0.477453	0.454092	0.434961	0.418953
0.2000	0.466928	0.439344	0.417738	0.400270	0.385793	0.373557
0.3000	0.420747	0.398740	0.381307	0.367081	0.355198	0.345087
0.4000	0.386570	0.368355	0.353796	0.341826	0.331766	0.323159
0.5000	0.358315	0.343002	0.330666	0.320458	0.311832	0.304418
0.6000	0.333105	0.320192	0.309715	0.300992	0.293583	0.287186
0.7000	0.309030	0.298236	0.289414	0.282025	0.275716	0.270245
0.8000	0.284104	0.275316	0.268076	0.261973	0.256733	0.252167
0.9000	0.254366	0.247707	0.242169	0.237462	0.233392	0.229824
0.9500	0.233301	0.227972	0.223507	0.219690	0.216373	0.213451
0.9750	0.217162	0.212747	0.209029	0.205834	0.203046	0.200582
0.9900	0.200514	0.196947	0.193925	0.191316	0.189029	0.187000
0.9950	0.190311	0.187215	0.184583	0.182303	0.180300	0.178518
0.9990	0.171741	0.169408	0.167411	0.165673	0.164137	0.162766
0.9995	0.165296	0.163200	0.161401	0.159832	0.158444	0.157202
0.9999	0.152776	0.151098	0.149652	0.148385	0.147261	0.146253

COEFFICIENTS OF w_0 IN EXACT LOWER CONFIDENCE BOUNDS FOR σ

P\n	14	15	16	17	18	19
0.0001	0.839449	0.787813	0.744835	0.708486	0.677326	0.650298
0.0005	0.725832	0.686491	0.653410	0.625178	0.600780	0.579466
0.0010	0.680094	0.645380	0.616056	0.590931	0.569140	0.550041
0.0050	0.579911	0.554599	0.532982	0.514278	0.497916	0.483462
0.0100	0.538805	0.517032	0.498344	0.482104	0.467839	0.455194
0.0250	0.485642	0.468140	0.453013	0.439786	0.428103	0.417696
0.0500	0.445740	0.431199	0.418560	0.407453	0.397599	0.388785
0.1000	0.405323	0.393548	0.383252	0.374155	0.366045	0.358758
0.2000	0.363047	0.353897	0.345842	0.338681	0.332262	0.326466
0.3000	0.336352	0.328708	0.321947	0.315913	0.310484	0.305565
0.4000	0.315689	0.309125	0.303298	0.298080	0.293372	0.289094
0.5000	0.297955	0.292257	0.287181	0.282623	0.278499	0.274743
0.6000	0.281589	0.276636	0.272212	0.268227	0.264612	0.261313
0.7000	0.265439	0.261172	0.257348	0.253894	0.250753	0.247880
0.8000	0.248139	0.244548	0.241320	0.238395	0.235728	0.233281
0.9000	0.226659	0.223825	0.221265	0.218937	0.216807	0.214846
0.9500	0.210849	0.208510	0.206392	0.204459	0.202686	0.201050
0.9750	0.198381	0.196396	0.194594	0.192947	0.191431	0.190031
0.9900	0.185182	0.183538	0.182040	0.180668	0.179403	0.178231
0.9950	0.176918	0.175469	0.174147	0.172933	0.171812	0.170773
0.9990	0.161529	0.160405	0.159376	0.158429	0.157553	0.156738
0.9995	0.156081	0.155061	0.154126	0.153265	0.152467	0.151725
0.9999	0.145340	0.144507	0.143742	0.143036	0.142381	0.141770

P\n	20	22	24	26	28	30
0.0001	0.626618	0.587038	0.555208	0.528998	0.506993	0.488223
0.0005	0.560669	0.528984	0.503248	0.481873	0.463795	0.448275
0.0010	0.533149	0.504567	0.481247	0.461805	0.445308	0.431102
0.0050	0.470588	0.448602	0.430466	0.415204	0.402148	0.390825
0.0100	0.443894	0.424514	0.408447	0.394868	0.383208	0.373063
0.0250	0.408353	0.392234	0.378777	0.367334	0.357456	0.348821
0.0500	0.380843	0.367075	0.355514	0.345634	0.337069	0.329553
0.1000	0.352166	0.340677	0.330969	0.322627	0.315361	0.308957
0.2000	0.321198	0.311963	0.304102	0.297306	0.291354	0.286083
0.3000	0.301081	0.293187	0.286436	0.280574	0.275421	0.270842
0.4000	0.285185	0.278280	0.272351	0.267186	0.262631	0.258573
0.5000	0.271303	0.265209	0.259958	0.255369	0.251311	0.247688
0.6000	0.258285	0.252906	0.248254	0.244178	0.240564	0.237329
0.7000	0.245237	0.240528	0.236443	0.232851	0.229659	0.226794
0.8000	0.231026	0.226995	0.223484	0.220387	0.217627	0.215143
0.9000	0.213033	0.209780	0.206932	0.204410	0.202153	0.200116
0.9500	0.199534	0.196804	0.194407	0.192276	0.190364	0.188633
0.9750	0.188730	0.186384	0.184316	0.182474	0.180817	0.179314
0.9900	0.177141	0.175169	0.173426	0.171869	0.170464	0.169188
0.9950	0.169805	0.168051	0.166498	0.165107	0.163851	0.162708
0.9990	0.155977	0.154594	0.153365	0.152261	0.151261	0.150348
0.9995	0.151031	0.149768	0.148645	0.147634	0.146718	0.145881
0.9999	0.141199	0.140157	0.139227	0.138388	0.137627	0.136929

COEFFICIENTS OF w_0 IN EXACT LOWER CONFIDENCE BOUNDS FOR σ

P\n	32	34	36	38	40	50
0.0001	0.471994	0.457801	0.445267	0.434101	0.424080	0.386012
0.0005	0.434779	0.422916	0.412390	0.402975	0.394493	0.361984
0.0010	0.418718	0.407808	0.398107	0.389413	0.381567	0.351376
0.0050	0.380891	0.372089	0.364223	0.357141	0.350723	0.325774
0.0100	0.364136	0.356205	0.349101	0.342691	0.336870	0.314137
0.0250	0.341192	0.334388	0.328274	0.322739	0.317699	0.297885
0.0500	0.322889	0.316928	0.311556	0.306682	0.302232	0.284641
0.1000	0.303259	0.298144	0.293520	0.289312	0.285462	0.270144
0.2000	0.281372	0.277127	0.273275	0.269759	0.266532	0.253601
0.3000	0.266738	0.263030	0.259657	0.256571	0.253733	0.242304
0.4000	0.254927	0.251625	0.248616	0.245858	0.243317	0.233042
0.5000	0.244424	0.241464	0.238761	0.236279	0.233989	0.224695
0.6000	0.234409	0.231756	0.229329	0.227097	0.225034	0.216634
0.7000	0.224204	0.221844	0.219683	0.217692	0.215849	0.208315
0.8000	0.212891	0.210836	0.208950	0.207209	0.205595	0.198969
0.9000	0.198263	0.196567	0.195006	0.193562	0.192220	0.186683
0.9500	0.187056	0.185609	0.184274	0.183037	0.181886	0.177116
0.9750	0.177942	0.176681	0.175516	0.174434	0.173427	0.169236
0.9900	0.168019	0.166944	0.165948	0.165023	0.164160	0.160555
0.9950	0.161660	0.160695	0.159800	0.158968	0.158190	0.154936
0.9990	0.149510	0.148735	0.148016	0.147345	0.146718	0.144080
0.9995	0.145111	0.144399	0.143738	0.143121	0.142544	0.140112
0.9999	0.136287	0.135692	0.135139	0.134622	0.134137	0.132090

P\n	60	70	80	90	100
0.0001	0.360319	0.341555	0.327104	0.315542	0.306022
0.0005	0.339770	0.323400	0.310704	0.300489	0.292039
0.0010	0.330630	0.315278	0.303334	0.293700	0.285712
0.0050	0.308389	0.295393	0.285200	0.276924	0.270027
0.0100	0.298194	0.286216	0.276787	0.269108	0.262691
0.0250	0.283857	0.273247	0.264848	0.257978	0.252217
0.0500	0.272091	0.262542	0.254950	0.248717	0.243473
0.1000	0.259120	0.250679	0.243933	0.238371	0.233676
0.2000	0.244201	0.236949	0.231118	0.226287	0.222192
0.3000	0.233939	0.227450	0.222211	0.217856	0.214154
0.4000	0.225478	0.219586	0.214812	0.210832	0.207440
0.5000	0.217817	0.212437	0.208066	0.204411	0.201290
0.6000	0.210385	0.205479	0.201480	0.198129	0.195261
0.7000	0.202682	0.198242	0.194611	0.191560	0.188943
0.8000	0.193986	0.190040	0.186801	0.184072	0.181725
0.9000	0.182486	0.179144	0.176387	0.174055	0.172043
0.9500	0.173479	0.170569	0.168160	0.166117	0.164348
0.9750	0.166026	0.163447	0.161307	0.159486	0.157907
0.9900	0.157779	0.155541	0.153677	0.152086	0.150704
0.9950	0.152422	0.150390	0.148693	0.147243	0.145981
0.9990	0.142029	0.140363	0.138966	0.137768	0.136722
0.9995	0.138216	0.136673	0.135378	0.134266	0.133295
0.9999	0.130487	0.129178	0.128075	0.127126	0.126295

COEFFICIENTS OF w_1 IN EXACT LOWER CONFIDENCE BOUNDS FOR σ

P\n	2	3	4	5	6	7
0.0001			13263.45	107.2697	20.03927	8.362937
0.0005			2652.358	47.81845	11.62765	5.528619
0.0010			1325.971	33.73029	9.182028	4.616604
0.0050			264.8610	14.92544	5.270218	3.014327
0.0100			132.2216	10.46674	4.130091	2.496233
0.0250			52.63614	6.505659	2.970485	1.930831
0.0500			26.10466	4.503533	2.295105	1.576274
0.1000			12.83304	3.079548	1.751542	1.271567
0.2000			6.184397	2.058588	1.308101	1.005074
0.3000			3.955115	1.594628	1.082953	0.861369
0.4000			2.828671	1.308899	0.933499	0.761934
0.5000			2.141038	1.105445	0.820502	0.684183
0.6000			1.669805	0.946451	0.727444	0.618215
0.7000			1.317825	0.812568	0.645143	0.558203
0.8000			1.032687	0.690636	0.566407	0.499117
0.9000			0.773664	0.564930	0.480641	0.432615
0.9500			0.631329	0.487419	0.424959	0.388075
0.9750			0.540450	0.433792	0.384970	0.355343
0.9900			0.460300	0.383212	0.346011	0.322801
0.9950			0.417004	0.354396	0.323225	0.303447
0.9990			0.347796	0.305815	0.283748	0.269318
0.9995			0.326310	0.290033	0.270616	0.257789
0.9999			0.287802	0.260826	0.245892	0.235830

P\n	8	9	10	11	12	13
0.0001	4.853985	3.336432	2.531693	2.046209	1.726311	1.501639
0.0005	3.468233	2.509941	1.975546	1.640925	1.413981	1.250869
0.0010	2.994160	2.215237	1.771176	1.488527	1.294376	1.153399
0.0050	2.111314	1.643930	1.363090	1.177181	1.045520	0.947538
0.0100	1.807107	1.438332	1.211477	1.058644	0.948907	0.866330
0.0250	1.460168	1.196576	1.029122	0.913545	0.828970	0.764344
0.0500	1.232293	1.032595	0.902445	0.810865	0.742824	0.690188
0.1000	1.028139	0.881344	0.783036	0.712422	0.659097	0.617296
0.2000	0.841460	0.738630	0.667684	0.615552	0.575470	0.543584
0.3000	0.736815	0.656405	0.599835	0.557643	0.524814	0.498441
0.4000	0.662425	0.596814	0.549940	0.514561	0.486772	0.464272
0.5000	0.602957	0.548414	0.508920	0.478801	0.454946	0.435498
0.6000	0.551496	0.505927	0.472517	0.446789	0.426250	0.409397
0.7000	0.503789	0.465997	0.437941	0.416124	0.398571	0.384073
0.8000	0.455899	0.425339	0.402345	0.384276	0.369613	0.357416
0.9000	0.400780	0.377765	0.360156	0.346134	0.334631	0.324977
0.9500	0.363063	0.344686	0.330451	0.319004	0.309538	0.301539
0.9750	0.334897	0.319683	0.307783	0.298140	0.290114	0.283295
0.9900	0.306488	0.294188	0.284469	0.276528	0.269873	0.264186
0.9950	0.289391	0.278705	0.270208	0.263228	0.257355	0.252317
0.9990	0.258852	0.250776	0.244277	0.238888	0.234316	0.230368
0.9995	0.248420	0.241151	0.235279	0.230392	0.226235	0.222637
0.9999	0.228381	0.222543	0.217788	0.213805	0.210398	0.207434

COEFFICIENTS OF w_1 IN EXACT LOWER CONFIDENCE BOUNDS FOR σ

P\n	14	15	16	17	18	19
0.0001	1.336085	1.209452	1.109664	1.029102	0.962744	0.907156
0.0005	1.128371	1.033168	0.957120	0.895000	0.843305	0.799604
0.0010	1.046631	0.963064	0.895908	0.840762	0.794661	0.755533
0.0050	0.871811	0.811518	0.762354	0.721468	0.686906	0.657282
0.0100	0.801921	0.750242	0.707820	0.672339	0.642192	0.616237
0.0250	0.713291	0.671886	0.637583	0.608660	0.583913	0.562472
0.0500	0.648179	0.613812	0.585127	0.560783	0.539834	0.521592
0.1000	0.583564	0.555709	0.532270	0.512239	0.494893	0.479705
0.2000	0.517534	0.495796	0.477339	0.461439	0.447574	0.435357
0.3000	0.476719	0.458465	0.442871	0.429367	0.417535	0.407066
0.4000	0.445618	0.429854	0.416321	0.404551	0.394199	0.385008
0.5000	0.419280	0.405506	0.393631	0.383262	0.374113	0.365964
0.6000	0.395266	0.383208	0.372769	0.363621	0.355523	0.348289
0.7000	0.371850	0.361369	0.352258	0.344245	0.337127	0.330750
0.8000	0.347069	0.338151	0.330363	0.323485	0.317353	0.311842
0.9000	0.316723	0.309561	0.303270	0.297685	0.292682	0.288167
0.9500	0.294661	0.288662	0.283369	0.278651	0.274410	0.270571
0.9750	0.277404	0.272245	0.267677	0.263593	0.259911	0.256568
0.9900	0.259249	0.254907	0.251046	0.247582	0.244450	0.241598
0.9950	0.247929	0.244060	0.240611	0.237510	0.234700	0.232137
0.9990	0.226909	0.223842	0.221096	0.218617	0.216362	0.214298
0.9995	0.219477	0.216671	0.214154	0.211878	0.209805	0.207905
0.9999	0.204821	0.202491	0.200395	0.198494	0.196757	0.195162

P\n	20	22	24	26	28	30
0.0001	0.859918	0.783937	0.725455	0.678999	0.641154	0.609687
0.0005	0.762167	0.701327	0.653934	0.615907	0.584662	0.558490
0.0010	0.721892	0.666966	0.623951	0.589281	0.560684	0.536651
0.0050	0.631588	0.589164	0.555501	0.528068	0.505227	0.485874
0.0100	0.593634	0.556122	0.526178	0.501651	0.481142	0.463699
0.0250	0.543696	0.512309	0.487042	0.466199	0.448663	0.433669
0.0500	0.505544	0.478558	0.456685	0.438535	0.423189	0.410011
0.1000	0.466276	0.443551	0.424990	0.409492	0.396315	0.384945
0.2000	0.424495	0.405978	0.390725	0.377895	0.366918	0.357394
0.3000	0.397722	0.381714	0.368451	0.357239	0.347606	0.339216
0.4000	0.376780	0.362626	0.350844	0.340843	0.332221	0.324688
0.5000	0.358649	0.346020	0.335463	0.326470	0.318691	0.311877
0.6000	0.341778	0.330500	0.321033	0.312940	0.305919	0.299752
0.7000	0.324995	0.314992	0.306559	0.299325	0.293030	0.287485
0.8000	0.306854	0.298148	0.290776	0.284426	0.278881	0.273981
0.9000	0.284064	0.276868	0.270737	0.265429	0.260772	0.256642
0.9500	0.267071	0.260909	0.255635	0.251050	0.247014	0.243423
0.9750	0.253514	0.248119	0.243485	0.239443	0.235875	0.232692
0.9900	0.238987	0.234357	0.230363	0.226868	0.223772	0.221003
0.9950	0.229786	0.225609	0.221997	0.218828	0.216015	0.213495
0.9990	0.212399	0.209010	0.206065	0.203469	0.201157	0.199078
0.9995	0.206155	0.203028	0.200304	0.197901	0.195757	0.193827
0.9999	0.193689	0.191049	0.188742	0.186699	0.184872	0.183222

COEFFICIENTS OF w_1 IN EXACT LOWER CONFIDENCE BOUNDS FOR σ

P\n	32	34	36	38	40	50
0.0001	0.583077	0.560249	0.540427	0.523035	0.507635	0.450909
0.0005	0.536213	0.516994	0.500220	0.485435	0.472289	0.423401
0.0010	0.516136	0.498390	0.482867	0.469155	0.456942	0.411322
0.0050	0.469235	0.454751	0.442010	0.430698	0.420574	0.382355
0.0100	0.448652	0.435516	0.423929	0.413618	0.404371	0.369284
0.0250	0.420675	0.409284	0.399198	0.390194	0.382093	0.351137
0.0500	0.398545	0.388460	0.379504	0.371485	0.364253	0.336452
0.1000	0.375011	0.366240	0.358424	0.351404	0.345056	0.320491
0.2000	0.349033	0.341618	0.334984	0.329006	0.323582	0.302431
0.3000	0.331825	0.325252	0.319356	0.314029	0.309184	0.290196
0.4000	0.318034	0.312101	0.306768	0.301940	0.297542	0.280225
0.5000	0.305843	0.300451	0.295594	0.291189	0.287170	0.271283
0.6000	0.294278	0.289377	0.284953	0.280934	0.277260	0.262684
0.7000	0.282551	0.278123	0.274119	0.270474	0.267138	0.253846
0.8000	0.269610	0.265677	0.262113	0.258863	0.255881	0.243949
0.9000	0.252943	0.249605	0.246570	0.243795	0.241243	0.230971
0.9500	0.240199	0.237281	0.234622	0.232186	0.229942	0.220864
0.9750	0.229827	0.227229	0.224858	0.222682	0.220673	0.212519
0.9900	0.218504	0.216233	0.214156	0.212246	0.210480	0.203280
0.9950	0.211217	0.209144	0.207246	0.205497	0.203879	0.197263
0.9990	0.197193	0.195473	0.193893	0.192435	0.191082	0.185522
0.9995	0.192074	0.190474	0.189002	0.187643	0.186381	0.181183
0.9999	0.181722	0.180348	0.179082	0.177911	0.176822	0.172320

P\n	60	70	80	90	100	
0.0001	0.414163	0.388086	0.368430	0.352964	0.340401	
0.0005	0.391314	0.368329	0.350877	0.337068	0.325798	
0.0010	0.381202	0.359533	0.343027	0.329932	0.319222	
0.0050	0.356746	0.338125	0.323824	0.312404	0.303014	
0.0100	0.345612	0.328312	0.314976	0.304292	0.295485	
0.0250	0.330046	0.314523	0.302489	0.292807	0.284795	
0.0500	0.317356	0.303216	0.292205	0.283312	0.275932	
0.1000	0.303463	0.290771	0.280836	0.272778	0.266068	
0.2000	0.287615	0.276485	0.267719	0.260576	0.254603	
0.3000	0.276797	0.266677	0.258674	0.252129	0.246640	
0.4000	0.267931	0.258605	0.251202	0.245132	0.240029	
0.5000	0.259941	0.251302	0.244423	0.238767	0.234002	
0.6000	0.252223	0.244223	0.237833	0.232565	0.228118	
0.7000	0.244253	0.236887	0.230983	0.226104	0.221976	
0.8000	0.235284	0.228598	0.223221	0.218763	0.214981	
0.9000	0.223449	0.217609	0.212888	0.208959	0.205614	
0.9500	0.214174	0.208954	0.204718	0.201181	0.198162	
0.9750	0.206476	0.201742	0.197888	0.194661	0.191900	
0.9900	0.197913	0.193689	0.190237	0.187338	0.184853	
0.9950	0.192312	0.188404	0.185203	0.182509	0.180196	
0.9990	0.181330	0.178001	0.175262	0.172948	0.170954	
0.9995	0.177253	0.174126	0.171549	0.169368	0.167487	
0.9999	0.168896	0.166160	0.163896	0.161976	0.160315	

COEFFICIENTS OF w_2 IN EXACT LOWER CONFIDENCE BOUNDS FOR σ

P\n	2	3	4	5	6	7
0.0001					21104.27	162.7241
0.0005					4220.195	72.46820
0.0010					2109.686	51.08054
0.0050					421.2787	22.53321
0.0100					210.2269	15.76524
0.0250					83.59335	9.753771
0.0500					41.37813	6.716586
0.1000					20.26252	4.558164
0.2000					9.687144	3.013246
0.3000					6.143900	2.313047
0.4000					4.355698	1.883072
0.5000					3.266039	1.577890
0.6000					2.521167	1.340271
0.7000					1.966745	1.141049
0.8000					1.519878	0.960599
0.9000					1.117210	0.775989
0.9500					0.898264	0.663178
0.9750					0.759836	0.585742
0.9900					0.638989	0.513279
0.9950					0.574330	0.472295
0.9990					0.472127	0.403781
0.9995					0.440744	0.381702
0.9999					0.384981	0.341106

P\n	8	9	10	11	12	13
0.0001	29.24531	11.81345	6.670399	4.479327	3.332267	2.648011
0.0005	16.92600	7.780905	4.745924	3.355042	2.589228	2.115042
0.0010	13.34462	6.483722	4.087915	2.954466	2.316449	1.914863
0.0050	7.617417	4.205997	2.863600	2.178824	1.772519	1.506533
0.0100	5.948988	3.470211	2.442320	1.900167	1.570829	1.351395
0.0250	4.253156	2.668129	1.962559	1.573055	1.328685	1.161859
0.0500	3.266581	2.166003	1.648092	1.351684	1.160877	1.028062
0.1000	2.473870	1.735423	1.367065	1.148037	1.003126	0.900135
0.2000	1.828967	1.360095	1.111006	0.956581	0.851282	0.774699
0.3000	1.502712	1.158496	0.968041	0.846706	0.762312	0.699990
0.4000	1.286883	1.019493	0.866759	0.767339	0.697090	0.644580
0.5000	1.124272	0.911174	0.786058	0.703077	0.643629	0.598716
0.6000	0.990854	0.819598	0.716455	0.646842	0.596324	0.557772
0.7000	0.873353	0.736611	0.652160	0.594165	0.551532	0.518666
0.8000	0.761504	0.655276	0.587886	0.540733	0.505582	0.478185
0.9000	0.640493	0.564281	0.514308	0.478519	0.451366	0.429908
0.9500	0.562527	0.503740	0.464254	0.435488	0.413376	0.395720
0.9750	0.506899	0.459497	0.427057	0.403103	0.384499	0.369521
0.9900	0.453053	0.415749	0.389716	0.370218	0.354909	0.342474
0.9950	0.421744	0.389859	0.367337	0.350319	0.336865	0.325877
0.9990	0.367865	0.344457	0.327553	0.314569	0.304173	0.295592
0.9995	0.350057	0.329199	0.314020	0.302295	0.292862	0.285050
0.9999	0.316697	0.300257	0.288114	0.278627	0.270927	0.264502

COEFFICIENTS OF w_2 IN EXACT LOWER CONFIDENCE BOUNDS FOR σ

P\n	14	15	16	17	18	19
0.0001	2.201625	1.890924	1.663833	1.491414	1.356464	1.248195
0.0005	1.796639	1.569833	1.400878	1.270535	1.167121	1.083169
0.0010	1.641749	1.445202	1.297547	1.182828	1.091257	1.016528
0.0050	1.320015	1.182436	1.076954	0.993580	0.926047	0.870229
0.0100	1.195382	1.079014	0.988966	0.917237	0.858744	0.810118
0.0250	1.040965	0.949387	0.877609	0.819809	0.772235	0.732363
0.0500	0.930329	0.855368	0.796003	0.747777	0.707783	0.674042
0.1000	0.823089	0.763197	0.715232	0.675896	0.643005	0.615058
0.2000	0.716350	0.670308	0.632970	0.602020	0.575901	0.553528
0.3000	0.651927	0.613620	0.582293	0.556140	0.533932	0.514805
0.4000	0.603687	0.570833	0.543784	0.521070	0.501687	0.484919
0.5000	0.563438	0.534894	0.511252	0.491298	0.474193	0.459339
0.6000	0.527244	0.502377	0.481664	0.464097	0.448974	0.435791
0.7000	0.492428	0.470911	0.452885	0.437521	0.424237	0.412613
0.8000	0.456118	0.437885	0.422514	0.409342	0.397898	0.387843
0.9000	0.412427	0.397846	0.385455	0.374761	0.365413	0.357155
0.9500	0.381214	0.369028	0.358608	0.349568	0.341629	0.334585
0.9750	0.357132	0.346664	0.337669	0.329831	0.322922	0.316771
0.9900	0.332113	0.323304	0.315695	0.309034	0.303138	0.297870
0.9950	0.316678	0.308827	0.302022	0.296047	0.290745	0.285996
0.9990	0.288347	0.282118	0.276683	0.271885	0.267605	0.263755
0.9995	0.278432	0.272727	0.267738	0.263325	0.259381	0.255828
0.9999	0.259027	0.254282	0.250114	0.246411	0.243090	0.240088

P\n	20	22	24	26	28	30
0.0001	1.159528	1.023180	0.923368	0.847175	0.787090	0.738464
0.0005	1.013705	0.905508	0.825128	0.763025	0.713555	0.673173
0.0010	0.954411	0.857101	0.784332	0.727805	0.682572	0.645506
0.0050	0.823310	0.748774	0.692134	0.647552	0.611479	0.581638
0.0100	0.769034	0.703352	0.653070	0.613251	0.580868	0.553962
0.0250	0.698435	0.643703	0.601370	0.567560	0.539865	0.516713
0.0500	0.645165	0.598238	0.561636	0.532198	0.507943	0.487562
0.1000	0.590987	0.551559	0.520521	0.495367	0.474506	0.456881
0.2000	0.534121	0.502043	0.476525	0.455663	0.438235	0.423415
0.3000	0.498134	0.470407	0.448195	0.429928	0.414590	0.401490
0.4000	0.470246	0.445722	0.425961	0.409632	0.395864	0.384063
0.5000	0.446295	0.424396	0.406658	0.391937	0.379480	0.368768
0.6000	0.424176	0.404594	0.388654	0.375370	0.364088	0.354357
0.7000	0.402337	0.384936	0.370700	0.358785	0.348628	0.339840
0.8000	0.378920	0.363736	0.351243	0.340738	0.331746	0.323938
0.9000	0.349791	0.337181	0.326729	0.317885	0.310275	0.303636
0.9500	0.328281	0.317434	0.308392	0.300705	0.294062	0.288247
0.9750	0.311250	0.301712	0.293725	0.286907	0.280996	0.275805
0.9900	0.293125	0.284893	0.277964	0.272025	0.266856	0.262302
0.9950	0.281710	0.274253	0.267957	0.262544	0.257822	0.253653
0.9990	0.260266	0.254163	0.248978	0.244496	0.240568	0.237086
0.9995	0.252603	0.246951	0.242138	0.237970	0.234310	0.231061
0.9999	0.237355	0.232548	0.228435	0.224860	0.221709	0.218903

COEFFICIENTS OF w_2 IN EXACT LOWER CONFIDENCE BOUNDS FOR σ

P\n	32	34	36	38	40	50
0.0001	0.698270	0.664460	0.635599	0.610652	0.588853	0.510872
0.0005	0.639546	0.611075	0.586631	0.565394	0.546752	0.479365
0.0010	0.614536	0.588239	0.565604	0.545891	0.528552	0.465579
0.0050	0.556500	0.535000	0.516376	0.500064	0.485643	0.432652
0.0100	0.531210	0.511687	0.494725	0.479830	0.466629	0.417859
0.0250	0.497029	0.480060	0.465254	0.452205	0.440600	0.397396
0.0500	0.470160	0.455098	0.441912	0.430254	0.419858	0.380902
0.1000	0.441757	0.428612	0.417060	0.406812	0.397644	0.363047
0.2000	0.410628	0.399459	0.389600	0.380819	0.372937	0.342942
0.3000	0.390145	0.380202	0.371398	0.363536	0.356461	0.329383
0.4000	0.373811	0.364801	0.356804	0.349646	0.343192	0.318372
0.5000	0.359436	0.351213	0.343899	0.337339	0.331413	0.308526
0.6000	0.345856	0.338348	0.331654	0.325640	0.320197	0.299088
0.7000	0.332142	0.325325	0.319235	0.313751	0.308779	0.289413
0.8000	0.317077	0.310984	0.305527	0.300602	0.296128	0.278614
0.9000	0.297778	0.292556	0.287864	0.283617	0.279748	0.264504
0.9500	0.283098	0.278496	0.274350	0.270588	0.267153	0.253553
0.9750	0.271197	0.267069	0.263342	0.259954	0.256855	0.244531
0.9900	0.258248	0.254606	0.251310	0.248307	0.245555	0.234561
0.9950	0.249934	0.246588	0.243555	0.240788	0.238249	0.228073
0.9990	0.233968	0.231154	0.228595	0.226254	0.224101	0.215418
0.9995	0.228148	0.225515	0.223119	0.220925	0.218904	0.210739
0.9999	0.216381	0.214096	0.212011	0.210098	0.208333	0.201169

P\n	60	70	80	90	100	
0.0001	0.462261	0.428652	0.403794	0.384519	0.369045	
0.0005	0.436760	0.407007	0.384835	0.367541	0.353590	
0.0010	0.425509	0.397398	0.376378	0.359939	0.346647	
0.0050	0.398392	0.374085	0.355754	0.341319	0.329584	
0.0100	0.386094	0.363437	0.346281	0.332730	0.321684	
0.0250	0.368956	0.348518	0.332953	0.320602	0.310496	
0.0500	0.355034	0.336327	0.322011	0.310607	0.301248	
0.1000	0.339848	0.322952	0.309953	0.299554	0.290988	
0.2000	0.322599	0.307662	0.296097	0.286799	0.279108	
0.3000	0.310875	0.297206	0.286578	0.278001	0.270887	
0.4000	0.301296	0.288626	0.278737	0.270735	0.264080	
0.5000	0.292687	0.280884	0.271641	0.264141	0.257891	
0.6000	0.284394	0.273399	0.264760	0.257732	0.251863	
0.7000	0.275853	0.265660	0.257626	0.251071	0.245584	
0.8000	0.266267	0.256941	0.249561	0.243521	0.238451	
0.9000	0.253660	0.245415	0.238856	0.233466	0.228926	
0.9500	0.243810	0.236363	0.230415	0.225510	0.221367	
0.9750	0.235651	0.228833	0.223368	0.218850	0.215025	
0.9900	0.226586	0.220433	0.215483	0.211377	0.207891	
0.9950	0.220661	0.214923	0.210296	0.206449	0.203178	
0.9990	0.209040	0.204072	0.200045	0.196683	0.193815	
0.9995	0.204722	0.200025	0.196210	0.193021	0.190296	
0.9999	0.195856	0.191688	0.188290	0.185440	0.182999	

COEFFICIENTS OF w_3 IN EXACT LOWER CONFIDENCE BOUNDS FOR σ

P\n	2	3	4	5	6	7
0.0001						
0.0005						
0.0010						
0.0050						
0.0100						
0.0250						
0.0500						
0.1000						
0.2000						
0.3000						
0.4000						
0.5000						
0.6000						
0.7000						
0.8000						
0.9000						
0.9500						
0.9750						
0.9900						
0.9950						
0.9990						
0.9995						
0.9999						

P\n	8	9	10	11	12	13
0.0001	29010.40	218.4575	38.46300	15.25459	8.474352	5.609558
0.0005	5801.099	97.24362	22.23121	10.02663	6.014164	4.190003
0.0010	2899.937	68.52000	17.51261	8.345120	5.173190	3.684413
0.0050	579.0060	30.18139	9.967431	5.393214	3.609063	2.705979
0.0100	288.8885	21.09242	7.769797	4.440033	3.071202	2.354764
0.0250	114.8149	13.01990	5.536630	3.401472	2.459102	1.942835
0.0500	56.78504	8.942003	4.238039	2.751798	2.058295	1.664391
0.1000	27.75966	6.044806	3.195337	2.195260	1.700558	1.408606
0.2000	13.22393	3.972383	2.348069	1.710915	1.375207	1.168609
0.3000	8.354876	3.034064	1.920136	1.451271	1.193944	1.031183
0.4000	5.898497	2.458535	1.637501	1.272572	1.065774	0.932105
0.5000	4.402546	2.050596	1.424917	1.133575	0.963838	0.852030
0.6000	3.380811	1.733480	1.250825	1.016289	0.876089	0.782090
0.7000	2.621271	1.468134	1.097835	0.910237	0.795206	0.716709
0.8000	2.010254	1.228413	0.952597	0.806567	0.714549	0.650548
0.9000	1.461488	0.984110	0.796054	0.691001	0.622530	0.573757
0.9500	1.164484	0.835528	0.695642	0.614425	0.560167	0.520830
0.9750	0.977573	0.733981	0.624280	0.558664	0.513972	0.481117
0.9900	0.815234	0.639388	0.555482	0.503725	0.467752	0.440913
0.9950	0.728811	0.586117	0.515628	0.471320	0.440133	0.416650
0.9990	0.593056	0.497526	0.447354	0.414717	0.391208	0.373200
0.9995	0.551631	0.469124	0.424886	0.395767	0.374622	0.358326
0.9999	0.478410	0.417122	0.382948	0.359935	0.342957	0.329717

COEFFICIENTS OF w_3 IN EXACT LOWER CONFIDENCE BOUNDS FOR σ

P\n	14	15	16	17	18	19
0.0001	4.120597	3.238121	2.665801	2.269588	1.981432	1.763655
0.0005	3.192738	2.579187	2.169603	1.879410	1.664305	1.499118
0.0010	2.852278	2.331843	1.979959	1.728080	1.539791	1.394168
0.0050	2.173869	1.827728	1.586402	1.409341	1.274254	1.167965
0.0100	1.922568	1.636414	1.434136	1.284054	1.168485	1.076839
0.0250	1.621163	1.402936	1.245695	1.127206	1.034783	0.960690
0.0500	1.412557	1.238346	1.110874	1.013610	0.936947	0.874944
0.1000	1.216749	1.081223	0.980403	0.902427	0.840272	0.789516
0.2000	1.028663	0.927485	0.850814	0.790613	0.742014	0.701899
0.3000	0.918704	0.836123	0.772771	0.722523	0.681613	0.647599
0.4000	0.838249	0.768489	0.714441	0.671222	0.635793	0.606164
0.5000	0.772420	0.712604	0.665855	0.628203	0.597149	0.571043
0.6000	0.714278	0.662802	0.622238	0.589344	0.562058	0.539005
0.7000	0.659331	0.615326	0.580359	0.551807	0.527983	0.507753
0.8000	0.603092	0.566286	0.536771	0.512487	0.492094	0.474679
0.9000	0.536935	0.507967	0.484466	0.464941	0.448407	0.434187
0.9500	0.490731	0.466797	0.447209	0.430815	0.416845	0.404764
0.9750	0.455708	0.435329	0.418534	0.404393	0.392281	0.381761
0.9900	0.419918	0.402926	0.388815	0.376858	0.366560	0.357572
0.9950	0.398149	0.383088	0.370520	0.359827	0.350586	0.342494
0.9990	0.358821	0.346987	0.337023	0.328479	0.321045	0.314498
0.9995	0.345252	0.334451	0.325326	0.317480	0.310637	0.304597
0.9999	0.318994	0.310066	0.302475	0.295912	0.290160	0.285061

P\n	20	22	24	26	28	30
0.0001	1.593935	1.347658	1.178353	1.055190	0.961698	0.888340
0.0005	1.368618	1.176111	1.041299	0.941770	0.865298	0.804685
0.0010	1.278423	1.106426	0.984976	0.894718	0.824992	0.769474
0.0050	1.082230	0.952532	0.859087	0.788511	0.733263	0.688779
0.0100	1.002418	0.888912	0.806376	0.743576	0.694113	0.654084
0.0250	0.899956	0.806259	0.737242	0.684180	0.642028	0.607666
0.0500	0.823733	0.743987	0.684626	0.638598	0.601775	0.571579
0.1000	0.747248	0.680756	0.630687	0.591502	0.559911	0.533836
0.2000	0.668177	0.614517	0.573581	0.541202	0.514868	0.492968
0.3000	0.618826	0.572682	0.537167	0.508875	0.485724	0.466374
0.4000	0.580972	0.540314	0.508797	0.483541	0.462774	0.445343
0.5000	0.548746	0.512556	0.484322	0.461577	0.442792	0.426966
0.6000	0.519229	0.486959	0.461627	0.441118	0.424107	0.409723
0.7000	0.490323	0.461723	0.439131	0.420744	0.405427	0.392428
0.8000	0.459601	0.434710	0.414908	0.398700	0.385132	0.373569
0.9000	0.421798	0.401181	0.384630	0.370980	0.359481	0.349628
0.9500	0.394187	0.376478	0.362163	0.350288	0.340235	0.331584
0.9750	0.372514	0.356956	0.344307	0.333764	0.324802	0.317064
0.9900	0.349638	0.336217	0.325236	0.316036	0.308182	0.301372
0.9950	0.335333	0.323176	0.313189	0.304794	0.297606	0.291358
0.9990	0.308673	0.298718	0.290477	0.283504	0.277500	0.272256
0.9995	0.299214	0.289990	0.282334	0.275839	0.270236	0.265334
0.9999	0.280498	0.272644	0.266086	0.260497	0.255654	0.251402

COEFFICIENTS OF w_3 IN EXACT LOWER CONFIDENCE BOUNDS FOR σ

P\n	32	34	36	38	40	50
0.0001	0.829242	0.780598	0.739835	0.705160	0.675285	0.571599
0.0005	0.755431	0.714585	0.680133	0.650659	0.625132	0.535525
0.0010	0.724185	0.686502	0.654623	0.627279	0.603544	0.519790
0.0050	0.652145	0.621409	0.595220	0.572612	0.552874	0.482332
0.0100	0.620971	0.593085	0.569244	0.548601	0.530532	0.465566
0.0250	0.579065	0.554848	0.534046	0.515959	0.500067	0.442442
0.0500	0.546317	0.524832	0.506305	0.490140	0.475892	0.423865
0.1000	0.511899	0.493149	0.476912	0.462689	0.450111	0.403819
0.2000	0.474424	0.458486	0.444614	0.432411	0.421574	0.381331
0.3000	0.449917	0.435718	0.423318	0.412375	0.402631	0.366221
0.4000	0.430465	0.417587	0.406309	0.396331	0.387428	0.353982
0.5000	0.413414	0.401650	0.391321	0.382163	0.373973	0.343066
0.6000	0.397368	0.386613	0.377147	0.368735	0.361199	0.332624
0.7000	0.381225	0.371446	0.362817	0.355132	0.348232	0.321947
0.8000	0.363567	0.354809	0.347059	0.340138	0.333911	0.310060
0.9000	0.341064	0.333533	0.326844	0.320851	0.315442	0.294576
0.9500	0.324037	0.317379	0.311447	0.306119	0.301298	0.282598
0.9750	0.310292	0.304301	0.298952	0.294136	0.289770	0.272754
0.9900	0.295394	0.290089	0.285339	0.281053	0.277158	0.261900
0.9950	0.285862	0.280974	0.276591	0.272628	0.269023	0.254851
0.9990	0.267622	0.263487	0.259765	0.256391	0.253311	0.241126
0.9995	0.260995	0.257117	0.253623	0.250451	0.247553	0.236059
0.9999	0.247627	0.244243	0.241186	0.238404	0.235858	0.225705

P\n	60	70	80	90	100	
0.0001	0.509443	0.467559	0.437145	0.413889	0.395422	
0.0005	0.480975	0.443821	0.416628	0.395706	0.379008	
0.0010	0.468446	0.433306	0.407494	0.387578	0.371647	
0.0050	0.438333	0.407856	0.385266	0.367713	0.353592	
0.0100	0.424718	0.396263	0.375083	0.358570	0.345250	
0.0250	0.405791	0.380055	0.360782	0.345684	0.333458	
0.0500	0.390458	0.366841	0.349067	0.335087	0.323730	
0.1000	0.373778	0.352382	0.336186	0.323392	0.312960	
0.2000	0.354893	0.335898	0.321425	0.309931	0.300518	
0.3000	0.342096	0.324657	0.311308	0.300669	0.291930	
0.4000	0.331665	0.315451	0.302993	0.293033	0.284831	
0.5000	0.322310	0.307162	0.295481	0.286116	0.278387	
0.6000	0.313316	0.299161	0.288208	0.279404	0.272120	
0.7000	0.304071	0.290905	0.280681	0.272438	0.265603	
0.8000	0.293719	0.281620	0.272187	0.264556	0.258212	
0.9000	0.280142	0.269378	0.260939	0.254084	0.248364	
0.9500	0.269562	0.259786	0.252089	0.245815	0.240564	
0.9750	0.260816	0.251822	0.244715	0.238904	0.234029	
0.9900	0.251119	0.242954	0.236475	0.231161	0.226690	
0.9950	0.244790	0.237144	0.231061	0.226060	0.221845	
0.9990	0.232395	0.225715	0.220372	0.215960	0.212229	
0.9995	0.227795	0.221456	0.216376	0.212174	0.208616	
0.9999	0.218354	0.212685	0.208123	0.204338	0.201123	

COEFFICIENTS OF w_4 IN EXACT LOWER CONFIDENCE BOUNDS FOR σ

P\n	2	3	4	5	6	7
0.0001						
0.0005						
0.0010						
0.0050						
0.0100						
0.0250						
0.0500						
0.1000						
0.2000						
0.3000						
0.4000						
0.5000						
0.6000						
0.7000						
0.8000						
0.9000						
0.9500						
0.9750						
0.9900						
0.9950						
0.9990						
0.9995						
0.9999						

P\n	8	9	10	11	12	13
0.0001			36944.39	274.3310	47.69050	18.69343
0.0005			7387.576	122.0821	27.54226	12.27073
0.0010			3692.975	86.00431	21.68525	10.20508
0.0050			737.2920	37.84996	12.32008	6.579210
0.0100			367.8302	26.43412	9.592589	5.408652
0.0250			146.1492	16.29525	6.821338	4.133563
0.0500			72.24900	11.17387	5.210209	3.336250
0.1000			35.28583	7.535774	3.916998	2.653619
0.2000			16.77549	4.934126	2.866825	2.060063
0.3000			10.57565	3.756771	2.336867	1.742232
0.4000			7.448393	3.035043	1.987156	1.523719
0.5000			5.544359	2.523826	1.724371	1.353939
0.6000			4.244399	2.126758	1.509398	1.210847
0.7000			3.278589	1.794864	1.320722	1.081636
0.8000			2.502348	1.495449	1.141893	0.955536
0.9000			1.806331	1.190976	0.949593	0.815291
0.9500			1.430542	1.006317	0.826590	0.722614
0.9750			1.194647	0.880448	0.739395	0.655291
0.9900			0.990359	0.763534	0.655558	0.589127
0.9950			0.881927	0.697874	0.607117	0.550191
0.9990			0.712247	0.589060	0.524390	0.482377
0.9995			0.660676	0.554295	0.497250	0.459737
0.9999			0.569832	0.490831	0.446724	0.417027

COEFFICIENTS OF w_4 IN EXACT LOWER CONFIDENCE BOUNDS FOR σ

P\n	14	15	16	17	18	19
0.0001	10.27356	6.734413	4.903411	3.822746	3.124589	2.642986
0.0005	7.278790	5.020637	3.791639	3.038626	2.537855	2.184314
0.0010	6.255208	4.410383	3.383814	2.744397	2.313705	2.006504
0.0050	4.351851	3.229772	2.571504	2.145023	1.848806	1.632231
0.0100	3.697567	2.806189	2.270786	1.917720	1.669075	1.485237
0.0250	2.953260	2.309631	1.910320	1.640502	1.446809	1.301358
0.0500	2.466158	1.974217	1.661039	1.445250	1.287936	1.168313
0.1000	2.031714	1.666357	1.427270	1.259043	1.134348	1.038234
0.2000	1.637031	1.377855	1.203016	1.077096	0.982013	0.907602
0.3000	1.417423	1.212881	1.072101	0.969130	0.890409	0.828172
0.4000	1.262322	1.094090	0.976434	0.889303	0.822027	0.768401
0.5000	1.139110	0.998195	0.898250	0.823423	0.765136	0.718336
0.6000	1.033177	0.914542	0.829280	0.764784	0.714125	0.673167
0.7000	0.935665	0.836448	0.764188	0.708956	0.665207	0.629586
0.8000	0.838586	0.757551	0.697671	0.651378	0.614369	0.584002
0.9000	0.728082	0.666175	0.619587	0.583045	0.553482	0.528985
0.9500	0.653385	0.603352	0.565181	0.534913	0.510206	0.489579
0.9750	0.598180	0.556315	0.524025	0.498196	0.476960	0.459122
0.9900	0.543073	0.508800	0.482055	0.460460	0.442567	0.427439
0.9950	0.510215	0.480183	0.456574	0.437398	0.421431	0.407873
0.9990	0.452162	0.429060	0.410646	0.395521	0.382808	0.371928
0.9995	0.432532	0.411600	0.394835	0.381008	0.369348	0.359339
0.9999	0.395136	0.378083	0.364292	0.352826	0.343092	0.334688

P\n	20	22	24	26	28	30
0.0001	2.293899	1.826538	1.531169	1.329131	1.182818	1.072204
0.0005	1.923121	1.565665	1.334243	1.172984	1.054462	0.963767
0.0010	1.777620	1.461300	1.254299	1.108859	1.001252	0.918461
0.0050	1.467540	1.234458	1.077897	0.965651	0.881249	0.815441
0.0100	1.344137	1.142267	1.005045	0.905750	0.830529	0.771518
0.0250	1.188271	1.024012	0.910478	0.827253	0.763544	0.713131
0.0500	1.074330	0.936130	0.839305	0.767573	0.712190	0.668055
0.1000	0.961862	0.848043	0.767115	0.706460	0.659191	0.621226
0.2000	0.847718	0.757109	0.691597	0.641848	0.602660	0.570906
0.3000	0.777655	0.700435	0.643969	0.600705	0.566376	0.538391
0.4000	0.724570	0.657016	0.607162	0.568685	0.537974	0.512813
0.5000	0.679850	0.620092	0.575629	0.541090	0.513374	0.490565
0.6000	0.639287	0.586310	0.546582	0.515528	0.490481	0.469781
0.7000	0.599946	0.553264	0.517976	0.490215	0.467707	0.449023
0.8000	0.558568	0.518189	0.487394	0.462994	0.443095	0.426494
0.9000	0.508293	0.475102	0.449496	0.429020	0.412192	0.398062
0.9500	0.472042	0.443692	0.421625	0.403853	0.389160	0.376762
0.9750	0.443878	0.419075	0.399631	0.383878	0.370792	0.359704
0.9900	0.414436	0.393133	0.376299	0.362575	0.351112	0.341355
0.9950	0.396179	0.376933	0.361647	0.349133	0.338646	0.329692
0.9990	0.362478	0.346790	0.334210	0.323828	0.315069	0.307547
0.9995	0.350624	0.336112	0.324433	0.314766	0.306591	0.299556
0.9999	0.327334	0.315011	0.305022	0.296705	0.289636	0.283527

COEFFICIENTS OF w_4 IN EXACT LOWER CONFIDENCE BOUNDS FOR σ

P\n	32	34	36	38	40	50
0.0001	0.985737	0.916316	0.859349	0.811754	0.771375	0.635715
0.0005	0.892147	0.834144	0.786192	0.745863	0.711453	0.594378
0.0010	0.852782	0.799384	0.755091	0.717729	0.685766	0.576400
0.0050	0.762647	0.719313	0.683068	0.652272	0.625757	0.533741
0.0100	0.723932	0.684700	0.651760	0.623678	0.599427	0.514714
0.0250	0.672182	0.638210	0.609532	0.584966	0.563661	0.488543
0.0500	0.631991	0.601916	0.576415	0.554484	0.535398	0.467581
0.1000	0.589999	0.563811	0.541495	0.522220	0.505381	0.445028
0.2000	0.544589	0.522375	0.503338	0.486813	0.472312	0.419816
0.3000	0.515078	0.495312	0.478307	0.463494	0.450456	0.402929
0.4000	0.491764	0.473853	0.458393	0.444889	0.432972	0.389285
0.5000	0.471412	0.455059	0.440905	0.428509	0.417544	0.377142
0.6000	0.452333	0.437390	0.424419	0.413031	0.402936	0.365551
0.7000	0.433215	0.419631	0.407806	0.397398	0.388151	0.353724
0.8000	0.412390	0.400225	0.389601	0.380224	0.371870	0.340585
0.9000	0.385990	0.375528	0.366352	0.358221	0.350953	0.323523
0.9500	0.366123	0.356868	0.348723	0.341485	0.334998	0.310361
0.9750	0.350155	0.341822	0.334469	0.327918	0.322034	0.299571
0.9900	0.332920	0.325533	0.318995	0.313154	0.307894	0.287701
0.9950	0.321932	0.315121	0.309080	0.303674	0.298798	0.280007
0.9990	0.300996	0.295221	0.290079	0.285461	0.281282	0.265061
0.9995	0.293417	0.287997	0.283164	0.278817	0.274880	0.259553
0.9999	0.278176	0.273436	0.269197	0.265375	0.261904	0.248316

P\n	60	70	80	90	100	
0.0001	0.557679	0.506461	0.469950	0.442415	0.420783	
0.0005	0.525890	0.480420	0.447735	0.422924	0.403330	
0.0010	0.511930	0.468906	0.437860	0.414224	0.395514	
0.0050	0.478465	0.441097	0.413875	0.392997	0.376371	
0.0100	0.463375	0.428459	0.402909	0.383244	0.367541	
0.0250	0.442444	0.410822	0.387533	0.369520	0.355077	
0.0500	0.425528	0.396473	0.374961	0.358252	0.344810	
0.1000	0.407169	0.380802	0.361163	0.345837	0.333461	
0.2000	0.386442	0.362980	0.345383	0.331576	0.320375	
0.3000	0.372433	0.350855	0.334592	0.321782	0.311358	
0.4000	0.361037	0.340942	0.325735	0.313719	0.303916	
0.5000	0.350835	0.332029	0.317745	0.306425	0.297167	
0.6000	0.341044	0.323440	0.310020	0.299355	0.290612	
0.7000	0.330996	0.314589	0.302035	0.292027	0.283804	
0.8000	0.319767	0.304654	0.293039	0.283749	0.276094	
0.9000	0.305074	0.291580	0.281149	0.272768	0.265836	
0.9500	0.293653	0.281358	0.271812	0.264113	0.257725	
0.9750	0.284231	0.272886	0.264044	0.256890	0.250940	
0.9900	0.273804	0.263467	0.255376	0.248808	0.243329	
0.9950	0.267009	0.257305	0.249687	0.243489	0.238310	
0.9990	0.253725	0.245202	0.238472	0.232972	0.228358	
0.9995	0.248803	0.240697	0.234283	0.229032	0.224622	
0.9999	0.238711	0.231428	0.225640	0.220885	0.216878	

COEFFICIENTS OF w_5 IN EXACT LOWER CONFIDENCE BOUNDS FOR σ

P\n	2	3	4	5	6	7
0.0001						
0.0005						
0.0010						
0.0050						
0.0100						
0.0250						
0.0500						
0.1000						
0.2000						
0.3000						
0.4000						
0.5000						
0.6000						
0.7000						
0.8000						
0.9000						
0.9500						
0.9750						
0.9900						
0.9950						
0.9990						
0.9995						
0.9999						

P\n	8	9	10	11	12	13
0.0001					44892.77	330.2846
0.0005					8976.932	146.9567
0.0010					4487.452	103.5142
0.0050					895.8665	45.53020
0.0100					446.9166	31.78418
0.0250					177.5421	19.57594
0.0500					87.74250	13.40949
0.1000					42.82706	9.029311
0.2000					20.33477	5.897479
0.3000					12.80156	4.480578
0.4000					9.002069	3.612285
0.5000					6.689040	2.997495
0.6000					5.110160	2.520211
0.7000					3.937487	2.121519
0.8000					2.995458	1.762152
0.9000					2.151568	1.397209
0.9500					1.696579	1.176270
0.9750					1.411404	1.025932
0.9900					1.164884	0.886556
0.9950					1.034284	0.808431
0.9990					0.830423	0.679269
0.9995					0.768631	0.638105
0.9999					0.660036	0.563122

COEFFICIENTS OF w_5 IN EXACT LOWER CONFIDENCE BOUNDS FOR σ

P\n	14	15	16	17	18	19
0.0001	56.92501	22.13185	12.07066	7.856579	5.683315	4.404377
0.0005	32.85750	14.51453	8.541791	5.849097	4.388101	3.495485
0.0010	25.86122	12.06474	7.335753	5.134349	3.913069	3.154517
0.0050	14.67468	7.764895	5.093402	3.751849	2.967147	2.460159
0.0100	11.41689	6.376918	4.322751	3.255983	2.617099	2.196954
0.0250	8.107054	4.865220	3.446269	2.674869	2.197668	1.876097
0.0500	6.183047	3.920180	2.872870	2.282516	1.907761	1.650241
0.1000	4.639002	3.111338	2.361690	1.922593	1.636065	1.434996
0.2000	3.385582	2.408425	1.897613	1.585573	1.375658	1.224873
0.3000	2.753382	2.032300	1.639611	1.393036	1.223789	1.100318
0.4000	2.336423	1.773883	1.457533	1.254511	1.112906	1.008307
0.5000	2.023290	1.573240	1.313003	1.142779	1.022364	0.932436
0.6000	1.767298	1.404268	1.188845	1.045394	0.942561	0.864963
0.7000	1.542802	1.251823	1.074664	0.954567	0.867318	0.800787
0.8000	1.330242	1.103213	0.961120	0.862910	0.790514	0.734671
0.9000	1.102019	0.938196	0.832078	0.756927	0.700494	0.656325
0.9500	0.956312	0.829355	0.745013	0.684190	0.637882	0.601234
0.9750	0.853204	0.750426	0.680775	0.629818	0.590590	0.559269
0.9900	0.754255	0.672995	0.616760	0.574984	0.542440	0.516207
0.9950	0.697184	0.627508	0.578653	0.542010	0.513250	0.489925
0.9990	0.599941	0.548450	0.511461	0.483214	0.460729	0.442284
0.9995	0.568112	0.522112	0.488785	0.463171	0.442679	0.425800
0.9999	0.508971	0.472516	0.445658	0.424755	0.407864	0.393836

P\n	20	22	24	26	28	30
0.0001	3.580358	2.603383	2.056236	1.711868	1.477192	1.307820
0.0005	2.903456	2.179249	1.760049	1.489764	1.302124	1.164679
0.0010	2.644931	2.012869	1.641606	1.399643	1.230265	1.105370
0.0050	2.108943	1.658464	1.384308	1.200899	1.069886	0.971702
0.0100	1.901839	1.517509	1.279809	1.118879	1.002852	0.915248
0.0250	1.645848	1.339575	1.145848	1.012481	0.915066	0.840741
0.0500	1.462984	1.209593	1.046369	0.932463	0.848376	0.783665
0.1000	1.286334	1.081392	0.946740	0.851369	0.780140	0.724806
0.2000	1.111298	0.951415	0.843997	0.766626	0.708071	0.662092
0.3000	1.006153	0.871720	0.780032	0.713236	0.662229	0.621881
0.4000	0.927735	0.811391	0.731071	0.672012	0.626582	0.590430
0.5000	0.862549	0.760611	0.689469	0.636724	0.595884	0.563210
0.6000	0.804151	0.714590	0.651437	0.604244	0.567471	0.537898
0.7000	0.748202	0.669997	0.614268	0.572283	0.539358	0.512738
0.8000	0.690121	0.623144	0.574855	0.538148	0.509153	0.485571
0.9000	0.620663	0.566297	0.526505	0.495902	0.471500	0.451500
0.9500	0.571375	0.525370	0.491308	0.464875	0.443645	0.426140
0.9750	0.533564	0.493616	0.463759	0.440419	0.421563	0.405937
0.9900	0.494506	0.460465	0.434762	0.414508	0.398037	0.384314
0.9950	0.470534	0.439933	0.416675	0.398253	0.383208	0.370630
0.9990	0.426801	0.402089	0.383069	0.367853	0.355325	0.344778
0.9995	0.411583	0.388797	0.371179	0.357033	0.345351	0.335493
0.9999	0.381939	0.362712	0.347708	0.335573	0.325491	0.316939

COEFFICIENTS OF w_5 IN EXACT LOWER CONFIDENCE BOUNDS FOR σ

P\n	32	34	36	38	40	50
0.0001	1.180174	1.080678	1.001006	0.935792	0.881427	0.705179
0.0005	1.059829	0.977267	0.910578	0.855580	0.809431	0.657715
0.0010	1.009575	0.933796	0.872348	0.821500	0.778706	0.637132
0.0050	0.895380	0.834325	0.784339	0.742630	0.707272	0.588452
0.0100	0.846730	0.791633	0.746324	0.708373	0.676090	0.566815
0.0250	0.782103	0.734604	0.695299	0.662196	0.633902	0.537133
0.0500	0.732248	0.690347	0.655494	0.626009	0.600709	0.513430
0.1000	0.680495	0.644145	0.613737	0.587884	0.565602	0.488002
0.2000	0.624941	0.594232	0.568374	0.546264	0.527112	0.459671
0.3000	0.589080	0.561827	0.538775	0.518987	0.501786	0.440754
0.4000	0.560893	0.536246	0.515323	0.497303	0.481595	0.425506
0.5000	0.536394	0.513932	0.494799	0.478273	0.463831	0.411964
0.6000	0.513523	0.493028	0.475516	0.460347	0.447057	0.399063
0.7000	0.490698	0.472097	0.456149	0.442296	0.430126	0.385925
0.8000	0.465949	0.449317	0.435004	0.422530	0.411541	0.371364
0.9000	0.434749	0.420470	0.408122	0.397314	0.387755	0.352505
0.9500	0.411404	0.398787	0.387835	0.378216	0.369683	0.338001
0.9750	0.392728	0.381377	0.371493	0.362787	0.355046	0.326137
0.9900	0.372659	0.362604	0.353817	0.346054	0.339131	0.313116
0.9950	0.359914	0.350646	0.342528	0.335341	0.328920	0.304692
0.9990	0.335742	0.327886	0.320973	0.314830	0.309321	0.288366
0.9995	0.327027	0.319654	0.313156	0.307372	0.302179	0.282363
0.9999	0.309562	0.303113	0.297410	0.292319	0.287735	0.270135

P\n	60	70	80	90	100	
0.0001	0.608218	0.546309	0.503007	0.470802	0.445773	
0.0005	0.572688	0.517728	0.478942	0.449899	0.427203	
0.0010	0.557121	0.505113	0.468262	0.440581	0.418897	
0.0050	0.519893	0.474706	0.442363	0.417880	0.398580	
0.0100	0.503150	0.460916	0.430543	0.407467	0.389223	
0.0250	0.479975	0.441704	0.413995	0.392831	0.376030	
0.0500	0.461286	0.426103	0.400485	0.380833	0.365177	
0.1000	0.441050	0.409097	0.385683	0.367633	0.353196	
0.2000	0.418262	0.389798	0.368786	0.352494	0.339403	
0.3000	0.402898	0.376694	0.357251	0.342115	0.329912	
0.4000	0.390423	0.365998	0.347797	0.333580	0.322088	
0.5000	0.379273	0.356395	0.339278	0.325868	0.315001	
0.6000	0.368589	0.347151	0.331052	0.318401	0.308123	
0.7000	0.357643	0.337641	0.322559	0.310670	0.300988	
0.8000	0.345431	0.326979	0.313003	0.301946	0.292915	
0.9000	0.329488	0.312977	0.300395	0.290393	0.282191	
0.9500	0.317123	0.302051	0.290510	0.281300	0.273724	
0.9750	0.306942	0.293010	0.282297	0.273722	0.266649	
0.9900	0.295694	0.282973	0.273147	0.265252	0.258721	
0.9950	0.288377	0.276415	0.267148	0.259685	0.253498	
0.9990	0.274098	0.263555	0.255338	0.248688	0.243154	
0.9995	0.268815	0.258775	0.250932	0.244574	0.239275	
0.9999	0.257999	0.248951	0.241850	0.236071	0.231240	

COEFFICIENTS OF w_6 IN EXACT LOWER CONFIDENCE BOUNDS FOR σ

P\n	14	15	16	17	18	19
0.0001	52849.55	386.2881	66.16455	25.57044	13.86669	8.977252
0.0005	10567.97	171.8538	38.17574	16.75843	9.803985	6.676346
0.0010	5282.768	121.0402	30.03962	13.92448	8.415563	5.857194
0.0050	1054.609	53.21776	17.03069	8.950564	5.834322	4.272956
0.0100	526.0874	37.13946	13.24229	7.345121	4.947322	3.704846
0.0250	208.9688	22.85998	9.393532	5.596735	3.938667	3.039209
0.0500	103.2532	15.64749	7.156420	4.503896	3.278952	2.589924
0.1000	50.37710	10.52449	5.361321	3.568758	2.691000	2.177931
0.2000	23.89857	6.861888	3.904432	2.756377	2.157477	1.792371
0.3000	15.03052	5.205126	3.169845	2.321880	1.861036	1.572246
0.4000	10.55799	4.190045	2.685524	2.023493	1.651943	1.413968
0.5000	7.835441	3.471498	2.321941	1.791927	1.486060	1.286378
0.6000	5.977240	2.913836	2.024838	1.597017	1.343641	1.175240
0.7000	4.597362	2.448186	1.764428	1.421280	1.212755	1.071659
0.8000	3.489217	2.028700	1.518037	1.250099	1.082708	0.967220
0.9000	2.497079	1.603087	1.253770	1.060234	0.935083	0.846598
0.9500	1.962637	1.345726	1.085274	0.935176	0.835616	0.763930
0.9750	1.627991	1.170815	0.966190	0.844600	0.762319	0.702208
0.9900	1.339052	1.008875	0.852065	0.755865	0.689374	0.640042
0.9950	1.186176	0.918224	0.786329	0.703804	0.646006	0.602703
0.9990	0.947954	0.768616	0.674513	0.613469	0.569655	0.536224
0.9995	0.875884	0.721024	0.637977	0.583424	0.543927	0.513596
0.9999	0.749440	0.634469	0.570193	0.526926	0.495063	0.470279

P\n	20	22	24	26	28	30
0.0001	6.461522	4.034273	2.910975	2.284052	1.890711	1.623435
0.0005	4.983134	3.267424	2.433665	1.952701	1.643561	1.429560
0.0010	4.440987	2.974602	2.246474	1.820238	1.543312	1.350012
0.0050	3.361610	2.367673	1.847876	1.532598	1.322330	1.172555
0.0100	2.962282	2.133244	1.689416	1.415836	1.231183	1.098426
0.0250	2.483928	1.843581	1.489468	1.266223	1.113003	1.001399
0.0500	2.153416	1.636761	1.343482	1.155186	1.024177	0.927730
0.1000	1.843802	1.437073	1.199580	1.044047	0.934212	0.852404
0.2000	1.547240	1.239353	1.053797	0.929527	0.840277	0.772911
0.3000	1.374409	1.120676	0.964483	0.858289	0.781144	0.722387
0.4000	1.248300	1.032225	0.896921	0.803798	0.735517	0.683127
0.5000	1.145389	0.958746	0.840089	0.757528	0.696485	0.649339
0.6000	1.054743	0.892962	0.788617	0.715255	0.660581	0.618084
0.7000	0.969337	0.829982	0.738778	0.673970	0.625276	0.587179
0.8000	0.882234	0.764658	0.686455	0.630229	0.587597	0.554000
0.9000	0.780262	0.686625	0.623041	0.576624	0.541012	0.512681
0.9500	0.709432	0.631326	0.577444	0.537648	0.506838	0.482146
0.9750	0.655998	0.588950	0.542103	0.507172	0.479928	0.457962
0.9900	0.601660	0.545228	0.505248	0.475127	0.451443	0.432220
0.9950	0.568756	0.518422	0.482444	0.455158	0.433589	0.416009
0.9990	0.509640	0.469584	0.440465	0.418097	0.400235	0.385556
0.9995	0.489352	0.452611	0.425738	0.404998	0.388377	0.374675
0.9999	0.450266	0.419584	0.396866	0.379167	0.364877	0.353023

COEFFICIENTS OF w_6 IN EXACT LOWER CONFIDENCE BOUNDS FOR σ

P\n	32	34	36	38	40	50
0.0001	1.431046	1.286410	1.173928	1.084049	1.010624	0.781900
0.0005	1.273220	1.154247	1.060774	0.985434	0.923424	0.727222
0.0010	1.207853	1.099081	1.013229	0.943762	0.886390	0.703585
0.0050	1.060603	0.973789	0.904494	0.847879	0.800732	0.647864
0.0100	0.998449	0.920442	0.857854	0.806489	0.763549	0.623187
0.0250	0.916463	0.849615	0.795583	0.750962	0.713454	0.589427
0.0500	0.853695	0.795008	0.747286	0.707672	0.674222	0.562549
0.1000	0.789009	0.738360	0.696900	0.662286	0.632912	0.533800
0.2000	0.720141	0.677600	0.642510	0.613021	0.587851	0.501877
0.3000	0.676021	0.638411	0.607226	0.580901	0.558342	0.480629
0.4000	0.641535	0.607628	0.579392	0.555466	0.534898	0.463543
0.5000	0.611706	0.580888	0.555124	0.533221	0.514336	0.448400
0.6000	0.583986	0.555939	0.532405	0.512332	0.494975	0.434002
0.7000	0.556448	0.531057	0.509669	0.491365	0.475492	0.419370
0.8000	0.526736	0.504095	0.484941	0.468488	0.454172	0.403188
0.9000	0.489509	0.470138	0.453657	0.439429	0.426996	0.382288
0.9500	0.461828	0.444755	0.430165	0.417520	0.406432	0.366260
0.9750	0.439796	0.424466	0.411318	0.399886	0.389833	0.353181
0.9900	0.416235	0.402683	0.391011	0.380827	0.371843	0.338858
0.9950	0.401337	0.388860	0.378086	0.368664	0.360335	0.329612
0.9990	0.373220	0.362667	0.353507	0.345460	0.338317	0.311732
0.9995	0.363130	0.353232	0.344625	0.337050	0.330317	0.305172
0.9999	0.342984	0.334338	0.326790	0.320125	0.314182	0.291833

P\n	60	70	80	90	100	
0.0001	0.662088	0.587799	0.536854	0.499504	0.470790	
0.0005	0.622319	0.556406	0.510774	0.477077	0.451025	
0.0010	0.604934	0.542574	0.499216	0.467093	0.442193	
0.0050	0.563457	0.509297	0.471233	0.442802	0.420619	
0.0100	0.544853	0.494236	0.458484	0.431676	0.410696	
0.0250	0.519153	0.473289	0.440659	0.416057	0.396719	
0.0500	0.498475	0.456309	0.426130	0.403270	0.385236	
0.1000	0.476134	0.437832	0.410233	0.389220	0.372574	
0.2000	0.451039	0.416907	0.392119	0.373131	0.358016	
0.3000	0.434160	0.402726	0.379773	0.362116	0.348013	
0.4000	0.420480	0.391169	0.369668	0.353070	0.339774	
0.5000	0.408273	0.380805	0.360572	0.344903	0.332319	
0.6000	0.396592	0.370843	0.351798	0.337003	0.325090	
0.7000	0.384644	0.360605	0.342749	0.328832	0.317597	
0.8000	0.371337	0.349144	0.332580	0.319621	0.309127	
0.9000	0.354001	0.334119	0.319184	0.307439	0.297891	
0.9500	0.340587	0.322417	0.308698	0.297866	0.289030	
0.9750	0.329561	0.312748	0.299997	0.289897	0.281635	
0.9900	0.317402	0.302029	0.290315	0.281000	0.273356	
0.9950	0.309504	0.295036	0.283975	0.275158	0.267907	
0.9990	0.294121	0.281341	0.271509	0.263631	0.257127	
0.9995	0.288439	0.276258	0.266865	0.259324	0.253088	
0.9999	0.276822	0.265823	0.257298	0.250428	0.244728	

COEFFICIENTS OF w_7 IN EXACT LOWER CONFIDENCE BOUNDS FOR σ

P\n	14	15	16	17	18	19
0.0001			60811.62	442.3250	75.40772	29.00936
0.0005			12160.06	196.7659	43.49614	19.00255
0.0010			6078.602	138.5768	34.21975	15.78441
0.0050			1213.459	60.91019	19.38773	10.13631
0.0100			605.3116	42.49824	15.06850	8.313361
0.0250			240.4171	26.14628	10.68058	6.328223
0.0500			118.7749	17.88709	8.130208	5.087531
0.1000			57.93272	12.02079	6.083904	4.026029
0.2000			27.46526	7.827020	4.423390	3.104100
0.3000			17.26143	5.930192	3.586316	2.611170
0.4000			12.11536	4.768179	3.034553	2.272765
0.5000			8.982955	3.945754	2.620450	2.010231
0.6000			6.845181	3.307603	2.282168	1.789339
0.7000			5.257882	2.774889	1.985778	1.590266
0.8000			3.983411	2.295171	1.705484	1.396466
0.9000			2.842787	1.808749	1.405081	1.181692
0.9500			2.228723	1.514866	1.213732	1.040373
0.9750			1.844481	1.315305	1.078622	0.938118
0.9900			1.512994	1.130723	0.949271	0.838045
0.9950			1.337764	1.027500	0.874841	0.779392
0.9990			1.065049	0.857369	0.748401	0.677747
0.9995			0.982657	0.803324	0.707143	0.643984
0.9999			0.838288	0.705152	0.630688	0.580567

P\n	20	22	24	26	28	30
0.0001	15.66219	7.238669	4.486973	3.217294	2.510574	2.068261
0.0005	11.06575	5.577270	3.630315	2.686924	2.144159	1.796150
0.0010	9.494982	4.968062	3.303246	2.478964	1.997715	1.685806
0.0050	6.574896	3.755325	2.625468	2.036257	1.679809	1.442656
0.0100	5.571549	3.306742	2.363745	1.860322	1.550813	1.342410
0.0250	4.430714	2.769489	2.040441	1.638394	1.385582	1.212480
0.0500	3.684664	2.398380	1.809683	1.476426	1.263005	1.114869
0.1000	3.019911	2.050849	1.586969	1.316840	1.140379	1.016055
0.2000	2.416894	1.718120	1.366572	1.155266	1.014100	0.912946
0.3000	2.081979	1.524315	1.234364	1.056341	0.935600	0.848084
0.4000	1.845839	1.382969	1.135877	0.981550	0.875587	0.798062
0.5000	1.658573	1.267678	1.054103	0.918668	0.824653	0.755293
0.6000	1.497866	1.166177	0.980930	0.861747	0.778145	0.715970
0.7000	1.350245	1.070596	0.910915	0.806662	0.732747	0.677326
0.8000	1.203661	0.973179	0.838343	0.748871	0.684681	0.636108
0.9000	1.037413	0.859239	0.751733	0.678892	0.625826	0.585193
0.9500	0.925515	0.780179	0.690417	0.628624	0.583075	0.547876
0.9750	0.843138	0.720592	0.643474	0.589697	0.549674	0.518514
0.9900	0.761240	0.660057	0.595085	0.549138	0.514584	0.487459
0.9950	0.712596	0.623436	0.565443	0.524064	0.492735	0.468009
0.9990	0.627065	0.557715	0.511498	0.477952	0.452223	0.431708
0.9995	0.598281	0.535187	0.492771	0.461791	0.437918	0.418812
0.9999	0.543669	0.491828	0.456363	0.430136	0.409731	0.393277

COEFFICIENTS OF w_7 IN EXACT LOWER CONFIDENCE BOUNDS FOR σ

P\n	32	34	36	38	40	50
0.0001	1.768395	1.553007	1.391401	1.265955	1.165891	0.868016
0.0005	1.555790	1.380563	1.247480	1.143116	1.059142	0.804752
0.0010	1.468582	1.309165	1.187426	1.091519	1.014046	0.777490
0.0050	1.274111	1.148391	1.051090	0.973566	0.910336	0.713453
0.0100	1.192913	1.080563	0.993070	0.922998	0.865590	0.685198
0.0250	1.086673	0.991126	0.916068	0.855510	0.805583	0.646656
0.0500	1.006050	0.922688	0.856732	0.803194	0.758826	0.616067
0.1000	0.923654	0.852195	0.795210	0.748643	0.709830	0.583449
0.2000	0.836758	0.777195	0.729264	0.689796	0.656680	0.547357
0.3000	0.781566	0.729177	0.686758	0.651645	0.622049	0.523413
0.4000	0.738701	0.691665	0.653387	0.621564	0.594641	0.504206
0.5000	0.701829	0.659235	0.624413	0.595351	0.570681	0.487219
0.6000	0.667740	0.629112	0.597393	0.570822	0.548192	0.471102
0.7000	0.634049	0.599203	0.570460	0.546287	0.525630	0.454755
0.8000	0.597902	0.566952	0.541292	0.519618	0.501026	0.436718
0.9000	0.552924	0.526576	0.504585	0.485903	0.469798	0.413489
0.9500	0.519715	0.496580	0.477170	0.460606	0.446271	0.395726
0.9750	0.493432	0.472722	0.455272	0.440325	0.427348	0.381267
0.9900	0.465479	0.447228	0.431778	0.418489	0.406910	0.365469
0.9950	0.447888	0.431119	0.416880	0.404600	0.393874	0.355291
0.9990	0.414871	0.400742	0.388671	0.378206	0.369024	0.335659
0.9995	0.403084	0.389850	0.378518	0.368676	0.360025	0.328472
0.9999	0.379646	0.368116	0.358199	0.349551	0.341923	0.313885

P\n	60	70	80	90	100	
0.0001	0.720286	0.631533	0.571920	0.528858	0.496121	
0.0005	0.675679	0.597012	0.543637	0.504787	0.475076	
0.0010	0.656223	0.581829	0.531121	0.494084	0.465682	
0.0050	0.609921	0.545370	0.500864	0.468076	0.442762	
0.0100	0.589208	0.528904	0.487103	0.456182	0.432233	
0.0250	0.560653	0.506038	0.467887	0.439503	0.417417	
0.0500	0.537731	0.487535	0.452248	0.425865	0.405258	
0.1000	0.513019	0.467436	0.435161	0.410899	0.391865	
0.2000	0.485334	0.444720	0.415723	0.393785	0.376488	
0.3000	0.466756	0.429355	0.402495	0.382085	0.365934	
0.4000	0.451726	0.416850	0.391681	0.372485	0.357250	
0.5000	0.438335	0.405650	0.381959	0.363827	0.349399	
0.6000	0.425542	0.394898	0.372589	0.355459	0.341791	
0.7000	0.412476	0.383861	0.362936	0.346812	0.333912	
0.8000	0.397948	0.371524	0.352101	0.337075	0.325015	
0.9000	0.379062	0.355376	0.337847	0.324213	0.313224	
0.9500	0.364481	0.342823	0.326706	0.314119	0.303938	
0.9750	0.352517	0.332465	0.317475	0.305725	0.296194	
0.9900	0.339348	0.321001	0.307214	0.296365	0.287535	
0.9950	0.330806	0.313530	0.300503	0.290224	0.281840	
0.9990	0.314201	0.298922	0.287323	0.278122	0.270585	
0.9995	0.308079	0.293508	0.282418	0.273604	0.266372	
0.9999	0.295578	0.282405	0.272326	0.264282	0.257659	

COEFFICIENTS OF w_8 IN EXACT LOWER CONFIDENCE BOUNDS FOR σ

P\n	14	15	16	17	18	19
0.0001					68777.27	498.3850
0.0005					13752.88	221.6885
0.0010					6874.822	156.1209
0.0050					1372.380	68.60602
0.0100					684.5717	47.85945
0.0250					271.8799	29.43414
0.0500					134.3038	20.12781
0.1000					65.49211	13.51787
0.2000					31.03389	8.792672
0.3000					19.49364	6.655637
0.4000					13.67371	5.346588
0.5000					10.13123	4.420203
0.6000					7.713713	3.701486
0.7000					5.918850	3.101635
0.8000					4.477914	2.561604
0.9000					3.188640	2.014271
0.9500					2.494839	1.683793
0.9750					2.060910	1.459523
0.9900					1.686782	1.252240
0.9950					1.489142	1.136411
0.9990					1.181835	0.945693
0.9995					1.089089	0.885174
0.9999					0.926733	0.775349

P\n	20	22	24	26	28	30
0.0001	84.65363	17.45739	8.015114	4.938835	3.522713	2.736167
0.0005	48.81819	12.32728	6.170813	3.992464	2.939364	2.334758
0.0010	38.40118	10.57419	5.494579	3.631182	2.710666	2.174357
0.0050	21.74556	7.315270	4.148539	2.882616	2.223907	1.826240
0.0100	16.89534	6.195577	3.650718	2.593617	2.030516	1.685029
0.0250	11.96807	4.922548	3.054576	2.236691	1.786629	1.504202
0.0500	9.104317	4.090146	2.642873	1.982002	1.608690	1.370101
0.1000	6.806700	3.348564	2.257418	1.736268	1.433432	1.235999
0.2000	4.942449	2.676018	1.888512	1.493193	1.256073	1.097972
0.3000	4.002815	2.302599	1.673722	1.347450	1.147540	1.012214
0.4000	3.383553	2.039388	1.517128	1.238925	1.065519	0.946683
0.5000	2.918880	1.830715	1.389446	1.148853	0.996588	0.891088
0.6000	2.539371	1.651695	1.277079	1.068287	0.934217	0.840346
0.7000	2.206950	1.487317	1.171310	0.991234	0.873887	0.790838
0.8000	1.892698	1.324170	1.063569	0.911408	0.810626	0.738448
0.9000	1.556092	1.139262	0.937642	0.816212	0.734080	0.674346
0.9500	1.341838	1.014908	0.850339	0.748874	0.679140	0.627819
0.9750	1.190663	0.923431	0.784589	0.697359	0.636626	0.591495
0.9900	1.046045	0.832560	0.717848	0.644299	0.592363	0.553361
0.9950	0.962897	0.778630	0.677503	0.611819	0.565018	0.529631
0.9990	0.821791	0.683899	0.605169	0.552764	0.514773	0.485670
0.9995	0.775796	0.652051	0.580398	0.532281	0.497179	0.470161
0.9999	0.690647	0.591680	0.532761	0.492491	0.462740	0.439619

COEFFICIENTS OF w_8 IN EXACT LOWER CONFIDENCE BOUNDS FOR σ

P\n	32	34	36	38	40	50
0.0001	2.244870	1.912413	1.674029	1.495462	1.357062	0.966132
0.0005	1.947860	1.681131	1.487017	1.339829	1.224579	0.892521
0.0010	1.827445	1.586285	1.409606	1.274905	1.168946	0.860910
0.0050	1.562176	1.374846	1.235352	1.127562	1.041811	0.786934
0.0100	1.452848	1.286595	1.161864	1.064885	0.987330	0.754424
0.0250	1.311190	1.171166	1.064999	0.981729	0.914645	0.710215
0.0500	1.204808	1.083602	0.990904	0.917676	0.858323	0.675245
0.1000	1.097159	0.994151	0.914619	0.851293	0.799621	0.638075
0.2000	0.984891	0.899865	0.833499	0.780175	0.736329	0.597100
0.3000	0.914304	0.840012	0.781593	0.734362	0.695319	0.570008
0.4000	0.859893	0.793549	0.741061	0.698409	0.662999	0.548332
0.5000	0.813390	0.753598	0.706035	0.667207	0.634845	0.529206
0.6000	0.770654	0.716677	0.673514	0.638121	0.608511	0.511097
0.7000	0.728672	0.680206	0.641238	0.609140	0.582181	0.492770
0.8000	0.683920	0.641095	0.606452	0.577771	0.553575	0.472596
0.9000	0.628676	0.592461	0.562932	0.538319	0.517435	0.446690
0.9500	0.588218	0.556581	0.530624	0.508875	0.490337	0.426941
0.9750	0.556407	0.528202	0.504943	0.485370	0.468625	0.410905
0.9900	0.522785	0.498040	0.477520	0.460169	0.445263	0.393427
0.9950	0.501740	0.479071	0.460202	0.444197	0.430410	0.382191
0.9990	0.462493	0.443494	0.427568	0.413976	0.402206	0.360571
0.9995	0.448562	0.430802	0.415874	0.403107	0.392029	0.352676
0.9999	0.420993	0.405580	0.392556	0.381365	0.371616	0.336682

P\n	60	70	80	90	100	
0.0001	0.783889	0.678095	0.608586	0.559147	0.521993	
0.0005	0.733718	0.640079	0.577887	0.533296	0.499576	
0.0010	0.711887	0.623388	0.564321	0.521815	0.489581	
0.0050	0.660067	0.583387	0.531578	0.493954	0.465218	
0.0100	0.636948	0.565358	0.516710	0.481230	0.454040	
0.0250	0.605146	0.540362	0.495978	0.463408	0.438327	
0.0500	0.579677	0.520173	0.479127	0.448852	0.425445	
0.1000	0.552281	0.498280	0.460744	0.432898	0.411271	
0.2000	0.521668	0.473586	0.439865	0.414680	0.395016	
0.3000	0.501176	0.456915	0.425679	0.402241	0.383873	
0.4000	0.484628	0.443366	0.414095	0.392046	0.374712	
0.5000	0.469909	0.431246	0.403691	0.382859	0.366436	
0.6000	0.455868	0.419625	0.393674	0.373987	0.358423	
0.7000	0.441550	0.407711	0.383365	0.364828	0.350130	
0.8000	0.425657	0.394411	0.371807	0.354523	0.340774	
0.9000	0.405041	0.377033	0.356623	0.340929	0.328388	
0.9500	0.389160	0.363546	0.344773	0.330273	0.318645	
0.9750	0.376153	0.352435	0.334966	0.321422	0.310528	
0.9900	0.361862	0.340155	0.324079	0.311562	0.301460	
0.9950	0.352607	0.332161	0.316964	0.305099	0.295501	
0.9990	0.334649	0.316557	0.303012	0.292377	0.283735	
0.9995	0.328040	0.310781	0.297825	0.287632	0.279334	
0.9999	0.314564	0.298951	0.287164	0.277849	0.270240	

Table A8

MOST EFFECTIVE (EFFICIENT) INTERVAL ESTIMATORS FOR σ, BASED ON ONE QUASI-RANGE OF A SAMPLE FROM A NORMAL POPULATION

[Upper and lower $(1-P)$ confidence bounds, B_{ur} and B_{tr}, of central $(1-2P)$ confidence interval, based on rth quasi-range, w_r, of sample size n]

F_u = **Effectiveness of Upper Confidence Bound**

F_i = **Effectiveness of Central Confidence Interval**

E_u = **Efficiency of Upper Confidence Bound**

E_i = **Efficiency of Central Confidence Interval**

A. VALUE OF r, CHOSEN TO MAXIMIZE EFFECTIVENESS F_u OF UPPER CONFIDENCE BOUND, ALSO MAXIMIZES EFFECTIVENESS F_i OF CENTRAL CONFIDENCE INTERVAL, UNLESS MARKED *

n	r	1-P	B_{ur}/w_r	$B_{\ell r}/w_r$	1-2P	F_u (%)	F_i (%)
2	0	0.9999	5641.895	0.181748	0.9998	100.0	100.0
2	0	0.9995	1128.379	0.203147	0.9990	100.0	100.0
2	0	0.9990	564.1894	0.214892	0.9980	100.0	100.0
2	0	0.9950	112.8372	0.251905	0.9900	100.0	100.0
2	0	0.9900	56.41748	0.274516	0.9800	100.0	100.0
2	0	0.9750	22.56389	0.315475	0.9500	100.0	100.0
2	0	0.9500	11.27640	0.360775	0.9000	100.0	100.0
2	0	0.9000	5.627083	0.429890	0.8000	100.0	100.0
2	0	0.8000	2.791059	0.551758	0.6000	100.0	100.0
2	0	0.7000	1.835113	0.682250	0.4000	100.0	100.0
2	0	0.6000	1.348410	0.840172	0.2000	100.0	100.0
2	0	0.5000	1.048358			100.0	
3	0	0.9999	52.50243	0.170527	0.9998	99.7	99.7
3	0	0.9995	23.47744	0.188097	0.9990	99.7	99.7
3	0	0.9990	16.59896	0.197494	0.9980	99.7	99.7
3	0	0.9950	7.415786	0.226028	0.9900	99.7	99.7
3	0	0.9900	5.237111	0.242701	0.9800	99.7	99.7
3	0	0.9750	3.299561	0.271572	0.9500	99.7	99.7
3	0	0.9500	2.318021	0.301705	0.9000	99.7	99.7
3	0	0.9000	1.617201	0.344545	0.8000	99.7	99.6
3	0	0.8000	1.110998	0.412621	0.6000	99.8	99.6
3	0	0.7000	0.878535	0.477420	0.4000	99.8	99.6
3	0	0.6000	0.733893	0.547549	0.2000	99.8	99.6
3	0	0.5000	0.629807			99.9	
4	0	0.9999	10.82318	0.164396	0.9998	99.2	99.1
4	0	0.9995	6.322927	0.180088	0.9990	99.2	99.1
4	0	0.9990	5.013886	0.188366	0.9980	99.2	99.1
4	0	0.9950	2.917986	0.213034	0.9900	99.2	99.0
4	0	0.9900	2.305867	0.227128	0.9800	99.2	99.0
4	0	0.9750	1.681682	0.251003	0.9500	99.2	98.9
4	0	0.9500	1.316598	0.275242	0.9000	99.3	98.9
4	0	0.9000	1.021069	0.308599	0.8000	99.3	98.8
4	0	0.8000	0.777803	0.359227	0.6000	99.4	98.8
4	0	0.7000	0.652961	0.405055	0.4000	99.5	98.7
4	0	0.6000	0.569305	0.452431	0.2000	99.6	98.7
4	0	0.5000	0.505479			99.7	
5	0	0.9999	4.866452	0.160264	0.9998	98.5	98.3
5	0	0.9995	3.244414	0.174771	0.9990	98.5	98.3
5	0	0.9990	2.721889	0.182357	0.9980	98.5	98.2
5	0	0.9950	1.802113	0.204684	0.9900	98.6	98.1
5	0	0.9900	1.503725	0.217258	0.9800	98.6	98.0
5	0	0.9750	1.176925	0.238264	0.9500	98.7	97.9
5	0	0.9500	0.970930	0.259225	0.9000	98.7	97.8
5	0	0.9000	0.792771	0.287498	0.8000	98.9	97.8
5	0	0.8000	0.635550	0.329238	0.6000	99.0	97.7
5	0	0.7000	0.549920	0.365913	0.4000	99.2	97.6
5	0	0.6000	0.490173	0.402832	0.2000	99.4	97.6
5	0	0.5000	0.443089			99.6	

A. VALUE OF r, CHOSEN TO MAXIMIZE EFFECTIVENESS F_u OF UPPER CONFIDENCE BOUND, ALSO MAXIMIZES EFFECTIVENESS F_i OF CENTRAL CONFIDENCE INTERVAL, UNLESS MARKED *

n	r	1-P	B_{ur}/w_r	$B_{\ell r}/w_r$	1-2P	F_u (%)	F_i (%)
6	0	0.9999	2.992513	0.157190	0.9998	97.9	97.5
6	0	0.9995	2.156566	0.170857	0.9990	97.9	97.3
6	0	0.9990	1.870081	0.177957	0.9980	97.9	97.3
6	0	0.9950	1.335144	0.198670	0.9900	98.0	97.1
6	0	0.9900	1.150066	0.210214	0.9800	98.0	97.0
6	0	0.9750	0.938129	0.229310	0.9500	98.2	96.8
6	0	0.9500	0.798157	0.248133	0.9000	98.3	96.7
6	0	0.9000	0.671955	0.273170	0.8000	98.5	96.6
6	0	0.8000	0.555557	0.309431	0.6000	98.7	96.5
6	0	0.7000	0.489709	0.340651	0.4000	99.0	96.5
6	0	0.6000	0.442548	0.371518	0.2000	99.2	96.4
6	0	0.5000	0.404588			99.5	
7	0	0.9999	2.152779	0.154765	0.9998	97.2	96.5
7	0	0.9995	1.632416	0.167793	0.9990	97.3	96.3
7	0	0.9990	1.446451	0.174528	0.9980	97.3	96.3
7	0	0.9950	1.084804	0.194039	0.9900	97.4	96.0
7	0	0.9900	0.954067	0.204827	0.9800	97.5	95.9
7	0	0.9750	0.799680	0.222538	0.9500	97.7	95.7
7	0	0.9500	0.694376	0.239834	0.9000	97.9	95.6
7	0	0.9000	0.596641	0.262598	0.8000	98.1	95.5
7	0	0.8000	0.503665	0.295100	0.6000	98.5	95.4
7	0	0.7000	0.449642	0.322664	0.4000	98.8	95.3
7	0	0.6000	0.410222	0.349561	0.2000	99.1	95.3
7	0	0.5000	0.378007			99.4	
8	0	0.9999	1.694382	0.152776	0.9998	96.6	95.6
8	0	0.9995	1.331535	0.165296	0.9990	96.7	95.4
8	0	0.9990	1.197855	0.171741	0.9980	96.8	95.2
8	0	0.9950	0.929989	0.190311	0.9900	96.9	95.0
8	0	0.9900	0.830000	0.200514	0.9800	97.1	94.8
8	0	0.9750	0.709210	0.217162	0.9500	97.3	94.7
8	0	0.9500	0.624839	0.233301	0.9000	97.5	94.5
8	0	0.9000	0.544826	0.254366	0.8000	97.8	94.4
8	0	0.8000	0.466928	0.284104	0.6000	98.2	94.2
8	0	0.7000	0.420747	0.309030	0.4000	98.6	94.2
8	0	0.6000	0.386570	0.333105	0.2000	99.0	94.1
8	0	0.5000	0.358315			99.3	
9	0	0.9999	1.411017	0.151098	0.9998	96.1	94.6
9	0	0.9995	1.138489	0.163200	0.9990	96.2	94.4
9	0	0.9990	1.035725	0.169408	0.9980	96.3	94.2
9	0	0.9950	0.825004	0.187215	0.9900	96.5	93.9
9	0	0.9900	0.744388	0.196947	0.9800	96.6	93.8
9	0	0.9750	0.645278	0.212747	0.9500	96.9	93.6
9	0	0.9500	0.574762	0.227972	0.9000	97.2	93.4
9	0	0.9000	0.506758	0.247707	0.8000	97.5	93.3
9	0	0.8000	0.439344	0.275316	0.6000	98.0	93.1
9	0	0.7000	0.398740	0.298236	0.4000	98.5	93.1
9	0	0.6000	0.368355	0.320192	0.2000	98.9	93.0
9	0	0.5000	0.343002			99.3	

A. VALUE OF r, CHOSEN TO MAXIMIZE EFFECTIVENESS F_u OF UPPER CONFIDENCE BOUND, ALSO MAXIMIZES EFFECTIVENESS F_i OF CENTRAL CONFIDENCE INTERVAL, UNLESS MARKED *

n	r	1-P	B_{ur}/w_r	$B_{\ell r}/w_r$	1-2P	F_u (%)	F_i (%)
10	0	0.9999	1.220356	0.149652	0.9998	95.6	93.7
10	0	0.9995	1.004804	0.161401	0.9990	95.7	93.4
10	0	0.9990	0.922013	0.167411	0.9980	95.8	93.3
10	0	0.9950	0.749104	0.184583	0.9900	96.1	92.9
10	0	0.9900	0.681648	0.193925	0.9800	96.3	92.8
10	0	0.9750	0.597544	0.209029	0.9500	96.6	92.6
10	0	0.9500	0.536814	0.223507	0.9000	96.9	92.4
10	0	0.9000	0.477453	0.242169	0.8000	97.3	92.2
10	0	0.8000	0.417738	0.268076	0.6000	97.9	92.1
10	0	0.7000	0.381307	0.289414	0.4000	98.3	92.0
10	0	0.6000	0.353796	0.309715	0.2000	98.8	92.0
10	0	0.5000	0.330666			99.3	
11	0	0.9999	1.083995	0.148385	0.9998	95.1	92.8
11	0	0.9995	0.906960	0.159832	0.9990	95.3	92.5
11	0	0.9990	0.837939	0.165673	0.9980	95.4	92.3
11	0	0.9950	0.691601	0.182303	0.9900	95.7	92.0
11	0	0.9900	0.633592	0.191316	0.9800	95.9	91.8
11	0	0.9750	0.560427	0.205834	0.9500	96.3	91.6
11	0	0.9500	0.506950	0.219690	0.9000	96.6	91.4
11	0	0.9000	0.454092	0.237462	0.8000	97.1	91.2
11	0	0.8000	0.400270	0.261973	0.6000	97.7	91.1
11	0	0.7000	0.367081	0.282025	0.4000	98.2	91.0
11	0	0.6000	0.341826	0.300992	0.2000	98.7	90.9
11	0	0.5000	0.320458			99.2	
12	0	0.9999	0.981891	0.147261	0.9998	94.7	91.9
12	0	0.9995	0.832298	0.158444	0.9990	94.9	91.6
12	0	0.9990	0.773245	0.164137	0.9980	95.0	91.4
12	0	0.9950	0.646455	0.180300	0.9900	95.4	91.0
12	0	0.9900	0.595520	0.189029	0.9800	95.6	90.9
12	0	0.9750	0.530652	0.203046	0.9500	96.0	90.6
12	0	0.9500	0.482752	0.216373	0.9000	96.4	90.4
12	0	0.9000	0.434961	0.233392	0.8000	96.9	90.2
12	0	0.8000	0.385793	0.256733	0.6000	97.6	90.1
12	0	0.7000	0.355198	0.275716	0.4000	98.1	90.0
12	0	0.6000	0.331766	0.293583	0.2000	98.7	90.0
12	1	0.5000	0.454946			99.3	
13	0	0.9999	0.902675	0.146253	0.9998	94.3	91.0
13	0	0.9995	0.773444	0.157202	0.9990	94.5	90.7
13	0	0.9990	0.721890	0.162766	0.9980	94.7	90.5
13	0	0.9950	0.610006	0.178518	0.9900	95.1	90.1
13	0	0.9900	0.564547	0.187000	0.9800	95.3	89.9
13	0	0.9750	0.506172	0.200582	0.9500	95.7	89.7
13	0	0.9500	0.462689	0.213451	0.9000	96.2	89.5
13	0	0.9000	0.418953	0.229824	0.8000	96.7	89.3
13	0	0.8000	0.373557	0.252167	0.6000	97.4	89.1
13	0	0.7000	0.345087	0.270245	0.4000	98.1	89.1
13	0	0.6000	0.323159	0.287186	0.2000	98.6	89.0
13	1	0.5000	0.435498			99.4	

A. VALUE OF r, CHOSEN TO MAXIMIZE EFFECTIVENESS F_u OF UPPER CONFIDENCE BOUND, ALSO MAXIMIZES EFFECTIVENESS F_i OF CENTRAL CONFIDENCE INTERVAL, UNLESS MARKED *

n	r	1-P	B_{ur}/W_r	$B_{\ell r}/W_r$	1-2P	F_u (%)	F_i (%)
14	0	0.9999	0.839449	0.145340	0.9998	93.9	90.2
14	0	0.9995	0.725832	0.156081	0.9990	94.2	89.9
14	0	0.9990	0.680094	0.161529	0.9980	94.4	89.7
14	0	0.9950	0.579911	0.176918	0.9900	94.8	89.3
14	0	0.9900	0.538805	0.185182	0.9800	95.1	89.1
14	0	0.9750	0.485642	0.198381	0.9500	95.5	88.8
14	0	0.9500	0.445740	0.210849	0.9000	96.0	88.6
14	0	0.9000	0.405323	0.226659	0.8000	96.5	88.4
14	0	0.8000	0.363047	0.248139	0.6000	97.3	88.3
14	0	0.7000	0.336352	0.265439	0.4000	98.0	88.2
14	0	0.6000	0.315689	0.281589	0.2000	98.6	88.1
14	1	0.5000	0.419280			99.4	
15	0	0.9999	0.787813	0.144507	0.9998	93.6	89.4
15	0	0.9995	0.686491	0.155061	0.9990	93.9	89.0
15	0	0.9990	0.645380	0.160405	0.9980	94.1	88.9
15	0	0.9950	0.554599	0.175469	0.9900	94.6	88.4
15	0	0.9900	0.517032	0.183538	0.9800	94.8	88.2
15	0	0.9750	0.468140	0.196396	0.9500	95.3	88.0
15	0	0.9500	0.431199	0.208510	0.9000	95.8	87.8
15	0	0.9000	0.393548	0.223825	0.8000	96.4	87.6
15	0	0.8000	0.353897	0.244548	0.6000	97.2	87.4
15	0	0.7000	0.328708	0.261172	0.4000	97.9	87.3
15	1	0.6000	0.429854	0.383208	0.2000	98.6	84.6*
15	1	0.5000	0.405506			99.4	
16	0	0.9999	0.744835	0.143742	0.9998	93.3	88.6
16	0	0.9995	0.653410	0.154126	0.9990	93.6	88.3
16	0	0.9990	0.616056	0.159376	0.9980	93.8	88.1
16	0	0.9950	0.532982	0.174147	0.9900	94.3	87.6
16	0	0.9900	0.498344	0.182040	0.9800	94.6	87.4
16	0	0.9750	0.453013	0.194594	0.9500	95.1	87.2
16	0	0.9500	0.418560	0.206392	0.9000	95.6	87.0
16	0	0.9000	0.383252	0.221265	0.8000	96.3	86.8
16	0	0.8000	0.345842	0.241320	0.6000	97.1	86.6
16	0	0.7000	0.321947	0.257348	0.4000	97.8	86.5
16	1	0.6000	0.416321	0.372769	0.2000	98.6	84.8*
16	1	0.5000	0.393631			99.5	
17	0	0.9999	0.708486	0.143036	0.9998	93.0	87.9
17	0	0.9995	0.625178	0.153265	0.9990	93.4	87.5
17	0	0.9990	0.590931	0.158429	0.9980	93.6	87.3
17	0	0.9950	0.514278	0.172933	0.9900	94.1	86.9
17	0	0.9900	0.482104	0.180668	0.9800	94.4	86.7
17	0	0.9750	0.439786	0.192947	0.9500	95.0	86.4
17	0	0.9500	0.407453	0.204459	0.9000	95.5	86.2
17	0	0.9000	0.374155	0.218937	0.8000	96.1	86.0
17	0	0.8000	0.338681	0.238395	0.6000	97.0	85.8
17	1	0.7000	0.429367	0.344245	0.4000	97.9	85.0*
17	1	0.6000	0.404551	0.363621	0.2000	98.7	85.0*
17	1	0.5000	0.383262			99.5	

A. VALUE OF r, CHOSEN TO MAXIMIZE EFFECTIVENESS F_u OF UPPER CONFIDENCE BOUND, ALSO MAXIMIZES EFFECTIVENESS F_i OF CENTRAL CONFIDENCE INTERVAL, UNLESS MARKED *

n	r	1-P	B_{ur}/W_r	$B_{\ell r}/W_r$	1-2P	F_u (%)	F_i (%)
18	0	0.9999	0.677326	0.142381	0.9998	92.8	87.2
18	0	0.9995	0.600780	0.152467	0.9990	93.1	86.8
18	0	0.9990	0.569140	0.157553	0.9980	93.3	86.6
18	0	0.9950	0.497916	0.171812	0.9900	93.9	86.1
18	0	0.9900	0.467839	0.179403	0.9800	94.3	85.9
18	0	0.9750	0.428103	0.191431	0.9500	94.8	85.6
18	0	0.9500	0.397599	0.202686	0.9000	95.3	85.4
18	0	0.9000	0.366045	0.216807	0.8000	96.0	85.2
18	1	0.8000	0.447574	0.317353	0.6000	97.0	85.0*
18	1	0.7000	0.417535	0.337127	0.4000	97.9	85.1
18	1	0.6000	0.394199	0.355523	0.2000	98.7	85.1
18	1	0.5000	0.374113			99.5	
19	0	0.9999	0.650298	0.141770	0.9998	92.5	86.5
19	0	0.9995	0.579466	0.151725	0.9990	92.9	86.1
19	0	0.9990	0.550041	0.156738	0.9980	93.1	85.9
19	0	0.9950	0.483462	0.170773	0.9900	93.7	85.4
19	0	0.9900	0.455194	0.178231	0.9800	94.1	85.2
19	0	0.9750	0.417696	0.190031	0.9500	94.7	84.9
19	0	0.9500	0.388785	0.201050	0.9000	95.2	84.7
19	0	0.9000	0.358758	0.214846	0.8000	95.9	84.5
19	1	0.8000	0.435357	0.311842	0.6000	97.1	85.0
19	1	0.7000	0.407066	0.330750	0.4000	98.0	85.1
19	1	0.6000	0.385008	0.348289	0.2000	98.8	85.1
19	1	0.5000	0.365964			99.5	
20	0	0.9999	0.626618	0.141199	0.9998	92.3	85.8
20	0	0.9995	0.560669	0.151031	0.9990	92.7	85.4
20	0	0.9990	0.533149	0.155977	0.9980	92.9	85.2
20	0	0.9950	0.470588	0.169805	0.9900	93.6	84.7
20	0	0.9900	0.443894	0.177141	0.9800	93.9	84.5
20	0	0.9750	0.408353	0.188730	0.9500	94.5	84.2*
20	0	0.9500	0.380843	0.199534	0.9000	95.1	84.0*
20	1	0.9000	0.466276	0.284064	0.8000	96.0	84.9
20	1	0.8000	0.424495	0.306854	0.6000	97.2	85.0
20	1	0.7000	0.397722	0.324995	0.4000	98.1	85.1
20	1	0.6000	0.376780	0.341778	0.2000	98.8	85.1
20	2	0.5000	0.446295			99.5	
22	0	0.9999	0.587038	0.140157	0.9998	91.9	84.5
22	0	0.9995	0.528984	0.149768	0.9990	92.3	84.1
22	0	0.9990	0.504567	0.154594	0.9980	92.5	83.9*
22	0	0.9950	0.448602	0.168051	0.9900	93.2	83.4*
22	0	0.9900	0.424514	0.175169	0.9800	93.6	83.2*
22	1	0.9750	0.512309	0.248119	0.9500	94.5	84.6
22	1	0.9500	0.478558	0.260909	0.9000	95.3	84.7
22	1	0.9000	0.443551	0.276868	0.8000	96.2	84.8
22	1	0.8000	0.405978	0.298148	0.6000	97.3	84.9
22	1	0.7000	0.381714	0.314992	0.4000	98.1	85.0
22	1	0.6000	0.362626	0.330500	0.2000	98.8	85.0
22	2	0.5000	0.424396			99.6	

A. VALUE OF r, CHOSEN TO MAXIMIZE EFFECTIVENESS F_u OF UPPER CONFIDENCE BOUND, ALSO MAXIMIZES EFFECTIVENESS F_i OF CENTRAL CONFIDENCE INTERVAL, UNLESS MARKED *

n	r	1-P	B_{ur}/w_r	$B_{\ell r}/w_r$	1-2P	F_u(%)	F_i (%)
24	0	0.9999	0.555208	0.139227	0.9998	91.5	83.3*
24	0	0.9995	0.503248	0.148645	0.9990	92.0	82.9*
24	0	0.9990	0.481247	0.153365	0.9980	92.2	82.7*
24	1	0.9950	0.555501	0.221997	0.9900	93.3	84.3
24	1	0.9900	0.526178	0.230363	0.9800	93.9	84.4
24	1	0.9750	0.487042	0.243485	0.9500	94.8	84.5
24	1	0.9500	0.456685	0.255635	0.9000	95.5	84.6
24	1	0.9000	0.424990	0.270737	0.8000	96.3	84.7
24	1	0.8000	0.390725	0.290776	0.6000	97.4	84.7
24	1	0.7000	0.368451	0.306559	0.4000	98.2	84.7
24	1	0.6000	0.350844	0.321033	0.2000	98.9	84.8
24	2	0.5000	0.406658			99.6	
26	1	0.9999	0.678999	0.186699	0.9998	91.2	83.7
26	1	0.9995	0.615907	0.197901	0.9990	92.1	83.9
26	1	0.9990	0.589281	0.203469	0.9980	92.5	84.0
26	1	0.9950	0.528068	0.218828	0.9900	93.6	84.1
26	1	0.9900	0.501651	0.226868	0.9800	94.1	84.2
26	1	0.9750	0.466199	0.239443	0.9500	94.9	84.3
26	1	0.9500	0.438535	0.251050	0.9000	95.6	84.3
26	1	0.9000	0.409492	0.265429	0.8000	96.4	84.4
26	1	0.8000	0.377895	0.284426	0.6000	97.5	84.4
26	1	0.7000	0.357239	0.299325	0.4000	98.2	84.4
26	1	0.6000	0.340843	0.312940	0.2000	98.9	84.4
26	2	0.5000	0.391937			99.6	
28	1	0.9999	0.641154	0.184872	0.9998	91.5	83.7
28	1	0.9995	0.584662	0.195757	0.9990	92.3	83.8
28	1	0.9990	0.560684	0.201157	0.9980	92.7	83.8
28	1	0.9950	0.505227	0.216015	0.9900	93.8	83.9
28	1	0.9900	0.481142	0.223772	0.9800	94.3	83.9
28	1	0.9750	0.448663	0.235875	0.9500	95.0	84.0
28	1	0.9500	0.423189	0.247014	0.9000	95.7	84.0
28	1	0.9000	0.396315	0.260772	0.8000	96.5	84.0
28	1	0.8000	0.366918	0.278881	0.6000	97.5	84.1
28	1	0.7000	0.347606	0.293030	0.4000	98.2	84.1
28	2	0.6000	0.395864	0.364088	0.2000	98.9	83.0*
28	3	0.5000	0.442792			99.6	
30	1	0.9999	0.609687	0.183222	0.9998	91.7	83.5
30	1	0.9995	0.558490	0.193827	0.9990	92.5	83.5
30	1	0.9990	0.536651	0.199078	0.9980	92.9	83.6
30	1	0.9950	0.485874	0.213495	0.9900	93.9	83.6
30	1	0.9900	0.463699	0.221003	0.9800	94.4	83.6
30	1	0.9750	0.433669	0.232692	0.9500	95.1	83.6
30	1	0.9500	0.410011	0.243423	0.9000	95.8	83.7
30	1	0.9000	0.384945	0.256642	0.8000	96.5	83.7
30	1	0.8000	0.357394	0.273981	0.6000	97.5	83.7
30	2	0.7000	0.401490	0.339840	0.4000	98.3	83.1*
30	2	0.6000	0.384063	0.354357	0.2000	99.0	83.1*
30	3	0.5000	0.426966			99.7	

A. VALUE OF r, CHOSEN TO MAXIMIZE EFFECTIVENESS F_u OF UPPER CONFIDENCE BOUND, ALSO MAXIMIZES EFFECTIVENESS F_i OF CENTRAL CONFIDENCE INTERVAL, UNLESS MARKED *

n	r	1-P	B_{ur}/W_r	$B_{\ell r}/W_r$	1-2P	F_u (%)	F_i (%)
32	1	0.9999	0.583077	0.181722	0.9998	91.9	83.2
32	1	0.9995	0.536213	0.192074	0.9990	92.7	83.3
32	1	0.9990	0.516136	0.197193	0.9980	93.0	83.3
32	1	0.9950	0.469235	0.211217	0.9900	94.0	83.3
32	1	0.9900	0.448652	0.218504	0.9800	94.4	83.3
32	1	0.9750	0.420675	0.229827	0.9500	95.2	83.3
32	1	0.9500	0.398545	0.240199	0.9000	95.8	83.3
32	1	0.9000	0.375011	0.252943	0.8000	96.6	83.3
32	2	0.8000	0.410628	0.317077	0.6000	97.5	83.2*
32	2	0.7000	0.390145	0.332142	0.4000	98.3	83.2*
32	2	0.6000	0.373811	0.345856	0.2000	99.0	83.2*
32	3	0.5000	0.413414			99.7	
34	1	0.9999	0.560249	0.180348	0.9998	92.0	83.0
34	1	0.9995	0.516994	0.190474	0.9990	92.7	82.9
34	1	0.9990	0.498390	0.195473	0.9980	93.1	82.9
34	1	0.9950	0.454751	0.209144	0.9900	94.0	82.9
34	1	0.9900	0.435516	0.216233	0.9800	94.5	82.9
34	1	0.9750	0.409284	0.227229	0.9500	95.2	82.9*
34	1	0.9500	0.388460	0.237281	0.9000	95.8	82.9*
34	2	0.9000	0.428612	0.292556	0.8000	96.6	83.1
34	2	0.8000	0.399459	0.310984	0.6000	97.6	83.2
34	2	0.7000	0.380202	0.325325	0.4000	98.4	83.3
34	2	0.6000	0.364801	0.338348	0.2000	99.0	83.3
34	3	0.5000	0.401650			99.7	
36	1	0.9999	0.540427	0.179082	0.9998	92.1	82.6
36	1	0.9995	0.500220	0.189002	0.9990	92.8	82.6
36	1	0.9990	0.482867	0.193893	0.9980	93.2	82.6
36	1	0.9950	0.442010	0.207246	0.9900	94.1	82.5*
36	1	0.9900	0.423929	0.214156	0.9800	94.5	82.5*
36	1	0.9750	0.399198	0.224858	0.9500	95.2	82.5*
36	2	0.9500	0.441912	0.274350	0.9000	95.9	83.0
36	2	0.9000	0.417060	0.287864	0.8000	96.7	83.1
36	2	0.8000	0.389600	0.305527	0.6000	97.7	83.2
36	2	0.7000	0.371398	0.319235	0.4000	98.4	83.2
36	2	0.6000	0.356804	0.331654	0.2000	99.1	83.2
36	4	0.5000	0.440905			99.7	
38	1	0.9999	0.523035	0.177911	0.9998	92.2	82.3
38	1	0.9995	0.485435	0.187643	0.9990	92.9	82.2*
38	1	0.9990	0.469155	0.192435	0.9980	93.2	82.2*
38	1	0.9950	0.430698	0.205497	0.9900	94.1	82.1*
38	1	0.9900	0.413618	0.212246	0.9800	94.5	82.1*
38	2	0.9750	0.452205	0.259954	0.9500	95.3	82.9
38	2	0.9500	0.430254	0.270588	0.9000	96.0	83.0
38	2	0.9000	0.406812	0.283617	0.8000	96.8	83.1
38	2	0.8000	0.380819	0.300602	0.6000	97.8	83.1
38	2	0.7000	0.363536	0.313751	0.4000	98.5	83.1
38	2	0.6000	0.349646	0.325640	0.2000	99.1	83.2
38	4	0.5000	0.428509			99.7	

A. VALUE OF r, CHOSEN TO MAXIMIZE EFFECTIVENESS F_u OF UPPER CONFIDENCE BOUND, ALSO MAXIMIZES EFFECTIVENESS F_i OF CENTRAL CONFIDENCE INTERVAL, UNLESS MARKED *

n	n	1-P	B_{ur}/W_r	$B_{\ell r}/W_r$	1-2P	F_u (%)	F_i (%)
40	1	0.9999	0.507635	0.176822	0.9998	92.2	81.9*
40	1	0.9995	0.472289	0.186381	0.9990	92.9	81.8*
40	1	0.9990	0.456942	0.191082	0.9980	93.2	81.8*
40	2	0.9950	0.485643	0.238249	0.9900	94.2	82.7
40	2	0.9900	0.466629	0.245555	0.9800	94.7	82.8
40	2	0.9750	0.440600	0.256855	0.9500	95.4	82.9
40	2	0.9500	0.419858	0.267153	0.9000	96.1	82.9
40	2	0.9000	0.397644	0.279748	0.8000	96.9	83.0
40	2	0.8000	0.372937	0.296128	0.6000	97.8	83.0
40	2	0.7000	0.356461	0.308779	0.4000	98.5	83.0
40	3	0.6000	0.387428	0.361199	0.2000	99.1	82.2*
40	4	0.5000	0.417544			99.7	
50	2	0.9999	0.510872	0.201169	0.9998	92.7	82.0
50	2	0.9995	0.479365	0.210739	0.9990	93.4	82.0
50	2	0.9990	0.465579	0.215418	0.9980	93.8	82.0
50	2	0.9950	0.432652	0.228073	0.9900	94.7	82.0*
50	2	0.9900	0.417859	0.234561	0.9800	95.1	82.1*
50	2	0.9750	0.397396	0.244531	0.9500	95.8	82.1*
50	2	0.9500	0.380902	0.253553	0.9000	96.4	82.1*
50	3	0.9000	0.403819	0.294576	0.8000	97.1	82.4
50	3	0.8000	0.381331	0.310060	0.6000	98.0	82.5
50	3	0.7000	0.366221	0.321947	0.4000	98.6	82.5
50	3	0.6000	0.353982	0.332624	0.2000	99.2	82.5
50	5	0.5000	0.411964			99.8	
60	2	0.9999	0.462261	0.195856	0.9998	93.1	81.0*
60	3	0.9995	0.480975	0.227795	0.9990	93.8	81.8
60	3	0.9990	0.468446	0.232395	0.9980	94.1	81.9
60	3	0.9950	0.438333	0.244790	0.9900	95.0	82.0
60	3	0.9900	0.424718	0.251119	0.9800	95.4	82.0
60	3	0.9750	0.405791	0.260816	0.9500	96.1	82.0
60	3	0.9500	0.390458	0.269562	0.9000	96.7	82.1
60	3	0.9000	0.373778	0.280142	0.8000	97.3	82.1
60	3	0.8000	0.354893	0.293719	0.6000	98.1	82.1
60	4	0.7000	0.372433	0.330996	0.4000	98.8	82.1*
60	4	0.6000	0.361037	0.341044	0.2000	99.3	82.1*
60	6	0.5000	0.408273			99.8	
70	3	0.9999	0.467559	0.212685	0.9998	93.5	81.3*
70	3	0.9995	0.443821	0.221456	0.9990	94.2	81.4*
70	3	0.9990	0.433306	0.225715	0.9980	94.5	81.4*
70	3	0.9950	0.407856	0.237144	0.9900	95.3	81.4*
70	4	0.9900	0.428459	0.263467	0.9800	95.7	81.8
70	4	0.9750	0.410822	0.272886	0.9500	96.3	81.9
70	4	0.9500	0.396473	0.281358	0.9000	96.9	81.9
70	4	0.9000	0.380802	0.291580	0.8000	97.5	82.0
70	4	0.8000	0.362980	0.304654	0.6000	98.3	82.0
70	4	0.7000	0.350855	0.314589	0.4000	98.8	82.0
70	5	0.6000	0.365998	0.347151	0.2000	99.3	81.8*
70	8	0.5000	0.431246			99.8	

A. VALUE OF **r**, CHOSEN TO MAXIMIZE EFFECTIVENESS F_u OF UPPER CONFIDENCE BOUND, ALSO MAXIMIZES EFFECTIVENESS F_i OF CENTRAL CONFIDENCE INTERVAL, UNLESS MARKED *

n	r	1-P	B_{ur}/w_r	$B_{\ell r}/w_r$	1-2P	F_u (%)	F_i (%)
80	4	0.9999	0.469950	0.225640	0.9998	93.8	81.4
80	4	0.9995	0.447735	0.234283	0.9990	94.5	81.4
80	4	0.9990	0.437860	0.238472	0.9980	94.8	81.5*
80	4	0.9950	0.413875	0.249687	0.9900	95.5	81.5*
80	4	0.9900	0.402909	0.255376	0.9800	95.9	81.5*
80	4	0.9750	0.387533	0.264044	0.9500	96.5	81.6*
80	4	0.9500	0.374961	0.271812	0.9000	97.0	81.6*
80	5	0.9000	0.385683	0.300395	0.8000	97.6	81.8
80	5	0.8000	0.368786	0.313003	0.6000	98.4	81.8
80	5	0.7000	0.357251	0.322559	0.4000	98.9	81.8
80	6	0.6000	0.369668	0.351798	0.2000	99.4	81.5*
80	8	0.5000	0.403691			99.9	
90	4	0.9999	0.442415	0.220885	0.9998	94.1	81.0*
90	4	0.9995	0.422924	0.229032	0.9990	94.7	81.0*
90	5	0.9990	0.440581	0.248688	0.9980	95.0	81.4
90	5	0.9950	0.417880	0.259685	0.9900	95.8	81.5
90	5	0.9900	0.407467	0.265252	0.9800	96.1	81.5
90	5	0.9750	0.392831	0.273722	0.9500	96.7	81.6
90	5	0.9500	0.380833	0.281300	0.9000	97.2	81.6
90	5	0.9000	0.367633	0.290393	0.8000	97.8	81.6
90	6	0.8000	0.373131	0.319621	0.6000	98.5	81.7*
90	6	0.7000	0.362116	0.328832	0.4000	99.0	81.7
90	6	0.6000	0.353070	0.337003	0.2000	99.4	81.7
90	8	0.5000	0.382859			99.9	
100	5	0.9999	0.445773	0.231240	0.9998	94.3	81.1*
100	5	0.9995	0.427203	0.239275	0.9990	94.9	81.2*
100	5	0.9990	0.418897	0.243154	0.9980	95.2	81.2*
100	5	0.9950	0.398580	0.253498	0.9900	95.9	81.2*
100	6	0.9900	0.410696	0.273356	0.9800	96.3	81.5
100	6	0.9750	0.396719	0.281635	0.9500	96.8	81.5
100	6	0.9500	0.385236	0.289030	0.9000	97.3	81.5
100	6	0.9000	0.372574	0.297891	0.8000	97.9	81.6
100	6	0.8000	0.358016	0.309127	0.6000	98.5	81.6
100	7	0.7000	0.365934	0.333912	0.4000	99.0	81.5*
100	7	0.6000	0.357250	0.341791	0.2000	99.5	81.5*
100	8	0.5000	0.366436			99.9	

B. VALUE OF $\mathbf{r}$, CHOSEN TO MAXIMIZE EFFECTIVENESS F_i OF CENTRAL CONFIDENCE INTERVAL, DOES NOT MAXIMIZE EFFECTIVENESS F_u OF UPPER CONFIDENCE BOUND--SEE TABLE A9A

n	r	1-P	B_{ur}/w_r	$B_{\ell r}/w_r$	1-2P	F_u(%)	F_i(%)
15	0	0.6000	0.309125	0.276636	0.2000	98.5	87.3
16	0	0.6000	0.303298	0.272212	0.2000	98.5	86.5
17	0	0.7000	0.315913	0.253894	0.4000	97.8	85.7
17	0	0.6000	0.298080	0.268227	0.2000	98.5	85.7
18	0	0.8000	0.332262	0.235728	0.6000	97.0	85.0
20	1	0.9750	0.543696	0.253514	0.9500	94.2	84.6
20	1	0.9500	0.505544	0.267071	0.9000	95.0	84.7
22	1	0.9990	0.666966	0.209010	0.9980	91.7	84.0
22	1	0.9950	0.589164	0.225609	0.9900	93.0	84.3
22	1	0.9900	0.556122	0.234357	0.9800	93.6	84.4
24	1	0.9999	0.725455	0.188742	0.9998	90.7	83.7
24	1	0.9995	0.653934	0.200304	0.9990	91.7	84.0
24	1	0.9990	0.623951	0.206065	0.9980	92.2	84.1
28	1	0.6000	0.332221	0.305919	0.2000	98.9	84.1
30	1	0.7000	0.339216	0.287485	0.4000	98.3	83.7
30	1	0.6000	0.324688	0.299752	0.2000	98.9	83.7
32	1	0.8000	0.349033	0.269610	0.6000	97.5	83.3
32	1	0.7000	0.331825	0.282551	0.4000	98.3	83.3
32	1	0.6000	0.318034	0.294278	0.2000	98.9	83.3
34	2	0.9750	0.480060	0.267069	0.9500	95.0	82.9
34	2	0.9500	0.455098	0.278496	0.9000	95.7	83.0
36	2	0.9950	0.516376	0.243555	0.9900	93.8	82.7
36	2	0.9900	0.494725	0.251310	0.9800	94.4	82.8
36	2	0.9750	0.465254	0.263342	0.9500	95.2	82.9
38	2	0.9995	0.565394	0.220925	0.9990	92.5	82.5
38	2	0.9990	0.545891	0.226254	0.9980	92.9	82.5
38	2	0.9950	0.500064	0.240788	0.9900	94.0	82.7
38	2	0.9900	0.479830	0.248307	0.9800	94.5	82.8
40	2	0.9999	0.588853	0.208333	0.9998	91.9	82.3
40	2	0.9995	0.546752	0.218904	0.9990	92.7	82.5
40	2	0.9990	0.528552	0.224101	0.9980	93.1	82.5
40	2	0.6000	0.343192	0.320197	0.2000	99.1	83.0
50	3	0.9950	0.482332	0.254851	0.9900	94.5	82.1
50	3	0.9900	0.465566	0.261900	0.9800	95.0	82.2
50	3	0.9750	0.442442	0.272754	0.9500	95.7	82.3
50	3	0.9500	0.423865	0.282598	0.9000	96.4	82.3
60	3	0.9999	0.509443	0.218354	0.9998	93.1	81.7
60	3	0.7000	0.342096	0.304071	0.4000	98.7	82.1
60	3	0.6000	0.331665	0.313316	0.2000	99.3	82.1

B. VALUE OF r, CHOSEN TO MAXIMIZE EFFECTIVENESS F_i OF CENTRAL CONFIDENCE INTERVAL, DOES NOT MAXIMIZE EFFECTIVENESS F_u OF UPPER CONFIDENCE BOUND--SEE TABLE A9A

n	r	1-P	B_{ur}/W_r	$B_{\ell r}/W_r$	1-2P	F_u (%)	F_i (%)
70	4	0.9999	0.506461	0.231428	0.9998	93.3	81.5
70	4	0.9995	0.480420	0.240697	0.9990	94.1	81.6
70	4	0.9990	0.468906	0.245202	0.9980	94.4	81.7
70	4	0.9950	0.441097	0.257305	0.9900	95.3	81.8
70	4	0.6000	0.340942	0.323440	0.2000	99.3	82.0
80	5	0.9990	0.468262	0.255338	0.9980	94.6	81.5
80	5	0.9950	0.442363	0.267148	0.9900	95.5	81.6
80	5	0.9900	0.430543	0.273147	0.9800	95.9	81.6
80	5	0.9750	0.413995	0.282297	0.9500	96.5	81.7
80	5	0.9500	0.400485	0.290510	0.9000	97.0	81.7
80	5	0.6000	0.347797	0.331052	0.2000	99.4	81.8
90	5	0.9999	0.470802	0.236071	0.9998	94.0	81.3
90	5	0.9995	0.449899	0.244574	0.9990	94.7	81.4
90	5	0.8000	0.352494	0.301946	0.6000	98.5	81.7
100	6	0.9999	0.470790	0.244728	0.9998	94.3	81.2
100	6	0.9995	0.451025	0.253088	0.9990	94.9	81.3
100	6	0.9990	0.442193	0.257127	0.9980	95.2	81.4
100	6	0.9950	0.420619	0.267907	0.9900	95.9	81.4
100	6	0.7000	0.348013	0.317597	0.4000	99.0	81.6
100	6	0.6000	0.339774	0.325090	0.2000	99.5	81.6

C. VALUE OF r, CHOSEN TO MAXIMIZE EFFICIENCY E_u OF UPPER CONFIDENCE BOUND, ALSO MAXIMIZES EFFICIENCY E_i OF CENTRAL CONFIDENCE INTERVAL, UNLESS MARKED *

n	r	1-P	B_{ur}/w_r	$B_{\ell r}/w_r$	1-2P	E_u (%)	E_i (%)
2	0	0.9999	5641.895	0.181748	0.9998	100.0	100.0
2	0	0.9995	1128.379	0.203147	0.9990	100.0	100.0
2	0	0.9990	564.1894	0.214892	0.9980	100.0	100.0
2	0	0.9950	112.8372	0.251905	0.9900	100.0	100.0
2	0	0.9900	56.41748	0.274516	0.9800	100.0	100.0
2	0	0.9750	22.56389	0.315475	0.9500	100.0	100.0
2	0	0.9500	11.27640	0.360775	0.9000	100.0	100.0
2	0	0.9000	5.627083	0.429890	0.8000	100.0	100.0
2	0	0.8000	2.791059	0.551758	0.6000	100.0	100.0
2	0	0.7000	1.835113	0.682250	0.4000	100.0	100.0
2	0	0.6000	1.348410	0.840172	0.2000	100.0	100.0
2	0	0.5000	1.048358			100.0	
3	0	0.9999	52.50243	0.170527	0.9998	99.3	99.3
3	0	0.9995	23.47744	0.188097	0.9990	99.3	99.3
3	0	0.9990	16.59896	0.197494	0.9980	99.3	99.3
3	0	0.9950	7.415786	0.226028	0.9900	99.2	99.2
3	0	0.9900	5.237111	0.242701	0.9800	99.2	99.2
3	0	0.9750	3.299561	0.271572	0.9500	99.1	99.1
3	0	0.9500	2.318021	0.301705	0.9000	99.1	99.1
3	0	0.9000	1.617201	0.344545	0.8000	99.0	99.0
3	0	0.8000	1.110998	0.412621	0.6000	98.9	98.9
3	0	0.7000	0.878535	0.477420	0.4000	98.8	98.9
3	0	0.6000	0.733893	0.547549	0.2000	98.8	98.9
3	0	0.5000	0.629807			98.9	
4	0	0.9999	10.82318	0.164396	0.9998	97.9	97.9
4	0	0.9995	6.322927	0.180088	0.9990	97.8	97.8
4	0	0.9990	5.013886	0.188366	0.9980	97.8	97.8
4	0	0.9950	2.917986	0.213034	0.9900	97.6	97.6
4	0	0.9900	2.305867	0.227128	0.9800	97.5	97.5
4	0	0.9750	1.681682	0.251003	0.9500	97.3	97.4
4	0	0.9500	1.316598	0.275242	0.9000	97.1	97.2
4	0	0.9000	1.021069	0.308599	0.8000	96.9	97.0
4	0	0.8000	0.777803	0.359227	0.6000	96.6	96.9
4	0	0.7000	0.652961	0.405055	0.4000	96.4	96.8
4	0	0.6000	0.569305	0.452431	0.2000	96.5	96.8
4	0	0.5000	0.505479			96.8	
5	0	0.9999	4.866452	0.160264	0.9998	96.2	96.2
5	0	0.9995	3.244414	0.174771	0.9990	96.0	96.0
5	0	0.9990	2.721889	0.182357	0.9980	95.9	95.9
5	0	0.9950	1.802113	0.204684	0.9900	95.6	95.6
5	0	0.9900	1.503725	0.217258	0.9800	95.4	95.5
5	0	0.9750	1.176925	0.238264	0.9500	95.1	95.2
5	0	0.9500	0.970930	0.259225	0.9000	94.8	94.9
5	0	0.9000	0.792771	0.287498	0.8000	94.4	94.7
5	0	0.8000	0.635550	0.329238	0.6000	93.9	94.5
5	0	0.7000	0.549920	0.365913	0.4000	93.7	94.4
5	0	0.6000	0.490173	0.402832	0.2000	93.9	94.5
5	0	0.5000	0.443089			94.5	

C. VALUE OF r, CHOSEN TO MAXIMIZE EFFICIENCY E_u OF UPPER CONFIDENCE BOUND, ALSO MAXIMIZES EFFICIENCY E_i OF CENTRAL CONFIDENCE INTERVAL, UNLESS MARKED *

n	r	1-P	B_{ur}/w_r	$B_{\ell r}/w_r$	1-2P	E_u(%)	E_i(%)
6	0	0.9999	2.992513	0.157190	0.9998	94.4	94.4
6	0	0.9995	2.156566	0.170857	0.9990	94.1	94.1
6	0	0.9990	1.870081	0.177957	0.9980	94.0	94.0
6	0	0.9950	1.335144	0.198670	0.9900	93.5	93.5
6	0	0.9900	1.150066	0.210214	0.9800	93.2	93.3
6	0	0.9750	0.938129	0.229310	0.9500	92.8	92.9
6	0	0.9500	0.798157	0.248133	0.9000	92.3	92.6
6	0	0.9000	0.671955	0.273170	0.8000	91.8	92.3
6	0	0.8000	0.555557	0.309431	0.6000	91.2	92.1
6	0	0.7000	0.489709	0.340651	0.4000	91.0	92.1
6	0	0.6000	0.442548	0.371518	0.2000	91.3	92.1
6	0	0.5000	0.404588			92.2	
7	0	0.9999	2.152779	0.154765	0.9998	92.5	92.5
7	0	0.9995	1.632416	0.167793	0.9990	92.2	92.1
7	0	0.9990	1.446451	0.174528	0.9980	92.0	92.0
7	0	0.9950	1.084804	0.194039	0.9900	91.3	91.4
7	0	0.9900	0.954067	0.204827	0.9800	91.0	91.1
7	0	0.9750	0.799680	0.222538	0.9500	90.4	90.7
7	0	0.9500	0.694376	0.239834	0.9000	89.9	90.3
7	0	0.9000	0.596641	0.262598	0.8000	89.2	90.0
7	0	0.8000	0.503665	0.295100	0.6000	88.5	89.7
7	0	0.7000	0.449642	0.322664	0.4000	88.4	89.7
7	0	0.6000	0.410222	0.349561	0.2000	88.8	89.8
7	0	0.5000	0.378007			89.9	
8	0	0.9999	1.694382	0.152776	0.9998	90.7	90.7
8	0	0.9995	1.331535	0.165296	0.9990	90.2	90.2
8	0	0.9990	1.197855	0.171741	0.9980	90.0	90.0
8	0	0.9950	0.929989	0.190311	0.9900	89.2	89.3
8	0	0.9900	0.830000	0.200514	0.9800	88.8	89.0
8	0	0.9750	0.709210	0.217162	0.9500	88.2	88.5
8	0	0.9500	0.624839	0.233301	0.9000	87.6	88.1
8	0	0.9000	0.544826	0.254366	0.8000	86.8	87.7
8	0	0.8000	0.466928	0.284104	0.6000	86.0	87.5
8	0	0.7000	0.420747	0.309030	0.4000	85.9	87.5
8	0	0.6000	0.386570	0.333105	0.2000	86.4	87.6
8	0	0.5000	0.358315			87.6	
9	0	0.9999	1.411017	0.151098	0.9998	88.9	88.8
9	0	0.9995	1.138489	0.163200	0.9990	88.3	88.3
9	0	0.9990	1.035725	0.169408	0.9980	88.0	88.0
9	0	0.9950	0.825004	0.187215	0.9900	87.2	87.3
9	0	0.9900	0.744388	0.196947	0.9800	86.7	86.9
9	0	0.9750	0.645278	0.212747	0.9500	86.0	86.4
9	0	0.9500	0.574762	0.227972	0.9000	85.3	86.0
9	0	0.9000	0.506758	0.247707	0.8000	84.5	85.6
9	0	0.8000	0.439344	0.275316	0.6000	83.7	85.4
9	0	0.7000	0.398740	0.298236	0.4000	83.6	85.4
9	0	0.6000	0.368355	0.320192	0.2000	84.2	85.5
9	0	0.5000	0.343002			85.5	

C. VALUE OF r , CHOSEN TO MAXIMIZE EFFICIENCY E_u OF UPPER CONFIDENCE BOUND, ALSO MAXIMIZES EFFICIENCY E_i OF CENTRAL CONFIDENCE INTERVAL, UNLESS MARKED *

n	r	1-P	B_{ur}/w_r	$B_{\ell r}/w_r$	1-2P	E_u(%)	E_i(%)
10	0	0.9999	1.220356	0.149652	0.9998	87.1	87.1
10	0	0.9995	1.004804	0.161401	0.9990	86.5	86.5
10	0	0.9990	0.922013	0.167411	0.9980	86.2	86.2
10	0	0.9950	0.749104	0.184583	0.9900	85.3	85.4
10	0	0.9900	0.681648	0.193925	0.9800	84.8	85.0
10	0	0.9750	0.597544	0.209029	0.9500	84.0	84.4
10	0	0.9500	0.536814	0.223507	0.9000	83.2	84.0
10	0	0.9000	0.477453	0.242169	0.8000	82.4	83.6
10	0	0.8000	0.417738	0.268076	0.6000	81.5	83.4
10	0	0.7000	0.381307	0.289414	0.4000	81.4	83.4
10	0	0.6000	0.353796	0.309715	0.2000	82.1	83.5
10	0	0.5000	0.330666			83.5	
11	0	0.9999	1.083995	0.148385	0.9998	85.4	85.4
11	0	0.9995	0.906960	0.159832	0.9990	84.8	84.7
11	0	0.9990	0.837939	0.165673	0.9980	84.4	84.4
11	0	0.9950	0.691601	0.182303	0.9900	83.4	83.5
11	0	0.9900	0.633592	0.191316	0.9800	82.9	83.1
11	0	0.9750	0.560427	0.205834	0.9500	82.0	82.6
11	0	0.9500	0.506950	0.219690	0.9000	81.3	82.1
11	0	0.9000	0.454092	0.237462	0.8000	80.3	81.7
11	0	0.8000	0.400270	0.261973	0.6000	79.4	81.5
11	0	0.7000	0.367081	0.282025	0.4000	79.3	81.5
11	0	0.6000	0.341826	0.300992	0.2000	80.1	81.6
11	0	0.5000	0.320458			81.7	
12	0	0.9999	0.981891	0.147261	0.9998	83.8	83.8
12	0	0.9995	0.832298	0.158444	0.9990	83.1	83.1
12	0	0.9990	0.773245	0.164137	0.9980	82.7	82.7
12	0	0.9950	0.646455	0.180300	0.9900	81.7	81.8
12	0	0.9900	0.595520	0.189029	0.9800	81.1	81.4
12	0	0.9750	0.530652	0.203046	0.9500	80.2	80.8
12	0	0.9500	0.482752	0.216373	0.9000	79.4	80.3
12	0	0.9000	0.434961	0.233392	0.8000	78.4	79.9
12	0	0.8000	0.385793	0.256733	0.6000	77.5	79.7
12	0	0.7000	0.355198	0.275716	0.4000	77.4	79.7
12	0	0.6000	0.331766	0.293583	0.2000	78.2	79.8
12	0	0.5000	0.311832			79.9	
13	0	0.9999	0.902675	0.146253	0.9998	82.3	82.2
13	0	0.9995	0.773444	0.157202	0.9990	81.5	81.5
13	0	0.9990	0.721890	0.162766	0.9980	81.1	81.1
13	0	0.9950	0.610006	0.178518	0.9900	80.0	80.2
13	0	0.9900	0.564547	0.187000	0.9800	79.4	79.7
13	0	0.9750	0.506172	0.200582	0.9500	78.5	79.1
13	0	0.9500	0.462689	0.213451	0.9000	77.6	78.7
13	0	0.9000	0.418953	0.229824	0.8000	76.6	78.3
13	0	0.8000	0.373557	0.252167	0.6000	75.7	78.0
13	0	0.7000	0.345087	0.270245	0.4000	75.6	78.0
13	0	0.6000	0.323159	0.287186	0.2000	76.4	78.1
13	0	0.5000	0.304418			78.2	

C. VALUE OF r, CHOSEN TO MAXIMIZE EFFICIENCY E_u OF UPPER CONFIDENCE BOUND, ALSO MAXIMIZES EFFICIENCY E_i OF CENTRAL CONFIDENCE INTERVAL, UNLESS MARKED *

n	r	1-P	B_{ur}/W_r	$B_{\ell r}/W_r$	1-2P	E_u(%)	E_i(%)
14	0	0.9999	0.839449	0.145340	0.9998	80.8	80.7
14	0	0.9995	0.725832	0.156081	0.9990	80.0	79.9
14	0	0.9990	0.680094	0.161529	0.9980	79.6	79.6
14	0	0.9950	0.579911	0.176918	0.9900	78.4	78.6
14	0	0.9900	0.538805	0.185182	0.9800	77.8	78.2
14	0	0.9750	0.485642	0.198381	0.9500	76.8	77.6
14	0	0.9500	0.445740	0.210849	0.9000	76.0	77.1
14	0	0.9000	0.405323	0.226659	0.8000	75.0	76.7
14	0	0.8000	0.363047	0.248139	0.6000	74.0	76.4
14	0	0.7000	0.336352	0.265439	0.4000	73.9	76.5
14	0	0.6000	0.315689	0.281589	0.2000	74.8	76.5
14	0	0.5000	0.297955			76.6	
15	0	0.9999	0.787813	0.144507	0.9998	79.4	79.3
15	0	0.9995	0.686491	0.155061	0.9990	78.6	78.5
15	0	0.9990	0.645380	0.160405	0.9980	78.1	78.1
15	0	0.9950	0.554599	0.175469	0.9900	76.9	77.1
15	0	0.9900	0.517032	0.183538	0.9800	76.3	76.7
15	0	0.9750	0.468140	0.196396	0.9500	75.3	76.1
15	0	0.9500	0.431199	0.208510	0.9000	74.4	75.6
15	0	0.9000	0.393548	0.223825	0.8000	73.4	75.2
15	0	0.8000	0.353897	0.244548	0.6000	72.4	74.9
15	0	0.7000	0.328708	0.261172	0.4000	72.3	75.0
15	0	0.6000	0.309125	0.276636	0.2000	73.2	75.0
15	0	0.5000	0.292257			75.1	
16	0	0.9999	0.744835	0.143742	0.9998	78.1	77.9
16	0	0.9995	0.653410	0.154126	0.9990	77.2	77.1
16	0	0.9990	0.616056	0.159376	0.9980	76.7	76.7
16	0	0.9950	0.532982	0.174147	0.9900	75.5	75.7
16	0	0.9900	0.498344	0.182040	0.9800	74.9	75.3
16	0	0.9750	0.453013	0.194594	0.9500	73.8	74.6
16	0	0.9500	0.418560	0.206392	0.9000	72.9	74.2
16	0	0.9000	0.383252	0.221265	0.8000	71.9	73.8
16	0	0.8000	0.345842	0.241320	0.6000	70.9	73.5
16	0	0.7000	0.321947	0.257348	0.4000	70.8	73.5
16	0	0.6000	0.303298	0.272212	0.2000	71.8	73.6
16	0	0.5000	0.287181			73.6	
17	0	0.9999	0.708486	0.143036	0.9998	76.8	76.6
17	0	0.9995	0.625178	0.153265	0.9990	75.9	75.8
17	0	0.9990	0.590931	0.158429	0.9980	75.4	75.4
17	0	0.9950	0.514278	0.172933	0.9900	74.2	74.4
17	0	0.9900	0.482104	0.180668	0.9800	73.5	73.9
17	0	0.9750	0.439786	0.192947	0.9500	72.5	73.3
17	0	0.9500	0.407453	0.204459	0.9000	71.5	72.8
17	0	0.9000	0.374155	0.218937	0.8000	70.5	72.4
17	0	0.8000	0.338681	0.238395	0.6000	69.4	72.2
17	0	0.7000	0.315913	0.253894	0.4000	69.4	72.2
17	0	0.6000	0.298080	0.268227	0.2000	70.4	72.2
17	0	0.5000	0.282623			72.3	

C. VALUE OF r, CHOSEN TO MAXIMIZE EFFICIENCY E_u OF UPPER CONFIDENCE BOUND, ALSO MAXIMIZES EFFICIENCY E_i OF CENTRAL CONFIDENCE INTERVAL, UNLESS MARKED *

n	r	1-P	B_{ur}/w_r	$B_{\ell r}/w_r$	1-2P	E_u(%)	E_i(%)
18	0	0.9999	0.677326	0.142381	0.9998	75.6	75.4
18	0	0.9995	0.600780	0.152467	0.9990	74.6	74.5
18	0	0.9990	0.569140	0.157553	0.9980	74.2	74.1
18	0	0.9950	0.497916	0.171812	0.9900	72.9	73.1
18	0	0.9900	0.467839	0.179403	0.9800	72.2	72.6
18	0	0.9750	0.428103	0.191431	0.9500	71.1	72.0
18	0	0.9500	0.397599	0.202686	0.9000	70.2	71.6
18	0	0.9000	0.366045	0.216807	0.8000	69.1	71.1
18	1	0.8000	0.447574	0.317353	0.6000	68.6	71.2
18	1	0.7000	0.417535	0.337127	0.4000	69.3	71.7
18	1	0.6000	0.394199	0.355523	0.2000	70.5	72.0
18	1	0.5000	0.374113			72.1	
19	0	0.9999	0.650298	0.141770	0.9998	74.4	74.2
19	0	0.9995	0.579466	0.151725	0.9990	73.4	73.3
19	0	0.9990	0.550041	0.156738	0.9980	73.0	72.9
19	0	0.9950	0.483462	0.170773	0.9900	71.6	71.9
19	0	0.9900	0.455194	0.178231	0.9800	71.0	71.4
19	0	0.9750	0.417696	0.190031	0.9500	69.9	70.8
19	0	0.9500	0.388785	0.201050	0.9000	68.9	70.3
19	1	0.9000	0.479705	0.288167	0.8000	68.4	70.8
19	1	0.8000	0.435357	0.311842	0.6000	68.7	71.4
19	1	0.7000	0.407066	0.330750	0.4000	69.4	71.8
19	1	0.6000	0.385008	0.348289	0.2000	70.5	72.0
19	1	0.5000	0.365964			72.1	
20	0	0.9999	0.626618	0.141199	0.9998	73.3	73.1
20	0	0.9995	0.560669	0.151031	0.9990	72.3	72.2
20	0	0.9990	0.533149	0.155977	0.9980	71.8	71.8
20	0	0.9950	0.470588	0.169805	0.9900	70.5	70.7
20	0	0.9900	0.443894	0.177141	0.9800	69.8	70.3
20	0	0.9750	0.408353	0.188730	0.9500	68.7	69.6*
20	1	0.9500	0.505544	0.267071	0.9000	68.3	70.4
20	1	0.9000	0.466276	0.284064	0.8000	68.5	70.9
20	1	0.8000	0.424495	0.306854	0.6000	68.8	71.4
20	1	0.7000	0.397722	0.324995	0.4000	69.4	71.7
20	1	0.6000	0.376780	0.341778	0.2000	70.5	72.0
20	1	0.5000	0.358649			72.1	
22	0	0.9999	0.587038	0.140157	0.9998	71.1	70.9
22	0	0.9995	0.528984	0.149768	0.9990	70.1	70.0
22	0	0.9990	0.504567	0.154594	0.9980	69.6	69.6
22	0	0.9950	0.448602	0.168051	0.9900	68.3	68.6*
22	1	0.9900	0.556122	0.234357	0.9800	68.4	69.9
22	1	0.9750	0.512309	0.248119	0.9500	68.5	70.2
22	1	0.9500	0.478558	0.260909	0.9000	68.5	70.5
22	1	0.9000	0.443551	0.276868	0.8000	68.5	70.8
22	1	0.8000	0.405978	0.298148	0.6000	68.7	71.2
22	1	0.7000	0.381714	0.314992	0.4000	69.1	71.5
22	1	0.6000	0.362626	0.330500	0.2000	70.2	71.7
22	1	0.5000	0.346020			71.8	

C. VALUE OF r, CHOSEN TO MAXIMIZE EFFICIENCY E_u OF UPPER CONFIDENCE BOUND, ALSO MAXIMIZES EFFICIENCY E_i OF CENTRAL CONFIDENCE INTERVAL, UNLESS MARKED *

n	r	1-P	B_{ur}/W_r	$B_{\ell r}/W_r$	1-2P	E_u(%)	E_i(%)
24	0	0.9999	0.555208	0.139227	0.9998	69.2	68.9
24	0	0.9995	0.503248	0.148645	0.9990	68.2	68.0*
24	1	0.9990	0.623951	0.206065	0.9980	68.2	69.2
24	1	0.9950	0.555501	0.221997	0.9900	68.5	69.7
24	1	0.9900	0.526178	0.230363	0.9800	68.5	69.9
24	1	0.9750	0.487042	0.243485	0.9500	68.5	70.2
24	1	0.9500	0.456685	0.255635	0.9000	68.4	70.4
24	1	0.9000	0.424990	0.270737	0.8000	68.3	70.6
24	1	0.8000	0.390725	0.290776	0.6000	68.3	70.9
24	1	0.7000	0.368451	0.306559	0.4000	68.7	71.1
24	1	0.6000	0.350844	0.321033	0.2000	69.7	71.2
24	1	0.5000	0.335463			71.3	
26	1	0.9999	0.678999	0.186699	0.9998	68.1	68.8
26	1	0.9995	0.615907	0.197901	0.9990	68.3	69.2
26	1	0.9990	0.589281	0.203469	0.9980	68.4	69.3
26	1	0.9950	0.528068	0.218828	0.9900	68.5	69.6
26	1	0.9900	0.501651	0.226868	0.9800	68.4	69.7
26	1	0.9750	0.466199	0.239443	0.9500	68.3	69.9
26	1	0.9500	0.438535	0.251050	0.9000	68.2	70.0
26	1	0.9000	0.409492	0.265429	0.8000	67.9	70.2
26	1	0.8000	0.377895	0.284426	0.6000	67.8	70.4
26	1	0.7000	0.357239	0.299325	0.4000	68.2	70.5
26	1	0.6000	0.340843	0.312940	0.2000	69.2	70.7
26	1	0.5000	0.326470			70.7	
28	1	0.9999	0.641154	0.184872	0.9998	68.3	68.9
28	1	0.9995	0.584662	0.195757	0.9990	68.4	69.1
28	1	0.9990	0.560684	0.201157	0.9980	68.4	69.2
28	1	0.9950	0.505227	0.216015	0.9900	68.3	69.3
28	1	0.9900	0.481142	0.223772	0.9800	68.2	69.4
28	1	0.9750	0.448663	0.235875	0.9500	68.0	69.5
28	1	0.9500	0.423189	0.247014	0.9000	67.7	69.5
28	1	0.9000	0.396315	0.260772	0.8000	67.4	69.6
28	1	0.8000	0.366918	0.278881	0.6000	67.2	69.8
28	1	0.7000	0.347606	0.293030	0.4000	67.6	69.9
28	1	0.6000	0.332221	0.305919	0.2000	68.5	70.0
28	1	0.5000	0.318691			70.1	
30	1	0.9999	0.609687	0.183222	0.9998	68.2	68.7
30	1	0.9995	0.558490	0.193827	0.9990	68.2	68.8
30	1	0.9990	0.536651	0.199078	0.9980	68.2	68.9
30	1	0.9950	0.485874	0.213495	0.9900	68.0	68.9
30	1	0.9900	0.463699	0.221003	0.9800	67.8	68.9
30	1	0.9750	0.433669	0.232692	0.9500	67.5	69.0
30	1	0.9500	0.410011	0.243423	0.9000	67.2	69.0
30	1	0.9000	0.384945	0.256642	0.8000	66.9	69.0
30	1	0.8000	0.357394	0.273981	0.6000	66.6	69.1
30	1	0.7000	0.339216	0.287485	0.4000	66.9	69.2
30	1	0.6000	0.324688	0.299752	0.2000	67.8	69.3
30	1	0.5000	0.311877			69.4	

C. VALUE OF r, CHOSEN TO MAXIMIZE EFFICIENCY E_u OF UPPER CONFIDENCE BOUND, ALSO MAXIMIZES EFFICIENCY E_i OF CENTRAL CONFIDENCE INTERVAL, UNLESS MARKED *

n	r	1-P	B_{ur}/w_r	$B_{\ell r}/w_r$	1-2P	E_u(%)	E_i(%)
32	1	0.9999	0.583077	0.181722	0.9998	68.1	68.5
32	1	0.9995	0.536213	0.192074	0.9990	68.0	68.5
32	1	0.9990	0.516136	0.197193	0.9980	67.9	68.5
32	1	0.9950	0.469235	0.211217	0.9900	67.6	68.4
32	1	0.9900	0.448652	0.218504	0.9800	67.4	68.4
32	1	0.9750	0.420675	0.229827	0.9500	67.0	68.4
32	1	0.9500	0.398545	0.240199	0.9000	66.7	68.4
32	1	0.9000	0.375011	0.252943	0.8000	66.2	68.4
32	2	0.8000	0.410628	0.317077	0.6000	66.1	68.5
32	2	0.7000	0.390145	0.332142	0.4000	66.6	68.8
32	2	0.6000	0.373811	0.345856	0.2000	67.6	68.9
32	2	0.5000	0.359436			69.0	
34	1	0.9999	0.560249	0.180348	0.9998	67.8	68.1
34	1	0.9995	0.516994	0.190474	0.9990	67.6	68.0
34	1	0.9990	0.498390	0.195473	0.9980	67.5	68.0
34	1	0.9950	0.454751	0.209144	0.9900	67.1	67.9
34	1	0.9900	0.435516	0.216233	0.9800	66.9	67.9
34	1	0.9750	0.409284	0.227229	0.9500	66.5	67.8
34	1	0.9500	0.388460	0.237281	0.9000	66.1	67.7*
34	2	0.9000	0.428612	0.292556	0.8000	66.0	68.3
34	2	0.8000	0.399459	0.310984	0.6000	66.2	68.6
34	2	0.7000	0.380202	0.325325	0.4000	66.7	68.8
34	2	0.6000	0.364801	0.338348	0.2000	67.6	69.0
34	2	0.5000	0.351213			69.0	
36	1	0.9999	0.540427	0.179082	0.9998	67.4	67.7
36	1	0.9995	0.500220	0.189002	0.9990	67.2	67.5
36	1	0.9990	0.482867	0.193893	0.9980	67.0	67.5
36	1	0.9950	0.442010	0.207246	0.9900	66.6	67.3*
36	1	0.9900	0.423929	0.214156	0.9800	66.3	67.2*
36	2	0.9750	0.465254	0.263342	0.9500	66.0	67.8
36	2	0.9500	0.441912	0.274350	0.9000	66.0	68.1
36	2	0.9000	0.417060	0.287864	0.8000	66.1	68.3
36	2	0.8000	0.389600	0.305527	0.6000	66.2	68.6
36	2	0.7000	0.371398	0.319235	0.4000	66.6	68.8
36	2	0.6000	0.356804	0.331654	0.2000	67.6	68.9
36	2	0.5000	0.343899			68.9	
38	1	0.9999	0.523035	0.177911	0.9998	67.0	67.2
38	1	0.9995	0.485435	0.187643	0.9990	66.7	67.0
38	1	0.9990	0.469155	0.192435	0.9980	66.5	66.9*
38	2	0.9950	0.500064	0.240788	0.9900	66.0	67.5
38	2	0.9900	0.479830	0.248307	0.9800	66.1	67.6
38	2	0.9750	0.452205	0.259954	0.9500	66.1	67.9
38	2	0.9500	0.430254	0.270588	0.9000	66.1	68.0
38	2	0.9000	0.406812	0.283617	0.8000	66.0	68.2
38	2	0.8000	0.380819	0.300602	0.6000	66.1	68.5
38	2	0.7000	0.363536	0.313751	0.4000	66.5	68.6
38	2	0.6000	0.349646	0.325640	0.2000	67.4	68.7
38	2	0.5000	0.337339			68.8	

C. VALUE OF r, CHOSEN TO MAXIMIZE EFFICIENCY E_u OF UPPER CONFIDENCE BOUND, ALSO MAXIMIZES EFFICIENCY E_i OF CENTRAL CONFIDENCE INTERVAL, UNLESS MARKED *

n	r	1-P	B_{ur}/W_r	$B_{\ell r}/W_r$	1-2P	E_u (%)	E_i (%)
40	1	0.9999	0.507635	0.176822	0.9998	66.6	66.6*
40	1	0.9995	0.472289	0.186381	0.9990	66.2	66.4*
40	2	0.9990	0.528552	0.224101	0.9980	66.0	67.2
40	2	0.9950	0.485643	0.238249	0.9900	66.2	67.5
40	2	0.9900	0.466629	0.245555	0.9800	66.2	67.6
40	2	0.9750	0.440600	0.256855	0.9500	66.1	67.8
40	2	0.9500	0.419858	0.267153	0.9000	66.1	67.9
40	2	0.9000	0.397644	0.279748	0.8000	65.9	68.1
40	2	0.8000	0.372937	0.296128	0.6000	66.0	68.3
40	2	0.7000	0.356461	0.308779	0.4000	66.3	68.4
40	2	0.6000	0.343192	0.320197	0.2000	67.2	68.5
40	2	0.5000	0.331413			68.6	
50	2	0.9999	0.510872	0.201169	0.9998	66.0	66.5
50	2	0.9995	0.479365	0.210739	0.9990	66.0	66.6
50	2	0.9990	0.465579	0.215418	0.9980	65.9	66.6
50	2	0.9950	0.432652	0.228073	0.9900	65.7	66.7
50	2	0.9900	0.417859	0.234561	0.9800	65.6	66.7*
50	2	0.9750	0.397396	0.244531	0.9500	65.3	66.7*
50	3	0.9500	0.423865	0.282598	0.9000	65.2	67.1
50	3	0.9000	0.403819	0.294576	0.8000	65.2	67.3
50	3	0.8000	0.381331	0.310060	0.6000	65.3	67.5
50	3	0.7000	0.366221	0.321947	0.4000	65.8	67.7
50	3	0.6000	0.353982	0.332624	0.2000	66.6	67.8
50	3	0.5000	0.343066			67.8	
60	3	0.9999	0.509443	0.218354	0.9998	65.3	66.1
60	3	0.9995	0.480975	0.227795	0.9990	65.4	66.3
60	3	0.9990	0.468446	0.232395	0.9980	65.4	66.4
60	3	0.9950	0.438333	0.244790	0.9900	65.4	66.6
60	3	0.9900	0.424718	0.251119	0.9800	65.3	66.6
60	3	0.9750	0.405791	0.260816	0.9500	65.2	66.7
60	3	0.9500	0.390458	0.269562	0.9000	65.1	66.8
60	3	0.9000	0.373778	0.280142	0.8000	64.9	66.9
60	3	0.8000	0.354893	0.293719	0.6000	64.9	67.0
60	4	0.7000	0.372433	0.330996	0.4000	65.3	67.1*
60	4	0.6000	0.361037	0.341044	0.2000	66.1	67.2
60	4	0.5000	0.350835			67.2	
70	3	0.9999	0.467559	0.212685	0.9998	65.2	65.7*
70	3	0.9995	0.443821	0.221456	0.9990	65.2	65.8*
70	3	0.9990	0.433306	0.225715	0.9980	65.1	65.8*
70	4	0.9950	0.441097	0.257305	0.9900	65.0	66.3
70	4	0.9900	0.428459	0.263467	0.9800	65.0	66.4
70	4	0.9750	0.410822	0.272886	0.9500	65.0	66.5
70	4	0.9500	0.396473	0.281358	0.9000	64.9	66.6
70	4	0.9000	0.380802	0.291580	0.8000	64.8	66.7
70	4	0.8000	0.362980	0.304654	0.6000	64.9	66.9
70	4	0.7000	0.350855	0.314589	0.4000	65.2	67.0
70	4	0.6000	0.340942	0.323440	0.2000	66.0	67.0
70	4	0.5000	0.332029			67.1	

C. VALUE OF r, CHOSEN TO MAXIMIZE EFFICIENCY E_u OF UPPER CONFIDENCE BOUND, ALSO MAXIMIZES EFFICIENCY E_i OF CENTRAL CONFIDENCE INTERVAL, UNLESS MARKED *

n	r	1-P	B_{ur}/W_r	$B_{\ell r}/r$	1-2P	E_u (%)	E_i (%)
80	4	0.9999	0.469950	0.225640	0.9998	65.0	65.7
80	4	0.9995	0.447735	0.234283	0.9990	65.0	65.9
80	4	0.9990	0.437860	0.238472	0.9980	65.0	65.9
80	4	0.9950	0.413875	0.249687	0.9900	65.0	66.0*
80	4	0.9900	0.402909	0.255376	0.9800	64.9	66.1*
80	4	0.9750	0.387533	0.264044	0.9500	64.8	66.1*
80	5	0.9500	0.400485	0.290510	0.9000	64.7	66.4
80	5	0.9000	0.385683	0.300395	0.8000	64.7	66.5
80	5	0.8000	0.368786	0.313003	0.6000	64.8	66.7
80	5	0.7000	0.357251	0.322559	0.4000	65.1	66.8
80	5	0.6000	0.347797	0.331052	0.2000	65.8	66.8
80	5	0.5000	0.339278			66.9	
90	4	0.9999	0.442415	0.220885	0.9998	64.8	65.3*
90	5	0.9995	0.449899	0.244574	0.9990	64.8	65.8
90	5	0.9990	0.440581	0.248688	0.9980	64.8	65.9
90	5	0.9950	0.417880	0.259685	0.9900	64.8	66.0
90	5	0.9900	0.407467	0.265252	0.9800	64.8	66.1
90	5	0.9750	0.392831	0.273722	0.9500	64.7	66.2
90	5	0.9500	0.380833	0.281300	0.9000	64.7	66.2
90	5	0.9000	0.367633	0.290393	0.8000	64.6	66.3
90	6	0.8000	0.373131	0.319621	0.6000	64.6	66.4
90	6	0.7000	0.362116	0.328832	0.4000	65.0	66.5
90	6	0.6000	0.353070	0.337003	0.2000	65.7	66.6
90	6	0.5000	0.344903			66.6	
100	5	0.9999	0.445773	0.231240	0.9998	64.8	65.5*
100	5	0.9995	0.427203	0.239275	0.9990	64.8	65.5*
100	5	0.9990	0.418897	0.243154	0.9980	64.8	65.6*
100	6	0.9950	0.420619	0.267907	0.9900	64.7	65.9
100	6	0.9900	0.410696	0.273356	0.9800	64.7	66.0
100	6	0.9750	0.396719	0.281635	0.9500	64.7	66.1
100	6	0.9500	0.385236	0.289030	0.9000	64.6	66.2
100	6	0.9000	0.372574	0.297891	0.8000	64.6	66.3
100	6	0.8000	0.358016	0.309127	0.6000	64.6	66.4
100	6	0.7000	0.348013	0.317597	0.4000	64.9	66.4
100	6	0.6000	0.339774	0.325090	0.2000	65.6	66.5
100	6	0.5000	0.332319			66.5	

D. VALUE OF r, CHOSEN TO MAXIMIZE EFFICIENCY E_i OF CENTRAL CONFIDENCE INTERVAL, DOES NOT MAXIMIZE EFFICIENCY E_u OF UPPER CONFIDENCE BOUND--SEE TABLE A9C

n	r	1-P	B_{ur}/w_r	$B_{\ell r}/w_r$	1-2P	E_u (%)	E_i (%)
20	1	0.9750	0.543696	0.253514	0.9500	68.2	70.0
22	1	0.9950	0.589164	0.225609	0.9900	68.2	69.6
24	1	0.9995	0.653934	0.200304	0.9990	68.1	69.0
34	2	0.9500	0.455098	0.278496	0.9000	65.9	68.0
36	2	0.9950	0.516376	0.243555	0.9900	65.8	67.3
36	2	0.9900	0.494725	0.251310	0.9800	65.9	67.6
38	2	0.9990	0.545891	0.226254	0.9980	65.8	67.1
40	2	0.9999	0.588853	0.208333	0.9998	65.7	66.7
40	2	0.9995	0.546752	0.218904	0.9990	66.0	67.0
50	3	0.9900	0.465566	0.261900	0.9800	65.1	66.7
50	3	0.9750	0.442442	0.272754	0.9500	65.2	67.0
60	3	0.7000	0.342096	0.304071	0.4000	65.2	67.1
70	4	0.9999	0.506461	0.231428	0.9998	64.7	65.8
70	4	0.9995	0.480420	0.240697	0.9990	64.9	66.0
70	4	0.9990	0.468906	0.245202	0.9980	64.9	66.1
80	5	0.9950	0.442363	0.267148	0.9900	64.6	66.0
80	5	0.9900	0.430543	0.273147	0.9800	64.7	66.1
80	5	0.9750	0.413995	0.282297	0.9500	64.7	66.3
90	5	0.9999	0.470802	0.236071	0.9998	64.8	65.6
100	6	0.9999	0.470790	0.244728	0.9998	64.5	65.5
100	6	0.9995	0.451025	0.253088	0.9990	64.6	65.7
100	6	0.9990	0.442193	0.257127	0.9980	64.6	65.7

Appendix B

TABLES BASED ON THE RANGE OF SAMPLES FROM A RECTANGULAR POPULATION

SOURCES OF TABLES

Tables B1 WADC TR 58-200 (Harter)
Tables B2-B3 ARL 31, Part I (Harter)

Table B1

EFFICIENT POINT ESTIMATORS FOR σ, BASED ON RANGE

ESTIMATES OF STANDARD DEVIATION OF RECTANGULAR POPULATION

Sample size, n	Estimate based on the range	Bias (%) of estimates which assume normality		
		One quasi-range	Two adjacent quasi-ranges	Any Two quasi-ranges*
2	.866025 w_o	2.33		
3	.577350 w_o	2.33		
4	.481125 w_o	0.96	1.98	1.98
5	.433013 w_o	-0.71	1.61	1.61
6	.404145 w_o	-2.37	1.13	1.13
7	.384900 w_o	-3.93	0.54	0.54
8	.371154 w_o	-5.37	-0.11	-0.11
9	.360844 w_o	-6.69	-0.80	-0.80
10	.352825 w_o	-7.90	-1.51	-1.51
11	.346410 w_o	-9.02	-2.22	-2.22
12	.341162 w_o	-10.04	-2.92	-1.47
13	.336788 w_o	-10.99	-3.60	-1.72
14	.333087 w_o	-11.87	-4.27	-2.01
15	.329914 w_o	-12.69	-4.92	-2.33
16	.327165 w_o	-13.46	-5.54	-2.66
17	.324760 w_o	-14.18	-6.14	-3.01
18	.322637 w_o	1.26	-6.72	-3.37
19	.320750 w_o	0.41	-7.28	-3.74
20	.319062 w_o	-0.39	-7.82	-4.11
21	.317543 w_o	-1.15	-8.33	-4.48
22	.316168 w_o	-1.87	-8.83	-3.11
23	.314918 w_o	-2.56	-9.31	-3.32
24	.313777 w_o	-3.22	-9.78	-3.53
25	.312731 w_o	-3.85	-10.23	-3.75
26	.311769 w_o	-4.45	-10.66	-3.98
27	.310881 w_o	-5.03	-11.08	-4.21
28	.310058 w_o	-5.58	-11.48	-4.45
29	.309295 w_o	-6.11	-11.87	-4.69
30	.308584 w_o	-6.63	-12.25	-4.94
31	.307920 w_o	-7.12	-12.61	-5.18
32	.307299 w_o	1.23	-12.97	-5.43
33	.306717 w_o	.0.71	-13.31	-3.95
34	.306171 w_o	0.20	-4.02	-4.11
35	.305656 w_o	-0.29	-4.43	-4.28

* Quasi-ranges 0 through 8 only

ESTIMATES OF STANDARD DEVIATION OF RECTANGULAR POPULATION

Sample size, n	Estimate based on the range	Bias (%) of estimates which assume normality		
		One quasi-range	Two adjacent quasi-ranges	Any Two quasi-ranges*
36	.305171 w_o	-0.77	-4.83	-4.45
37	.304713 w_o	-1.23	-5.22	-4.63
38	.304279 w_o	-1.68	-5.60	-4.81
39	.303869 w_o	-2.11	-5.96	-4.99
40	.303479 w_o	-2.53	-6.32	-5.17
41	.303109 w_o	-2.94	-6.67	-5.36
42	.302757 w_o	-3.34	-7.00	-5.54
43	.302422 w_o	-3.73	-7.33	-5.72
44	.302102 w_o	-4.10	-7.65	-5.91
45	.301797 w_o	-4.47	-7.97	-4.45
46	.301505 w_o	1.24	-8.27	-4.59
47	.301226 w_o	0.86	-8.57	-4.73
48	.300959 w_o	0.49	-2.46	-4.88
49	.300703 w_o	0.12	-2.79	-5.02
50	.300458 w_o	-0.23	-3.10	-5.17
51	.300222 w_o	-0.58	-3.41	-5.32
52	.299996 w_o	-0.92	-3.71	-5.46
53	.299778 w_o	-1.25	-4.01	-5.61
54	.299569 w_o	-1.57	-4.30	-5.76
55	.299367 w_o	-1.89	-4.58	-5.91
56	.299172 w_o	-2.20	-4.86	-0.63
57	.298985 w_o	-2.50	-5.13	-0.76
58	.298804 w_o	-2.79	-5.40	-0.89
59	.298629 w_o	-3.08	-5.66	-1.02
60	.298461 w_o	-3.37	-5.91	-1.15
61	.298298 w_o	0.95	-6.16	-1.28
62	.298140 w_o	0.66	-1.61	-1.41
63	.297987 w_o	0.37	-1.87	-1.55
64	.297839 w_o	0.09	-2.13	-1.67
65	.297696 w_o	-0.19	-2.39	-1.81
66	.297557 w_o	-0.46	-2.64	-1.94
67	.297423 w_o	-0.73	-2.88	-2.07
68	.297292 w_o	-0.99	-3.12	-2.20
69	.297166 w_o	-1.25	-3.36	-2.34
70	.297043 w_o	-1.50	-3.59	-2.47

* Quasi-ranges 0 through 8 only

ESTIMATES OF STANDARD DEVIATION OF RECTANGULAR POPULATION

Sample size, n	Estimate based on the range	Bias (%) of estimates which assume normality		
		One quasi-range	Two adjacent quasi-ranges	Any Two quasi-ranges*
71	.296923 w_o	-1.75	-3.82	-2.59
72	.296807 w_o	-1.99	-4.05	-2.73
73	.296694 w_o	-2.23	-4.27	-2.85
74	.296584 w_o	-2.47	-4.49	-2.98
75	.296477 w_o	1.01	-4.70	-3.11
76	.296373 w_o	0.77	-4.91	-3.24
77	.296272 w_o	0.53	-1.29	-3.36
78	.296173 w_o	0.30	-1.51	-3.49
79	.296077 w_o	0.07	-1.73	-3.62
80	.295983 w_o	-0.16	-1.94	-3.74
81	.295892 w_o	-0.38	-2.15	-3.87
82	.295803 w_o	-0.60	-2.36	-3.99
83	.295716 w_o	-0.82	-2.56	-4.11
84	.295631 w_o	-1.04	-2.76	-4.24
85	.295548 w_o	-1.25	-2.96	-4.36
86	.295467 w_o	-1.45	-3.15	-4.48
87	.295389 w_o	-1.66	-3.34	-4.60
88	.295311 w_o	-1.86	-3.53	-4.72
89	.295236 w_o	1.06	-3.72	-4.83
90	.295162 w_o	0.85	-3.91	-4.95
91	.295090 w_o	0.65	-0.89	-5.07
92	.295020 w_o	0.45	-1.08	-5.19
93	.294951 w_o	0.25	-1.27	-5.30
94	.294883 w_o	0.05	-1.45	-5.42
95	.294817 w_o	-0.14	-1.63	-5.53
96	.294753 w_o	-0.33	-1.81	-5.64
97	.294689 w_o	-0.52	-1.99	-5.76
98	.294627 w_o	-0.70	-2.17	-5.87
99	.294566 w_o	-0.88	-2.34	-5.98
100	.294507 w_o	-1.07	-2.51	-6.09

* Quasi-ranges 0 through 8 only

Table B2

PERCENTAGE POINTS OF THE RANGE W FOR SAMPLES OF n FROM $R(\mu, 1)$

[P = cumulative probability = Prob($W \leq$ tabular value)]

PERCENTAGE POINTS OF THE RANGE

P\n	2	3	4	5	6	7
0.0001	0.000173	0.020039	0.102048	0.234911	0.391335	0.553000
0.0005	0.000866	0.044916	0.175456	0.353873	0.544437	0.729418
0.0010	0.001732	0.063636	0.221835	0.422668	0.628340	0.822649
0.0050	0.008671	0.143414	0.384117	0.641196	0.879856	1.091498
0.0100	0.017364	0.204046	0.487979	0.769281	1.019532	1.235421
0.0250	0.043575	0.326662	0.672453	0.982357	1.242800	1.459346
0.0500	0.087713	0.468867	0.861192	1.186772	1.448675	1.660334
0.1000	0.177766	0.678271	1.110108	1.441446	1.696314	1.896371
0.2000	0.365715	0.994685	1.446423	1.766026	2.000700	2.179400
0.3000	0.565826	1.258361	1.702936	2.002222	2.215786	2.375394
0.4000	0.780820	1.499717	1.924308	2.199756	2.392155	2.533946
0.5000	1.014612	1.732051	2.127902	2.377031	2.548020	2.672589
0.6000	1.273211	1.964384	2.323835	2.544145	2.693050	2.800441
0.7000	1.566735	2.205741	2.520536	2.708834	2.834316	2.923976
0.8000	1.914908	2.469417	2.728614	2.880023	2.979541	3.050006
0.9000	2.368656	2.785830	2.970262	3.075308	3.143342	3.191049
0.9500	2.689505	2.995234	3.125966	3.199304	3.246383	3.279203
0.9750	2.916379	3.137439	3.229977	3.281388	3.314203	3.336995
0.9900	3.117691	3.260055	3.318614	3.350887	3.371391	3.385588
0.9950	3.219153	3.320687	3.362102	3.384839	3.399251	3.409215
0.9990	3.354557	3.400465	3.418989	3.429108	3.435504	3.439917
0.9995	3.386642	3.419186	3.432284	3.439431	3.443946	3.447059
0.9999	3.429461	3.444063	3.449921	3.453112	3.455126	3.456515

P\n	8	9	10	11	12	13
0.0001	0.710259	0.858661	0.996538	1.123634	1.240378	1.347496
0.0005	0.901647	1.058999	1.201592	1.330456	1.446929	1.552377
0.0010	1.000168	1.160095	1.303461	1.431916	1.547209	1.650981
0.0050	1.276462	1.437708	1.578636	1.702382	1.811634	1.908634
0.0100	1.420483	1.579514	1.716973	1.836611	1.941478	2.034029
0.0250	1.640219	1.792685	1.922521	2.034190	2.131129	2.215995
0.0500	1.833620	1.977532	2.098680	2.201923	2.290877	2.368269
0.1000	2.056826	2.188057	2.297227	2.389387	2.468180	2.536292
0.2000	2.319684	2.432597	2.525371	2.602919	2.668685	2.725154
0.3000	2.499043	2.597588	2.677938	2.744690	2.801019	2.849184
0.4000	2.642707	2.728747	2.798499	2.856182	2.904674	2.946009
0.5000	2.767363	2.841881	2.902007	2.951541	2.993055	3.028350
0.6000	2.881563	2.945007	2.995986	3.037845	3.072831	3.102508
0.7000	2.991258	3.043619	3.085531	3.119840	3.148444	3.172657
0.8000	3.102544	3.143233	3.175681	3.202163	3.224187	3.242793
0.9000	3.226374	3.253591	3.275209	3.292795	3.307383	3.319680
0.9500	3.303404	3.321993	3.336722	3.348680	3.358585	3.366922
0.9750	3.353756	3.366605	3.376769	3.385012	3.391832	3.397568
0.9900	3.396005	3.403977	3.410276	3.415379	3.419597	3.423142
0.9950	3.416519	3.422104	3.426514	3.430085	3.433036	3.435515
0.9990	3.443147	3.445615	3.447562	3.449138	3.450439	3.451532
0.9995	3.449338	3.451078	3.452451	3.453561	3.454478	3.455248
0.9999	3.457530	3.458306	3.458917	3.459412	3.459820	3.460163

PERCENTAGE POINTS OF THE RANGE

P\n	14	15	16	17	18	19
0.0001	1.445820	1.536183	1.619372	1.696111	1.767053	1.832778
0.0005	1.648078	1.735184	1.814706	1.887527	1.954413	2.016028
0.0010	1.744701	1.829645	1.906912	1.977443	2.042044	2.101406
0.0050	1.995231	2.072946	2.143034	2.206534	2.264314	2.317096
0.0100	2.116235	2.189691	2.255690	2.315290	2.369363	2.418632
0.0250	2.290864	2.357375	2.416832	2.470288	2.518597	2.562463
0.0500	2.436184	2.496245	2.549726	2.597645	2.640820	2.679918
0.1000	2.595739	2.648067	2.694475	2.735909	2.773126	2.806734
0.2000	2.774161	2.817088	2.854998	2.888720	2.918910	2.946095
0.3000	2.890836	2.927211	2.959250	2.987685	3.013090	3.035926
0.4000	2.981660	3.012723	3.040031	3.064225	3.085808	3.105182
0.5000	3.058725	3.085142	3.108328	3.128840	3.147117	3.163504
0.6000	3.128001	3.150135	3.169534	3.186675	3.201930	3.215595
0.7000	3.193418	3.211418	3.227172	3.241077	3.253439	3.264503
0.8000	3.258719	3.272505	3.284557	3.295181	3.304618	3.313056
0.9000	3.330186	3.339266	3.347192	3.354172	3.360365	3.365897
0.9500	3.374037	3.380181	3.385540	3.390255	3.394436	3.398169
0.9750	3.402460	3.406681	3.410361	3.413597	3.416466	3.419026
0.9900	3.426164	3.428771	3.431042	3.433038	3.434807	3.436386
0.9950	3.437628	3.439449	3.441036	3.442431	3.443667	3.444769
0.9990	3.452463	3.453265	3.453964	3.454578	3.455122	3.455607
0.9995	3.455904	3.456470	3.456962	3.457395	3.457778	3.458120
0.9999	3.460455	3.460707	3.460926	3.461118	3.461289	3.461441

P\n	20	22	24	26	28	30
0.0001	1.893805	2.003546	2.099345	2.183619	2.258275	2.324839
0.C005	2.072945	2.174619	2.262704	2.339698	2.407538	2.467744
0.0010	2.156120	2.253577	2.337728	2.411080	2.475560	2.532667
0.0050	2.365490	2.451110	2.524464	2.587988	2.643520	2.692469
0.0100	2.463702	2.543203	2.611082	2.669697	2.720812	2.765774
0.0250	2.602466	2.672747	2.732478	2.783857	2.828517	2.867689
0.0500	2.715488	2.777786	2.830544	2.875789	2.915017	2.949350
0.1000	2.837234	2.890483	2.935410	2.973822	3.007037	3.036043
0.2000	2.970701	3.013517	3.049502	3.080169	3.106614	3.129653
0.3000	3.056562	3.092395	3.122438	3.147991	3.169988	3.189125
0.4000	3.122668	3.152985	3.178358	3.199906	3.218432	3.234531
0.5000	3.178280	3.203864	3.225246	3.243380	3.258956	3.272478
0.6000	3.227906	3.249199	3.266970	3.282026	3.294945	3.306153
0.7000	3.274462	3.291670	3.306014	3.318154	3.328563	3.337586
0.8000	3.320646	3.333746	3.344653	3.353875	3.361775	3.368619
0.9000	3.370868	3.379440	3.386568	3.392589	3.397742	3.402202
0.9500	3.401522	3.407299	3.412099	3.416151	3.419617	3.422616
0.9750	3.421325	3.425284	3.428573	3.431347	3.433720	3.435772
0.9900	3.437803	3.440242	3.442267	3.443976	3.445436	3.446698
0.9950	3.445759	3.447462	3.448876	3.450068	3.451087	3.451968
0.9990	3.456043	3.456792	3.457414	3.457938	3.458386	3.458774
0.9995	3.458427	3.458954	3.459392	3.459762	3.460077	3.460350
0.9999	3.461577	3.461812	3.462007	3.462171	3.462312	3.462433

PERCENTAGE POINTS OF THE RANGE

P\n	32	34	36	38	40	50
0.0001	2.384535	2.438357	2.487122	2.531503	2.572058	2.731610
0.0005	2.521520	2.569834	2.613471	2.653075	2.689175	2.830392
0.0010	2.583586	2.629264	2.670463	2.707809	2.741814	2.874506
0.0050	2.735934	2.774783	2.809711	2.841282	2.869955	2.981192
0.0100	2.805626	2.841189	2.873118	2.901942	2.928091	3.029275
0.0250	2.902325	2.933167	2.960805	2.985713	3.008275	3.095280
0.0500	2.979649	3.006584	3.030686	3.052379	3.072006	3.147489
0.1000	3.061590	3.084262	3.104519	3.122726	3.139179	3.202283
0.2000	3.149904	3.167844	3.183846	3.198209	3.211172	3.260746
0.3000	3.205923	3.220788	3.234035	3.245914	3.256627	3.297523
0.4000	3.248649	3.261132	3.272248	3.282210	3.291188	3.325417
0.5000	3.284328	3.294798	3.304115	3.312460	3.319978	3.348608
0.6000	3.315967	3.324634	3.332342	3.339243	3.345456	3.369097
0.7000	3.345482	3.352451	3.358646	3.364190	3.369180	3.388147
0.8000	3.374604	3.379883	3.384574	3.388770	3.392546	3.406883
0.9000	3.406100	3.409536	3.412588	3.415317	3.417771	3.427081
0.9500	3.425236	3.427545	3.429594	3.431427	3.433074	3.439320
0.9750	3.437564	3.439143	3.440545	3.441797	3.442923	3.447192
0.9900	3.447801	3.448772	3.449634	3.450405	3.451097	3.453721
0.9950	3.452737	3.453415	3.454016	3.454554	3.455036	3.456866
0.9990	3.459112	3.459409	3.459674	3.459910	3.460122	3.460926
0.9995	3.460588	3.460798	3.460984	3.461150	3.461300	3.461866
0.9999	3.462539	3.462632	3.462715	3.462789	3.462856	3.463107

P\n	60	70	80	90	100	
0.0001	2.842924	2.924944	2.987868	3.037659	3.078035	
0.0005	2.928185	2.999877	3.054674	3.097913	3.132898	
0.0010	2.966099	3.033098	3.084225	3.124518	3.157089	
0.0050	3.057391	3.112840	3.154993	3.188117	3.214833	
0.0100	3.098356	3.148511	3.186575	3.216449	3.240519	
0.0250	3.154414	3.197214	3.229624	3.255017	3.275449	
0.0500	3.198613	3.235527	3.263431	3.285264	3.302814	
0.1000	3.244868	3.275540	3.298685	3.316769	3.331290	
0.2000	3.294075	3.318018	3.336050	3.350121	3.361405	
0.3000	3.324953	3.344627	3.359426	3.370964	3.380211	
0.4000	3.348334	3.364751	3.377091	3.386704	3.394404	
0.5000	3.367748	3.381446	3.391734	3.399745	3.406159	
0.6000	3.384882	3.396169	3.404640	3.411234	3.416510	
0.7000	3.400796	3.409834	3.416613	3.421887	3.426107	
0.8000	3.416434	3.423253	3.428365	3.432340	3.435519	
0.9000	3.433276	3.437695	3.441005	3.443578	3.445636	
0.9500	3.443473	3.446434	3.448651	3.450374	3.451752	
0.9750	3.450028	3.452050	3.453564	3.454739	3.455679	
0.9900	3.455463	3.456705	3.457634	3.458356	3.458933	
0.9950	3.458081	3.458947	3.459595	3.460098	3.460500	
0.9990	3.461459	3.461839	3.462124	3.462345	3.462521	
0.9995	3.462241	3.462509	3.462709	3.462865	3.462989	
0.9999	3.463274	3.463393	3.463483	3.463552	3.463607	

Table B3

COEFFICIENTS OF SAMPLE RANGE w IN EXACT LOWER CONFIDENCE BOUNDS FOR STANDARD DEVIATION σ OF RECTANGULAR POPULATION

[P = confidence level; n = sample size]

COEFFICIENTS OF w IN EXACT LOWER CONFIDENCE BOUNDS FOR σ

P\n	2	3	4	5	6	7
0.0001	5773.358	49.90350	9.799300	4.256924	2.555358	1.808319
0.0005	1154.556	22.26383	5.699450	2.825874	1.836760	1.370956
0.0010	577.2059	15.71427	4.507863	2.365921	1.591495	1.215585
0.0050	115.3255	6.972804	2.603374	1.559586	1.136549	0.916172
0.0100	57.59033	4.900845	2.049266	1.299915	0.980842	0.809440
0.0250	22.94876	3.061264	1.487093	1.017960	0.804635	0.685239
0.0500	11.40082	2.132799	1.161182	0.842622	0.690286	0.602289
0.1000	5.625364	1.474336	0.900813	0.693748	0.589513	0.527323
0.2000	2.734370	1.005344	0.691361	0.566243	0.499825	0.458842
0.3000	1.767327	0.794685	0.587221	0.499445	0.451307	0.420983
0.4000	1.280705	0.666792	0.519667	0.454596	0.418033	0.394641
0.5000	0.985599	0.577350	0.469946	0.420693	0.392462	0.374169
0.6000	0.785416	0.509065	0.430323	0.393059	0.371326	0.357087
0.7000	0.638270	0.453362	0.396741	0.369162	0.352819	0.342000
0.8000	0.522218	0.404954	0.366486	0.347219	0.335622	0.327868
0.9000	0.422180	0.358959	0.336671	0.325171	0.318133	0.313377
0.9500	0.371816	0.333864	0.319901	0.312568	0.308035	0.304952
0.9750	0.342891	0.318731	0.309600	0.304749	0.301732	0.299671
0.9900	0.320750	0.306743	0.301331	0.298428	0.296613	0.295370
0.9950	0.310641	0.301142	0.297433	0.295435	0.294182	0.293323
0.9990	0.298102	0.294077	0.292484	0.291621	0.291078	0.290705
0.9995	0.295278	0.292467	0.291351	0.290746	0.290365	0.290102
0.9999	0.291591	0.290355	0.289862	0.289594	0.289425	0.289309

P\n	8	9	10	11	12	13
0.0001	1.407938	1.164604	1.003474	0.889969	0.806206	0.742117
0.0005	1.109082	0.944288	0.832229	0.751622	0.691119	0.644174
0.0010	0.999832	0.861999	0.767188	0.698365	0.646325	0.605700
0.0050	0.783415	0.695552	0.633458	0.587412	0.551988	0.523935
0.0100	0.703986	0.633106	0.582420	0.544481	0.515071	0.491635
0.0250	0.609675	0.557823	0.520150	0.491596	0.469235	0.451265
0.0500	0.545369	0.505681	0.476490	0.454148	0.436514	0.422249
0.1000	0.486186	0.457026	0.435307	0.418517	0.405157	0.394276
0.2000	0.431093	0.411083	0.395981	0.384184	0.374716	0.366952
0.3000	0.400153	0.384973	0.373422	0.364340	0.357013	0.350978
0.4000	0.378400	0.366469	0.357334	0.350118	0.344273	0.339442
0.5000	0.361355	0.351880	0.344589	0.338806	0.334107	0.330213
0.6000	0.347034	0.339558	0.333780	0.329181	0.325433	0.322320
0.7000	0.334307	0.328556	0.324093	0.320529	0.317617	0.315193
0.8000	0.322316	0.318144	0.314893	0.312289	0.310156	0.308376
0.9000	0.309945	0.307353	0.305324	0.303693	0.302354	0.301234
0.9500	0.302718	0.301024	0.299695	0.298625	0.297744	0.297007
0.9750	0.298173	0.297035	0.296141	0.295420	0.294826	0.294328
0.9900	0.294464	0.293774	0.293231	0.292793	0.292432	0.292129
0.9950	0.292696	0.292218	0.291842	0.291538	0.291287	0.291077
0.9990	0.290432	0.290224	0.290060	0.289928	0.289818	0.289726
0.9995	0.289911	0.289765	0.289649	0.289556	0.289479	0.289415
0.9999	0.289224	0.289159	0.289108	0.289066	0.289032	0.289004

COEFFICIENTS OF w IN EXACT LOWER CONFIDENCE BOUNDS FOR σ

P\n	14	15	16	17	18	19
0.0001	0.691649	0.650964	0.617523	0.589584	0.565914	0.545620
0.0005	0.606767	0.576308	0.551054	0.529794	0.511663	0.496025
0.0010	0.573164	0.546554	0.524408	0.505704	0.489705	0.475872
0.0050	0.501195	0.482405	0.466628	0.453199	0.441635	0.431575
0.0100	0.472537	0.456685	0.443323	0.431911	0.422054	0.413457
0.0250	0.436517	0.424201	0.413765	0.404811	0.397046	0.390250
0.0500	0.410478	0.400602	0.392199	0.384964	0.378670	0.373146
0.1000	0.385247	0.377634	0.371130	0.365509	0.360604	0.356286
0.2000	0.360469	0.354977	0.350263	0.346174	0.342594	0.339432
0.3000	0.345921	0.341622	0.337923	0.334707	0.331885	0.329389
0.4000	0.335384	0.331926	0.328944	0.326347	0.324064	0.322042
0.5000	0.326934	0.324134	0.321716	0.319607	0.317751	0.316105
0.6000	0.319693	0.317447	0.315504	0.313807	0.312312	0.310984
0.7000	0.313144	0.311389	0.309869	0.308539	0.307367	0.306325
0.8000	0.306869	0.305576	0.304455	0.303473	0.302607	0.301836
0.9000	0.300284	0.299467	0.298758	0.298136	0.297587	0.297098
0.9500	0.296381	0.295842	0.295374	0.294963	0.294600	0.294276
0.9750	0.293905	0.293541	0.293224	0.292946	0.292700	0.292481
0.9900	0.291872	0.291650	0.291457	0.291287	0.291137	0.291003
0.9950	0.290898	0.290744	0.290610	0.290492	0.290388	0.290295
0.9990	0.289648	0.289581	0.289522	0.289471	0.289425	0.289385
0.9995	0.289360	0.289313	0.289271	0.289235	0.289203	0.289174
0.9999	0.288979	0.288958	0.288940	0.288924	0.288910	0.288897

P\n	20	22	24	26	28	30
0.0001	0.528037	0.499115	0.476339	0.457955	0.442816	0.430137
0.0005	0.482405	0.459851	0.441949	0.427406	0.415362	0.405229
0.0010	0.463796	0.443739	0.427766	0.414752	0.403949	0.394841
0.0050	0.422745	0.407978	0.396124	0.386400	0.378283	0.371406
0.0100	0.405893	0.393205	0.382983	0.374574	0.367537	0.361562
0.0250	0.384251	0.374147	0.365968	0.359214	0.353542	0.348713
0.0500	0.368258	0.359999	0.353289	0.347731	0.343051	0.339058
0.1000	0.352456	0.345963	0.340668	0.336268	0.332553	0.329376
0.2000	0.336621	0.331838	0.327922	0.324658	0.321894	0.319524
0.3000	0.327165	0.323374	0.320263	0.317663	0.315459	0.313566
0.4000	0.320239	0.317160	0.314628	0.312509	0.310710	0.309164
0.5000	0.314636	0.312123	0.310054	0.308320	0.306847	0.305579
0.6000	0.309798	0.307768	0.306094	0.304690	0.303495	0.302466
0.7000	0.305394	0.303797	0.302479	0.301372	0.300430	0.299618
0.8000	0.301146	0.299963	0.298985	0.298163	0.297462	0.296858
0.9000	0.296659	0.295907	0.295284	0.294760	0.294313	0.293927
0.9500	0.293986	0.293488	0.293075	0.292727	0.292430	0.292174
0.9750	0.292284	0.291947	0.291667	0.291431	0.291229	0.291055
0.9900	0.290883	0.290677	0.290506	0.290362	0.290239	0.290133
0.9950	0.290212	0.290068	0.289950	0.289849	0.289764	0.289690
0.9990	0.289348	0.289286	0.289234	0.289190	0.289152	0.289120
0.9995	0.289149	0.289105	0.289068	0.289037	0.289011	0.288988
0.9999	0.288886	0.288866	0.288850	0.288836	0.288824	0.288814

COEFFICIENTS OF w IN EXACT LOWER CONFIDENCE BOUNDS FOR σ

P\n	32	34	36	38	40	50
0.0001	0.419369	0.410112	0.402071	0.395022	0.388794	0.366084
0.0005	0.396586	0.389130	0.382633	0.376921	0.371861	0.353308
0.0010	0.387059	0.380335	0.374467	0.369302	0.364722	0.347886
0.0050	0.365506	0.360389	0.355908	0.351954	0.348438	0.335436
0.0100	0.356427	0.351965	0.348054	0.344597	0.341519	0.330112
0.0250	0.344551	0.340928	0.337746	0.334928	0.332416	0.323073
0.0500	0.335610	0.332603	0.329958	0.327613	0.325520	0.317714
0.1000	0.326628	0.324227	0.322111	0.320233	0.318555	0.312277
0.2000	0.317470	0.315672	0.314086	0.312675	0.311413	0.306678
0.3000	0.311923	0.310483	0.309211	0.308080	0.307066	0.303258
0.4000	0.307820	0.306642	0.305600	0.304673	0.303842	0.300714
0.5000	0.304476	0.303509	0.302653	0.301890	0.301207	0.298632
0.6000	0.301571	0.300785	0.300089	0.299469	0.298913	0.296815
0.7000	0.298911	0.298289	0.297739	0.297248	0.296808	0.295147
0.8000	0.296331	0.295868	0.295458	0.295092	0.294764	0.293523
0.9000	0.293591	0.293295	0.293033	0.292799	0.292588	0.291793
0.9500	0.291951	0.291754	0.291580	0.291424	0.291284	0.290755
0.9750	0.290904	0.290770	0.290652	0.290546	0.290451	0.290091
0.9900	0.290040	0.289958	0.289886	0.289821	0.289763	0.289543
0.9950	0.289625	0.289568	0.289518	0.289473	0.289433	0.289279
0.9990	0.289092	0.289067	0.289045	0.289025	0.289007	0.288940
0.9995	0.288968	0.288951	0.288935	0.288921	0.288909	0.288862
0.9999	0.288805	0.288798	0.288791	0.288785	0.288779	0.288758

P\n	60	70	80	90	100	
0.0001	0.351751	0.341887	0.334687	0.329201	0.324883	
0.0005	0.341509	0.333347	0.327367	0.322798	0.319193	
0.0010	0.337143	0.329696	0.324231	0.320049	0.316747	
0.0050	0.327076	0.321250	0.316958	0.313665	0.311058	
0.0100	0.322752	0.317610	0.313817	0.310902	0.308593	
0.0250	0.317016	0.312772	0.309634	0.307218	0.305302	
0.0500	0.312636	0.309069	0.306426	0.304390	0.302772	
0.1000	0.308179	0.305293	0.303151	0.301498	0.300184	
0.2000	0.303575	0.301385	0.299756	0.298497	0.297495	
0.3000	0.300756	0.298987	0.297670	0.296651	0.295840	
0.4000	0.298656	0.297199	0.296113	0.295272	0.294603	
0.5000	0.296934	0.295731	0.294834	0.294140	0.293586	
0.6000	0.295431	0.294449	0.293717	0.293149	0.292696	
0.7000	0.294049	0.293269	0.292687	0.292236	0.291877	
0.8000	0.292703	0.292120	0.291684	0.291346	0.291077	
0.9000	0.291267	0.290893	0.290613	0.290396	0.290222	
0.9500	0.290404	0.290155	0.289968	0.289824	0.289708	
0.9750	0.289853	0.289683	0.289556	0.289457	0.289379	
0.9900	0.289397	0.289293	0.289215	0.289155	0.289106	
0.9950	0.289178	0.289105	0.289051	0.289009	0.288976	
0.9990	0.288896	0.288864	0.288840	0.288822	0.288807	
0.9995	0.288830	0.288808	0.288791	0.288778	0.288768	
0.9999	0.288744	0.288734	0.288727	0.288721	0.288716	

Appendix C

TABLES OF EXPECTED VALUES OF ORDER STATISTICS OF SAMPLES FROM VARIOUS POPULATIONS

SOURCES OF TABLES

Table C1.1	ARL TR 60-292 (Harter)
Table C1.2	*Ann. Math. Statist.* **27** (1956), 427-451 (Sarhan & Greenberg)
Table C2.1	ARL 64-31 (Harter)
Table C2.2	Freshly computed (Harter)
Table C3.1	ARL 64-31 (Harter)
Table C3.2	AFIT M.S. thesis GRE/Math/64-3 (Clark)
Table C4.1	ARL 64-31 (Harter) [values for $m = 1$, ALPHA = 0.5 freshly computed]
Table C4.2	Computed in connection with AFIT M.S. theses GRE/Math/64-2 (Bruce) and GRE/Math/65-5 (Hill)
Table C5	ARL 64-31 (Harter) [last two lines freshly computed]
Tables C6.1,6.2	*Sankhyā B* **36**, 230-260 (Gupta, McDonald & Galarneau)
Tables C7.1,7.2	*J. Indust. Math. Soc.* **1** (1964), 21-60 (White) UCLS doctoral dissertation (1965) (Mann)
Tables C8.1,8.2	McMaster Ph.D. thesis (1993) (Chan)
Tables C9.1,9.2	*J. Amer. Statist. Assoc.* **61** (1966), 1205-1218 (Barnett)
Tables C10.1,10.2	*J. Amer. Statist. Assoc.* **61** (1966), 248-258 (Govindarajulu)
Tables C11.1,11.2	*Rep. Statist. Appl. Res., JUSE* **12** (1965), 149-164 (Govindarajulu & Eisenstat)
Tables C12.2,12.2	*Selected Tables in Math. Statist.* **12** (1992), (Balakrishnan & Malik)
Tables C13.1,13.2	*Skand. Akturarietidskr.* **49** (1966), 144-157 (Malik)
Tables C14.1,14.2	*Selected Tables in Math. Statist.* **8** (1985) (Tiku & Kumra)
Tables C15.1,15.2	*J. Statist. Comput. and Simul.* **20** (1985), 287-309 (Balakrishnan)
Tables C16.1,16.2	*Commun. Statist. B: Simul. and Comput.* **17** (1988), 51-84 (Balakrishnan & Leung)

Table C1.1

EXPECTED VALUES OF NORMAL ORDER STATISTICS

$$E(x_{k|n}) = \frac{n!}{(n-k)!(k-1)!}\int_{-\infty}^{\infty} x\left[\frac{1}{2}-\Phi(x)\right]^{k-1}\left[\frac{1}{2}+\Phi(x)\right]^{n-k}\phi(x)dx\ ,$$

where $\phi(x) = (2\pi)^{-1/2}e^{-x^2/2}$ and $\Phi(x) = \int_0^x \phi(x)dx$.

[Tabular values are the expected values of the kth largest normal deviate for a sample of size n from $N(0,1)$; or when preceded by a minus sign, they are the expected values of the kth smallest normal deviate.]

A. EXPECTED VALUES OF NORMAL ORDER STATISTICS

k \ n	2	3	4	5	6	7	8
1	0.56419	0.84628	1.02938	1.16296	1.26721	1.35218	1.42360
2		0.00000	0.29701	0.49502	0.64176	0.75737	0.85222
3				0.00000	0.20155	0.35271	0.47282
4						0.00000	0.15251

k \ n	9	10	11	12	13	14	15
1	1.48501	1.53875	1.58644	1.62923	1.66799	1.70338	1.73591
2	0.93230	1.00136	1.06192	1.11573	1.16408	1.20790	1.24794
3	0.57197	0.65606	0.72884	0.79284	0.84983	0.90113	0.94769
4	0.27453	0.37576	0.46198	0.53684	0.60285	0.66176	0.71488
5	0.00000	0.12267	0.22489	0.31225	0.38833	0.45557	0.51570
6			0.00000	0.10259	0.19052	0.26730	0.33530
7					0.00000	0.08816	0.16530
8							0.00000

k \ n	16	17	18	19	20	21	22
1	1.76599	1.79394	1.82003	1.84448	1.86748	1.88917	1.90969
2	1.28474	1.31878	1.35041	1.37994	1.40760	1.43362	1.45816
3	0.99027	1.02946	1.06573	1.09945	1.13095	1.16047	1.18824
4	0.76317	0.80738	0.84812	0.88586	0.92098	0.95380	0.98459
5	0.57001	0.61946	0.66479	0.70661	0.74538	0.78150	0.81527
6	0.39622	0.45133	0.50158	0.54771	0.59030	0.62982	0.66667
7	0.23375	0.29519	0.35084	0.40164	0.44833	0.49148	0.53157
8	0.07729	0.14599	0.20774	0.26374	0.31493	0.36203	0.40559
9		0.00000	0.06880	0.13072	0.18696	0.23841	0.28579
10				0.00000	0.06200	0.11836	0.16997
11						0.00000	0.05642

k \ n	23	24	25	26	27	28	29
1	1.92916	1.94767	1.96531	1.98216	1.99827	2.01371	2.02852
2	1.48137	1.50338	1.52430	1.54423	1.56326	1.58145	1.59888
3	1.21445	1.23924	1.26275	1.28511	1.30641	1.32674	1.34619
4	1.01356	1.04091	1.06679	1.09135	1.11471	1.13697	1.15822
5	0.84697	0.87682	0.90501	0.93171	0.95705	0.98115	1.00414
6	0.70115	0.73354	0.76405	0.79289	0.82021	0.84615	0.87084
7	0.56896	0.60399	0.63690	0.66794	0.69727	0.72508	0.75150
8	0.44609	0.48391	0.51935	0.55267	0.58411	0.61385	0.64205
9	0.32965	0.37047	0.40860	0.44436	0.47801	0.50977	0.53982
10	0.21755	0.26163	0.30268	0.34105	0.37706	0.41096	0.44298
11	0.10813	0.15583	0.20006	0.24128	0.27983	0.31603	0.35013
12	0.00000	0.05176	0.09953	0.14387	0.18520	0.22389	0.26023
13			0.00000	0.04781	0.09220	0.13361	0.17240
14					0.00000	0.04442	0.08588
15							0.00000

EXPECTED VALUES OF NORMAL ORDER STATISTICS

k \ n	30	31	32	33	34	35	36
1	2.04276	2.05646	2.06967	2.08241	2.09471	2.10661	2.11812
2	1.61560	1.63166	1.64712	1.66200	1.67636	1.69023	1.70362
3	1.36481	1.38268	1.39985	1.41637	1.43228	1.44762	1.46244
4	1.17855	1.19803	1.21672	1.23468	1.25196	1.26860	1.28466
5	1.02609	1.04709	1.06721	1.08652	1.10509	1.12295	1.14016
6	0.89439	0.91688	0.93841	0.95905	0.97886	0.99790	1.01624
7	0.77666	0.80066	0.82359	0.84555	0.86660	0.88681	0.90625
8	0.66885	0.69438	0.71875	0.74204	0.76435	0.78574	0.80629
9	0.56834	0.59545	0.62129	0.64596	0.66954	0.69214	0.71382
10	0.47329	0.50206	0.52943	0.55552	0.58043	0.60427	0.62710
11	0.38235	0.41287	0.44185	0.46942	0.49572	0.52084	0.54488
12	0.29449	0.32686	0.35755	0.38669	0.41444	0.44091	0.46620
13	0.20885	0.24322	0.27573	0.30654	0.33582	0.36371	0.39032
14	0.12473	0.16126	0.19572	0.22832	0.25924	0.28863	0.31663
15	0.04148	0.08037	0.11695	0.15147	0.18415	0.21515	0.24463
16		0.00000	0.03890	0.07552	0.11009	0.14282	0.17388
17				0.00000	0.03663	0.07123	0.10399
18						0.00000	0.03461

k \ n	37	38	39	40	41	42	43
1	2.12928	2.14009	2.15059	2.16078	2.17068	2.18032	2.18969
2	1.71659	1.72914	1.74131	1.75312	1.76458	1.77571	1.78654
3	1.47676	1.49061	1.50402	1.51702	1.52964	1.54188	1.55377
4	1.30016	1.31514	1.32964	1.34368	1.35728	1.37048	1.38329
5	1.15677	1.17280	1.18830	1.20330	1.21782	1.23190	1.24556
6	1.03390	1.05095	1.06741	1.08332	1.09872	1.11364	1.12810
7	0.92496	0.94300	0.96041	0.97722	0.99348	1.00922	1.02446
8	0.82605	0.84508	0.86343	0.88114	0.89825	0.91480	0.93082
9	0.73465	0.75468	0.77398	0.79259	0.81056	0.82792	0.84472
10	0.64902	0.67009	0.69035	0.70988	0.72871	0.74690	0.76448
11	0.56793	0.59005	0.61131	0.63177	0.65149	0.67052	0.68889
12	0.49042	0.51363	0.53592	0.55736	0.57799	0.59788	0.61707
13	0.41576	0.44012	0.46348	0.48591	0.50749	0.52827	0.54830
14	0.34336	0.36892	0.39340	0.41688	0.43944	0.46114	0.48204
15	0.27272	0.29954	0.32520	0.34978	0.37337	0.39604	0.41784
16	0.20342	0.23159	0.25849	0.28423	0.30890	0.33257	0.35533
17	0.13509	0.16469	0.19292	0.21988	0.24569	0.27043	0.29418
18	0.06739	0.09853	0.12817	0.15644	0.18345	0.20931	0.23411
19	0.00000	0.03280	0.06395	0.09362	0.12192	0.14897	0.17488
20			0.00000	0.03117	0.06085	0.08917	0.11625
21					0.00000	0.02969	0.05803
22							0.00000

EXPECTED VALUES OF NORMAL ORDER STATISTICS

k\n	44	45	46	47	48	49	50
1	2.19882	2.20772	2.21639	2.22486	2.23312	2.24119	2.24907
2	1.79707	1.80733	1.81732	1.82706	1.83655	1.84582	1.85487
3	1.56533	1.57658	1.58754	1.59820	1.60860	1.61874	1.62863
4	1.39574	1.40784	1.41962	1.43108	1.44224	1.45312	1.46374
5	1.25881	1.27170	1.28422	1.29641	1.30827	1.31983	1.33109
6	1.14213	1.15576	1.16899	1.18186	1.19439	1.20658	1.21846
7	1.03924	1.05358	1.06751	1.08104	1.09420	1.10701	1.11948
8	0.94634	0.96139	0.97599	0.99018	1.00396	1.01737	1.03042
9	0.86097	0.87673	0.89201	0.90684	0.92125	0.93525	0.94887
10	0.78148	0.79795	0.81391	0.82939	0.84442	0.85902	0.87321
11	0.70666	0.72385	0.74049	0.75663	0.77228	0.78748	0.80225
12	0.63561	0.65353	0.67088	0.68768	0.70397	0.71978	0.73513
13	0.56763	0.58631	0.60438	0.62186	0.63881	0.65523	0.67117
14	0.50220	0.52166	0.54046	0.55865	0.57625	0.59331	0.60986
15	0.43885	0.45912	0.47868	0.49759	0.51588	0.53360	0.55077
16	0.37723	0.39833	0.41868	0.43834	0.45734	0.47573	0.49354
17	0.31701	0.33898	0.36016	0.38060	0.40034	0.41942	0.43789
18	0.25792	0.28081	0.30285	0.32410	0.34460	0.36441	0.38357
19	0.19972	0.22358	0.24652	0.26862	0.28992	0.31049	0.33036
20	0.14219	0.16707	0.19097	0.21396	0.23610	0.25746	0.27807
21	0.08513	0.11109	0.13600	0.15993	0.18296	0.20514	0.22653
22	0.02835	0.05546	0.08144	0.10637	0.13033	0.15338	0.17559
23		0.00000	0.02712	0.05311	0.07805	0.10203	0.12511
24				0.00000	0.02599	0.05095	0.07494
25						0.00000	0.02496

EXPECTED VALUES OF NORMAL ORDER STATISTICS

k\n	51	52	53	54	55	56	57
1	2.25678	2.26432	2.27169	2.27891	2.28598	2.29291	2.29970
2	1.86371	1.87235	1.88080	1.88906	1.89715	1.90506	1.91282
3	1.63829	1.64773	1.65695	1.66596	1.67478	1.68340	1.69185
4	1.47409	1.48420	1.49407	1.50372	1.51315	1.52237	1.53140
5	1.34207	1.35279	1.36326	1.37348	1.38346	1.39323	1.40278
6	1.23003	1.24132	1.25234	1.26310	1.27361	1.28387	1.29391
7	1.13162	1.14347	1.15502	1.16629	1.17729	1.18804	1.19855
8	1.04312	1.05550	1.06757	1.07934	1.09083	1.10205	1.11300
9	0.96213	0.97504	0.98762	0.99988	1.01185	1.02352	1.03493
10	0.88701	0.90045	0.91354	0.92629	0.93873	0.95086	0.96271
11	0.81661	0.83058	0.84417	0.85742	0.87033	0.88292	0.89520
12	0.75004	0.76455	0.77866	0.79240	0.80578	0.81883	0.83155
13	0.68666	0.70170	0.71633	0.73057	0.74444	0.75794	0.77111
14	0.62592	0.64152	0.65668	0.67143	0.68578	0.69976	0.71337
15	0.56742	0.58358	0.59928	0.61455	0.62940	0.64385	0.65793
16	0.51080	0.52755	0.54380	0.55960	0.57495	0.58989	0.60444
17	0.45578	0.47312	0.48995	0.50629	0.52217	0.53761	0.55263
18	0.40211	0.42007	0.43749	0.45439	0.47080	0.48675	0.50226
19	0.34957	0.36818	0.38621	0.40369	0.42065	0.43713	0.45314
20	0.29799	0.31726	0.33592	0.35400	0.37154	0.38856	0.40510
21	0.24719	0.26716	0.28648	0.30518	0.32331	0.34090	0.35797
22	0.19702	0.21772	0.23772	0.25708	0.27583	0.29400	0.31163
23	0.14735	0.16880	0.18953	0.20957	0.22896	0.24774	0.26595
24	0.09803	0.12029	0.14177	0.16252	0.18259	0.20201	0.22082
25	0.04896	0.07206	0.09434	0.11584	0.13661	0.15669	0.17614
26	0.00000	0.02400	0.04712	0.06940	0.09091	0.11170	0.13180
27			0.00000	0.02312	0.04541	0.06693	0.08773
28					0.00000	0.02229	0.04382
29							0.00000

EXPECTED VALUES OF NORMAL ORDER STATISTICS

k\n	58	59	60	61	62	63	64
1	2.30635	2.31288	2.31928	2.32556	2.33173	2.33778	2.34373
2	1.92041	1.92786	1.93516	1.94232	1.94934	1.95624	1.96301
3	1.70012	1.70822	1.71616	1.72394	1.73158	1.73906	1.74641
4	1.54024	1.54889	1.55736	1.56567	1.57381	1.58180	1.58963
5	1.41212	1.42127	1.43023	1.43900	1.44760	1.45603	1.46430
6	1.30373	1.31334	1.32274	1.33195	1.34097	1.34982	1.35848
7	1.20882	1.21886	1.22869	1.23832	1.24774	1.25698	1.26603
8	1.12371	1.13419	1.14443	1.15445	1.16427	1.17388	1.18329
9	1.04607	1.05695	1.06760	1.07802	1.08821	1.09819	1.10797
10	0.97427	0.98557	0.99662	1.00742	1.01799	1.02833	1.03846
11	0.90719	0.91890	0.93034	0.94153	0.95247	0.96317	0.97365
12	0.84397	0.85609	0.86793	0.87950	0.89081	0.90187	0.91270
13	0.78396	0.79649	0.80873	0.82068	0.83237	0.84379	0.85496
14	0.72665	0.73960	0.75224	0.76459	0.77665	0.78843	0.79996
15	0.67164	0.68502	0.69807	0.71081	0.72324	0.73540	0.74727
16	0.61860	0.63241	0.64587	0.65901	0.67183	0.68436	0.69659
17	0.56725	0.58150	0.59538	0.60893	0.62214	0.63504	0.64764
18	0.51736	0.53205	0.54637	0.56033	0.57395	0.58723	0.60020
19	0.46872	0.48388	0.49864	0.51303	0.52705	0.54073	0.55408
20	0.42117	0.43681	0.45202	0.46685	0.48129	0.49537	0.50911
21	0.37456	0.39068	0.40637	0.42164	0.43652	0.45101	0.46515
22	0.32875	0.34538	0.36155	0.37729	0.39260	0.40752	0.42207
23	0.28362	0.30078	0.31745	0.33366	0.34944	0.36480	0.37976
24	0.23906	0.25677	0.27396	0.29066	0.30691	0.32272	0.33812
25	0.19498	0.21325	0.23098	0.24820	0.26494	0.28122	0.29706
26	0.15127	0.17013	0.18842	0.20618	0.22343	0.24019	0.25650
27	0.10785	0.12733	0.14621	0.16452	0.18230	0.19957	0.21636
28	0.06463	0.08476	0.10425	0.12315	0.14148	0.15927	0.17656
29	0.02153	0.04234	0.06248	0.08198	0.10089	0.11923	0.13704
30		0.00000	0.02081	0.04096	0.06047	0.07938	0.09774
31				0.00000	0.02014	0.03966	0.05858
32						0.00000	0.01952

EXPECTED VALUES OF NORMAL ORDER STATISTICS

k \ n	65	66	67	68	69	70	71
1	2.34958	2.35532	2.36097	2.36652	2.37199	2.37736	2.38265
2	1.96965	1.97618	1.98260	1.98891	1.99510	2.00120	2.00720
3	1.75363	1.76071	1.76767	1.77451	1.78122	1.78783	1.79432
4	1.59732	1.60487	1.61228	1.61955	1.62670	1.63373	1.64063
5	1.47241	1.48036	1.48817	1.49584	1.50338	1.51078	1.51805
6	1.36698	1.37532	1.38351	1.39154	1.39942	1.40717	1.41478
7	1.27490	1.28360	1.29213	1.30051	1.30873	1.31680	1.32473
8	1.19252	1.20157	1.21044	1.21915	1.22769	1.23608	1.24431
9	1.11754	1.12693	1.13613	1.14516	1.15401	1.16270	1.17123
10	1.04838	1.05810	1.06762	1.07696	1.08612	1.09511	1.10393
11	0.98391	0.99395	1.00380	1.01345	1.02291	1.03220	1.04130
12	0.92329	0.93367	0.94383	0.95379	0.96355	0.97313	0.98252
13	0.86590	0.87660	0.88708	0.89735	0.90741	0.91728	0.92695
14	0.81123	0.82226	0.83306	0.84364	0.85400	0.86416	0.87412
15	0.75889	0.77025	0.78138	0.79226	0.80293	0.81338	0.82362
16	0.70856	0.72025	0.73170	0.74290	0.75387	0.76462	0.77514
17	0.65996	0.67200	0.68377	0.69529	0.70657	0.71761	0.72843
18	0.61288	0.62526	0.63737	0.64921	0.66080	0.67214	0.68325
19	0.56712	0.57985	0.59230	0.60447	0.61638	0.62803	0.63943
20	0.52252	0.53561	0.54841	0.56091	0.57314	0.58510	0.59681
21	0.47894	0.49240	0.50555	0.51839	0.53095	0.54323	0.55525
22	0.43625	0.45009	0.46360	0.47680	0.48969	0.50230	0.51463
23	0.39435	0.40857	0.42245	0.43601	0.44925	0.46219	0.47484
24	0.35312	0.36775	0.38201	0.39594	0.40953	0.42281	0.43579
25	0.31249	0.32753	0.34219	0.35649	0.37045	0.38408	0.39739
26	0.27237	0.28784	0.30290	0.31759	0.33192	0.34591	0.35958
27	0.23269	0.24859	0.26408	0.27917	0.29389	0.30825	0.32227
28	0.19337	0.20973	0.22565	0.24116	0.25627	0.27102	0.28540
29	0.15435	0.17118	0.18755	0.20349	0.21902	0.23416	0.24893
30	0.11556	0.13288	0.14972	0.16611	0.18207	0.19762	0.21277
31	0.07694	0.09478	0.11211	0.12896	0.14536	0.16134	0.17690
32	0.03844	0.05681	0.07465	0.09199	0.10885	0.12527	0.14125
33	0.00000	0.01893	0.03730	0.05514	0.07249	0.08936	0.10579
34			0.00000	0.01837	0.03622	0.05357	0.07045
35					0.00000	0.01785	0.03520
36							0.00000

EXPECTED VALUES OF NORMAL ORDER STATISTICS

k\n	72	73	74	75	76	77	78
1	2.38785	2.39298	2.39802	2.40299	2.40789	2.41271	2.41747
2	2.01310	2.01890	2.02462	2.03024	2.03578	2.04124	2.04662
3	1.80071	1.80699	1.81317	1.81926	1.82525	1.83115	1.83696
4	1.64742	1.65410	1.66067	1.66714	1.67350	1.67976	1.68592
5	1.52520	1.53223	1.53914	1.54594	1.55263	1.55921	1.56569
6	1.42226	1.42961	1.43684	1.44395	1.45094	1.45782	1.46459
7	1.33252	1.34017	1.34770	1.35510	1.36237	1.36953	1.37657
8	1.25240	1.26034	1.26815	1.27583	1.28338	1.29080	1.29810
9	1.17961	1.18784	1.19592	1.20387	1.21168	1.21936	1.22691
10	1.11259	1.12110	1.12945	1.13766	1.14572	1.15365	1.16145
11	1.05024	1.05902	1.06764	1.07610	1.08442	1.09260	1.10063
12	0.99173	1.00078	1.00966	1.01838	1.02695	1.03537	1.04364
13	0.93644	0.94576	0.95490	0.96387	0.97269	0.98135	0.98986
14	0.88388	0.89346	0.90286	0.91209	0.92115	0.93005	0.93880
15	0.83366	0.84351	0.85317	0.86265	0.87196	0.88110	0.89008
16	0.78546	0.79558	0.80550	0.81524	0.82480	0.83418	0.84339
17	0.73903	0.74942	0.75960	0.76960	0.77940	0.78903	0.79848
18	0.69413	0.70480	0.71526	0.72551	0.73557	0.74544	0.75512
19	0.65060	0.66155	0.67227	0.68279	0.69310	0.70322	0.71314
20	0.60827	0.61950	0.63050	0.64128	0.65185	0.66222	0.67239
21	0.56701	0.57852	0.58980	0.60085	0.61168	0.62230	0.63272
22	0.52669	0.53850	0.55006	0.56138	0.57248	0.58336	0.59403
23	0.48721	0.49932	0.51117	0.52277	0.53414	0.54528	0.55621
24	0.44848	0.46089	0.47304	0.48493	0.49657	0.50798	0.51917
25	0.41041	0.42313	0.43558	0.44777	0.45970	0.47138	0.48283
26	0.37292	0.38597	0.39873	0.41122	0.42343	0.43540	0.44711
27	0.33596	0.34934	0.36242	0.37521	0.38772	0.39997	0.41196
28	0.29945	0.31317	0.32657	0.33968	0.35250	0.36504	0.37731
29	0.26333	0.27740	0.29114	0.30457	0.31770	0.33055	0.34311
30	0.22756	0.24199	0.25608	0.26984	0.28329	0.29645	0.30931
31	0.19208	0.20688	0.22133	0.23543	0.24922	0.26269	0.27586
32	0.15683	0.17202	0.18684	0.20130	0.21543	0.22923	0.24272
33	0.12178	0.13737	0.15257	0.16740	0.18188	0.19602	0.20983
34	0.08688	0.10289	0.11848	0.13370	0.14854	0.16303	0.17718
35	0.05209	0.06852	0.08453	0.10014	0.11536	0.13021	0.14471
36	0.01736	0.03424	0.05068	0.06670	0.08231	0.09754	0.11240
37		0.00000	0.01689	0.03333	0.04935	0.06497	0.08020
38				0.00000	0.01644	0.03247	0.04809
39						0.00000	0.01602

EXPECTED VALUES OF NORMAL ORDER STATISTICS

k \ n	79	80	81	82	83	84	85
1	2.42215	2.42677	2.43133	2.43582	2.44026	2.44463	2.44894
2	2.05191	2.05714	2.06228	2.06735	2.07236	2.07729	2.08216
3	1.84268	1.84832	1.85387	1.85935	1.86475	1.87007	1.87532
4	1.69200	1.69798	1.70387	1.70968	1.71540	1.72104	1.72660
5	1.57207	1.57836	1.58455	1.59065	1.59665	1.60258	1.60841
6	1.47125	1.47781	1.48428	1.49064	1.49691	1.50309	1.50918
7	1.38350	1.39032	1.39704	1.40366	1.41017	1.41659	1.42292
8	1.30529	1.31236	1.31932	1.32617	1.33292	1.33957	1.34611
9	1.23434	1.24165	1.24884	1.25593	1.26290	1.26977	1.27653
10	1.16912	1.17666	1.18409	1.19139	1.19859	1.20567	1.21264
11	1.10854	1.11631	1.12396	1.13148	1.13889	1.14618	1.15336
12	1.05178	1.05978	1.06764	1.07539	1.08300	1.09050	1.09788
13	0.99822	1.00644	1.01453	1.02249	1.03031	1.03802	1.04560
14	0.94739	0.95584	0.96414	0.97231	0.98034	0.98825	0.99603
15	0.89890	0.90757	0.91609	0.92447	0.93271	0.94082	0.94880
16	0.85244	0.86134	0.87007	0.87867	0.88711	0.89542	0.90360
17	0.80776	0.81687	0.82583	0.83464	0.84329	0.85180	0.86017
18	0.76463	0.77398	0.78315	0.79217	0.80103	0.80975	0.81832
19	0.72289	0.73246	0.74186	0.75109	0.76016	0.76908	0.77785
20	0.68237	0.69217	0.70179	0.71124	0.72053	0.72965	0.73862
21	0.64294	0.65297	0.66282	0.67249	0.68199	0.69133	0.70050
22	0.60449	0.61476	0.62484	0.63473	0.64445	0.65399	0.66337
23	0.56692	0.57742	0.58773	0.59785	0.60779	0.61755	0.62714
24	0.53013	0.54088	0.55143	0.56178	0.57193	0.58191	0.59171
25	0.49404	0.50504	0.51583	0.52641	0.53680	0.54700	0.55701
26	0.45859	0.46985	0.48088	0.49170	0.50232	0.51274	0.52297
27	0.42371	0.43522	0.44651	0.45757	0.46842	0.47907	0.48952
28	0.38934	0.40111	0.41265	0.42397	0.43506	0.44594	0.45662
29	0.35542	0.36747	0.37927	0.39084	0.40218	0.41330	0.42421
30	0.32190	0.33423	0.34630	0.35813	0.36972	0.38108	0.39223
31	0.28875	0.30136	0.31371	0.32580	0.33765	0.34926	0.36065
32	0.25591	0.26881	0.28144	0.29381	0.30592	0.31779	0.32943
33	0.22334	0.23655	0.24947	0.26212	0.27450	0.28664	0.29852
34	0.19101	0.20453	0.21775	0.23069	0.24335	0.25576	0.26790
35	0.15888	0.17272	0.18625	0.19949	0.21244	0.22512	0.23753
36	0.12691	0.14108	0.15493	0.16848	0.18172	0.19469	0.20738
37	0.09507	0.10959	0.12377	0.13763	0.15118	0.16444	0.17741
38	0.06333	0.07820	0.09272	0.10691	0.12078	0.13434	0.14761
39	0.03165	0.04689	0.06177	0.07629	0.09049	0.10436	0.11793
40	0.00000	0.01562	0.03087	0.04575	0.06028	0.07448	0.08836
41			0.00000	0.01524	0.03013	0.04466	0.05886
42					0.00000	0.01488	0.02942
43							0.00000

EXPECTED VALUES OF NORMAL ORDER STATISTICS

k \ n	86	87	88	89	90	91	92
1	2.45320	2.45741	2.46156	2.46565	2.46970	2.47370	2.47764
2	2.08696	2.09170	2.09637	2.10099	2.10554	2.11004	2.11448
3	1.88049	1.88560	1.89064	1.89561	1.90052	1.90536	1.91015
4	1.73209	1.73750	1.74283	1.74810	1.75329	1.75842	1.76348
5	1.61417	1.61984	1.62544	1.63096	1.63641	1.64178	1.64709
6	1.51518	1.52110	1.52693	1.53269	1.53836	1.54396	1.54949
7	1.42915	1.43529	1.44135	1.44732	1.45321	1.45903	1.46476
8	1.35257	1.35893	1.36520	1.37138	1.37747	1.38348	1.38941
9	1.28320	1.28976	1.29624	1.30262	1.30891	1.31511	1.32123
10	1.21951	1.22628	1.23295	1.23952	1.24600	1.25239	1.25869
11	1.16043	1.16740	1.17426	1.18102	1.18769	1.19426	1.20073
12	1.10515	1.11231	1.11936	1.12631	1.13316	1.13990	1.14656
13	1.05306	1.06041	1.06765	1.07478	1.08181	1.08873	1.09555
14	1.00369	1.01122	1.01865	1.02596	1.03316	1.04026	1.04726
15	0.95665	0.96437	0.97198	0.97948	0.98686	0.99413	1.00129
16	0.91164	0.91956	0.92735	0.93502	0.94258	0.95002	0.95735
17	0.86841	0.87651	0.88449	0.89234	0.90007	0.90769	0.91519
18	0.82675	0.83504	0.84320	0.85123	0.85914	0.86693	0.87460
19	0.78647	0.79496	0.80330	0.81152	0.81960	0.82756	0.83540
20	0.74744	0.75611	0.76465	0.77304	0.78131	0.78944	0.79745
21	0.70952	0.71838	0.72710	0.73568	0.74412	0.75243	0.76061
22	0.67259	0.68165	0.69056	0.69932	0.70795	0.71643	0.72478
23	0.63656	0.64581	0.65492	0.66387	0.67267	0.68134	0.68986
24	0.60133	0.61079	0.62009	0.62923	0.63822	0.64706	0.65576
25	0.56684	0.57650	0.58600	0.59533	0.60451	0.61353	0.62241
26	0.53301	0.54288	0.55258	0.56210	0.57147	0.58068	0.58974
27	0.49979	0.50986	0.51976	0.52949	0.53905	0.54845	0.55769
28	0.46710	0.47739	0.48750	0.49743	0.50718	0.51677	0.52620
29	0.43491	0.44542	0.45574	0.46587	0.47582	0.48561	0.49522
30	0.40316	0.41389	0.42443	0.43477	0.44493	0.45491	0.46472
31	0.37182	0.38278	0.39353	0.40409	0.41445	0.42463	0.43464
32	0.34084	0.35203	0.36300	0.37378	0.38436	0.39474	0.40495
33	0.31018	0.32161	0.33281	0.34381	0.35461	0.36520	0.37561
34	0.27981	0.29148	0.30292	0.31415	0.32517	0.33598	0.34660
35	0.24970	0.26162	0.27330	0.28476	0.29601	0.30704	0.31787
36	0.21981	0.23199	0.24392	0.25562	0.26710	0.27835	0.28940
37	0.19012	0.20256	0.21475	0.22669	0.23841	0.24990	0.26117
38	0.16059	0.17330	0.18576	0.19796	0.20991	0.22164	0.23314
39	0.13121	0.14420	0.15692	0.16938	0.18159	0.19356	0.20530
40	0.10193	0.11521	0.12821	0.14094	0.15341	0.16563	0.17761
41	0.07275	0.08633	0.09961	0.11262	0.12536	0.13783	0.15006
42	0.04362	0.05751	0.07110	0.08439	0.09740	0.11014	0.12262
43	0.01454	0.02874	0.04263	0.05622	0.06952	0.08253	0.09528
44		0.00000	0.01421	0.02810	0.04169	0.05499	0.06801
45				0.00000	0.01389	0.02748	0.04078
46						0.00000	0.01359

EXPECTED VALUES OF NORMAL ORDER STATISTICS

k\n	93	94	95	96	97	98	99
1	2.48154	2.48540	2.48920	2.49297	2.49669	2.50036	2.50400
2	2.11887	2.12321	2.12749	2.13172	2.13590	2.14003	2.14411
3	1.91487	1.91953	1.92414	1.92869	1.93318	1.93763	1.94201
4	1.76848	1.77341	1.77828	1.78309	1.78784	1.79254	1.79718
5	1.65232	1.65749	1.66259	1.66763	1.67261	1.67752	1.68238
6	1.55494	1.56033	1.56564	1.57089	1.57607	1.58118	1.58624
7	1.47042	1.47600	1.48151	1.48695	1.49232	1.49762	1.50286
8	1.39526	1.40103	1.40673	1.41235	1.41790	1.42338	1.42879
9	1.32726	1.33321	1.33909	1.34489	1.35061	1.35626	1.36183
10	1.26491	1.27104	1.27708	1.28305	1.28894	1.29475	1.30049
11	1.20712	1.21342	1.21964	1.22577	1.23182	1.23779	1.24368
12	1.15311	1.15958	1.16596	1.17226	1.17847	1.18459	1.19064
13	1.10228	1.10891	1.11546	1.12191	1.12827	1.13455	1.14075
14	1.05415	1.06095	1.06765	1.07426	1.08078	1.08721	1.09356
15	1.00835	1.01531	1.02217	1.02894	1.03561	1.04219	1.04868
16	0.96458	0.97170	0.97872	0.98564	0.99246	0.99919	1.00583
17	0.92258	0.92986	0.93704	0.94411	0.95109	0.95797	0.96475
18	0.88215	0.88959	0.89693	0.90416	0.91129	0.91831	0.92524
19	0.84312	0.85072	0.85822	0.86560	0.87288	0.88006	0.88713
20	0.80533	0.81310	0.82075	0.82829	0.83572	0.84305	0.85027
21	0.76866	0.77659	0.78441	0.79210	0.79968	0.80716	0.81452
22	0.73300	0.74110	0.74907	0.75692	0.76466	0.77228	0.77980
23	0.69825	0.70651	0.71464	0.72266	0.73055	0.73832	0.74598
24	0.66432	0.67275	0.68105	0.68922	0.69727	0.70519	0.71301
25	0.63115	0.63974	0.64821	0.65654	0.66474	0.67282	0.68079
26	0.59865	0.60742	0.61605	0.62454	0.63291	0.64115	0.64926
27	0.56678	0.57572	0.58452	0.59318	0.60170	0.61010	0.61837
28	0.53547	0.54459	0.55356	0.56239	0.57108	0.57963	0.58805
29	0.50468	0.51398	0.52312	0.53212	0.54097	0.54969	0.55827
30	0.47436	0.48384	0.49316	0.50233	0.51136	0.52024	0.52898
31	0.44447	0.45414	0.46364	0.47299	0.48218	0.49123	0.50013
32	0.41498	0.42483	0.43452	0.44404	0.45341	0.46263	0.47170
33	0.38584	0.39588	0.40576	0.41547	0.42501	0.43440	0.44364
34	0.35702	0.36727	0.37733	0.38722	0.39695	0.40652	0.41593
35	0.32850	0.33895	0.34921	0.35929	0.36920	0.37895	0.38853
36	0.30025	0.31090	0.32136	0.33163	0.34173	0.35166	0.36142
37	0.27223	0.28309	0.29375	0.30423	0.31452	0.32464	0.33458
38	0.24443	0.25550	0.26637	0.27705	0.28754	0.29785	0.30797
39	0.21681	0.22810	0.23919	0.25008	0.26077	0.27127	0.28159
40	0.18936	0.20088	0.21219	0.22328	0.23418	0.24488	0.25539
41	0.16205	0.17380	0.18533	0.19665	0.20776	0.21866	0.22937
42	0.13486	0.14685	0.15861	0.17015	0.18148	0.19259	0.20351
43	0.10777	0.12001	0.13201	0.14378	0.15533	0.16666	0.17778
44	0.08076	0.09325	0.10550	0.11750	0.12928	0.14083	0.15217
45	0.05381	0.06656	0.07906	0.09131	0.10332	0.11510	0.12666
46	0.02689	0.03992	0.05267	0.06518	0.07743	0.08944	0.10123
47	0.00000	0.01330	0.02633	0.03909	0.05159	0.06385	0.07586
48			0.00000	0.01303	0.02579	0.03829	0.05055
49					0.00000	0.01276	0.02527
50							0.00000

EXPECTED VALUES OF NORMAL ORDER STATISTICS

k\n	100	105	108	112	120	125	126
1	2.50759	2.52498	2.53498	2.54784	2.57208	2.58634	2.58912
2	2.14814	2.16766	2.17886	2.19326	2.22038	2.23630	2.23940
3	1.94635	1.96731	1.97934	1.99478	2.02385	2.04090	2.04421
4	1.80176	1.82389	1.83658	1.85288	1.88351	1.90146	1.90495
5	1.68718	1.71033	1.72361	1.74064	1.77263	1.79137	1.79501
6	1.59123	1.61531	1.62911	1.64681	1.68002	1.69947	1.70324
7	1.50803	1.53298	1.54726	1.56557	1.59992	1.62002	1.62392
8	1.43414	1.45990	1.47464	1.49353	1.52895	1.54966	1.55368
9	1.36734	1.39388	1.40906	1.42851	1.46494	1.48623	1.49036
10	1.30615	1.33344	1.34905	1.36902	1.40644	1.42828	1.43252
11	1.24950	1.27752	1.29354	1.31403	1.35239	1.37477	1.37911
12	1.19661	1.22534	1.24175	1.26275	1.30203	1.32493	1.32937
13	1.14687	1.17630	1.19311	1.21460	1.25477	1.27819	1.28273
14	1.09982	1.12994	1.14713	1.16911	1.21017	1.23409	1.23872
15	1.05509	1.08590	1.10347	1.12593	1.16786	1.19226	1.19700
16	1.01238	1.04387	1.06182	1.08475	1.12754	1.15243	1.15726
17	0.97145	1.00360	1.02193	1.04533	1.08898	1.11435	1.11927
18	0.93208	0.96491	0.98361	1.00748	1.05197	1.07783	1.08283
19	0.89411	0.92761	0.94668	0.97102	1.01636	1.04268	1.04778
20	0.85739	0.89156	0.91100	0.93581	0.98199	1.00879	1.01398
21	0.82179	0.85663	0.87645	0.90172	0.94874	0.97601	0.98129
22	0.78720	0.82272	0.84292	0.86866	0.91651	0.94426	0.94962
23	0.75353	0.78974	0.81031	0.83651	0.88521	0.91342	0.91888
24	0.72070	0.75759	0.77854	0.80521	0.85475	0.88344	0.88899
25	0.68863	0.72621	0.74753	0.77469	0.82507	0.85423	0.85986
26	0.65725	0.69552	0.71724	0.74486	0.79610	0.82573	0.83146
27	0.62651	0.66548	0.68758	0.71569	0.76778	0.79789	0.80370
28	0.59635	0.63603	0.65852	0.68711	0.74007	0.77065	0.77656
29	0.56672	0.60713	0.63001	0.65909	0.71291	0.74398	0.74997
30	0.53758	0.57872	0.60200	0.63157	0.68627	0.71782	0.72391
31	0.50890	0.55077	0.57445	0.60453	0.66011	0.69215	0.69833
32	0.48062	0.52324	0.54734	0.57792	0.63439	0.66692	0.67320
33	0.45273	0.49611	0.52062	0.55171	0.60909	0.64212	0.64849
34	0.42518	0.46934	0.49427	0.52588	0.58417	0.61770	0.62416
35	0.39796	0.44290	0.46826	0.50040	0.55961	0.59365	0.60021
36	0.37102	0.41677	0.44256	0.47523	0.53539	0.56993	0.57659
37	0.34436	0.39092	0.41715	0.45037	0.51147	0.54653	0.55329
38	0.31793	0.36533	0.39201	0.42578	0.48784	0.52343	0.53029
39	0.29173	0.33997	0.36712	0.40144	0.46449	0.50061	0.50756
40	0.26572	0.31483	0.34245	0.37734	0.44138	0.47804	0.48509
41	0.23990	0.28989	0.31798	0.35346	0.41851	0.45571	0.46287
42	0.21423	0.26513	0.29371	0.32978	0.39585	0.43361	0.44087
43	0.18870	0.24054	0.26961	0.30628	0.37340	0.41172	0.41909
44	0.16330	0.21608	0.24566	0.28295	0.35113	0.39002	0.39750
45	0.13800	0.19176	0.22186	0.25977	0.32904	0.36851	0.37609
46	0.11279	0.16754	0.19818	0.23674	0.30710	0.34717	0.35486
47	0.08765	0.14343	0.17461	0.21382	0.28531	0.32598	0.33379
48	0.06257	0.11940	0.15113	0.19102	0.26366	0.30494	0.31286
49	0.03753	0.09544	0.12774	0.16832	0.24213	0.28403	0.29207
50	0.01251	0.07153	0.10442	0.14571	0.22071	0.26325	0.27140

EXPECTED VALUES OF NORMAL ORDER STATISTICS

k\n	100	105	108	112	120	125	126
51		0.04766	0.08116	0.12317	0.19940	0.24258	0.25086
52		0.02383	0.05794	0.10069	0.17817	0.22201	0.23041
53		0.00000	0.03475	0.07826	0.15703	0.20154	0.21006
54			0.01158	0.05587	0.13595	0.18115	0.18980
55				0.03351	0.11493	0.16084	0.16962
56				0.01117	0.09397	0.14059	0.14950
57					0.07304	0.12040	0.12945
58					0.05215	0.10026	0.10945
59					0.03128	0.08016	0.08949
60					0.01043	0.06009	0.06956
61						0.04005	0.04967
62						0.02002	0.02979
63						0.00000	0.00993

EXPECTED VALUES OF NORMAL ORDER STATISTICS

k \ n	128	135	140	144	147	150	160
1	2.59460	2.61305	2.62559	2.63527	2.64234	2.64925	2.67122
2	2.24551	2.26609	2.28006	2.29083	2.29869	2.30638	2.33078
3	2.05075	2.07275	2.08768	2.09919	2.10758	2.11578	2.14181
4	1.91184	1.93498	1.95068	1.96277	1.97158	1.98019	2.00751
5	1.80219	1.82632	1.84267	1.85527	1.86445	1.87341	1.90184
6	1.71069	1.73571	1.75265	1.76570	1.77521	1.78448	1.81390
7	1.63162	1.65745	1.67494	1.68840	1.69820	1.70777	1.73809
8	1.56160	1.58820	1.60619	1.62004	1.63013	1.63997	1.67114
9	1.49850	1.52583	1.54430	1.55852	1.56887	1.57896	1.61093
10	1.44087	1.46889	1.48783	1.50239	1.51299	1.52333	1.55606
11	1.38767	1.41635	1.43574	1.45064	1.46148	1.47206	1.50551
12	1.33813	1.36746	1.38728	1.40250	1.41358	1.42438	1.45855
13	1.29167	1.32164	1.34188	1.35742	1.36873	1.37975	1.41460
14	1.24786	1.27845	1.29909	1.31494	1.32647	1.33771	1.37323
15	1.20631	1.23751	1.25856	1.27471	1.28646	1.29791	1.33408
16	1.16676	1.19856	1.22000	1.23645	1.24842	1.26007	1.29689
17	1.12895	1.16134	1.18317	1.19992	1.21210	1.22396	1.26140
18	1.09269	1.12567	1.14789	1.16493	1.17731	1.18937	1.22744
19	1.05782	1.09138	1.11398	1.13131	1.14390	1.15616	1.19484
20	1.02419	1.05832	1.08130	1.09892	1.11172	1.12417	1.16347
21	0.99168	1.02639	1.04975	1.06765	1.08065	1.09330	1.13320
22	0.96019	0.99547	1.01920	1.03739	1.05059	1.06344	1.10393
23	0.92962	0.96548	0.98958	1.00805	1.02145	1.03449	1.07558
24	0.89990	0.93633	0.96081	0.97955	0.99316	1.00639	1.04808
25	0.87095	0.90795	0.93281	0.95183	0.96564	0.97907	1.02134
26	0.84272	0.88029	0.90552	0.92483	0.93883	0.95245	0.99532
27	0.81515	0.85329	0.87889	0.89848	0.91268	0.92650	0.96995
28	0.78818	0.82690	0.85288	0.87274	0.88715	0.90115	0.94520
29	0.76177	0.80107	0.82742	0.84757	0.86218	0.87638	0.92101
30	0.73589	0.77577	0.80250	0.82293	0.83773	0.85212	0.89734
31	0.71048	0.75095	0.77806	0.79877	0.81378	0.82836	0.87418
32	0.68553	0.72659	0.75408	0.77508	0.79029	0.80506	0.85147
33	0.66101	0.70265	0.73052	0.75181	0.76722	0.78219	0.82919
34	0.63687	0.67911	0.70737	0.72894	0.74456	0.75973	0.80732
35	0.61310	0.65594	0.68459	0.70645	0.72227	0.73764	0.78582
36	0.58967	0.63312	0.66216	0.68431	0.70034	0.71590	0.76469
37	0.56656	0.61062	0.64005	0.66250	0.67874	0.69450	0.74389
38	0.54374	0.58843	0.61826	0.64100	0.65745	0.67341	0.72341
39	0.52121	0.56653	0.59676	0.61979	0.63645	0.65261	0.70322
40	0.49894	0.54489	0.57552	0.59886	0.61573	0.63210	0.68332
41	0.47692	0.52350	0.55455	0.57819	0.59528	0.61185	0.66369
42	0.45512	0.50236	0.53382	0.55776	0.57507	0.59184	0.64431
43	0.43354	0.48143	0.51331	0.53757	0.55509	0.57208	0.62517
44	0.41216	0.46072	0.49302	0.51759	0.53533	0.55253	0.60626
45	0.39097	0.44020	0.47293	0.49781	0.51578	0.53319	0.58756
46	0.36995	0.41986	0.45303	0.47823	0.49643	0.51405	0.56906
47	0.34909	0.39970	0.43330	0.45883	0.47725	0.49509	0.55076
48	0.32839	0.37970	0.41375	0.43960	0.45826	0.47632	0.53264
49	0.30782	0.35984	0.39435	0.42054	0.43942	0.45770	0.51469
50	0.28739	0.34013	0.37510	0.40162	0.42075	0.43925	0.49691

EXPECTED VALUES OF NORMAL ORDER STATISTICS

k \ n	128	135	140	144	147	150	160
51	0.26707	0.32055	0.35598	0.38285	0.40221	0.42094	0.47928
52	0.24686	0.30110	0.33700	0.36421	0.38382	0.40278	0.46181
53	0.22675	0.28176	0.31814	0.34570	0.36555	0.38475	0.44447
54	0.20674	0.26252	0.29938	0.32730	0.34741	0.36684	0.42726
55	0.18681	0.24338	0.28074	0.30902	0.32937	0.34904	0.41018
56	0.16695	0.22432	0.26219	0.29084	0.31145	0.33136	0.39322
57	0.14715	0.20535	0.24373	0.27275	0.29362	0.31378	0.37637
58	0.12742	0.18645	0.22536	0.25476	0.27589	0.29630	0.35963
59	0.10773	0.16762	0.20705	0.23684	0.25825	0.27891	0.34299
60	0.08809	0.14885	0.18882	0.21900	0.24068	0.26160	0.32644
61	0.06848	0.13013	0.17066	0.20123	0.22319	0.24437	0.30998
62	0.04889	0.11146	0.15254	0.18353	0.20577	0.22721	0.29361
63	0.02933	0.09282	0.13448	0.16588	0.18840	0.21012	0.27731
64	0.00978	0.07422	0.11646	0.14828	0.17110	0.19309	0.26109
65		0.05564	0.09848	0.13073	0.15385	0.17612	0.24493
66		0.03708	0.08053	0.11322	0.13664	0.15919	0.22884
67		0.01854	0.06261	0.09574	0.11947	0.14232	0.21281
68		0.00000	0.04471	0.07829	0.10234	0.12548	0.19683
69			0.02682	0.06087	0.08524	0.10868	0.18090
70			0.00894	0.04347	0.06816	0.09191	0.16502
71				0.02607	0.05110	0.07516	0.14918
72				0.00869	0.03406	0.05844	0.13338
73					0.01703	0.04173	0.11761
74					0.00000	0.02503	0.10187
75						0.00834	0.08616
76							0.07046
77							0.05479
78							0.03912
79							0.02347
80							0.00782

EXPECTED VALUES OF NORMAL ORDER STATISTICS

k \ n	162	168	175	180	189	192	196
1	2.67543	2.68773	2.70148	2.71093	2.72724	2.73248	2.73934
2	2.33546	2.34910	2.36434	2.37481	2.39286	2.39867	2.40625
3	2.14679	2.16133	2.17755	2.18870	2.20790	2.21406	2.22212
4	2.01274	2.02799	2.04500	2.05667	2.07678	2.08324	2.09167
5	1.90728	1.92313	1.94081	1.95294	1.97382	1.98052	1.98928
6	1.81953	1.83592	1.85419	1.86673	1.88830	1.89522	1.90426
7	1.74389	1.76077	1.77959	1.79250	1.81470	1.82182	1.83111
8	1.67709	1.69444	1.71376	1.72701	1.74979	1.75709	1.76663
9	1.61703	1.63481	1.65462	1.66819	1.69151	1.69899	1.70876
10	1.56231	1.58050	1.60075	1.61463	1.63848	1.64612	1.65610
11	1.51190	1.53049	1.55118	1.56535	1.58969	1.59749	1.60767
12	1.46506	1.48404	1.50514	1.51960	1.54442	1.55238	1.56275
13	1.42124	1.44059	1.46210	1.47683	1.50211	1.51022	1.52078
14	1.38000	1.39970	1.42161	1.43660	1.46234	1.47058	1.48133
15	1.34098	1.36103	1.38333	1.39858	1.42476	1.43314	1.44407
16	1.30390	1.32430	1.34697	1.36248	1.38908	1.39760	1.40871
17	1.26853	1.28927	1.31232	1.32808	1.35510	1.36375	1.37503
18	1.23469	1.25576	1.27917	1.29518	1.32261	1.33139	1.34284
19	1.20220	1.22361	1.24738	1.26362	1.29147	1.30038	1.31199
20	1.17094	1.19267	1.21680	1.23328	1.26152	1.27056	1.28234
21	1.14078	1.16284	1.18731	1.20403	1.23268	1.24184	1.25377
22	1.11163	1.13401	1.15883	1.17578	1.20482	1.21410	1.22620
23	1.08340	1.10609	1.13126	1.14844	1.17787	1.18727	1.19953
24	1.05600	1.07901	1.10452	1.12194	1.15175	1.16127	1.17368
25	1.02937	1.05270	1.07855	1.09619	1.12639	1.13604	1.14860
26	1.00346	1.02710	1.05329	1.07116	1.10173	1.11150	1.12422
27	0.97820	1.00215	1.02868	1.04678	1.07773	1.08762	1.10049
28	0.95356	0.97782	1.00469	1.02301	1.05434	1.06435	1.07737
29	0.92947	0.95405	0.98125	0.99980	1.03151	1.04163	1.05481
30	0.90592	0.93081	0.95835	0.97712	1.00921	1.01945	1.03277
31	0.88286	0.90806	0.93594	0.95493	0.98739	0.99775	1.01123
32	0.86026	0.88577	0.91399	0.93321	0.96604	0.97652	0.99014
33	0.83809	0.86391	0.89247	0.91191	0.94512	0.95571	0.96949
34	0.81633	0.84246	0.87135	0.89102	0.92460	0.93531	0.94924
35	0.79495	0.82140	0.85062	0.87052	0.90447	0.91530	0.92938
36	0.77392	0.80069	0.83025	0.85037	0.88470	0.89564	0.90987
37	0.75324	0.78031	0.81022	0.83056	0.86527	0.87633	0.89071
38	0.73287	0.76026	0.79051	0.81108	0.84616	0.85734	0.87187
39	0.71279	0.74051	0.77110	0.79190	0.82735	0.83865	0.85333
40	0.69301	0.72104	0.75197	0.77300	0.80883	0.82025	0.83508
41	0.67349	0.70185	0.73312	0.75438	0.79059	0.80212	0.81711
42	0.65422	0.68290	0.71453	0.73601	0.77261	0.78426	0.79939
43	0.63520	0.66420	0.69618	0.71789	0.75487	0.76664	0.78193
44	0.61640	0.64574	0.67806	0.70001	0.73737	0.74925	0.76469
45	0.59782	0.62748	0.66016	0.68234	0.72009	0.73209	0.74769
46	0.57944	0.60944	0.64247	0.66489	0.70302	0.71515	0.73089
47	0.56125	0.59159	0.62498	0.64763	0.68615	0.69840	0.71431
48	0.54325	0.57393	0.60768	0.63057	0.66948	0.68185	0.69791
49	0.52543	0.55644	0.59056	0.61369	0.65300	0.66549	0.68170
50	0.50777	0.53913	0.57361	0.59698	0.63668	0.64930	0.66567

EXPECTED VALUES OF NORMAL ORDER STATISTICS

k\n	162	168	175	180	189	192	196
51	0.49026	0.52197	0.55682	0.58044	0.62054	0.63328	0.64981
52	0.47291	0.50497	0.54019	0.56405	0.60456	0.61742	0.63411
53	0.45570	0.48811	0.52371	0.54782	0.58873	0.60171	0.61856
54	0.43862	0.47139	0.50737	0.53172	0.57304	0.58615	0.60316
55	0.42167	0.45480	0.49116	0.51577	0.55750	0.57074	0.58791
56	0.40484	0.43834	0.47508	0.49994	0.54209	0.55545	0.57279
57	0.38812	0.42199	0.45913	0.48424	0.52680	0.54030	0.55780
58	0.37151	0.40576	0.44329	0.46866	0.51164	0.52527	0.54293
59	0.35500	0.38963	0.42756	0.45319	0.49660	0.51036	0.52819
60	0.33859	0.37360	0.41193	0.43783	0.48167	0.49556	0.51355
61	0.32227	0.35767	0.39641	0.42257	0.46684	0.48086	0.49903
62	0.30604	0.34182	0.38098	0.40741	0.45212	0.46627	0.48461
63	0.28988	0.32607	0.36564	0.39234	0.43749	0.45178	0.47029
64	0.27380	0.31039	0.35039	0.37736	0.42296	0.43739	0.45607
65	0.25779	0.29479	0.33521	0.36247	0.40851	0.42308	0.44194
66	0.24185	0.27926	0.32012	0.34765	0.39415	0.40886	0.42790
67	0.22597	0.26380	0.30510	0.33291	0.37988	0.39472	0.41394
68	0.21014	0.24840	0.29014	0.31825	0.36568	0.38066	0.40006
69	0.19437	0.23306	0.27525	0.30365	0.35155	0.36668	0.38626
70	0.17865	0.21778	0.26042	0.28912	0.33749	0.35276	0.37253
71	0.16297	0.20254	0.24565	0.27464	0.32350	0.33892	0.35887
72	0.14733	0.18736	0.23093	0.26023	0.30957	0.32514	0.34528
73	0.13173	0.17221	0.21626	0.24586	0.29570	0.31142	0.33175
74	0.11615	0.15711	0.20164	0.23155	0.28189	0.29776	0.31828
75	0.10061	0.14204	0.18706	0.21729	0.26813	0.28416	0.30487
76	0.08509	0.12700	0.17252	0.20307	0.25442	0.27060	0.29151
77	0.06959	0.11199	0.15802	0.18889	0.24076	0.25710	0.27821
78	0.05411	0.09701	0.14355	0.17475	0.22715	0.24364	0.26495
79	0.03864	0.08205	0.12911	0.16064	0.21357	0.23023	0.25175
80	0.02318	0.06711	0.11470	0.14657	0.20004	0.21686	0.23858
81	0.00773	0.05218	0.10031	0.13252	0.18654	0.20353	0.22546
82		0.03726	0.08594	0.11850	0.17308	0.19023	0.21237
83		0.02235	0.07159	0.10451	0.15965	0.17697	0.19932
84		0.00745	0.05725	0.09053	0.14624	0.16374	0.18631
85			0.04293	0.07657	0.13286	0.15053	0.17332
86			0.02862	0.06263	0.11951	0.13736	0.16037
87			0.01431	0.04870	0.10618	0.12421	0.14744
88			0.00000	0.03478	0.09287	0.11107	0.13454
89				0.02087	0.07957	0.09796	0.12166
90				0.00695	0.06629	0.08487	0.10880
91					0.05301	0.07179	0.09596
92					0.03975	0.05872	0.08313
93					0.02650	0.04566	0.07032
94					0.01325	0.03261	0.05752
95					0.00000	0.01956	0.04473
96						0.00652	0.03194
97							0.01916
98							0.00639

EXPECTED VALUES OF NORMAL ORDER STATISTICS

k\n	200	210	216	224	225	240	243
1	2.74604	2.76217	2.77145	2.78339	2.78485	2.80592	2.80997
2	2.41365	2.43147	2.44171	2.45488	2.45649	2.47971	2.48416
3	2.22999	2.24892	2.25979	2.27376	2.27547	2.30009	2.30480
4	2.09991	2.11970	2.13107	2.14567	2.14746	2.17317	2.17810
5	1.99783	2.01836	2.03015	2.04529	2.04713	2.07378	2.07888
6	1.91308	1.93427	1.94643	1.96204	1.96395	1.99141	1.99667
7	1.84019	1.86198	1.87447	1.89052	1.89247	1.92068	1.92608
8	1.77594	1.79828	1.81109	1.82753	1.82953	1.85843	1.86395
9	1.71828	1.74114	1.75424	1.77105	1.77310	1.80264	1.80829
10	1.66583	1.68918	1.70256	1.71972	1.72182	1.75196	1.75772
11	1.61760	1.64142	1.65506	1.67256	1.67470	1.70542	1.71129
12	1.57287	1.59714	1.61104	1.62886	1.63103	1.66231	1.66828
13	1.53109	1.55579	1.56993	1.58806	1.59027	1.62208	1.62816
14	1.49182	1.51694	1.53132	1.54975	1.55200	1.58433	1.59050
15	1.45472	1.48025	1.49487	1.51359	1.51588	1.54870	1.55497
16	1.41953	1.44546	1.46030	1.47931	1.48163	1.51495	1.52130
17	1.38602	1.41234	1.42740	1.44669	1.44904	1.48283	1.48928
18	1.35399	1.38070	1.39597	1.41553	1.41792	1.45218	1.45872
19	1.32330	1.35038	1.36587	1.38570	1.38812	1.42284	1.42946
20	1.29381	1.32126	1.33696	1.35705	1.35950	1.39467	1.40137
21	1.26540	1.29322	1.30912	1.32948	1.33195	1.36757	1.37436
22	1.23798	1.26616	1.28227	1.30288	1.30539	1.34144	1.34831
23	1.21146	1.24000	1.25631	1.27717	1.27971	1.31619	1.32314
24	1.18577	1.21466	1.23117	1.25228	1.25485	1.29176	1.29879
25	1.16084	1.19008	1.20678	1.22814	1.23074	1.26807	1.27518
26	1.13661	1.16620	1.18310	1.20470	1.20733	1.24508	1.25227
27	1.11303	1.14297	1.16006	1.18191	1.18457	1.22273	1.22999
28	1.09005	1.12034	1.13762	1.15971	1.16240	1.20097	1.20831
29	1.06763	1.09827	1.11574	1.13807	1.14079	1.17977	1.18719
30	1.04574	1.07672	1.09438	1.11695	1.11970	1.15908	1.16658
31	1.02434	1.05566	1.07351	1.09632	1.09909	1.13888	1.14645
32	1.00340	1.03506	1.05310	1.07614	1.07895	1.11914	1.12678
33	0.98290	1.01488	1.03311	1.05640	1.05923	1.09982	1.10754
34	0.96279	0.99512	1.01354	1.03706	1.03992	1.08090	1.08870
35	0.94307	0.97573	0.99434	1.01809	1.02098	1.06237	1.07023
36	0.92371	0.95671	0.97550	0.99949	1.00241	1.04419	1.05213
37	0.90469	0.93803	0.95701	0.98123	0.98418	1.02635	1.03436
38	0.88599	0.91967	0.93883	0.96329	0.96626	1.00883	1.01692
39	0.86760	0.90161	0.92096	0.94565	0.94866	0.99162	0.99978
40	0.84950	0.88384	0.90338	0.92831	0.93134	0.97469	0.98293
41	0.83167	0.86635	0.88608	0.91124	0.91429	0.95804	0.96635
42	0.81410	0.84912	0.86904	0.89443	0.89751	0.94166	0.95003
43	0.79678	0.83214	0.85224	0.87787	0.88098	0.92552	0.93397
44	0.77969	0.81539	0.83568	0.86155	0.86469	0.90962	0.91814
45	0.76283	0.79887	0.81935	0.84545	0.84862	0.89394	0.90254
46	0.74619	0.78257	0.80324	0.82957	0.83277	0.87848	0.88715
47	0.72975	0.76647	0.78733	0.81390	0.81712	0.86323	0.87198
48	0.71350	0.75057	0.77162	0.79842	0.80168	0.84818	0.85700
49	0.69744	0.73486	0.75609	0.78314	0.78642	0.83332	0.84221
50	0.68156	0.71932	0.74075	0.76803	0.77134	0.81864	0.82760

EXPECTED VALUES OF NORMAL ORDER STATISTICS

k\n	200	210	216	224	225	240	243
51	0.66585	0.70396	0.72558	0.75310	0.75644	0.80413	0.81317
52	0.65030	0.68876	0.71058	0.73833	0.74170	0.78980	0.79890
53	0.63490	0.67372	0.69573	0.72373	0.72712	0.77562	0.78480
54	0.61966	0.65883	0.68103	0.70927	0.71270	0.76160	0.77085
55	0.60456	0.64408	0.66648	0.69497	0.69842	0.74772	0.75705
56	0.58959	0.62948	0.65207	0.68080	0.68428	0.73399	0.74339
57	0.57476	0.61501	0.63780	0.66677	0.67028	0.72039	0.72987
58	0.56005	0.60066	0.62365	0.65287	0.65641	0.70693	0.71648
59	0.54546	0.58644	0.60963	0.63909	0.64267	0.69359	0.70322
60	0.53099	0.57233	0.59573	0.62544	0.62904	0.68038	0.69008
61	0.51663	0.55834	0.58194	0.61190	0.61553	0.66728	0.67706
62	0.50237	0.54446	0.56826	0.59847	0.60213	0.65430	0.66416
63	0.48822	0.53068	0.55468	0.58515	0.58884	0.64143	0.65136
64	0.47416	0.51700	0.54121	0.57193	0.57566	0.62866	0.63867
65	0.46020	0.50342	0.52784	0.55882	0.56257	0.61599	0.62608
66	0.44632	0.48993	0.51456	0.54579	0.54958	0.60343	0.61359
67	0.43253	0.47652	0.50136	0.53286	0.53668	0.59095	0.60119
68	0.41882	0.46321	0.48826	0.52002	0.52386	0.57857	0.58889
69	0.40519	0.44997	0.47524	0.50726	0.51114	0.56628	0.57667
70	0.39164	0.43682	0.46230	0.49459	0.49850	0.55407	0.56454
71	0.37816	0.42373	0.44944	0.48199	0.48593	0.54194	0.55250
72	0.36474	0.41072	0.43665	0.46947	0.47344	0.52989	0.54053
73	0.35139	0.39778	0.42393	0.45702	0.46103	0.51792	0.52864
74	0.33811	0.38491	0.41128	0.44465	0.44869	0.50603	0.51682
75	0.32488	0.37210	0.39869	0.43234	0.43641	0.49420	0.50508
76	0.31171	0.35935	0.38617	0.42010	0.42420	0.48244	0.49340
77	0.29859	0.34666	0.37371	0.40792	0.41205	0.47075	0.48180
78	0.28553	0.33402	0.36131	0.39580	0.39997	0.45912	0.47025
79	0.27251	0.32144	0.34896	0.38373	0.38794	0.44756	0.45877
80	0.25954	0.30891	0.33666	0.37173	0.37596	0.43605	0.44735
81	0.24661	0.29643	0.32442	0.35977	0.36404	0.42461	0.43599
82	0.23373	0.28399	0.31222	0.34787	0.35218	0.41321	0.42468
83	0.22088	0.27160	0.30007	0.33602	0.34036	0.40188	0.41343
84	0.20807	0.25924	0.28797	0.32421	0.32859	0.39059	0.40223
85	0.19529	0.24693	0.27590	0.31245	0.31686	0.37935	0.39108
86	0.18254	0.23466	0.26388	0.30073	0.30518	0.36816	0.37998
87	0.16983	0.22242	0.25189	0.28906	0.29354	0.35701	0.36892
88	0.15714	0.21021	0.23994	0.27742	0.28194	0.34591	0.35791
89	0.14448	0.19803	0.22803	0.26582	0.27038	0.33486	0.34694
90	0.13184	0.18589	0.21615	0.25425	0.25885	0.32384	0.33602
91	0.11922	0.17377	0.20429	0.24272	0.24736	0.31286	0.32513
92	0.10662	0.16168	0.19247	0.23123	0.23590	0.30192	0.31429
93	0.09404	0.14961	0.18067	0.21976	0.22447	0.29102	0.30348
94	0.08147	0.13756	0.16890	0.20832	0.21307	0.28015	0.29270
95	0.06891	0.12553	0.15715	0.19691	0.20170	0.26931	0.28196
96	0.05637	0.11352	0.14543	0.18552	0.19035	0.25851	0.27125
97	0.04383	0.10153	0.13372	0.17416	0.17903	0.24773	0.26058
98	0.03130	0.08955	0.12203	0.16282	0.16773	0.23699	0.24993
99	0.01878	0.07758	0.11036	0.15150	0.15645	0.22627	0.23931
100	0.00626	0.06563	0.09870	0.14020	0.14520	0.21557	0.22872

EXPECTED VALUES OF NORMAL ORDER STATISTICS

k\n	200	210	216	224	225	240	243
101		0.05368	0.08706	0.12892	0.13396	0.20491	0.21815
102		0.04175	0.07543	0.11766	0.12274	0.19426	0.20761
103		0.02981	0.06381	0.10641	0.11153	0.18364	0.19709
104		0.01789	0.05219	0.09517	0.10034	0.17304	0.18659
105		0.00596	0.04059	0.08394	0.08916	0.16245	0.17611
106			0.02899	0.07273	0.07799	0.15189	0.16565
107			0.01739	0.06153	0.06683	0.14134	0.15521
108			0.00580	0.05033	0.05568	0.13081	0.14479
109				0.03914	0.04453	0.12029	0.13438
110				0.02795	0.03340	0.10979	0.12399
111				0.01677	0.02226	0.09929	0.11361
112				0.00559	0.01113	0.08881	0.10325
113					0.00000	0.07834	0.09289
114						0.06788	0.08254
115						0.05742	0.07221
116						0.04697	0.06188
117						0.03653	0.05155
118						0.02609	0.04124
119						0.01565	0.03092
120						0.00522	0.02061
121							0.01031
122							0.00000

EXPECTED VALUES OF NORMAL ORDER STATISTICS

k \ n	245	250	252	256	270	280	288
1	2.81263	2.81918	2.82177	2.82686	2.84404	2.85572	2.86474
2	2.48709	2.49431	2.49715	2.50276	2.52164	2.53447	2.54437
3	2.30791	2.31555	2.31856	2.32450	2.34449	2.35807	2.36854
4	2.18134	2.18932	2.19246	2.19866	2.21951	2.23367	2.24458
5	2.08224	2.09050	2.09375	2.10017	2.12175	2.13639	2.14769
6	2.00013	2.00864	2.01199	2.01860	2.04082	2.05590	2.06752
7	1.92963	1.93837	1.94181	1.94859	1.97139	1.98686	1.99877
8	1.86759	1.87654	1.88006	1.88700	1.91034	1.92616	1.93835
9	1.81201	1.82115	1.82474	1.83184	1.85568	1.87183	1.88428
10	1.76152	1.77084	1.77451	1.78174	1.80605	1.82252	1.83521
11	1.71516	1.72466	1.72839	1.73576	1.76052	1.77729	1.79020
12	1.67222	1.68189	1.68569	1.69319	1.71837	1.73542	1.74855
13	1.63216	1.64199	1.64585	1.65348	1.67907	1.69640	1.70974
14	1.59456	1.60455	1.60848	1.61622	1.64221	1.65980	1.67334
15	1.55910	1.56923	1.57322	1.58108	1.60745	1.62530	1.63903
16	1.52549	1.53577	1.53982	1.54779	1.57454	1.59263	1.60656
17	1.49352	1.50395	1.50805	1.51613	1.54325	1.56159	1.57569
18	1.46302	1.47359	1.47774	1.48593	1.51340	1.53197	1.54626
19	1.43382	1.44452	1.44873	1.45702	1.48484	1.50365	1.51811
20	1.40579	1.41663	1.42089	1.42929	1.45745	1.47648	1.49112
21	1.37882	1.38980	1.39411	1.40261	1.43111	1.45037	1.46517
22	1.35283	1.36393	1.36830	1.37690	1.40573	1.42520	1.44017
23	1.32772	1.33895	1.34337	1.35206	1.38122	1.40091	1.41605
24	1.30342	1.31478	1.31924	1.32804	1.35751	1.37742	1.39272
25	1.27986	1.29135	1.29586	1.30476	1.33455	1.35467	1.37012
26	1.25700	1.26861	1.27317	1.28216	1.31227	1.33259	1.34821
27	1.23477	1.24651	1.25112	1.26020	1.29062	1.31115	1.32692
28	1.21314	1.22500	1.22966	1.23884	1.26956	1.29030	1.30622
29	1.19207	1.20405	1.20875	1.21802	1.24905	1.26999	1.28606
30	1.17151	1.18361	1.18836	1.19772	1.22906	1.25019	1.26642
31	1.15143	1.16365	1.16845	1.17790	1.20954	1.23087	1.24725
32	1.13181	1.14415	1.14899	1.15854	1.19047	1.21200	1.22853
33	1.11262	1.12507	1.12996	1.13960	1.17183	1.19356	1.21023
34	1.09382	1.10640	1.11133	1.12106	1.15359	1.17551	1.19233
35	1.07541	1.08810	1.09308	1.10290	1.13572	1.15783	1.17480
36	1.05735	1.07016	1.07519	1.08509	1.11821	1.14051	1.15763
37	1.03963	1.05256	1.05763	1.06762	1.10103	1.12353	1.14079
38	1.02224	1.03528	1.04040	1.05048	1.08418	1.10687	1.12427
39	1.00514	1.01830	1.02347	1.03363	1.06762	1.09050	1.10805
40	0.98834	1.00161	1.00682	1.01708	1.05136	1.07443	1.09211
41	0.97181	0.98520	0.99045	1.00080	1.03536	1.05862	1.07645
42	0.95554	0.96905	0.97435	0.98478	1.01963	1.04308	1.06105
43	0.93952	0.95314	0.95849	0.96901	1.00415	1.02779	1.04590
44	0.92374	0.93748	0.94287	0.95348	0.98890	1.01273	1.03098
45	0.90819	0.92204	0.92748	0.93817	0.97389	0.99790	1.01629
46	0.89285	0.90682	0.91230	0.92308	0.95908	0.98328	1.00182
47	0.87772	0.89180	0.89733	0.90820	0.94449	0.96887	0.98755
48	0.86279	0.87699	0.88256	0.89352	0.93009	0.95466	0.97348
49	0.84805	0.86236	0.86798	0.87902	0.91589	0.94065	0.95960
50	0.83349	0.84792	0.85358	0.86471	0.90186	0.92681	0.94591

EXPECTED VALUES OF NORMAL ORDER STATISTICS

k \ n	245	250	252	256	270	280	288
51	0.81910	0.83365	0.83936	0.85058	0.88801	0.91315	0.93239
52	0.80489	0.81955	0.82530	0.83661	0.87433	0.89966	0.91903
53	0.79083	0.80561	0.81141	0.82281	0.86082	0.88632	0.90584
54	0.77693	0.79183	0.79767	0.80915	0.84745	0.87315	0.89281
55	0.76318	0.77819	0.78408	0.79565	0.83424	0.86013	0.87992
56	0.74957	0.76470	0.77063	0.78230	0.82117	0.84724	0.86718
57	0.73610	0.75135	0.75732	0.76908	0.80824	0.83450	0.85458
58	0.72276	0.73812	0.74415	0.75599	0.79545	0.82190	0.84212
59	0.70954	0.72503	0.73110	0.74303	0.78278	0.80942	0.82978
60	0.69645	0.71206	0.71817	0.73020	0.77024	0.79706	0.81757
61	0.68348	0.69921	0.70537	0.71748	0.75781	0.78483	0.80548
62	0.67063	0.68647	0.69268	0.70488	0.74551	0.77272	0.79350
63	0.65788	0.67384	0.68010	0.69239	0.73331	0.76071	0.78164
64	0.64524	0.66132	0.66763	0.68001	0.72123	0.74882	0.76989
65	0.63270	0.64891	0.65525	0.66773	0.70924	0.73703	0.75824
66	0.62026	0.63659	0.64298	0.65555	0.69736	0.72534	0.74670
67	0.60791	0.62437	0.63081	0.64347	0.68558	0.71375	0.73525
68	0.59566	0.61224	0.61873	0.63148	0.67389	0.70225	0.72390
69	0.58350	0.60020	0.60674	0.61958	0.66229	0.69085	0.71264
70	0.57142	0.58824	0.59483	0.60777	0.65079	0.67954	0.70148
71	0.55943	0.57637	0.58301	0.59604	0.63936	0.66831	0.69039
72	0.54751	0.56458	0.57127	0.58440	0.62802	0.65717	0.67940
73	0.53567	0.55287	0.55961	0.57283	0.61676	0.64610	0.66848
74	0.52391	0.54124	0.54802	0.56134	0.60558	0.63512	0.65764
75	0.51222	0.52967	0.53651	0.54992	0.59447	0.62421	0.64688
76	0.50060	0.51818	0.52506	0.53858	0.58344	0.61338	0.63620
77	0.48904	0.50676	0.51369	0.52730	0.57247	0.60262	0.62559
78	0.47755	0.49540	0.50238	0.51609	0.56158	0.59192	0.61504
79	0.46613	0.48410	0.49114	0.50494	0.55075	0.58130	0.60457
80	0.45476	0.47287	0.47995	0.49386	0.53999	0.57074	0.59416
81	0.44345	0.46169	0.46883	0.48284	0.52928	0.56024	0.58381
82	0.43220	0.45058	0.45777	0.47187	0.51864	0.54980	0.57353
83	0.42101	0.43952	0.44676	0.46096	0.50806	0.53943	0.56331
84	0.40986	0.42851	0.43580	0.45011	0.49753	0.52911	0.55314
85	0.39877	0.41755	0.42490	0.43931	0.48706	0.51885	0.54304
86	0.38772	0.40665	0.41404	0.42856	0.47664	0.50864	0.53298
87	0.37673	0.39579	0.40324	0.41786	0.46627	0.49848	0.52299
88	0.36577	0.38498	0.39248	0.40721	0.45595	0.48838	0.51304
89	0.35487	0.37421	0.38177	0.39660	0.44569	0.47832	0.50314
90	0.34400	0.36349	0.37110	0.38604	0.43546	0.46832	0.49330
91	0.33317	0.35280	0.36048	0.37552	0.42529	0.45836	0.48350
92	0.32239	0.34216	0.34989	0.36504	0.41515	0.44844	0.47374
93	0.31164	0.33156	0.33934	0.35460	0.40506	0.43857	0.46404
94	0.30092	0.32099	0.32883	0.34420	0.39501	0.42875	0.45437
95	0.29025	0.31046	0.31836	0.33384	0.38500	0.41896	0.44475
96	0.27960	0.29997	0.30792	0.32352	0.37503	0.40921	0.43517
97	0.26899	0.28951	0.29752	0.31322	0.36510	0.39951	0.42563
98	0.25840	0.27907	0.28714	0.30296	0.35520	0.38984	0.41612
99	0.24785	0.26867	0.27680	0.29274	0.34533	0.38020	0.40666
100	0.23732	0.25830	0.26649	0.28254	0.33550	0.37060	0.39723

EXPECTED VALUES OF NORMAL ORDER STATISTICS

k \ n	245	250	252	256	270	280	288
101	0.22682	0.24796	0.25621	0.27237	0.32571	0.36104	0.38783
102	0.21634	0.23764	0.24595	0.26223	0.31594	0.35151	0.37847
103	0.20589	0.22735	0.23572	0.25212	0.30620	0.34201	0.36915
104	0.19546	0.21708	0.22551	0.24203	0.29650	0.33254	0.35985
105	0.18505	0.20683	0.21533	0.23197	0.28682	0.32310	0.35059
106	0.17466	0.19661	0.20517	0.22193	0.27716	0.31369	0.34135
107	0.16429	0.18641	0.19503	0.21192	0.26754	0.30430	0.33215
108	0.15394	0.17622	0.18491	0.20192	0.25793	0.29495	0.32297
109	0.14360	0.16606	0.17481	0.19195	0.24836	0.28562	0.31382
110	0.13328	0.15591	0.16473	0.18199	0.23880	0.27631	0.30470
111	0.12297	0.14577	0.15466	0.17205	0.22927	0.26703	0.29560
112	0.11268	0.13566	0.14461	0.16213	0.21975	0.25777	0.28653
113	0.10240	0.12555	0.13457	0.15223	0.21026	0.24853	0.27748
114	0.09213	0.11546	0.12455	0.14234	0.20078	0.23932	0.26845
115	0.08187	0.10538	0.11454	0.13246	0.19133	0.23012	0.25945
116	0.07162	0.09531	0.10454	0.12260	0.18189	0.22095	0.25046
117	0.06137	0.08526	0.09456	0.11275	0.17246	0.21179	0.24150
118	0.05113	0.07520	0.08458	0.10291	0.16306	0.20265	0.23255
119	0.04090	0.06516	0.07461	0.09308	0.15366	0.19353	0.22363
120	0.03067	0.05513	0.06464	0.08325	0.14428	0.18442	0.21472
121	0.02045	0.04510	0.05469	0.07344	0.13492	0.17533	0.20583
122	0.01022	0.03507	0.04474	0.06363	0.12556	0.16625	0.19695
123	0.00000	0.02505	0.03479	0.05383	0.11622	0.15719	0.18809
124		0.01503	0.02485	0.04404	0.10688	0.14814	0.17925
125		0.00501	0.01491	0.03425	0.09756	0.13910	0.17041
126			0.00497	0.02446	0.08824	0.13007	0.16160
127				0.01468	0.07893	0.12106	0.15279
128				0.00489	0.06963	0.11205	0.14400
129					0.06033	0.10305	0.13522
130					0.05104	0.09406	0.12644
131					0.04176	0.08508	0.11768
132					0.03247	0.07611	0.10893
133					0.02319	0.06714	0.10018
134					0.01392	0.05818	0.09145
135					0.00464	0.04922	0.08272
136						0.04026	0.07399
137						0.03131	0.06527
138						0.02237	0.05656
139						0.01342	0.04785
140						0.00447	0.03915
141							0.03044
142							0.02174
143							0.01305
144							0.00435

EXPECTED VALUES OF NORMAL ORDER STATISTICS

k \ n	294	300	315	320	324	336	343
1	2.87133	2.87777	2.89327	2.89826	2.90219	2.91367	2.92016
2	2.55160	2.55867	2.57567	2.58114	2.58544	2.59802	2.60512
3	2.37618	2.38365	2.40162	2.40739	2.41194	2.42521	2.43271
4	2.25255	2.26033	2.27904	2.28506	2.28979	2.30361	2.31141
5	2.15592	2.16397	2.18331	2.18952	2.19441	2.20869	2.21675
6	2.07599	2.08427	2.10416	2.11055	2.11558	2.13025	2.13854
7	2.00747	2.01595	2.03634	2.04289	2.04804	2.06307	2.07156
8	1.94724	1.95592	1.97676	1.98345	1.98872	2.00408	2.01276
9	1.89335	1.90220	1.92347	1.93030	1.93567	1.95134	1.96018
10	1.84445	1.85348	1.87515	1.88210	1.88757	1.90353	1.91253
11	1.79961	1.80879	1.83084	1.83792	1.84348	1.85971	1.86887
12	1.75812	1.76746	1.78986	1.79705	1.80271	1.81920	1.82850
13	1.71945	1.72894	1.75169	1.75899	1.76473	1.78147	1.79091
14	1.68320	1.69283	1.71591	1.72332	1.72915	1.74612	1.75570
15	1.64904	1.65880	1.68221	1.68972	1.69562	1.71283	1.72254
16	1.61670	1.62659	1.65031	1.65792	1.66391	1.68134	1.69117
17	1.58597	1.59599	1.62002	1.62772	1.63378	1.65143	1.66138
18	1.55666	1.56681	1.59114	1.59893	1.60507	1.62293	1.63300
19	1.52864	1.53891	1.56353	1.57142	1.57762	1.59570	1.60588
20	1.50177	1.51216	1.53706	1.54504	1.55132	1.56959	1.57989
21	1.47595	1.48645	1.51163	1.51970	1.52605	1.54452	1.55493
22	1.45107	1.46169	1.48715	1.49530	1.50171	1.52038	1.53090
23	1.42706	1.43780	1.46352	1.47176	1.47824	1.49710	1.50773
24	1.40385	1.41470	1.44068	1.44901	1.45555	1.47460	1.48533
25	1.38137	1.39233	1.41858	1.42698	1.43359	1.45283	1.46366
26	1.35956	1.37063	1.39714	1.40563	1.41230	1.43172	1.44266
27	1.33839	1.34957	1.37633	1.38490	1.39163	1.41124	1.42227
28	1.31780	1.32908	1.35610	1.36475	1.37154	1.39132	1.40246
29	1.29775	1.30914	1.33641	1.34513	1.35199	1.37195	1.38319
30	1.27822	1.28971	1.31722	1.32603	1.33295	1.35308	1.36441
31	1.25916	1.27076	1.29851	1.30739	1.31437	1.33468	1.34611
32	1.24054	1.25225	1.28024	1.28920	1.29624	1.31672	1.32824
33	1.22235	1.23415	1.26239	1.27143	1.27853	1.29917	1.31079
34	1.20455	1.21646	1.24494	1.25405	1.26120	1.28202	1.29373
35	1.18713	1.19914	1.22785	1.23704	1.24425	1.26524	1.27704
36	1.17006	1.18217	1.21112	1.22038	1.22765	1.24880	1.26070
37	1.15332	1.16553	1.19472	1.20405	1.21138	1.23270	1.24469
38	1.13691	1.14921	1.17863	1.18804	1.19543	1.21691	1.22899
39	1.12079	1.13320	1.16285	1.17233	1.17977	1.20142	1.21359
40	1.10496	1.11746	1.14735	1.15690	1.16440	1.18621	1.19847
41	1.08940	1.10200	1.13212	1.14174	1.14930	1.17127	1.18362
42	1.07410	1.08680	1.11714	1.12684	1.13445	1.15658	1.16903
43	1.05905	1.07185	1.10242	1.11219	1.11986	1.14215	1.15468
44	1.04423	1.05713	1.08793	1.09777	1.10549	1.12795	1.14056
45	1.02964	1.04264	1.07366	1.08357	1.09136	1.11397	1.12667
46	1.01527	1.02836	1.05961	1.06960	1.07743	1.10020	1.11300
47	1.00110	1.01429	1.04577	1.05582	1.06372	1.08664	1.09953
48	0.98714	1.00042	1.03213	1.04225	1.05020	1.07328	1.08625
49	0.97336	0.98674	1.01867	1.02886	1.03687	1.06011	1.07317
50	0.95976	0.97324	1.00540	1.01566	1.02372	1.04712	1.06026

EXPECTED VALUES OF NORMAL ORDER STATISTICS

k \ n	294	300	315	320	324	336	343
51	0.94634	0.95991	0.99230	1.00263	1.01074	1.03430	1.04753
52	0.93309	0.94676	0.97937	0.98977	0.99794	1.02165	1.03497
53	0.92000	0.93376	0.96660	0.97707	0.98530	1.00917	1.02257
54	0.90706	0.92093	0.95398	0.96453	0.97281	0.99683	1.01032
55	0.89428	0.90824	0.94152	0.95214	0.96047	0.98465	0.99823
56	0.88164	0.89570	0.92920	0.93989	0.94828	0.97261	0.98628
57	0.86914	0.88329	0.91702	0.92778	0.93622	0.96072	0.97446
58	0.85678	0.87102	0.90498	0.91581	0.92430	0.94895	0.96279
59	0.84454	0.85888	0.89306	0.90396	0.91251	0.93732	0.95124
60	0.83243	0.84687	0.88127	0.89224	0.90085	0.92581	0.93982
61	0.82044	0.83498	0.86961	0.88064	0.88930	0.91442	0.92852
62	0.80857	0.82320	0.85806	0.86916	0.87788	0.90316	0.91733
63	0.79681	0.81154	0.84662	0.85780	0.86657	0.89200	0.90626
64	0.78516	0.79998	0.83529	0.84654	0.85537	0.88095	0.89530
65	0.77361	0.78854	0.82407	0.83539	0.84427	0.87001	0.88445
66	0.76217	0.77719	0.81295	0.82434	0.83328	0.85918	0.87370
67	0.75082	0.76595	0.80193	0.81340	0.82238	0.84844	0.86305
68	0.73958	0.75480	0.79101	0.80254	0.81159	0.83780	0.85250
69	0.72842	0.74374	0.78018	0.79179	0.80089	0.82726	0.84204
70	0.71736	0.73277	0.76945	0.78112	0.79027	0.81680	0.83167
71	0.70638	0.72189	0.75880	0.77054	0.77975	0.80644	0.82139
72	0.69548	0.71110	0.74823	0.76005	0.76931	0.79616	0.81120
73	0.68467	0.70039	0.73775	0.74964	0.75896	0.78596	0.80109
74	0.67394	0.68976	0.72735	0.73931	0.74869	0.77585	0.79106
75	0.66329	0.67920	0.71703	0.72906	0.73849	0.76581	0.78111
76	0.65271	0.66872	0.70678	0.71889	0.72837	0.75585	0.77124
77	0.64220	0.65831	0.69661	0.70879	0.71833	0.74597	0.76145
78	0.63176	0.64798	0.68651	0.69876	0.70836	0.73616	0.75172
79	0.62139	0.63771	0.67647	0.68880	0.69845	0.72641	0.74207
80	0.61109	0.62751	0.66651	0.67891	0.68862	0.71674	0.73248
81	0.60085	0.61738	0.65661	0.66908	0.67885	0.70713	0.72296
82	0.59068	0.60730	0.64678	0.65932	0.66915	0.69759	0.71351
83	0.58056	0.59729	0.63701	0.64962	0.65951	0.68811	0.70412
84	0.57051	0.58734	0.62729	0.63998	0.64993	0.67870	0.69479
85	0.56051	0.57745	0.61764	0.63041	0.64040	0.66934	0.68552
86	0.55057	0.56761	0.60804	0.62088	0.63094	0.66004	0.67631
87	0.54068	0.55783	0.59851	0.61142	0.62153	0.65080	0.66716
88	0.53085	0.54810	0.58902	0.60201	0.61218	0.64161	0.65806
89	0.52106	0.53842	0.57959	0.59265	0.60288	0.63248	0.64902
90	0.51133	0.52879	0.57020	0.58335	0.59364	0.62339	0.64003
91	0.50164	0.51922	0.56087	0.57409	0.58444	0.61436	0.63109
92	0.49200	0.50968	0.55159	0.56488	0.57529	0.60538	0.62220
93	0.48241	0.50020	0.54235	0.55572	0.56619	0.59645	0.61336
94	0.47286	0.49076	0.53316	0.54661	0.55714	0.58757	0.60457
95	0.46335	0.48136	0.52402	0.53754	0.54813	0.57873	0.59583
96	0.45389	0.47201	0.51492	0.52852	0.53917	0.56994	0.58713
97	0.44446	0.46269	0.50586	0.51954	0.53025	0.56119	0.57847
98	0.43508	0.45342	0.49684	0.51060	0.52137	0.55248	0.56986
99	0.42573	0.44419	0.48786	0.50170	0.51253	0.54382	0.56129
100	0.41642	0.43499	0.47893	0.49284	0.50374	0.53519	0.55276

EXPECTED VALUES OF NORMAL ORDER STATISTICS

k \ n	294	300	315	320	324	336	343
101	0.40715	0.42583	0.47003	0.48402	0.49498	0.52661	0.54427
102	0.39791	0.41670	0.46116	0.47524	0.48626	0.51806	0.53582
103	0.38870	0.40761	0.45234	0.46649	0.47757	0.50956	0.52740
104	0.37953	0.39856	0.44354	0.45778	0.46892	0.50109	0.51903
105	0.37039	0.38953	0.43479	0.44911	0.46031	0.49265	0.51069
106	0.36128	0.38054	0.42606	0.44047	0.45173	0.48425	0.50239
107	0.35220	0.37158	0.41737	0.43186	0.44319	0.47588	0.49412
108	0.34315	0.36265	0.40871	0.42328	0.43467	0.46755	0.48588
109	0.33413	0.35375	0.40008	0.41473	0.42619	0.45925	0.47768
110	0.32513	0.34487	0.39148	0.40622	0.41774	0.45098	0.46951
111	0.31616	0.33602	0.38291	0.39773	0.40932	0.44274	0.46137
112	0.30722	0.32720	0.37436	0.38927	0.40093	0.43453	0.45326
113	0.29830	0.31841	0.36585	0.38084	0.39256	0.42635	0.44518
114	0.28940	0.30963	0.35736	0.37244	0.38422	0.41820	0.43713
115	0.28053	0.30089	0.34889	0.36406	0.37591	0.41008	0.42911
116	0.27168	0.29216	0.34046	0.35571	0.36763	0.40198	0.42112
117	0.26285	0.28346	0.33204	0.34738	0.35937	0.39391	0.41315
118	0.25404	0.27478	0.32365	0.33908	0.35113	0.38587	0.40521
119	0.24525	0.26612	0.31528	0.33080	0.34292	0.37785	0.39729
120	0.23648	0.25748	0.30693	0.32254	0.33473	0.36985	0.38940
121	0.22773	0.24885	0.29861	0.31431	0.32657	0.36188	0.38153
122	0.21899	0.24025	0.29030	0.30609	0.31842	0.35393	0.37369
123	0.21027	0.23167	0.28202	0.29790	0.31030	0.34601	0.36587
124	0.20157	0.22310	0.27375	0.28972	0.30220	0.33810	0.35807
125	0.19288	0.21455	0.26551	0.28157	0.29411	0.33022	0.35030
126	0.18421	0.20601	0.25728	0.27343	0.28605	0.32235	0.34254
127	0.17555	0.19749	0.24907	0.26532	0.27800	0.31451	0.33481
128	0.16691	0.18898	0.24087	0.25722	0.26998	0.30669	0.32709
129	0.15827	0.18049	0.23270	0.24913	0.26197	0.29888	0.31940
130	0.14965	0.17201	0.22453	0.24107	0.25397	0.29109	0.31172
131	0.14104	0.16354	0.21638	0.23302	0.24600	0.28333	0.30406
132	0.13244	0.15508	0.20825	0.22498	0.23804	0.27557	0.29642
133	0.12385	0.14664	0.20013	0.21696	0.23009	0.26784	0.28880
134	0.11527	0.13820	0.19202	0.20895	0.22216	0.26012	0.28119
135	0.10670	0.12978	0.18393	0.20096	0.21424	0.25241	0.27360
136	0.09813	0.12136	0.17585	0.19298	0.20634	0.24472	0.26603
137	0.08958	0.11296	0.16778	0.18501	0.19845	0.23705	0.25847
138	0.08103	0.10456	0.15972	0.17705	0.19057	0.22939	0.25092
139	0.07248	0.09617	0.15167	0.16910	0.18270	0.22174	0.24339
140	0.06394	0.08778	0.14363	0.16117	0.17484	0.21410	0.23588
141	0.05541	0.07940	0.13560	0.15324	0.16700	0.20648	0.22837
142	0.04688	0.07103	0.12758	0.14533	0.15916	0.19887	0.22088
143	0.03835	0.06266	0.11957	0.13742	0.15134	0.19127	0.21341
144	0.02982	0.05430	0.11156	0.12952	0.14352	0.18369	0.20594
145	0.02130	0.04594	0.10356	0.12163	0.13571	0.17611	0.19849
146	0.01278	0.03758	0.09557	0.11375	0.12792	0.16854	0.19104
147	0.00426	0.02923	0.08759	0.10587	0.12012	0.16098	0.18361
148		0.02088	0.07961	0.09801	0.11234	0.15344	0.17619
149		0.01252	0.07163	0.09014	0.10456	0.14590	0.16878
150		0.00417	0.06366	0.08229	0.09679	0.13837	0.16137

EXPECTED VALUES OF NORMAL ORDER STATISTICS

k \ n	294	300	315	320	324	336	343
151			0.05569	0.07443	0.08903	0.13084	0.15398
152			0.04773	0.06659	0.08127	0.12333	0.14659
153			0.03977	0.05874	0.07351	0.11582	0.13921
154			0.03181	0.05090	0.06576	0.10832	0.13184
155			0.02386	0.04307	0.05802	0.10082	0.12448
156			0.01590	0.03523	0.05028	0.09333	0.11712
157			0.00795	0.02740	0.04254	0.08584	0.10977
158			0.00000	0.01957	0.03480	0.07836	0.10243
159				0.01174	0.02706	0.07089	0.09509
160				0.00391	0.01933	0.06341	0.08775
161					0.01160	0.05594	0.08042
162					0.00387	0.04848	0.07310
163						0.04102	0.06578
164						0.03356	0.05846
165						0.02610	0.05115
166						0.01864	0.04383
167						0.01118	0.03652
168						0.00373	0.02922
169							0.02191
170							0.01461
171							0.00730
172							0.00000

EXPECTED VALUES OF NORMAL ORDER STATISTICS

k \ n	350	360	375	378	384	392	400
1	2.92651	2.93534	2.94810	2.95059	2.95549	2.96191	2.96818
2	2.61207	2.62173	2.63568	2.63840	2.64376	2.65077	2.65761
3	2.44004	2.45024	2.46495	2.46781	2.47346	2.48084	2.48806
4	2.31904	2.32964	2.34494	2.34792	2.35379	2.36147	2.36897
5	2.22462	2.23557	2.25136	2.25443	2.26049	2.26841	2.27615
6	2.14663	2.15788	2.17410	2.17725	2.18348	2.19161	2.19955
7	2.07985	2.09137	2.10797	2.11120	2.11757	2.12589	2.13402
8	2.02122	2.03299	2.04995	2.05325	2.05975	2.06825	2.07654
9	1.96882	1.98082	1.99810	2.00146	2.00809	2.01675	2.02521
10	1.92133	1.93354	1.95113	1.95455	1.96130	1.97011	1.97871
11	1.87781	1.89022	1.90811	1.91158	1.91844	1.92739	1.93614
12	1.83758	1.85019	1.86835	1.87188	1.87885	1.88793	1.89681
13	1.80013	1.81292	1.83135	1.83493	1.84199	1.85121	1.86021
14	1.76504	1.77802	1.79670	1.80033	1.80749	1.81683	1.82595
15	1.73201	1.74515	1.76408	1.76776	1.77501	1.78448	1.79371
16	1.70076	1.71407	1.73324	1.73696	1.74431	1.75389	1.76324
17	1.67109	1.68457	1.70396	1.70773	1.71516	1.72486	1.73432
18	1.64283	1.65646	1.67608	1.67989	1.68741	1.69721	1.70678
19	1.61582	1.62961	1.64945	1.65330	1.66090	1.67081	1.68048
20	1.58994	1.60388	1.62393	1.62783	1.63551	1.64552	1.65530
21	1.56508	1.57917	1.59943	1.60337	1.61113	1.62125	1.63112
22	1.54116	1.55539	1.57586	1.57983	1.58767	1.59789	1.60786
23	1.51809	1.53246	1.55313	1.55714	1.56506	1.57537	1.58544
24	1.49580	1.51031	1.53118	1.53523	1.54322	1.55363	1.56379
25	1.47423	1.48888	1.50994	1.51403	1.52209	1.53260	1.54285
26	1.45332	1.46811	1.48936	1.49349	1.50162	1.51222	1.52257
27	1.43303	1.44795	1.46940	1.47356	1.48176	1.49246	1.50289
28	1.41332	1.42837	1.45000	1.45420	1.46247	1.47326	1.48378
29	1.39414	1.40932	1.43114	1.43537	1.44371	1.45459	1.46520
30	1.37546	1.39077	1.41277	1.41703	1.42545	1.43641	1.44711
31	1.35725	1.37268	1.39486	1.39917	1.40765	1.41870	1.42948
32	1.33947	1.35504	1.37740	1.38173	1.39028	1.40142	1.41228
33	1.32211	1.33780	1.36034	1.36471	1.37332	1.38455	1.39550
34	1.30515	1.32096	1.34367	1.34807	1.35675	1.36807	1.37910
35	1.28854	1.30448	1.32737	1.33180	1.34055	1.35195	1.36306
36	1.27229	1.28835	1.31141	1.31588	1.32469	1.33617	1.34736
37	1.25637	1.27255	1.29578	1.30028	1.30915	1.32072	1.33199
38	1.24076	1.25706	1.28045	1.28499	1.29393	1.30558	1.31693
39	1.22544	1.24186	1.26543	1.27000	1.27900	1.29073	1.30216
40	1.21041	1.22695	1.25068	1.25529	1.26435	1.27616	1.28767
41	1.19565	1.21230	1.23621	1.24084	1.24997	1.26186	1.27344
42	1.18114	1.19792	1.22198	1.22665	1.23584	1.24781	1.25947
43	1.16688	1.18377	1.20800	1.21270	1.22195	1.23400	1.24574
44	1.15285	1.16986	1.19425	1.19898	1.20830	1.22043	1.23225
45	1.13904	1.15617	1.18073	1.18549	1.19486	1.20707	1.21897
46	1.12545	1.14269	1.16742	1.17221	1.18164	1.19393	1.20590
47	1.11207	1.12942	1.15431	1.15913	1.16863	1.18099	1.19304
48	1.09888	1.11635	1.14139	1.14625	1.15580	1.16825	1.18037
49	1.08587	1.10346	1.12867	1.13355	1.14317	1.15569	1.16789
50	1.07305	1.09075	1.11612	1.12104	1.13071	1.14332	1.15559

EXPECTED VALUES OF NORMAL ORDER STATISTICS

k \ n	350	360	375	378	384	392	400
51	1.06041	1.07822	1.10375	1.10870	1.11843	1.13111	1.14346
52	1.04793	1.06586	1.09155	1.09652	1.10632	1.11907	1.13149
53	1.03561	1.05365	1.07950	1.08451	1.09436	1.10720	1.11969
54	1.02345	1.04160	1.06761	1.07265	1.08256	1.09547	1.10804
55	1.01144	1.02970	1.05587	1.06094	1.07091	1.08390	1.09654
56	0.99957	1.01795	1.04427	1.04937	1.05940	1.07247	1.08518
57	0.98784	1.00633	1.03281	1.03794	1.04804	1.06117	1.07396
58	0.97624	0.99485	1.02149	1.02665	1.03680	1.05001	1.06287
59	0.96478	0.98350	1.01030	1.01548	1.02569	1.03898	1.05192
60	0.95344	0.97227	0.99923	1.00444	1.01471	1.02808	1.04108
61	0.94222	0.96116	0.98828	0.99353	1.00386	1.01730	1.03037
62	0.93112	0.95017	0.97745	0.98272	0.99311	1.00663	1.01978
63	0.92013	0.93930	0.96673	0.97204	0.98248	0.99608	1.00930
64	0.90925	0.92853	0.95612	0.96146	0.97197	0.98563	0.99893
65	0.89848	0.91788	0.94562	0.95099	0.96155	0.97529	0.98866
66	0.88782	0.90732	0.93522	0.94062	0.95124	0.96506	0.97850
67	0.87725	0.89687	0.92493	0.93035	0.94103	0.95493	0.96844
68	0.86678	0.88651	0.91473	0.92018	0.93092	0.94489	0.95848
69	0.85640	0.87625	0.90462	0.91011	0.92091	0.93495	0.94861
70	0.84612	0.86607	0.89461	0.90012	0.91098	0.92510	0.93883
71	0.83592	0.85599	0.88468	0.89023	0.90114	0.91534	0.92914
72	0.82581	0.84599	0.87484	0.88042	0.89139	0.90566	0.91954
73	0.81579	0.83608	0.86509	0.87069	0.88173	0.89607	0.91002
74	0.80584	0.82625	0.85541	0.86105	0.87214	0.88656	0.90058
75	0.79598	0.81650	0.84582	0.85149	0.86264	0.87713	0.89122
76	0.78619	0.80682	0.83630	0.84200	0.85321	0.86778	0.88194
77	0.77648	0.79722	0.82686	0.83259	0.84386	0.85850	0.87274
78	0.76684	0.78770	0.81749	0.82325	0.83458	0.84930	0.86361
79	0.75727	0.77824	0.80820	0.81398	0.82537	0.84017	0.85455
80	0.74777	0.76885	0.79897	0.80479	0.81623	0.83110	0.84556
81	0.73833	0.75953	0.78981	0.79566	0.80716	0.82211	0.83663
82	0.72896	0.75028	0.78071	0.78659	0.79816	0.81318	0.82778
83	0.71966	0.74109	0.77168	0.77759	0.78922	0.80432	0.81899
84	0.71041	0.73196	0.76272	0.76866	0.78034	0.79551	0.81026
85	0.70123	0.72289	0.75381	0.75978	0.77152	0.78677	0.80159
86	0.69211	0.71389	0.74496	0.75096	0.76276	0.77809	0.79298
87	0.68304	0.70494	0.73617	0.74221	0.75406	0.76947	0.78443
88	0.67403	0.69604	0.72744	0.73350	0.74542	0.76090	0.77594
89	0.66507	0.68720	0.71876	0.72486	0.73684	0.75239	0.76750
90	0.65617	0.67842	0.71014	0.71626	0.72830	0.74394	0.75912
91	0.64732	0.66968	0.70157	0.70772	0.71982	0.73554	0.75079
92	0.63852	0.66100	0.69305	0.69923	0.71139	0.72718	0.74252
93	0.62976	0.65236	0.68458	0.69080	0.70302	0.71888	0.73429
94	0.62106	0.64378	0.67616	0.68241	0.69469	0.71063	0.72611
95	0.61240	0.63524	0.66778	0.67406	0.68640	0.70243	0.71798
96	0.60379	0.62675	0.65946	0.66577	0.67817	0.69427	0.70990
97	0.59522	0.61830	0.65118	0.65752	0.66998	0.68616	0.70186
98	0.58670	0.60990	0.64294	0.64931	0.66184	0.67809	0.69387
99	0.57822	0.60154	0.63475	0.64115	0.65374	0.67007	0.68593
100	0.56978	0.59322	0.62659	0.63303	0.64568	0.66209	0.67802

EXPECTED VALUES OF NORMAL ORDER STATISTICS

k\n	350	360	375	378	384	392	400
101	0.56138	0.58494	0.61849	0.62495	0.63766	0.65416	0.67016
102	0.55302	0.57670	0.61042	0.61692	0.62969	0.64626	0.66234
103	0.54470	0.56850	0.60239	0.60892	0.62175	0.63841	0.65456
104	0.53641	0.56034	0.59440	0.60096	0.61386	0.63059	0.64682
105	0.52817	0.55222	0.58644	0.59304	0.60600	0.62281	0.63912
106	0.51996	0.54413	0.57853	0.58515	0.59818	0.61507	0.63145
107	0.51178	0.53608	0.57065	0.57731	0.59039	0.60737	0.62383
108	0.50364	0.52806	0.56280	0.56950	0.58264	0.59970	0.61624
109	0.49553	0.52008	0.55499	0.56172	0.57493	0.59206	0.60868
110	0.48745	0.51213	0.54721	0.55397	0.56725	0.58447	0.60116
111	0.47941	0.50421	0.53947	0.54626	0.55960	0.57690	0.59367
112	0.47139	0.49632	0.53176	0.53858	0.55199	0.56937	0.58622
113	0.46341	0.48847	0.52408	0.53094	0.54441	0.56187	0.57880
114	0.45545	0.48064	0.51643	0.52332	0.53685	0.55440	0.57141
115	0.44753	0.47285	0.50881	0.51574	0.52933	0.54696	0.56405
116	0.43963	0.46508	0.50122	0.50818	0.52184	0.53955	0.55672
117	0.43176	0.45734	0.49366	0.50065	0.51438	0.53218	0.54942
118	0.42392	0.44962	0.48613	0.49316	0.50695	0.52483	0.54215
119	0.41610	0.44194	0.47862	0.48568	0.49954	0.51751	0.53491
120	0.40831	0.43428	0.47115	0.47824	0.49217	0.51021	0.52770
121	0.40054	0.42664	0.46369	0.47082	0.48482	0.50295	0.52051
122	0.39280	0.41903	0.45627	0.46343	0.47749	0.49571	0.51335
123	0.38508	0.41145	0.44887	0.45606	0.47019	0.48849	0.50622
124	0.37738	0.40389	0.44149	0.44872	0.46292	0.48130	0.49911
125	0.36970	0.39635	0.43414	0.44140	0.45567	0.47414	0.49203
126	0.36205	0.38883	0.42681	0.43411	0.44844	0.46700	0.48497
127	0.35442	0.38134	0.41950	0.42684	0.44124	0.45989	0.47794
128	0.34681	0.37386	0.41222	0.41959	0.43406	0.45279	0.47093
129	0.33922	0.36641	0.40495	0.41236	0.42690	0.44572	0.46394
130	0.33164	0.35898	0.39771	0.40516	0.41976	0.43868	0.45698
131	0.32409	0.35157	0.39049	0.39797	0.41265	0.43165	0.45004
132	0.31656	0.34417	0.38329	0.39081	0.40556	0.42464	0.44312
133	0.30904	0.33680	0.37611	0.38367	0.39848	0.41766	0.43622
134	0.30154	0.32945	0.36895	0.37654	0.39143	0.41070	0.42934
135	0.29406	0.32211	0.36181	0.36944	0.38439	0.40375	0.42248
136	0.28659	0.31479	0.35469	0.36235	0.37738	0.39683	0.41564
137	0.27914	0.30748	0.34758	0.35528	0.37038	0.38992	0.40883
138	0.27171	0.30019	0.34049	0.34823	0.36340	0.38304	0.40203
139	0.26429	0.29292	0.33342	0.34120	0.35644	0.37617	0.39524
140	0.25689	0.28567	0.32637	0.33418	0.34950	0.36932	0.38848
141	0.24950	0.27843	0.31933	0.32718	0.34257	0.36248	0.38174
142	0.24212	0.27120	0.31231	0.32020	0.33566	0.35567	0.37501
143	0.23475	0.26399	0.30530	0.31323	0.32877	0.34887	0.36830
144	0.22740	0.25679	0.29831	0.30628	0.32189	0.34208	0.36160
145	0.22006	0.24960	0.29133	0.29934	0.31503	0.33532	0.35492
146	0.21274	0.24243	0.28437	0.29241	0.30818	0.32856	0.34826
147	0.20542	0.23527	0.27742	0.28550	0.30134	0.32182	0.34161
148	0.19812	0.22812	0.27048	0.27861	0.29452	0.31510	0.33498
149	0.19082	0.22098	0.26356	0.27172	0.28772	0.30839	0.32836
150	0.18354	0.21386	0.25665	0.26485	0.28092	0.30170	0.32176

EXPECTED VALUES OF NORMAL ORDER STATISTICS

k \ n	350	360	375	378	384	392	400
151	0.17626	0.20675	0.24975	0.25799	0.27414	0.29501	0.31517
152	0.16900	0.19964	0.24287	0.25115	0.26738	0.28834	0.30860
153	0.16174	0.19255	0.23599	0.24431	0.26062	0.28169	0.30203
154	0.15450	0.18546	0.22913	0.23749	0.25388	0.27505	0.29548
155	0.14726	0.17839	0.22227	0.23068	0.24714	0.26841	0.28895
156	0.14003	0.17132	0.21543	0.22388	0.24042	0.26179	0.28242
157	0.13280	0.16427	0.20860	0.21709	0.23371	0.25519	0.27591
158	0.12558	0.15722	0.20178	0.21031	0.22701	0.24859	0.26941
159	0.11837	0.15017	0.19496	0.20354	0.22032	0.24200	0.26292
160	0.11117	0.14314	0.18816	0.19677	0.21365	0.23543	0.25644
161	0.10397	0.13611	0.18136	0.19002	0.20698	0.22886	0.24998
162	0.09678	0.12909	0.17458	0.18328	0.20031	0.22231	0.24352
163	0.08959	0.12208	0.16780	0.17654	0.19366	0.21576	0.23707
164	0.08240	0.11507	0.16103	0.16981	0.18702	0.20922	0.23064
165	0.07522	0.10807	0.15426	0.16309	0.18038	0.20269	0.22421
166	0.06805	0.10107	0.14750	0.15638	0.17376	0.19618	0.21779
167	0.06088	0.09408	0.14075	0.14968	0.16714	0.18966	0.21138
168	0.05371	0.08709	0.13401	0.14298	0.16053	0.18316	0.20498
169	0.04654	0.08011	0.12727	0.13628	0.15392	0.17667	0.19859
170	0.03938	0.07313	0.12054	0.12960	0.14732	0.17018	0.19220
171	0.03221	0.06616	0.11381	0.12292	0.14073	0.16370	0.18583
172	0.02505	0.05918	0.10709	0.11624	0.13414	0.15722	0.17946
173	0.01789	0.05221	0.10038	0.10957	0.12756	0.15076	0.17310
174	0.01074	0.04525	0.09366	0.10291	0.12099	0.14430	0.16674
175	0.00358	0.03828	0.08696	0.09625	0.11442	0.13784	0.16040
176		0.03132	0.08025	0.08959	0.10786	0.13139	0.15406
177		0.02436	0.07355	0.08294	0.10130	0.12495	0.14772
178		0.01740	0.06685	0.07629	0.09474	0.11851	0.14139
179		0.01044	0.06016	0.06965	0.08819	0.11208	0.13507
180		0.00348	0.05347	0.06300	0.08164	0.10565	0.12875
181			0.04678	0.05636	0.07510	0.09922	0.12244
182			0.04009	0.04973	0.06856	0.09280	0.11613
183			0.03341	0.04309	0.06202	0.08639	0.10983
184			0.02673	0.03646	0.05548	0.07997	0.10353
185			0.02004	0.02983	0.04895	0.07356	0.09723
186			0.01336	0.02320	0.04242	0.06716	0.09094
187			0.00668	0.01657	0.03589	0.06075	0.08465
188			0.00000	0.00994	0.02936	0.05435	0.07837
189				0.00331	0.02284	0.04795	0.07209
190					0.01631	0.04155	0.06581
191					0.00979	0.03516	0.05954
192					0.00326	0.02876	0.05326
193						0.02237	0.04699
194						0.01598	0.04072
195						0.00959	0.03445
196						0.00320	0.02819
197							0.02192
198							0.01566
199							0.00939
200							0.00313

Blom's Approximation

$$[\Phi(x) = \int_{-\infty}^{x} \phi(x)dx, \text{ with } \phi(x) = (2\pi)^{-1/2} e^{-x^2/2}]$$

B. Values of $\alpha_{i,n}$ such that $E(x_i) = \Phi^{-1}[(i - \alpha_{i,n}) / (n - 2\alpha_{i,n} + 1)]$

i	n=25	n=50	n=100	n=200	n=400	i	n=100	n=200	n=400
1	0.377	0.384	0.391	0.396	0.401	30	0.394	0.404	0.414
2	0.394	0.403	0.412	0.419	0.426	35	0.393	0.402	0.412
3	0.395	0.405	0.415	0.423	0.430	40	0.392	0.400	0.410
4	0.394	0.405	0.415	0.424	0.431	45	0.391	0.398	0.408
5	0.392	0.403	0.414	0.423	0.431	50	0.391	0.397	0.407
6	0.391	0.402	0.412	0.422	0.430	55		0.396	0.405
7	0.390	0.400	0.411	0.421	0.429	60		0.395	0.404
8	0.389	0.399	0.410	0.420	0.429	65		0.394	0.403
9	0.388	0.398	0.408	0.418	0.428	70		0.394	0.402
10	0.388	0.397	0.407	0.417	0.427	75		0.393	0.401
11	0.387	0.396	0.406	0.416	0.426	80		0.393	0.400
12	0.387	0.395	0.405	0.415	0.425	85		0.392	0.399
13		0.394	0.404	0.414	0.424	90		0.392	0.399
14		0.393	0.403	0.414	0.423	95		0.391	0.398
15		0.393	0.402	0.413	0.423	100		0.391	0.398
16		0.392	0.402	0.412	0.422	110			0.396
17		0.392	0.401	0.411	0.421	120			0.396
18		0.391	0.400	0.410	0.420	130			0.395
19		0.391	0.399	0.410	0.420	140			0.394
20		0.391	0.399	0.409	0.419	150			0.394
21		0.390	0.398	0.408	0.419	160			0.393
22		0.390	0.398	0.408	0.418	170			0.393
23		0.390	0.397	0.407	0.417	180			0.392
24		0.390	0.397	0.407	0.417	190			0.392
25		0.390	0.396	0.406	0.416	200			0.391

C. Compromise Values of α

n	α_n	$\alpha_{1,n}$	$\alpha_{2,n}$
2	0.330	0.330	
4	0.349	0.347	0.359
6	0.359	0.355	0.368
8	0.364	0.360	0.374
10	0.368	0.364	0.378
15	0.374	0.370	0.385
20	0.378	0.374	0.390
25	0.381	0.377	0.394
50	0.389	0.384	0.403
100	0.396	0.391	0.412
200	0.402	0.396	0.419
400	0.407	0.401	0.426

To estimate α for intermediate values of n, use the following equations:

$$\alpha_n = .314195 + .063336X - .010895X^2$$

$$\alpha_{1,n} = .315065 + .057974X - .009776X^2$$

$$\alpha_{2,n} = .327511 + .058212X - .007909X^2$$

where $X = \log_{10} n$

Table C1.2

VARIANCES AND COVARIANCES OF NORMAL ORDER STATISTICS

[When $I = J$, tabular values are variances of the Ith order statistic of a sample of size N from $N(0,1)$; when $I \neq J$, they are covariances of the Ith and Jth order statistics.]

Variances and Covariances of Normal Order Statistics

N	*I*	*J*	*COV*	*N*	*I*	*J*	*COV*	*N*	*I*	*J*	*COV*	*N*	*I*	*J*	*COV*
1	1	1	1.00000	8	1	3	0.12597	10	1	10	0.02670	11	4	8	0.07255
2	1	1	0.68169	8	1	4	0.09472	10	2	2	0.21452	11	5	5	0.13964
2	1	2	0.31831	8	1	5	0.07477	10	2	3	0.14662	11	5	6	0.11674
3	1	1	0.55947	8	1	6	0.06021	10	2	4	0.11170	11	5	7	0.09919
3	1	2	0.27566	8	1	7	0.04830	10	2	5	0.08974	11	6	6	0.13716
3	1	3	0.16487	8	1	8	0.03684	10	2	6	0.07420	12	1	1	0.32364
3	2	2	0.44867	8	2	2	0.23940	10	2	7	0.06223	12	1	2	0.16024
4	1	1	0.49172	8	2	3	0.16320	10	2	8	0.05231	12	1	3	0.10893
4	1	2	0.24559	8	2	4	0.12326	10	2	9	0.04337	12	1	4	0.08307
4	1	3	0.15801	8	2	5	0.09756	10	3	3	0.17500	12	1	5	0.06709
4	1	4	0.10468	8	2	6	0.07872	10	3	4	0.13380	12	1	6	0.05599
4	2	2	0.36046	8	2	7	0.06325	10	3	5	0.10774	12	1	7	0.04766
4	2	3	0.23594	8	3	3	0.20077	10	3	6	0.08923	12	1	8	0.04102
5	1	1	0.44753	8	3	4	0.15236	10	3	7	0.07492	12	1	9	0.03544
5	1	2	0.22433	8	3	5	0.12096	10	3	8	0.06303	12	1	10	0.03050
5	1	3	0.14815	8	3	6	0.09782	10	4	4	0.15794	12	1	11	0.02579
5	1	4	0.10577	8	4	4	0.18719	10	4	5	0.12751	12	1	12	0.02062
5	1	5	0.07422	8	4	5	0.14918	10	4	6	0.10579	12	2	2	0.19726
5	2	2	0.31152	9	1	1	0.35735	10	4	7	0.08895	12	2	3	0.13490
5	2	3	0.20844	9	1	2	0.17814	10	5	5	0.15105	12	2	4	0.10320
5	2	4	0.14994	9	1	3	0.12075	10	5	6	0.12560	12	2	5	0.08350
5	3	3	0.28683	9	1	4	0.09131	11	1	1	0.33325	12	2	6	0.06979
6	1	1	0.41593	9	1	5	0.07274	11	1	2	0.16536	12	2	7	0.05946
6	1	2	0.20850	9	1	6	0.05948	11	1	3	0.11236	12	2	8	0.05121
6	1	3	0.13944	9	1	7	0.04908	11	1	4	0.08552	12	2	9	0.04427
6	1	4	0.10243	9	1	8	0.04009	11	1	5	0.06885	12	2	10	0.03812
6	1	5	0.07736	9	1	9	0.03106	11	1	6	0.05720	12	2	11	0.03225
6	1	6	0.05634	9	2	2	0.22570	11	1	7	0.04838	12	3	3	0.15798
6	2	2	0.27958	9	2	3	0.15412	11	1	8	0.04124	12	3	4	0.12121
6	2	3	0.18899	9	2	4	0.11701	11	1	9	0.03511	12	3	5	0.09826
6	2	4	0.13966	9	2	5	0.09345	11	1	10	0.02942	12	3	6	0.08222
6	2	5	0.10591	9	2	6	0.07655	11	1	11	0.02332	12	3	7	0.07012
6	3	3	0.24621	9	2	7	0.06324	11	2	2	0.20520	12	3	8	0.06044
6	3	4	0.18327	9	2	8	0.05171	11	2	3	0.14031	12	3	9	0.05228
7	1	1	0.39192	9	3	3	0.18638	11	2	4	0.10715	12	3	10	0.04504
7	1	2	0.19620	9	3	4	0.14208	11	2	5	0.08644	12	4	4	0.13981
7	1	3	0.13212	9	3	5	0.11377	11	2	6	0.07193	12	4	5	0.11357
7	1	4	0.09849	9	3	6	0.09336	11	2	7	0.06089	12	4	6	0.09516
7	1	5	0.07656	9	3	7	0.07724	11	2	8	0.05195	12	4	7	0.08124
7	1	6	0.05992	9	4	4	0.17056	11	2	9	0.04425	12	4	8	0.07008
7	1	7	0.04480	9	4	5	0.13699	11	2	10	0.03710	12	4	9	0.06066
7	2	2	0.25673	9	4	6	0.11267	11	3	3	0.16572	12	5	5	0.13061
7	2	3	0.17448	9	5	5	0.16610	11	3	4	0.12697	12	5	6	0.10962
7	2	4	0.13073	10	1	1	0.34434	11	3	5	0.10264	12	5	7	0.09370
7	2	5	0.10196	10	1	2	0.17126	11	3	6	0.08552	12	5	8	0.08090
7	2	6	0.07998	10	1	3	0.11626	11	3	7	0.07247	12	6	6	0.12664
7	3	3	0.21972	10	1	4	0.08825	11	3	8	0.06189	12	6	7	0.10839
7	3	4	0.16556	10	1	5	0.07074	11	3	9	0.05276	13	1	1	0.31521
7	3	5	0.12960	10	1	6	0.05840	11	4	4	0.14795	13	1	2	0.15573
7	4	4	0.21045	10	1	7	0.04892	11	4	5	0.11988	13	1	3	0.10589
8	1	1	0.37290	10	1	8	0.04108	11	4	6	0.10003	13	1	4	0.08086
8	1	2	0.18631	10	1	9	0.03404	11	4	7	0.08488	13	1	5	0.06546

Variances and Covariances of Normal Order Statistics

N	I	J	COV	N	I	J	COV	N	I	J	COV	N	I	J	COV
13	1	6	0.05482	14	1	9	0.03521	15	1	5	0.06258	15	6	7	0.09147
13	1	7	0.04688	14	1	10	0.03104	15	1	6	0.05265	15	6	8	0.08014
13	1	8	0.04061	14	1	11	0.02734	15	1	7	0.04531	15	6	9	0.07086
13	1	9	0.03542	14	1	12	0.02391	15	1	8	0.03957	15	6	10	0.06298
13	1	10	0.03093	14	1	13	0.02051	15	1	9	0.03490	15	7	7	0.10269
13	1	11	0.02685	14	1	14	0.01663	15	1	10	0.03096	15	7	8	0.09005
13	1	12	0.02289	14	2	2	0.18442	15	1	11	0.02752	15	7	9	0.07967
13	1	13	0.01843	14	2	3	0.12608	15	1	12	0.02441	15	8	8	0.10169
13	2	2	0.19041	14	2	4	0.09665	15	1	13	0.02148	16	1	1	0.29501
13	2	3	0.13021	14	2	5	0.07852	15	1	14	0.01853	16	1	2	0.14489
13	2	4	0.09973	14	2	6	0.06600	15	1	15	0.01511	16	1	3	0.09850
13	2	5	0.08088	14	2	7	0.05669	15	2	2	0.17912	16	1	4	0.07540
13	2	6	0.06781	14	2	8	0.04937	15	2	3	0.12242	16	1	5	0.06131
13	2	7	0.05805	14	2	9	0.04336	15	2	4	0.09391	16	1	6	0.05166
13	2	8	0.05032	14	2	10	0.03823	15	2	5	0.07639	16	1	7	0.04455
13	2	9	0.04391	14	2	11	0.03369	15	2	6	0.06434	16	1	8	0.03902
13	2	10	0.03836	14	2	12	0.02947	15	2	7	0.05541	16	1	9	0.03454
13	2	11	0.03331	14	2	13	0.02529	15	2	8	0.04842	16	1	10	0.03078
13	2	12	0.02840	14	3	3	0.14570	15	2	9	0.04273	16	1	11	0.02754
13	3	3	0.15139	14	3	4	0.11198	15	2	10	0.03792	16	1	12	0.02465
13	3	4	0.11627	14	3	5	0.09112	15	2	11	0.03372	16	1	13	0.02200
13	3	5	0.09446	14	3	6	0.07668	15	2	12	0.02992	16	1	14	0.01946
13	3	6	0.07929	14	3	7	0.06591	15	2	13	0.02633	16	1	15	0.01687
13	3	7	0.06793	14	3	8	0.05743	15	2	14	0.02272	16	1	16	0.01383
13	3	8	0.05892	14	3	9	0.05047	15	3	3	0.14073	16	2	2	0.17439
13	3	9	0.05145	14	3	10	0.04452	15	3	4	0.10821	16	2	3	0.11914
13	3	10	0.04496	14	3	11	0.03924	15	3	5	0.08816	16	2	4	0.09144
13	3	11	0.03906	14	3	12	0.03433	15	3	6	0.07433	16	2	5	0.07446
13	4	4	0.13301	14	4	4	0.12723	15	3	7	0.06406	16	2	6	0.06281
13	4	5	0.10825	14	4	5	0.10369	15	3	8	0.05601	16	2	7	0.05420
13	4	6	0.09099	14	4	6	0.08736	15	3	9	0.04945	16	2	8	0.04750
13	4	7	0.07802	14	4	7	0.07515	15	3	10	0.04390	16	2	9	0.04206
13	4	8	0.06772	14	4	8	0.06553	15	3	11	0.03904	16	2	10	0.03750
13	4	9	0.05916	14	4	9	0.05761	15	3	12	0.03465	16	2	11	0.03356
13	4	10	0.05173	14	4	10	0.05084	15	3	13	0.03051	16	2	12	0.03005
13	5	5	0.12325	14	4	11	0.04482	15	4	4	0.12223	16	2	13	0.02682
13	5	6	0.10374	14	5	5	0.11710	15	4	5	0.09973	16	2	14	0.02373
13	5	7	0.08904	14	5	6	0.09877	15	4	6	0.08417	16	2	15	0.02058
13	5	8	0.07736	14	5	7	0.08505	15	4	7	0.07259	16	3	3	0.13634
13	5	9	0.06762	14	5	8	0.07422	15	4	8	0.06352	16	3	4	0.10487
13	6	6	0.11832	14	5	9	0.06529	15	4	9	0.05610	16	3	5	0.08552
13	6	7	0.10168	14	5	10	0.05764	15	4	10	0.04982	16	3	6	0.07221
13	6	8	0.08842	14	6	6	0.11153	15	4	11	0.04432	16	3	7	0.06236
13	7	7	0.11680	14	6	7	0.09614	15	4	12	0.03935	16	3	8	0.05467
14	1	1	0.30773	14	6	8	0.08396	15	5	5	0.11187	16	3	9	0.04844
14	1	2	0.15172	14	6	9	0.07391	15	5	6	0.09452	16	3	10	0.04320
14	1	3	0.10317	14	7	7	0.10903	15	5	7	0.08159	16	3	11	0.03867
14	1	4	0.07887	14	7	8	0.09531	15	5	8	0.07143	16	3	12	0.03463
14	1	5	0.06397	15	1	1	0.30104	15	5	9	0.06312	16	3	13	0.03091
14	1	6	0.05371	15	1	2	0.14813	15	5	10	0.05608	16	3	14	0.02736
14	1	7	0.04609	15	1	3	0.10072	15	5	11	0.04991	16	4	4	0.11787
14	1	8	0.04011	15	1	4	0.07706	15	6	6	0.10587	16	4	5	0.09625

Variances and Covariances of Normal Order Statistics

N	I	J	COV	N	I	J	COV	N	I	J	COV	N	I	J	COV
16	4	6	0.08135	17	2	8	0.04661	17	7	9	0.07282	18	3	13	0.03092
16	4	7	0.07030	17	2	9	0.04139	17	7	10	0.06527	18	3	14	0.02799
16	4	8	0.06167	17	2	10	0.03703	17	7	11	0.05876	18	3	15	0.02522
16	4	9	0.05466	17	2	11	0.03329	17	8	8	0.09074	18	3	16	0.02252
16	4	10	0.04876	17	2	12	0.03000	17	8	9	0.08080	18	4	4	0.11057
16	4	11	0.04366	17	2	13	0.02702	17	8	10	0.07246	18	4	5	0.09040
16	4	12	0.03911	17	2	14	0.02424	17	9	9	0.09005	18	4	6	0.07656
16	4	13	0.03493	17	2	15	0.02155	18	1	1	0.28453	18	4	7	0.06635
16	5	5	0.10735	17	2	16	0.01877	18	1	2	0.13925	18	4	8	0.05843
16	5	6	0.09082	17	3	3	0.13242	18	1	3	0.09462	18	4	9	0.05204
16	5	7	0.07855	17	3	4	0.10188	18	1	4	0.07249	18	4	10	0.04672
16	5	8	0.06895	17	3	5	0.08314	18	1	5	0.05903	18	4	11	0.04217
16	5	9	0.06114	17	3	6	0.07029	18	1	6	0.04986	18	4	12	0.03819
16	5	10	0.05456	17	3	7	0.06080	18	1	7	0.04313	18	4	13	0.03462
16	5	11	0.04887	17	3	8	0.05342	18	1	8	0.03793	18	4	14	0.03135
16	5	12	0.04379	17	3	9	0.04746	18	1	9	0.03374	18	4	15	0.02825
16	6	6	0.10105	17	3	10	0.04247	18	1	10	0.03026	18	5	5	0.09991
16	6	7	0.08746	17	3	11	0.03819	18	1	11	0.02729	18	5	6	0.08469
16	6	8	0.07682	17	3	12	0.03442	18	1	12	0.02470	18	5	7	0.07345
16	6	9	0.06815	17	3	13	0.03100	18	1	13	0.02238	18	5	8	0.06471
16	6	10	0.06085	17	3	14	0.02782	18	1	14	0.02025	18	5	9	0.05765
16	6	11	0.05452	17	3	15	0.02473	18	1	15	0.01825	18	5	10	0.05178
16	7	7	0.09740	17	4	4	0.11401	18	1	16	0.01629	18	5	11	0.04675
16	7	8	0.08562	17	4	5	0.09316	18	1	17	0.01424	18	5	12	0.04234
16	7	9	0.07600	17	4	6	0.07783	18	1	18	0.01177	18	5	13	0.03839
16	7	10	0.06789	17	4	7	0.06823	18	2	2	0.16629	18	5	14	0.03477
16	8	8	0.09572	17	4	8	0.05998	18	2	3	0.11351	18	6	6	0.09324
16	8	9	0.08503	17	4	9	0.05331	18	2	4	0.08716	18	6	7	0.08092
17	1	1	0.28953	17	4	10	0.04772	18	2	5	0.07108	18	6	8	0.07133
17	1	2	0.14194	17	4	11	0.04293	18	2	6	0.06010	18	6	9	0.06358
17	1	3	0.09647	17	4	12	0.03869	18	2	7	0.05202	18	6	10	0.05712
17	1	4	0.07388	17	4	13	0.03486	18	2	8	0.04577	18	6	11	0.05159
17	1	5	0.06013	17	4	14	0.03129	18	2	9	0.04073	18	6	12	0.04674
17	1	6	0.05073	17	5	5	0.10340	18	2	10	0.03655	18	6	13	0.04239
17	1	7	0.04382	17	5	6	0.08757	18	2	11	0.03297	18	7	7	0.08902
17	1	8	0.03847	17	5	7	0.07585	18	2	12	0.02984	18	7	8	0.07852
17	1	9	0.03414	17	5	8	0.06672	18	2	13	0.02705	18	7	9	0.07002
17	1	10	0.03054	17	5	9	0.05932	18	2	14	0.02448	18	7	10	0.06293
17	1	11	0.02745	17	5	10	0.05313	18	2	15	0.02206	18	7	11	0.05685
17	1	12	0.02472	17	5	11	0.04780	18	2	16	0.01969	18	7	12	0.05152
17	1	13	0.02226	17	5	12	0.04310	18	2	17	0.01722	18	8	8	0.08650
17	1	14	0.01997	17	5	13	0.03884	18	3	3	0.12890	18	8	9	0.07718
17	1	15	0.01775	17	6	6	0.09688	18	3	4	0.09918	18	8	10	0.06939
17	1	16	0.01546	17	6	7	0.08398	18	3	5	0.08099	18	8	11	0.06271
17	1	17	0.01273	17	6	8	0.07391	18	3	6	0.06853	18	9	9	0.08531
17	2	2	0.17014	17	6	9	0.06594	18	3	7	0.05936	18	9	10	0.07674
17	2	3	0.11619	17	6	10	0.05890	18	3	8	0.05225	19	1	1	0.27994
17	2	4	0.08920	17	6	11	0.05301	18	3	9	0.04652	19	1	2	0.13678
17	2	5	0.07270	17	6	12	0.04781	18	3	10	0.04175	19	1	3	0.09291
17	2	6	0.06140	17	7	7	0.09290	18	3	11	0.03767	19	1	4	0.07119
17	2	7	0.05308	17	7	8	0.08182	18	3	12	0.03411	19	1	5	0.05801

Variances and Covariances of Normal Order Statistics

N	I	J	COV	N	I	J	COV	N	I	J	COV	N	I	J	COV
19	1	6	0.04904	19	4	9	0.05086	20	1	8	0.03689	20	4	8	0.05569
19	1	7	0.04247	19	4	10	0.04576	20	1	9	0.03293	20	4	9	0.04975
19	1	8	0.03740	19	4	11	0.04142	20	1	10	0.02966	20	4	10	0.04485
19	1	9	0.03333	19	4	12	0.03764	20	1	11	0.02688	20	4	11	0.04068
19	1	10	0.02996	19	4	13	0.03428	20	1	12	0.02448	20	4	12	0.03707
19	1	11	0.02710	19	4	14	0.03123	20	1	13	0.02236	20	4	13	0.03388
19	1	12	0.02461	19	4	15	0.02839	20	1	14	0.02046	20	4	14	0.03100
19	1	13	0.02240	19	4	16	0.02569	20	1	15	0.01871	20	4	15	0.02837
19	1	14	0.02040	19	5	5	0.09679	20	1	16	0.01707	20	4	16	0.02589
19	1	15	0.01854	19	5	6	0.08211	20	1	17	0.01550	20	4	17	0.02351
19	1	16	0.01677	19	5	7	0.07128	20	1	18	0.01392	20	5	5	0.09400
19	1	17	0.01502	19	5	8	0.06289	20	1	19	0.01225	20	5	6	0.07978
19	1	18	0.01318	19	5	9	0.05613	20	1	20	0.01020	20	5	7	0.06932
19	1	19	0.01094	19	5	10	0.05051	20	2	2	0.15957	20	5	8	0.06123
19	2	2	0.16279	19	5	11	0.04573	20	2	3	0.10881	20	5	9	0.05472
19	2	3	0.11106	19	5	12	0.04157	20	2	4	0.08358	20	5	10	0.04934
19	2	4	0.08529	19	5	13	0.03786	20	2	5	0.06822	20	5	11	0.04477
19	2	5	0.06960	19	5	14	0.03450	20	2	6	0.05777	20	5	12	0.04080
19	2	6	0.05889	19	5	15	0.03138	20	2	7	0.05011	20	5	13	0.03729
19	2	7	0.05104	19	6	6	0.09002	20	2	8	0.04420	20	5	14	0.03414
19	2	8	0.04497	19	6	7	0.07820	20	2	9	0.03947	20	5	15	0.03123
19	2	9	0.04009	19	6	8	0.06903	20	2	10	0.03556	20	5	16	0.02851
19	2	10	0.03605	19	6	9	0.06163	20	2	11	0.03224	20	6	6	0.08715
19	2	11	0.03261	19	6	10	0.05549	20	2	12	0.02937	20	6	7	0.07577
19	2	12	0.02963	19	6	11	0.05025	20	2	13	0.02683	20	6	8	0.06696
19	2	13	0.02697	19	6	12	0.04568	20	2	14	0.02455	20	6	9	0.05987
19	2	14	0.02456	19	6	13	0.04162	20	2	15	0.02245	20	6	10	0.05399
19	2	15	0.02233	19	6	14	0.03793	20	2	16	0.02049	20	6	11	0.04900
19	2	16	0.02020	19	7	7	0.08562	20	2	17	0.01860	20	6	12	0.04467
19	2	17	0.01810	19	7	8	0.07562	20	2	18	0.01671	20	6	13	0.04084
19	2	18	0.01588	19	7	9	0.06754	20	2	19	0.01471	20	6	14	0.03738
19	3	3	0.12571	19	7	10	0.06083	20	3	3	0.12281	20	6	15	0.03421
19	3	4	0.09674	19	7	11	0.05510	20	3	4	0.09450	20	7	7	0.08261
19	3	5	0.07903	19	7	12	0.05011	20	3	5	0.07724	20	7	8	0.07304
19	3	6	0.06693	19	7	13	0.04566	20	3	6	0.06545	20	7	9	0.06533
19	3	7	0.05803	19	8	8	0.08283	20	3	7	0.05681	20	7	10	0.05894
19	3	8	0.05115	19	8	9	0.07403	20	3	8	0.05013	20	7	11	0.05351
19	3	9	0.04562	19	8	10	0.06670	20	3	9	0.04478	20	7	12	0.04879
19	3	10	0.04104	19	8	11	0.06044	20	3	10	0.04035	20	7	13	0.04461
19	3	11	0.03713	19	8	12	0.05498	20	3	11	0.03659	20	7	14	0.04085
19	3	12	0.03374	19	9	9	0.08129	20	3	12	0.03334	20	8	8	0.07963
19	3	13	0.03072	19	9	10	0.07327	20	3	13	0.03046	20	8	9	0.07126
19	3	14	0.02798	19	9	11	0.06642	20	3	14	0.02788	20	8	10	0.06431
19	3	15	0.02544	19	10	10	0.08079	20	3	15	0.02550	20	8	11	0.05840
19	3	16	0.02302	20	1	1	0.27570	20	3	16	0.02327	20	8	12	0.05326
19	3	17	0.02062	20	1	2	0.13449	20	3	17	0.02113	20	8	13	0.04872
19	4	4	0.10747	20	1	3	0.09132	20	3	18	0.01899	20	9	9	0.07781
19	4	5	0.08791	20	1	4	0.06999	20	4	4	0.10468	20	9	10	0.07025
19	4	6	0.07450	20	1	5	0.05706	20	4	5	0.08564	20	9	11	0.06382
19	4	7	0.06464	20	1	6	0.04827	20	4	6	0.07263	20	9	12	0.05822
19	4	8	0.05700	20	1	7	0.04184	20	4	7	0.06307	20	10	10	0.07695
												20	10	11	0.06993

Table C2.1

EXPECTED VALUES OF EXPONENTIAL ORDER STATISTICS

[$E(x_{M,N})$, where $x_{M,N}$ is the Mth order statistic of a sample of size N from an exponential population with location parameter 0 and scale parameter 1]

EXPECTED VALUES OF EXPONENTIAL ORDER STATISTICS

M \ N	1	2	3	4	5	6	7	8
1	1.00000	0.50000	0.33333	0.25000	0.20000	0.16667	0.14286	0.12500
2		1.50000	0.83333	0.58333	0.45000	0.36667	0.30952	0.26786
3			1.83333	1.08333	0.78333	0.61667	0.50952	0.43452
4				2.08333	1.28333	0.95000	0.75952	0.63452
5					2.28333	1.45000	1.09286	0.88452
6						2.45000	1.59286	1.21786
7							2.59286	1.71786
8								2.71786

M \ N	9	10	11	12	13	14	15	16
1	0.11111	0.10000	0.09091	0.08333	0.07692	0.07143	0.06667	0.06250
2	0.23611	0.21111	0.19091	0.17424	0.16026	0.14835	0.13810	0.12917
3	0.37897	0.33611	0.30202	0.27424	0.25117	0.23168	0.21502	0.20060
4	0.54563	0.47897	0.42702	0.38535	0.35117	0.32259	0.29835	0.27752
5	0.74563	0.64563	0.56988	0.51035	0.46228	0.42259	0.38926	0.36085
6	0.99563	0.84563	0.73654	0.65321	0.58728	0.53371	0.48926	0.45176
7	1.32897	1.09563	0.93654	0.81988	0.73013	0.65871	0.60037	0.55176
8	1.82897	1.42897	1.18654	1.01988	0.89680	0.80156	0.72537	0.66287

M \ N	9	10	11	12	13	14	15	16
9	2.82897	1.92897	1.51988	1.26988	1.09680	0.96823	0.86823	0.78787
10		2.92897	2.01988	1.60321	1.34680	1.16823	1.03490	0.93073
11			3.01988	2.10321	1.68013	1.41823	1.23490	1.09740
12				3.10321	2.18013	1.75156	1.48490	1.29740
13					3.18013	2.25156	1.81823	1.54740
14						3.25156	2.31823	1.88073
15							3.31823	2.38073
16								3.38073

M \ N	17	18	19	20	21	22	23	24
1	0.05882	0.05556	0.05263	0.05000	0.04762	0.04545	0.04348	0.04167
2	0.12132	0.11438	0.10819	0.10263	0.09762	0.09307	0.08893	0.08514
3	0.18799	0.17688	0.16701	0.15819	0.15025	0.14307	0.13655	0.13060
4	0.25942	0.24355	0.22951	0.21701	0.20581	0.19571	0.18655	0.17822
5	0.33634	0.31497	0.29618	0.27951	0.26463	0.25126	0.23918	0.22822
6	0.41968	0.39190	0.36761	0.34618	0.32713	0.31008	0.29474	0.28085
7	0.51058	0.47523	0.44453	0.41761	0.39380	0.37258	0.35356	0.33641
8	0.61058	0.56614	0.52786	0.49453	0.46522	0.43925	0.41606	0.39523

M \ N	17	18	19	20	21	22	23	24
9	0.72170	0.66614	0.61877	0.57786	0.54215	0.51068	0.48273	0.45773
10	0.84670	0.77725	0.71877	0.66877	0.62548	0.58760	0.55416	0.52440
11	0.98955	0.90225	0.82988	0.76877	0.71639	0.67094	0.63108	0.59582
12	1.15622	1.04511	0.95488	0.87988	0.81639	0.76184	0.71441	0.67275
13	1.35622	1.21177	1.09774	1.00488	0.92750	0.86184	0.80532	0.75608
14	1.60622	1.41177	1.26441	1.14774	1.05250	0.97296	0.90532	0.84699
15	1.93955	1.66177	1.46441	1.31441	1.19536	1.09796	1.01643	0.94699
16	2.43955	1.99511	1.71441	1.51441	1.36203	1.24081	1.14143	1.05810

EXPECTED VALUES OF EXPONENTIAL ORDER STATISTICS

M \ N	17	18	19	20	21	22	23	24
17	3.43955	2.49511	2.04774	1.76441	1.56203	1.40748	1.28429	1.18310
18		3.49511	2.54774	2.09774	1.81203	1.60748	1.45096	1.32596
19			3.54774	2.59774	2.14536	1.85748	1.65096	1.49262
20				3.59774	2.64536	2.19081	1.90096	1.69262
21					3.64536	2.69081	2.23429	1.94262
22						3.69081	2.73429	2.27596
23							3.73429	2.77596
24								3.77596

M \ N	25	26	27	28	29	30	31	32
1	0.04000	0.03846	0.03704	0.03571	0.03448	0.03333	0.03226	0.03125
2	0.08167	0.07846	0.07550	0.07275	0.07020	0.06782	0.06559	0.06351
3	0.12514	0.12013	0.11550	0.11121	0.10723	0.10353	0.10007	0.09684
4	0.17060	0.16361	0.15717	0.15121	0.14570	0.14057	0.13579	0.13132
5	0.21822	0.20906	0.20064	0.19288	0.18570	0.17903	0.17283	0.16704
6	0.26822	0.25668	0.24610	0.23636	0.22736	0.21903	0.21129	0.20408
7	0.32085	0.30668	0.29372	0.28181	0.27084	0.26070	0.25129	0.24254
8	0.37641	0.35931	0.34372	0.32943	0.31630	0.30417	0.29295	0.28254

M \ N	25	26	27	28	29	30	31	32
9	0.43523	0.41487	0.39635	0.37943	0.36391	0.34963	0.33643	0.32420
10	0.49773	0.47369	0.45190	0.43206	0.41391	0.39725	0.38189	0.36768
11	0.56440	0.53619	0.51073	0.48762	0.46655	0.44725	0.42951	0.41314
12	0.63582	0.60286	0.57323	0.54644	0.52210	0.49988	0.47951	0.46076
13	0.71275	0.67429	0.63989	0.60894	0.58092	0.55543	0.53214	0.51076
14	0.79608	0.75121	0.71132	0.67561	0.64342	0.61426	0.58769	0.56339
15	0.88699	0.83454	0.78825	0.74704	0.71009	0.67676	0.64652	0.61894
16	0.98699	0.92545	0.87158	0.82396	0.78152	0.74342	0.70902	0.67777

M \ N	25	26	27	28	29	30	31	32
17	1.09810	1.02545	0.96249	0.90729	0.85844	0.81485	0.77568	0.74027
18	1.22310	1.13656	1.06249	0.99820	0.94178	0.89178	0.84711	0.80693
19	1.36596	1.26156	1.17360	1.09820	1.03269	0.97511	0.92403	0.87836
20	1.53262	1.40442	1.29860	1.20931	1.13269	1.06602	1.00737	0.95528
21	1.73262	1.57109	1.44146	1.33431	1.24380	1.16602	1.09828	1.03862
22	1.98262	1.77109	1.60812	1.47717	1.36880	1.27713	1.19828	1.12953
23	2.31596	2.02109	1.80812	1.64384	1.51165	1.40213	1.30939	1.22953
24	2.81596	2.35442	2.05812	1.84384	1.67832	1.54499	1.43439	1.34064

M \ N	25	26	27	28	29	30	31	32
25	3.81596	2.85442	2.39146	2.09384	1.87832	1.71165	1.57725	1.46564
26		3.85442	2.89146	2.42717	2.12832	1.91165	1.74391	1.60850
27			3.89146	2.92717	2.46165	2.16165	1.94391	1.77516
28				3.92717	2.96165	2.49499	2.19391	1.97516
29					3.96165	2.99499	2.52725	2.22516
30						3.99499	3.02725	2.55850
31							4.02725	3.05850
32								4.05850

EXPECTED VALUES OF EXPONENTIAL ORDER STATISTICS

M \ N	33	34	35	36	37	38	39	40
1	0.03030	0.02941	0.02857	0.02778	0.02703	0.02632	0.02564	0.02500
2	0.06155	0.05971	0.05798	0.05635	0.05480	0.05334	0.05196	0.05064
3	0.09381	0.09096	0.08829	0.08576	0.08338	0.08112	0.07898	0.07696
4	0.12714	0.12322	0.11954	0.11606	0.11279	0.10969	0.10676	0.10398
5	0.16163	0.15656	0.15179	0.14731	0.14309	0.13910	0.13533	0.13176
6	0.19734	0.19104	0.18513	0.17957	0.17434	0.16941	0.16474	0.16033
7	0.23438	0.22675	0.21961	0.21291	0.20660	0.20066	0.19505	0.18974
8	0.27284	0.26379	0.25532	0.24739	0.23993	0.23291	0.22630	0.22005
9	0.31284	0.30225	0.29236	0.28310	0.27442	0.26625	0.25856	0.25130
10	0.35451	0.34225	0.33082	0.32014	0.31013	0.30073	0.29189	0.28356
11	0.39798	0.38392	0.37082	0.35860	0.34717	0.33645	0.32637	0.31689
12	0.44344	0.42740	0.41249	0.39860	0.38563	0.37348	0.36209	0.35137
13	0.49106	0.47285	0.45597	0.44027	0.42563	0.41194	0.39912	0.38709
14	0.54106	0.52047	0.50142	0.48375	0.46729	0.45194	0.43758	0.42412
15	0.59369	0.57047	0.54904	0.52920	0.51077	0.49361	0.47758	0.46258
16	0.64925	0.62310	0.59904	0.57682	0.55623	0.53709	0.51925	0.50258
17	0.70807	0.67866	0.65167	0.62682	0.60385	0.58254	0.56273	0.54425
18	0.77057	0.73748	0.70723	0.67945	0.65385	0.63016	0.60818	0.58773
19	0.83724	0.79998	0.76605	0.73501	0.70648	0.68016	0.65580	0.63318
20	0.90866	0.86665	0.82855	0.79383	0.76203	0.73279	0.70580	0.68080
21	0.98559	0.93808	0.89522	0.85633	0.82086	0.78835	0.75843	0.73080
22	1.06892	1.01500	0.96665	0.92300	0.88336	0.84717	0.81399	0.78343
23	1.15983	1.09833	1.04357	0.99443	0.95002	0.90967	0.87281	0.83899
24	1.25983	1.18924	1.12690	1.07135	1.02145	0.97634	0.93531	0.89781
25	1.37094	1.28924	1.21781	1.15468	1.09838	1.04777	1.00198	0.96031
26	1.49594	1.40035	1.31781	1.24559	1.18171	1.12469	1.07341	1.02698
27	1.63880	1.52535	1.42892	1.34559	1.27262	1.20802	1.15033	1.09841
28	1.80546	1.66821	1.55392	1.45670	1.37262	1.29893	1.23367	1.17533
29	2.00546	1.83488	1.69678	1.58170	1.48373	1.39893	1.32457	1.25867
30	2.25546	2.03488	1.86345	1.72456	1.60873	1.51004	1.42457	1.34957
31	2.58880	2.28488	2.06345	1.89123	1.75159	1.63504	1.53569	1.44957
32	3.08880	2.61821	2.31345	2.09123	1.91825	1.77790	1.66069	1.56069
33	4.08880	3.11821	2.64678	2.34123	2.11825	1.94457	1.80354	1.68569
34		4.11821	3.14678	2.67456	2.36825	2.14457	1.97021	1.82854
35			4.14678	3.17456	2.70159	2.39457	2.17021	1.99521
36				4.17456	3.20159	2.72790	2.42021	2.19521
37					4.20159	3.22790	2.75354	2.44521
38						4.22790	3.25354	2.77854
39							4.25354	3.27854
40								4.27854

EXPECTED VALUES OF EXPONENTIAL ORDER STATISTICS

M \ N	41	42	43	44	45	46	47	48
1	0.02439	0.02381	0.02326	0.02273	0.02222	0.02174	0.02128	0.02083
2	0.04939	0.04820	0.04707	0.04598	0.04495	0.04396	0.04302	0.04211
3	0.07503	0.07320	0.07146	0.06979	0.06821	0.06669	0.06524	0.06385
4	0.10135	0.09884	0.09646	0.09418	0.09201	0.08994	0.08797	0.08607
5	0.12837	0.12516	0.12210	0.11918	0.11641	0.11375	0.11122	0.10880
6	0.15615	0.15218	0.14841	0.14482	0.14141	0.13814	0.13503	0.13205
7	0.18472	0.17996	0.17544	0.17114	0.16705	0.16314	0.15942	0.15586
8	0.21414	0.20853	0.20322	0.19817	0.19336	0.18879	0.18442	0.18025
M \ N	41	42	43	44	45	46	47	48
9	0.24444	0.23794	0.23179	0.22594	0.22039	0.21510	0.21006	0.20525
10	0.27569	0.26825	0.26120	0.25452	0.24817	0.24213	0.23638	0.23090
11	0.30795	0.29950	0.29150	0.28393	0.27674	0.26991	0.26340	0.25721
12	0.34128	0.33176	0.32275	0.31423	0.30615	0.29848	0.29118	0.28424
13	0.37576	0.36509	0.35501	0.34548	0.33645	0.32789	0.31975	0.31202
14	0.41148	0.39957	0.38834	0.37774	0.36770	0.35819	0.34917	0.34059
15	0.44851	0.43529	0.42283	0.41107	0.39996	0.38944	0.37947	0.37000
16	0.48698	0.47232	0.45854	0.44555	0.43329	0.42170	0.41072	0.40030
M \ N	41	42	43	44	45	46	47	48
17	0.52698	0.51078	0.49558	0.48127	0.46778	0.45503	0.44298	0.43155
18	0.56864	0.55078	0.53404	0.51831	0.50349	0.48952	0.47631	0.46381
19	0.61212	0.59245	0.57404	0.55677	0.54053	0.52523	0.51079	0.49714
20	0.65757	0.63593	0.61571	0.59677	0.57899	0.56227	0.54651	0.53163
21	0.70519	0.68138	0.65919	0.63843	0.61899	0.60073	0.58354	0.56734
22	0.75519	0.72900	0.70464	0.68191	0.66066	0.64073	0.62201	0.60438
23	0.80783	0.77900	0.75226	0.72737	0.70413	0.68240	0.66201	0.64284
24	0.86338	0.83163	0.80226	0.77499	0.74959	0.72587	0.70367	0.68284
M \ N	41	42	43	44	45	46	47	48
25	0.92220	0.88719	0.85489	0.82499	0.79721	0.77133	0.74715	0.72451
26	0.98470	0.94601	0.91045	0.87762	0.84721	0.81895	0.79261	0.76798
27	1.05137	1.00851	0.96927	0.93317	0.89984	0.86895	0.84022	0.81344
28	1.12280	1.07518	1.03177	0.99200	0.95540	0.92158	0.89022	0.86106
29	1.19972	1.14661	1.09844	1.05450	1.01422	0.97713	0.94286	0.91106
30	1.28306	1.22353	1.16986	1.12116	1.07672	1.03596	0.99841	0.96369
31	1.37397	1.30687	1.24679	1.19259	1.14339	1.09846	1.05723	1.01924
32	1.47397	1.39777	1.33012	1.26952	1.21481	1.16512	1.11973	1.07807
M \ N	41	42	43	44	45	46	47	48
33	1.58508	1.49777	1.42103	1.35285	1.29174	1.23655	1.18640	1.14057
34	1.71008	1.60889	1.52103	1.44376	1.37507	1.31348	1.25783	1.20723
35	1.85293	1.73389	1.63214	1.54376	1.46598	1.39681	1.33475	1.27866
36	2.01960	1.87674	1.75714	1.65487	1.56598	1.48772	1.41809	1.35559
37	2.21960	2.04341	1.90000	1.77987	1.67709	1.58772	1.50900	1.43892
38	2.46960	2.24341	2.06667	1.92273	1.80209	1.69883	1.60900	1.52983
39	2.80293	2.49341	2.26667	2.08939	1.94495	1.82383	1.72011	1.62983
40	3.30293	2.82674	2.51667	2.28939	2.11161	1.96669	1.84511	1.74094

EXPECTED VALUES OF EXPONENTIAL ORDER STATISTICS

M \ N	41	42	43	44	45	46	47	48
41	4.30293	3.32674	2.85000	2.53939	2.31161	2.13335	1.98796	1.86594
42		4.32674	3.35000	2.87273	2.56161	2.33335	2.15463	2.00880
43			4.35000	3.37273	2.89495	2.58335	2.35463	2.17546
44				4.37273	3.39495	2.91669	2.60463	2.37546
45					4.39495	3.41669	2.93796	2.62546
46						4.41669	3.43796	2.95880
47							4.43796	3.45880
48								4.45880

M \ N	49	50	51	52	53	54	55	56
1	0.02041	0.02000	0.01961	0.01923	0.01887	0.01852	0.01818	0.01786
2	0.04124	0.04041	0.03961	0.03884	0.03810	0.03739	0.03670	0.03604
3	0.06252	0.06124	0.06002	0.05884	0.05771	0.05662	0.05557	0.05456
4	0.08426	0.08252	0.08085	0.07925	0.07771	0.07623	0.07480	0.07343
5	0.10648	0.10426	0.10213	0.10008	0.09811	0.09623	0.09441	0.09266
6	0.12921	0.12648	0.12387	0.12136	0.11895	0.11663	0.11441	0.11226
7	0.15246	0.14921	0.14609	0.14310	0.14022	0.13747	0.13482	0.13226
8	0.17627	0.17246	0.16881	0.16532	0.16196	0.15874	0.15565	0.15267
9	0.20066	0.19627	0.19207	0.18805	0.18419	0.18048	0.17692	0.17351
10	0.22566	0.22066	0.21588	0.21130	0.20691	0.20270	0.19866	0.19478
11	0.25130	0.24566	0.24027	0.23511	0.23017	0.22543	0.22089	0.21652
12	0.27762	0.27130	0.26527	0.25950	0.25398	0.24869	0.24361	0.23874
13	0.30465	0.29762	0.29091	0.28450	0.27837	0.27250	0.26687	0.26147
14	0.33242	0.32465	0.31723	0.31014	0.30337	0.29689	0.29068	0.28473
15	0.36100	0.35242	0.34425	0.33646	0.32901	0.32189	0.31507	0.30854
16	0.39041	0.38100	0.37203	0.36348	0.35533	0.34753	0.34007	0.33293
17	0.42071	0.41041	0.40060	0.39126	0.38235	0.37384	0.36571	0.35793
18	0.45196	0.44071	0.43001	0.41983	0.41013	0.40087	0.39203	0.38357
19	0.48422	0.47196	0.46032	0.44925	0.43870	0.42865	0.41905	0.40988
20	0.51755	0.50422	0.49157	0.47955	0.46811	0.45722	0.44683	0.43691
21	0.55203	0.53755	0.52383	0.51080	0.49842	0.48663	0.47540	0.46469
22	0.58775	0.57203	0.55716	0.54306	0.52967	0.51694	0.50481	0.49326
23	0.62479	0.60775	0.59164	0.57639	0.56192	0.54819	0.53512	0.52267
24	0.66325	0.64479	0.62736	0.61087	0.59526	0.58044	0.56637	0.55297
25	0.70325	0.68325	0.66439	0.64659	0.62974	0.61378	0.59863	0.58422
26	0.74491	0.72325	0.70285	0.68362	0.66546	0.64826	0.63196	0.61648
27	0.78839	0.76491	0.74285	0.72209	0.70249	0.68397	0.66644	0.64982
28	0.83385	0.80839	0.78452	0.76209	0.74095	0.72101	0.70216	0.68430
29	0.88147	0.85385	0.82800	0.80375	0.78095	0.75947	0.73919	0.72001
30	0.93147	0.90147	0.87345	0.84723	0.82262	0.79947	0.77765	0.75705
31	0.98410	0.95147	0.92107	0.89269	0.86610	0.84114	0.81765	0.79551
32	1.03965	1.00410	0.97107	0.94030	0.91155	0.88462	0.85932	0.83551

EXPECTED VALUES OF EXPONENTIAL ORDER STATISTICS

M \ N	49	50	51	52	53	54	55	56
33	1.09848	1.05965	1.02371	0.99030	0.95917	0.93007	0.90280	0.87718
34	1.16098	1.11848	1.07926	1.04294	1.00917	0.97769	0.94825	0.92066
35	1.22764	1.18098	1.13808	1.09849	1.06180	1.02769	0.99587	0.96611
36	1.29907	1.24764	1.20058	1.15731	1.11736	1.08032	1.04587	1.01373
37	1.37599	1.31907	1.26725	1.21981	1.17618	1.13588	1.09850	1.06373
38	1.45933	1.39599	1.33868	1.28648	1.23868	1.19470	1.15406	1.11636
39	1.55024	1.47933	1.41560	1.35791	1.30535	1.25720	1.21288	1.17192
40	1.65024	1.57024	1.49894	1.43483	1.37678	1.32387	1.27538	1.23074

M \ N	49	50	51	52	53	54	55	56
41	1.76135	1.67024	1.58984	1.51817	1.45370	1.39530	1.34205	1.29324
42	1.88635	1.78135	1.68984	1.60908	1.53703	1.47222	1.41348	1.35991
43	2.02921	1.90635	1.80096	1.70908	1.62794	1.55555	1.49040	1.43134
44	2.19587	2.04921	1.92596	1.82019	1.72794	1.64646	1.57373	1.50826
45	2.39587	2.21587	2.06881	1.94519	1.83905	1.74646	1.66464	1.59159
46	2.64587	2.41587	2.23548	2.08804	1.96405	1.85757	1.76464	1.68250
47	2.97921	2.66587	2.43548	2.25471	2.10691	1.98257	1.87576	1.78250
48	3.47921	2.99921	2.68548	2.45471	2.27358	2.12543	2.00076	1.89361

M \ N	49	50	51	52	53	54	55	56
49	4.47921	3.49921	3.01881	2.70471	2.47358	2.29210	2.14361	2.01861
50		4.49921	3.51881	3.03804	2.72358	2.49210	2.31028	2.16147
51			4.51881	3.53804	3.05691	2.74210	2.51028	2.32814
52				4.53804	3.55691	3.07543	2.76028	2.52814
53					4.55691	3.57543	3.09361	2.77814
54						4.57543	3.59361	3.11147
55							4.59361	3.61147
56								4.61147

M \ N	57	58	59	60	61	62	63	64
1	0.01754	0.01724	0.01695	0.01667	0.01639	0.01613	0.01587	0.01563
2	0.03540	0.03479	0.03419	0.03362	0.03306	0.03252	0.03200	0.03150
3	0.05358	0.05264	0.05173	0.05086	0.05001	0.04919	0.04840	0.04763
4	0.07210	0.07082	0.06959	0.06840	0.06725	0.06614	0.06506	0.06402
5	0.09097	0.08934	0.08777	0.08626	0.08479	0.08338	0.08201	0.08069
6	0.11020	0.10821	0.10629	0.10444	0.10265	0.10092	0.09925	0.09764
7	0.12981	0.12744	0.12516	0.12296	0.12083	0.11878	0.11680	0.11488
8	0.14981	0.14705	0.14439	0.14183	0.13935	0.13696	0.13465	0.13242

M \ N	57	58	59	60	61	62	63	64
9	0.17022	0.16705	0.16400	0.16106	0.15822	0.15548	0.15284	0.15028
10	0.19105	0.18746	0.18400	0.18067	0.17745	0.17435	0.17135	0.16846
11	0.21233	0.20829	0.20441	0.20067	0.19706	0.19358	0.19022	0.18698
12	0.23407	0.22957	0.22524	0.22107	0.21706	0.21319	0.20945	0.20585
13	0.25629	0.25131	0.24652	0.24191	0.23747	0.23319	0.22906	0.22508
14	0.27901	0.27353	0.26826	0.26318	0.25830	0.25360	0.24906	0.24469
15	0.30227	0.29626	0.29048	0.28492	0.27958	0.27443	0.26947	0.26469
16	0.32608	0.31951	0.31321	0.30714	0.30132	0.29571	0.29030	0.28509

EXPECTED VALUES OF EXPONENTIAL ORDER STATISTICS

M \ N	57	58	59	60	61	62	63	64
17	0.35047	0.34332	0.33646	0.32987	0.32354	0.31744	0.31158	0.30593
18	0.37547	0.36771	0.36027	0.35313	0.34627	0.33967	0.33332	0.32720
19	0.40111	0.39271	0.38466	0.37694	0.36952	0.36239	0.35554	0.34894
20	0.42743	0.41835	0.40966	0.40133	0.39333	0.38565	0.37827	0.37117
21	0.45445	0.44467	0.43530	0.42633	0.41772	0.40946	0.40152	0.39389
22	0.48223	0.47170	0.46162	0.45197	0.44272	0.43385	0.42533	0.41715
23	0.51080	0.49947	0.48864	0.47828	0.46836	0.45885	0.44972	0.44096
24	0.54021	0.52804	0.51642	0.50531	0.49468	0.48449	0.47472	0.46535
25	0.57052	0.55746	0.54499	0.53309	0.52170	0.51081	0.50036	0.49035
26	0.60177	0.58776	0.57441	0.56166	0.54948	0.53783	0.52668	0.51599
27	0.63403	0.61901	0.60471	0.59107	0.57805	0.56561	0.55371	0.54230
28	0.66736	0.65127	0.63596	0.62138	0.60747	0.59418	0.58148	0.56933
29	0.70184	0.68460	0.66822	0.65263	0.63777	0.62359	0.61006	0.59711
30	0.73756	0.71908	0.70155	0.68488	0.66902	0.65390	0.63947	0.62568
31	0.77459	0.75480	0.73603	0.71822	0.70128	0.68515	0.66977	0.65509
32	0.81306	0.79183	0.77175	0.75270	0.73461	0.71741	0.70102	0.68540
33	0.85306	0.83030	0.80878	0.78841	0.76909	0.75074	0.73328	0.71665
34	0.89472	0.87030	0.84725	0.82545	0.80481	0.78522	0.76661	0.74890
35	0.93820	0.91196	0.88725	0.86391	0.84184	0.82094	0.80109	0.78224
36	0.98365	0.95544	0.92891	0.90391	0.88031	0.85797	0.83681	0.81672
37	1.03127	1.00090	0.97239	0.94558	0.92031	0.89643	0.87385	0.85243
38	1.08127	1.04851	1.01785	0.98906	0.96197	0.93643	0.91231	0.88947
39	1.13391	1.09851	1.06546	1.03451	1.00545	0.97810	0.95231	0.92793
40	1.18946	1.15115	1.11546	1.08213	1.05091	1.02158	0.99397	0.96793
41	1.24828	1.20670	1.16810	1.13213	1.09852	1.06703	1.03745	1.00960
42	1.31078	1.26553	1.22365	1.18476	1.14852	1.11465	1.08291	1.05308
43	1.37745	1.32803	1.28247	1.24032	1.20116	1.16465	1.13053	1.09853
44	1.44888	1.39469	1.34497	1.29914	1.25671	1.21728	1.18053	1.14615
45	1.52580	1.46612	1.41164	1.36164	1.31553	1.27284	1.23316	1.19615
46	1.60914	1.54304	1.48307	1.42831	1.37803	1.33166	1.28871	1.24878
47	1.70004	1.62638	1.55999	1.49974	1.44470	1.39416	1.34754	1.30434
48	1.80004	1.71729	1.64333	1.57666	1.51613	1.46083	1.41004	1.36316
49	1.91116	1.81729	1.73424	1.65999	1.59305	1.53226	1.47670	1.42566
50	2.03616	1.92840	1.83424	1.75090	1.67639	1.60918	1.54813	1.49233
51	2.17901	2.05340	1.94535	1.85090	1.76730	1.69252	1.62506	1.56376
52	2.34568	2.19625	2.07035	1.96201	1.86730	1.78342	1.70839	1.64068
53	2.54568	2.36292	2.21320	2.08701	1.97841	1.88342	1.79930	1.72401
54	2.79568	2.56292	2.37987	2.22987	2.10341	1.99454	1.89930	1.81492
55	3.12901	2.81292	2.57987	2.39654	2.24626	2.11954	2.01041	1.91492
56	3.62901	3.14625	2.82987	2.59654	2.41293	2.26239	2.13541	2.02603

EXPECTED VALUES OF EXPONENTIAL ORDER STATISTICS

M \ N	57	58	59	60	61	62	63	64
57	4.62901	3.64625	3.16320	2.84654	2.61293	2.42906	2.27827	2.15103
58		4.64625	3.66320	3.17987	2.86293	2.62906	2.44493	2.29389
59			4.66320	3.67987	3.19626	2.87906	2.64493	2.46056
60				4.67987	3.69626	3.21239	2.89493	2.66056
61					4.69626	3.71239	3.22827	2.91056
62						4.71239	3.72827	3.24389
63							4.72827	3.74389
64								4.74389

M \ N	65	66	67	68	69	70	71	72
1	0.01538	0.01515	0.01493	0.01471	0.01449	0.01429	0.01408	0.01389
2	0.03101	0.03054	0.03008	0.02963	0.02920	0.02878	0.02837	0.02797
3	0.04688	0.04616	0.04546	0.04478	0.04412	0.04348	0.04286	0.04226
4	0.06301	0.06203	0.06109	0.06017	0.05928	0.05841	0.05757	0.05675
5	0.07941	0.07816	0.07696	0.07579	0.07466	0.07356	0.07249	0.07146
6	0.09607	0.09456	0.09309	0.09167	0.09029	0.08895	0.08765	0.08638
7	0.11302	0.11122	0.10948	0.10779	0.10616	0.10457	0.10303	0.10153
8	0.13026	0.12817	0.12615	0.12419	0.12229	0.12044	0.11866	0.11692

M \ N	65	66	67	68	69	70	71	72
9	0.14781	0.14541	0.14310	0.14085	0.13868	0.13657	0.13453	0.13254
10	0.16566	0.16296	0.16034	0.15780	0.15535	0.15297	0.15066	0.14842
11	0.18385	0.18081	0.17788	0.17505	0.17230	0.16963	0.16705	0.16455
12	0.20236	0.19900	0.19574	0.19259	0.18954	0.18658	0.18372	0.18094
13	0.22123	0.21752	0.21392	0.21045	0.20708	0.20382	0.20067	0.19761
14	0.24046	0.23638	0.23244	0.22863	0.22494	0.22137	0.21791	0.21456
15	0.26007	0.25561	0.25131	0.24715	0.24312	0.23922	0.23545	0.23180
16	0.28007	0.27522	0.27054	0.26601	0.26164	0.25741	0.25331	0.24934

M \ N	65	66	67	68	69	70	71	72
17	0.30048	0.29522	0.29015	0.28525	0.28051	0.27592	0.27149	0.26720
18	0.32131	0.31563	0.31015	0.30485	0.29974	0.29479	0.29001	0.28538
19	0.34259	0.33646	0.33056	0.32485	0.31935	0.31402	0.30888	0.30390
20	0.36433	0.35774	0.35139	0.34526	0.33935	0.33363	0.32811	0.32277
21	0.38655	0.37948	0.37267	0.36609	0.35975	0.35363	0.34772	0.34200
22	0.40928	0.40170	0.39440	0.38737	0.38059	0.37404	0.36772	0.36160
23	0.43253	0.42443	0.41663	0.40911	0.40186	0.39487	0.38812	0.38160
24	0.45634	0.44768	0.43935	0.43133	0.42360	0.41615	0.40896	0.40201

M \ N	65	66	67	68	69	70	71	72
25	0.48073	0.47149	0.46261	0.45406	0.44583	0.43789	0.43023	0.42285
26	0.50573	0.49588	0.48642	0.47732	0.46855	0.46011	0.45197	0.44412
27	0.53137	0.52088	0.51081	0.50113	0.49181	0.48284	0.47420	0.46586
28	0.55769	0.54653	0.53581	0.52552	0.51562	0.50609	0.49692	0.48808
29	0.58472	0.57284	0.56145	0.55052	0.54001	0.52990	0.52018	0.51081
30	0.61249	0.59987	0.58777	0.57616	0.56501	0.55429	0.54399	0.53407
31	0.64107	0.62765	0.61479	0.60247	0.59065	0.57929	0.56838	0.55788
32	0.67048	0.65622	0.64257	0.62950	0.61696	0.60493	0.59338	0.58227

EXPECTED VALUES OF EXPONENTIAL ORDER STATISTICS

M \ N	65	66	67	68	69	70	71	72
33	0.70078	0.68563	0.67114	0.65728	0.64399	0.63125	0.61902	0.60727
34	0.73203	0.71593	0.70055	0.68585	0.67177	0.65828	0.64534	0.63291
35	0.76429	0.74718	0.73086	0.71526	0.70034	0.68606	0.67236	0.65922
36	0.79762	0.77944	0.76211	0.74556	0.72975	0.71463	0.70014	0.68625
37	0.83210	0.81277	0.79437	0.77681	0.76006	0.74404	0.72871	0.71403
38	0.86782	0.84726	0.82770	0.80907	0.79131	0.77434	0.75812	0.74260
39	0.90486	0.88297	0.86218	0.84240	0.82356	0.80559	0.78843	0.77201
40	0.94332	0.92001	0.89790	0.87689	0.85690	0.83785	0.81968	0.80231
M \ N	65	66	67	68	69	70	71	72
41	0.98332	0.95847	0.93493	0.91260	0.89138	0.87118	0.85193	0.83356
42	1.02498	0.99847	0.97339	0.94964	0.92709	0.90567	0.88527	0.86582
43	1.06846	1.04014	1.01339	0.98810	0.96413	0.94138	0.91975	0.89916
44	1.11392	1.08361	1.05506	1.02810	1.00259	0.97842	0.95546	0.93364
45	1.16154	1.12907	1.09854	1.06977	1.04259	1.01688	0.99250	0.96935
46	1.21154	1.17669	1.14399	1.11325	1.08426	1.05688	1.03096	1.00639
47	1.26417	1.22669	1.19161	1.15870	1.12774	1.09855	1.07096	1.04485
48	1.31972	1.27932	1.24161	1.20632	1.17319	1.14202	1.11263	1.08485
M \ N	65	66	67	68	69	70	71	72
49	1.37855	1.33487	1.29424	1.25632	1.22081	1.18748	1.15611	1.12652
50	1.44105	1.39370	1.34980	1.30895	1.27081	1.23510	1.20156	1.17000
51	1.50771	1.45620	1.40862	1.36451	1.32344	1.28510	1.24918	1.21545
52	1.57914	1.52286	1.47112	1.42333	1.37900	1.33773	1.29918	1.26307
53	1.65606	1.59429	1.53779	1.48583	1.43782	1.39328	1.35181	1.31307
54	1.73940	1.67122	1.60922	1.55250	1.50032	1.45211	1.40737	1.36570
55	1.83031	1.75455	1.68614	1.62392	1.56699	1.51461	1.46619	1.42126
56	1.93031	1.84546	1.76948	1.70085	1.63842	1.58127	1.52869	1.48008
M \ N	65	66	67	68	69	70	71	72
57	2.04142	1.94546	1.86038	1.78418	1.71534	1.65270	1.59536	1.54258
58	2.16642	2.05657	1.96038	1.87509	1.79867	1.72963	1.66679	1.60925
59	2.30928	2.18157	2.07150	1.97509	1.88958	1.81296	1.74371	1.68068
60	2.47594	2.32443	2.19650	2.08620	1.98958	1.90387	1.82704	1.75760
61	2.67594	2.49109	2.33935	2.21120	2.10069	2.00387	1.91795	1.84093
62	2.92594	2.69109	2.50602	2.35406	2.22569	2.11498	2.01795	1.93184
63	3.25928	2.94109	2.70602	2.52072	2.36855	2.23998	2.12906	2.03184
64	3.75928	3.27443	2.95602	2.72072	2.53522	2.38284	2.25406	2.14295
M \ N	65	66	67	68	69	70	71	72
65	4.75928	3.77443	3.28935	2.97072	2.73522	2.54950	2.39692	2.26795
66		4.77443	3.78935	3.30406	2.98522	2.74950	2.56359	2.41081
67			4.78935	3.80406	3.31855	2.99950	2.76359	2.57748
68				4.80406	3.81855	3.33284	3.01359	2.77748
69					4.81855	3.83284	3.34692	3.02748
70						4.83284	3.84692	3.36081
71							4.84692	3.86081
72								4.86081

EXPECTED VALUES OF EXPONENTIAL ORDER STATISTICS

M \ N	73	74	75	76	77	78	79	80
1	0.01370	0.01351	0.01333	0.01316	0.01299	0.01282	0.01266	0.01250
2	0.02759	0.02721	0.02685	0.02649	0.02614	0.02581	0.02548	0.02516
3	0.04167	0.04110	0.04055	0.04000	0.03948	0.03897	0.03847	0.03798
4	0.05596	0.05519	0.05443	0.05370	0.05299	0.05230	0.05162	0.05097
5	0.07045	0.06947	0.06852	0.06759	0.06669	0.06581	0.06496	0.06412
6	0.08516	0.08396	0.08280	0.08168	0.08058	0.07951	0.07847	0.07746
7	0.10008	0.09867	0.09730	0.09596	0.09466	0.09340	0.09217	0.09097
8	0.11523	0.11360	0.11200	0.11046	0.10895	0.10748	0.10606	0.10467
9	0.13062	0.12875	0.12693	0.12516	0.12344	0.12177	0.12014	0.11856
10	0.14624	0.14413	0.14208	0.14009	0.13815	0.13626	0.13443	0.13264
11	0.16212	0.15976	0.15746	0.15524	0.15307	0.15097	0.14892	0.14693
12	0.17824	0.17563	0.17309	0.17062	0.16823	0.16589	0.16363	0.16142
13	0.19464	0.19176	0.18896	0.18625	0.18361	0.18105	0.17855	0.17613
14	0.21131	0.20815	0.20509	0.20212	0.19923	0.19643	0.19370	0.19105
15	0.22825	0.22482	0.22149	0.21825	0.21511	0.21206	0.20909	0.20620
16	0.24550	0.24177	0.23815	0.23464	0.23124	0.22793	0.22471	0.22159
17	0.26304	0.25901	0.25510	0.25131	0.24763	0.24406	0.24059	0.23721
18	0.28090	0.27655	0.27234	0.26826	0.26430	0.26045	0.25672	0.25309
19	0.29908	0.29441	0.28989	0.28550	0.28125	0.27712	0.27311	0.26922
20	0.31760	0.31259	0.30774	0.30304	0.29849	0.29407	0.28978	0.28561
21	0.33646	0.33111	0.32593	0.32090	0.31603	0.31131	0.30672	0.30228
22	0.35570	0.34998	0.34444	0.33908	0.33389	0.32885	0.32397	0.31922
23	0.37530	0.36921	0.36331	0.35760	0.35207	0.34671	0.34151	0.33647
24	0.39530	0.38882	0.38254	0.37647	0.37059	0.36489	0.35937	0.35401
25	0.41571	0.40882	0.40215	0.39570	0.38946	0.38341	0.37755	0.37187
26	0.43654	0.42923	0.42215	0.41531	0.40869	0.40228	0.39607	0.39005
27	0.45782	0.45006	0.44256	0.43531	0.42830	0.42151	0.41494	0.40857
28	0.47956	0.47134	0.46339	0.45572	0.44830	0.44112	0.43417	0.42744
29	0.50178	0.49307	0.48467	0.47655	0.46870	0.46112	0.45377	0.44667
30	0.52451	0.51530	0.50641	0.49783	0.48954	0.48152	0.47377	0.46627
31	0.54777	0.53802	0.52863	0.51957	0.51081	0.50236	0.49418	0.48627
32	0.57158	0.56128	0.55136	0.54179	0.53255	0.52363	0.51502	0.50668
33	0.59597	0.58509	0.57461	0.56451	0.55477	0.54537	0.53629	0.52752
34	0.62097	0.60948	0.59842	0.58777	0.57750	0.56760	0.55803	0.54879
35	0.64661	0.63448	0.62281	0.61158	0.60076	0.59032	0.58025	0.57053
36	0.67292	0.66012	0.64781	0.63597	0.62457	0.61358	0.60298	0.59275
37	0.69995	0.68644	0.67345	0.66097	0.64896	0.63739	0.62624	0.61548
38	0.72773	0.71346	0.69977	0.68661	0.67396	0.66178	0.65005	0.63874
39	0.75630	0.74124	0.72680	0.71293	0.69960	0.68678	0.67444	0.66255
40	0.78571	0.76981	0.75457	0.73995	0.72591	0.71242	0.69944	0.68694

EXPECTED VALUES OF EXPONENTIAL ORDER STATISTICS

M \ N	73	74	75	76	77	78	79	80
41	0.81601	0.79922	0.78315	0.76773	0.75294	0.73873	0.72508	0.71194
42	0.84726	0.82953	0.81256	0.79630	0.78072	0.76576	0.75139	0.73758
43	0.87952	0.86078	0.84286	0.82572	0.80929	0.79354	0.77842	0.76389
44	0.91285	0.89304	0.87411	0.85602	0.83870	0.82211	0.80620	0.79092
45	0.94734	0.92637	0.90637	0.88727	0.86901	0.85152	0.83477	0.81870
46	0.98305	0.96085	0.93970	0.91953	0.90026	0.88183	0.86418	0.84727
47	1.02009	0.99657	0.97418	0.95286	0.93251	0.91308	0.89448	0.87668
48	1.05855	1.03360	1.00990	0.98734	0.96585	0.94533	0.92573	0.90698
49	1.09855	1.07206	1.04694	1.02306	1.00033	0.97867	0.95799	0.93823
50	1.14022	1.11206	1.08540	1.06009	1.03604	1.01315	0.99133	0.97049
51	1.18370	1.15373	1.12540	1.09856	1.07308	1.04886	1.02581	1.00383
52	1.22915	1.19721	1.16706	1.13856	1.11154	1.08590	1.06152	1.03831
53	1.27677	1.24266	1.21054	1.18022	1.15154	1.12436	1.09856	1.07402
54	1.32677	1.29028	1.25600	1.22370	1.19321	1.16436	1.13702	1.11106
55	1.37940	1.34028	1.30362	1.26915	1.23669	1.20603	1.17702	1.14952
56	1.43496	1.39291	1.35362	1.31677	1.28214	1.24951	1.21869	1.18952
57	1.49378	1.44847	1.40625	1.36677	1.32976	1.29496	1.26217	1.23119
58	1.55628	1.50729	1.46180	1.41941	1.37976	1.34258	1.30762	1.27467
59	1.62295	1.56979	1.52063	1.47496	1.43239	1.39258	1.35524	1.32012
60	1.69438	1.63646	1.58313	1.53378	1.48795	1.44521	1.40524	1.36774
61	1.77130	1.70789	1.64979	1.59628	1.54677	1.50077	1.45787	1.41774
62	1.85463	1.78481	1.72122	1.66295	1.60927	1.55959	1.51343	1.47037
63	1.94554	1.86814	1.79814	1.73438	1.67594	1.62209	1.57225	1.52593
64	2.04554	1.95905	1.88148	1.81130	1.74737	1.68876	1.63475	1.58475
65	2.15665	2.05905	1.97239	1.89464	1.82429	1.76019	1.70142	1.64725
66	2.28165	2.17017	2.07239	1.98555	1.90762	1.83711	1.77285	1.71392
67	2.42451	2.29517	2.18350	2.08555	1.99853	1.92044	1.84977	1.78535
68	2.59118	2.43802	2.30850	2.19666	2.09853	2.01135	1.93310	1.86227
69	2.79118	2.60469	2.45136	2.32166	2.20964	2.11135	2.02401	1.94560
70	3.04118	2.80469	2.61802	2.46451	2.33464	2.22246	2.12401	2.03651
71	3.37451	3.05469	2.81802	2.63118	2.47750	2.34746	2.23512	2.13651
72	3.87451	3.38802	3.06802	2.83118	2.64417	2.49032	2.36012	2.24762
73	4.87451	3.88802	3.40136	3.08118	2.84417	2.65699	2.50298	2.37262
74		4.88802	3.90136	3.41451	3.09417	2.85699	2.66965	2.51548
75			4.90136	3.91451	3.42750	3.10699	2.86965	2.68215
76				4.91451	3.92750	3.44032	3.11965	2.88215
77					4.92750	3.94032	3.45298	3.13215
78						4.94032	3.95298	3.46548
79							4.95298	3.96548
80								4.96548

EXPECTED VALUES OF EXPONENTIAL ORDER STATISTICS

M \ N	81	82	83	84	85	86	87	88
1	0.01235	0.01220	0.01205	0.01190	0.01176	0.01163	0.01149	0.01136
2	0.02485	0.02454	0.02424	0.02395	0.02367	0.02339	0.02312	0.02286
3	0.03750	0.03704	0.03659	0.03615	0.03572	0.03530	0.03489	0.03449
4	0.05032	0.04970	0.04909	0.04849	0.04791	0.04735	0.04679	0.04625
5	0.06331	0.06252	0.06175	0.06099	0.06026	0.05954	0.05884	0.05816
6	0.07647	0.07551	0.07457	0.07365	0.07276	0.07189	0.07103	0.07020
7	0.08980	0.08866	0.08755	0.08647	0.08542	0.08439	0.08338	0.08240
8	0.10332	0.10200	0.10071	0.09946	0.09824	0.09704	0.09588	0.09474
9	0.11701	0.11551	0.11405	0.11262	0.11122	0.10987	0.10854	0.10724
10	0.13090	0.12921	0.12756	0.12595	0.12438	0.12285	0.12136	0.11990
11	0.14499	0.14310	0.14126	0.13946	0.13772	0.13601	0.13435	0.13272
12	0.15927	0.15718	0.15515	0.15316	0.15123	0.14934	0.14750	0.14571
13	0.17377	0.17147	0.16923	0.16705	0.16493	0.16286	0.16084	0.15887
14	0.18847	0.18596	0.18352	0.18114	0.17882	0.17656	0.17435	0.17220
15	0.20340	0.20067	0.19801	0.19542	0.19290	0.19044	0.18805	0.18571
16	0.21855	0.21559	0.21272	0.20991	0.20719	0.20453	0.20194	0.19941
17	0.23393	0.23074	0.22764	0.22462	0.22168	0.21881	0.21602	0.21330
18	0.24956	0.24613	0.24279	0.23955	0.23639	0.23331	0.23031	0.22739
19	0.26543	0.26175	0.25818	0.25470	0.25131	0.24801	0.24480	0.24167
20	0.28156	0.27763	0.27380	0.27008	0.26646	0.26294	0.25951	0.25617
21	0.29795	0.29376	0.28968	0.28571	0.28185	0.27809	0.27443	0.27087
22	0.31462	0.31015	0.30580	0.30158	0.29747	0.29347	0.28958	0.28580
23	0.33157	0.32682	0.32220	0.31771	0.31334	0.30910	0.30497	0.30095
24	0.34881	0.34377	0.33886	0.33410	0.32947	0.32497	0.32059	0.31633
25	0.36636	0.36101	0.35581	0.35077	0.34587	0.34110	0.33647	0.33196
26	0.38421	0.37855	0.37306	0.36772	0.36253	0.35750	0.35260	0.34783
27	0.40239	0.39641	0.39060	0.38496	0.37948	0.37416	0.36899	0.36396
28	0.42091	0.41459	0.40846	0.40250	0.39672	0.39111	0.38566	0.38035
29	0.43978	0.43311	0.42664	0.42036	0.41427	0.40835	0.40261	0.39702
30	0.45901	0.45198	0.44516	0.43854	0.43213	0.42590	0.41985	0.41397
31	0.47862	0.47121	0.46402	0.45706	0.45031	0.44375	0.43739	0.43121
32	0.49862	0.49081	0.48326	0.47593	0.46883	0.46194	0.45525	0.44875
33	0.51903	0.51081	0.50286	0.49516	0.48769	0.48045	0.47343	0.46661
34	0.53986	0.53122	0.52286	0.51477	0.50692	0.49932	0.49195	0.48479
35	0.56114	0.55206	0.54327	0.53477	0.52653	0.51855	0.51082	0.50331
36	0.58288	0.57333	0.56410	0.55518	0.54653	0.53816	0.53005	0.52218
37	0.60510	0.59507	0.58538	0.57601	0.56694	0.55816	0.54965	0.54141
38	0.62783	0.61729	0.60712	0.59729	0.58777	0.57857	0.56965	0.56102
39	0.65108	0.64002	0.62934	0.61902	0.60905	0.59940	0.59006	0.58102
40	0.67489	0.66328	0.65207	0.64125	0.63079	0.62068	0.61090	0.60143

EXPECTED VALUES OF EXPONENTIAL ORDER STATISTICS

M \ N	81	82	83	84	85	86	87	88
41	0.69928	0.68709	0.67533	0.66397	0.65301	0.64242	0.63217	0.62226
42	0.72428	0.71148	0.69913	0.68723	0.67574	0.66464	0.65391	0.64354
43	0.74992	0.73648	0.72353	0.71104	0.69899	0.68737	0.67613	0.66528
44	0.77624	0.76212	0.74853	0.73543	0.72280	0.71062	0.69886	0.68750
45	0.80327	0.78843	0.77417	0.76043	0.74719	0.73443	0.72212	0.71022
46	0.83104	0.81546	0.80048	0.78607	0.77219	0.75882	0.74593	0.73348
47	0.85961	0.84324	0.82751	0.81239	0.79784	0.78382	0.77032	0.75729
48	0.88903	0.87181	0.85529	0.83941	0.82415	0.80946	0.79532	0.78168
M \ N	81	82	83	84	85	86	87	88
49	0.91933	0.90122	0.88386	0.86719	0.85118	0.83578	0.82096	0.80668
50	0.95058	0.93152	0.91327	0.89576	0.87896	0.86281	0.84727	0.83232
51	0.98284	0.96277	0.94357	0.92517	0.90753	0.89058	0.87430	0.85864
52	1.01617	0.99503	0.97482	0.95548	0.93694	0.91916	0.90208	0.88566
53	1.05065	1.02837	1.00708	0.98673	0.96724	0.94857	0.93065	0.91344
54	1.08637	1.06285	1.04041	1.01899	0.99849	0.97887	0.96006	0.94201
55	1.12341	1.09856	1.07490	1.05232	1.03075	1.01012	0.99036	0.97143
56	1.16187	1.13560	1.11061	1.08680	1.06408	1.04238	1.02161	1.00173
M \ N	81	82	83	84	85	86	87	88
57	1.20187	1.17406	1.14765	1.12252	1.09857	1.07571	1.05387	1.03298
58	1.24353	1.21406	1.18611	1.15955	1.13428	1.11019	1.08721	1.06524
59	1.28701	1.25573	1.22611	1.19801	1.17132	1.14591	1.12169	1.09857
60	1.33247	1.29921	1.26778	1.23801	1.20978	1.18295	1.15740	1.13305
61	1.38009	1.34466	1.31126	1.27968	1.24978	1.22141	1.19444	1.16877
62	1.43009	1.39228	1.35671	1.32316	1.29145	1.26141	1.23290	1.20580
63	1.48272	1.44228	1.40433	1.36861	1.33492	1.30307	1.27290	1.24427
64	1.53827	1.49491	1.45433	1.41623	1.38038	1.34655	1.31457	1.28427
M \ N	81	82	83	84	85	86	87	88
65	1.59710	1.55047	1.50696	1.46623	1.42800	1.39201	1.35805	1.32593
66	1.65960	1.60929	1.56252	1.51886	1.47800	1.43963	1.40350	1.36941
67	1.72626	1.67179	1.62134	1.57442	1.53063	1.48963	1.45112	1.41486
68	1.79769	1.73846	1.68384	1.63324	1.58619	1.54226	1.50112	1.46248
69	1.87461	1.80989	1.75051	1.69574	1.64501	1.59781	1.55375	1.51248
70	1.95795	1.88681	1.82193	1.76241	1.70751	1.65664	1.60931	1.56512
71	2.04886	1.97014	1.89886	1.83384	1.77418	1.71914	1.66813	1.62067
72	2.14886	2.06105	1.98219	1.91076	1.84560	1.78580	1.73063	1.67949
M \ N	81	82	83	84	85	86	87	88
73	2.25997	2.16105	2.07310	1.99410	1.92253	1.85723	1.79730	1.74199
74	2.38497	2.27216	2.17310	2.08500	2.00586	1.93415	1.86873	1.80866
75	2.52782	2.39716	2.28421	2.18500	2.09677	2.01749	1.94565	1.88009
76	2.69449	2.54002	2.40921	2.29612	2.19677	2.10840	2.02898	1.95701
77	2.89449	2.70669	2.55207	2.42112	2.30788	2.20840	2.11989	2.04035
78	3.14449	2.90669	2.71873	2.56397	2.43288	2.31951	2.21989	2.13126
79	3.47782	3.15669	2.91873	2.73064	2.57574	2.44451	2.33100	2.23126
80	3.97782	3.49002	3.16873	2.93064	2.74240	2.58737	2.45600	2.34237

EXPECTED VALUES OF EXPONENTIAL ORDER STATISTICS

M \ N	81	82	83	84	85	86	87	88
81	4.97782	3.99002	3.50207	3.18064	2.94240	2.75403	2.59886	2.46737
82		4.99002	4.00207	3.51397	3.19240	2.95403	2.76553	2.61022
83			5.00207	4.01397	3.52574	3.20403	2.96553	2.77689
84				5.01397	4.02574	3.53737	3.21553	2.97689
85					5.02574	4.03737	3.54886	3.22689
86						5.03737	4.04886	3.56022
87							5.04886	4.06022
88								5.06022

M \ N	89	90	91	92	93	94	95	96
1	0.01124	0.01111	0.01099	0.01087	0.01075	0.01064	0.01053	0.01042
2	0.02260	0.02235	0.02210	0.02186	0.02162	0.02139	0.02116	0.02094
3	0.03409	0.03371	0.03334	0.03297	0.03261	0.03226	0.03192	0.03158
4	0.04572	0.04520	0.04470	0.04421	0.04372	0.04325	0.04279	0.04233
5	0.05749	0.05683	0.05619	0.05557	0.05496	0.05436	0.05378	0.05320
6	0.06939	0.06860	0.06782	0.06706	0.06632	0.06560	0.06489	0.06419
7	0.08144	0.08050	0.07959	0.07869	0.07782	0.07696	0.07612	0.07530
8	0.09363	0.09255	0.09149	0.09046	0.08944	0.08845	0.08749	0.08654
9	0.10598	0.10475	0.10354	0.10236	0.10121	0.10008	0.09898	0.09790
10	0.11848	0.11709	0.11573	0.11441	0.11311	0.11185	0.11061	0.10940
11	0.13114	0.12959	0.12808	0.12660	0.12516	0.12375	0.12237	0.12103
12	0.14396	0.14225	0.14058	0.13895	0.13736	0.13580	0.13428	0.13279
13	0.15695	0.15507	0.15324	0.15145	0.14970	0.14800	0.14633	0.14469
14	0.17010	0.16806	0.16606	0.16411	0.16220	0.16034	0.15852	0.15674
15	0.18344	0.18121	0.17905	0.17693	0.17486	0.17284	0.17087	0.16894
16	0.19695	0.19455	0.19220	0.18992	0.18768	0.18550	0.18337	0.18128
17	0.21065	0.20806	0.20554	0.20307	0.20067	0.19832	0.19603	0.19378
18	0.22454	0.22176	0.21905	0.21641	0.21383	0.21131	0.20885	0.20644
19	0.23862	0.23565	0.23275	0.22992	0.22716	0.22446	0.22183	0.21926
20	0.25291	0.24973	0.24664	0.24362	0.24067	0.23780	0.23499	0.23225
21	0.26740	0.26402	0.26072	0.25751	0.25437	0.25131	0.24832	0.24541
22	0.28211	0.27851	0.27501	0.27159	0.26826	0.26501	0.26184	0.25874
23	0.29703	0.29322	0.28950	0.28588	0.28235	0.27890	0.27554	0.27225
24	0.31218	0.30814	0.30421	0.30037	0.29663	0.29298	0.28943	0.28595
25	0.32757	0.32330	0.31913	0.31508	0.31112	0.30727	0.30351	0.29984
26	0.34319	0.33868	0.33428	0.33000	0.32583	0.32176	0.31780	0.31393
27	0.35907	0.35430	0.34967	0.34515	0.34075	0.33647	0.33229	0.32821
28	0.37520	0.37018	0.36529	0.36054	0.35591	0.35139	0.34699	0.34270
29	0.39159	0.38631	0.38117	0.37616	0.37129	0.36654	0.36192	0.35741
30	0.40826	0.40270	0.39730	0.39204	0.38692	0.38193	0.37707	0.37234
31	0.42520	0.41937	0.41369	0.40817	0.40279	0.39755	0.39246	0.38749
32	0.44245	0.43632	0.43036	0.42456	0.41892	0.41343	0.40808	0.40287

EXPECTED VALUES OF EXPONENTIAL ORDER STATISTICS

M \ N	89	90	91	92	93	94	95	96
33	0.45999	0.45356	0.44731	0.44123	0.43531	0.42956	0.42395	0.41850
34	0.47785	0.47110	0.46455	0.45817	0.45198	0.44595	0.44008	0.43437
35	0.49603	0.48896	0.48209	0.47542	0.46893	0.46262	0.45648	0.45050
36	0.51455	0.50714	0.49995	0.49296	0.48617	0.47957	0.47314	0.46689
37	0.53342	0.52566	0.51813	0.51082	0.50371	0.49681	0.49009	0.48356
38	0.55265	0.54453	0.53665	0.52900	0.52157	0.51435	0.50733	0.50051
39	0.57225	0.56376	0.55552	0.54752	0.53975	0.53221	0.52488	0.51775
40	0.59225	0.58337	0.57475	0.56639	0.55827	0.55039	0.54273	0.53529
41	0.61266	0.60337	0.59435	0.58562	0.57714	0.56891	0.56092	0.55315
42	0.63350	0.62377	0.61435	0.60522	0.59637	0.58778	0.57943	0.57133
43	0.65477	0.64461	0.63476	0.62522	0.61598	0.60701	0.59830	0.58985
44	0.67651	0.66588	0.65560	0.64563	0.63598	0.62661	0.61753	0.60872
45	0.69873	0.68762	0.67687	0.66647	0.65638	0.64661	0.63714	0.62795
46	0.72146	0.70984	0.69861	0.68774	0.67722	0.66702	0.65714	0.64756
47	0.74472	0.73257	0.72083	0.70948	0.69849	0.68786	0.67755	0.66756
48	0.76853	0.75583	0.74356	0.73170	0.72023	0.70913	0.69838	0.68797
49	0.79292	0.77964	0.76682	0.75443	0.74246	0.73087	0.71966	0.70880
50	0.81792	0.80403	0.79063	0.77769	0.76518	0.75309	0.74140	0.73008
51	0.84356	0.82903	0.81502	0.80150	0.78844	0.77582	0.76362	0.75182
52	0.86987	0.85467	0.84002	0.82589	0.81225	0.79908	0.78635	0.77404
53	0.89690	0.88098	0.86566	0.85089	0.83664	0.82289	0.80960	0.79676
54	0.92468	0.90801	0.89197	0.87653	0.86164	0.84728	0.83341	0.82002
55	0.95325	0.93579	0.91900	0.90284	0.88728	0.87228	0.85780	0.84383
56	0.98266	0.96436	0.94678	0.92987	0.91360	0.89792	0.88280	0.86822
57	1.01296	0.99377	0.97535	0.95765	0.94062	0.92423	0.90844	0.89322
58	1.04421	1.02408	1.00476	0.98622	0.96840	0.95126	0.93476	0.91886
59	1.07647	1.05533	1.03506	1.01563	0.99697	0.97904	0.96179	0.94518
60	1.10981	1.08758	1.06631	1.04593	1.02638	1.00761	0.98957	0.97220
61	1.14429	1.12092	1.09857	1.07718	1.05669	1.03702	1.01814	0.99998
62	1.18000	1.15540	1.13191	1.10944	1.08794	1.06732	1.04755	1.02855
63	1.21704	1.19111	1.16639	1.14278	1.12019	1.09857	1.07785	1.05796
64	1.25550	1.22815	1.20210	1.17726	1.15353	1.13083	1.10910	1.08827
65	1.29550	1.26661	1.23914	1.21297	1.18801	1.16417	1.14136	1.11952
66	1.33717	1.30661	1.27760	1.25001	1.22373	1.19865	1.17469	1.15178
67	1.38065	1.34828	1.31760	1.28847	1.26076	1.23436	1.20918	1.18511
68	1.42610	1.39176	1.35927	1.32847	1.29922	1.27140	1.24489	1.21959
69	1.47372	1.43721	1.40275	1.37014	1.33922	1.30986	1.28193	1.25531
70	1.52372	1.48483	1.44820	1.41362	1.38089	1.34986	1.32039	1.29234
71	1.57635	1.53483	1.49582	1.45907	1.42437	1.39153	1.36039	1.33080
72	1.63191	1.58746	1.54582	1.50669	1.46982	1.43501	1.40205	1.37080

EXPECTED VALUES OF EXPONENTIAL ORDER STATISTICS

M \ N	89	90	91	92	93	94	95	96
73	1.69073	1.64302	1.59845	1.55669	1.51744	1.48046	1.44553	1.41247
74	1.75323	1.70184	1.65401	1.60932	1.56744	1.52808	1.49099	1.45595
75	1.81990	1.76434	1.71283	1.66488	1.62007	1.57808	1.53861	1.50140
76	1.89133	1.83101	1.77533	1.72370	1.67563	1.63071	1.58861	1.54902
77	1.96825	1.90244	1.84200	1.78620	1.73445	1.68627	1.64124	1.59902
78	2.05158	1.97936	1.91343	1.85287	1.79695	1.74509	1.69679	1.65166
79	2.14249	2.06269	1.99035	1.92430	1.86362	1.80759	1.75562	1.70721
80	2.24249	2.15360	2.07368	2.00122	1.93505	1.87426	1.81812	1.76603
81	2.35360	2.25360	2.16459	2.08455	2.01197	1.94569	1.88478	1.82853
82	2.47860	2.36471	2.26459	2.17546	2.09530	2.02261	1.95621	1.89520
83	2.62146	2.48971	2.37570	2.27546	2.18621	2.10594	2.03314	1.96663
84	2.78813	2.63257	2.50070	2.38657	2.28621	2.19685	2.11647	2.04355
85	2.98813	2.79924	2.64356	2.51157	2.39732	2.29685	2.20738	2.12689
86	3.23813	2.99924	2.81023	2.65443	2.52232	2.40796	2.30738	2.21779
87	3.57146	3.24924	3.01023	2.82110	2.66518	2.53296	2.41849	2.31779
88	4.07146	3.58257	3.26023	3.02110	2.83185	2.67582	2.54349	2.42891
89	5.07146	4.08257	3.59356	3.27110	3.03185	2.84249	2.68635	2.55391
90		5.08257	4.09356	3.60443	3.28185	3.04249	2.85301	2.69676
91			5.09356	4.10443	3.61518	3.29249	3.05301	2.86343
92				5.10443	4.11518	3.62582	3.30301	3.06343
93					5.11518	4.12582	3.63635	3.31343
94						5.12582	4.13635	3.64676
95							5.13635	4.14676
96								5.14676

M \ N	97	98	99	100	101	102	103	104
1	0.01031	0.01020	0.01010	0.01000	0.00990	0.00980	0.00971	0.00962
2	0.02073	0.02051	0.02031	0.02010	0.01990	0.01970	0.01951	0.01932
3	0.03125	0.03093	0.03061	0.03031	0.03000	0.02970	0.02941	0.02913
4	0.04189	0.04146	0.04103	0.04061	0.04021	0.03981	0.03941	0.03903
5	0.05264	0.05209	0.05156	0.05103	0.05052	0.05001	0.04951	0.04903
6	0.06351	0.06285	0.06220	0.06156	0.06093	0.06032	0.05972	0.05913
7	0.07450	0.07372	0.07295	0.07220	0.07146	0.07074	0.07003	0.06933
8	0.08561	0.08471	0.08382	0.08295	0.08210	0.08126	0.08044	0.07964
9	0.09685	0.09582	0.09481	0.09382	0.09285	0.09190	0.09097	0.09006
10	0.10821	0.10705	0.10592	0.10481	0.10372	0.10265	0.10161	0.10059
11	0.11971	0.11842	0.11715	0.11592	0.11471	0.11352	0.11236	0.11122
12	0.13133	0.12991	0.12852	0.12715	0.12582	0.12451	0.12323	0.12198
13	0.14310	0.14154	0.14001	0.13852	0.13705	0.13562	0.13422	0.13285
14	0.15500	0.15330	0.15164	0.15001	0.14842	0.14686	0.14533	0.14384
15	0.16705	0.16521	0.16340	0.16164	0.15991	0.15822	0.15657	0.15495
16	0.17925	0.17726	0.17531	0.17340	0.17154	0.16972	0.16793	0.16618

EXPECTED VALUES OF EXPONENTIAL ORDER STATISTICS

M \ N	97	98	99	100	101	102	103	104
17	0.19159	0.18945	0.18736	0.18531	0.18331	0.18134	0.17943	0.17755
18	0.20409	0.20180	0.19955	0.19736	0.19521	0.19311	0.19105	0.18904
19	0.21675	0.21430	0.21190	0.20955	0.20726	0.20501	0.20282	0.20067
20	0.22957	0.22696	0.22440	0.22190	0.21945	0.21706	0.21472	0.21243
21	0.24256	0.23978	0.23706	0.23440	0.23180	0.22926	0.22677	0.22434
22	0.25572	0.25276	0.24988	0.24706	0.24430	0.24160	0.23897	0.23639
23	0.26905	0.26592	0.26286	0.25988	0.25696	0.25410	0.25131	0.24858
24	0.28256	0.27925	0.27602	0.27286	0.26978	0.26676	0.26381	0.26093
25	0.29626	0.29277	0.28936	0.28602	0.28276	0.27958	0.27647	0.27343
26	0.31015	0.30647	0.30287	0.29936	0.29592	0.29257	0.28929	0.28609
27	0.32424	0.32036	0.31657	0.31287	0.30926	0.30573	0.30228	0.29891
28	0.33852	0.33444	0.33046	0.32657	0.32277	0.31906	0.31544	0.31189
29	0.35301	0.34873	0.34454	0.34046	0.33647	0.33257	0.32877	0.32505
30	0.36772	0.36322	0.35883	0.35454	0.35036	0.34627	0.34228	0.33838
31	0.38265	0.37792	0.37332	0.36883	0.36444	0.36016	0.35598	0.35190
32	0.39780	0.39285	0.38803	0.38332	0.37873	0.37425	0.36987	0.36560
33	0.41318	0.40800	0.40295	0.39803	0.39322	0.38853	0.38395	0.37949
34	0.42881	0.42339	0.41810	0.41295	0.40793	0.40302	0.39824	0.39357
35	0.44468	0.43901	0.43349	0.42810	0.42285	0.41773	0.41273	0.40786
36	0.46081	0.45488	0.44911	0.44349	0.43800	0.43266	0.42744	0.42235
37	0.47720	0.47101	0.46498	0.45911	0.45339	0.44781	0.44236	0.43705
38	0.49387	0.48741	0.48111	0.47498	0.46901	0.46319	0.45752	0.45198
39	0.51082	0.50407	0.49751	0.49111	0.48489	0.47882	0.47290	0.46713
40	0.52806	0.52102	0.51417	0.50751	0.50101	0.49469	0.48853	0.48252
41	0.54560	0.53826	0.53112	0.52417	0.51741	0.51082	0.50440	0.49814
42	0.56346	0.55581	0.54836	0.54112	0.53407	0.52721	0.52053	0.51401
43	0.58164	0.57366	0.56591	0.55836	0.55102	0.54388	0.53692	0.53014
44	0.60016	0.59185	0.58377	0.57591	0.56827	0.56083	0.55359	0.54654
45	0.61903	0.61036	0.60195	0.59377	0.58581	0.57807	0.57054	0.56320
46	0.63826	0.62923	0.62047	0.61195	0.60367	0.59561	0.58778	0.58015
47	0.65787	0.64846	0.63933	0.63047	0.62185	0.61347	0.60532	0.59739
48	0.67787	0.66807	0.65856	0.64933	0.64037	0.63165	0.62318	0.61494
49	0.69828	0.68807	0.67817	0.66856	0.65923	0.65017	0.64136	0.63279
50	0.71911	0.70848	0.69817	0.68817	0.67847	0.66904	0.65988	0.65098
51	0.74039	0.72931	0.71858	0.70817	0.69807	0.68827	0.67875	0.66949
52	0.76212	0.75059	0.73941	0.72858	0.71807	0.70788	0.69798	0.68836
53	0.78435	0.77233	0.76069	0.74941	0.73848	0.72788	0.71759	0.70759
54	0.80707	0.79455	0.78243	0.77069	0.75931	0.74829	0.73759	0.72720
55	0.83033	0.81728	0.80465	0.79243	0.78059	0.76912	0.75799	0.74720
56	0.85414	0.84053	0.82738	0.81465	0.80233	0.79040	0.77883	0.76761

EXPECTED VALUES OF EXPONENTIAL ORDER STATISTICS

M \ N	97	98	99	100	101	102	103	104
57	0.87853	0.86434	0.85063	0.83738	0.82455	0.81213	0.80010	0.78844
58	0.90353	0.88873	0.87444	0.86063	0.84728	0.83436	0.82184	0.80972
59	0.92917	0.91373	0.89883	0.88444	0.87054	0.85708	0.84407	0.83146
60	0.95549	0.93937	0.92383	0.90883	0.89435	0.88034	0.86679	0.85368
61	0.98251	0.96569	0.94948	0.93383	0.91874	0.90415	0.89005	0.87641
62	1.01029	0.99272	0.97579	0.95948	0.94374	0.92854	0.91386	0.89966
63	1.03886	1.02050	1.00282	0.98579	0.96938	0.95354	0.93825	0.92347
64	1.06827	1.04907	1.03060	1.01282	0.99569	0.97918	0.96325	0.94786
M \ N	97	98	99	100	101	102	103	104
65	1.09858	1.07848	1.05917	1.04060	1.02272	1.00550	0.98889	0.97286
66	1.12983	1.10878	1.08858	1.06917	1.05050	1.03252	1.01520	0.99850
67	1.16209	1.14003	1.11888	1.09858	1.07907	1.06030	1.04223	1.02482
68	1.19542	1.17229	1.15013	1.12888	1.10848	1.08887	1.07001	1.05185
69	1.22990	1.20562	1.18239	1.16013	1.13878	1.11828	1.09858	1.07963
70	1.26562	1.24011	1.21572	1.19239	1.17003	1.14859	1.12799	1.10820
71	1.30265	1.27582	1.25021	1.22572	1.20229	1.17984	1.15830	1.13761
72	1.34111	1.31286	1.28592	1.26021	1.23562	1.21210	1.18955	1.16791
M \ N	97	98	99	100	101	102	103	104
73	1.38111	1.35132	1.32296	1.29592	1.27011	1.24543	1.22180	1.19916
74	1.42278	1.39132	1.36142	1.33296	1.30582	1.27991	1.25514	1.23142
75	1.46626	1.43298	1.40142	1.37142	1.34286	1.31563	1.28962	1.26475
76	1.51171	1.47646	1.44309	1.41142	1.38132	1.35266	1.32533	1.29924
77	1.55933	1.52192	1.48656	1.45309	1.42132	1.39112	1.36237	1.33495
78	1.60933	1.56954	1.53202	1.49656	1.46299	1.43112	1.40083	1.37199
79	1.66196	1.61954	1.57964	1.54202	1.50647	1.47279	1.44083	1.41045
80	1.71752	1.67217	1.62964	1.58964	1.55192	1.51627	1.48250	1.45045
M \ N	97	98	99	100	101	102	103	104
81	1.77634	1.72772	1.68227	1.63964	1.59954	1.56172	1.52598	1.49212
82	1.83884	1.78655	1.73782	1.69227	1.64954	1.60934	1.57143	1.53559
83	1.90551	1.84905	1.79665	1.74782	1.70217	1.65934	1.61905	1.58105
84	1.97694	1.91571	1.85915	1.80665	1.75773	1.71197	1.66905	1.62867
85	2.05386	1.98714	1.92582	1.86915	1.81655	1.76753	1.72168	1.67867
86	2.13720	2.06407	1.99724	1.93582	1.87905	1.82635	1.77724	1.73130
87	2.22810	2.14740	2.07417	2.00724	1.94572	1.88885	1.83606	1.78685
88	2.32810	2.23831	2.15750	2.08417	2.01714	1.95552	1.89856	1.84568
M \ N	97	98	99	100	101	102	103	104
89	2.43922	2.33831	2.24841	2.16750	2.09407	2.02695	1.96523	1.90818
90	2.56422	2.44942	2.34841	2.25841	2.17740	2.10387	2.03666	1.97484
91	2.70707	2.57442	2.45952	2.35841	2.26831	2.18721	2.11358	2.04627
92	2.87374	2.71728	2.58452	2.46952	2.36831	2.27811	2.19691	2.12320
93	3.07374	2.88394	2.72738	2.59452	2.47942	2.37811	2.28782	2.20653
94	3.32374	3.08394	2.89404	2.73738	2.60442	2.48923	2.38782	2.29744
95	3.65707	3.33394	3.09404	2.90404	2.74728	2.61423	2.49893	2.39744
96	4.15707	3.66728	3.34404	3.10404	2.91395	2.75708	2.62393	2.50855

EXPECTED VALUES OF EXPONENTIAL ORDER STATISTICS

M \ N	97	98	99	100	101	102	103	104
97	5.15707	4.16728	3.67738	3.35404	3.11395	2.92375	2.76679	2.63355
98		5.16728	4.17738	3.68738	3.36395	3.12375	2.93346	2.77641
99			5.17738	4.18738	3.69728	3.37375	3.13346	2.94307
100				5.18738	4.19728	3.70708	3.38346	3.14307
101					5.19728	4.20708	3.71679	3.39307
102						5.20708	4.21679	3.72641
103							5.21679	4.22641
104								5.22641

M \ N	105	106	107	108	109	110	111	112
1	0.00952	0.00943	0.00935	0.00926	0.00917	0.00909	0.00901	0.00893
2	0.01914	0.01896	0.01878	0.01861	0.01843	0.01827	0.01810	0.01794
3	0.02885	0.02857	0.02830	0.02804	0.02778	0.02752	0.02727	0.02703
4	0.03865	0.03828	0.03792	0.03756	0.03721	0.03687	0.03653	0.03620
5	0.04855	0.04809	0.04763	0.04718	0.04674	0.04630	0.04588	0.04546
6	0.05855	0.05799	0.05743	0.05689	0.05635	0.05583	0.05531	0.05481
7	0.06865	0.06799	0.06733	0.06669	0.06606	0.06544	0.06484	0.06424
8	0.07886	0.07809	0.07733	0.07659	0.07587	0.07515	0.07445	0.07377

M \ N	105	106	107	108	109	110	111	112
9	0.08917	0.08829	0.08743	0.08659	0.08577	0.08496	0.08416	0.08338
10	0.09958	0.09860	0.09764	0.09669	0.09577	0.09486	0.09397	0.09309
11	0.11011	0.10902	0.10795	0.10690	0.10587	0.10486	0.10387	0.10289
12	0.12075	0.11954	0.11836	0.11721	0.11607	0.11496	0.11387	0.11279
13	0.13150	0.13018	0.12889	0.12762	0.12638	0.12516	0.12397	0.12279
14	0.14237	0.14094	0.13953	0.13815	0.13680	0.13547	0.13417	0.13290
15	0.15336	0.15180	0.15028	0.14879	0.14732	0.14589	0.14448	0.14310
16	0.16447	0.16279	0.16115	0.15954	0.15796	0.15641	0.15490	0.15341

M \ N	105	106	107	108	109	110	111	112
17	0.17571	0.17390	0.17214	0.17041	0.16871	0.16705	0.16542	0.16383
18	0.18707	0.18514	0.18325	0.18140	0.17958	0.17781	0.17606	0.17435
19	0.19856	0.19650	0.19449	0.19251	0.19057	0.18867	0.18681	0.18499
20	0.21019	0.20800	0.20585	0.20375	0.20168	0.19966	0.19768	0.19574
21	0.22196	0.21963	0.21734	0.21511	0.21292	0.21078	0.20867	0.20661
22	0.23386	0.23139	0.22897	0.22660	0.22428	0.22201	0.21978	0.21760
23	0.24591	0.24330	0.24074	0.23823	0.23578	0.23337	0.23102	0.22871
24	0.25811	0.25534	0.25264	0.25000	0.24741	0.24487	0.24238	0.23995

M \ N	105	106	107	108	109	110	111	112
25	0.27045	0.26754	0.26469	0.26190	0.25917	0.25650	0.25388	0.25131
26	0.28295	0.27989	0.27689	0.27395	0.27108	0.26826	0.26551	0.26281
27	0.29561	0.29239	0.28923	0.28614	0.28312	0.28017	0.27727	0.27443
28	0.30843	0.30504	0.30173	0.29849	0.29532	0.29221	0.28918	0.28620
29	0.32142	0.31786	0.31439	0.31099	0.30766	0.30441	0.30122	0.29810
30	0.33457	0.33085	0.32721	0.32365	0.32016	0.31676	0.31342	0.31015
31	0.34791	0.34401	0.34020	0.33647	0.33282	0.32926	0.32576	0.32235
32	0.36142	0.35734	0.35335	0.34946	0.34564	0.34191	0.33826	0.33469

EXPECTED VALUES OF EXPONENTIAL ORDER STATISTICS

M \ N	105	106	107	108	109	110	111	112
33	0.37512	0.37086	0.36669	0.36261	0.35863	0.35473	0.35092	0.34719
34	0.38901	0.38455	0.38020	0.37595	0.37179	0.36772	0.36374	0.35985
35	0.40309	0.39844	0.39390	0.38946	0.38512	0.38088	0.37673	0.37267
36	0.41738	0.41253	0.40779	0.40316	0.39863	0.39421	0.38989	0.38566
37	0.43187	0.42681	0.42187	0.41705	0.41233	0.40773	0.40322	0.39882
38	0.44658	0.44131	0.43616	0.43113	0.42622	0.42142	0.41673	0.41215
39	0.46150	0.45601	0.45065	0.44542	0.44031	0.43531	0.43043	0.42566
40	0.47665	0.47094	0.46536	0.45991	0.45459	0.44940	0.44432	0.43936
41	0.49204	0.48609	0.48028	0.47462	0.46909	0.46368	0.45841	0.45325
42	0.50766	0.50147	0.49543	0.48954	0.48379	0.47818	0.47269	0.46734
43	0.52354	0.51710	0.51082	0.50469	0.49872	0.49288	0.48719	0.48162
44	0.53967	0.53297	0.52644	0.52008	0.51387	0.50781	0.50189	0.49611
45	0.55606	0.54910	0.54232	0.53570	0.52925	0.52296	0.51682	0.51082
46	0.57273	0.56549	0.55845	0.55158	0.54488	0.53834	0.53197	0.52575
47	0.58968	0.58216	0.57484	0.56771	0.56075	0.55397	0.54735	0.54090
48	0.60692	0.59911	0.59151	0.58410	0.57688	0.56984	0.56298	0.55628
49	0.62446	0.61635	0.60846	0.60077	0.59327	0.58597	0.57885	0.57191
50	0.64232	0.63389	0.62570	0.61771	0.60994	0.60236	0.59498	0.58778
51	0.66050	0.65175	0.64324	0.63496	0.62689	0.61903	0.61137	0.60391
52	0.67902	0.66993	0.66110	0.65250	0.64413	0.63598	0.62804	0.62030
53	0.69789	0.68845	0.67928	0.67036	0.66167	0.65322	0.64499	0.63697
54	0.71712	0.70732	0.69780	0.68854	0.67953	0.67077	0.66223	0.65392
55	0.73673	0.72655	0.71667	0.70706	0.69771	0.68862	0.67977	0.67116
56	0.75673	0.74616	0.73590	0.72593	0.71623	0.70680	0.69763	0.68870
57	0.77713	0.76616	0.75550	0.74516	0.73510	0.72532	0.71581	0.70656
58	0.79797	0.78657	0.77550	0.76476	0.75433	0.74419	0.73433	0.72474
59	0.81924	0.80740	0.79591	0.78476	0.77394	0.76342	0.75320	0.74326
60	0.84098	0.82868	0.81675	0.80517	0.79394	0.78303	0.77243	0.76213
61	0.86320	0.85042	0.83802	0.82601	0.81435	0.80303	0.79204	0.78136
62	0.88593	0.87264	0.85976	0.84728	0.83518	0.82344	0.81204	0.80097
63	0.90919	0.89537	0.88198	0.86902	0.85646	0.84427	0.83245	0.82097
64	0.93300	0.91862	0.90471	0.89124	0.87820	0.86555	0.85328	0.84138
65	0.95739	0.94243	0.92797	0.91397	0.90042	0.88729	0.87456	0.86221
66	0.98239	0.96682	0.95178	0.93723	0.92315	0.90951	0.89630	0.88348
67	1.00803	0.99182	0.97617	0.96104	0.94640	0.93224	0.91852	0.90522
68	1.03434	1.01746	1.00117	0.98543	0.97021	0.95549	0.94124	0.92745
69	1.06137	1.04378	1.02681	1.01043	0.99460	0.97930	0.96450	0.95017
70	1.08915	1.07081	1.05312	1.03607	1.01960	1.00369	0.98831	0.97343
71	1.11772	1.09858	1.08015	1.06238	1.04524	1.02869	1.01270	0.99724
72	1.14713	1.12715	1.10793	1.08941	1.07156	1.05433	1.03770	1.02163

EXPECTED VALUES OF EXPONENTIAL ORDER STATISTICS

M \ N	105	106	107	108	109	110	111	112
73	1.17744	1.15657	1.13650	1.11719	1.09858	1.08065	1.06334	1.04663
74	1.20869	1.18687	1.16591	1.14576	1.12636	1.10768	1.08966	1.07227
75	1.24094	1.21812	1.19621	1.17517	1.15493	1.13545	1.11668	1.09859
76	1.27428	1.25038	1.22746	1.20547	1.18435	1.16402	1.14446	1.12561
77	1.30876	1.28371	1.25972	1.23672	1.21465	1.19344	1.17303	1.15339
78	1.34447	1.31819	1.29306	1.26898	1.24590	1.22374	1.20245	1.18196
79	1.38151	1.35391	1.32754	1.30232	1.27816	1.25499	1.23275	1.21137
80	1.41997	1.39094	1.36325	1.33680	1.31149	1.28725	1.26400	1.24168
M \ N	105	106	107	108	109	110	111	112
81	1.45997	1.42941	1.40029	1.37251	1.34597	1.32058	1.29626	1.27293
82	1.50164	1.46941	1.43875	1.40955	1.38169	1.35506	1.32959	1.30519
83	1.54512	1.51107	1.47875	1.44801	1.41872	1.39078	1.36407	1.33852
84	1.59057	1.55455	1.52042	1.48801	1.45719	1.42781	1.39979	1.37300
85	1.63819	1.60001	1.56390	1.52968	1.49719	1.46628	1.43682	1.40872
86	1.68819	1.64762	1.60935	1.57316	1.53885	1.50628	1.47529	1.44575
87	1.74082	1.69762	1.65697	1.61861	1.58233	1.54794	1.51529	1.48421
88	1.79638	1.75026	1.70697	1.66623	1.62778	1.59142	1.55695	1.52421
M \ N	105	106	107	108	109	110	111	112
89	1.85520	1.80581	1.75960	1.71623	1.67540	1.63688	1.60043	1.56588
90	1.91770	1.86464	1.81516	1.76886	1.72540	1.68449	1.64588	1.60936
91	1.98437	1.92714	1.87398	1.82442	1.77804	1.73449	1.69350	1.65481
92	2.05580	1.99380	1.93648	1.88324	1.83359	1.78713	1.74350	1.70243
93	2.13272	2.06523	2.00315	1.94574	1.89241	1.84268	1.79614	1.75243
94	2.21605	2.14215	2.07458	2.01241	1.95491	1.90151	1.85169	1.80506
95	2.30696	2.22549	2.15150	2.08384	2.02158	1.96401	1.91051	1.86062
96	2.40696	2.31640	2.23483	2.16076	2.09301	2.03067	1.97301	1.91944
M \ N	105	106	107	108	109	110	111	112
97	2.51807	2.41640	2.32574	2.24409	2.16993	2.10210	2.03968	1.98194
98	2.64307	2.52751	2.42574	2.33500	2.25327	2.17902	2.11111	2.04861
99	2.78593	2.65251	2.53685	2.43500	2.34418	2.26236	2.18803	2.12004
100	2.95260	2.79536	2.66185	2.54611	2.44418	2.35327	2.27137	2.19696
101	3.15260	2.96203	2.80471	2.67111	2.55529	2.45327	2.36228	2.28029
102	3.40260	3.16203	2.97138	2.81397	2.68029	2.56438	2.46228	2.37120
103	3.73593	3.41203	3.17138	2.98064	2.82314	2.68938	2.57339	2.47120
104	4.23593	3.74536	3.42138	3.18064	2.98981	2.83223	2.69839	2.58232
M \ N	105	106	107	108	109	110	111	112
105	5.23593	4.24536	3.75471	3.43064	3.18981	2.99890	2.84124	2.70732
106		5.24536	4.25471	3.76397	3.43981	3.19890	3.00791	2.85017
107			5.25471	4.26397	3.77314	3.44890	3.20791	3.01684
108				5.26397	4.27314	3.78223	3.45791	3.21684
109					5.27314	4.28223	3.79124	3.46684
110						5.28223	4.29124	3.80017
111							5.29124	4.30017
112								5.30017

EXPECTED VALUES OF EXPONENTIAL ORDER STATISTICS

M \ N	113	114	115	116	117	118	119	120
1	0.00885	0.00877	0.00870	0.00862	0.00855	0.00847	0.00840	0.00833
2	0.01778	0.01762	0.01747	0.01732	0.01717	0.01702	0.01688	0.01674
3	0.02679	0.02655	0.02632	0.02609	0.02586	0.02564	0.02542	0.02521
4	0.03588	0.03556	0.03525	0.03494	0.03464	0.03434	0.03405	0.03376
5	0.04505	0.04465	0.04425	0.04387	0.04348	0.04311	0.04274	0.04238
6	0.05431	0.05382	0.05335	0.05288	0.05241	0.05196	0.05151	0.05107
7	0.06366	0.06308	0.06252	0.06197	0.06142	0.06089	0.06036	0.05985
8	0.07309	0.07243	0.07178	0.07114	0.07051	0.06990	0.06929	0.06870
M \ N	113	114	115	116	117	118	119	120
9	0.08262	0.08186	0.08112	0.08040	0.07969	0.07899	0.07830	0.07762
10	0.09223	0.09139	0.09056	0.08975	0.08895	0.08816	0.08739	0.08663
11	0.10194	0.10100	0.10008	0.09918	0.09829	0.09742	0.09657	0.09572
12	0.11174	0.11071	0.10970	0.10870	0.10773	0.10677	0.10582	0.10490
13	0.12164	0.12052	0.11941	0.11832	0.11725	0.11620	0.11517	0.11416
14	0.13164	0.13042	0.12921	0.12803	0.12687	0.12573	0.12460	0.12350
15	0.14175	0.14042	0.13911	0.13783	0.13657	0.13534	0.13413	0.13294
16	0.15195	0.15052	0.14911	0.14773	0.14638	0.14505	0.14374	0.14246
M \ N	113	114	115	116	117	118	119	120
17	0.16226	0.16072	0.15921	0.15773	0.15628	0.15485	0.15345	0.15208
18	0.17268	0.17103	0.16942	0.16783	0.16628	0.16475	0.16326	0.16179
19	0.18320	0.18145	0.17973	0.17804	0.17638	0.17475	0.17316	0.17159
20	0.19384	0.19197	0.19014	0.18835	0.18658	0.18486	0.18316	0.18149
21	0.20459	0.20261	0.20067	0.19876	0.19689	0.19506	0.19326	0.19149
22	0.21546	0.21336	0.21131	0.20929	0.20731	0.20537	0.20346	0.20159
23	0.22645	0.22423	0.22206	0.21993	0.21784	0.21579	0.21377	0.21180
24	0.23756	0.23522	0.23293	0.23068	0.22848	0.22631	0.22419	0.22211
M \ N	113	114	115	116	117	118	119	120
25	0.24880	0.24633	0.24392	0.24155	0.23923	0.23695	0.23471	0.23252
26	0.26016	0.25757	0.25503	0.25254	0.25010	0.24770	0.24535	0.24305
27	0.27166	0.26893	0.26627	0.26365	0.26109	0.25857	0.25611	0.25369
28	0.28328	0.28043	0.27763	0.27489	0.27220	0.26956	0.26698	0.26444
29	0.29505	0.29206	0.28912	0.28625	0.28343	0.28067	0.27796	0.27531
30	0.30695	0.30382	0.30075	0.29774	0.29480	0.29191	0.28908	0.28630
31	0.31900	0.31573	0.31252	0.30937	0.30629	0.30327	0.30031	0.29741
32	0.33120	0.32777	0.32442	0.32114	0.31792	0.31477	0.31168	0.30864
M \ N	113	114	115	116	117	118	119	120
33	0.34354	0.33997	0.33647	0.33304	0.32968	0.32639	0.32317	0.32001
34	0.35604	0.35231	0.34866	0.34509	0.34159	0.33816	0.33480	0.33150
35	0.36870	0.36481	0.36101	0.35729	0.35364	0.35006	0.34656	0.34313
36	0.38152	0.37747	0.37351	0.36963	0.36583	0.36211	0.35847	0.35490
37	0.39451	0.39029	0.38617	0.38213	0.37818	0.37431	0.37051	0.36680
38	0.40767	0.40328	0.39899	0.39479	0.39068	0.38665	0.38271	0.37885
39	0.42100	0.41644	0.41198	0.40761	0.40334	0.39915	0.39506	0.39104
40	0.43451	0.42977	0.42513	0.42060	0.41616	0.41181	0.40756	0.40339

EXPECTED VALUES OF EXPONENTIAL ORDER STATISTICS

M \ N	113	114	115	116	117	118	119	120
41	0.44821	0.44328	0.43847	0.43375	0.42914	0.42463	0.42021	0.41589
42	0.46210	0.45698	0.45198	0.44709	0.44230	0.43762	0.43303	0.42855
43	0.47618	0.47087	0.46568	0.46060	0.45563	0.45078	0.44602	0.44137
44	0.49047	0.48496	0.47957	0.47430	0.46915	0.46411	0.45918	0.45435
45	0.50496	0.49924	0.49365	0.48819	0.48285	0.47762	0.47251	0.46751
46	0.51967	0.51374	0.50794	0.50227	0.49674	0.49132	0.48603	0.48085
47	0.53459	0.52844	0.52243	0.51656	0.51082	0.50521	0.49972	0.49436
48	0.54975	0.54337	0.53714	0.53105	0.52511	0.51929	0.51361	0.50806
M \ N	113	114	115	116	117	118	119	120
49	0.56513	0.55852	0.55206	0.54576	0.53960	0.53358	0.52770	0.52195
50	0.58076	0.57390	0.56721	0.56068	0.55430	0.54807	0.54198	0.53603
51	0.59663	0.58953	0.58260	0.57583	0.56923	0.56278	0.55648	0.55032
52	0.61276	0.60540	0.59822	0.59122	0.58438	0.57770	0.57118	0.56481
53	0.62915	0.62153	0.61410	0.60684	0.59977	0.59286	0.58611	0.57952
54	0.64582	0.63792	0.63023	0.62272	0.61539	0.60824	0.60126	0.59444
55	0.66277	0.65459	0.64662	0.63885	0.63126	0.62387	0.61664	0.60959
56	0.68001	0.67154	0.66329	0.65524	0.64739	0.63974	0.63227	0.62498
M \ N	113	114	115	116	117	118	119	120
57	0.69755	0.68878	0.68023	0.67191	0.66379	0.65587	0.64814	0.64060
58	0.71541	0.70632	0.69748	0.68886	0.68045	0.67226	0.66427	0.65648
59	0.73359	0.72418	0.71502	0.70610	0.69740	0.68893	0.68066	0.67260
60	0.75211	0.74236	0.73288	0.72364	0.71464	0.70588	0.69733	0.68900
61	0.77098	0.76088	0.75106	0.74150	0.73219	0.72312	0.71428	0.70566
62	0.79021	0.77975	0.76958	0.75968	0.75004	0.74066	0.73152	0.72261
63	0.80982	0.79898	0.78845	0.77820	0.76823	0.75852	0.74907	0.73986
64	0.82982	0.81859	0.80768	0.79707	0.78675	0.77670	0.76692	0.75740
M \ N	113	114	115	116	117	118	119	120
65	0.85022	0.83859	0.82728	0.81630	0.80561	0.79522	0.78510	0.77526
66	0.87106	0.85900	0.84728	0.83590	0.82484	0.81409	0.80362	0.79344
67	0.89233	0.87983	0.86769	0.85590	0.84445	0.83332	0.82249	0.81196
68	0.91407	0.90111	0.88853	0.87631	0.86445	0.85293	0.84172	0.83082
69	0.93630	0.92285	0.90980	0.89715	0.88486	0.87293	0.86133	0.85006
70	0.95902	0.94507	0.93154	0.91842	0.90569	0.89333	0.88133	0.86966
71	0.98228	0.96780	0.95376	0.94016	0.92697	0.91417	0.90174	0.88966
72	1.00609	0.99105	0.97649	0.96238	0.94871	0.93544	0.92257	0.91007
M \ N	113	114	115	116	117	118	119	120
73	1.03048	1.01486	0.99975	0.98511	0.97093	0.95718	0.94385	0.93090
74	1.05548	1.03925	1.02356	1.00837	0.99366	0.97941	0.96559	0.95218
75	1.08112	1.06425	1.04795	1.03218	1.01691	1.00213	0.98781	0.97392
76	1.10744	1.08989	1.07295	1.05657	1.04072	1.02539	1.01054	0.99614
77	1.13446	1.11621	1.09859	1.08157	1.06511	1.04920	1.03379	1.01887
78	1.16224	1.14323	1.12490	1.10721	1.09011	1.07359	1.05760	1.04213
79	1.19081	1.17101	1.15193	1.13352	1.11575	1.09859	1.08199	1.06594
80	1.22022	1.19958	1.17971	1.16055	1.14207	1.12423	1.10699	1.09033

EXPECTED VALUES OF EXPONENTIAL ORDER STATISTICS

M \ N	113	114	115	116	117	118	119	120
81	1.25053	1.22900	1.20828	1.18833	1.16910	1.15055	1.13263	1.11533
82	1.28178	1.25930	1.23769	1.21690	1.19688	1.17757	1.15895	1.14097
83	1.31403	1.29055	1.26799	1.24631	1.22545	1.20535	1.18598	1.16728
84	1.34737	1.32281	1.29924	1.27661	1.25486	1.23392	1.21375	1.19431
85	1.38185	1.35614	1.33150	1.30786	1.28516	1.26333	1.24232	1.22209
86	1.41756	1.39062	1.36484	1.34012	1.31641	1.29364	1.27174	1.25066
87	1.45460	1.42634	1.39932	1.37346	1.34867	1.32489	1.30204	1.28007
88	1.49306	1.46337	1.43503	1.40794	1.38200	1.35714	1.33329	1.31037
89	1.53306	1.50184	1.47207	1.44365	1.41649	1.39048	1.36555	1.34162
90	1.57473	1.54184	1.51053	1.48069	1.45220	1.42496	1.39888	1.37388
91	1.61821	1.58350	1.55053	1.51915	1.48924	1.46067	1.43336	1.40721
92	1.66366	1.62698	1.59220	1.55915	1.52770	1.49771	1.46908	1.44170
93	1.71128	1.67243	1.63568	1.60082	1.56770	1.53617	1.50612	1.47741
94	1.76128	1.72005	1.68113	1.64430	1.60937	1.57617	1.54458	1.51445
95	1.81391	1.77005	1.72875	1.68975	1.65284	1.61784	1.58458	1.55291
96	1.86947	1.82269	1.77875	1.73737	1.69830	1.66132	1.62624	1.59291
97	1.92829	1.87824	1.83138	1.78737	1.74592	1.70677	1.66972	1.63458
98	1.99079	1.93706	1.88694	1.84000	1.79592	1.75439	1.71518	1.67806
99	2.05746	1.99956	1.94576	1.89556	1.84855	1.80439	1.76280	1.72351
100	2.12889	2.06623	2.00826	1.95438	1.90410	1.85702	1.81280	1.77113
101	2.20581	2.13766	2.07493	2.01688	1.96293	1.91258	1.86543	1.82113
102	2.28914	2.21458	2.14636	2.08355	2.02543	1.97140	1.92098	1.87376
103	2.38005	2.29792	2.22328	2.15498	2.09209	2.03390	1.97981	1.92932
104	2.48005	2.38883	2.30661	2.23190	2.16352	2.10057	2.04231	1.98814
105	2.59116	2.48883	2.39752	2.31523	2.24045	2.17200	2.10897	2.05064
106	2.71616	2.59994	2.49752	2.40614	2.32378	2.24892	2.18040	2.11731
107	2.85902	2.72494	2.60863	2.50614	2.41469	2.33225	2.25732	2.18873
108	3.02569	2.86779	2.73363	2.61725	2.51469	2.42316	2.34066	2.26566
109	3.22569	3.03446	2.87649	2.74225	2.62580	2.52316	2.43157	2.34899
110	3.47569	3.23446	3.04316	2.88511	2.75080	2.63427	2.53157	2.43990
111	3.80902	3.48446	3.24316	3.05178	2.89366	2.75927	2.64268	2.53990
112	4.30902	3.81779	3.49316	3.25178	3.06032	2.90213	2.76768	2.65101
113	5.30902	4.31779	3.82649	3.50178	3.26032	3.06880	2.91053	2.77601
114		5.31779	4.32649	3.83511	3.51032	3.26880	3.07720	2.91887
115			5.32649	4.33511	3.84366	3.51880	3.27720	3.08553
116				5.33511	4.34366	3.85213	3.52720	3.28553
117					5.34366	4.35213	3.86053	3.53553
118						5.35213	4.36053	3.86887
119							5.36053	4.36887
120								5.36887

Table C2.2

VARIANCES AND COVARIANCES OF EXPONENTIAL ORDER STATISTICS

[When $I = J$, tabular values are variances of the Ith order statistics of a sample of size N from an exponential population with location parameter 0 and scale parameter 1; when $I < J \leq N$, they are covariances of the Ith and Jth order statistics.]

Variances and Covariances of Exponential Order Statistics

N	$I \leq J$	COV	N	$I \leq J$	COV	N	$I \leq J$	COV	N	$I \leq J$	COV
2	1	0.25000	10	9	0.54977	15	1	0.00444	18	5	0.02000
2	2	1.25000	10	10	1.54977	15	2	0.00955	18	6	0.02592
3	1	0.11111	11	1	0.00826	15	3	0.01546	18	7	0.03286
3	2	0.36111	11	2	0.01826	15	4	0.02241	18	8	0.04113
3	3	1.36111	11	3	0.03061	15	5	0.03067	18	9	0.05113
4	1	0.06250	11	4	0.04624	15	6	0.04067	18	10	0.06347
4	2	0.17361	11	5	0.06664	15	7	0.05302	18	11	0.07910
4	3	0.42361	11	6	0.09442	15	8	0.06864	18	12	0.09950
4	4	1.42361	11	7	0.13442	15	9	0.08905	18	13	0.12728
5	1	0.04000	11	8	0.19692	15	10	0.11683	18	14	0.16728
5	2	0.10250	11	9	0.30803	15	11	0.15683	18	15	0.22978
5	3	0.21361	11	10	0.55803	15	12	0.21933	18	16	0.34089
5	4	0.46361	11	11	1.55803	15	13	0.33044	18	17	0.59089
5	5	1.46361	12	1	0.00694	15	14	0.58044	18	18	1.59089
6	1	0.02778	12	2	0.01521	15	15	1.58044	19	1	0.00277
6	2	0.06778	12	3	0.02521	16	1	0.00391	19	2	0.00586
6	3	0.13028	12	4	0.03755	16	2	0.00835	19	3	0.00932
6	4	0.24139	12	5	0.05318	16	3	0.01345	19	4	0.01322
6	5	0.49139	12	6	0.07359	16	4	0.01937	19	5	0.01767
6	6	1.49139	12	7	0.10137	16	5	0.02631	19	6	0.02277
7	1	0.02041	12	8	0.14137	16	6	0.03458	19	7	0.02869
7	2	0.04819	12	9	0.20387	16	7	0.04458	19	8	0.03563
7	3	0.08819	12	10	0.31498	16	8	0.05692	19	9	0.04390
7	4	0.15069	12	11	0.56498	16	9	0.07255	19	10	0.05390
7	5	0.26180	12	12	1.56498	16	10	0.09296	19	11	0.06624
7	6	0.51180	13	1	0.00592	16	11	0.12074	19	12	0.08187
7	7	1.51180	13	2	0.01286	16	12	0.16074	19	13	0.10227
8	1	0.01562	13	3	0.02113	16	13	0.22324	19	14	0.13005
8	2	0.03603	13	4	0.03113	16	14	0.33435	19	15	0.17005
8	3	0.06381	13	5	0.04347	16	15	0.58435	19	16	0.23255
8	4	0.10381	13	6	0.05910	16	16	1.58435	19	17	0.34366
8	5	0.16631	13	7	0.07950	17	1	0.00346	19	18	0.59366
8	6	0.27742	13	8	0.10728	17	2	0.00737	19	19	1.59366
8	7	0.52742	13	9	0.14728	17	3	0.01181	20	1	0.00250
8	8	1.52742	13	10	0.20978	17	4	0.01691	20	2	0.00527
9	1	0.01235	13	11	0.32089	17	5	0.02283	20	3	0.00836
9	2	0.02797	13	12	0.57089	17	6	0.02977	20	4	0.01182
9	3	0.04838	13	13	1.57089	17	7	0.03804	20	5	0.01572
9	4	0.07616	14	1	0.00510	17	8	0.04804	20	6	0.02017
9	5	0.11616	14	2	0.01102	17	9	0.06038	20	7	0.02527
9	6	0.17866	14	3	0.01796	17	10	0.07601	20	8	0.03119
9	7	0.28977	14	4	0.02623	17	11	0.09642	20	9	0.03813
9	8	0.53977	14	5	0.03623	17	12	0.12420	20	10	0.04640
9	9	1.53977	14	6	0.04857	17	13	0.16420	20	11	0.05640
10	1	0.01000	14	7	0.06420	17	14	0.22670	20	12	0.06874
10	2	0.02235	14	8	0.08461	17	15	0.33781	20	13	0.08437
10	3	0.03797	14	9	0.11238	17	16	0.58781	20	14	0.10477
10	4	0.05838	14	10	0.15238	17	17	1.58781	20	15	0.13255
10	5	0.08616	14	11	0.21488	18	1	0.00309	20	16	0.17255
10	6	0.12616	14	12	0.32600	18	2	0.00655	20	17	0.23505
10	7	0.18866	14	13	0.57600	18	3	0.01045	20	18	0.34616
10	8	0.29977	14	14	1.57600	18	4	0.01490	20	19	0.59616
									20	20	1.59616

Table C3.1

EXPECTED VALUES OF WEIBULL ORDER STATISTICS

[$E(x_{M,N})$, where $x_{M,N}$ is the Mth order statistic of a sample of size N from a Weibull population with location parameter 0, scale parameter 1, and shape parameter $K = 0.5,\ 1.5(0.5)4.0(1.0)8.0$; when $K = 1.0$, see Table C2.1]

EXPECTED VALUES OF WEIBULL ORDER STATISTICS

SHAPE PARAMETER K=0.5

M \ N	1	2	3	4	5	6	7	8
1	2.00000	.50000	.22222	.12500	.08000	.05556	.04082	.03125
2		3.50000	1.05556	.51389	.30500	.20222	.14399	.10778
3			4.72222	1.59722	.82722	.51056	.34780	.25262
4				5.76389	2.11056	1.14389	.72756	.50643
5					6.67722	2.59389	1.45613	.94869
6						7.49389	3.04899	1.76060
7							8.23471	3.47846
8								8.91417

M \ N	9	10	11	12	13	14	15	16
1	.02469	.02000	.01653	.01389	.01183	.01020	.00889	.00781
2	.08372	.06691	.05471	.04557	.03854	.03303	.02862	.02503
3	.19200	.15094	.12183	.10042	.08421	.07164	.06170	.05369
4	.37387	.28779	.22858	.18605	.15444	.13030	.11142	.09639
5	.67213	.50300	.39140	.31364	.25717	.21481	.18220	.15653
6	1.16995	.84126	.63692	.50027	.40399	.33342	.28005	.23867
7	2.05592	1.38907	1.01154	.77356	.61260	.49809	.41346	.34902
8	3.88489	2.34172	1.60481	1.18152	.91153	.72711	.59481	.49632

M \ N	9	10	11	12	13	14	15	16
9	9.54283	4.27069	2.61806	1.81645	1.35025	1.04985	.84287	.69329
10		10.12862	4.63794	2.88526	2.02365	1.51714	1.18784	.95921
11			10.67769	4.98847	3.14374	2.22626	1.68180	1.32501
12				11.19489	5.32388	3.39397	2.42424	1.84397
13					11.68414	5.64553	3.63640	2.61767
14						12.14865	5.95463	3.87149
15							12.59108	6.25222
16								13.01368

M \ N	17	18	19	20	21	22	23	24
1	.00692	.00617	.00554	.00500	.00454	.00413	.00378	.00347
2	.02209	.01963	.01756	.01580	.01430	.01300	.01187	.01088
3	.04715	.04174	.03721	.03338	.03011	.02730	.02487	.02275
4	.08421	.07421	.06590	.05891	.05298	.04790	.04353	.03972
5	.13596	.11921	.10539	.09385	.08411	.07582	.06870	.06254
6	.20590	.17950	.15790	.14001	.12500	.11230	.10145	.09211
7	.29874	.25871	.22629	.19966	.17751	.15888	.14305	.12949
8	.42085	.36164	.31427	.27575	.24397	.21744	.19506	.17598

M \ N	17	18	19	20	21	22	23	24
9	.58123	.49487	.42677	.37206	.32738	.29040	.25942	.23320
10	.79290	.66759	.57053	.49365	.43163	.38080	.33858	.30312
11	1.07563	.89315	.75495	.64740	.56188	.49262	.43567	.38824
12	1.46104	1.19176	.99367	.84293	.72516	.63114	.55474	.49174
13	2.00353	1.59568	1.30731	1.09416	.93127	.80351	.70117	.61775
14	2.80664	2.16039	1.72878	1.42208	1.19439	1.01972	.88223	.77175
15	4.09967	2.99128	2.31454	1.86022	1.53592	1.29421	1.10810	.96115
16	6.53922	4.32135	3.17174	2.46598	1.98993	1.64873	1.39346	1.19628

EXPECTED VALUES OF WEIBULL ORDER STATISTICS

SHAPE PARAMETER K=0.5

M \ N	17	18	19	20	21	22	23	24
17	13.41833	6.81646	4.53690	3.34818	2.61474	2.11789	1.76040	1.49205
18		13.80667	7.08464	4.74667	3.52076	2.76088	2.24406	1.87090
19			14.18012	7.34441	4.95099	3.68962	2.90444	2.36844
20				14.53989	7.59635	5.15016	3.85492	3.04549
21					14.88707	7.84097	5.34445	4.01680
22						15.22260	8.07874	5.53411
23							15.54732	8.31007
24								15.86198

M \ N	25	26	27	28	29	30	31	32
1	.00320	.00296	.00274	.00255	.00238	.00222	.00208	.00195
2	.01001	.00924	.00855	.00794	.00739	.00690	.00645	.00605
3	.02089	.01925	.01779	.01649	.01534	.01429	.01336	.01251
4	.03640	.03347	.03089	.02859	.02654	.02471	.02305	.02156
5	.05718	.05248	.04834	.04467	.04140	.03848	.03586	.03349
6	.08400	.07692	.07071	.06522	.06035	.05600	.05211	.04861
7	.11778	.10759	.09868	.09084	.08390	.07773	.07221	.06727
8	.15960	.14541	.13305	.12221	.11265	.10417	.09663	.08987
9	.21080	.19151	.17477	.16015	.14731	.13596	.12588	.11689
10	.27302	.24724	.22499	.20564	.18870	.17379	.16060	.14886
11	.34827	.31426	.28507	.25982	.23781	.21852	.20150	.18642
12	.43910	.39464	.35672	.32410	.29582	.27114	.24945	.23030
13	.54876	.49097	.44204	.40022	.36417	.33285	.30547	.28138
14	.68144	.60654	.54366	.49030	.44459	.40512	.37077	.34068
15	.84271	.74563	.66493	.59702	.53927	.48971	.44683	.40945
16	1.04010	.91390	.81019	.72378	.65092	.58883	.53545	.48919
17	1.28413	1.11899	.98519	.87500	.78299	.70524	.63888	.58172
18	1.58990	1.37156	1.19769	1.05649	.93995	.84244	.75989	.68931
19	1.98018	1.68695	1.45849	1.27613	1.12771	1.00496	.90205	.81479
20	2.49105	2.08821	1.78314	1.54487	1.35425	1.19878	1.06995	.96176
21	3.18410	2.61190	2.19498	1.87845	1.63065	1.43198	1.26964	1.13486
22	4.17541	3.32034	2.73102	2.30050	1.97285	1.71579	1.50929	1.34023
23	5.71939	4.33088	3.45427	2.84844	2.40475	2.06632	1.80027	1.58614
24	8.53534	5.90049	4.48334	3.58598	2.96419	2.50775	2.15886	1.88406
25	16.16726	8.75491	6.07764	4.63290	3.71552	3.07830	2.60950	2.25047
26		16.46375	8.96910	6.25101	4.77968	3.84296	3.19081	2.71004
27			16.75201	9.17818	6.42078	4.92379	3.96837	3.30176
28				17.03252	9.38243	6.58711	5.06533	4.09182
29					17.30574	9.58210	6.75016	5.20440
30						17.57207	9.77740	6.91006
31							17.83189	9.96856
32								18.08555

EXPECTED VALUES OF WEIBULL ORDER STATISTICS

SHAPE PARAMETER K=0.5

M \ N	33	34	35	36	37	38	39	40
1	.00184	.00173	.00163	.00154	.00146	.00139	.00131	.00125
2	.00568	.00535	.00504	.00476	.00451	.00427	.00405	.00385
3	.01174	.01103	.01039	.00981	.00927	.00878	.00832	.00790
4	.02021	.01898	.01787	.01684	.01590	.01504	.01425	.01352
5	.03136	.02942	.02766	.02605	.02458	.02323	.02198	.02084
6	.04545	.04260	.04000	.03763	.03547	.03349	.03167	.03000
7	.06282	.05879	.05515	.05183	.04880	.04603	.04350	.04116
8	.08380	.07833	.07338	.06889	.06480	.06106	.05764	.05450
9	.10883	.10158	.09504	.08911	.08372	.07881	.07432	.07020
10	.13837	.12896	.12049	.11283	.10587	.09955	.09378	.08850
11	.17298	.16096	.15015	.14041	.13159	.12358	.11629	.10962
12	.21329	.19812	.18453	.17230	.16125	.15125	.14215	.13386
13	.26006	.24111	.22418	.20899	.19530	.18294	.17172	.16151
14	.31417	.29068	.26976	.25105	.23425	.21909	.20538	.19292
15	.37666	.34772	.32205	.29916	.27866	.26023	.24358	.22851
16	.44880	.41331	.38195	.35410	.32923	.30693	.28685	.26871
17	.53210	.48872	.45055	.41678	.38674	.35989	.33579	.31407
18	.62842	.57548	.52913	.48830	.45212	.41990	.39108	.36517
19	.74005	.67548	.61926	.56997	.52649	.48792	.45353	.42274
20	.86986	.79103	.72283	.66336	.61116	.56506	.52411	.48757
21	1.02149	.92504	.84219	.77040	.70773	.65265	.60395	.56065
22	1.19965	1.08120	.98028	.89347	.81815	.75232	.69439	.64312
23	1.41052	1.26425	1.14083	1.03553	.94482	.86603	.79708	.73634
24	1.66249	1.48048	1.32865	1.20035	1.09074	.99620	.91399	.84197
25	1.96714	1.73833	1.55007	1.39280	1.25972	1.14589	1.04759	.96201
26	2.34113	2.04952	1.81363	1.61927	1.45667	1.31892	1.20093	1.09894
27	2.80936	2.43085	2.13117	1.88839	1.68806	1.52025	1.37791	1.25585
28	3.41118	2.90749	2.51965	2.21210	1.96258	1.75642	1.58352	1.43667
29	4.21336	3.51911	3.00445	2.60752	2.29230	2.03621	1.82435	1.64645
30	5.34110	4.33306	3.62559	3.10025	2.69448	2.37177	2.10926	1.89183
31	7.06696	5.47550	4.45097	3.73066	3.19493	2.78054	2.45053	2.18174
32	10.15576	7.22097	5.60770	4.56715	3.83435	3.28851	2.86570	2.52856
33	18.33336	10.33918	7.37222	5.73777	4.68165	3.93670	3.38100	2.94998
34		18.57560	10.51900	7.52081	5.86578	4.79453	4.03773	3.47242
35			18.81256	10.69536	7.66684	5.99181	4.90582	4.13749
36				19.04448	10.86842	7.81041	6.11592	5.01558
37					19.27160	11.03831	7.95162	6.23818
38						19.49412	11.20516	8.09055
39							19.71225	11.36909
40								19.92617

EXPECTED VALUES OF WEIBULL ORDER STATISTICS

SHAPE PARAMETER K=1.5

M \ N	1	2	3	4	5	6	7	8
1	.90275	.56869	.43399	.35825	.30873	.27340	.24670	.22569
2		1.23680	.83809	.66121	.55634	.48541	.43361	.39378
3			1.43615	1.01497	.81853	.69819	.61492	.55307
4				1.57654	1.14593	.93888	.80921	.71800
5					1.68420	1.24945	1.03613	.90041
6						1.77114	1.33478	1.11757
7							1.84387	1.40718
8								1.90626

M \ N	9	10	11	12	13	14	15	16
1	.20864	.19449	.18252	.17223	.16328	.15541	.14842	.14217
2	.36203	.33601	.31423	.29566	.27963	.26560	.25322	.24218
3	.50491	.46612	.43406	.40703	.38387	.36377	.34611	.33046
4	.64939	.59543	.55160	.51514	.48422	.45760	.43439	.41394
5	.80377	.73034	.67212	.62452	.58470	.55077	.52143	.49576
6	.97773	.87719	.80021	.73876	.68824	.64578	.60944	.57792
7	1.18749	1.04475	.94135	.86165	.79770	.74487	.70027	.66199
8	1.46995	1.24866	1.10384	.99827	.91647	.85052	.79583	.74950

M \ N	9	10	11	12	13	14	15	16
9	1.96079	1.52528	1.30297	1.15663	1.04940	.96594	.89838	.84216
10		2.00918	1.57468	1.35175	1.20428	1.09576	1.01097	.94211
11			2.05263	1.61927	1.39598	1.24769	1.13816	1.05229
12				2.09203	1.65986	1.43643	1.28752	1.17719
13					2.12805	1.69710	1.47365	1.32430
14						2.16120	1.73148	1.50812
15							2.19189	1.76339
16								2.22046

M \ N	17	18	19	20	21	22	23	24
1	.13654	.13144	.12678	.12252	.11860	.11498	.11162	.10850
2	.23228	.22333	.21520	.20776	.20094	.19466	.18884	.18344
3	.31647	.30387	.29247	.28207	.27256	.26381	.25573	.24825
4	.39574	.37944	.36472	.35136	.33916	.32797	.31767	.30813
5	.47306	.45282	.43462	.41817	.40320	.38950	.37692	.36532
6	.55024	.52570	.50375	.48399	.46608	.44975	.43479	.42103
7	.62866	.59932	.57324	.54987	.52877	.50962	.49213	.47609
8	.70960	.67477	.64403	.61664	.59205	.56982	.54960	.53110

M \ N	17	18	19	20	21	22	23	24
9	.79439	.75314	.71704	.68510	.65660	.63096	.60774	.58658
10	.88462	.83565	.79325	.75606	.72311	.69364	.66708	.64300
11	.98235	.92380	.87381	.83043	.79231	.75847	.72816	.70080
12	1.09044	1.01960	.96017	.90930	.86508	.82616	.79154	.76049
13	1.21333	1.12587	1.05427	.99408	.94246	.89752	.85789	.82259
14	1.35845	1.24697	1.15891	1.08668	1.02584	.97358	.92800	.88775
15	1.54019	1.39030	1.27842	1.18987	1.11710	1.05570	1.00288	.95675
16	1.79314	1.57017	1.42013	1.30793	1.21897	1.14575	1.08387	1.03056

EXPECTED VALUES OF WEIBULL ORDER STATISTICS

SHAPE PARAMETER K=1.5

M \ N	17	18	19	20	21	22	23	24
17	2.24716	1.82102	1.59830	1.44818	1.33573	1.24643	1.17282	1.11053
18		2.27223	1.84722	1.62480	1.47464	1.36200	1.27242	1.19847
19			2.29584	1.87193	1.64982	1.49967	1.38688	1.29707
20				2.31815	1.89531	1.67353	1.52342	1.41052
21					2.33930	1.91749	1.69605	1.54600
22						2.35938	1.93858	1.71748
23							2.37851	1.95868
24								2.39676

M \ N	25	26	27	28	29	30	31	32
1	.10559	.10286	.10031	.09790	.09564	.09350	.09148	.08956
2	.17841	.17371	.16931	.16518	.16129	.15762	.15415	.15087
3	.24129	.23480	.22872	.22303	.21768	.21264	.20788	.20338
4	.29929	.29106	.28337	.27617	.26941	.26305	.25705	.25139
5	.35457	.34458	.33527	.32657	.31841	.31074	.30352	.29671
6	.40831	.39652	.38555	.37531	.36573	.35674	.34828	.34031
7	.46130	.44762	.43492	.42309	.41204	.40169	.39197	.38282
8	.51411	.49843	.48391	.47041	.45782	.44605	.43502	.42465

M \ N	25	26	27	28	29	30	31	32
9	.56721	.54939	.53292	.51765	.50345	.49019	.47778	.46613
10	.62102	.60088	.58232	.56515	.54922	.53438	.52052	.50754
11	.67596	.65326	.63243	.61321	.59542	.57890	.56349	.54909
12	.73243	.70691	.68357	.66212	.64232	.62397	.60691	.59099
13	.79089	.76220	.73608	.71217	.69017	.66984	.65099	.63344
14	.85186	.81958	.79033	.76367	.73924	.71675	.69594	.67663
15	.91595	.87953	.84673	.81699	.78985	.76495	.74201	.72077
16	.98394	.94267	.90577	.87251	.84232	.81474	.78943	.76608

M \ N	25	26	27	28	29	30	31	32
17	1.05679	1.00974	.96803	.93071	.89704	.86644	.83848	.81278
18	1.13582	1.08170	1.03427	.99218	.95448	.92044	.88948	.86115
19	1.22283	1.15987	1.10542	1.05765	1.01522	.97718	.94280	.91150
20	1.32051	1.24603	1.18279	1.12805	1.07997	1.03724	.99890	.96421
21	1.43302	1.34285	1.26817	1.20468	1.14969	1.10134	1.05833	1.01971
22	1.56752	1.45449	1.36419	1.28933	1.22563	1.17041	1.12182	1.07856
23	1.73793	1.58807	1.47501	1.38461	1.30959	1.24572	1.19029	1.14148
24	1.97787	1.75748	1.60773	1.49467	1.40418	1.32903	1.26500	1.20939

M \ N	25	26	27	28	29	30	31	32
25	2.41422	1.99624	1.77619	1.62658	1.51352	1.42297	1.34771	1.28353
26		2.43094	2.01384	1.79415	1.64467	1.53163	1.44103	1.36568
27			2.44698	2.03074	1.81140	1.66206	1.54905	1.45842
28				2.46239	2.04699	1.82799	1.67880	1.56584
29					2.47723	2.06263	1.84397	1.69494
30						2.49153	2.07771	1.85939
31							2.50532	2.09227
32								2.51864

EXPECTED VALUES OF WEIBULL ORDER STATISTICS

SHAPE PARAMETER K=1.5

M \ N	33	34	35	36	37	38	39	40
1	.08775	.08602	.08437	.08280	.08130	.07987	.07850	.07718
2	.14776	.14480	.14199	.13931	.13675	.13431	.13197	.12974
3	.19911	.19507	.19122	.18755	.18406	.18073	.17754	.17449
4	.24603	.24094	.23611	.23152	.22714	.22297	.21898	.21516
5	.29027	.28416	.27837	.27287	.26763	.26263	.25787	.25331
6	.33278	.32566	.31891	.31250	.30640	.30059	.29505	.28976
7	.37418	.36602	.35830	.35097	.34400	.33738	.33106	.32503
8	.41488	.40565	.39693	.38866	.38081	.37335	.36625	.35948
9	.45518	.44485	.43510	.42586	.41711	.40879	.40088	.39335
10	.49534	.48387	.47304	.46280	.45310	.44390	.43516	.42684
11	.53558	.52289	.51094	.49965	.48898	.47886	.46926	.46012
12	.57609	.56212	.54898	.53659	.52489	.51381	.50331	.49333
13	.61706	.60171	.58731	.57376	.56097	.54889	.53744	.52658
14	.65865	.64184	.62609	.61129	.59736	.58421	.57177	.55999
15	.70105	.68266	.66546	.64934	.63418	.61990	.60642	.59366
16	.74445	.72434	.70558	.68803	.67157	.65608	.64148	.62768
17	.78906	.76707	.74661	.72752	.70964	.69286	.67707	.66217
18	.83511	.81105	.78873	.76795	.74855	.73037	.71330	.69722
19	.88286	.85649	.83212	.80950	.78843	.76875	.75029	.73295
20	.93261	.90367	.87702	.85236	.82946	.80812	.78817	.76946
21	.98474	.95288	.92366	.89674	.87182	.84867	.82708	.80688
22	1.03969	1.00447	.97235	.94289	.91572	.89057	.86718	.84535
23	1.09800	1.05890	1.02345	.99110	.96141	.93402	.90864	.88503
24	1.16039	1.11670	1.07739	1.04174	1.00918	.97927	.95167	.92609
25	1.22776	1.17859	1.13472	1.09522	1.05937	1.02662	.99652	.96873
26	1.30138	1.24547	1.19613	1.15210	1.11243	1.07640	1.04347	1.01320
27	1.38299	1.31858	1.26254	1.21307	1.16889	1.12906	1.09287	1.05977
28	1.47518	1.39969	1.33519	1.27903	1.22943	1.18511	1.14514	1.10880
29	1.58203	1.49136	1.41581	1.35123	1.29498	1.24526	1.20082	1.16071
30	1.71051	1.59766	1.50699	1.43140	1.36675	1.31041	1.26059	1.21603
31	1.87428	1.72556	1.61277	1.52210	1.44648	1.38177	1.32535	1.27544
32	2.10633	1.88867	1.74011	1.62739	1.53674	1.46109	1.39633	1.33984
33	2.53153	2.11994	1.90260	1.75420	1.64156	1.55093	1.47526	1.41046
34		2.54400	2.13311	1.91609	1.76786	1.65529	1.56468	1.48901
35			2.55609	2.14588	1.92917	1.78110	1.66861	1.57804
36				2.56781	2.15826	1.94186	1.79395	1.68155
37					2.57918	2.17028	1.95418	1.80644
38						2.59023	2.18196	1.96616
39							2.60098	2.19332
40								2.61143

EXPECTED VALUES OF WEIBULL ORDER STATISTICS

SHAPE PARAMETER K=2.0

M \ N	1	2	3	4	5	6	7	8
1	.88623	.62666	.51166	.44311	.39633	.36180	.33496	.31333
2		1.14580	.85664	.71731	.63024	.56899	.52283	.48640
3			1.29037	.99598	.84793	.75272	.68440	.63213
4				1.38851	1.09467	.94313	.84382	.77152
5					1.46196	1.17045	1.01762	.91612
6						1.52027	1.23158	1.07852
7							1.56838	1.28260
8								1.60921

M \ N	9	10	11	12	13	14	15	16
1	.29541	.28025	.26721	.25583	.24580	.23685	.22882	.22156
2	.45669	.43184	.41067	.39234	.37627	.36203	.34929	.33781
3	.59040	.55605	.52712	.50232	.48073	.46172	.44481	.42964
4	.71559	.67054	.63319	.60154	.57427	.55043	.52936	.51056
5	.84144	.78316	.73590	.69648	.66291	.63386	.60837	.58577
6	.97586	.89971	.83989	.79108	.75018	.71521	.68483	.65809
7	1.12985	1.02663	.94957	.88869	.83880	.79681	.76078	.72939
8	1.32624	1.17408	1.07066	.99305	.93146	.88078	.83799	.80115

M \ N	9	10	11	12	13	14	15	16
9	1.64458	1.36428	1.21286	1.10947	1.03154	.96947	.91822	.87482
10		1.67572	1.39792	1.24733	1.14410	1.06603	1.00364	.95198
11			1.70350	1.42804	1.27830	1.17533	1.09722	1.03464
12				1.72855	1.45527	1.30638	1.20373	1.12567
13					1.75132	1.48009	1.33204	1.22975
14						1.77218	1.50286	1.35564
15							1.79142	1.52389
16								1.80926

M \ N	17	18	19	20	21	22	23	24
1	.21494	.20889	.20331	.19817	.19339	.18894	.18479	.18090
2	.32740	.31789	.30917	.30113	.29368	.28676	.28031	.27428
3	.41593	.40345	.39204	.38155	.37185	.36286	.35450	.34669
4	.49364	.47830	.46432	.45151	.43970	.42878	.41863	.40918
5	.56555	.54730	.53073	.51559	.50168	.48885	.47696	.46591
6	.63432	.61299	.59370	.57615	.56009	.54531	.53165	.51898
7	.70168	.67698	.65477	.63466	.61631	.59950	.58400	.56967
8	.76896	.74049	.71506	.69214	.67134	.65235	.63491	.61882

M \ N	17	18	19	20	21	22	23	24
9	.83736	.80455	.77547	.74943	.72594	.70458	.68504	.66709
10	.90812	.87017	.83687	.80729	.78077	.75679	.73496	.71497
11	.98268	.93847	.90015	.86645	.83646	.80954	.78516	.76294
12	1.06297	1.01082	.96634	.92772	.89370	.86339	.83613	.81142
13	1.15180	1.08905	1.03676	.99209	.95324	.91896	.88838	.86084
14	1.25374	1.17593	1.11319	1.06081	1.01600	.97697	.94249	.91168
15	1.37748	1.27597	1.19834	1.13564	1.08322	1.03831	.99913	.96449
16	1.54341	1.39778	1.29668	1.21924	1.15661	1.10417	1.05920	1.01992

EXPECTED VALUES OF WEIBULL ORDER STATISTICS

SHAPE PARAMETER K=2.0

M \ N	17	18	19	20	21	22	23	24
17	1.82587	1.56162	1.41674	1.31604	1.23881	1.17627	1.12385	1.07884
18		1.84141	1.57866	1.43451	1.33421	1.25720	1.19477	1.14239
19			1.85601	1.59468	1.45123	1.35132	1.27454	1.21224
20				1.86977	1.60978	1.46700	1.36749	1.29094
21					1.88277	1.62406	1.48193	1.38280
22						1.89508	1.63759	1.49609
23							1.90679	1.65046
24								1.91793

M \ N	25	26	27	28	29	30	31	32
1	.17725	.17380	.17055	.16748	.16457	.16180	.15917	.15666
2	.26862	.26330	.25828	.25353	.24904	.24478	.24074	.23688
3	.33937	.33250	.32603	.31993	.31415	.30868	.30348	.29854
4	.40034	.39205	.38426	.37691	.36997	.36340	.35717	.35125
5	.45559	.44593	.43686	.42833	.42028	.41267	.40546	.39861
6	.50718	.49615	.48582	.47612	.46697	.45834	.45016	.44242
7	.55635	.54393	.53231	.52142	.51117	.50151	.49238	.48374
8	.60392	.59005	.57711	.56500	.55362	.54292	.53281	.52326

M \ N	25	26	27	28	29	30	31	32
9	.65050	.63511	.62078	.60740	.59486	.58307	.57196	.56148
10	.69658	.67956	.66376	.64904	.63527	.62236	.61021	.59876
11	.74257	.72380	.70642	.69027	.67520	.66110	.64786	.63540
12	.78887	.76817	.74908	.73138	.71492	.69956	.68516	.67164
13	.83585	.81302	.79204	.77267	.75470	.73798	.72235	.70770
14	.88391	.85868	.83561	.81439	.79478	.77658	.75962	.74376
15	.93350	.90554	.88011	.85683	.83540	.81558	.79717	.78001
16	.98514	.95401	.92588	.90028	.87683	.85522	.83522	.81663

M \ N	25	26	27	28	29	30	31	32
17	1.03948	1.00460	.97335	.94508	.91934	.89573	.87397	.85381
18	1.09735	1.05795	1.02299	.99163	.96326	.93739	.91365	.89176
19	1.15990	1.11487	1.07543	1.04041	1.00897	.98051	.95453	.93068
20	1.22876	1.17649	1.13148	1.09202	1.05695	1.02546	.99691	.97085
21	1.30648	1.24444	1.19225	1.14726	1.10779	1.07270	1.04116	1.01255
22	1.39733	1.32125	1.25936	1.20724	1.16229	1.12283	1.08772	1.05614
23	1.50956	1.41117	1.33531	1.27357	1.22154	1.17664	1.13720	1.10208
24	1.66271	1.52239	1.42436	1.34874	1.28714	1.23521	1.19036	1.15094

M \ N	25	26	27	28	29	30	31	32
25	1.92857	1.67440	1.53465	1.43696	1.36157	1.30013	1.24829	1.20350
26		1.93873	1.68558	1.54637	1.44902	1.37386	1.31257	1.26083
27			1.94847	1.69629	1.55760	1.46059	1.38564	1.32451
28				1.95781	1.70657	1.56838	1.47169	1.39697
29					1.96678	1.71644	1.57874	1.48236
30						1.97542	1.72593	1.58871
31							1.98373	1.73508
32								1.99175

EXPECTED VALUES OF WEIBULL ORDER STATISTICS

SHAPE PARAMETER K=2.0

M \ N	33	34	35	36	37	38	39	40
1	.15427	.15199	.14980	.14770	.14569	.14376	.14191	.14012
2	.23321	.22970	.22634	.22313	.22005	.21710	.21426	.21153
3	.29384	.28934	.28505	.28095	.27701	.27324	.26962	.26614
4	.34562	.34025	.33512	.33022	.32552	.32103	.31671	.31256
5	.39210	.38590	.37999	.37434	.36894	.36376	.35880	.35403
6	.43506	.42806	.42138	.41501	.40892	.40310	.39751	.39216
7	.47553	.46773	.46031	.45323	.44647	.44000	.43381	.42787
8	.51420	.50561	.49743	.48964	.48221	.47511	.46831	.46180
9	.55155	.54214	.53320	.52470	.51658	.50884	.50144	.49435
10	.58794	.57770	.56797	.55873	.54993	.54153	.53352	.52585
11	.62365	.61253	.60200	.59200	.58249	.57343	.56478	.55652
12	.65891	.64689	.63551	.62473	.61448	.60473	.59544	.58657
13	.69393	.68095	.66869	.65708	.64607	.63561	.62564	.61614
14	.72888	.71489	.70170	.68923	.67741	.66620	.65554	.64538
15	.76394	.74887	.73468	.72129	.70863	.69663	.68524	.67440
16	.79928	.78304	.76779	.75342	.73986	.72703	.71487	.70331
17	.83506	.81755	.80115	.78574	.77122	.75751	.74452	.73220
18	.87146	.85257	.83492	.81838	.80282	.78816	.77431	.76119
19	.90867	.88825	.86924	.85146	.83479	.81911	.80433	.79035
20	.94690	.92479	.90427	.88514	.86725	.85047	.83468	.81978
21	.98641	.96239	.94018	.91956	.90034	.88236	.86547	.84958
22	1.02748	1.00128	.97719	.95491	.93421	.91490	.89683	.87986
23	1.07047	1.04177	1.01552	.99137	.96902	.94825	.92887	.91072
24	1.11582	1.08420	1.05547	1.02918	1.00497	.98257	.96173	.94229
25	1.16411	1.12900	1.09736	1.06861	1.04229	1.01804	.99559	.97470
26	1.21611	1.17675	1.14165	1.11002	1.08125	1.05490	1.03061	1.00812
27	1.27287	1.22821	1.18890	1.15382	1.12219	1.09341	1.06704	1.04273
28	1.33598	1.28445	1.23986	1.20059	1.16554	1.13391	1.10513	1.07874
29	1.40786	1.34702	1.29560	1.25108	1.21186	1.17683	1.14522	1.11644
30	1.49264	1.41835	1.35766	1.30635	1.26190	1.22273	1.18774	1.15614
31	1.59832	1.50255	1.42846	1.36792	1.31672	1.27235	1.23323	1.19827
32	1.74391	1.60759	1.51211	1.43822	1.37783	1.32673	1.28244	1.24338
33	1.99950	1.75243	1.61654	1.52134	1.44766	1.38742	1.33642	1.29221
34		2.00699	1.76066	1.62519	1.53027	1.45679	1.39669	1.34580
35			2.01423	1.76863	1.63356	1.53892	1.46563	1.40567
36				2.02125	1.77635	1.64168	1.54730	1.47419
37					2.02805	1.78383	1.64954	1.55542
38						2.03465	1.79109	1.65717
39							2.04106	1.79814
40								2.04729

EXPECTED VALUES OF WEIBULL ORDER STATISTICS

SHAPE PARAMETER K=2.5

M \ N	1	2	3	4	5	6	7	8
1	.88726	.67242	.57175	.50960	.46608	.43330	.40739	.38620
2		1.10211	.87377	.75819	.68366	.62999	.58877	.55572
3			1.21628	.98934	.86999	.79099	.73305	.68792
4				1.29192	1.06890	.94899	.86826	.80826
5					1.34768	1.12886	1.00955	.92825
6						1.39144	1.17658	1.05832
7							1.42726	1.21600
8								1.45744

M \ N	9	10	11	12	13	14	15	16
1	.36843	.35323	.34001	.32838	.31804	.30875	.30034	.29269
2	.52839	.50527	.48536	.46795	.45255	.43880	.42640	.41516
3	.65135	.62087	.59489	.57239	.55263	.53508	.51935	.50513
4	.76105	.72249	.69012	.66240	.63826	.61698	.59802	.58096
5	.86726	.81890	.77913	.74558	.71670	.69146	.66914	.64918
6	.97705	.91562	.86662	.82611	.79178	.76212	.73612	.71304
7	1.09896	1.01800	.95646	.90712	.86617	.83132	.80113	.77458
8	1.24944	1.13366	1.05317	.99170	.94223	.90101	.86584	.83527

M \ N	9	10	11	12	13	14	15	16
9	1.48343	1.27838	1.16384	1.08390	1.02263	.97314	.93179	.89641
10		1.50622	1.30384	1.19049	1.11113	1.05012	1.00070	.95931
11			1.52646	1.32651	1.21430	1.13553	1.07483	1.02554
12				1.54463	1.34691	1.23578	1.15761	1.09723
13					1.56111	1.36543	1.25532	1.17774
14						1.57616	1.38237	1.27323
15							1.59001	1.39796
16								1.60281

M \ N	17	18	19	20	21	22	23	24
1	.28568	.27922	.27324	.26770	.26252	.25768	.25314	.24887
2	.40488	.39545	.38675	.37868	.37117	.36416	.35759	.35142
3	.49219	.48035	.46944	.45936	.45001	.44128	.43313	.42547
4	.56551	.55142	.53849	.52656	.51552	.50525	.49566	.48669
5	.63119	.61484	.59991	.58618	.57350	.56173	.55078	.54054
6	.69236	.67368	.65667	.64110	.62676	.61349	.60117	.58968
7	.75095	.72974	.71053	.69301	.67694	.66213	.64840	.63564
8	.80832	.78429	.76267	.74306	.72514	.70869	.69350	.67940

M \ N	17	18	19	20	21	22	23	24
9	.86558	.83835	.81403	.79209	.77217	.75394	.73718	.72168
10	.92381	.89281	.86538	.84083	.81866	.79849	.78002	.76301
11	.98416	.94860	.91750	.88993	.86521	.84287	.82251	.80384
12	1.04811	1.00679	.97122	.94006	.91239	.88756	.86508	.84457
13	1.11770	1.06877	1.02754	.99200	.96081	.93308	.90817	.88559
14	1.19621	1.13652	1.08780	1.04669	1.01119	.98000	.95225	.92728
15	1.28973	1.21326	1.15392	1.10541	1.06443	1.02901	.99785	.97008
16	1.41239	1.30503	1.22909	1.17009	1.12181	1.08097	1.04562	1.01451

EXPECTED VALUES OF WEIBULL ORDER STATISTICS

SHAPE PARAMETER K=2.5

M \ N	17	18	19	20	21	22	23	24
17	1.61471	1.42581	1.31926	1.24384	1.18518	1.13712	1.09643	1.06118
18		1.62582	1.43834	1.33257	1.25764	1.19931	1.15148	1.11095
19			1.63624	1.45010	1.34506	1.27060	1.21260	1.16499
20				1.64603	1.46115	1.35682	1.28282	1.22512
21					1.65528	1.47159	1.36792	1.29435
22						1.66403	1.48146	1.37843
23							1.67232	1.49083
24								1.68022

M \ N	25	26	27	28	29	30	31	32
1	.24484	.24103	.23741	.23399	.23072	.22762	.22465	.22182
2	.34560	.34011	.33492	.32999	.32531	.32085	.31659	.31253
3	.41828	.41149	.40507	.39899	.39322	.38773	.38250	.37751
4	.47826	.47032	.46283	.45574	.44901	.44262	.43654	.43074
5	.53094	.52192	.51341	.50537	.49776	.49054	.48367	.47712
6	.57893	.56884	.55935	.55039	.54192	.53388	.52626	.51900
7	.62372	.61256	.60208	.59220	.58287	.57404	.56567	.55771
8	.66628	.65402	.64252	.63171	.62152	.61188	.60276	.59409

M \ N	25	26	27	28	29	30	31	32
9	.70729	.69387	.68132	.66955	.65847	.64801	.63812	.62874
10	.74726	.73263	.71897	.70619	.69418	.68287	.67219	.66208
11	.78663	.77068	.75584	.74199	.72900	.71680	.70529	.69443
12	.82574	.80837	.79226	.77726	.76323	.75008	.73771	.72605
13	.86496	.84601	.82851	.81226	.79712	.78296	.76967	.75716
14	.90462	.88391	.86487	.84726	.83090	.81564	.80136	.78795
15	.94508	.92237	.90160	.88248	.86478	.84833	.83298	.81860
16	.98675	.96174	.93899	.91817	.89899	.88123	.86471	.84928

M \ N	25	26	27	28	29	30	31	32
17	1.03012	1.00239	.97737	.95461	.93376	.91453	.89672	.88014
18	1.07580	1.04480	1.01710	.99210	.96933	.94846	.92920	.91135
19	1.12462	1.08958	1.05865	1.03099	1.00601	.98325	.96236	.94309
20	1.17774	1.13753	1.10260	1.07175	1.04414	1.01919	.99644	.97555
21	1.23697	1.18981	1.14975	1.11494	1.08417	1.05662	1.03170	1.00897
22	1.30529	1.24820	1.20125	1.16135	1.12666	1.09598	1.06849	1.04361
23	1.38840	1.31567	1.25887	1.21213	1.17239	1.13782	1.10722	1.07980
24	1.49973	1.39789	1.32554	1.26902	1.22250	1.18292	1.14846	1.11795

M \ N	25	26	27	28	29	30	31	32
25	1.68774	1.50822	1.40693	1.33496	1.27872	1.23240	1.19297	1.15863
26		1.69492	1.51632	1.41557	1.34396	1.28798	1.24186	1.20258
27			1.70179	1.52407	1.42383	1.35258	1.29685	1.25092
28				1.70837	1.53150	1.43175	1.36083	1.30535
29					1.71468	1.53862	1.43935	1.36876
30						1.72075	1.54547	1.44665
31							1.72660	1.55206
32								1.73223

EXPECTED VALUES OF WEIBULL ORDER STATISTICS

SHAPE PARAMETER K=2.5

M \ N	33	34	35	36	37	38	39	40
1	.21910	.21650	.21401	.21161	.20930	.20708	.20494	.20288
2	.30865	.30493	.30136	.29793	.29464	.29147	.28842	.28547
3	.37274	.36818	.36380	.35960	.35557	.35169	.34795	.34436
4	.42521	.41991	.41484	.40998	.40531	.40083	.39651	.39235
5	.47088	.46492	.45921	.45374	.44849	.44345	.43860	.43394
6	.51208	.50548	.49916	.49312	.48732	.48176	.47641	.47127
7	.55013	.54290	.53600	.52939	.52306	.51699	.51117	.50556
8	.58586	.57801	.57052	.56336	.55651	.54994	.54364	.53758
9	.61984	.61136	.60328	.59557	.58820	.58113	.57436	.56786
10	.65249	.64338	.63470	.62642	.61852	.61095	.60371	.59676
11	.68413	.67436	.66507	.65622	.64777	.63970	.63197	.62456
12	.71501	.70456	.69463	.68519	.67618	.66759	.65937	.65150
13	.74535	.73418	.72359	.71352	.70394	.69481	.68608	.67773
14	.77532	.76339	.75210	.74139	.73121	.72151	.71226	.70342
15	.80509	.79236	.78033	.76894	.75812	.74784	.73803	.72868
16	.83482	.82122	.80840	.79628	.78480	.77389	.76352	.75363
17	.86464	.85011	.83645	.82355	.81136	.79979	.78881	.77835
18	.89472	.87917	.86459	.85086	.83790	.82564	.81401	.80296
19	.92520	.90854	.89295	.87832	.86454	.85152	.83920	.82752
20	.95626	.93836	.92167	.90604	.89137	.87755	.86449	.85212
21	.98809	.96880	.95088	.93417	.91852	.90381	.88995	.87686
22	1.02090	1.00003	.98074	.96282	.94609	.93042	.91569	.90180
23	1.05496	1.03228	1.01143	.99215	.97422	.95749	.94180	.92705
24	1.09059	1.06581	1.04316	1.02233	1.00306	.98514	.96840	.95270
25	1.12821	1.10092	1.07619	1.05358	1.03277	1.01351	.99560	.97886
26	1.16836	1.13804	1.11082	1.08613	1.06357	1.04279	1.02355	1.00564
27	1.21180	1.17769	1.14746	1.12031	1.09568	1.07315	1.05241	1.03319
28	1.25962	1.22064	1.18665	1.15651	1.12943	1.10486	1.08238	1.06166
29	1.31352	1.26797	1.22914	1.19526	1.16521	1.13821	1.11369	1.09125
30	1.37638	1.32138	1.27601	1.23731	1.20355	1.17359	1.14666	1.12220
31	1.45368	1.38371	1.32894	1.28374	1.24519	1.21154	1.18167	1.15481
32	1.55840	1.46045	1.39078	1.33623	1.29121	1.25279	1.21925	1.18947
33	1.73766	1.56453	1.46698	1.39760	1.34326	1.29841	1.26013	1.22669
34		1.74291	1.57044	1.47329	1.40418	1.35006	1.30537	1.26722
35			1.74798	1.57615	1.47939	1.41055	1.35663	1.31210
36				1.75289	1.58168	1.48529	1.41671	1.36299
37					1.75764	1.58704	1.49100	1.42268
38						1.76226	1.59223	1.49654
39							1.76673	1.59726
40								1.77107

EXPECTED VALUES OF WEIBULL ORDER STATISTICS

SHAPE PARAMETER K=3.0

M \ N	1	2	3	4	5	6	7	8
1	.89298	.70876	.61916	.56254	.52222	.49143	.46681	.44649
2		1.07720	.88796	.78900	.72384	.67618	.63911	.60907
3			1.17182	.98692	.88675	.81916	.76886	.72922
4				1.23346	1.05369	.95434	.88623	.83492
5					1.27840	1.10337	1.00542	.93755
6						1.31340	1.14254	1.04615
7							1.34188	1.17467
8								1.36577

M \ N	9	10	11	12	13	14	15	16
1	.42930	.41448	.40152	.39004	.37978	.37051	.36209	.35438
2	.58401	.56264	.54410	.52779	.51327	.50024	.48844	.47768
3	.69678	.66949	.64608	.62565	.60760	.59148	.57695	.56376
4	.79410	.76043	.73194	.70736	.68582	.66671	.64959	.63412
5	.88595	.84460	.81029	.78111	.75582	.73358	.71379	.69601
6	.97882	.92730	.88577	.85115	.82157	.79585	.77316	.75292
7	1.07982	1.01317	.96190	.92039	.88565	.85587	.82989	.80691
8	1.20178	1.10838	1.04247	.99155	.95018	.91543	.88556	.85943

M \ N	9	10	11	12	13	14	15	16
9	1.38626	1.22513	1.13310	1.06792	1.01741	.97624	.94156	.91168
10		1.40417	1.24558	1.15482	1.09037	1.04029	.99936	.96480
11			1.42003	1.26373	1.17415	1.11041	1.06076	1.02009
12				1.43424	1.28001	1.19154	1.12846	1.07924
13					1.44709	1.29476	1.20731	1.14487
14						1.45881	1.30821	1.22172
15							1.46956	1.32057
16								1.47950

M \ N	17	18	19	20	21	22	23	24
1	.34729	.34074	.33465	.32898	.32367	.31869	.31400	.30958
2	.46781	.45871	.45028	.44244	.43512	.42826	.42181	.41573
3	.55170	.54061	.53037	.52086	.51200	.50372	.49595	.48864
4	.62003	.60713	.59524	.58424	.57401	.56446	.55553	.54713
5	.67990	.66520	.65170	.63925	.62771	.61696	.60692	.59750
6	.73468	.71812	.70298	.68906	.67619	.66425	.65311	.64270
7	.78635	.76780	.75092	.73546	.72123	.70805	.69580	.68437
8	.83627	.81551	.79673	.77963	.76393	.74946	.73605	.72356

M \ N	17	18	19	20	21	22	23	24
9	.88549	.86222	.84133	.82240	.80512	.78926	.77461	.76102
10	.93497	.90876	.88544	.86446	.84543	.82804	.81205	.79726
11	.98569	.95593	.92975	.90641	.88539	.86630	.84883	.83275
12	1.03885	1.00463	.97497	.94884	.92552	.90449	.88536	.86784
13	1.09607	1.05596	1.02193	.99239	.96634	.94305	.92202	.90288
14	1.15989	1.11149	1.07167	1.03783	1.00843	.98246	.95922	.93822
15	1.23496	1.17371	1.12572	1.08617	1.05253	1.02327	.99739	.97422
16	1.33198	1.24721	1.18651	1.13890	1.09963	1.06619	1.03707	1.01130

EXPECTED VALUES OF WEIBULL ORDER STATISTICS

SHAPE PARAMETER K=3.0

M \ N	17	18	19	20	21	22	23	24
17	1.48872	1.34258	1.25860	1.19842	1.15117	1.11217	1.07893	1.04996
18		1.49731	1.35246	1.26922	1.20953	1.16264	1.12390	1.09086
19			1.50536	1.36171	1.27916	1.21995	1.17340	1.13492
20				1.51292	1.37040	1.28851	1.22975	1.18353
21					1.52005	1.37859	1.29733	1.23900
22						1.52678	1.38633	1.30566
23							1.53317	1.39366
24								1.53923

M \ N	25	26	27	28	29	30	31	32
1	.30539	.30143	.29766	.29407	.29065	.28739	.28426	.28127
2	.40999	.40456	.39940	.39450	.38983	.38537	.38111	.37703
3	.48174	.47521	.46903	.46316	.45757	.45224	.44715	.44228
4	.53922	.53175	.52468	.51798	.51160	.50553	.49973	.49420
5	.58865	.58030	.57241	.56493	.55783	.55107	.54464	.53849
6	.63292	.62371	.61503	.60681	.59901	.59161	.58455	.57783
7	.67366	.66360	.65412	.64517	.63669	.62864	.62099	.61370
8	.71190	.70097	.69068	.68098	.67182	.66313	.65487	.64702

M \ N	25	26	27	28	29	30	31	32
9	.74835	.73650	.72539	.71493	.70505	.69571	.68685	.67843
10	.78353	.77073	.75874	.74748	.73687	.72685	.71737	.70837
11	.81786	.80402	.79111	.77900	.76763	.75691	.74677	.73717
12	.85169	.83673	.82281	.80981	.79762	.78615	.77533	.76510
13	.88533	.86914	.85413	.84015	.82708	.81482	.80328	.79238
14	.91908	.90152	.88530	.87026	.85624	.84312	.83080	.81920
15	.95326	.93414	.91657	.90035	.88528	.87123	.85808	.84572
16	.98820	.96728	.94819	.93064	.91441	.89933	.88526	.87208

M \ N	25	26	27	28	29	30	31	32
17	1.02429	1.00127	.98041	.96135	.94382	.92760	.91252	.89845
18	1.06203	1.03648	1.01354	.99274	.97372	.95622	.94002	.92495
19	1.10206	1.07339	1.04795	1.02510	1.00436	.98539	.96792	.95174
20	1.14529	1.11263	1.08410	1.05878	1.03601	1.01534	.99642	.97899
21	1.19309	1.15509	1.12261	1.09423	1.06902	1.04634	1.02575	1.00688
22	1.24774	1.20214	1.16437	1.13207	1.10384	1.07874	1.05615	1.03563
23	1.31356	1.25603	1.21072	1.17318	1.14106	1.11296	1.08798	1.06548
24	1.40063	1.32106	1.26391	1.21888	1.18156	1.14961	1.12165	1.09678

M \ N	25	26	27	28	29	30	31	32
25	1.54501	1.40726	1.32820	1.27142	1.22666	1.18955	1.15777	1.12994
26		1.55052	1.41358	1.33502	1.27858	1.23408	1.19717	1.16556
27			1.55578	1.41963	1.34153	1.28542	1.24118	1.20447
28				1.56083	1.42541	1.34777	1.29198	1.24798
29					1.56566	1.43096	1.35374	1.29826
30						1.57031	1.43628	1.35948
31							1.57478	1.44140
32								1.57908

EXPECTED VALUES OF WEIBULL ORDER STATISTICS

SHAPE PARAMETER K=3.0

M \ N	33	34	35	36	37	38	39	40
1	.27840	.27564	.27299	.27044	.26798	.26561	.26332	.26111
2	.37312	.36937	.36576	.36230	.35896	.35573	.35263	.34962
3	.43762	.43315	.42886	.42473	.42075	.41692	.41323	.40967
4	.48890	.48382	.47895	.47426	.46976	.46542	.46124	.45721
5	.53261	.52698	.52159	.51641	.51142	.50663	.50201	.49756
6	.57140	.56526	.55937	.55372	.54829	.54307	.53804	.53319
7	.60674	.60009	.59372	.58762	.58176	.57613	.57071	.56550
8	.63954	.63239	.62555	.61901	.61273	.60670	.60090	.59532
9	.67041	.66277	.65546	.64847	.64177	.63534	.62917	.62323
10	.69981	.69165	.68387	.67643	.66931	.66249	.65593	.64963
11	.72805	.71938	.71111	.70321	.69566	.68843	.68149	.67483
12	.75540	.74619	.73742	.72905	.72106	.71341	.70609	.69906
13	.78207	.77230	.76300	.75415	.74570	.73763	.72990	.72249
14	.80825	.79787	.78803	.77866	.76974	.76123	.75308	.74529
15	.83407	.82306	.81264	.80274	.79332	.78434	.77577	.76756
16	.85969	.84801	.83697	.82650	.81655	.80709	.79806	.78944
17	.88525	.87284	.86113	.85005	.83955	.82957	.82006	.81100
18	.91087	.89766	.88523	.87351	.86241	.85188	.84187	.83233
19	.93668	.92261	.90940	.89696	.88522	.87410	.86356	.85352
20	.96284	.94780	.93373	.92052	.90808	.89634	.88521	.87464
21	.98949	.97337	.95835	.94429	.93109	.91866	.90691	.89577
22	1.01682	.99947	.98338	.96839	.95435	.94116	.92873	.91698
23	1.04503	1.02628	1.00898	.99293	.97796	.96394	.95077	.93835
24	1.07438	1.05400	1.03531	1.01806	1.00204	.98710	.97311	.95995
25	1.10518	1.08287	1.06256	1.04393	1.02673	1.01076	.99585	.98188
26	1.13787	1.11321	1.09099	1.07076	1.05219	1.03504	1.01910	1.00423
27	1.17301	1.14545	1.12091	1.09877	1.07861	1.06011	1.04301	1.02711
28	1.21146	1.18016	1.15272	1.12828	1.10624	1.08615	1.06771	1.05066
29	1.25450	1.21817	1.18702	1.15971	1.13537	1.11341	1.09340	1.07502
30	1.30430	1.26076	1.22461	1.19361	1.16642	1.14219	1.12031	1.10037
31	1.36500	1.31010	1.26679	1.23081	1.19995	1.17288	1.14875	1.12696
32	1.44633	1.37031	1.31569	1.27259	1.23679	1.20607	1.17911	1.15507
33	1.58323	1.45108	1.37544	1.32108	1.27818	1.24255	1.21196	1.18512
34		1.58723	1.45567	1.38038	1.32628	1.28358	1.24811	1.21766
35			1.59110	1.46010	1.38515	1.33130	1.28880	1.25348
36				1.59484	1.46438	1.38977	1.33616	1.29385
37					1.59847	1.46852	1.39423	1.34086
38						1.60198	1.47254	1.39856
39							1.60539	1.47643
40								1.60869

EXPECTED VALUES OF WEIBULL ORDER STATISTICS

SHAPE PARAMETER K=3.5

M \ N	1	2	3	4	5	6	7	8
1	.89975	.73809	.65736	.60548	.56809	.53925	.51602	.49670
2		1.06140	.89957	.81297	.75508	.71226	.67866	.65123
3			1.14231	.98618	.89980	.84071	.79627	.76095
4				1.19436	1.04377	.95889	.89996	.85512
5					1.23201	1.08620	1.00309	.94479
6						1.26117	1.11945	1.03807
7							1.28479	1.14657
8								1.30453

M \ N	9	10	11	12	13	14	15	16
1	.48026	.46602	.45350	.44237	.43237	.42331	.41504	.40746
2	.62820	.60844	.59121	.57599	.56239	.55013	.53898	.52879
3	.73185	.70722	.68596	.66733	.65080	.63597	.62255	.61033
4	.81917	.78931	.76390	.74186	.72245	.70517	.68962	.67552
5	.90007	.86395	.83379	.80799	.78552	.76566	.74793	.73193
6	.98058	.93618	.90014	.86991	.84394	.82125	.80114	.78313
7	1.06682	1.01017	.96622	.93038	.90020	.87420	.85141	.83116
8	1.16935	1.09110	1.03529	.99182	.95625	.92620	.90024	.87744

M \ N	9	10	11	12	13	14	15	16
9	1.32143	1.18892	1.11203	1.05702	1.01405	.97878	.94892	.92305
10		1.33615	1.20600	1.13037	1.07612	1.03364	.99869	.96903
11			1.34917	1.22113	1.14664	1.09312	1.05111	1.01649
12				1.36081	1.23467	1.16124	1.10839	1.06685
13					1.37132	1.24691	1.17445	1.12224
14						1.38089	1.25805	1.18650
15							1.38967	1.26828
16								1.39776

M \ N	17	18	19	20	21	22	23	24
1	.40046	.39398	.38794	.38229	.37700	.37202	.36733	.36289
2	.51941	.51074	.50268	.49517	.48814	.48153	.47531	.46944
3	.59913	.58879	.57922	.57031	.56198	.55418	.54684	.53993
4	.66263	.65080	.63986	.62971	.62025	.61140	.60309	.59527
5	.71738	.70407	.69181	.68046	.66992	.66007	.65085	.64219
6	.76684	.75200	.73839	.72584	.71421	.70338	.69327	.68378
7	.81298	.79651	.78148	.76768	.75492	.74309	.73205	.72173
8	.85713	.83886	.82228	.80713	.79318	.78029	.76830	.75712

M \ N	17	18	19	20	21	22	23	24
9	.90028	.87997	.86166	.84502	.82978	.81575	.80276	.79067
10	.94330	.92059	.90031	.88200	.86533	.85005	.83597	.82290
11	.98705	.96146	.93885	.91862	.90033	.88367	.86837	.85425
12	1.03255	1.00333	.97790	.95540	.93524	.91700	.90035	.88506
13	1.08114	1.04716	1.01817	.99290	.97052	.95044	.93226	.91564
14	1.13489	1.09421	1.06053	1.03178	1.00668	.98442	.96443	.94631
15	1.19756	1.14651	1.10623	1.07286	1.04432	1.01940	.99727	.97738
16	1.27770	1.20777	1.15726	1.11736	1.08427	1.05596	1.03120	1.00920

EXPECTED VALUES OF WEIBULL ORDER STATISTICS

SHAPE PARAMETER K=3.5

M \ N	17	18	19	20	21	22	23	24
17	1.40526	1.28644	1.21724	1.16723	1.12770	1.09489	1.06679	1.04220
18		1.41225	1.29459	1.22607	1.17653	1.13735	1.10481	1.07692
19			1.41879	1.30220	1.23433	1.18524	1.14639	1.11410
20				1.42492	1.30934	1.24208	1.19341	1.15489
21					1.43070	1.31607	1.24938	1.20112
22						1.43616	1.32242	1.25627
23							1.44133	1.32843
24								1.44624

M \ N	25	26	27	28	29	30	31	32
1	.35868	.35468	.35088	.34725	.34379	.34048	.33730	.33425
2	.46388	.45860	.45359	.44881	.44425	.43989	.43572	.43172
3	.53339	.52719	.52130	.51570	.51036	.50526	.50039	.49571
4	.58789	.58090	.57427	.56798	.56198	.55625	.55078	.54555
5	.63403	.62631	.61900	.61206	.60546	.59917	.59317	.58743
6	.67485	.66643	.65847	.65092	.64374	.63691	.63040	.62418
7	.71204	.70292	.69430	.68615	.67841	.67106	.66405	.65736
8	.74665	.73681	.72753	.71877	.71047	.70258	.69508	.68793

M \ N	25	26	27	28	29	30	31	32
9	.77937	.76879	.75883	.74944	.74056	.73214	.72414	.71653
10	.81074	.79937	.78870	.77866	.76918	.76021	.75170	.74361
11	.84115	.82894	.81751	.80678	.79667	.78712	.77808	.76950
12	.87093	.85780	.84556	.83409	.82332	.81316	.80356	.79446
13	.90037	.88624	.87311	.86085	.84936	.83855	.82836	.81872
14	.92974	.91450	.90038	.88725	.87499	.86348	.85266	.84245
15	.95933	.94281	.92760	.91351	.90040	.88813	.87663	.86579
16	.98941	.97144	.95498	.93981	.92575	.91266	.90041	.88890

M \ N	25	26	27	28	29	30	31	32
17	1.02033	1.00065	.98275	.96636	.95124	.93721	.92414	.91191
18	1.05249	1.03075	1.01117	.99336	.97703	.96196	.94798	.93494
19	1.08641	1.06215	1.04054	1.02107	1.00334	.98708	.97206	.95812
20	1.12285	1.09535	1.07125	1.04977	1.03039	1.01275	.99656	.98160
21	1.16290	1.13110	1.10379	1.07984	1.05848	1.03922	1.02166	1.00554
22	1.20840	1.17047	1.13890	1.11178	1.08797	1.06674	1.04758	1.03010
23	1.26280	1.21529	1.17765	1.14629	1.11935	1.09569	1.07458	1.05552
24	1.33414	1.26900	1.22184	1.18446	1.15332	1.12655	1.10304	1.08204

M \ N	25	26	27	28	29	30	31	32
25	1.45091	1.33957	1.27489	1.22807	1.19095	1.16001	1.13341	1.11003
26		1.45537	1.34474	1.28051	1.23401	1.19714	1.16640	1.13996
27			1.45962	1.34969	1.28588	1.23968	1.20305	1.17250
28				1.46369	1.35441	1.29101	1.24511	1.20870
29					1.46759	1.35894	1.29593	1.25031
30						1.47134	1.36329	1.30065
31							1.47494	1.36746
32								1.47841

EXPECTED VALUES OF WEIBULL ORDER STATISTICS

SHAPE PARAMETER K=3.5

M \ N	33	34	35	36	37	38	39	40
1	.33133	.32851	.32581	.32319	.32067	.31824	.31589	.31361
2	.42788	.42419	.42064	.41722	.41392	.41073	.40766	.40468
3	.49123	.48693	.48279	.47880	.47496	.47125	.46767	.46421
4	.54053	.53572	.53109	.52663	.52235	.51821	.51422	.51036
5	.58193	.57665	.57159	.56672	.56203	.55751	.55315	.54894
6	.61823	.61252	.60705	.60179	.59673	.59185	.58715	.58262
7	.65097	.64485	.63898	.63335	.62793	.62272	.61770	.61285
8	.68111	.67458	.66832	.66232	.65656	.65102	.64568	.64053
9	.70927	.70233	.69569	.68932	.68321	.67734	.67170	.66626
10	.73590	.72854	.72151	.71478	.70833	.70213	.69617	.69043
11	.76133	.75355	.74612	.73901	.73221	.72568	.71941	.71337
12	.78582	.77760	.76976	.76227	.75510	.74823	.74164	.73531
13	.80958	.80090	.79263	.78474	.77720	.76998	.76307	.75642
14	.83278	.82361	.81489	.80659	.79866	.79108	.78382	.77686
15	.85557	.84588	.83669	.82795	.81961	.81165	.80404	.79675
16	.87807	.86783	.85814	.84893	.84017	.83181	.82383	.81620
17	.90042	.88959	.87935	.86965	.86043	.85165	.84329	.83529
18	.92272	.91125	.90042	.89019	.88049	.87127	.86249	.85411
19	.94511	.93293	.92147	.91066	.90043	.89073	.88151	.87273
20	.96771	.95474	.94258	.93114	.92035	.91013	.90044	.89122
21	.99064	.97679	.96385	.95173	.94031	.92954	.91934	.90966
22	1.01405	.99921	.98541	.97252	.96042	.94904	.93828	.92810
23	1.03813	1.02215	1.00737	.99361	.98076	.96870	.95734	.94662
24	1.06308	1.04577	1.02986	1.01514	1.00144	.98863	.97660	.96528
25	1.08916	1.07029	1.05306	1.03722	1.02256	1.00891	.99614	.98415
26	1.11671	1.09595	1.07717	1.06003	1.04426	1.02966	1.01606	1.00334
27	1.14622	1.12310	1.10245	1.08377	1.06671	1.05100	1.03646	1.02291
28	1.17834	1.15221	1.12922	1.10868	1.09009	1.07310	1.05747	1.04298
29	1.21413	1.18394	1.15796	1.13509	1.11465	1.09615	1.07925	1.06368
30	1.25530	1.21933	1.18932	1.16348	1.14073	1.12039	1.10198	1.08515
31	1.30518	1.26010	1.22433	1.19449	1.16878	1.14616	1.12592	1.10759
32	1.37148	1.30954	1.26472	1.22914	1.19946	1.17389	1.15138	1.13124
33	1.48175	1.37535	1.31375	1.26916	1.23378	1.20426	1.17882	1.15641
34		1.48498	1.37908	1.31780	1.27345	1.23826	1.20888	1.18357
35			1.48809	1.38269	1.32171	1.27759	1.24258	1.21335
36				1.49110	1.38617	1.32550	1.28159	1.24675
37					1.49402	1.38954	1.32915	1.28546
38						1.49684	1.39281	1.33270
39							1.49958	1.39597
40								1.50223

EXPECTED VALUES OF WEIBULL ORDER STATISTICS

SHAPE PARAMETER K=4.0

M \ N	1	2	3	4	5	6	7	8
1	.90640	.76219	.68872	.64092	.60615	.57914	.55725	.53895
2		1.05061	.90914	.83210	.78003	.74119	.71050	.68531
3			1.12135	.98618	.91021	.85770	.81790	.78608
4				1.16641	1.03683	.96272	.91076	.87094
5					1.19881	1.07388	1.00168	.95058
6						1.22379	1.10276	1.03235
7							1.24396	1.12623
8								1.26078

M \ N	9	10	11	12	13	14	15	16
1	.52331	.50971	.49771	.48700	.47735	.46859	.46057	.45320
2	.66406	.64575	.62972	.61551	.60278	.59126	.58077	.57115
3	.75971	.73729	.71787	.70078	.68556	.67187	.65946	.64812
4	.83882	.81201	.78909	.76913	.75150	.73575	.72153	.70861
5	.91109	.87903	.85212	.82901	.80880	.79089	.77484	.76032
6	.98216	.94316	.91132	.88448	.86134	.84104	.82300	.80678
7	1.05744	1.00817	.96969	.93816	.91148	.88840	.86811	.85002
8	1.14588	1.07856	1.03015	.99222	.96102	.93455	.91160	.89136

M \ N	9	10	11	12	13	14	15	16
9	1.27515	1.16271	1.09671	1.04912	1.01172	.98087	.95464	.93184
10		1.28764	1.17738	1.11258	1.06574	1.02885	.99836	.97237
11			1.29866	1.19034	1.12663	1.08050	1.04409	1.01395
12				1.30851	1.20193	1.13920	1.09374	1.05780
13					1.31739	1.21238	1.15057	1.10572
14						1.32547	1.22189	1.16092
15							1.33287	1.23060
16								1.33969

M \ N	17	18	19	20	21	22	23	24
1	.44638	.44005	.43414	.42861	.42342	.41852	.41389	.40951
2	.56227	.55405	.54640	.53924	.53254	.52623	.52027	.51464
3	.63769	.62806	.61911	.61077	.60296	.59563	.58873	.58220
4	.69677	.68586	.67577	.66638	.65761	.64939	.64167	.63438
5	.74708	.73493	.72372	.71333	.70365	.69459	.68609	.67810
6	.79208	.77866	.76632	.75491	.74431	.73443	.72518	.71649
7	.83373	.81894	.80540	.79293	.78140	.77067	.76065	.75125
8	.87329	.85698	.84214	.82855	.81601	.80439	.79357	.78346

M \ N	17	18	19	20	21	22	23	24
9	.91169	.89367	.87738	.86254	.84892	.83634	.82468	.81380
10	.94974	.92972	.91177	.89552	.88070	.86708	.85450	.84280
11	.98822	.96577	.94587	.92801	.91183	.89705	.88345	.87087
12	1.02799	1.00250	.98024	.96048	.94272	.92661	.91188	.89832
13	1.07022	1.04073	1.01549	.99341	.97379	.95615	.94012	.92545
14	1.11664	1.08155	1.05238	1.02738	1.00548	.98601	.96847	.95253
15	1.17041	1.12667	1.09197	1.06310	1.03833	1.01661	.99728	.97986
16	1.23863	1.17916	1.13592	1.10160	1.07301	1.04846	1.02692	1.00773

EXPECTED VALUES OF WEIBULL ORDER STATISTICS

SHAPE PARAMETER K=4.0

M \ N	17	18	19	20	21	22	23	24
17	1.34601	1.24606	1.18727	1.14450	1.11053	1.08222	1.05788	1.03652
18		1.35188	1.25298	1.19482	1.15249	1.11886	1.09081	1.06668
19			1.35738	1.25944	1.20187	1.15997	1.12665	1.09885
20				1.36253	1.26550	1.20849	1.16698	1.13397
21					1.36739	1.27120	1.21471	1.17358
22						1.37197	1.27658	1.22059
23							1.37630	1.28167
24								1.38042

M \ N	25	26	27	28	29	30	31	32
1	.40536	.40140	.39763	.39403	.39059	.38729	.38413	.38110
2	.50931	.50423	.49941	.49480	.49040	.48619	.48215	.47827
3	.57603	.57017	.56460	.55928	.55421	.54937	.54472	.54027
4	.62749	.62096	.61476	.60885	.60322	.59785	.59270	.58777
5	.67055	.66340	.65662	.65018	.64404	.63818	.63258	.62722
6	.70830	.70056	.69323	.68627	.67964	.67333	.66730	.66154
7	.74242	.73409	.72621	.71875	.71165	.70489	.69845	.69229
8	.77397	.76503	.75660	.74862	.74105	.73385	.72699	.72044

M \ N	25	26	27	28	29	30	31	32
9	.80362	.79406	.78506	.77655	.76850	.76085	.75357	.74663
10	.83189	.82168	.81207	.80302	.79446	.78634	.77864	.77130
11	.85917	.84824	.83800	.82837	.81928	.81069	.80253	.79478
12	.88575	.87406	.86314	.85289	.84324	.83413	.82551	.81733
13	.91192	.89939	.88772	.87680	.86656	.85690	.84779	.83915
14	.93793	.92446	.91196	.90032	.88942	.87918	.86953	.86041
15	.96401	.94948	.93606	.92361	.91199	.90112	.89090	.88126
16	.99043	.97467	.96021	.94685	.93445	.92287	.91202	.90182

M \ N	25	26	27	28	29	30	31	32
17	1.01747	1.00028	.98461	.97023	.95693	.94458	.93304	.92222
18	1.04548	1.02657	1.00949	.99392	.97961	.96638	.95408	.94258
19	1.07492	1.05389	1.03511	1.01814	1.00266	.98843	.97526	.96302
20	1.10640	1.08267	1.06180	1.04315	1.02629	1.01090	.99675	.98365
21	1.14086	1.11352	1.08998	1.06926	1.05074	1.03398	1.01869	1.00461
22	1.17982	1.14737	1.12025	1.09688	1.07631	1.05791	1.04127	1.02606
23	1.22615	1.18572	1.15353	1.12662	1.10343	1.08300	1.06472	1.04818
24	1.28650	1.23142	1.19131	1.15938	1.13268	1.10965	1.08936	1.07120

M \ N	25	26	27	28	29	30	31	32
25	1.38433	1.29109	1.23643	1.19664	1.16494	1.13843	1.11556	1.09541
26		1.38806	1.29546	1.24121	1.20171	1.17024	1.14392	1.12121
27			1.39162	1.29963	1.24577	1.20655	1.17530	1.14916
28				1.39503	1.30362	1.25012	1.21118	1.18015
29					1.39829	1.30744	1.25430	1.21561
30						1.40143	1.31111	1.25830
31							1.40444	1.31463
32								1.40733

EXPECTED VALUES OF WEIBULL ORDER STATISTICS

SHAPE PARAMETER K=4.0

M \ N	33	34	35	36	37	38	39	40
1	.37817	.37536	.37265	.37004	.36751	.36507	.36271	.36042
2	.47455	.47097	.46752	.46419	.46097	.45787	.45486	.45196
3	.53599	.53188	.52792	.52411	.52043	.51687	.51343	.51011
4	.58304	.57849	.57411	.56990	.56583	.56191	.55812	.55445
5	.62208	.61714	.61240	.60783	.60343	.59918	.59508	.59111
6	.65601	.65071	.64562	.64072	.63601	.63146	.62707	.62283
7	.68640	.68075	.67532	.67011	.66509	.66026	.65560	.65110
8	.71418	.70819	.70244	.69692	.69161	.68650	.68157	.67682
9	.74000	.73366	.72759	.72176	.71616	.71078	.70559	.70059
10	.76430	.75761	.75121	.74507	.73918	.73352	.72807	.72282
11	.78740	.78035	.77361	.76716	.76098	.75504	.74932	.74382
12	.80955	.80213	.79505	.78827	.78179	.77556	.76958	.76383
13	.83095	.82314	.81570	.80859	.80179	.79527	.78902	.78300
14	.85176	.84355	.83574	.82828	.82115	.81433	.80779	.80151
15	.87214	.86350	.85528	.84745	.83999	.83285	.82601	.81945
16	.89220	.88309	.87445	.86623	.85841	.85093	.84379	.83694
17	.91205	.90244	.89335	.88472	.87651	.86868	.86121	.85406
18	.93180	.92165	.91207	.90299	.89438	.88618	.87836	.87088
19	.95157	.94082	.93070	.92115	.91209	.90349	.89530	.88749
20	.97145	.96005	.94934	.93926	.92972	.92069	.91211	.90393
21	.99157	.97943	.96808	.95741	.94736	.93786	.92885	.92028
22	1.01206	.99909	.98700	.97570	.96507	.95505	.94558	.93659
23	1.03306	1.01913	1.00623	.99420	.98294	.97236	.96237	.95293
24	1.05475	1.03972	1.02587	1.01302	1.00105	.98984	.97930	.96935
25	1.07737	1.06102	1.04607	1.03229	1.01951	1.00759	.99643	.98593
26	1.10118	1.08325	1.06700	1.05213	1.03842	1.02571	1.01384	1.00273
27	1.12660	1.10670	1.08888	1.07272	1.05793	1.04429	1.03164	1.01983
28	1.15418	1.13176	1.11198	1.09426	1.07819	1.06349	1.04992	1.03732
29	1.18478	1.15898	1.13670	1.11704	1.09943	1.08345	1.06882	1.05532
30	1.21986	1.18923	1.16359	1.14145	1.12190	1.10439	1.08849	1.07394
31	1.26214	1.22394	1.19351	1.16802	1.14601	1.12658	1.10915	1.09334
32	1.31802	1.26584	1.22787	1.19762	1.17228	1.15039	1.13107	1.11374
33	1.41012	1.32128	1.26940	1.23165	1.20157	1.17639	1.15462	1.13540
34		1.41282	1.32442	1.27283	1.23530	1.20539	1.18034	1.15870
35			1.41542	1.32746	1.27614	1.23882	1.20907	1.18416
36				1.41793	1.33039	1.27934	1.24222	1.21263
37					1.42036	1.33322	1.28244	1.24550
38						1.42272	1.33597	1.28543
39							1.42500	1.33863
40								1.42721

EXPECTED VALUES OF WEIBULL ORDER STATISTICS

SHAPE PARAMETER K=5.0

M \ N	1	2	3	4	5	6	7	8
1	.91817	.79931	.73705	.69584	.66547	.64164	.62216	.60577
2		1.03703	.92383	.86068	.81733	.78461	.75852	.73693
3			1.09362	.98698	.92572	.88276	.84985	.82329
4				1.12917	1.02783	.96867	.92664	.89410
5					1.15450	1.05740	1.00020	.95918
6						1.17392	1.08028	1.02480
7							1.18953	1.09878
8								1.20250

M \ N	9	10	11	12	13	14	15	16
1	.59166	.57933	.56839	.55858	.54971	.54162	.53420	.52735
2	.71859	.70270	.68871	.67625	.66503	.65485	.64553	.63697
3	.80113	.78216	.76563	.75102	.73794	.72613	.71538	.70552
4	.86763	.84538	.82624	.80948	.79460	.78125	.76915	.75811
5	.92719	.90100	.87887	.85975	.84295	.82799	.81451	.80228
6	.98478	.95338	.92755	.90564	.88664	.86989	.85493	.84143
7	1.04482	1.00571	.97490	.94946	.92781	.90898	.89233	.87743
8	1.11420	1.06157	1.02331	.99307	.96803	.94665	.92800	.91149

M \ N	9	10	11	12	13	14	15	16
9	1.21353	1.12735	1.07592	1.03844	1.00873	.98406	.96296	.94452
10		1.22311	1.13878	1.08842	1.05164	1.02243	.99813	.97730
11			1.23154	1.14885	1.09945	1.06332	1.03458	1.01063
12				1.23906	1.15783	1.10930	1.07377	1.04547
13					1.24583	1.16592	1.11819	1.08321
14						1.25197	1.17327	1.12626
15							1.25760	1.17998
16								1.26277

M \ N	17	18	19	20	21	22	23	24
1	.52099	.51507	.50953	.50433	.49943	.49481	.49043	.48627
2	.62904	.62167	.61479	.60834	.60228	.59656	.59116	.58604
3	.69643	.68800	.68015	.67281	.66592	.65943	.65331	.64751
4	.74796	.73858	.72987	.72175	.71414	.70699	.70025	.69388
5	.79109	.78078	.77124	.76237	.75408	.74630	.73899	.73209
6	.82914	.81788	.80749	.79786	.78889	.78050	.77263	.76521
7	.86395	.85167	.84039	.82997	.82029	.81127	.80282	.79488
8	.89668	.88326	.87100	.85974	.84931	.83963	.83058	.82210

M \ N	17	18	19	20	21	22	23	24
9	.92815	.91345	.90011	.88791	.87667	.86627	.85659	.84754
10	.95907	.94286	.92827	.91502	.90288	.89170	.88133	.87167
11	.99006	.97203	.95598	.94152	.92837	.91631	.90518	.89485
12	1.02185	1.00154	.98371	.96781	.95348	.94043	.92845	.91738
13	1.05531	1.03200	1.01194	.99430	.97857	.96436	.95141	.93951
14	1.09179	1.06428	1.04126	1.02143	1.00399	.98840	.97432	.96147
15	1.13365	1.09965	1.07250	1.04976	1.03015	1.01289	.99746	.98350
16	1.18616	1.14044	1.10689	1.08008	1.05761	1.03821	1.02112	1.00583

EXPECTED VALUES OF WEIBULL ORDER STATISTICS

SHAPE PARAMETER K=5.0

M \ N	17	18	19	20	21	22	23	24
17	1.26756	1.19187	1.14673	1.11360	1.08710	1.06488	1.04569	1.02877
18		1.27201	1.19719	1.15258	1.11983	1.09363	1.07165	1.05266
19			1.27617	1.20214	1.15804	1.12566	1.09974	1.07798
20				1.28006	1.20678	1.16315	1.13111	1.10546
21					1.28373	1.21115	1.16796	1.13624
22						1.28718	1.21526	1.17249
23							1.29045	1.21915
24								1.29355

M \ N	25	26	27	28	29	30	31	32
1	.48232	.47855	.47495	.47151	.46821	.46505	.46201	.45908
2	.58117	.57653	.57211	.56789	.56384	.55997	.55624	.55266
3	.64201	.63678	.63180	.62704	.62249	.61813	.61395	.60993
4	.68785	.68212	.67666	.67146	.66649	.66173	.65717	.65279
5	.72557	.71937	.71349	.70788	.70253	.69741	.69251	.68781
6	.75820	.75157	.74527	.73928	.73357	.72811	.72289	.71789
7	.78739	.78032	.77361	.76724	.76118	.75539	.74986	.74457
8	.81412	.80660	.79948	.79273	.78631	.78019	.77435	.76877

M \ N	25	26	27	28	29	30	31	32
9	.83905	.83106	.82351	.81636	.80958	.80313	.79698	.79110
10	.86263	.85415	.84615	.83860	.83144	.82464	.81816	.81199
11	.88522	.87621	.86774	.85975	.85220	.84504	.83823	.83175
12	.90711	.89752	.88853	.88008	.87211	.86457	.85741	.85061
13	.92851	.91829	.90875	.89980	.89138	.88343	.87590	.86875
14	.94966	.93874	.92857	.91908	.91017	.90177	.89385	.88634
15	.97075	.95903	.94817	.93807	.92862	.91975	.91140	.90350
16	.99199	.97935	.96771	.95693	.94689	.93749	.92867	.92035

M \ N	25	26	27	28	29	30	31	32
17	1.01362	.99990	.98735	.97580	.96509	.95511	.94577	.93699
18	1.03590	1.02088	1.00727	.99483	.98336	.97272	.96280	.95351
19	1.05917	1.04257	1.02769	1.01419	1.00184	.99045	.97988	.97003
20	1.08392	1.06529	1.04884	1.03408	1.02069	1.00843	.99712	.98663
21	1.11085	1.08951	1.07105	1.05474	1.04010	1.02682	1.01465	1.00342
22	1.14108	1.11592	1.09479	1.07649	1.06031	1.04580	1.03261	1.02053
23	1.17677	1.14566	1.12073	1.09978	1.08164	1.06559	1.05119	1.03810
24	1.22283	1.18083	1.14999	1.12528	1.10451	1.08652	1.07060	1.05631

M \ N	25	26	27	28	29	30	31	32
25	1.29650	1.22633	1.18469	1.15411	1.12961	1.10901	1.09116	1.07537
26		1.29931	1.22967	1.18835	1.15803	1.13373	1.11330	1.09558
27			1.30198	1.23284	1.19185	1.16177	1.13766	1.11738
28				1.30454	1.23588	1.19520	1.16534	1.14141
29					1.30700	1.23879	1.19839	1.16876
30						1.30935	1.24157	1.20146
31							1.31161	1.24425
32								1.31378

EXPECTED VALUES OF WEIBULL ORDER STATISTICS

SHAPE PARAMETER K=5.0

M \ N	33	34	35	36	37	38	39	40
1	.45627	.45355	.45093	.44840	.44595	.44357	.44128	.43905
2	.54922	.54590	.54269	.53960	.53661	.53372	.53091	.52820
3	.60606	.60234	.59875	.59529	.59194	.58870	.58557	.58253
4	.64858	.64453	.64063	.63686	.63323	.62971	.62631	.62302
5	.68330	.67896	.67478	.67075	.66686	.66310	.65947	.65595
6	.71309	.70848	.70404	.69977	.69564	.69166	.68781	.68409
7	.73949	.73462	.72993	.72542	.72107	.71688	.71283	.70891
8	.76342	.75829	.75336	.74862	.74405	.73965	.73540	.73130
9	.78548	.78009	.77492	.76995	.76517	.76056	.75612	.75183
10	.80609	.80044	.79503	.78983	.78483	.78002	.77538	.77090
11	.82556	.81965	.81398	.80855	.80333	.79831	.79347	.78881
12	.84412	.83793	.83201	.82633	.82088	.81565	.81061	.80576
13	.86195	.85547	.84928	.84336	.83768	.83222	.82698	.82194
14	.87921	.87242	.86595	.85976	.85384	.84816	.84271	.83746
15	.89602	.88891	.88213	.87567	.86949	.86358	.85790	.85245
16	.91248	.90502	.89794	.89118	.88473	.87857	.87266	.86699
17	.92871	.92087	.91344	.90638	.89964	.89321	.88706	.88116
18	.94478	.93654	.92874	.92134	.91430	.90759	.90118	.89504
19	.96079	.95210	.94390	.93614	.92877	.92176	.91507	.90868
20	.97683	.96765	.95901	.95085	.94312	.93578	.92880	.92213
21	.99299	.98326	.97413	.96553	.95742	.94973	.94242	.93546
22	1.00938	.99902	.98934	.98027	.97172	.96364	.95599	.94872
23	1.02611	1.01503	1.00474	.99512	.98609	.97759	.96956	.96194
24	1.04332	1.03141	1.02040	1.01017	1.00061	.99164	.98318	.97519
25	1.06118	1.04828	1.03645	1.02552	1.01535	1.00585	.99692	.98851
26	1.07991	1.06582	1.05301	1.04126	1.03040	1.02029	1.01084	1.00197
27	1.09980	1.08424	1.07026	1.05753	1.04586	1.03506	1.02502	1.01562
28	1.12129	1.10384	1.08839	1.07450	1.06186	1.05026	1.03953	1.02954
29	1.14501	1.12503	1.10770	1.09236	1.07856	1.06600	1.05447	1.04381
30	1.17204	1.14845	1.12861	1.11140	1.09616	1.08245	1.06998	1.05852
31	1.20440	1.17518	1.15176	1.13205	1.11496	1.09982	1.08620	1.07380
32	1.24682	1.20723	1.17820	1.15494	1.13536	1.11838	1.10333	1.08980
33	1.31587	1.24929	1.20995	1.18111	1.15800	1.13855	1.12167	1.10672
34		1.31789	1.25167	1.21257	1.18391	1.16094	1.14161	1.12484
35			1.31984	1.25397	1.21510	1.18661	1.16378	1.14457
36				1.32172	1.25620	1.21754	1.18922	1.16653
37					1.32354	1.25834	1.21990	1.19174
38						1.32530	1.26042	1.22219
39							1.32701	1.26243
40								1.32867

EXPECTED VALUES OF WEIBULL ORDER STATISTICS

SHAPE PARAMETER K=6.0

M \ N	1	2	3	4	5	6	7	8
1	.92772	.82650	.77250	.73633	.70945	.68822	.67076	.65600
2		1.02893	.93452	.88099	.84386	.81562	.79295	.77410
3			1.07614	.98805	.93669	.90034	.87228	.84951
4				1.10551	1.02229	.97305	.93775	.91023
5					1.12631	1.04690	.99952	.96526
6						1.14219	1.06586	1.02008
7							1.15492	1.08112
8								1.16546

M \ N	9	10	11	12	13	14	15	16
1	.64324	.63205	.62209	.61313	.60501	.59758	.59075	.58443
2	.75801	.74402	.73166	.72061	.71063	.70155	.69323	.68555
3	.83041	.81400	.79964	.78690	.77547	.76511	.75566	.74697
4	.88772	.86871	.85229	.83786	.82501	.81344	.80293	.79331
5	.93837	.91623	.89744	.88115	.86677	.85393	.84234	.83178
6	.98678	.96050	.93877	.92026	.90414	.88989	.87712	.86556
7	1.03672	1.00430	.97861	.95729	.93906	.92315	.90904	.89638
8	1.09380	1.05062	1.01898	.99384	.97291	.95497	.93928	.92533

M \ N	9	10	11	12	13	14	15	16
9	1.17442	1.10460	1.06248	1.03156	1.00692	.98636	.96871	.95323
10		1.18217	1.11396	1.07279	1.04251	1.01834	.99813	.98075
11			1.18900	1.12219	1.08187	1.05218	1.02844	1.00856
12				1.19507	1.12953	1.08997	1.06081	1.03747
13					1.20053	1.13612	1.09726	1.06860
14						1.20549	1.14210	1.10387
15							1.21001	1.14756
16								1.21418

M \ N	17	18	19	20	21	22	23	24
1	.57855	.57307	.56793	.56309	.55853	.55422	.55013	.54624
2	.67843	.67180	.66560	.65978	.65430	.64912	.64421	.63956
3	.73894	.73148	.72451	.71799	.71186	.70608	.70061	.69543
4	.78445	.77624	.76861	.76147	.75478	.74848	.74253	.73690
5	.82210	.81317	.80488	.79715	.78992	.78313	.77673	.77069
6	.85502	.84533	.83637	.82805	.82029	.81301	.80617	.79972
7	.88489	.87439	.86473	.85579	.84747	.83969	.83240	.82553
8	.91278	.90139	.89096	.88134	.87243	.86413	.85636	.84907

M \ N	17	18	19	20	21	22	23	24
9	.93944	.92703	.91573	.90538	.89582	.88696	.87869	.87095
10	.96548	.95186	.93958	.92839	.91812	.90863	.89982	.89159
11	.99143	.97637	.96292	.95077	.93968	.92950	.92009	.91133
12	1.01791	1.00102	.98615	.97286	.96084	.94987	.93977	.93043
13	1.04562	1.02635	1.00969	.99501	.98188	.96998	.95912	.94911
14	1.07566	1.05304	1.03403	1.01760	1.00310	.99011	.97834	.96758
15	1.10992	1.08213	1.05982	1.04108	1.02485	1.01052	.99768	.98603
16	1.15258	1.11547	1.08808	1.06607	1.04757	1.03154	1.01737	1.00466

EXPECTED VALUES OF WEIBULL ORDER STATISTICS

SHAPE PARAMETER K=6.0

M \ N	17	18	19	20	21	22	23	24
17	1.21803	1.15722	1.12061	1.09358	1.07186	1.05358	1.03774	1.02373
18		1.22160	1.16152	1.12538	1.09869	1.07723	1.05917	1.04350
19			1.22494	1.16554	1.12983	1.10346	1.08225	1.06439
20				1.22807	1.16930	1.13399	1.10792	1.08695
21					1.23101	1.17283	1.13790	1.11211
22						1.23378	1.17615	1.14159
23							1.23640	1.17930
24								1.23888

M \ N	25	26	27	28	29	30	31	32
1	.54253	.53900	.53562	.53238	.52928	.52630	.52343	.52066
2	.63513	.63091	.62687	.62301	.61931	.61576	.61235	.60906
3	.69050	.68581	.68133	.67706	.67296	.66903	.66525	.66162
4	.73155	.72647	.72162	.71700	.71257	.70833	.70426	.70035
5	.76496	.75951	.75433	.74939	.74466	.74014	.73581	.73164
6	.79361	.78782	.78231	.77707	.77206	.76727	.76269	.75829
7	.81905	.81291	.80709	.80154	.79626	.79121	.78638	.78175
8	.84220	.83571	.82956	.82371	.81815	.81284	.80777	.80292

M \ N	25	26	27	28	29	30	31	32
9	.86367	.85681	.85032	.84417	.83832	.83275	.82743	.82234
10	.88388	.87663	.86979	.86331	.85716	.85132	.84575	.84043
11	.90315	.89548	.88826	.88144	.87499	.86886	.86302	.85745
12	.92174	.91361	.90599	.89880	.89201	.88558	.87946	.87364
13	.93985	.93122	.92315	.91557	.90842	.90166	.89526	.88917
14	.95767	.94848	.93991	.93190	.92436	.91726	.91053	.90416
15	.97537	.96554	.95643	.94793	.93997	.93248	.92542	.91873
16	.99313	.98258	.97284	.96380	.95536	.94746	.94002	.93300

M \ N	25	26	27	28	29	30	31	32
17	1.01115	.99973	.98927	.97962	.97065	.96228	.95443	.94704
18	1.02964	1.01719	1.00589	.99552	.98595	.97705	.96874	.96095
19	1.04889	1.03518	1.02285	1.01165	1.00137	.99188	.98305	.97480
20	1.06928	1.05395	1.04037	1.02816	1.01705	1.00687	.99745	.98869
21	1.09137	1.07388	1.05870	1.04525	1.03315	1.02215	1.01204	1.00270
22	1.11607	1.09553	1.07822	1.06319	1.04986	1.03787	1.02696	1.01694
23	1.14507	1.11980	1.09947	1.08232	1.06743	1.05422	1.04233	1.03151
24	1.18227	1.14836	1.12333	1.10319	1.08621	1.07145	1.05835	1.04657

M \ N	25	26	27	28	29	30	31	32
25	1.24124	1.18510	1.15149	1.12669	1.10673	1.08990	1.07526	1.06228
26		1.24348	1.18779	1.15447	1.12988	1.11010	1.09341	1.07890
27			1.24563	1.19035	1.15731	1.13293	1.11331	1.09676
28				1.24767	1.19280	1.16002	1.13583	1.11638
29					1.24963	1.19514	1.16261	1.13861
30						1.25151	1.19738	1.16509
31							1.25332	1.19953
32								1.25505

EXPECTED VALUES OF WEIBULL ORDER STATISTICS

SHAPE PARAMETER K=6.0

M \ N	33	34	35	36	37	38	39	40
1	.51800	.51543	.51295	.51054	.50822	.50596	.50378	.50166
2	.60590	.60284	.59989	.59704	.59428	.59161	.58902	.58651
3	.65813	.65476	.65150	.64836	.64532	.64238	.63953	.63676
4	.69658	.69295	.68946	.68608	.68281	.67965	.67659	.67363
5	.72764	.72378	.72007	.71648	.71302	.70967	.70643	.70329
6	.75406	.74999	.74608	.74230	.73865	.73513	.73172	.72841
7	.77731	.77304	.76893	.76497	.76114	.75745	.75388	.75043
8	.79826	.79379	.78948	.78534	.78135	.77749	.77377	.77017
9	.81747	.81279	.80830	.80398	.79982	.79580	.79193	.78818
10	.83533	.83045	.82577	.82127	.81693	.81276	.80873	.80483
11	.85214	.84705	.84216	.83748	.83297	.82863	.82444	.82040
12	.86809	.86278	.85770	.85282	.84813	.84362	.83928	.83509
13	.88336	.87782	.87253	.86745	.86258	.85790	.85339	.84905
14	.89809	.89231	.88679	.88151	.87644	.87158	.86691	.86241
15	.91239	.90635	.90059	.89509	.88983	.88478	.87993	.87527
16	.92635	.92003	.91403	.90829	.90281	.89757	.89253	.88770
17	.94006	.93345	.92717	.92119	.91548	.91003	.90480	.89979
18	.95361	.94667	.94010	.93385	.92790	.92222	.91679	.91159
19	.96706	.95977	.95288	.94635	.94013	.93422	.92856	.92315
20	.98051	.97282	.96558	.95873	.95223	.94605	.94017	.93454
21	.99401	.98589	.97825	.97106	.96425	.95779	.95165	.94579
22	1.00767	.99904	.99097	.98339	.97624	.96948	.96306	.95695
23	1.02157	1.01237	1.00381	.99580	.98827	.98116	.97444	.96806
24	1.03583	1.02597	1.01684	1.00834	1.00038	.99290	.98584	.97916
25	1.05059	1.03994	1.03016	1.02109	1.01265	1.00475	.99731	.99030
26	1.06602	1.05443	1.04386	1.03414	1.02515	1.01676	1.00891	1.00152
27	1.08237	1.06959	1.05808	1.04759	1.03795	1.02901	1.02069	1.01289
28	1.09996	1.08568	1.07300	1.06158	1.05117	1.04159	1.03272	1.02444
29	1.11931	1.10302	1.08885	1.07626	1.06493	1.05459	1.04508	1.03626
30	1.14128	1.12212	1.10595	1.09188	1.07939	1.06813	1.05787	1.04842
31	1.16747	1.14383	1.12481	1.10876	1.09480	1.08240	1.07121	1.06102
32	1.20160	1.16976	1.14629	1.12740	1.11146	1.09760	1.08528	1.07417
33	1.25672	1.20359	1.17196	1.14865	1.12989	1.11406	1.10029	1.08806
34		1.25833	1.20551	1.17408	1.15092	1.13229	1.11657	1.10289
35			1.25989	1.20736	1.17612	1.15311	1.13460	1.11898
36				1.26139	1.20914	1.17809	1.15523	1.13684
37					1.26284	1.21087	1.18000	1.15727
38						1.26424	1.21254	1.18184
39							1.26560	1.21415
40								1.26692

EXPECTED VALUES OF WEIBULL ORDER STATISTICS

SHAPE PARAMETER K=7.0

M \ N	1	2	3	4	5	6	7	8
1	.93544	.84725	.79957	.76737	.74330	.72419	.70841	.69503
2		1.02363	.94261	.89615	.86368	.83885	.81883	.80211
3			1.06414	.98907	.94486	.91334	.88889	.86897
4				1.08916	1.01855	.97637	.94594	.92210
5					1.10681	1.03964	.99919	.96978
6						1.12024	1.05582	1.01685
7							1.13098	1.06881
8								1.13986

M \ N	9	10	11	12	13	14	15	16
1	.68343	.67322	.66412	.65591	.64846	.64163	.63533	.62950
2	.78780	.77532	.76426	.75436	.74540	.73723	.72973	.72279
3	.85220	.83774	.82506	.81378	.80363	.79443	.78601	.77825
4	.90252	.88593	.87156	.85890	.84759	.83739	.82811	.81960
5	.94657	.92740	.91108	.89688	.88433	.87309	.86292	.85364
6	.98834	.96574	.94699	.93096	.91697	.90456	.89343	.88333
7	1.03110	1.00341	.98137	.96302	.94728	.93351	.92127	.91026
8	1.07958	1.04297	1.01600	.99448	.97650	.96105	.94750	.93543

M \ N	9	10	11	12	13	14	15	16
9	1.14739	1.08874	1.05308	1.02676	1.00571	.98809	.97291	.95957
10		1.15391	1.09666	1.06185	1.03612	1.01550	.99821	.98329
11			1.15964	1.10363	1.06957	1.04437	1.02414	1.00716
12				1.16473	1.10982	1.07644	1.05173	1.03186
13					1.16930	1.11538	1.08262	1.05835
14						1.17345	1.12042	1.08822
15							1.17724	1.12502
16								1.18072

M \ N	17	18	19	20	21	22	23	24
1	.62407	.61900	.61424	.60975	.60552	.60151	.59770	.59408
2	.71636	.71035	.70473	.69944	.69446	.68974	.68527	.68103
3	.77108	.76440	.75816	.75230	.74679	.74159	.73667	.73199
4	.81175	.80447	.79768	.79133	.78536	.77974	.77442	.76939
5	.84512	.83724	.82992	.82309	.81669	.81067	.80498	.79961
6	.87410	.86560	.85774	.85042	.84358	.83716	.83112	.82541
7	.90025	.89109	.88264	.87481	.86752	.86069	.85428	.84824
8	.92455	.91465	.90557	.89718	.88940	.88215	.87535	.86895

M \ N	17	18	19	20	21	22	23	24
9	.94767	.93692	.92713	.91814	.90983	.90210	.89489	.88813
10	.97015	.95842	.94781	.93812	.92922	.92099	.91332	.90616
11	.99248	.97954	.96797	.95749	.94791	.93910	.93095	.92335
12	1.01516	1.00072	.98796	.97654	.96619	.95672	.94801	.93992
13	1.03882	1.02239	1.00816	.99558	.98430	.97408	.96472	.95609
14	1.06435	1.04514	1.02896	1.01493	1.00252	.99138	.98128	.97202
15	1.09334	1.06984	1.05092	1.03497	1.02113	1.00888	.99788	.98789
16	1.12925	1.09804	1.07489	1.05624	1.04050	1.02685	1.01475	1.00388

EXPECTED VALUES OF WEIBULL ORDER STATISTICS

SHAPE PARAMETER K=7.0

M \ N	17	18	19	20	21	22	23	24
17	1.18394	1.13315	1.10238	1.07955	1.06115	1.04563	1.03214	1.02019
18		1.18692	1.13677	1.10640	1.08388	1.06572	1.05039	1.03706
19			1.18971	1.14014	1.11016	1.08792	1.06998	1.05483
20				1.19232	1.14330	1.11367	1.09169	1.07397
21					1.19477	1.14626	1.11696	1.09524
22						1.19708	1.14905	1.12007
23							1.19926	1.15169
24								1.20133

M \ N	25	26	27	28	29	30	31	32
1	.59062	.58732	.58416	.58114	.57823	.57544	.57275	.57016
2	.67698	.67312	.66943	.66589	.66250	.65925	.65611	.65309
3	.72755	.72331	.71926	.71539	.71167	.70811	.70468	.70139
4	.76460	.76005	.75570	.75154	.74756	.74375	.74008	.73656
5	.79451	.78966	.78504	.78063	.77641	.77237	.76849	.76476
6	.82001	.81487	.80999	.80533	.80088	.79663	.79254	.78862
7	.84252	.83711	.83197	.82707	.82239	.81792	.81364	.80953
8	.86293	.85722	.85181	.84666	.84176	.83708	.83260	.82831

M \ N	25	26	27	28	29	30	31	32
9	.88177	.87576	.87007	.86467	.85954	.85464	.84996	.84548
10	.89944	.89311	.88714	.88147	.87609	.87097	.86608	.86141
11	.91624	.90957	.90328	.89733	.89169	.88634	.88123	.87636
12	.93239	.92534	.91872	.91247	.90656	.90095	.89562	.89053
13	.94808	.94062	.93362	.92705	.92084	.91497	.90939	.90409
14	.96348	.95555	.94815	.94121	.93469	.92852	.92269	.91715
15	.97873	.97027	.96242	.95508	.94821	.94173	.93561	.92981
16	.99399	.98493	.97656	.96878	.96150	.95468	.94825	.94218

M \ N	25	26	27	28	29	30	31	32
17	1.00944	.99966	.99069	.98239	.97468	.96747	.96071	.95433
18	1.02524	1.01461	1.00494	.99605	.98784	.98020	.97305	.96633
19	1.04166	1.02997	1.01945	1.00987	1.00107	.99293	.98536	.97827
20	1.05899	1.04596	1.03440	1.02399	1.01450	1.00579	.99772	.99021
21	1.07771	1.06290	1.05001	1.03856	1.02825	1.01886	1.01022	1.00223
22	1.09858	1.08124	1.06658	1.05382	1.04249	1.03228	1.02297	1.01441
23	1.12300	1.10173	1.08457	1.07006	1.05743	1.04621	1.03609	1.02686
24	1.15418	1.12577	1.10472	1.08772	1.07336	1.06084	1.04972	1.03970

M \ N	25	26	27	28	29	30	31	32
25	1.20330	1.15655	1.12840	1.10755	1.09072	1.07648	1.06409	1.05307
26		1.20517	1.15880	1.13091	1.11024	1.09356	1.07946	1.06717
27			1.20695	1.16095	1.13329	1.11281	1.09628	1.08229
28				1.20865	1.16300	1.13557	1.11526	1.09887
29					1.21028	1.16496	1.13774	1.11760
30						1.21185	1.16683	1.13983
31							1.21335	1.16863
32								1.21479

EXPECTED VALUES OF WEIBULL ORDER STATISTICS

SHAPE PARAMETER K=7.0

M \ N	33	34	35	36	37	38	39	40
1	.56765	.56524	.56290	.56064	.55845	.55633	.55427	.55227
2	.65018	.64737	.64466	.64203	.63949	.63702	.63463	.63231
3	.69821	.69514	.69218	.68932	.68655	.68386	.68126	.67873
4	.73316	.72989	.72673	.72368	.72072	.71786	.71509	.71240
5	.76117	.75771	.75438	.75116	.74804	.74503	.74211	.73928
6	.78485	.78122	.77772	.77435	.77109	.76793	.76488	.76192
7	.80559	.80179	.79813	.79461	.79120	.78791	.78473	.78165
8	.82419	.82023	.81642	.81274	.80920	.80578	.80247	.79926
9	.84118	.83706	.83309	.82927	.82559	.82203	.81860	.81528
10	.85693	.85264	.84851	.84455	.84073	.83704	.83348	.83004
11	.87170	.86723	.86295	.85883	.85486	.85105	.84736	.84380
12	.88568	.88103	.87658	.87231	.86820	.86424	.86043	.85674
13	.89903	.89420	.88957	.88513	.88087	.87677	.87282	.86902
14	.91187	.90684	.90203	.89742	.89300	.88875	.88467	.88073
15	.92431	.91906	.91406	.90927	.90468	.90028	.89605	.89198
16	.93642	.93095	.92574	.92076	.91599	.91143	.90705	.90284
17	.94830	.94258	.93714	.93196	.92701	.92227	.91773	.91337
18	.96000	.95402	.94834	.94293	.93778	.93286	.92815	.92363
19	.97161	.96533	.95938	.95374	.94837	.94325	.93836	.93367
20	.98318	.97657	.97033	.96443	.95883	.95349	.94840	.94354
21	.99478	.98780	.98124	.97505	.96919	.96363	.95833	.95327
22	1.00648	.99910	.99218	.98567	.97952	.97370	.96817	.96290
23	1.01838	1.01051	1.00318	.99632	.98986	.98375	.97797	.97248
24	1.03055	1.02214	1.01434	1.00707	1.00025	.99384	.98778	.98203
25	1.04313	1.03406	1.02571	1.01797	1.01076	1.00399	.99762	.99160
26	1.05625	1.04639	1.03740	1.02912	1.02144	1.01428	1.00756	1.00123
27	1.07011	1.05928	1.04950	1.04058	1.03237	1.02475	1.01763	1.01096
28	1.08500	1.07292	1.06218	1.05248	1.04363	1.03547	1.02791	1.02084
29	1.10134	1.08759	1.07561	1.06495	1.05532	1.04654	1.03844	1.03093
30	1.11984	1.10371	1.09007	1.07818	1.06760	1.05805	1.04933	1.04129
31	1.14182	1.12199	1.10599	1.09245	1.08065	1.07015	1.06067	1.05201
32	1.17036	1.14374	1.12406	1.10817	1.09473	1.08302	1.07259	1.06318
33	1.21618	1.17203	1.14559	1.12604	1.11027	1.09693	1.08530	1.07495
34		1.21752	1.17363	1.14737	1.12795	1.11229	1.09904	1.08749
35			1.21881	1.17517	1.14908	1.12980	1.11424	1.10108
36				1.22005	1.17666	1.15073	1.13157	1.11612
37					1.22126	1.17811	1.15233	1.13329
38						1.22242	1.17950	1.15387
39							1.22355	1.18085
40								1.22465

EXPECTED VALUES OF WEIBULL ORDER STATISTICS

SHAPE PARAMETER K=8.0

M \ N	1	2	3	4	5	6	7	8
1	.94174	.86358	.82090	.79191	.77012	.75277	.73841	.72618
2		1.01990	.94894	.90789	.87904	.85689	.83897	.82396
3			1.05539	.98999	.95116	.92335	.90169	.88398
4				1.07719	1.01587	.97898	.95223	.93120
5					1.09251	1.03432	.99904	.97326
6						1.10415	1.04843	1.01450
7							1.11344	1.05974
8								1.12111

M \ N	9	10	11	12	13	14	15	16
1	.71557	.70621	.69784	.69029	.68342	.67712	.67131	.66591
2	.81109	.79983	.78984	.78088	.77276	.76534	.75852	.75221
3	.86903	.85612	.84477	.83466	.82555	.81726	.80968	.80269
4	.91388	.89916	.88639	.87511	.86502	.85591	.84760	.83998
5	.95285	.93595	.92152	.90894	.89780	.88781	.87875	.87048
6	.98958	.96976	.95326	.93913	.92677	.91579	.90592	.89695
7	1.02696	1.00280	.98350	.96739	.95355	.94141	.93060	.92086
8	1.06911	1.03732	1.01382	.99501	.97926	.96569	.95376	.94312

M \ N	9	10	11	12	13	14	15	16
9	1.12761	1.07705	1.04613	1.02323	1.00485	.98943	.97613	.96441
10		1.13323	1.08392	1.05377	1.03140	1.01342	.99831	.98524
11			1.13816	1.08995	1.06048	1.03859	1.02097	1.00615
12				1.14254	1.09531	1.06645	1.04500	1.02771
13					1.14648	1.10012	1.07181	1.05076
14						1.15004	1.10448	1.07667
15							1.15330	1.10845
16								1.15629

M \ N	17	18	19	20	21	22	23	24
1	.66089	.65618	.65176	.64760	.64366	.63993	.63638	.63300
2	.74635	.74087	.73574	.73091	.72635	.72203	.71793	.71404
3	.79621	.79017	.78452	.77922	.77422	.76950	.76503	.76078
4	.83293	.82639	.82029	.81457	.80919	.80412	.79932	.79477
5	.86287	.85583	.84928	.84316	.83742	.83201	.82691	.82208
6	.88874	.88118	.87417	.86764	.86153	.85580	.85039	.84528
7	.91200	.90387	.89637	.88941	.88292	.87683	.87112	.86572
8	.93352	.92477	.91673	.90930	.90239	.89595	.88990	.88422

M \ N	17	18	19	20	21	22	23	24
9	.95393	.94446	.93582	.92787	.92052	.91368	.90728	.90128
10	.97372	.96340	.95406	.94553	.93768	.93040	.92362	.91728
11	.99331	.98197	.97181	.96260	.95417	.94641	.93921	.93250
12	1.01315	1.00052	.98936	.97934	.97025	.96193	.95426	.94714
13	1.03378	1.01946	1.00704	.99604	.98616	.97719	.96897	.96138
14	1.05598	1.03929	1.02520	1.01296	1.00211	.99237	.98351	.97539
15	1.08110	1.06075	1.04432	1.03044	1.01838	1.00768	.99806	.98931
16	1.11210	1.08517	1.06514	1.04895	1.03527	1.02337	1.01281	1.00331

EXPECTED VALUES OF WEIBULL ORDER STATISTICS

SHAPE PARAMETER K=8.0

M \ N	17	18	19	20	21	22	23	24
17	1.15905	1.11547	1.08893	1.06918	1.05322	1.03973	1.02799	1.01756
18		1.16161	1.11859	1.09241	1.07294	1.05719	1.04387	1.03228
19			1.16400	1.12150	1.09566	1.07644	1.06089	1.04774
20				1.16624	1.12422	1.09869	1.07971	1.06435
21					1.16834	1.12677	1.10154	1.08278
22						1.17032	1.12917	1.10422
23							1.17219	1.13144
24								1.17396

M \ N	25	26	27	28	29	30	31	32
1	.62978	.62670	.62375	.62092	.61820	.61559	.61307	.61064
2	.71033	.70678	.70339	.70014	.69702	.69402	.69113	.68835
3	.75673	.75287	.74919	.74565	.74227	.73901	.73588	.73287
4	.79044	.78632	.78238	.77862	.77501	.77155	.76822	.76502
5	.81749	.81312	.80896	.80498	.80117	.79752	.79401	.79064
6	.84043	.83583	.83144	.82726	.82326	.81943	.81575	.81222
7	.86062	.85578	.85117	.84679	.84260	.83859	.83475	.83106
8	.87884	.87376	.86893	.86434	.85995	.85577	.85176	.84792

M \ N	25	26	27	28	29	30	31	32
9	.89563	.89029	.88523	.88042	.87584	.87147	.86729	.86329
10	.91133	.90572	.90041	.89538	.89059	.88603	.88168	.87752
11	.92622	.92031	.91474	.90947	.90447	.89971	.89517	.89084
12	.94049	.93427	.92841	.92289	.91765	.91268	.90796	.90345
13	.95433	.94776	.94159	.93578	.93030	.92511	.92017	.91547
14	.96789	.96091	.95440	.94829	.94253	.93709	.93194	.92704
15	.98128	.97386	.96696	.96051	.95445	.94875	.94335	.93824
16	.99467	.98673	.97938	.97255	.96616	.96016	.95450	.94915

M \ N	25	26	27	28	29	30	31	32
17	1.00818	.99963	.99178	.98451	.97774	.97141	.96546	.95985
18	1.02198	1.01270	1.00425	.99648	.98928	.98258	.97631	.97041
19	1.03628	1.02611	1.01693	1.00857	1.00087	.99375	.98711	.98090
20	1.05136	1.04003	1.02997	1.02089	1.01261	1.00500	.99794	.99137
21	1.06760	1.05476	1.04356	1.03360	1.02462	1.01642	1.00888	1.00189
22	1.08567	1.07066	1.05795	1.04688	1.03702	1.02813	1.02001	1.01254
23	1.10675	1.08840	1.07355	1.06098	1.05001	1.04026	1.03145	1.02341
24	1.13359	1.10914	1.09098	1.07628	1.06384	1.05298	1.04332	1.03459

M \ N	25	26	27	28	29	30	31	32
25	1.17565	1.13562	1.11142	1.09343	1.07888	1.06655	1.05580	1.04623
26		1.17725	1.13756	1.11357	1.09576	1.08134	1.06913	1.05848
27			1.17877	1.13940	1.11563	1.09797	1.08369	1.07159
28				1.18023	1.14117	1.11759	1.10009	1.08593
29					1.18163	1.14285	1.11947	1.10211
30						1.18297	1.14446	1.12126
31							1.18425	1.14601
32								1.18548

EXPECTED VALUES OF WEIBULL ORDER STATISTICS

SHAPE PARAMETER K=8.0

M \ N	33	34	35	36	37	38	39	40
1	.60830	.60603	.60384	.60172	.59966	.59767	.59573	.59385
2	.68566	.68307	.68056	.67813	.67578	.67350	.67129	.66914
3	.72996	.72716	.72445	.72182	.71928	.71682	.71444	.71212
4	.76193	.75895	.75608	.75330	.75061	.74800	.74547	.74302
5	.78739	.78426	.78124	.77832	.77550	.77276	.77011	.76754
6	.80882	.80555	.80239	.79934	.79639	.79354	.79078	.78810
7	.82751	.82410	.82081	.81764	.81457	.81161	.80874	.80595
8	.84422	.84067	.83726	.83396	.83078	.82770	.82473	.82184
9	.85945	.85576	.85221	.84879	.84550	.84231	.83923	.83625
10	.87353	.86970	.86602	.86247	.85906	.85576	.85258	.84950
11	.88669	.88272	.87890	.87523	.87169	.86829	.86499	.86181
12	.89913	.89501	.89105	.88725	.88359	.88006	.87666	.87338
13	.91099	.90670	.90260	.89865	.89487	.89122	.88771	.88432
14	.92237	.91791	.91365	.90957	.90565	.90188	.89825	.89475
15	.93337	.92874	.92431	.92007	.91601	.91211	.90836	.90474
16	.94407	.93924	.93464	.93024	.92603	.92199	.91811	.91438
17	.95455	.94951	.94471	.94014	.93577	.93158	.92757	.92371
18	.96485	.95958	.95458	.94982	.94528	.94094	.93678	.93279
19	.97505	.96953	.96431	.95934	.95462	.95011	.94579	.94166
20	.98520	.97941	.97393	.96875	.96382	.95913	.95465	.95036
21	.99537	.98926	.98351	.97808	.97294	.96804	.96338	.95893
22	1.00561	.99915	.99309	.98739	.98200	.97690	.97204	.96741
23	1.01601	1.00914	1.00273	.99672	.99107	.98572	.98065	.97583
24	1.02663	1.01929	1.01248	1.00613	1.00017	.99455	.98925	.98421
25	1.03758	1.02969	1.02241	1.01566	1.00936	1.00344	.99787	.99260
26	1.04899	1.04042	1.03260	1.02538	1.01868	1.01243	1.00656	1.00103
27	1.06103	1.05163	1.04313	1.03537	1.02822	1.02157	1.01537	1.00954
28	1.07394	1.06347	1.05415	1.04572	1.03802	1.03092	1.02433	1.01817
29	1.08808	1.07618	1.06580	1.05655	1.04820	1.04056	1.03351	1.02697
30	1.10405	1.09013	1.07833	1.06803	1.05886	1.05057	1.04299	1.03599
31	1.12298	1.10590	1.09209	1.08039	1.07017	1.06107	1.05284	1.04532
32	1.14749	1.12464	1.10769	1.09398	1.08237	1.07223	1.06319	1.05503
33	1.18667	1.14892	1.12623	1.10940	1.09579	1.08427	1.07420	1.06524
34		1.18781	1.15030	1.12776	1.11105	1.09754	1.08610	1.07610
35			1.18892	1.15162	1.12923	1.11264	1.09922	1.08786
36				1.18998	1.15290	1.13065	1.11417	1.10085
37					1.19101	1.15414	1.13203	1.11565
38						1.19201	1.15534	1.13335
39							1.19297	1.15649
40								1.19391

Table C3.2

VARIANCES AND COVARIANCES OF WEIBULL ORDER STATISTICS

[When $I = J$, tabular values are variances of the Ith order statistics of a sample of size N from a Weibull population with location parameter 0, scale parameter 1, and shape parameter $K = 0.5, 1.5(0.5)4.0$ (when $K = 1.0$, see Table C2.2); when $I \neq J$, they are covariances of the Ith and Jth order statistics.]

Variances and Covariances of Weibull Order Statistics (Shape Parameter K)

N	I	J	$K=0.5$	$K=1.5$	$K=2.0$	$K=2.5$	$K=3.0$	$K=3.5$	$K=4.0$
1	1	1	20.00000	0.37569	0.21460	0.14415	0.10533	0.08107	0.06466
2	1	1	1.25000	0.14909	0.10730	0.08279	0.06636	0.05456	0.04572
2	1	2	2.25000	0.11159	0.06738	0.04616	0.03394	0.02613	0.02080
2	2	2	34.25000	0.37911	0.18715	0.11319	0.07644	0.05532	0.04201
3	1	1	0.24691	0.08683	0.07513	0.05986	0.05064	0.04327	0.03733
3	1	2	0.39506	0.06775	0.04783	0.03600	0.02818	0.02269	0.01868
3	1	3	0.69136	0.05391	0.03355	0.02329	0.01724	0.01332	0.01063
3	2	2	2.79321	0.16476	0.09949	0.06785	0.04962	0.03801	0.03011
3	2	3	4.64506	0.13255	0.07087	0.04470	0.03095	0.02278	0.01750
3	3	3	45.49691	0.36706	0.16827	0.09675	0.06298	0.04434	0.03294
4	1	1	0.07812	0.05917	0.05365	0.04755	0.04180	0.03671	0.03233
4	1	2	0.11979	0.04691	0.03675	0.02944	0.02403	0.01994	0.01678
4	1	3	0.18229	0.03904	0.02757	0.02065	0.01608	0.01288	0.01055
4	1	4	0.30729	0.03210	0.02044	0.01435	0.01069	0.00830	0.00664
4	2	2	0.63985	0.10098	0.06880	0.05042	0.03869	0.03067	0.02493
4	2	3	0.94772	0.08458	0.05207	0.03573	0.02618	0.02005	0.01587
4	2	4	1.56346	0.06986	0.03884	0.02502	0.01755	0.01302	0.01006
4	3	3	4.35976	0.16596	0.09137	0.05856	0.04097	0.03035	0.02343
4	3	4	6.90606	0.13821	0.06894	0.04154	0.02785	0.02001	0.01508
4	4	4	54.86902	0.35525	0.15539	0.08659	0.05513	0.03817	0.02799
5	1	1	0.03200	0.04394	0.04292	0.03978	0.03602	0.03232	0.02892
5	1	2	0.04800	0.03513	0.02978	0.02501	0.02107	0.01788	0.01531
5	1	3	0.06933	0.02976	0.02294	0.01811	0.01460	0.01200	0.01002
5	1	4	0.10133	0.02558	0.01817	0.01362	0.01060	0.00849	0.00695
5	1	5	0.16533	0.02148	0.01393	0.00988	0.00741	0.00577	0.00463
5	2	2	0.22212	0.07103	0.05280	0.04078	0.03239	0.02632	0.02180
5	2	3	0.31512	0.06043	0.04093	0.02974	0.02263	0.01781	0.01439
5	2	4	0.45462	0.05210	0.03254	0.02247	0.01651	0.01266	0.01003
5	2	5	0.73362	0.04385	0.02504	0.01636	0.01158	0.00865	0.00671
5	3	3	1.10282	0.10465	0.06435	0.04406	0.03220	0.02462	0.01945
5	3	4	1.55880	0.09062	0.05149	0.03353	0.02368	0.01765	0.01367
5	3	5	2.47076	0.07653	0.03981	0.02455	0.01671	0.01213	0.00921
5	4	4	5.87228	0.16396	0.08502	0.05240	0.03566	0.02589	0.01967
5	4	5	8.99480	0.13939	0.06634	0.03878	0.02545	0.01800	0.01341
5	5	5	62.94732	0.34513	0.14600	0.07960	0.04990	0.03415	0.02482
6	1	1	0.01543	0.03446	0.03577	0.03438	0.03190	0.02912	0.02640
6	1	2	0.02284	0.02769	0.02501	0.02183	0.01886	0.01630	0.01414
6	1	3	0.03210	0.02368	0.01954	0.01608	0.01333	0.01116	0.00946
6	1	4	0.04444	0.02072	0.01589	0.01248	0.01002	0.00820	0.00682
6	1	5	0.06296	0.01817	0.01300	0.00978	0.00762	0.00610	0.00500
6	1	6	0.10000	0.01548	0.01020	0.00730	0.00550	0.00430	0.00346
6	2	2	0.09691	0.05390	0.04291	0.03453	0.02819	0.02336	0.01963
6	2	3	0.13440	0.04265	0.03369	0.02558	0.02004	0.01611	0.01322
6	2	4	0.18437	0.04055	0.02748	0.01992	0.01512	0.01187	0.00957
6	2	5	0.25933	0.03561	0.02252	0.01564	0.01153	0.00887	0.00703
6	2	6	0.40926	0.03037	0.01771	0.01171	0.00835	0.00626	0.00488
6	3	3	0.40917	0.07511	0.05008	0.03598	0.02716	0.02125	0.01708
6	3	4	0.55396	0.06605	0.04101	0.02817	0.02060	0.01575	0.01244
6	3	5	0.77114	0.05812	0.03371	0.02219	0.01577	0.01180	0.00917
6	3	6	1.20551	0.04967	0.02658	0.01666	0.01145	0.00837	0.00639
6	4	4	1.59591	0.10523	0.06050	0.03965	0.02811	0.02101	0.01631
6	4	5	2.18513	0.09292	0.04997	0.03142	0.02164	0.01584	0.01211
6	4	6	3.36358	0.07965	0.03956	0.02369	0.01580	0.01129	0.00848
6	5	5	7.30963	0.16117	0.08006	0.04799	0.03203	0.02292	0.01723
6	5	6	10.92085	0.13893	0.06387	0.03652	0.02360	0.01650	0.01219

Variances and Covariances of Weibull Order Statistics (Shape Parameter K)

N	I	J	$K = 0.5$	$K = 1.5$	$K = 2.0$	$K = 2.5$	$K = 3.0$	$K = 3.5$	$K = 4.0$
6	6	6	70.07319	0.33656	0.13879	0.07442	0.04612	0.03129	0.02260
7	1	1	0.00833	0.02806	0.03066	0.03039	0.02878	0.02667	0.02444
7	1	2	0.01222	0.02263	0.02155	0.01942	0.01714	0.01504	0.01320
7	1	3	0.01688	0.01947	0.01699	0.01446	0.01226	0.01044	0.00895
7	1	4	0.02271	0.01720	0.01402	0.01142	0.00940	0.00783	0.00660
7	1	5	0.03049	0.01534	0.01177	0.00923	0.00740	0.00604	0.00502
7	1	6	0.04215	0.01363	0.00983	0.00742	0.00580	0.00465	0.00381
7	1	7	0.06547	0.01174	0.00785	0.00566	0.00429	0.00337	0.00271
7	2	2	0.04892	0.04293	0.03617	0.03011	0.02515	0.02118	0.01801
7	2	3	0.06689	0.03703	0.02863	0.02252	0.01807	0.01478	0.01228
7	2	4	0.08935	0.03277	0.02367	0.01784	0.01389	0.01111	0.00908
7	2	5	0.11930	0.02926	0.01990	0.01444	0.01096	0.00860	0.00693
7	2	6	0.16421	0.02602	0.01665	0.01163	0.00860	0.00663	0.00527
7	2	7	0.25405	0.02243	0.01331	0.00888	0.00637	0.00480	0.00375
7	3	3	0.18722	0.05784	0.04112	0.03071	0.02378	0.01894	0.01544
7	3	4	0.24770	0.05129	0.03411	0.02441	0.01835	0.01430	0.01146
7	3	5	0.32834	0.04588	0.02873	0.01980	0.01451	0.01110	0.00877
7	3	6	0.44930	0.04085	0.02408	0.01599	0.01142	0.00858	0.00668
7	3	7	0.69121	0.03525	0.01927	0.01223	0.00847	0.00623	0.00477
7	4	4	0.62268	0.07656	0.04749	0.03257	0.02380	0.01817	0.01434
7	4	5	0.81684	0.06863	0.04013	0.02652	0.01889	0.01416	0.01102
7	4	6	1.10808	0.06122	0.03371	0.02147	0.01491	0.01098	0.00842
7	4	7	1.69056	0.05291	0.02705	0.01647	0.01110	0.00799	0.00603
7	5	5	2.09834	0.10466	0.05731	0.03641	0.02526	0.01857	0.01425
7	5	6	2.80696	0.09362	0.04833	0.02961	0.02003	0.01447	0.01095
7	5	7	4.22422	0.08112	0.03892	0.02280	0.01497	0.01059	0.00788
7	6	6	8.66924	0.15830	0.07608	0.04466	0.02937	0.02079	0.01550
7	6	7	12.70294	0.13784	0.06168	0.03466	0.02214	0.01534	0.01126
7	7	7	76.46552	0.32924	0.13304	0.07041	0.04323	0.02914	0.02093
8	1	1	0.00488	0.02348	0.02683	0.02731	0.02633	0.02471	0.02286
8	1	2	0.00711	0.01898	0.01893	0.01754	0.01577	0.01402	0.01243
8	1	3	0.00972	0.01640	0.01502	0.01316	0.01137	0.00982	0.00850
8	1	4	0.01284	0.01458	0.01251	0.01051	0.00882	0.00745	0.00636
8	1	5	0.01675	0.01313	0.01064	0.00863	0.00707	0.00587	0.00494
8	1	6	0.02196	0.01186	0.00912	0.00716	0.00574	0.00469	0.00389
8	1	7	0.02977	0.01064	0.00774	0.00587	0.00460	0.00369	0.00303
8	1	8	0.04540	0.00924	0.00626	0.00455	0.00346	0.00273	0.00220
8	2	2	0.02733	0.03536	0.03127	0.02680	0.02282	0.01949	0.01674
8	2	3	0.03701	0.03062	0.02489	0.02018	0.01652	0.01370	0.01151
8	2	4	0.04863	0.02725	0.02076	0.01615	0.01285	0.01043	0.00862
8	2	5	0.06315	0.02456	0.01770	0.01329	0.01032	0.00824	0.00672
8	2	6	0.08251	0.02221	0.01517	0.01103	0.00838	0.00658	0.00530
8	2	7	0.11155	0.01994	0.01289	0.00905	0.00672	0.00519	0.00413
8	2	8	0.16963	0.01732	0.01043	0.00702	0.00507	0.00383	0.00300
8	3	3	0.09795	0.04659	0.03494	0.02694	0.02131	0.01724	0.01421
8	3	4	0.12773	0.04154	0.02922	0.02162	0.01663	0.01317	0.01068
8	3	5	0.16495	0.03748	0.02494	0.01783	0.01338	0.01042	0.00834
8	3	6	0.21459	0.03393	0.02142	0.01482	0.01088	0.00834	0.00659
8	3	7	0.28904	0.03048	0.01820	0.01218	0.00874	0.00658	0.00514
8	3	8	0.43794	0.02650	0.01475	0.00946	0.00660	0.00487	0.00375
8	4	4	0.29576	0.05959	0.03928	0.02794	0.02091	0.01624	0.01298
8	4	5	0.37913	0.05386	0.03361	0.02309	0.01688	0.01289	0.01017
8	4	6	0.49028	0.04881	0.02891	0.01923	0.01375	0.01033	0.00805
8	4	7	0.65702	0.04390	0.02461	0.01583	0.01107	0.00818	0.00630
8	4	8	0.99050	0.03821	0.01997	0.01231	0.00837	0.00606	0.00459

Variances and Covariances of Weibull Order Statistics (Shape Parameter K)

N	I	J	$K = 0.5$	$K = 1.5$	$K = 2.0$	$K = 2.5$	$K = 3.0$	$K = 3.5$	$K = 4.0$
8	5	5	0.85181	0.07690	0.04525	0.03001	0.02142	0.01608	0.01252
8	5	6	1.09212	0.06981	0.03902	0.02507	0.01751	0.01293	0.00995
8	5	7	1.45257	0.06288	0.03329	0.02068	0.01412	0.01026	0.00780
8	5	8	2.17349	0.05480	0.02706	0.01612	0.01070	0.00762	0.00571
8	6	6	2.59906	0.10364	0.05466	0.03390	0.02313	0.01681	0.01278
8	6	7	3.41510	0.09357	0.04678	0.02808	0.01874	0.01340	0.01006
8	6	8	5.04718	0.08175	0.03816	0.02197	0.01426	0.00999	0.00739
8	7	7	9.95487	0.15555	0.07281	0.04202	0.02732	0.01918	0.01420
8	7	8	14.35966	0.13650	0.05975	0.03311	0.02095	0.01442	0.01052
8	8	8	82.27366	0.32292	0.12831	0.06718	0.04094	0.02744	0.01963
9	1	1	0.00305	0.02007	0.02384	0.02485	0.02434	0.02310	0.02155
9	1	2	0.00442	0.01626	0.01688	0.01602	0.01463	0.01316	0.01177
9	1	3	0.00599	0.01409	0.01345	0.01208	0.01062	0.00928	0.00811
9	1	4	0.00782	0.01257	0.01127	0.00972	0.00831	0.00711	0.00612
9	1	5	0.01001	0.01139	0.00968	0.00807	0.00674	0.00567	0.00482
9	1	6	0.01276	0.01039	0.00841	0.00681	0.00557	0.00462	0.00388
9	1	7	0.01641	0.00947	0.00731	0.00575	0.00461	0.00377	0.00313
9	1	8	0.02190	0.00856	0.00628	0.00478	0.00375	0.00302	0.00248
9	1	9	0.03287	0.00749	0.00513	0.00375	0.00287	0.00227	0.00183
9	2	2	0.01646	0.02987	0.02755	0.02421	0.02097	0.01812	0.01571
9	2	3	0.02214	0.02593	0.02202	0.01832	0.01527	0.01282	0.01087
9	2	4	0.02875	0.02317	0.01848	0.01477	0.01197	0.00984	0.00822
9	2	5	0.03670	0.02100	0.01589	0.01229	0.00973	0.00787	0.00649
9	2	6	0.04663	0.01917	0.01381	0.01036	0.00804	0.00641	0.00522
9	2	7	0.05987	0.01749	0.01202	0.00876	0.00666	0.00524	0.00422
9	2	8	0.07972	0.01582	0.01032	0.00729	0.00543	0.00420	0.00335
9	2	9	0.11944	0.01383	0.00844	0.00573	0.00415	0.00315	0.00248
9	3	3	0.05627	0.03872	0.03040	0.02409	0.01941	0.01591	0.01324
9	3	4	0.07265	0.03465	0.02556	0.01947	0.01526	0.01225	0.01004
9	3	5	0.09231	0.03144	0.02201	0.01621	0.01242	0.00981	0.00793
9	3	6	0.11689	0.02872	0.01915	0.01370	0.01028	0.00800	0.00641
9	3	7	0.14965	0.02622	0.01668	0.01158	0.00852	0.00654	0.00517
9	3	8	0.19880	0.02372	0.01433	0.00965	0.00695	0.00526	0.00411
9	3	9	0.29710	0.02076	0.01173	0.00759	0.00532	0.00394	0.00304
9	4	4	0.15925	0.04841	0.03357	0.02461	0.01880	0.01482	0.01197
9	4	5	0.20119	0.04399	0.02895	0.02054	0.01534	0.01189	0.00949
9	4	6	0.25361	0.04021	0.02523	0.01738	0.01271	0.00971	0.00766
9	4	7	0.32351	0.03674	0.02199	0.01472	0.01056	0.00796	0.00621
9	4	8	0.42836	0.03326	0.01892	0.01227	0.00862	0.00639	0.00493
9	4	9	0.63805	0.02912	0.01550	0.00966	0.00661	0.00586	0.00366
9	5	5	0.41697	0.06033	0.03762	0.02582	0.01886	0.01439	0.01135
9	5	6	0.52245	0.05522	0.03284	0.02189	0.01567	0.01178	0.00919
9	5	7	0.66309	0.05051	0.02867	0.01857	0.01303	0.00966	0.00745
9	5	8	0.87405	0.04577	0.02469	0.01551	0.01066	0.00779	0.00594
9	5	9	1.29598	0.04011	0.02025	0.01222	0.00818	0.00586	0.00441
9	6	6	1.08954	0.07670	0.04333	0.02800	0.01963	0.01455	0.01122
9	6	7	1.37270	0.07026	0.03790	0.02381	0.01637	0.01196	0.00913
9	6	8	1.79744	0.06375	0.03270	0.01992	0.01342	0.00966	0.00729
9	6	9	2.64693	0.05594	0.02687	0.01573	0.01032	0.00729	0.00542
9	7	7	3.09216	0.10244	0.05242	0.03190	0.02148	0.01546	0.01166
9	7	8	4.00541	0.09314	0.04536	0.02678	0.01767	0.01253	0.00935
9	7	9	5.83191	0.08191	0.03738	0.02122	0.01364	0.00949	0.00698
9	8	8	11.17229	0.15299	0.07006	0.03989	0.02569	0.01791	0.01319
9	8	9	15.90728	0.13509	0.05806	0.03180	0.01996	0.01366	0.00992
9	9	9	87.60442	0.31739	0.12433	0.06450	0.03906	0.02606	0.01858

Variances and Covariances of Weibull Order Statistics (Shape Parameter K)

N	I	J	$K = 0.5$	$K = 1.5$	$K = 2.0$	$K = 2.5$	$K = 3.0$	$K = 3.5$	$K = 4.0$
10	1	1	0.00200	0.01744	0.02146	0.02285	0.02269	0.02175	0.02045
10	1	2	0.00289	0.01415	0.01523	0.01476	0.01369	0.01244	0.01121
10	1	3	0.00389	0.01229	0.01218	0.01118	0.00998	0.00881	0.00776
10	1	4	0.00503	0.01100	0.01025	0.00905	0.00785	0.00680	0.00590
10	1	5	0.00637	0.01001	0.00886	0.00757	0.00643	0.00547	0.00469
10	1	6	0.00797	0.00918	0.00776	0.00645	0.00537	0.00451	0.00383
10	1	7	0.00997	0.00845	0.00684	0.00553	0.00452	0.00375	0.00315
10	1	8	0.01263	0.00776	0.00601	0.00474	0.00380	0.00311	0.00259
10	1	9	0.01663	0.00706	0.00521	0.00399	0.00314	0.00253	0.00208
10	1	10	0.02463	0.00620	0.00430	0.00316	0.00243	0.00192	0.00156
10	2	2	0.01050	0.02571	0.02462	0.02213	0.01945	0.01699	0.01485
10	2	3	0.01405	0.02237	0.01974	0.01681	0.01422	0.01208	0.01032
10	2	4	0.01810	0.02005	0.01664	0.01362	0.01122	0.00933	0.00786
10	2	5	0.02282	0.01825	0.01439	0.01142	0.00919	0.00753	0.00626
10	2	6	0.02849	0.01675	0.01262	0.00973	0.00768	0.00620	0.00510
10	2	7	0.03558	0.01542	0.01113	0.00836	0.00648	0.00517	0.00421
10	2	8	0.04504	0.01417	0.00979	0.00715	0.00545	0.00429	0.00346
10	2	9	0.05922	0.01289	0.00848	0.00602	0.00450	0.00349	0.00279
10	2	10	0.08757	0.01133	0.00700	0.00478	0.00348	0.00265	0.00209
10	3	3	0.03465	0.03294	0.02692	0.02185	0.01789	0.01483	0.01245
10	3	4	0.04441	0.02955	0.02273	0.01775	0.01414	0.01149	0.00950
10	3	5	0.05581	0.02692	0.01968	0.01488	0.01160	0.00927	0.00757
10	3	6	0.06948	0.02473	0.01728	0.01271	0.00972	0.00767	0.00620
10	3	7	0.08656	0.02278	0.01524	0.01091	0.00819	0.00637	0.00509
10	3	8	0.10935	0.02094	0.01342	0.00936	0.00691	0.00531	0.00421
10	3	9	0.14352	0.01905	0.01163	0.00788	0.00570	0.00431	0.00338
10	3	10	0.21187	0.01676	0.00960	0.00626	0.00441	0.00328	0.00254
10	4	4	0.09361	0.04051	0.02935	0.02209	0.01717	0.01370	0.01117
10	4	5	0.11708	0.03695	0.02545	0.01856	0.01412	0.01109	0.00893
10	4	6	0.14524	0.03396	0.02236	0.01585	0.01182	0.00915	0.00730
10	4	7	0.18044	0.03130	0.01976	0.01365	0.01001	0.00766	0.00605
10	4	8	0.22738	0.02879	0.01739	0.01169	0.00841	0.00634	0.00495
10	4	9	0.29778	0.02620	0.01509	0.00986	0.00696	0.00518	0.00401
10	4	10	0.43859	0.02306	0.01246	0.00783	0.00539	0.00394	0.00300
10	5	5	0.22992	0.04933	0.03229	0.02282	0.01699	0.01315	0.01047
10	5	6	0.28392	0.04539	0.02841	0.01953	0.01427	0.01089	0.00858
10	5	7	0.35141	0.04187	0.02511	0.01681	0.01206	0.00908	0.00709
10	5	8	0.44141	0.03853	0.02214	0.01444	0.01019	0.00758	0.00587
10	5	9	0.57640	0.03509	0.01922	0.01218	0.00842	0.00617	0.00472
10	5	10	0.84638	0.03090	0.01589	0.00969	0.00653	0.00470	0.00355
10	6	6	0.54682	0.06055	0.03615	0.02415	0.01731	0.01303	0.01017
10	6	7	0.67337	0.05591	0.03201	0.02084	0.01467	0.01090	0.00843
10	6	8	0.84211	0.05150	0.02825	0.01791	0.01239	0.00908	0.00695
10	6	9	1.09521	0.04694	0.02456	0.01513	0.01027	0.00743	0.00563
10	6	10	1.60142	0.04137	0.02032	0.01205	0.00797	0.00566	0.00423
10	7	7	1.33131	0.07623	0.04167	0.02638	0.01823	0.01337	0.01023
10	7	8	1.65423	0.07031	0.03684	0.02273	0.01544	0.01118	0.00847
10	7	9	2.13862	0.06416	0.03208	0.01923	0.01281	0.00915	0.00686
10	7	10	3.10740	0.05661	0.02658	0.01534	0.00997	0.00699	0.00517
10	8	8	3.57455	0.10120	0.05050	0.03025	0.02015	0.01439	0.01079
10	8	9	4.57633	0.09253	0.04409	0.02568	0.01678	0.01182	0.00877
10	8	10	6.57989	0.08181	0.03663	0.02055	0.01310	0.00906	0.00664
10	9	9	12.32754	0.15063	0.06772	0.03810	0.02434	0.01687	0.01237
10	9	10	17.35960	0.13368	0.05656	0.03068	0.01912	0.01302	0.00942
10	10	10	92.53698	0.31251	0.12092	0.06225	0.03749	0.02492	0.01771

Table C4.1

EXPECTED VALUES OF GAMMA ORDER STATISTICS

[$E(x_{M,N})$, where $x_{M,N}$ is the Mth order statistic of a sample of size N from a Gamma population with location parameter 0, scale parameter 1, and shape parameter $\alpha = 0.5, 1.5(0.5)4.0(1.0)6.0$; when $\alpha = 1.0$, see Table C2.1]

EXPECTED VALUES OF GAMMA ORDER STATISTICS

SHAPE PARAMETER ALPHA=0.5

M \ N	1	2	3	4	5	6	7	8
1	0.50000	0.18169	0.09640	0.06035	0.04154	0.03042	0.02327	0.01840
2		0.81831	0.35227	0.20454	0.13560	0.09714	0.07329	0.05738
3			1.05133	0.50000	0.30796	0.21251	0.15679	0.12100
4				1.23511	0.62803	0.40340	0.28681	0.21645
5					1.38687	0.74034	0.49085	0.35717
6						1.51618	0.84014	0.57106
7							1.62886	0.92983
8								1.72872

M \ N	9	10	11	12	13	14	15	16
1	0.01492	0.01235	0.01040	0.00887	0.00766	0.00669	0.00589	0.00522
2	0.04622	0.03806	0.03190	0.02715	0.02339	0.02036	0.01790	0.01585
3	0.09647	0.07885	0.06574	0.05570	0.04782	0.04153	0.03641	0.03219
4	0.17005	0.13757	0.11382	0.09588	0.08195	0.07090	0.06199	0.05468
5	0.27445	0.21877	0.17912	0.14972	0.12721	0.10956	0.09542	0.08392
6	0.42335	0.33013	0.26634	0.22029	0.18572	0.15899	0.13783	0.12074
7	0.64491	0.48549	0.38328	0.31240	0.26062	0.22137	0.19074	0.16630
8	1.01123	0.71324	0.54390	0.43391	0.35677	0.29988	0.25637	0.22216

M \ N	9	10	11	12	13	14	15	16
9	1.81840	1.08573	0.77674	0.59890	0.48212	0.39944	0.33795	0.29058
10		1.89981	1.15440	0.83601	0.65081	0.52805	0.44044	0.37480
11			1.97435	1.21807	0.89158	0.69991	0.57186	0.47982
12				2.04310	1.27744	0.94385	0.74647	0.61369
13					2.10691	1.33304	0.99319	0.79074
14						2.16644	1.38532	1.03991
15							2.22223	1.43466
16								2.27474

M \ N	17	18	19	20	21	22	23	24
1	0.00466	0.00419	0.00379	0.00344	0.00314	0.00287	0.00264	0.00244
2	0.01414	0.01270	0.01146	0.01040	0.00948	0.00868	0.00797	0.00735
3	0.02868	0.02571	0.02319	0.02102	0.01914	0.01751	0.01608	0.01482
4	0.04861	0.04351	0.03918	0.03547	0.03227	0.02949	0.02705	0.02491
5	0.07441	0.06646	0.05974	0.05401	0.04907	0.04479	0.04105	0.03777
6	0.10673	0.09508	0.08528	0.07695	0.06980	0.06362	0.05824	0.05353
7	0.14643	0.13003	0.11632	0.10472	0.09481	0.08628	0.07887	0.07239
8	0.19468	0.17220	0.15354	0.13786	0.12453	0.11310	0.10321	0.09460

M \ N	17	18	19	20	21	22	23	24
9	0.25308	0.22278	0.19785	0.17707	0.15951	0.14453	0.13164	0.12044
10	0.32390	0.28339	0.25047	0.22326	0.20047	0.18115	0.16460	0.15030
11	0.41043	0.35632	0.31302	0.27767	0.24834	0.22366	0.20266	0.18462
12	0.51766	0.44487	0.38781	0.34194	0.30434	0.27301	0.24657	0.22399
13	0.65370	0.55406	0.47815	0.41838	0.37014	0.33044	0.29725	0.26914
14	0.83290	0.69203	0.58910	0.51033	0.44807	0.39763	0.35597	0.32104
15	1.08427	0.87315	0.72879	0.62285	0.54147	0.47689	0.42440	0.38093
16	1.48138	1.12650	0.91164	0.76410	0.65540	0.57160	0.50489	0.45049

EXPECTED VALUES OF GAMMA ORDER STATISTICS

SHAPE PARAMETER ALPHA=0.5

M \ N	17	18	19	20	21	22	23	24
17	2.32432	1.52574	1.16678	0.94853	0.79807	0.68683	0.60079	0.53208
18		2.37130	1.56797	1.20530	0.98393	0.83079	0.71720	0.62908
19			2.41593	1.60827	1.24219	1.01796	0.86234	0.74657
20				2.45843	1.64680	1.27760	1.05072	0.89281
21					2.49902	1.68373	1.31163	1.08230
22						2.53784	1.71916	1.34439
23							2.57505	1.75323
24								2.61078

M \ N	25	26	27	28	29	30	31	32
1	0.00226	0.00209	0.00195	0.00182	0.00170	0.00159	0.00150	0.00141
2	0.00680	0.00631	0.00587	0.00547	0.00511	0.00479	0.00450	0.00423
3	0.01370	0.01270	0.01181	0.01101	0.01029	0.00964	0.00905	0.00851
4	0.02302	0.02133	0.01983	0.01847	0.01726	0.01616	0.01516	0.01425
5	0.03486	0.03229	0.02999	0.02793	0.02608	0.02441	0.02289	0.02151
6	0.04937	0.04568	0.04240	0.03946	0.03682	0.03444	0.03229	0.03033
7	0.06670	0.06166	0.05718	0.05317	0.04958	0.04635	0.04342	0.04077
8	0.08704	0.08037	0.07446	0.06918	0.06446	0.06021	0.05637	0.05290

M \ N	25	26	27	28	29	30	31	32
9	0.11066	0.10204	0.09442	0.08764	0.08158	0.07614	0.07124	0.06680
10	0.13784	0.12692	0.11729	0.10874	0.10111	0.09428	0.08813	0.08258
11	0.16898	0.15531	0.14330	0.13268	0.12322	0.11477	0.10719	0.10035
12	0.20452	0.18761	0.17279	0.15973	0.14814	0.13782	0.12857	0.12025
13	0.24507	0.22426	0.20613	0.19020	0.17613	0.16363	0.15246	0.14244
14	0.29136	0.26588	0.24380	0.22450	0.20752	0.19248	0.17910	0.16711
15	0.34435	0.31320	0.28639	0.26309	0.24269	0.22470	0.20874	0.19450
16	0.40531	0.36720	0.33466	0.30658	0.28213	0.26068	0.24173	0.22488

M \ N	25	26	27	28	29	30	31	32
17	0.47591	0.42912	0.38957	0.35572	0.32644	0.30090	0.27845	0.25857
18	0.55852	0.50068	0.45239	0.41147	0.37638	0.34597	0.31939	0.29599
19	0.65652	0.58423	0.52483	0.47512	0.43292	0.39665	0.36517	0.33760
20	0.77501	0.68315	0.60924	0.54837	0.49733	0.45392	0.41654	0.38403
21	0.92226	0.80257	0.70902	0.63358	0.57134	0.51904	0.47447	0.43604
22	1.11279	0.95075	0.82930	0.73416	0.65730	0.59375	0.54027	0.49461
23	1.37597	1.14225	0.97836	0.85525	0.75862	0.68041	0.61563	0.56102
24	1.78604	1.40645	1.17075	1.00512	0.88046	0.78243	0.70294	0.63699

M \ N	25	26	27	28	29	30	31	32
25	2.64515	1.81767	1.43592	1.19836	1.03109	0.90496	0.80561	0.72492
26		2.67825	1.84821	1.46442	1.22512	1.05631	0.92881	0.82821
27			2.71017	1.87773	1.49204	1.25109	1.08083	0.95203
28				2.74100	1.90630	1.51881	1.27632	1.10469
29					2.77081	1.93398	1.54479	1.30084
30						2.79967	1.96082	1.57002
31							2.82763	1.98688
32								2.85475

EXPECTED VALUES OF GAMMA ORDER STATISTICS

SHAPE PARAMETER ALPHA=0.5

M \ N	33	34	35	36	37	38	39	40
1	0.00133	0.00125	0.00118	0.00112	0.00106	0.00101	0.00096	0.00092
2	0.00398	0.00376	0.00356	0.00337	0.00319	0.00303	0.00288	0.00274
3	0.00802	0.00757	0.00716	0.00678	0.00643	0.00610	0.00580	0.00552
4	0.01343	0.01267	0.01197	0.01134	0.01075	0.01020	0.00970	0.00923
5	0.02025	0.01911	0.01805	0.01708	0.01619	0.01537	0.01461	0.01390
6	0.02855	0.02692	0.02543	0.02405	0.02279	0.02163	0.02055	0.01955
7	0.03835	0.03615	0.03413	0.03228	0.03058	0.02900	0.02755	0.02621
8	0.04974	0.04686	0.04422	0.04180	0.03958	0.03754	0.03564	0.03389
M \ N	33	34	35	36	37	38	39	40
9	0.06277	0.05910	0.05575	0.05268	0.04986	0.04726	0.04486	0.04265
10	0.07755	0.07297	0.06879	0.06496	0.06146	0.05823	0.05525	0.05250
11	0.09416	0.08854	0.08341	0.07873	0.07444	0.07049	0.06686	0.06350
12	0.11273	0.10591	0.09971	0.09405	0.08887	0.08412	0.07974	0.07571
13	0.13340	0.12522	0.11780	0.11103	0.10485	0.09918	0.09396	0.08916
14	0.15634	0.14661	0.13779	0.12977	0.12245	0.11575	0.10960	0.10394
15	0.18174	0.17024	0.15984	0.15040	0.14180	0.13394	0.12673	0.12011
16	0.20982	0.19630	0.18410	0.17305	0.16301	0.15385	0.14546	0.13777
M \ N	33	34	35	36	37	38	39	40
17	0.24088	0.22504	0.21079	0.19791	0.18624	0.17561	0.16590	0.15700
18	0.27523	0.25672	0.24012	0.22518	0.21165	0.19937	0.18817	0.17793
19	0.31328	0.29168	0.27239	0.25507	0.23945	0.22530	0.21243	0.20068
20	0.35552	0.33033	0.30793	0.28789	0.26987	0.25360	0.23885	0.22541
21	0.40256	0.37315	0.34713	0.32396	0.30320	0.28452	0.26762	0.25228
22	0.45517	0.42077	0.39050	0.36369	0.33977	0.31833	0.29900	0.28150
23	0.51433	0.47393	0.43865	0.40757	0.37999	0.35537	0.33326	0.31332
24	0.58132	0.53364	0.49234	0.45622	0.42435	0.39605	0.37075	0.34801
M \ N	33	34	35	36	37	38	39	40
25	0.65787	0.60118	0.55257	0.51041	0.47348	0.44086	0.41186	0.38591
26	0.74637	0.67828	0.62063	0.57113	0.52813	0.49043	0.45711	0.42743
27	0.85024	0.76732	0.69824	0.63967	0.58932	0.54553	0.50710	0.47308
28	0.97464	0.87174	0.78779	0.71776	0.65832	0.60715	0.56262	0.52348
29	1.12791	0.99670	0.89272	0.80780	0.73687	0.67659	0.62465	0.57939
30	1.32469	1.15053	1.01821	0.91322	0.82737	0.75558	0.69450	0.64182
31	1.59456	1.34791	1.17258	1.03921	0.93325	0.84651	0.77390	0.71206
32	2.01219	1.61843	1.37053	1.19410	1.05971	0.95284	0.86525	0.79185
M \ N	33	34	35	36	37	38	39	40
33	2.88108	2.03680	1.64167	1.39259	1.21509	1.07975	0.97200	0.88360
34		2.90666	2.06075	1.66431	1.41410	1.23560	1.09934	0.99075
35			2.93154	2.08407	1.68639	1.43510	1.25564	1.11851
36				2.95576	2.10679	1.70792	1.45562	1.27523
37					2.97934	2.12895	1.72895	1.47566
38						3.00232	2.15057	1.74949
39							3.02474	2.17168
40								3.04661

EXPECTED VALUES OF GAMMA ORDER STATISTICS

SHAPE PARAMETER ALPHA=1.5

M \ N	1	2	3	4	5	6	7	8
1	1.50000	0.86338	0.63154	0.50801	0.43003	0.37579	0.33559	0.30445
2		2.13662	1.32706	1.00214	0.81991	0.70125	0.61697	0.55359
3			2.54140	1.65198	1.27548	1.05725	0.91195	0.80711
4				2.83787	1.90298	1.49371	1.25099	1.08669
5					3.07159	2.10762	1.67575	1.41528
6						3.26439	2.28037	1.83203
7							3.42839	2.42981
8								3.57104

M \ N	9	10	11	12	13	14	15	16
1	0.27951	0.25902	0.24185	0.22722	0.21458	0.20354	0.19379	0.18511
2	0.50395	0.46386	0.43073	0.40280	0.37891	0.35819	0.34003	0.32397
3	0.72732	0.66427	0.61299	0.57034	0.53423	0.50321	0.47622	0.45250
4	0.96667	0.87446	0.80102	0.74093	0.69070	0.64799	0.61116	0.57902
5	1.23672	1.10498	1.00297	0.92121	0.85394	0.79747	0.74927	0.70757
6	1.55813	1.36846	1.22739	1.11744	1.02883	0.95560	0.89387	0.84101
7	1.96898	1.68458	1.48602	1.33735	1.22082	1.12647	1.04819	0.98197
8	2.56148	2.09087	1.79804	1.59221	1.43723	1.31517	1.21594	1.13332

M \ N	9	10	11	12	13	14	15	16
9	3.69724	2.67913	2.20068	1.90095	1.68907	1.52877	1.40200	1.29857
10		3.81036	2.78546	2.30059	1.99512	1.77813	1.61328	1.48244
11			3.91285	2.88243	2.39224	2.08191	1.86056	1.69178
12				4.00653	2.97155	2.47687	2.16240	1.93728
13					4.09278	3.05400	2.55549	2.23744
14						4.17268	3.13069	2.62889
15							4.24711	3.20238
16								4.31676

M \ N	17	18	19	20	21	22	23	24
1	0.17732	0.17030	0.16392	0.15810	0.15276	0.14785	0.14330	0.13909
2	0.30963	0.29676	0.28511	0.27453	0.26486	0.25598	0.24779	0.24022
3	0.43146	0.41265	0.39572	0.38038	0.36642	0.35364	0.34190	0.33107
4	0.55069	0.52551	0.50295	0.48261	0.46416	0.44733	0.43191	0.41773
5	0.67108	0.63884	0.61011	0.58433	0.56103	0.53988	0.52056	0.50283
6	0.79514	0.75491	0.71927	0.68746	0.65886	0.63298	0.60943	0.58790
7	0.92510	0.87562	0.83211	0.79350	0.75897	0.72787	0.69969	0.67401
8	1.06322	1.00285	0.95020	0.90381	0.86257	0.82561	0.79228	0.76204

M \ N	17	18	19	20	21	22	23	24
9	1.21218	1.13869	1.07524	1.01979	0.97083	0.92724	0.88811	0.85277
10	1.37536	1.28567	1.20918	1.14301	1.08506	1.03381	0.98809	0.94701
11	1.55740	1.44711	1.35451	1.27536	1.20675	1.14656	1.09324	1.04561
12	1.76508	1.62759	1.51446	1.41926	1.33774	1.26693	1.20472	1.14953
13	2.00903	1.83382	1.69358	1.57793	1.48041	1.39674	1.32396	1.25992
14	2.30772	2.07642	1.89855	1.75585	1.63794	1.53833	1.45273	1.37815
15	2.69771	2.37380	2.13995	1.95970	1.81481	1.69486	1.59336	1.50600
16	3.26967	2.76249	2.43617	2.20003	2.01766	1.87078	1.74899	1.64577

EXPECTED VALUES OF GAMMA ORDER STATISTICS

SHAPE PARAMETER ALPHA=1.5

M \ N	17	18	19	20	21	22	23	24
17	4.38220	3.33307	2.82368	2.49520	2.25703	2.07273	1.92407	1.80060
18		4.44392	3.39300	2.88164	2.55124	2.31123	2.12520	1.97491
19			4.50230	3.44981	2.93671	2.60457	2.36290	2.17530
20				4.55769	3.50382	2.98915	2.65545	2.41227
21					4.61039	3.55529	3.03921	2.70409
22						4.66063	3.60444	3.08708
23							4.70864	3.65148
24								4.75460

M \ N	25	26	27	28	29	30	31	32
1	0.13517	0.13151	0.12808	0.12487	0.12185	0.11900	0.11632	0.11377
2	0.23320	0.22665	0.22054	0.21482	0.20945	0.20440	0.19964	0.19514
3	0.32104	0.31172	0.30304	0.29492	0.28732	0.28018	0.27346	0.26712
4	0.40463	0.39249	0.38120	0.37067	0.36082	0.35158	0.34291	0.33474
5	0.48651	0.47141	0.45741	0.44438	0.43221	0.42083	0.41015	0.40011
6	0.56813	0.54990	0.53303	0.51736	0.50277	0.48913	0.47637	0.46438
7	0.65051	0.62890	0.60896	0.59048	0.57330	0.55729	0.54233	0.52830
8	0.73445	0.70916	0.68589	0.66439	0.64445	0.62591	0.60860	0.59242

M \ N	25	26	27	28	29	30	31	32
9	0.82066	0.79134	0.76444	0.73965	0.71672	0.69545	0.67565	0.65716
10	0.90985	0.87605	0.84515	0.81676	0.79059	0.76636	0.74386	0.72290
11	1.00275	0.96393	0.92859	0.89624	0.86650	0.83905	0.81362	0.78998
12	1.10016	1.05568	1.01535	0.97858	0.94490	0.91391	0.88528	0.85874
13	1.20302	1.15206	1.10609	1.06437	1.02630	0.99139	0.95924	0.92952
14	1.31244	1.25398	1.20156	1.15423	1.11123	1.07195	1.03590	1.00268
15	1.42978	1.36254	1.30265	1.24890	1.20030	1.15611	1.11572	1.07862
16	1.55681	1.47910	1.41045	1.34925	1.29425	1.24449	1.19920	1.15777

M \ N	25	26	27	28	29	30	31	32
17	1.69581	1.60539	1.52629	1.45635	1.39393	1.33779	1.28695	1.24064
18	1.84991	1.74369	1.65191	1.57155	1.50041	1.43687	1.37966	1.32781
19	2.02352	1.89712	1.78958	1.69656	1.61502	1.54278	1.47818	1.41998
20	2.22323	2.07009	1.94240	1.83364	1.73947	1.65685	1.58357	1.51800
21	2.45953	2.26918	2.11478	1.98590	1.87602	1.78078	1.69715	1.62291
22	2.75067	2.50486	2.31329	2.15774	2.02776	1.91684	1.82061	1.73603
23	3.13295	2.79536	2.54839	2.35571	2.19909	2.06810	1.95620	1.85905
24	3.69657	3.17699	2.83832	2.59028	2.39657	2.23896	2.10702	1.99422

M \ N	25	26	27	28	29	30	31	32
25	4.79869	3.73986	3.21932	2.87965	2.63064	2.43597	2.27744	2.14462
26		4.84104	3.78151	3.26008	2.91950	2.66957	2.47402	2.31463
27			4.88179	3.82162	3.29938	2.95795	2.70718	2.51080
28				4.92106	3.86030	3.33732	2.99510	2.74354
29					4.95894	3.89766	3.37398	3.03104
30						4.99554	3.93377	3.40946
31							5.03093	3.96873
32								5.06519

EXPECTED VALUES OF GAMMA ORDER STATISTICS

SHAPE PARAMETER ALPHA=1.5

M \ N	33	34	35	36	37	38	39	40
1	0.11136	0.10908	0.10690	0.10483	0.10286	0.10097	0.09917	0.09745
2	0.19088	0.18685	0.18301	0.17937	0.17590	0.17259	0.16943	0.16641
3	0.26112	0.25545	0.25007	0.24496	0.24010	0.23547	0.23106	0.22684
4	0.32703	0.31973	0.31283	0.30628	0.30005	0.29413	0.28848	0.28309
5	0.39064	0.38171	0.37325	0.36524	0.35764	0.35041	0.34353	0.33697
6	0.45310	0.44247	0.43243	0.42292	0.41390	0.40535	0.39720	0.38945
7	0.51513	0.50272	0.49102	0.47996	0.46949	0.45955	0.45012	0.44114
8	0.57724	0.56297	0.54953	0.53685	0.52485	0.51348	0.50270	0.49245
9	0.63985	0.62361	0.60833	0.59394	0.58034	0.56747	0.55528	0.54370
10	0.70331	0.68497	0.66774	0.65153	0.63624	0.62179	0.60812	0.59515
11	0.76794	0.74734	0.72803	0.70989	0.69281	0.67669	0.66145	0.64702
12	0.83405	0.81102	0.78947	0.76926	0.75027	0.73237	0.71548	0.69950
13	0.90194	0.87628	0.85232	0.82989	0.80884	0.78904	0.77038	0.75276
14	0.97194	0.94341	0.91683	0.89200	0.86874	0.84691	0.82636	0.80699
15	1.04440	1.01271	0.98328	0.95584	0.93020	0.90617	0.88360	0.86235
16	1.11968	1.08453	1.05196	1.02168	0.99345	0.96704	0.94229	0.91902
17	1.19823	1.15923	1.12321	1.08981	1.05874	1.02975	1.00263	0.97719
18	1.28054	1.23724	1.19737	1.16053	1.12636	1.09455	1.06485	1.03705
19	1.36720	1.31904	1.27489	1.23422	1.19661	1.16170	1.12919	1.09883
20	1.45888	1.40521	1.35622	1.31127	1.26985	1.23152	1.19592	1.16275
21	1.55644	1.49645	1.44196	1.39218	1.34649	1.30435	1.26533	1.22908
22	1.66090	1.59357	1.53277	1.47751	1.42700	1.38060	1.33779	1.29813
23	1.77359	1.69763	1.62950	1.56794	1.51195	1.46074	1.41368	1.37023
24	1.89620	1.80993	1.73317	1.66430	1.60202	1.54534	1.49348	1.44579
25	2.03098	1.93215	1.84510	1.76761	1.69803	1.63508	1.57776	1.52528
26	2.18099	2.06655	1.96697	1.87920	1.80101	1.73077	1.66717	1.60924
27	2.35062	2.21620	2.10102	2.00073	1.91227	1.83344	1.76256	1.69837
28	2.54639	2.38546	2.25033	2.13445	2.03349	1.94439	1.86494	1.79347
29	2.77875	2.58088	2.41925	2.28343	2.16691	2.06531	1.97561	1.89557
30	3.06583	2.81286	2.61432	2.45203	2.31558	2.19843	2.09625	2.00597
31	3.44382	3.09956	2.84595	2.64678	2.48387	2.34682	2.22909	2.12634
32	4.00259	3.47714	3.13229	2.87808	2.67831	2.51482	2.37720	2.25892
33	5.09840	4.03543	3.50947	3.16407	2.90929	2.70896	2.54492	2.40677
34		5.13061	4.06731	3.54087	3.19495	2.93964	2.73879	2.57423
35			5.16188	4.09827	3.57139	3.22498	2.96918	2.76783
36				5.19227	4.12838	3.60108	3.25422	2.99794
37					5.22183	4.15768	3.62999	3.28269
38						5.25059	4.18620	3.65815
39							5.27860	4.21399
40								5.30589

EXPECTED VALUES OF GAMMA ORDER STATISTICS

SHAPE PARAMETER ALPHA=2.0

M \ N	1	2	3	4	5	6	7	8
1	2.00000	1.25000	0.96296	0.80469	0.70208	0.62912	0.57402	0.53063
2		2.75000	1.82407	1.43779	1.21512	1.06690	0.95969	0.87777
3			3.21296	2.21036	1.77180	1.51154	1.33494	1.20545
4				3.54716	2.50273	2.03205	1.74701	1.55077
5					3.80827	2.73808	2.24583	1.94325
6						4.02231	2.93497	2.42737
7							4.20353	3.10417
8								4.36058

M \ N	9	10	11	12	13	14	15	16
1	0.49537	0.46602	0.44112	0.41967	0.40095	0.38443	0.36972	0.35652
2	0.81270	0.75949	0.71499	0.67710	0.64435	0.61570	0.59037	0.56779
3	1.10551	1.02553	0.95974	0.90446	0.85721	0.81624	0.78032	0.74849
4	1.40532	1.29214	1.20097	1.12558	1.06197	1.00740	0.95996	0.91824
5	1.73259	1.57509	1.45169	1.35173	1.26871	1.19839	1.13787	1.08511
6	2.11178	1.89009	1.72316	1.59163	1.48457	1.39530	1.31942	1.25394
7	2.58517	2.25957	2.02920	1.85470	1.71653	1.60360	1.50911	1.42856
8	3.25246	2.72471	2.39122	2.15384	1.97314	1.82945	1.71160	1.61268

M \ N	9	10	11	12	13	14	15	16
9	4.49910	3.38440	2.84977	2.50991	2.26678	2.08091	1.93257	1.81052
10		4.62296	3.50321	2.96306	2.61797	2.37004	2.17980	2.02749
11			4.73493	3.61124	3.06658	2.71714	2.46515	2.27119
12				4.83708	3.71027	3.16189	2.80877	2.55332
13					4.93099	3.80166	3.25016	2.89392
14						5.01786	3.88651	3.33237
15							5.09867	3.96567
16								5.17420

M \ N	17	18	19	20	21	22	23	24
1	0.34458	0.33373	0.32380	0.31468	0.30626	0.29846	0.29121	0.28445
2	0.54748	0.52911	0.51239	0.49709	0.48302	0.47003	0.45799	0.44679
3	0.72005	0.69445	0.67125	0.65011	0.63074	0.61293	0.59646	0.58119
4	0.88121	0.84805	0.81816	0.79104	0.76629	0.74360	0.72270	0.70337
5	1.03861	0.99724	0.96015	0.92665	0.89622	0.86841	0.84289	0.81935
6	1.19671	1.14616	1.10111	1.06064	1.02404	0.99075	0.96030	0.93231
7	1.35886	1.29781	1.24378	1.19554	1.15214	1.11283	1.07703	1.04425
8	1.52813	1.45480	1.39043	1.33337	1.28233	1.23636	1.19466	1.15664

M \ N	17	18	19	20	21	22	23	24
9	1.70780	1.61979	1.54330	1.47603	1.41630	1.36279	1.31453	1.27071
10	1.90183	1.79581	1.70477	1.62551	1.55568	1.49357	1.43787	1.38756
11	2.11545	1.98665	1.87774	1.78404	1.70231	1.63021	1.56599	1.50831
12	2.35613	2.19742	2.06586	1.95440	1.85834	1.77442	1.70027	1.63414
13	2.63548	2.43549	2.27416	2.14017	2.02644	1.92827	1.84239	1.76640
14	2.97344	2.71240	2.50995	2.34631	2.21015	2.09440	1.99434	1.90668
15	3.40929	3.04802	2.78471	2.58008	2.41438	2.27630	2.15873	2.05696
16	4.03985	3.48154	3.11824	2.85292	2.64636	2.47883	2.33900	2.21979

EXPECTED VALUES OF GAMMA ORDER STATISTICS

SHAPE PARAMETER ALPHA=2.0

M \ N	17	18	19	20	21	22	23	24
17	5.24510	4.10964	3.54966	3.18457	2.91747	2.70918	2.54000	2.39860
18		5.31189	4.17552	3.61409	3.24742	2.97873	2.76889	2.59822
19			5.37502	4.23791	3.67520	3.30713	3.03701	2.82579
20				5.43487	4.29714	3.73332	3.36399	3.09260
21					5.49175	4.35352	3.78872	3.41827
22						5.54596	4.40731	3.84164
23							5.59771	4.45873
24								5.64723

M \ N	25	26	27	28	29	30	31	32
1	0.27812	0.27218	0.26660	0.26133	0.25636	0.25165	0.24718	0.24294
2	0.43634	0.42656	0.41738	0.40875	0.40061	0.39292	0.38564	0.37873
3	0.56697	0.55370	0.54128	0.52961	0.51864	0.50829	0.49850	0.48924
4	0.68542	0.66871	0.65310	0.63847	0.62474	0.61181	0.59961	0.58807
5	0.79757	0.77734	0.75848	0.74085	0.72432	0.70880	0.69417	0.68036
6	0.90649	0.88256	0.86032	0.83957	0.82017	0.80196	0.78485	0.76873
7	1.01410	0.98624	0.96041	0.93638	0.91396	0.89297	0.87327	0.85473
8	1.12179	1.08970	1.06003	1.03250	1.00687	0.98293	0.96051	0.93945

M \ N	25	26	27	28	29	30	31	32
9	1.23070	1.19399	1.16016	1.12886	1.09979	1.07270	1.04739	1.02367
10	1.34183	1.30004	1.26165	1.22625	1.19346	1.16299	1.13458	1.10801
11	1.45616	1.40870	1.36529	1.32538	1.28855	1.25440	1.22265	1.19303
12	1.57469	1.52087	1.47185	1.42696	1.38567	1.34752	1.31213	1.27921
13	1.69855	1.63748	1.58214	1.53170	1.48546	1.44289	1.40354	1.36701
14	1.82903	1.75962	1.69708	1.64036	1.58860	1.54113	1.49739	1.45692
15	1.96769	1.88853	1.81769	1.75381	1.69581	1.64285	1.59424	1.54942
16	2.11647	2.02574	1.94520	1.87306	1.80794	1.74877	1.69470	1.64504

M \ N	25	26	27	28	29	30	31	32
17	2.27791	2.17318	2.08111	1.99931	1.92597	1.85971	1.79946	1.74437
18	2.45540	2.33335	2.22733	2.13404	2.05108	1.97663	1.90933	1.84808
19	2.65376	2.50965	2.38636	2.27916	2.18475	2.10071	2.02524	1.95696
20	2.88011	2.70686	2.56156	2.43713	2.32885	2.23340	2.14837	2.07196
21	3.14572	2.93208	2.75771	2.61133	2.48586	2.37658	2.28017	2.19422
22	3.47018	3.19659	2.98190	2.80651	2.65913	2.53270	2.42249	2.32519
23	3.89229	3.51993	3.24538	3.02974	2.85340	2.70510	2.57778	2.46672
24	4.50799	3.94086	3.56768	3.29226	3.07574	2.89853	2.74939	2.62124

M \ N	25	26	27	28	29	30	31	32
25	5.69470	4.55525	3.98751	3.61358	3.33737	3.12004	2.94203	2.79210
26		5.74028	4.60067	4.03238	3.65777	3.38083	3.16276	2.98402
27			5.78411	4.64439	4.07560	3.70038	3.42277	3.20401
28				5.82632	4.68652	4.11730	3.74150	3.46328
29					5.86703	4.72718	4.15756	3.78125
30						5.90634	4.76646	4.19649
31							5.94433	4.80446
32								5.98110

EXPECTED VALUES OF GAMMA ORDER STATISTICS

SHAPE PARAMETER ALPHA=2.0

M \ N	33	34	35	36	37	38	39	40
1	0.23890	0.23505	0.23138	0.22787	0.22450	0.22128	0.21819	0.21523
2	0.37217	0.36592	0.35997	0.35429	0.34886	0.34366	0.33868	0.33390
3	0.48045	0.47209	0.46414	0.45656	0.44932	0.44241	0.43579	0.42944
4	0.57714	0.56677	0.55691	0.54752	0.53857	0.53003	0.52186	0.51403
5	0.66730	0.65493	0.64318	0.63200	0.62136	0.61122	0.60152	0.59225
6	0.75350	0.73908	0.72542	0.71245	0.70010	0.68834	0.67713	0.66641
7	0.83726	0.82075	0.80512	0.79030	0.77622	0.76282	0.75005	0.73786
8	0.91963	0.90094	0.88327	0.86653	0.85065	0.83556	0.82120	0.80750
9	1.00138	0.98039	0.96058	0.94185	0.92410	0.90725	0.89123	0.87598
10	1.08310	1.05968	1.03762	1.01678	0.99707	0.97839	0.96064	0.94377
11	1.16531	1.13931	1.11485	1.09179	1.07001	1.04939	1.02984	1.01127
12	1.24847	1.21969	1.19267	1.16725	1.14327	1.12061	1.09916	1.07880
13	1.33300	1.30123	1.27147	1.24352	1.21720	1.19237	1.16889	1.14665
14	1.41934	1.38432	1.35160	1.32092	1.29210	1.26495	1.23932	1.21508
15	1.50792	1.46936	1.43341	1.39980	1.36827	1.33864	1.31071	1.28434
16	1.59921	1.55676	1.51729	1.48048	1.44603	1.41372	1.38332	1.35467
17	1.69372	1.64697	1.60363	1.56332	1.52569	1.49047	1.45741	1.42630
18	1.79203	1.74048	1.69286	1.64869	1.60758	1.56920	1.53325	1.49949
19	1.89479	1.83785	1.78545	1.73702	1.69208	1.65023	1.61114	1.57451
20	2.00278	1.93974	1.88197	1.82879	1.77960	1.73393	1.69139	1.65162
21	2.11693	2.04690	1.98306	1.92452	1.87060	1.82070	1.77435	1.73115
22	2.23839	2.16028	2.08947	2.02487	1.96561	1.91099	1.86043	1.81344
23	2.36859	2.28099	2.20212	2.13057	2.06527	2.00534	1.95006	1.89887
24	2.50938	2.41048	2.32214	2.24256	2.17032	2.10436	2.04379	1.98790
25	2.66319	2.55059	2.45097	2.36194	2.28168	2.20880	2.14222	2.08104
26	2.83335	2.70373	2.59043	2.49015	2.40046	2.31958	2.24609	2.17892
27	3.02458	2.87324	2.74295	2.62900	2.52809	2.43780	2.35632	2.28226
28	3.24389	3.06381	2.91184	2.78093	2.66638	2.56487	2.47401	2.39197
29	3.50246	3.28248	3.10181	2.94924	2.81776	2.70263	2.60057	2.50917
30	3.81970	3.54039	3.31986	3.13863	2.98552	2.85348	2.73783	2.63524
31	4.23417	3.85695	3.57714	3.35610	3.17436	3.02073	2.88818	2.77202
32	4.84125	4.27067	3.89305	3.61279	3.39128	3.20905	3.05493	2.92190
33	6.01672	4.87691	4.30607	3.92808	3.64740	3.42544	3.24276	3.08819
34		6.05126	4.91151	4.34044	3.96210	3.68103	3.45866	3.27555
35			6.08478	4.94510	4.37382	3.99517	3.71374	3.49097
36				6.11735	4.97775	4.40627	4.02734	3.74556
37					6.14900	5.00949	4.43785	4.05864
38						6.17980	5.04039	4.46860
39							6.20978	5.07049
40								6.23900

EXPECTED VALUES OF GAMMA ORDER STATISTICS

SHAPE PARAMETER ALPHA=2.5

M \ N	1	2	3	4	5	6	7	8
1	2.50000	1.65117	1.31558	1.12646	1.00178	0.91192	0.84327	0.78866
2		3.34883	2.32235	1.88296	1.62514	1.45113	1.32380	1.22554
3			3.86206	2.76175	2.26969	1.97316	1.76947	1.61857
4				4.22883	3.08978	2.56623	2.24475	2.02097
5					4.51360	3.35156	2.80733	2.46853
6						4.74600	3.56925	3.01061
7							4.94213	3.75546
8								5.11165

M \ N	9	10	11	12	13	14	15	16
1	0.74389	0.70633	0.67423	0.64640	0.62195	0.60026	0.58084	0.56332
2	1.14681	1.08193	1.02729	0.98046	0.93975	0.90394	0.87212	0.84361
3	1.50109	1.40632	1.32782	1.26144	1.20436	1.15461	1.11075	1.07172
4	1.85355	1.72221	1.61566	1.52697	1.45170	1.38677	1.33003	1.27991
5	2.23024	2.05054	1.90869	1.79302	1.69635	1.61401	1.54279	1.48041
6	2.65917	2.40994	2.22077	2.07064	1.94768	1.84456	1.75645	1.68004
7	3.18633	2.82531	2.56759	2.37089	2.21409	2.08518	1.97672	1.88380
8	3.91807	3.34105	2.97259	2.70809	2.50529	2.34300	2.20914	2.09619

M \ N	9	10	11	12	13	14	15	16
9	5.26085	4.06233	3.47923	3.10484	2.83484	2.62701	2.46013	2.32209
10		5.39402	4.19191	3.60403	3.22484	2.95030	2.73826	2.56750
11			5.51423	4.30948	3.71778	3.33465	3.05632	2.84071
12				5.62375	4.41706	3.82228	3.43586	3.15432
13					5.72431	4.51620	3.91888	3.52971
14						5.81724	4.60809	4.00869
15							5.90361	4.69372
16								5.98427

M \ N	17	18	19	20	21	22	23	24
1	0.54741	0.53288	0.51954	0.50722	0.49582	0.48521	0.47532	0.46605
2	0.81787	0.79447	0.77309	0.75345	0.73533	0.71854	0.70292	0.68835
3	1.03668	1.00501	0.97620	0.94985	0.92562	0.90325	0.88251	0.86322
4	1.23521	1.19503	1.15866	1.12554	1.09521	1.06730	1.04151	1.01759
5	1.42518	1.37584	1.33142	1.29115	1.25444	1.22079	1.18980	1.16113
6	1.61296	1.55346	1.50021	1.45221	1.40863	1.36886	1.33236	1.29872
7	1.80303	1.73197	1.66882	1.61223	1.56113	1.51470	1.47227	1.43329
8	1.99919	1.91469	1.84021	1.77392	1.71443	1.66063	1.61170	1.56692

M \ N	17	18	19	20	21	22	23	24
9	2.20532	2.10481	2.01709	1.93965	1.87061	1.80856	1.75239	1.70124
10	2.42588	2.30582	2.20228	2.11175	2.03170	1.96024	1.89593	1.83765
11	2.66663	2.52193	2.39901	2.29281	2.19981	2.11745	2.04384	1.97752
12	2.93566	2.75871	2.61133	2.48590	2.37736	2.28216	2.19776	2.12222
13	3.24543	3.02414	2.84469	2.69495	2.56730	2.45668	2.35954	2.27329
14	3.61718	3.33054	3.10696	2.92532	2.77350	2.64389	2.53141	2.43251
15	4.09258	3.69908	3.41040	3.18480	3.00123	2.84756	2.71619	2.60205
16	4.77387	4.17128	3.77606	3.48559	3.25823	3.07294	2.91762	2.78468

EXPECTED VALUES OF GAMMA ORDER STATISTICS

SHAPE PARAMETER ALPHA=2.5

M \ N	17	18	19	20	21	22	23	24
17	6.05992	4.84920	4.24538	3.84868	3.55664	3.32772	3.14089	2.98409
18		6.13114	4.92023	4.31539	3.91739	3.62398	3.39365	3.20546
19			6.19841	4.98744	4.38173	3.98259	3.68796	3.45639
20				6.26215	5.05120	4.44475	4.04462	3.74889
21					6.32269	5.11184	4.50477	4.10377
22						6.38035	5.16966	4.56205
23							6.43538	5.22490
24								6.48801

M \ N	25	26	27	28	29	30	31	32
1	0.45736	0.44918	0.44146	0.43416	0.42725	0.42069	0.41445	0.40850
2	0.67471	0.66190	0.64986	0.63849	0.62775	0.61757	0.60791	0.59872
3	0.84521	0.82834	0.81251	0.79760	0.78354	0.77024	0.75764	0.74568
4	0.99531	0.97451	0.95502	0.93672	0.91948	0.90321	0.88783	0.87324
5	1.13452	1.10973	1.08656	1.06485	1.04444	1.02522	1.00707	0.98990
6	1.26758	1.23864	1.21167	1.18645	1.16280	1.14056	1.11960	1.09980
7	1.39733	1.36402	1.33305	1.30415	1.27712	1.25175	1.22789	1.20539
8	1.52576	1.48775	1.45251	1.41973	1.38912	1.36047	1.33356	1.30825

M \ N	25	26	27	28	29	30	31	32
9	1.65439	1.61128	1.57144	1.53448	1.50007	1.46792	1.43781	1.40952
10	1.78452	1.73582	1.69096	1.64948	1.61096	1.57507	1.54153	1.51009
11	1.91735	1.86244	1.81207	1.76564	1.72266	1.68273	1.64551	1.61070
12	2.05409	1.99223	1.93572	1.88383	1.83597	1.79163	1.75042	1.71197
13	2.19603	2.12627	2.06286	2.00490	1.95163	1.90246	1.85689	1.81450
14	2.34461	2.26579	2.19455	2.12975	2.07045	2.01593	1.96557	1.91885
15	2.50157	2.41218	2.33193	2.25935	2.19327	2.13277	2.07709	2.02562
16	2.66903	2.56713	2.47638	2.39484	2.32103	2.25378	2.19216	2.13542

M \ N	25	26	27	28	29	30	31	32
17	2.84973	2.73272	2.62952	2.53753	2.45481	2.37987	2.31155	2.24890
18	3.04733	2.91167	2.79343	2.68904	2.59592	2.51212	2.43614	2.36682
19	3.26695	3.10761	2.97079	2.85143	2.74595	2.65179	2.56699	2.49006
20	3.51621	3.32566	3.16522	3.02734	2.90694	2.80047	2.70535	2.61963
21	3.80707	3.57337	3.38181	3.22038	3.08152	2.96017	2.85279	2.75678
22	4.16028	3.86271	3.62810	3.43562	3.27328	3.13352	3.01131	2.90308
23	4.61684	4.21438	3.91603	3.68059	3.48728	3.32410	3.18352	3.06050
24	5.27777	4.66933	4.26627	3.96721	3.73103	3.53694	3.37299	3.23166

M \ N	25	26	27	28	29	30	31	32
25	6.53844	5.32847	4.71971	4.31612	4.01642	3.77955	3.58476	3.42011
26		6.58684	5.37717	4.76815	4.36407	4.06379	3.82630	3.63086
27			6.63337	5.42402	4.81477	4.41026	4.10946	3.87140
28				6.67816	5.46915	4.85972	4.45483	4.15355
29					6.72133	5.51268	4.90310	4.49787
30						6.76301	5.55472	4.94502
31							6.80329	5.59537
32								6.84225

EXPECTED VALUES OF GAMMA ORDER STATISTICS

SHAPE PARAMETER ALPHA=2.5

M \ N	33	34	35	36	37	38	39	40
1	0.40283	0.39741	0.39223	0.38726	0.38250	0.37793	0.37353	0.36930
2	0.58998	0.58163	0.57366	0.56604	0.55873	0.55173	0.54501	0.53854
3	0.73431	0.72348	0.71315	0.70328	0.69384	0.68479	0.67612	0.66779
4	0.85940	0.84623	0.83369	0.82172	0.81028	0.79934	0.78886	0.77881
5	0.97362	0.95816	0.94345	0.92943	0.91606	0.90327	0.89104	0.87932
6	1.08107	1.06330	1.04641	1.03035	1.01503	1.00042	0.98645	0.97307
7	1.18413	1.16399	1.14489	1.12674	1.10946	1.09299	1.07726	1.06222
8	1.28436	1.26178	1.24039	1.22009	1.20080	1.18242	1.16489	1.14816
9	1.38288	1.35775	1.33397	1.31144	1.29005	1.26971	1.25033	1.23184
10	1.48055	1.45271	1.42643	1.40156	1.37799	1.35560	1.33430	1.31400
11	1.57805	1.54735	1.51842	1.49109	1.46521	1.44068	1.41737	1.39519
12	1.67599	1.64223	1.61048	1.58053	1.55224	1.52544	1.50002	1.47585
13	1.77492	1.73788	1.70310	1.67037	1.63949	1.61030	1.58264	1.55639
14	1.87537	1.83477	1.79673	1.76101	1.72737	1.69563	1.66561	1.63715
15	1.97786	1.93338	1.89182	1.85287	1.81627	1.78179	1.74924	1.71845
16	2.08294	2.03421	1.98880	1.94635	1.90655	1.86914	1.83388	1.80057
17	2.19118	2.13776	2.08813	2.04186	1.99859	1.95799	1.91982	1.88383
18	2.30323	2.24459	2.19031	2.13984	2.09277	2.04873	2.00740	1.96851
19	2.41982	2.35534	2.29587	2.24077	2.18953	2.14171	2.09695	2.05492
20	2.54181	2.47072	2.40543	2.34517	2.28931	2.23734	2.18883	2.14339
21	2.67021	2.59157	2.51969	2.45364	2.39264	2.33608	2.28344	2.23426
22	2.80625	2.71889	2.63949	2.56687	2.50011	2.43843	2.38121	2.32793
23	2.95149	2.85391	2.76581	2.68569	2.61239	2.54497	2.48265	2.42481
24	3.10789	2.99816	2.89987	2.81109	2.73031	2.65637	2.58832	2.52540
25	3.27807	3.15362	3.04321	2.94426	2.85484	2.77344	2.69890	2.63027
26	3.46556	3.32287	3.19778	3.08675	2.98719	2.89717	2.81519	2.74008
27	3.67537	3.50946	3.36617	3.24049	3.12887	3.02873	2.93815	2.85563
28	3.91496	3.71838	3.55191	3.40807	3.28183	3.16966	3.06899	2.97789
29	4.19615	3.95708	3.76000	3.59301	3.44865	3.32189	3.20921	3.10804
30	4.53948	4.23737	3.99786	3.80030	3.63284	3.48799	3.36074	3.24758
31	4.98557	4.57976	4.27729	4.03737	3.83938	3.67147	3.52616	3.39846
32	5.63471	5.02484	4.61879	4.31599	4.07569	3.87730	3.70896	3.56323
33	6.87999	5.67283	5.06291	4.65664	4.35354	4.11289	3.91412	3.74540
34		6.91657	5.70979	5.09984	4.69338	4.39000	4.14903	3.94991
35			6.95206	5.74567	5.13571	4.72907	4.42544	4.18416
36				6.98653	5.78053	5.17056	4.76377	4.45991
37					7.02003	5.81441	5.20446	4.79754
38						7.05262	5.84738	5.23746
39							7.08433	5.87949
40								7.11523

EXPECTED VALUES OF GAMMA ORDER STATISTICS

SHAPE PARAMETER ALPHA=3.0

M \ N	1	2	3	4	5	6	7	8
1	3.00000	2.06250	1.68313	1.46603	1.32126	1.21592	1.13483	1.06987
2		3.93750	2.82124	2.33441	2.04512	1.84796	1.70251	1.58951
3			4.49563	3.30808	2.76834	2.43945	2.21156	2.04151
4				4.89148	3.66791	3.09724	2.74330	2.49496
5					5.19737	3.95324	3.36269	2.99164
6						5.44619	4.18946	3.58532
7							5.65565	4.39084
8								5.83633

M \ N	9	10	11	12	13	14	15	16
1	1.01630	0.97111	0.93231	0.89851	0.86870	0.84216	0.81831	0.79672
2	1.49843	1.42297	1.35912	1.30415	1.25618	1.21383	1.17606	1.14211
3	1.90830	1.80026	1.71033	1.63394	1.56800	1.51031	1.45928	1.41372
4	2.30794	2.16040	2.04008	1.93949	1.85375	1.77952	1.71443	1.65674
5	2.72874	2.52926	2.37096	2.24125	2.13240	2.03932	1.95852	1.88751
6	3.20195	2.92822	2.71923	2.55255	2.41542	2.29994	2.20091	2.11474
7	3.77701	3.38444	3.10237	2.88591	2.71253	2.56939	2.44848	2.34453
8	4.56622	3.94525	3.54563	3.25699	3.03452	2.85568	2.70756	2.58213

M \ N	9	10	11	12	13	14	15	16
9	5.99510	4.72147	4.09511	3.68995	3.39603	3.16865	2.98528	2.83300
10		6.13661	4.86066	4.23016	3.82058	3.52235	3.29090	3.10373
11			6.26421	4.98676	4.35304	3.93987	3.63807	3.40320
12				6.38034	5.10198	4.46572	4.04961	3.74483
13					6.48687	5.20802	4.56975	4.15121
14						6.58524	5.30622	4.66633
15							6.67660	5.39763
16								6.76186

M \ N	17	18	19	20	21	22	23	24
1	0.77705	0.75904	0.74244	0.72710	0.71285	0.69956	0.68714	0.67548
2	1.11137	1.08335	1.05767	1.03403	1.01215	0.99184	0.97290	0.95520
3	1.37270	1.33551	1.30160	1.27049	1.24183	1.21530	1.19065	1.16767
4	1.60514	1.55862	1.51640	1.47785	1.44247	1.40984	1.37962	1.35152
5	1.82445	1.76795	1.71695	1.67060	1.62824	1.58931	1.55339	1.52009
6	2.03885	1.97134	1.91076	1.85599	1.80616	1.76057	1.71864	1.67991
7	2.25388	2.17388	2.10260	2.03854	1.98056	1.92774	1.87937	1.83484
8	2.47403	2.37958	2.29608	2.22157	2.15451	2.09374	2.03832	1.98751

M \ N	17	18	19	20	21	22	23	24
9	2.70373	2.59211	2.49438	2.40786	2.33053	2.26086	2.19764	2.13994
10	2.94791	2.81536	2.70068	2.60013	2.51097	2.43118	2.35920	2.29382
11	3.21280	3.05395	2.91856	2.80124	2.69821	2.60672	2.52475	2.45072
12	3.50706	3.31388	3.15241	3.01456	2.89491	2.78969	2.69615	2.61224
13	3.84390	3.60364	3.40807	3.24432	3.10429	2.98259	2.87543	2.78006
14	4.24576	3.93631	3.69391	3.49625	3.33049	3.18855	3.06502	2.95613
15	4.75646	4.33418	4.02288	3.77862	3.57913	3.41160	3.26795	3.14280
16	5.48312	4.84091	4.41719	4.10431	3.85841	3.65731	3.48821	3.34304

EXPECTED VALUES OF GAMMA ORDER STATISTICS

SHAPE PARAMETER ALPHA=3.0

M \ N	17	18	19	20	21	22	23	24
17	6.84179	5.56340	4.92036	4.49541	4.18115	3.93383	3.73130	3.56079
18		6.91698	5.63905	4.99535	4.56936	4.25389	4.00531	3.80151
19			6.98798	5.71057	5.06635	4.63946	4.32294	4.07324
20				7.05521	5.77839	5.13375	4.70610	4.38865
21					7.11905	5.84285	5.19790	4.76959
22						7.17983	5.90428	5.25908
23							7.23781	5.96293
24								7.29323

M \ N	25	26	27	28	29	30	31	32
1	0.66452	0.65418	0.64441	0.63515	0.62636	0.61801	0.61005	0.60246
2	0.93859	0.92297	0.90825	0.89433	0.88115	0.86865	0.85676	0.84544
3	1.14618	1.12601	1.10704	1.08915	1.07225	1.05623	1.04104	1.02659
4	1.32531	1.30078	1.27776	1.25610	1.23567	1.21636	1.19806	1.18069
5	1.48912	1.46021	1.43314	1.40772	1.38380	1.36122	1.33987	1.31964
6	1.64399	1.61055	1.57931	1.55005	1.52257	1.49668	1.47224	1.44913
7	1.79367	1.75546	1.71986	1.68660	1.65542	1.62611	1.59850	1.57242
8	1.94070	1.89739	1.85716	1.81966	1.78459	1.75171	1.72078	1.69163

M \ N	25	26	27	28	29	30	31	32
9	2.08698	2.03815	1.99294	1.95091	1.91171	1.87503	1.84062	1.80824
10	2.23408	2.17921	2.12858	2.08166	2.03802	1.99729	1.95916	1.92336
11	2.38342	2.32187	2.26529	2.21303	2.16458	2.11948	2.07737	2.03792
12	2.53638	2.46735	2.40417	2.34604	2.29232	2.24248	2.19605	2.15268
13	2.69441	2.61692	2.54633	2.48167	2.42215	2.36709	2.31598	2.26835
14	2.85911	2.77191	2.69293	2.62094	2.55494	2.49414	2.43787	2.38559
15	3.03236	2.93386	2.84525	2.76492	2.69165	2.62443	2.56246	2.50508
16	3.21643	3.10459	3.00475	2.91486	2.83331	2.75887	2.69053	2.62749

M \ N	25	26	27	28	29	30	31	32
17	3.41426	3.28634	3.17322	3.07217	2.98111	2.89845	2.82294	2.75357
18	3.62974	3.48199	3.35287	3.23861	3.13645	3.04433	2.96064	2.88414
19	3.86830	3.69541	3.54655	3.41635	3.30104	3.19787	3.10477	3.02014
20	4.13796	3.93200	3.75809	3.60822	3.47704	3.36077	3.25667	3.16267
21	4.45133	4.19974	3.99287	3.81803	3.66725	3.53517	3.41803	3.31307
22	4.83021	4.51123	4.25885	4.05115	3.87547	3.72386	3.59095	3.47300
23	5.31757	4.88820	4.56859	4.31550	4.10705	3.93061	3.77823	3.64457
24	6.01905	5.37357	4.94379	4.62361	4.36987	4.16075	3.98361	3.83053

M \ N	25	26	27	28	29	30	31	32
25	7.34633	6.07284	5.42730	4.99715	4.67647	4.42216	4.21241	4.03464
26		7.39727	6.12448	5.47891	5.04846	4.72733	4.47250	4.26219
27			7.44622	6.17414	5.52858	5.09786	4.77634	4.52103
28				7.49333	6.22196	5.57644	5.14550	4.82361
29					7.53874	6.26807	5.62261	5.19148
30						7.58255	6.31258	5.66721
31							7.62489	6.35561
32								7.66583

EXPECTED VALUES OF GAMMA ORDER STATISTICS

SHAPE PARAMETER ALPHA=3.0

M \ N	33	34	35	36	37	38	39	40
1	0.59520	0.58826	0.58161	0.57523	0.56909	0.56320	0.55752	0.55206
2	0.83464	0.82432	0.81445	0.80499	0.79593	0.78722	0.77885	0.77079
3	1.01283	0.99971	0.98717	0.97518	0.96369	0.95267	0.94209	0.93192
4	1.16418	1.14845	1.13344	1.11910	1.10538	1.09224	1.07964	1.06753
5	1.30043	1.28216	1.26475	1.24814	1.23227	1.21708	1.20253	1.18857
6	1.42721	1.40640	1.38660	1.36773	1.34972	1.33250	1.31603	1.30024
7	1.54774	1.52434	1.50210	1.48095	1.46078	1.44152	1.42311	1.40549
8	1.66409	1.63801	1.61328	1.58977	1.56739	1.54605	1.52567	1.50619
9	1.77770	1.74884	1.72150	1.69556	1.67089	1.64741	1.62501	1.60361
10	1.88967	1.85787	1.82781	1.79933	1.77229	1.74657	1.72207	1.69870
11	2.00086	1.96597	1.93303	1.90187	1.87234	1.84429	1.81761	1.79219
12	2.11203	2.07382	2.03783	2.00384	1.97168	1.94118	1.91221	1.88464
13	2.22382	2.18207	2.14281	2.10581	2.07085	2.03776	2.00637	1.97654
14	2.33686	2.29127	2.24850	2.20827	2.17034	2.13449	2.10054	2.06832
15	2.45174	2.40198	2.35542	2.31171	2.27059	2.23179	2.19512	2.16037
16	2.56909	2.51477	2.46407	2.41660	2.37203	2.33007	2.29048	2.25303
17	2.68955	2.63020	2.57497	2.52340	2.47510	2.42972	2.38699	2.34665
18	2.81383	2.74890	2.68867	2.63261	2.58023	2.53115	2.48503	2.44157
19	2.94273	2.87155	2.80577	2.74474	2.68789	2.63477	2.58496	2.53814
20	3.07717	2.99893	2.92694	2.86038	2.79860	2.74102	2.68719	2.63671
21	3.21824	3.13194	3.05292	2.98018	2.91290	2.85041	2.79216	2.73768
22	3.36726	3.27167	3.18462	3.10488	3.03144	2.96349	2.90034	2.84146
23	3.52587	3.41941	3.32310	3.23537	3.15496	3.08086	3.01228	2.94852
24	3.69617	3.57679	3.46965	3.37269	3.28431	3.20328	3.12858	3.05940
25	3.88091	3.74591	3.62590	3.51813	3.42056	3.33158	3.24996	3.17470
26	4.08383	3.92951	3.79392	3.67332	3.56497	3.46682	3.37729	3.29513
27	4.31020	4.13131	3.97645	3.84031	3.71916	3.61027	3.51159	3.42154
28	4.56788	4.35658	4.17720	4.02183	3.88518	3.76352	3.65413	3.55495
29	4.86928	4.61316	4.40143	4.22159	4.06575	3.92863	3.80649	3.69663
30	5.23592	4.91344	4.65696	4.44484	4.26458	4.10831	3.97074	3.84817
31	5.71034	5.27892	4.95619	4.69939	4.48690	4.30625	4.14958	4.01160
32	6.39724	5.75209	5.32057	4.99760	4.74051	4.52769	4.34668	4.18963
33	7.70548	6.43756	5.79254	5.36094	5.03778	4.78042	4.56729	4.38595
34		7.74390	6.47665	5.83178	5.40011	5.07677	4.81917	4.60576
35			7.78117	6.51459	5.86987	5.43815	5.11465	4.85683
36				7.81736	6.55143	5.90687	5.47512	5.15148
37					7.85252	6.58724	5.94285	5.51108
38						7.88672	6.62207	5.97786
39							7.92000	6.65597
40								7.95241

EXPECTED VALUES OF GAMMA ORDER STATISTICS

SHAPE PARAMETER ALPHA=3.5

M \ N	1	2	3	4	5	6	7	8
1	3.50000	2.48141	2.06188	1.81904	1.65572	1.53606	1.44340	1.36882
2		4.51859	3.32047	2.79037	2.47235	2.25401	2.09199	1.96547
3			5.11765	3.85057	3.26740	2.90903	2.65907	2.47156
4				5.54001	4.23935	3.62578	3.24229	2.97160
5					5.86518	4.54613	3.91340	3.51299
6						6.12898	4.79923	4.15364
7							6.35061	5.01442
8								6.54149

M \ N	9	10	11	12	13	14	15	16
1	1.30705	1.25473	1.20966	1.17026	1.13542	1.10430	1.07628	1.05085
2	1.86304	1.77785	1.70550	1.64303	1.58835	1.53994	1.49667	1.45768
3	2.32398	2.20380	2.10340	2.01785	1.94378	1.87880	1.82118	1.76961
4	2.76672	2.60439	2.47152	2.36006	2.26476	2.18203	2.10930	2.04467
5	3.22770	3.01022	2.83693	2.69443	2.57447	2.47159	2.38205	2.30316
6	3.74122	3.44518	3.21817	3.03641	2.88638	2.75965	2.65067	2.55561
7	4.35986	3.93858	3.63436	3.39992	3.21145	3.05535	2.92311	2.80912
8	5.20144	4.54041	4.11242	3.80181	3.56146	3.36756	3.20647	3.06967

M \ N	9	10	11	12	13	14	15	16
9	6.70900	5.36670	4.70090	4.26772	3.95204	3.70688	3.50851	3.34328
10		6.85815	5.51466	4.84529	4.40803	4.08823	3.83913	3.63703
11			6.99249	5.64853	4.97647	4.53594	4.21279	3.96039
12				7.11467	5.77072	5.09662	4.65346	4.32751
13					7.22667	5.88307	5.20741	4.76210
14						7.33002	5.98702	5.31017
15							7.42595	6.08371
16								7.51543

M \ N	17	18	19	20	21	22	23	24
1	1.02763	1.00632	0.98665	0.96843	0.95147	0.93564	0.92081	0.90687
2	1.42230	1.38998	1.36031	1.33293	1.30757	1.28397	1.26194	1.24130
3	1.72308	1.68081	1.64218	1.60669	1.57393	1.54356	1.51530	1.48890
4	1.98674	1.93441	1.88682	1.84329	1.80326	1.76628	1.73198	1.70004
5	2.23294	2.16990	2.11288	2.06096	2.01342	1.96966	1.92921	1.89167
6	2.47168	2.39685	2.32957	2.26863	2.21309	2.16218	2.11528	2.07189
7	2.70947	2.62133	2.54263	2.47176	2.40750	2.34886	2.29505	2.24545
8	2.95149	2.84796	2.75626	2.67424	2.60029	2.53316	2.47183	2.41552

M \ N	17	18	19	20	21	22	23	24
9	3.20263	3.08089	2.97406	2.87928	2.79440	2.71778	2.64814	2.58447
10	3.46829	3.32438	3.19958	3.08990	2.99246	2.90508	2.82611	2.75427
11	3.75514	3.58342	3.43670	3.30926	3.19710	3.09731	3.00773	2.92669
12	4.07234	3.86441	3.69013	3.54097	3.41122	3.29688	3.19503	3.10351
13	4.43383	4.17631	3.96607	3.78957	3.63828	3.50651	3.39024	3.28656
14	4.86311	4.53288	4.27335	4.06111	3.88267	3.72951	3.59595	3.47796
15	5.40597	4.95746	4.62557	4.36430	4.15033	3.97019	3.81538	3.68022
16	6.17408	5.49567	5.04597	4.71266	4.44989	4.23440	4.05275	3.89648

EXPECTED VALUES OF GAMMA ORDER STATISTICS

SHAPE PARAMETER ALPHA=3.5

M \ N	17	18	19	20	21	22	23	24
17	7.59927	6.25888	5.57999	5.12929	4.79477	4.53070	4.31387	4.13089
18		7.67812	6.33874	5.65953	5.20801	4.87244	4.60722	4.38922
19			7.75253	6.41421	5.73478	5.28258	4.94611	4.67989
20				7.82296	6.48573	5.80618	5.35341	5.01617
21					7.88983	6.55369	5.87410	5.42086
22						7.95345	6.61841	5.93885
23							8.01413	6.68019
24								8.07213

M \ N	25	26	27	28	29	30	31	32
1	0.89374	0.88135	0.86961	0.85848	0.84791	0.83783	0.82823	0.81905
2	1.22192	1.20367	1.18644	1.17013	1.15467	1.13998	1.12599	1.11266
3	1.46418	1.44095	1.41908	1.39842	1.37887	1.36034	1.34272	1.32596
4	1.67021	1.64225	1.61597	1.59121	1.56783	1.54570	1.52471	1.50476
5	1.85669	1.82399	1.79334	1.76452	1.73735	1.71169	1.68739	1.66434
6	2.03158	1.99401	1.95887	1.92591	1.89491	1.86567	1.83804	1.81187
7	2.19952	2.15683	2.11700	2.07974	2.04476	2.01185	1.98080	1.95144
8	2.36355	2.31540	2.27061	2.22881	2.18966	2.15290	2.11830	2.08564
M \ N	25	26	27	28	29	30	31	32
9	2.52594	2.47189	2.42178	2.37513	2.33156	2.29074	2.25240	2.21628
10	2.68851	2.62803	2.57213	2.52026	2.47195	2.42680	2.38448	2.34470
11	2.85289	2.78529	2.72306	2.66550	2.61205	2.56224	2.51567	2.47199
12	3.02062	2.94507	2.87582	2.81201	2.75296	2.69809	2.64692	2.59905
13	3.19330	3.10876	3.03164	2.96090	2.89567	2.83526	2.77910	2.72670
14	3.37265	3.27783	3.19181	3.11328	3.04118	2.97466	2.91302	2.85568
15	3.56071	3.45393	3.35771	3.27034	3.19053	3.11720	3.04951	2.98674
16	3.75989	3.63901	3.53091	3.43342	3.34484	3.26386	3.18941	3.12064
M \ N	25	26	27	28	29	30	31	32
17	3.97330	3.83544	3.71332	3.60403	3.50539	3.41570	3.33365	3.25818
18	4.20505	4.04629	3.90728	3.78404	3.67367	3.57397	3.48327	3.40024
19	4.46084	4.27561	4.11579	3.97575	3.85149	3.74013	3.63948	3.54785
20	4.74907	4.52908	4.34290	4.18213	4.04114	3.91596	3.80370	3.70217
21	5.08294	4.81506	4.59425	4.40721	4.24558	4.10373	3.97771	3.86462
22	5.48522	5.14672	4.87816	4.65659	4.46878	4.30637	4.16374	4.03695
23	6.00071	5.54677	5.20776	4.93858	4.71635	4.52784	4.36472	4.22138
24	6.73927	6.05992	5.60573	5.26628	4.99656	4.77372	4.58458	4.42081
M \ N	25	26	27	28	29	30	31	32
25	8.12767	6.79589	6.11669	5.66230	5.32247	5.05227	4.82889	4.63917
26		8.18094	6.85022	6.17122	5.71667	5.37651	5.10588	4.88201
27			8.23212	6.90245	6.22366	5.76901	5.42856	5.15754
28				8.28137	6.95273	6.27418	5.81944	5.47875
29					8.32882	7.00120	6.32290	5.86811
30						8.37460	7.04798	6.36995
31							8.41882	7.09318
32								8.46158

EXPECTED VALUES OF GAMMA ORDER STATISTICS

SHAPE PARAMETER ALPHA=3.5

M \ N	33	34	35	36	37	38	39	40
1	0.81028	0.80187	0.79380	0.78605	0.77861	0.77144	0.76453	0.75787
2	1.09993	1.08776	1.07610	1.06492	1.05419	1.04387	1.03394	1.02438
3	1.30998	1.29472	1.28013	1.26615	1.25275	1.23989	1.22753	1.21563
4	1.48578	1.46767	1.45038	1.43385	1.41801	1.40283	1.38825	1.37424
5	1.64242	1.62156	1.60167	1.58266	1.56449	1.54708	1.53038	1.51435
6	1.78704	1.76342	1.74094	1.71949	1.69900	1.67939	1.66061	1.64260
7	1.92363	1.89722	1.87211	1.84819	1.82537	1.80355	1.78268	1.76269
8	2.05475	2.02547	1.99767	1.97122	1.94601	1.92195	1.89896	1.87696
9	2.18217	2.14990	2.11931	2.09024	2.06258	2.03622	2.01105	1.98699
10	2.30722	2.27181	2.23829	2.20650	2.17629	2.14754	2.12012	2.09393
11	2.43092	2.39219	2.35560	2.32095	2.28807	2.25681	2.22705	2.19867
12	2.55414	2.51188	2.47203	2.43435	2.39866	2.36478	2.33257	2.30189
13	2.67765	2.63161	2.58827	2.54738	2.50871	2.47206	2.43727	2.40417
14	2.80216	2.75203	2.70495	2.66062	2.61877	2.57919	2.54166	2.50601
15	2.92832	2.87376	2.82265	2.77461	2.72937	2.68664	2.64620	2.60786
16	3.05684	2.99744	2.94192	2.88989	2.84098	2.79488	2.75134	2.71011
17	3.18842	3.12368	3.06336	3.00696	2.95408	2.90436	2.85748	2.81317
18	3.32383	3.25317	3.18755	3.12638	3.06918	3.01551	2.96503	2.91741
19	3.46392	3.38664	3.31514	3.24871	3.18677	3.12880	3.07441	3.02322
20	3.60969	3.52494	3.44685	3.37458	3.30740	3.24473	3.18606	3.13099
21	3.76228	3.66901	3.58350	3.50467	3.43168	3.36380	3.30045	3.24114
22	3.92310	3.82002	3.72603	3.63980	3.56029	3.48663	3.41810	3.35412
23	4.09387	3.97932	3.87556	3.78089	3.69402	3.61386	3.53958	3.47045
24	4.27681	4.14866	4.03346	3.92906	3.83378	3.74629	3.66554	3.59067
25	4.47480	4.33021	4.20146	4.08567	3.98067	3.88481	3.79676	3.71545
26	4.69177	4.52686	4.38171	4.25240	4.13606	4.03052	3.93412	3.84554
27	4.93323	4.74251	4.57710	4.43145	4.30162	4.18477	4.07872	3.98182
28	5.20738	4.98267	4.79152	4.62565	4.47953	4.34923	4.23190	4.12538
29	5.52721	5.25554	5.03046	4.83891	4.67262	4.52606	4.39533	4.27755
30	5.91513	5.57405	5.30210	5.07670	4.88478	4.71811	4.57114	4.44000
31	6.41543	5.96061	5.61937	5.34718	5.12148	4.92923	4.76219	4.61486
32	7.13691	6.45945	6.00464	5.66327	5.39087	5.16489	4.97233	4.80497
33	8.50298	7.17925	6.50209	6.04731	5.70584	5.43324	5.20702	5.01417
34		8.54309	7.22029	6.54343	6.08870	5.74714	5.47437	5.24792
35			8.58200	7.26010	6.58355	6.12889	5.78725	5.51433
36				8.61977	7.29876	6.62252	6.16793	5.82624
37					8.65646	7.33633	6.66041	6.20590
38						8.69214	7.37287	6.69726
39							8.72686	7.40843
40								8.76066

EXPECTED VALUES OF GAMMA ORDER STATISTICS

SHAPE PARAMETER ALPHA=4.0

M \ N	1	2	3	4	5	6	7	8
1	4.00000	2.90625	2.44942	2.18264	2.00200	1.86896	1.76549	1.68188
2		5.09375	3.81990	3.24979	2.90517	2.66721	2.48981	2.35073
3			5.73067	4.39002	3.75671	3.38108	3.11073	2.90705
4				6.17756	4.80556	4.15235	3.74156	3.45020
5					6.52056	5.13216	4.46044	4.03291
6						6.79824	5.40085	4.71696
7							7.03114	5.62881
8								7.23147

M \ N	9	10	11	12	13	14	15	16
1	1.61240	1.55338	1.50240	1.45773	1.41813	1.38270	1.35072	1.32166
2	2.23775	2.14350	2.06325	1.99378	1.93284	1.87878	1.83036	1.78666
3	2.74616	2.61473	2.50464	2.41058	2.32896	2.25721	2.19346	2.13629
4	3.22882	3.05282	2.90832	2.78680	2.68266	2.59204	2.51222	2.44117
5	3.72693	3.49282	3.30568	3.15137	3.02113	2.90919	2.81155	2.72537
6	4.27769	3.96105	3.71739	3.52171	3.35975	3.22263	3.10445	3.00116
7	4.93659	4.48878	4.16410	3.91306	3.71066	3.54259	3.39989	3.27662
8	5.82659	5.12851	4.67431	4.34342	4.08654	3.87874	3.70568	3.55838

M \ N	9	10	11	12	13	14	15	16
9	7.40708	6.00111	5.29883	4.83976	4.50397	4.24239	4.03017	3.85297
10		7.56329	6.15718	5.45185	4.98900	4.64929	4.38387	4.16798
11			7.70391	6.29824	5.59071	5.12489	4.78199	4.51341
12				7.83169	6.42688	5.71775	5.24957	4.90408
13					7.94876	6.54507	5.83480	5.36474
14						8.05674	6.65434	5.94328
15							8.15691	6.75593
16								8.25031

M \ N	17	18	19	20	21	22	23	24
1	1.29508	1.27064	1.24806	1.22710	1.20757	1.18931	1.17219	1.15608
2	1.74693	1.71059	1.67718	1.64630	1.61766	1.59097	1.56603	1.54265
3	2.08463	2.03763	1.99461	1.95504	1.91845	1.88449	1.85285	1.82327
4	2.37737	2.31964	2.26706	2.21890	2.17455	2.13353	2.09543	2.05992
5	2.64853	2.57942	2.51681	2.45972	2.40738	2.35914	2.31449	2.27300
6	2.90980	2.82821	2.75472	2.68807	2.62723	2.57138	2.51988	2.47217
7	3.16865	3.07298	2.98742	2.91025	2.84017	2.77614	2.71731	2.66301
8	3.43087	3.31897	3.21967	3.13072	3.05041	2.97739	2.91060	2.84919

M \ N	17	18	19	20	21	22	23	24
9	3.70183	3.57074	3.45550	3.35309	3.26124	3.17820	3.10262	3.03343
10	3.98733	3.83292	3.69877	3.58067	3.47557	3.38118	3.29576	3.21794
11	4.29444	4.11085	3.95366	3.81687	3.69628	3.58883	3.49223	3.40471
12	4.63284	4.41128	4.22517	4.06558	3.92650	3.80373	3.69422	3.59566
13	5.01709	4.74363	4.51984	4.33157	4.16988	4.02880	3.90412	3.79278
14	5.47171	5.12227	4.84691	4.62122	4.43106	4.26755	4.12471	3.99834
15	6.04433	5.57155	5.22061	4.94364	4.71629	4.52450	4.35938	4.21498
16	6.85081	6.13888	5.66513	5.31294	5.03458	4.80580	4.61256	4.44602

EXPECTED VALUES OF GAMMA ORDER STATISTICS

SHAPE PARAMETER ALPHA=4.0

M \ N	17	18	19	20	21	22	23	24
17	8.33777	6.93980	6.22771	5.75318	5.39992	5.12037	4.89034	4.69583
18		8.42001	7.02357	6.31145	5.83630	5.48214	5.20156	4.97043
19			8.49759	7.10270	6.39064	5.91500	5.56009	5.27860
20				8.57100	7.17765	6.46575	5.98972	5.63416
21					8.64067	7.24884	6.53715	6.06083
22						8.70695	7.31662	6.60520
23							8.77015	7.38129
24								8.83053

M \ N	25	26	27	28	29	30	31	32
1	1.14089	1.12653	1.11292	1.10000	1.08771	1.07599	1.06481	1.05412
2	1.52066	1.49992	1.48033	1.46177	1.44416	1.42741	1.41145	1.39622
3	1.79553	1.76944	1.74484	1.72159	1.69957	1.67867	1.65879	1.63986
4	2.02670	1.99554	1.96623	1.93858	1.91245	1.88769	1.86418	1.84183
5	2.23429	2.19808	2.16409	2.13210	2.10192	2.07339	2.04634	2.02066
6	2.42780	2.38640	2.34763	2.31123	2.27696	2.24461	2.21402	2.18501
7	2.61267	2.56583	2.52208	2.48110	2.44261	2.40635	2.37211	2.33971
8	2.79245	2.73982	2.69081	2.64501	2.60209	2.56175	2.52373	2.48782

M \ N	25	26	27	28	29	30	31	32
9	2.96975	2.91088	2.85622	2.80530	2.75768	2.71304	2.67105	2.63147
10	3.14663	3.08095	3.02018	2.96373	2.91110	2.86186	2.81567	2.77221
11	3.32491	3.25172	3.18426	3.12180	3.06374	3.00957	2.95887	2.91128
12	3.50628	3.42471	3.34984	3.28078	3.21680	3.15729	3.10174	3.04973
13	3.69248	3.60144	3.51829	3.44192	3.37142	3.30607	3.24525	3.18844
14	3.88537	3.78351	3.69099	3.60642	3.52868	3.45689	3.39029	3.32827
15	4.08710	3.97268	3.86942	3.77556	3.68971	3.61074	3.53776	3.47002
16	4.30023	4.17100	4.05528	3.95077	3.85570	3.76867	3.68858	3.61452

M \ N	25	26	27	28	29	30	31	32
17	4.52803	4.38100	4.25056	4.13366	4.02801	3.93184	3.84376	3.76264
18	4.77480	4.60587	4.45773	4.32620	4.20824	4.10156	4.00438	3.91533
19	5.04650	4.84988	4.67994	4.53080	4.39829	4.27936	4.17174	4.07365
20	5.35190	5.11894	4.92143	4.75059	4.60054	4.46714	4.34733	4.23885
21	5.70473	5.42178	5.18807	4.98976	4.81811	4.66725	4.53303	4.41242
22	6.12866	5.77210	5.48855	5.25418	5.05515	4.88277	4.73116	4.59620
23	6.67018	6.19348	5.83654	5.55248	5.31750	5.11784	4.94479	4.79251
24	7.44313	6.73236	6.25556	5.89829	5.61377	5.37827	5.17803	5.00438

M \ N	25	26	27	28	29	30	31	32
25	8.88834	7.50236	6.79196	6.31511	5.95756	5.67265	5.43667	5.23591
26		8.94378	7.55919	6.84918	6.37232	6.01454	5.72928	5.49288
27			8.99703	7.61381	6.90420	6.42736	6.06940	5.78384
28				9.04826	7.66638	6.95718	6.48039	6.12228
29					9.09762	7.71703	7.00827	6.53155
30						9.14522	7.76591	7.05758
31							9.19120	7.81313
32								9.23565

EXPECTED VALUES OF GAMMA ORDER STATISTICS

SHAPE PARAMETER ALPHA=4.0

M \ N	33	34	35	36	37	38	39	40
1	1.04389	1.03407	1.02465	1.01560	1.00688	0.99849	0.99040	0.98259
2	1.38167	1.36774	1.35439	1.34158	1.32927	1.31743	1.30603	1.29504
3	1.62179	1.60453	1.58800	1.57217	1.55697	1.54238	1.52834	1.51482
4	1.82053	1.80020	1.78078	1.76219	1.74437	1.72728	1.71085	1.69505
5	1.99624	1.97296	1.95074	1.92951	1.90918	1.88969	1.87100	1.85303
6	2.15746	2.13125	2.10626	2.08241	2.05960	2.03777	2.01684	1.99675
7	2.30898	2.27979	2.25201	2.22552	2.20023	2.17604	2.15288	2.13067
8	2.45383	2.42158	2.39093	2.36175	2.33392	2.30734	2.28192	2.25757
9	2.59406	2.55863	2.52502	2.49306	2.46262	2.43359	2.40585	2.37931
10	2.73122	2.69246	2.65575	2.62090	2.58775	2.55618	2.52605	2.49726
11	2.86649	2.82422	2.78425	2.74636	2.71039	2.67616	2.64355	2.61242
12	3.00087	2.95487	2.91144	2.87035	2.83140	2.79439	2.75918	2.72561
13	3.13522	3.08521	3.03810	2.99361	2.95151	2.91157	2.87362	2.83750
14	3.27032	3.21600	3.16494	3.11682	3.07135	3.02831	2.98747	2.94865
15	3.40692	3.34792	3.29259	3.24056	3.19151	3.14514	3.10123	3.05956
16	3.54575	3.48165	3.42169	3.36543	3.31251	3.26260	3.21540	3.17069
17	3.68760	3.61787	3.55285	3.49200	3.43489	3.38115	3.33043	3.28247
18	3.83328	3.75732	3.68672	3.62085	3.55919	3.50129	3.44678	3.39532
19	3.98371	3.90079	3.82400	3.75259	3.68594	3.62352	3.56489	3.50966
20	4.13992	4.04916	3.96546	3.88790	3.81574	3.74836	3.68523	3.62592
21	4.30315	4.20346	4.11194	4.02750	3.94923	3.87638	3.80833	3.74455
22	4.47486	4.36487	4.26446	4.17226	4.08714	4.00821	3.93471	3.86603
23	4.65687	4.53486	4.42420	4.32314	4.23029	4.14455	4.06500	3.99090
24	4.85148	4.71522	4.59259	4.48132	4.37966	4.28621	4.19988	4.11977
25	5.06172	4.90826	4.77143	4.64823	4.53639	4.43417	4.34017	4.25330
26	5.29166	5.11696	4.96299	4.82564	4.70191	4.58955	4.48680	4.39229
27	5.54706	5.34541	5.17026	5.01581	4.87798	4.75377	4.64092	4.53769
28	5.83646	5.59934	5.39730	5.22174	5.06686	4.92859	4.80393	4.69063
29	6.17333	5.88727	5.64985	5.44747	5.27152	5.11625	4.97756	4.85248
30	6.58096	6.22265	5.93639	5.69870	5.49600	5.31971	5.16407	5.02500
31	7.10525	6.62873	6.27035	5.98393	5.74600	5.54301	5.36641	5.21042
32	7.85880	7.15136	6.67497	6.31655	6.02998	5.79183	5.58859	5.41169
33	9.27868	7.90302	7.19602	6.71977	6.35133	6.07463	5.83629	5.63281
34		9.32037	7.94587	7.23932	6.76322	6.40477	6.11796	5.87946
35			9.36079	7.98743	7.28133	6.80539	6.44695	6.16005
36				9.40003	8.02778	7.32212	6.84636	6.48793
37					9.43815	8.06698	7.36177	6.88618
38						9.47521	8.10510	7.40033
39							9.51126	8.14219
40								9.54637

EXPECTED VALUES OF GAMMA ORDER STATISTICS

SHAPE PARAMETER ALPHA = 5.0

M \ N	1	2	3	4	5	6	7	8
1	5.00000	3.76953	3.24473	2.93419	2.72188	2.56430	2.44095	2.34074
2		6.23047	4.81292	4.17637	3.78343	3.50979	3.30438	3.14242
3			6.93614	5.46188	4.76577	4.33070	4.02332	3.79027
4				7.42756	5.92596	5.20084	4.74055	4.41173
5					7.80296	6.28851	5.54606	5.06937
6						8.10585	6.58549	5.83208
7							8.35924	6.83663
8								8.57676

M \ N	9	10	11	12	13	14	15
1	2.25706	2.18568	1.12378	2.06936	2.02097	1.97753	1.93823
2	3.01020	2.89942	2.80471	2.72245	3.65005	2.58563	2.52778
3	3.60521	3.45333	3.32558	3.21604	3.12066	3.03656	2.96162
4	4.16040	3.95959	3.79400	3.65420	3.53396	3.42902	3.33630
5	4.72589	4.46162	4.24936	4.07361	3.92472	3.79632	3.68399
6	5.34415	4.99017	4.71632	4.49541	4.31183	4.15585	4.02098
7	6.07605	5.58013	5.21837	4.93723	4.70958	4.51981	4.35814
8	7.05393	6.28859	5.78686	5.41918	5.13236	4.89936	4.70459
9	8.76711	7.24527	6.47674	5.97069	5.59845	5.30711	5.06978
10		8.93621	7.41606	6.64542	6.13614	5.76030	5.46533
11			9.08822	7.57019	6.79820	6.28647	5.90779
12				9.22623	7.71055	6.93776	6.42417
13					9.35253	7.83934	7.06616
14						9.46893	7.95830
15							9.57684

SHAPE PARAMETER ALPHA = 6.0

M \ N	1	2	3	4	5	6	7	8
1	6.00000	4.64648	4.06042	3.71036	3.46939	3.28956	3.14818	3.03288
2		7.35352	5.81862	5.11060	4.67423	4.36850	4.13787	3.95529
3			8.12096	6.52663	5.76516	5.28570	4.94506	4.68563
4				8.65241	7.03428	6.24461	5.73989	5.37745
5					9.05694	7.42911	6.62316	6.10234
6						9.38251	7.75149	6.93565
7							9.65435	8.02343
8								9.88734

M \ N	9	10	11	12	13	14	15
1	2.93627	2.85364	2.78178	2.71845	2.66202	2.61126	2.56525
2	3.80571	3.67999	3.57224	3.47840	3.39563	3.32183	3.25544
3	4.47883	4.30855	4.16491	4.04142	3.93364	3.83841	3.75338
4	5.09923	4.87614	4.69161	4.53537	4.40067	4.28284	4.17851
5	5.72521	5.43387	5.19908	5.00408	4.83845	4.69526	4.56972
6	6.40404	6.01655	5.71563	5.47207	5.26909	5.09618	4.94634
7	7.20146	6.66236	6.26732	5.95918	5.70888	5.49965	5.32094
8	8.25829	7.43250	6.88809	6.48742	6.17373	5.91812	5.70388
9	10.09097	8.46473	7.63665	7.08843	6.68347	6.36543	6.10558
10		10.27166	8.64875	7.81939	7.26842	6.86016	6.53867
11			10.43395	8.81462	7.98468	7.43172	7.02091
12				10.58116	8.96552	8.13549	7.58110
13					10.71580	9.10386	8.27408
14						10.83979	9.23152
15							10.95467

Table C4.2

VARIANCES AND COVARIANCES OF GAMMA ORDER STATISTICS

[When $I = J$, tabular values are variances of the Ith order statistic of a sample of size N from a Gamma population with location parameter 0, scale parameter 1, and shape parameter $\alpha = 2.0(1.0)6.0$ (when $\alpha = 1.0$, see Table C2.2); when $I \neq J$, they are covariances of the Ith and Jth order statistics.]

Variances and Covariances of Gamma Order Statistics (Shape Parameter α)

N	I	J	$\alpha = 2.0$	$\alpha = 3.0$	$\alpha = 4.0$	$\alpha = 5.0$	$\alpha = 6.0$
1	1	1	2.00000	3.00000	4.00000	5.00000	6.00000
2	1	1	0.68750	1.18359	1.70996	2.25548	2.81448
2	1	2	0.56250	0.87891	1.19629	1.51405	1.83200
2	2	2	2.18750	3.05859	3.89746	4.71642	5.52151
3	1	1	0.38134	0.71715	1.08834	1.48227	1.89235
3	1	2	0.31139	0.53278	0.76395	1.00101	1.24210
3	1	3	0.27023	0.43320	0.59714	0.76146	0.92597
3	2	2	0.80547	1.25294	1.70107	2.14943	2.59790
3	2	3	0.70722	1.03553	1.35489	1.66869	1.97862
3	3	3	2.23551	3.02690	3.77864	4.50599	5.21638
4	1	1	0.25482	0.51168	0.80468	1.12125	1.45468
4	1	2	0.20689	0.37770	0.56145	0.75316	0.95039
4	1	3	0.18113	0.31191	0.44795	0.58706	0.72823
4	1	4	0.16184	0.26474	0.36855	0.47272	0.57706
4	2	2	0.46030	0.76801	1.08519	1.40807	1.73484
4	2	3	0.40607	0.64104	0.87654	1.11222	1.34800
4	2	4	0.36453	0.54766	0.72661	0.90291	1.07737
4	3	3	0.85221	1.26385	1.66687	2.06451	2.45839
4	3	4	0.77094	1.09128	1.39866	1.69809	1.99201
4	4	4	2.24985	2.98780	3.68374	4.35384	5.00598
5	1	1	0.18792	0.39759	0.64299	0.91198	1.19794
5	1	2	0.15172	0.29149	0.44551	0.60843	0.77756
5	1	3	0.13298	0.24149	0.35722	0.47729	0.60032
5	1	4	0.12018	0.20887	0.30106	0.39523	0.49072
5	1	5	0.10928	0.18183	0.25522	0.32895	0.40285
5	2	2	0.31188	0.54887	0.79888	1.05685	1.32030
5	2	3	0.27495	0.45844	0.64662	0.83754	1.03033
5	2	4	0.24933	0.39838	0.54792	0.69762	0.84741
5	2	5	0.22724	0.34794	0.46624	0.58299	0.69863
5	3	3	0.49699	0.78289	1.06930	1.35590	1.64258
5	3	4	0.45281	0.68473	0.91276	1.13833	1.36217
5	3	5	0.41406	0.60079	0.78081	0.95672	1.12976
5	4	4	0.87532	1.26081	1.63358	1.99851	2.35800
5	4	5	0.80509	1.11512	1.41025	1.69626	1.97597
5	5	5	2.25260	2.95170	3.60804	4.23805	4.84973
6	1	1	0.14725	0.32543	0.53855	0.77500	1.02833
6	1	2	0.11829	0.23710	0.37070	0.51365	0.66315
6	1	3	0.10361	0.19643	0.29749	0.40357	0.51311
6	1	4	0.09394	0.17091	0.25271	0.33737	0.42395
6	1	5	0.08647	0.15173	0.21961	0.28894	0.35921
6	1	6	0.07955	0.13433	0.18990	0.24578	0.30183
6	2	2	0.23152	0.42549	0.63417	0.85190	1.07592
6	2	3	0.20374	0.35489	0.51289	0.67499	0.83990
6	2	4	0.18523	0.30995	0.43761	0.56694	0.69740
6	2	5	0.17079	0.27585	0.38137	0.48705	0.59281
6	2	6	0.15733	0.24466	0.33048	0.41526	0.49932
6	3	3	0.34080	0.56237	0.78857	1.01748	1.24823
6	3	4	0.31091	0.49347	0.67642	0.85951	1.04267
6	3	5	0.28733	0.44057	0.59162	0.74126	0.88990
6	3	6	0.26515	0.39170	0.51409	0.63389	0.75190
6	4	4	0.51771	0.78706	1.05261	1.31574	1.57718
6	4	5	0.48007	0.70593	0.92547	1.14109	1.35399
6	4	6	0.44418	0.62991	0.80753	0.98020	1.14943
6	5	5	0.88796	1.25343	1.60405	1.94555	2.28074
6	5	6	0.82546	1.12572	1.41004	1.68463	1.95246

Variances and Covariances of Gamma Order Statistics (Shape Parameter α)

N	I	J	$\alpha = 2.0$	$\alpha = 3.0$	$\alpha = 4.0$	$\alpha = 5.0$	$\alpha = 6.0$
6	6	6	2.25065	2.91986	3.54620	4.14609	4.72757
7	1	1	0.12022	0.27582	0.46547	0.67808	0.90740
7	1	2	0.09616	0.19983	0.31849	0.44670	0.58166
7	1	3	0.08411	0.16538	0.25543	0.35090	0.45013
7	1	4	0.07633	0.14421	0.21768	0.29453	0.37367
7	1	5	0.07054	0.12891	0.19088	0.25491	0.32034
7	1	6	0.06570	0.11639	0.16917	0.22309	0.27772
7	1	7	0.06096	0.10429	0.14837	0.19724	0.23726
7	2	2	0.18195	0.34689	0.52736	0.71747	0.91432
7	2	3	0.15980	0.28876	0.42579	0.56772	0.71301
7	2	4	0.14536	0.25261	0.36423	0.47844	0.59441
7	2	5	0.13453	0.22629	0.32013	0.41515	0.51095
7	2	6	0.12543	0.20461	0.28421	0.36398	0.44382
7	2	7	0.11646	0.18354	0.24959	0.31492	0.37973
7	3	3	0.25485	0.43690	0.62580	0.81877	1.01454
7	3	4	0.23245	0.38364	0.53760	0.69317	0.84982
7	3	5	0.21552	0.34451	0.47383	0.60328	0.73278
7	3	6	0.20120	0.31205	0.42151	0.53006	0.63795
7	3	7	0.18701	0.28032	0.37077	0.45942	0.54684
7	4	4	0.35838	0.56810	0.77821	0.98847	1.19880
7	4	5	0.33308	0.51181	0.68841	0.86366	1.03794
7	4	6	0.31149	0.46469	0.61406	0.76104	0.90636
7	4	7	0.28992	0.41826	0.54135	0.66123	0.77889
7	5	5	0.53058	0.78687	1.03693	1.28311	1.52661
7	5	6	0.49750	0.71699	0.92869	1.13559	1.33918
7	5	7	0.46408	0.64731	0.82157	0.99035	1.15536
7	6	6	0.89522	1.24476	1.57822	1.90184	2.21864
7	6	7	0.83844	1.12998	1.40499	1.66990	1.92782
7	7	7	2.24666	2.89195	3.49451	4.07067	4.62845
8	1	1	0.10109	0.23966	0.41140	0.60570	0.81649
8	1	2	0.08054	0.17277	0.27999	0.39684	0.52052
8	1	3	0.07035	0.14277	0.22427	0.31143	0.40253
8	1	4	0.06384	0.12458	0.19139	0.26188	0.33491
8	1	5	0.05910	0.11172	0.16852	0.22780	0.28876
8	1	6	0.05529	0.10161	0.15076	0.20153	0.25337
8	1	7	0.05193	0.09283	0.13547	0.17905	0.22321
8	1	8	0.04848	0.08393	0.12008	0.15651	0.19308
8	2	2	0.14869	0.29265	0.45255	0.62241	0.79925
8	2	3	0.13033	0.24309	0.36467	0.49163	0.62231
8	2	4	0.11852	0.21273	0.31222	0.41488	0.51970
8	2	5	0.10985	0.19111	0.27549	0.36170	0.44913
8	2	6	0.10286	0.17403	0.24681	0.32048	0.39473
8	2	7	0.09667	0.15914	0.22202	0.28506	0.34817
8	2	8	0.09031	0.14400	0.19698	0.24942	0.30148
8	3	3	0.20121	0.35637	0.51967	0.68788	0.85947
8	3	4	0.18340	0.31284	0.44653	0.58273	0.72063
8	3	5	0.17024	0.28161	0.39491	0.50928	0.62438
8	3	6	0.15956	0.25681	0.35437	0.45204	0.54976
8	3	7	0.15008	0.23509	0.31917	0.40261	0.48558
8	3	8	0.14029	0.21292	0.28346	0.35268	0.42097
8	4	4	0.26973	0.44262	0.61828	0.79554	0.97387
8	4	5	0.25084	0.39944	0.54836	0.69739	0.84647
8	4	6	0.23542	0.36492	0.49307	0.62035	0.74700
8	4	7	0.22165	0.33452	0.44479	0.55345	0.66097
8	4	8	0.20737	0.30333	0.39556	0.48552	0.57389

Variances and Covariances of Gamma Order Statistics (Shape Parameter α)

N	I	J	$\alpha = 2.0$	$\alpha = 3.0$	$\alpha = 4.0$	$\alpha = 5.0$	$\alpha = 6.0$
8	5	5	0.37002	0.57023	0.76837	0.96517	1.16100
8	5	6	0.34792	0.52223	0.69282	0.86109	1.02775
8	5	7	0.32803	0.47964	0.62635	0.77003	0.91161
8	5	8	0.30727	0.43565	0.55809	0.67691	0.79323
8	6	6	0.53902	0.78468	1.02259	1.25573	1.48557
8	6	7	0.50931	0.72281	0.92756	1.12696	1.32265
8	6	8	0.47799	0.65824	0.82897	0.99389	1.15482
8	7	7	0.89944	1.23590	1.55556	1.86493	2.16718
8	7	8	0.84708	1.13091	1.39789	1.65456	1.90407
8	8	8	2.24180	2.86735	3.45043	4.00727	4.54580
9	1	1	0.08691	0.21215	0.36971	0.54944	0.74544
9	1	2	0.06901	0.15226	0.25041	0.35821	0.47287
9	1	3	0.06018	0.12562	0.20028	0.28075	0.36527
9	1	4	0.05460	0.10962	0.17099	0.23626	0.30422
9	1	5	0.05058	0.09845	0.15089	0.20607	0.26312
9	1	6	0.04741	0.08986	0.13561	0.18330	0.23229
9	1	7	0.04473	0.08271	0.12302	0.16464	0.20712
9	1	8	0.04227	0.07622	0.11167	0.14791	0.18464
9	1	9	0.03967	0.06941	0.09982	0.13048	0.16128
9	2	2	0.12500	0.25307	0.39724	0.55155	0.71297
9	2	3	0.10937	0.20978	0.31946	0.43483	0.55416
9	2	4	0.09940	0.18352	0.27354	0.36710	0.46310
9	2	5	0.09218	0.16509	0.24183	0.32083	0.40137
9	2	6	0.08648	0.15085	0.21763	0.28578	0.35486
9	2	7	0.08164	0.13896	0.19759	0.25694	0.31674
9	2	8	0.07718	0.12815	0.19751	0.23102	0.28260
9	2	9	0.07245	0.11676	0.16055	0.20394	0.24703
9	3	3	0.16490	0.30051	0.44508	0.59507	0.74881
9	3	4	0.15017	0.26362	0.38228	0.50405	0.62794
9	3	5	0.13943	0.23756	0.33864	0.44146	0.54544
9	3	6	0.13092	0.21733	0.30517	0.39381	0.48298
9	3	7	0.12368	0.20038	0.27736	0.35446	0.43160
9	3	8	0.11699	0.18492	0.25218	0.31897	0.38543
9	3	9	0.10987	0.16859	0.22571	0.28181	0.33719
9	4	4	0.21391	0.36161	0.51357	0.66802	0.82418
9	4	5	0.19892	0.32653	0.45600	0.58653	0.71776
9	4	6	0.18699	0.29916	0.41161	0.52415	0.63674
9	4	7	0.17678	0.27613	0.37457	0.47240	0.56979
9	4	8	0.16733	0.25505	0.34089	0.42556	0.50941
9	4	9	0.15723	0.23270	0.30538	0.37634	0.44610
9	5	5	0.27999	0.44550	0.61133	0.77728	0.94329
9	5	6	0.26356	0.40892	0.55297	0.69617	0.83876
9	5	7	0.24943	0.37797	0.50402	0.62852	0.75192
9	5	8	0.23630	0.34951	0.45931	0.56700	0.67322
9	5	9	0.22220	0.31922	0.41194	0.50204	0.59034
9	6	6	0.37813	0.57050	0.75919	0.94559	1.13038
9	6	7	0.35839	0.52836	0.69352	0.85572	1.01587
9	6	8	0.33992	0.48937	0.63316	0.77349	0.91142
9	6	9	0.31998	0.44761	0.56882	0.68613	0.80074
9	7	7	0.54477	0.78154	1.00957	1.23224	1.45120
9	7	8	0.51765	0.72568	0.92433	1.11723	1.30617
9	7	9	0.48810	0.66529	0.83261	0.99390	1.15103
9	8	8	0.90182	1.22730	1.53554	1.83320	2.12354
9	8	9	0.85299	1.13002	1.39001	1.63955	1.88186
9	9	9	2.23662	2.84550	3.41225	3.95292	4.47539

Variances and Covariances of Gamma Order Statistics (Shape Parameter α)

N	I	J	$\alpha = 2.0$	$\alpha = 3.0$	$\alpha = 4.0$	$\alpha = 5.0$	$\alpha = 6.0$
10	1	1	0.07603	0.19053	0.33656	0.50437	0.68823
10	1	2	0.06018	0.13619	0.22696	0.32735	0.43461
10	1	3	0.05241	0.11219	0.18125	0.25620	0.33528
10	1	4	0.04752	0.09786	0.15473	0.21563	0.27935
10	1	5	0.04402	0.08795	0.13669	0.18836	0.24204
10	1	6	0.04131	0.08043	0.12318	0.16808	0.21444
10	1	7	0.03906	0.07431	0.11228	0.15182	0.19241
10	1	8	0.03709	0.06900	0.10289	0.13788	0.17360
10	1	9	0.03522	0.06402	0.09414	0.12494	0.15617
10	1	10	0.03318	0.05863	0.08471	0.11104	0.13750
10	2	2	0.10738	0.22297	0.35469	0.49663	0.64576
10	2	3	0.09378	0.18446	0.28467	0.39080	0.50103
10	2	4	0.08517	0.16128	0.24367	0.32989	0.41873
10	2	5	0.07899	0.14516	0.21564	0.28869	0.36350
10	2	6	0.07417	0.13288	0.19454	0.25793	0.32248
10	2	7	0.07018	0.12286	0.17748	0.23319	0.28963
10	2	8	0.06665	0.11414	0.16274	0.21193	0.26150
10	2	9	0.06331	0.10595	0.14898	0.19215	0.23539
10	2	10	0.05967	0.09708	0.13413	0.17086	0.20736
10	3	3	0.13887	0.25961	0.38980	0.52577	0.66574
10	3	4	0.12635	0.22754	0.33458	0.44514	0.55811
10	3	5	0.11731	0.20511	0.29661	0.39029	0.48545
10	3	6	0.11024	0.18796	0.26791	0.34915	0.43124
10	3	7	0.10436	0.17392	0.24463	0.31596	0.38770
10	3	8	0.09916	0.16168	0.22447	0.28737	0.35031
10	3	9	0.09422	0.15016	0.20561	0.26071	0.31555
10	3	10	0.08882	0.13764	0.18520	0.23195	0.27813
10	4	4	0.17586	0.30517	0.43971	0.57736	0.71713
10	4	5	0.16349	0.27558	0.39058	0.50729	0.62514
10	4	6	0.15378	0.25285	0.35328	0.45449	0.55620
10	4	7	0.14567	0.23417	0.32292	0.41175	0.50062
10	4	8	0.13849	0.21784	0.29655	0.37481	0.45275
10	4	9	0.13165	0.20244	0.27181	0.34029	0.40813
10	4	10	0.12415	0.18566	0.24498	0.30295	0.35998
10	5	5	0.22294	0.36463	0.50819	0.65280	0.79812
10	5	6	0.20993	0.33505	0.46044	0.58591	0.71144
10	5	7	0.19903	0.31066	0.42139	0.53154	0.64126
10	5	8	0.18935	0.28925	0.38738	0.48438	0.58059
10	5	9	0.18009	0.26899	0.35536	0.44016	0.52387
10	5	10	0.16992	0.24687	0.32053	0.39219	0.46247
10	6	6	0.28743	0.44678	0.60485	0.76208	0.91864
10	6	7	0.27280	0.41486	0.55448	0.69257	0.82959
10	6	8	0.25975	0.38672	0.51039	0.63202	0.75223
10	6	9	0.24724	0.36000	0.46874	0.57501	0.67959
10	6	10	0.23342	0.33069	0.42324	0.51293	0.60065
10	7	7	0.38399	0.56972	0.75069	0.92870	1.10467
10	7	8	0.36607	0.53196	0.69229	0.84919	1.00374
10	7	9	0.34879	0.49589	0.63679	0.77392	0.90844
10	7	10	0.32692	0.45611	0.57585	0.69148	0.80429
10	8	8	0.54877	0.77796	0.99774	1.21176	1.42178
10	8	9	0.52371	0.72680	0.92006	1.10731	1.29040
10	8	10	0.49567	0.66990	0.83400	0.99193	1.14559
10	9	9	0.90304	1.21914	1.51769	1.80551	2.08588
10	9	10	0.85711	1.12809	1.38193	1.62527	1.86131
10	10	10	2.23139	2.82594	3.37871	3.90560	4.41439

Table C5

MOMENTS OF EXPONENTIAL, WEIBULL, AND GAMMA POPULATIONS

[Location parameter = 0; scale parameter = 1.]

MOMENTS OF EXPONENTIAL, WEIBULL, AND GAMMA POPULATIONS

Population	*Shapeparameter*	*Mean*	*Variance*	*Skewness*	*Kurtosis*
EXPONENTIAL		1.00000000	1.00000000	2.00000000	9.00000000
WEIBULL	0.5	2.00000000	20.00000000	6.61876121	87.72000000
WEIBULL	1.0	1.00000000	1.00000000	2.00000000	9.00000000
WEIBULL	1.5	0.90274529	0.37569028	1.07198657	4.39040356
WEIBULL	2.0	0.88622693	0.21460184	0.63111066	3.24508930
WEIBULL	2.5	0.88726382	0.14414669	0.35863184	2.85678309
WEIBULL	3.0	0.89297951	0.10533288	0.16810284	2.72946363
WEIBULL	3.5	0.89974718	0.08107275	0.02510816	2.71273189
WEIBULL	4.0	0.90640248	0.06446148	-0.08723697	2.74782953
WEIBULL	5.0	0.91816874	0.04422998	-0.25410959	2.88029006
WEIBULL	6.0	0.92771933	0.03231635	-0.37326156	3.03545528
WEIBULL	7.0	0.93543756	0.02470374	-0.46318962	3.18718296
WEIBULL	8.0	0.94174270	0.01952316	-0.53372638	3.32767551
GAMMA	0.5	0.50000000	0.50000000	2.82842712	15.00000000
GAMMA	1.0	1.00000000	1.00000000	2.00000000	9.00000000
GAMMA	1.5	1.50000000	1.50000000	1.63299316	7.00000000
GAMMA	2.0	2.00000000	2.00000000	1.41421356	6.00000000
GAMMA	2.5	2.50000000	2.50000000	1.26491106	5.40000000
GAMMA	3.0	3.00000000	3.00000000	1.15470054	5.00000000
GAMMA	3.5	3.50000000	3.50000000	1.06904497	4.71428571
GAMMA	4.0	4.00000000	4.00000000	1.00000000	4.50000000
GAMMA	5.0	5.00000000	5.00000000	0.89442719	4.20000000
GAMMA	6.0	6.00000000	6.00000000	0.81649658	4.00000000

Table C6.1

EXPECTED VALUES OF LOGNORMAL ORDER STATISTICS

[Tabular values are expected values of the Mth order statistics of a sample of size N from a lognormal population with location parameter 0 and scale parameter 1.]

Expected Values of Lognormal Order Statistics

N	*M*	*Mean*	*N*	*M*	*Mean*	*N*	*M*	*Mean*	*N*	*M*	*Mean*
1	1	1.64872	10	9	3.04158	15	1	0.20255	18	5	0.54046
2	1	0.79056	10	10	5.62108	15	2	0.31302	18	6	0.63427
2	2	2.50688	11	1	0.23885	15	3	0.41530	18	7	0.73600
3	1	0.55922	11	2	0.38191	15	4	0.51968	18	8	0.84824
3	2	1.25185	11	3	0.52340	15	5	0.63114	18	9	0.97416
3	3	3.13439	11	4	0.67794	15	6	0.75380	18	10	1.11795
4	1	0.45059	11	5	0.85611	15	7	0.89219	18	11	1.28544
4	2	0.88788	11	6	1.07100	15	8	1.05217	18	12	1.48521
4	3	1.61582	11	7	1.34318	15	9	1.24207	18	13	1.73072
4	4	3.64058	11	8	1.71039	15	10	1.47479	18	14	2.04475
5	1	0.38568	11	9	2.25527	15	11	1.77204	18	15	2.47011
5	2	0.71024	11	10	3.21632	15	12	2.17460	18	16	3.10095
5	3	1.15434	11	11	5.86156	15	13	2.77214	18	17	4.20799
5	4	1.92348	12	1	0.22779	15	14	3.82304	18	18	7.21613
5	5	4.06986	12	2	0.36048	15	15	6.69229	19	1	0.17990
6	1	0.34213	12	3	0.48905	16	1	0.19598	19	2	0.27212
6	2	0.60342	12	4	0.62646	16	2	0.30101	19	3	0.35417
6	3	0.92389	12	5	0.78090	16	3	0.39712	19	4	0.43478
6	4	1.38479	12	6	0.96139	16	4	0.49410	19	5	0.51752
6	5	2.19283	12	7	1.18062	16	5	0.59642	19	6	0.60470
6	6	4.44526	12	8	1.45930	16	6	0.70752	19	7	0.69834
7	1	0.31059	12	9	1.83593	16	7	0.83093	19	8	0.80055
7	2	0.53137	12	10	2.39506	16	8	0.97096	19	9	0.91380
7	3	0.78354	12	11	3.38057	16	9	1.13338	19	10	1.04123
7	4	1.11101	12	12	6.08710	16	10	1.32660	19	11	1.18701
7	5	1.59012	13	1	0.21825	16	11	1.56370	19	12	1.35703
7	6	2.43392	13	2	0.34231	16	12	1.86675	19	13	1.55999
7	7	4.78049	13	3	0.46044	16	13	2.27722	19	14	1.80952
8	1	0.28652	13	4	0.58444	16	14	2.88636	19	15	2.12876
8	2	0.47910	13	5	0.72100	16	15	3.95685	19	16	2.56113
8	3	0.68819	13	6	0.87674	16	16	6.87465	19	17	3.20216
8	4	0.94247	13	7	1.06015	17	1	0.19008	19	18	4.32632
8	5	1.27956	13	8	1.28388	17	2	0.29033	19	19	7.37667
8	6	1.77645	13	9	1.56894	17	3	0.38112	20	1	0.17546
8	7	2.65307	13	10	1.95460	17	4	0.47181	20	2	0.26428
8	8	5.08441	13	11	2.52719	17	5	0.56652	20	3	0.34270
9	1	0.26743	13	12	3.53573	17	6	0.66818	20	4	0.41919
9	2	0.43922	13	13	6.29971	17	7	0.77965	20	5	0.49714
9	3	0.61870	14	1	0.20991	17	8	0.90420	20	6	0.57865
9	4	0.82715	14	2	0.32666	17	9	1.04606	20	7	0.66548
9	5	1.08663	14	3	0.43618	17	10	1.21100	20	8	0.75937
9	6	1.43390	14	4	0.54940	17	11	1.40752	20	9	0.86232
9	7	1.94773	14	5	0.67203	17	12	1.64889	20	10	0.97673
9	8	2.85460	14	6	0.80916	17	13	1.95752	20	11	1.10572
9	9	5.36313	14	7	0.96685	17	14	2.37558	20	12	1.25351
10	1	0.25186	14	8	1.15345	17	15	2.99581	20	13	1.42604
10	2	0.40764	14	9	1.38170	17	16	4.08499	20	14	1.63211
10	3	0.56555	14	10	1.67296	17	17	7.04901	20	15	1.88556
10	4	0.74273	14	11	2.06725	18	1	0.18475	20	16	2.20983
10	5	0.95379	14	12	2.65263	18	2	0.28076	20	17	2.64896
10	6	1.21947	14	13	3.68292	18	3	0.36690	20	18	3.29979
10	7	1.57686	14	14	6.50101	18	4	0.45220	20	19	4.44038
10	8	2.10667							20	20	7.53121

Table C6.2

VARIANCES AND COVARIANCES OF LOGNORMAL ORDER STATISTICS

[When $I = J$, tabular values are variances of the Ith order statistic of a sample of size N from a lognormal population with location parameter 0 and scale parameter 1; when $I \neq J$, they are covariances of the Ith and Jth order statistics.]

Variances and Covariances of Lognormal Order Statistics

N	I	J	COV	N	I	J	COV	N	I	J	COV	N	I	J	COV
1	1	1	4.67077	6	6	6	13.86507	8	5	5	0.34249	10	1	1	0.02051
2	1	1	0.53730	7	1	1	0.03690	8	5	6	0.38054	10	1	2	0.01810
2	1	2	0.73644	7	1	2	0.03429	8	5	7	0.44740	10	1	3	0.01753
2	2	2	7.33137	7	1	3	0.03477	8	5	8	0.61663	10	1	4	0.01768
3	1	1	0.19781	7	1	4	0.03688	8	6	6	0.73703	10	1	5	0.01826
3	1	2	0.22222	7	1	5	0.04078	8	6	7	0.86578	10	1	6	0.01928
3	1	3	0.30896	7	1	6	0.04796	8	6	8	1.19033	10	1	7	0.02081
3	2	2	0.89711	7	1	7	0.06649	8	7	7	2.13018	10	1	8	0.02315
3	2	3	1.24394	7	2	2	0.07612	8	7	8	2.91207	10	1	9	0.02722
3	3	3	9.36718	7	2	3	0.07753	8	8	8	16.17767	10	1	10	0.03747
4	1	1	0.10605	7	2	4	0.08240	9	1	1	0.02425	10	2	2	0.03606
4	1	2	0.10982	7	2	5	0.09117	9	1	2	0.02171	10	2	3	0.03507
4	1	3	0.12711	7	2	6	0.10723	9	1	3	0.02129	10	2	4	0.03543
4	1	4	0.17719	7	2	7	0.14857	9	1	4	0.02172	10	2	5	0.03665
4	2	2	0.32969	7	3	3	0.14635	9	1	5	0.02278	10	2	6	0.03869
4	2	3	0.38296	7	3	4	0.15584	9	1	6	0.02452	10	2	7	0.04178
4	2	4	0.53299	7	3	5	0.17257	9	1	7	0.02726	10	2	8	0.04649
4	3	3	1.19958	7	3	6	0.20295	9	1	8	0.03206	10	2	9	0.05466
4	3	4	1.66142	7	3	7	0.28094	9	1	9	0.04424	10	2	10	0.07519
4	4	4	11.06480	7	4	4	0.29019	9	2	2	0.04448	10	3	3	0.05819
5	1	1	0.06807	7	4	5	0.32158	9	2	3	0.04378	10	3	4	0.05887
5	1	2	0.06714	7	4	6	0.37812	9	2	4	0.04478	10	3	5	0.06097
5	1	3	0.07257	7	4	7	0.52274	9	2	5	0.04701	10	3	6	0.06440
5	1	4	0.08487	7	5	5	0.64695	9	2	6	0.05061	10	3	7	0.06955
5	1	5	0.11821	7	5	6	0.76023	9	2	7	0.05627	10	3	8	0.07740
5	2	2	0.17368	7	5	7	1.04843	9	2	8	0.06619	10	3	9	0.09099
5	2	3	0.18860	7	6	6	1.92598	9	2	9	0.09128	10	3	10	0.12514
5	2	4	0.22078	7	6	7	2.64088	9	3	3	0.07495	10	4	4	0.09210
5	2	5	0.30713	7	7	7	15.06829	9	3	4	0.07680	10	4	5	0.09547
5	3	3	0.44537	8	1	1	0.02941	9	3	5	0.08070	10	4	6	0.10091
5	3	4	0.52178	8	1	2	0.02678	9	3	6	0.08692	10	4	7	0.10899
5	3	5	0.72432	8	1	3	0.02664	9	3	7	0.09667	10	4	8	0.12130
5	4	4	1.46575	8	1	4	0.02764	9	3	8	0.11369	10	4	9	0.14258
5	4	5	2.02350	8	1	5	0.02959	9	3	9	0.15669	10	4	10	0.19600
5	5	5	12.54318	8	1	6	0.03284	9	4	4	0.12491	10	5	5	0.14741
6	1	1	0.04847	8	1	7	0.03864	9	4	5	0.13141	10	5	6	0.15590
6	1	2	0.04622	8	1	8	0.05343	9	4	6	0.14160	10	5	7	0.16842
6	1	3	0.04808	8	2	2	0.05682	9	4	7	0.15750	10	5	8	0.18747
6	1	4	0.05278	8	2	3	0.05678	9	4	8	0.18520	10	5	9	0.22031
6	1	5	0.06198	8	2	4	0.05901	9	4	9	0.25509	10	5	10	0.30266
6	1	6	0.08613	8	2	5	0.06324	9	5	5	0.21388	10	6	6	0.24505
6	2	2	0.10917	8	2	6	0.07019	9	5	6	0.23061	10	6	7	0.26482
6	2	3	0.11408	8	2	7	0.08258	9	5	7	0.25651	10	6	8	0.29475
6	2	4	0.12546	8	2	8	0.11414	9	5	8	0.30156	10	6	9	0.34626
6	2	5	0.14737	8	3	3	0.10126	9	5	9	0.41500	10	6	10	0.47529
6	2	6	0.20464	8	3	4	0.10544	9	6	6	0.39179	10	7	7	0.43852
6	3	3	0.23426	8	3	5	0.11311	9	6	7	0.43583	10	7	8	0.48805
6	3	4	0.25808	8	3	6	0.12559	9	6	8	0.51210	10	7	9	0.57301
6	3	5	0.30318	8	3	7	0.14774	9	6	9	0.70382	10	7	10	0.78549
6	3	6	0.42047	8	3	8	0.20408	9	7	7	0.82164	10	8	8	0.90162
6	4	4	0.55026	8	4	4	0.18108	9	7	8	0.96448	10	8	9	1.05746
6	4	5	0.64632	8	4	5	0.19442	9	7	9	1.32233	10	8	10	1.44609
6	4	6	0.89425	8	4	6	0.21595	9	8	8	2.32129	10	9	9	2.50139
6	5	5	1.70585	8	4	7	0.25400	9	8	9	3.16463	10	9	10	3.40168
6	5	6	2.34677	8	4	8	0.35055	9	9	9	17.21053	10	10	10	18.17950

Table C7.1

EXPECTED VALUES OF LOG-WEIBULL (TYPE I SMALLEST EXTREME-VALUE) ORDER STATISTICS

[Tabular values are expected values of the Mth order statistic of a sample of size N from a log-Weibull population with location parameter 0 and scale parameter 1.]

Expected Values of Log-Weibull (Type I Smallest Extreme-Value) Order Statistics

N	*M*	*Mean*	*N*	*M*	*Mean*	*N*	*M*	*Mean*	*N*	*M*	*Mean*
1	1	-0.57722	10	9	0.58456	15	1	-3.28527	18	5	-1.25925
2	1	-1.27036	10	10	0.98987	15	2	-2.25037	18	6	-1.02329
2	2	0.11593	11	1	-2.97511	15	3	-1.71329	18	7	-0.81823
3	1	-1.67583	11	2	-1.92670	15	4	-1.34035	18	8	-0.63418
3	2	-0.45943	11	3	-1.37393	15	5	-1.04786	18	9	-0.46468
3	3	0.40361	11	4	-0.98252	15	6	-0.80189	18	10	-0.30511
4	1	-1.96351	11	5	-0.66781	15	7	-0.58518	18	11	-0.15185
4	2	-0.81278	11	6	-0.39457	15	8	-0.38728	18	12	-0.00167
4	3	-0.10608	11	7	-0.14318	15	9	-0.20102	18	13	0.14866
4	4	0.57351	11	8	0.10075	15	10	-0.02061	18	14	0.30297
5	1	-2.18665	11	9	0.35229	15	11	0.15946	18	15	0.46660
5	2	-1.07094	11	10	0.63617	15	12	0.34581	18	16	0.64857
5	3	-0.42555	11	11	1.02524	15	13	0.54853	18	17	0.86840
5	4	0.10689	12	1	-3.06212	15	14	0.78837	18	18	1.19098
5	5	0.69017	12	2	-2.01799	15	15	1.13268	19	1	-3.52165
6	1	-2.36898	12	3	-1.47026	16	1	-3.34980	19	2	-2.49438
6	2	-1.27505	12	4	-1.08493	16	2	-2.31719	19	3	-1.96578
6	3	-0.66272	12	5	-0.77771	16	3	-1.78267	19	4	-1.60239
6	4	-0.18839	12	6	-0.51397	16	4	-1.41265	19	5	-1.32070
6	5	0.25453	12	7	-0.27517	16	5	-1.12348	19	6	-1.08717
6	6	0.77729	12	8	-0.04891	16	6	-0.88143	19	7	-0.88488
7	1	-2.52313	12	9	0.17557	16	7	-0.66932	19	8	-0.70399
7	2	-1.44407	12	10	0.41120	16	8	-0.47699	19	9	-0.53819
7	3	-0.85248	12	11	0.68117	16	9	-0.29758	19	10	-0.38300
7	4	-0.40969	12	12	1.05652	16	10	-0.12591	19	11	-0.23501
7	5	-0.02240	13	1	-3.14216	16	11	0.04256	19	12	-0.09136
7	6	0.36531	13	2	-2.10161	16	12	0.21259	19	13	0.05065
7	7	0.84596	13	3	-1.55805	16	13	0.39021	19	14	0.19390
8	1	-2.65666	13	4	-1.17764	16	14	0.58507	19	15	0.34193
8	2	-1.58841	13	5	-0.87633	16	15	0.81742	19	16	0.49985
8	3	-1.01107	13	6	-0.61990	16	16	1.15370	19	17	0.67646
8	4	-0.58818	13	7	-0.39038	17	1	-3.41043	19	18	0.89098
8	5	-0.23121	13	8	-0.17642	17	2	-2.37981	19	19	1.20765
8	6	0.10288	13	9	0.03079	17	3	-1.84752	20	1	-3.57295
8	7	0.45279	13	10	0.23992	17	4	-1.48001	20	2	-2.54708
8	8	0.90212	13	11	0.46259	17	5	-1.19370	20	3	-2.02004
9	1	-2.77444	13	12	0.72091	17	6	-0.95494	20	4	-1.65835
9	2	-1.71439	13	13	1.08499	17	7	-0.74666	20	5	-1.37856
9	3	-1.14745	14	1	-3.21627	17	8	-0.55885	20	6	-1.14714
9	4	-0.73829	14	2	-2.17876	17	9	-0.38489	20	7	-0.94723
9	5	-0.40053	14	3	-1.63870	17	10	-0.21997	20	8	-0.76907
9	6	-0.09575	14	4	-1.26235	17	11	-0.06007	20	9	-0.60638
9	7	0.20220	14	5	-0.96585	17	12	0.09855	20	10	-0.45483
9	8	0.52438	14	6	-0.71520	17	13	0.26011	20	11	-0.31116
9	9	0.94934	14	7	-0.49283	17	14	0.43024	20	12	-0.17271
10	1	-2.87980	14	8	-0.28794	17	15	0.61825	20	13	-0.03713
10	2	-1.82620	14	9	-0.09277	17	16	0.84397	20	14	0.09791
10	3	-1.26718	14	10	0.09943	17	17	1.17306	20	15	0.23503
10	4	-0.86808	14	11	0.29611	18	1	-3.46759	20	16	0.37756
10	5	-0.54361	14	12	0.50799	18	2	-2.43874	20	17	0.53042
10	6	-0.25745	14	13	0.75640	18	3	-1.90841	20	18	0.70223
10	7	0.01204	14	14	1.10973	18	4	-1.54309	20	19	0.91195
10	8	0.28369							20	20	1.22321

Table C7.2

VARIANCES AND COVARIANCES OF LOG-WEIBULL (TYPE I SMALLEST EXTREME-VALUE) ORDER STATISTICS

[When $I = J$, tabular values are variances of the Ith order statistic of a sample of size N from a log-Weibull population with location parameter 0 and scale parameter 1; when $I \neq J$, they are covariances of the Ith and Jth order statistics.]

Variances and Covariances of Log-Weibull (Type I Smallest Extreme-Value) Order Statistics

N	*I*	*J*	*COV*	*N*	*I*	*J*	*COV*	*N*	*I*	*J*	*COV*	*N*	*I*	*J*	*COV*
1	1	1	1.64493	6	6	6	0.24658	8	5	5	0.23166	10	1	1	1.64493
2	1	1	1.64493	7	1	1	1.64493	8	5	6	0.16868	10	1	2	0.61876
2	1	2	0.48045	7	1	2	0.60675	8	5	7	0.12103	10	1	3	0.35919
2	2	2	0.68043	7	1	3	0.34211	8	5	8	0.07953	10	1	4	0.24260
3	1	1	1.64493	7	1	4	0.22190	8	6	6	0.19837	10	1	5	0.17615
3	1	2	0.54629	7	1	5	0.15195	8	6	7	0.14306	10	1	6	0.13282
3	1	3	0.24376	7	1	6	0.10427	8	6	8	0.09447	10	1	7	0.10185
3	2	2	0.65852	7	1	7	0.06601	8	7	7	0.18355	10	1	8	0.07803
3	2	3	0.30137	7	2	2	0.64691	8	7	8	0.12233	10	1	9	0.05824
3	3	3	0.44850	7	2	3	0.36717	8	8	8	0.19956	10	1	10	0.03962
4	1	1	1.64493	7	2	4	0.23906	9	1	1	1.64493	10	2	2	0.64586
4	1	2	0.57432	7	2	5	0.16412	9	1	2	0.61569	10	2	3	0.37650
4	1	3	0.29271	7	2	6	0.11283	9	1	3	0.35488	10	2	4	0.25489
4	1	4	0.15389	7	2	7	0.07155	9	1	4	0.23745	10	2	5	0.18536
4	2	2	0.65180	7	3	3	0.39969	9	1	5	0.17027	10	2	6	0.13991
4	2	3	0.33721	7	3	4	0.26155	9	1	6	0.12615	10	2	7	0.10738
4	2	4	0.17903	7	3	5	0.18017	9	1	7	0.09419	10	2	8	0.08232
4	3	3	0.41553	7	3	6	0.12419	9	1	8	0.06886	10	2	9	0.06148
4	3	4	0.22455	7	3	7	0.07893	9	1	9	0.04597	10	2	10	0.04184
4	4	4	0.34402	7	4	4	0.29271	9	2	2	0.64609	10	3	3	0.39702
5	1	1	1.64493	7	4	5	0.20262	9	2	3	0.37418	10	3	4	0.26954
5	1	2	0.58992	7	4	6	0.14020	9	2	4	0.25105	10	3	5	0.19637
5	1	3	0.31716	7	4	7	0.08941	9	2	5	0.18033	10	3	6	0.14842
5	1	4	0.18967	7	5	5	0.23701	9	2	6	0.13377	10	3	7	0.11403
5	1	5	0.10901	7	5	6	0.16497	9	2	7	0.09998	10	3	8	0.08749
5	2	2	0.64907	7	5	7	0.10578	9	2	8	0.07315	10	3	9	0.06538
5	2	3	0.35267	7	6	6	0.21021	9	2	9	0.04887	10	3	10	0.04453
5	2	4	0.21227	7	6	7	0.13618	9	3	3	0.39758	10	4	4	0.28739
5	2	5	0.12258	7	7	7	0.21964	9	3	4	0.26763	10	4	5	0.20986
5	3	3	0.40598	8	1	1	1.64493	9	3	5	0.19267	10	4	6	0.15888
5	3	4	0.24677	8	1	2	0.61182	9	3	6	0.14315	10	4	7	0.12221
5	3	5	0.14359	8	1	3	0.34937	9	3	7	0.10712	10	4	8	0.09387
5	4	4	0.30850	8	1	4	0.23081	9	3	8	0.07846	10	4	9	0.07021
5	4	5	0.18203	8	1	5	0.16256	9	3	9	0.05246	10	4	10	0.04786
5	5	5	0.28486	8	1	6	0.11719	9	4	4	0.28844	10	5	5	0.22686
6	1	1	1.64493	8	1	7	0.08340	9	4	5	0.20823	10	5	6	0.17211
6	1	2	0.59986	8	1	8	0.05440	9	4	6	0.15503	10	5	7	0.13262
6	1	3	0.33205	8	2	2	0.64642	9	4	7	0.11620	10	5	8	0.10200
6	1	4	0.20925	8	2	3	0.37119	9	4	8	0.08522	10	5	9	0.07639
6	1	5	0.13619	8	2	4	0.24600	9	4	9	0.05706	10	5	10	0.05213
6	1	6	0.08285	8	2	5	0.17362	9	5	5	0.22869	10	6	6	0.18958
6	2	2	0.64770	8	2	6	0.12534	9	5	6	0.17074	10	6	7	0.14641
6	2	3	0.36146	8	2	7	0.08931	9	5	7	0.12825	10	6	8	0.11282
6	2	4	0.22888	8	2	8	0.05832	9	5	8	0.09424	10	6	9	0.08463
6	2	5	0.14945	8	3	3	0.39840	9	5	9	0.06322	10	6	10	0.05785
6	2	6	0.09116	8	3	4	0.26509	9	6	6	0.19275	10	7	7	0.16581
6	3	3	0.40186	8	3	5	0.18759	9	6	7	0.14525	10	7	8	0.12812
6	3	4	0.25617	8	3	6	0.13570	9	6	8	0.10703	10	7	9	0.09635
6	3	5	0.16806	8	3	7	0.09685	9	6	9	0.07199	10	7	10	0.06603
6	3	6	0.10292	8	3	8	0.06332	9	7	7	0.17158	10	8	8	0.15192
6	4	4	0.29762	8	4	4	0.29005	9	7	8	0.12699	10	8	9	0.11471
6	4	5	0.19671	8	4	5	0.20599	9	7	9	0.08580	10	8	10	0.07893
6	4	6	0.12122	8	4	6	0.14941	9	8	8	0.16390	10	9	9	0.14879
6	5	5	0.24855	8	4	7	0.10686	9	8	9	0.11167	10	9	10	0.10319
6	5	6	0.15497	8	4	8	0.07001	9	9	9	0.18395	10	10	10	0.17143

Table C8.1

EXPECTED VALUES OF LOG-GAMMA ORDER STATISTICS

[Tabular values are expected values of the rth order statistic of a sample of size n from a log-Gamma population with location parameter 0, scale parameter 1, and shape parameter
$\kappa = 0.5(0.5)2.0(1.0)4.0(2.0)12.0.$]

Expected Values of Log-Gamma Order Statistics
Shape Parameter κ

		κ									
n	r	0.5	1.0	1.5	2.0	3.0	4.0	6.0	8.0	10.0	12.0
1	1	-1.96351	-0.57722	0.03649	0.42278	0.92278	1.25612	1.70612	2.01564	2.25175	2.44266
2	1	-3.12975	-1.27036	-0.49313	-0.02036	0.57339	0.95880	1.46759	1.81088	2.06957	2.27693
2	2	-0.79727	0.11593	0.56611	0.86593	1.27218	1.55343	1.94464	2.22041	2.43394	2.60839
3	1	-3.84691	-1.67583	-0.79467	-0.26842	0.38179	0.79776	1.34027	1.70253	1.97373	2.19014
3	2	-1.69544	-0.45943	0.10995	0.47575	0.95659	1.28089	1.72223	2.02758	2.26123	2.45052
3	3	-0.34818	0.40361	0.79420	1.06102	1.42998	1.68970	2.05585	2.31682	2.52029	2.68733
4	1	-4.36924	-1.96351	-1.00538	-0.44007	0.25077	0.68843	1.25456	1.62995	1.90976	2.13234
4	2	-2.27992	-0.81278	-0.16255	0.24654	0.77483	1.12577	1.59741	1.92027	2.16566	2.36353
4	3	-1.11095	-0.10608	0.38244	0.70497	1.13835	1.43601	1.84705	2.13489	2.35680	2.53751
4	4	-0.09392	0.57351	0.93145	1.17970	1.52719	1.77427	2.12544	2.37746	2.57479	2.73727
5	1	-4.78116	-2.18665	-1.16715	-0.57097	0.15168	0.60614	1.19043	1.57583	1.86217	2.08942
5	2	-2.72156	-1.07094	-0.35831	0.08353	0.64713	1.01756	1.51107	1.84640	2.10011	2.30401
5	3	-1.61747	-0.42555	0.13110	0.49104	0.96639	1.28809	1.72693	2.03106	2.26400	2.45282
5	4	-0.77328	0.10689	0.55000	0.84759	1.25299	1.53462	1.92714	2.20411	2.41866	2.59397
5	5	0.07592	0.69017	1.02681	1.26273	1.59574	1.83418	2.17502	2.42080	2.61382	2.77310
6	1	-5.12163	-2.36898	-1.29833	-0.67659	0.07223	0.54042	1.13943	1.53292	1.82450	2.05548
6	2	-3.07879	-1.27505	-0.51126	-0.04290	0.54893	0.93477	1.44541	1.79043	2.05055	2.25909
6	3	-2.00710	-0.66272	-0.05241	0.33639	0.84353	1.18313	1.64240	1.95835	2.19923	2.39385
6	4	-1.22783	-0.18839	0.31461	0.64569	1.08925	1.39304	1.81147	2.10378	2.32878	2.51179
6	5	-0.54600	0.25453	0.66770	0.94854	1.33486	1.60541	1.98497	2.25428	2.46361	2.63506
6	6	0.20030	0.77729	1.09863	1.32557	1.64791	1.87993	2.21303	2.45410	2.64387	2.80070
7	1	-5.41197	-2.52313	-1.40859	-0.76501	0.00604	0.48583	1.09723	1.49747	1.79343	2.02753
7	2	-3.37959	-1.44407	-0.63678	-0.14607	0.46933	0.86793	1.39264	1.74556	2.01090	2.22320
7	3	-2.32680	-0.85248	-0.19747	0.21503	0.74793	1.10187	1.57733	1.90258	2.14967	2.34881
7	4	-1.58084	-0.40969	0.14101	0.49821	0.97099	1.29147	1.72914	2.03271	2.26531	2.45390
7	5	-0.96307	-0.02240	0.44480	0.75629	1.17794	1.46922	1.87321	2.15709	2.37638	2.55521
7	6	-0.37918	0.36531	0.75686	1.02544	1.39763	1.65989	2.02967	2.29315	2.49850	2.66700
7	7	0.29688	0.84596	1.15560	1.37559	1.68962	1.91661	2.24359	2.48092	2.66809	2.82299
8	1	-5.66515	-2.65666	-1.50365	-0.84099	-0.05060	0.43924	1.06132	1.46736	1.76707	2.00383
8	2	-3.63976	-1.58841	-0.74318	-0.23314	0.40252	0.81200	1.34865	1.70825	1.97797	2.19343
8	3	-2.59909	-1.01107	-0.31755	0.11514	0.66977	1.03570	1.52459	1.85750	2.10967	2.31251
8	4	-1.87299	-0.58818	0.00267	0.38152	0.87820	1.21216	1.66523	1.97772	2.21632	2.40930
8	5	-1.28868	-0.23121	0.27935	0.61490	1.06378	1.37078	1.79305	2.08769	2.31429	2.49850
8	6	-0.76770	0.10288	0.54408	0.84113	1.24644	1.52829	1.92131	2.19872	2.41363	2.58923
8	7	-0.24967	0.45279	0.82779	1.08687	1.44802	1.70376	2.06579	2.32463	2.52679	2.69292
8	8	0.37496	0.90212	1.20242	1.41684	1.72414	1.94701	2.26899	2.50325	2.68828	2.84157
9	1	-5.88964	-2.77444	-1.58717	-0.90756	-0.10005	0.39864	1.03010	1.44123	1.74421	1.98330
9	2	-3.86918	-1.71439	-0.83551	-0.30841	0.34503	0.76401	1.31102	1.67639	1.94990	2.16807
9	3	-2.83679	-1.14745	-0.42004	0.03028	0.60374	0.97998	1.48035	1.81976	2.07625	2.28221
9	4	-2.12368	-0.73829	-0.11258	0.28485	0.80183	1.14715	1.61308	1.93297	2.17652	2.37311
9	5	-1.55964	-0.40053	0.14674	0.50236	0.97366	1.29343	1.73043	2.03366	2.26606	2.45453
9	6	-1.07192	-0.09575	0.38543	0.70493	1.13589	1.43266	1.84316	2.13092	2.35288	2.53368
9	7	-0.61559	0.20220	0.62340	0.90923	1.30171	1.57610	1.96039	2.23262	2.44400	2.61700
9	8	-0.14512	0.52438	0.88618	1.13763	1.48983	1.74024	2.09591	2.35092	2.55044	2.71461
9	9	0.43997	0.94934	1.24196	1.45174	1.75343	1.97286	2.29062	2.52229	2.70551	2.85744
10	1	-6.09133	-2.87980	-1.66163	-0.96677	-0.14390	0.36271	1.00254	1.41819	1.72407	1.96522
10	2	-4.07447	-1.82620	-0.91703	-0.37465	0.29463	0.72204	1.27820	1.64865	1.92547	2.14602
10	3	-3.04802	-1.26718	-0.50943	-0.04344	0.54664	0.93192	1.44232	1.78738	2.04761	2.25627
10	4	-2.34392	-0.86808	-0.21146	0.20231	0.73698	1.09211	1.56910	1.89531	2.14309	2.34274
10	5	-1.79331	-0.54361	0.03572	0.40867	0.89911	1.22970	1.67905	1.98945	2.22668	2.41867
10	6	-1.32597	-0.25745	0.25775	0.59605	1.04820	1.35716	1.78180	2.07787	2.30545	2.49039
10	7	-0.90256	0.01204	0.47055	0.77752	1.19434	1.48298	1.88406	2.16629	2.38450	2.56255
10	8	-0.49261	0.28369	0.68891	0.96567	1.34773	1.61601	1.99310	2.26105	2.46950	2.64033
10	9	-0.05825	0.58456	0.93550	1.18062	1.52535	1.77129	2.12161	2.37339	2.57067	2.73318
10	10	0.49533	0.98987	1.27601	1.48187	1.77877	1.99526	2.30940	2.53884	2.72049	2.87125

Table C8.2

VARIANCES AND COVARIANCES OF LOG-GAMMA ORDER STATISTICS

[When $r = s$, tabular values are variances of teh rth order statistic of a sample of size n from a log-Gamma population with location parameter 0, scale parameter 1, and shape parameter $\kappa = 0.5(0.5)2.0(1.0)4.0(2.0)12.0$; when $r \neq s$, they are the covariances of the rth and sth order statistics.]

Variances and Covariances of Log-Gamma Order Statistics
Shape Parameter κ

n	r	s	κ = 0.5	1.0	1.5	2.0	3.0	4.0	6.0	8.0	10.0	12.0
1	1	1	4.93480	1.64493	0.93480	0.64493	0.39493	0.28382	0.18132	0.13314	0.10517	0.08690
2	1	1	5.46327	1.64493	0.87896	0.58243	0.33927	0.23651	0.14570	0.10468	0.08147	0.06658
2	1	2	1.36012	0.48045	0.28050	0.19638	0.12208	0.08840	0.05689	0.04193	0.03319	0.02747
2	2	2	1.68609	0.68403	0.42965	0.31468	0.20644	0.15434	0.10316	0.07774	0.06248	0.05229
3	1	1	5.72305	1.64493	0.85222	0.55303	0.31377	0.21524	0.13007	0.09239	0.07134	0.05798
3	1	2	1.69021	0.54629	0.30419	0.20676	0.12403	0.08791	0.05517	0.04005	0.03138	0.02577
3	1	3	0.67166	0.24376	0.14352	0.10085	0.06290	0.04561	0.02940	0.02168	0.01717	0.01421
3	2	2	1.85783	0.65852	0.38688	0.27205	0.17001	0.12345	0.07970	0.05882	0.04661	0.03859
3	2	3	0.75230	0.30137	0.18747	0.13634	0.08860	0.06582	0.04364	0.03272	0.02620	0.02186
3	3	3	0.99518	0.44850	0.29496	0.22181	0.14996	0.11408	0.07779	0.05931	0.04805	0.04044
4	1	1	5.88131	1.64493	0.83573	0.53501	0.29832	0.20247	0.12080	0.08515	0.06542	0.05296
4	1	2	1.86835	0.57432	0.31077	0.20743	0.12172	0.08514	0.05260	0.03783	0.02946	0.02408
4	1	3	0.87229	0.29271	0.16565	0.11360	0.06881	0.04903	0.03095	0.02254	0.01770	0.01456
4	1	4	0.41735	0.15389	0.09093	0.06398	0.03993	0.02896	0.01867	0.01377	0.01090	0.00902
4	2	2	1.97432	0.65180	0.36889	0.25351	0.15413	0.11011	0.06973	0.05088	0.04000	0.03293
4	2	3	0.93069	0.33721	0.19988	0.14120	0.08863	0.06451	0.04173	0.03084	0.02445	0.02025
4	2	4	0.44843	0.17903	0.11087	0.08037	0.05200	0.03852	0.02545	0.01904	0.01522	0.01269
4	3	3	1.05808	0.41553	0.25636	0.18552	0.11981	0.08867	0.05851	0.04373	0.03495	0.02912
4	3	4	0.51782	0.22455	0.14477	0.10750	0.07154	0.05389	0.03631	0.02748	0.02215	0.01857
4	4	4	0.71563	0.34402	0.23248	0.17756	0.12221	0.09394	0.06484	0.04979	0.04053	0.03424
5	1	1	5.98885	1.64493	0.82427	0.52251	0.28767	0.19370	0.11449	0.08025	0.06142	0.04958
5	1	2	1.98346	0.58992	0.31291	0.20618	0.11909	0.08251	0.05041	0.03602	0.02792	0.02275
5	1	3	0.99183	0.31716	0.17496	0.11809	0.07017	0.04943	0.03078	0.02224	0.01737	0.01423
5	1	4	0.55197	0.18967	0.10826	0.07456	0.04536	0.03240	0.02050	0.01495	0.01175	0.00968
5	1	5	0.29250	0.10901	0.06453	0.04542	0.02835	0.02056	0.01325	0.00977	0.00773	0.00640
5	2	2	2.05761	0.64907	0.35818	0.24230	0.14455	0.10212	0.06381	0.04620	0.03613	0.02963
5	2	3	1.03511	0.35267	0.20268	0.14052	0.08627	0.06196	0.03947	0.02890	0.02277	0.01878
5	2	4	0.57839	0.21227	0.12630	0.08937	0.05619	0.04092	0.02649	0.01958	0.01553	0.01286
5	2	5	0.30749	0.12258	0.07567	0.05473	0.03530	0.02610	0.01721	0.01286	0.01027	0.00856
5	3	3	1.11797	0.40598	0.24124	0.17068	0.10733	0.07819	0.05064	0.03743	0.02969	0.02460
5	3	4	0.62960	0.24677	0.15186	0.10967	0.07062	0.05216	0.03433	0.02562	0.02045	0.01702
5	3	5	0.33688	0.14359	0.09169	0.06768	0.04471	0.03353	0.02247	0.01695	0.01363	0.01141
5	4	4	0.73310	0.30850	0.19625	0.14455	0.09527	0.07134	0.04772	0.03596	0.02889	0.02417
5	4	5	0.39750	0.18203	0.12015	0.09044	0.06115	0.04649	0.03166	0.02411	0.01951	0.01641
5	5	5	0.56703	0.28486	0.19607	0.15134	0.10545	0.08164	0.05683	0.04386	0.03583	0.03034
6	1	1	6.06705	1.64493	0.81572	0.51319	0.27974	0.18720	0.10983	0.07664	0.05848	0.04711
6	1	2	2.06502	0.59986	0.31344	0.20450	0.11669	0.08026	0.04861	0.03456	0.02669	0.02170
6	1	3	1.07393	0.33205	0.17978	0.11992	0.07025	0.04906	0.03025	0.02173	0.01690	0.01380
6	1	4	0.63722	0.20925	0.11670	0.07924	0.04738	0.03350	0.02094	0.01517	0.01186	0.00973
6	1	5	0.38978	0.13619	0.07814	0.05395	0.03290	0.02353	0.01491	0.01088	0.00856	0.00705
6	1	6	0.22056	0.08285	0.04910	0.03455	0.02156	0.01563	0.01007	0.00742	0.00587	0.00486
6	2	2	2.12021	0.64770	0.35080	0.23451	0.13793	0.09661	0.05977	0.04302	0.03351	0.02740
6	2	3	1.10723	0.36146	0.20312	0.13890	0.08390	0.05968	0.03759	0.02734	0.02144	0.01762
6	2	4	0.65876	0.22888	0.13257	0.09230	0.05692	0.04099	0.02618	0.01920	0.01514	0.01249
6	2	5	0.40375	0.14945	0.08908	0.06308	0.03967	0.02890	0.01871	0.01383	0.01097	0.00909
6	2	6	0.22886	0.09116	0.05614	0.04052	0.02608	0.01925	0.01267	0.00946	0.00755	0.00629
6	3	3	1.16672	0.40186	0.23258	0.16196	0.09995	0.07201	0.04603	0.03376	0.02664	0.02199
6	3	4	0.69753	0.25617	0.15287	0.10839	0.06831	0.04981	0.03229	0.02388	0.01895	0.01570
6	3	5	0.42909	0.16806	0.10323	0.07444	0.04784	0.03529	0.02319	0.01729	0.01379	0.01147
6	3	6	0.24400	0.10292	0.06531	0.04802	0.03158	0.02361	0.01577	0.01187	0.00953	0.00797
6	4	4	0.76558	0.29762	0.18255	0.13157	0.08452	0.06234	0.04095	0.03052	0.02435	0.02025
6	4	5	0.47404	0.19671	0.12419	0.09104	0.05964	0.04450	0.02963	0.02226	0.01785	0.01491
6	4	6	0.27113	0.12122	0.07907	0.05909	0.03961	0.02995	0.02027	0.01538	0.01242	0.01042
6	5	5	0.56190	0.24855	0.16154	0.12047	0.08054	0.06081	0.04107	0.03113	0.02511	0.02106
6	5	6	0.32515	0.15497	0.10405	0.07910	0.05409	0.04140	0.02841	0.02174	0.01765	0.01488

Variances and Covariances of Log-Gamma Order Statistics
Shape Parameter κ

			κ									
n	r	s	0.5	1.0	1.5	2.0	3.0	4.0	6.0	8.0	10.0	12.0
6	6	6	0.47523	0.24658	0.17202	0.13383	0.09410	0.07325	0.05131	0.03975	0.03255	0.02762
7	1	1	6.12663	1.64493	0.80902	0.50588	0.27354	0.18213	0.10620	0.07385	0.05621	0.04520
7	1	2	2.12622	0.60675	0.31332	0.20281	0.11459	0.07835	0.04712	0.03336	0.02570	0.02084
7	1	3	1.13463	0.34211	0.18256	0.12065	0.06988	0.04848	0.02965	0.02120	0.01643	0.01339
7	1	4	0.69833	0.22190	0.12163	0.08169	0.04822	0.03382	0.02095	0.01509	0.01176	0.00962
7	1	5	0.45384	0.15195	0.08535	0.05816	0.03491	0.02473	0.01550	0.01124	0.00880	0.00722
7	1	6	0.29470	0.10427	0.06003	0.04151	0.02535	0.01814	0.01150	0.00840	0.00661	0.00544
7	1	7	0.17465	0.06601	0.03914	0.02754	0.01718	0.01245	0.00801	0.00591	0.00467	0.00387
7	2	2	2.16908	0.64691	0.34529	0.22867	0.13297	0.09251	0.05678	0.04068	0.03158	0.02577
7	2	3	1.16107	0.36717	0.20278	0.13718	0.08180	0.05775	0.03605	0.02608	0.02038	0.01671
7	2	4	0.71599	0.23906	0.13570	0.09332	0.05672	0.04049	0.02560	0.01866	0.01466	0.01206
7	2	5	0.46597	0.16412	0.09550	0.06664	0.04120	0.02970	0.01900	0.01394	0.01100	0.00908
7	2	6	0.30291	0.11283	0.06731	0.04767	0.02998	0.02183	0.01413	0.01045	0.00828	0.00686
7	2	7	0.17970	0.07155	0.04397	0.03169	0.02035	0.01501	0.00986	0.00736	0.00587	0.00489
7	3	3	1.20635	0.39969	0.22673	0.15598	0.09488	0.06777	0.04289	0.03128	0.02458	0.02023
7	3	4	0.74642	0.26155	0.15255	0.10669	0.06616	0.04779	0.03063	0.02251	0.01778	0.01468
7	3	5	0.48696	0.18017	0.10775	0.07647	0.04823	0.03519	0.02282	0.01688	0.01339	0.01110
7	3	6	0.31717	0.12419	0.07616	0.05485	0.03520	0.02594	0.01702	0.01268	0.01011	0.00841
7	3	7	0.18849	0.07893	0.04986	0.03655	0.02395	0.01787	0.01191	0.00895	0.00718	0.00600
7	4	4	0.79591	0.29271	0.17491	0.12412	0.07828	0.05711	0.03704	0.02740	0.02174	0.01802
7	4	5	0.52134	0.20262	0.12416	0.08941	0.05736	0.04226	0.02773	0.02065	0.01646	0.01369
7	4	6	0.34068	0.14020	0.08809	0.06438	0.04202	0.03127	0.02076	0.01558	0.01248	0.01041
7	4	7	0.20307	0.08941	0.05787	0.04304	0.02869	0.02162	0.01457	0.01103	0.00889	0.00746
7	5	5	0.57927	0.23701	0.14872	0.10862	0.07085	0.05271	0.03498	0.02624	0.02101	0.01753
7	5	6	0.38067	0.16497	0.10615	0.07868	0.05220	0.03923	0.02635	0.01990	0.01601	0.01341
7	5	7	0.22812	0.10578	0.07012	0.05289	0.03583	0.02727	0.01859	0.01417	0.01147	0.00965
7	6	6	0.45755	0.21021	0.13884	0.10451	0.07063	0.05367	0.03651	0.02779	0.02248	0.01890
7	6	7	0.27706	0.13618	0.09264	0.07096	0.04896	0.03767	0.02601	0.01998	0.01626	0.01373
7	7	7	0.41288	0.21964	0.15484	0.12120	0.08583	0.06709	0.04724	0.03671	0.03012	0.02560
8	1	1	6.17360	1.64493	0.80360	0.49995	0.26852	0.17802	0.10327	0.07159	0.05438	0.04366
8	1	2	2.17402	0.61182	0.31289	0.20121	0.11276	0.07672	0.04587	0.03236	0.02487	0.02014
8	1	3	1.18167	0.34937	0.18425	0.12085	0.06935	0.04784	0.02907	0.02070	0.01601	0.01302
8	1	4	0.74497	0.23081	0.12478	0.08308	0.04853	0.03383	0.02080	0.01492	0.01159	0.00946
8	1	5	0.50115	0.16256	0.08984	0.06061	0.03595	0.02528	0.01571	0.01133	0.00884	0.00724
8	1	6	0.34482	0.11719	0.06615	0.04519	0.02720	0.01929	0.01211	0.00879	0.00688	0.00565
8	1	7	0.23341	0.08340	0.04814	0.03332	0.02037	0.01458	0.00925	0.00676	0.00532	0.00438
8	1	8	0.14320	0.05440	0.03227	0.02270	0.01415	0.01025	0.00660	0.00486	0.00385	0.00318
8	2	2	2.20836	0.64642	0.34095	0.22406	0.12907	0.08929	0.05445	0.03885	0.03009	0.02450
8	2	3	1.20318	0.37119	0.20215	0.13556	0.07999	0.05612	0.03478	0.02505	0.01952	0.01597
8	2	4	0.75966	0.24600	0.13742	0.09357	0.05621	0.03985	0.02500	0.01813	0.01420	0.01165
8	2	5	0.51157	0.17362	0.09918	0.06844	0.04175	0.02987	0.01893	0.01382	0.01086	0.00894
8	2	6	0.35227	0.12534	0.07315	0.05112	0.03165	0.02283	0.01462	0.01073	0.00847	0.00700
8	2	7	0.23860	0.08931	0.05331	0.03775	0.02373	0.01728	0.01118	0.00826	0.00655	0.00543
8	2	8	0.14648	0.05832	0.03577	0.02575	0.01651	0.01216	0.00798	0.00595	0.00475	0.00395
8	3	3	1.23902	0.39840	0.22242	0.15151	0.09109	0.06462	0.04057	0.02944	0.02306	0.01893
8	3	4	0.78423	0.26509	0.15186	0.10506	0.06431	0.04610	0.02929	0.02141	0.01685	0.01388
8	3	5	0.52905	0.18759	0.10992	0.07706	0.04791	0.03465	0.02225	0.01636	0.01293	0.01068
8	3	6	0.36480	0.13570	0.08125	0.05769	0.03639	0.02655	0.01722	0.01274	0.01011	0.00838
8	3	7	0.24738	0.09685	0.05930	0.04267	0.02734	0.02013	0.01320	0.00983	0.00783	0.00651
8	3	8	0.15203	0.06332	0.03986	0.02915	0.01905	0.01419	0.00944	0.00709	0.00568	0.00474
8	4	4	0.82238	0.29005	0.16984	0.11907	0.07404	0.05356	0.03440	0.02530	0.02000	0.01652
8	4	5	0.55633	0.20599	0.12339	0.08767	0.05536	0.04042	0.02623	0.01941	0.01540	0.01277
8	4	6	0.38444	0.14941	0.09146	0.06581	0.04217	0.03105	0.02036	0.01515	0.01207	0.01004
8	4	7	0.26117	0.10686	0.06691	0.04878	0.03175	0.02359	0.01563	0.01171	0.00937	0.00782
8	4	8	0.16078	0.07001	0.04506	0.03340	0.02217	0.01667	0.01120	0.00846	0.00681	0.00571

Variances and Covariances of Log-Gamma Order Statistics
Shape Parameter κ

			κ									
n	r	s	0.5	1.0	1.5	2.0	3.0	4.0	6.0	8.0	10.0	12.0
8	5	5	0.59873	0.23166	0.14170	0.10193	0.06531	0.04808	0.03152	0.02346	0.01869	0.01554
8	5	6	0.41517	0.16868	0.10543	0.07681	0.04994	0.03709	0.02455	0.01839	0.01471	0.01226
8	5	7	0.28288	0.12103	0.07738	0.05712	0.03772	0.02827	0.01892	0.01426	0.01146	0.00958
8	5	8	0.17464	0.07953	0.05226	0.03922	0.02641	0.02003	0.01359	0.01033	0.00835	0.00702
8	6	6	0.46581	0.19837	0.12666	0.09344	0.06166	0.04619	0.03090	0.02328	0.01870	0.01564
8	6	7	0.31896	0.14306	0.09342	0.06983	0.04680	0.03538	0.02392	0.01814	0.01464	0.01228
8	6	8	0.19788	0.09447	0.06341	0.04818	0.03292	0.02518	0.01727	0.01320	0.01072	0.00903
8	7	7	0.38770	0.18355	0.12278	0.09311	0.06346	0.04846	0.03317	0.02533	0.02054	0.01730
8	7	8	0.24283	0.12233	0.08409	0.06481	0.04503	0.03479	0.02415	0.01860	0.01517	0.01283
8	8	8	0.36771	0.19956	0.14188	0.11160	0.07950	0.06236	0.04409	0.03434	0.02823	0.02402
9	1	1	6.21164	1.64493	0.79910	0.49502	0.26434	0.17461	0.10085	0.06972	0.05287	0.04239
9	1	2	2.21247	0.61569	0.31234	0.19975	0.11116	0.07530	0.04480	0.03151	0.02417	0.01954
9	1	3	1.21933	0.35488	0.18533	0.12078	0.06876	0.04721	0.02853	0.02025	0.01562	0.01269
9	1	4	0.78202	0.23745	0.12692	0.08390	0.04858	0.03368	0.02059	0.01471	0.01140	0.00929
9	1	5	0.53813	0.17027	0.09288	0.06215	0.03650	0.02552	0.01575	0.01132	0.00880	0.00719
9	1	6	0.38266	0.12615	0.07014	0.04746	0.02824	0.01990	0.01239	0.00895	0.00699	0.00572
9	1	7	0.27387	0.09419	0.05337	0.03652	0.02202	0.01564	0.00982	0.00713	0.00559	0.00459
9	1	8	0.19117	0.06886	0.03982	0.02758	0.01687	0.01208	0.00767	0.00560	0.00441	0.00363
9	1	9	0.12051	0.04597	0.02728	0.01919	0.01195	0.00865	0.00557	0.00410	0.00324	0.00268
9	2	2	2.24064	0.64609	0.33742	0.22031	0.12590	0.08668	0.05256	0.03738	0.02888	0.02348
9	2	3	1.23720	0.37418	0.20142	0.13408	0.07842	0.05473	0.03371	0.02419	0.01881	0.01536
9	2	4	0.79441	0.25105	0.13839	0.09347	0.05561	0.03920	0.02442	0.01764	0.01378	0.01129
9	2	5	0.54710	0.18033	0.10149	0.06940	0.04188	0.02977	0.01873	0.01361	0.01067	0.00876
9	2	6	0.38927	0.13377	0.07675	0.05308	0.03246	0.02325	0.01476	0.01078	0.00848	0.00699
9	2	7	0.27874	0.09998	0.05847	0.04089	0.02534	0.01829	0.01171	0.00860	0.00679	0.00561
9	2	8	0.19465	0.07315	0.04367	0.03091	0.01943	0.01414	0.00915	0.00676	0.00536	0.00444
9	2	9	0.12275	0.04887	0.02993	0.02152	0.01378	0.01014	0.00665	0.00495	0.00395	0.00329
9	3	3	1.26639	0.39758	0.21904	0.14799	0.08812	0.06215	0.03875	0.02802	0.02189	0.01793
9	3	4	0.81471	0.26763	0.15106	0.10357	0.06273	0.04469	0.02819	0.02051	0.01610	0.01323
9	3	5	0.56184	0.19267	0.11104	0.07708	0.04736	0.03403	0.02167	0.01586	0.01249	0.01030
9	3	6	0.40016	0.14315	0.08412	0.05906	0.03678	0.02662	0.01711	0.01259	0.00995	0.00823
9	3	7	0.28678	0.10712	0.06417	0.04556	0.02875	0.02097	0.01360	0.01006	0.00798	0.00662
9	3	8	0.20041	0.07846	0.04798	0.03448	0.02207	0.01624	0.01064	0.00792	0.00631	0.00524
9	3	9	0.12647	0.05246	0.03292	0.02403	0.01567	0.01166	0.00774	0.00581	0.00465	0.00388
9	4	4	0.84526	0.28844	0.16615	0.11534	0.07089	0.05093	0.03245	0.02375	0.01871	0.01543
9	4	5	0.58411	0.20823	0.12250	0.08610	0.05369	0.03890	0.02502	0.01842	0.01457	0.01204
9	4	6	0.41667	0.15503	0.09300	0.06612	0.04178	0.03050	0.01980	0.01465	0.01163	0.00964
9	4	7	0.29899	0.11620	0.07106	0.05109	0.03271	0.02407	0.01577	0.01173	0.00934	0.00777
9	4	8	0.20918	0.08522	0.05320	0.03872	0.02515	0.01866	0.01235	0.00924	0.00739	0.00616
9	4	9	0.13214	0.05706	0.03655	0.02702	0.01787	0.01341	0.00899	0.00679	0.00546	0.00457
9	5	5	0.61704	0.22869	0.13710	0.09746	0.06158	0.04496	0.02919	0.02160	0.01715	0.01421
9	5	6	0.44120	0.17074	0.10438	0.07505	0.04805	0.03536	0.02316	0.01723	0.01372	0.01141
9	5	7	0.31720	0.12825	0.07994	0.05813	0.03771	0.02796	0.01848	0.01382	0.01105	0.00921
9	5	8	0.22229	0.09424	0.05996	0.04414	0.02904	0.02172	0.01450	0.01091	0.00876	0.00732
9	5	9	0.14066	0.06322	0.04127	0.03085	0.02068	0.01564	0.01058	0.00803	0.00648	0.00544
9	6	6	0.47837	0.19275	0.12005	0.08727	0.05660	0.04196	0.02773	0.02074	0.01658	0.01382
9	6	7	0.34495	0.14525	0.09223	0.06781	0.04455	0.03329	0.02220	0.01669	0.01339	0.01119
9	6	8	0.24239	0.10703	0.06938	0.05163	0.03441	0.02593	0.01746	0.01321	0.01064	0.00892
9	6	9	0.15379	0.07199	0.04788	0.03619	0.02457	0.01872	0.01278	0.00975	0.00789	0.00664
9	7	7	0.39011	0.17158	0.11109	0.08261	0.05502	0.04144	0.02790	0.02110	0.01699	0.01424
9	7	8	0.27534	0.12700	0.08393	0.06318	0.04268	0.03242	0.02204	0.01677	0.01356	0.01140
9	7	9	0.17549	0.08580	0.05818	0.04447	0.03060	0.02350	0.01619	0.01242	0.01010	0.00852
9	8	8	0.33783	0.16390	0.11078	0.08451	0.05801	0.04448	0.03059	0.02343	0.01904	0.01606
9	8	9	0.21722	0.11167	0.07743	0.05997	0.04192	0.03250	0.02265	0.01749	0.01429	0.01210
9	9	9	0.33341	0.18395	0.13170	0.10402	0.07446	0.05858	0.04156	0.03244	0.02671	0.02275

Variances and Covariances of Log-Gamma Order Statistics
Shape Parameter κ

			κ									
n	r	s	0.5	1.0	1.5	2.0	3.0	4.0	6.0	8.0	10.0	12.0
10	1	1	6.24308	1.64493	0.79528	0.49083	0.26078	0.17171	0.09879	0.06814	0.05159	0.04132
10	1	2	2.24412	0.61876	0.31174	0.19841	0.10974	0.07407	0.04387	0.03078	0.02356	0.01903
10	1	3	1.25026	0.35919	0.18603	0.12056	0.06817	0.04662	0.02803	0.01984	0.01528	0.01239
10	1	4	0.81229	0.24260	0.12843	0.08438	0.04849	0.03347	0.02035	0.01449	0.01121	0.00912
10	1	5	0.56808	0.17615	0.09505	0.06317	0.03679	0.02560	0.01570	0.01124	0.00873	0.00712
10	1	6	0.41276	0.13282	0.07294	0.04898	0.02888	0.02023	0.01252	0.00901	0.00701	0.00573
10	1	7	0.30491	0.10185	0.05689	0.03859	0.02302	0.01624	0.01013	0.00732	0.00572	0.00469
10	1	8	0.22465	0.07803	0.04435	0.03039	0.01834	0.01303	0.00819	0.00595	0.00467	0.00383
10	1	9	0.16060	0.05824	0.03373	0.02338	0.01430	0.01024	0.00650	0.00475	0.00374	0.00308
10	1	10	0.10347	0.03962	0.02351	0.01653	0.01029	0.00745	0.00479	0.00353	0.00279	0.00231
10	2	2	2.26768	0.64586	0.33448	0.21716	0.12325	0.08450	0.05099	0.03616	0.02789	0.02264
10	2	3	1.26535	0.37650	0.20067	0.13274	0.07705	0.05353	0.03279	0.02346	0.01820	0.01484
10	2	4	0.82288	0.25489	0.13895	0.09320	0.05498	0.03857	0.02389	0.01720	0.01340	0.01096
10	2	5	0.57585	0.18536	0.10302	0.06991	0.04181	0.02956	0.01848	0.01338	0.01046	0.00858
10	2	6	0.41861	0.13991	0.07916	0.05428	0.03286	0.02340	0.01475	0.01073	0.00842	0.00692
10	2	7	0.30936	0.10738	0.06180	0.04281	0.02622	0.01880	0.01194	0.00873	0.00687	0.00566
10	2	8	0.22800	0.08232	0.04821	0.03374	0.02091	0.01510	0.00967	0.00711	0.00561	0.00463
10	2	9	0.16303	0.06148	0.03670	0.02597	0.01632	0.01187	0.00768	0.00567	0.00450	0.00373
10	2	10	0.10507	0.04184	0.02559	0.01838	0.01175	0.00864	0.00566	0.00422	0.00336	0.00280
10	3	3	1.28963	0.39702	0.21630	0.14513	0.08569	0.06014	0.03729	0.02686	0.02094	0.01713
10	3	4	0.83996	0.26954	0.15025	0.10223	0.06136	0.04348	0.02726	0.01976	0.01547	0.01269
10	3	5	0.58844	0.19637	0.11163	0.07685	0.04676	0.03340	0.02113	0.01540	0.01210	0.00995
10	3	6	0.42810	0.14842	0.08590	0.05976	0.03681	0.02648	0.01689	0.01237	0.00975	0.00804
10	3	7	0.31657	0.11403	0.06714	0.04718	0.02941	0.02130	0.01370	0.01008	0.00797	0.00659
10	3	8	0.23343	0.08749	0.05242	0.03722	0.02348	0.01712	0.01110	0.00821	0.00652	0.00540
10	3	9	0.16700	0.06538	0.03993	0.02868	0.01833	0.01348	0.00882	0.00656	0.00523	0.00434
10	3	10	0.10768	0.04453	0.02787	0.02031	0.01321	0.00982	0.00651	0.00488	0.00391	0.00326
10	4	4	0.86511	0.28739	0.16328	0.11241	0.06842	0.04888	0.03093	0.02255	0.01772	0.01458
10	4	5	0.60701	0.20986	0.12161	0.08471	0.05227	0.03764	0.02403	0.01762	0.01389	0.01146
10	4	6	0.44214	0.15888	0.09374	0.06599	0.04122	0.02989	0.01925	0.01418	0.01122	0.00928
10	4	7	0.32726	0.12221	0.07337	0.05217	0.03297	0.02408	0.01563	0.01157	0.00918	0.00761
10	4	8	0.24152	0.09387	0.05735	0.04120	0.02635	0.01938	0.01269	0.00944	0.00751	0.00624
10	4	9	0.17290	0.07021	0.04373	0.03177	0.02060	0.01527	0.01009	0.00755	0.00603	0.00503
10	4	10	0.11156	0.04786	0.03054	0.02252	0.01486	0.01113	0.00745	0.00562	0.00452	0.00378
10	5	5	0.63359	0.22686	0.13379	0.09417	0.05882	0.04266	0.02748	0.02024	0.01601	0.01324
10	5	6	0.46230	0.17211	0.10335	0.07353	0.04649	0.03396	0.02205	0.01633	0.01296	0.01074
10	5	7	0.34265	0.13262	0.08103	0.05823	0.03726	0.02740	0.01794	0.01334	0.01062	0.00883
10	5	8	0.25317	0.10200	0.06342	0.04605	0.02982	0.02209	0.01458	0.01090	0.00871	0.00725
10	5	9	0.18144	0.07639	0.04842	0.03556	0.02333	0.01742	0.01161	0.00873	0.00700	0.00585
10	5	10	0.11719	0.05213	0.03386	0.02523	0.01685	0.01272	0.00858	0.00650	0.00524	0.00440
10	6	6	0.49128	0.18958	0.11577	0.08319	0.05322	0.03914	0.02562	0.01905	0.01517	0.01261
10	6	7	0.36488	0.14641	0.09097	0.06603	0.04274	0.03165	0.02089	0.01561	0.01247	0.01038
10	6	8	0.27006	0.11282	0.07134	0.05232	0.03427	0.02556	0.01701	0.01277	0.01024	0.00855
10	6	9	0.19386	0.08463	0.05456	0.04046	0.02686	0.02019	0.01356	0.01024	0.00824	0.00690
10	6	10	0.12541	0.05785	0.03822	0.02876	0.01943	0.01477	0.01005	0.00765	0.00619	0.00520
10	7	7	0.39805	0.16581	0.10479	0.07682	0.05031	0.03752	0.02496	0.01874	0.01502	0.01254
10	7	8	0.29540	0.12812	0.08240	0.06103	0.04045	0.03038	0.02038	0.01537	0.01236	0.01035
10	7	9	0.21256	0.09635	0.06317	0.04732	0.03178	0.02406	0.01629	0.01236	0.00998	0.00838
10	7	10	0.13786	0.06603	0.04436	0.03372	0.02305	0.01763	0.01209	0.00925	0.00751	0.00633
10	8	8	0.33629	0.15192	0.09948	0.07448	0.04999	0.03782	0.02559	0.01941	0.01567	0.01315
10	8	9	0.24297	0.11471	0.07656	0.05797	0.03942	0.03006	0.02054	0.01567	0.01269	0.01068
10	8	10	0.15825	0.07893	0.05398	0.04147	0.02870	0.02212	0.01531	0.01177	0.00959	0.00810
10	9	9	0.30048	0.14879	0.10144	0.07778	0.05370	0.04132	0.02854	0.02191	0.01783	0.01506
10	9	10	0.19733	0.10319	0.07207	0.05606	0.03938	0.03063	0.02142	0.01658	0.01356	0.01150
10	10	10	0.30642	0.17143	0.12347	0.09787	0.07035	0.05548	0.03948	0.03088	0.02545	0.02170

Table C9.1

EXPECTED VALUES OF CAUCHY ORDER STATISTICS

[Tabular values are expected values of the Mth largest value in a sample of size N from a Cauchy population with location parameter 0 and scale parameter 1.]

Expected Values of Cauchy Order Statistics

N	*M*	*Mean*	*N*	*M*	*Mean*	*N*	*M*	*Mean*	*N*	*M*	*Mean*
3	2	0.0000	11	4	0.7680	15	4	1.3099	18	5	1.1357
4	2	0.6978	11	5	0.3421	15	5	0.8298	18	6	0.7799
5	2	1.1630	11	6	0.0000	15	6	0.4998	18	7	0.5134
5	3	0.0000	12	2	3.6518	15	7	0.2366	18	8	0.2930
6	2	1.5637	12	3	1.6502	15	8	0.0000	18	9	0.0954
6	3	0.3618	12	4	0.9126	16	2	4.9667	19	2	5.9413
7	2	1.9359	12	5	0.4790	16	3	2.3538	19	3	2.8620
7	3	0.6331	12	6	0.1505	16	4	1.4348	19	4	1.7966
7	4	0.0000	13	2	3.9830	16	5	0.9352	19	5	1.2322
8	2	2.2928	13	3	1.8303	16	6	0.5981	19	6	0.8655
8	3	0.8654	13	4	1.0499	16	7	0.3361	19	7	0.5945
8	4	0.2459	13	5	0.6037	16	8	0.1086	19	8	0.3743
9	2	2.6402	13	6	0.2795	17	2	5.2923	19	9	0.1812
9	3	1.0768	13	7	0.0000	17	3	2.5245	19	10	0.0000
9	4	0.4425	14	2	4.3123	17	4	1.5573	20	2	6.2648
9	5	0.0000	14	3	2.0072	17	5	1.0369	20	3	3.0293
10	2	2.9814	14	4	1.1819	17	6	0.6911	20	4	1.9140
10	3	1.2755	14	5	0.7198	17	7	0.4277	20	5	1.3268
10	4	0.6132	14	6	0.3945	17	8	0.2052	20	6	0.9484
10	5	0.1866	14	7	0.1262	17	9	0.0000	20	7	0.6720
11	2	3.3182	15	2	4.6401	18	2	5.6171	20	8	0.4506
11	3	1.4658	15	3	2.1815	18	3	2.6938	20	9	0.2599
						18	4	1.6778	20	10	0.0850

Table C9.2

VARIANCES AND COVARIANCES OF CAUCHY ORDER STATISTICS

[When $I = J$, tabular values are variances of the Ith order statistic of a sample of size N from a Cauchy population with location parameter 0 and scale parameter 1; when $I \neq J$, they are covariances of the Ith and Jth order statistics.]

Variances and Covariances of Cauchy Order Statistics

N	I	J	COV	N	I	J	COV	N	I	J	COV	N	I	J	COV
5	3	3	1.2213	11	3	9	0.3959	13	4	5	0.5051	14	6	7	0.2206
6	3	3	1.0904	11	4	4	0.7293	13	4	6	0.3341	14	6	8	0.1875
6	3	4	0.6309	11	4	5	0.4103	13	4	7	0.2569	14	6	9	0.1753
7	3	3	1.2774	11	4	6	0.2937	13	4	8	0.2234	14	7	7	0.2297
7	3	4	0.5713	11	4	7	0.2521	13	4	9	0.2183	14	7	8	0.1981
7	3	5	0.4732	11	4	8	0.2566	13	4	10	0.2424	15	3	3	5.2472
7	4	4	0.6121	11	5	5	0.3765	13	5	5	0.4298	15	3	4	1.7460
8	3	3	1.5689	11	5	6	0.2775	13	5	6	0.2894	15	3	5	0.8914
8	3	4	0.6206	11	5	7	0.2436	13	5	7	0.2255	15	3	6	0.5614
8	3	5	0.4204	11	6	6	0.3068	13	5	8	0.1982	15	3	7	0.4056
8	3	6	0.4156	12	3	3	3.3524	13	5	9	0.1953	15	3	8	0.3255
8	4	4	0.5516	12	3	4	1.1442	13	6	6	0.2811	15	3	9	0.2859
8	4	5	0.4035	12	3	5	0.6096	13	6	7	0.2231	15	3	10	0.2736
9	3	3	1.9297	12	3	6	0.4100	13	6	8	0.1990	15	3	11	0.2864
9	3	4	0.7142	12	3	7	0.3256	13	7	7	0.2456	15	3	12	0.3339
9	3	5	0.4343	12	3	8	0.2986	14	3	3	4.5642	15	3	13	0.4553
9	3	6	0.3590	12	3	9	0.3173	14	3	4	1.5277	15	4	4	1.2245
9	3	7	0.3946	12	3	10	0.4066	14	3	5	0.7878	15	4	5	0.6324
9	4	4	0.5790	12	4	4	0.8325	14	3	6	0.5039	15	4	6	0.4017
9	4	5	0.3706	12	4	5	0.4530	14	3	7	0.3722	15	4	7	0.2921
9	4	6	0.3182	12	4	6	0.3095	14	3	8	0.3080	15	4	8	0.2356
9	5	5	0.4087	12	4	7	0.2488	14	3	9	0.2821	15	4	9	0.2078
10	3	3	2.3496	12	4	8	0.2303	14	3	10	0.2859	15	4	10	0.1995
10	3	4	0.8360	12	4	9	0.2464	14	3	11	0.3257	15	4	11	0.2094
10	3	5	0.4767	12	5	5	0.3975	14	3	12	0.4371	15	4	12	0.2446
10	3	6	0.3561	12	5	6	0.2777	14	4	4	1.0811	15	5	5	0.5176
10	3	7	0.3314	12	5	7	0.2272	14	4	5	0.5652	15	5	6	0.3326
10	3	8	0.3907	12	5	8	0.2131	14	4	6	0.3653	15	5	7	0.2441
10	4	4	0.6430	12	6	6	0.2842	14	4	7	0.2719	15	5	8	0.1983
10	4	5	0.3803	12	6	7	0.2379	14	4	8	0.2264	15	5	9	0.1760
10	4	6	0.2920	13	3	3	3.9325	14	4	9	0.2084	15	5	10	0.1697
10	4	7	0.2773	13	3	4	1.3269	14	4	10	0.2120	15	5	11	0.1788
10	5	5	0.3738	13	3	5	0.6937	14	4	11	0.2422	15	6	6	0.3034
10	5	6	0.2986	13	3	6	0.4531	14	5	5	0.4704	15	6	7	0.2253
11	3	3	2.8244	13	3	7	0.3450	14	5	6	0.3083	15	6	8	0.1848
11	3	4	0.9802	13	3	8	0.2970	14	5	7	0.2321	15	6	9	0.1652
11	3	5	0.5366	13	3	9	0.2893	14	5	8	0.1950	15	6	10	0.1603
11	3	6	0.3763	13	3	10	0.3198	14	5	9	0.1807	15	7	7	0.2254
11	3	7	0.3180	13	3	11	0.4207	14	5	10	0.1848	15	7	8	0.1871
11	3	8	0.3200	13	4	4	0.9501	14	6	6	0.2888	15	7	9	0.1689
												15	8	8	0.2048

Table C10.1

EXPECTED VALUES OF DOUBLE EXPONENTIAL ORDER STATISTICS

[Tabular values are expected values of the Mth largest value in a sample of size N from a double exponential population with location parameter 0 and scale parameter 1.]

Expected Values of Double Exponential Order Statistics

N	M	$Mean$	N	M	$Mean$	N	M	$Mean$	N	M	$Mean$
1	1	0.00000	10	3	0.73412	14	6	0.26714	18	2	1.80196
2	1	0.75000	10	4	0.39447	14	7	0.08639	18	3	1.30196
3	1	1.12500	10	5	0.12461	15	1	2.62508	18	4	0.96859
3	2	0.00000	11	1	2.32672	15	2	1.62508	18	5	0.71843
4	1	1.38542	11	2	1.32663	15	3	1.12504	18	6	0.51782
4	2	0.34375	11	3	0.82592	15	4	0.79147	18	7	0.34926
5	1	1.58854	11	4	0.48933	15	5	0.54048	18	8	0.20154
5	2	0.57292	11	5	0.22845	15	6	0.33724	18	9	0.06586
5	3	0.00000	11	6	0.00000	15	7	0.16198	19	1	2.85459
6	1	1.75625	12	1	2.41006	15	8	0.00000	19	2	1.85459
6	2	0.75000	12	2	1.41002	16	1	2.68758	19	3	1.35459
6	3	0.21875	12	3	0.90968	16	2	1.68758	19	4	1.02124
7	1	1.89948	12	4	0.57466	16	3	1.18756	19	5	0.77115
7	2	0.89688	12	5	0.31868	16	4	0.85410	19	6	0.57080
7	3	0.38281	12	6	0.10213	16	5	0.60356	19	7	0.40301
7	4	0.00000	13	1	2.48698	16	6	0.40169	19	8	0.25711
8	1	2.02462	13	2	1.48696	16	7	0.22981	19	9	0.12513
8	2	1.02350	13	3	0.98680	16	8	0.07477	19	10	0.00000
8	3	0.51699	13	4	0.65259	17	1	2.74640	20	1	2.90459
8	4	0.15918	13	5	0.39930	17	2	1.74640	20	2	1.90459
9	1	2.13578	13	6	0.18967	17	3	1.24640	20	3	1.40459
9	2	1.13530	13	7	0.00000	17	4	0.91300	20	4	1.07125
9	3	0.63223	14	1	2.55841	17	5	0.66270	20	5	0.82120
9	4	0.28652	14	2	1.55840	17	6	0.46163	20	6	0.62100
9	5	0.00000	14	3	1.05832	17	7	0.29181	20	7	0.45367
10	1	2.23581	14	4	0.72454	17	8	0.14124	20	8	0.30893
10	2	1.23559	14	5	0.47273	17	9	0.00000	20	9	0.17940
						18	1	2.80196	20	10	0.05881

Table C10.2

VARIANCES AND COVARIANCES OF DOUBLE EXPONENTIAL ORDER STATISTICS

[When $I = J$, tabular values are variances of the Ith order statistic of a sample of size N from a double exponential population with location parameter 0 and scale parameter 1; when $I \neq J$, they are covariances of the Ith and Jth order statistics.]

Variances and Covariances of Double Exponential Order Statistics

N	I	J	COV	N	I	J	COV	N	I	J	COV	N	I	J	COV
1	1	1	2.00000	7	3	3	0.29128	9	4	5	0.15065	11	2	4	0.20014
2	1	1	1.47350	7	3	4	0.19676	9	4	6	0.12757	11	2	5	0.14278
2	1	2	0.56250	7	3	5	0.16033	9	5	5	0.17507	11	2	6	0.11174
3	1	1	1.41493	7	4	4	0.23565	10	1	1	1.54984	11	2	7	0.09626
3	1	2	0.47222	8	1	1	1.52785	10	1	2	0.55013	11	2	8	0.08974
3	1	3	0.32118	8	1	2	0.52913	10	1	3	0.30148	11	2	9	0.08758
3	2	2	0.63889	8	1	3	0.28401	10	1	4	0.19429	11	2	10	0.08707
4	1	1	1.44173	8	1	4	0.18429	10	1	5	0.13992	11	3	3	0.30984
4	1	2	0.46821	8	1	5	0.14048	10	1	6	0.11266	11	3	4	0.20081
4	1	3	0.27485	8	1	6	0.12356	10	1	7	0.10066	11	3	5	0.14337
4	1	4	0.23188	8	1	7	0.11861	10	1	8	0.09644	11	3	6	0.11228
4	2	2	0.52072	8	1	8	0.11768	10	1	9	0.09537	11	3	7	0.09678
4	2	3	0.31955	8	2	2	0.53073	10	1	10	0.09519	11	3	8	0.09025
5	1	1	1.47025	8	2	3	0.28523	10	2	2	0.55047	11	3	9	0.08809
5	1	2	0.48257	8	2	4	0.18533	10	2	3	0.30174	11	4	4	0.20369
5	1	3	0.26403	8	2	5	0.14145	10	2	4	0.19450	11	4	5	0.14587
5	1	4	0.19948	8	2	6	0.12450	10	2	5	0.14011	11	4	6	0.11458
5	1	5	0.18471	8	2	7	0.11955	10	2	6	0.11284	11	4	7	0.09899
5	2	2	0.50246	8	3	3	0.29186	10	2	7	0.10084	11	4	8	0.09242
5	2	3	0.28017	8	3	4	0.19093	10	2	8	0.09662	11	5	5	0.15354
5	2	4	0.21448	8	3	5	0.14660	10	2	9	0.09554	11	5	6	0.12157
5	3	3	0.35118	8	3	6	0.12950	10	3	3	0.30337	11	5	7	0.10567
6	1	1	1.49395	8	4	4	0.21031	10	3	4	0.19584	11	6	6	0.13828
6	1	2	0.49972	8	4	5	0.16431	10	3	5	0.14130	12	1	1	1.56499
6	1	3	0.26684	9	1	1	1.53995	10	3	6	0.11396	12	1	2	0.56506
6	1	4	0.18590	9	1	2	0.54055	10	3	7	0.10193	12	1	3	0.31542
6	1	5	0.16031	9	1	3	0.29313	10	3	8	0.09770	12	1	4	0.20558
6	1	6	0.15473	9	1	4	0.18876	10	4	4	0.20120	12	1	5	0.14628
6	2	2	0.50795	9	1	5	0.13877	10	4	5	0.14601	12	1	6	0.11240
6	2	3	0.27332	9	1	6	0.11628	10	4	6	0.11835	12	1	7	0.09388
6	2	4	0.19172	9	1	7	0.10792	10	4	7	0.10619	12	1	8	0.08493
6	2	5	0.16594	9	1	8	0.10563	10	5	5	0.15954	12	1	9	0.08135
6	3	3	0.30333	9	1	9	0.10523	10	5	6	0.13091	12	1	10	0.08024
6	3	4	0.21848	9	2	2	0.54129	11	1	1	1.55806	12	1	11	0.08000
7	1	1	1.51282	9	2	3	0.29368	11	1	2	0.55820	12	1	12	0.07996
7	1	2	0.51554	9	2	4	0.18923	11	1	3	0.30891	12	2	2	0.56513
7	1	3	0.27473	9	2	5	0.13920	11	1	4	0.20004	12	2	3	0.31548
7	1	4	0.18242	9	2	6	0.11668	11	1	5	0.14270	12	2	4	0.20563
7	1	5	0.14681	9	2	7	0.10832	11	1	6	0.11166	12	2	5	0.14632
7	1	6	0.13581	9	2	8	0.10603	11	1	7	0.09618	12	2	6	0.11244
7	1	7	0.13357	9	3	3	0.29694	11	1	8	0.08966	12	2	7	0.09392
7	2	2	0.51912	9	3	4	0.19194	11	1	9	0.08750	12	2	8	0.08496
7	2	3	0.27749	9	3	5	0.14165	11	1	10	0.08699	12	2	9	0.08139
7	2	4	0.18483	9	3	6	0.11902	11	1	11	0.08691	12	2	10	0.08028
7	2	5	0.14910	9	3	7	0.11062	11	2	2	0.55836	12	2	11	0.08003
7	2	6	0.13806	9	4	4	0.20202	11	2	3	0.30902	12	3	3	0.31589

Variances and Covariances of Double Exponential Order Statistics

N	*I*	*J*	*COV*	*N*	*I*	*J*	*COV*	*N*	*I*	*J*	*COV*	*N*	*I*	*J*	*COV*
12	3	4	0.20597	13	3	7	0.09336	14	3	3	0.32623	15	2	4	0.21961
12	3	5	0.14662	13	3	8	0.08207	14	3	4	0.21549	15	2	5	0.15781
12	3	6	0.11271	13	3	9	0.07685	14	3	5	0.15415	15	2	6	0.11951
12	3	7	0.09417	13	3	10	0.07489	14	3	6	0.11680	15	2	7	0.09504
12	3	8	0.08521	13	3	11	0.07431	14	3	7	0.09380	15	2	8	0.07991
12	3	9	0.08163	13	4	4	0.21176	14	3	8	0.08043	15	2	9	0.07132
12	3	10	0.08053	13	4	5	0.15106	14	3	9	0.07350	15	2	10	0.06706
12	4	4	0.20752	13	4	6	0.11505	14	3	10	0.07045	15	2	11	0.06528
12	4	5	0.14795	13	4	7	0.09397	14	3	11	0.06937	15	2	12	0.06468
12	4	6	0.11392	13	4	8	0.08266	14	3	12	0.06907	15	2	13	0.06453
12	4	7	0.09532	13	4	9	0.07743	14	4	4	0.21595	15	2	14	0.06450
12	4	8	0.08634	13	4	10	0.07546	14	4	5	0.15454	15	3	3	0.33056
12	4	9	0.08275	13	5	5	0.15358	14	4	6	0.11715	15	3	4	0.21966
12	5	5	0.15234	13	5	6	0.11729	14	4	7	0.09412	15	3	5	0.15784
12	5	6	0.11787	13	5	7	0.09606	14	4	8	0.08074	15	3	6	0.11954
12	5	7	0.09904	13	5	8	0.08467	14	4	9	0.07380	15	3	7	0.09507
12	5	8	0.08996	13	5	9	0.07941	14	4	10	0.07075	15	3	8	0.07994
12	6	6	0.12785	13	6	6	0.12332	14	4	11	0.06966	15	3	9	0.07135
12	6	7	0.10840	13	6	7	0.10163	14	5	5	0.15599	15	3	10	0.06708
13	1	1	1.57090	13	6	8	0.09002	14	5	6	0.11843	15	3	11	0.06531
13	1	2	0.57093	13	7	7	0.11377	14	5	7	0.09530	15	3	12	0.06471
13	1	3	0.32112	14	1	1	1.57600	14	5	8	0.08186	15	3	13	0.06456
13	1	4	0.21072	14	1	2	0.57601	14	5	9	0.07490	15	4	4	0.21990
13	1	5	0.15018	14	1	3	0.32611	14	5	10	0.07184	15	4	5	0.15805
13	1	6	0.11425	14	1	4	0.21540	14	6	6	0.12208	15	4	6	0.11973
13	1	7	0.09322	14	1	5	0.15407	14	6	7	0.09864	15	4	7	0.09524
13	1	8	0.08194	14	1	6	0.11673	14	6	8	0.08503	15	4	8	0.08010
13	1	9	0.07672	14	1	7	0.09373	14	6	9	0.07799	15	4	9	0.07150
13	1	10	0.07475	14	1	8	0.08036	14	7	7	0.10630	15	4	10	0.06724
13	1	11	0.07418	14	1	9	0.07343	14	7	8	0.09227	15	4	11	0.06546
13	1	12	0.07406	14	1	10	0.07038	15	1	1	1.58044	15	4	12	0.06487
13	1	13	0.07404	14	1	11	0.06930	15	1	2	0.58045	15	5	5	0.15888
13	2	2	0.57097	14	1	12	0.06900	15	1	3	0.33050	15	5	6	0.12046
13	2	3	0.32115	14	1	13	0.06895	15	1	4	0.21961	15	5	7	0.09591
13	2	4	0.21074	14	1	14	0.06894	15	1	5	0.15780	15	5	8	0.08073
13	2	5	0.15019	14	2	2	0.57603	15	1	6	0.11950	15	5	9	0.07211
13	2	6	0.11427	14	2	3	0.32612	15	1	7	0.09504	15	5	10	0.06784
13	2	7	0.09324	14	2	4	0.21541	15	1	8	0.07990	15	5	11	0.06606
13	2	8	0.08195	14	2	5	0.15408	15	1	9	0.07132	15	6	6	0.12266
13	2	9	0.07673	14	2	6	0.11674	15	1	10	0.06705	15	6	7	0.09791
13	2	10	0.07477	14	2	7	0.09374	15	1	11	0.06528	15	6	8	0.08261
13	2	11	0.07419	14	2	8	0.08037	15	1	12	0.06468	15	6	9	0.07393
13	2	12	0.07407	14	2	9	0.07344	15	1	13	0.06453	15	6	10	0.06963
13	3	3	0.32136	14	2	10	0.07039	15	1	14	0.06450	15	7	7	0.10277
13	3	4	0.21092	14	2	11	0.06931	15	1	15	0.06449	15	7	8	0.08715
13	3	5	0.15034	14	2	12	0.06901	15	2	2	0.58046	15	7	9	0.07831
13	3	6	0.11440	14	2	13	0.06895	15	2	3	0.33051	15	8	8	0.09635

Table C11.1

EXPECTED VALUES OF HALF NORMAL [CHI (1 d.f.)] ORDER STATISTICS

[Tabular values are expected values of the Mth order statistic of a sample of size N from a half normal [chi (1 d.f.)] population with location parameter 0 and scale parameter 1.]

Expected Values of Half Normal [Chi (1 d.f.)] Order Statistics

N	*M*	*Mean*	*N*	*M*	*Mean*	*N*	*M*	*Mean*	*N*	*M*	*Mean*
1	1	0.79788	10	9	1.42436	15	1	0.07875	18	5	0.33923
2	1	0.46739	10	10	1.88072	15	2	0.15835	18	6	0.41120
2	2	1.12838	11	1	0.10539	15	3	0.23929	18	7	0.48552
3	1	0.33490	11	2	0.21278	15	4	0.32213	18	8	0.56276
3	2	0.73236	11	3	0.32339	15	5	0.40751	18	9	0.64366
3	3	1.32639	11	4	0.43876	15	6	0.49622	18	10	0.72914
4	1	0.26208	11	5	0.56088	15	7	0.58926	18	11	0.82046
4	2	0.55336	11	6	0.69259	15	8	0.68794	18	12	0.91933
4	3	0.91136	11	7	0.83814	15	9	0.79405	18	13	1.02828
4	4	1.46473	11	8	1.00467	15	10	0.91021	18	14	1.15119
5	1	0.21569	11	9	1.20569	15	11	1.04042	18	15	1.29469
5	2	0.44764	11	10	1.47296	15	12	1.19144	18	16	1.47145
5	3	0.71195	11	11	1.92149	15	13	1.37627	18	17	1.71154
5	4	1.04430	12	1	0.09716	15	14	1.62555	18	18	2.12468
5	5	1.56983	12	2	0.19590	15	15	2.05090	19	1	0.06288
6	1	0.18344	12	3	0.29716	16	1	0.07407	19	2	0.12621
6	2	0.37695	12	4	0.40208	16	2	0.14886	19	3	0.19022
6	3	0.58903	12	5	0.51210	16	3	0.22477	19	4	0.25519
6	4	0.83487	12	6	0.62917	16	4	0.30224	19	5	0.32139
6	5	1.14902	12	7	0.75601	16	5	0.38180	19	6	0.38918
6	6	1.65400	12	8	0.89682	16	6	0.46406	19	7	0.45893
7	1	0.15967	12	9	1.05860	16	7	0.54980	19	8	0.53111
7	2	0.32605	12	10	1.25472	16	8	0.63998	19	9	0.60629
7	3	0.50420	12	11	1.51660	16	9	0.73589	19	10	0.68518
7	4	0.70212	12	12	1.95830	16	10	0.83929	19	11	0.76871
7	5	0.93444	13	1	0.09013	16	11	0.95276	19	12	0.85810
7	6	1.23485	13	2	0.18153	16	12	1.08026	19	13	0.95505
7	7	1.72385	13	3	0.27494	16	13	1.22851	19	14	1.06207
8	1	0.14141	13	4	0.37123	16	14	1.41037	19	15	1.18302
8	2	0.28752	13	5	0.47149	16	15	1.65629	19	16	1.32447
8	3	0.44163	13	6	0.57709	16	16	2.07720	19	17	1.49901
8	4	0.60849	13	7	0.68992	17	1	0.06992	19	18	1.73654
8	5	0.79575	13	8	0.81265	17	2	0.14045	19	19	2.14624
8	6	1.01765	13	9	0.94942	17	3	0.21192	20	1	0.05987
8	7	1.30726	13	10	1.10712	17	4	0.28470	20	2	0.12013
8	8	1.78336	13	11	1.29899	17	5	0.35922	20	3	0.18097
9	1	0.12693	13	12	1.55617	17	6	0.43598	20	4	0.24264
9	2	0.25728	13	13	1.99181	17	7	0.51556	20	5	0.30537
9	3	0.39335	14	1	0.08405	17	8	0.59872	20	6	0.36945
9	4	0.53820	14	2	0.16914	17	9	0.68640	20	7	0.43520
9	5	0.69636	14	3	0.25586	17	10	0.77988	20	8	0.50299
9	6	0.87526	14	4	0.34490	17	11	0.88088	20	9	0.57328
9	7	1.08884	14	5	0.43708	17	12	0.99196	20	10	0.64663
9	8	1.36966	14	6	0.53343	17	13	1.11705	20	11	0.72374
9	9	1.83508	14	7	0.63531	17	14	1.26280	20	12	0.80551
10	1	0.11515	14	8	0.74453	17	15	1.44199	20	13	0.89316
10	2	0.23289	14	9	0.86374	17	16	1.68486	20	14	0.98838
10	3	0.35485	14	10	0.99701	17	17	2.10172	20	15	1.09366
10	4	0.48316	14	11	1.15117	18	1	0.06622	20	16	1.21281
10	5	0.62075	14	12	1.33931	18	2	0.13295	20	17	1.35238
10	6	0.77198	14	13	1.59231	18	3	0.20048	20	18	1.52489
10	7	0.94412	14	14	2.02254	18	4	0.26913	20	19	1.76006
10	8	1.15086							20	20	2.16657

Table C11.2

VARIANCES AND COVARIANCES OF HALF NORMAL [CHI (1 d.f.)] ORDER STATISTICS

[When $I = J$, tabular values are variances of the Ith order statistic of a sample of size N from a half normal [chi (1 d.f.)] population with location parameter 0 and scale parameter 1; when $I \neq J$, they are covariances of the Ith and Jth order statistics.]

Variances and Covariances of Half Normal [Chi (1 d.f.)] Order Statistics

N	*I*	*J*	*COV*	*N*	*I*	*J*	*COV*	*N*	*I*	*J*	*COV*	*N*	*I*	*J*	*COV*
1	1	1	0.36338	6	6	6	0.29666	8	5	5	0.08112	10	1	1	0.01144
2	1	1	0.14493	7	1	1	0.02105	8	5	6	0.07282	10	1	2	0.01061
2	1	2	0.10923	7	1	2	0.01904	8	5	7	0.06401	10	1	3	0.00984
2	2	2	0.36338	7	1	3	0.01718	8	5	8	0.05321	10	1	4	0.00911
3	1	1	0.08064	7	1	4	0.01543	8	6	6	0.10651	10	1	5	0.00843
3	1	2	0.06630	7	1	5	0.01372	8	6	7	0.09379	10	1	6	0.00776
3	1	3	0.05142	7	1	6	0.01194	8	6	8	0.07813	10	1	7	0.00710
3	2	2	0.16819	7	1	7	0.00982	8	7	7	0.15074	10	1	8	0.00643
3	2	3	0.13126	7	2	2	0.04027	8	7	8	0.12611	10	1	9	0.00570
3	3	3	0.34336	7	2	3	0.03636	8	8	8	0.27704	10	1	10	0.00479
4	1	1	0.05202	7	2	4	0.03267	9	1	1	0.01373	10	2	2	0.02188
4	1	2	0.04453	7	2	5	0.02906	9	1	2	0.01265	10	2	3	0.02029
4	1	3	0.03739	7	2	6	0.02530	9	1	3	0.01165	10	2	4	0.01880
4	1	4	0.02962	7	2	7	0.02082	9	1	4	0.01071	10	2	5	0.01738
4	2	2	0.10287	7	3	3	0.05936	9	1	5	0.00981	10	2	6	0.01601
4	2	3	0.08660	7	3	4	0.05337	9	1	6	0.00893	10	2	7	0.01466
4	2	4	0.06877	7	3	5	0.04750	9	1	7	0.00805	10	2	8	0.01327
4	3	3	0.16942	7	3	6	0.04139	9	1	8	0.00710	10	2	9	0.01177
4	3	4	0.13531	7	3	7	0.03408	9	1	9	0.00592	10	2	10	0.00988
4	4	4	0.32478	7	4	4	0.08064	9	2	2	0.02624	10	3	3	0.03179
5	1	1	0.03655	7	4	5	0.07185	9	2	3	0.02416	10	3	4	0.02946
5	1	2	0.03208	7	4	6	0.06268	9	2	4	0.02222	10	3	5	0.02724
5	1	3	0.02792	7	4	7	0.05168	9	2	5	0.02036	10	3	6	0.02510
5	1	4	0.02381	7	5	5	0.10853	9	2	6	0.01854	10	3	7	0.02299
5	1	5	0.01915	7	5	6	0.09487	9	2	7	0.01671	10	3	8	0.02082
5	2	2	0.07082	7	5	7	0.07839	9	2	8	0.01474	10	3	9	0.01846
5	2	3	0.06171	7	6	6	0.15541	9	2	9	0.01230	10	3	10	0.01550
5	2	4	0.05268	7	6	7	0.12900	9	3	3	0.03822	10	4	4	0.04169
5	2	5	0.04244	7	7	7	0.28604	9	3	4	0.03515	10	4	5	0.03857
5	3	3	0.10904	8	1	1	0.01680	9	3	5	0.03222	10	4	6	0.03555
5	3	4	0.09331	8	1	2	0.01535	9	3	6	0.02935	10	4	7	0.03256
5	3	5	0.07536	8	1	3	0.01401	9	3	7	0.02646	10	4	8	0.02950
5	4	4	0.16548	8	1	4	0.01275	9	3	8	0.02335	10	4	9	0.02616
5	4	5	0.13436	8	1	5	0.01154	9	3	9	0.01949	10	4	10	0.02197
5	5	5	0.30937	8	1	6	0.01034	9	4	4	0.05044	10	5	5	0.05221
6	1	1	0.02718	8	1	7	0.00906	9	4	5	0.04626	10	5	6	0.04814
6	1	2	0.02427	8	1	8	0.00751	9	4	6	0.04216	10	5	7	0.04411
6	1	3	0.02158	8	2	2	0.03210	9	4	7	0.03802	10	5	8	0.03998
6	1	4	0.01901	8	2	3	0.02931	9	4	8	0.03357	10	5	9	0.03548
6	1	5	0.01640	8	2	4	0.02668	9	4	9	0.02804	10	5	10	0.02981
6	1	6	0.01336	8	2	5	0.02416	9	5	5	0.06397	10	6	6	0.06430
6	2	2	0.05220	8	2	6	0.02164	9	5	6	0.05835	10	6	7	0.05895
6	2	3	0.04645	8	2	7	0.01898	9	5	7	0.05265	10	6	8	0.05346
6	2	4	0.04094	8	2	8	0.01574	9	5	8	0.04651	10	6	9	0.04747
6	2	5	0.03534	8	3	3	0.04695	9	5	9	0.03588	10	6	10	0.03992
6	2	6	0.02880	8	3	4	0.04277	9	6	6	0.08061	10	7	7	0.07963
6	3	3	0.07807	8	3	5	0.03874	9	6	7	0.07281	10	7	8	0.07228
6	3	4	0.06890	8	3	6	0.03472	9	6	8	0.06439	10	7	9	0.06425
6	3	5	0.05954	8	3	7	0.03046	9	6	9	0.05389	10	7	10	0.05409
6	3	6	0.04859	8	3	8	0.02527	9	7	7	0.10426	10	8	8	0.10198
6	4	4	0.10979	8	4	4	0.06263	9	7	8	0.09236	10	8	9	0.09080
6	4	5	0.09509	8	4	5	0.05676	9	7	9	0.07746	10	8	10	0.07659
6	4	6	0.07779	8	4	6	0.05091	9	8	8	0.14649	10	9	9	0.14266
6	5	5	0.16043	8	4	7	0.04470	9	8	9	0.12337	10	9	10	0.12080
6	5	6	0.13189	8	4	8	0.03711	9	9	9	0.26929	10	10	10	0.26253

Table C12.1

EXPECTED VALUES OF LOGISTIC ORDER STATISTICS

[Tabular values are the expected values of the kth largest logistic deviate for a sample of size N; or when preceded by a minus sign, they are the expected values of the kth smallest logistic deviate.]

Expected Values of Logistic Order Statistics

N	k	$Mean$	N	k	$Mean$	N	k	$Mean$	N	k	$Mean$
1	1	0.00000	10	3	0.60252	14	6	0.23957	18	2	1.31256
2	1	0.55133	10	4	0.33999	14	7	0.07876	18	3	1.00244
3	1	0.82699	10	5	0.11027	15	1	1.79268	18	4	0.78191
3	2	0.00000	11	1	1.61482	15	2	1.20197	18	5	0.60470
4	1	1.01077	11	2	1.00836	15	3	0.88390	18	6	0.45202
4	2	0.27566	11	3	0.67144	15	4	0.65418	18	7	0.31419
5	1	1.14860	11	4	0.41875	15	5	0.46622	18	8	0.18531
5	2	0.45944	11	5	0.20215	15	6	0.30082	18	9	0.06126
5	3	0.00000	11	6	0.00000	15	7	0.14768	19	1	1.92695
6	1	1.25887	12	1	1.66495	15	8	0.00000	19	2	1.34500
6	2	0.59727	12	2	1.06350	16	1	1.82944	19	3	1.03690
6	3	0.18378	12	3	0.73270	16	2	1.24135	19	4	0.81867
7	1	1.35076	12	4	0.48766	16	3	0.92631	19	5	0.64408
7	2	0.70754	12	5	0.28092	16	4	0.70012	19	6	0.49443
7	3	0.32161	12	6	0.09189	16	5	0.51634	19	7	0.36013
7	4	0.00000	13	1	1.71089	16	6	0.35596	19	8	0.23543
8	1	1.42952	13	2	1.11362	16	7	0.20894	19	9	0.11639
8	2	0.79943	13	3	0.78783	16	8	0.06892	19	10	0.00000
8	3	0.43187	13	4	0.54892	17	1	1.86389	20	1	1.95597
8	4	0.13783	13	5	0.34983	17	2	1.27811	20	2	1.37563
9	1	1.49843	13	6	0.17065	17	3	0.96569	20	3	1.06933
9	2	0.87819	13	7	0.00000	17	4	0.74253	20	4	0.85312
9	3	0.52376	14	1	1.75330	17	5	0.56229	20	5	0.68083
9	4	0.24810	14	2	1.15956	17	6	0.40608	20	6	0.53381
9	5	0.00000	14	3	0.83795	17	7	0.26407	20	7	0.40254
10	1	1.55969	14	4	0.60406	17	8	0.13017	20	8	0.28137
10	2	0.94710	14	5	0.41109	17	9	0.00000	20	9	0.16651
						18	1	1.89632	20	10	0.05513

Table C12.2

VARIANCES AND COVARIANCES OF LOGISTIC ORDER STATISTICS

[When $I = J$, tabular values are variances of the Ith order statistic of a sample of size N from logistic distribution; when $I \neq J$, they are covariances of the Ith and Jth order statistics.]

Variances and Covariances of Logistic Order Statistics

N	I	J	COV	N	I	J	COV	N	I	J	COV	N	I	J	COV
1	1	1	1.00000	8	1	3	0.13399	10	1	10	0.03531	11	4	8	0.06130
2	1	1	0.69604	8	1	4	0.09680	10	2	2	0.23176	11	5	5	0.11395
2	1	2	0.30396	8	1	5	0.07574	10	2	3	0.14420	11	5	6	0.09441
3	1	1	0.62005	8	1	6	0.06219	10	2	4	0.10455	11	5	7	0.08063
3	1	2	0.25991	8	1	7	0.05276	10	2	5	0.08198	11	6	6	0.11023
3	1	3	0.16410	8	1	8	0.04580	10	2	6	0.06743	12	1	1	0.52642
3	2	2	0.39207	8	2	2	0.24271	10	2	7	0.05726	12	1	2	0.20943
4	1	1	0.58627	8	2	3	0.15174	10	2	8	0.04976	12	1	3	0.12904
4	1	2	0.24118	8	2	4	0.11032	10	2	9	0.04399	12	1	4	0.09305
4	1	3	0.15089	8	2	5	0.08667	10	3	3	0.16051	12	1	5	0.07271
4	1	4	0.10976	8	2	6	0.07137	10	3	4	0.11701	12	1	6	0.05966
4	2	2	0.31608	8	2	7	0.06067	10	3	5	0.09209	12	1	7	0.05057
4	2	3	0.20374	8	3	3	0.17516	10	3	6	0.07592	12	1	8	0.04389
5	1	1	0.56727	8	3	4	0.12832	10	3	7	0.06460	12	1	9	0.03876
5	1	2	0.23090	8	3	5	0.10132	10	3	8	0.05621	12	1	10	0.03471
5	1	3	0.14374	8	3	6	0.08374	10	4	4	0.13294	12	1	11	0.03142
5	1	4	0.10425	8	4	4	0.15355	10	4	5	0.10507	12	1	12	0.02870
5	1	5	0.08178	8	4	5	0.12202	10	4	6	0.08689	12	2	2	0.22496
5	2	2	0.28231	9	1	1	0.53572	10	4	7	0.07410	12	2	3	0.13956
5	2	3	0.17950	9	1	2	0.21425	10	5	5	0.12239	12	2	4	0.10101
5	2	4	0.13181	9	1	3	0.13231	10	5	6	0.10160	12	2	5	0.07912
5	3	3	0.24009	9	1	4	0.09552	11	1	1	0.52893	12	2	6	0.06502
6	1	1	0.55512	9	1	5	0.07471	11	1	2	0.21073	12	2	7	0.05518
6	1	2	0.22442	9	1	6	0.06133	11	1	3	0.12992	12	2	8	0.04794
6	1	3	0.13927	9	1	7	0.05201	11	1	4	0.09371	12	2	9	0.04237
6	1	4	0.10083	9	1	8	0.04515	11	1	5	0.07325	12	2	10	0.03796
6	1	5	0.07899	9	1	9	0.03988	11	1	6	0.06011	12	2	11	0.03438
6	1	6	0.06493	9	2	2	0.23651	11	1	7	0.05096	12	3	3	0.15201
6	2	2	0.26331	9	2	3	0.14746	11	1	8	0.04422	12	3	4	0.11049
6	2	3	0.16609	9	2	4	0.10704	11	1	9	0.03906	12	3	5	0.08679
6	2	4	0.12138	9	2	5	0.08401	11	1	10	0.03498	12	3	6	0.07146
6	2	5	0.09569	9	2	6	0.06913	11	1	11	0.03167	12	3	7	0.06073
6	3	3	0.20632	9	2	7	0.05873	11	2	2	0.22800	12	3	8	0.05281
6	3	4	0.15268	9	2	8	0.05105	11	2	3	0.14163	12	3	9	0.04672
7	1	1	0.54667	9	3	3	0.16672	11	2	4	0.10259	12	3	10	0.04188
7	1	2	0.21997	9	3	4	0.12179	11	2	5	0.08040	12	4	4	0.12199
7	1	3	0.13621	9	3	5	0.09598	11	2	6	0.06610	12	4	5	0.09612
7	1	4	0.09849	9	3	6	0.07922	11	2	7	0.05611	12	4	6	0.07932
7	1	5	0.07711	9	3	7	0.06745	11	2	8	0.04875	12	4	7	0.06753
7	1	6	0.06334	9	4	4	0.14139	11	2	9	0.04309	12	4	8	0.05879
7	1	7	0.05374	9	4	5	0.11200	11	2	10	0.03861	12	4	9	0.05206
7	2	2	0.25115	9	4	6	0.09278	11	3	3	0.15576	12	5	5	0.10774
7	2	3	0.15760	9	5	5	0.13455	11	3	4	0.11337	12	5	6	0.08915
7	2	4	0.11482	10	1	1	0.53197	11	3	5	0.08912	12	5	7	0.07604
7	2	5	0.09033	10	1	2	0.21230	11	3	6	0.07342	12	5	8	0.06631
7	2	6	0.07446	10	1	3	0.13098	11	3	7	0.06243	12	6	6	0.10179
7	3	3	0.18732	10	1	4	0.09452	11	3	8	0.05431	12	6	7	0.08704
7	3	4	0.13778	10	1	5	0.07390	11	3	9	0.04805	13	1	1	0.52430
7	3	5	0.10908	10	1	6	0.06065	11	4	4	0.12674	13	1	2	0.20835
7	4	4	0.17254	10	1	7	0.05143	11	4	5	0.09999	13	1	3	0.12831
8	1	1	0.54047	10	1	8	0.04463	11	4	6	0.08259	13	1	4	0.09250
8	1	2	0.21672	10	1	9	0.03943	11	4	7	0.07037	13	1	5	0.07227

Variances and Covariances of Logistic Order Statistics

N	*I*	*J*	*COV*	*N*	*I*	*J*	*COV*	*N*	*I*	*J*	*COV*	*N*	*I*	*J*	*COV*
13	1	6	0.05928	14	1	9	0.03830	15	1	5	0.07156	15	6	7	0.07425
13	1	7	0.05025	14	1	10	0.03429	15	1	6	0.05870	15	6	8	0.06472
13	1	8	0.04360	14	1	11	0.03104	15	1	7	0.04975	15	6	9	0.05736
13	1	9	0.03851	14	1	12	0.02835	15	1	8	0.04316	15	6	10	0.05151
13	1	10	0.03448	14	1	13	0.02609	15	1	9	0.03811	15	7	7	0.08239
13	1	11	0.03121	14	1	14	0.02417	15	1	10	0.03412	15	7	8	0.07192
13	1	12	0.02851	14	2	2	0.22034	15	1	11	0.03089	15	7	9	0.06382
13	1	13	0.02624	14	2	3	0.13641	15	1	12	0.02821	15	8	8	0.08094
13	2	2	0.22245	14	2	4	0.09862	15	1	13	0.02597	16	1	1	0.51960
13	2	3	0.13784	14	2	5	0.07719	15	1	14	0.02405	16	1	2	0.20594
13	2	4	0.09971	14	2	6	0.06340	15	1	15	0.02240	16	1	3	0.12669
13	2	5	0.07807	14	2	7	0.05379	15	2	2	0.21854	16	1	4	0.09127
13	2	6	0.06414	14	2	8	0.04671	15	2	3	0.13519	16	1	5	0.07128
13	2	7	0.05443	14	2	9	0.04127	15	2	4	0.09769	16	1	6	0.05846
13	2	8	0.04727	14	2	10	0.03697	15	2	5	0.07644	16	1	7	0.04954
13	2	9	0.04177	14	2	11	0.03348	15	2	6	0.06278	16	1	8	0.04298
13	2	10	0.03742	14	2	12	0.03059	15	2	7	0.05325	16	1	9	0.03796
13	2	11	0.03389	14	2	13	0.02816	15	2	8	0.04624	16	1	10	0.03398
13	2	12	0.03097	14	3	3	0.14646	15	2	9	0.04085	16	1	11	0.03076
13	3	3	0.14897	14	3	4	0.10625	15	2	10	0.03659	16	1	12	0.02809
13	3	4	0.10817	14	3	5	0.08335	15	2	11	0.03314	16	1	13	0.02586
13	3	5	0.08490	14	3	6	0.06857	15	2	12	0.03028	16	1	14	0.02395
13	3	6	0.06987	14	3	7	0.05824	15	2	13	0.02787	16	1	15	0.02230
13	3	7	0.05937	14	3	8	0.05062	15	2	14	0.02582	16	1	16	0.02087
13	3	8	0.05161	14	3	9	0.04476	15	3	3	0.14485	16	2	2	0.21699
13	3	9	0.04564	14	3	10	0.04011	15	3	4	0.10464	16	2	3	0.13413
13	3	10	0.04091	14	3	11	0.03634	15	3	5	0.08205	16	2	4	0.09689
13	3	11	0.03707	14	3	12	0.03322	15	3	6	0.06748	16	2	5	0.07580
13	4	4	0.11824	14	4	4	0.11520	15	3	7	0.05730	16	2	6	0.06224
13	4	5	0.09306	14	4	5	0.09059	15	3	8	0.04979	16	2	7	0.05279
13	4	6	0.07674	14	4	6	0.07466	15	3	9	0.04402	16	2	8	0.04583
13	4	7	0.06530	14	4	7	0.06349	15	3	10	0.03945	16	2	9	0.04049
13	4	8	0.05682	14	4	8	0.05524	15	3	11	0.03574	16	2	10	0.03627
13	4	9	0.05030	14	4	9	0.04888	15	3	12	0.03266	16	2	11	0.03284
13	4	10	0.04512	14	4	10	0.04384	15	3	13	0.03008	16	2	12	0.03000
13	5	5	0.10299	14	4	11	0.03974	15	4	4	0.11269	16	2	13	0.02762
13	5	6	0.08512	14	5	5	0.09924	15	4	5	0.08856	16	2	14	0.02558
13	5	7	0.07255	14	5	6	0.08195	15	4	6	0.07294	16	2	15	0.02383
13	5	8	0.06322	14	5	7	0.06980	15	4	7	0.06201	16	3	3	0.14255
13	5	9	0.05602	14	5	8	0.06079	15	4	8	0.05393	16	3	4	0.10327
13	6	6	0.09558	14	5	9	0.05385	15	4	9	0.04772	16	3	5	0.08094
13	6	7	0.08163	14	5	10	0.04833	15	4	10	0.04279	16	3	6	0.06655
13	6	8	0.07125	14	6	6	0.09083	15	4	11	0.03878	16	3	7	0.05650
13	7	7	0.09334	14	6	7	0.07750	15	4	12	0.03546	16	3	8	0.04908
14	1	1	0.52251	14	6	8	0.06760	15	5	5	0.09620	16	3	9	0.04339
14	1	2	0.20743	14	6	9	0.05994	15	5	6	0.07938	16	3	10	0.03888
14	1	3	0.12769	14	7	7	0.08714	15	5	7	0.06757	16	3	11	0.03522
14	1	4	0.09203	14	7	8	0.07613	15	5	8	0.05883	16	3	12	0.03219
14	1	5	0.07189	15	1	1	0.52095	15	5	9	0.05209	16	3	13	0.02964
14	1	6	0.05897	15	1	2	0.20663	15	5	10	0.04674	16	3	14	0.02746
14	1	7	0.04998	15	1	3	0.12715	15	5	11	0.04239	16	4	4	0.11058
14	1	8	0.04337	15	1	4	0.09162	15	6	6	0.08708	16	4	5	0.08684

Variances and Covariances of Logistic Order Statistics

N	I	J	COV	N	I	J	COV	N	I	J	COV	N	I	J	COV
16	4	6	0.07150	17	2	8	0.04548	17	7	9	0.05843	18	3	13	0.02893
16	4	7	0.06077	17	2	9	0.04018	17	7	10	0.05247	18	3	14	0.02681
16	4	8	0.05284	17	2	10	0.03598	17	7	11	0.04762	18	3	15	0.02497
16	4	9	0.04674	17	2	11	0.03258	17	8	8	0.07244	18	3	16	0.02337
16	4	10	0.04190	17	2	12	0.02977	17	8	9	0.06428	18	4	4	0.10723
16	4	11	0.03797	17	2	13	0.02740	17	8	10	0.05778	18	4	5	0.08413
16	4	12	0.03472	17	2	14	0.02538	17	9	9	0.07144	18	4	6	0.06922
16	4	13	0.03198	17	2	15	0.02364	18	1	1	0.51736	18	4	7	0.05880
16	5	5	0.09369	17	2	16	0.02212	18	1	2	0.20480	18	4	8	0.05111
16	5	6	0.07726	17	3	3	0.14100	18	1	3	0.12592	18	4	9	0.04519
16	5	7	0.06574	17	3	4	0.10209	18	1	4	0.09069	18	4	10	0.04051
16	5	8	0.05721	17	3	5	0.07999	18	1	5	0.07081	18	4	11	0.03670
16	5	9	0.05065	17	3	6	0.06575	18	1	6	0.05807	18	4	12	0.03355
16	5	10	0.04543	17	3	7	0.05581	18	1	7	0.04921	18	4	13	0.03090
16	5	11	0.04120	17	3	8	0.04848	18	1	8	0.04269	18	4	14	0.02863
16	5	12	0.03768	17	3	9	0.04285	18	1	9	0.03769	18	4	15	0.02668
16	6	6	0.08404	17	3	10	0.03839	18	1	10	0.03374	18	5	5	0.08978
16	6	7	0.07161	17	3	11	0.03478	18	1	11	0.03054	18	5	6	0.07397
16	6	8	0.06239	17	3	12	0.03178	18	1	12	0.02790	18	5	7	0.06289
16	6	9	0.05527	17	3	13	0.02926	18	1	13	0.02567	18	5	8	0.05471
16	6	10	0.04962	17	3	14	0.02711	18	1	14	0.02378	18	5	9	0.04841
16	6	11	0.04502	17	3	15	0.02526	18	1	15	0.02214	18	5	10	0.04341
16	7	7	0.07864	17	4	4	0.10878	18	1	16	0.02072	18	5	11	0.03935
16	7	8	0.06860	17	4	5	0.08539	18	1	17	0.01946	18	5	12	0.03598
16	7	9	0.06084	17	4	6	0.07028	18	1	18	0.01835	18	5	13	0.03314
16	7	10	0.05466	17	4	7	0.05971	18	2	2	0.21445	18	5	14	0.03072
16	8	8	0.07619	17	4	8	0.05191	18	2	3	0.13242	18	6	6	0.07942
16	8	9	0.06765	17	4	9	0.04591	18	2	4	0.09559	18	6	7	0.06761
17	1	1	0.51842	17	4	10	0.04115	18	2	5	0.07475	18	6	8	0.05886
17	1	2	0.20534	17	4	11	0.03729	18	2	6	0.06136	18	6	9	0.05211
17	1	3	0.12628	17	4	12	0.03409	18	2	7	0.05203	18	6	10	0.04676
17	1	4	0.09096	17	4	13	0.03140	18	2	8	0.04517	18	6	11	0.04240
17	1	5	0.07103	17	4	14	0.02910	18	2	9	0.03990	18	6	12	0.03879
17	1	6	0.05825	17	5	5	0.09158	18	2	10	0.03573	18	6	13	0.03575
17	1	7	0.04936	17	5	6	0.07548	18	2	11	0.03235	18	7	7	0.07309
17	1	8	0.04283	17	5	7	0.06420	18	2	12	0.02956	18	7	8	0.06369
17	1	9	0.03782	17	5	8	0.05586	18	2	13	0.02721	18	7	9	0.05644
17	1	10	0.03386	17	5	9	0.04944	18	2	14	0.02520	18	7	10	0.05067
17	1	11	0.03064	17	5	10	0.04434	18	2	15	0.02347	18	7	11	0.04598
17	1	12	0.02799	17	5	11	0.04020	18	2	16	0.02196	18	7	12	0.04208
17	1	13	0.02576	17	5	12	0.03676	18	2	17	0.02064	18	8	8	0.06940
17	1	14	0.02386	17	5	13	0.03387	18	3	3	0.13965	18	8	9	0.06155
17	1	15	0.02222	17	6	6	0.08153	18	3	4	0.10107	18	8	10	0.05530
17	1	16	0.02079	17	6	7	0.06943	18	3	5	0.07916	18	8	11	0.05021
17	1	17	0.01953	17	6	8	0.06047	18	3	6	0.06505	18	9	9	0.06769
17	2	2	0.21564	17	6	9	0.05355	18	3	7	0.05521	18	9	10	0.06087
17	2	3	0.13322	17	6	10	0.04806	18	3	8	0.04795	19	1	1	0.51643
17	2	4	0.09620	17	6	11	0.04359	18	3	9	0.04238	19	1	2	0.20432
17	2	5	0.07524	17	6	12	0.03989	18	3	10	0.03797	19	1	3	0.12560
17	2	6	0.06177	17	7	7	0.07560	18	3	11	0.03439	19	1	4	0.09045
17	2	7	0.05239	17	7	8	0.06591	18	3	12	0.03143	19	1	5	0.07062

Variances and Covariances of Logistic Order Statistics

N	*I*	*J*	*COV*	*N*	*I*	*J*	*COV*	*N*	*I*	*J*	*COV*	*N*	*I*	*J*	*COV*
19	1	6	0.05791	19	4	9	0.04457	20	1	8	0.04246	20	4	8	0.04980
19	1	7	0.04907	19	4	10	0.03995	20	1	9	0.03749	20	4	9	0.04403
19	1	8	0.04257	19	4	11	0.03619	20	1	10	0.03356	20	4	10	0.03946
19	1	9	0.03758	19	4	12	0.03308	20	1	11	0.03037	20	4	11	0.03574
19	1	10	0.03365	19	4	13	0.03046	20	1	12	0.02774	20	4	12	0.03267
19	1	11	0.03045	19	4	14	0.02823	20	1	13	0.02553	20	4	13	0.03008
19	1	12	0.02782	19	4	15	0.02630	20	1	14	0.02364	20	4	14	0.02788
19	1	13	0.02560	19	4	16	0.02462	20	1	15	0.02202	20	4	15	0.02597
19	1	14	0.02371	19	5	5	0.08823	20	1	16	0.02060	20	4	16	0.02431
19	1	15	0.02208	19	5	6	0.07266	20	1	17	0.01935	20	4	17	0.02285
19	1	16	0.02066	19	5	7	0.06177	20	1	18	0.01825	20	5	5	0.08688
19	1	17	0.01941	19	5	8	0.05371	20	1	19	0.01727	20	5	6	0.07153
19	1	18	0.01830	19	5	9	0.04752	20	1	20	0.01638	20	5	7	0.06079
19	1	19	0.01731	19	5	10	0.04261	20	2	2	0.21246	20	5	8	0.05285
19	2	2	0.21340	19	5	11	0.03862	20	2	3	0.13107	20	5	9	0.04675
19	2	3	0.13170	19	5	12	0.03531	20	2	4	0.09457	20	5	10	0.04191
19	2	4	0.09505	19	5	13	0.03252	20	2	5	0.07393	20	5	11	0.03798
19	2	5	0.07432	19	5	14	0.03015	20	2	6	0.06068	20	5	12	0.03473
19	2	6	0.06100	19	5	15	0.02809	20	2	7	0.05145	20	5	13	0.03198
19	2	7	0.05172	19	6	6	0.07762	20	2	8	0.04465	20	5	14	0.02964
19	2	8	0.04489	19	6	7	0.06605	20	2	9	0.03944	20	5	15	0.02762
19	2	9	0.03966	19	6	8	0.05748	20	2	10	0.03532	20	5	16	0.02586
19	2	10	0.03551	19	6	9	0.05089	20	2	11	0.03198	20	6	6	0.07607
19	2	11	0.03215	19	6	10	0.04565	20	2	12	0.02921	20	6	7	0.06471
19	2	12	0.02937	19	6	11	0.04139	20	2	13	0.02689	20	6	8	0.05630
19	2	13	0.02704	19	6	12	0.03786	20	2	14	0.02490	20	6	9	0.04983
19	2	14	0.02504	19	6	13	0.03488	20	2	15	0.02319	20	6	10	0.04470
19	2	15	0.02333	19	6	14	0.03234	20	2	16	0.02170	20	6	11	0.04052
19	2	16	0.02183	19	7	7	0.07098	20	2	17	0.02039	20	6	12	0.03706
19	2	17	0.02051	19	7	8	0.06183	20	2	18	0.01923	20	6	13	0.03414
19	2	18	0.01934	19	7	9	0.05477	20	2	19	0.01820	20	6	14	0.03165
19	3	3	0.13846	19	7	10	0.04916	20	3	3	0.13741	20	6	15	0.02950
19	3	4	0.10017	19	7	11	0.04460	20	3	4	0.09937	20	7	7	0.06918
19	3	5	0.07843	19	7	12	0.04081	20	3	5	0.07779	20	7	8	0.06024
19	3	6	0.06444	19	7	13	0.03761	20	3	6	0.06390	20	7	9	0.05335
19	3	7	0.05468	19	8	8	0.06688	20	3	7	0.05422	20	7	10	0.04788
19	3	8	0.04749	19	8	9	0.05930	20	3	8	0.04708	20	7	11	0.04342
19	3	9	0.04197	19	8	10	0.05326	20	3	9	0.04161	20	7	12	0.03973
19	3	10	0.03760	19	8	11	0.04834	20	3	10	0.03727	20	7	13	0.03661
19	3	11	0.03405	19	8	12	0.04426	20	3	11	0.03375	20	7	14	0.03395
19	3	12	0.03112	19	9	9	0.06465	20	3	12	0.03084	20	8	8	0.06477
19	3	13	0.02865	19	9	10	0.05811	20	3	13	0.02839	20	8	9	0.05741
19	3	14	0.02654	19	9	11	0.05278	20	3	14	0.02631	20	8	10	0.05155
19	3	15	0.02472	19	10	10	0.06393	20	3	15	0.02450	20	8	11	0.04678
19	3	16	0.02314	20	1	1	0.51558	20	3	16	0.02293	20	8	12	0.04281
19	3	17	0.02174	20	1	2	0.20390	20	3	17	0.02155	20	8	13	0.03947
19	4	4	0.10588	20	1	3	0.12531	20	3	18	0.02033	20	9	9	0.06213
19	4	5	0.08304	20	1	4	0.09023	20	4	4	0.10469	20	9	10	0.05583
19	4	6	0.06830	20	1	5	0.07045	20	4	5	0.08208	20	9	11	0.05069
19	4	7	0.05801	20	1	6	0.05776	20	4	6	0.06750	20	9	12	0.04642
19	4	8	0.05041	20	1	7	0.04894	20	4	7	0.05731	20	10	10	0.06089
												20	10	11	0.05532

Table C13.1

EXPECTED VALUES OF PARETO ORDER STATISTICS

[Tabular values are expected values of the Mth order statistic of a sample of size N from a Pareto population with location parameter 0, scale parameter 1, and shape parameter $v = 2.5(0.5)5.0$.]

Expected Values of Pareto Order Statistics (Shape Parameter ν)

N	M	$\nu = 2.5$	$\nu = 3.0$	$\nu = 3.5$	$\nu = 4.0$	$\nu = 4.5$	$\nu = 5.0$
1	1	1.6667	1.5000	1.4000	1.3333	1.2857	1.2500
2	1	1.2500	1.2000	1.1667	1.1429	1.1250	1.1111
2	2	2.0833	1.8000	1.6333	1.5238	1.4464	1.3889
3	1	1.1538	1.1250	1.1053	1.0909	1.0800	1.0714
3	2	1.4423	1.3500	1.2895	1.2468	1.2150	1.1905
3	3	2.4038	2.0250	1.8053	1.6623	1.5621	1.4881
4	1	1.1111	1.0909	1.0769	1.0667	1.0588	1.0526
4	2	1.2821	1.2273	1.1903	1.1636	1.1435	1.1278
4	3	1.6026	1.4727	1.3887	1.3299	1.2865	1.2531
4	4	2.6709	2.2091	1.9441	1.7732	1.6540	1.5664
5	1	1.0870	1.0714	1.0606	1.0526	1.0465	1.0417
5	2	1.2077	1.1688	1.1422	1.1228	1.1081	1.0965
5	3	1.3935	1.3149	1.2624	1.2249	1.1967	1.1748
5	4	1.7419	1.5779	1.4728	1.3999	1.3463	1.3053
5	5	2.9032	2.3669	2.0620	1.8665	1.7310	1.6317
6	1	1.0714	1.0588	1.0500	1.0435	1.0385	1.0345
6	2	1.1616	1.1345	1.1136	1.0984	1.0868	1.0776
6	3	1.2940	1.2376	1.1993	1.1716	1.1507	1.1343
6	4	1.4931	1.3923	1.3255	1.2781	1.2427	1.2153
6	5	1.8663	1.6707	1.5464	1.4607	1.3981	1.3504
6	6	3.1106	2.5061	2.1651	1.9476	1.7975	1.6879
7	1	1.0606	1.0500	1.0426	1.0370	1.0328	1.0294
7	2	1.1364	1.1118	1.0947	1.0821	1.0725	1.0649
7	3	1.2352	1.1912	1.1610	1.1391	1.1224	1.1093
7	4	1.3724	1.2995	1.2503	1.2150	1.1884	1.1677
7	5	1.5836	1.4619	1.3819	1.3255	1.2835	1.2511
7	6	1.9795	1.7543	1.6123	1.5148	1.4439	1.3901
7	7	3.2991	2.6314	2.2572	2.0198	1.8565	1.7376
8	1	1.0526	1.0435	1.0370	1.0323	1.0286	1.0256
8	2	1.1164	1.0957	1.0812	1.0705	1.0623	1.0558
8	3	1.1962	1.1601	1.1352	1.1170	1.1032	1.0922
8	4	1.3002	1.2430	1.2042	1.1758	1.1545	1.1377
8	5	1.4447	1.3560	1.2966	1.2542	1.2224	1.1976
8	6	1.6669	1.5255	1.4331	1.3682	1.3202	1.2831
8	7	2.0836	1.8306	1.6720	1.5637	1.4852	1.4257
8	8	3.4727	2.7458	2.3408	2.0849	1.9095	1.7821

Table C13.2

VARIANCES AND COVARIANCES OF PARETO ORDER STATISTICS

[When $I = J$, tabular values are variances of the Ith order statistic of a sample of size N from a Pareto population with location parameter 0, shape parameter 1, and shape parameter $v = 2.5(0.5)4.0$; when $I \neq J$, they are covariances of the Ith and Jth order statistics.]

Variances and Covariances of Pareto Order Statistics (Shape Parameter ν)

N	I	J	$\nu = 2.5$	$\nu = 3.0$	$\nu = 3.5$	$\nu = 4.0$	$\nu = 4.5$	$\nu = 5.0$
1	1	1	2.2222	0.7500	0.3733	0.2222	0.1469	0.1042
2	1	1	0.1042	0.0600	0.0389	0.0272	0.0201	0.0154
2	1	2	0.1736	0.0900	0.0544	0.0363	0.0258	0.0193
2	2	2	3.9931	1.2600	0.5989	0.3447	0.2221	0.1543
3	1	1	0.0323	0.0201	0.0137	0.0099	0.0075	0.0059
3	1	2	0.0403	0.0241	0.0160	0.0113	0.0085	0.0065
3	1	3	0.0672	0.0362	0.0224	0.0151	0.0109	0.0082
3	2	2	0.1925	0.1061	0.0667	0.0456	0.0331	0.0251
3	2	3	0.3208	0.1591	0.0933	0.0608	0.0425	0.0313
3	3	3	5.5852	1.6851	0.7763	0.4366	0.2765	0.1894
4	1	1	0.0154	0.0099	0.0069	0.0051	0.0039	0.0031
4	1	2	0.0178	0.0112	0.0076	0.0055	0.0042	0.0033
4	1	3	0.0223	0.0134	0.0089	0.0063	0.0047	0.0037
4	1	4	0.0371	0.0201	0.0125	0.0084	0.0061	0.0046
4	2	2	0.0609	0.0367	0.0244	0.0174	0.0130	0.0101
4	2	3	0.0761	0.0440	0.0285	0.0199	0.0146	0.0112
4	2	4	0.1269	0.0660	0.0399	0.0265	0.0188	0.0140
4	3	3	0.2727	0.1454	0.0893	0.0600	0.0430	0.0322
4	3	4	0.4545	0.2180	0.1250	0.0800	0.0553	0.0403
4	4	4	7.0706	2.0628	0.9282	0.5130	0.3205	0.2173
5	1	1	0.0090	0.0059	0.0041	0.0031	0.0024	0.0019
5	1	2	0.0100	0.0064	0.0045	0.0033	0.0025	0.0020
5	1	3	0.0115	0.0072	0.0049	0.0036	0.0027	0.0021
5	1	4	0.0144	0.0087	0.0058	0.0041	0.0031	0.0024
5	1	5	0.0240	0.0130	0.0081	0.0055	0.0039	0.0030
5	2	2	0.0295	0.0184	0.0126	0.0091	0.0069	0.0054
5	2	3	0.0340	0.0208	0.0139	0.0100	0.0075	0.0058
5	2	4	0.0425	0.0249	0.0163	0.0114	0.0084	0.0065
5	2	5	0.0709	0.0374	0.0228	0.0152	0.0108	0.0081
5	3	3	0.0873	0.0512	0.0334	0.0235	0.0174	0.0134
5	3	4	0.1091	0.0614	0.0390	0.0268	0.0195	0.0148
5	3	5	0.1818	0.0921	0.0546	0.0358	0.0251	0.0185
5	4	4	0.3478	0.1805	0.1088	0.0721	0.0511	0.0380
5	4	5	0.5796	0.2707	0.1523	0.0962	0.0657	0.0475
5	5	5	8.4816	2.4089	1.0636	0.5797	0.3583	0.2408
6	1	1	0.0059	0.0039	0.0028	0.0021	0.0016	0.0013
6	1	2	0.0064	0.0042	0.0029	0.0022	0.0017	0.0013
6	1	3	0.0071	0.0045	0.0032	0.0023	0.0018	0.0014
6	1	4	0.0082	0.0051	0.0035	0.0025	0.0019	0.0015
6	1	5	0.0103	0.0061	0.0041	0.0029	0.0022	0.0017

Variances and Covariances of Pareto Order Statistics (Shape Parameter ν)

N	I	J	$\nu = 2.5$	$\nu = 3.0$	$\nu = 3.5$	$\nu = 4.0$	$\nu = 4.5$	$\nu = 5.0$
6	1	6	0.0171	0.0092	0.0057	0.0038	0.0028	0.0021
6	2	2	0.0173	0.0111	0.0077	0.0056	0.0043	0.0034
6	2	3	0.0193	0.0121	0.0083	0.0060	0.0046	0.0036
6	2	4	0.0222	0.0136	0.0092	0.0066	0.0049	0.0038
6	2	5	0.0278	0.0163	0.0107	0.0075	0.0055	0.0043
6	2	6	0.0463	0.0245	0.0150	0.0100	0.0071	0.0053
6	3	3	0.0426	0.0261	0.0175	0.0126	0.0095	0.0074
6	3	4	0.0492	0.0293	0.0194	0.0137	0.0102	0.0079
6	3	5	0.0615	0.0352	0.0226	0.0157	0.0115	0.0088
6	3	6	0.1024	0.0528	0.0317	0.0209	0.0148	0.0109
6	4	4	0.1121	0.0643	0.0413	0.0287	0.0210	0.0161
6	4	5	0.1402	0.0772	0.0482	0.0328	0.0237	0.0178
6	4	6	0.2336	0.1157	0.0675	0.0437	0.0304	0.0223
6	5	5	0.4191	0.2127	0.1262	0.0827	0.0581	0.0429
6	5	6	0.6985	0.3191	0.1767	0.1103	0.0747	0.0536
6	6	6	9.8361	2.7318	1.1874	0.6396	0.3918	0.2614
7	1	1	0.0041	0.0028	0.0020	0.0015	0.0011	0.0009
7	1	2	0.0044	0.0029	0.0021	0.0015	0.0012	0.0009
7	1	3	0.0048	0.0031	0.0022	0.0016	0.0012	0.0010
7	1	4	0.0054	0.0034	0.0024	0.0017	0.0013	0.0010
7	1	5	0.0062	0.0038	0.0026	0.0019	0.0014	0.0011
7	1	6	0.0077	0.0046	0.0030	0.0022	0.0016	0.0012
7	1	7	0.0129	0.0069	0.0043	0.0029	0.0021	0.0015
7	2	2	0.0114	0.0074	0.0052	0.0038	0.0029	0.0023
7	2	3	0.0124	0.0079	0.0055	0.0040	0.0031	0.0024
7	2	4	0.0138	0.0086	0.0059	0.0043	0.0033	0.0026
7	2	5	0.0159	0.0097	0.0065	0.0047	0.0035	0.0027
7	2	6	0.0199	0.0117	0.0076	0.0054	0.0040	0.0030
7	2	7	0.0331	0.0175	0.0107	0.0071	0.0051	0.0038
7	3	3	0.0252	0.0158	0.0108	0.0079	0.0060	0.0047
7	3	4	0.0280	0.0173	0.0117	0.0084	0.0063	0.0049
7	3	5	0.0323	0.0194	0.0129	0.0091	0.0068	0.0053
7	3	6	0.0404	0.0233	0.0150	0.0105	0.0077	0.0059
7	3	7	0.0673	0.0349	0.0210	0.0139	0.0099	0.0073
7	4	4	0.0550	0.0331	0.0219	0.0156	0.0116	0.0090
7	4	5	0.0635	0.0372	0.0242	0.0170	0.0125	0.0096
7	4	6	0.0794	0.0446	0.0283	0.0194	0.0141	0.0107
7	4	7	0.1323	0.0669	0.0396	0.0259	0.0181	0.0134
7	5	5	0.1359	0.0764	0.0485	0.0333	0.0242	0.0184
7	5	6	0.1698	0.0917	0.0566	0.0381	0.0273	0.0204

Variances and Covariances of Pareto Order Statistics (Shape Parameter ν)

N	I	J	$\nu = 2.5$	$\nu = 3.0$	$\nu = 3.5$	$\nu = 4.0$	$\nu = 4.5$	$\nu = 5.0$
7	5	7	0.2830	0.1375	0.0792	0.0508	0.0351	0.0255
7	6	6	0.4876	0.2429	0.1422	0.0922	0.0643	0.0471
7	6	7	0.8127	0.3643	0.1990	0.1230	0.0826	0.0589
7	7	7	11.1454	3.0367	1.3021	0.6944	0.4220	0.2799
8	1	1	0.0031	0.0021	0.0015	0.0011	0.0009	0.0007
8	1	2	0.0033	0.0022	0.0015	0.0012	0.0009	0.0007
8	1	3	0.0035	0.0023	0.0016	0.0012	0.0009	0.0007
8	1	4	0.0038	0.0025	0.0017	0.0013	0.0010	0.0008
8	1	5	0.0042	0.0027	0.0018	0.0013	0.0010	0.0008
8	1	6	0.0049	0.0030	0.0020	0.0015	0.0011	0.0009
8	1	7	0.0061	0.0036	0.0024	0.0017	0.0012	0.0010
8	1	8	0.0102	0.0054	0.0033	0.0022	0.0016	0.0012
8	2	2	0.0081	0.0053	0.0037	0.0028	0.0021	0.0017
8	2	3	0.0086	0.0056	0.0039	0.0029	0.0022	0.0018
8	2	4	0.0094	0.0060	0.0042	0.0030	0.0023	0.0018
8	2	5	0.0104	0.0065	0.0045	0.0032	0.0025	0.0019
8	2	6	0.0120	0.0074	0.0049	0.0035	0.0027	0.0021
8	2	7	0.0151	0.0088	0.0058	0.0040	0.0030	0.0023
8	2	8	0.0251	0.0133	0.0081	0.0054	0.0038	0.0029
8	3	3	0.0166	0.0106	0.0074	0.0054	0.0041	0.0032
8	3	4	0.0181	0.0114	0.0078	0.0057	0.0043	0.0034
8	3	5	0.0201	0.0124	0.0084	0.0060	0.0046	0.0036
8	3	6	0.0232	0.0140	0.0093	0.0066	0.0049	0.0038
8	3	7	0.0290	0.0168	0.0108	0.0075	0.0055	0.0042
8	3	8	0.0483	0.0251	0.0152	0.0100	0.0071	0.0053
8	4	4	0.0327	0.0202	0.0136	0.0098	0.0074	0.0058
8	4	5	0.0363	0.0220	0.0147	0.0105	0.0078	0.0061
8	4	6	0.0419	0.0248	0.0162	0.0114	0.0085	0.0065
8	4	7	0.0524	0.0297	0.0189	0.0131	0.0095	0.0072
8	4	8	0.0873	0.0446	0.0265	0.0174	0.0122	0.0090
8	5	5	0.0670	0.0395	0.0259	0.0182	0.0135	0.0104
8	5	6	0.0773	0.0445	0.0287	0.0199	0.0146	0.0111
8	5	7	0.0966	0.0534	0.0334	0.0228	0.0164	0.0124
8	5	8	0.1610	0.0801	0.0468	0.0303	0.0211	0.0155
8	6	6	0.1587	0.0878	0.0550	0.0375	0.0271	0.0204
8	6	7	0.1983	0.1053	0.0642	0.0429	0.0305	0.0227
8	6	8	0.3306	0.1580	0.0899	0.0571	0.0392	0.0284
8	7	7	0.5539	0.2713	0.1569	0.1009	0.0698	0.0510
8	7	8	0.9232	0.4069	0.2197	0.1346	0.0898	0.0637
8	8	8	12.4173	3.3270	1.4098	0.7452	0.4498	0.2967

Table C14.1

EXPECTED VALUES OF STUDENT'S t ORDER STATISTICS

[Tabular values are expected values of the Mth largest value in a sample of size N from a Student's t population with location parameter 0, scale parameter 1, and degrees of freedom
3(1)19.]

Expected Values of Student's t Order Statistics

N	M	$d.f. = 3$	$d.f. = 4$	$d.f. = 5$	$d.f. = 6$	$d.f. = 7$	$d.f. = 8$	$d.f. = 9$	$d.f. = 10$
1	1	0.00000	0.00000	0.00000	0.00000	0.00000	0.00000	0.00000	0.00000
2	1	0.47746	0.52065	0.53601	0.54360	0.54806	0.55096	0.55300	0.55450
3	1	0.71620	0.78098	0.80402	0.81541	0.82209	0.82644	0.82949	0.83174
3	2	0.00000	0.00000	0.00000	0.00000	0.00000	0.00000	0.00000	0.00000
4	1	0.88438	0.95975	0.98558	0.99800	1.00514	1.00972	1.01288	1.01519
4	2	0.21165	0.24466	0.25934	0.26762	0.27293	0.27662	0.27933	0.28141
5	1	1.01729	1.09774	1.12392	1.13600	1.14270	1.14689	1.14971	1.15173
5	2	0.35275	0.40777	0.43223	0.44603	0.45488	0.46103	0.46555	0.46901
5	3	0.00000	0.00000	0.00000	0.00000	0.00000	0.00000	0.00000	0.00000
6	1	1.12862	1.21103	1.23620	1.24719	1.25299	1.25646	1.25871	1.26026
6	2	0.46060	0.53130	0.56251	0.58005	0.59126	0.59904	0.60475	0.60911
6	3	0.13705	0.16072	0.17167	0.17800	0.18211	0.18501	0.18716	0.18881
7	1	1.22524	1.30764	1.33101	1.34047	1.34511	1.34768	1.34922	1.35020
7	2	0.54891	0.63132	0.66735	0.68747	0.70028	0.70915	0.71564	0.72059
7	3	0.23984	0.28126	0.30042	0.31149	0.31870	0.32377	0.32752	0.33041
7	4	0.00000	0.00000	0.00000	0.00000	0.00000	0.00000	0.00000	0.00000
8	1	1.31110	1.39220	1.41326	1.42094	1.42426	1.42582	1.42659	1.42695
8	2	0.62425	0.71571	0.75525	0.77718	0.79108	0.80066	0.80766	0.81299
8	3	0.32286	0.37815	0.40365	0.41835	0.42790	0.43461	0.43958	0.44341
8	4	0.10147	0.11976	0.12838	0.13340	0.13670	0.13902	0.14075	0.14209
9	1	1.38869	1.46761	1.48604	1.49178	1.49368	1.49419	1.49413	1.49384
9	2	0.69033	0.78893	0.83105	0.85424	0.86886	0.87890	0.88622	0.89178
9	3	0.39297	0.45944	0.48993	0.50746	0.51883	0.52680	0.53269	0.53723
9	4	0.18265	0.21557	0.23108	0.24012	0.24606	0.25024	0.25336	0.25577
9	5	0.00000	0.00000	0.00000	0.00000	0.00000	0.00000	0.00000	0.00000
10	1	1.45972	1.53582	1.55140	1.55511	1.55555	1.55496	1.55407	1.55312
10	2	0.74943	0.85375	0.89777	0.92182	0.93690	0.94722	0.95471	0.96039
10	3	0.45395	0.52964	0.56415	0.58390	0.59669	0.60564	0.61225	0.61733
10	4	0.25069	0.29565	0.31677	0.32908	0.33714	0.34283	0.34706	0.35033
10	5	0.08059	0.09547	0.10254	0.10669	0.10942	0.11136	0.11280	0.11391
11	1	1.52539	1.59820	1.61080	1.61242	1.61137	1.60968	1.60794	1.60632
11	2	0.80306	0.91202	0.95743	0.98203	0.99738	1.00783	1.01539	1.02111
11	3	0.50811	0.59154	0.62932	0.65086	0.66476	0.67447	0.68163	0.68713
11	4	0.30951	0.36457	0.39036	0.40535	0.41516	0.42208	0.42722	0.43118
11	5	0.14775	0.17502	0.18799	0.19560	0.20061	0.20416	0.20680	0.20844
11	6	0.00000	0.00000	0.00000	0.00000	0.00000	0.00000	0.00000	0.00000
12	1	1.58658	1.65576	1.66529	1.66479	1.66224	1.65944	1.65685	1.65457
12	2	0.85227	0.96503	1.01143	1.03635	1.05181	1.06229	1.06985	1.07555
12	3	0.55698	0.64699	0.68746	0.71044	0.72523	0.73553	0.74312	0.74894
12	4	0.36150	0.42519	0.45489	0.47212	0.48337	0.49130	0.49718	0.50172
12	5	0.20553	0.24334	0.26129	0.27182	0.27874	0.28364	0.28729	0.29011
12	6	0.06685	0.07937	0.08537	0.08890	0.09123	0.09288	0.09411	0.09507
13	1	1.64398	1.70926	1.71567	1.71303	1.70898	1.70508	1.70166	1.69871
13	2	0.89784	1.01371	1.06077	1.08585	1.10130	1.11173	1.11922	1.12486
13	3	0.60162	0.69728	0.73999	0.76412	0.77961	0.79038	0.79829	0.80435
13	4	0.40820	0.47936	0.51239	0.53150	0.54396	0.55272	0.55921	0.56422
13	5	0.25640	0.30330	0.32551	0.33852	0.34706	0.35311	0.35761	0.36109
13	6	0.12414	0.14741	0.15854	0.16509	0.16942	0.17249	0.17478	0.17655
13	7	0.00000	0.00000	0.00000	0.00000	0.00000	0.00000	0.00000	0.00000
14	1	1.69810	1.75930	1.76255	1.75778	1.75223	1.74724	1.74298	1.73939
14	2	0.94036	1.05877	1.10625	1.13133	1.14668	1.15699	1.16437	1.16991
14	3	0.64278	0.74334	0.78791	0.81298	0.82902	0.84014	0.84831	0.85455
14	4	0.45070	0.52839	0.56427	0.58497	0.59844	0.60789	0.61490	0.62029
14	5	0.30195	0.35679	0.38269	0.39782	0.40776	0.41478	0.42000	0.42404
14	6	0.17442	0.20702	0.22260	0.23177	0.23782	0.24211	0.24531	0.24779

Expected Values of Student's t Order Statistics

N	M	$d.f. = 3$	$d.f. = 4$	$d.f. = 5$	$d.f. = 6$	$d.f. = 7$	$d.f. = 8$	$d.f. = 9$	$d.f. = 10$
14	7	0.05711	0.06793	0.07312	0.07619	0.07822	0.07966	0.08074	0.08157
15	1	1.74938	1.80634	1.80641	1.79952	1.79249	1.78642	1.78134	1.77711
15	2	0.98025	1.10076	1.14844	1.17341	1.18858	1.19873	1.20596	1.21137
15	3	0.68103	0.78588	0.83200	0.85783	0.87429	0.88569	0.89404	0.90041
15	4	0.48977	0.57321	0.61156	0.63361	0.64793	0.65797	0.66539	0.67111
15	5	0.34327	0.40513	0.43423	0.45121	0.46234	0.47019	0.47604	0.48055
15	6	0.21931	0.26012	0.27959	0.29105	0.29859	0.30394	0.30793	0.31102
15	7	0.10709	0.12736	0.13711	0.14286	0.14666	0.14936	0.15138	0.15294
15	8	0.00000	0.00000	0.00000	0.00000	0.00000	0.00000	0.00000	0.00000
16	1	1.79814	1.85076	1.84765	1.83865	1.83016	1.82302	1.81713	1.81226
16	2	1.01789	1.14009	1.18782	1.21257	1.22752	1.23745	1.24451	1.24976
16	3	0.71681	0.82542	0.87284	0.89927	0.91607	0.92766	0.93614	0.94261
16	4	0.52597	0.61453	0.65502	0.67823	0.69326	0.70379	0.71157	0.71755
16	5	0.38116	0.44925	0.48118	0.49976	0.51192	0.52050	0.52687	0.53179
16	6	0.25993	0.30805	0.33095	0.34440	0.35325	0.35953	0.36420	0.36782
16	7	0.15160	0.18025	0.19400	0.20212	0.20749	0.21129	0.21414	0.21634
16	8	0.04986	0.05937	0.06395	0.06667	0.06846	0.06974	0.07069	0.07143
17	1	1.84468	1.89286	1.88658	1.87549	1.86555	1.85736	1.85068	1.84518
17	2	1.05355	1.17712	1.22474	1.24920	1.26387	1.27357	1.28042	1.28551
17	3	0.75046	0.86239	0.91090	0.93781	0.95485	0.96659	0.97516	0.98168
17	4	0.55975	0.65288	0.69524	0.71944	0.73508	0.74602	0.75409	0.76029
17	5	0.41619	0.48988	0.52429	0.54428	0.55734	0.56654	0.57337	0.57864
17	6	0.29709	0.35175	0.37770	0.39291	0.40292	0.41000	0.41527	0.41935
17	7	0.19180	0.22793	0.24525	0.25546	0.26221	0.26699	0.27057	0.27334
17	8	0.09417	0.11214	0.12080	0.12592	0.12931	0.13172	0.13353	0.13492
17	9	0.00000	0.00000	0.00000	0.00000	0.00000	0.00000	0.00000	0.00000
18	1	1.88922	1.93290	1.92347	1.91031	1.89894	1.88971	1.88224	1.87614
18	2	1.08746	1.21212	1.25951	1.28362	1.29798	1.30740	1.31404	1.31894
18	3	0.78226	0.89713	0.94654	0.97382	0.99104	1.00287	1.01149	1.01804
18	4	0.59145	0.68869	0.73268	0.75774	0.77390	0.78517	0.79348	0.79986
18	5	0.44880	0.52754	0.56417	0.58539	0.59924	0.60898	0.61620	0.62177
18	6	0.33139	0.39195	0.42061	0.43739	0.44841	0.45620	0.46201	0.46650
18	7	0.22850	0.27136	0.29187	0.30396	0.31193	0.31758	0.32180	0.32507
18	8	0.13412	0.15967	0.17198	0.17926	0.18407	0.18749	0.19005	0.19203
18	9	0.04424	0.05272	0.05683	0.05926	0.06087	0.06201	0.06287	0.06353
19	1	1.93197	1.97110	1.95853	1.94332	1.93054	1.92029	1.91206	1.90535
19	2	1.11981	1.24532	1.29239	1.31609	1.33009	1.33924	1.34563	1.35034
19	3	0.81244	0.92992	0.98007	1.00763	1.02497	1.03685	1.04549	1.05205
19	4	0.62135	0.72230	0.76772	0.79351	0.81010	0.82166	0.83017	0.83670
19	5	0.47935	0.56267	0.60128	0.62359	0.63812	0.64833	0.65589	0.66172
19	6	0.36326	0.42918	0.46028	0.47845	0.49038	0.49880	0.50507	0.50991
19	7	0.26232	0.31128	0.33466	0.34841	0.35748	0.36391	0.36871	0.37242
19	8	0.17054	0.20294	0.21853	0.22774	0.23384	0.23816	0.24140	0.24391
19	9	0.08407	0.10018	0.10797	0.11259	0.11565	0.11782	0.11945	0.12071
19	10	0.00000	0.00000	0.00000	0.00000	0.00000	0.00000	0.00000	0.00000
20	1	1.97309	2.00764	1.99195	1.97471	1.96055	1.94929	1.94030	1.93300
20	2	1.15077	1.27691	1.32357	1.34683	1.36045	1.36928	1.37543	1.37994
20	3	0.84116	0.96097	1.01173	1.03949	1.05690	1.06879	1.07743	1.08397
20	4	0.64966	0.75397	0.80066	0.82708	0.84404	0.85583	0.86450	0.87114
20	5	0.50811	0.59561	0.63598	0.65925	0.67438	0.68500	0.69286	0.69892
20	6	0.39307	0.46387	0.49718	0.51660	0.52932	0.53830	0.54498	0.55014
20	7	0.29370	0.34822	0.37419	0.38946	0.39951	0.40663	0.41194	0.41605
20	8	0.20404	0.24267	0.26123	0.27219	0.27943	0.28458	0.28842	0.29140
20	9	0.12029	0.14334	0.15448	0.16107	0.16544	0.16854	0.17087	0.17267
20	10	0.03975	0.04742	0.05113	0.05333	0.05479	0.05583	0.05660	0.05721

Expected Values of Student's t Order Statistics

N	M	$d.f. = 11$	$d.f. = 12$	$d.f. = 13$	$d.f. = 14$	$d.f. = 15$	$d.f. = 16$	$d.f. = 17$	$d.f. = 18$	$d.f. = 19$
1	1	0.00000	0.00000	0.00000	0.00000	0.00000	0.00000	0.00000	0.00000	0.00000
2	1	0.55565	0.55656	0.55729	0.55790	0.55841	0.55884	0.55922	0.55954	0.55983
3	1	0.83347	0.83483	0.83594	0.83685	0.83762	0.83827	0.83883	0.83932	0.83974
3	2	0.00000	0.00000	0.00000	0.00000	0.00000	0.00000	0.00000	0.00000	0.00000
4	1	1.01694	1.01832	1.01943	1.02033	1.02109	1.02173	1.02228	1.02276	1.02318
4	2	0.28305	0.28438	0.28548	0.28640	0.28719	0.28786	0.28846	0.28897	0.28943
5	1	1.15324	1.15441	1.15533	1.15608	1.15670	1.15722	1.15767	1.15805	1.15838
5	2	0.47175	0.47396	0.47580	0.47733	0.47864	0.47977	0.48076	0.48162	0.48239
5	3	0.00000	0.00000	0.00000	0.00000	0.00000	0.00000	0.00000	0.00000	0.00000
6	1	1.26138	1.26222	1.26287	1.26338	1.26380	1.26414	1.26442	1.26466	1.26486
6	2	0.61256	0.61535	0.61766	0.61959	0.62124	0.62266	0.62389	0.62498	0.62594
6	3	0.19012	0.19119	0.19207	0.19282	0.19346	0.19401	0.19449	0.19491	0.19529
7	1	1.35086	1.35131	1.35164	1.35187	1.35204	1.35217	1.35227	1.35235	1.35241
7	2	0.72450	0.72766	0.73027	0.73245	0.73431	0.73591	0.73731	0.73853	0.73962
7	3	0.33271	0.33458	0.33613	0.33744	0.33855	0.33951	0.34036	0.34110	0.34175
7	4	0.00000	0.00000	0.00000	0.00000	0.00000	0.00000	0.00000	0.00000	0.00000
8	1	1.42710	1.42713	1.42710	1.42704	1.42695	1.42685	1.42675	1.42665	1.42655
8	2	0.81719	0.82057	0.82337	0.82571	0.82770	0.82941	0.83090	0.83221	0.83336
8	3	0.44645	0.44892	0.45096	0.45269	0.45416	0.45543	0.45654	0.45752	0.45838
8	4	0.14316	0.14402	0.14474	0.14535	0.14587	0.14632	0.14671	0.14706	0.14737
9	1	1.49347	1.49307	1.49267	1.49229	1.49194	1.49161	1.49130	1.49102	1.49076
9	2	0.89614	0.89967	0.90256	0.90499	0.90705	0.90882	0.91036	0.91171	0.91291
9	3	0.54083	0.54375	0.54617	0.54821	0.54996	0.55146	0.55277	0.55392	0.55495
9	4	0.25768	0.25924	0.26054	0.26163	0.26257	0.26338	0.26408	0.26471	0.26526
9	5	0.00000	0.00000	0.00000	0.00000	0.00000	0.00000	0.00000	0.00000	0.00000
10	1	1.55220	1.55136	1.55059	1.54990	1.54927	1.54871	1.54819	1.54773	1.54730
10	2	0.96484	0.96843	0.97137	0.97384	0.97593	0.97772	0.97928	0.98064	0.98186
10	3	0.62135	0.62462	0.62733	0.62961	0.63155	0.63323	0.63469	0.63598	0.63712
10	4	0.35293	0.35505	0.35681	0.35829	0.35956	0.36066	0.36162	0.36246	0.36321
10	5	0.11480	0.11553	0.11613	0.11664	0.11708	0.11745	0.11778	0.11807	0.11833
11	1	1.60487	1.60358	0.60244	1.60143	1.60053	1.59973	1.59901	1.59836	1.59777
11	2	1.02559	1.02918	1.03213	1.03460	1.03669	1.03848	1.04004	1.04140	1.04260
11	3	0.69149	0.69502	0.69794	0.70040	0.70249	0.70430	0.70588	0.70726	0.70849
11	4	0.43434	0.43691	0.43904	0.44084	0.44237	0.44370	0.44486	0.44589	0.44679
11	5	0.21047	0.21180	0.21291	0.21384	0.21464	0.21533	0.21593	0.21647	0.21694
11	6	0.00000	0.00000	0.00000	0.00000	0.00000	0.00000	0.00000	0.00000	0.00000
12	1	1.65258	1.65085	1.64934	1.64802	1.64685	1.64582	1.64489	1.64406	1.64332
12	2	1.08000	1.08357	1.08649	1.08893	1.09100	1.09277	1.09430	1.09564	1.09683
12	3	0.75354	0.75727	0.76035	0.76295	0.76516	0.76706	0.76872	0.77018	0.77148
12	4	0.50533	0.50826	0.51070	0.51275	0.51451	0.51602	0.51735	0.51851	0.51955
12	5	0.29236	0.29420	0.29573	0.29701	0.29811	0.29907	0.29990	0.30063	0.30129
12	6	0.09583	0.09645	0.09697	0.09740	0.09778	0.09810	0.09838	0.09863	0.09885
13	1	1.69619	1.69403	1.69215	1.69052	1.68909	1.68782	1.68669	1.68569	1.68478
13	2	1.12925	1.13276	1.13563	1.13802	1.14005	1.14178	1.14328	1.14459	1.14575
13	3	0.80914	0.81302	0.81622	0.81892	0.82121	0.82319	0.82491	0.82642	0.82776
13	4	0.56820	0.57143	0.57412	0.57638	0.57831	0.57998	0.58144	0.58272	0.58386
13	5	0.36386	0.36613	0.36800	0.36959	0.37095	0.37212	0.37314	0.37405	0.37485
13	6	0.17797	0.17912	0.18008	0.18089	0.18159	0.18219	0.18271	0.18317	0.18359
13	7	0.00000	0.00000	0.00000	0.00000	0.00000	0.00000	0.00000	0.00000	0.00000
14	1	1.73635	1.73375	1.73151	1.72958	1.72788	1.72639	1.72506	1.72388	1.72282
14	2	1.17421	1.17764	1.18044	1.18277	1.18474	1.18643	1.18788	1.18916	1.19028
14	3	0.85948	0.86346	0.86676	0.86952	0.87187	0.87390	0.87566	0.87721	0.87859
14	4	0.62457	0.62806	0.63094	0.63337	0.63545	0.63724	0.63881	0.64018	0.64141
14	5	0.42725	0.42987	0.43205	0.43389	0.43546	0.43682	0.43800	0.43905	0.43998
14	6	0.24976	0.25138	0.25272	0.25385	0.25482	0.25566	0.25639	0.25704	0.25761

Expected Values of Student's t Order Statistics

N	M	$d.f. = 11$	$d.f. = 12$	$d.f. = 13$	$d.f. = 14$	$d.f. = 15$	$d.f. = 16$	$d.f. = 17$	$d.f. = 18$	$d.f. = 19$
14	7	0.08224	0.08278	0.08323	0.08361	0.08394	0.08422	0.08447	0.08469	0.08488
15	1	1.77354	1.77052	1.76794	1.76570	1.76375	1.76203	1.76051	1.75915	1.75794
15	2	1.21556	1.21889	1.22161	1.22387	1.22577	1.22740	1.22880	1.23003	1.23111
15	3	0.90544	0.90950	0.91285	0.91566	0.91805	0.92012	0.92191	0.92348	0.92488
15	4	0.67564	0.67932	0.68238	0.68495	0.68714	0.68903	0.69069	0.69214	0.69343
15	5	0.48414	0.48707	0.48950	0.49155	0.49330	0.49482	0.49614	0.49731	0.49834
15	6	0.31348	0.31549	0.31716	0.31857	0.31977	0.32082	0.32173	0.32254	0.32325
15	7	0.15419	0.15521	0.15606	0.15677	0.15739	0.15792	0.15838	0.15879	0.15916
15	8	0.00000	0.00000	0.00000	0.00000	0.00000	0.00000	0.00000	0.00000	0.00000
16	1	1.80819	1.80476	1.80182	1.79929	1.79708	1.79515	1.79344	1.79191	1.79055
16	2	1.25382	1.25705	1.25967	1.26184	1.26367	1.26524	1.26658	1.26776	1.26879
16	3	0.94770	0.95181	0.95520	0.95804	0.96046	0.96254	0.96435	0.96594	0.96734
16	4	0.72229	0.72613	0.72932	0.73201	0.73429	0.73627	0.73799	0.73951	0.74085
16	5	0.53571	0.53889	0.54154	0.54377	0.54568	0.54733	0.54877	0.55003	0.55116
16	6	0.37070	0.37306	0.37501	0.37666	0.37808	0.37930	0.38037	0.38131	0.38214
16	7	0.21810	0.21954	0.22073	0.22174	0.22261	0.22336	0.22401	0.22459	0.22510
16	8	0.07202	0.07250	0.07290	0.07324	0.07353	0.07379	0.07401	0.07420	0.07437
17	1	1.84062	1.83677	1.83349	1.83067	1.82822	1.82607	1.82418	1.82249	1.82097
17	2	1.28942	1.29253	1.29504	1.29713	1.29888	1.30037	1.30165	1.30277	1.30376
17	3	0.98681	0.99095	0.99436	0.99721	0.99964	1.00173	1.00355	1.00514	1.00655
17	4	0.76520	0.76918	0.77248	0.77525	0.77762	0.77966	0.78144	0.78301	0.78439
17	5	0.58283	0.58624	0.58907	0.59146	0.59349	0.59526	0.59679	0.59815	0.59935
17	6	0.42261	0.42526	0.42746	0.42932	0.43091	0.43229	0.43350	0.43456	0.43550
17	7	0.27555	0.27735	0.27885	0.28012	0.28120	0.28214	0.28296	0.28369	0.28433
17	8	0.13604	0.13695	0.13771	0.13835	0.13890	0.13937	0.13979	0.14016	0.14048
17	9	0.00000	0.00000	0.00000	0.00000	0.00000	0.00000	0.00000	0.00000	0.00000
18	1	1.87108	1.86683	1.86322	1.86012	1.85743	1.85507	1.85299	1.85114	1.84948
18	2	1.32270	1.32567	1.32808	1.33007	1.33173	1.33315	1.33437	1.33543	1.33636
18	3	1.02319	1.02734	1.03075	1.03361	1.03604	1.03812	1.03994	1.04153	1.04293
18	4	0.80491	0.80900	0.81239	0.81524	0.81766	0.81976	0.82158	0.82319	0.82461
18	5	0.62620	0.62980	0.63279	0.63530	0.63746	0.63931	0.64094	0.64236	0.64363
18	6	0.47007	0.47298	0.47540	0.47745	0.47919	0.48071	0.48203	0.48319	0.48422
18	7	0.32768	0.32981	0.33158	0.33308	0.33436	0.33546	0.33643	0.33729	0.33804
18	8	0.19362	0.19491	0.19599	0.19690	0.19768	0.19835	0.19894	0.19946	0.19992
18	9	0.06406	0.06450	0.06486	0.06516	0.06542	0.06565	0.06585	0.06602	0.06618
19	1	1.89981	1.89517	1.89123	1.88785	1.88492	1.88236	1.88010	1.87809	1.87629
19	2	1.35394	1.35677	1.35906	1.36095	1.36252	1.36386	1.36501	1.36601	1.36689
19	3	1.05719	1.06133	1.06474	1.06759	1.07001	1.07208	1.07389	1.07547	1.07687
19	4	0.84186	0.84604	0.84950	0.85240	0.85488	0.85701	0.85887	0.86051	0.86196
19	5	0.66635	0.67012	0.67324	0.67587	0.67811	0.68005	0.68175	0.68324	0.68456
19	6	0.51377	0.51691	0.51952	0.52173	0.52361	0.52524	0.52666	0.52792	0.52903
19	7	0.37538	0.37780	0.37981	0.38151	0.38296	0.38422	0.38531	0.38628	0.38714
19	8	0.24591	0.24754	0.24890	0.25005	0.25104	0.25189	0.25264	0.25329	0.25388
19	9	0.12171	0.12254	0.12323	0.12381	0.12430	0.12473	0.12511	0.12544	0.12574
19	10	0.00000	0.00000	0.00000	0.00000	0.00000	0.00000	0.00000	0.00000	0.00000
20	1	1.92699	1.92197	1.91771	1.91406	1.91089	1.90813	1.90569	1.90353	1.90159
20	2	1.38336	1.38606	1.38822	1.39000	1.39149	1.39275	1.39383	1.39476	1.39558
20	3	1.08910	1.09322	1.09661	1.09944	1.10184	1.10390	1.10569	1.10726	1.10865
20	4	0.87639	0.88064	0.88415	0.88710	0.88962	0.89178	0.89367	0.89533	0.89680
20	5	0.70372	0.70762	0.71086	0.71359	0.71591	0.71792	0.71968	0.72122	0.72259
20	6	0.55425	0.55759	0.56037	0.56271	0.56471	0.56645	0.56796	0.56929	0.57047
20	7	0.41933	0.42200	0.42422	0.42610	0.42770	0.42909	0.43031	0.43138	0.43233
20	8	0.29378	0.29572	0.29733	0.29870	0.29987	0.30088	0.30176	0.30254	0.30323
20	9	0.17411	0.17528	0.17626	0.17709	0.17780	0.17841	0.17895	0.17942	0.17984
20	10	0.05769	0.05808	0.05841	0.05869	0.05892	0.05913	0.05931	0.05947	0.05961

Table C14.2

VARIANCES AND COVARIANCES OF STUDENT'S t ORDER STATISTICS

[When $I = J$, tabular values are variances of the Ith order statistic of a sample of size N from a Student's t population with location parameter 0, scale parameter 1, and degrees of freedom $= 3(1)19$; when $I \neq J$, they are covariances of the Ith and Jth order statistics.]

Variances and Covariances of Student's t Order Statistics

N	I	J	$d.f. = 3$	$d.f. = 4$	$d.f. = 5$	$d.f. = 6$	$d.f. = 7$	$d.f. = 8$	$d.f. = 9$	$d.f. = 10$
1	1	1	1.00000	1.00000	1.00000	1.00000	1.00000	1.00000	1.00000	1.00000
2	1	1	0.77203	0.72892	0.71269	0.70449	0.69963	0.69644	0.69420	0.69253
2	1	2	0.22797	0.27108	0.28731	0.29551	0.30037	0.30356	0.30580	0.30747
3	1	1	0.86702	0.73293	0.67862	0.64980	0.63206	0.62010	0.61149	0.60501
3	1	2	0.18998	0.22857	0.24380	0.25175	0.25657	0.25980	0.26211	0.26383
3	1	3	0.13298	0.15278	0.15884	0.16139	0.16268	0.16340	0.16385	0.16414
3	2	2	0.24009	0.31429	0.34986	0.37063	0.38422	0.39380	0.40090	0.40638
4	1	1	0.97778	0.76460	0.67877	0.63336	0.60548	0.58668	0.57317	0.56301
4	1	2	0.19277	0.22233	0.23201	0.23641	0.23882	0.24030	0.24128	0.24198
4	1	3	0.11119	0.13376	0.14232	0.14661	0.14913	0.15077	0.15192	0.15276
4	1	4	0.09821	0.10788	0.10944	0.10949	0.10920	0.10885	0.10852	0.10821
4	2	2	0.19530	0.25442	0.28260	0.29901	0.30973	0.31728	0.32288	0.32719
4	2	3	0.12079	0.16092	0.18054	0.19209	0.19969	0.20505	0.20904	0.21212
5	1	1	1.08754	0.80196	0.68873	0.62936	0.59313	0.56881	0.55140	0.53833
5	1	2	0.20236	0.22370	0.22815	0.22919	0.22930	0.22911	0.22884	0.22855
5	1	3	0.11008	0.13022	0.13730	0.14065	0.14253	0.14371	0.14450	0.14507
5	1	4	0.08095	0.09516	0.09991	0.10204	0.10317	0.10385	0.10429	0.10459
5	1	5	0.08028	0.08463	0.08390	0.08273	0.08168	0.08083	0.08013	0.07956
5	2	2	0.18544	0.23432	0.25620	0.26851	0.27637	0.28181	0.28580	0.28885
5	2	3	0.10607	0.14151	0.15891	0.16919	0.17596	0.18074	0.18430	0.18705
5	2	4	0.08012	0.10551	0.11758	0.12456	0.12909	0.13226	0.13459	0.13639
5	3	3	0.13541	0.18482	0.21010	0.22540	0.23563	0.24295	0.24845	0.25272
6	1	1	1.19369	0.84011	0.70221	0.63065	0.58727	0.55830	0.53763	0.52216
6	1	2	0.21389	0.22771	0.22743	0.22547	0.22354	0.22190	0.22053	0.21939
6	1	3	0.11290	0.13019	0.13536	0.13745	0.13845	0.13899	0.13929	0.13948
6	1	4	0.07889	0.09261	0.09713	0.09913	0.10018	0.10080	0.10120	0.10147
6	1	5	0.06509	0.07451	0.07706	0.07794	0.07828	0.07841	0.07844	0.07843
6	1	6	0.06928	0.07043	0.06841	0.06657	0.06514	0.06403	0.06315	0.06245
6	2	2	0.18496	0.22621	0.24307	0.25202	0.25751	0.26121	0.26386	0.26585
6	2	3	0.10159	0.13324	0.14828	0.15699	0.16265	0.16662	0.16956	0.17181
6	2	4	0.07245	0.09626	0.10779	0.11453	0.11894	0.12204	0.12434	0.12611
6	2	5	0.06057	0.07823	0.08624	0.09073	0.09358	0.09554	0.09697	0.09806
6	3	3	0.11663	0.15898	0.18063	0.19372	0.20247	0.20873	0.21342	0.21708
6	3	4	0.08525	0.11698	0.13333	0.14325	0.14990	0.15467	0.15824	0.16102
7	1	1	1.29593	0.87759	0.71702	0.63451	0.58484	0.55182	0.52834	0.51083
7	1	2	0.22603	0.23280	0.22822	0.22357	0.21984	0.21692	0.21459	0.21272
7	1	3	0.11705	0.13152	0.13477	0.13559	0.13571	0.13561	0.13543	0.13523
7	1	4	0.07979	0.09232	0.09604	0.09751	0.09819	0.09854	0.09873	0.09883
7	1	5	0.06265	0.07237	0.07519	0.07627	0.07674	0.07697	0.07708	0.07713
7	1	6	0.05534	0.06172	0.06289	0.06301	0.06287	0.06266	0.06246	0.06227
7	1	7	0.06178	0.06085	0.05803	0.05581	0.05417	0.05293	0.05196	0.05120
7	2	2	0.18813	0.22316	0.23583	0.24198	0.24550	0.24773	0.24927	0.25037
7	2	3	0.10072	0.12930	0.14223	0.14950	0.15413	0.15732	0.15966	0.16144
7	2	4	0.06982	0.09194	0.10247	0.10855	0.11250	0.11527	0.11731	0.11888
7	2	5	0.05539	0.07266	0.08077	0.08542	0.08842	0.09051	0.09204	0.09322
7	2	6	0.04928	0.06232	0.06789	0.07088	0.07272	0.07396	0.07485	0.07551
7	3	3	0.10880	0.14629	0.16500	0.17616	0.18356	0.18882	0.19274	0.19579
7	3	4	0.07685	0.10551	0.12029	0.12927	0.13529	0.13960	0.14284	0.14537
7	3	5	0.06170	0.08415	0.09556	0.10243	0.10700	0.11027	0.11271	0.11461
7	4	4	0.09419	0.13070	0.14989	0.16168	0.16964	0.17537	0.17970	0.18307

Variances and Covariances of Student's t Order Statistics

N	I	J	d.f. = 3	d.f. = 4	d.f. = 5	d.f. = 6	d.f. = 7	d.f. = 8	d.f. = 9	d.f. = 10
8	1	1	1.39452	0.91395	0.73223	0.63972	0.58438	0.54776	0.52182	0.50251
8	1	2	0.23830	0.23837	0.22980	0.22272	0.21738	0.21332	0.21015	0.20764
8	1	3	0.12175	0.13347	0.13487	0.13349	0.13379	0.13308	0.13245	0.13190
8	1	4	0.08175	0.09291	0.09568	0.09652	0.09676	0.09678	0.09672	0.09664
8	1	5	0.06271	0.07197	0.07448	0.07535	0.07568	0.07580	0.07584	0.07583
8	1	6	0.05270	0.05978	0.06146	0.06191	0.06199	0.06195	0.06187	0.06178
8	1	7	0.04871	0.05303	0.05328	0.05290	0.05246	0.05204	0.05169	0.05139
8	1	8	0.05631	0.05392	0.05059	0.04815	0.04639	0.04508	0.04407	0.04328
8	2	2	0.19300	0.22265	0.23169	0.23542	0.23725	0.23824	0.23883	0.23919
8	2	3	0.10150	0.12747	0.13854	0.14451	0.14821	0.15071	0.15251	0.15386
8	2	4	0.06913	0.08973	0.09921	0.10458	0.10802	0.11040	0.11214	0.11347
8	2	5	0.05349	0.06998	0.07768	0.08207	0.08490	0.08686	0.08830	0.08940
8	2	6	0.04522	0.05841	0.06436	0.06767	0.06977	0.07121	0.07226	0.07305
8	2	7	0.04199	0.05199	0.05597	0.05799	0.05919	0.05996	0.06050	0.06090
8	3	3	0.10537	0.13922	0.15554	0.16508	0.17132	0.17571	0.17897	0.18148
8	3	4	0.07288	0.09915	0.11249	0.12052	0.12587	0.12968	0.13254	0.13476
8	3	5	0.05693	0.07789	0.08862	0.09510	0.09943	0.10253	0.10484	0.10665
8	3	6	0.04846	0.06535	0.07375	0.07873	0.08202	0.08434	0.08608	0.08742
8	4	4	0.08389	0.11636	0.13341	0.14388	0.15095	0.15604	0.15989	0.16288
8	4	5	0.06631	0.09224	0.10591	0.11431	0.12000	0.12409	0.12718	0.12960
9	1	1	1.48981	0.94907	0.74742	0.64566	0.58513	0.54526	0.51711	0.49621
9	1	2	0.25048	0.24412	0.23183	0.22252	0.21572	0.21064	0.20672	0.20363
9	1	3	0.12668	0.13573	0.13538	0.13387	0.13239	0.13113	0.13008	0.12921
9	1	4	0.08419	0.09393	0.09571	0.09590	0.09569	0.09538	0.09508	0.09479
9	1	5	0.06369	0.07224	0.07425	0.07478	0.07489	0.07484	0.07474	0.07463
9	1	6	0.05231	0.05930	0.06092	0.06134	0.06140	0.06134	0.06125	0.06114
9	1	7	0.04598	0.05120	0.05208	0.05210	0.05192	0.05170	0.05149	0.05130
9	1	8	0.04390	0.04673	0.04635	0.04563	0.04498	0.04444	0.04399	0.04362
9	1	9	0.05211	0.04866	0.04499	0.04241	0.04059	0.03925	0.03823	0.03743
9	2	2	0.19875	0.22356	0.22936	0.23095	0.23130	0.23122	0.23098	0.23070
9	2	3	0.10312	0.12681	0.13620	0.14102	0.14389	0.14577	0.14709	0.14806
9	2	4	0.06940	0.08863	0.09711	0.10179	0.10472	0.10672	0.10817	0.10926
9	2	5	0.05290	0.06857	0.07572	0.07974	0.08230	0.08407	0.08535	0.08633
9	2	6	0.04367	0.05652	0.06235	0.06561	0.06767	0.06909	0.07013	0.07091
9	2	7	0.03853	0.04895	0.05343	0.05585	0.05735	0.05835	0.05907	0.05961
9	2	8	0.03690	0.04477	0.04765	0.04902	0.04977	0.05024	0.05054	0.05076
9	3	3	0.10410	0.13503	0.14936	0.15752	0.16276	0.16641	0.16909	0.17113
9	3	4	0.07096	0.09532	0.10739	0.11455	0.11927	0.12262	0.12511	0.12704
9	3	5	0.05453	0.07420	0.08417	0.09016	0.09415	0.09698	0.09911	0.10076
9	3	6	0.04526	0.06143	0.06956	0.07442	0.07765	0.07994	0.08165	0.08298
9	3	7	0.04010	0.05337	0.05978	0.06352	0.06595	0.06766	0.06892	0.06989
9	4	4	0.07843	0.10795	0.12325	0.13257	0.13884	0.14334	0.14672	0.14936
9	4	5	0.06083	0.08464	0.09720	0.10492	0.11015	0.11391	0.11675	0.11897
9	4	6	0.05083	0.07043	0.08068	0.08695	0.09117	0.09421	0.09650	0.09828
9	5	5	0.07219	0.10106	0.11645	0.12598	0.13246	0.13714	0.14068	0.14345
10	1	1	1.58211	0.98296	0.76240	0.65197	0.58667	0.54382	0.51366	0.49133
10	1	2	0.26249	0.24992	0.23413	0.22274	0.21460	0.20859	0.20401	0.20040
10	1	3	0.13169	0.13816	0.13614	0.13357	0.13137	0.12960	0.12816	0.12699
10	1	4	0.08685	0.09517	0.09598	0.09553	0.09488	0.09425	0.09370	0.09323
10	1	5	0.06510	0.07283	0.07427	0.07443	0.07427	0.07402	0.07378	0.07355

Variances and Covariances of Student's t Order Statistics

N	I	J	$d.f.=3$	$d.f.=4$	$d.f.=5$	$d.f.=6$	$d.f.=7$	$d.f.=8$	$d.f.=9$	$d.f.=10$
10	1	6	0.05276	0.05939	0.06075	0.06100	0.06094	0.06080	0.06064	0.06048
10	1	7	0.04532	0.05067	0.05164	0.05171	0.05157	0.05138	0.05119	0.05102
10	1	8	0.04112	0.04498	0.04527	0.04499	0.04462	0.04428	0.04398	0.04373
10	1	9	0.04024	0.04195	0.04111	0.04016	0.03937	0.03874	0.03823	0.03782
10	1	10	0.04877	0.04452	0.04062	0.03796	0.03611	0.03476	0.03374	0.03295
10	2	2	0.20498	0.22532	0.22816	0.22784	0.22686	0.22582	0.22486	0.22402
10	2	3	0.10524	0.12684	0.13472	0.13849	0.14061	0.14194	0.14282	0.14345
10	2	4	0.07019	0.08816	0.09572	0.09974	0.10219	0.10384	0.10501	0.10588
10	2	5	0.05297	0.06784	0.07441	0.07803	0.08030	0.08185	0.08297	0.08381
10	2	6	0.04312	0.05551	0.06106	0.06413	0.06606	0.06739	0.06835	0.06907
10	2	7	0.03715	0.04749	0.05201	0.05448	0.05601	0.05705	0.05780	0.05836
10	2	8	0.03380	0.04224	0.04569	0.04747	0.04854	0.04924	0.04973	0.05009
10	2	9	0.03314	0.03946	0.04154	0.04243	0.04288	0.04312	0.04327	0.04336
10	3	3	0.10402	0.13251	0.14511	0.15207	0.15645	0.15945	0.16162	0.16327
10	3	4	0.07016	0.09291	0.10386	0.11024	0.11440	0.11732	0.11948	0.12114
10	3	5	0.05331	0.07188	0.08111	0.08660	0.09023	0.09280	0.09472	0.09620
10	3	6	0.04361	0.05904	0.06677	0.07138	0.07443	0.07659	0.07821	0.07946
10	3	7	0.03771	0.05064	0.05701	0.06076	0.06323	0.06497	0.06626	0.06726
10	3	8	0.03439	0.04514	0.05017	0.05303	0.05487	0.05615	0.05708	0.05780
10	4	4	0.07536	0.10259	0.11643	0.12477	0.13034	0.13431	0.13729	0.13960
10	4	5	0.05771	0.07984	0.09140	0.09847	0.10324	0.10667	0.10925	0.11126
10	4	6	0.04746	0.06585	0.07550	0.08142	0.08541	0.08829	0.09045	0.09214
10	4	7	0.04120	0.05667	0.06464	0.06948	0.07273	0.07505	0.07679	0.07814
10	5	5	0.06569	0.09194	0.10594	0.11460	0.12048	0.12474	0.12795	0.13047
10	5	6	0.05438	0.07623	0.08790	0.09513	0.10004	0.10359	0.10628	0.10839

N	I	J	$d.f.=11$	$d.f.=12$	$d.f.=13$	$d.f.=14$	$d.f.=15$	$d.f.=16$	$d.f.=17$	$d.f.=18$	$d.f.=19$
1	1	1	1.00000	1.00000	1.00000	1.00000	1.00000	1.00000	1.00000	1.00000	1.00000
2	1	1	0.69126	0.69025	0.68943	0.68875	0.68818	0.68769	0.68728	0.68691	0.68659
2	1	2	0.30874	0.30975	0.31057	0.31125	0.31182	0.31231	0.31272	0.31309	0.31341
3	1	1	0.59996	0.59591	0.59260	0.58984	0.58750	0.58549	0.58375	0.58223	0.58089
3	1	1	0.59996	0.59591	0.59260	0.58984	0.58750	0.58549	0.58375	0.58223	0.58089
3	1	2	0.26517	0.26624	0.26711	0.26783	0.26845	0.26897	0.26942	0.26982	0.27017
3	1	3	0.16433	0.16447	0.16457	0.16465	0.16471	0.16475	0.16479	0.16481	0.16483
3	2	2	0.41073	0.41428	0.41722	0.41969	0.42181	0.42364	0.42523	0.42664	0.42788
4	1	1	0.55509	0.54875	0.54356	0.53923	0.53556	0.53242	0.52970	0.52732	0.52522
4	1	2	0.24249	0.24289	0.24320	0.24344	0.24365	0.24382	0.24396	0.24409	0.24419
4	1	3	0.15340	0.15390	0.15431	0.15464	0.15492	0.15516	0.15537	0.15554	0.15570
4	1	4	0.10795	0.10771	0.10750	0.10732	0.10716	0.10702	0.10689	0.10678	0.10667
4	2	2	0.33062	0.33341	0.33572	0.33767	0.33933	0.34077	0.34202	0.34313	0.34411
4	2	3	0.21456	0.21656	0.21821	0.21960	0.22080	0.22182	0.22272	0.22351	0.22422
5	1	1	0.52816	0.52003	0.51339	0.50786	0.50318	0.49917	0.49570	0.49267	0.49000
5	1	2	0.22828	0.22803	0.22780	0.22760	0.22741	0.22724	0.22709	0.22696	0.22683
5	1	3	0.14549	0.14582	0.14608	0.14629	0.14646	0.14660	0.14672	0.14683	0.14692
5	1	4	0.10480	0.10496	0.10508	0.10517	0.10525	0.10531	0.10536	0.10540	0.10544
5	1	5	0.07908	0.07868	0.07834	0.07804	0.07779	0.07756	0.07737	0.07719	0.07703
5	2	2	0.29125	0.29319	0.29480	0.29615	0.29729	0.29828	0.29914	0.29989	0.30056
5	2	3	0.18924	0.19103	0.19251	0.19376	0.19482	0.19575	0.19655	0.19726	0.19789
5	2	4	0.13781	0.13896	0.13991	0.14071	0.14140	0.14199	0.14250	0.14295	0.14335
5	3	3	0.25614	0.25894	0.26127	0.26324	0.26493	0.26640	0.26768	0.26881	0.26981
6	1	1	0.51016	0.50058	0.49276	0.48626	0.48076	0.47607	0.47200	0.46845	0.46532
6	1	2	0.21844	0.21763	0.21694	0.21635	0.21583	0.21538	0.21497	0.21462	0.21429
6	1	3	0.13959	0.13966	0.13971	0.13974	0.13976	0.13977	0.13977	0.13977	0.13977

Variances and Covariances of Student's t Order Statistics

N	I	J	$d.f.=11$	$d.f.=12$	$d.f.=13$	$d.f.=14$	$d.f.=15$	$d.f.=16$	$d.f.=17$	$d.f.=18$	$d.f.=19$
6	1	4	0.10165	0.10179	0.10190	0.10198	0.10204	0.10209	0.10213	0.10217	0.10220
6	1	5	0.07840	0.07837	0.07833	0.07829	0.07825	0.07821	0.07817	0.07814	0.07811
6	1	6	0.06188	0.06140	0.06100	0.06065	0.06036	0.06010	0.05987	0.05967	0.05949
6	2	2	0.26739	0.26863	0.26964	0.27048	0.27119	0.27179	0.27232	0.27278	0.27318
6	2	3	0.17360	0.17505	0.17626	0.17727	0.17813	0.17888	0.17952	0.18010	0.18060
6	2	4	0.12752	0.12866	0.12960	0.13040	0.13108	0.13167	0.13218	0.13263	0.13303
6	2	5	0.09891	0.09960	0.10016	0.10063	0.10103	0.10138	0.10168	0.10194	0.10217
6	3	3	0.22000	0.22239	0.22438	0.22606	0.22751	0.22876	0.22985	0.23081	0.23167
6	3	4	0.16325	0.16507	0.16659	0.16788	0.16898	0.16993	0.17077	0.17151	0.17216
7	1	1	0.49727	0.48647	0.47766	0.47035	0.46419	0.45892	0.45436	0.45038	0.44688
7	1	2	0.21118	0.20990	0.20882	0.20789	0.20709	0.20639	0.20577	0.20523	0.20474
7	1	3	0.13504	0.13486	0.13470	0.13455	0.13442	0.13429	0.13418	0.13408	0.13399
7	1	4	0.09889	0.09892	0.09894	0.09894	0.09894	0.09894	0.09893	0.09892	0.09891
7	1	5	0.07715	0.07715	0.07714	0.07713	0.07712	0.07710	0.07708	0.07707	0.07705
7	1	6	0.06210	0.06195	0.06181	0.06169	0.06159	0.06149	0.06141	0.06133	0.06126
7	1	7	0.05058	0.05007	0.04964	0.04928	0.04896	0.04869	0.04845	0.04824	0.04805
7	2	2	0.25121	0.25185	0.25237	0.25279	0.25314	0.25343	0.25369	0.25390	0.25409
7	2	3	0.16284	0.16397	0.16490	0.16568	0.16634	0.16691	0.16741	0.16784	0.16823
7	2	4	0.12012	0.12113	0.12196	0.12266	0.12326	0.12377	0.12422	0.12461	0.12496
7	2	5	0.09414	0.09489	0.09551	0.09603	0.09647	0.09685	0.09719	0.09748	0.09773
7	2	6	0.07603	0.07644	0.07677	0.07705	0.07728	0.07748	0.07765	0.07780	0.07793
7	3	3	0.19821	0.20020	0.20184	0.20323	0.20443	0.20546	0.20636	0.20715	0.20785
7	3	4	0.14739	0.14904	0.15042	0.15158	0.15258	0.15345	0.15421	0.15487	0.15547
7	3	5	0.11612	0.11736	0.11839	0.11926	0.12000	0.12065	0.12121	0.12171	0.12215
7	4	4	0.18578	0.18801	0.18986	0.19143	0.19278	0.19395	0.19498	0.19588	0.19669
8	1	1	0.48760	0.47575	0.46610	0.45810	0.45136	0.44560	0.44063	0.43629	0.43247
8	1	2	0.20559	0.20389	0.20247	0.20125	0.20021	0.19930	0.19850	0.19779	0.19716
8	1	3	0.13142	0.13101	0.13065	0.13034	0.13007	0.12982	0.12961	0.12941	0.12924
8	1	4	0.09654	0.09644	0.09635	0.09627	0.09619	0.09612	0.09605	0.09599	0.09593
8	1	5	0.07580	0.07576	0.07572	0.07568	0.07564	0.07560	0.07557	0.07553	0.07550
8	1	6	0.06169	0.06160	0.06152	0.06145	0.06138	0.06132	0.06127	0.06122	0.06117
8	1	7	0.05113	0.05092	0.05073	0.05056	0.05042	0.05029	0.05018	0.05007	0.04998
8	1	8	0.04265	0.04212	0.04169	0.04131	0.04099	0.04072	0.04047	0.04026	0.04007
8	2	2	0.23942	0.23957	0.23968	0.23975	0.23979	0.23982	0.23984	0.23986	0.23986
8	2	3	0.15491	0.15575	0.15644	0.15701	0.15749	0.15791	0.15827	0.15858	0.15886
8	2	4	0.11452	0.11537	0.11606	0.11665	0.11715	0.11757	0.11795	0.11827	0.11856
8	2	5	0.09027	0.09097	0.09155	0.09204	0.09245	0.09281	0.09312	0.09339	0.09363
8	2	6	0.07367	0.07417	0.07458	0.07493	0.07522	0.07547	0.07568	0.07587	0.07604
8	2	7	0.06120	0.06143	0.06162	0.06177	0.06190	0.06201	0.06210	0.06218	0.06225
8	3	3	0.18347	0.18510	0.18644	0.18757	0.18854	0.18937	0.19010	0.19074	0.19131
8	3	4	0.13654	0.13799	0.13919	0.14021	0.14109	0.14184	0.14250	0.14309	0.14361
8	3	5	0.10809	0.10926	0.11025	0.11107	0.11179	0.11240	0.11294	0.11341	0.11383
8	3	6	0.08848	0.08935	0.09007	0.09068	0.09120	0.09165	0.09204	0.09239	0.09269
8	4	4	0.16529	0.16726	0.16891	0.17031	0.17151	0.17254	0.17345	0.17426	0.17497
8	4	5	0.13153	0.13312	0.13445	0.13557	0.13654	0.13737	0.13811	0.13876	0.13933
9	1	1	0.48011	0.46732	0.45693	0.44833	0.44109	0.43491	0.42957	0.42492	0.42083
9	1	2	0.20114	0.19908	0.19735	0.19589	0.19463	0.19354	0.19258	0.19173	0.19098
9	1	3	0.12847	0.12785	0.12732	0.12686	0.12646	0.12611	0.12579	0.12551	0.12526
9	1	4	0.09454	0.09431	0.09412	0.09394	0.09378	0.09364	0.09352	0.09340	0.09330
9	1	5	0.07452	0.07441	0.07431	0.07423	0.07415	0.07407	0.07401	0.07395	0.07389

Variances and Covariances of Student's t Order Statistics

N	I	J	$d.f.=11$	$d.f.=12$	$d.f.=13$	$d.f.=14$	$d.f.=15$	$d.f.=16$	$d.f.=17$	$d.f.=18$	$d.f.=19$
9	1	6	0.06104	0.06095	0.06086	0.06078	0.06071	0.06065	0.06059	0.06054	0.06049
9	1	7	0.05113	0.05099	0.05086	0.05075	0.05065	0.05056	0.05047	0.05040	0.05034
9	1	8	0.04331	0.04305	0.04283	0.04263	0.04247	0.04232	0.04219	0.04207	0.04197
9	1	9	0.03679	0.03627	0.03583	0.03546	0.03514	0.03487	0.03463	0.03442	0.03423
9	2	2	0.23041	0.23014	0.22989	0.22966	0.22945	0.22926	0.22908	0.22892	0.22877
9	2	3	0.14880	0.14938	0.14985	0.15024	0.15056	0.15084	0.15107	0.15128	0.15146
9	2	4	0.11012	0.11081	0.11137	0.11184	0.11224	0.11258	0.11288	0.11314	0.11336
9	2	5	0.08710	0.08772	0.08823	0.08866	0.08902	0.08934	0.08961	0.08985	0.09006
9	2	6	0.07153	0.07202	0.07243	0.07277	0.07306	0.07331	0.07352	0.07371	0.07388
9	2	7	0.06002	0.06036	0.06063	0.06085	0.06104	0.06121	0.06135	0.06147	0.06157
9	2	8	0.05091	0.05103	0.05112	0.05119	0.05125	0.05130	0.05134	0.05137	0.05140
9	3	3	0.17275	0.17406	0.17514	0.17605	0.17682	0.17748	0.17806	0.17857	0.17902
9	3	4	0.12858	0.12983	0.13087	0.13175	0.13250	0.13315	0.13372	0.13421	0.13466
9	3	5	0.10207	0.10315	0.10404	0.10480	0.10544	0.10600	0.10649	0.10692	0.10731
9	3	6	0.08403	0.08490	0.08561	0.08622	0.08673	0.08718	0.08757	0.08792	0.08822
9	3	7	0.07066	0.07128	0.07180	0.07223	0.07260	0.07292	0.07320	0.07344	0.07366
9	4	4	0.15147	0.15320	0.15465	0.15587	0.15692	0.15782	0.15862	0.15932	0.15994
9	4	5	0.12075	0.12222	0.12344	0.12447	0.12536	0.12613	0.12680	0.12740	0.12793
9	4	6	0.09971	0.10089	0.10186	0.10269	0.10340	0.10402	0.10456	0.10503	0.10546
9	5	5	0.14567	0.14750	0.14903	0.15033	0.15144	0.15241	0.15326	0.15400	0.15467
10	1	1	0.47414	0.46053	0.44948	0.44034	0.43265	0.42610	0.42044	0.41552	0.41119
10	1	2	0.19750	0.19512	0.19313	0.19144	0.19000	0.18874	0.18765	0.18668	0.18582
10	1	3	0.12602	0.12521	0.12452	0.12392	0.12341	0.12296	0.12256	0.12220	0.12189
10	1	4	0.09283	0.09248	0.09218	0.09191	0.09168	0.09148	0.09129	0.09113	0.09098
10	1	5	0.07335	0.07317	0.07301	0.07287	0.07274	0.07263	0.07253	0.07243	0.07235
10	1	6	0.06034	0.06021	0.06009	0.05999	0.05990	0.05982	0.05974	0.05967	0.05961
10	1	7	0.05087	0.05073	0.05061	0.05050	0.05041	0.05033	0.05025	0.05018	0.05012
10	1	8	0.04351	0.04333	0.04317	0.04303	0.04290	0.04279	0.04270	0.04261	0.04253
10	1	9	0.03747	0.03719	0.03694	0.03673	0.03655	0.03639	0.03625	0.03613	0.03602
10	1	10	0.03231	0.03179	0.03136	0.03100	0.03068	0.03041	0.03018	0.02997	0.02978
10	2	2	0.22328	0.22264	0.22208	0.22159	0.22115	0.22077	0.22042	0.22011	0.21983
10	2	3	0.14392	0.14427	0.14455	0.14478	0.14496	0.14511	0.14524	0.14535	0.14544
10	2	4	0.10656	0.10710	0.10753	0.10790	0.10820	0.10846	0.10869	0.10888	0.10905
10	2	5	0.08447	0.08500	0.08543	0.08579	0.08610	0.08636	0.08659	0.08679	0.08696
10	2	6	0.06964	0.07009	0.07046	0.07078	0.07104	0.07127	0.07146	0.07163	0.07179
10	2	7	0.05880	0.05915	0.05944	0.05967	0.05987	0.06005	0.06019	0.06032	0.06044
10	2	8	0.05036	0.05058	0.05075	0.05090	0.05102	0.05112	0.05120	0.05128	0.05135
10	2	9	0.04342	0.04345	0.04348	0.04349	0.04350	0.04351	0.04352	0.04352	0.04352
10	3	3	0.16455	0.16559	0.16644	0.16715	0.16775	0.16827	0.16872	0.16911	0.16946
10	3	4	0.12246	0.12353	0.12441	0.12516	0.12579	0.12634	0.12682	0.12724	0.12762
10	3	5	0.09738	0.09834	0.09914	0.09981	0.10039	0.10089	0.10132	0.10171	0.10205
10	3	6	0.08045	0.08127	0.08194	0.08251	0.08300	0.08342	0.08379	0.08411	0.08440
10	3	7	0.06805	0.06869	0.06922	0.06967	0.07006	0.07039	0.07068	0.07093	0.07116
10	3	8	0.05836	0.05881	0.05918	0.05950	0.05976	0.05999	0.06019	0.06037	0.06052
10	4	4	0.14144	0.14295	0.14421	0.14527	0.14618	0.14697	0.14766	0.14827	0.14881
10	4	5	0.11288	0.11420	0.11530	0.11624	0.11704	0.11774	0.11835	0.11888	0.11936
10	4	6	0.09349	0.09461	0.09553	0.09632	0.09699	0.09757	0.09809	0.09854	0.09894
10	4	7	0.07923	0.08011	0.08085	0.08148	0.08201	0.08248	0.08288	0.08324	0.08356
10	5	5	0.13249	0.13416	0.13554	0.13672	0.13773	0.13861	0.13938	0.14006	0.14067
10	5	6	0.11008	0.11147	0.11263	0.11361	0.11446	0.11519	0.11584	0.11641	0.11691

Table C15.1

EXPECTED VALUES OF HALF LOGISTIC ORDER STATISTICS

[Tabular values are expected values of the Mth order statistic of a sample of size N from a half logistic population with location parameter 0 and scale parameter 1.]

Expected Values of Half Logistic Order Statistics

N	M	*Mean*	N	M	*Mean*	N	M	*Mean*	N	M	*Mean*	N	M	*Mean*
1	1	1.38629	7	4	1.15872	10	4	0.78398	12	7	1.24686	14	6	0.86635
2	1	0.77259	7	5	1.57326	10	5	1.01542	12	8	1.49614	14	7	1.03806
2	2	2.00000	7	6	2.14890	10	6	1.27655	12	9	1.79324	14	8	1.22566
3	1	0.54518	7	7	3.21963	10	7	1.58441	12	10	2.17166	14	9	1.43525
3	2	1.22741	8	1	0.22663	10	8	1.97264	12	11	2.71490	14	10	1.67650
3	3	2.38629	8	2	0.46297	10	9	2.52485	12	12	3.75642	14	11	1.96618
4	1	0.42369	8	3	0.71605	10	10	3.57460	13	1	0.14410	14	12	2.33776
4	2	0.90965	8	4	0.99630	11	1	0.16860	13	2	0.29079	14	13	2.87464
4	3	1.54518	8	5	1.32114	11	2	0.34130	13	3	0.44162	14	14	3.91027
4	4	2.66667	8	6	1.72454	11	3	0.52066	13	4	0.59843	15	1	0.12584
5	1	0.34738	8	7	2.29035	11	4	0.70995	13	5	0.76353	15	2	0.25344
5	2	0.72894	8	8	3.35238	11	5	0.91352	13	6	0.93994	15	3	0.38378
5	3	1.18071	9	1	0.20326	11	6	1.13769	13	7	1.13186	15	4	0.51804
5	4	1.78816	9	2	0.41359	11	7	1.39227	13	8	1.34542	15	5	0.65757
5	5	2.88629	9	3	0.63579	11	8	1.69421	13	9	1.59034	15	6	0.80408
6	1	0.29475	9	4	0.87655	11	9	2.07706	13	10	1.88342	15	7	0.95976
6	2	0.61049	9	5	1.14598	11	10	2.62436	13	11	2.25813	15	8	1.12754
6	3	0.96582	9	6	1.46127	11	11	3.66963	13	12	2.79795	15	9	1.31151
6	4	1.39560	9	7	1.85617	12	1	0.15538	13	13	3.83629	15	10	1.51774
6	5	1.98443	9	8	2.41441	12	2	0.31400	14	1	0.13435	15	11	1.75587
6	6	3.06667	9	9	3.46963	12	3	0.47781	14	2	0.27082	15	12	2.04266
7	1	0.25617	10	1	0.18430	12	4	0.69423	14	3	0.41063	15	13	2.41153
7	2	0.52624	10	2	0.37391	12	5	0.83138	14	4	0.55524	15	14	2.94589
7	3	0.82114	10	3	0.57229	12	6	1.02852	14	5	0.70641	15	15	3.97915

Table C15.2

VARIANCES AND COVARIANCES OF HALF LOGISTIC ORDER STATISTICS

[When $I = J$, tabular values are variances of the Ith order statistic of a sample of size N from a half logistic population with location parameter 0 and scale parameter 1; when $I \neq j$, they are covariances of the Ith and Jth order statistics.]

Variances and Covariances of Half Logistic Order Statistics

N	*I*	*J*	*Cov*	*N*	*I*	*J*	*Cov*	*N*	*I*	*J*	*Cov*	*N*	*I*	*J*	*Cov*	*N*	*I*	*J*	*Cov*
1	1	1	1.36806	5	4	1	0.07313	6	6	3	0.18412	7	7	2	0.07664	8	6	6	0.37550
2	1	1	0.43767	5	4	2	0.16723	6	6	4	0.31242	7	7	3	0.12890	8	7	1	0.02911
2	2	1	0.37663	5	4	3	0.31284	6	6	5	0.58461	7	7	4	0.20339	8	7	2	0.06183
2	2	2	1.54518	5	4	4	0.61088	6	6	6	1.63149	7	7	5	0.32763	8	7	3	0.10130
3	1	1	0.22672	5	5	1	0.06729	7	1	1	0.05514	7	7	6	0.59546	8	7	4	0.15306
3	2	1	0.19878	5	5	2	0.15414	7	2	1	0.05064	7	7	7	1.63498	8	7	5	0.22850
3	2	2	0.54926	5	5	3	0.28923	7	2	2	0.10893	8	1	1	0.04377	8	7	6	0.35595
3	3	1	0.17628	5	5	4	0.56827	7	3	1	0.04676	8	2	1	0.04049	8	7	7	0.63203
3	3	2	0.49130	5	5	5	1.62584	7	3	2	0.10066	8	2	2	0.08579	8	8	1	0.02753
3	3	3	1.59547	6	1	1	0.07175	7	3	3	0.16863	8	3	1	0.03762	8	8	2	0.05848
4	1	1	0.14147	6	2	1	0.06536	7	4	1	0.04339	8	3	2	0.07975	8	8	3	0.09585
4	2	1	0.12598	6	2	2	0.14374	7	4	2	0.09346	8	3	3	0.13032	8	8	4	0.14496
4	2	2	0.30538	6	3	1	0.05991	7	4	3	0.15674	8	4	1	0.03509	8	8	5	0.21668
4	3	1	0.11323	6	3	2	0.13191	7	4	4	0.24619	8	4	2	0.07444	8	8	6	0.33828
4	3	2	0.27549	6	3	3	0.22978	7	5	1	0.04043	8	4	3	0.12172	8	8	7	0.60313
4	3	3	0.59119	6	4	1	0.05523	7	5	2	0.08715	8	4	4	0.18340	8	8	8	1.63727
4	4	1	0.10259	6	4	2	0.12173	7	5	3	0.14631	8	5	1	0.03286	9	1	1	0.03564
4	4	2	0.25048	6	4	3	0.21243	7	5	4	0.23018	8	5	2	0.06974	9	2	1	0.03316
4	4	3	0.54143	6	4	4	0.35840	7	5	5	0.36890	8	5	3	0.11412	9	2	2	0.06952
4	4	4	1.61579	6	5	1	0.05117	7	6	1	0.03782	8	5	4	0.17212	9	3	1	0.03097
5	1	1	0.09760	6	5	2	0.11289	7	6	2	0.08158	8	5	5	0.25621	9	3	2	0.06495
5	2	1	0.08801	6	5	3	0.19734	7	6	3	0.13709	8	6	1	0.03088	9	3	3	0.10435
5	2	2	0.20047	6	5	4	0.33392	7	6	4	0.21601	8	6	2	0.06556	9	4	1	0.02903
5	3	1	0.07996	6	5	5	0.62154	7	6	5	0.34710	8	6	3	0.10735	9	4	2	0.06091
5	3	2	0.18250	6	6	1	0.04762	7	6	6	0.62792	8	6	4	0.16207	9	4	3	0.09790
5	3	3	0.34027	6	6	2	0.10516	7	7	1	0.03551	8	6	5	0.24160	9	4	4	0.14361

Variances and Covariances of Half Logistic Order Statistics

N	*I*	*J*	*Cov*	*N*	*I*	*J*	*Cov*	*N*	*I*	*J*	*Cov*	*N*	*I*	*J*	*Cov*	*N*	*I*	*J*	*Cov*
9	5	1	0.02730	10	6	6	0.19915	11	6	6	0.15845	12	4	4	0.08177	12	11	10	0.37327
9	5	2	0.05730	10	7	1	0.02073	11	7	1	0.01788	12	5	1	0.01717	12	11	11	0.63934
9	5	3	0.09214	10	7	2	0.04317	11	7	2	0.03700	12	5	2	0.03535	12	12	1	0.01264
9	5	4	0.13526	10	7	3	0.06861	11	7	3	0.05824	12	5	3	0.05522	12	12	2	0.02604
9	5	5	0.19282	10	7	4	0.09894	11	7	4	0.08285	12	5	4	0.07772	12	12	3	0.04073
9	6	1	0.02576	10	7	5	0.13730	11	7	5	0.11272	12	5	5	0.10417	12	12	4	0.05740
9	6	2	0.05407	10	7	6	0.18943	11	7	6	0.15103	12	6	1	0.01635	12	12	5	0.07708
9	6	3	0.08699	10	7	7	0.26723	11	7	7	0.20361	12	6	2	0.03366	12	12	6	0.10137
9	6	4	0.12777	10	8	1	0.01972	11	8	1	0.01705	12	6	3	0.05259	12	12	7	0.13299
9	6	5	0.18231	10	8	2	0.04107	11	8	2	0.03529	12	6	4	0.07403	12	12	8	0.17697
9	6	6	0.26275	10	8	3	0.06528	11	8	3	0.05555	12	6	5	0.09925	12	12	9	0.24390
9	7	1	0.02437	10	8	4	0.09418	11	8	4	0.07905	12	6	6	0.13025	12	12	10	0.36061
9	7	2	0.05117	10	8	5	0.13078	11	8	5	0.10759	12	7	1	0.01559	12	12	11	0.61941
9	7	3	0.08235	10	8	6	0.18057	11	8	6	0.14424	12	7	2	0.03211	12	12	12	1.64149
9	7	4	0.12103	10	8	7	0.25505	11	8	7	0.19461	12	7	3	0.05018	13	1	1	0.01854
9	7	5	0.17284	10	8	8	0.38296	11	8	8	0.27043	12	7	4	0.07065	13	2	1	0.01753
9	7	6	0.24943	10	9	1	0.01880	11	9	1	0.01629	12	7	5	0.09476	13	2	2	0.03595
9	7	7	0.37989	10	9	2	0.03916	11	9	2	0.03372	12	7	6	0.12441	13	3	1	0.01662
9	8	1	0.02311	10	9	3	0.06225	11	9	3	0.05309	12	7	7	0.16281	13	3	2	0.03409
9	8	2	0.04854	10	9	4	0.08984	11	9	4	0.07557	12	8	1	0.01490	13	3	3	0.05294
9	8	3	0.07815	10	9	5	0.12482	11	9	5	0.10289	12	8	2	0.03069	13	4	1	0.01580
9	8	4	0.11492	10	9	6	0.17248	11	9	6	0.13801	12	8	3	0.04797	13	4	2	0.03240
9	8	5	0.16426	10	9	7	0.24389	11	9	7	0.18635	12	8	4	0.06756	13	4	3	0.05033
9	8	6	0.23734	10	9	8	0.36689	11	9	8	0.25923	12	8	5	0.09063	13	4	4	0.07027
9	8	7	0.36224	10	9	9	0.63679	11	9	9	0.38518	12	8	6	0.11904	13	5	1	0.01505
9	8	8	0.63482	10	10	1	0.01795	11	10	1	0.01559	12	8	7	0.15587	13	5	2	0.03086
9	9	1	0.02197	10	10	2	0.03740	11	10	2	0.03227	12	8	8	0.20686	13	5	3	0.04795
9	9	2	0.04616	10	10	3	0.05948	11	10	3	0.05083	12	9	1	0.01427	13	5	4	0.06695
9	9	3	0.07434	10	10	4	0.08587	11	10	4	0.07237	12	9	2	0.02939	13	5	5	0.08878
9	9	4	0.10938	10	10	5	0.11935	11	10	5	0.09857	12	9	3	0.04594	13	6	1	0.01436
9	9	5	0.15645	10	10	6	0.16505	11	10	6	0.13228	12	9	4	0.06471	13	6	2	0.02945
9	9	6	0.22633	10	10	7	0.23364	11	10	7	0.17873	12	9	5	0.08684	13	6	3	0.04577
9	9	7	0.34611	10	10	8	0.35210	11	10	8	0.24890	12	9	6	0.11409	13	6	4	0.06392
9	9	8	0.60881	10	10	9	0.61318	11	10	9	0.37046	12	9	7	0.14947	13	6	5	0.08478
9	9	9	1.63885	10	10	10	1.63999	11	10	10	0.63825	12	9	8	0.19851	13	6	6	0.10963
10	1	1	0.02961	11	1	1	0.02500	11	11	1	0.01495	12	9	9	0.27280	13	7	1	0.01373
10	2	1	0.02768	11	2	1	0.02348	11	11	2	0.03095	12	10	1	0.01368	13	7	2	0.02816
10	2	2	0.05759	11	2	2	0.04855	11	11	3	0.04875	12	10	2	0.02819	13	7	3	0.04376
10	3	1	0.02597	11	3	1	0.02212	11	11	4	0.06941	12	10	3	0.04407	13	7	4	0.06113
10	3	2	0.05404	11	3	2	0.04574	11	11	5	0.09458	12	10	4	0.06208	13	7	5	0.08111
10	3	3	0.08577	11	3	3	0.07193	11	11	6	0.12698	12	10	5	0.08333	13	7	6	0.10491
10	4	1	0.02445	11	4	1	0.02089	11	11	7	0.17169	12	10	6	0.10953	13	7	7	0.13448
10	4	2	0.05088	11	4	2	0.04322	11	11	8	0.23934	12	10	7	0.14356	13	8	1	0.01315
10	4	3	0.08078	11	4	3	0.06798	11	11	9	0.35681	12	10	8	0.19079	13	8	2	0.02697
10	4	4	0.11633	11	4	4	0.09661	11	11	10	0.61662	12	10	9	0.26245	13	8	3	0.04192
10	5	1	0.02308	11	5	1	0.01979	11	11	11	1.64084	12	10	10	0.38684	13	8	4	0.05857
10	5	2	0.04804	11	5	2	0.04094	12	1	1	0.02140	12	11	1	0.01314	13	8	5	0.07772
10	5	3	0.07630	11	5	3	0.06441	12	2	1	0.02018	12	11	2	0.02707	13	8	6	0.10056
10	5	4	0.10994	11	5	4	0.09157	12	2	2	0.04152	12	11	3	0.04233	13	8	7	0.12896
10	5	5	0.15238	11	5	5	0.12448	12	3	1	0.01907	12	11	4	0.05965	13	8	8	0.16605
10	6	1	0.02185	11	6	1	0.01879	12	3	2	0.03926	12	11	5	0.08009	13	9	1	0.01261
10	6	2	0.04549	11	6	2	0.03888	12	3	3	0.06131	12	11	6	0.10530	13	9	2	0.02588
10	6	3	0.07226	11	6	3	0.06118	12	4	1	0.01807	12	11	7	0.13808	13	9	3	0.04022
10	6	4	0.10416	11	6	4	0.08700	12	4	2	0.03721	12	11	8	0.18363	13	9	4	0.05620
10	6	5	0.14447	11	6	5	0.11833	12	4	3	0.05812	12	11	9	0.25285	13	9	5	0.07460

Variances and Covariances of Half Logistic Order Statistics

N	*I*	*J*	*Cov*	*N*	*I*	*J*	*Cov*	*N*	*I*	*J*	*Cov*	*N*	*I*	*J*	*Cov*	*N*	*I*	*J*	*Cov*
13	9	6	0.09655	14	3	3	0.04623	14	11	6	0.07719	15	5	1	0.01185	15	11	11	0.21264
13	9	7	0.12385	14	4	1	0.01393	14	11	7	0.09755	15	5	2	0.02417	15	12	1	0.00906
13	9	8	0.15955	14	4	2	0.02849	14	11	8	0.12310	15	5	3	0.03724	15	12	2	0.01848
13	9	9	0.20930	14	4	3	0.04406	14	11	9	0.15675	15	5	4	0.05143	15	12	3	0.02848
13	10	1	0.01212	14	4	4	0.06113	14	11	10	0.20392	15	5	5	0.06719	15	12	4	0.03936
13	10	2	0.02486	14	5	1	0.01330	14	11	11	0.27598	15	6	1	0.01136	15	12	5	0.05147
13	10	3	0.03865	14	5	2	0.02720	14	12	1	0.01005	15	6	2	0.02316	15	12	6	0.06531
13	10	4	0.05401	14	5	3	0.04207	14	12	2	0.02056	15	6	3	0.03569	15	12	7	0.08158
13	10	5	0.07170	14	5	4	0.05838	14	12	3	0.03182	15	6	4	0.04929	15	12	8	0.10135
13	10	6	0.09282	14	5	5	0.07678	14	12	4	0.04420	15	6	5	0.06441	15	12	9	0.12631
13	10	7	0.11912	14	6	1	0.01272	14	12	5	0.05819	15	6	6	0.08165	15	12	10	0.15935
13	10	8	0.15353	14	6	2	0.02602	14	12	6	0.07450	15	7	1	0.01090	15	12	11	0.20584
13	10	9	0.20153	14	6	3	0.04024	14	12	7	0.09418	15	7	2	0.02223	15	12	12	0.27708
13	10	10	0.27459	14	6	4	0.05585	14	12	8	0.11888	15	7	3	0.03426	15	13	1	0.00876
13	11	1	0.01166	14	6	5	0.07347	14	12	9	0.15144	15	7	4	0.04732	15	13	2	0.01787
13	11	2	0.02392	14	6	6	0.09394	14	12	10	0.19713	15	7	5	0.06184	15	13	3	0.02755
13	11	3	0.03719	14	7	1	0.01219	14	12	11	0.26703	15	7	6	0.07841	15	13	4	0.03807
13	11	4	0.05198	14	7	2	0.02492	14	12	12	0.38910	15	7	7	0.09784	15	13	5	0.04979
13	11	5	0.06902	14	7	3	0.03856	14	13	1	0.00971	15	8	1	0.01048	15	13	6	0.06318
13	11	6	0.08937	14	7	4	0.05352	14	13	2	0.01986	15	8	2	0.02137	15	13	7	0.07893
13	11	7	0.11472	14	7	5	0.07042	14	13	3	0.03074	15	8	3	0.03293	15	13	8	0.09809
13	11	8	0.14793	14	7	6	0.09006	14	13	4	0.04270	15	8	4	0.04549	15	13	9	0.12228
13	11	9	0.19430	14	7	7	0.11370	14	13	5	0.05622	15	8	5	0.05946	15	13	10	0.15433
13	11	10	0.26499	14	8	1	0.01169	14	13	6	0.07199	15	8	6	0.07540	15	13	11	0.19946
13	11	11	0.38811	14	8	2	0.02392	14	13	7	0.09102	15	8	7	0.09412	15	13	12	0.26871
13	12	1	0.01123	14	8	3	0.03700	14	13	8	0.11494	15	8	8	0.11681	15	13	13	0.38990
13	12	2	0.02305	14	8	4	0.05137	14	13	9	0.14648	15	9	1	0.01008	15	14	1	0.00848
13	12	3	0.03583	14	8	5	0.06760	14	13	10	0.19077	15	9	2	0.02057	15	14	2	0.01730
13	12	4	0.05009	14	8	6	0.08647	14	13	11	0.25864	15	9	3	0.03170	15	14	3	0.02667
13	12	5	0.06652	14	8	7	0.10920	14	13	12	0.37738	15	9	4	0.04380	15	14	4	0.03686
13	12	6	0.08615	14	8	8	0.13766	14	13	13	0.64086	15	9	5	0.05725	15	14	5	0.04821
13	12	7	0.11062	14	9	1	0.01124	14	14	1	0.00939	15	9	6	0.07261	15	14	6	0.06118
13	12	8	0.14271	14	9	2	0.02298	14	14	2	0.01920	15	9	7	0.09065	15	14	7	0.07645
13	12	9	0.18756	14	9	3	0.03556	14	14	3	0.02972	15	9	8	0.11254	15	14	8	0.09502
13	12	10	0.25603	14	9	4	0.04938	14	14	4	0.04129	15	9	9	0.14012	15	14	9	0.11849
13	12	11	0.37553	14	9	5	0.06498	14	14	5	0.05438	15	10	1	0.00972	15	14	10	0.14960
13	12	12	0.64019	14	9	6	0.08315	14	14	6	0.06964	15	10	2	0.01982	15	14	11	0.19345
13	13	1	0.01083	14	9	7	0.10503	14	14	7	0.08807	15	10	3	0.03055	15	14	12	0.26081
13	13	2	0.02223	14	9	8	0.13246	14	14	8	0.11124	15	10	4	0.04222	15	14	13	0.37893
13	13	3	0.03456	14	9	9	0.16851	14	14	9	0.14182	15	10	5	0.05519	15	14	14	0.64140
13	13	4	0.04832	14	10	1	0.01081	14	14	10	0.18480	15	10	6	0.07001	15	15	1	0.00821
13	13	5	0.06418	14	10	2	0.02212	14	14	11	0.25074	15	10	7	0.08742	15	15	2	0.01676
13	13	6	0.08315	14	10	3	0.03422	14	14	12	0.36534	15	10	8	0.10855	15	15	3	0.02584
13	13	7	0.10680	14	10	4	0.04753	14	14	13	0.62363	15	10	9	0.13521	15	15	4	0.03572
13	13	8	0.13783	14	10	5	0.06256	14	14	14	1.64240	15	10	10	0.17043	15	15	5	0.04672
13	13	9	0.18126	14	10	6	0.08006	15	1	1	0.01430	15	11	1	0.00937	15	15	6	0.05930
13	13	10	0.24764	14	10	7	0.10116	15	2	1	0.01361	15	11	2	0.01913	15	15	7	0.07411
13	13	11	0.36373	14	10	8	0.12761	15	2	2	0.02774	15	11	3	0.02948	15	15	8	0.09213
13	13	12	0.62171	14	10	9	0.16243	15	3	1	0.01297	15	11	4	0.04074	15	15	9	0.11492
13	13	13	1.64200	14	10	10	0.21117	15	3	2	0.02645	15	11	5	0.05327	15	15	10	0.14514
14	1	1	0.01621	14	11	1	0.01042	15	3	3	0.04075	15	11	6	0.06758	15	15	11	0.18778
14	2	1	0.01538	14	11	2	0.02131	15	4	1	0.01239	15	11	7	0.08440	15	15	12	0.25337
14	2	2	0.03144	14	11	3	0.03298	15	4	2	0.02526	15	11	8	0.10483	15	15	13	0.36856
14	3	1	0.01463	14	11	4	0.04580	15	4	3	0.03892	15	11	9	0.13061	15	15	14	0.62527
14	3	2	0.02990	14	11	5	0.06030	15	4	4	0.05374	15	11	10	0.16471	15	15	15	1.64272

Table C16.1

EXPECTED VALUES OF TYPE I GENERALIZED LOGISTIC ORDER STATISTICS

[Tabular values are expected values of the Rth order statistic of a sample of size N from a Type I generalized logistic population with location parameter 0, scale parameter 1, and shape parameter $B = 1.0(0.5)5.0(1.0)8.0$.]

Expected Values of Type I Generalized Logistic Order Statistics

N	R	B = 1.0	B = 1.5	B = 2.0	B = 2.5	B = 3.0	B = 3.5	B = 4.0	B = 4.5	B = 5.0	B = 6.0	B = 7.0	B = 8.0
1	1	0.00000	0.61371	1.00000	1.28037	1.50000	1.68037	1.83333	1.96609	2.08333	2.28333	2.45000	2.59286
2	1	-1.00000	-0.27259	0.16667	0.47741	0.71667	0.91074	1.07381	1.21432	1.33770	1.54679	1.71987	1.86749
2	2	1.00000	1.50000	1.83333	2.08333	2.28333	2.45000	2.59286	2.71786	2.82897	3.01988	3.18013	3.31823
3	1	-1.50000	-0.69280	-0.21667	0.11509	0.36785	0.57133	0.74131	0.88710	1.01466	1.22992	1.40734	1.55818
3	2	0.00000	0.56783	0.93333	1.20205	1.41429	1.58957	1.73882	1.86875	1.98378	2.18053	2.34492	2.48610
3	3	1.50000	1.96609	2.28333	2.52397	2.71786	2.88022	3.01988	3.14241	3.25156	3.43955	3.59774	3.73429
4	1	-1.83333	-0.96416	-0.45952	-0.11159	0.15155	0.36222	0.53747	0.68730	0.81803	1.03799	1.21870	1.37198
4	2	-0.50000	0.12131	0.51190	0.79512	1.01677	1.19867	1.35281	1.48651	1.60453	1.80572	1.97326	2.11678
4	3	0.50000	1.01435	1.35476	1.60899	1.81180	1.98047	2.12482	2.25099	2.36303	2.55534	2.71659	2.85543
4	4	1.83333	2.28333	2.59286	2.82897	3.01988	3.18013	3.31823	3.43955	3.54774	3.73429	3.89146	4.02725
5	1	-2.08333	-1.16330	-0.63532	-0.27417	-0.00259	0.21393	0.39346	0.54656	0.67987	0.90361	1.08699	1.24223
5	2	-0.83333	-0.16762	0.24365	0.53875	0.76810	0.95538	1.11350	1.25026	1.37070	1.57548	1.74556	1.89097
5	3	0.00000	0.55470	0.91429	1.17968	1.38979	1.56360	1.71178	1.84089	1.95528	2.15108	2.31481	2.45550
5	4	0.83333	1.32077	1.64841	1.89520	2.09314	2.25838	2.40019	2.52439	2.63487	2.82484	2.98444	3.12205
5	5	2.08333	2.52397	2.82897	3.06241	3.25156	3.41057	3.54774	3.66834	3.77596	3.96165	4.11821	4.25354
6	1	-2.28333	-1.32005	-0.77226	-0.39992	-0.12120	0.10026	0.28340	0.43925	0.57472	0.80166	0.98726	1.14415
6	2	-1.08333	-0.37753	0.04939	0.35459	0.59046	0.78229	0.94377	1.08311	1.20559	1.41339	1.58560	1.73259
6	3	-0.33333	0.25619	0.63218	0.90708	1.12338	1.30155	1.45295	1.58456	1.70092	1.89965	2.06547	2.20771
6	4	0.33333	0.85322	1.19539	1.45228	1.65619	1.82565	1.97060	2.09722	2.20964	2.40250	2.56414	2.70328
6	5	1.08333	1.55455	1.87442	2.11666	2.31161	2.47474	2.61498	2.73797	2.84748	3.03602	3.19459	3.33144
6	6	2.28333	2.71786	3.01988	3.25156	3.43955	3.59774	3.73429	3.85442	3.96165	4.14678	4.30293	4.43796
7	1	-2.45000	-1.44904	-0.88401	-0.50195	-0.21703	0.00871	0.19497	0.35320	0.49055	0.72024	0.90777	1.06608
7	2	-1.28333	-0.54611	-0.10176	0.21224	0.45379	0.64957	0.81396	0.95552	1.07977	1.29019	1.46423	1.51259
7	3	-0.58333	0.03690	0.42726	0.71045	0.93215	1.11410	1.26831	1.40206	1.52013	1.72141	1.88902	2.03259
7	4	0.00000	0.54859	0.90541	1.16924	1.37835	1.55148	1.69915	1.82788	1.94197	2.13732	2.30073	2.44119
7	5	0.58333	1.08169	1.41463	1.66455	1.86457	2.03127	2.17418	2.29923	2.41039	2.60138	2.76170	2.89985
7	6	1.28333	1.74370	2.05834	2.29750	2.49043	2.65213	2.79130	2.91346	3.02232	3.20987	3.36775	3.50407
7	7	2.45000	2.88022	3.18013	3.41057	3.59774	3.75534	3.89146	4.01125	4.11821	4.30293	4.45880	4.59361
8	1	-2.59286	-1.55849	-0.97817	-0.58750	-0.29710	-0.06759	0.12144	0.28176	0.42076	0.65287	0.84210	1.00166
8	2	-1.45000	-0.68294	-0.22488	0.09693	0.34350	0.54279	0.70975	0.85328	0.97908	1.19180	1.36746	1.51703
8	3	-0.78333	-0.13559	0.26759	0.55815	0.78453	0.96993	1.12661	1.26226	1.38184	1.58536	1.75454	1.89930
8	4	-0.25000	0.32438	0.69338	0.96429	1.17802	1.35439	1.50447	1.63506	1.75062	1.94816	2.11313	2.25475
8	5	0.25000	0.77279	1.11743	1.37419	1.57809	1.74857	1.89384	2.02071	2.13332	2.32647	2.48833	2.62763
8	6	0.78333	1.26702	1.59295	1.83877	2.03609	2.20090	2.34239	2.46634	2.57663	2.76532	2.92572	3.06319
8	7	1.45000	1.90259	2.21347	2.45041	2.64188	2.80254	2.94094	3.06250	3.17088	3.35772	3.51509	3.65104
8	8	2.59286	3.01988	3.31823	3.54774	3.73429	3.89146	4.02725	4.14678	4.25354	4.43796	4.59361	4.72827
9	1	-2.71786	-1.65344	-1.05939	-0.66100	-0.36568	-0.13278	0.05871	0.22092	0.36139	0.59567	0.78641	0.94709
9	2	-1.59286	-0.79884	-0.32840	0.00044	0.25152	0.45394	0.62321	0.76850	0.89570	1.11047	1.28759	1.43822
9	3	-0.95000	-0.27731	0.13745	0.43466	0.66543	0.85374	1.01263	1.14999	1.27092	1.47644	1.64704	1.79285
9	4	-0.45000	0.14784	0.52786	0.80514	1.02302	1.20231	1.35455	1.48680	1.60368	1.80319	1.96955	2.11221
9	5	0.00000	0.54506	0.90029	1.16322	1.37175	1.54449	1.69187	1.82038	1.93429	2.12938	2.29261	2.43293
9	6	0.45000	0.95498	1.29114	1.54297	1.74424	1.91184	2.05541	2.18098	2.29255	2.48415	2.64491	2.78339
9	7	0.95000	1.42305	1.74386	1.98667	2.18201	2.34542	2.48588	2.60903	2.71867	2.90741	3.06613	3.20308
9	8	1.59286	2.03961	2.34764	2.58290	2.77327	2.93314	3.07096	3.19206	3.30008	3.48638	3.64337	3.77902
9	9	2.71786	3.14241	3.43955	3.66834	3.85442	4.01125	4.14678	4.26612	4.37273	4.55691	4.71239	4.84692
10	1	-2.82897	-1.73724	-1.13071	-0.72530	-0.42552	-0.18955	0.00418	0.16808	0.30989	0.54613	0.73824	0.89993
10	2	-1.71786	-0.89922	-0.41750	-0.08227	0.17292	0.37818	0.54953	0.69643	0.82488	1.04152	1.21994	1.37155
10	3	-1.09286	-0.39732	0.02802	0.33128	0.56596	0.75699	0.91790	1.05680	1.17895	1.38628	1.55816	1.70492
10	4	-0.61667	0.00270	0.39279	0.67589	0.89755	1.07948	1.23368	1.36744	1.48552	1.68681	1.85443	1.99802
10	5	-0.20000	0.36554	0.73045	0.99902	1.21123	1.38655	1.53585	1.66584	1.78092	1.97777	2.14224	2.28349
10	6	0.20000	0.72458	1.07012	1.32743	1.53229	1.70243	1.84789	1.97492	2.08765	2.28099	2.44298	2.58237
10	7	0.61667	1.10857	1.43849	1.68666	1.88553	2.05144	2.19376	2.31835	2.42914	2.61959	2.77954	2.91740
10	8	1.09286	1.55782	1.87473	2.11524	2.30908	2.47142	2.61107	2.73360	2.84276	3.03076	3.18896	3.32552
10	9	1.71786	2.16005	2.46587	2.69982	2.88931	3.04857	3.18593	3.30668	3.41441	3.60029	3.75697	3.89240
10	10	2.82897	3.25156	3.54774	3.77596	3.96165	4.11821	4.25354	4.37273	4.47921	4.66320	4.81355	4.95298

Table C16.2

VARIANCES AND COVARIANCES OF TYPE I GENERALIZED LOGISTIC ORDER STATISTICS

[**When $R = S$, tabular values are variances of the Rth order statistic of a sample of size N from a Type I generalized logistic population with location parameter 0, scale parameter 1, and shape parameter $B = 1.0(0.5)5.0(1.0)8.0$; when $R \neq S$, they are covariances of the Rth and Sth order statistics.**]

Variances and Covariances of Type I Generalized Logistic Order Statistics

N	S	R	B = 1.0	B = 1.5	B = 2.0	B = 2.5	B = 3.0	B = 3.5	B = 4.0	B = 4.5	B = 5.0	B = 6.0	B = 7.0	B = 8.0
1	1	1	3.28987	2.57974	2.28987	2.13529	2.03987	1.97529	1.92876	1.89366	1.86626	1.82626	1.79848	1.77807
2	1	1	2.28987	1.54857	1.26209	1.11483	1.02626	0.96745	0.92569	0.89455	0.87047	0.83568	0.81179	0.79439
2	2	1	1.00000	0.78552	0.69444	0.64475	0.61361	0.59233	0.57688	0.56516	0.55597	0.54250	0.53310	0.52516
2	2	2	2.28987	2.03987	1.92876	1.86626	1.82626	1.79848	1.77807	1.76245	1.75010	1.73184	1.71897	1.70943
3	1	1	2.03987	1.27870	0.99292	0.84908	0.76387	0.70797	0.66863	0.63951	0.61712	0.58500	0.56310	0.54722
3	2	1	0.85507	0.61356	0.51562	0.46365	0.43170	0.41014	0.39466	0.38300	0.37392	0.36070	0.35154	0.34483
3	2	2	1.28987	1.02887	0.91876	0.85869	0.82101	0.79522	0.77646	0.76222	0.75104	0.73462	0.72315	0.71467
3	3	1	0.53987	0.41988	0.36792	0.33932	0.32132	0.30897	0.30000	0.29318	0.28783	0.27996	0.27450	0.27046
3	3	2	0.85507	0.73555	0.68229	0.65231	0.63312	0.61980	0.61002	0.60253	0.59661	0.58812	0.58176	0.57715
3	3	3	2.03987	1.89366	1.82626	1.78755	1.76245	1.74485	1.73184	1.72182	1.71387	1.70206	1.69370	1.68748
4	1	1	1.92876	1.15342	0.86759	0.72572	0.64257	0.58846	0.55063	0.52278	0.50145	0.47100	0.45035	0.43545
4	2	1	0.79347	0.53656	0.43536	0.38266	0.35068	0.32931	0.31406	0.30265	0.29380	0.28098	0.27215	0.26570
4	2	2	1.03987	0.77085	0.66115	0.60256	0.56631	0.54175	0.52403	0.51065	0.50020	0.48493	0.47432	0.46551
4	3	1	0.49640	0.36170	0.30561	0.27543	0.25672	0.24402	0.23486	0.22795	0.22255	0.21467	0.20920	0.20518
4	3	2	0.67026	0.53474	0.47675	0.44487	0.42478	0.41098	0.40092	0.39327	0.38726	0.37841	0.37222	0.36765
4	3	3	1.03987	0.88812	0.82115	0.78363	0.75968	0.74308	0.73090	0.72158	0.71422	0.70335	0.69570	0.69003
4	4	1	0.36111	0.27790	0.24148	0.22134	0.20864	0.19993	0.19359	0.18878	0.18500	0.17945	0.17558	0.17273
4	4	2	0.49640	0.41764	0.38244	0.36262	0.34994	0.34114	0.33468	0.32974	0.32584	0.32075	0.31615	0.31303
4	4	3	0.79347	0.71160	0.67409	0.65265	0.63877	0.62907	0.62190	0.61640	0.61206	0.60555	0.60097	0.59756
4	4	4	1.92876	1.82626	1.77807	1.75010	1.73184	1.71897	1.70943	1.70206	1.69620	1.68748	1.68129	1.67568
5	1	1	1.86626	1.08011	0.79398	0.65339	0.57164	0.51877	0.48199	0.45502	0.43443	0.40516	0.38539	0.37115
5	2	1	0.75964	0.49235	0.38920	0.33621	0.30437	0.28325	0.26826	0.25710	0.24847	0.23602	0.22748	0.22126
5	2	2	0.92876	0.65353	0.54399	0.48639	0.45114	0.42743	0.41043	0.39766	0.38771	0.37324	0.36323	0.35590
5	3	1	0.47287	0.32860	0.27013	0.23918	0.22019	0.20742	0.19826	0.19137	0.18602	0.17824	0.17286	0.16892
5	3	2	0.59053	0.44565	0.38542	0.35288	0.33260	0.31879	0.30879	0.30122	0.29528	0.28660	0.28054	0.27509
5	3	3	0.78987	0.63378	0.56705	0.53034	0.50718	0.49127	0.47967	0.47064	0.46389	0.45368	0.44652	0.44123
5	4	1	0.34298	0.25091	0.21194	0.19080	0.17764	0.16868	0.16221	0.15732	0.15350	0.14791	0.14403	0.14117
5	4	2	0.43365	0.34438	0.30583	0.28454	0.27110	0.26185	0.25510	0.24997	0.24593	0.23999	0.23582	0.23275
5	4	3	0.59053	0.49768	0.45662	0.43360	0.41889	0.40870	0.40122	0.39550	0.39098	0.38432	0.37962	0.37513
5	4	4	0.92876	0.82293	0.77498	0.74771	0.73013	0.71786	0.70882	0.70187	0.69637	0.68822	0.68246	0.67318
5	5	1	0.26903	0.20511	0.17693	0.16131	0.15146	0.14470	0.13978	0.13604	0.13311	0.12881	0.12581	0.12359
5	5	2	0.34298	0.28368	0.25714	0.24220	0.23265	0.22602	0.22116	0.21744	0.21451	0.21146	0.20742	0.20494
5	5	3	0.47287	0.41441	0.38770	0.37247	0.36263	0.35576	0.35069	0.34679	0.34371	0.33914	0.33590	0.33350
5	5	4	0.75964	0.69762	0.66871	0.65202	0.64116	0.63353	0.62789	0.62352	0.62006	0.61491	0.61126	0.60854
5	5	5	1.86626	1.78755	1.75010	1.72822	1.71387	1.70374	1.69620	1.69038	1.68574	1.67883	1.67392	1.67025
6	1	1	1.82626	1.03150	0.74495	0.60526	0.52454	0.47259	0.43659	0.41027	0.39025	0.36186	0.34275	0.32903
6	2	1	0.73832	0.46335	0.35887	0.30577	0.27411	0.25322	0.23847	0.22752	0.21908	0.20695	0.19865	0.19263
6	2	2	0.86626	0.58601	0.47652	0.41963	0.38511	0.36205	0.34558	0.33326	0.32369	0.30982	0.30025	0.29327
6	3	1	0.45816	0.30701	0.24697	0.21558	0.19650	0.18374	0.17463	0.16782	0.16253	0.15487	0.14960	0.14575
6	3	2	0.54641	0.39506	0.33352	0.30071	0.28046	0.26675	0.25688	0.24943	0.24361	0.23513	0.22923	0.22490
6	3	3	0.67876	0.51913	0.45252	0.41641	0.39385	0.37845	0.36728	0.35882	0.35218	0.34245	0.33565	0.33065
6	4	1	0.33170	0.23335	0.19272	0.17100	0.15761	0.14856	0.14205	0.13715	0.13333	0.12778	0.12393	0.12111
6	4	2	0.39933	0.30314	0.26265	0.24063	0.22686	0.21745	0.21062	0.20545	0.20139	0.19544	0.19129	0.18823
6	4	3	0.50230	0.40300	0.36025	0.33664	0.32172	0.31145	0.30395	0.29824	0.29375	0.28713	0.28249	0.27906
6	4	4	0.67876	0.57021	0.52240	0.49564	0.47858	0.46675	0.45808	0.45145	0.44621	0.43848	0.43304	0.42901
6	5	1	0.25988	0.19013	0.16027	0.14399	0.13383	0.12690	0.12189	0.11810	0.11514	0.11081	0.10780	0.10559
6	5	2	0.31480	0.24849	0.21968	0.20373	0.19364	0.18670	0.18163	0.17778	0.17474	0.17028	0.16715	0.16484
6	5	3	0.39933	0.33288	0.30343	0.28691	0.27636	0.26905	0.26368	0.25958	0.25634	0.25155	0.24819	0.24569
6	5	4	0.54641	0.47616	0.44441	0.42539	0.41478	0.40669	0.40073	0.39616	0.39253	0.38717	0.38338	0.38056
6	5	5	0.86626	0.78534	0.74803	0.72661	0.71272	0.70298	0.69578	0.69023	0.68584	0.67930	0.67468	0.67123
6	6	1	0.21361	0.16154	0.13847	0.12568	0.11760	0.11205	0.10802	0.10496	0.10256	0.09903	0.09658	0.09476
6	6	2	0.25988	0.21200	0.19054	0.17847	0.17076	0.16542	0.16150	0.15850	0.15614	0.15472	0.15067	0.14849
6	6	3	0.33170	0.28555	0.26450	0.25252	0.24479	0.23940	0.23543	0.23238	0.22997	0.22642	0.22387	0.22199
6	6	4	0.45816	0.41174	0.39022	0.37782	0.36978	0.36413	0.36002	0.35676	0.35420	0.35040	0.34771	0.34572
6	6	5	0.73832	0.68847	0.66497	0.65131	0.64242	0.63610	0.63143	0.62783	0.62496	0.62069	0.61766	0.61540
6	6	6	1.82626	1.76245	1.73184	1.71387	1.70206	1.69370	1.68748	1.68267	1.67883	1.67310	1.66903	1.66597
7	1	1	1.79848	0.99664	0.70964	0.57061	0.49067	0.43944	0.40405	0.37825	0.35866	0.33096	0.31237	0.29906
7	2	1	0.72366	0.44271	0.33723	0.28409	0.25260	0.23194	0.21739	0.20663	0.19835	0.18648	0.17840	0.17254
7	2	2	0.82626	0.54184	0.43232	0.37597	0.34203	0.31946	0.30342	0.29145	0.28218	0.26878	0.25957	0.25283
7	3	1	0.44811	0.29168	0.23051	0.19886	0.17976	0.16705	0.15802	0.15128	0.14606	0.13853	0.13336	0.12960
7	3	2	0.51847	0.36225	0.29981	0.26691	0.24674	0.23317	0.22343	0.21611	0.21041	0.20212	0.19637	0.19217
7	3	3	0.61626	0.45367	0.38712	0.35146	0.32937	0.31438	0.30355	0.29536	0.28897	0.27961	0.27311	0.26833
7	4	1	0.32403	0.22090	0.17911	0.15702	0.14350	0.13442	0.12792	0.12304	0.11925	0.11375	0.10995	0.10718
7	4	2	0.37775	0.27653	0.23477	0.21233	0.19841	0.18896	0.18213	0.17697	0.17293	0.16704	0.16294	0.15992
7	4	3	0.45329	0.34950	0.30575	0.28190	0.26695	0.25672	0.24928	0.24364	0.23921	0.23270	0.22816	0.22481
7	4	4	0.56765	0.45680	0.40908	0.38272	0.36605	0.35457	0.34619	0.33981	0.33478	0.32738	0.32219	0.31835
7	5	1	0.25367	0.17953	0.14849	0.13178	0.12144	0.11443	0.10939	0.10559	0.10262	0.09831	0.09531	0.09312
7	5	2	0.29717	0.22586	0.19557	0.17902	0.16865	0.16156	0.15640	0.15249	0.14943	0.14493	0.14179	0.13947
7	5	3	0.35886	0.28715	0.25611	0.23893	0.22806	0.22056	0.21509	0.21092	0.20764	0.20281	0.19942	0.19691
7	5	4	0.45329	0.37822	0.34511	0.32557	0.31474	0.30654	0.30053	0.29593	0.29230	0.28694	0.28317	0.28037
7	5	5	0.61626	0.53347	0.49626	0.47519	0.46165	0.45222	0.44528	0.43996	0.43575	0.42951	0.42511	0.42184
7	6	1	0.20839	0.15223	0.12798	0.11472	0.10643	0.10077	0.09668	0.09358	0.09116	0.08762	0.08516	0.08335
7	6	2	0.24497	0.19217	0.16912	0.15634	0.14825	0.14269	0.13862	0.13553	0.13310	0.12952	0.12701	0.12516

Variances and Covariances of Type I Generalized Logistic Order Statistics

N	S	R	B = 1.0	B = 1.5	B = 2.0	B = 2.5	B = 3.0	B = 3.5	B = 4.0	B = 4.5	B = 5.0	B = 6.0	B = 7.0	B = 8.0
7	6	3	0.29717	0.24534	0.22233	0.20941	0.20117	0.19546	0.19127	0.18806	0.18553	0.18180	0.17917	0.17722
7	6	4	0.37775	0.32495	0.30112	0.28760	0.27891	0.27285	0.26839	0.26496	0.26226	0.25825	0.25542	0.25331
7	6	5	0.51847	0.46210	0.43624	0.42142	0.41183	0.40512	0.40016	0.39635	0.39332	0.38883	0.38565	0.38329
7	6	6	0.82626	0.76087	0.73036	0.71272	0.70123	0.69315	0.68717	0.68255	0.57889	0.67343	0.66956	0.66668
7	7	1	0.17581	0.13278	0.11321	0.10233	0.09547	0.09076	0.08734	0.08474	0.08270	0.07971	0.07763	0.07509
7	7	2	0.20839	0.16804	0.14994	0.13977	0.13328	0.12878	0.12549	0.12297	0.12099	0.12111	0.11670	0.11464
7	7	3	0.25367	0.21519	0.19768	0.18772	0.18131	0.17684	0.17355	0.17103	0.16903	0.16613	0.16400	0.16244
7	7	4	0.32403	0.28623	0.26876	0.25873	0.25223	0.24768	0.24431	0.24177	0.23968	0.23662	0.23445	0.23285
7	7	5	0.44811	0.40964	0.39161	0.38117	0.37436	0.36959	0.36602	0.36328	0.36110	0.35786	0.35557	0.35385
7	7	6	0.72366	0.68202	0.66223	0.65071	0.64310	0.63775	0.63378	0.63070	0.62826	0.62461	0.62202	0.62009
7	7	7	1.79848	1.74485	1.71897	1.70374	1.59370	1.58659	1.68129	1.67719	1.67392	1.66903	1.66555	1.66295
8	1	1	1.77807	0.97027	0.68282	0.54428	0.46497	0.41430	0.37941	0.35403	0.33480	0.30766	0.28950	0.27551
8	2	1	0.71297	0.42716	0.32091	0.26777	0.23644	0.21596	0.20160	0.19099	0.18286	0.17122	0.16331	0.15757
8	2	2	0.79848	0.51051	0.40092	0.34501	0.31153	0.28936	0.27366	0.26197	0.25294	0.23993	0.23100	0.22451
8	3	1	0.44081	0.28016	0.21814	0.18632	0.16723	0.15458	0.14562	0.13896	0.13382	0.12640	0.12133	0.11764
8	3	2	0.49921	0.33910	0.27603	0.24309	0.22303	0.20960	0.19999	0.19279	0.18720	0.17907	0.17347	0.16937
8	3	3	0.57626	0.41113	0.34460	0.30932	0.28760	0.27293	0.26237	0.25442	0.24822	0.23917	0.23290	0.22329
8	4	1	0.31847	0.21156	0.16889	0.14655	0.13297	0.12389	0.11741	0.11257	0.10881	0.10337	0.09963	0.09590
8	4	2	0.36294	0.25782	0.21516	0.19247	0.17849	0.16904	0.16224	0.15711	0.15312	0.14729	0.14325	0.14029
8	4	3	0.42217	0.31496	0.27055	0.24660	0.23169	0.22154	0.21419	0.20862	0.20427	0.19791	0.19345	0.19018
8	4	4	0.50515	0.39233	0.34466	0.31862	0.30227	0.29108	0.28293	0.27674	0.27189	0.26475	0.25976	0.25608
8	5	1	0.24917	0.17157	0.13965	0.12265	0.11220	0.10516	0.10011	0.09631	0.09336	0.08907	0.08611	0.08394
8	5	2	0.28512	0.20999	0.17866	0.16173	0.15120	0.14403	0.13884	0.13492	0.13185	0.12736	0.12424	0.12194
8	5	3	0.33333	0.25776	0.22568	0.20812	0.19709	0.18953	0.18403	0.17986	0.17658	0.17176	0.16839	0.16590
8	5	4	0.40143	0.32300	0.28909	0.27032	0.25844	0.25025	0.24426	0.23970	0.23611	0.23081	0.22710	0.22435
8	5	5	0.50515	0.42074	0.38360	0.36282	0.34956	0.34038	0.33365	0.32851	0.32445	0.31845	0.31423	0.31110
8	6	1	0.20461	0.14524	0.12012	0.10654	0.09811	0.09239	0.08827	0.08516	0.08274	0.07920	0.07675	0.07496
8	6	2	0.23480	0.17828	0.15411	0.14088	0.13256	0.12687	0.12274	0.11960	0.11713	0.11352	0.11100	0.10913
8	6	3	0.27548	0.21959	0.19530	0.18183	0.17330	0.16742	0.16313	0.15985	0.15728	0.15348	0.15082	0.14885
8	6	4	0.33333	0.27633	0.25115	0.23705	0.22805	0.22181	0.21723	0.21374	0.21098	0.20690	0.20403	0.20190
8	6	5	0.42217	0.36200	0.33500	0.31972	0.30991	0.30308	0.29806	0.29420	0.29116	0.28664	0.28346	0.28110
8	6	6	0.57626	0.50951	0.47906	0.46168	0.45045	0.44260	0.43681	0.43236	0.42883	0.42359	0.41990	0.41714
8	7	1	0.17356	0.12652	0.10608	0.09487	0.08785	0.08306	0.07959	0.07697	0.07492	0.07192	0.06983	0.06330
8	7	2	0.19958	0.15563	0.13637	0.12569	0.11892	0.11427	0.11087	0.10828	0.10625	0.10325	0.10116	0.09961
8	7	3	0.23480	0.19216	0.17321	0.16258	0.15580	0.15109	0.14765	0.14501	0.14293	0.13986	0.13771	0.13510
8	7	4	0.28512	0.24259	0.22340	0.21253	0.20555	0.20068	0.19710	0.19436	0.19219	0.18897	0.18670	0.18502
8	7	5	0.36294	0.31919	0.29917	0.28773	0.28033	0.27515	0.27134	0.26841	0.26608	0.26262	0.26018	0.25837
8	7	6	0.49921	0.45220	0.43038	0.41780	0.40964	0.40389	0.39964	0.39637	0.39378	0.38991	0.38718	[illegible]
8	7	7	0.79848	0.74367	0.71786	0.70287	0.69307	0.68617	0.68105	0.67710	0.67395	0.66927	0.66595	0.66347
8	8	1	0.15069	0.11248	0.09544	0.08578	0.08000	0.07590	0.07292	0.07066	0.06888	0.06629	0.06448	0.06314
8	8	2	0.17356	0.13858	0.12289	0.11407	0.10845	0.10456	0.10171	0.09953	0.09781	0.09951	0.09448	0.09242
8	8	3	0.20461	0.17144	0.15636	0.14780	0.14230	0.13847	0.13565	0.13348	0.13177	0.12933	0.12747	0.12614
8	8	4	0.24917	0.21696	0.20212	0.19362	0.18812	0.18427	0.18143	0.17933	0.17753	0.17494	0.17312	0.17177
8	6	5	0.31847	0.28646	0.27152	0.26289	0.25728	0.25341	0.25042	0.24816	0.24637	0.24371	0.24183	0.24043
8	8	6	0.44081	0.40798	0.39247	0.38344	0.37756	0.37338	0.37029	0.36790	0.36600	0.36318	0.36117	0.35967
8	8	7	0.71297	0.67724	0.66014	0.65013	0.64355	0.63890	0.63544	0.63276	0.63063	0.62744	0.62518	0.62349
8	8	8	1.77807	1.73184	1.70943	1.69620	1.58748	1.58129	1.67668	1.67310	1.67025	1.56599	1.66295	1.66368
9	1	1	1.76245	0.94953	0.66163	0.52348	0.44467	0.39449	0.36000	0.33497	0.31603	0.28937	0.27155	0.25384
9	2	1	0.70484	0.41498	0.30809	0.25496	0.22378	0.20347	0.18926	0.17880	0.17079	0.15934	0.15158	0.14598
9	2	2	0.77807	0.48703	0.37736	0.32179	0.28869	0.26686	0.25144	0.23999	0.23116	0.21846	0.20977	0.20346
9	3	1	0.43527	0.27114	0.20844	0.17651	0.15744	0.14487	0.13599	0.12939	0.12431	0.11701	0.11202	0.10840
9	3	2	0.48514	0.32183	0.25825	0.22532	0.20537	0.19207	0.18258	0.17549	0.16999	0.16203	0.15654	0.15254
9	3	3	0.54845	0.38114	0.31461	0.27963	0.25822	0.24382	0.23349	0.22573	0.21969	0.21090	0.20482	0.20036
9	4	1	0.31426	0.20424	0.16089	0.13838	0.12477	0.11571	0.10926	0.10445	0.10073	0.09535	0.09166	0.08893
9	4	2	0.35216	0.24389	0.20055	0.17769	0.16369	0.15427	0.14752	0.14244	0.13848	0.13273	0.12875	0.12584
9	4	3	0.40068	0.29072	0.24584	0.22185	0.20701	0.19695	0.18969	0.18421	0.17992	0.17367	0.16933	0.16513
9	4	4	0.46515	0.35060	0.30296	0.27719	0.26111	0.25014	0.24219	0.23617	0.23145	0.22454	0.21972	0.21616
9	5	1	0.24577	0.16534	0.13274	0.11553	0.10502	0.09796	0.09292	0.08914	0.08621	0.08195	0.07902	0.07533
9	5	2	0.27637	0.19818	0.16608	0.14889	0.13827	0.13107	0.12588	0.12196	0.11890	0.11444	0.11134	0.10906
9	5	3	0.31576	0.23720	0.20438	0.18660	0.17549	0.16791	0.16242	0.15826	0.15500	0.15022	0.14689	0.14443
9	5	4	0.36845	0.28745	0.25302	0.23415	0.22228	0.21413	0.20820	0.20369	0.20015	0.19494	0.19130	0.18861
9	5	5	0.44265	0.35683	0.31972	0.29918	0.28616	0.27719	0.27064	0.26564	0.26170	0.25590	0.25183	0.24882
9	6	1	0.20176	0.13977	0.11398	0.10017	0.09165	0.08590	0.08177	0.07866	0.07624	0.07272	0.07029	0.06351
9	6	2	0.22742	0.16797	0.14297	0.12941	0.12096	0.11519	0.11102	0.10786	0.10539	0.10177	0.09925	0.09740
9	6	3	0.26061	0.20161	0.17642	0.16260	0.15390	0.14793	0.14359	0.14029	0.13770	0.13389	0.13123	0.12925
9	6	4	0.30522	0.24515	0.21910	0.20465	0.19550	0.18918	0.18457	0.18105	0.17828	0.17420	0.17133	0.16921
9	6	5	0.36845	0.30564	0.27797	0.26248	0.25260	0.24576	0.24074	0.23690	0.23388	0.22940	0.22625	0.22393
9	6	6	0.46515	0.39719	0.36680	0.34963	0.33862	0.33096	0.32533	0.32101	0.31760	0.31254	0.30898	0.30634
9	7	1	0.17110	0.12162	0.10051	0.08906	0.08193	0.07710	0.07361	0.07098	0.05892	0.06593	0.06385	0.06233
9	7	2	0.19321	0.14642	0.12630	0.11526	0.10832	0.10356	0.10010	0.09748	0.09542	0.09240	0.09029	0.08373
9	7	3	0.22189	0.17612	0.15616	0.14508	0.13807	0.13323	0.12969	0.12700	0.12488	0.12175	0.11956	0.11794
9	7	4	0.26061	0.21470	0.19439	0.18302	0.17576	0.17073	0.16704	0.16422	0.16199	0.15870	0.15639	0.15468
9	7	5	0.31576	0.26855	0.24737	0.23539	0.22771	0.22236	0.21843	0.21542	0.21303	0.20950	0.20701	0.20517
9	7	6	0.40068	0.35056	0.32775	0.31477	0.30638	0.30053	0.29621	0.29289	0.29026	0.28636	0.28361	0.28156
9	7	7	0.54845	0.49263	0.46687	0.45208	0.44248	0.43576	0.43078	0.42596	0.42392	0.41940	0.41621	0.41383
9	8	1	0.14853	0.10802	0.09033	0.08060	0.07451	0.07035	0.06734	0.06506	0.06328	0.06068	0.05887	0.05753

Variances and Covariances of Type I Generalized Logistic Order Statistics

N	S	R	B = 1.0	B = 1.5	B = 2.0	B = 2.5	B = 3.0	B = 3.5	B = 4.0	B = 4.5	B = 5.0	B = 6.0	B = 7.0	B = 8.0
9	8	2	0.16795	0.13023	0.11366	0.10445	0.09863	0.09462	0.09169	0.08946	0.08771	0.08514	0.08333	0.08200
9	8	3	0.19321	0.15689	0.14073	0.13167	0.12589	0.12188	0.11894	0.11670	0.11493	0.11232	0.11048	0.10912
9	8	4	0.22742	0.19164	0.17551	0.16638	0.16051	0.15643	0.15343	0.15113	0.14931	0.14662	0.14472	0.14331
9	8	5	0.27637	0.24032	0.22386	0.21447	0.20840	0.20416	0.20103	0.19863	0.19673	0.19390	0.19191	0.19043
9	8	6	0.35216	0.31483	0.29758	0.28765	0.28122	0.27671	0.27336	0.27079	0.26875	0.26572	0.26357	0.26197
9	8	7	0.48514	0.44484	0.42597	0.41505	0.40793	0.40292	0.39920	0.39634	0.39406	0.39067	0.38827	0.38548
9	8	8	0.77807	0.73092	0.70856	0.69553	0.68699	0.68096	0.67649	0.67303	0.67027	0.66617	0.66326	0.66108
9	9	1	0.13121	0.09743	0.08233	0.07393	0.06863	0.06499	0.06235	0.06035	0.05878	0.05648	0.05488	0.09107
9	9	2	0.14853	0.11759	0.10370	0.09590	0.09093	0.08749	0.08498	0.08306	0.08155	0.08492	0.07904	0.07689
9	9	3	0.17110	0.14184	0.12855	0.12102	0.11619	0.11282	0.11035	0.10845	0.10695	0.10487	0.10319	0.10202
9	9	4	0.20176	0.17352	0.16055	0.15313	0.14833	0.14498	0.14251	0.14077	0.13914	0.13687	0.13530	0.13412
9	9	5	0.24578	0.21806	0.20516	0.19774	0.19291	0.18971	0.18705	0.18510	0.18356	0.18128	0.17967	0.17347
9	9	6	0.31425	0.28650	0.27345	0.26587	0.26100	0.25746	0.25487	0.25287	0.25129	0.24893	0.24726	0.24502
9	9	7	0.43527	0.40664	0.39303	0.38511	0.37987	0.37619	0.37345	0.37134	0.36965	0.36715	0.36537	0.36404
9	9	8	0.70484	0.67355	0.65852	0.64967	0.64385	0.63974	0.63667	0.63430	0.63241	0.62958	0.62758	0.62608
9	9	9	1.76245	1.72182	1.70206	1.69038	1.68267	1.67719	1.67310	1.66993	1.66741	1.66363	1.66093	1.65392
10	1	1	1.75010	0.93272	0.64439	0.50555	0.42817	0.37838	0.34425	0.31951	0.30082	0.27456	0.25704	0.24456
10	2	1	0.69844	0.40513	0.29771	0.24459	0.21355	0.19339	0.17932	0.16898	0.16107	0.14981	0.14218	0.13568
10	2	2	0.76245	0.46871	0.35894	0.30366	0.27087	0.24933	0.23414	0.22289	0.21424	0.20180	0.19332	0.18716
10	3	1	0.43079	0.26386	0.20061	0.16859	0.14957	0.13706	0.12824	0.12172	0.11670	0.10949	0.10458	0.10102
10	3	2	0.47441	0.30839	0.24441	0.21149	0.19166	0.17847	0.16910	0.16211	0.15669	0.14887	0.14349	0.13957
10	3	3	0.52807	0.35879	0.29224	0.25750	0.23636	0.22219	0.21205	0.20445	0.19855	0.18998	0.18406	0.17973
10	4	1	0.31096	0.19834	0.15443	0.13179	0.11817	0.10914	0.10273	0.09795	0.09427	0.08895	0.08531	0.08267
10	4	2	0.34395	0.23307	0.18919	0.16623	0.15224	0.14286	0.13614	0.13111	0.12720	0.12153	0.11761	0.11474
10	4	3	0.38495	0.27270	0.22747	0.20347	0.18871	0.17874	0.17157	0.16616	0.16195	0.15581	0.15155	0.14842
10	4	4	0.43737	0.32130	0.27367	0.24812	0.23226	0.22149	0.21371	0.20783	0.20323	0.19650	0.19182	0.18338
10	5	1	0.24311	0.16031	0.12716	0.10980	0.09925	0.09219	0.08716	0.08340	0.08049	0.07628	0.07338	0.07126
10	5	2	0.26972	0.18902	0.15631	0.13895	0.12827	0.12106	0.11588	0.11198	0.10894	0.10451	0.10145	0.09920
10	5	3	0.30295	0.22196	0.18859	0.17065	0.15951	0.15195	0.14648	0.14234	0.13911	0.13438	0.13110	0.12867
10	5	4	0.34566	0.26258	0.22777	0.20886	0.19703	0.18895	0.18308	0.17863	0.17514	0.17002	0.16645	0.16381
10	5	5	0.40265	0.31557	0.27849	0.25815	0.24534	0.23654	0.23013	0.22526	0.22143	0.21580	0.21186	0.20894
10	6	1	0.19953	0.13536	0.10903	0.09504	0.08647	0.08070	0.07657	0.07347	0.07106	0.06757	0.06515	0.06339
10	6	2	0.22183	0.15997	0.13433	0.12054	0.11199	0.10619	0.10200	0.09834	0.09637	0.09276	0.09025	0.08341
10	6	3	0.24978	0.18830	0.16244	0.14838	0.13958	0.13357	0.12921	0.12590	0.12331	0.11950	0.11685	0.11490
10	6	4	0.28586	0.22340	0.19672	0.18207	0.17283	0.16649	0.16187	0.15836	0.15559	0.15153	0.14869	0.14559
10	6	5	0.33426	0.26942	0.24131	0.22572	0.21584	0.20902	0.20404	0.20024	0.19725	0.19283	0.18974	0.18745
10	6	6	0.40265	0.33363	0.30326	0.28628	0.27545	0.26795	0.26246	0.25825	0.25494	0.25004	0.24659	0.24404
10	7	1	0.16888	0.11766	0.09602	0.08439	0.07719	0.07233	0.06883	0.06619	0.06414	0.06116	0.05910	0.05753
10	7	2	0.18838	0.13928	0.11850	0.10719	0.10013	0.09532	0.09183	0.08918	0.08711	0.08408	0.08197	0.08042
10	7	3	0.21251	0.16425	0.14354	0.13216	0.12499	0.12007	0.11649	0.11376	0.11162	0.10848	0.10628	0.10465
10	7	4	0.24377	0.19528	0.17418	0.16248	0.15505	0.14993	0.14619	0.14333	0.14108	0.13777	0.13544	0.13372
10	7	5	0.28586	0.23612	0.21418	0.20190	0.19406	0.18863	0.18465	0.18161	0.17921	0.17566	0.17317	0.17132
10	7	6	0.34566	0.29337	0.27000	0.25681	0.24835	0.24248	0.23815	0.23483	0.23221	0.22834	0.22560	0.22358
10	7	7	0.43737	0.38059	0.35487	0.34025	0.33082	0.32424	0.31939	0.31567	0.31272	0.30835	0.30527	0.30297
10	8	1	0.14647	0.10442	0.08620	0.07629	0.07011	0.06592	0.06289	0.06060	0.05882	0.05622	0.05441	0.05309
10	8	2	0.15370	0.12376	0.10651	0.09702	0.09106	0.08697	0.08400	0.08174	0.07997	0.07737	0.07555	0.07422
10	8	3	0.18493	0.14614	0.12919	0.11977	0.11380	0.10968	0.10668	0.10438	0.10258	0.09992	0.09806	0.09668
10	8	4	0.21251	0.17403	0.15700	0.14745	0.14136	0.13714	0.13405	0.13168	0.12981	0.12706	0.12512	0.12368
10	8	5	0.24978	0.21086	0.19341	0.18355	0.17722	0.17282	0.16959	0.16711	0.16515	0.16225	0.16020	0.15869
10	8	6	0.30295	0.26269	0.24441	0.23401	0.22730	0.22262	0.21917	0.21652	0.21442	0.21131	0.20911	0.20748
10	8	7	0.38496	0.34204	0.32232	0.31102	0.30370	0.29857	0.29478	0.29186	0.28955	0.28612	0.28369	0.28188
10	8	8	0.52807	0.48010	0.45776	0.44490	0.43652	0.43064	0.42628	0.42292	0.42025	0.41628	0.41347	0.41137
10	9	1	0.12971	0.09411	0.07849	0.06990	0.06451	0.06083	0.05817	0.05615	0.05458	0.05227	0.05067	0.04949
10	9	2	0.14474	0.11165	0.09708	0.08898	0.08386	0.08033	0.07776	0.07580	0.07426	0.07199	0.07041	0.06924
10	9	3	0.16370	0.13199	0.11787	0.10995	0.10490	0.10140	0.09884	0.09688	0.09534	0.09306	0.09146	0.09027
10	9	4	0.18838	0.15738	0.14342	0.13551	0.13044	0.12692	0.12432	0.12234	0.12077	0.11845	0.11681	0.11559
10	9	5	0.22183	0.19100	0.17694	0.16893	0.16376	0.16015	0.15749	0.15545	0.15383	0.15142	0.14973	0.14347
10	9	6	0.26972	0.23845	0.22404	0.21576	0.21040	0.20664	0.20387	0.20173	0.20003	0.19751	0.19573	0.19441
10	9	7	0.34396	0.31143	0.29626	0.28750	0.28181	0.27779	0.27483	0.27254	0.27072	0.26802	0.26610	0.26468
10	9	8	0.47441	0.43916	0.42254	0.41289	0.40658	0.40213	0.39883	0.39628	0.39425	0.39123	0.38909	0.38750
10	9	9	0.76245	0.72109	0.70137	0.68984	0.68227	0.67692	0.67295	0.66987	0.66742	0.66377	0.66118	0.65924
10	10	1	0.11616	0.08584	0.07227	0.06472	0.05995	0.05668	0.05431	0.05251	0.05110	0.04904	0.04760	0.08906
10	10	2	0.12971	0.10192	0.08945	0.08244	0.07799	0.07490	0.07265	0.07093	0.06957	0.07475	0.06783	0.06552
10	10	3	0.14684	0.12060	0.10869	0.10195	0.09763	0.09462	0.09241	0.09072	0.08938	0.08760	0.08604	0.08499
10	10	4	0.16918	0.14395	0.13238	0.12577	0.12151	0.11853	0.11633	0.11493	0.11338	0.11134	0.10994	0.10390
10	10	5	0.19953	0.17493	0.16352	0.15696	0.15270	0.14972	0.14758	0.14583	0.14447	0.14247	0.14106	0.14001
10	10	6	0.24311	0.21878	0.20738	0.20078	0.19666	0.19348	0.19122	0.18949	0.18811	0.18607	0.18463	0.18355
10	10	7	0.31096	0.28646	0.27486	0.26823	0.26370	0.26058	0.25827	0.25648	0.25506	0.25294	0.25144	0.25032
10	10	8	0.43092	0.40555	0.39353	0.38632	0.38165	0.37836	0.37591	0.37401	0.37250	0.37025	0.36864	0.36745
10	10	9	0.69844	0.67072	0.65719	0.64927	0.64407	0.64038	0.63762	0.63549	0.63380	0.63126	0.69245	0.62811
10	10	10	1.75010	1.71387	1.69620	1.68574	1.67883	1.67392	1.67025	1.66741	1.66514	1.66174	1.65932	1.65751

Appendix D

TABLES OF ONE- AND TWO-ORDER STATISTIC ESTIMATORS FOR EXPONENTIAL POPULATIONS

SOURCES OF TABLES

Tables D1,D3	*Ann. Math. Statist.* 32(1961), 1078-1090 (Harter)
Table D2	IEEE Trans. Reliability 14(1965), 100-106; 15(1966), 120-126 (Moore and Harter)
Tables D4 A,B	*Rider Anniversary Volume* (1963), 77-100 (Harter) [with additional columns freshly computed (Harter)]
Tables D4 C,D	*Technometrics* 6(1964), 301-317 (Harter)

Table D1

MOST EFFICIENT UNBIASED POINT ESTIMATORS FOR σ, BASED ON ONE OR TWO ORDER STATISTICS OF A SAMPLE FROM A ONE-PARAMETER EXPONENTIAL POPULATION

[One-order statistics estimator: $\tilde{\sigma}_k = c_k x_k$; two-order-statistic estimator: $\tilde{\sigma}_{lm} = c_l x_l + c_m x_m$]

Best Estimators of Parameter σ of 1-Parameter Exponential Population

n	From 1 Order Statistic, x_k				From 2 Order Statistics, x_ℓ and x_m					
	k	c_k	V_k/σ^2	$E_k(\%)$	ℓ	m	c_ℓ	c_m	$V_{\ell m}/\sigma^2$	$E_{\ell m}(\%)$
1	1	1.00000	1.000000	100.00						
2	2	.666667	.5555556	90.000	1	2	.500000	.500000	.5000000	100.00
3	3	.545455	.4049587	82.313	2	3	.447368	.342105	.3421053	97.436
4	4	.480000	.3280000	76.220	3	4	.413043	.265217	.2652174	94.262
5	5	.437956	.2807289	71.243	3	5	.527997	.256818	.2140154	93.451
6	5	.689655	.2337165	71.311	4	6	.493892	.216654	.1805452	92.313
7	6	.627803	.2017178	70.820	5	7	.467528	.188618	.1571815	90.887
8	7	.582121	.1787245	69.940	5	8	.553626	.187760	.1393976	89.672
9	8	.546756	.1613595	68.859	6	9	.527062	.167990	.1247200	89.088
10	8	.699806	.1468046	68.118	7	10	.505032	.152501	.1132201	88.324
11	9	.657948	.1333457	68.175	8	11	.486391	.140031	.1039622	87.444
12	10	.623748	.1225454	68.002	8	12	.552632	.140623	.0960924	86.722
13	11	.595191	.1136773	67.668	9	13	.533452	.130469	.0891540	86.281
14	12	.570919	.1062578	67.222	10	14	.516702	.121903	.0833001	85.748
15	12	.673448	.0994728	67.020	11	15	.501916	.114575	.0782930	85.150
16	13	.646247	.0932310	67.038	10	15	.519329	.217012	.0735428	84.984
17	14	.622580	.0878686	66.945	11	16	.506584	.204426	.0692778	84.910
18	15	.601766	.0832093	66.766	12	17	.495052	.193425	.0655497	84.753
19	16	.583292	.0791212	66.520	12	18	.531320	.193368	.0621307	84.711
20	16	.660325	.0752377	66.456	13	19	.519816	.183870	.0590788	84.633
21	17	.640194	.0716497	66.461	14	20	.509272	.175398	.0563567	84.496
22	18	.622092	.0684545	66.401	14	21	.542190	.175587	.0539094	84.317
23	19	.605709	.0655900	66.288	15	22	.531650	.168092	.0516082	84.247
24	20	.590798	.0630065	66.131	16	23	.521895	.161307	.0495251	84.132
25	20	.652475	.0605006	66.115	17	24	.512834	.155136	.0476304	83.980
26	21	.636502	.0581740	66.115	17	25	.542341	.155498	.0458907	83.811
27	22	.621843	.0560556	66.072	18	26	.533236	.149905	.0442401	83.718
28	23	.608333	.0541184	65.993	19	27	.524718	.144765	.0427233	83.594
29	24	.595834	.0523395	65.883	18	27	.529588	.203622	.0412679	83.558
30	24	.647255	.0505915	65.887	19	28	.521922	.196822	.0398898	83.564
31	25	.634017	.0489616	65.884	20	29	.514693	.190530	.0386145	83.539
32	26	.621699	.0474550	65.852	21	30	.507861	.184689	.0374308	83.487
33	27	.610203	.0460582	65.793	21	31	.529266	.184782	.0363033	83.471
34	28	.599445	.0447593	65.711	22	32	.522441	.179406	.0352472	83.444
35	28	.643532	.0434715	65.724	23	33	.515963	.174384	.0342605	83.395
36	29	.632230	.0422665	65.721	23	34	.536163	.174543	.0333365	83.325
37	30	.621609	.0411405	65.694	24	35	.529688	.169881	.0324462	83.298
38	31	.611604	.0400859	65.649	25	36	.523519	.165502	.0316097	83.252
39	31	.651175	.0390941	65.588	26	37	.517635	.161380	.0308224	83.189
40	32	.640744	.0381083	65.603	26	38	.536491	.161608	.0300780	83.117
41	33	.630885	.0371813	65.598	27	39	.530595	.157745	.0293590	83.076
42	34	.621548	.0363080	65.577	27	39	.511693	.194092	.0286687	83.051
43	35	.612692	.0354837	65.539	27	40	.527739	.194098	.0280016	83.052
44	35	.647770	.0346995	65.497	28	41	.522497	.189685	.0273649	83.053
45	36	.638578	.0339230	65.508	29	42	.517464	.185500	.0267612	83.039
46	37	.629834	.0331879	65.503	30	43	.512624	.181525	.0261877	83.013
47	38	.621506	.0324908	65.485	30	44	.527790	.181619	.0256330	83.005
48	39	.613561	.0318289	65.454	31	45	.522953	.177867	.0251035	82.990
49	39	.645063	.0311934	65.425	32	46	.518293	.174292	.0245990	82.963
50	40	.636847	.0305660	65.432	32	47	.532844	.174417	.0241175	82.927

Best Estimators of Parameter σ of 1-Parameter Exponential Population

n	From 1 Order Statistic, x_k				From 2 Order Statistics, x_ℓ and x_m					
	k	c_k	V_k/σ^2	$E_k(\%)$	ℓ	m	c_ℓ	c_m	$V_{\ell m}/\sigma^2$	$E_{\ell m}(\%)$
51	41	.628992	.0299688	65.427	33	48	.528185	.171029	.0236490	82.912
52	42	.621475	.0293996	65.412	34	49	.523688	.167791	.0232014	82.886
53	43	.614272	.0288564	65.386	35	50	.519344	.164695	.0227732	82.851
54	43	.642858	.0283310	65.365	35	51	.533183	.164857	.0223621	82.812
55	44	.635431	.0278136	65.370	35	51	.518358	.192720	.0219586	82.800
56	45	.628302	.0273189	65.366	36	52	.514374	.189295	.0215684	82.793
57	46	.621452	.0268453	65.352	36	53	.526631	.189331	.0211897	82.794
58	47	.614863	.0263915	65.329	37	54	.522652	.186069	.0208246	82.793
59	47	.641028	.0259499	65.315	38	55	.518794	.182935	.0204738	82.784
60	48	.634252	.0255159	65.319	39	56	.515050	.179922	.0201366	82.768
61	49	.627725	.0250994	65.314	39	57	.526789	.180005	.0198076	82.763
62	50	.621434	.0246992	65.302	40	58	.523047	.177123	.0194905	82.753
63	51	.615363	.0243145	65.282	41	59	.519411	.174347	.0191850	82.737
64	51	.639486	.0239381	65.273	41	60	.530779	.174447	.0188902	82.715
65	52	.633255	.0235689	65.275	42	61	.527144	.171785	.0186019	82.704
66	53	.627237	.0232133	65.271	43	62	.523609	.169216	.0183238	82.688
67	54	.621420	.0228707	65.260	44	63	.520168	.166736	.0180552	82.665
68	55	.615792	.0225404	65.242	43	63	.522477	.191905	.0177944	82.643
69	55	.638167	.0222158	65.236	44	64	.519185	.189124	.0175365	82.644
70	56	.632401	.0218979	65.238	45	65	.515976	.186434	.0172871	82.638
71	57	.626818	.0215909	65.234	45	66	.525891	.186478	.0170434	82.639
72	58	.621409	.0212943	65.224	46	67	.522686	.183891	.0168070	82.638
73	59	.616163	.0210076	65.208	47	68	.519559	.181386	.0165781	82.631
74	59	.637027	.0207248	65.204	48	69	.516508	.178961	.0163564	82.619
75	60	.631661	.0204481	65.206	48	70	.526081	.179033	.0161388	82.617
76	61	.626455	.0201804	65.201	49	71	.523031	.176693	.0159279	82.609
77	62	.621399	.0199211	65.192	50	72	.520052	.174423	.0157233	82.597
78	63	.616488	.0196699	65.178	50	73	.529377	.174507	.0155244	82.583
79	63	.636031	.0194214	65.177	51	74	.526398	.172314	.0153294	82.575
80	64	.631014	.0191784	65.177	52	75	.523487	.170185	.0151399	82.563
81	65	.626136	.0189428	65.173	53	76	.520640	.168116	.0149559	82.547
82	66	.621392	.0187142	65.165	52	76	.522467	.189026	.0147749	82.540
83	67	.616774	.0184924	65.152	53	77	.519723	.186749	.0145969	82.539
84	67	.635154	.0182722	65.152	54	78	.517038	.184536	.0144240	82.535
85	68	.630443	.0180572	65.153	54	79	.525362	.184580	.0142541	82.536
86	69	.625856	.0178483	65.149	55	80	.522678	.182437	.0140886	82.534
87	70	.621385	.0176452	65.141	56	81	.520050	.180352	.0139276	82.529
88	71	.617028	.0174478	65.129	56	82	.528185	.180407	.0137707	82.520
89	71	.634376	.0172514	65.131	57	83	.525558	.178385	.0136163	82.518
90	72	.629936	.0170598	65.130	58	84	.522984	.176415	.0134660	82.512
91	73	.625605	.0168733	65.127	59	85	.520461	.174496	.0133195	82.503
92	74	.621380	.0166917	65.119	59	86	.528366	.174567	.0131763	82.493
93	75	.617256	.0165150	65.109	60	87	.525843	.172703	.0130356	82.487
94	75	.633681	.0163387	65.111	61	88	.523368	.170884	.0128984	82.478
95	76	.629482	.0161668	65.111	62	89	.520941	.169111	.0127645	82.466
96	77	.625382	.0159993	65.107	61	89	.522452	.186991	.0126315	82.466
97	78	.621376	.0158360	65.100	62	90	.520100	.185065	.0125015	82.465
98	78	.637131	.0156767	65.091	63	91	.517792	.183185	.0123745	82.461
99	79	.633057	.0155177	65.093	63	92	.524964	.183228	.0122493	82.462
100	80	.629074	.0153627	65.093	64	93	.522657	.181399	.0121271	82.460

Table D2

UNBIASED POINT ESTIMATORS FOR σ, BASED ON ONE ORDER STATISTIC OF A CENSORED SAMPLE FROM A ONE-PARAMETER EXPONENTIAL POPULATION

[Estimator: $\tilde{\sigma}_k = c_k c_k$, where $c_k = C(K, N)$ and the Kth order statistic is optimal when the first M order statistics of a sample of size N are known (single censoring from above).]

N	K	M	C(K,N)	EFF.	N	K	M	C(K,N)	EFF.
1	1	1	1.00000	100.00	10	6	6	1.18254	94.47
2	1	1	2.00000	100.00	10	7	7	0.91271	90.90
2	2	2	0.66667	90.00	10	8	8	0.69981	85.15
3	1	1	3.00000	100.00	10	8	9	0.69981	75.69
3	2	2	1.20000	96.15	10	8	10	0.69981	68.12
3	3	3	0.54545	82.31	11	1	1	11.00000	100.00
4	1	1	4.00000	100.00	11	2	2	5.23810	99.77
4	2	2	1.71429	98.00	11	3	3	3.31104	99.33
4	3	3	0.92308	92.35	11	4	4	2.34181	98.60
4	4	4	0.48000	76.22	11	5	5	1.75476	97.46
5	1	1	5.00000	100.00	11	6	6	1.35769	95.76
5	2	2	2.22222	98.78	11	7	7	1.06776	93.22
5	3	3	1.27660	95.75	11	8	8	0.84278	89.37
5	4	4	0.77922	88.81	11	9	9	0.65795	83.33
5	5	5	0.43796	71.24	11	9	10	0.65795	74.99
6	1	1	6.00000	100.00	11	9	11	0.65795	68.18
6	2	2	2.72727	99.18	12	1	1	12.00000	100.00
6	3	3	1.62162	97.30	12	2	2	5.73913	99.81
6	4	4	1.05263	93.47	12	3	3	3.64641	99.45
6	5	5	0.68966	85.57	12	4	4	2.59502	98.85
6	5	6	0.68966	71.31	12	5	5	1.95943	97.96
7	1	1	7.00000	100.00	12	6	6	1.53090	96.64
7	2	2	3.23077	99.41	12	7	7	1.21969	94.73
7	3	3	1.96262	98.13	12	8	8	0.98051	91.97
7	4	4	1.31661	95.71	12	9	9	0.78748	87.89
7	5	5	0.91503	91.24	12	10	10	0.62375	81.60
7	6	6	0.62780	82.62	12	10	11	0.62375	74.18
7	6	7	0.62780	70.82	12	10	12	0.62375	68.00
8	1	1	8.00000	100.00	13	1	1	13.00000	100.00
8	2	2	3.73333	99.56	13	2	2	6.24000	99.84
8	3	3	2.30137	98.63	13	3	3	3.98144	99.54
8	4	4	1.57598	96.96	13	4	4	2.84766	99.05
8	5	5	1.13055	94.09	13	5	5	2.16321	98.32
8	6	6	0.82111	89.10	13	6	6	1.70278	97.27
8	7	7	0.58212	79.93	13	7	7	1.36961	95.79
8	7	8	0.58212	69.94	13	8	8	1.11508	93.71
9	1	1	9.00000	100.00	13	9	9	0.91174	90.75
9	2	2	4.23529	99.66	13	10	10	0.74250	86.46
9	3	3	2.63874	98.95	13	11	11	0.59519	79.97
9	4	4	1.83273	97.73	13	11	12	0.59519	73.31
9	5	5	1.34114	95.73	13	11	13	0.59519	67.67
9	6	6	1.00438	92.48	14	1	1	14.00000	100.00
9	7	7	0.75246	87.07	14	2	2	6.74074	99.86
9	8	8	0.54676	77.47	14	3	3	4.31621	99.60
9	8	9	0.54676	68.86	14	4	4	3.09987	99.19
10	1	1	10.00000	100.00	14	5	5	2.36634	98.59
10	2	2	4.73684	99.72	14	6	6	1.87369	97.73
10	3	3	2.97521	99.17	14	7	7	1.51813	96.55
10	4	4	2.08782	98.24	14	8	8	1.24756	94.92
10	5	5	1.54886	96.76	14	9	9	1.03281	92.68

N	K	M	C(K,N)	EFF.	N	K	M	C(K,N)	EFF.
14	10	10	0.85600	89.56	17	14	15	0.62258	75.87
14	11	11	0.70510	85.09	17	14	16	0.62258	71.13
14	12	12	0.57092	78.43	17	14	17	0.62258	66.94
14	12	13	0.57092	72.39	18	1	1	18.00000	100.00
14	12	14	0.57092	67.22	18	2	2	8.74286	99.92
15	1	1	15.00000	100.00	18	3	3	5.65358	99.77
15	2	2	7.24138	99.88	18	4	4	4.10600	99.54
15	3	3	4.65077	99.66	18	5	5	3.17486	99.21
15	4	4	3.35175	99.31	18	6	6	2.55169	98.77
15	5	5	2.56897	98.80	18	7	7	2.10424	98.18
15	6	6	2.04390	98.09	18	8	8	1.76635	97.42
15	7	7	1.66563	97.12	18	9	9	1.50119	96.44
15	8	8	1.37860	95.82	18	10	10	1.28659	95.18
15	9	9	1.15177	94.06	18	11	11	1.10834	93.56
15	10	10	0.96628	91.67	18	12	12	0.95684	91.47
15	11	11	0.80979	88.40	18	13	13	0.82524	88.74
15	12	12	0.67345	83.77	18	14	14	0.70833	85.10
15	12	13	0.67345	77.33	18	15	15	0.60177	80.12
15	12	14	0.67345	71.81	18	15	16	0.60177	75.11
15	12	15	0.67345	67.02	18	15	17	0.60177	70.69
16	1	1	16.00000	100.00	18	15	18	0.60177	66.77
16	2	2	7.74194	99.90	19	1	1	19.00000	100.00
16	3	3	4.98516	99.70	19	2	2	9.24324	99.93
16	4	4	3.60337	99.40	19	3	3	5.98764	99.79
16	5	5	2.77122	98.97	19	4	4	4.35710	99.59
16	6	6	2.21356	98.37	19	5	5	3.37636	99.30
16	7	7	1.81238	97.56	19	6	6	2.72030	98.91
16	8	8	1.50859	96.49	19	7	7	2.24957	98.41
16	9	9	1.26924	95.07	19	8	8	1.89443	97.75
16	10	10	1.07443	93.19	19	9	9	1.61611	96.92
16	11	11	0.91125	90.68	19	10	10	1.39126	95.86
16	12	12	0.77077	87.27	19	11	11	1.20499	94.52
16	13	13	0.64625	82.51	19	12	12	1.04725	92.81
16	13	14	0.64625	76.61	19	13	13	0.91096	90.63
16	13	15	0.64625	71.51	19	14	14	0.79089	87.81
16	13	16	0.64625	67.04	19	15	15	0.68287	84.07
17	1	1	17.00000	100.00	19	16	16	0.58329	78.99
17	2	2	8.24242	99.91	19	16	17	0.58329	74.35
17	3	3	5.31943	99.74	19	16	18	0.58329	70.22
17	4	4	3.85477	99.48	19	16	19	0.58329	66.52
17	5	5	2.97317	99.10	20	1	1	20.00000	100.00
17	6	6	2.38280	98.59	20	2	2	9.74359	99.93
17	7	7	1.95854	97.91	20	3	3	6.32163	99.82
17	8	8	1.63778	97.01	20	4	4	4.60807	99.63
17	9	9	1.38563	95.84	20	5	5	3.57768	99.38
17	10	10	1.18106	94.32	20	6	6	2.88869	99.04
17	11	11	1.01056	92.33	20	7	7	2.39460	98.59
17	12	12	0.86489	89.70	20	8	8	2.02213	98.02
17	13	13	0.73734	86.17	20	9	9	1.73052	97.30
17	14	14	0.62258	81.29	20	10	10	1.49528	96.40

N	K	M	C(K,N)	EFF.	N	K	M	C(K,N)	EFF.
20	11	11	1.30078	95.27	22	18	20	0.62209	73.04
20	12	12	1.13652	93.85	22	18	21	0.62209	69.56
20	13	13	0.99514	92.07	22	18	22	0.62209	66.40
20	14	14	0.87128	89.81	23	1	1	23.00000	100.00
20	15	15	0.76080	86.89	23	2	2	11.24444	99.95
20	16	16	0.66032	83.07	23	3	3	7.32323	99.86
20	16	17	0.66032	78.18	23	4	4	5.36044	99.73
20	16	18	0.66032	73.84	23	5	5	4.18089	99.54
20	16	19	0.66032	69.95	23	6	6	3.39283	99.30
20	16	20	0.66032	66.46	23	7	7	2.82835	98.99
21	1	1	21.00000	100.00	23	8	8	2.40349	98.59
21	2	2	10.24390	99.94	23	9	9	2.07155	98.11
21	3	3	6.65555	99.83	23	10	10	1.80454	97.51
21	4	4	4.85894	99.67	23	11	11	1.58458	96.78
21	5	5	3.77887	99.44	23	12	12	1.39975	95.89
21	6	6	3.05689	99.14	23	13	13	1.24174	94.81
21	7	7	2.53938	98.75	23	14	14	1.10458	93.49
21	8	8	2.14950	98.25	23	15	15	0.98383	91.88
21	9	9	1.84451	97.62	23	16	16	0.87609	89.89
21	10	10	1.59877	96.84	23	17	17	0.77864	87.41
21	11	11	1.39589	95.88	23	18	18	0.68920	84.28
21	12	12	1.22490	94.68	23	19	19	0.60571	80.24
21	13	13	1.07817	93.19	23	19	20	0.60571	76.23
21	14	14	0.95012	91.33	23	19	21	0.60571	72.60
21	15	15	0.83657	88.99	23	19	22	0.60571	69.30
21	16	16	0.73420	86.00	23	19	23	0.60571	66.29
21	17	17	0.64019	82.10	24	1	1	24.00000	100.00
21	17	18	0.64019	77.54	24	2	2	11.74468	99.95
21	17	19	0.64019	73.46	24	3	3	7.65700	99.87
21	17	20	0.64019	69.78	24	4	4	5.61109	99.75
21	17	21	0.64019	66.46	24	5	5	4.38177	99.58
22	1	1	22.00000	100.00	24	6	6	3.56062	99.36
22	2	2	10.74419	99.95	24	7	7	2.97260	99.08
22	3	3	6.98941	99.85	24	8	8	2.53018	98.73
22	4	4	5.10973	99.70	24	9	9	2.18470	98.30
22	5	5	3.97993	99.50	24	10	10	1.90696	97.77
22	6	6	3.22493	99.23	24	11	11	1.67835	97.12
22	7	7	2.68396	98.88	24	12	12	1.48644	96.34
22	8	8	2.27660	98.44	24	13	13	1.32261	95.41
22	9	9	1.95818	97.89	24	14	14	1.18065	94.27
22	10	10	1.70183	97.21	24	15	15	1.05598	92.90
22	11	11	1.49046	96.37	24	16	16	0.94509	91.23
22	12	12	1.31260	95.34	24	17	17	0.84524	89.18
22	13	13	1.16030	94.08	24	18	18	0.75417	86.64
22	14	14	1.02780	92.53	24	19	19	0.66996	83.45
22	15	15	0.91078	90.61	24	20	20	0.59080	79.36
22	16	16	0.80592	88.19	24	20	21	0.59080	75.58
22	17	17	0.71049	85.13	24	20	22	0.59080	72.14
22	18	18	0.62209	81.16	24	20	23	0.59080	69.01
22	18	19	0.62209	76.89	24	20	24	0.59080	66.13

N	K	M	C(K,N)	EFF.	N	K	M	C(K,N)	EFF.
25	1	1	25.00000	100.00	26	21	26	0.63650	66.11
25	2	2	12.24490	99.96	27	1	1	27.00000	100.00
25	3	3	7.99074	99.88	27	2	2	13.24528	99.96
25	4	4	5.86168	99.77	27	3	3	8.65812	99.90
25	5	5	4.58256	99.62	27	4	4	6.36273	99.81
25	6	6	3.72830	99.42	27	5	5	4.98396	99.68
25	7	7	3.11672	99.16	27	6	6	4.06342	99.51
25	8	8	2.65671	98.85	27	7	7	3.40464	99.30
25	9	9	2.29764	98.46	27	8	8	2.90937	99.04
25	10	10	2.00912	97.98	27	9	9	2.52303	98.72
25	11	11	1.77181	97.41	27	10	10	2.21286	98.33
25	12	12	1.57276	96.73	27	11	11	1.95799	97.87
25	13	13	1.40302	95.90	27	12	12	1.74451	97.33
25	14	14	1.25615	94.92	27	13	13	1.56276	96.68
25	15	15	1.12741	93.74	27	14	14	1.40583	95.92
25	16	16	1.01318	92.31	27	15	15	1.26864	95.01
25	17	17	0.91066	90.59	27	16	16	1.14734	93.94
25	18	18	0.81759	88.48	27	17	17	1.03897	92.66
25	19	19	0.73209	85.89	27	18	18	0.94119	91.15
25	20	20	0.65248	82.64	27	19	19	0.85208	89.33
25	20	21	0.65248	78.71	27	20	20	0.77006	87.13
25	20	22	0.65248	75.13	27	21	21	0.69374	84.43
25	20	23	0.65248	71.86	27	22	22	0.62184	81.09
25	20	24	0.65248	68.87	27	22	23	0.62184	77.56
25	20	25	0.65248	66.12	27	22	24	0.62184	74.33
26	1	1	26.00000	100.00	27	22	25	0.62184	71.36
26	2	2	12.74510	99.96	27	22	26	0.62184	68.61
26	3	3	8.32444	99.89	27	22	27	0.62184	66.07
26	4	4	6.11223	99.79	28	1	1	28.00000	100.00
26	5	5	4.78329	99.65	28	2	2	13.74545	99.97
26	6	6	3.89590	99.47	28	3	3	8.99177	99.91
26	7	7	3.26073	99.24	28	4	4	6.61319	99.82
26	8	8	2.78310	98.95	28	5	5	5.18458	99.70
26	9	9	2.41041	98.60	28	6	6	4.23087	99.55
26	10	10	2.11108	98.17	28	7	7	3.54846	99.35
26	11	11	1.86501	97.66	28	8	8	3.03553	99.12
26	12	12	1.65877	97.05	28	9	9	2.63552	98.82
26	13	13	1.48305	96.32	28	10	10	2.31448	98.48
26	14	14	1.33119	95.46	28	11	11	2.05078	98.06
26	15	15	1.19826	94.43	28	12	12	1.83002	97.57
26	16	16	1.08055	93.20	28	13	13	1.64219	96.99
26	17	17	0.97518	91.73	28	14	14	1.48015	96.31
26	18	18	0.87985	89.95	28	15	15	1.33862	95.50
26	19	19	0.79267	87.80	28	16	16	1.21365	94.56
26	20	20	0.71204	85.15	28	17	17	1.10218	93.45
26	21	21	0.63650	81.86	28	18	18	1.00180	92.13
26	21	22	0.63650	78.14	28	19	19	0.91058	90.57
26	21	23	0.63650	74.74	28	20	20	0.82692	88.71
26	21	24	0.63650	71.62	28	21	21	0.74945	86.46
26	21	25	0.63650	68.76	28	22	22	0.67697	83.73

N	K	M	C(K,N)	EFF.	N	K	M	C(K,N)	EFF.
28	23	23	0.60833	80.34	30	16	16	1.34513	95.54
28	23	24	0.60833	76.99	30	17	17	1.22721	94.67
28	23	25	0.60833	73.91	30	18	18	1.12136	93.66
28	23	26	0.60833	71.07	30	19	19	1.02553	92.47
28	23	27	0.60833	68.44	30	20	20	0.93807	91.08
28	23	28	0.60833	65.99	30	21	21	0.85762	89.45
29	1	1	29.00000	100.00	30	22	22	0.78301	87.50
29	2	2	14.24561	99.97	30	23	23	0.71320	85.18
29	3	3	9.32539	99.91	30	24	24	0.64725	82.36
29	4	4	6.86362	99.83	30	24	25	0.64725	79.06
29	5	5	5.38516	99.72	30	24	26	0.64725	76.02
29	6	6	4.39827	99.58	30	24	27	0.64725	73.21
29	7	7	3.69221	99.40	30	24	28	0.64725	70.59
29	8	8	3.16160	99.18	30	24	29	0.64725	68.16
29	9	9	2.74790	98.92	30	24	30	0.64725	65.89
29	10	10	2.41596	98.60	31	1	1	31.00000	100.00
29	11	11	2.14341	98.22	31	2	2	15.24590	99.97
29	12	12	1.91534	97.78	31	3	3	9.99259	99.93
29	13	13	1.72139	97.25	31	4	4	7.36440	99.86
29	14	14	1.55418	96.64	31	5	5	5.78618	99.76
29	15	15	1.40827	95.93	31	6	6	4.73290	99.64
29	16	16	1.27956	95.09	31	7	7	3.97951	99.49
29	17	17	1.16490	94.11	31	8	8	3.41351	99.30
29	18	18	1.06182	92.96	31	9	9	2.97237	99.07
29	19	19	0.96835	91.61	31	10	10	2.61858	98.81
29	20	20	0.88286	90.01	31	11	11	2.32826	98.49
29	21	21	0.80399	88.10	31	12	12	2.08548	98.12
29	22	22	0.73057	85.81	31	13	13	1.87921	97.69
29	23	23	0.66153	83.04	31	14	14	1.70157	97.19
29	24	24	0.59583	79.61	31	15	15	1.54675	96.61
29	24	25	0.59583	76.42	31	16	16	1.41041	95.93
29	24	26	0.59583	73.48	31	17	17	1.28919	95.16
29	24	27	0.59583	70.76	31	18	18	1.18048	94.25
29	24	28	0.59583	68.24	31	19	19	1.08221	93.21
29	24	29	0.59583	65.88	31	20	20	0.99269	91.99
30	1	1	30.00000	100.00	31	21	21	0.91052	90.56
30	2	2	14.74576	99.97	31	22	22	0.83453	88.89
30	3	3	9.65900	99.92	31	23	23	0.76372	86.91
30	4	4	7.11402	99.85	31	24	24	0.69716	84.55
30	5	5	5.58569	99.74	31	25	25	0.63402	81.70
30	6	6	4.56561	99.61	31	25	26	0.63402	78.55
30	7	7	3.83589	99.45	31	25	27	0.63402	75.65
30	8	8	3.28759	99.25	31	25	28	0.63402	72.94
30	9	9	2.86018	99.00	31	25	29	0.63402	70.43
30	10	10	2.51732	98.71	31	25	30	0.63402	68.08
30	11	11	2.23590	98.36	31	25	31	0.63402	65.88
30	12	12	2.00048	97.96	32	1	1	32.00000	100.00
30	13	13	1.80039	97.49	32	2	2	15.74603	99.97
30	14	14	1.62798	96.93	32	3	3	10.32616	99.93
30	15	15	1.47763	96.29	32	4	4	7.61475	99.87

N	K	M	C(K,N)	EFF.	N	K	M	C(K,N)	EFF.
32	5	5	5.98665	99.78	33	23	23	0.86220	89.54
32	6	6	4.90015	99.66	33	24	24	0.79376	87.80
32	7	7	4.12308	99.52	33	25	25	0.72943	85.76
32	8	8	3.53936	99.35	33	26	26	0.66848	83.33
32	9	9	3.08448	99.14	33	27	27	0.61020	80.41
32	10	10	2.71974	98.89	33	27	28	0.61020	77.54
32	11	11	2.42051	98.60	33	27	29	0.61020	74.87
32	12	12	2.17035	98.26	33	27	30	0.61020	72.37
32	13	13	1.95788	97.87	33	27	31	0.61020	70.04
32	14	14	1.77498	97.41	33	27	32	0.61020	67.85
32	15	15	1.61566	96.88	33	27	33	0.61020	65.79
32	16	16	1.47544	96.28	34	1	1	34.00000	100.00
32	17	17	1.35087	95.58	34	2	2	16.74627	99.98
32	18	18	1.23926	94.77	34	3	3	10.99326	99.94
32	19	19	1.13848	93.84	34	4	4	8.11538	99.88
32	20	20	1.04681	92.76	34	5	5	6.38748	99.80
32	21	21	0.96282	91.51	34	6	6	5.23453	99.71
32	22	22	0.88533	90.05	34	7	7	4.41008	99.58
32	23	23	0.81332	88.34	34	8	8	3.79089	99.43
32	24	24	0.74591	86.33	34	9	9	3.30850	99.25
32	25	25	0.68230	83.93	34	10	10	2.92183	99.04
32	26	26	0.62170	81.05	34	11	11	2.60472	98.79
32	26	27	0.62170	78.05	34	12	12	2.33975	98.50
32	26	28	0.62170	75.26	34	13	13	2.11483	98.17
32	26	29	0.62170	72.66	34	14	14	1.92134	97.79
32	26	30	0.62170	70.24	34	15	15	1.75294	97.35
32	26	31	0.62170	67.98	34	16	16	1.60487	96.84
32	26	32	0.62170	65.85	34	17	17	1.47350	96.27
33	1	1	33.00000	100.00	34	18	18	1.35597	95.61
33	2	2	16.24615	99.98	34	19	19	1.25003	94.85
33	3	3	10.65972	99.93	34	20	20	1.15387	93.99
33	4	4	7.86507	99.87	34	21	21	1.06601	93.00
33	5	5	6.18708	99.79	34	22	22	0.98522	91.87
33	6	6	5.06736	99.69	34	23	23	0.91047	90.56
33	7	7	4.26660	99.55	34	24	24	0.84087	89.04
33	8	8	3.66515	99.39	34	25	25	0.77565	87.27
33	9	9	3.19652	99.20	34	26	26	0.71411	85.20
33	10	10	2.82082	98.97	34	27	27	0.65559	82.74
33	11	11	2.51266	98.70	34	28	28	0.59944	79.79
33	12	12	2.25510	98.39	34	28	29	0.59944	77.04
33	13	13	2.03642	98.03	34	28	30	0.59944	74.47
33	14	14	1.84823	97.61	34	28	31	0.59944	72.07
33	15	15	1.68438	97.13	34	28	32	0.59944	69.82
33	16	16	1.54025	96.58	34	28	33	0.59944	67.70
33	17	17	1.41229	95.94	34	28	34	0.59944	65.71
33	18	18	1.29774	95.22	35	1	1	35.00000	100.00
33	19	19	1.19441	94.38	35	2	2	17.24638	99.98
33	20	20	1.10052	93.42	35	3	3	11.32680	99.94
33	21	21	1.01462	92.31	35	4	4	8.36566	99.89
33	22	22	0.93552	91.03	35	5	5	6.58786	99.82

N	K	M	C(K,N)	EFF.	N	K	M	C(K,N)	EFF.
35	6	6	5.40168	99.72	36	21	21	1.16777	94.13
35	7	7	4.55352	99.61	36	22	22	1.08343	93.21
35	8	8	3.91658	99.47	36	23	23	1.00561	92.18
35	9	9	3.42042	99.30	36	24	24	0.93340	90.99
35	10	10	3.02276	99.10	36	25	25	0.86604	89.62
35	11	11	2.69670	98.87	36	26	26	0.80283	88.05
35	12	12	2.42430	98.60	36	27	27	0.74317	86.23
35	13	13	2.19314	98.30	36	28	28	0.68648	84.10
35	14	14	1.99433	97.94	36	29	29	0.63223	81.58
35	15	15	1.82136	97.54	36	29	30	0.63223	78.86
35	16	16	1.66933	97.08	36	29	31	0.63223	76.32
35	17	17	1.53451	96.55	36	29	32	0.63223	73.94
35	18	18	1.41397	95.95	36	29	33	0.63223	71.70
35	19	19	1.30539	95.27	36	29	34	0.63223	69.59
35	20	20	1.20692	94.49	36	29	35	0.63223	67.60
35	21	21	1.11705	93.60	36	29	36	0.63223	65.72
35	22	22	1.03450	92.59	37	1	1	37.00000	100.00
35	23	23	0.95825	91.43	37	2	2	18.24658	99.98
35	24	24	0.88739	90.09	37	3	3	11.99383	99.95
35	25	25	0.82114	88.54	37	4	4	8.86619	99.90
35	26	26	0.75883	86.75	37	5	5	6.98856	99.84
35	27	27	0.69983	84.64	37	6	6	5.73588	99.75
35	28	28	0.64353	82.16	37	7	7	4.84029	99.65
35	28	29	0.64353	79.32	37	8	8	4.16784	99.53
35	28	30	0.64353	76.68	37	9	9	3.64411	99.38
35	28	31	0.64353	74.21	37	10	10	3.22446	99.21
35	28	32	0.64353	71.89	37	11	11	2.88046	99.01
35	28	33	0.64353	69.71	37	12	12	2.59317	98.78
35	28	34	0.64353	67.66	37	13	13	2.34947	98.51
35	28	35	0.64353	65.72	37	14	14	2.13998	98.21
36	1	1	36.00000	100.00	37	15	15	1.95782	97.86
36	2	2	17.74648	99.98	37	16	16	1.79783	97.47
36	3	3	11.66032	99.95	37	17	17	1.65605	97.03
36	4	4	8.61594	99.89	37	18	18	1.52941	96.53
36	5	5	6.78822	99.83	37	19	19	1.41547	95.96
36	6	6	5.56879	99.74	37	20	20	1.31228	95.31
36	7	7	4.69692	99.63	37	21	21	1.21824	94.59
36	8	8	4.04223	99.50	37	22	22	1.13204	93.76
36	9	9	3.53229	99.34	37	23	23	1.05261	92.83
36	10	10	3.12364	99.16	37	24	24	0.97900	91.76
36	11	11	2.78861	98.94	37	25	25	0.91044	90.55
36	12	12	2.50877	98.70	37	26	26	0.84623	89.16
36	13	13	2.27135	98.41	37	27	27	0.78578	87.57
36	14	14	2.06720	98.08	37	28	28	0.72853	85.72
36	15	15	1.88964	97.71	37	29	29	0.67398	83.56
36	16	16	1.73364	97.29	37	30	30	0.62161	81.02
36	17	17	1.59536	96.80	37	30	31	0.62161	78.41
36	18	18	1.47178	96.25	37	30	32	0.62161	75.96
36	19	19	1.36053	95.63	37	30	33	0.62161	73.66
36	20	20	1.25972	94.93	37	30	34	0.62161	71.49

N	K	M	C(K,N)	EFF.	N	K	M	C(K,N)	EFF.
37	30	35	0.62161	69.45	39	10	10	3.42596	99.30
37	30	36	0.62161	67.52	39	11	11	3.06399	99.12
37	30	37	0.62161	65.69	39	12	12	2.76177	98.92
38	1	1	38.00000	100.00	39	13	13	2.50549	98.69
38	2	2	18.74667	99.98	39	14	14	2.28527	98.43
38	3	3	12.32733	99.95	39	15	15	2.09387	98.13
38	4	4	9.11643	99.91	39	16	16	1.92585	97.79
38	5	5	7.18888	99.85	39	17	17	1.77705	97.41
38	6	6	5.90295	99.77	39	18	18	1.64424	96.99
38	7	7	4.98363	99.67	39	19	19	1.52485	96.50
38	8	8	4.29341	99.56	39	20	20	1.41683	95.96
38	9	9	3.75589	99.42	39	21	21	1.31850	95.36
38	10	10	3.32523	99.26	39	22	22	1.22852	94.67
38	11	11	2.97225	99.07	39	23	23	1.14572	93.90
38	12	12	2.67750	98.85	39	24	24	1.06916	93.04
38	13	13	2.42752	98.61	39	25	25	0.99802	92.06
38	14	14	2.21266	98.32	39	26	26	0.93161	90.95
38	15	15	2.02589	98.00	39	27	27	0.86931	89.69
38	16	16	1.86189	97.64	39	28	28	0.81059	88.26
38	17	17	1.71661	97.23	39	29	29	0.75496	86.61
38	18	18	1.58689	96.77	39	30	30	0.70196	84.72
38	19	19	1.47024	96.25	39	31	31	0.65117	82.51
38	20	20	1.36464	95.66	39	31	32	0.65117	79.94
38	21	21	1.26847	94.99	39	31	33	0.65117	77.51
38	22	22	1.18040	94.24	39	31	34	0.65117	75.23
38	23	23	1.09930	93.40	39	31	35	0.65117	73.08
38	24	24	1.02423	92.44	39	31	36	0.65117	71.05
38	25	25	0.95441	91.36	39	31	37	0.65117	69.13
38	26	26	0.88913	90.12	39	31	38	0.65117	67.31
38	27	27	0.82780	88.71	39	31	39	0.65117	65.59
38	28	28	0.76986	87.09	40	1	1	40.00000	100.00
38	29	29	0.71483	85.21	40	2	2	19.74684	99.98
38	30	30	0.66223	83.03	40	3	3	12.99430	99.96
38	31	31	0.61160	80.47	40	4	4	9.61688	99.92
38	31	32	0.61160	77.96	40	5	5	7.58946	99.86
38	31	33	0.61160	75.60	40	6	6	6.23702	99.79
38	31	34	0.61160	73.37	40	7	7	5.27024	99.71
38	31	35	0.61160	71.28	40	8	8	4.54447	99.60
38	31	36	0.61160	69.30	40	9	9	3.97934	99.48
38	31	37	0.61160	67.42	40	10	10	3.52664	99.34
38	31	38	0.61160	65.65	40	11	11	3.15568	99.17
39	1	1	39.00000	100.00	40	12	12	2.84599	98.99
39	2	2	19.24675	99.98	40	13	13	2.58340	98.77
39	3	3	12.66082	99.95	40	14	14	2.35780	98.52
39	4	4	9.36666	99.91	40	15	15	2.16177	98.24
39	5	5	7.38918	99.85	40	16	16	1.98971	97.93
39	6	6	6.06999	99.78	40	17	17	1.83739	97.58
39	7	7	5.12695	99.69	40	18	18	1.70146	97.18
39	8	8	4.41896	99.58	40	19	19	1.57932	96.74
39	9	9	3.86764	99.45	40	20	20	1.46885	96.24

N	K	M	C(K,N)	EFF.	N	K	M	C(K,N)	EFF.
40	21	21	1.36836	95.68	40	31	31	0.68986	84.23
40	22	22	1.27643	95.05	40	32	32	0.64074	82.00
40	23	23	1.19191	94.35	40	32	33	0.64074	79.52
40	24	24	1.11382	93.56	40	32	34	0.64074	77.18
40	25	25	1.04133	92.68	40	32	35	0.64074	74.97
40	26	26	0.97373	91.68	40	32	36	0.64074	72.89
40	27	27	0.91041	90.55	40	32	37	0.64074	70.92
40	28	28	0.85082	89.27	40	32	38	0.64074	69.06
40	29	29	0.79449	87.81	40	32	39	0.64074	67.28
40	30	30	0.74097	86.14	40	32	40	0.64074	65.60

Table D3

MOST EFFICIENT UNBIASED POINT ESTIMATORS FOR TWO PARAMETERS, BASED ON TWO ORDER STATISTICS OF A TWO-PARAMETER EXPONENTIAL POPULATION

[Estimator of location parameter α : $\tilde{\alpha} = c_{\alpha l}x_l + c_{\alpha m}x_m$; **estimator of scale parameter** σ : $\tilde{\sigma} = c_{\sigma l}x_l + c_{\sigma m}x_m$; **estimator of mean** μ : $\tilde{\mu} = c_{\mu l}x_l + c_{\mu m}x_m$, **where** $c_{\alpha l} = 1 + c_\alpha$, $c_{\alpha m} = -c_\alpha$; $c_{\sigma l} = -c_\alpha$, $c_{\sigma m} = c_\sigma$; $c_{\mu l} = 1 + c_\alpha - c_\sigma$, $c_{\mu m} = c_\sigma - c_\alpha$.]

Best Estimators from 2 Order Statistics x_1, x_m of Parameters α, σ and Mean μ of 2-Parameter Exponential Population

n	m	c_α	c_σ	$V_{\tilde{\alpha}}/\sigma^2$	$E_{\tilde{\alpha}}(\%)$	$V_{\tilde{\sigma}}/\sigma^2$	$E_{\tilde{\sigma}}(\%)$	$V_{\tilde{\mu}}/\sigma^2$	$E_{\tilde{\mu}}(\%)$
2	2	.500000	1.00000	.5000000	100.00	1.000000	100.00	.5000000	100.00
3	3	.222222	.666667	.1728395	96.429	.5555556	90.000	.3580247	93.103
4	4	.136364	.545455	.08780992	94.902	.4049587	82.313	.2902893	86.121
5	5	.096000	.480000	.05312000	94.127	.3280000	76.220	.2499200	80.026
6	6	.072993	.437956	.03557580	93.697	.2807289	71.243	.2227284	74.830
7	6	.098522	.689655	.02517789	94.565	.2337165	71.311	.1921182	74.359
8	7	.078475	.627803	.01877684	95.102	.2017178	70.820	.1700652	73.501
9	8	.064680	.582121	.01455215	95.442	.1787246	69.940	.1535601	72.357
10	9	.054676	.546756	.01161360	95.673	.1613595	68.859	.1407012	71.073
11	9	.063619	.699806	.009477724	95.919	.1468046	68.118	.1295906	70.151
12	10	.054829	.657948	.007870456	96.256	.1333457	68.175	.1189919	70.033
13	11	.047981	.623748	.006642280	96.507	.1225454	68.002	.1103346	69.718
14	12	.042514	.595191	.005682027	96.700	.1136773	67.668	.1031197	69.268
15	13	.038061	.570919	.004916701	96.852	.1062578	67.222	.0970068	68.724
16	13	.042091	.673448	.004294816	97.016	.0994728	67.020	.0913336	68.430
17	14	.038015	.646247	.003782806	97.189	.0932310	67.038	.0860455	68.363
18	15	.034588	.622580	.003357619	97.330	.0878686	66.945	.0814630	68.197
19	16	.031672	.601766	.003000580	97.447	.0832093	66.766	.0774510	67.955
20	17	.029165	.583292	.002697803	97.545	.0791212	66.520	.0739069	67.653
21	17	.031444	.660325	.002438181	97.653	.0752377	66.456	.0705104	67.535
22	18	.029100	.640195	.002214152	97.758	.0716497	66.461	.0673502	67.490
23	19	.027047	.622092	.002019763	97.847	.0684545	66.401	.0645217	67.385
24	20	.025238	.605709	.001849983	97.925	.0655900	66.288	.0619741	67.232
25	21	.023632	.590798	.001700810	97.993	.0630065	66.131	.0596668	67.039
26	21	.025095	.652475	.001568788	98.067	.0605006	66.115	.0574155	66.988
27	22	.023574	.636502	.001451542	98.137	.0581740	66.115	.0553164	66.955
28	23	.022209	.621843	.001347010	98.199	.0560557	66.072	.0533987	66.882
29	24	.020977	.608333	.001253411	98.254	.0541184	65.993	.0516395	66.776
30	25	.019861	.595834	.001169266	98.303	.0523395	65.883	.0500195	66.641
31	25	.020879	.647255	.001093227	98.357	.0505915	65.887	.0484208	66.620
32	26	.019813	.634017	.001024377	98.408	.0489616	65.884	.0469259	66.594
33	27	.018839	.621699	.000961850	98.453	.0474551	65.852	.0455408	66.540
34	28	.017947	.610203	.000904895	98.494	.0460582	65.793	.0442538	66.462
35	29	.017127	.599445	.000852865	98.531	.0447593	65.711	.0430545	66.361
36	29	.017876	.643532	.000805148	98.572	.0434715	65.724	.0418616	66.356
37	30	.017087	.632230	.000761334	98.610	.0422665	65.721	.0407431	66.335
38	31	.016358	.621609	.000721011	98.644	.0411405	65.694	.0396962	66.293
39	32	.015682	.611604	.000683817	98.676	.0400859	65.649	.0387140	66.232
40	32	.016279	.651175	.000649434	98.705	.0390941	65.588	.0377889	66.157
41	33	.015628	.640744	.000617554	98.737	.0381083	65.602	.0368669	66.158
42	34	.015021	.630885	.000587971	98.767	.0371813	65.598	.0359988	66.140
43	35	.014455	.621548	.000560469	98.794	.0363080	65.577	.0351797	66.106
44	36	.013925	.612692	.000534857	98.819	.0354837	65.539	.0344057	66.057
45	36	.014395	.647770	.000510963	98.843	.0346995	65.497	.0336683	66.003
46	37	.013882	.638578	.000488621	98.868	.0339230	65.508	.0329368	66.003
47	38	.013401	.629835	.000467717	98.892	.0331879	65.503	.0322434	65.987
48	39	.012948	.621506	.000448130	98.914	.0324909	65.485	.0315852	65.959
49	40	.012522	.613562	.000429750	98.934	.0318289	65.454	.0309596	65.919
50	40	.012901	.645063	.000412477	98.954	.0311934	65.425	.0303581	65.880

Best Estimators from 2 Order Statistics x_1, x_m of Parameters α, σ and Mean μ of 2-Parameter Exponential Population

n	m	c_α	c_σ	$V_{\tilde\alpha}/\sigma^2$	$E_{\tilde\alpha}(\%)$	$V_{\tilde\sigma}/\sigma^2$	$E_{\tilde\sigma}(\%)$	$V_{\tilde\mu}/\sigma^2$	$E_{\tilde\mu}(\%)$
51	41	.012487	.636847	.000396219	98.975	.0305661	65.432	.0297636	65.879
52	42	.012096	.628992	.000380906	98.994	.0299689	65.427	.0291971	65.865
53	43	.011726	.621475	.000366465	99.012	.0293996	65.412	.0286567	65.841
54	44	.011375	.614272	.000352831	99.029	.0288564	65.385	.0281405	65.807
55	44	.011688	.642858	.000339944	99.046	.0283310	65.365	.0276407	65.779
56	45	.011347	.635431	.000327747	99.063	.0278136	65.370	.0271480	65.777
57	46	.011023	.628302	.000316195	99.079	.0273189	65.366	.0266765	65.765
58	47	.010715	.621452	.000305245	99.094	.0268453	65.352	.0262249	65.744
59	48	.010421	.614864	.000294855	99.109	.0263915	65.329	.0257918	65.715
60	48	.010684	.641029	.000284986	99.123	.0259499	65.315	.0253699	65.695
61	49	.010398	.634253	.000275602	99.137	.0255159	65.319	.0249549	65.692
62	50	.010125	.627726	.000266675	99.151	.0250994	65.314	.0245564	65.682
63	51	.009864	.621434	.000258176	99.164	.0246992	65.302	.0241733	65.663
64	52	.009615	.615364	.000250077	99.176	.0243145	65.282	.0238047	65.638
65	52	.009838	.639486	.000242352	99.188	.0239381	65.272	.0234439	65.623
66	53	.009595	.633256	.000234979	99.200	.0235689	65.275	.0230896	65.620
67	54	.009362	.627237	.000227938	99.212	.0232133	65.271	.0227483	65.611
68	55	.009139	.621420	.000221209	99.223	.0228708	65.260	.0224193	65.595
69	56	.008925	.615792	.000214774	99.234	.0225404	65.242	.0221018	65.573
70	56	.009117	.638167	.000208615	99.244	.0222159	65.236	.0217897	65.562
71	57	.008907	.632402	.000202717	99.255	.0218979	65.238	.0214838	65.559
72	58	.008706	.626818	.000197066	99.265	.0215909	65.234	.0211882	65.550
73	59	.008512	.621409	.000191648	99.275	.0212943	65.224	.0209025	65.536
74	60	.008327	.616164	.000186451	99.284	.0210076	65.208	.0206263	65.516
75	60	.008494	.637027	.000181462	99.294	.0207248	65.204	.0203536	65.508
76	61	.008311	.631662	.000176670	99.303	.0204481	65.206	.0200867	65.505
77	62	.008136	.626455	.000172066	99.312	.0201804	65.201	.0198283	65.497
78	63	.007967	.621399	.000167640	99.320	.0199211	65.192	.0195779	65.484
79	64	.007804	.616488	.000163382	99.328	.0196699	65.178	.0193353	65.467
80	64	.007950	.636031	.000159285	99.337	.0194214	65.177	.0190951	65.462
81	65	.007790	.631015	.000155339	99.345	.0191784	65.177	.0188602	65.459
82	66	.007636	.626137	.000151538	99.353	.0189428	65.173	.0186324	65.451
83	67	.007487	.621392	.000147875	99.360	.0187143	65.165	.0184112	65.440
84	68	.007343	.616774	.000144344	99.367	.0184924	65.152	.0181964	65.424
85	68	.007472	.635155	.000140937	99.375	.0182722	65.152	.0179832	65.420
86	69	.007331	.630444	.000137650	99.382	.0180572	65.152	.0177749	65.418
87	70	.007194	.625856	.000134476	99.389	.0178483	65.149	.0175725	65.411
88	71	.007061	.621386	.000131411	99.396	.0176453	65.141	.0173756	65.400
89	72	.006933	.617029	.000128449	99.402	.0174478	65.129	.0171842	65.385
90	72	.007049	.634377	.000125587	99.409	.0172515	65.130	.0169937	65.384
91	73	.006922	.629936	.000122818	99.415	.0170598	65.130	.0168077	65.381
92	74	.006800	.625606	.000120141	99.421	.0168733	65.127	.0166266	65.374
93	75	.006682	.621380	.000117550	99.427	.0166918	65.119	.0164503	65.364
94	76	.006567	.617256	.000115042	99.433	.0165150	65.109	.0162786	65.351
95	76	.006670	.633682	.000112614	99.439	.0163387	65.111	.0161074	65.351
96	77	.006557	.629483	.000110261	99.445	.0161668	65.111	.0159403	65.348
97	78	.006447	.625382	.000107982	99.451	.0159993	65.107	.0157774	65.342
98	79	.006341	.621376	.000105772	99.456	.0158360	65.100	.0156186	65.333
99	79	.006436	.637131	.000103630	99.461	.0156767	65.091	.0154636	65.321
100	80	.006331	.633057	.000101552	99.467	.0155177	65.093	.0153089	65.321

Table D4

MOST EFFECTIVE (EFFICIENT) INTERVAL ESTIMATORS FOR σ, BASED ON ONE ORDER STATISTIC OF A SAMPLE FROM A ONE-PARAMETER EXPONENTIAL POPULATION

[Upper and lower $(1-P)$ confidence bounds, B_{um} and B_{lm}, of central $(1-2P)$ confidence interval, based on mth order statistic, x_m, of sample of size n]

F_u = Effectiveness of Upper Confidence Bound Relative to Conventional Bound

F_i = Effectiveness of Central Confidence Interval Relative to Conventional Interval

S_u = Effectiveness of Upper Confidence Bound Relative to Censored Bound

S_i = Effectiveness of Central Confidence Interval Relative to Censored Interval

E_u = Efficiency of Upper Confidence Bound Relative to Conventional Bound

E_i = Efficiency of Central Confidence Interval Relative to Conventional Interval

R_u = Efficiency of Upper Confidence Bound Relative to Censored Bound

R_i = Efficiency of Central Confidence Interval Relative to Censored Interval

A. VALUE OF m, CHOSEN TO MAXIMIZE EFFECTIVENESS F_u OF UPPER CONFIDENCE BOUND, ALSO MAXIMIZES EFFECTIVENESS F_i OF CENTRAL CONFIDENCE INTERVAL UNLESS MARKED *

n	m	1−P	B_{um}/x_m	$B_{\ell m}/x_m$	1−2P	F_u (%)	F_i (%)	S_u (%)	S_i (%)
1	1	0.9999	9999.500	0.108574	0.9998	100.0	100.0	100.0	100.0
1	1	0.9995	1999.500	0.131563	0.9990	100.0	100.0	100.0	100.0
1	1	0.9990	999.4999	0.144765	0.9980	100.0	100.0	100.0	100.0
1	1	0.9950	199.4996	0.188739	0.9900	100.0	100.0	100.0	100.0
1	1	0.9900	99.49916	0.217147	0.9800	100.0	100.0	100.0	100.0
1	1	0.9750	39.49789	0.271085	0.9500	100.0	100.0	100.0	100.0
1	1	0.9500	19.49573	0.333808	0.9000	100.0	100.0	100.0	100.0
1	1	0.9000	9.491222	0.434294	0.8000	100.0	100.0	100.0	100.0
1	1	0.8000	4.481420	0.621335	0.6000	100.0	100.0	100.0	100.0
1	1	0.7000	2.803673	0.830584	0.4000	100.0	100.0	100.0	100.0
1	1	0.6000	1.957615	1.091357	0.2000	100.0	100.0	100.0	100.0
1	1	0.5000	1.442695			100.0		100.0	
2	2	0.9999	99.49916	0.100975	0.9998	94.3	94.3	94.3	94.3
2	2	0.9995	44.21947	0.120570	0.9990	94.3	94.3	94.3	94.3
2	2	0.9990	31.12010	0.131568	0.9980	94.4	94.3	94.4	94.3
2	2	0.9950	13.63602	0.166939	0.9900	94.5	94.3	94.5	94.3
2	2	0.9900	9.491222	0.188829	0.9800	94.6	94.3	94.6	94.3
2	2	0.9750	5.810219	0.228534	0.9500	94.7	94.3	94.7	94.3
2	2	0.9500	3.951067	0.272025	0.9000	95.0	94.3	95.0	94.3
2	2	0.9000	2.630676	0.336730	0.8000	95.3	94.4	95.3	94.4
2	2	0.8000	1.686956	0.444770	0.6000	95.9	94.4	95.9	94.4
2	2	0.7000	1.260304	0.551900	0.4000	96.4	94.4	96.4	94.4
2	2	0.6000	0.999090	0.671202	0.2000	97.0	94.4	97.0	94.4
2	2	0.5000	0.814367			97.6		97.6	
3	3	0.9999	21.04039	0.097003	0.9998	90.2	90.1	90.2	90.1
3	3	0.9995	12.09232	0.114951	0.9990	90.4	90.1	90.4	90.1
3	3	0.9990	9.491222	0.124906	0.9980	90.5	90.1	90.5	90.1
3	3	0.9950	5.332417	0.156366	0.9900	90.8	90.2	90.8	90.2
3	3	0.9900	4.121388	0.175425	0.9800	91.1	90.2	91.1	90.2
3	3	0.9750	2.891186	0.209245	0.9500	91.5	90.2	91.5	90.2
3	3	0.9500	2.176260	0.245258	0.9000	92.0	90.2	92.0	90.2
3	3	0.9000	1.602776	0.297045	0.8000	92.6	90.2	92.6	90.2
3	3	0.8000	1.137652	0.379433	0.6000	93.7	90.2	93.7	90.2
3	3	0.7000	0.903386	0.456955	0.4000	94.6	90.1	94.6	90.1
3	3	0.6000	0.749139	0.539296	0.2000	95.6	90.1	95.6	90.1
3	3	0.5000	0.633542			96.6		96.6	
4	4	0.9999	9.491222	0.094370	0.9998	87.3	86.9	87.3	86.9
4	4	0.9995	6.173910	0.111272	0.9990	87.6	86.9	87.6	86.9
4	4	0.9990	5.107108	0.120574	0.9980	87.7	86.9	87.7	86.9
4	4	0.9950	3.234883	0.149639	0.9900	88.3	86.9	88.3	86.9
4	4	0.9900	2.630676	0.167009	0.9800	88.7	86.9	88.7	86.9
4	4	0.9750	1.972805	0.197406	0.9500	89.3	86.9	89.3	86.9
4	4	0.9500	1.561745	0.229206	0.9000	90.0	86.9	90.0	86.9
4	4	0.9000	1.210191	0.273987	0.8000	90.9	86.8	90.9	86.8
4	4	0.8000	0.905098	0.343171	0.6000	92.4	86.8	92.4	86.8
4	4	0.7000	0.742174	0.406260	0.4000	93.6	86.8	93.6	86.8
4	4	0.6000	0.630491	0.471432	0.2000	94.8	86.8	94.8	86.8
4	4	0.5000	0.544011			96.1		96.1	

A. VALUE OF m, CHOSEN TO MAXIMIZE EFFECTIVENESS F_u OF UPPER CONFIDENCE BOUND, ALSO MAXIMIZES EFFECTIVENESS F_i OF CENTRAL CONFIDENCE INTERVAL UNLESS MARKED *

n	m	1-P	B_{um}/x_m	$B_{\ell m}/x_m$	1-2P	F_u(%)	F_i(%)	S_u(%)	S_i(%)
5	5	0.9999	5.795202	0.092424	0.9998	85.0	84.2	85.0	84.2
5	5	0.9995	4.052508	0.108576	0.9990	85.4	84.3	85.4	84.3
5	5	0.9990	3.457000	0.117415	0.9980	85.7	84.3	85.7	84.3
5	5	0.9950	2.350046	0.144807	0.9900	86.4	84.2	86.4	84.2
5	5	0.9900	1.969761	0.161015	0.9800	86.9	84.2	86.9	84.2
5	5	0.9750	1.537456	0.189100	0.9500	87.7	84.2	87.7	84.2
5	5	0.9500	1.254847	0.218114	0.9000	88.6	84.1	88.6	84.1
5	5	0.9000	1.003167	0.258376	0.8000	89.7	84.1	89.7	84.1
5	5	0.8000	0.775084	0.319325	0.6000	91.4	84.0	91.4	84.0
5	5	0.7000	0.648595	0.373717	0.4000	92.9	84.0	92.9	84.0
5	4	0.6000	0.992163	0.754215	0.2000	94.7	82.1*	97.8	94.2
5	4	0.5000	0.862837			96.7		98.4	
6	6	0.9999	4.121388	0.090892	0.9998	83.3	82.0	83.3	82.0
6	6	0.9995	3.022012	0.106468	0.9990	83.8	82.0	83.8	82.0
6	6	0.9990	2.630676	0.114954	0.9980	84.1	82.0	84.1	82.0
6	6	0.9950	1.874013	0.141084	0.9900	85.0	82.0	85.0	82.0
6	6	0.9900	1.602776	0.156427	0.9800	85.6	81.9	85.6	81.9
6	6	0.9750	1.285111	0.182811	0.9500	86.5	81.9	86.5	81.9
6	6	0.9500	1.070836	0.209810	0.9000	87.5	81.8	87.5	81.8
6	6	0.9000	0.874523	0.246859	0.8000	88.8	81.7	88.8	81.7
6	5	0.8000	1.160510	0.508397	0.6000	91.3	82.3	96.2	92.5
6	5	0.7000	0.979760	0.586569	0.4000	93.5	82.8	96.9	92.5
6	5	0.6000	0.852532	0.665578	0.2000	95.3	83.0	97.5	92.5
6	5	0.5000	0.751821			97.1		98.2	
7	7	0.9999	3.201607	0.089636	0.9998	81.8	80.1	81.8	80.1
7	7	0.9995	2.427707	0.104749	0.9990	82.5	80.1	82.5	80.1
7	7	0.9990	2.143967	0.112953	0.9980	82.8	80.1	82.8	80.1
7	7	0.9950	1.579244	0.138082	0.9900	83.9	80.0	83.9	80.0
7	7	0.9900	1.370422	0.152747	0.9800	84.5	79.9	84.5	79.9
7	7	0.9750	1.120405	0.177810	0.9500	85.6	79.8*	85.6	79.8
7	6	0.9500	1.532390	0.341250	0.9000	87.3	81.1	94.1	90.9
7	6	0.9000	1.261314	0.393616	0.8000	89.5	81.9	94.7	90.9
7	6	0.8000	1.008139	0.470785	0.6000	92.1	82.7	95.7	90.9
7	6	0.7000	0.863959	0.538095	0.4000	94.0	83.0	96.5	90.9
7	6	0.6000	0.760539	0.605166	0.2000	95.7	83.1	97.3	90.9
7	6	0.5000	0.677386			97.3		98.1	
8	8	0.9999	2.630676	0.088576	0.9998	80.7	78.4	80.7	78.4
8	8	0.9995	2.045421	0.103304	0.9990	81.4	78.4	81.4	78.4
8	8	0.9990	1.825962	0.111275	0.9980	81.8	78.4	81.8	78.4
8	7	0.9950	2.175662	0.233223	0.9900	83.3	79.3	91.9	89.6
8	7	0.9900	1.894905	0.254500	0.9800	84.6	79.9	92.3	89.6
8	7	0.9750	1.558882	0.290142	0.9500	86.5	80.8	92.9	89.5
8	7	0.9500	1.327002	0.325660	0.9000	88.2	81.5	93.5	89.5
8	7	0.9000	1.110127	0.373262	0.8000	90.1	82.0	94.2	89.5
8	7	0.8000	0.902888	0.442520	0.6000	92.5	82.6	95.3	89.5
8	7	0.7000	0.782460	0.502148	0.4000	94.3	82.8	96.2	89.4
8	7	0.6000	0.694828	0.560911	0.2000	95.9	82.9	97.1	89.4
8	6	0.5000	0.878878			97.5		98.9	

A. VALUE OF m, CHOSEN TO MAXIMIZE EFFECTIVENESS F_u OF UPPER CONFIDENCE BOUND, ALSO MAXIMIZES EFFECTIVENESS F_i OF CENTRAL CONFIDENCE INTERVAL UNLESS MARKED *

n	m	1−P	B_{um}/x_m	$B_{\ell m}/x_m$	1−2P	F_u (%)	F_i (%)	S_u (%)	S_i (%)
9	9	0.9999	2.245571	0.087661	0.9998	79.7	76.9	79.7	76.9
9	8	0.9995	2.740770	0.179101	0.9990	80.9	77.9	90.3	88.3
9	8	0.9990	2.452161	0.191088	0.9980	81.8	78.5	90.5	88.3
9	8	0.9950	1.864991	0.226624	0.9900	84.2	79.7	91.2	88.3
9	8	0.9900	1.642726	0.246683	0.9800	85.4	80.2	91.6	88.2
9	8	0.9750	1.372153	0.280082	0.9500	87.1	80.9	92.3	88.2
9	8	0.9500	1.182057	0.313124	0.9000	88.7	81.4	93.0	88.2
9	8	0.9000	1.001314	0.357057	0.8000	90.5	81.8	93.8	88.1
9	8	0.8000	0.825468	0.420313	0.6000	92.7	82.2	95.0	88.1
9	8	0.7000	0.721641	0.474202	0.4000	94.4	82.4	96.0	88.1
9	7	0.6000	0.887721	0.721851	0.2000	96.0	81.0*	98.2	93.3
9	7	0.5000	0.799401			97.7		98.8	
10	9	0.9999	2.947923	0.153739	0.9998	80.0	77.3	89.0	87.1
10	9	0.9995	2.347536	0.175598	0.9990	81.8	78.4	89.5	87.1
10	9	0.9990	2.118425	0.187109	0.9980	82.7	78.8	89.8	87.1
10	9	0.9950	1.643984	0.221066	0.9900	84.8	79.8	90.6	87.0
10	9	0.9900	1.460964	0.240127	0.9800	85.9	80.2	91.1	87.0
10	9	0.9750	1.235109	0.271703	0.9500	87.5	80.7	91.8	87.0
10	9	0.9500	1.074125	0.302752	0.9000	89.0	81.1	92.5	86.9
10	9	0.9000	0.919013	0.343757	0.8000	90.7	81.4	93.5	86.9
10	9	0.8000	0.765884	0.402287	0.6000	92.9	81.7	94.8	86.9
10	8	0.7000	0.909329	0.607133	0.4000	94.6	81.4*	97.5	92.3
10	8	0.6000	0.816001	0.671425	0.2000	96.3	81.5*	98.1	92.3
10	8	0.5000	0.739333			97.9		98.7	
11	10	0.9999	2.550353	0.151406	0.9998	80.8	77.7	88.3	86.0
11	10	0.9995	2.063467	0.172563	0.9990	82.4	78.5	88.9	86.0
11	10	0.9990	1.874971	0.183669	0.9980	83.2	78.9	89.2	86.0
11	10	0.9950	1.478934	0.216293	0.9900	85.2	79.7	90.1	85.9
11	10	0.9900	1.323778	0.234518	0.9800	86.2	80.0	90.6	85.8
11	10	0.9750	1.130147	0.264576	0.9500	87.8	80.4	91.4	85.8
11	10	0.9500	0.990477	0.293977	0.9000	89.1	80.6	92.1	85.8
11	10	0.9000	0.854410	0.332584	0.8000	90.8	80.9	93.1	85.7
11	9	0.8000	0.952715	0.514005	0.6000	93.1	81.4	96.7	91.4
11	9	0.7000	0.842195	0.573666	0.4000	95.0	81.7	97.4	91.3
11	9	0.6000	0.760095	0.631482	0.2000	96.5	81.8	98.0	91.3
11	9	0.5000	0.692122			98.0		98.7	
12	11	0.9999	2.255961	0.149346	0.9998	81.3	77.8	87.7	85.0
12	11	0.9995	1.849261	0.169894	0.9990	82.8	78.5	88.3	84.9
12	11	0.9990	1.689910	0.180650	0.9980	83.5	78.7	88.6	84.9
12	11	0.9950	1.351016	0.212129	0.9900	85.4	79.4	89.6	84.8
12	11	0.9900	1.216517	0.229641	0.9800	86.4	79.6	90.1	84.8
12	11	0.9750	1.047075	0.258411	0.9500	87.9	79.9*	91.0	84.7
12	11	0.9500	0.923619	0.286423	0.9000	89.2	80.1*	91.8	84.7
12	10	0.9000	1.049144	0.426071	0.8000	91.1	81.1	95.6	90.4
12	10	0.8000	0.886772	0.491465	0.6000	93.5	81.5	96.5	90.4
12	10	0.7000	0.788618	0.546209	0.4000	95.2	81.7	97.3	90.4
12	10	0.6000	0.715134	0.598927	0.2000	96.7	81.8	98.0	90.4
12	9	0.5000	0.825023			98.2		99.1	

A. VALUE OF m, CHOSEN TO MAXIMIZE EFFECTIVENESS F_u OF UPPER CONFIDENCE BOUND, ALSO MAXIMIZES EFFECTIVENESS F_i OF CENTRAL CONFIDENCE INTERVAL UNLESS MARKED *

n	m	1-P	B_{um}/x_m	$B_{\ell m}/x_m$	1-2P	F_u (%)	F_i (%)	S_u (%)	S_i (%)
13	12	0.9999	2.029881	0.147507	0.9998	81.6	77.7	87.1	84.0
13	12	0.9995	1.682235	0.167520	0.9990	83.0	78.3	87.8	83.9
13	12	0.9990	1.544630	0.177970	0.9980	83.7	78.5	88.2	83.9
13	12	0.9950	1.248943	0.208452	0.9900	85.6	79.0*	89.2	83.8
13	12	0.9900	1.130291	0.225345	0.9800	86.5	79.2*	89.7	83.7
13	11	0.9750	1.268036	0.334656	0.9500	88.2	80.2	94.0	89.6
13	11	0.9500	1.121884	0.368006	0.9000	89.7	80.7	94.6	89.6
13	11	0.9000	0.978212	0.411445	0.8000	91.5	81.1	95.3	89.6
13	11	0.8000	0.833211	0.472453	0.6000	93.7	81.5	96.3	89.5
13	11	0.7000	0.744749	0.523193	0.4000	95.3	81.7	97.1	89.5
13	11	0.6000	0.678079	0.571790	0.2000	96.8	81.8	97.9	89.5
13	10	0.5000	0.775205			98.3		99.1	
14	13	0.9999	1.851142	0.145849	0.9998	81.8	77.5	86.6	83.1
14	13	0.9995	1.548457	0.165387	0.9990	83.2	78.0	87.3	83.0
14	13	0.9990	1.427597	0.175566	0.9980	83.8	78.1*	87.7	82.9
14	12	0.9950	1.495147	0.273156	0.9900	85.8	79.3	92.7	88.8
14	12	0.9900	1.355763	0.293296	0.9800	86.9	79.8	93.1	88.8
14	12	0.9750	1.178786	0.326094	0.9500	88.6	80.3	93.7	88.8
14	12	0.9500	1.048767	0.357757	0.9000	90.0	80.7	94.3	88.7
14	12	0.9000	0.919932	0.398814	0.8000	91.8	81.1	95.1	88.7
14	12	0.8000	0.788749	0.456148	0.6000	93.9	81.4	96.2	88.7
14	12	0.7000	0.708079	0.503560	0.4000	95.5	81.5	97.0	88.7
14	11	0.6000	0.798285	0.676158	0.2000	97.0	81.0*	98.5	92.3
14	11	0.5000	0.734080			98.4		99.0	
15	14	0.9999	1.706449	0.144344	0.9998	81.9	77.2*	86.2	82.2
15	13	0.9995	1.833879	0.220180	0.9990	83.3	78.2	91.3	88.1
15	13	0.9990	1.692871	0.232450	0.9980	84.1	78.6	91.6	88.1
15	13	0.9950	1.387217	0.267782	0.9900	86.3	79.4	92.4	88.0
15	13	0.9900	1.263430	0.287130	0.9800	87.3	79.8	92.8	88.0
15	13	0.9750	1.105156	0.318540	0.9500	88.9	80.3	93.5	87.9
15	13	0.9500	0.988009	0.348749	0.9000	90.3	80.6	94.1	87.9
15	13	0.9000	0.871119	0.387765	0.8000	91.9	80.9	94.9	87.9
15	13	0.8000	0.751176	0.441972	0.6000	94.0	81.2	96.0	87.9
15	12	0.7000	0.828123	0.594566	0.4000	95.6	81.1*	97.9	91.6
15	12	0.6000	0.758287	0.646176	0.2000	97.1	81.2*	98.4	91.6
15	11	0.5000	0.841029			98.5		99.3	
16	14	0.9999	2.005152	0.193769	0.9998	82.1	77.6	90.4	87.4
16	14	0.9995	1.695043	0.216892	0.9990	83.8	78.4	91.0	87.3
16	14	0.9990	1.570296	0.228792	0.9980	84.6	78.7	91.2	87.3
16	14	0.9950	1.297775	0.262963	0.9900	86.6	79.5	92.1	87.2
16	14	0.9900	1.186488	0.281616	0.9800	87.6	79.8	92.5	87.2
16	14	0.9750	1.043321	0.311808	0.9500	89.2	80.2	93.2	87.2
16	14	0.9500	0.936661	0.340749	0.9000	90.5	80.4	93.9	87.1
16	14	0.9000	0.829582	0.377993	0.8000	92.1	80.7	94.8	87.1
16	13	0.8000	0.872865	0.521712	0.6000	94.2	81.1	97.1	90.9
16	13	0.7000	0.788504	0.572361	0.4000	95.8	81.3	97.8	90.9
16	13	0.6000	0.724240	0.620435	0.2000	97.2	81.4	98.4	90.9
16	12	0.5000	0.798618			98.6		99.3	

A. VALUE OF m, CHOSEN TO MAXIMIZE EFFECTIVENESS F_u OF UPPER CONFIDENCE BOUND, ALSO MAXIMIZES EFFECTIVENESS F_i OF CENTRAL CONFIDENCE INTERVAL UNLESS MARKED *

n	m	1-P	B_{um}/x_m	$B_{\ell m}/x_m$	1-2P	F_u (%)	F_i (%)	S_u (%)	S_i (%)
17	15	0.9999	1.855646	0.191376	0.9998	82.6	77.8	90.0	86.6
17	15	0.9995	1.579822	0.213903	0.9990	84.2	78.5	90.6	86.6
17	15	0.9990	1.468123	0.225474	0.9980	84.9	78.7	90.9	86.6
17	15	0.9950	1.222420	0.258609	0.9900	86.9	79.4	91.8	86.5
17	15	0.9900	1.121347	0.276643	0.9800	87.9	79.6	92.2	86.5
17	15	0.9750	0.990611	0.305758	0.9500	89.3	80.0*	93.0	86.4
17	15	0.9500	0.892645	0.333581	0.9000	90.6	80.2*	93.7	86.4
17	14	0.9000	0.957399	0.447099	0.8000	92.3	80.8	96.1	90.3
17	14	0.8000	0.832352	0.505315	0.6000	94.4	81.1	97.0	90.3
17	14	0.7000	0.754460	0.552980	0.4000	95.9	81.3	97.7	90.3
17	14	0.6000	0.694857	0.598052	0.2000	97.3	81.4	98.3	90.3
17	13	0.5000	0.762453			98.6		99.2	
18	16	0.9999	1.730698	0.189181	0.9998	83.0	77.9	89.7	86.0
18	16	0.9995	1.482686	0.211170	0.9990	84.5	78.4	90.3	85.9
18	16	0.9990	1.381653	0.222443	0.9980	85.2	78.7	90.6	85.9
18	16	0.9950	1.158040	0.254647	0.9900	87.1	79.2	91.5	85.8
18	16	0.9900	1.065450	0.272127	0.9800	88.1	79.5*	92.0	85.8
18	15	0.9750	1.134481	0.363976	0.9500	89.5	80.1	94.8	89.7
18	15	0.9500	1.024211	0.395328	0.9000	90.9	80.5	95.3	89.7
18	15	0.9000	0.912925	0.435503	0.8000	92.5	80.8	96.0	89.6
18	15	0.8000	0.797279	0.490796	0.6000	94.6	81.1	96.9	89.6
18	15	0.7000	0.724847	0.535883	0.4000	96.1	81.3	97.6	89.6
18	14	0.6000	0.787487	0.679805	0.2000	97.4	80.8*	98.7	92.3
18	14	0.5000	0.731197			98.7		99.2	
19	17	0.9999	1.624765	0.187158	0.9998	83.3	77.8	89.4	85.3
19	17	0.9995	1.399692	0.208657	0.9990	84.7	78.4	90.0	85.2
19	17	0.9990	1.307514	0.219659	0.9980	85.4	78.6	90.3	85.2
19	17	0.9950	1.102371	0.251020	0.9900	87.3	79.0*	91.3	85.1
19	16	0.9900	1.213796	0.324908	0.9800	88.3	79.7	94.0	89.1
19	16	0.9750	1.078882	0.356544	0.9500	89.8	80.2	94.6	89.1
19	16	0.9500	0.977280	0.386644	0.9000	91.1	80.5	95.2	89.0
19	16	0.9000	0.874242	0.425100	0.8000	92.7	80.8	95.9	89.0
19	16	0.8000	0.766582	0.477829	0.6000	94.7	81.1	96.8	89.0
19	15	0.7000	0.818087	0.608666	0.4000	96.1	80.9*	98.2	91.8
19	15	0.6000	0.756412	0.655755	0.2000	97.5	81.0*	98.7	91.8
19	14	0.5000	0.815108			98.8		99.4	
20	18	0.9999	1.533835	0.185284	0.9998	83.5	77.8	89.1	84.6
20	18	0.9995	1.327954	0.206334	0.9990	84.9	78.2*	89.7	84.6
20	18	0.9990	1.243230	0.217089	0.9980	85.6	78.4*	90.1	84.5
20	17	0.9950	1.250475	0.300207	0.9900	87.6	79.5	93.4	88.5
20	17	0.9900	1.155046	0.319249	0.9800	88.6	79.8	93.8	88.5
20	17	0.9750	1.030490	0.349790	0.9500	90.0	80.2	94.4	88.5
20	17	0.9500	0.936249	0.378773	0.9000	91.3	80.5	95.0	88.4
20	17	0.9000	0.840256	0.415701	0.8000	92.9	80.7	95.7	88.4
20	17	0.8000	0.739461	0.466161	0.6000	94.8	81.0	96.7	88.4
20	16	0.7000	0.786774	0.590012	0.4000	96.3	81.0*	98.1	91.2
20	16	0.6000	0.729065	0.634461	0.2000	97.6	81.1*	98.7	91.2
20	15	0.5000	0.782936			98.8		99.4	

A. VALUE OF m, CHOSEN TO MAXIMIZE EFFECTIVENESS F_u OF UPPER CONFIDENCE BOUND, ALSO MAXIMIZES EFFECTIVENESS F_i OF CENTRAL CONFIDENCE INTERVAL UNLESS MARKED *

n	m	1-P	B_{um}/x_m	$B_{\ell m}/x_m$	1-2P	F_u (%)	F_i (%)	S_u (%)	S_i (%)
22	19	0.9999	1.633258	0.222895	0.9998	83.9	78.0	91.3	87.5
22	19	0.9995	1.422825	0.246148	0.9990	85.5	78.6	91.9	87.4
22	19	0.9990	1.335863	0.257958	0.9980	86.2	78.9	92.2	87.4
22	19	0.9950	1.140511	0.291350	0.9900	88.1	79.5	93.0	87.4
22	19	0.9900	1.058338	0.309288	0.9800	89.0	79.7	93.4	87.3
22	19	0.9750	0.950269	0.337944	0.9500	90.4	80.0*	94.1	87.3
22	19	0.9500	0.867838	0.365016	0.9000	91.6	80.3*	94.7	87.3
22	18	0.9000	0.904556	0.462181	0.8000	93.1	80.7	96.5	90.2
22	18	0.8000	0.800424	0.515348	0.6000	95.1	81.0	97.4	90.2
22	18	0.7000	0.734334	0.558256	0.4000	96.5	81.1	98.0	90.2
22	17	0.6000	0.780078	0.682688	0.2000	97.7	80.6*	98.9	92.3
22	17	0.5000	0.729377			98.9		99.4	
24	21	0.9999	1.480046	0.218376	0.9998	84.5	78.1	90.9	86.4
24	21	0.9995	1.299520	0.240663	0.9990	86.0	78.6	91.5	86.4
24	21	0.9990	1.224348	0.251950	0.9980	86.6	78.8	91.8	86.3
24	21	0.9950	1.054133	0.283750	0.9900	88.4	79.3*	92.6	86.3
24	21	0.9900	0.981925	0.300767	0.9800	89.3	79.5*	93.1	86.3
24	20	0.9750	1.016340	0.378702	0.9500	90.7	80.2	95.2	89.2
24	20	0.9500	0.931525	0.407375	0.9000	92.0	80.4	95.7	89.2
24	20	0.9000	0.844192	0.443635	0.8000	93.4	80.7	96.4	89.2
24	20	0.8000	0.751366	0.492735	0.6000	95.3	80.9	97.2	89.2
24	19	0.7000	0.784638	0.602885	0.4000	96.6	80.8*	98.3	91.4
24	19	0.6000	0.731876	0.644422	0.2000	97.8	80.9*	98.8	91.4
24	18	0.5000	0.772545			99.0		99.5	
26	23	0.9999	1.359356	0.214399	0.9998	84.9	78.1*	90.5	85.4
26	23	0.9995	1.201441	0.235854	0.9990	86.3	78.5*	91.1	85.4
26	23	0.9990	1.135260	0.246693	0.9980	86.9	78.7*	91.4	85.3
26	22	0.9950	1.121519	0.320166	0.9900	88.8	79.6	93.9	88.3
26	22	0.9900	1.047435	0.338253	0.9800	89.7	79.8	94.3	88.3
26	22	0.9750	0.949123	0.366976	0.9500	91.1	80.1	94.9	88.3
26	22	0.9500	0.873405	0.393935	0.9000	92.3	80.4	95.5	88.3
26	21	0.9000	0.896438	0.482693	0.8000	93.6	80.6*	96.9	90.6
26	21	0.8000	0.801204	0.533849	0.6000	95.4	80.8	97.7	90.6
26	21	0.7000	0.740143	0.574802	0.4000	96.8	81.0	98.3	90.5
26	21	0.6000	0.692415	0.612826	0.2000	97.9	81.0	98.8	90.5
26	19	0.5000	0.810547			99.1		99.5	
28	24	0.9999	1.431924	0.245342	0.9998	85.3	78.4	92.0	87.5
28	24	0.9995	1.270901	0.268334	0.9990	86.7	78.9	92.6	87.5
28	24	0.9990	1.203231	0.279909	0.9980	87.4	79.1	92.9	87.5
28	24	0.9950	1.048515	0.312305	0.9900	89.2	79.6	93.7	87.4
28	24	0.9900	0.982206	0.329524	0.9800	90.0	79.8*	94.1	87.4
28	24	0.9750	0.893769	0.356792	0.9500	91.3	80.0*	94.7	87.4
28	23	0.9500	0.925437	0.429916	0.9000	92.5	80.4	96.2	89.7
28	23	0.9000	0.845026	0.465383	0.8000	93.9	80.6	96.8	89.7
28	23	0.8000	0.758737	0.513062	0.6000	95.6	80.9	97.6	89.7
28	23	0.7000	0.703098	0.551063	0.4000	96.9	80.9	98.2	89.7
28	22	0.6000	0.733611	0.651911	0.2000	98.0	80.8*	99.0	91.6
28	21	0.5000	0.765155			99.1		99.5	

A. VALUE OF m, CHOSEN TO MAXIMIZE EFFECTIVENESS F_u OF UPPER CONFIDENCE BOUND, ALSO MAXIMIZES EFFECTIVENESS F_i OF CENTRAL CONFIDENCE INTERVAL UNLESS MARKED *

n	m	1-P	B_{um}/X_m	$B_{\ell m}/X_m$	1-2P	F_u (%)	F_i(%)	S_u(%)	S_i (%)
30	26	0.9999	1.331245	0.240983	0.9998	85.7	78.4	91.7	86.7
30	26	0.9995	1.187769	0.263146	0.9990	87.1	78.8*	92.3	86.7
30	26	0.9990	1.127161	0.274279	0.9980	87.7	79.0*	92.6	86.6
30	26	0.9950	0.987834	0.305356	0.9900	89.4	79.4*	93.4	86.6
30	25	0.9900	1.035766	0.361724	0.9800	90.3	79.9	94.9	89.0
30	25	0.9750	0.945340	0.390291	0.9500	91.6	80.2	95.5	89.0
30	25	0.9500	0.875155	0.416965	0.9000	92.7	80.4	96.0	88.9
30	25	0.9000	0.801920	0.450391	0.8000	94.1	80.6	96.7	88.9
30	24	0.8000	0.800851	0.548396	0.6000	95.8	80.7*	97.9	90.8
30	24	0.7000	0.743887	0.587613	0.4000	97.0	80.8*	98.5	90.8
30	24	0.6000	0.699093	0.623834	0.2000	98.1	80.9*	98.9	90.8
30	22	0.5000	0.798347			99.2		99.6	
32	28	0.9999	1.247588	0.237062	0.9998	86.1	78.3*	91.4	85.9
32	28	0.9995	1.118229	0.258492	0.9990	87.4	78.7*	92.0	85.8
32	27	0.9990	1.182691	0.303082	0.9980	88.0	79.2	93.6	88.3
32	27	0.9950	1.040594	0.335733	0.9900	89.7	79.7	94.4	88.2
32	27	0.9900	0.979187	0.352999	0.9800	90.6	79.9	94.7	88.2
32	27	0.9750	0.896772	0.380224	0.9500	91.8	80.1*	95.3	88.2
32	27	0.9500	0.832521	0.405574	0.9000	92.9	80.3*	95.9	88.2
32	26	0.9000	0.844328	0.482919	0.8000	94.3	80.6	97.1	90.1
32	26	0.8000	0.763488	0.529213	0.6000	95.9	80.8	97.8	90.1
32	26	0.7000	0.710993	0.565898	0.4000	97.1	80.9	98.4	90.1
32	25	0.6000	0.734721	0.657776	0.2000	98.2	80.7*	99.1	91.7
32	24	0.5000	0.759630			99.2		99.6	
34	29	0.9999	1.304359	0.263372	0.9998	86.4	78.6	92.6	87.6
34	29	0.9995	1.172722	0.285958	0.9990	87.7	79.0	93.1	87.5
34	29	0.9990	1.116753	0.297258	0.9980	88.4	79.2	93.4	87.5
34	29	0.9950	0.987200	0.328650	0.9900	90.0	79.6*	94.2	87.5
34	29	0.9900	0.930945	0.345207	0.9800	90.8	79.8*	94.5	87.5
34	28	0.9750	0.940470	0.409567	0.9500	92.0	80.2	95.9	89.4
34	28	0.9500	0.874911	0.435872	0.9000	93.1	80.4	96.4	89.4
34	28	0.9000	0.806087	0.468692	0.8000	94.4	80.6	97.0	89.4
34	28	0.8000	0.731322	0.512404	0.6000	96.0	80.8	97.8	89.4
34	27	0.7000	0.746381	0.597874	0.4000	97.2	80.8*	98.6	91.0
34	27	0.6000	0.704060	0.632517	0.2000	98.3	80.8*	99.1	91.0
34	25	0.5000	0.789085			99.3		99.6	
36	31	0.9999	1.231646	0.259203	0.9998	86.7	78.6*	92.3	86.9
36	31	0.9995	1.111570	0.281069	0.9990	88.0	79.0*	92.9	86.8
36	31	0.9990	1.060322	0.291989	0.9980	88.6	79.1*	93.2	86.8
36	30	0.9950	1.031931	0.355450	0.9900	90.2	79.8	94.9	88.8
36	30	0.9900	0.974643	0.372669	0.9800	91.0	79.9	95.2	88.8
36	30	0.9750	0.897353	0.399724	0.9500	92.3	80.2	95.8	88.8
36	30	0.9500	0.836759	0.424820	0.9000	93.3	80.4	96.3	88.8
36	30	0.9000	0.772922	0.456054	0.8000	94.6	80.5*	96.9	88.7
36	29	0.8000	0.766625	0.542415	0.6000	96.1	80.7	98.0	90.4
36	29	0.7000	0.716834	0.577885	0.4000	97.3	80.8	98.6	90.4
36	28	0.6000	0.735442	0.662513	0.2000	98.3	80.6*	99.2	91.8
36	27	0.5000	0.755342			99.3		99.6	

A. VALUE OF m, CHOSEN TO MAXIMIZE EFFECTIVENESS F_u OF UPPER CONFIDENCE BOUND, ALSO MAXIMIZES EFFECTIVENESS F_i OF CENTRAL CONFIDENCE INTERVAL UNLESS MARKED *

n	m	1-P	B_{um}/x_m	$B_{\ell m}/x_m$	1-2P	F_u(%)	F_i(%)	S_u(%)	S_i(%)
38	33	0.9999	1.169119	0.255393	0.9998	87.0	78.6*	92.1	86.2
38	32	0.9995	1.157737	0.305457	0.9990	88.2	79.2	93.8	88.2
38	32	0.9990	1.105670	0.316842	0.9980	88.9	79.3	94.0	88.2
38	32	0.9950	0.984449	0.348350	0.9900	90.5	79.7	94.7	88.2
38	32	0.9900	0.931487	0.364904	0.9800	91.3	79.9*	95.1	88.1
38	32	0.9750	0.859813	0.390866	0.9500	92.4	80.1*	95.6	88.1
38	31	0.9500	0.873587	0.451754	0.9000	93.5	80.4	96.7	89.8
38	31	0.9000	0.808551	0.483943	0.8000	94.7	80.6	97.2	89.8
38	31	0.8000	0.737490	0.526622	0.6000	96.3	80.7	98.0	89.8
38	30	0.7000	0.748075	0.606308	0.4000	97.4	80.7*	98.7	91.2
38	30	0.6000	0.707875	0.639558	0.2000	98.4	80.7*	99.1	91.2
38	28	0.5000	0.781813			99.3		99.7	
40	34	0.9999	1.215558	0.278301	0.9998	87.3	78.8	93.0	87.6
40	34	0.9995	1.103409	0.300435	0.9990	88.5	79.2	93.6	87.6
40	34	0.9990	1.055310	0.311453	0.9980	89.1	79.3*	93.8	87.6
40	34	0.9950	0.942943	0.341893	0.9900	90.7	79.7*	94.5	87.5
40	34	0.9900	0.893670	0.357852	0.9800	91.4	79.8*	94.9	87.5
40	33	0.9750	0.896554	0.416292	0.9500	92.6	80.2	96.1	89.2
40	33	0.9500	0.839140	0.441081	0.9000	93.6	80.4	96.6	89.2
40	33	0.9000	0.778377	0.471827	0.8000	94.9	80.6	97.2	89.2
40	32	0.8000	0.768717	0.553447	0.6000	96.4	80.7*	98.2	90.6
40	32	0.7000	0.721278	0.587802	0.4000	97.5	80.8*	98.7	90.6
40	31	0.6000	0.735911	0.666431	0.2000	98.4	80.5*	99.2	91.8
40	29	0.5000	0.806226			99.4		99.7	

B. VALUE OF m, CHOSEN TO MAXIMIZE EFFECTIVENESS F_i OF CENTRAL CONFIDENCE INTERVAL, DOES NOT MAXIMIZE EFFECTIVENESS F_u OF UPPER CONFIDENCE BOUND--SEE TABLE D4A

n	m	1-P	B_{um}/x_m	$B_{\ell m}/x_m$	1-2P	F_u (%)	F_i (%)	S_u (%)	S_i (%)
5	5	0.6000	0.559569	0.428851	0.2000	94.3	84.0	94.3	84.0
7	6	0.9750	1.828384	0.302555	0.9500	85.4	80.3	93.6	91.0
9	8	0.6000	0.645221	0.526836	0.2000	96.0	82.5	97.0	88.1
10	9	0.7000	0.674289	0.451711	0.4000	94.5	81.8	95.8	86.8
10	9	0.6000	0.606238	0.499627	0.2000	96.0	81.9	96.8	86.8
12	10	0.9750	1.378577	0.344476	0.9500	87.6	80.0	94.4	90.5
12	10	0.9500	1.211687	0.379810	0.9000	89.2	80.5	94.9	90.5
13	11	0.9950	1.627988	0.279206	0.9900	85.2	79.0	93.1	89.6
13	11	0.9900	1.468685	0.300255	0.9800	86.4	79.6	93.4	89.6
14	12	0.9990	1.842593	0.236511	0.9980	83.5	78.3	92.0	88.9
14	12	0.6000	0.646930	0.548755	0.2000	96.9	81.6	97.8	88.7
15	13	0.9999	2.187085	0.196396	0.9998	81.5	77.3	90.8	88.1
15	13	0.7000	0.676905	0.486571	0.4000	95.6	81.3	96.9	87.9
15	13	0.6000	0.620320	0.528907	0.2000	96.9	81.4	97.8	87.9
17	14	0.9750	1.199072	0.372209	0.9500	89.1	80.0	95.0	90.3
17	14	0.9500	1.078461	0.404975	0.9000	90.6	80.4	95.5	90.3
18	15	0.9900	1.281664	0.331117	0.9800	87.9	79.6	94.2	89.7
18	15	0.6000	0.669201	0.578375	0.2000	97.4	81.3	98.3	89.6
19	16	0.9950	1.317612	0.305226	0.9900	87.2	79.4	93.6	89.1
19	16	0.7000	0.698820	0.520667	0.4000	96.1	81.2	97.6	89.0
19	16	0.6000	0.646573	0.560915	0.2000	97.4	81.3	98.3	89.0
20	17	0.9995	1.582632	0.252492	0.9990	84.7	78.4	92.4	88.6
20	17	0.9990	1.479642	0.264921	0.9980	85.5	78.7	92.6	88.5
20	17	0.7000	0.675737	0.507016	0.4000	96.2	81.1	97.5	88.4
20	17	0.6000	0.626441	0.545295	0.2000	97.5	81.1	98.2	88.4
22	18	0.9750	1.099866	0.392402	0.9500	90.3	80.0	95.4	90.2
22	18	0.9500	1.003299	0.423138	0.9000	91.6	80.4	95.9	90.2
22	18	0.6000	0.683061	0.598356	0.2000	97.7	81.2	98.6	90.2
24	20	0.9950	1.211151	0.329163	0.9900	88.3	79.5	94.2	89.3
24	20	0.9900	1.127156	0.348266	0.9800	89.3	79.8	94.6	89.3
24	20	0.7000	0.692030	0.532142	0.4000	96.6	81.0	97.9	89.2
24	20	0.6000	0.645753	0.568803	0.2000	97.8	81.1	98.6	89.2
26	22	0.9999	1.555456	0.250229	0.9998	84.7	78.2	92.4	88.4
26	22	0.9995	1.372122	0.274169	0.9990	86.2	78.8	92.9	88.4
26	22	0.9990	1.295533	0.286250	0.9980	86.9	79.0	93.2	88.4
26	21	0.9000	0.896438	0.482693	0.8000	93.6	80.6	96.9	90.6
28	23	0.9900	1.103136	0.371585	0.9800	89.9	79.8	95.2	89.8
28	23	0.9750	1.002869	0.401705	0.9500	91.3	80.2	95.7	89.8
28	23	0.6000	0.659424	0.586217	0.2000	98.0	81.0	98.8	89.7

B. VALUE OF m, CHOSEN TO MAXIMIZE EFFECTIVENESS F_i OF CENTRAL CONFIDENCE INTERVAL, DOES NOT MAXIMIZE EFFECTIVENESS F_u OF UPPER CONFIDENCE BOUND--SEE TABLE D4A

n	m	1-P	B_{um}/x_m	$B_{\ell m}/x_m$	1-2P	F_u(%)	F_i(%)	S_u(%)	S_i(%)
30	25	0.9995	1.329477	0.297361	0.9990	86.9	78.9	93.7	89.0
30	25	0.9990	1.260820	0.309568	0.9980	87.6	79.1	93.9	89.0
30	25	0.9950	1.103420	0.343651	0.9900	89.4	79.7	94.6	89.0
30	25	0.8000	0.722905	0.495148	0.6000	95.8	80.8	97.5	88.9
30	25	0.7000	0.671704	0.530678	0.4000	97.0	80.9	98.1	88.9
30	25	0.6000	0.631367	0.563442	0.2000	98.1	80.9	98.7	88.9
32	27	0.9999	1.390009	0.267963	0.9998	86.0	78.5	92.9	88.3
32	27	0.9995	1.244357	0.291357	0.9990	87.4	79.0	93.4	88.3
32	26	0.9750	0.990566	0.420589	0.9500	91.8	80.1	96.1	90.1
32	26	0.9500	0.919078	0.448278	0.9000	92.9	80.4	96.5	90.1
32	26	0.6000	0.669571	0.599678	0.2000	98.2	80.9	98.9	90.1
34	28	0.9950	1.086829	0.363308	0.9900	89.9	79.7	95.1	89.5
34	28	0.9900	1.024403	0.381275	0.9800	90.8	79.9	95.4	89.5
34	28	0.7000	0.682572	0.546928	0.4000	97.2	80.8	98.8	90.4
34	28	0.6000	0.643987	0.578633	0.2000	98.2	80.9	98.9	89.4
36	30	0.9999	1.353606	0.287349	0.9998	86.5	78.6	93.5	88.9
36	30	0.9995	1.220349	0.310958	0.9990	87.9	79.1	94.0	88.8
36	30	0.9990	1.163585	0.322749	0.9980	88.6	79.3	94.2	88.8
36	29	0.9000	0.842850	0.497436	0.8000	94.6	80.6	97.3	90.4
36	29	0.6000	0.677373	0.610420	0.2000	98.3	80.9	99.0	90.4
38	32	0.9999	1.279531	0.282624	0.9998	86.9	78.8	93.2	88.2
38	31	0.9900	1.013717	0.397906	0.9800	91.2	79.9	95.8	89.8
38	31	0.9750	0.935225	0.425858	0.9500	92.4	80.2	96.2	89.8
38	31	0.7000	0.690910	0.560183	0.4000	97.4	80.8	98.5	89.8
38	31	0.6000	0.653898	0.590896	0.2000	98.4	80.8	99.0	89.8
40	33	0.9990	1.146099	0.339744	0.9980	89.0	79.4	94.7	89.3
40	33	0.9950	1.023220	0.372365	0.9900	90.6	79.8	95.3	89.2
40	33	0.9900	0.969437	0.389479	0.9800	91.4	80.0	95.6	89.2
40	33	0.8000	0.711757	0.512485	0.6000	96.3	80.7	97.9	89.2
40	33	0.7000	0.667951	0.544375	0.4000	97.4	80.8	98.5	89.2
40	32	0.6000	0.683540	0.619208	0.2000	98.4	80.8	99.1	90.6

C. VALUE OF m, CHOSEN TO MAXIMIZE EFFICIENCY E_u OF UPPER CONFIDENCE BOUND, ALSO MAXIMIZES EFFICIENCY E_i OF CENTRAL CONFIDENCE INTERVAL, UNLESS MARKED *

n	m	1-P	B_{um}/m	$B_{\ell m}/m$	1-2P	E_u (%)	E_i (%)	R_u (%)	R_i (%)
1	1	0.9999	9999.500	0.108574	0.9998	100.0	100.0	100.0	100.0
1	1	0.9995	1999.500	0.131563	0.9990	100.0	100.0	100.0	100.0
1	1	0.9990	999.4999	0.144765	0.9980	100.0	100.0	100.0	100.0
1	1	0.9950	199.4996	0.188739	0.9900	100.0	100.0	100.0	100.0
1	1	0.9900	99.49916	0.217147	0.9800	100.0	100.0	100.0	100.0
1	1	0.9750	39.49789	0.271085	0.9500	100.0	100.0	100.0	100.0
1	1	0.9500	19.49573	0.333808	0.9000	100.0	100.0	100.0	100.0
1	1	0.9000	9.491222	0.434294	0.8000	100.0	100.0	100.0	100.0
1	1	0.8000	4.481420	0.621335	0.6000	100.0	100.0	100.0	100.0
1	1	0.7000	2.803673	0.830584	0.4000	100.0	100.0	100.0	100.0
1	1	0.6000	1.957615	1.091357	0.2000	100.0	100.0	100.0	100.0
1	1	0.5000	1.442695			100.0		100.0	
2	2	0.9999	99.49916	0.100975	0.9998	85.7	85.7	85.7	85.7
2	2	0.9995	44.21947	0.120570	0.9990	85.7	85.7	85.7	85.7
2	2	0.9990	31.12010	0.131568	0.9980	85.6	85.6	85.6	85.6
2	2	0.9950	13.63602	0.166939	0.9900	85.5	85.5	85.5	85.5
2	2	0.9900	9.491222	0.188829	0.9800	85.5	85.5	85.5	85.5
2	2	0.9750	5.810219	0.228534	0.9500	85.3	85.3	85.3	85.3
2	2	0.9500	3.951067	0.272025	0.9000	85.1	85.2	85.1	85.2
2	2	0.9000	2.630676	0.336730	0.8000	84.8	85.0	84.8	85.0
2	2	0.8000	1.686956	0.444770	0.6000	84.4	84.9	84.4	84.9
2	2	0.7000	1.260304	0.551900	0.4000	84.1	84.9	84.1	84.9
2	2	0.6000	0.999090	0.671202	0.2000	84.3	85.0	84.3	85.0
2	2	0.5000	0.814367			85.0		85.0	
3	3	0.9999	21.04039	0.097003	0.9998	76.8	76.8	76.8	76.8
3	3	0.9995	12.09232	0.114951	0.9990	76.7	76.7	76.7	76.7
3	3	0.9990	9.491222	0.124906	0.9980	76.6	76.6	76.6	76.6
3	3	0.9950	5.332417	0.156366	0.9900	76.4	76.4	76.4	76.4
3	3	0.9900	4.121388	0.175425	0.9800	76.2	76.3	76.2	76.3
3	3	0.9750	2.891186	0.209245	0.9500	75.8	76.1	75.8	76.1
3	3	0.9500	2.176260	0.245258	0.9000	75.5	75.8	75.5	75.8
3	3	0.9000	1.602776	0.297045	0.8000	74.9	75.6	74.9	75.6
3	3	0.8000	1.137652	0.379433	0.6000	74.3	75.5	74.3	75.5
3	3	0.7000	0.903386	0.456955	0.4000	74.0	75.6	74.0	75.6
3	3	0.6000	0.749139	0.539296	0.2000	74.4	75.8	74.4	75.8
3	3	0.5000	0.633542			75.9		75.9	
4	4	0.9999	9.491222	0.094370	0.9998	70.5	70.5	70.5	70.5
4	4	0.9995	6.173910	0.111272	0.9990	70.3	70.3	70.3	70.3
4	4	0.9990	5.107108	0.120574	0.9980	70.1	70.2	70.1	70.2
4	4	0.9950	3.234883	0.149639	0.9900	69.7	69.9	69.7	69.9
4	4	0.9900	2.630676	0.167009	0.9800	69.5	69.7	69.5	69.7
4	4	0.9750	1.972805	0.197406	0.9500	69.0	69.5	69.0	69.5
4	4	0.9500	1.561745	0.229206	0.9000	68.5	69.2	68.5	69.2
4	4	0.9000	1.210191	0.273987	0.8000	67.8	69.0	67.8	69.0
4	4	0.8000	0.905098	0.343171	0.6000	67.1	68.9	67.1	68.9
4	4	0.7000	0.742174	0.406260	0.4000	66.9	69.0	66.9	69.0
4	4	0.6000	0.630491	0.471432	0.2000	67.5	69.3	67.5	69.3
4	4	0.5000	0.544011			69.4		69.4	

C. VALUE OF m, CHOSEN TO MAXIMIZE EFFICIENCY E_u OF UPPER CONFIDENCE BOUND, ALSO MAXIMIZES EFFICIENCY E_i OF CENTRAL CONFIDENCE INTERVAL, UNLESS MARKED *

n	m	1-P	B_{um}/x_m	$B_{\ell m}/x_m$	1-2P	E_u (%)	E_i (%)	R_u (%)	R_i (%)
5	5	0.9999	5.795202	0.092424	0.9998	65.6	65.6	65.6	65.6
5	5	0.9995	4.052508	0.108576	0.9990	65.3	65.4	65.3	65.4
5	5	0.9990	3.457000	0.117415	0.9980	65.1	65.3	65.1	65.3
5	5	0.9950	2.350046	0.144807	0.9900	64.6	64.9	64.6	64.9
5	5	0.9900	1.969761	0.161015	0.9800	64.3	64.7	64.3	64.7
5	5	0.9750	1.537456	0.189100	0.9500	63.7	64.4	63.7	64.4
5	5	0.9500	1.254847	0.218114	0.9000	63.2	64.2	63.2	64.2
5	5	0.9000	1.003167	0.258376	0.8000	62.4	64.0	62.4	64.0
5	5	0.8000	0.775084	0.319325	0.6000	61.6	63.9	61.6	63.9
5	5	0.7000	0.648595	0.373717	0.4000	61.4	64.0	61.4	64.0
5	4	0.6000	0.992163	0.754215	0.2000	62.7	65.0	84.4	85.4
5	4	0.5000	0.862837			65.6		85.4	
6	6	0.9999	4.121388	0.090892	0.9998	61.6	61.7	61.6	61.7
6	6	0.9995	3.022012	0.106468	0.9990	61.3	61.5	61.3	61.5
6	6	0.9990	2.630676	0.114954	0.9980	61.1	61.4	61.1	61.4
6	6	0.9950	1.874013	0.141084	0.9900	60.5	61.0	60.5	61.0
6	6	0.9900	1.602776	0.156427	0.9800	60.1	60.8	60.1	60.8
6	6	0.9750	1.285111	0.182811	0.9500	59.5	60.4	59.5	60.4
6	6	0.9500	1.070836	0.209810	0.9000	58.9	60.2	58.9	60.2
6	5	0.9000	1.486492	0.420253	0.8000	58.7	60.7	80.9	81.9
6	5	0.8000	1.160510	0.508397	0.6000	60.5	63.2	80.3	81.8
6	5	0.7000	0.979760	0.586569	0.4000	61.9	64.8	80.2	81.8
6	5	0.6000	0.852532	0.665578	0.2000	63.8	66.0	80.6	81.9
6	5	0.5000	0.751821			66.5		81.9	
7	7	0.9999	3.201607	0.089636	0.9998	58.4	58.5	58.4	58.5
7	7	0.9995	2.427707	0.104749	0.9990	58.0	58.2	58.0	58.2
7	7	0.9990	2.143967	0.112953	0.9980	57.8	58.1	57.8	58.1
7	7	0.9950	1.579244	0.138082	0.9900	57.1	57.7	57.1	57.7
7	7	0.9900	1.370422	0.152747	0.9800	56.7	57.5	56.7	57.5
7	6	0.9750	1.828384	0.302555	0.9500	57.7	59.0	78.5	79.1
7	6	0.9500	1.532390	0.341250	0.9000	58.8	60.5	78.1	78.9
7	6	0.9000	1.261314	0.393616	0.8000	59.9	62.1	77.5	78.7
7	6	0.8000	1.008139	0.470785	0.6000	61.0	63.9	76.9	78.7
7	6	0.7000	0.863959	0.538095	0.4000	62.2	65.1	76.8	78.7
7	6	0.6000	0.760539	0.605166	0.2000	63.8	66.0	77.3	78.8
7	6	0.5000	0.677386			66.3		78.8	
8	8	0.9999	2.630676	0.088576	0.9998	55.6	55.8	55.6	55.8
8	8	0.9995	2.045421	0.103304	0.9990	55.1	55.5	55.1	55.5
8	8	0.9990	1.825962	0.111275	0.9980	54.9	55.4	54.9	55.4
8	7	0.9950	2.175662	0.233223	0.9900	56.7	57.6	76.3	76.6
8	7	0.9900	1.894905	0.254500	0.9800	57.6	58.7	76.0	76.5
8	7	0.9750	1.558882	0.290142	0.9500	58.7	60.1	75.6	76.3
8	7	0.9500	1.327002	0.325660	0.9000	59.4	61.2	75.1	76.1
8	7	0.9000	1.110127	0.373262	0.8000	60.1	62.4	74.5	76.0
8	7	0.8000	0.902888	0.442520	0.6000	60.8	63.8	73.8	75.9
8	7	0.7000	0.782460	0.502148	0.4000	61.7	64.7	73.7	75.9
8	7	0.6000	0.694828	0.560911	0.2000	63.2	65.4	74.4	76.0
8	7	0.5000	0.623520			65.6		76.1	

C. VALUE OF m, CHOSEN TO MAXIMIZE EFFICIENCY E_u OF UPPER CONFIDENCE BOUND, ALSO MAXIMIZES EFFICIENCY E_i OF CENTRAL CONFIDENCE INTERVAL, UNLESS MARKED *

n	m	1-P	B_{um}/x_m	$B_{\ell m}/x_m$	1-2P	E_u (%)	E_i (%)	R_u (%)	R_i (%)
9	8	0.9999	3.510602	0.156420	0.9998	53.7	54.1	74.6	74.8
9	8	0.9995	2.740770	0.179101	0.9990	55.4	55.9	74.3	74.6
9	8	0.9990	2.452161	0.191088	0.9980	56.0	56.7	74.2	74.5
9	8	0.9950	1.864991	0.226624	0.9900	57.6	58.5	73.7	74.2
9	8	0.9900	1.642726	0.246683	0.9800	58.2	59.3	73.4	74.0
9	8	0.9750	1.372153	0.280082	0.9500	58.9	60.4	72.9	73.8
9	8	0.9500	1.182057	0.313124	0.9000	59.3	61.3	72.4	73.7
9	8	0.9000	1.001314	0.357057	0.8000	59.7	62.2	71.8	73.5
9	8	0.8000	0.825468	0.420313	0.6000	60.2	63.2	71.1	73.4
9	8	0.7000	0.721641	0.474202	0.4000	60.9	64.0	71.0	73.5
9	8	0.6000	0.645221	0.526836	0.2000	62.3	64.5	71.7	73.6
9	8	0.5000	0.582440			64.7		73.6	
10	9	0.9999	2.947923	0.153739	0.9998	54.9	55.3	72.4	72.6
10	9	0.9995	2.347536	0.175598	0.9990	56.1	56.8	72.1	72.4
10	9	0.9990	2.118425	0.187109	0.9980	56.7	57.4	71.9	72.3
10	9	0.9950	1.643984	0.221066	0.9900	57.8	58.8	71.4	72.0
10	9	0.9900	1.460964	0.240127	0.9800	58.2	59.5	71.1	71.8
10	9	0.9750	1.235109	0.271703	0.9500	58.6	60.3	70.6	71.6
10	9	0.9500	1.074125	0.302752	0.9000	58.9	61.0	70.0	71.4
10	9	0.9000	0.919013	0.343757	0.8000	59.1	61.7	69.4	71.3
10	9	0.8000	0.765884	0.402287	0.6000	59.3	62.5	68.6	71.2
10	8	0.7000	0.909329	0.607133	0.4000	60.4	63.5	80.7	82.5
10	8	0.6000	0.816001	0.671425	0.2000	62.2	64.4	81.2	82.5
10	8	0.5000	0.739333			64.7		82.5	
11	10	0.9999	2.550353	0.151406	0.9998	55.5	55.9	70.3	70.6
11	10	0.9995	2.063467	0.172563	0.9990	56.4	57.1	70.0	70.4
11	10	0.9990	1.874971	0.183669	0.9980	56.8	57.6	69.8	70.2
11	10	0.9950	1.478934	0.216293	0.9900	57.6	58.7	69.3	69.9
11	10	0.9900	1.323778	0.234518	0.9800	57.9	59.2	68.9	69.8
11	10	0.9750	1.130147	0.264576	0.9500	58.1	59.9	68.4	69.6
11	10	0.9500	0.990477	0.293977	0.9000	58.3	60.4	67.8	69.4
11	9	0.9000	1.137529	0.443272	0.8000	58.5	61.3	79.2	80.6
11	9	0.8000	0.952715	0.514005	0.6000	59.7	62.9	78.7	80.6
11	9	0.7000	0.842195	0.573666	0.4000	60.9	64.0	78.6	80.6
11	9	0.6000	0.760095	0.631482	0.2000	62.5	64.7	79.2	80.6
11	9	0.5000	0.692122			65.0		80.7	
12	11	0.9999	2.255961	0.149346	0.9998	55.6	56.2	68.5	68.7
12	11	0.9995	1.849261	0.169894	0.9990	56.4	57.1	68.1	68.5
12	11	0.9990	1.689910	0.180650	0.9980	56.6	57.5	67.9	68.4
12	11	0.9950	1.351016	0.212129	0.9900	57.2	58.4	67.3	68.1
12	11	0.9900	1.216517	0.229641	0.9800	57.4	58.8	67.0	67.9
12	10	0.9750	1.378577	0.344476	0.9500	57.5	59.6	78.2	79.1
12	10	0.9500	1.211687	0.379810	0.9000	58.3	60.7	77.8	79.0
12	10	0.9000	1.049144	0.426071	0.8000	59.1	61.9	77.3	78.9
12	10	0.8000	0.886772	0.491465	0.6000	60.0	63.2	76.7	78.8
12	10	0.7000	0.788618	0.546209	0.4000	61.0	64.1	76.7	78.9
12	10	0.6000	0.715134	0.598927	0.2000	62.5	64.7	77.4	78.9
12	10	0.5000	0.653892			64.9		78.9	

C. VALUE OF m, CHOSEN TO MAXIMIZE EFFICIENCY E_u OF UPPER CONFIDENCE BOUND, ALSO MAXIMIZES EFFICIENCY E_i OF CENTRAL CONFIDENCE INTERVAL, UNLESS MARKED *

n	m	1-P	B_{um}/x_m	$B_{\ell m}/x_m$	1-2P	E_u (%)	E_i (%)	R_u (%)	R_i (%)
13	12	0.9999	2.029881	0.147507	0.9998	55.5	56.1	66.7	67.1
13	12	0.9995	1.682235	0.167520	0.9990	56.1	56.9	66.4	66.8
13	12	0.9990	1.544630	0.177970	0.9980	56.3	57.2	66.2	66.7
13	12	0.9950	1.248943	0.208452	0.9900	56.6	57.9*	65.5	66.4
13	11	0.9900	1.468685	0.300255	0.9800	57.3	59.0	76.9	77.6
13	11	0.9750	1.268036	0.334656	0.9500	58.1	60.2	76.4	77.4
13	11	0.9500	1.121884	0.368006	0.9000	58.7	61.1	76.0	77.3
13	11	0.9000	0.978212	0.411445	0.8000	59.3	62.1	75.5	77.2
13	11	0.8000	0.833211	0.472453	0.6000	60.0	63.2	74.9	77.2
13	11	0.7000	0.744749	0.523193	0.4000	60.9	64.0	74.9	77.2
13	11	0.6000	0.678079	0.571790	0.2000	62.3	64.5	75.7	77.3
13	11	0.5000	0.622201			64.7		77.3	
14	13	0.9999	1.851142	0.145849	0.9998	55.2	55.9	65.1	65.5
14	13	0.9995	1.548457	0.165387	0.9990	55.7	56.5	64.7	65.3
14	12	0.9990	1.842593	0.236511	0.9980	55.9	57.0	76.0	76.4
14	12	0.9950	1.495147	0.273156	0.9900	57.2	58.7	75.5	76.2
14	12	0.9900	1.355763	0.293296	0.9800	57.7	59.5	75.3	76.0
14	12	0.9750	1.178786	0.326094	0.9500	58.4	60.5	74.8	75.9
14	12	0.9500	1.048767	0.357757	0.9000	58.8	61.3	74.4	75.8
14	12	0.9000	0.919932	0.398814	0.8000	59.3	62.1	73.8	75.7
14	12	0.8000	0.788749	0.456148	0.6000	59.8	63.1	73.3	75.7
14	12	0.7000	0.708079	0.503560	0.4000	60.6	63.7	73.3	75.7
14	12	0.6000	0.646930	0.548755	0.2000	62.0	64.2	74.0	75.7
14	12	0.5000	0.595428			64.3		75.8	
15	14	0.9999	1.706449	0.144344	0.9998	54.8	55.5*	63.6	64.0
15	13	0.9995	1.833879	0.220180	0.9990	55.9	57.0	74.6	75.0
15	13	0.9990	1.692871	0.232450	0.9980	56.4	57.6	74.5	74.9
15	13	0.9950	1.387217	0.267782	0.9900	57.5	59.1	74.0	74.7
15	13	0.9900	1.263430	0.287130	0.9800	57.9	59.7	73.7	74.6
15	13	0.9750	1.105156	0.318540	0.9500	58.5	60.6	73.3	74.5
15	13	0.9500	0.988009	0.348749	0.9000	58.8	61.3	72.8	74.4
15	13	0.9000	0.871119	0.387765	0.8000	59.1	62.0	72.3	74.3
15	12	0.8000	0.920263	0.540418	0.6000	59.5	62.9	79.7	81.7
15	12	0.7000	0.828123	0.594566	0.4000	60.7	63.7	79.8	81.7
15	12	0.6000	0.758287	0.646176	0.2000	62.2	64.3	80.4	81.7
15	12	0.5000	0.699473			64.5		81.7	
16	14	0.9999	2.005152	0.193769	0.9998	55.4	56.2	73.5	73.8
16	14	0.9995	1.695043	0.216892	0.9990	56.4	57.4	73.2	73.7
16	14	0.9990	1.570296	0.228792	0.9980	56.8	58.0	73.1	73.6
16	14	0.9950	1.297775	0.262963	0.9900	57.7	59.2	72.6	73.4
16	14	0.9900	1.186488	0.281616	0.9800	58.0	59.8	72.3	73.3
16	14	0.9750	1.043321	0.311808	0.9500	58.4	60.5	71.8	73.1
16	14	0.9500	0.936661	0.340749	0.9000	58.6	61.1	71.4	73.0
16	13	0.9000	1.009128	0.460131	0.8000	59.0	62.0	78.8	80.4
16	13	0.8000	0.872865	0.521712	0.6000	59.9	63.2	78.3	80.4
16	13	0.7000	0.788504	0.572361	0.4000	60.9	63.9	78.4	80.4
16	13	0.6000	0.724240	0.620435	0.2000	62.4	64.5	79.0	80.5
16	13	0.5000	0.669886			64.6		80.5	

C. VALUE OF m, CHOSEN TO MAXIMIZE EFFICIENCY E_u OF UPPER CONFIDENCE BOUND, ALSO MAXIMIZES EFFICIENCY E_i OF CENTRAL CONFIDENCE INTERVAL, UNLESS MARKED *

n	m	1-P	B_{um}/X_m	$B_{\ell m}/X_m$	1-2P	E_u (%)	E_i (%)	R_u (%)	R_i (%)
17	15	0.9999	1.855646	0.191376	0.9998	55.7	56.6	72.2	72.6
17	15	0.9995	1.579822	0.213903	0.9990	56.6	57.7	71.9	72.4
17	15	0.9990	1.468123	0.225474	0.9980	56.9	58.2	71.7	72.3
17	15	0.9950	1.222420	0.258609	0.9900	57.6	59.2	71.2	72.1
17	15	0.9900	1.121347	0.276643	0.9800	57.9	59.7	71.0	72.0
17	15	0.9750	0.990611	0.305758	0.9500	58.2	60.4*	70.5	71.8
17	14	0.9500	1.078461	0.404975	0.9000	58.6	61.3	77.9	79.2
17	14	0.9000	0.957399	0.447099	0.8000	59.3	62.3	77.5	79.2
17	14	0.8000	0.832352	0.505315	0.6000	60.0	63.3	77.0	79.2
17	14	0.7000	0.754460	0.552980	0.4000	61.0	64.0	77.1	79.2
17	14	0.6000	0.694857	0.598052	0.2000	62.4	64.5	77.8	79.2
17	14	0.5000	0.644254			64.6		79.3	
18	16	0.9999	1.730698	0.189181	0.9998	55.9	56.9	71.0	71.4
18	16	0.9995	1.482686	0.211170	0.9990	56.6	57.8	70.7	71.2
18	16	0.9990	1.381653	0.222443	0.9980	56.9	58.2	70.5	71.1
18	16	0.9950	1.158040	0.254647	0.9900	57.5	59.1	70.0	70.9
18	15	0.9900	1.281664	0.331117	0.9800	57.7	59.8	77.4	78.3
18	15	0.9750	1.134481	0.363976	0.9500	58.4	60.8	77.1	78.2
18	15	0.9500	1.024211	0.395328	0.9000	58.9	61.6	76.7	78.1
18	15	0.9000	0.912925	0.435503	0.8000	59.4	62.4	76.2	78.0
18	15	0.8000	0.797279	0.490796	0.6000	60.1	63.4	75.7	78.0
18	15	0.7000	0.724847	0.535883	0.4000	60.9	64.0	75.8	78.0
18	15	0.6000	0.669201	0.578375	0.2000	62.3	64.4	76.5	78.1
18	15	0.5000	0.621795			64.5		78.1	
19	17	0.9999	1.624765	0.187158	0.9998	56.0	57.0	69.8	70.2
19	17	0.9995	1.399692	0.208657	0.9990	56.6	57.8	69.5	70.0
19	17	0.9990	1.307514	0.219659	0.9980	56.8	58.1	69.3	69.9
19	16	0.9950	1.317612	0.305226	0.9900	57.6	59.4	76.5	77.2
19	16	0.9900	1.213796	0.324908	0.9800	58.1	60.1	76.3	77.2
19	16	0.9750	1.078882	0.356544	0.9500	58.6	61.0	75.9	77.1
19	16	0.9500	0.977280	0.386644	0.9000	59.1	61.7	75.5	77.0
19	16	0.9000	0.874242	0.425100	0.8000	59.4	62.5	75.0	76.9
19	16	0.8000	0.766582	0.477829	0.6000	60.0	63.3	74.5	76.9
19	16	0.7000	0.698820	0.520667	0.4000	60.8	63.8	74.6	76.9
19	15	0.6000	0.756412	0.655755	0.2000	62.2	64.2	81.2	82.4
19	15	0.5000	0.703873			64.4		82.4	
20	18	0.9999	1.533835	0.185284	0.9998	56.0	56.9	68.7	69.1
20	18	0.9995	1.327954	0.206334	0.9990	56.5	57.7*	68.3	68.9
20	17	0.9990	1.479642	0.264921	0.9980	56.9	58.4	75.8	76.3
20	17	0.9950	1.250475	0.300207	0.9900	57.9	59.7	75.4	76.2
20	17	0.9900	1.155046	0.319249	0.9800	58.3	60.3	75.2	76.1
20	17	0.9750	1.030490	0.349790	0.9500	58.7	61.1	74.7	76.0
20	17	0.9500	0.936249	0.378773	0.9000	59.1	61.7	74.3	75.9
20	17	0.9000	0.840256	0.415701	0.8000	59.4	62.4	73.8	75.8
20	16	0.8000	0.861627	0.542702	0.6000	59.9	63.2	79.4	81.3
20	16	0.7000	0.786774	0.590012	0.4000	60.9	63.9	79.4	81.4
20	16	0.6000	0.729065	0.634461	0.2000	62.4	64.3	80.1	81.4
20	16	0.5000	0.679753			64.5		81.4	

C. VALUE OF m, CHOSEN TO MAXIMIZE EFFICIENCY E_u OF UPPER CONFIDENCE BOUND, ALSO MAXIMIZES EFFICIENCY E_i OF CENTRAL CONFIDENCE INTERVAL, UNLESS MARKED *

n	m	1-P	B_{um}/x_m	$B_{\ell m}/x_m$	1-2P	E_u (%)	E_i (%)	R_u (%)	R_i (%)
22	19	0.9999	1.633258	0.222895	0.9998	56.2	57.4	74.2	74.6
22	19	0.9995	1.422825	0.246148	0.9990	57.0	58.4	73.9	74.4
22	19	0.9990	1.335863	0.257958	0.9980	57.4	58.9	73.7	74.3
22	19	0.9950	1.140511	0.291350	0.9900	58.1	60.0	73.3	74.2
22	19	0.9900	1.058338	0.309288	0.9800	58.4	60.4	73.0	74.1
22	19	0.9750	0.950269	0.337944	0.9500	58.7	61.1*	72.6	74.0
22	18	0.9500	1.003299	0.423138	0.9000	59.0	61.8	78.1	79.5
22	18	0.9000	0.904556	0.462181	0.8000	59.6	62.6	77.7	79.5
22	18	0.8000	0.800424	0.515348	0.6000	60.2	63.5	77.3	79.4
22	18	0.7000	0.734334	0.558256	0.4000	61.1	64.0	77.4	79.5
22	18	0.6000	0.683061	0.598356	0.2000	62.5	64.4	78.1	79.5
22	18	0.5000	0.639014			64.5		79.5	
24	21	0.9999	1.480046	0.218376	0.9998	56.6	57.8	72.3	72.7
24	21	0.9995	1.299520	0.240663	0.9990	57.2	58.6	72.0	72.6
24	21	0.9990	1.224348	0.251950	0.9980	57.5	59.0	71.8	72.5
24	20	0.9950	1.211151	0.329163	0.9900	58.0	60.1	77.1	77.9
24	20	0.9900	1.127156	0.348266	0.9800	58.5	60.7	76.9	77.9
24	20	0.9750	1.016340	0.378702	0.9500	59.0	61.5	76.6	77.8
24	20	0.9500	0.931525	0.407375	0.9000	59.3	62.1	76.2	77.7
24	20	0.9000	0.844192	0.443635	0.8000	59.7	62.8	75.8	77.7
24	20	0.8000	0.751366	0.492735	0.6000	60.2	63.5	75.4	77.7
24	19	0.7000	0.784638	0.602885	0.4000	61.0	63.9*	80.2	82.0
24	19	0.6000	0.731876	0.644422	0.2000	62.4	64.3	80.8	82.1
24	19	0.5000	0.686446			64.4		82.1	
26	23	0.9999	1.359356	0.214399	0.9998	56.7	57.9	70.5	71.0
26	22	0.9995	1.372122	0.274169	0.9990	57.3	58.9	75.9	76.4
26	22	0.9990	1.295533	0.286250	0.9980	57.6	59.3	75.8	76.4
26	22	0.9950	1.121519	0.320166	0.9900	58.4	60.5	75.4	76.3
26	22	0.9900	1.047435	0.338253	0.9800	58.7	60.9	75.2	76.2
26	22	0.9750	0.949123	0.366976	0.9500	59.1	61.6	74.8	76.1
26	22	0.9500	0.873405	0.393935	0.9000	59.4	62.1	74.4	76.1
26	21	0.9000	0.896438	0.482693	0.8000	59.7	62.8	78.7	80.4
26	21	0.8000	0.801204	0.533849	0.6000	60.4	63.6	78.4	80.4
26	21	0.7000	0.740143	0.574802	0.4000	61.2	64.1	78.5	80.4
26	21	0.6000	0.692415	0.612826	0.2000	62.6	64.4	79.2	80.5
26	21	0.5000	0.651148			64.5		80.5	
28	24	0.9999	1.431924	0.245342	0.9998	56.9	58.3	74.6	75.0
28	24	0.9995	1.270901	0.268334	0.9990	57.6	59.2	74.3	74.9
28	24	0.9990	1.203231	0.279909	0.9980	57.9	59.6	74.2	74.8
28	24	0.9950	1.048515	0.312305	0.9900	58.5	60.6	73.8	74.7
28	24	0.9900	0.982206	0.329524	0.9800	58.8	61.0*	73.5	74.6
28	23	0.9750	1.002869	0.401705	0.9500	59.2	61.8	77.8	79.0
28	23	0.9500	0.925437	0.429916	0.9000	59.5	62.4	77.5	78.9
28	23	0.9000	0.845026	0.465383	0.8000	59.9	63.0	77.1	78.9
28	23	0.8000	0.758737	0.513062	0.6000	60.5	63.7	76.7	78.9
28	23	0.7000	0.703098	0.551063	0.4000	61.3	64.1	76.9	78.9
28	23	0.6000	0.659424	0.586217	0.2000	62.6	64.4	77.6	79.0
28	23	0.5000	0.621528			64.5		79.0	

C. VALUE OF m, CHOSEN TO MAXIMIZE EFFICIENCY E_u OF UPPER CONFIDENCE BOUND, ALSO MAXIMIZES EFFICIENCY E_i OF CENTRAL CONFIDENCE INTERVAL, UNLESS MARKED *

n	m	1-P	B_{um}/x_m	$B_{\ell m}/x_m$	1-2P	E_u (%)	E_i (%)	R_u (%)	R_i (%)
30	26	0.9999	1.331245	0.240983	0.9998	57.2	58.6	73.1	73.5
30	26	0.9995	1.187769	0.263146	0.9990	57.8	59.3	72.8	73.4
30	26	0.9990	1.127161	0.274279	0.9980	58.0	59.7*	72.7	73.4
30	25	0.9950	1.103420	0.343651	0.9900	58.6	60.8	76.9	77.7
30	25	0.9900	1.035766	0.361724	0.9800	59.0	61.3	76.6	77.6
30	25	0.9750	0.945340	0.390291	0.9500	59.4	62.0	76.3	77.6
30	25	0.9500	0.875155	0.416965	0.9000	59.7	62.5	76.0	77.5
30	25	0.9000	0.801920	0.450391	0.8000	60.0	63.0	75.6	77.5
30	24	0.8000	0.800851	0.548396	0.6000	60.5	63.6	79.2	81.1
30	24	0.7000	0.743887	0.587613	0.4000	61.4	64.1	79.3	81.1
30	24	0.6000	0.699093	0.623834	0.2000	62.7	64.4	79.9	81.2
30	24	0.5000	0.660162			64.5		81.2	
32	28	0.9999	1.247588	0.237062	0.9998	57.3	58.6*	71.7	72.2
32	27	0.9995	1.244357	0.291357	0.9990	57.9	59.6	75.9	76.5
32	27	0.9990	1.182691	0.303082	0.9980	58.2	60.0	75.8	76.5
32	27	0.9950	1.040594	0.335733	0.9900	58.8	61.0	75.4	76.3
32	27	0.9900	0.979187	0.352999	0.9800	59.1	61.4	75.2	76.3
32	27	0.9750	0.896772	0.380224	0.9500	59.4	62.0	74.9	76.2
32	26	0.9500	0.919078	0.448278	0.9000	59.7	62.5	78.4	79.8
32	26	0.9000	0.844328	0.482919	0.8000	60.1	63.1	78.1	79.8
32	26	0.8000	0.763488	0.529213	0.6000	60.7	63.8	77.8	79.8
32	26	0.7000	0.710993	0.565898	0.4000	61.5	64.2	77.9	79.8
32	26	0.6000	0.669571	0.599678	0.2000	62.7	64.4	78.6	79.9
32	26	0.5000	0.633464			64.5		79.9	
34	29	0.9999	1.304359	0.263372	0.9998	57.5	59.0	74.9	75.3
34	29	0.9995	1.172722	0.285958	0.9990	58.1	59.8	74.6	75.2
34	29	0.9990	1.116753	0.297258	0.9980	58.4	60.2	74.5	75.2
34	29	0.9950	0.987200	0.328650	0.9900	58.9	61.0*	74.1	75.1
34	28	0.9900	1.024403	0.381275	0.9800	59.1	61.6	77.7	78.7
34	28	0.9750	0.940470	0.409567	0.9500	59.6	62.2	77.4	78.6
34	28	0.9500	0.874911	0.435872	0.9000	59.9	62.7	77.1	78.6
34	28	0.9000	0.806087	0.468692	0.8000	60.2	63.2	76.7	78.6
34	28	0.8000	0.731322	0.512404	0.6000	60.7	63.8	76.4	78.6
34	27	0.7000	0.746381	0.597874	0.4000	61.5	64.1	79.9	81.7
34	27	0.6000	0.704060	0.632517	0.2000	62.7	64.4	80.5	81.7
34	27	0.5000	0.667123			64.5		81.7	
36	31	0.9999	1.231646	0.259203	0.9998	57.7	59.2	73.6	74.1
36	31	0.9995	1.111570	0.281069	0.9990	58.2	59.9*	73.4	74.0
36	31	0.9990	1.060322	0.291989	0.9980	58.4	60.2*	73.2	74.0
36	30	0.9950	1.031931	0.355450	0.9900	59.1	61.3	76.7	77.5
36	30	0.9900	0.974643	0.372669	0.9800	59.3	61.7	76.5	77.5
36	30	0.9750	0.897353	0.399724	0.9500	59.7	62.3	76.2	77.4
36	30	0.9500	0.836759	0.424820	0.9000	59.9	62.7	75.8	77.4
36	29	0.9000	0.842850	0.497436	0.8000	60.2	63.2	78.8	80.5
36	29	0.8000	0.766625	0.542415	0.6000	60.8	63.8	78.5	80.5
36	29	0.7000	0.716834	0.577885	0.4000	61.6	64.2	78.7	80.5
36	29	0.6000	0.677372	0.610420	0.2000	62.8	64.5	79.3	80.5
36	29	0.5000	0.642844			64.5		80.5	

C. VALUE OF m, CHOSEN TO MAXIMIZE EFFICIENCY E_u OF UPPER CONFIDENCE BOUND, ALSO MAXIMIZES EFFICIENCY E_i OF CENTRAL CONFIDENCE INTERVAL, UNLESS MARKED *

n	m	1-P	B_{um}/x_m	$B_{\ell m}/x_m$	1-2P	E_u (%)	E_i(%)	R_u(%)	R_i (%)
38	33	0.9999	1.169119	0.255393	0.9998	57.7	59.2*	72.4	73.0
38	32	0.9995	1.157737	0.305457	0.9990	58.4	60.2	76.0	76.5
38	32	0.9990	1.105670	0.316842	0.9980	58.6	60.6	75.9	76.5
38	32	0.9950	0.984449	0.348350	0.9900	59.2	61.4	75.5	76.4
38	32	0.9900	0.931487	0.364904	0.9800	59.4	61.8*	75.3	76.4
38	31	0.9750	0.935225	0.425858	0.9500	59.7	62.4	78.3	79.4
38	31	0.9500	0.873587	0.451754	0.9000	60.0	62.9	78.0	79.4
38	31	0.9000	0.808551	0.483943	0.8000	60.4	63.3	77.6	79.4
38	31	0.8000	0.737490	0.526622	0.6000	60.9	63.9	77.4	79.4
38	31	0.7000	0.690910	0.560183	0.4000	61.6	64.2	77.5	79.4
38	31	0.6000	0.653898	0.590896	0.2000	62.8	64.4	78.2	79.4
38	31	0.5000	0.621440			64.5		79.4	
40	34	0.9999	1.215558	0.278301	0.9998	58.0	59.6	75.1	75.6
40	34	0.9995	1.103409	0.300435	0.9990	58.5	60.3	74.9	75.5
40	34	0.9990	1.055310	0.311453	0.9980	58.7	60.6	74.7	75.4
40	33	0.9950	1.023220	0.372365	0.9900	59.2	61.5	77.6	78.5
40	33	0.9900	0.969437	0.389479	0.9800	59.5	62.0	77.4	78.4
40	33	0.9750	0.896554	0.416292	0.9500	59.9	62.5	77.1	78.4
40	33	0.9500	0.839140	0.441081	0.9000	60.2	62.9	76.9	78.4
40	33	0.9000	0.778377	0.471827	0.8000	60.4	63.4	76.5	78.3
40	32	0.8000	0.768717	0.553447	0.6000	60.9	63.9	79.1	81.0
40	32	0.7000	0.721278	0.587802	0.4000	61.7	64.2	79.3	81.0
40	32	0.6000	0.683540	0.619208	0.2000	62.9	64.5	79.9	81.1
40	32	0.5000	0.650412			64.5		81.1	

D. VALUE OF m, CHOSEN TO MAXIMIZE EFFICIENCY E_i OF CENTRAL CONFIDENCE INTERVAL, DOES NOT MAXIMIZE EFFICIENCY E_u OF UPPER CONFIDENCE BOUND--SEE TABLE D4C

n	m	1-P	B_{um}/x_m	$B_{\ell m}/x_m$	1-2P	E_u (%)	E_i (%)	R_u (%)	R_i (%)
13	11	0.9950	1.627988	0.279206	0.9900	56.6	58.1	77.1	77.7
15	13	0.9999	2.187085	0.196396	0.9998	54.8	55.6	74.9	75.2
17	14	0.9750	1.199072	0.372209	0.9500	58.0	60.4	78.3	79.3
20	17	0.9995	1.582632	0.252492	0.9990	56.4	57.8	76.0	76.4
22	18	0.9750	1.099866	0.392402	0.9500	58.5	61.1	78.4	79.5
24	20	0.7000	0.692030	0.532142	0.4000	61.0	63.9	75.5	77.7
28	23	0.9900	1.103136	0.371585	0.9800	58.6	61.0	78.1	79.1
30	25	0.9990	1.260820	0.309568	0.9980	57.8	59.7	77.2	77.8
32	27	0.9999	1.390009	0.267963	0.9998	57.2	58.7	76.2	76.6
34	28	0.9950	1.086829	0.363308	0.9900	58.8	61.1	77.9	78.7
36	30	0.9995	1.220349	0.310958	0.9990	58.1	60.0	77.1	77.7
36	30	0.9990	1.163585	0.322749	0.9980	58.4	60.4	77.0	77.6
38	32	0.9999	1.279531	0.282624	0.9998	57.7	59.4	76.2	76.6
38	31	0.9900	1.013717	0.397906	0.9800	59.3	61.8	78.5	79.5

Appendix E

TABLES OF CONDITIONAL MAXIMUM-LIKELIHOOD ESTIMATORS FROM SINGLY CENSORED SAMPLES

SOURCES OF TABLES

Table E1	*Technometrics* 7(1965), 405-422 (Harter and Moore)
Table E2	*Technometrics* 9(1967), 325-331 (Harter and Moore)
Table E3	*Technometrics* 10(1968), 349-259 (Harter and Moore) [with additional values not previously published (Harter and Moore)]

Table E1

WEIBULL POPULATION - UNBIASING FACTORS AND VARIANCES OF UNBIASED ESTIMATORS

[$\hat{\theta}$ = maximum-likelihood estimator and $\tilde{\theta}$ = unbiased estimator of scale parameter θ; shape parameter $K = 0.5(0.5)4.0(1.0)8.0$.]

UNBIASING FACTORS FOR MAXIMUM LIKELIHOOD ESTIMATORS AND VARIANCES OF UNBIASED ESTIMATORS FROM m ORDER STATISTICS OF SCALE PARAMETER θ OF WEIBULL POPULATION

SHAPE PARAMETER K=0.5

m	$\tilde{\theta}/\hat{\theta}$	Var $\tilde{\theta}/\theta^2$	m	$\tilde{\theta}/\hat{\theta}$	Var $\tilde{\theta}/\theta^2$
1	.500000	5.00000000	51	.980769	.07918552
2	.666667	2.33333333	52	.981132	.07764877
3	.750000	1.50000000	53	.981481	.07617051
4	.800000	1.10000000	54	.981818	.07474747
5	.833333	.86666667	55	.982143	.07337662
6	.857143	.71428571	56	.982456	.07205514
7	.875000	.60714286	57	.982758	.07078040
8	.888889	.52777778	58	.983051	.06954997
9	.900000	.46666667	59	.983333	.06836158
10	.909091	.41818182	60	.983606	.06721311
11	.916667	.37878788	61	.983871	.06610259
12	.923077	.34615385	62	.984127	.06502816
13	.928571	.31868132	63	.984375	.06398810
14	.933333	.29523810	64	.984615	.06298077
15	.937500	.27500000	65	.984848	.06200466
16	.941176	.25735294	66	.985075	.06105834
17	.944444	.24183007	67	.985294	.06014047
18	.947368	.22807018	68	.985507	.05924979
19	.950000	.21578947	69	.985714	.05838509
20	.952381	.20476190	70	.985915	.05754527
21	.954545	.19480519	71	.986111	.05672926
22	.956522	.18577075	72	.986302	.05593607
23	.958333	.17753623	73	.986486	.05516475
24	.960000	.17000000	74	.986667	.05441441
25	.961538	.16307692	75	.986842	.05368421
26	.962963	.15669516	76	.987013	.05297334
27	.964286	.15079365	77	.987180	.05228105
28	.965517	.14532020	78	.987342	.05160662
29	.966667	.14022989	79	.987500	.05094937
30	.967742	.13548387	80	.987654	.05030864
31	.968750	.13104839	81	.987805	.04968383
32	.969697	.12689394	82	.987952	.04907435
33	.970588	.12299465	83	.988095	.04847963
34	.971429	.11932773	84	.988235	.04789916
35	.972222	.11587302	85	.988372	.04733242
36	.972973	.11261261	86	.988506	.04677894
37	.973684	.10953058	87	.988636	.04623824
38	.974359	.10661269	88	.988764	.04570991
39	.975000	.10384615	89	.988889	.04519351
40	.975610	.10121951	90	.989011	.04468864
41	.976190	.09872242	91	.989131	.04419494
42	.976744	.09634551	92	.989247	.04371201
43	.977273	.09408034	93	.989362	.04323953
44	.977778	.09191919	94	.989474	.04277716
45	.978261	.08985507	95	.989583	.04232456
46	.978723	.08788159	96	.989691	.04188144
47	.979167	.08599291	97	.989796	.04144751
48	.979592	.08418367	98	.989899	.04102247
49	.980000	.08244898	99	.990000	.04060606
50	.980392	.08078431	100	.990099	.04019802

UNBIASING FACTORS FOR MAXIMUM LIKELIHOOD ESTIMATORS AND VARIANCES OF UNBIASED ESTIMATORS FROM m ORDER STATISTICS OF SCALE PARAMETER θ OF WEIBULL POPULATION

SHAPE PARAMETER K=1.0

m	$\tilde{\theta}/\hat{\theta}$	Var $\tilde{\theta}/\theta^2$	m	$\tilde{\theta}/\hat{\theta}$	Var $\tilde{\theta}/\theta^2$
1	1.000000	1.00000000	51	1.000000	.01960784
2	1.000000	.50000000	52	1.000000	.01923077
3	1.000000	.33333333	53	1.000000	.01886792
4	1.000000	.25000000	54	1.000000	.01851852
5	1.000000	.20000000	55	1.000000	.01818182
6	1.000000	.16666667	56	1.000000	.01785714
7	1.000000	.14285714	57	1.000000	.01754386
8	1.000000	.12500000	58	1.000000	.01724138
9	1.000000	.11111111	59	1.000000	.01694915
10	1.000000	.10000000	60	1.000000	.01666667
11	1.000000	.09090909	61	1.000000	.01639344
12	1.000000	.08333333	62	1.000000	.01612903
13	1.000000	.07692308	63	1.000000	.01587302
14	1.000000	.07142857	64	1.000000	.01562500
15	1.000000	.06666667	65	1.000000	.01538462
16	1.000000	.06250000	66	1.000000	.01515152
17	1.000000	.05882353	67	1.000000	.01492537
18	1.000000	.05555556	68	1.000000	.01470588
19	1.000000	.05263158	69	1.000000	.01449275
20	1.000000	.05000000	70	1.000000	.01428571
21	1.000000	.04761905	71	1.000000	.01408451
22	1.000000	.04545455	72	1.000000	.01388889
23	1.000000	.04347826	73	1.000000	.01369863
24	1.000000	.04166667	74	1.000000	.01351351
25	1.000000	.04000000	75	1.000000	.01333333
26	1.000000	.03846154	76	1.000000	.01315789
27	1.000000	.03703704	77	1.000000	.01298701
28	1.000000	.03571429	78	1.000000	.01282051
29	1.000000	.03448276	79	1.000000	.01265823
30	1.000000	.03333333	80	1.000000	.01250000
31	1.000000	.03225806	81	1.000000	.01234568
32	1.000000	.03125000	82	1.000000	.01219512
33	1.000000	.03030303	83	1.000000	.01204819
34	1.000000	.02941176	84	1.000000	.01190476
35	1.000000	.02857143	85	1.000000	.01176471
36	1.000000	.02777778	86	1.000000	.01162791
37	1.000000	.02702703	87	1.000000	.01149425
38	1.000000	.02631579	88	1.000000	.01136364
39	1.000000	.02564103	89	1.000000	.01123596
40	1.000000	.02500000	90	1.000000	.01111111
41	1.000000	.02439024	91	1.000000	.01098901
42	1.000000	.02380952	92	1.000000	.01086957
43	1.000000	.02325581	93	1.000000	.01075269
44	1.000000	.02272727	94	1.000000	.01063830
45	1.000000	.02222222	95	1.000000	.01052632
46	1.000000	.02173913	96	1.000000	.01041667
47	1.000000	.02127660	97	1.000000	.01030928
48	1.000000	.02083333	98	1.000000	.01020408
49	1.000000	.02040816	99	1.000000	.01010101
50	1.000000	.02000000	100	1.000000	.01000000

UNBIASING FACTORS FOR MAXIMUM LIKELIHOOD ESTIMATORS AND VARIANCES OF UNBIASED ESTIMATORS FROM m ORDER STATISTICS OF SCALE PARAMETER θ OF WEIBULL POPULATION

SHAPE PARAMETER K=1.5

m	$\tilde{\theta}/\hat{\theta}$	Var $\tilde{\theta}/\theta^2$	m	$\tilde{\theta}/\hat{\theta}$	Var $\tilde{\theta}/\theta^2$
1	1.107732	.46099849	51	1.002179	.00872401
2	1.055049	.22723873	52	1.002137	.00855607
3	1.036879	.15053631	53	1.002096	.00839447
4	1.027710	.11250205	54	1.002058	.00823886
5	1.022187	.08979793	55	1.002020	.00808891
6	1.018498	.07471422	56	1.001984	.00794432
7	1.015860	.06396708	57	1.001949	.00780482
8	1.013880	.05592196	58	1.001916	.00767012
9	1.012340	.04967390	59	1.001883	.00754000
10	1.011107	.04468139	60	1.001852	.00741422
11	1.010098	.04060061	61	1.001821	.00729257
12	1.009257	.03720273	62	1.001792	.00717484
13	1.008545	.03432959	63	1.001764	.00706085
14	1.007935	.03186837	64	1.001736	.00695043
15	1.007406	.02973641	65	1.001709	.00684341
16	1.006943	.02787179	66	1.001683	.00673964
17	1.006535	.02622719	67	1.001658	.00663897
18	1.006172	.02476585	68	1.001634	.00654126
19	1.005847	.02345875	69	1.001610	.00644638
20	1.005555	.02228270	70	1.001587	.00635422
21	1.005291	.02121893	71	1.001565	.00626465
22	1.005050	.02025209	72	1.001543	.00617758
23	1.004831	.01936952	73	1.001522	.00609289
24	1.004629	.01856066	74	1.001501	.00601049
25	1.004444	.01781664	75	1.001481	.00593029
26	1.004273	.01712997	76	1.001462	.00585221
27	1.004115	.01649427	77	1.001443	.00577615
28	1.003968	.01590405	78	1.001424	.00570204
29	1.003831	.01535462	79	1.001406	.00562981
30	1.003704	.01484188	80	1.001389	.00555939
31	1.003584	.01436227	81	1.001372	.00549071
32	1.003472	.01391270	82	1.001355	.00542371
33	1.003367	.01349041	83	1.001339	.00535832
34	1.003268	.01309300	84	1.001323	.00529449
35	1.003174	.01271833	85	1.001307	.00523216
36	1.003086	.01236452	86	1.001292	.00517128
37	1.003003	.01202985	87	1.001277	.00511180
38	1.002924	.01171282	88	1.001263	.00505368
39	1.002849	.01141208	89	1.001248	.00499686
40	1.002778	.01112639	90	1.001235	.00494131
41	1.002710	.01085465	91	1.001221	.00488697
42	1.002645	.01059587	92	1.001208	.00483382
43	1.002584	.01034914	93	1.001195	.00478181
44	1.002525	.01011365	94	1.001182	.00473091
45	1.002469	.00988863	95	1.001170	.00468109
46	1.002415	.00967340	96	1.001157	.00463230
47	1.002364	.00946735	97	1.001145	.00458451
48	1.002315	.00926989	98	1.001134	.00453771
49	1.002268	.00908049	99	1.001122	.00449185
50	1.002222	.00889869	100	1.001111	.00444690

UNBIASING FACTORS FOR MAXIMUM LIKELIHOOD ESTIMATORS AND VARIANCES OF UNBIASED ESTIMATORS FROM m ORDER STATISTICS OF SCALE PARAMETER θ OF WEIBULL POPULATION

SHAPE PARAMETER K=2.0

m	$\tilde{\theta}/\hat{\theta}$	Var $\tilde{\theta}/\theta^2$
1	1.128379	.27323954
2	1.063846	.13176848
3	1.042352	.08649774
4	1.031661	.06432432
5	1.025273	.05118452
6	1.021027	.04249704
7	1.018002	.03632842
8	1.015737	.03172251
9	1.013979	.02815254
10	1.012573	.02530447
11	1.011424	.02297952
12	1.010468	.02104572
13	1.009659	.01941205
14	1.008967	.01801368
15	1.008367	.01680320
16	1.007842	.01574513
17	1.007379	.01481240
18	1.006968	.01398398
19	1.006600	.01324330
20	1.006269	.01257713
21	1.005970	.01197477
22	1.005697	.01142746
23	1.005449	.01092799
24	1.005222	.01047035
25	1.005012	.01004949
26	1.004819	.00966116
27	1.004640	.00930172
28	1.004474	.00896807
29	1.004319	.00865752
30	1.004175	.00836776
31	1.004040	.00809677
32	1.003914	.00784278
33	1.003795	.00760423
34	1.003683	.00737977
35	1.003578	.00716818
36	1.003478	.00696839
37	1.003384	.00677943
38	1.003295	.00660045
39	1.003210	.00643067
40	1.003130	.00626941
41	1.003053	.00611604
42	1.002981	.00596999
43	1.002911	.00583076
44	1.002845	.00569787
45	1.002782	.00557090
46	1.002721	.00544947
47	1.002663	.00533322
48	1.002608	.00522183
49	1.002554	.00511499
50	1.002503	.00501244
51	1.002454	.00491392
52	1.002407	.00481919
53	1.002361	.00472805
54	1.002317	.00464030
55	1.002275	.00455574
56	1.002235	.00447421
57	1.002195	.00439554
58	1.002157	.00431959
59	1.002121	.00424623
60	1.002085	.00417531
61	1.002051	.00410672
62	1.002018	.00404035
63	1.001986	.00397610
64	1.001955	.00391385
65	1.001925	.00385352
66	1.001896	.00379503
67	1.001867	.00373828
68	1.001840	.00368320
69	1.001813	.00362973
70	1.001787	.00357778
71	1.001762	.00352730
72	1.001738	.00347823
73	1.001714	.00343050
74	1.001691	.00338407
75	1.001668	.00333887
76	1.001646	.00329487
77	1.001625	.00325201
78	1.001604	.00321025
79	1.001584	.00316955
80	1.001564	.00312987
81	1.001544	.00309117
82	1.001526	.00305341
83	1.001507	.00301657
84	1.001489	.00298061
85	1.001472	.00294549
86	1.001455	.00291119
87	1.001438	.00287768
88	1.001421	.00284493
89	1.001405	.00281292
90	1.001390	.00278163
91	1.001375	.00275102
92	1.001360	.00272107
93	1.001345	.00269178
94	1.001331	.00266310
95	1.001317	.00263503
96	1.001303	.00260755
97	1.001289	.00258063
98	1.001276	.00255427
99	1.001263	.00252843
100	1.001251	.00250312

UNBIASING FACTORS FOR MAXIMUM LIKELIHOOD ESTIMATORS AND VARIANCES OF UNBIASED ESTIMATORS FROM m ORDER STATISTICS OF SCALE PARAMETER θ OF WEIBULL POPULATION

SHAPE PARAMETER K=2.5

m	$\tilde{\theta}/\hat{\theta}$	Var $\tilde{\theta}/\theta^2$
1	1.127060	.18310455
2	1.062261	.08652458
3	1.041086	.05634335
4	1.030634	.04172268
5	1.024414	.03311340
6	1.020292	.02744474
7	1.017359	.02343128
8	1.015167	.02044098
9	1.013466	.01812705
10	1.012108	.01628345
11	1.010999	.01478008
12	1.010075	.01353073
13	1.009295	.01247607
14	1.008627	.01157389
15	1.008049	.01079335
16	1.007543	.01011142
17	1.007097	.00951052
18	1.006701	.00897702
19	1.006346	.00850019
20	1.006027	.00807145
21	1.005739	.00768388
22	1.005477	.00733182
23	1.005238	.00701061
24	1.005019	.00671635
25	1.004818	.00644580
26	1.004632	.00619620
27	1.004460	.00596521
28	1.004300	.00575082
29	1.004151	.00555131
30	1.004012	.00536517
31	1.003882	.00519111
32	1.003761	.00502799
33	1.003647	.00487481
34	1.003539	.00473069
35	1.003438	.00459484
36	1.003342	.00446657
37	1.003251	.00434528
38	1.003166	.00423039
39	1.003084	.00412143
40	1.003007	.00401793
41	1.002933	.00391951
42	1.002863	.00382579
43	1.002797	.00373645
44	1.002733	.00365119
45	1.002672	.00356973
46	1.002614	.00349183
47	1.002558	.00341725
48	1.002505	.00334579
49	1.002454	.00327726
50	1.002404	.00321149
51	1.002357	.00314830
52	1.002312	.00308754
53	1.002268	.00302909
54	1.002226	.00297281
55	1.002185	.00291859
56	1.002146	.00286630
57	1.002109	.00281586
58	1.002072	.00276716
59	1.002037	.00272012
60	1.002003	.00267465
61	1.001970	.00263067
62	1.001938	.00258812
63	1.001908	.00254692
64	1.001878	.00250701
65	1.001849	.00246834
66	1.001821	.00243084
67	1.001794	.00239446
68	1.001767	.00235916
69	1.001741	.00232488
70	1.001717	.00229158
71	1.001692	.00225922
72	1.001669	.00222777
73	1.001646	.00219717
74	1.001624	.00216741
75	1.001602	.00213844
76	1.001581	.00211024
77	1.001560	.00208277
78	1.001540	.00205601
79	1.001521	.00202992
80	1.001502	.00200449
81	1.001483	.00197969
82	1.001465	.00195549
83	1.001447	.00193188
84	1.001430	.00190884
85	1.001413	.00188633
86	1.001397	.00186435
87	1.001381	.00184288
88	1.001365	.00182189
89	1.001350	.00180138
90	1.001335	.00178133
91	1.001320	.00176171
92	1.001306	.00174253
93	1.001292	.00172375
94	1.001278	.00170538
95	1.001264	.00168740
96	1.001251	.00166979
97	1.001238	.00165254
98	1.001226	.00163565
99	1.001213	.00161910
100	1.001201	.00160288

UNBIASING FACTORS FOR MAXIMUM LIKELIHOOD ESTIMATORS AND VARIANCES OF UNBIASED ESTIMATORS FROM m ORDER STATISTICS OF SCALE PARAMETER θ OF WEIBULL POPULATION

SHAPE PARAMETER K=3.0

m	$\tilde{\theta}/\hat{\theta}$	Var $\tilde{\theta}/\theta^2$	m	$\tilde{\theta}/\hat{\theta}$	Var $\tilde{\theta}/\theta^2$
1	1.119847	.13209336	51	1.002183	.00218813
2	1.058189	.06133753	52	1.002141	.00214587
3	1.038277	.03967758	53	1.002101	.00210521
4	1.028495	.02928080	54	1.002062	.00206607
5	1.022689	.02319038	55	1.002024	.00202835
6	1.018846	.01919354	56	1.001988	.00199199
7	1.016115	.01637029	57	1.001953	.00195691
8	1.014075	.01427035	58	1.001919	.00192304
9	1.012494	.01264752	59	1.001887	.00189032
10	1.011231	.01135588	60	1.001855	.00185870
11	1.010201	.01030348	61	1.001825	.00182812
12	1.009343	.00942952	62	1.001795	.00179853
13	1.008619	.00869217	63	1.001767	.00176988
14	1.007998	.00806173	64	1.001739	.00174213
15	1.007461	.00751654	65	1.001712	.00171524
16	1.006992	.00704040	66	1.001686	.00168916
17	1.006578	.00662097	67	1.001661	.00166387
18	1.006210	.00624870	68	1.001637	.00163932
19	1.005882	.00591606	69	1.001613	.00161549
20	1.005586	.00561704	70	1.001590	.00159234
21	1.005319	.00534678	71	1.001567	.00156984
22	1.005076	.00510133	72	1.001546	.00154797
23	1.004854	.00487743	73	1.001524	.00152670
24	1.004651	.00467235	74	1.001504	.00150601
25	1.004464	.00448383	75	1.001484	.00148587
26	1.004292	.00430992	76	1.001464	.00146626
27	1.004132	.00414900	77	1.001445	.00144716
28	1.003984	.00399966	78	1.001427	.00142856
29	1.003846	.00386070	79	1.001408	.00141042
30	1.003717	.00373107	80	1.001391	.00139274
31	1.003597	.00360986	81	1.001374	.00137550
32	1.003484	.00349628	82	1.001357	.00135868
33	1.003378	.00338962	83	1.001340	.00134227
34	1.003279	.00328928	84	1.001324	.00132625
35	1.003185	.00319471	85	1.001309	.00131060
36	1.003096	.00310543	86	1.001294	.00129533
37	1.003012	.00302100	87	1.001279	.00128040
38	1.002932	.00294104	88	1.001264	.00126581
39	1.002857	.00286520	89	1.001250	.00125155
40	1.002785	.00279318	90	1.001236	.00123761
41	1.002717	.00272469	91	1.001222	.00122398
42	1.002652	.00265947	92	1.001209	.00121064
43	1.002591	.00259731	93	1.001196	.00119760
44	1.002532	.00253798	94	1.001183	.00118483
45	1.002475	.00248131	95	1.001171	.00117232
46	1.002421	.00242711	96	1.001159	.00116008
47	1.002370	.00237523	97	1.001147	.00114810
48	1.002320	.00232551	98	1.001135	.00113636
49	1.002273	.00227784	99	1.001124	.00112485
50	1.002227	.00223208	100	1.001112	.00111358

UNBIASING FACTORS FOR MAXIMUM LIKELIHOOD ESTIMATORS AND VARIANCES OF UNBIASED ESTIMATORS FROM m ORDER STATISTICS OF SCALE PARAMETER θ OF WEIBULL POPULATION

SHAPE PARAMETER K=3.5

m	$\tilde{\theta}/\hat{\theta}$	Var $\tilde{\theta}/\theta^2$	m	$\tilde{\theta}/\hat{\theta}$	Var $\tilde{\theta}/\theta^2$
1	1.111423	.10014607	51	1.002006	.00160864
2	1.053765	.04581787	52	1.001967	.00157756
3	1.035293	.02947697	53	1.001930	.00154765
4	1.026247	.02169264	54	1.001894	.00151885
5	1.020886	.01715179	55	1.001859	.00149111
6	1.017341	.01417983	56	1.001826	.00146436
7	1.014824	.01208442	57	1.001794	.00143856
8	1.012945	.01052796	58	1.001763	.00141365
9	1.011488	.00932639	59	1.001733	.00138958
10	1.010326	.00837081	60	1.001704	.00136633
11	1.009378	.00759275	61	1.001676	.00134383
12	1.008589	.00694696	62	1.001649	.00132207
13	1.007922	.00640237	63	1.001623	.00130100
14	1.007351	.00593692	64	1.001597	.00128059
15	1.006857	.00553455	65	1.001573	.00126081
16	1.006426	.00518324	66	1.001549	.00124164
17	1.006045	.00487386	67	1.001526	.00122304
18	1.005707	.00459932	68	1.001503	.00120498
19	1.005405	.00435406	69	1.001481	.00118745
20	1.005133	.00413362	70	1.001460	.00117043
21	1.004887	.00393443	71	1.001440	.00115389
22	1.004664	.00375355	72	1.001420	.00113780
23	1.004460	.00358857	73	1.001400	.00112216
24	1.004273	.00343748	74	1.001381	.00110695
25	1.004101	.00329859	75	1.001363	.00109214
26	1.003943	.00317049	76	1.001345	.00107772
27	1.003796	.00305197	77	1.001327	.00106368
28	1.003660	.00294199	78	1.001310	.00104999
29	1.003533	.00283966	79	1.001294	.00103666
30	1.003415	.00274421	80	1.001277	.00102366
31	1.003305	.00265496	81	1.001262	.00101098
32	1.003201	.00257134	82	1.001246	.00099862
33	1.003104	.00249283	83	1.001231	.00098655
34	1.003012	.00241896	84	1.001217	.00097477
35	1.002926	.00234935	85	1.001202	.00096327
36	1.002844	.00228363	86	1.001188	.00095203
37	1.002767	.00222149	87	1.001175	.00094106
38	1.002694	.00216264	88	1.001161	.00093033
39	1.002625	.00210683	89	1.001148	.00091985
40	1.002559	.00205382	90	1.001135	.00090960
41	1.002496	.00200342	91	1.001123	.00089958
42	1.002437	.00195543	92	1.001111	.00088977
43	1.002380	.00190969	93	1.001099	.00088018
44	1.002326	.00186604	94	1.001087	.00087079
45	1.002274	.00182434	95	1.001075	.00086160
46	1.002224	.00178446	96	1.001064	.00085260
47	1.002177	.00174629	97	1.001053	.00084379
48	1.002131	.00170971	98	1.001043	.00083515
49	1.002088	.00167464	99	1.001032	.00082670
50	1.002046	.00164098	100	1.001022	.00081841

UNBIASING FACTORS FOR MAXIMUM LIKELIHOOD ESTIMATORS AND VARIANCES OF UNBIASED ESTIMATORS FROM m ORDER STATISTICS OF SCALE PARAMETER θ OF WEIBULL POPULATION

SHAPE PARAMETER K=4.0

m	$\tilde{\theta}/\hat{\theta}$	Var $\tilde{\theta}/\theta^2$	m	$\tilde{\theta}/\hat{\theta}$	Var $\tilde{\theta}/\theta^2$
1	1.103263	.07870520	51	1.001843	.00123225
2	1.049606	.03555699	52	1.001807	.00120843
3	1.032516	.02277234	53	1.001773	.00118551
4	1.024163	.01672043	54	1.001740	.00116344
5	1.019220	.01320237	55	1.001709	.00114218
6	1.015954	.01090486	56	1.001678	.00112168
7	1.013635	.00928741	57	1.001648	.00110190
8	1.011905	.00808731	58	1.001620	.00108281
9	1.010564	.00716161	59	1.001592	.00106437
10	1.009495	.00642591	60	1.001566	.00104655
11	1.008622	.00582721	61	1.001540	.00102932
12	1.007896	.00533050	62	1.001515	.00101264
13	1.007283	.00491179	63	1.001491	.00099649
14	1.006758	.00455404	64	1.001468	.00098086
15	1.006304	.00424485	65	1.001445	.00096570
16	1.005907	.00397497	66	1.001423	.00095101
17	1.005557	.00373734	67	1.001402	.00093675
18	1.005246	.00352652	68	1.001381	.00092292
19	1.004968	.00333820	69	1.001361	.00090949
20	1.004718	.00316898	70	1.001342	.00089645
21	1.004492	.00301608	71	1.001323	.00088377
22	1.004286	.00287725	72	1.001304	.00087145
23	1.004099	.00275064	73	1.001287	.00085946
24	1.003927	.00263470	74	1.001269	.00084781
25	1.003769	.00252814	75	1.001252	.00083646
26	1.003624	.00242987	76	1.001236	.00082541
27	1.003489	.00233894	77	1.001220	.00081465
28	1.003364	.00225458	78	1.001204	.00080417
29	1.003247	.00217609	79	1.001189	.00079396
30	1.003138	.00210288	80	1.001174	.00078400
31	1.003037	.00203443	81	1.001159	.00077428
32	1.002942	.00197030	82	1.001145	.00076481
33	1.002852	.00191009	83	1.001131	.00075556
34	1.002768	.00185345	84	1.001118	.00074654
35	1.002688	.00180007	85	1.001105	.00073773
36	1.002614	.00174968	86	1.001092	.00072912
37	1.002543	.00170204	87	1.001079	.00072071
38	1.002476	.00165692	88	1.001067	.00071250
39	1.002412	.00161413	89	1.001055	.00070447
40	1.002351	.00157349	90	1.001043	.00069662
41	1.002294	.00153485	91	1.001032	.00068894
42	1.002239	.00149806	92	1.001020	.00068143
43	1.002187	.00146300	93	1.001009	.00067408
44	1.002137	.00142954	94	1.000999	.00066688
45	1.002089	.00139757	95	1.000988	.00065984
46	1.002044	.00136701	96	1.000978	.00065295
47	1.002000	.00133775	97	1.000968	.00064620
48	1.001958	.00130972	98	1.000958	.00063959
49	1.001918	.00128283	99	1.000948	.00063311
50	1.001880	.00125703	100	1.000939	.00062676

UNBIASING FACTORS FOR MAXIMUM LIKELIHOOD ESTIMATORS AND VARIANCES OF UNBIASED ESTIMATORS FROM m ORDER STATISTICS OF SCALE PARAMETER θ OF WEIBULL POPULATION

SHAPE PARAMETER K=5.0

m	$\tilde{\theta}/\hat{\theta}$	Var $\tilde{\theta}/\theta^2$	m	$\tilde{\theta}/\hat{\theta}$	Var $\tilde{\theta}/\theta^2$
1	1.089124	.05246525	51	1.001573	.00078924
2	1.042563	.02323010	52	1.001543	.00077397
3	1.027845	.01477365	53	1.001513	.00075928
4	1.020674	.01080969	54	1.001485	.00074514
5	1.016435	.00851761	55	1.001458	.00073151
6	1.013637	.00702572	56	1.001432	.00071837
7	1.011653	.00597782	57	1.001407	.00070570
8	1.010172	.00520161	58	1.001383	.00069347
9	1.009025	.00460363	59	1.001359	.00068165
10	1.008110	.00412886	60	1.001336	.00067023
11	1.007364	.00374281	61	1.001314	.00065918
12	1.006744	.00342274	62	1.001293	.00064850
13	1.006219	.00315307	63	1.001273	.00063815
14	1.005771	.00292278	64	1.001253	.00062813
15	1.005383	.00272383	65	1.001233	.00061842
16	1.005043	.00255023	66	1.001215	.00060900
17	1.004744	.00239742	67	1.001197	.00059987
18	1.004479	.00226189	68	1.001179	.00059101
19	1.004241	.00214086	69	1.001162	.00058240
20	1.004028	.00203212	70	1.001145	.00057404
21	1.003835	.00193389	71	1.001129	.00056592
22	1.003659	.00184472	72	1.001113	.00055803
23	1.003499	.00176341	73	1.001098	.00055035
24	1.003353	.00168896	74	1.001083	.00054288
25	1.003218	.00162054	75	1.001069	.00053561
26	1.003093	.00155745	76	1.001055	.00052853
27	1.002978	.00149909	77	1.001041	.00052164
28	1.002871	.00144494	78	1.001027	.00051493
29	1.002772	.00139457	79	1.001014	.00050838
30	1.002679	.00134759	80	1.001002	.00050200
31	1.002592	.00130367	81	1.000989	.00049578
32	1.002511	.00126253	82	1.000977	.00048971
33	1.002434	.00122390	83	1.000965	.00048379
34	1.002363	.00118757	84	1.000954	.00047801
35	1.002295	.00115333	85	1.000943	.00047236
36	1.002231	.00112101	86	1.000932	.00046685
37	1.002170	.00109045	87	1.000921	.00046146
38	1.002113	.00106151	88	1.000911	.00045620
39	1.002059	.00103407	89	1.000900	.00045106
40	1.002007	.00100802	90	1.000890	.00044603
41	1.001958	.00098324	91	1.000880	.00044111
42	1.001911	.00095965	92	1.000871	.00043630
43	1.001867	.00093717	93	1.000862	.00043159
44	1.001824	.00091571	94	1.000852	.00042698
45	1.001783	.00089522	95	1.000843	.00042247
46	1.001744	.00087562	96	1.000835	.00041806
47	1.001707	.00085687	97	1.000826	.00041373
48	1.001672	.00083890	98	1.000817	.00040950
49	1.001637	.00082167	99	1.000809	.00040535
50	1.001604	.00080513	100	1.000801	.00040128

UNBIASING FACTORS FOR MAXIMUM LIKELIHOOD ESTIMATORS AND VARIANCES OF UNBIASED ESTIMATORS FROM m ORDER STATISTICS OF SCALE PARAMETER θ OF WEIBULL POPULATION

SHAPE PARAMETER K=6.0

m	$\tilde{\theta}/\hat{\theta}$	Var $\tilde{\theta}/\theta^2$	m	$\tilde{\theta}/\hat{\theta}$	Var $\tilde{\theta}/\theta^2$
1	1.077912	.03754820	51	1.001366	.00054838
2	1.037070	.01637374	52	1.001339	.00053776
3	1.024223	.01035970	53	1.001314	.00052755
4	1.017974	.00756092	54	1.001289	.00051772
5	1.014284	.00594882	55	1.001266	.00050825
6	1.011850	.00490205	56	1.001243	.00049911
7	1.010124	.00416801	57	1.001221	.00049030
8	1.008837	.00362492	58	1.001200	.00048180
9	1.007840	.00320692	59	1.001180	.00047359
10	1.007045	.00287528	60	1.001160	.00046565
11	1.006396	.00260576	61	1.001141	.00045797
12	1.005857	.00238241	62	1.001123	.00045054
13	1.005401	.00219431	63	1.001105	.00044335
14	1.005012	.00203373	64	1.001088	.00043639
15	1.004674	.00189504	65	1.001071	.00042964
16	1.004380	.00177405	66	1.001055	.00042309
17	1.004120	.00166758	67	1.001039	.00041675
18	1.003889	.00157317	68	1.001023	.00041059
19	1.003683	.00148887	69	1.001009	.00040461
20	1.003497	.00141314	70	1.000994	.00039880
21	1.003330	.00134474	71	1.000980	.00039315
22	1.003177	.00128266	72	1.000966	.00038767
23	1.003038	.00122605	73	1.000953	.00038233
24	1.002911	.00117423	74	1.000940	.00037714
25	1.002794	.00112661	75	1.000928	.00037209
26	1.002686	.00108271	76	1.000915	.00036717
27	1.002586	.00104209	77	1.000904	.00036238
28	1.002493	.00100442	78	1.000892	.00035771
29	1.002407	.00096937	79	1.000881	.00035317
30	1.002326	.00093668	80	1.000870	.00034873
31	1.002251	.00090613	81	1.000859	.00034441
32	1.002180	.00087751	82	1.000848	.00034019
33	1.002114	.00085064	83	1.000838	.00033607
34	1.002051	.00082537	84	1.000828	.00033206
35	1.001992	.00080155	85	1.000818	.00032813
36	1.001937	.00077907	86	1.000809	.00032430
37	1.001884	.00075782	87	1.000800	.00032056
38	1.001834	.00073769	88	1.000790	.00031690
39	1.001787	.00071861	89	1.000782	.00031333
40	1.001742	.00070049	90	1.000773	.00030983
41	1.001700	.00068326	91	1.000764	.00030642
42	1.001659	.00066686	92	1.000756	.00030307
43	1.001620	.00065123	93	1.000748	.00029980
44	1.001584	.00063631	94	1.000740	.00029660
45	1.001548	.00062206	95	1.000732	.00029347
46	1.001514	.00060843	96	1.000724	.00029040
47	1.001482	.00059539	97	1.000717	.00028740
48	1.001451	.00058290	98	1.000710	.00028445
49	1.001421	.00057092	99	1.000702	.00028157
50	1.001393	.00055942	100	1.000695	.00027874

UNBIASING FACTORS FOR MAXIMUM LIKELIHOOD ESTIMATORS AND VARIANCES OF UNBIASED ESTIMATORS FROM **m** ORDER STATISTICS OF SCALE PARAMETER θ OF WEIBULL POPULATION

SHAPE PARAMETER K=7.0

m	$\tilde{\theta}/\hat{\theta}$	Var $\tilde{\theta}/\theta^2$	m	$\tilde{\theta}/\hat{\theta}$	Var $\tilde{\theta}/\theta^2$
1	1.069018	.02823145	51	1.001204	.00040305
2	1.032756	.01216533	52	1.001181	.00039525
3	1.021387	.00766682	53	1.001158	.00038774
4	1.015863	.00558486	54	1.001137	.00038051
5	1.012604	.00438916	55	1.001116	.00037354
6	1.010454	.00361417	56	1.001096	.00036683
7	1.008931	.00307138	57	1.001077	.00036035
8	1.007794	.00267015	58	1.001058	.00035410
9	1.006915	.00236154	59	1.001040	.00034806
10	1.006213	.00211683	60	1.001023	.00034222
11	1.005641	.00191803	61	1.001006	.00033658
12	1.005165	.00175335	62	1.000990	.00033112
13	1.004763	.00161470	63	1.000974	.00032583
14	1.004420	.00149636	64	1.000959	.00032071
15	1.004122	.00139418	65	1.000944	.00031575
16	1.003862	.00130506	66	1.000930	.00031094
17	1.003633	.00122664	67	1.000916	.00030627
18	1.003430	.00115711	68	1.000902	.00030174
19	1.003248	.00109504	69	1.000889	.00029735
20	1.003084	.00103929	70	1.000877	.00029308
21	1.002936	.00098893	71	1.000864	.00028893
22	1.002802	.00094324	72	1.000852	.00028490
23	1.002679	.00090157	73	1.000840	.00028097
24	1.002567	.00086343	74	1.000829	.00027716
25	1.002464	.00082839	75	1.000818	.00027344
26	1.002368	.00079608	76	1.000807	.00026983
27	1.002280	.00076620	77	1.000797	.00026631
28	1.002198	.00073848	78	1.000786	.00026288
29	1.002122	.00071269	79	1.000776	.00025953
30	1.002051	.00068864	80	1.000767	.00025628
31	1.001984	.00066617	81	1.000757	.00025310
32	1.001922	.00064511	82	1.000748	.00025000
33	1.001864	.00062534	83	1.000739	.00024697
34	1.001809	.00060675	84	1.000730	.00024402
35	1.001757	.00058924	85	1.000722	.00024114
36	1.001708	.00057270	86	1.000713	.00023832
37	1.001661	.00055707	87	1.000705	.00023557
38	1.001618	.00054227	88	1.000697	.00023288
39	1.001576	.00052823	89	1.000689	.00023025
40	1.001536	.00051491	90	1.000681	.00022768
41	1.001499	.00050224	91	1.000674	.00022517
42	1.001463	.00049017	92	1.000667	.00022272
43	1.001429	.00047868	93	1.000659	.00022031
44	1.001396	.00046771	94	1.000652	.00021796
45	1.001365	.00045723	95	1.000645	.00021565
46	1.001335	.00044721	96	1.000639	.00021340
47	1.001307	.00043762	97	1.000632	.00021119
48	1.001279	.00042843	98	1.000626	.00020903
49	1.001253	.00041962	99	1.000619	.00020691
50	1.001228	.00041117	100	1.000613	.00020483

UNBIASING FACTORS FOR MAXIMUM LIKELIHOOD ESTIMATORS AND VARIANCES OF UNBIASED ESTIMATORS FROM m ORDER STATISTICS OF SCALE PARAMETER θ OF WEIBULL POPULATION

SHAPE PARAMETER K=8.0

m	$\tilde{\theta}/\hat{\theta}$	Var $\tilde{\theta}/\theta^2$
1	1.061861	.02201333
2	1.029305	.00939588
3	1.019123	.00590316
4	1.014180	.00429371
5	1.011265	.00337150
6	1.009343	.00277461
7	1.007980	.00235696
8	1.006965	.00204845
9	1.006178	.00181128
10	1.005551	.00162328
11	1.005040	.00147062
12	1.004615	.00134419
13	1.004256	.00123776
14	1.003949	.00114695
15	1.003683	.00106854
16	1.003450	.00100017
17	1.003246	.00094002
18	1.003064	.00088669
19	1.002901	.00083908
20	1.002755	.00079633
21	1.002623	.00075772
22	1.002503	.00072268
23	1.002393	.00069074
24	1.002293	.00066150
25	1.002201	.00063463
26	1.002116	.00060987
27	1.002037	.00058696
28	1.001964	.00056571
29	1.001896	.00054595
30	1.001832	.00052752
31	1.001773	.00051029
32	1.001717	.00049415
33	1.001665	.00047901
34	1.001616	.00046476
35	1.001569	.00045133
36	1.001526	.00043866
37	1.001484	.00042669
38	1.001445	.00041534
39	1.001408	.00040459
40	1.001372	.00039438
41	1.001339	.00038467
42	1.001307	.00037543
43	1.001276	.00036662
44	1.001247	.00035822
45	1.001219	.00035019
46	1.001193	.00034251
47	1.001167	.00033516
48	1.001143	.00032813
49	1.001120	.00032138
50	1.001097	.00031490
51	1.001076	.00030868
52	1.001055	.00030270
53	1.001035	.00029695
54	1.001016	.00029141
55	1.000997	.00028607
56	1.000979	.00028093
57	1.000962	.00027597
58	1.000945	.00027118
59	1.000929	.00026655
60	1.000914	.00026208
61	1.000899	.00025776
62	1.000884	.00025358
63	1.000870	.00024953
64	1.000857	.00024560
65	1.000843	.00024180
66	1.000831	.00023812
67	1.000818	.00023454
68	1.000806	.00023108
69	1.000794	.00022771
70	1.000783	.00022444
71	1.000772	.00022126
72	1.000761	.00021817
73	1.000751	.00021517
74	1.000741	.00021224
75	1.000731	.00020940
76	1.000721	.00020663
77	1.000712	.00020393
78	1.000702	.00020131
79	1.000694	.00019875
80	1.000685	.00019625
81	1.000676	.00019381
82	1.000668	.00019144
83	1.000660	.00018912
84	1.000652	.00018686
85	1.000645	.00018465
86	1.000637	.00018250
87	1.000630	.00018039
88	1.000623	.00017833
89	1.000616	.00017632
90	1.000609	.00017435
91	1.000602	.00017243
92	1.000595	.00017054
93	1.000589	.00016870
94	1.000583	.00016690
95	1.000577	.00016514
96	1.000571	.00016341
97	1.000565	.00016172
98	1.000559	.00016006
99	1.000553	.00015844
100	1.000548	.00015685

Table E2

TYPE I EXTREME-VALUE POPULATION - BIASES, AND VARIANCES OF UNBIASED ESTIMATORS

[b = scale parameter (unknown); u = location parameter (unknown); $\hat{u}_{mn}$ = maximum-likelihood estimator of u from first m order statistics of a sample of size $n \geq m$.]

VALUES OF $E[(\hat{u}_{mn}|b)-u]/b$ AND $V(\hat{u}_{mn}|b)/b^2$

m	E/b	V/b	m	E/b	V/b
1	-0.577216	1.644934	51	-0.009836	0.019801
2	-0.270363	0.644934	52	-0.009646	0.019417
3	-0.175828	0.394934	53	-0.009464	0.019047
4	-0.130177	0.283823	54	-0.009288	0.018691
5	-0.103320	0.221323	55	-0.009118	0.018348
6	-0.085642	0.181323	56	-0.008955	0.018018
7	-0.073126	0.153545	57	-0.008798	0.017699
8	-0.063800	0.133137	58	-0.008645	0.017391
9	-0.056583	0.117512	59	-0.008499	0.017094
10	-0.050833	0.105166	60	-0.008356	0.016806
11	-0.046143	0.095166	61	-0.008219	0.016529
12	-0.042245	0.086902	62	-0.008086	0.016260
13	-0.038954	0.079957	63	-0.007958	0.016000
14	-0.036139	0.074040	64	-0.007833	0.015748
15	-0.033704	0.068938	65	-0.007712	0.015504
16	-0.031575	0.064494	66	-0.007595	0.015267
17	-0.029700	0.060588	67	-0.007481	0.015037
18	-0.028035	0.057127	68	-0.007371	0.014815
19	-0.026547	0.054041	69	-0.007264	0.014598
20	-0.025208	0.051271	70	-0.007160	0.014388
21	-0.023998	0.048771	71	-0.007059	0.014184
22	-0.022899	0.046503	72	-0.006961	0.013986
23	-0.021897	0.044437	73	-0.006865	0.013793
24	-0.020978	0.042547	74	-0.006772	0.013605
25	-0.020133	0.040811	75	-0.006681	0.013423
26	-0.019354	0.039211	76	-0.006593	0.013245
27	-0.018633	0.037731	77	-0.006508	0.013072
28	-0.017963	0.036360	78	-0.006424	0.012903
29	-0.017340	0.035084	79	-0.006342	0.012739
30	-0.016759	0.033895	80	-0.006263	0.012578
31	-0.016216	0.032784	81	-0.006186	0.012422
32	-0.015706	0.031743	82	-0.006110	0.012270
33	-0.015228	0.030767	83	-0.006036	0.012121
34	-0.014778	0.029849	84	-0.005964	0.011976
35	-0.014354	0.028983	85	-0.005894	0.011834
36	-0.013953	0.028167	86	-0.005825	0.011696
37	-0.013574	0.027396	87	-0.005758	0.011561
38	-0.013216	0.026665	88	-0.005693	0.011428
39	-0.012875	0.025973	89	-0.005628	0.011299
40	-0.012552	0.025315	90	-0.005566	0.011173
41	-0.012245	0.024690	91	-0.005505	0.011050
42	-0.011952	0.024095	92	-0.005445	0.010929
43	-0.011673	0.023528	93	-0.005386	0.010811
44	-0.011407	0.0229[illegible]7	94	-0.005329	0.010695
45	-0.011152	0.022471	95	-0.005272	0.010582
46	-0.010909	0.021977	96	-0.005217	0.010471
47	-0.010676	0.021505	97	-0.005163	0.010363
48	-0.010453	0.021052	98	-0.005111	0.010256
49	-0.010239	0.020618	99	-0.005059	0.010152
50	-0.010033	0.020201	100	-0.005008	0.010050

Table E3

TYPE II EXTREME-VALUE POPULATION - UNBIASING FACTOR, VARIANCE, AND EFFICIENCY

[Scale parameter $v = v_n$ for distribution of largest values; scale parameter $v = v_1$ for distribution of smallest values; $\hat{v} =$ maximum-likelihood estimator and $\tilde{v} =$ unbiased estimator of v; sample size $n \geq m$; shape parameter $K = 0.5(0.5)4.0(1.0)8.0$.]

UNBIASING FACTORS FOR ML ESTIMATORS AND VARIANCES AND EFFICIENCIES OF UNBIASED ESTIMATORS FROM m ORDER STATISTICS OF SCALE PARAMETER v OF TYPE 2 E-V POPULATION

SHAPE PARAMETER K=0.5

m	$\tilde{v}/\hat{v}$	$var(\tilde{v})/v^2$	Eff $(\tilde{v})$	m	$\tilde{v}/\hat{v}$	$var(\tilde{v})/v^2$	Eff $(\hat{v})$
1				51	0.941945	0.08599290	91.21
2				52	0.943047	0.08418367	91.38
3	0.222222			53	0.944108	0.08244897	91.54
4	0.375000			54	0.945130	0.08078431	91.69
5	0.480000	5.00000000	16.00	55	0.946116	0.07918552	91.84
6	0.555556	2.33333333	28.57	56	0.947066	0.07764876	91.99
7	0.612245	1.50000000	38.10	57	0.947984	0.07617050	92.13
8	0.656250	1.10000000	45.45	58	0.948870	0.07474747	92.26
9	0.691358	0.86666667	51.28	59	0.949727	0.07337662	92.40
10	0.720000	0.71428571	56.00	60	0.950556	0.07205513	92.52
11	0.743802	0.60714286	59.89	61	0.951357	0.07078040	92.64
12	0.763889	0.52777778	63.16	62	0.952133	0.06954997	92.76
13	0.781065	0.46666667	65.93	63	0.952885	0.06836158	92.88
14	0.795918	0.41818182	68.32	64	0.953613	0.06721311	92.99
15	0.808889	0.37878788	70.40	65	0.954320	0.06610259	93.10
16	0.820313	0.34615384	72.22	66	0.955005	0.06502816	93.20
17	0.830450	0.31868131	73.83	67	0.955669	0.06398809	93.30
18	0.839506	0.29523809	75.27	68	0.956315	0.06298076	93.40
19	0.847645	0.27500000	76.56	69	0.956942	0.06200466	93.49
20	0.855000	0.25735294	77.71	70	0.957551	0.06105834	93.59
21	0.861678	0.24183006	78.76	71	0.958143	0.06014047	93.68
22	0.867769	0.22807017	79.72	72	0.958719	0.05924978	93.76
23	0.873346	0.21578947	80.59	73	0.959279	0.05838509	93.85
24	0.878472	0.20476190	81.40	74	0.959825	0.05754527	93.93
25	0.883200	0.19480519	82.13	75	0.960356	0.05672926	94.01
26	0.887574	0.18577075	82.82	76	0.960873	0.05593607	94.09
27	0.891632	0.17753623	83.45	77	0.961376	0.05516475	94.17
28	0.895408	0.17000000	84.03	78	0.961867	0.05441441	94.24
29	0.898930	0.16307692	84.58	79	0.962346	0.05368420	94.32
30	0.902222	0.15669515	85.09	80	0.962812	0.05297334	94.39
31	0.905307	0.15079365	85.57	81	0.963268	0.05228105	94.46
32	0.908203	0.14532019	86.02	82	0.963712	0.05160562	94.52
33	0.910927	0.14022988	86.44	83	0.964146	0.05094936	94.59
34	0.913495	0.13548387	86.83	84	0.964569	0.05030864	94.65
35	0.915918	0.13104838	87.21	85	0.964983	0.04968382	94.72
36	0.918210	0.12689394	87.56	86	0.965387	0.04907434	94.78
37	0.920380	0.12299465	87.90	87	0.965781	0.04847963	94.84
38	0.922438	0.11932772	88.21	88	0.966167	0.04789916	94.90
39	0.924392	0.11587301	88.51	89	0.966545	0.04733242	94.95
40	0.926250	0.11261261	88.80	90	0.966914	0.04677893	95.01
41	0.928019	0.10953058	89.07	91	0.967274	0.04623824	95.06
42	0.929705	0.10661268	89.33	92	0.967628	0.04570991	95.12
43	0.931314	0.10384615	89.58	93	0.967973	0.04519350	95.17
44	0.932851	0.10121951	89.81	94	0.968311	0.04468864	95.22
45	0.934321	0.09872241	90.04	95	0.968643	0.04419493	95.27
46	0.935728	0.09634551	90.25	96	0.968967	0.04371201	95.32
47	0.937076	0.09408034	90.46	97	0.969285	0.04323953	95.37
48	0.938368	0.09191919	90.66	98	0.969596	0.04277715	95.42
49	0.939608	0.08985507	90.85	99	0.969901	0.04232456	95.46
50	0.940800	0.08788159	91.03	100	0.970200	0.04188144	95.51

UNBIASING FACTORS FOR ML ESTIMATORS AND VARIANCES AND EFFICIENCIES OF UNBIASED ESTIMATORS FROM **m** ORDER STATISTICS OF SCALE PARAMETER **v** OF TYPE 2 E-V POPULATION

SHAPE PARAMETER K=1.0

m	$\tilde{v}/\hat{v}$	var$(\tilde{v})/v^2$	Eff$(\tilde{v})$	m	$\tilde{v}/\hat{v}$	var$(\tilde{v})/v^2$	Eff$(\tilde{v})$
1				51	0.980392	0.02040816	96.08
2	0.500000			52	0.980769	0.02000000	96.15
3	0.666667	1.00000000	33.33	53	0.981132	0.01960784	96.23
4	0.750000	0.50000000	50.00	54	0.981481	0.01923077	96.30
5	0.800000	0.33333333	60.00	55	0.981818	0.01886792	96.36
6	0.833333	0.25000000	66.67	56	0.982143	0.01851851	96.43
7	0.857143	0.20000000	71.43	57	0.982456	0.01818182	96.49
8	0.875000	0.16666667	75.00	58	0.982759	0.01785714	96.55
9	0.888889	0.14285714	77.78	59	0.983051	0.01754385	96.61
10	0.900000	0.12500000	80.00	60	0.983333	0.01724137	96.67
11	0.909091	0.11111111	81.82	61	0.983607	0.01694915	96.72
12	0.916667	0.10000000	83.33	62	0.983871	0.01666667	96.77
13	0.923077	0.09090909	84.62	63	0.984127	0.01639344	96.83
14	0.928571	0.08333333	85.71	64	0.984375	0.01612903	96.88
15	0.933333	0.07692307	86.67	65	0.984615	0.01587301	96.92
16	0.937500	0.07142857	87.50	66	0.984848	0.01562500	96.97
17	0.941176	0.06666667	88.24	67	0.985075	0.01538461	97.01
18	0.944444	0.06250000	88.89	68	0.985294	0.01515151	97.06
19	0.947368	0.05882353	89.47	69	0.985507	0.01492537	97.10
20	0.950000	0.05555556	90.00	70	0.985714	0.01470588	97.14
21	0.952381	0.05263157	90.48	71	0.985915	0.01449275	97.18
22	0.954545	0.05000000	90.91	72	0.986111	0.01428571	97.22
23	0.956522	0.04761904	91.30	73	0.986301	0.01408450	97.26
24	0.958333	0.04545455	91.67	74	0.986486	0.01388889	97.30
25	0.960000	0.04347826	92.00	75	0.986667	0.01369863	97.33
26	0.961538	0.04166667	92.31	76	0.986842	0.01351351	97.37
27	0.962963	0.04000000	92.59	77	0.987013	0.01333333	97.40
28	0.964286	0.03846154	92.86	78	0.987179	0.01315789	97.44
29	0.965517	0.03703703	93.10	79	0.987342	0.01298701	97.47
30	0.966667	0.03571428	93.33	80	0.987500	0.01282051	97.50
31	0.967742	0.03448275	93.55	81	0.987654	0.01265822	97.53
32	0.968750	0.03333333	93.75	82	0.987805	0.01250000	97.56
33	0.969697	0.03225806	93.94	83	0.987952	0.01234568	97.59
34	0.970588	0.03125000	94.12	84	0.988095	0.01219512	97.62
35	0.971429	0.03030303	94.29	85	0.988235	0.01204819	97.65
36	0.972222	0.02941176	94.44	86	0.988372	0.01190476	97.67
37	0.972973	0.02857143	94.59	87	0.988506	0.01176471	97.70
38	0.973684	0.02777778	94.74	88	0.988636	0.01162791	97.73
39	0.974359	0.02702703	94.87	89	0.988764	0.01149425	97.75
40	0.975000	0.02631579	95.00	90	0.988889	0.01136364	97.78
41	0.975610	0.02564102	95.12	91	0.989011	0.01123595	97.80
42	0.976190	0.02500000	95.24	92	0.989130	0.01111111	97.83
43	0.976744	0.02439024	95.35	93	0.989247	0.01098901	97.85
44	0.977273	0.02380952	95.45	94	0.989362	0.01086956	97.87
45	0.977778	0.02325581	95.56	95	0.989474	0.01075269	97.89
46	0.978261	0.02272727	95.65	96	0.989583	0.01063830	97.92
47	0.978723	0.02222222	95.74	97	0.989691	0.01052631	97.94
48	0.979167	0.02173913	95.83	98	0.989796	0.01041667	97.96
49	0.979592	0.02127659	95.92	99	0.989899	0.01030927	97.98
50	0.980000	0.02083333	96.00	100	0.990000	0.01020408	98.00

UNBIASING FACTORS FOR ML ESTIMATORS AND VARIANCES AND EFFICIENCIES OF UNBIASED ESTIMATORS FROM m ORDER STATISTICS OF SCALE PARAMETER v OF TYPE 2 E-V POPULATION

SHAPE PARAMETER K=1.5

m	$\tilde{v}/\hat{v}$	$\mathrm{var}(\tilde{v})/v^2$	Eff($\tilde{v}$)	m	$\tilde{v}/\hat{v}$	$\mathrm{var}(\tilde{v})/v^2$	Eff($\tilde{v}$)
1	0.373282			51	0.989083	0.00895847	97.28
2	0.705459	0.69814004	31.83	52	0.989293	0.00878147	97.33
3	0.807549	0.27360503	54.15	53	0.989496	0.00861132	97.38
4	0.857079	0.16963727	65.50	54	0.989691	0.00844765	97.43
5	0.886330	0.12285178	72.35	55	0.989879	0.00829007	97.48
6	0.905641	0.09627540	76.94	56	0.990060	0.00813828	97.52
7	0.919342	0.07914610	80.22	57	0.990234	0.00799194	97.56
8	0.929569	0.06718881	82.69	58	0.990403	0.00785077	97.61
9	0.937494	0.05836906	84.60	59	0.990566	0.00771450	97.65
10	0.943816	0.05159550	86.14	60	0.990724	0.00758288	97.69
11	0.948976	0.04623022	87.40	61	0.990876	0.00745568	97.72
12	0.953268	0.04187546	88.45	62	0.991023	0.00733267	97.76
13	0.956895	0.03827035	89.33	63	0.991166	0.00721365	97.80
14	0.959998	0.03523669	90.09	64	0.991304	0.00709844	97.83
15	0.962685	0.03264860	90.75	65	0.991438	0.00698685	97.86
16	0.965034	0.03041464	91.33	66	0.991568	0.00687872	97.90
17	0.967104	0.02846679	91.84	67	0.991694	0.00677387	97.93
18	0.968943	0.02675339	92.29	68	0.991817	0.00667218	97.96
19	0.970588	0.02523452	92.70	69	0.991935	0.00657350	97.99
20	0.972066	0.02387884	93.06	70	0.992051	0.00647769	98.02
21	0.973404	0.02266139	93.39	71	0.992163	0.00638464	98.04
22	0.974619	0.02156205	93.69	72	0.992272	0.00629422	98.07
23	0.975728	0.02056443	93.97	73	0.992378	0.00620633	98.10
24	0.976744	0.01965503	94.22	74	0.992481	0.00612085	98.12
25	0.977678	0.01882266	94.45	75	0.992582	0.00603770	98.15
26	0.978541	0.01805793	94.66	76	0.992679	0.00595678	98.17
27	0.979339	0.01735290	94.86	77	0.992775	0.00587800	98.20
28	0.980079	0.01670086	95.04	78	0.992867	0.00580128	98.22
29	0.980769	0.01609604	95.21	79	0.992958	0.00572653	98.24
30	0.981412	0.01553349	95.37	80	0.993046	0.00565368	98.26
31	0.982014	0.01500894	95.52	81	0.993132	0.00558267	98.29
32	0.982578	0.01451866	95.66	82	0.993216	0.00551341	98.31
33	0.983108	0.01405939	95.79	83	0.993298	0.00544586	98.33
34	0.983606	0.01362829	95.92	84	0.993377	0.00537994	98.35
35	0.984076	0.01322284	96.03	85	0.993455	0.00531559	98.37
36	0.984520	0.01284081	96.14	86	0.993532	0.00525277	98.39
37	0.984940	0.01248024	96.25	87	0.993606	0.00519142	98.40
38	0.985337	0.01213937	96.35	88	0.993679	0.00513148	98.42
39	0.985714	0.01181662	96.44	89	0.993750	0.00507290	98.44
40	0.986072	0.01151059	96.53	90	0.993820	0.00501566	98.46
41	0.986413	0.01122001	96.61	91	0.993888	0.00495968	98.47
42	0.986737	0.01094374	96.69	92	0.993954	0.00490495	98.49
43	0.987047	0.01068075	96.77	93	0.994019	0.00485141	98.51
44	0.987342	0.01043010	96.84	94	0.994083	0.00479902	98.52
45	0.987624	0.01019094	96.91	95	0.994145	0.00474776	98.54
46	0.987893	0.00996251	96.98	96	0.994206	0.00469758	98.55
47	0.988152	0.00974409	97.05	97	0.994266	0.00464845	98.57
48	0.988399	0.00953504	97.11	98	0.994325	0.00460033	98.58
49	0.988636	0.00933478	97.17	99	0.994382	0.00455320	98.60
50	0.988864	0.00914276	97.22	100	0.994438	0.00450703	98.61

UNBIASING FACTORS FOR ML ESTIMATORS AND VARIANCES AND EFFICIENCIES OF UNBIASED ESTIMATORS FROM m ORDER STATISTICS OF SCALE PARAMETER v OF TYPE 2 E-V POPULATION

SHAPE PARAMETER K=2.0

m	$\tilde{v}/\hat{v}$	$\mathrm{var}(\tilde{v})/v^2$	$\mathrm{Eff}(\tilde{v})$	m	$\tilde{v}/\hat{v}$	$\mathrm{var}(\tilde{v})/v^2$	$\mathrm{Eff}(\tilde{v})$
1	0.564190			51	0.992626	0.00501243	97.80
2	0.797885	0.27323954	45.75	52	0.992768	0.00491391	97.84
3	0.868627	0.13176848	63.24	53	0.992905	0.00481919	97.88
4	0.902703	0.08649774	72.26	54	0.993037	0.00472805	97.92
5	0.922746	0.06432432	77.73	55	0.993164	0.00464030	97.96
6	0.935942	0.05118451	81.40	56	0.993286	0.00455573	97.99
7	0.945288	0.04249704	84.04	57	0.993404	0.00447420	98.03
8	0.952254	0.03632841	86.02	58	0.993518	0.00439554	98.06
9	0.957646	0.03172251	87.56	59	0.993628	0.00431959	98.09
10	0.961945	0.02815253	88.80	60	0.993735	0.00424622	98.13
11	0.965451	0.02530447	89.82	61	0.993838	0.00417531	98.16
12	0.968365	0.02297951	90.66	62	0.993937	0.00410672	98.19
13	0.970826	0.02104571	91.38	63	0.994034	0.00404035	98.22
14	0.972932	0.01941204	91.99	64	0.994127	0.00397609	98.24
15	0.974754	0.01801367	92.52	65	0.994218	0.00391384	98.27
16	0.976347	0.01680319	92.99	66	0.994306	0.00385351	98.30
17	0.977750	0.01574513	93.40	67	0.994391	0.00379502	98.32
18	0.978996	0.01481239	93.77	68	0.994473	0.00373828	98.35
19	0.980110	0.01398397	94.09	69	0.994554	0.00368320	98.37
20	0.981112	0.01324330	94.39	70	0.994632	0.00362972	98.39
21	0.982018	0.01257713	94.65	71	0.994707	0.00357778	98.42
22	0.982841	0.01197477	94.90	72	0.994781	0.00352730	98.44
23	0.983592	0.01142745	95.12	73	0.994853	0.00347823	98.46
24	0.984279	0.01092798	95.32	74	0.994922	0.00343050	98.48
25	0.984912	0.01047035	95.51	75	0.994990	0.00338406	98.50
26	0.985496	0.01004949	95.68	76	0.995056	0.00333887	98.52
27	0.986036	0.00966116	95.84	77	0.995121	0.00329486	98.54
28	0.986537	0.00930172	95.99	78	0.995183	0.00325200	98.56
29	0.987004	0.00896807	96.13	79	0.995244	0.00321025	98.58
30	0.987439	0.00865752	96.26	80	0.995304	0.00316954	98.59
31	0.987846	0.00836776	96.38	81	0.995362	0.00312986	98.61
32	0.988228	0.00809677	96.49	82	0.995419	0.00309116	98.63
33	0.988586	0.00784277	96.60	83	0.995474	0.00305341	98.65
34	0.988923	0.00760423	96.70	84	0.995528	0.00301657	98.66
35	0.989241	0.00737977	96.79	85	0.995581	0.00298060	98.68
36	0.989541	0.00716818	96.88	86	0.995632	0.00294548	98.69
37	0.989825	0.00696839	96.96	87	0.995682	0.00291119	98.71
38	0.990094	0.00677942	97.04	88	0.995732	0.00287768	98.72
39	0.990349	0.00660044	97.12	89	0.995780	0.00284493	98.74
40	0.990591	0.00643066	97.19	90	0.995827	0.00281292	98.75
41	0.990821	0.00626940	97.26	91	0.995873	0.00278162	98.76
42	0.991040	0.00611603	97.32	92	0.995917	0.00275102	98.78
43	0.991249	0.00596999	97.39	93	0.995961	0.00272107	98.79
44	0.991449	0.00583075	97.45	94	0.996004	0.00269177	98.80
45	0.991640	0.00569786	97.50	95	0.996047	0.00266310	98.82
46	0.991822	0.00557090	97.56	96	0.996088	0.00263503	98.83
47	0.991996	0.00544947	97.61	97	0.996128	0.00260755	98.84
48	0.992164	0.00533322	97.66	98	0.996168	0.00258063	98.85
49	0.992324	0.00522182	97.71	99	0.996207	0.00255426	98.86
50	0.992478	0.00511499	97.75	100	0.996245	0.00252843	98.88

UNBIASING FACTORS FOR ML ESTIMATORS AND VARIANCES AND EFFICIENCIES OF UNBIASED ESTIMATORS FROM **m** ORDER STATISTICS OF SCALE PARAMETER **v** OF TYPE 2 E-V POPULATION

SHAPE PARAMETER K=2.5

m	$\tilde{v}/\hat{v}$	$\mathrm{var}(\tilde{v})/v^2$	$\mathrm{Eff}(\tilde{v})$	m	$\tilde{v}/\hat{v}$	$\mathrm{var}(\tilde{v})/v^2$	$\mathrm{Eff}(\tilde{v})$
1	0.671505	1.07009831	14.95	51	0.994493	0.00319864	98.08
2	0.848176	0.15005462	53.31	52	0.994599	0.00313595	98.12
3	0.901487	0.07817621	68.22	53	0.994701	0.00307567	98.15
4	0.927111	0.05265725	75.96	54	0.994799	0.00301766	98.19
5	0.942162	0.03966147	80.68	55	0.994894	0.00296180	98.22
6	0.952062	0.03180014	83.86	56	0.994986	0.00290798	98.25
7	0.959068	0.02653585	86.14	57	0.995074	0.00285607	98.28
8	0.964288	0.02276529	87.85	58	0.995159	0.00280598	98.31
9	0.968328	0.01993214	89.19	59	0.995241	0.00275762	98.34
10	0.971546	0.01772569	90.26	60	0.995321	0.00271089	98.37
11	0.974171	0.01595881	91.14	61	0.995398	0.00266573	98.40
12	0.976352	0.01451209	91.88	62	0.995472	0.00262205	98.42
13	0.978194	0.01330578	92.50	63	0.995544	0.00257977	98.45
14	0.979769	0.01228455	93.03	64	0.995614	0.00253884	98.47
15	0.981132	0.01140887	93.49	65	0.995682	0.00249918	98.49
16	0.982323	0.01064970	93.90	66	0.995747	0.00246074	98.52
17	0.983373	0.00998524	94.26	67	0.995811	0.00242347	98.54
18	0.984305	0.00939880	94.57	68	0.995873	0.00238732	98.56
19	0.985138	0.00887742	94.86	69	0.995933	0.00235222	98.58
20	0.985887	0.00841083	95.12	70	0.995991	0.00231814	98.60
21	0.986564	0.00799084	95.35	71	0.996047	0.00228503	98.62
22	0.987180	0.00761078	95.56	72	0.996102	0.00225285	98.64
23	0.987741	0.00726524	95.75	73	0.996156	0.00222158	98.66
24	0.988255	0.00694971	95.93	74	0.996208	0.00219116	98.68
25	0.988728	0.00666044	96.09	75	0.996259	0.00216155	98.69
26	0.989164	0.00639428	96.24	76	0.996308	0.00213274	98.71
27	0.989568	0.00614858	96.38	77	0.996356	0.00210468	98.73
28	0.989943	0.00592106	96.51	78	0.996403	0.00207736	98.74
29	0.990291	0.00570977	96.63	79	0.996449	0.00205074	98.76
30	0.990617	0.00551305	96.74	80	0.996493	0.00202478	98.78
31	0.990921	0.00532943	96.85	81	0.996536	0.00199948	98.79
32	0.991206	0.00515764	96.94	82	0.996579	0.00197480	98.81
33	0.991474	0.00499659	97.04	83	0.996620	0.00195073	98.82
34	0.991726	0.00484528	97.12	84	0.996660	0.00192723	98.83
35	0.991963	0.00470287	97.20	85	0.996700	0.00190429	98.85
36	0.992188	0.00456860	97.28	86	0.996738	0.00188189	98.86
37	0.992400	0.00444178	97.36	87	0.996776	0.00186002	98.87
38	0.992600	0.00432180	97.43	88	0.996812	0.00183864	98.89
39	0.992791	0.00420814	97.49	89	0.996848	0.00181775	98.90
40	0.992972	0.00410030	97.55	90	0.996883	0.00179733	98.91
41	0.993144	0.00399785	97.61	91	0.996918	0.00177737	98.92
42	0.993308	0.00390039	97.67	92	0.996951	0.00175784	98.94
43	0.993464	0.00380758	97.72	93	0.996984	0.00173873	98.95
44	0.993613	0.00371908	97.78	94	0.997016	0.00172005	98.96
45	0.993756	0.00363459	97.83	95	0.997048	0.00170175	98.97
46	0.993892	0.00355387	97.87	96	0.997078	0.00168384	98.98
47	0.994022	0.00347665	97.92	97	0.997109	0.00166631	98.99
48	0.994147	0.00340271	97.96	98	0.997138	0.00164913	99.00
49	0.994267	0.00333185	98.00	99	0.997167	0.00163230	99.01
50	0.994382	0.00326389	98.04	100	0.997196	0.00161582	99.02

UNBIASING FACTORS FOR ML ESTIMATORS AND VARIANCES AND EFFICIENCIES OF UNBIASED
Estimators from m ORDER STATISTICS OF SCALE PARAMETER **v** OF TYPE 2 E-V POPULATION

SHAPE PARAMETER K=3.0

m	$\tilde{v}/\hat{v}$	$\mathrm{var}(\tilde{v})/v^2$	$\mathrm{Eff}(\tilde{v})$	m	$\tilde{v}/\hat{v}$	$\mathrm{var}(\tilde{v})/v^2$	$\mathrm{Eff}(\tilde{v})$
1	0.738488	0.46099848	24.10	51	0.995628	0.00221723	98.26
2	0.879208	0.09574886	58.02	52	0.995713	0.00217386	98.29
3	0.921670	0.05191891	71.34	53	0.995794	0.00213214	98.33
4	0.942067	0.03548267	78.29	54	0.995872	0.00209200	98.36
5	0.954041	0.02692497	82.53	55	0.995947	0.00205334	98.39
6	0.961915	0.02168556	85.40	56	0.996020	0.00201608	98.41
7	0.967486	0.01815031	87.45	57	0.996090	0.00198016	98.44
8	0.971635	0.01560494	89.00	58	0.996158	0.00194548	98.47
9	0.974845	0.01368508	90.21	59	0.996223	0.00191201	98.50
10	0.977403	0.01218554	91.18	60	0.996286	0.00187966	98.52
11	0.979489	0.01098199	91.98	61	0.996347	0.00184839	98.54
12	0.981222	0.00999470	92.64	62	0.996406	0.00181814	98.57
13	0.982685	0.00917022	93.20	63	0.996463	0.00178888	98.59
14	0.983936	0.00847135	93.69	64	0.996519	0.00176053	98.61
15	0.985019	0.00787142	94.11	65	0.996572	0.00173307	98.63
16	0.985965	0.00735082	94.47	66	0.996624	0.00170645	98.65
17	0.986799	0.00689480	94.80	67	0.996675	0.00168064	98.67
18	0.987539	0.00649205	95.08	68	0.996724	0.00165560	98.69
19	0.988201	0.00613374	95.34	69	0.996772	0.00163130	98.71
20	0.988796	0.00581291	95.57	70	0.996818	0.00160769	98.73
21	0.989334	0.00552396	95.78	71	0.996863	0.00158476	98.75
22	0.989822	0.00526238	95.97	72	0.996906	0.00156248	98.77
23	0.990268	0.00502445	96.15	73	0.996949	0.00154081	98.78
24	0.990676	0.00480710	96.31	74	0.996990	0.00151973	98.80
25	0.991052	0.00460777	96.46	75	0.997030	0.00149923	98.82
26	0.991398	0.00442431	96.59	76	0.997070	0.00147927	98.83
27	0.991719	0.00425491	96.72	77	0.997108	0.00145984	98.85
28	0.992016	0.00409799	96.83	78	0.997145	0.00144091	98.86
29	0.992293	0.00395223	96.94	79	0.997181	0.00142246	98.88
30	0.992551	0.00381649	97.04	80	0.997216	0.00140448	98.89
31	0.992793	0.00368977	97.14	81	0.997251	0.00138695	98.90
32	0.993019	0.00357118	97.23	82	0.997284	0.00136985	98.92
33	0.993232	0.00345998	97.31	83	0.997317	0.00135316	98.93
34	0.993432	0.00335550	97.39	84	0.997349	0.00133688	98.94
35	0.993620	0.00325714	97.47	85	0.997380	0.00132099	98.96
36	0.993798	0.00316439	97.54	86	0.997411	0.00130547	98.97
37	0.993967	0.00307676	97.60	87	0.997441	0.00129031	98.98
38	0.994126	0.00299387	97.67	88	0.997470	0.00127549	98.99
39	0.994278	0.00291532	97.73	89	0.997498	0.00126102	99.00
40	0.994421	0.00284078	97.78	90	0.997526	0.00124687	99.01
41	0.994558	0.00276997	97.84	91	0.997554	0.00123303	99.02
42	0.994688	0.00270259	97.89	92	0.997580	0.00121950	99.03
43	0.994812	0.00263842	97.94	93	0.997606	0.00120626	99.04
44	0.994930	0.00257722	97.98	94	0.997632	0.00119331	99.06
45	0.995043	0.00251880	98.03	95	0.997657	0.00118063	99.07
46	0.995152	0.00246297	98.07	96	0.997681	0.00116821	99.07
47	0.995255	0.00240956	98.11	97	0.997705	0.00115605	99.08
48	0.995354	0.00235841	98.15	98	0.997729	0.00114416	99.09
49	0.995449	0.00230940	98.19	99	0.997752	0.00113249	99.10
50	0.995541	0.00226238	98.22	100	0.997774	0.00112107	99.11

UNBIASING FACTORS FOR ML ESTIMATORS AND VARIANCES AND EFFICIENCIES OF UNBIASED ESTIMATORS FROM m ORDER STATISTICS OF SCALE PARAMETER v OF TYPE 2 E-V POPULATION

SHAPE PARAMETER K=3.5

m	$\tilde{v}/\hat{v}$	$var(\tilde{v})/v^2$	$Eff(\tilde{v})$	m	$\tilde{v}/\hat{v}$	$var(\tilde{v})/v^2$	$Eff(\tilde{v})$
1	0.783704	0.26984765	30.25	51	0.996387	0.00162696	98.38
2	0.900060	0.06667202	61.22	52	0.996456	0.00159517	98.41
3	0.935204	0.03704224	73.46	53	0.996523	0.00156459	98.44
4	0.952084	0.02555147	79.87	54	0.996588	0.00153517	98.47
5	0.961992	0.01948312	83.80	55	0.996650	0.00150684	98.50
6	0.968506	0.01573846	86.45	56	0.996710	0.00147953	98.53
7	0.973114	0.01319911	88.35	57	0.996768	0.00145319	98.55
8	0.976547	0.01136444	89.79	58	0.996824	0.00142777	98.58
9	0.979202	0.00997711	90.91	59	0.996878	0.00140323	98.60
10	0.981317	0.00889140	91.81	60	0.996930	0.00137952	98.62
11	0.983042	0.00801866	92.55	61	0.996981	0.00135659	98.65
12	0.984475	0.00730184	93.16	62	0.997029	0.00133441	98.67
13	0.985685	0.00670262	93.69	63	0.997077	0.00131296	98.69
14	0.986720	0.00619425	94.13	64	0.997122	0.00129217	98.71
15	0.987615	0.00575753	94.52	65	0.997167	0.00127204	98.73
16	0.988398	0.00537832	94.86	66	0.997210	0.00125252	98.75
17	0.989087	0.00504597	95.16	67	0.997252	0.00123359	98.77
18	0.989699	0.00475229	95.43	68	0.997292	0.00121523	98.79
19	0.990246	0.00449090	95.67	69	0.997331	0.00119741	98.80
20	0.990738	0.00425677	95.89	70	0.997370	0.00118010	98.82
21	0.991182	0.00404584	96.08	71	0.997407	0.00116327	98.84
22	0.991586	0.00385482	96.26	72	0.997443	0.00114693	98.85
23	0.991955	0.00368102	96.42	73	0.997478	0.00113104	98.87
24	0.992292	0.00352222	96.57	74	0.997512	0.00111558	98.88
25	0.992603	0.00337654	96.71	75	0.997545	0.00110055	98.90
26	0.992889	0.00324244	96.83	76	0.997578	0.00108591	98.91
27	0.993154	0.00311858	96.95	77	0.997609	0.00107165	98.93
28	0.993400	0.00300384	97.06	78	0.997640	0.00105777	98.94
29	0.993629	0.00289724	97.16	79	0.997670	0.00104424	98.95
30	0.993843	0.00279795	97.25	80	0.997699	0.00103105	98.97
31	0.994042	0.00270523	97.34	81	0.997728	0.00101819	98.98
32	0.994230	0.00261846	97.42	82	0.997755	0.00100564	98.99
33	0.994405	0.00253709	97.50	83	0.997783	0.00099341	99.01
34	0.994571	0.00246062	97.58	84	0.997809	0.00098146	99.02
35	0.994727	0.00238863	97.64	85	0.997835	0.00096980	99.03
36	0.994874	0.00232072	97.71	86	0.997860	0.00095841	99.04
37	0.995013	0.00225657	97.77	87	0.997885	0.00094730	99.05
38	0.995145	0.00219588	97.83	88	0.997909	0.00093643	99.06
39	0.995270	0.00213836	97.89	89	0.997932	0.00092581	99.07
40	0.995389	0.00208378	97.94	90	0.997955	0.00091542	99.08
41	0.995502	0.00203191	97.99	91	0.997978	0.00090528	99.09
42	0.995609	0.00198257	98.04	92	0.998000	0.00089534	99.10
43	0.995712	0.00193556	98.08	93	0.998021	0.00088563	99.11
44	0.995809	0.00189073	98.13	94	0.998042	0.00087613	99.12
45	0.995903	0.00184793	98.17	95	0.998063	0.00086682	99.13
46	0.995992	0.00180703	98.21	96	0.998083	0.00085771	99.14
47	0.996078	0.00176790	98.24	97	0.998103	0.00084879	99.15
48	0.996160	0.00173042	98.28	98	0.998123	0.00084006	99.16
49	0.996239	0.00169450	98.32	99	0.998142	0.00083151	99.17
50	0.996314	0.00166004	98.35	100	0.998160	0.00082312	99.17

UNBIASING FACTORS FOR ML ESTIMATORS AND VARIANCES AND EFFICIENCIES OF UNBIASED ESTIMATORS FROM m ORDER STATISTICS OF SCALE PARAMETER v OF TYPE 2 E-V POPULATION

SHAPE PARAMETER K=4.0

m	$\tilde{v}/\hat{v}$	$\mathrm{var}(\tilde{v})/v^2$	$\mathrm{Eff}(\tilde{v})$	m	$\tilde{v}/\hat{v}$	$\mathrm{var}(\tilde{v})/v^2$	$\mathrm{Eff}(\tilde{v})$
1	0.816049	0.18034060	34.66	51	0.996926	0.00124452	98.47
2	0.914950	0.04919164	63.53	52	0.996985	0.00122022	98.50
3	0.944858	0.02777956	75.00	53	0.997042	0.00119685	98.53
4	0.959225	0.01928552	81.02	54	0.997097	0.00117436	98.56
5	0.967658	0.01475536	84.71	55	0.997150	0.00115270	98.58
6	0.973201	0.01194441	87.21	56	0.997201	0.00113183	98.61
7	0.977124	0.01003147	89.01	57	0.997250	0.00111170	98.63
8	0.980044	0.00864597	90.36	58	0.997298	0.00109227	98.66
9	0.982304	0.00759638	91.42	59	0.997344	0.00107351	98.68
10	0.984104	0.00677386	92.27	60	0.997388	0.00105538	98.70
11	0.985572	0.00611194	92.96	61	0.997431	0.00103786	98.72
12	0.986791	0.00556780	93.54	62	0.997473	0.00102091	98.74
13	0.987821	0.00511259	94.04	63	0.997513	0.00100449	98.76
14	0.988701	0.00472616	94.46	64	0.997552	0.00098861	98.78
15	0.989463	0.00439402	94.83	65	0.997590	0.00097322	98.80
16	0.990129	0.00410548	95.15	66	0.997626	0.00095829	98.82
17	0.990715	0.00385249	95.43	67	0.997662	0.00094382	98.84
18	0.991236	0.00362886	95.68	68	0.997696	0.00092978	98.85
19	0.991702	0.00342977	95.91	69	0.997730	0.00091615	98.87
20	0.992120	0.00325139	96.11	70	0.997762	0.00090291	98.89
21	0.992498	0.00309063	96.30	71	0.997794	0.00089006	98.90
22	0.992842	0.00294503	96.46	72	0.997825	0.00087756	98.92
23	0.993156	0.00281252	96.62	73	0.997855	0.00086541	98.93
24	0.993443	0.00269142	96.76	74	0.997884	0.00085359	98.95
25	0.993707	0.00258032	96.89	75	0.997912	0.00084209	98.96
26	0.993951	0.00247803	97.01	76	0.997939	0.00083090	98.97
27	0.994176	0.00238353	97.12	77	0.997966	0.00082000	98.99
28	0.994385	0.00229598	97.22	78	0.997992	0.00080938	99.00
29	0.994580	0.00221463	97.31	79	0.998018	0.00079903	99.01
30	0.994762	0.00213885	97.40	80	0.998043	0.00078894	99.02
31	0.994932	0.00206808	97.49	81	0.998067	0.00077911	99.04
32	0.995091	0.00200184	97.57	82	0.998091	0.00076951	99.05
33	0.995240	0.00193972	97.64	83	0.998114	0.00076015	99.06
34	0.995381	0.00188134	97.71	84	0.998136	0.00075102	99.07
35	0.995514	0.00182637	97.77	85	0.998158	0.00074210	99.08
36	0.995639	0.00177451	97.84	86	0.998180	0.00073340	99.09
37	0.995757	0.00172552	97.89	87	0.998200	0.00072489	99.10
38	0.995870	0.00167917	97.95	88	0.998221	0.00071658	99.11
39	0.995976	0.00163524	98.00	89	0.998241	0.00070845	99.12
40	0.996077	0.00159355	98.05	90	0.998261	0.00070052	99.13
41	0.996173	0.00155393	98.10	91	0.998280	0.00069275	99.14
42	0.996264	0.00151623	98.14	92	0.998298	0.00068516	99.15
43	0.996352	0.00148032	98.19	93	0.998317	0.00067773	99.16
44	0.996435	0.00144607	98.23	94	0.998335	0.00067046	99.17
45	0.996514	0.00141337	98.27	95	0.998352	0.00066334	99.18
46	0.996591	0.00138212	98.31	96	0.998369	0.00065637	99.19
47	0.996663	0.00135221	98.34	97	0.998386	0.00064955	99.20
48	0.996733	0.00132358	98.38	98	0.998403	0.00064287	99.20
49	0.996800	0.00129613	98.41	99	0.998419	0.00063632	99.21
50	0.996864	0.00126980	98.44	100	0.998435	0.00062991	99.22

UNBIASING FACTORS FOR ML ESTIMATORS AND VARIANCES AND EFFICIENCIES OF UNBIASED ESTIMATORS FROM m ORDER STATISTICS OF SCALE PARAMETER v OF TYPE 2 E-V POPULATION

SHAPE PARAMETER K=5.0

m	$\tilde{v}/\hat{v}$	$\mathrm{var}(\tilde{v})/v^2$	$\mathrm{Eff}(\tilde{v})$	m	$\tilde{v}/\hat{v}$	$\mathrm{var}(\tilde{v})/v^2$	$\mathrm{Eff}(\tilde{v})$
1	0.858937	0.09868553	40.53	51	0.997639	0.00079551	98.59
2	0.934685	0.03001769	66.63	52	0.997685	0.00078000	98.62
3	0.957645	0.01730143	77.06	53	0.997728	0.00076509	98.64
4	0.968679	0.01211111	82.57	54	0.997771	0.00075073	98.67
5	0.975157	0.00930747	85.95	55	0.997811	0.00073690	98.69
6	0.979415	0.00755521	88.24	56	0.997851	0.00072357	98.72
7	0.982428	0.00635716	89.89	57	0.997888	0.00071071	98.74
8	0.984672	0.00548661	91.13	58	0.997925	0.00069831	98.76
9	0.986408	0.00482554	92.10	59	0.997960	0.00068633	98.78
10	0.987790	0.00430652	92.88	60	0.997994	0.00067475	98.80
11	0.988918	0.00388823	93.52	61	0.998027	0.00066355	98.82
12	0.989855	0.00354397	94.06	62	0.998059	0.00065272	98.84
13	0.990645	0.00325567	94.51	63	0.998090	0.00064225	98.86
14	0.991322	0.00301073	94.90	64	0.998120	0.00063209	98.88
15	0.991907	0.00280006	95.24	65	0.998149	0.00062227	98.89
16	0.992418	0.00261694	95.53	66	0.998177	0.00061273	98.91
17	0.992869	0.00245629	95.79	67	0.998204	0.00060349	98.93
18	0.993269	0.00231422	96.02	68	0.998231	0.00059452	98.94
19	0.993626	0.00218768	96.23	69	0.998256	0.00058581	98.96
20	0.993948	0.00207426	96.42	70	0.998281	0.00057735	98.97
21	0.994238	0.00197201	96.59	71	0.998306	0.00056914	98.99
22	0.994502	0.00187938	96.74	72	0.998329	0.00056116	99.00
23	0.994743	0.00179505	96.88	73	0.998352	0.00055339	99.02
24	0.994964	0.00171797	97.01	74	0.998375	0.00054584	99.03
25	0.995167	0.00164723	97.13	75	0.998396	0.00053849	99.04
26	0.995354	0.00158209	97.24	76	0.998417	0.00053134	99.05
27	0.995527	0.00152190	97.34	77	0.998438	0.00052437	99.07
28	0.995688	0.00146612	97.44	78	0.998458	0.00051758	99.08
29	0.995837	0.00141428	97.53	79	0.998478	0.00051098	99.09
30	0.995977	0.00136600	97.61	80	0.998497	0.00050453	99.10
31	0.996107	0.00132089	97.69	81	0.998515	0.00049825	99.11
32	0.996230	0.00127867	97.76	82	0.998533	0.00049212	99.12
33	0.996344	0.00123906	97.83	83	0.998551	0.00048614	99.13
34	0.996453	0.00120184	97.89	84	0.998568	0.00048030	99.14
35	0.996554	0.00116678	97.95	85	0.998585	0.00047460	99.15
36	0.996651	0.00113371	98.01	86	0.998602	0.00046904	99.16
37	0.996742	0.00110247	98.06	87	0.998618	0.00046360	99.17
38	0.996828	0.00107290	98.11	88	0.998634	0.00045829	99.18
39	0.996909	0.00104488	98.16	89	0.998649	0.00045309	99.19
40	0.996987	0.00101828	98.20	90	0.998664	0.00044802	99.20
41	0.997061	0.00099300	98.25	91	0.998679	0.00044306	99.21
42	0.997131	0.00096895	98.29	92	0.998693	0.00043821	99.22
43	0.997198	0.00094603	98.33	93	0.998707	0.00043345	99.23
44	0.997262	0.00092417	98.37	94	0.998721	0.00042881	99.23
45	0.997323	0.00090330	98.40	95	0.998735	0.00042426	99.24
46	0.997381	0.00088336	98.44	96	0.998748	0.00041981	99.25
47	0.997437	0.00086427	98.47	97	0.998761	0.00041544	99.26
48	0.997491	0.00084599	98.50	98	0.998773	0.00041118	99.27
49	0.997542	0.00082847	98.53	99	0.998786	0.00040700	99.27
50	0.997592	0.00081166	98.56	100	0.998798	0.00040290	99.28

UNBIASING FACTORS FOR ML ESTIMATORS AND VARIANCES AND EFFICIENCIES OF UNBIASED ESTIMATORS FROM m ORDER STATISTICS OF SCALE PARAMETER v OF TYPE 2 E-V POPULATION

SHAPE PARAMETER K=6.0

m	$\tilde{v}/\hat{v}$	$\mathrm{var}(\tilde{v})/v^2$	Eff$(\tilde{v})$	m	$\tilde{v}/\hat{v}$	$\mathrm{var}(\tilde{v})/v^2$	Eff$(\tilde{v})$
1	0.885907	0.06275332	44.27	51	0.998087	0.00055201	98.67
2	0.947104	0.02024318	68.61	52	0.998124	0.00054125	98.69
3	0.965690	0.01181142	78.39	53	0.998160	0.00053091	98.72
4	0.974626	0.00831035	83.56	54	0.998194	0.00052095	98.74
5	0.979873	0.00640427	86.75	55	0.998227	0.00051136	98.77
6	0.983323	0.00520760	88.90	56	0.998258	0.00050212	98.79
7	0.985764	0.00438702	90.45	57	0.998289	0.00049320	98.81
8	0.987581	0.00378953	91.63	58	0.998319	0.00048460	98.83
9	0.988988	0.00333512	92.54	59	0.998347	0.00047629	98.85
10	0.990108	0.00297793	93.28	60	0.998375	0.00046826	98.87
11	0.991021	0.00268980	93.88	61	0.998402	0.00046050	98.89
12	0.991780	0.00245248	94.39	62	0.998428	0.00045299	98.90
13	0.992421	0.00225362	94.81	63	0.998453	0.00044572	98.92
14	0.992969	0.00208458	95.18	64	0.998477	0.00043868	98.94
15	0.993443	0.00193912	95.50	65	0.998500	0.00043186	98.95
16	0.993857	0.00181263	95.78	66	0.998523	0.00042525	98.97
17	0.994222	0.00170162	96.03	67	0.998545	0.00041883	98.99
18	0.994546	0.00160342	96.24	68	0.998567	0.00041261	99.00
19	0.994836	0.00151594	96.44	69	0.998587	0.00040658	99.02
20	0.995097	0.00143751	96.62	70	0.998608	0.00040071	99.03
21	0.995332	0.00136679	96.78	71	0.998627	0.00039501	99.04
22	0.995546	0.00130270	96.92	72	0.998646	0.00038947	99.06
23	0.995741	0.00124435	97.06	73	0.998665	0.00038409	99.07
24	0.995920	0.00119101	97.18	74	0.998683	0.00037885	99.08
25	0.996084	0.00114205	97.29	75	0.998701	0.00037375	99.09
26	0.996236	0.00109696	97.39	76	0.998718	0.00036879	99.11
27	0.996376	0.00105529	97.49	77	0.998735	0.00036396	99.12
28	0.996506	0.00101667	97.58	78	0.998751	0.00035925	99.13
29	0.996627	0.00098077	97.66	79	0.998767	0.00035466	99.14
30	0.996740	0.00094733	97.74	80	0.998782	0.00035019	99.15
31	0.996846	0.00091609	97.81	81	0.998797	0.00034583	99.16
32	0.996945	0.00088684	97.88	82	0.998812	0.00034158	99.17
33	0.997038	0.00085941	97.95	83	0.998826	0.00033743	99.18
34	0.997126	0.00083362	98.01	84	0.998840	0.00033338	99.19
35	0.997208	0.00080933	98.06	85	0.998854	0.00032943	99.20
36	0.997286	0.00078642	98.12	86	0.998867	0.00032557	99.21
37	0.997360	0.00076477	98.17	87	0.998880	0.00032180	99.22
38	0.997430	0.00074428	98.21	88	0.998893	0.00031811	99.23
39	0.997496	0.00072486	98.26	89	0.998905	0.00031451	99.24
40	0.997559	0.00070643	98.30	90	0.998918	0.00031099	99.24
41	0.997619	0.00068890	98.34	91	0.998930	0.00030755	99.25
42	0.997676	0.00067224	98.38	92	0.998941	0.00030418	99.26
43	0.997730	0.00065635	98.42	93	0.998953	0.00030088	99.27
44	0.997782	0.00064120	98.46	94	0.998964	0.00029766	99.28
45	0.997831	0.00062674	98.49	95	0.998975	0.00029450	99.28
46	0.997878	0.00061291	98.52	96	0.998985	0.00029141	99.29
47	0.997924	0.00059967	98.56	97	0.998996	0.00028839	99.30
48	0.997967	0.00058700	98.59	98	0.999006	0.00028542	99.31
49	0.998009	0.00057486	98.61	99	0.999016	0.00028252	99.31
50	0.998049	0.00056320	98.64	100	0.999026	0.00027967	99.32

UNBIASING FACTORS FOR ML ESTIMATORS AND VARIANCES AND EFFICIENCIES OF UNBIASED ESTIMATORS FROM m ORDER STATISTICS OF SCALE PARAMETER v OF TYPE 2 E-V POPULATION

SHAPE PARAMETER K=7.0

m	$\tilde{v}/\hat{v}$	$\mathrm{var}(\tilde{v})/v^2$	Eff($\tilde{v}$)	m	$\tilde{v}/\hat{v}$	$\mathrm{var}(\tilde{v})/v^2$	Eff($\tilde{v}$)
1	0.904350	0.04356829	46.84	51	0.998394	0.00040533	98.72
2	0.955606	0.01458028	69.99	52	0.998425	0.00039744	98.75
3	0.971198	0.00857685	79.31	53	0.998455	0.00038985	98.77
4	0.978698	0.00605541	84.26	54	0.998483	0.00038254	98.79
5	0.983102	0.00467536	87.30	55	0.998511	0.00037550	98.82
6	0.985998	0.00380626	89.36	56	0.998538	0.00036872	98.84
7	0.988047	0.00320911	90.85	57	0.998563	0.00036217	98.86
8	0.989573	0.00277369	91.97	58	0.998588	0.00035586	98.88
9	0.990754	0.00244220	92.85	59	0.998612	0.00034976	98.90
10	0.991694	0.00218141	93.55	60	0.998636	0.00034387	98.91
11	0.992461	0.00197092	94.13	61	0.998658	0.00033817	98.93
12	0.993098	0.00179745	94.62	62	0.998680	0.00033265	98.95
13	0.993636	0.00165202	95.03	63	0.998701	0.00032732	98.97
14	0.994097	0.00152837	95.38	64	0.998721	0.00032216	98.98
15	0.994495	0.00142192	95.68	65	0.998741	0.00031715	99.00
16	0.994842	0.00132933	95.95	66	0.998760	0.00031230	99.01
17	0.995149	0.00124806	96.19	67	0.998778	0.00030759	99.03
18	0.995421	0.00117616	96.40	68	0.998796	0.00030302	99.04
19	0.995664	0.00111208	96.59	69	0.998814	0.00029859	99.06
20	0.995883	0.00105462	96.76	70	0.998831	0.00029428	99.07
21	0.996081	0.00100281	96.91	71	0.998847	0.00029010	99.08
22	0.996260	0.00095585	97.05	72	0.998863	0.00028604	99.09
23	0.996424	0.00091309	97.18	73	0.998879	0.00028208	99.11
24	0.996574	0.00087399	97.29	74	0.998894	0.00027823	99.12
25	0.996712	0.00083811	97.40	75	0.998909	0.00027449	99.13
26	0.996839	0.00080505	97.50	76	0.998923	0.00027085	99.14
27	0.996957	0.00077450	97.59	77	0.998937	0.00026730	99.15
28	0.997066	0.00074618	97.68	78	0.998951	0.00026385	99.16
29	0.997168	0.00071987	97.76	79	0.998964	0.00026048	99.17
30	0.997263	0.00069534	97.83	80	0.998977	0.00025719	99.18
31	0.997352	0.00067243	97.90	81	0.998990	0.00025399	99.19
32	0.997435	0.00065099	97.97	82	0.999002	0.00025087	99.20
33	0.997513	0.00063086	98.03	83	0.999014	0.00024782	99.21
34	0.997587	0.00061195	98.09	84	0.999026	0.00024485	99.22
35	0.997656	0.00059413	98.14	85	0.999038	0.00024195	99.23
36	0.997721	0.00057732	98.19	86	0.999049	0.00023911	99.24
37	0.997783	0.00056145	98.24	87	0.999060	0.00023635	99.25
38	0.997842	0.00054641	98.29	88	0.999071	0.00023364	99.26
39	0.997898	0.00053217	98.33	89	0.999081	0.00023100	99.27
40	0.997950	0.00051864	98.37	90	0.999091	0.00022841	99.28
41	0.998001	0.00050579	98.41	91	0.999101	0.00022588	99.28
42	0.998048	0.00049356	98.45	92	0.999111	0.00022341	99.29
43	0.998094	0.00048190	98.49	93	0.999121	0.00022099	99.30
44	0.998137	0.00047079	98.52	94	0.999130	0.00021862	99.31
45	0.998179	0.00046017	98.55	95	0.999139	0.00021631	99.31
46	0.998219	0.00045002	98.58	96	0.999148	0.00021403	99.32
47	0.998257	0.00044031	98.61	97	0.999157	0.00021181	99.33
48	0.998293	0.00043102	98.64	98	0.999166	0.00020964	99.33
49	0.998328	0.00042210	98.67	99	0.999174	0.00020751	99.34
50	0.998362	0.00041354	98.70	100	0.999182	0.00020542	99.35

UNBIASING FACTORS FOR ML ESTIMATORS AND VARIANCES AND EFFICIENCIES OF UNBIASED ESTIMATORS FROM m ORDER STATISTICS OF SCALE PARAMETER v OF TYPE 2 E-V POPULATION

SHAPE PARAMETER K=8.0

m	$\tilde{v}/\hat{v}$	$var(\tilde{v})/v^2$	Eff $(\tilde{v})$	m	$\tilde{v}/\hat{v}$	$var(\tilde{v})/v^2$	Eff $(\tilde{v})$
1	0.917724	0.03206694	48.73	51	0.998617	0.00031020	98.76
2	0.961779	0.01100435	70.99	52	0.998643	0.00030417	98.79
3	0.975197	0.00651100	79.99	53	0.998669	0.00029836	98.81
4	0.981654	0.00460833	84.76	54	0.998694	0.00029277	98.83
5	0.985447	0.00356296	87.71	55	0.998718	0.00028738	98.85
6	0.987941	0.00290315	89.70	56	0.998741	0.00028220	98.87
7	0.989705	0.00244914	91.14	57	0.998763	0.00027719	98.89
8	0.991020	0.00211775	92.23	58	0.998784	0.00027236	98.91
9	0.992036	0.00186527	93.08	59	0.998805	0.00026769	98.93
10	0.992846	0.00166652	93.76	60	0.998825	0.00026318	98.95
11	0.993507	0.00150603	94.32	61	0.998844	0.00025883	98.96
12	0.994056	0.00137371	94.79	62	0.998863	0.00025461	98.98
13	0.994519	0.00126275	95.18	63	0.998881	0.00025053	99.00
14	0.994915	0.00116838	95.52	64	0.998898	0.00024657	99.01
15	0.995258	0.00108712	95.82	65	0.998915	0.00024274	99.03
16	0.995558	0.00101642	96.08	66	0.998932	0.00023903	99.04
17	0.995822	0.00095436	96.31	67	0.998948	0.00023542	99.06
18	0.996056	0.00089944	96.51	68	0.998963	0.00023193	99.07
19	0.996266	0.00085049	96.69	69	0.998978	0.00022854	99.08
20	0.996454	0.00080660	96.86	70	0.998993	0.00022525	99.10
21	0.996624	0.00076701	97.01	71	0.999007	0.00022204	99.11
22	0.996779	0.00073113	97.14	72	0.999021	0.00021893	99.12
23	0.996920	0.00069845	97.26	73	0.999035	0.00021590	99.13
24	0.997049	0.00066857	97.38	74	0.999048	0.00021296	99.15
25	0.997168	0.00064114	97.48	75	0.999060	0.00021010	99.16
26	0.997278	0.00061587	97.58	76	0.999073	0.00020731	99.17
27	0.997379	0.00059252	97.67	77	0.999085	0.00020460	99.18
28	0.997473	0.00057087	97.75	78	0.999097	0.00020196	99.19
29	0.997561	0.00055075	97.83	79	0.999108	0.00019938	99.20
30	0.997643	0.00053200	97.90	80	0.999119	0.00019687	99.21
31	0.997719	0.00051448	97.97	81	0.999130	0.00019442	99.22
32	0.997791	0.00049809	98.03	82	0.999141	0.00019202	99.23
33	0.997858	0.00048270	98.09	83	0.999151	0.00018969	99.24
34	0.997921	0.00046824	98.15	84	0.999161	0.00018742	99.25
35	0.997981	0.00045461	98.20	85	0.999171	0.00018520	99.26
36	0.998037	0.00044176	98.25	86	0.999181	0.00018302	99.27
37	0.998091	0.00042962	98.30	87	0.999190	0.00018091	99.27
38	0.998141	0.00041812	98.34	88	0.999199	0.00017884	99.28
39	0.998189	0.00040723	98.38	89	0.999208	0.00017682	99.29
40	0.998235	0.00039688	98.42	90	0.999217	0.00017484	99.30
41	0.998278	0.00038705	98.46	91	0.999226	0.00017290	99.31
42	0.998319	0.00037769	98.50	92	0.999234	0.00017101	99.31
43	0.998358	0.00036878	98.53	93	0.999243	0.00016916	99.32
44	0.998396	0.00036027	98.57	94	0.999251	0.00016735	99.33
45	0.998431	0.00035216	98.60	95	0.999259	0.00016557	99.33
46	0.998466	0.00034440	98.63	96	0.999266	0.00016384	99.34
47	0.998498	0.00033697	98.66	97	0.999274	0.00016213	99.35
48	0.998530	0.00032985	98.69	98	0.999281	0.00016047	99.36
49	0.998560	0.00032303	98.71	99	0.999289	0.00015884	99.36
50	0.998589	0.00031649	98.74	100	0.999296	0.00015724	99.37

Appendix F

TABLES OF ASYMPTOTIC VARIANCES AND COVARIANCES OF MAXIMUM-LIKELIHOOD ESTIMATORS FROM DOUBLY CENSORED SAMPLES

SOURCES OF TABLES

Table F1	*Biometrika* 53(1966), 205-213 (Harter and Moore)
Table F2	*J. Amer. Statist. Assoc.* 61(1966), 842-851: Corrigendum, 1247 (Harter and Moore)
Tables F3,F4	*Ann. Math. Statist.* 38(1967), 557-570 (Harter and Moore)
Table F5	*J. Amer. Statist. Assoc.* 62(1967), 675-684 (Harter and Moore)
Table F6	*J. Amer. Statist. Assoc.* 63(1968), 889-901 (Harter and Moore)
Table F7	McMaster Ph.D. thesis (1993) (Chan)

Table F1

COEFFICIENTS IN ASYMPTOTIC VARIANCES AND COVARIANCES OF MAXIMUM-LIKELIHOOD ESTIMATORS OF PARAMETERS μ AND σ OF NORMAL POPULATION FROM SAMPLES OF SIZE n WITH PROPORTIONS q_1 CENSORED FROM BELOW AND q_2 FROM ABOVE

COEFFICIENTS OF σ^2/N IN ASYMPTOTIC VARIANCES AND COVARIANCES OF MAXIMUM-LIKELIHOOD ESTIMATORS OF PARAMETERS μ AND σ OF NORMAL POPULATION FROM SAMPLES OF SIZE N WITH PROPORTIONS Q1 CENSORED FROM BELOW AND Q2 FROM ABOVE

		(BOTH PARAMETERS UNKNOWN)			(σ KNOWN)	(μ KNOWN)
Q1	Q2	$N\ VAR(\hat{\mu})/\sigma^2$	$N\ COV(\hat{\mu},\hat{\sigma})/\sigma^2$	$N\ VAR(\hat{\sigma})/\sigma^2$	$N\ VAR(\hat{\mu})/\sigma^2$	$N\ VAR(\hat{\sigma})/\sigma^2$
0.0	0.0	1.000000	0.000000	0.500000	1.000000	0.500000
0.0	0.1	1.020092	0.041136	0.585925	1.017205	0.584266
0.0	0.2	1.062323	0.106905	0.688692	1.045728	0.677934
0.0	0.3	1.138257	0.206568	0.819749	1.086204	0.782262
0.0	0.4	1.272656	0.359824	0.994759	1.142501	0.893025
0.0	0.5	1.517094	0.605233	1.241453	1.222031	1.000000
0.0	0.6	1.990850	1.025933	1.615494	1.339322	1.086805
0.0	0.7	3.019940	1.832190	2.247997	1.526647	1.136413
0.0	0.8	5.780392	3.717327	3.537484	1.874080	1.146899
0.0	0.9	17.794599	10.620022	7.513928	2.784491	1.175776
0.1	0.1	1.035011	0.000000	0.702692	1.035011	0.702692
0.1	0.2	1.070615	0.071658	0.847527	1.064557	0.842731
0.1	0.3	1.140391	0.187749	1.041120	1.106533	1.010210
0.1	0.4	1.274494	0.379562	1.315918	1.165014	1.202879
0.1	0.5	1.542208	0.715075	1.736943	1.247822	1.405385
0.1	0.6	2.128202	1.364988	2.458565	1.370365	1.583087
0.1	0.7	3.665653	2.880735	3.954475	1.567111	1.690586
0.1	0.8	9.774446	8.237227	8.655663	1.935427	1.713898
0.2	0.2	1.095839	0.000000	1.052478	1.095839	1.052478
0.2	0.3	1.152548	0.127812	1.341466	1.140370	1.327293
0.2	0.4	1.275501	0.360575	1.783003	1.202582	1.681071
0.2	0.5	1.556437	0.820702	2.537708	1.291020	2.104956
0.2	0.6	2.301737	1.897104	4.093984	1.422641	2.530381
0.2	0.7	5.184839	5.628780	8.927385	1.635853	2.816652
0.3	0.3	1.188673	0.000000	1.796338	1.188673	1.796338
0.3	0.4	1.285467	0.273191	2.569779	1.256424	2.511720
0.3	0.5	1.565414	0.938941	4.155856	1.353277	3.592676
0.3	0.6	2.689726	3.281978	9.043125	1.498614	5.038488
0.4	0.4	1.332365	0.000000	4.173987	1.332365	4.173987
0.4	0.5	1.569895	1.079093	9.089706	1.441790	8.347974

(INTERCHANGING Q1 AND Q2 LEAVES VARIANCES AND ABSOLUTE VALUE OF COVARIANCE UNCHANGED, BUT CHANGES SIGN OF COVARIANCE)

Table F2

COEFFICIENTS IN ASYMPTOTIC VARIANCES AND COVARIANCES OF MAXIMUM-LIKELIHOOD ESTIMATORS OF PARAMETERS μ, σ, and τ OF LOGNORMAL POPULATION FROM SAMPLES OF SIZE n WITH PROPORTIONS q_1 CENSORED FROM BELOW AND q_2 FROM ABOVE

A. $\sigma = 1$

Q1	Q2	N VAR($\hat{\mu}$)	N VAR($\hat{\sigma}$)	N VAR($\hat{\tau}$) / EXP(2μ−4)	N COV($\hat{\mu}$, $\hat{\sigma}$)	N COV($\hat{\mu}$, $\hat{\tau}$) / EXP(μ−2)	N COV($\hat{\sigma}$, $\hat{\tau}$) / EXP(μ−2)
0.00	0.0	1.410414	0.910414	8.243387	-0.410414	-1.839348	1.839348
0.00	0.1	1.412588	1.146076	8.901992	-0.427754	-1.869223	2.233038
0.00	0.2	1.414626	1.434485	9.568442	-0.405682	-1.836026	2.671344
0.00	0.3	1.429019	1.818560	10.330370	-0.332334	-1.733111	3.212178
0.00	0.4	1.478482	2.358012	11.249300	-0.169886	-1.521643	3.916075
0.00	0.5	1.618193	3.163315	12.414950	0.164441	-1.120328	4.884651
0.00	0.6	1.999588	4.467645	13.987247	0.868068	-0.349597	6.316150
0.00	0.7	3.115817	6.853719	16.302302	2.496707	1.250207	8.665095
0.00	0.8	7.115683	12.229846	20.249630	7.124206	5.199917	13.267144
0.00	0.9	31.407288	31.443446	29.512080	28.668431	20.043421	26.574609
0.01	0.0	1.777448	1.188926	16.582264	-0.729769	-3.588245	3.359194
0.01	0.1	1.808473	1.568272	18.764757	-0.836072	-3.844798	4.268600
0.01	0.2	1.818528	2.051113	21.113066	-0.903977	-3.995210	5.333216
0.01	0.3	1.819907	2.727682	23.991118	-0.927218	-4.043905	6.728391
0.01	0.4	1.825317	3.737812	27.755774	-0.859376	-3.914049	8.678105
0.01	0.5	1.874389	5.363918	33.027651	-0.581240	-3.415193	11.605384
0.01	0.6	2.109547	8.268538	41.100989	0.239625	-2.050745	16.446600
0.01	0.7	3.135404	14.362138	55.265556	2.729248	1.734025	25.733631
0.01	0.8	8.554546	31.444398	87.020906	12.318465	14.763655	49.010430
0.01	0.9	68.049468	129.28688	218.08099	88.370847	102.33117	162.11335
0.02	0.0	1.982316	1.317602	21.791553	-0.892116	-4.621259	4.177695
0.02	0.1	2.041723	1.773131	25.129949	-1.054644	-5.063234	5.410237
0.02	0.2	2.070846	2.362711	28.795805	-1.184340	-5.387484	6.880097
0.02	0.3	2.082860	3.206757	33.399397	-1.282091	-5.616773	8.850969
0.02	0.4	2.084254	4.500022	39.598571	-1.303527	-5.665201	11.681932
0.02	0.5	2.101512	6.650618	48.598897	-1.121588	-5.295680	16.080598
0.02	0.6	2.254557	10.661227	63.053861	-0.348631	-3.834051	23.692701
0.02	0.7	3.149650	19.617860	90.209795	2.464325	1.047586	39.283045
0.02	0.8	9.115507	47.545382	158.47117	15.315516	21.066673	82.921684
0.02	0.9	102.58354	257.10173	531.39495	154.77808	206.21759	362.18770
0.05	0.0	2.522554	1.613960	37.010394	-1.292131	-7.488117	6.299471
0.05	0.1	2.689855	2.268699	44.400142	-1.621216	-8.596862	8.498017
0.05	0.2	2.811150	3.148450	52.854782	-1.946760	-9.607481	11.224782
0.05	0.3	2.907809	4.469320	64.020049	-2.302199	-10.642625	15.064459
0.05	0.4	2.973894	6.614480	80.004008	-2.674170	-11.660722	20.919094
0.05	0.5	2.999411	10.457280	105.09468	-2.968286	-12.417451	30.736531
0.05	0.6	3.011875	18.391847	149.92022	-2.761610	-11.938437	49.591423
0.05	0.7	3.439670	39.084595	248.86461	0.133599	-5.646706	94.826016
0.05	0.8	10.105184	125.22396	583.44947	23.867431	40.910795	264.51545
0.05	0.9	384.34840	1821.0122	5380.7591	817.36166	1370.3050	3114.8512
0.10	0.0	3.452578	2.044686	67.248898	-1.924859	-12.789917	9.904549
0.10	0.1	3.897738	3.047913	84.899904	-2.591086	-15.589908	14.110599
0.10	0.2	4.305458	4.469161	106.20027	-3.351121	-18.534864	19.611692
0.10	0.3	4.741352	6.750278	136.26962	-4.346395	-22.151787	27.892379
0.10	0.4	5.219343	10.777561	183.07459	-5.729836	-26.873808	41.619543
0.10	0.5	5.730231	18.838010	265.21225	-7.747765	-33.327389	67.345471
0.10	0.6	6.200764	38.384658	437.74036	-10.730930	-42.222325	125.40455
0.10	0.7	6.397504	105.14187	937.06877	-13.745415	-50.595771	307.92783
0.10	0.8	10.476445	620.11852	4059.9720	28.955482	53.386278	1575.6021

Λ,σ = 1

Q1	Q2	N VAR($\hat{\mu}$|τ)	N VAR($\hat{\sigma}$|τ)	N COV($\hat{\mu}$,$\hat{\sigma}$|τ)	N VAR($\hat{\mu}$|σ)	N VAR($\hat{\tau}$|σ) / EXP(2μ−4)	N COV($\hat{\mu}$,$\hat{\tau}$|σ) / EXP(μ−2)
0.00	0.0	1.000000	0.500000	0.000000	1.225400	4.527273	-1.010171
0.00	0.1	1.020092	0.585925	0.041136	1.252937	4.551094	-1.035779
0.00	0.2	1.062323	0.688692	0.106905	1.299896	4.593782	-1.080552
0.00	0.3	1.138257	0.819749	0.206568	1.368286	4.656604	-1.146099
0.00	0.4	1.272656	0.994759	0.359824	1.466243	4.745670	-1.239504
0.00	0.5	1.517094	1.241453	0.605233	1.609644	4.872287	-1.374250
0.00	0.6	1.990850	1.615494	1.025933	1.830922	5.057767	-1.576831
0.00	0.7	3.019940	2.247997	1.832190	2.206304	5.347102	-1.906357
0.00	0.8	5.780392	3.537484	3.717327	2.965645	5.857208	-2.528543
0.00	0.9	17.794599	7.513928	10.620022	5.268965	7.052396	-4.185868
0.01	0.0	1.000986	0.508429	-0.002871	1.329512	7.091192	-1.526355
0.01	0.1	1.020695	0.597252	0.038541	1.362749	7.146269	-1.569136
0.01	0.2	1.062517	0.703928	0.105223	1.420122	7.245865	-1.644727
0.01	0.3	1.138273	0.840682	0.206909	1.504718	7.394152	-1.756729
0.01	0.4	1.273368	1.024521	0.364388	1.627734	7.607756	-1.918830
0.01	0.5	1.521244	1.285973	0.618804	1.811405	7.918218	-2.157621
0.01	0.6	2.007225	1.687415	1.060233	2.102603	8.387750	-2.527374
0.01	0.7	3.080997	2.379633	1.921823	2.616763	9.156838	-3.156156
0.01	0.8	6.049798	3.841579	4.003531	3.728740	10.631390	-4.436371
0.01	0.9	20.032136	8.777806	12.301642	7.645760	14.806396	-8.477405
0.02	0.0	1.002301	0.516689	-0.006167	1.378286	8.545415	-1.792644
0.02	0.1	1.021572	0.608359	0.035421	1.414429	8.622050	-1.845267
0.02	0.2	1.062887	0.718869	0.102876	1.477180	8.761301	-1.938745
0.02	0.3	1.138288	0.861216	0.206376	1.570268	8.969843	-2.078075
0.02	0.4	1.273758	1.053747	0.367758	1.706660	9.272597	-2.281281
0.02	0.5	1.524457	1.329805	0.630667	1.912362	9.717450	-2.583779
0.02	0.6	2.021424	1.758614	1.092026	2.243157	10.401000	-3.059279
0.02	0.7	3.137485	2.511540	2.008140	2.840090	11.548939	-3.887010
0.02	0.8	6.314968	4.155751	4.292161	4.182009	13.851328	-5.644404
0.02	0.9	22.557009	10.242164	14.224477	9.405429	21.169185	-11.823398
0.05	0.0	1.007523	0.541739	-0.017592	1.488079	12.422837	-2.444784
0.05	0.1	1.025310	0.642211	0.024190	1.531331	12.568559	-2.524166
0.05	0.2	1.064786	0.764640	0.093583	1.607424	12.836445	-2.666938
0.05	0.3	1.138591	0.924525	0.202101	1.721920	13.243212	-2.882746
0.05	0.4	1.274324	1.144648	0.374824	1.892753	13.844841	-3.203336
0.05	0.5	1.532228	1.467916	0.663386	2.156867	14.752414	-3.692924
0.05	0.6	2.061194	1.987727	1.187450	2.597208	16.202857	-4.492084
0.05	0.7	3.311547	2.952606	2.285190	3.439213	18.800224	-5.970842
0.05	0.8	7.236567	5.301972	5.319917	5.556101	24.703201	-9.505307
0.05	0.9	35.376143	17.865569	24.110026	17.475472	52.789498	-27.796554
0.10	0.0	1.020092	0.585925	-0.041136	1.640523	19.270832	-3.465814
0.10	0.1	1.035011	0.702692	0.000000	1.695009	19.573561	-3.594233
0.10	0.2	1.070615	0.847527	0.071658	1.792679	20.139713	-3.829383
0.10	0.3	1.140391	1.041120	0.187749	1.942779	21.017367	-4.192337
0.10	0.4	1.274494	1.315918	0.379562	2.173104	22.353037	-4.746987
0.10	0.5	1.542208	1.736943	0.715075	2.543702	24.453676	-5.629301
0.10	0.6	2.128202	2.458565	1.364988	3.200793	28.037583	-7.163854
0.10	0.7	3.665653	3.954475	2.880735	4.600538	35.243948	-10.339726
0.10	0.8	9.774446	8.655663	8.237227	9.124413	56.669411	-20.184043

A. σ = 1

Q1	Q2	N VAR($\hat{\sigma}$|μ)	N VAR($\hat{\tau}$|μ) / EXP(2μ−4)	N COV($\hat{\sigma}$,$\hat{\tau}$|μ) / EXP(μ−2)	N VAR($\hat{\mu}$|σ,τ)	N VAR($\hat{\sigma}$|μ,τ)	N VAR($\hat{\tau}$|μ,σ) / EXP(2μ−4)
0.00	0.0	0.790988	5.844657	1.304119	1.000000	0.500000	3.694528
0.00	0.1	1.016546	6.428522	1.667009	1.017205	0.584266	3.694834
0.00	0.2	1.318145	7.185487	2.144814	1.045728	0.677934	3.695562
0.00	0.3	1.741272	8.228456	2.809123	1.086204	0.782262	3.696613
0.00	0.4	2.338491	9.683236	3.741229	1.142501	0.893025	3.697842
0.00	0.5	3.146604	11.639310	4.998499	1.222031	1.000000	3.699007
0.00	0.6	4.090797	13.926126	6.467918	1.339322	1.086805	3.699765
0.00	0.7	4.853106	15.800662	7.663303	1.526647	1.136413	3.699914
0.00	0.8	5.097106	16.449694	8.060998	1.874080	1.146899	3.701344
0.00	0.9	5.275031	16.720821	8.279064	2.784491	1.175776	3.726982
0.01	0.0	0.889304	9.338449	1.885963	1.000969	0.508421	5.338851
0.01	0.1	1.181749	10.590751	2.491118	1.018208	0.595797	5.339490
0.01	0.2	1.601752	12.335798	3.347225	1.046788	0.693508	5.341010
0.01	0.3	2.255277	15.005401	4.668075	1.087348	0.803071	5.343205
0.01	0.4	3.333210	19.362832	6.835335	1.143767	0.920247	5.345773
0.01	0.5	5.183678	26.805068	10.546348	1.223479	1.034259	5.348209
0.01	0.6	8.241318	39.107408	16.679546	1.341062	1.127391	5.349793
0.01	0.7	11.986433	54.306559	24.224229	1.528907	1.180864	5.350104
0.01	0.8	13.705927	61.541413	27.750910	1.877487	1.192191	5.353095
0.01	0.9	14.526145	64.197828	29.223377	2.792021	1.223425	5.406887
0.02	0.0	0.916116	11.018277	2.097956	1.002228	0.516651	6.213843
0.02	0.1	1.228358	12.573722	2.794843	1.019510	0.607131	6.214710
0.02	0.2	1.685374	14.779802	3.798935	1.048165	0.708912	6.216769
0.02	0.3	2.417575	18.252847	5.393601	1.088833	0.823800	6.219743
0.02	0.4	3.684775	24.200015	8.138822	1.145410	0.947569	6.223223
0.02	0.5	6.052020	35.254113	13.254265	1.225360	1.068898	6.226524
0.02	0.6	10.607316	56.533758	23.099826	1.343321	1.168673	6.228671
0.02	0.7	17.689741	89.861364	38.463402	1.531845	1.226234	6.229093
0.02	0.8	21.812855	109.78439	47.526290	1.881920	1.238453	6.233148
0.02	0.9	23.572517	116.84799	51.046529	2.801834	1.272192	6.306202
0.05	0.0	0.952091	14.782170	2.463825	1.006952	0.541432	8.406271
0.05	0.1	1.291568	16.924294	3.316558	1.024398	0.641641	8.407857
0.05	0.2	1.800292	20.019934	4.571471	1.053333	0.756415	8.411626
0.05	0.3	2.646601	25.067883	6.638377	1.094411	0.888652	8.417072
0.05	0.4	4.209827	34.281989	10.433599	1.151584	1.034399	8.423447
0.05	0.5	7.519796	53.686887	18.447936	1.232429	1.180700	8.429495
0.05	0.6	15.859705	102.59877	38.644982	1.351822	1.303639	8.433431
0.05	0.7	39.079406	239.59474	95.045339	1.542908	1.375672	8.434204
0.05	0.8	68.851488	417.82229	167.88826	1.898645	1.391069	8.441641
0.05	0.9	82.797400	495.25509	200.73801	2.839069	1.433779	8.576192
0.10	0.0	0.971551	19.869238	2.773995	1.017205	0.584266	11.948858
0.10	0.1	1.325446	22.544450	3.746950	1.035011	0.702692	11.952063
0.10	0.2	1.860840	26.408260	5.185219	1.064557	0.842731	11.959681
0.10	0.3	2.765941	32.775586	7.585849	1.106533	1.010210	11.970692
0.10	0.4	4.487301	44.704382	12.117263	1.165014	1.202879	11.983590
0.10	0.5	8.362366	71.378019	22.283976	1.247822	1.405385	11.995836
0.10	0.6	19.813906	150.23955	52.335361	1.370365	1.583087	12.003808
0.10	0.7	75.609034	536.92334	199.21983	1.567111	1.690586	12.005375
0.10	0.8	540.08947	3787.9241	1428.0497	1.935427	1.713898	12.020448

B. $\sigma = 2$

Q1	Q2	N VAR($\hat{\mu}$)	N VAR($\hat{\sigma}$)	$\frac{\text{N VAR}(\hat{\tau})}{\text{EXP}(2\mu-8)}$	N COV($\hat{\mu},\hat{\sigma}$)	$\frac{\text{N COV}(\hat{\mu},\hat{\tau})}{\text{EXP}(\mu-4)}$	$\frac{\text{N COV}(\hat{\sigma},\hat{\tau})}{\text{EXP}(\mu-4)}$
0.00	0.0	4.015152	2.060608	0.827274	-0.030304	-0.111959	0.223919
0.00	0.1	4.091516	2.424300	0.830610	0.134569	-0.096219	0.258743
0.00	0.2	4.255492	2.861590	0.834228	0.401882	-0.071924	0.298517
0.00	0.3	4.554600	3.422938	0.838487	0.811235	-0.036291	0.347407
0.00	0.4	5.090952	4.178529	0.843716	1.447365	0.016596	0.410260
0.00	0.5	6.079645	5.254124	0.850419	2.477933	0.097897	0.495163
0.00	0.6	8.025234	6.905653	0.859489	4.269366	0.230531	0.617525
0.00	0.7	12.331795	9.746593	0.872746	7.764863	0.469002	0.811528
0.00	0.8	24.191903	15.687904	0.894783	16.152327	0.978630	1.173093
0.00	0.9	78.241470	34.877748	0.942970	48.316045	2.580748	2.132378
0.01	0.0	4.214417	2.587046	19.696570	-0.352750	-2.036080	3.301319
0.01	0.1	4.254732	3.144779	20.397153	-0.206331	-1.872792	3.926267
0.01	0.2	4.367371	3.844368	21.183865	0.073526	-1.576366	4.668071
0.01	0.3	4.607252	4.791560	22.153035	0.549449	-1.095368	5.626098
0.01	0.4	5.097602	6.152365	23.412547	1.365438	-0.310978	6.935128
0.01	0.5	6.126614	8.255096	25.148672	2.835131	1.023280	8.845486
0.01	0.6	8.465396	11.852013	27.738004	5.733298	3.479588	11.896600
0.01	0.7	14.663260	19.058555	32.095821	12.411349	8.664920	17.498423
0.01	0.8	36.871341	38.279034	41.223199	33.053883	22.855777	30.733298
0.01	0.9	208.68444	142.33940	74.231077	166.61531	97.687475	89.216941
0.02	0.0	4.332557	2.807065	38.013137	-0.513932	-3.505939	5.304855
0.02	0.1	4.359389	3.456637	39.797928	-0.386932	-3.296789	6.381326
0.02	0.2	4.448881	4.284758	41.827713	-0.115845	-2.872971	7.677695
0.02	0.3	4.657093	5.429276	44.370447	0.371341	-2.147542	9.383453
0.02	0.4	5.112406	7.115947	47.744882	1.246523	-0.910807	11.768853
0.02	0.5	6.129487	9.808393	52.524255	2.899656	1.289503	15.355472
0.02	0.6	8.600994	14.620511	59.923096	6.345242	5.556817	21.320879
0.02	0.7	15.716029	24.901344	73.091180	14.890612	15.212275	32.951222
0.02	0.8	44.613630	55.497021	103.51321	44.597795	44.759019	63.434806
0.02	0.9	343.90604	273.79291	245.22267	299.92546	249.41451	238.94306
0.05	0.0	4.673637	3.316628	118.29087	-0.930522	-8.724988	11.661725
0.05	0.1	4.679376	4.203784	126.70234	-0.875463	-8.558704	14.392725
0.05	0.2	4.720120	5.376500	136.55198	-0.659355	-7.933917	17.790985
0.05	0.3	4.850354	7.073618	149.36917	-0.191160	-6.649218	22.454357
0.05	0.4	5.200681	9.722999	167.21632	0.770007	-4.157878	29.329658
0.05	0.5	16.132521	14.281779	194.14986	2.827748	0.837008	40.408205
0.05	0.6	8.836394	23.321970	239.75342	7.765394	11.909761	60.706382
0.05	0.7	18.509544	45.968385	333.17349	22.549173	41.876142	106.67955
0.05	0.8	74.018607	136.57600	615.57130	93.390074	166.56902	266.49072
0.05	0.9	1689.5180	1854.2510	3883.3175	1757.6989	2451.8213	2631.1850
0.10	0.0	5.320614	4.059563	377.17265	-1.623340	-21.628369	25.439673
0.10	0.1	5.334682	5.356380	416.16023	-1.743869	-22.297094	32.548149
0.10	0.2	5.336669	7.160202	463.78029	-1.706974	-22.111550	41.815007
0.10	0.3	5.369733	9.946280	529.22447	-1.410643	-20.681001	55.316087
0.10	0.4	5.539794	14.673882	627.01619	-0.520768	-16.644112	76.813755
0.10	0.5	6.217687	23.776403	789.58411	1.953580	-6.210807	115.27194
0.10	0.6	9.008836	45.048380	1107.6571	9.639324	23.439865	197.49727
0.10	0.7	24.044058	115.23938	1943.6777	42.063357	135.03520	439.59450
0.10	0.8	204.29975	640.49046	6468.2685	349.31982	1033.7169	1979.6251

B. $\sigma = 2$

Q1	Q2	$N\,VAR(\hat{\sigma}\mid\mu)$	$\frac{N\,VAR(\hat{\tau}\mid\mu)}{EXP(2\mu-8)}$	$\frac{N\,COV(\hat{\sigma},\hat{\tau}\mid\mu)}{EXP(\mu-4)}$	$N\,VAR(\hat{\mu}\mid\sigma,\tau)$	$N\,VAR(\hat{\sigma}\mid\mu,\tau)$	$\frac{N\,VAR(\hat{\tau}\mid\mu,\sigma)}{EXP(2\mu-8)}$
0.00	0.0	2.060379	0.824152	0.223074	4.000000	2.000000	0.800000
0.00	0.1	2.419874	0.828347	0.261907	4.068818	2.337064	0.800000
0.00	0.2	2.823637	0.833012	0.305309	4.182913	2.711737	0.800000
0.00	0.3	3.278446	0.838197	0.353871	4.344818	3.129048	0.800001
0.00	0.4	3.767041	0.843662	0.405542	4.570005	3.572100	0.800003
0.00	0.5	4.244172	0.848843	0.455262	4.888124	4.000000	0.800008
0.00	0.6	4.634382	0.852867	0.494884	5.357289	4.347220	0.800020
0.00	0.7	4.857354	0.854909	0.516215	6.106587	4.545650	0.800049
0.00	0.8	4.903400	0.855195	0.519686	7.496319	4.587597	0.800116
0.00	0.9	5.041394	0.857845	0.538702	11.137966	4.703105	0.800282
0.01	0.0	2.557520	18.712894	3.130898	4.003878	2.033682	14.880071
0.01	0.1	3.134774	19.572812	3.835447	4.072830	2.383187	14.880079
0.01	0.2	3.843130	20.614889	4.694609	4.187153	2.774031	14.880149
0.01	0.3	4.726034	21.892612	5.756729	4.349393	3.212285	14.880406
0.01	0.4	5.786621	23.393576	7.018426	4.575067	3.680987	14.881130
0.01	0.5	6.943120	24.977762	8.371956	4.893916	4.137038	14.882926
0.01	0.6	7.969063	26.307766	9.540004	5.364247	4.509564	14.887141
0.01	0.7	8.553280	26.975484	10.164219	6.115629	4.723456	14.896919
0.01	0.8	8.647370	27.055378	10.243883	7.509949	4.768765	14.920228
0.01	0.9	9.312407	28.502499	11.222486	11.168083	4.893700	14.978156
0.02	0.0	2.746102	35.176104	4.888977	4.008911	2.066603	26.472094
0.02	0.1	3.422294	37.304731	6.088709	4.078039	2.428522	26.472121
0.02	0.2	4.281741	39.972423	7.602885	4.192659	2.835648	26.472341
0.02	0.3	5.399667	43.380144	9.554690	4.355334	3.295199	26.473155
0.02	0.4	6.812016	47.582616	11.990929	4.581641	3.790274	26.475446
0.02	0.5	8.436662	52.252974	14.745451	4.901438	4.275592	26.481132
0.02	0.6	9.939413	56.333021	17.221429	5.373286	4.674693	26.494478
0.02	0.7	10.792797	58.366512	18.537907	6.127381	4.904935	26.525464
0.02	0.8	10.915056	58.608332	18.691674	7.527678	4.953810	26.599457
0.02	0.9	12.223550	64.337223	21.425035	11.207335	5.088767	26.784130
0.05	0.0	3.131361	102.00261	9.924579	4.027807	2.165726	70.547514
0.05	0.1	4.039994	111.04825	12.791479	4.097594	2.566563	70.547706
0.05	0.2	5.284395	123.21609	16.682693	4.213331	3.025662	70.549269
0.05	0.3	7.066084	140.25394	22.192302	4.377646	3.554609	70.555049
0.05	0.4	9.608993	163.89214	29.945268	4.606338	4.137595	70.571326
0.05	0.5	12.977885	194.03562	40.022255	4.929714	4.722798	70.611744
0.05	0.6	16.497768	223.70135	50.240124	5.407287	5.214555	70.706714
0.05	0.7	18.497952	238.43257	55.664123	6.171634	5.502687	70.927836
0.05	0.8	18.744742	240.72990	56.328789	7.594580	5.564275	71.459367
0.05	0.9	25.619683	325.24493	80.419863	11.356275	5.735117	72.807989
0.10	0.0	3.564276	289.25304	18.840774	4.068818	2.337064	189.66066
0.10	0.1	4.786322	322.96623	25.259388	4.140045	2.810769	189.66205
0.10	0.2	6.614213	372.16498	34.742459	4.258228	3.370924	189.67334
0.10	0.3	9.575701	449.57362	49.883136	4.426133	4.040841	189.71513
0.10	0.4	14.624927	577.00955	75.249126	4.660054	4.811518	189.83286
0.10	0.5	23.162594	783.38018	117.22335	4.991288	5.621539	190.12560
0.10	0.6	34.734443	1046.6695	172.41695	5.481458	6.332349	190.81568
0.10	0.7	41.652553	1185.2989	203.36010	6.268444	6.762342	192.43470
0.10	0.8	43.209581	1237.8624	212.13494	7.741710	6.855594	196.39815

B. σ = 2

Q1	Q2	N VAR($\hat{\mu}\|\tau$)	N VAR($\hat{\sigma}\|\tau$)	N COV($\hat{\mu}, \hat{\sigma}\|\tau$)	N VAR($\hat{\mu}\|\sigma$)	$\frac{\text{N VAR}(\hat{\tau}\|\sigma)}{\text{EXP}(2\mu-8)}$	$\frac{\text{N COV}(\hat{\mu},\hat{\tau}\|\sigma)}{\text{EXP}(\mu-4)}$
0.00	0.0	4.000000	2.000000	0.000000	4.014706	0.802941	-0.108666
0.00	0.1	4.080370	2.343700	0.164542	4.084046	0.802994	-0.110582
0.00	0.2	4.249291	2.754769	0.427619	4.199052	0.803087	-0.113847
0.00	0.3	4.553029	3.278998	0.826272	4.362338	0.803227	-0.118627
0.00	0.4	5.090625	3.979038	1.439295	4.589612	0.803435	-0.125511
0.00	0.5	6.068376	4.965812	2.420932	4.911011	0.803754	-0.135630
0.00	0.6	7.963401	6.461974	4.103734	5.385732	0.804268	-0.151249
0.00	0.7	12.079760	8.991990	7.328760	6.145726	0.805176	-0.177522
0.00	0.8	23.121569	14.149937	14.869308	7.561405	0.807063	-0.229191
0.00	0.9	71.178397	30.055711	42.480088	11.309390	0.812599	-0.373228
0.01	0.0	4.003943	2.033715	-0.011485	4.166319	15.483768	-1.585937
0.01	0.1	4.082779	2.389009	0.154165	4.241194	15.495197	-1.615188
0.01	0.2	4.250068	2.815713	0.420893	4.365965	15.515604	-1.665646
0.01	0.3	4.553091	3.362728	0.827634	4.544246	15.547049	-1.740514
0.01	0.4	5.093471	4.098082	1.457554	4.794560	15.595067	-1.850139
0.01	0.5	6.084977	5.143893	2.475215	5.152916	15.670572	-2.014615
0.01	0.6	8.028899	6.749660	4.240931	5.691968	15.796649	-2.275278
0.01	0.7	12.323989	9.518534	7.687294	6.580718	16.029818	-2.730437
0.01	0.8	24.199191	15.366315	16.014123	8.329367	16.548189	-3.682374
0.01	0.9	80.128543	35.111225	49.206567	13.652963	18.310771	-6.745374
0.02	0.0	4.009206	2.066755	-0.024667	4.238464	27.987902	-2.534700
0.02	0.1	4.086289	2.433435	0.141686	4.316076	28.017309	-2.582470
0.02	0.2	4.251548	2.875477	0.411504	4.445748	28.070344	-2.665393
0.02	0.3	4.553151	3.444866	0.825502	4.631695	28.152966	-2.789332
0.02	0.4	5.095031	4.214989	1.471032	4.894049	28.280726	-2.972395
0.02	0.5	6.097829	5.319219	2.522670	5.272261	28.484587	-3.250036
0.02	0.6	8.085697	7.034457	4.368105	5.847185	28.831168	-3.696358
0.02	0.7	12.549938	10.046158	8.032561	6.811677	29.487789	-4.492037
0.02	0.8	25.259871	16.623002	17.168643	8.774530	31.005275	-6.217635
0.02	0.9	90.228035	40.968656	56.897908	15.353767	36.693585	-12.334807
0.05	0.0	4.030092	2.166955	-0.070367	4.412567	77.286626	-5.453144
0.05	0.1	4.101239	2.568846	0.096761	4.497056	77.425190	-5.561334
0.05	0.2	4.259145	3.058562	0.374333	4.639259	77.681135	-5.752093
0.05	0.3	4.554362	3.698101	0.808404	4.845188	78.090495	-6.042405
0.05	0.4	5.097294	4.578592	1.499296	5.139701	78.742707	-6.480621
0.05	0.5	6.128913	5.871662	2.653543	5.572636	79.820759	-7.163692
0.05	0.6	8.244776	7.950908	4.749802	6.250792	81.736547	-8.303324
0.05	0.7	13.246187	11.810424	9.140759	7.448351	85.600577	-10.454077
0.05	0.8	28.946268	21.207888	21.279667	10.158883	95.587561	-15.656165
0.05	0.9	141.50457	71.462275	96.440102	23.343612	149.66189	-42.356199
0.10	0.0	4.080370	2.343700	-0.164542	4.671473	217.75232	-11.455542
0.10	0.1	4.140045	2.810769	0.000000	4.766933	218.38077	-11.700437
0.10	0.2	4.282462	3.390109	0.286630	4.929731	219.58398	-12.142955
0.10	0.3	4.561563	4.164481	0.750996	5.169667	221.58489	-12.835734
0.10	0.4	5.097977	5.263674	1.518249	5.521312	224.91722	-13.918034
0.10	0.5	6.168833	6.947773	2.860300	6.057171	230.72669	-15.682085
0.10	0.6	8.512809	9.834259	5.459954	6.946241	241.80640	-18.820038
0.10	0.7	14.662614	15.817898	11.522939	8.690574	266.79158	-25.420542
0.10	0.8	39.097786	34.622654	32.948908	13.782729	349.65177	-45.959254

Table F3

COEFFICIENTS IN ASYMPTOTIC VARIANCES AND COVARIANCES OF MAXIMUM-LIKELIHOOD ESTIMATORS OF PARAMETERS c, θ, AND K OF WEIBULL POPULATION FROM SAMPLES OF SIZE n WITH PROPORTIONS q_1 CENSORED FROM BELOW AND q_2 FROM ABOVE

A. K=1

Q1	Q2	$\frac{V(\hat{\theta})}{\theta^2/N}$	$\frac{V(\hat{K})}{1/N}$	$\frac{V(\hat{c})}{\theta^2/N}$	$\frac{C(\hat{\theta},\hat{K})}{\theta/N}$	$\frac{C(\hat{\theta},\hat{c})}{\theta^2/N}$	$\frac{C(\hat{K},\hat{c})}{\theta/N}$	$\frac{V(\hat{K}\mid\theta)}{1/N}$	$\frac{V(\hat{c}\mid\theta)}{\theta^2/N}$	$\frac{C(\hat{K},\hat{c}\mid\theta)}{\theta/N}$
.005	.00	1.143	0.678	0.006	0.306	-0.014	-0.020	0.596	0.005	-0.016
.010	.00	1.173	0.725	0.012	0.343	-0.028	-0.037	0.624	0.011	-0.029
.015	.00	1.202	0.766	0.019	0.378	-0.042	-0.054	0.647	0.018	-0.041
.020	.00	1.231	0.805	0.027	0.411	-0.057	-0.072	0.667	0.025	-0.053
.025	.00	1.260	0.842	0.036	0.445	-0.073	-0.090	0.685	0.032	-0.064
.005	.25	1.342	1.196	0.006	-0.002	-0.007	-0.032	1.196	0.006	-0.032
.010	.25	1.352	1.317	0.013	0.033	-0.015	-0.061	1.316	0.013	-0.061
.015	.25	1.364	1.427	0.021	0.069	-0.025	-0.091	1.424	0.021	-0.090
.020	.25	1.377	1.532	0.031	0.106	-0.036	-0.123	1.524	0.030	-0.120
.025	.25	1.392	1.635	0.041	0.146	-0.049	-0.156	1.620	0.040	-0.151
.005	.50	2.554	2.157	0.006	-1.073	0.016	-0.052	1.706	0.006	-0.045
.010	.50	2.569	2.471	0.015	-1.143	0.027	-0.103	1.963	0.014	-0.091
.015	.50	2.579	2.766	0.025	-1.196	0.037	-0.157	2.211	0.024	-0.140
.020	.50	2.585	3.057	0.037	-1.239	0.045	-0.216	2.463	0.036	-0.194
.025	.50	2.589	3.350	0.050	-1.273	0.053	-0.280	2.724	0.049	-0.254
.005	.75	12.501	5.496	0.007	-6.803	0.118	-0.112	1.794	0.006	-0.048
.010	.75	13.780	6.880	0.019	-8.133	0.238	-0.236	2.080	0.015	-0.096
.015	.75	14.946	8.295	0.034	-9.417	0.371	-0.383	2.362	0.025	-0.150
.020	.75	16.080	9.806	0.054	-10.726	0.521	-0.556	2.651	0.037	-0.209
.025	.75	17.218	11.449	0.079	-12.093	0.690	-0.759	2.955	0.051	-0.275

Q1	Q2	$\frac{V(\hat{\theta}\mid K)}{\theta^2/N}$	$\frac{V(\hat{c}\mid K)}{\theta^2/N}$	$\frac{C(\hat{\theta},\hat{c}\mid K)}{\theta^2/N}$	$\frac{V(\hat{\theta}\mid c)}{\theta^2/N}$	$\frac{V(\hat{K}\mid c)}{1/N}$	$\frac{C(\hat{\theta},\hat{K}\mid c)}{\theta/N}$	$\frac{V(\hat{\theta}\mid K,c)}{\theta^2/N}$	$\frac{V(\hat{K}\mid\theta,c)}{1/N}$	$\frac{V(\hat{c}\mid\theta,K)}{\theta^2/N}$
.000	.00				1.109	0.608	0.257	1.000	0.548	
.005	.00	1.005	0.005	-0.005	1.109	0.610	0.258	1.000	0.550	0.005
.010	.00	1.010	0.010	-0.010	1.109	0.612	0.259	1.000	0.551	0.010
.015	.00	1.015	0.015	-0.015	1.110	0.614	0.260	1.000	0.553	0.015
.020	.00	1.020	0.021	-0.021	1.110	0.616	0.260	1.000	0.555	0.020
.025	.00	1.026	0.026	-0.026	1.111	0.618	0.261	1.000	0.556	0.026
.000	.25				1.335	1.020	-0.037	1.333	1.019	
.005	.25	1.342	0.005	-0.007	1.335	1.025	-0.038	1.333	1.024	0.005
.010	.25	1.351	0.010	-0.014	1.335	1.031	-0.038	1.333	1.030	0.010
.015	.25	1.361	0.016	-0.021	1.335	1.036	-0.038	1.333	1.035	0.015
.020	.25	1.370	0.021	-0.028	1.335	1.042	-0.038	1.333	1.041	0.020
.025	.25	1.379	0.027	-0.035	1.335	1.048	-0.038	1.333	1.047	0.026
.000	.50				2.510	1.716	-0.936	2.000	1.367	
.005	.50	2.020	0.005	-0.010	2.515	1.731	-0.944	2.000	1.377	0.005
.010	.50	2.041	0.010	-0.021	2.519	1.747	-0.952	2.000	1.387	0.010
.015	.50	2.062	0.016	-0.031	2.524	1.763	-0.961	2.000	1.397	0.015
.020	.50	2.083	0.021	-0.042	2.529	1.780	-0.970	2.000	1.408	0.020
.025	.50	2.105	0.027	-0.053	2.534	1.797	-0.979	2.000	1.418	0.026
.000	.75				10.498	3.736	-4.927	4.000	1.423	
.005	.75	4.082	0.005	-0.020	10.623	3.807	-5.022	4.000	1.434	0.005
.010	.75	4.167	0.011	-0.042	10.754	3.883	-5.121	4.000	1.444	0.010
.015	.75	4.255	0.016	-0.064	10.892	3.963	-5.226	4.000	1.456	0.015
.020	.75	4.348	0.022	-0.088	11.036	4.047	-5.336	4.000	1.467	0.020
.025	.75	4.444	0.028	-0.113	11.187	4.136	-5.452	4.000	1.479	0.026

B. K=2

Q1	Q2	$\frac{V(\hat{\theta})}{\theta^2/N}$	$\frac{V(\hat{K})}{1/N}$	$\frac{V(\hat{c})}{\theta^2/N}$	$\frac{C(\hat{\theta},\hat{K})}{\theta/N}$	$\frac{C(\hat{\theta},\hat{c})}{\theta^2/N}$	$\frac{C(\hat{K},\hat{c})}{\theta/N}$	$\frac{V(\hat{K}\mid\theta)}{1/N}$	$\frac{V(\hat{c}\mid\theta)}{\theta^2/N}$	$\frac{C(\hat{K},\hat{c}\mid\theta)}{\theta/N}$
.005	.00	0.886	5.897	0.434	1.709	-0.514	-1.225	2.602	0.136	-0.234
.010	.00	1.064	6.765	0.567	2.102	-0.668	-1.565	2.613	0.148	-0.246
.015	.00	1.214	7.459	0.681	2.424	-0.799	-1.846	2.618	0.156	-0.251
.020	.00	1.351	8.072	0.787	2.714	-0.919	-2.101	2.619	0.162	-0.254
.025	.00	1.481	8.639	0.888	2.986	-1.034	-2.340	2.620	0.166	-0.256
.005	.25	0.916	11.535	0.544	2.042	-0.563	-2.011	6.980	0.198	-0.756
.010	.25	1.133	13.758	0.746	2.738	-0.773	-2.682	7.144	0.220	-0.815
.015	.25	1.330	15.628	0.929	3.344	-0.962	-3.265	7.219	0.233	-0.847
.020	.25	1.520	17.347	1.105	3.916	-1.145	-3.816	7.259	0.243	-0.866
.025	.25	1.710	18.992	1.281	4.474	-1.327	-4.354	7.282	0.250	-0.880
.005	.50	0.969	22.484	0.691	1.358	-0.485	-3.279	20.581	0.448	-2.600
.010	.50	1.156	27.993	1.002	2.371	-0.726	-4.588	23.128	0.546	-3.099
.015	.50	1.344	32.883	1.300	3.332	-0.963	-5.795	24.626	0.610	-3.409
.020	.50	1.543	37.580	1.603	4.298	-1.209	-6.989	25.609	0.657	-3.622
.025	.50	1.756	42.248	1.918	5.295	-1.468	-8.201	26.283	0.692	-3.776
.005	.75	2.736	63.723	1.059	-6.992	0.291	-7.167	45.853	1.028	-6.422
.010	.75	2.736	86.664	1.716	-6.967	0.287	-11.048	68.923	1.686	-10.318
.015	.75	2.743	109.409	2.429	-6.593	0.220	-15.076	93.562	2.412	-14.547
.020	.75	2.762	133.401	3.234	-5.921	0.097	-19.469	120.707	3.230	-19.261
.025	.75	2.799	159.352	4.151	-4.948	-0.086	-24.349	150.605	4.149	-24.501

Q1	Q2	$\frac{V(\hat{\theta}\mid K)}{\theta^2/N}$	$\frac{V(\hat{c}\mid K)}{\theta^2/N}$	$\frac{C(\hat{\theta},\hat{c}\mid K)}{\theta^2/N}$	$\frac{V(\hat{\theta}\mid c)}{\theta^2/N}$	$\frac{V(\hat{K}\mid c)}{1/N}$	$\frac{C(\hat{\theta},\hat{K}\mid c)}{\theta/N}$	$\frac{V(\hat{\theta}\mid K,c)}{\theta^2/N}$	$\frac{V(\hat{K}\mid\theta,c)}{1/N}$	$\frac{V(\hat{c}\mid\theta,K)}{\theta^2/N}$
.000	.00				0.277	2.432	0.257	0.250	2.193	
.005	.00	0.391	0.179	-0.159	0.277	2.439	0.258	0.250	2.199	0.115
.010	.00	0.411	0.205	-0.182	0.277	2.447	0.259	0.250	2.206	0.125
.015	.00	0.426	0.224	-0.199	0.277	2.455	0.260	0.250	2.212	0.132
.020	.00	0.438	0.240	-0.213	0.278	2.463	0.260	0.250	2.219	0.137
.025	.00	0.449	0.254	-0.225	0.278	2.471	0.261	0.250	2.226	0.141
.000	.25				0.334	4.080	-0.037	0.333	4.075	
.005	.25	0.554	0.193	-0.207	0.334	4.101	-0.038	0.333	4.097	0.116
.010	.25	0.588	0.224	-0.239	0.334	4.123	-0.038	0.333	4.119	0.127
.015	.25	0.614	0.246	-0.263	0.334	4.145	-0.038	0.333	4.141	0.134
.020	.25	0.636	0.266	-0.284	0.334	4.169	-0.038	0.333	4.164	0.139
.025	.25	0.655	0.283	-0.302	0.334	4.192	-0.038	0.333	4.188	0.144
.000	.50				0.628	6.865	-0.936	0.500	5.469	
.005	.50	0.887	0.213	-0.287	0.629	6.925	-0.944	0.500	5.508	0.120
.010	.50	0.955	0.250	-0.337	0.630	6.987	-0.952	0.500	5.547	0.131
.015	.50	1.007	0.279	-0.376	0.631	7.052	-0.961	0.500	5.588	0.138
.020	.50	1.052	0.304	-0.409	0.632	7.119	-0.970	0.500	5.630	0.144
.025	.50	1.093	0.326	-0.440	0.633	7.187	-0.979	0.500	5.674	0.149
.000	.75				2.624	14.942	-4.927	1.000	5.693	
.005	.75	1.969	0.253	-0.495	2.656	15.230	-5.022	1.000	5.735	0.129
.010	.75	2.176	0.308	-0.601	2.688	15.534	-5.121	1.000	5.778	0.141
.015	.75	2.346	0.352	-0.688	2.723	15.853	-5.226	1.000	5.822	0.150
.020	.75	2.499	0.392	-0.767	2.759	16.190	-5.336	1.000	5.868	0.157
.025	.75	2.645	0.431	-0.842	2.797	16.544	-5.452	1.000	5.915	0.163

C. K=3

Q1	Q2	$V(\hat\theta)$ / θ^2/N	$V(\hat K)$ / $1/N$	$V(\hat c)$ / θ^2/N	$C(\hat\theta,\hat K)$ / θ/N	$C(\hat\theta,\hat c)$ / θ^2/N	$C(\hat K,\hat c)$ / θ/N	$V(\hat K\|\theta)$ / $1/N$	$V(\hat c\|\theta)$ / θ^2/N	$C(\hat K,\hat c\|\theta)$ / θ/N
.000	.00	1.361	19.832	1.072	4.474	-1.152	-3.924	5.130	0.097	-0.138
.005	.00	2.340	30.383	1.931	7.686	-2.069	-6.933	5.134	0.102	-0.137
.010	.00	2.735	34.326	2.283	8.934	-2.442	-8.111	5.139	0.103	-0.135
.015	.00	3.064	37.522	2.578	9.960	-2.753	-9.082	5.146	0.104	-0.132
.020	.00	3.364	40.367	2.847	10.884	-3.038	-9.957	5.154	0.104	-0.130
.025	.00	3.648	43.007	3.103	11.749	-3.307	-10.779	5.163	0.105	-0.128
.000	.25	1.503	35.308	1.317	5.913	-1.336	-5.867	12.052	0.130	-0.612
.005	.25	2.981	61.075	2.752	12.082	-2.792	-11.944	12.112	0.137	-0.629
.010	.25	3.660	71.795	3.409	14.779	-3.460	-14.599	12.115	0.138	-0.627
.015	.25	4.258	80.907	3.989	17.115	-4.049	-16.899	12.125	0.139	-0.625
.020	.25	4.828	89.336	4.541	19.305	-4.610	-19.055	12.140	0.139	-0.622
.025	.25	5.387	97.425	5.083	21.432	-5.160	-21.149	12.160	0.140	-0.619
.000	.50	1.528	60.538	1.581	6.570	-1.406	-8.444	32.294	0.289	-2.402
.005	.50	3.429	119.035	3.890	17.108	-3.500	-20.060	33.689	0.317	-2.598
.010	.50	4.447	146.176	5.084	22.364	-4.603	-25.754	33.715	0.320	-2.607
.015	.50	5.410	170.457	6.199	27.198	-5.639	-30.956	33.716	0.321	-2.607
.020	.50	6.376	193.877	7.308	31.955	-6.674	-36.052	33.726	0.322	-2.604
.025	.50	7.370	217.202	8.441	36.771	-7.735	-41.194	33.751	0.322	-2.601
.000	.75	1.768	133.333	2.061	2.815	-1.113	-14.335	128.849	1.360	-12.562
.005	.75	3.625	334.411	6.921	22.065	-4.113	-45.576	200.095	2.254	-20.539
.010	.75	5.069	449.338	10.083	34.945	-6.250	-64.640	208.423	2.377	-21.552
.015	.75	6.684	563.786	13.398	48.539	-8.564	-84.118	211.287	2.426	-21.927
.020	.75	8.539	684.737	17.038	63.516	-11.162	-105.098	212.260	2.447	-22.069
.025	.75	10.684	815.705	21.102	80.279	-14.115	-128.169	212.509	2.455	-22.114

Q1	Q2	$V(\hat\theta\|K)$ / θ^2/N	$V(\hat c\|K)$ / θ^2/N	$C(\hat\theta,\hat c\|K)$ / θ^2/N	$V(\hat\theta\|c)$ / θ^2/N	$V(\hat K\|c)$ / $1/N$	$C(\hat\theta,\hat K\|c)$ / θ/N	$V(\hat\theta\|K,c)$ / θ^2/N	$V(\hat K\|\theta,c)$ / $1/N$	$V(\hat c\|\theta,K)$ / θ^2/N
.000	.00	0.352	0.296	-0.267	0.123	5.471	0.257	0.111	4.935	0.093
.005	.00	0.395	0.349	-0.315	0.123	5.488	0.258	0.111	4.949	0.098
.010	.00	0.409	0.366	-0.330	0.123	5.506	0.259	0.111	4.963	0.099
.015	.00	0.420	0.379	-0.343	0.123	5.524	0.260	0.111	4.977	0.100
.020	.00	0.430	0.391	-0.353	0.123	5.542	0.260	0.111	4.992	0.101
.025	.00	0.438	0.401	-0.362	0.123	5.561	0.261	0.111	5.007	0.102
.000	.25	0.513	0.342	-0.354	0.148	9.179	-0.037	0.148	9.170	0.099
.005	.25	0.591	0.416	-0.429	0.148	9.227	-0.038	0.148	9.217	0.104
.010	.25	0.618	0.440	-0.455	0.148	9.276	-0.038	0.148	9.267	0.106
.015	.25	0.638	0.460	-0.475	0.148	9.327	-0.038	0.148	9.317	0.107
.020	.25	0.656	0.477	-0.492	0.148	9.379	-0.038	0.148	9.369	0.108
.025	.25	0.672	0.492	-0.508	0.148	9.433	-0.038	0.148	9.423	0.108
.000	.50	0.815	0.403	-0.489	0.279	15.446	-0.936	0.222	12.306	0.110
.005	.50	0.971	0.509	-0.617	0.279	15.581	-0.944	0.222	12.392	0.117
.010	.50	1.026	0.547	-0.663	0.280	15.722	-0.952	0.222	12.481	0.118
.015	.50	1.070	0.577	-0.699	0.280	15.867	-0.961	0.222	12.573	0.120
.020	.50	1.109	0.604	-0.732	0.281	16.017	-0.970	0.222	12.668	0.121
.025	.50	1.145	0.628	-0.762	0.282	16.172	-0.979	0.222	12.766	0.122
.000	.75	1.708	0.520	-0.810	1.166	33.620	-4.927	0.444	12.810	0.135
.005	.75	2.169	0.709	-1.106	1.180	34.267	-5.022	0.444	12.903	0.145
.010	.75	2.351	0.784	-1.223	1.195	34.951	-5.121	0.444	13.000	0.148
.015	.75	2.505	0.848	-1.322	1.210	35.670	-5.226	0.444	13.100	0.150
.020	.75	2.647	0.906	-1.413	1.226	36.427	-5.336	0.444	13.203	0.152
.025	.75	2.783	0.963	-1.501	1.243	37.223	-5.452	0.444	13.309	0.154

Table F4

COEFFICIENTS IN ASYMPTOTIC VARIANCES AND COVARIANCES OF MAXIMUM-LIKELIHOOD ESTIMATORS OF PARAMETERS c, θ, AND α OF GAMMA POPULATION FROM SAMPLES OF SIZE n WITH PROPORTIONS q_1 CENSORED FROM BELOW AND q_2 FROM ABOVE

A. α=1

Q1	Q2	$\frac{V(\hat{\theta})}{\theta^2/N}$	$\frac{V(\hat{\alpha})}{1/N}$	$\frac{V(\hat{c})}{\theta^2/N}$	$\frac{C(\hat{\theta},\hat{\alpha})}{\theta/N}$	$\frac{C(\hat{\theta},\hat{c})}{\theta^2/N}$	$\frac{C(\hat{\alpha},\hat{c})}{\theta/N}$	$\frac{V(\hat{\alpha}\mid\theta)}{1/N}$	$\frac{V(\hat{c}\mid\theta)}{\theta^2/N}$	$\frac{C(\hat{\alpha},\hat{c}\mid\theta)}{\theta/N}$
.005	.00	2.812	1.896	0.006	-1.851	0.039	-0.045	0.678	0.006	-0.019
.010	.00	2.970	2.131	0.014	-2.044	0.074	-0.088	0.725	0.012	-0.037
.015	.00	3.105	2.344	0.023	-2.213	0.109	-0.131	0.766	0.019	-0.054
.020	.00	3.228	2.546	0.033	-2.371	0.144	-0.177	0.805	0.027	-0.071
.025	.00	3.343	2.744	0.045	-2.522	0.181	-0.225	0.842	0.035	-0.089
.005	.25	4.956	2.556	0.006	-3.039	0.063	-0.058	0.692	0.006	-0.020
.010	.25	5.360	2.941	0.015	-3.433	0.120	-0.115	0.741	0.012	-0.038
.015	.25	5.714	3.297	0.025	-3.788	0.180	-0.175	0.785	0.019	-0.055
.020	.25	6.045	3.643	0.037	-4.127	0.243	-0.239	0.826	0.027	-0.073
.025	.25	6.363	3.989	0.050	-4.459	0.308	-0.307	0.865	0.035	-0.091
.005	.50	10.389	3.808	0.007	-5.645	0.110	-0.081	0.741	0.006	-0.021
.010	.50	11.625	4.540	0.016	-6.596	0.218	-0.164	0.797	0.012	-0.040
.015	.50	12.751	5.242	0.028	-7.485	0.334	-0.256	0.848	0.019	-0.060
.020	.50	13.839	5.946	0.043	-8.361	0.459	-0.357	0.895	0.027	-0.079
.025	.50	14.918	6.669	0.060	-9.244	0.595	-0.468	0.941	0.036	-0.099
.005	.75	35.945	7.938	0.008	-15.904	0.277	-0.148	0.901	0.006	-0.026
.010	.75	43.463	10.290	0.020	-20.109	0.584	-0.320	0.986	0.013	-0.050
.015	.75	50.925	12.746	0.038	-24.390	0.946	-0.528	1.065	0.020	-0.075
.020	.75	58.722	15.415	0.061	-28.952	1.372	-0.777	1.141	0.029	-0.101
.025	.75	67.055	18.365	0.091	-33.910	1.872	-1.075	1.217	0.039	-0.128

Q1	Q2	$\frac{V(\hat{\theta}\mid\alpha)}{\theta^2/N}$	$\frac{V(\hat{c}\mid\alpha)}{\theta^2/N}$	$\frac{C(\hat{\theta},\hat{c}\mid\alpha)}{\theta^2/N}$	$\frac{V(\hat{\theta}\mid c)}{\theta^2/N}$	$\frac{V(\hat{\alpha}\mid c)}{1/N}$	$\frac{C(\hat{\theta},\hat{\alpha}\mid c)}{\theta/N}$	$\frac{V(\hat{\theta}\mid\alpha,c)}{\theta^2/N}$	$\frac{V(\hat{\alpha}\mid\theta,c)}{1/N}$	$\frac{V(\hat{c}\mid\theta,\alpha)}{\theta^2/N}$
.000	.00				2.551	1.551	-1.551	1.000	0.608	
.005	.00	1.005	0.005	-0.005	2.563	1.563	-1.563	1.000	0.610	0.005
.010	.00	1.010	0.010	-0.010	2.575	1.575	-1.575	1.000	0.612	0.010
.015	.00	1.015	0.015	-0.015	2.587	1.587	-1.587	1.000	0.614	0.015
.020	.00	1.020	0.021	-0.021	2.600	1.600	-1.600	1.000	0.615	0.020
.025	.00	1.026	0.026	-0.026	2.612	1.613	-1.612	1.000	0.617	0.026
.000	.25				4.314	2.005	-2.445	1.333	0.620	
.005	.25	1.342	0.005	-0.007	4.344	2.025	-2.469	1.333	0.622	0.005
.010	.25	1.351	0.010	-0.014	4.375	2.046	-2.494	1.333	0.623	0.010
.015	.25	1.361	0.016	-0.021	4.406	2.067	-2.520	1.333	0.625	0.015
.020	.25	1.370	0.021	-0.028	4.437	2.088	-2.546	1.333	0.627	0.020
.025	.25	1.379	0.027	-0.035	4.469	2.110	-2.572	1.333	0.629	0.026
.000	.50				8.513	2.801	-4.271	2.000	0.658	
.005	.50	2.020	0.005	-0.010	8.605	2.840	-4.331	2.000	0.660	0.005
.010	.50	2.041	0.010	-0.021	8.699	2.881	-4.393	2.000	0.662	0.010
.015	.50	2.062	0.016	-0.031	8.795	2.922	-4.456	2.000	0.665	0.015
.020	.50	2.083	0.021	-0.042	8.893	2.965	-4.521	2.000	0.667	0.020
.025	.50	2.105	0.027	-0.053	8.993	3.009	-4.587	2.000	0.669	0.026
.000	.75				25.685	5.021	-10.435	4.000	0.782	
.005	.75	4.082	0.005	-0.020	26.241	5.150	-10.702	4.000	0.785	0.005
.010	.75	4.167	0.011	-0.042	26.820	5.284	-10.981	4.000	0.788	0.010
.015	.75	4.255	0.016	-0.064	27.424	5.425	-11.273	4.000	0.791	0.015
.020	.75	4.000	0.794	0.020	28.054	5.572	-11.577	4.000	0.794	0.020
.025	.75	4.444	0.028	-0.113	28.713	5.726	-11.895	4.000	0.798	0.026

B. $\alpha=2$

Q1	Q2	$\frac{V(\hat{\theta})}{\theta^2/N}$	$\frac{V(\hat{\alpha})}{1/N}$	$\frac{V(\hat{c})}{\theta^2/N}$	$\frac{C(\hat{\theta},\hat{\alpha})}{\theta/N}$	$\frac{C(\hat{\theta},\hat{c})}{\theta^2/N}$	$\frac{C(\hat{\alpha},\hat{c})}{\theta/N}$	$\frac{V(\hat{\alpha}\mid\theta)}{1/N}$	$\frac{V(\hat{c}\mid\theta)}{\theta^2/N}$	$\frac{C(\hat{\alpha},\hat{c}\mid\theta)}{\theta/N}$
.005	.00	3.637	18.996	1.226	-7.580	1.309	-3.841	3.196	0.755	-1.113
.010	.00	4.000	22.592	1.710	-8.722	1.728	-5.161	3.569	0.964	-1.393
.015	.00	4.286	25.571	2.149	-9.646	2.082	-6.305	3.861	1.138	-1.618
.020	.00	4.536	28.272	2.574	-10.468	2.408	-7.376	4.115	1.295	-1.818
.025	.00	4.765	30.820	2.996	-11.231	2.719	-8.413	4.346	1.445	-2.004
.005	.25	6.743	28.967	1.446	-13.141	2.134	-5.324	3.355	0.771	-1.164
.010	.25	7.687	35.606	2.096	-15.645	2.917	-7.400	3.764	0.989	-1.463
.015	.25	8.467	41.334	2.710	-17.759	3.609	-9.275	4.087	1.171	-1.705
.020	.25	9.175	46.702	3.323	-19.708	4.268	-11.089	4.369	1.338	-1.922
.025	.25	9.843	51.911	3.951	-21.573	4.915	-12.898	4.627	1.496	-2.124
.005	.50	14.483	48.691	1.788	-25.492	3.755	-7.916	3.824	0.814	-1.306
.010	.50	17.245	62.453	2.725	-31.657	5.364	-11.507	4.342	1.057	-1.660
.015	.50	19.652	74.952	3.661	-37.141	6.865	-14.927	4.757	1.263	-1.953
.020	.50	21.930	87.162	4.640	-42.415	8.358	-18.384	5.127	1.454	-2.219
.025	.50	24.165	99.456	5.680	-47.657	9.883	-21.960	5.470	1.638	-2.470
.005	.75	50.409	118.410	2.668	-75.497	9.356	-15.741	5.339	0.931	-1.728
.010	.75	65.498	166.105	4.525	-102.323	14.648	-25.150	6.254	1.249	-2.267
.015	.75	80.141	214.395	6.609	-128.914	20.171	-35.182	7.026	1.532	-2.734
.020	.75	95.331	266.111	9.013	-156.941	26.213	-46.330	7.741	1.805	-3.176
.025	.75	111.532	322.739	11.803	-187.230	32.936	-58.899	8.432	2.077	-3.609

Q1	Q2	$\frac{V(\hat{\theta}\mid\alpha)}{\theta^2/N}$	$\frac{V(\hat{c}\mid\alpha)}{\theta^2/N}$	$\frac{C(\hat{\theta},\hat{c}\mid\alpha)}{\theta^2/N}$	$\frac{V(\hat{\theta}\mid c)}{\theta^2/N}$	$\frac{V(\hat{\alpha}\mid c)}{1/N}$	$\frac{C(\hat{\theta},\hat{\alpha}\mid c)}{\theta/N}$	$\frac{V(\hat{\theta}\mid\alpha,c)}{\theta^2/N}$	$\frac{V(\hat{\alpha}\mid\theta,c)}{1/N}$	$\frac{V(\hat{c}\mid\theta,\alpha)}{\theta^2/N}$
.000	.00				2.225	6.900	-3.450	0.500	1.551	
.005	.00	0.612	0.449	-0.224	2.239	6.959	-3.479	0.500	1.554	0.367
.010	.00	0.632	0.531	-0.265	2.254	7.018	-3.508	0.500	1.557	0.420
.015	.00	0.647	0.595	-0.296	2.268	7.078	-3.538	0.500	1.560	0.460
.020	.00	0.660	0.650	-0.323	2.283	7.138	-3.567	0.500	1.563	0.492
.025	.00	0.672	0.700	-0.347	2.297	7.199	-3.597	0.500	1.567	0.521
.000	.25				3.561	9.269	-5.227	0.612	1.594	
.005	.25	0.781	0.468	-0.281	3.594	9.375	-5.287	0.612	1.597	0.367
.010	.25	0.813	0.558	-0.334	3.627	9.482	-5.347	0.612	1.601	0.421
.015	.25	0.837	0.629	-0.376	3.661	9.589	-5.406	0.612	1.604	0.460
.020	.25	0.858	0.690	-0.412	3.694	9.698	-5.467	0.612	1.608	0.492
.025	.25	0.877	0.747	-0.445	3.728	9.808	-5.528	0.612	1.611	0.521
.000	.50				6.500	13.417	-8.718	0.836	1.725	
.005	.50	1.138	0.501	-0.389	6.593	13.639	-8.861	0.836	1.729	0.368
.010	.50	1.199	0.605	-0.469	6.686	13.862	-9.005	0.836	1.733	0.422
.015	.50	1.247	0.688	-0.532	6.780	14.089	-9.151	0.836	1.737	0.461
.020	.50	1.290	0.762	-0.588	6.874	14.319	-9.299	0.836	1.741	0.494
.025	.50	1.329	0.831	-0.640	6.970	14.554	-9.449	0.836	1.745	0.523
.000	.75				17.124	24.788	-19.698	1.470	2.128	
.005	.75	2.273	0.576	-0.680	17.599	25.546	-20.298	1.470	2.134	0.372
.010	.75	2.466	0.717	-0.845	18.082	26.320	-20.910	1.470	2.140	0.427
.015	.75	2.626	0.836	-0.983	18.578	27.120	-21.540	1.470	2.146	0.468
.020	.75	2.773	0.947	-1.111	19.091	27.950	-22.193	1.470	2.153	0.502
.025	.75	2.914	1.054	-1.233	19.624	28.814	-22.871	1.470	2.159	0.532

C. $\alpha=3$

Q1	Q2	$\frac{V(\hat{\theta})}{\theta^2/N}$	$\frac{V(\hat{\alpha})}{1/N}$	$\frac{V(\hat{c})}{\theta^2/N}$	$\frac{C(\hat{\theta},\hat{\alpha})}{\theta/N}$	$\frac{C(\hat{\theta},\hat{c})}{\theta^2/N}$	$\frac{C(\hat{\alpha},\hat{c})}{\theta/N}$	$\frac{V(\hat{\alpha}\mid\theta)}{1/N}$	$\frac{V(\hat{c}\mid\theta)}{\theta^2/N}$	$\frac{C(\hat{\alpha},\hat{c}\mid\theta)}{\theta/N}$
.000	.00	3.635	50.165	4.635	-12.541	2.635	-12.541	6.900	2.725	-3.450
.005	.00	5.022	84.475	10.073	-19.437	5.376	-26.191	9.247	4.318	-5.383
.010	.00	5.517	97.990	12.554	-22.024	6.485	-31.981	10.073	4.933	-6.096
.015	.00	5.911	109.157	14.721	-24.121	7.408	-36.901	10.727	5.436	-6.669
.020	.00	6.256	119.241	16.761	-25.987	8.248	-41.435	11.298	5.888	-7.177
.025	.00	6.573	128.710	18.741	-27.719	9.039	-45.766	11.819	6.310	-7.646
.000	.25	6.376	73.928	5.328	-20.605	4.008	-16.594	7.336	2.809	-3.641
.005	.25	9.865	138.790	13.140	-35.646	9.219	-39.087	9.990	4.523	-5.774
.010	.25	11.249	166.799	17.040	-41.871	11.542	-49.538	10.942	5.196	-6.574
.015	.25	12.402	190.929	20.590	-47.146	13.566	-58.794	11.702	5.751	-7.224
.020	.25	13.452	213.467	24.043	-52.012	15.470	-67.615	12.370	6.253	-7.802
.025	.25	14.447	235.264	27.494	-56.668	17.323	-76.288	12.985	6.723	-8.340
.000	.50	12.355	115.244	6.203	-36.316	6.288	-22.598	8.501	3.003	-4.116
.005	.50	21.756	246.979	17.975	-71.499	16.789	-61.947	12.001	5.019	-6.772
.010	.50	25.920	310.255	24.544	-87.732	22.018	-82.334	13.310	5.840	-7.809
.015	.50	29.573	367.583	30.845	-102.203	26.816	-101.340	14.376	6.530	-8.666
.020	.50	33.044	423.385	37.241	-116.121	31.528	-120.232	15.328	7.161	-9.442
.025	.50	36.458	479.369	43.880	-129.944	36.288	-139.509	16.216	7.761	-10.172
.000	.75	34.127	230.795	7.847	-86.430	12.237	-36.352	11.901	3.459	-5.362
.005	.75	76.378	638.600	31.199	-217.665	43.580	-133.842	18.288	6.333	-9.645
.010	.75	99.404	879.632	47.419	-292.163	62.903	-196.364	20.925	7.614	-11.483
.015	.75	121.867	1122.696	64.835	-366.053	82.681	-261.425	23.178	8.740	-13.075
.020	.75	145.232	1381.914	84.276	-443.877	103.993	-332.414	25.278	9.812	-14.576
.025	.75	170.198	1664.652	106.274	-527.894	127.428	-411.278	27.313	10.869	-16.042

Q1	Q2	$\frac{V(\hat{\theta}\mid\alpha)}{\theta^2/N}$	$\frac{V(\hat{c}\mid\alpha)}{\theta^2/N}$	$\frac{C(\hat{\theta},\hat{c}\mid\alpha)}{\theta^2/N}$	$\frac{V(\hat{\theta}\mid c)}{\theta^2/N}$	$\frac{V(\hat{\alpha}\mid c)}{1/N}$	$\frac{C(\hat{\theta},\hat{\alpha}\mid c)}{\theta/N}$	$\frac{V(\hat{\theta}\mid\alpha,c)}{\theta^2/N}$	$\frac{V(\hat{\alpha}\mid\theta,c)}{1/N}$	$\frac{V(\hat{c}\mid\theta,\alpha)}{\theta^2/N}$
.000	.00	0.500	1.500	-0.500	2.137	16.234	-5.411	0.333	2.532	1.000
.005	.00	0.550	1.953	-0.650	2.153	16.376	-5.458	0.333	2.536	1.184
.010	.00	0.567	2.116	-0.703	2.168	16.519	-5.505	0.333	2.540	1.244
.015	.00	0.581	2.247	-0.746	2.183	16.661	-5.551	0.333	2.544	1.290
.020	.00	0.593	2.362	-0.783	2.198	16.805	-5.598	0.333	2.549	1.328
.025	.00	0.604	2.468	-0.817	2.213	16.950	-5.644	0.333	2.553	1.363
.000	.25	0.633	1.604	-0.617	3.361	22.251	-8.124	0.395	2.616	1.002
.005	.25	0.710	2.132	-0.819	3.396	22.516	-8.220	0.395	2.620	1.186
.010	.25	0.738	2.327	-0.893	3.430	22.781	-8.315	0.395	2.624	1.246
.015	.25	0.760	2.485	-0.952	3.464	23.046	-8.410	0.395	2.629	1.292
.020	.25	0.780	2.626	-1.005	3.498	23.314	-8.506	0.395	2.634	1.331
.025	.25	0.797	2.756	-1.053	3.533	23.585	-8.602	0.395	2.638	1.366
.000	.50	0.911	1.771	-0.833	5.981	32.914	-13.408	0.519	2.858	1.010
.005	.50	1.057	2.437	-1.145	6.075	33.489	-13.640	0.519	2.863	1.198
.010	.50	1.112	2.695	-1.264	6.168	34.061	-13.870	0.519	2.869	1.259
.015	.50	1.157	2.907	-1.361	6.260	34.638	-14.102	0.519	2.874	1.306
.020	.50	1.196	3.098	-1.448	6.354	35.223	-14.336	0.520	2.880	1.345
.025	.50	1.233	3.279	-1.530	6.449	35.817	-14.573	0.520	2.886	1.381
.000	.75	1.760	2.121	-1.377	15.045	62.390	-29.743	0.866	3.591	1.044
.005	.75	2.187	3.148	-2.039	15.503	64.428	-30.709	0.866	3.599	1.246
.010	.75	2.365	3.584	-2.317	15.960	66.480	-31.677	0.866	3.608	1.313
.015	.75	2.516	3.961	-2.556	16.428	68.586	-32.669	0.866	3.616	1.364
.020	.75	2.657	4.315	-2.780	16.909	70.762	-33.693	0.866	3.625	1.407
.025	.75	2.793	4.662	-2.997	17.406	73.021	-34.753	0.866	3.634	1.446

Table F5

COEFFICIENTS IN ASYMPTOTIC VARIANCES AND COVARIANCES OF MAXIMUM-LIKELIHOOD ESTIMATORS OF PARAMETERS μ AND σ OF LOGISTIC POPULATION FROM SAMPLES OF SIZE n WITH PROPORTIONS q_1 CENSORED FROM BELOW AND q FROM ABOVE

COEFFICIENTS OF σ^2/N IN ASYMPTOTIC VARIANCES AND COVARIANCES OF MAXIMUM-LIKELIHOOD ESTIMATORS OF PARAMETERS μ AND σ OF LOGISTIC DISTRIBUTION FROM SAMPLES OF SIZE N WITH PROPORTIONS Q1 CENSORED FROM BELOW AND Q2 FROM ABOVE

Q1	Q2	$N\ VAR(\hat{\mu})/\sigma^2$	$N\ VAR(\hat{\sigma})/\sigma^2$	$N\ COV(\hat{\mu},\hat{\sigma})/\sigma^2$	$N\ VAR(\hat{\mu}\mid\sigma)/\sigma^2$	$N\ VAR(\hat{\sigma}\mid\mu)/\sigma^2$
0.0	0.0	0.911891	0.699322	0.000000	0.911891	0.699322
0.0	0.1	0.912876	0.769054	0.007452	0.912803	0.768994
0.0	0.2	0.920737	0.876663	0.036169	0.919245	0.875242
0.0	0.3	0.946701	1.029899	0.098945	0.937195	1.019557
0.0	0.4	1.012343	1.246821	0.217955	0.974242	1.199896
0.0	0.5	1.163950	1.562093	0.436172	1.042161	1.398644
0.0	0.6	1.513872	2.046448	0.847222	1.163126	1.572310
0.0	0.7	2.386396	2.866066	1.691620	1.387961	1.666944
0.0	0.8	5.025345	4.520589	3.777594	1.868628	1.680940
0.0	0.9	18.049503	9.509748	11.817224	3.364910	1.772871
0.1	0.1	0.913718	0.854083	0.000000	0.913718	0.854083
0.1	0.2	0.921155	0.988235	0.031160	0.920172	0.987181
0.1	0.3	0.946972	1.185761	0.102228	0.938159	1.174725
0.1	0.4	1.015380	1.479166	0.243534	0.975284	1.420756
0.1	0.5	1.182939	1.936695	0.519938	1.043353	1.708167
0.1	0.6	1.604612	2.720513	1.094087	1.164611	1.974522
0.1	0.7	2.825894	4.322156	2.491149	1.390077	2.126099
0.1	0.8	8.034268	9.220459	7.537549	1.872465	2.148919
0.2	0.2	0.926718	1.169419	0.000000	0.926718	1.169419
0.2	0.3	0.949261	1.448721	0.078894	0.944964	1.442164
0.2	0.4	1.015735	1.893218	0.250308	0.982641	1.831535
0.2	0.5	1.198442	2.664977	0.625186	1.051777	2.338839
0.2	0.6	1.743912	4.257226	1.556113	1.175117	2.868688
0.2	0.7	4.020561	9.157141	4.893916	1.405070	3.200158
0.3	0.3	0.963944	1.880831	0.000000	0.963944	1.880831
0.3	0.4	1.018918	2.643180	0.203958	1.003180	2.602354
0.3	0.5	1.209178	4.229832	0.752396	1.075343	3.761663
0.3	0.6	2.055657	9.132028	2.787791	1.204611	5.351351
0.4	0.4	1.045746	4.221988	0.000000	1.045746	4.221988
0.4	0.5	1.214726	9.122277	0.907717	1.124403	8.443975

(INTERCHANGING Q1 AND Q2 LEAVES VARIANCES AND ABSOLUTE VALUE OF COVARIANCE UNCHANGED, BUT CHANGES SIGN OF COVARIANCE)

Table F6

COEFFICIENTS IN ASYMPTOTIC VARIANCES AND COVARIANCES OF MAXIMUM-LIKELIHOOD ESTIMATORS OF PARAMETERS u AND b OF TYPE I EXTREME-VALUE POPULATION FROM SAMPLES OF SIZE n WITH PROPORTIONS q_1 CENSORED FROM BELOW AND q_2 FROM ABOVE

COEFFICIENTS OF b^2/n IN ASYMPTOTIC VARIANCES AND COVARIANCES OF MAXIMUM-LIKELIHOOD ESTIMATORS OF PARAMETERS **u** AND **b** OF TYPE I EXTREME-VALUE DISTRIBUTION FROM SAMPLES OF SIZE **n** WITH PROPORTIONS q_1 CENSORED FROM BELOW AND q_2 FROM ABOVE

q_1	q_2	$n\ VAR(\hat{u})/b^2$	$n\ VAR(\hat{b})/b^2$	$n\ COV(\hat{u},\hat{b})/b^2$	$n\ VAR(\hat{u}\|b)/b^2$	$n\ VAR(\hat{b}\|u)/b^2$
0.0	0.0	1.108665	0.607927	-0.257022	1.000000	0.548342
0.0	0.1	1.151684	0.767044	-0.176413	1.111111	0.740022
0.0	0.2	1.252617	0.928191	-0.049288	1.250000	0.926252
0.0	0.3	1.447258	1.122447	0.144825	1.428571	1.107954
0.0	0.4	1.811959	1.372781	0.446603	1.666667	1.262704
0.0	0.5	2.510236	1.716182	0.935766	2.000000	1.367347
0.0	0.6	3.933022	2.224740	1.785525	2.500000	1.414141
0.0	0.7	7.190427	3.065515	3.438601	3.333333	1.421110
0.0	0.8	16.478771	4.738764	7.375310	5.000000	1.437839
0.0	0.9	60.517110	9.744662	22.187207	10.000000	1.610233
0.1	0.0	1.118727	0.654702	-0.278694	1.000092	0.585275
0.1	0.1	1.157024	0.842250	-0.196403	1.111225	0.808911
0.1	0.2	1.253484	1.039534	-0.058920	1.250144	1.036765
0.1	0.3	1.448869	1.287741	0.160918	1.428760	1.269869
0.1	0.4	1.834013	1.625480	0.521153	1.666924	1.477389
0.1	0.5	2.619087	2.124585	1.146524	2.000370	1.622686
0.1	0.6	4.368319	2.950577	2.347534	2.500578	1.689013
0.1	0.7	9.027691	4.599896	5.117492	3.334361	1.698963
0.1	0.8	27.763083	9.562344	14.752841	5.002312	1.722929
0.2	0.0	1.137715	0.721168	-0.314194	1.000829	0.634399
0.2	0.1	1.168880	0.952080	-0.232437	1.112134	0.905859
0.2	0.2	1.257014	1.207078	-0.083088	1.251295	1.201586
0.2	0.3	1.449409	1.546735	0.172088	1.430263	1.526303
0.2	0.4	1.859086	2.045511	0.623606	1.668969	1.836330
0.2	0.5	2.782556	2.870038	1.495475	2.003317	2.066300
0.2	0.6	5.203882	4.518143	3.491862	2.505185	2.175065
0.2	0.7	14.464850	9.484080	10.270576	3.342557	2.191594
0.3	0.0	1.171354	0.813416	-0.369870	1.003170	0.696625
0.3	0.1	1.192915	1.110818	-0.294142	1.115026	1.038290
0.3	0.2	1.267204	1.460390	-0.133736	1.254958	1.446276
0.3	0.3	1.449526	1.963716	0.168598	1.435050	1.944106
0.3	0.4	1.888035	2.790888	0.770185	1.675492	2.476707
0.3	0.5	3.067689	4.441586	2.164654	2.012722	2.914141
0.3	0.6	7.565384	9.412787	6.891442	2.519909	3.135250
0.4	0.0	1.230645	0.943493	-0.457647	1.008660	0.773305
0.4	0.1	1.241015	1.347812	-0.400827	1.121813	1.218351
0.4	0.2	1.293682	1.864610	-0.236991	1.263561	1.821195
0.4	0.3	1.452168	2.698811	0.125731	1.446311	2.687925
0.4	0.4	1.924836	4.354570	1.009384	1.690862	3.825249
0.4	0.5	3.813216	9.331397	4.073545	2.034943	4.979750
0.5	0.0	1.338098	1.134038	-0.600672	1.019937	0.864396
0.5	0.1	1.340734	1.724528	-0.594518	1.135779	1.460903
0.5	0.2	1.363071	2.576961	-0.459023	1.281307	2.422382
0.5	0.3	1.469964	4.242607	-0.038778	1.469609	4.241584
0.5	0.4	1.989349	9.226891	1.568276	1.722792	7.990561
0.6	0.0	1.545203	1.431430	-0.848735	1.041964	0.965244
0.6	0.1	1.566598	2.389556	-0.981653	1.163162	1.774188
0.6	0.2	1.568133	4.080712	-1.013806	1.316264	3.425281
0.6	0.3	1.581070	9.078796	-0.769896	1.515781	8.703898
0.7	0.0	1.991079	1.946111	-1.327539	1.085498	1.060983
0.7	0.1	2.178199	3.804751	-1.911687	1.217677	2.126966
0.7	0.2	2.432859	8.841320	-3.041565	1.386509	5.038750
0.8	0.0	3.170376	3.016820	-2.450529	1.179840	1.122695
0.8	0.1	4.683862	8.369955	-5.292215	1.337663	2.390373
0.9	0.0	8.177861	6.388191	-6.555257	1.451169	1.133590

Table F7

COEFFICIENTS IN ASYMPTOTIC VARIANCES AND COVARIANCES OF MAXIMUM-LIKELIHOOD ESTIMATORS OF PARAMETERS μ AND σ OF LOG-GAMMA POPULATION WITH SHAPE PARAMETER $\kappa = 0.5(0.5)2.0(1.0)4.0(2.0)12.0$ FROM SAMPLES OF SIZE n WITH PROPORTIONS q_1 CENSORED FROM BELOW AND q_2 FROM ABOVE

Coefficients of σ^2/n in Asymptotic Variances V_{11} and V_{22} and Covariances V_{12} of Maximum-Likelihood Estimators

q_1	q_2	κ	0.5	1.0	1.5	2.0	3.0	4.0	6.0	8.0	10.0	12.0
0.0	0.0	V_{11}	1.181	1.109	1.076	1.059	1.040	1.030	1.021	1.015	1.012	1.010
		V_{22}	0.681	0.608	0.575	0.559	0.540	0.530	0.520	0.515	0.512	0.510
		V_{12}	-0.352	-0.257	-0.210	-0.181	-0.147	-0.127	-0.103	-0.089	-0.080	-0.073
	0.1	V_{11}	1.249	1.152	1.113	1.093	1.071	1.059	1.048	1.042	1.038	1.035
		V_{22}	0.869	0.767	0.721	0.694	0.665	0.650	0.633	0.624	0.619	0.615
		V_{12}	-0.243	-0.176	-0.139	-0.115	-0.087	-0.069	-0.049	-0.037	-0.028	-0.022
	0.2	V_{11}	1.411	1.253	1.197	1.168	1.138	1.122	1.106	1.098	1.092	1.088
		V_{22}	1.051	0.928	0.871	0.837	0.800	0.780	0.757	0.745	0.737	0.732
		V_{12}	-0.074	-0.049	-0.027	-0.012	0.008	0.021	0.036	0.045	0.052	0.057
	0.3	V_{11}	1.727	1.447	1.356	1.311	1.264	1.239	1.214	1.200	1.191	1.185
		V_{22}	1.259	1.122	1.053	1.013	0.966	0.940	0.912	0.896	0.886	0.879
		V_{12}	0.184	0.145	0.143	0.146	0.153	0.158	0.165	0.170	0.173	0.176
	0.4	V_{11}	2.329	1.812	1.652	1.573	1.493	1.452	1.408	1.384	1.369	1.359
		V_{22}	1.526	1.373	1.291	1.241	1.184	1.151	1.115	1.095	1.082	1.073
		V_{12}	0.584	0.447	0.408	0.391	0.376	0.369	0.363	0.361	0.360	0.359
	0.5	V_{11}	3.488	2.510	2.214	2.069	1.923	1.847	1.768	1.725	1.698	1.678
		V_{22}	1.881	1.716	1.618	1.557	1.486	1.445	1.399	1.373	1.357	1.345
		V_{12}	1.226	0.936	0.836	0.786	0.736	0.710	0.683	0.669	0.660	0.654
0.1	0.0	V_{11}	1.195	1.119	1.086	1.070	1.052	1.043	1.034	1.030	1.027	1.026
		V_{22}	0.732	0.655	0.624	0.610	0.596	0.590	0.584	0.582	0.581	0.580
		V_{12}	-0.378	-0.279	-0.232	-0.205	-0.173	-0.154	-0.133	-0.120	-0.112	-0.105
	0.1	V_{11}	1.255	1.157	1.119	1.099	1.078	1.067	1.057	1.051	1.048	1.046
		V_{22}	0.953	0.842	0.798	0.774	0.750	0.738	0.726	0.720	0.717	0.715
		V_{12}	-0.267	-0.196	-0.160	-0.137	-0.111	-0.095	-0.077	-0.067	-0.060	-0.054
	0.2	V_{11}	1.411	1.253	1.198	1.170	1.140	1.125	1.110	1.102	1.097	1.093
		V_{22}	1.177	1.040	0.983	0.953	0.922	0.906	0.889	0.881	0.875	0.872
		V_{12}	-0.083	-0.059	-0.039	-0.025	-0.008	0.003	0.015	0.023	0.028	0.032
	0.3	V_{11}	1.731	1.449	1.357	1.311	1.264	1.240	1.214	1.200	1.192	1.186
		V_{22}	1.444	1.288	1.220	1.183	1.144	1.123	1.101	1.090	1.083	1.078
		V_{12}	0.211	0.161	0.155	0.155	0.158	0.160	0.164	0.166	0.168	0.170
	0.4	V_{11}	2.369	1.834	1.669	1.587	1.504	1.461	1.415	1.391	1.375	1.364
		V_{22}	1.805	1.625	1.544	1.499	1.450	1.425	1.397	1.383	1.373	1.367
		V_{12}	0.690	0.521	0.473	0.450	0.429	0.418	0.408	0.402	0.399	0.396
	0.5	V_{11}	3.674	2.619	2.300	2.144	1.986	1.904	1.818	1.771	1.741	1.720
		V_{22}	2.325	2.125	2.026	1.971	1.911	1.879	1.845	1.826	1.814	1.805
		V_{12}	1.513	1.147	1.024	0.963	0.900	0.867	0.832	0.813	0.800	0.791
0.2	0.0	V_{11}	1.214	1.138	1.108	1.093	1.078	1.071	1.064	1.061	1.060	1.059
		V_{22}	0.797	0.721	0.694	0.685	0.675	0.672	0.670	0.670	0.670	0.670
		V_{12}	-0.413	-0.314	-0.270	-0.246	-0.218	-0.202	-0.183	-0.173	-0.165	-0.160
	0.1	V_{11}	1.265	1.169	1.133	1.115	1.097	1.088	1.080	1.076	1.073	1.072
		V_{22}	1.064	0.952	0.912	0.892	0.873	0.864	0.855	0.851	0.849	0.848
		V_{12}	-0.300	-0.232	-0.200	-0.181	-0.159	-0.146	-0.132	-0.123	-0.117	-0.113
	0.2	V_{11}	1.413	1.257	1.204	1.177	1.150	1.136	1.123	1.116	1.112	1.109
		V_{22}	1.349	1.207	1.156	1.129	1.103	1.090	1.078	1.071	1.068	1.065
		V_{12}	-0.098	-0.083	-0.069	-0.060	-0.049	-0.042	-0.034	-0.029	-0.026	-0.024
	0.3	V_{11}	1.736	1.449	1.358	1.312	1.265	1.241	1.217	1.204	1.196	1.191
		V_{22}	1.709	1.547	1.484	1.451	1.418	1.401	1.384	1.375	1.369	1.366
		V_{12}	0.244	0.172	0.155	0.148	0.142	0.139	0.136	0.134	0.133	0.132
	0.4	V_{11}	2.425	1.859	1.684	1.598	1.510	1.465	1.417	1.392	1.376	1.365
		V_{22}	2.236	2.046	1.969	1.929	1.888	1.866	1.844	1.832	1.825	1.820
		V_{12}	0.846	0.624	0.553	0.518	0.481	0.461	0.439	0.427	0.419	0.413
	0.5	V_{11}	3.976	2.783	2.420	2.242	2.062	1.968	1.869	1.816	1.782	1.758
		V_{22}	3.082	2.870	2.778	2.729	2.677	2.650	2.622	2.606	2.597	2.590
		V_{12}	1.991	1.495	1.324	1.235	1.140	1.088	1.031	0.999	0.978	0.963

Appendix G

TABLES OF RESULTS OF MONTE CARLO STUDIES OF MAXIMUM-LIKELIHOOD ESTIMATORS FROM DOUBLY CENSORED SAMPLES

SOURCES OF TABLES

Table G1	*Biometrika* 53(1966), 205-213: corrigendum, 56(1969), 229 (Harter and Moore)
Table G2	*J. Amer. Statist. Assoc.* 61(1966), 842-851: corrigenda, 61(1966), 1247 and 62(1967), 1519-1520 (Harter and Moore)
Table G3	*Ann. Math. Statist.* 38(1967), 557, 570 (Harter and Moore)
Table G4	*J. Amer. Statist. Assoc.* 62(1967), 675-684 (Harter and Moore) [with additional values not previously published (Harter and Moore)
Table G5	*J. Amer. Statist. Assoc.* 63(1968), 889-901 (Harter and Moore)
Table G6	McMaster Ph.D. thesis (1993) (Chan)

Table G1

NORMAL POPULATION - SAMPLE SIZES 10 AND 20

A. MEANS, VARIANCES, AND COVARIANCES OF ML ESTIMATES OF MEAN AND STANDARD DEVIATION OF STANDARD NORMAL POPULATION (μ=0, σ=1) FROM 1000 SAMPLES OF SIZE N=10 WITH PROPORTIONS Q1 CENSORED FROM BELOW AND Q2 FROM ABOVE

Q1	Q2	$M(\hat{\mu})$	$M(\hat{\sigma})$	$V(\hat{\mu})$	$C(\hat{\mu},\hat{\sigma})$	$V(\hat{\sigma})$	$M(\hat{\mu}\mid\sigma)$	$M(\hat{\sigma}\mid\mu)$	$V(\hat{\mu}\mid\sigma)$	$V(\hat{\sigma}\mid\mu)$
0.0	0.0	-0.00	0.93	0.100	0.000	0.049	-0.00	0.98	0.100	0.048
0.0	0.1	-0.01	0.92	0.101	0.003	0.057	-0.00	0.98	0.101	0.057
0.0	0.2	-0.02	0.90	0.104	0.009	0.067	-0.00	0.97	0.103	0.067
0.0	0.3	-0.04	0.88	0.114	0.019	0.078	-0.01	0.97	0.109	0.077
0.0	0.4	-0.07	0.85	0.125	0.028	0.084	-0.01	0.96	0.116	0.081
0.0	0.5	-0.11	0.81	0.146	0.047	0.101	-0.02	0.95	0.124	0.091
0.0	0.6	-0.19	0.76	0.184	0.075	0.123	-0.03	0.94	0.138	0.101
0.0	0.7	-0.31	0.67	0.265	0.130	0.160	-0.03	0.93	0.160	0.105
0.0	0.8	-0.57	0.51	0.412	0.210	0.204	-0.05	0.93	0.197	0.108
0.1	0.1	-0.00	0.90	0.101	0.000	0.067	-0.00	0.97	0.101	0.067
0.1	0.2	-0.01	0.88	0.105	0.007	0.078	-0.00	0.97	0.104	0.079
0.1	0.3	-0.03	0.86	0.114	0.018	0.091	-0.01	0.96	0.111	0.093
0.1	0.4	-0.07	0.81	0.126	0.029	0.101	-0.01	0.95	0.117	0.101
0.1	0.5	-0.12	0.75	0.147	0.052	0.125	-0.02	0.93	0.126	0.117
0.1	0.6	-0.22	0.66	0.187	0.085	0.152	-0.03	0.92	0.140	0.134
0.1	0.7	-0.41	0.49	0.271	0.143	0.189	-0.04	0.90	0.163	0.139
0.2	0.2	-0.01	0.87	0.108	-0.001	0.096	-0.00	0.97	0.108	0.100
0.2	0.3	-0.03	0.83	0.116	0.011	0.114	-0.01	0.96	0.115	0.122
0.2	0.4	-0.06	0.77	0.127	0.024	0.129	-0.02	0.95	0.122	0.137
0.2	0.5	-0.13	0.68	0.147	0.051	0.164	-0.02	0.92	0.131	0.169
0.2	0.6	-0.27	0.49	0.184	0.084	0.188	-0.03	0.89	0.146	0.206
0.3	0.3	-0.01	0.79	0.119	0.003	0.145	-0.01	0.96	0.119	0.157
0.3	0.4	-0.05	0.69	0.127	0.020	0.176	-0.02	0.94	0.125	0.196
0.3	0.5	-0.13	0.52	0.148	0.050	0.217	-0.02	0.90	0.136	0.256
0.4	0.4	-0.01	0.51	0.134	0.000	0.218	-0.01	0.90	0.134	0.282

$M(\hat{\mu})$=MEAN VALUE OF MLE $\hat{\mu}$ IN 1000 SAMPLES (σ UNKNOWN)
$M(\hat{\sigma})$=MEAN VALUE OF MLE $\hat{\sigma}$ IN 1000 SAMPLES (μ UNKNOWN)
$V(\hat{\mu})$=VARIANCE OF MLE $\hat{\mu}$ IN 1000 SAMPLES (σ UNKNOWN)
$C(\hat{\mu},\hat{\sigma})$=COVARIANCE OF MLES $\hat{\mu}$ AND $\hat{\sigma}$ IN 1000 SAMPLES
$V(\hat{\sigma})$=VARIANCE OF MLE $\hat{\sigma}$ IN 1000 SAMPLES (μ UNKNOWN)
$M(\hat{\mu}\mid\sigma)$=MEAN VALUE OF MLE $\hat{\mu}$ IN 1000 SAMPLES (σ KNOWN)
$M(\hat{\sigma}\mid\mu)$=MEAN VALUE OF MLE $\hat{\sigma}$ IN 1000 SAMPLES (μ KNOWN)
$V(\hat{\mu}\mid\sigma)$=VARIANCE OF MLE $\hat{\mu}$ IN 1000 SAMPLES (σ KNOWN)
$V(\hat{\sigma}\mid\mu)$=VARIANCE OF MLE $\hat{\sigma}$ IN 1000 SAMPLES (μ KNOWN)

B. MEANS, VARIANCES, AND COVARIANCES OF ML ESTIMATES OF MEAN AND STANDARD DEVIATION OF STANDARD NORMAL POPULATION ($\mu=0$, $\sigma=1$) FROM 1000 SAMPLES OF SIZE N=20 WITH PROPORTIONS Q1 CENSORED FROM BELOW AND Q2 FROM ABOVE

Q1	Q2	$M(\hat{\mu})$	$M(\hat{\sigma})$	$V(\hat{\mu})$	$C(\hat{\mu},\hat{\sigma})$	$V(\hat{\sigma})$	$M(\hat{\mu}\mid\sigma)$	$M(\hat{\sigma}\mid\mu)$	$V(\hat{\mu}\mid\sigma)$	$V(\hat{\sigma}\mid\mu)$
0.0	0.0	0.01	0.97	0.048	0.000	0.023	0.01	0.99	0.048	0.022
0.0	0.1	0.01	0.96	0.048	0.002	0.027	0.01	0.99	0.048	0.027
0.0	0.2	-0.00	0.95	0.050	0.005	0.032	0.01	0.99	0.049	0.031
0.0	0.3	-0.01	0.94	0.054	0.010	0.038	0.00	0.98	0.052	0.036
0.0	0.4	-0.03	0.93	0.061	0.017	0.045	-0.00	0.97	0.055	0.040
0.0	0.5	-0.04	0.91	0.073	0.029	0.057	-0.00	0.97	0.059	0.046
0.0	0.6	-0.07	0.89	0.095	0.048	0.075	-0.00	0.97	0.064	0.052
0.0	0.7	-0.13	0.85	0.142	0.083	0.100	-0.00	0.96	0.074	0.055
0.0	0.8	-0.27	0.76	0.257	0.151	0.141	-0.02	0.96	0.095	0.055
0.0	0.9	-0.73	0.52	0.563	0.289	0.202	-0.05	0.97	0.150	0.059
0.1	0.1	0.01	0.96	0.049	-0.001	0.033	0.01	0.99	0.049	0.033
0.1	0.2	0.00	0.94	0.051	0.002	0.039	0.01	0.98	0.051	0.038
0.1	0.3	-0.01	0.92	0.054	0.008	0.048	0.00	0.98	0.053	0.045
0.1	0.4	-0.03	0.90	0.061	0.015	0.057	0.00	0.97	0.056	0.052
0.1	0.5	-0.05	0.88	0.074	0.031	0.076	0.00	0.96	0.061	0.063
0.1	0.6	-0.09	0.84	0.097	0.056	0.104	0.00	0.95	0.066	0.073
0.1	0.7	-0.18	0.76	0.152	0.104	0.146	-0.00	0.94	0.077	0.079
0.1	0.8	-0.51	0.49	0.274	0.183	0.193	-0.02	0.93	0.100	0.080
0.2	0.2	0.01	0.93	0.052	-0.001	0.048	0.01	0.98	0.052	0.047
0.2	0.3	-0.00	0.91	0.055	0.006	0.063	0.00	0.97	0.054	0.060
0.2	0.4	-0.02	0.87	0.061	0.014	0.076	0.00	0.96	0.058	0.070
0.2	0.5	-0.05	0.83	0.074	0.034	0.108	0.00	0.96	0.063	0.092
0.2	0.6	-0.11	0.75	0.101	0.072	0.160	0.00	0.94	0.069	0.116
0.2	0.7	-0.31	0.51	0.168	0.143	0.233	-0.00	0.91	0.081	0.133
0.3	0.3	0.01	0.88	0.057	-0.001	0.078	0.01	0.97	0.057	0.075
0.3	0.4	-0.02	0.83	0.062	0.009	0.101	0.00	0.96	0.061	0.095
0.3	0.5	-0.06	0.74	0.074	0.032	0.145	0.00	0.94	0.067	0.137
0.3	0.6	-0.18	0.50	0.098	0.073	0.214	0.00	0.89	0.073	0.190
0.4	0.4	0.00	0.74	0.064	-0.000	0.152	0.00	0.94	0.064	0.148
0.4	0.5	-0.06	0.49	0.074	0.029	0.220	0.00	0.90	0.070	0.258

SEE GLOSSARY OF COLUMN HEADINGS AT FOOT OF TABLE G1A

C. COMPARISON OF MEASURES OF PRECISION OF BEST LINEAR UNBIASED ESTIMATORS (BLUE) AND MAXIMUM-LIKELIHOOD ESTIMATORS (MLE) OF PARAMETERS OF NORMAL POPULATION FROM SAMPLES OF SIZE N=10 WITH PROPORTIONS Q1 CENSORED FROM BELOW AND Q2 FROM ABOVE

Q1	Q2	$V(\mu^*)$	$MSE(\hat{\mu})$	$AV(\hat{\mu})$	$MSE(\hat{\mu}\|\sigma)$	$AV(\hat{\mu}\|\sigma)$	$V(\sigma^*)$	$MSE(\hat{\sigma})$	$AV(\hat{\sigma})$	$MSE(\hat{\sigma}\|\mu)$	$AV(\hat{\sigma}\|\mu)$
0.0	0.0	0.100	0.100	0.100	0.100	0.100	0.058	0.053	0.050	0.049	0.050
0.0	0.1	0.102	0.101	0.102	0.101	0.102	0.068	0.064	0.059	0.057	0.058
0.0	0.2	0.107	0.105	0.106	0.103	0.105	0.081	0.076	0.069	0.068	0.068
0.0	0.3	0.117	0.115	0.114	0.109	0.109	0.099	0.092	0.082	0.078	0.078
0.0	0.4	0.134	0.130	0.127	0.116	0.114	0.124	0.106	0.099	0.082	0.089
0.0	0.5	0.166	0.159	0.152	0.125	0.122	0.161	0.136	0.124	0.094	0.100
0.0	0.6	0.237	0.220	0.199	0.139	0.134	0.225	0.184	0.161	0.105	0.109
0.0	0.7	0.417	0.359	0.302	0.161	0.153	0.354	0.269	0.225	0.110	0.114
0.0	0.8	1.127	0.733	0.578	0.199	0.187	0.749	0.444	0.354	0.114	0.115
0.1	0.1	0.104	0.101	0.103	0.101	0.103	0.082	0.076	0.070	0.068	0.070
0.1	0.2	0.108	0.105	0.107	0.104	0.106	0.101	0.091	0.085	0.080	0.084
0.1	0.3	0.117	0.115	0.114	0.111	0.111	0.128	0.112	0.104	0.094	0.101
0.1	0.4	0.134	0.130	0.127	0.117	0.116	0.168	0.136	0.132	0.103	0.120
0.1	0.5	0.171	0.162	0.154	0.126	0.125	0.237	0.185	0.174	0.122	0.140
0.1	0.6	0.267	0.237	0.213	0.141	0.137	0.377	0.271	0.246	0.141	0.158
0.1	0.7	0.630	0.440	0.367	0.164	0.157	0.807	0.453	0.395	0.148	0.169
0.2	0.2	0.111	0.108	0.110	0.108	0.110	0.129	0.114	0.105	0.101	0.105
0.2	0.3	0.118	0.117	0.115	0.115	0.114	0.171	0.144	0.134	0.123	0.133
0.2	0.4	0.134	0.131	0.128	0.122	0.120	0.243	0.183	0.178	0.140	0.168
0.2	0.5	0.177	0.164	0.156	0.132	0.129	0.388	0.269	0.254	0.175	0.210
0.2	0.6	0.340	0.256	0.230	0.147	0.142	0.832	0.448	0.409	0.219	0.253
0.3	0.3	0.122	0.119	0.119	0.119	0.119	0.244	0.191	0.180	0.159	0.180
0.3	0.4	0.134	0.130	0.128	0.125	0.126	0.392	0.272	0.257	0.200	0.251
0.3	0.5	0.187	0.165	0.156	0.136	0.135	0.843	0.450	0.416	0.266	0.359
0.4	0.4	0.138	0.134	0.133	0.134	0.133	0.846	0.457	0.417	0.292	0.417

$V(\mu^*)$=EXACT VARIANCE OF BLUE μ^*
$MSE(\hat{\mu})$=MEAN SQUARE ERROR OF MLE $\hat{\mu}$ IN 1000 SAMPLES (σ UNKNOWN)
$AV(\hat{\mu})$=VARIANCE OF MLE $\hat{\mu}$ AS GIVEN BY ASYMPTOTIC FORMULA (σ UNKNOWN)
$MSE(\hat{\mu}|\sigma)$=MEAN SQUARE ERROR OF MLE $\hat{\mu}$ IN 1000 SAMPLES (σ KNOWN)
$AV(\hat{\mu}|\sigma)$=VARIANCE OF MLE $\hat{\mu}$ AS GIVEN BY ASYMPTOTIC FORMULA (σ KNOWN)
$V(\sigma^*)$=EXACT VARIANCE OF BLUE σ^*
$MSE(\hat{\sigma})$=MEAN SQUARE ERROR OF MLE $\hat{\sigma}$ IN 1000 SAMPLES (μ UNKNOWN)
$AV(\hat{\sigma})$=VARIANCE OF MLE $\hat{\sigma}$ AS GIVEN BY ASYMPTOTIC FORMULA (μ UNKNOWN)
$MSE(\hat{\sigma}|\mu)$=MEAN SQUARE ERROR OF MLE $\hat{\sigma}$ IN 1000 SAMPLES (μ KNOWN)
$AV(\hat{\sigma}|\mu)$=VARIANCE OF MLE $\hat{\sigma}$ AS GIVEN BY ASYMPTOTIC FORMULA (μ KNOWN)

D. COMPARISON OF MEASURES OF PRECISION OF BEST LINEAR UNBIASED ESTIMATORS (BLUE) AND MAXIMUM-LIKELIHOOD ESTIMATORS (MLE) OF PARAMETERS OF NORMAL POPULATION FROM SAMPLES OF SIZE N=20 WITH PROPORTIONS Q1 CENSORED FROM BELOW AND Q2 FROM ABOVE

Q1	Q2	$V(\mu^*)$	$MSE(\hat{\mu})$	$AV(\hat{\mu})$	$MSE(\hat{\mu}\|\sigma)$	$AV(\hat{\mu}\|\sigma)$	$V(\sigma^*)$	$MSE(\hat{\sigma})$	$AV(\hat{\sigma})$	$MSE(\hat{\sigma}\|\mu)$	$AV(\hat{\sigma}\|\mu)$
0.0	0.0	0.050	0.048	0.050	0.048	0.050	0.027	0.024	0.025	0.022	0.025
0.0	0.1	0.051	0.048	0.051	0.048	0.051	0.032	0.029	0.029	0.027	0.029
0.0	0.2	0.053	0.050	0.053	0.049	0.052	0.037	0.034	0.034	0.031	0.034
0.0	0.3	0.058	0.054	0.057	0.052	0.054	0.045	0.042	0.041	0.036	0.039
0.0	0.4	0.065	0.061	0.064	0.055	0.057	0.055	0.050	0.050	0.041	0.045
0.0	0.5	0.079	0.075	0.076	0.059	0.061	0.070	0.065	0.062	0.047	0.050
0.0	0.6	0.108	0.100	0.099	0.064	0.067	0.094	0.087	0.081	0.053	0.054
0.0	0.7	0.175	0.158	0.151	0.074	0.076	0.139	0.122	0.112	0.056	0.057
0.0	0.8	0.383	0.331	0.289	0.095	0.094	0.244	0.198	0.177	0.057	0.057
0.0	0.9	1.871	1.089	0.890	0.152	0.139	0.786	0.430	0.376	0.060	0.059
0.1	0.1	0.052	0.049	0.052	0.049	0.052	0.038	0.035	0.035	0.033	0.035
0.1	0.2	0.054	0.051	0.053	0.051	0.053	0.046	0.042	0.042	0.038	0.042
0.1	0.3	0.058	0.054	0.057	0.053	0.055	0.057	0.054	0.052	0.046	0.050
0.1	0.4	0.065	0.061	0.064	0.056	0.058	0.074	0.067	0.066	0.053	0.060
0.1	0.5	0.081	0.076	0.077	0.061	0.062	0.100	0.091	0.087	0.064	0.070
0.1	0.6	0.118	0.105	0.106	0.066	0.068	0.149	0.130	0.123	0.075	0.079
0.1	0.7	0.228	0.185	0.183	0.077	0.078	0.266	0.205	0.198	0.082	0.084
0.1	0.8	0.934	0.531	0.489	0.100	0.097	0.873	0.456	0.433	0.084	0.086
0.2	0.2	0.055	0.052	0.055	0.052	0.055	0.058	0.053	0.053	0.047	0.053
0.2	0.3	0.058	0.055	0.058	0.055	0.057	0.075	0.071	0.067	0.061	0.066
0.2	0.4	0.065	0.061	0.064	0.058	0.060	0.103	0.092	0.089	0.072	0.084
0.2	0.5	0.082	0.077	0.078	0.063	0.065	0.154	0.136	0.127	0.094	0.105
0.2	0.6	0.134	0.114	0.115	0.069	0.071	0.275	0.220	0.205	0.120	0.126
0.2	0.7	0.456	0.265	0.259	0.081	0.082	0.898	0.470	0.446	0.142	0.141
0.3	0.3	0.060	0.058	0.059	0.058	0.059	0.104	0.092	0.090	0.076	0.090
0.3	0.4	0.066	0.062	0.064	0.061	0.063	0.156	0.130	0.128	0.097	0.126
0.3	0.5	0.084	0.077	0.078	0.067	0.068	0.279	0.211	0.208	0.140	0.180
0.3	0.6	0.201	0.131	0.134	0.073	0.075	0.908	0.462	0.452	0.202	0.252
0.4	0.4	0.068	0.064	0.067	0.064	0.067	0.280	0.220	0.209	0.152	0.209
0.4	0.5	0.087	0.077	0.078	0.070	0.072	0.913	0.478	0.454	0.268	0.417

SEE GLOSSARY OF COLUMN HEADINGS AT FOOT OF TABLE G1C

Table G2

LOGNORMAL POPULATION - SAMPLE SIZES 50, 100, AND 200

MEANS, VARIANCES, AND COVARIANCES OF ESTIMATES OF PARAMETERS FROM 500 RANDOM SAMPLES, EACH OF SIZE N, WITH PROPORTIONS Q1 CENSORED FROM BELOW AND Q2 FROM ABOVE, DRAWN FROM LOGNORMAL POPULATION WITH PARAMETERS $\mu=4$, $\sigma=2$, AND $\tau=10$

N	Q1	Q2	MEAN($\hat{\mu}$)	MEAN($\hat{\mu}\|\sigma$)	MEAN($\hat{\mu}\|\tau$)	MEAN($\hat{\mu}\|\sigma,\tau$)
50	0.00	0.0	3.9172	3.9295	3.9899	3.9899
100	0.00	0.0	3.9624	3.9682	4.0009	4.0009
200	0.00	0.0	3.9903	3.9917	4.0080	4.0080
100	0.00	0.5	3.9838	3.9490	3.9608	3.9912
100	0.01	0.0	3.9654	3.9707	4.0011	4.0010
100	0.01	0.5	3.9726	3.9513	3.9602	3.9912

N	Q1	Q2	MEAN($\hat{\sigma}$)	MEAN($\hat{\sigma}\|\mu$)	MEAN($\hat{\sigma}\|\tau$)	MEAN($\hat{\sigma}\|\mu,\tau$)
50	0.00	0.0	2.0758	2.0958	1.9570	1.9774
100	0.00	0.0	2.0450	2.0541	1.9810	1.9913
200	0.00	0.0	2.0129	2.0180	1.9828	1.9881
100	0.00	0.5	2.0943	2.1273	1.9383	1.9709
100	0.01	0.0	2.0330	2.0410	1.9806	1.9911
100	0.01	0.5	2.0613	2.1065	1.9361	1.9697

N	Q1	Q2	MEAN($\hat{\tau}$)	MEAN($\hat{\tau}\|\mu$)	MEAN($\hat{\tau}\|\sigma$)	MEAN($\hat{\tau}\|\mu,\sigma$)
50	0.00	0.0	10.5885	10.6103	10.6109	10.6046
100	0.00	0.0	10.3209	10.3252	10.3160	10.3131
200	0.00	0.0	10.1490	10.1520	10.1501	10.1499
100	0.00	0.5	10.3114	10.3349	10.3197	10.3133
100	0.01	0.0	10.3513	10.3620	10.3582	10.3551
100	0.01	0.5	10.3118	10.3691	10.3653	10.3557

N	Q1	Q2	VAR($\hat{\mu}$)	VAR($\hat{\sigma}$)	VAR($\hat{\tau}$)	COV($\hat{\mu},\hat{\sigma}$)	COV($\hat{\mu},\hat{\tau}$)	COV($\hat{\sigma},\hat{\tau}$)
50	0.00	0.0	0.0860	0.0693	0.6132	-0.0119	-0.0087	0.0663
100	0.00	0.0	0.0416	0.0312	0.1733	-0.0032	-0.0015	0.0232
200	0.00	0.0	0.0211	0.0121	0.0477	-0.0010	-0.0016	0.0072
100	0.00	0.5	0.0600	0.0961	0.2109	0.0296	0.0146	0.0628
100	0.01	0.0	0.0425	0.0314	0.3359	-0.0042	0.0108	0.0380
100	0.01	0.5	0.0591	0.1036	0.5098	0.0279	0.0194	0.1303

N	Q1	Q2	VAR($\hat{\mu}\|\tau$)	VAR($\hat{\sigma}\|\tau$)	COV($\hat{\mu},\hat{\sigma}\|\tau$)	VAR($\hat{\mu}\|\sigma$)	VAR($\hat{\tau}\|\sigma$)	COV($\hat{\mu},\hat{\tau}\|\sigma$)
50	0.00	0.0	0.0798	0.0408	0.0006	0.0837	0.5240	0.0059
100	0.00	0.0	0.0406	0.0218	0.0002	0.0413	0.1536	0.0015
200	0.00	0.0	0.0208	0.0096	-0.0001	0.0211	0.0433	-0.0010
100	0.00	0.5	0.0576	0.0460	0.0198	0.0511	0.1540	-0.0009
100	0.01	0.0	0.0406	0.0224	0.0000	0.0421	0.2831	-0.0048
100	0.01	0.5	0.0574	0.0473	0.0197	0.0516	0.2815	-0.0057

N	Q1	Q2	VAR($\hat{\sigma}\|\mu$)	VAR($\hat{\tau}\|\mu$)	COV($\hat{\sigma},\hat{\tau}\|\mu$)	VAR($\hat{\mu}\|\sigma,\tau$)	VAR($\hat{\sigma}\|\mu,\tau$)	VAR($\hat{\tau}\|\mu,\sigma$)
50	0.00	0.0	0.0692	0.6047	0.0603	0.0798	0.0407	0.5315
100	0.00	0.0	0.0308	0.1725	0.0221	0.0406	0.0216	0.1552
200	0.00	0.0	0.0120	0.0477	0.0069	0.0208	0.0096	0.0436
100	0.00	0.5	0.0789	0.1905	0.0469	0.0492	0.0393	0.1553
100	0.01	0.0	0.0312	0.3299	0.0363	0.0406	0.0222	0.2858
100	0.01	0.5	0.0872	0.4374	0.1011	0.0494	0.0407	0.2858

Table G3

WEIBULL POPULATION - SAMPLE SIZES 50 AND 100

Results of Monte Carlo Study of Estimates of Parameters of Weibull Population with Scale Parameter $\theta = 10$, Shape Parameter $K = 3$, and Location Parameter $c = 20$ from 500 Random Samples of Size N

	N=50	N=100		N=50	N=100		N=50	N=100
$M(\hat{\theta})$	9.92	9.83	$M(\hat{K})$	3.11	3.00	$M(\hat{c})$	20.03	20.11
$M(\hat{\theta}\|K)$	9.76	9.90	$M(\hat{K}\|\theta)$	3.10	3.04	$M(\hat{c}\|\theta)$	20.01	19.99
$M(\hat{\theta}\|c)$	10.01	9.98	$M(\hat{K}\|c)$	3.11	3.03	$M(\hat{c}\|K)$	20.25	20.08
$M(\hat{\theta}\|K,c)$	9.99	9.98	$M(\hat{K}\|\theta,c)$	3.07	3.02	$M(\hat{c}\|\theta,K)$	20.04	19.99
$V(\hat{\theta})$	11.15	3.44	$V(\hat{K}\|\theta)$	0.122	0.051	$V(\hat{\theta}\|c)$	0.241	0.132
$AV(\hat{\theta})$	2.72	1.36	$AV(\hat{K}\|\theta)$	0.103	0.051	$AV(\hat{\theta}\|c)$	0.246	0.123
$V(\hat{K})$	1.69	0.55	$V(\hat{c}\|\theta)$	0.190	0.108	$V(\hat{K}\|c)$	0.123	0.056
$AV(\hat{K})$	0.40	0.20	$AV(\hat{c}\|\theta)$	0.194	0.097	$AV(\hat{K}\|c)$	0.109	0.055
$V(\hat{c})$	9.66	2.99	$C(\hat{K},\hat{c}\|\theta)$	-0.033	-0.014	$C(\hat{\theta},\hat{K}\|c)$	0.045	0.031
$AV(\hat{c})$	2.14	1.07	$AC(\hat{K},\hat{c}\|\theta)$	-0.028	-0.014	$AC(\hat{\theta},\hat{K}\|c)$	0.051	0.026
$C(\hat{\theta},\hat{K})$	4.15	1.30	$V(\hat{\theta}\|K)$	0.787	0.359	$V(\hat{\theta}\|K,c)$	0.218	0.116
$AC(\hat{\theta},\hat{K})$	0.89	0.45	$AV(\hat{\theta}\|K)$	0.704	0.352	$AV(\hat{\theta}\|K,c)$	0.222	0.111
$C(\hat{\theta},\hat{c})$	-10.26	-3.15	$V(\hat{c}\|K)$	0.657	0.328	$V(\hat{K}\|\theta,c)$	0.112	0.049
$AC(\hat{\theta},\hat{c})$	-2.30	-1.15	$AV(\hat{c}\|K)$	0.592	0.296	$AV(\hat{K}\|\theta,c)$	0.099	0.049
$C(\hat{K},\hat{c})$	-3.87	-1.21	$C(\hat{\theta},\hat{c}\|K)$	-0.611	-0.282	$V(\hat{c}\|\theta,K)$	0.181	0.104
$AC(\hat{K},\hat{c})$	-0.78	-0.39	$AC(\hat{\theta},\hat{c}\|K)$	-0.534	-0.267	$AV(\hat{c}\|\theta,K)$	0.187	0.093

M = mean; V = variance; C = covariance; prefix A indicates asymptotic value

Table G4

LOGISTIC POPULATION - SAMPLE SIZES 10 AND 20

A. MEANS, VARIANCES AND COVARIANCES OF MAXIMUM-LIKELIHOOD ESTIMATES OF MEAN AND STANDARD DEVIATION OF STANDARD LOGISTIC DISTRIBUTION (μ=0, σ=1) FROM 1000 SAMPLES OF SIZE N=10 WITH PROPORTIONS Q1 CENSORED FROM BELOW AND Q2 FROM ABOVE

Q1	Q2	$M(\hat{\mu})$	$M(\hat{\sigma})$	$V(\hat{\mu})$	$C(\hat{\mu},\hat{\sigma})$	$V(\hat{\sigma})$	$M(\hat{\mu}\|\sigma)$	$M(\hat{\sigma}\|\mu)$	$V(\hat{\mu}\|\sigma)$	$V(\hat{\sigma}\|\mu)$
0.0	0.0	0.00	0.94	0.092	-0.001	0.071	0.00	0.99	0.092	0.073
0.0	0.1	-0.00	0.93	0.093	-0.000	0.079	-0.00	0.99	0.092	0.082
0.0	0.2	-0.01	0.92	0.094	0.002	0.085	-0.00	0.99	0.094	0.090
0.0	0.3	-0.02	0.90	0.097	0.008	0.095	-0.00	0.99	0.096	0.101
0.0	0.4	-0.05	0.87	0.105	0.017	0.106	-0.01	0.98	0.102	0.113
0.0	0.5	-0.08	0.84	0.113	0.028	0.121	-0.02	0.98	0.107	0.126
0.0	0.6	-0.14	0.78	0.147	0.056	0.146	-0.04	0.97	0.126	0.142
0.0	0.7	-0.26	0.69	0.231	0.110	0.184	-0.07	0.97	0.163	0.153
0.0	0.8	-0.54	0.52	0.394	0.194	0.224	-0.12	0.97	0.228	0.167
0.1	0.0	0.00	0.93	0.092	-0.002	0.080	0.00	0.99	0.092	0.082
0.1	0.1	-0.00	0.92	0.093	0.000	0.090	0.00	0.98	0.092	0.094
0.1	0.2	-0.01	0.90	0.094	0.003	0.099	-0.00	0.98	0.094	0.105
0.1	0.3	-0.02	0.88	0.097	0.009	0.113	-0.00	0.98	0.096	0.120
0.1	0.4	-0.05	0.84	0.105	0.022	0.126	-0.01	0.97	0.102	0.135
0.1	0.5	-0.09	0.79	0.115	0.034	0.153	-0.02	0.96	0.107	0.157
0.1	0.6	-0.17	0.70	0.153	0.071	0.187	-0.04	0.95	0.126	0.179
0.1	0.7	-0.35	0.52	0.242	0.132	0.221	-0.07	0.94	0.163	0.196
0.2	0.0	0.01	0.91	0.094	-0.006	0.088	0.00	0.99	0.094	0.091
0.2	0.1	0.01	0.90	0.095	-0.005	0.100	0.00	0.98	0.094	0.105
0.2	0.2	0.00	0.88	0.096	-0.001	0.111	0.00	0.98	0.096	0.120
0.2	0.3	-0.01	0.84	0.098	0.005	0.130	-0.00	0.97	0.098	0.142
0.2	0.4	-0.04	0.78	0.106	0.017	0.149	-0.01	0.96	0.104	0.168
0.2	0.5	-0.10	0.70	0.114	0.035	0.189	-0.02	0.94	0.108	0.203
0.2	0.6	-0.22	0.52	0.149	0.068	0.215	-0.04	0.91	0.128	0.241
0.3	0.0	0.03	0.89	0.096	-0.010	0.097	0.01	0.98	0.095	0.101
0.3	0.1	0.02	0.87	0.097	-0.008	0.109	0.01	0.98	0.096	0.116
0.3	0.2	0.02	0.84	0.097	-0.005	0.127	0.00	0.97	0.097	0.140
0.3	0.3	0.00	0.79	0.099	0.001	0.150	0.00	0.96	0.099	0.170
0.3	0.4	-0.03	0.69	0.106	0.016	0.173	-0.01	0.94	0.105	0.210
0.3	0.5	-0.10	0.52	0.114	0.028	0.204	-0.01	0.90	0.110	0.262
0.4	0.0	0.05	0.87	0.100	-0.017	0.114	0.01	0.98	0.098	0.117
0.4	0.1	0.05	0.83	0.100	-0.017	0.133	0.01	0.97	0.098	0.138
0.4	0.2	0.05	0.78	0.101	-0.015	0.159	0.01	0.96	0.099	0.173
0.4	0.3	0.04	0.69	0.102	-0.008	0.193	0.01	0.94	0.102	0.224
0.4	0.4	0.00	0.50	0.108	0.010	0.210	0.00	0.89	0.108	0.299
0.5	0.0	0.08	0.84	0.121	-0.037	0.133	0.02	0.98	0.110	0.131
0.5	0.1	0.09	0.78	0.122	-0.042	0.161	0.02	0.96	0.110	0.160
0.5	0.2	0.10	0.70	0.122	-0.045	0.197	0.02	0.95	0.112	0.208
0.5	0.3	0.10	0.53	0.122	-0.042	0.238	0.02	0.92	0.115	0.292
0.6	0.0	0.15	0.78	0.156	-0.068	0.162	0.04	0.97	0.127	0.143
0.6	0.1	0.18	0.69	0.159	-0.077	0.195	0.04	0.95	0.127	0.178
0.6	0.2	0.22	0.52	0.157	-0.078	0.222	0.04	0.92	0.130	0.234
0.7	0.0	0.27	0.69	0.229	-0.114	0.190	0.07	0.96	0.160	0.152
0.7	0.1	0.36	0.51	0.232	-0.124	0.218	0.07	0.93	0.161	0.193
0.8	0.0	0.53	0.52	0.399	-0.196	0.229	0.11	0.96	0.230	0.163

$M(\hat{\mu})$=MEAN VALUE OF MLE $\hat{\mu}$ IN 1000 SAMPLES (σ UNKNOWN)
$M(\hat{\sigma})$=MEAN VALUE OF MLE $\hat{\sigma}$ IN 1000 SAMPLES (μ UNKNOWN)
$V(\hat{\mu})$=VARIANCE OF MLE $\hat{\mu}$ IN 1000 SAMPLES (σ UNKNOWN)
$V(\hat{\sigma})$=VARIANCE OF MLE $\hat{\sigma}$ IN 1000 SAMPLES (μ UNKNOWN)
$C(\hat{\mu},\hat{\sigma})$=COVARIANCE OF MLES $\hat{\mu}$ AND $\hat{\sigma}$ IN 1000 SAMPLES
$M(\hat{\mu}|\sigma)$=MEAN VALUE OF MLE $\hat{\mu}$ IN 1000 SAMPLES (σ KNOWN)
$M(\hat{\sigma}|\mu)$=MEAN VALUE OF MLE $\hat{\sigma}$ IN 1000 SAMPLES (μ KNOWN)
$V(\hat{\mu}|\sigma)$=VARIANCE OF MLE $\hat{\mu}$ IN 1000 SAMPLES (σ KNOWN)
$V(\hat{\sigma}|\mu)$=VARIANCE OF MLE $\hat{\sigma}$ IN 1000 SAMPLES (μ KNOWN)

B. MEANS, VARIANCES AND COVARIANCES OF MAXIMUM-LIKELIHOOD ESTIMATES OF MEAN AND STANDARD DEVIATION OF STANDARD LOGISTIC DISTRIBUTION ($\mu=0$, $\sigma=1$) FROM 1000 SAMPLES OF SIZE N=20 WITH PROPORTIONS Q1 CENSORED FROM BELOW AND Q2 FROM ABOVE

Q1	Q2	$M(\hat{\mu})$	$M(\hat{\sigma})$	$V(\hat{\mu})$	$C(\hat{\mu},\hat{\sigma})$	$V(\hat{\sigma})$	$M(\hat{\mu}\|\sigma)$	$M(\hat{\sigma}\|\mu)$	$V(\hat{\mu}\|\sigma)$	$V(\hat{\sigma}\|\mu)$
0.0	0.0	0.00	0.97	0.047	-0.001	0.035	0.00	1.00	0.047	0.035
0.0	0.1	0.00	0.96	0.047	-0.001	0.038	0.00	1.00	0.047	0.038
0.0	0.2	-0.00	0.96	0.048	0.001	0.043	0.00	0.99	0.047	0.044
0.0	0.3	-0.01	0.95	0.048	0.004	0.050	0.00	0.99	0.048	0.052
0.0	0.4	-0.02	0.94	0.052	0.009	0.061	-0.00	1.00	0.051	0.061
0.0	0.5	-0.03	0.93	0.060	0.020	0.077	-0.00	1.00	0.055	0.074
0.0	0.6	-0.06	0.90	0.083	0.042	0.097	-0.02	1.00	0.065	0.082
0.0	0.7	-0.12	0.86	0.125	0.078	0.131	-0.03	0.99	0.078	0.088
0.0	0.8	-0.27	0.76	0.243	0.157	0.183	-0.06	0.98	0.111	0.091
0.0	0.9	-0.72	0.53	0.740	0.411	0.323	-0.12	1.00	0.214	0.100
0.1	0.0	0.00	0.97	0.047	-0.002	0.038	0.00	1.00	0.047	0.038
0.1	0.1	0.00	0.96	0.047	-0.001	0.041	0.00	0.99	0.047	0.042
0.1	0.2	-0.00	0.95	0.048	0.000	0.047	-0.00	0.99	0.047	0.049
0.1	0.3	-0.01	0.94	0.048	0.003	0.057	0.00	0.99	0.048	0.059
0.1	0.4	-0.02	0.93	0.052	0.009	0.070	-0.00	1.00	0.051	0.071
0.1	0.5	-0.03	0.91	0.060	0.021	0.091	-0.00	1.00	0.055	0.087
0.1	0.6	-0.08	0.86	0.085	0.047	0.119	-0.02	0.99	0.065	0.100
0.1	0.7	-0.16	0.78	0.134	0.099	0.172	-0.03	0.98	0.078	0.110
0.1	0.8	-0.47	0.50	0.264	0.188	0.232	-0.06	0.97	0.111	0.114
0.2	0.0	0.01	0.96	0.048	-0.003	0.042	0.00	1.00	0.047	0.042
0.2	0.1	0.01	0.95	0.048	-0.004	0.048	0.00	0.99	0.047	0.047
0.2	0.2	0.00	0.94	0.048	-0.001	0.055	0.00	0.99	0.048	0.056
0.2	0.3	-0.00	0.92	0.049	-0.001	0.067	0.00	0.99	0.049	0.069
0.2	0.4	-0.02	0.90	0.052	0.007	0.087	-0.00	0.99	0.051	0.087
0.2	0.5	-0.04	0.87	0.060	0.022	0.115	-0.00	0.99	0.056	0.111
0.2	0.6	-0.10	0.78	0.086	0.057	0.157	-0.02	0.97	0.066	0.133
0.2	0.7	-0.28	0.53	0.143	0.120	0.229	-0.03	0.94	0.079	0.152
0.3	0.0	0.01	0.95	0.049	-0.006	0.049	0.00	0.99	0.048	0.048
0.3	0.1	0.01	0.93	0.049	-0.007	0.057	0.00	0.99	0.048	0.056
0.3	0.2	0.01	0.91	0.049	-0.006	0.066	0.00	0.98	0.049	0.066
0.3	0.3	0.00	0.89	0.049	-0.002	0.082	0.00	0.98	0.049	0.085
0.3	0.4	-0.01	0.85	0.052	0.006	0.113	0.00	0.98	0.052	0.114
0.3	0.5	-0.04	0.77	0.060	0.025	0.162	-0.00	0.97	0.056	0.160
0.3	0.6	-0.16	0.51	0.090	0.072	0.218	-0.01	0.92	0.067	0.218
0.4	0.0	0.03	0.93	0.051	-0.011	0.059	0.01	0.99	0.049	0.056
0.4	0.1	0.03	0.91	0.052	-0.013	0.071	0.01	0.98	0.049	0.067
0.4	0.2	0.03	0.88	0.052	-0.012	0.082	0.01	0.97	0.050	0.081
0.4	0.3	0.02	0.83	0.052	-0.011	0.109	0.01	0.96	0.051	0.113
0.4	0.4	0.01	0.75	0.053	-0.003	0.165	0.01	0.95	0.053	0.173
0.4	0.5	-0.04	0.50	0.060	0.019	0.243	0.00	0.91	0.058	0.292
0.5	0.0	0.04	0.91	0.058	-0.020	0.071	0.01	0.99	0.052	0.064
0.5	0.1	0.05	0.88	0.059	-0.024	0.087	0.01	0.98	0.052	0.078
0.5	0.2	0.06	0.83	0.059	-0.026	0.109	0.01	0.96	0.053	0.101
0.5	0.3	0.06	0.74	0.060	-0.030	0.155	0.01	0.94	0.054	0.153
0.5	0.4	0.06	0.49	0.060	-0.022	0.209	0.01	0.90	0.057	0.264
0.6	0.0	0.07	0.89	0.075	-0.040	0.096	0.02	0.99	0.058	0.075
0.6	0.1	0.09	0.84	0.077	-0.048	0.119	0.02	0.97	0.058	0.093
0.6	0.2	0.11	0.74	0.080	-0.058	0.158	0.02	0.94	0.059	0.127
0.6	0.3	0.17	0.49	0.083	-0.072	0.228	0.02	0.90	0.060	0.219
0.7	0.0	0.12	0.85	0.117	-0.079	0.134	0.03	0.98	0.071	0.083
0.7	0.1	0.18	0.76	0.131	-0.105	0.181	0.03	0.96	0.071	0.104
0.7	0.2	0.31	0.48	0.130	-0.111	0.211	0.03	0.91	0.072	0.141
0.8	0.0	0.24	0.78	0.238	-0.160	0.191	0.05	0.98	0.103	0.086
0.8	0.1	0.45	0.51	0.270	-0.203	0.247	0.05	0.96	0.104	0.107
0.9	0.0	0.73	0.53	0.673	-0.356	0.281	0.13	1.01	0.221	0.103

C. COMPARISON OF MEASURES OF PRECISION OF BEST LINEAR UNBIASED ESTIMATORS AND MAXIMUM-LIKELIHOOD ESTIMATORS OF PARAMETERS OF LOGISTIC DISTRIBUTION FROM SAMPLES OF SIZE N=10 WITH PROPORTIONS Q1 CENSORED FROM BELOW AND Q2 FROM ABOVE

Q1	Q2	$V(\mu^*)$	$S(\hat{\mu})$	$A(\hat{\mu})$	$S(\hat{\mu}\mid\sigma)$	$A(\hat{\mu}\mid\sigma)$	$V(\sigma^*)$	$S(\hat{\sigma})$	$A(\hat{\sigma})$	$S(\hat{\sigma}\mid\mu)$	$A(\hat{\sigma}\mid\mu)$
0.0	0.0	0.093	0.092	0.091	0.092	0.091	0.077	0.075	0.070	0.073	0.070
0.0	0.1	0.094	0.093	0.091	0.092	0.091	0.086	0.084	0.077	0.082	0.077
0.0	0.2	0.095	0.094	0.092	0.094	0.092	0.100	0.092	0.088	0.090	0.088
0.0	0.3	0.099	0.098	0.095	0.096	0.094	0.120	0.105	0.103	0.101	0.102
0.0	0.4	0.109	0.107	0.101	0.102	0.097	0.149	0.123	0.125	0.113	0.120
0.0	0.5	0.132	0.119	0.116	0.107	0.104	0.195	0.147	0.156	0.127	0.140
0.0	0.6	0.189	0.168	0.151	0.127	0.116	0.273	0.194	0.205	0.142	0.157
0.0	0.7	0.358	0.299	0.239	0.167	0.139	0.430	0.278	0.287	0.154	0.167
0.0	0.8	1.102	0.689	0.503	0.243	0.187	0.904	0.458	0.452	0.168	0.168
0.1	0.0	0.094	0.092	0.091	0.092	0.091	0.086	0.085	0.077	0.082	0.077
0.1	0.1	0.094	0.093	0.091	0.092	0.091	0.098	0.097	0.085	0.094	0.085
0.1	0.2	0.095	0.094	0.092	0.094	0.092	0.116	0.109	0.099	0.105	0.099
0.1	0.3	0.099	0.098	0.095	0.096	0.094	0.144	0.128	0.119	0.120	0.117
0.1	0.4	0.109	0.108	0.102	0.102	0.098	0.187	0.152	0.148	0.136	0.142
0.1	0.5	0.136	0.122	0.118	0.107	0.104	0.261	0.197	0.194	0.158	0.171
0.1	0.6	0.213	0.183	0.160	0.127	0.116	0.413	0.278	0.272	0.181	0.197
0.1	0.7	0.535	0.367	0.283	0.168	0.139	0.875	0.451	0.432	0.200	0.213
0.2	0.0	0.095	0.094	0.092	0.094	0.092	0.100	0.096	0.088	0.091	0.088
0.2	0.1	0.095	0.095	0.092	0.094	0.092	0.116	0.110	0.099	0.105	0.099
0.2	0.2	0.096	0.096	0.093	0.096	0.093	0.142	0.126	0.117	0.120	0.117
0.2	0.3	0.099	0.098	0.095	0.098	0.094	0.183	0.155	0.145	0.143	0.144
0.2	0.4	0.109	0.108	0.102	0.104	0.098	0.256	0.197	0.189	0.169	0.183
0.2	0.5	0.141	0.123	0.120	0.109	0.105	0.405	0.280	0.266	0.206	0.234
0.2	0.6	0.274	0.197	0.174	0.129	0.118	0.862	0.447	0.426	0.250	0.287
0.3	0.0	0.099	0.097	0.095	0.095	0.094	0.120	0.108	0.103	0.101	0.102
0.3	0.1	0.099	0.097	0.095	0.096	0.094	0.144	0.125	0.119	0.116	0.117
0.3	0.2	0.099	0.098	0.095	0.097	0.094	0.183	0.153	0.145	0.141	0.144
0.3	0.3	0.101	0.099	0.096	0.099	0.096	0.254	0.195	0.188	0.172	0.188
0.3	0.4	0.109	0.107	0.102	0.105	0.100	0.402	0.270	0.264	0.214	0.260
0.3	0.5	0.149	0.124	0.121	0.110	0.108	0.856	0.431	0.423	0.272	0.376
0.4	0.0	0.109	0.103	0.101	0.098	0.097	0.149	0.132	0.125	0.117	0.120
0.4	0.1	0.109	0.103	0.102	0.098	0.098	0.187	0.161	0.148	0.140	0.142
0.4	0.2	0.109	0.103	0.102	0.099	0.098	0.256	0.206	0.189	0.175	0.183
0.4	0.3	0.109	0.103	0.102	0.102	0.100	0.402	0.288	0.264	0.228	0.260
0.4	0.4	0.112	0.108	0.105	0.108	0.105	0.855	0.457	0.422	0.310	0.422
0.5	0.0	0.132	0.127	0.116	0.111	0.104	0.195	0.160	0.156	0.132	0.140
0.5	0.1	0.136	0.130	0.118	0.111	0.104	0.261	0.208	0.194	0.161	0.171
0.5	0.2	0.141	0.132	0.120	0.112	0.105	0.405	0.287	0.266	0.211	0.234
0.5	0.3	0.149	0.133	0.121	0.115	0.108	0.856	0.462	0.423	0.299	0.376
0.6	0.0	0.189	0.179	0.151	0.128	0.116	0.273	0.212	0.205	0.144	0.157
0.6	0.1	0.213	0.191	0.160	0.129	0.116	0.413	0.294	0.272	0.181	0.197
0.6	0.2	0.274	0.207	0.174	0.131	0.118	0.862	0.455	0.426	0.241	0.287
0.7	0.0	0.358	0.301	0.239	0.164	0.139	0.430	0.288	0.287	0.154	0.167
0.7	0.1	0.535	0.364	0.283	0.165	0.139	0.875	0.461	0.432	0.198	0.213
0.8	0.0	1.102	0.681	0.503	0.243	0.187	0.904	0.456	0.452	0.165	0.168

$V(\mu^*)$=EXACT VARIANCE OF BLUE μ^*
$S(\hat{\mu})$=MEAN SQUARE ERROR OF MLE $\hat{\mu}$ IN 1000 SAMPLES (σ UNKNOWN)
$A(\hat{\mu})$=VARIANCE OF MLE $\hat{\mu}$ AS GIVEN BY ASYMPTOTIC FORMULA (σ UNKNOWN)
$S(\hat{\mu}\mid\sigma)$=MEAN SQUARE ERROR OF MLE $\hat{\mu}$ IN 1000 SAMPLES (σ KNOWN)
$A(\hat{\mu}\mid\sigma)$=VARIANCE OF MLE $\hat{\mu}$ AS GIVEN BY ASYMPTOTIC FORMULA (σ KNOWN)
$V(\sigma^*)$=EXACT VARIANCE OF BLUE σ^*
$S(\hat{\sigma})$=MEAN SQUARE ERROR OF MLE $\hat{\sigma}$ IN 1000 SAMPLES (μ UNKNOWN)
$A(\hat{\sigma})$=VARIANCE OF MLE $\hat{\sigma}$ AS GIVEN BY ASYMPTOTIC FORMULA (μ UNKNOWN)
$S(\hat{\sigma}\mid\mu)$=MEAN SQUARE ERROR OF MLE $\hat{\sigma}$ IN 1000 SAMPLES (μ KNOWN)
$A(\hat{\sigma}\mid\mu)$=VARIANCE OF MLE $\hat{\sigma}$ AS GIVEN BY ASYMPTOTIC FORMULA (μ KNOWN)

D. COMPARISON OF MEASURES OF PRECISION OF BEST LINEAR UNBIASED ESTIMATORS AND MAXIMUM-LIKELIHOOD ESTIMATORS OF PARAMETERS OF LOGISTIC DISTRIBUTION FROM SAMPLES OF SIZE N=20 WITH PROPORTIONS Q1 CENSORED FROM BELOW AND Q2 FROM ABOVE

Q1	Q2	$V(\mu^*)$	$S(\hat{\mu})$	$A(\hat{\mu})$	$S(\hat{\mu}\|\sigma)$	$A(\hat{\mu}\|\sigma)$	$V(\sigma^*)$	$S(\hat{\sigma})$	$A(\hat{\sigma})$	$S(\hat{\sigma}\|\mu)$	$A(\hat{\sigma}\|\mu)$
0.0	0.0	0.046	0.047	0.046	0.047	0.046	0.037	0.035	0.035	0.035	0.035
0.0	0.1	0.046	0.047	0.046	0.047	0.046	0.041	0.039	0.038	0.038	0.038
0.0	0.2	0.047	0.048	0.046	0.047	0.046	0.047	0.045	0.044	0.044	0.044
0.0	0.3	0.048	0.049	0.047	0.048	0.047	0.055	0.053	0.051	0.052	0.051
0.0	0.4	0.052	0.052	0.051	0.051	0.049	0.068	0.064	0.062	0.061	0.060
0.0	0.5	0.062	0.061	0.058	0.055	0.052	0.087	0.082	0.078	0.074	0.070
0.0	0.6	0.084	0.087	0.076	0.066	0.058	0.117	0.107	0.102	0.082	0.079
0.0	0.7	0.143	0.140	0.119	0.079	0.069	0.172	0.151	0.143	0.088	0.083
0.0	0.8	0.351	0.317	0.251	0.115	0.093	0.301	0.242	0.226	0.091	0.084
0.0	0.9	2.052	1.262	0.902	0.229	0.168	0.951	0.547	0.475	0.100	0.089
0.1	0.0	0.046	0.047	0.046	0.047	0.046	0.041	0.039	0.038	0.038	0.038
0.1	0.1	0.046	0.047	0.046	0.047	0.046	0.046	0.043	0.043	0.042	0.043
0.1	0.2	0.047	0.048	0.046	0.047	0.046	0.053	0.050	0.049	0.049	0.049
0.1	0.3	0.048	0.048	0.047	0.048	0.047	0.065	0.060	0.059	0.059	0.059
0.1	0.4	0.053	0.052	0.051	0.051	0.049	0.083	0.075	0.074	0.071	0.071
0.1	0.5	0.063	0.061	0.059	0.055	0.052	0.111	0.099	0.097	0.087	0.085
0.1	0.6	0.091	0.091	0.080	0.066	0.058	0.164	0.138	0.136	0.100	0.099
0.1	0.7	0.184	0.161	0.141	0.079	0.070	0.290	0.222	0.216	0.110	0.106
0.1	0.8	0.822	0.489	0.402	0.115	0.094	0.928	0.478	0.461	0.115	0.107
0.2	0.0	0.047	0.048	0.046	0.047	0.046	0.047	0.044	0.044	0.042	0.044
0.2	0.1	0.047	0.048	0.046	0.047	0.046	0.053	0.050	0.049	0.047	0.049
0.2	0.2	0.047	0.048	0.046	0.048	0.046	0.064	0.059	0.058	0.056	0.058
0.2	0.3	0.048	0.049	0.047	0.049	0.047	0.081	0.073	0.072	0.069	0.072
0.2	0.4	0.053	0.052	0.051	0.051	0.049	0.109	0.097	0.095	0.087	0.092
0.2	0.5	0.064	0.061	0.060	0.056	0.053	0.161	0.133	0.133	0.111	0.117
0.2	0.6	0.104	0.097	0.087	0.067	0.059	0.286	0.208	0.213	0.134	0.143
0.2	0.7	0.371	0.223	0.201	0.080	0.070	0.920	0.454	0.458	0.155	0.160
0.3	0.0	0.048	0.049	0.047	0.048	0.047	0.055	0.052	0.051	0.048	0.051
0.3	0.1	0.048	0.049	0.047	0.048	0.047	0.065	0.062	0.059	0.056	0.059
0.3	0.2	0.048	0.049	0.047	0.049	0.047	0.081	0.074	0.072	0.066	0.072
0.3	0.3	0.049	0.049	0.048	0.049	0.048	0.108	0.095	0.094	0.085	0.094
0.3	0.4	0.053	0.052	0.051	0.052	0.050	0.160	0.135	0.132	0.114	0.130
0.3	0.5	0.066	0.062	0.060	0.056	0.054	0.284	0.213	0.211	0.161	0.188
0.3	0.6	0.158	0.117	0.103	0.067	0.060	0.917	0.461	0.457	0.224	0.268
0.4	0.0	0.052	0.052	0.051	0.049	0.049	0.068	0.063	0.062	0.056	0.060
0.4	0.1	0.053	0.053	0.051	0.049	0.049	0.083	0.079	0.074	0.068	0.071
0.4	0.2	0.053	0.053	0.051	0.050	0.049	0.109	0.098	0.095	0.082	0.092
0.4	0.3	0.053	0.052	0.051	0.051	0.050	0.160	0.137	0.132	0.114	0.130
0.4	0.4	0.054	0.053	0.052	0.053	0.052	0.283	0.228	0.211	0.175	0.211
0.4	0.5	0.069	0.062	0.061	0.058	0.056	0.916	0.494	0.456	0.300	0.422
0.5	0.0	0.062	0.060	0.058	0.053	0.052	0.087	0.079	0.078	0.064	0.070
0.5	0.1	0.063	0.062	0.059	0.053	0.052	0.111	0.101	0.097	0.079	0.085
0.5	0.2	0.064	0.062	0.060	0.053	0.053	0.161	0.138	0.133	0.103	0.117
0.5	0.3	0.066	0.063	0.060	0.054	0.054	0.284	0.221	0.211	0.157	0.188
0.5	0.4	0.069	0.063	0.061	0.057	0.056	0.916	0.466	0.456	0.274	0.422
0.6	0.0	0.084	0.080	0.076	0.058	0.058	0.117	0.108	0.102	0.075	0.079
0.6	0.1	0.091	0.085	0.080	0.059	0.058	0.164	0.144	0.136	0.094	0.099
0.6	0.2	0.104	0.093	0.087	0.059	0.059	0.286	0.225	0.213	0.130	0.143
0.6	0.3	0.158	0.113	0.103	0.061	0.060	0.917	0.488	0.457	0.230	0.268
0.7	0.0	0.143	0.133	0.119	0.071	0.069	0.172	0.156	0.143	0.083	0.083
0.7	0.1	0.184	0.163	0.141	0.072	0.070	0.290	0.241	0.216	0.106	0.106
0.7	0.2	0.371	0.227	0.201	0.073	0.070	0.920	0.485	0.458	0.149	0.160
0.8	0.0	0.351	0.296	0.251	0.106	0.093	0.301	0.239	0.226	0.087	0.084
0.8	0.1	0.822	0.474	0.402	0.106	0.094	0.928	0.483	0.461	0.109	0.107
0.9	0.0	2.052	1.208	0.902	0.239	0.168	0.951	0.505	0.475	0.103	0.089

Table G5

TYPE I EXTREME-VALUE POPULATION - SAMPLE SIZES 10 AND 20

A. MEANS, VARIANCES AND COVARIANCES OF MAXIMUM-LIKELIHOOD ESTIMATES OF PARAMETERS OF STANDARD TYPE I ASYMPTOTIC SMALLEST-VALUE DISTRIBUTION ($u=0$, $b=1$) FROM 2000 SAMPLES OF SIZE $n=10$ WITH PROPORTIONS q_1 CENSORED FROM BELOW AND q_2 FROM ABOVE

q_1	q_2	$M(\hat{u})$	$M(\hat{b})$	$V(\hat{u})$	$C(\hat{u},\hat{b})$	$V(\hat{b})$	$M(\hat{u}\mid b)$	$M(\hat{b}\mid u)$	$V(\hat{u}\mid b)$	$V(\hat{b}\mid u)$
0.0	0.0	-0.04	0.93	0.112	-0.022	0.058	-0.05	0.98	0.104	0.053
0.0	0.1	-0.05	0.92	0.119	-0.012	0.070	-0.06	0.99	0.116	0.067
0.0	0.2	-0.08	0.90	0.132	0.000	0.084	-0.06	0.99	0.131	0.083
0.0	0.3	-0.11	0.88	0.154	0.017	0.098	-0.08	0.99	0.151	0.097
0.0	0.4	-0.16	0.85	0.200	0.046	0.118	-0.10	0.99	0.182	0.112
0.0	0.5	-0.22	0.82	0.267	0.084	0.140	-0.11	0.99	0.217	0.122
0.0	0.6	-0.34	0.77	0.400	0.153	0.179	-0.14	0.98	0.269	0.131
0.0	0.7	-0.56	0.68	0.685	0.249	0.211	-0.19	0.98	0.392	0.137
0.0	0.8	-1.02	0.53	1.321	0.430	0.264	-0.26	0.99	0.622	0.147
0.1	0.0	-0.04	0.92	0.114	-0.024	0.062	-0.05	0.97	0.104	0.058
0.1	0.1	-0.05	0.90	0.120	-0.015	0.077	-0.06	0.98	0.116	0.074
0.1	0.2	-0.07	0.88	0.132	-0.003	0.094	-0.06	0.98	0.131	0.093
0.1	0.3	-0.11	0.85	0.154	0.017	0.112	-0.08	0.98	0.151	0.112
0.1	0.4	-0.17	0.81	0.201	0.049	0.136	-0.09	0.98	0.182	0.129
0.1	0.5	-0.25	0.77	0.271	0.093	0.164	-0.10	0.98	0.218	0.143
0.1	0.6	-0.41	0.67	0.409	0.171	0.211	-0.13	0.97	0.270	0.156
0.1	0.7	-0.76	0.49	0.688	0.261	0.230	-0.18	0.96	0.393	0.165
0.2	0.0	-0.03	0.91	0.116	-0.027	0.069	-0.05	0.97	0.104	0.063
0.2	0.1	-0.05	0.89	0.121	-0.019	0.086	-0.06	0.98	0.116	0.083
0.2	0.2	-0.07	0.86	0.132	-0.005	0.108	-0.06	0.98	0.132	0.107
0.2	0.3	-0.11	0.82	0.154	0.018	0.132	-0.08	0.98	0.152	0.132
0.2	0.4	-0.18	0.76	0.202	0.055	0.162	-0.10	0.97	0.183	0.156
0.2	0.5	-0.28	0.69	0.277	0.107	0.200	-0.11	0.97	0.218	0.178
0.2	0.6	-0.52	0.51	0.427	0.201	0.258	-0.14	0.95	0.271	0.199
0.3	0.0	-0.02	0.89	0.119	-0.032	0.075	-0.05	0.96	0.105	0.068
0.3	0.1	-0.04	0.86	0.123	-0.024	0.098	-0.06	0.97	0.117	0.093
0.3	0.2	-0.06	0.83	0.133	-0.008	0.123	-0.06	0.97	0.132	0.123
0.3	0.3	-0.11	0.77	0.155	0.016	0.152	-0.08	0.97	0.153	0.157
0.3	0.4	-0.20	0.67	0.202	0.055	0.182	-0.09	0.95	0.184	0.189
0.3	0.5	-0.35	0.52	0.272	0.104	0.206	-0.10	0.94	0.220	0.219
0.4	0.0	-0.00	0.86	0.125	-0.041	0.087	-0.05	0.95	0.106	0.075
0.4	0.1	-0.02	0.83	0.128	-0.034	0.116	-0.05	0.96	0.118	0.108
0.4	0.2	-0.05	0.77	0.137	-0.018	0.149	-0.06	0.96	0.134	0.150
0.4	0.3	-0.10	0.68	0.155	0.009	0.190	-0.07	0.94	0.155	0.206
0.4	0.4	-0.22	0.49	0.200	0.049	0.205	-0.09	0.92	0.187	0.262
0.5	0.0	0.01	0.84	0.136	-0.052	0.102	-0.05	0.95	0.108	0.084
0.5	0.1	0.00	0.79	0.137	-0.047	0.140	-0.05	0.95	0.120	0.128
0.5	0.2	-0.02	0.70	0.143	-0.033	0.191	-0.06	0.95	0.137	0.197
0.5	0.3	-0.08	0.52	0.158	-0.003	0.242	-0.08	0.93	0.158	0.294
0.6	0.0	0.06	0.78	0.151	-0.070	0.121	-0.05	0.93	0.110	0.092
0.6	0.1	0.06	0.69	0.152	-0.068	0.168	-0.05	0.92	0.124	0.147
0.6	0.2	0.05	0.52	0.154	-0.052	0.221	-0.06	0.90	0.141	0.245
0.7	0.0	0.15	0.69	0.185	-0.102	0.151	-0.04	0.90	0.116	0.103
0.7	0.1	0.19	0.51	0.183	-0.103	0.203	-0.04	0.88	0.131	0.178
0.8	0.0	0.34	0.52	0.246	-0.146	0.180	-0.03	0.88	0.127	0.110

$M(\hat{u})$ = MEAN VALUE OF MLE $\hat{u}$ IN 2000 SAMPLES (b UNKNOWN)
$M(\hat{b})$ = MEAN VALUE OF MLE $\hat{b}$ IN 2000 SAMPLES (u UNKNOWN)
$V(\hat{u})$ = VARIANCE OF MLE $\hat{u}$ IN 2000 SAMPLES (b UNKNOWN)
$V(\hat{b})$ = VARIANCE OF MLE $\hat{b}$ IN 2000 SAMPLES (u UNKNOWN)
$C(\hat{u},\hat{b})$ = COVARIANCE OF MLES $\hat{u}$ AND $\hat{b}$ IN 2000 SAMPLES
$M(\hat{u}\mid b)$ = MEAN VALUE OF MLE $\hat{u}$ IN 2000 SAMPLES (b KNOWN)
$M(\hat{b}\mid u)$ = MEAN VALUE OF MLE $\hat{b}$ IN 2000 SAMPLES (u KNOWN)
$V(\hat{u}\mid b)$ = VARIANCE OF MLE $\hat{u}$ IN 2000 SAMPLES (b KNOWN)
$V(\hat{b}\mid u)$ = VARIANCE OF MLE $\hat{b}$ IN 2000 SAMPLES (u KNOWN)

B. MEANS, VARIANCES AND COVARIANCES OF MAXIMUM-LIKELIHOOD ESTIMATES OF PARAMETERS OF STANDARD TYPE I ASYMPTOTIC SMALLEST-VALUE DISTRIBUTION ($\mathbf{u}=0$, $\mathbf{b}=1$) FROM 2000 SAMPLES OF SIZE $\mathbf{n}=20$ WITH PROPORTIONS q_1 CENSORED FROM BELOW AND q_2 FROM ABOVE

q_1	q_2	$M(\hat{u})$	$M(\hat{b})$	$V(\hat{u})$	$C(\hat{u},\hat{b})$	$V(\hat{b})$	$M(\hat{u}\mid b)$	$M(\hat{b}\mid u)$	$V(\hat{u}\mid b)$	$V(\hat{b}\mid u)$
0.0	0.0	-0.02	0.96	0.056	-0.013	0.032	-0.02	0.99	0.050	0.029
0.0	0.1	-0.02	0.96	0.059	-0.007	0.040	-0.02	0.99	0.058	0.038
0.0	0.2	-0.04	0.95	0.065	-0.002	0.047	-0.03	0.99	0.065	0.047
0.0	0.3	-0.05	0.94	0.075	0.007	0.056	-0.03	0.99	0.074	0.055
0.0	0.4	-0.07	0.93	0.094	0.023	0.069	-0.04	1.00	0.087	0.063
0.0	0.5	-0.11	0.91	0.131	0.044	0.081	-0.05	0.99	0.107	0.069
0.0	0.6	-0.16	0.88	0.202	0.083	0.103	-0.06	0.99	0.136	0.073
0.0	0.7	-0.26	0.84	0.358	0.152	0.136	-0.08	0.99	0.186	0.074
0.0	0.8	-0.52	0.76	0.787	0.299	0.187	-0.13	0.99	0.307	0.077
0.0	0.9	-1.39	0.51	1.620	0.519	0.232	-0.27	1.02	0.660	0.095
0.1	0.0	-0.01	0.96	0.056	-0.014	0.034	-0.02	0.99	0.050	0.031
0.1	0.1	-0.02	0.95	0.060	-0.009	0.043	-0.02	0.99	0.058	0.042
0.1	0.2	-0.04	0.94	0.065	-0.002	0.052	-0.03	0.99	0.065	0.053
0.1	0.3	-0.05	0.93	0.075	0.008	0.063	-0.03	0.99	0.074	0.062
0.1	0.4	-0.08	0.91	0.096	0.027	0.080	-0.04	0.99	0.087	0.074
0.1	0.5	-0.12	0.88	0.137	0.054	0.099	-0.05	0.99	0.108	0.081
0.1	0.6	-0.20	0.84	0.219	0.104	0.130	-0.06	0.98	0.136	0.085
0.1	0.7	-0.36	0.76	0.419	0.208	0.185	-0.08	0.98	0.186	0.087
0.1	0.8	-0.92	0.49	0.907	0.374	0.232	-0.13	0.98	0.308	0.092
0.2	0.0	-0.01	0.95	0.058	-0.016	0.038	-0.02	0.98	0.050	0.033
0.2	0.1	-0.02	0.94	0.061	-0.012	0.049	-0.02	0.99	0.058	0.047
0.2	0.2	-0.03	0.93	0.066	-0.005	0.060	-0.03	0.99	0.065	0.061
0.2	0.3	-0.05	0.91	0.075	0.007	0.074	-0.03	0.99	0.075	0.075
0.2	0.4	-0.08	0.88	0.095	0.029	0.099	-0.04	0.99	0.087	0.092
0.2	0.5	-0.14	0.84	0.140	0.064	0.127	-0.05	0.98	0.108	0.104
0.2	0.6	-0.26	0.76	0.238	0.135	0.181	-0.06	0.97	0.136	0.111
0.2	0.7	-0.62	0.51	0.503	0.297	0.278	-0.08	0.96	0.187	0.114
0.3	0.0	-0.01	0.94	0.060	-0.019	0.042	-0.02	0.98	0.051	0.036
0.3	0.1	-0.02	0.93	0.062	-0.015	0.055	-0.02	0.98	0.058	0.052
0.3	0.2	-0.03	0.91	0.066	-0.006	0.071	-0.03	0.98	0.065	0.072
0.3	0.3	-0.05	0.88	0.075	0.007	0.089	-0.03	0.98	0.075	0.092
0.3	0.4	-0.09	0.84	0.096	0.032	0.123	-0.04	0.98	0.088	0.117
0.3	0.5	-0.18	0.75	0.147	0.081	0.171	-0.05	0.96	0.108	0.140
0.3	0.6	-0.43	0.50	0.253	0.164	0.231	-0.06	0.93	0.137	0.150
0.4	0.0	0.00	0.93	0.063	-0.023	0.049	-0.02	0.98	0.051	0.040
0.4	0.1	-0.01	0.91	0.064	-0.020	0.068	-0.02	0.97	0.059	0.062
0.4	0.2	-0.02	0.88	0.068	-0.011	0.090	-0.03	0.97	0.066	0.090
0.4	0.3	-0.05	0.84	0.076	0.003	0.119	-0.03	0.97	0.076	0.124
0.4	0.4	-0.10	0.75	0.096	0.035	0.174	-0.04	0.96	0.089	0.173
0.4	0.5	-0.27	0.50	0.147	0.093	0.232	-0.05	0.91	0.110	0.224
0.5	0.0	0.01	0.92	0.067	-0.028	0.055	-0.02	0.97	0.052	0.043
0.5	0.1	0.00	0.89	0.068	-0.025	0.080	-0.03	0.97	0.059	0.070
0.5	0.2	-0.01	0.84	0.070	-0.019	0.112	-0.03	0.96	0.067	0.110
0.5	0.3	-0.04	0.76	0.076	-0.001	0.160	-0.03	0.96	0.076	0.171
0.5	0.4	-0.12	0.51	0.096	0.038	0.229	-0.04	0.92	0.090	0.278
0.6	0.0	0.03	0.89	0.077	-0.040	0.069	-0.02	0.96	0.053	0.047
0.6	0.1	0.03	0.84	0.077	-0.040	0.105	-0.02	0.95	0.061	0.084
0.6	0.2	0.03	0.75	0.077	-0.037	0.157	-0.03	0.94	0.069	0.147
0.6	0.3	0.01	0.50	0.080	-0.018	0.220	-0.03	0.90	0.079	0.280
0.7	0.0	0.07	0.86	0.098	-0.063	0.093	-0.02	0.95	0.055	0.054
0.7	0.1	0.09	0.76	0.099	-0.073	0.152	-0.02	0.93	0.064	0.102
0.7	0.2	0.14	0.50	0.098	-0.077	0.235	-0.03	0.88	0.073	0.213
0.8	0.0	0.16	0.77	0.150	-0.108	0.132	-0.01	0.94	0.061	0.056
0.8	0.1	0.30	0.50	0.149	-0.124	0.200	-0.02	0.89	0.072	0.111
0.9	0.0	0.49	0.51	0.289	-0.199	0.189	-0.00	0.94	0.079	0.059

C. COMPARISON OF MEASURES OF PRECISION OF BEST LINEAR UNBIASED, BEST LINEAR INVARIANT, AND MAXIMUM-LIKELIHOOD ESTIMATORS OF PARAMETERS OF TYPE I ASYMPTOTIC DISTRIBUTION OF SMALLEST VALUES FROM SAMPLES OF SIZE $n=10$ WITH PROPORTIONS q_1 CENSORED FROM BELOW AND q_2 FROM ABOVE

q_1	q_2	$V(u^*)$	$S(\tilde{u})$	$S(\hat{u})$	$A(\hat{u})$	$S(\hat{u}\|b)$	$A(\hat{u}\|b)$	$V(b^*)$	$S(\tilde{b})$	$S(\hat{b})$	$A(\hat{b})$	$S(\hat{b}\|u)$	$A(\hat{b}\|u)$
0.0	0.0	0.113	0.113	0.114	0.111	0.107	0.100	0.072	0.067	0.063	0.061	0.054	0.055
0.0	0.1	0.120	0.120	0.122	0.115	0.119	0.111	0.088	0.081	0.077	0.077	0.068	0.074
0.0	0.2	0.134	0.134	0.137	0.125	0.135	0.125	0.107	0.097	0.094	0.093	0.083	0.093
0.0	0.3	0.162	0.161	0.166	0.145	0.157	0.143	0.132	0.117	0.113	0.112	0.097	0.111
0.0	0.4	0.214	0.210	0.226	0.181	0.191	0.167	0.166	0.142	0.141	0.137	0.112	0.126
0.0	0.5	0.321	0.303	0.315	0.251	0.229	0.200	0.215	0.177	0.172	0.172	0.122	0.137
0.0	0.6	0.559	0.496	0.515	0.393	0.288	0.250	0.298	0.229	0.233	0.222	0.131	0.141
0.0	0.7	1.204	0.949	0.999	0.719	0.426	0.333	0.461	0.315	0.313	0.307	0.137	0.142
0.0	0.8	3.904	2.317	2.354	1.648	0.691	0.500	0.949	0.487	0.488	0.474	0.147	0.144
0.1	0.0			0.115	0.112	0.107	0.100			0.069	0.065	0.058	0.059
0.1	0.1			0.122	0.116	0.119	0.111			0.086	0.084	0.075	0.081
0.1	0.2			0.138	0.125	0.135	0.125			0.108	0.104	0.093	0.104
0.1	0.3			0.167	0.145	0.157	0.143			0.134	0.129	0.112	0.127
0.1	0.4			0.231	0.183	0.191	0.167			0.172	0.163	0.130	0.148
0.1	0.5			0.332	0.262	0.229	0.200			0.219	0.212	0.144	0.162
0.1	0.6			0.578	0.437	0.288	0.250			0.318	0.295	0.157	0.169
0.1	0.7			1.269	0.903	0.427	0.333			0.485	0.460	0.167	0.170
0.2	0.0			0.117	0.114	0.107	0.100			0.077	0.072	0.064	0.063
0.2	0.1			0.123	0.117	0.120	0.111			0.099	0.095	0.083	0.091
0.2	0.2			0.138	0.126	0.136	0.125			0.127	0.121	0.108	0.120
0.2	0.3			0.167	0.145	0.158	0.143			0.163	0.155	0.133	0.153
0.2	0.4			0.236	0.186	0.192	0.167			0.218	0.205	0.157	0.184
0.2	0.5			0.355	0.278	0.229	0.200			0.296	0.287	0.179	0.207
0.2	0.6			0.700	0.520	0.289	0.251			0.494	0.452	0.202	0.218
0.3	0.0			0.119	0.117	0.107	0.100			0.087	0.081	0.069	0.070
0.3	0.1			0.124	0.119	0.120	0.112			0.116	0.111	0.094	0.104
0.3	0.2			0.138	0.127	0.136	0.125			0.153	0.146	0.124	0.145
0.3	0.3			0.167	0.145	0.159	0.144			0.205	0.196	0.158	0.194
0.3	0.4			0.241	0.189	0.193	0.168			0.290	0.279	0.191	0.248
0.3	0.5			0.394	0.307	0.231	0.201			0.439	0.444	0.223	0.291
0.4	0.0			0.125	0.123	0.108	0.101			0.105	0.094	0.078	0.077
0.4	0.1			0.129	0.124	0.121	0.112			0.146	0.135	0.110	0.122
0.4	0.2			0.139	0.129	0.138	0.126			0.201	0.186	0.152	0.182
0.4	0.3			0.166	0.145	0.160	0.145			0.294	0.270	0.209	0.269
0.4	0.4			0.248	0.192	0.196	0.169			0.468	0.435	0.269	0.383
0.5	0.0			0.136	0.134	0.110	0.102			0.127	0.113	0.087	0.086
0.5	0.1			0.137	0.134	0.123	0.114			0.186	0.172	0.131	0.146
0.5	0.2			0.143	0.136	0.141	0.128			0.282	0.258	0.200	0.242
0.5	0.3			0.165	0.147	0.163	0.147			0.470	0.424	0.299	0.424
0.6	0.0			0.155	0.155	0.112	0.104			0.168	0.143	0.098	0.097
0.6	0.1			0.156	0.157	0.126	0.116			0.262	0.239	0.154	0.177
0.6	0.2			0.156	0.157	0.145	0.132			0.453	0.408	0.255	0.343
0.7	0.0			0.207	0.199	0.117	0.109			0.245	0.195	0.112	0.106
0.7	0.1			0.219	0.218	0.133	0.122			0.440	0.380	0.192	0.213
0.8	0.0			0.361	0.317	0.129	0.118			0.414	0.302	0.124	0.112

$V(u^*)$=EXACT VARIANCE OF BLUE u^*, $S(\tilde{u})$=MEAN SQUARE ERROR OF BLIE $\tilde{u}$
$S(\hat{u})$=MEAN SQUARE ERROR OF MLE $\hat{u}$ IN 2000 SAMPLES (**b** UNKNOWN)
$A(\hat{u})$=VARIANCE OF MLE $\hat{u}$ AS GIVEN BY ASYMPTOTIC FORMULA (**b** UNKNOWN)
$S(\hat{u}|b)$=MEAN SQUARE ERROR OF MLE $\hat{u}$ IN 2000 SAMPLES (**b** KNOWN)
$A(\hat{u}|b)$=VARIANCE OF MLE $\hat{u}$ AS GIVEN BY ASYMPTOTIC FORMULA (**b** KNOWN)
$V(b^*)$=EXACT VARIANCE OF BLUE b^*, $S(\tilde{b})$=MEAN SQUARE ERROR OF BLIE $\tilde{b}$
$S(\hat{b})$=MEAN SQUARE ERROR OF MLE $\hat{b}$ IN 2000 SAMPLES (**u** UNKNOWN)
$A(\hat{b})$=VARIANCE OF MLE $\hat{b}$ AS GIVEN BY ASYMPTOTIC FORMULA (**u** UNKNOWN)
$S(\hat{b}|u)$=MEAN SQUARE ERROR OF MLE $\hat{b}$ IN 2000 SAMPLES (**u** KNOWN)
$A(\hat{b}|u)$=VARIANCE OF MLE $\hat{b}$ AS GIVEN BY ASYMPTOTIC FORMULA (**u** KNOWN)

D. COMPARISON OF MEASURES OF PRECISION OF BEST LINEAR UNBIASED, BEST LINEAR INVARIANT, AND MAXIMUM-LIKELIHOOD ESTIMATORS OF PARAMETERS OF TYPE I ASYMPTOTIC DISTRIBUTION OF SMALLEST VALUES FROM SAMPLES OF SIZE $n = 20$ WITH PROPORTIONS q_1 CENSORED FROM BELOW AND q_2 FROM ABOVE

q_1	q_2	$V(u^*)$	$S(\tilde{u})$	$S(\hat{u})$	$A(\hat{u})$	$S(\hat{u}\|b)$	$A(\hat{u}\|b)$	$V(b^*)$	$S(\tilde{b})$	$S(\hat{b})$	$A(\hat{b})$	$S(\hat{b}\|u)$	$A(\hat{b}\|u)$
0.0	0.0	0.056	0.056	0.056	0.055	0.051	0.050	0.033	0.032	0.033	0.030	0.029	0.027
0.0	0.1	0.059	0.059	0.060	0.058	0.058	0.056	0.041	0.039	0.042	0.038	0.038	0.037
0.0	0.2	0.065	0.065	0.066	0.063	0.066	0.062	0.050	0.047	0.050	0.046	0.047	0.046
0.0	0.3	0.076	0.076	0.078	0.072	0.075	0.071	0.061	0.057	0.060	0.056	0.055	0.055
0.0	0.4	0.098	0.097	0.099	0.091	0.088	0.083	0.075	0.070	0.074	0.069	0.063	0.063
0.0	0.5	0.141	0.138	0.143	0.126	0.110	0.100	0.096	0.087	0.090	0.086	0.069	0.068
0.0	0.6	0.232	0.221	0.228	0.197	0.139	0.125	0.127	0.113	0.117	0.111	0.073	0.071
0.0	0.7	0.456	0.413	0.428	0.360	0.193	0.167	0.184	0.155	0.160	0.153	0.074	0.071
0.0	0.8	1.197	0.978	1.055	0.824	0.324	0.250	0.316	0.240	0.247	0.237	0.077	0.072
0.0	0.9	7.033	3.880	3.749	3.026	0.731	0.500	0.975	0.494	0.471	0.487	0.096	0.081
0.1	0.0			0.056	0.056	0.051	0.050			0.036	0.033	0.031	0.029
0.1	0.1			0.060	0.058	0.058	0.056			0.046	0.042	0.042	0.040
0.1	0.2			0.066	0.063	0.066	0.063			0.056	0.052	0.053	0.052
0.1	0.3			0.078	0.072	0.075	0.071			0.068	0.064	0.062	0.063
0.1	0.4			0.101	0.092	0.088	0.083			0.089	0.081	0.074	0.074
0.1	0.5			0.151	0.131	0.110	0.100			0.113	0.106	0.081	0.081
0.1	0.6			0.257	0.218	0.139	0.125			0.156	0.148	0.086	0.084
0.1	0.7			0.549	0.451	0.193	0.167			0.245	0.230	0.088	0.085
0.1	0.8			1.751	1.388	0.325	0.250			0.490	0.478	0.093	0.086
0.2	0.0			0.058	0.057	0.051	0.050			0.040	0.036	0.033	0.032
0.2	0.1			0.061	0.058	0.059	0.056			0.052	0.048	0.047	0.045
0.2	0.2			0.067	0.063	0.066	0.063			0.065	0.060	0.061	0.060
0.2	0.3			0.078	0.072	0.076	0.072			0.082	0.077	0.075	0.076
0.2	0.4			0.102	0.093	0.089	0.083			0.112	0.102	0.092	0.092
0.2	0.5			0.159	0.139	0.111	0.100			0.153	0.144	0.105	0.103
0.2	0.6			0.303	0.260	0.140	0.125			0.241	0.226	0.112	0.109
0.2	0.7			0.885	0.723	0.194	0.167			0.520	0.474	0.115	0.110
0.3	0.0			0.060	0.059	0.051	0.050			0.045	0.041	0.036	0.035
0.3	0.1			0.062	0.060	0.059	0.056			0.060	0.056	0.053	0.052
0.3	0.2			0.067	0.063	0.066	0.053			0.079	0.073	0.072	0.072
0.3	0.3			0.078	0.072	0.076	0.072			0.103	0.098	0.092	0.097
0.3	0.4			0.104	0.094	0.089	0.084			0.148	0.140	0.118	0.124
0.3	0.5			0.178	0.153	0.111	0.101			0.234	0.222	0.142	0.146
0.3	0.6			0.438	0.378	0.141	0.126			0.484	0.471	0.155	0.157
0.4	0.0			0.063	0.062	0.051	0.050			0.053	0.047	0.040	0.039
0.4	0.1			0.064	0.062	0.059	0.056			0.076	0.067	0.063	0.061
0.4	0.2			0.068	0.065	0.067	0.063			0.104	0.093	0.091	0.091
0.4	0.3			0.078	0.073	0.077	0.072			0.145	0.135	0.125	0.134
0.4	0.4			0.106	0.096	0.090	0.085			0.234	0.218	0.174	0.191
0.4	0.5			0.221	0.191	0.113	0.102			0.487	0.467	0.232	0.249
0.5	0.0			0.067	0.067	0.052	0.051			0.061	0.057	0.043	0.043
0.5	0.1			0.068	0.067	0.060	0.057			0.092	0.086	0.071	0.073
0.5	0.2			0.070	0.068	0.068	0.064			0.138	0.129	0.111	0.121
0.5	0.3			0.077	0.073	0.077	0.073			0.220	0.212	0.173	0.212
0.5	0.4			0.111	0.099	0.091	0.086			0.466	0.461	0.284	0.400
0.6	0.0			0.078	0.077	0.053	0.052			0.080	0.072	0.049	0.048
0.6	0.1			0.078	0.078	0.061	0.058			0.130	0.119	0.086	0.089
0.6	0.2			0.078	0.078	0.069	0.066			0.219	0.204	0.151	0.171
0.6	0.3			0.080	0.079	0.080	0.076			0.474	0.454	0.289	0.435
0.7	0.0			0.103	0.100	0.055	0.054			0.114	0.097	0.056	0.053
0.7	0.1			0.108	0.109	0.065	0.061			0.209	0.190	0.107	0.106
0.7	0.2			0.118	0.122	0.073	0.069			0.483	0.442	0.227	0.252
0.8	0.0			0.177	0.159	0.061	0.059			0.185	0.151	0.060	0.056
0.8	0.1			0.236	0.234	0.073	0.067			0.451	0.418	0.123	0.120
0.9	0.0			0.531	0.409	0.079	0.073			0.430	0.319	0.063	0.057

Table G6

LOG-GAMMA POPULATION - SAMPLE SIZES 20 AND 40
SHAPE PARAMETER $\kappa = 0.5(0.5)2.0(1.0)4.0(2.0)12.0$
(1) Bias($\hat{\mu}$)/σ, (2) Bias($\hat{\sigma}$)/σ, (3) Var($\hat{\mu}$)σ^2, (4) Var($\hat{\sigma}$)/σ^2, (5) Cov($\hat{\mu}, \hat{\sigma}$)/σ^2
(PROPORTION q_1 CENSORED FROM BELOW AND q_2 FROM ABOVE)

Biases, Variances and Covariances in Monte Carlo Study

		$\kappa = 0.5$					$\kappa = 1.0$				
q_1	q_2	(1)	(2)	(3)	(4)	(5)	(1)	(2)	(3)	(4)	(5)
n= 20											
.0	.0	-.021	-.045	.064	.034	-.018	-.014	-.034	.056	.031	-.012
.0	.1	-.032	-.051	.069	.042	-.012	-.022	-.041	.059	.039	-.008
.0	.2	-.050	-.062	.080	.050	-.003	-.034	-.051	.064	.046	-.001
.0	.3	-.071	-.072	.098	.059	.010	-.049	-.060	.075	.055	.008
.0	.4	-.096	-.082	.127	.072	.028	-.069	-.070	.093	.066	.022
.0	.5	-.134	-.095	.187	.089	.059	-.106	-.090	.131	.081	.046
.1	.0	-.020	-.050	.065	.036	-.019	-.012	-.038	.056	.033	-.013
.1	.1	-.031	-.057	.069	.045	-.013	-.021	-.047	.059	.042	-.009
.1	.2	-.050	-.072	.080	.054	-.004	-.034	-.059	.065	.051	-.002
.1	.3	-.073	-.085	.098	.066	.010	-.051	-.071	.075	.063	.009
.1	.4	-.104	-.100	.128	.081	.031	-.074	-.087	.094	.078	.026
.1	.5	-.151	-.121	.194	.104	.069	-.120	-.115	.136	.100	.056
.2	.0	-.017	-.055	.066	.039	-.020	-.010	-.043	.057	.036	-.015
.2	.1	-.029	-.065	.069	.050	-.014	-.019	-.054	.059	.047	-.010
.2	.2	-.050	-.083	.080	.060	-.004	-.033	-.070	.065	.058	-.003
.2	.3	-.076	-.102	.099	.076	.013	-.051	-.087	.075	.074	.010
.2	.4	-.113	-.125	.130	.096	.037	-.080	-.110	.095	.095	.030
.2	.5	-.178	-.163	.205	.128	.086	-.138	-.156	.142	.127	.068
n= 40											
.0	.0	-.010	-.020	.030	.018	-.009	-.010	-.020	.028	.015	-.006
.0	.1	-.016	-.023	.031	.022	-.006	-.014	-.024	.029	.019	-.003
.0	.2	-.025	-.028	.036	.027	-.002	-.019	-.027	.030	.023	-.001
.0	.3	-.033	-.031	.045	.031	.004	-.026	-.031	.036	.028	.004
.0	.4	-.047	-.037	.061	.037	.014	-.040	-.040	.046	.034	.012
.0	.5	-.068	-.045	.091	.045	.029	-.058	-.049	.064	.041	.023
.1	.0	-.010	-.021	.030	.019	-.010	-.010	-.021	.028	.016	-.006
.1	.1	-.016	-.024	.032	.024	-.007	-.014	-.025	.029	.021	-.004
.1	.2	-.025	-.031	.036	.030	-.002	-.019	-.029	.030	.026	-.001
.1	.3	-.034	-.035	.045	.036	.005	-.027	-.034	.036	.032	.005
.1	.4	-.050	-.043	.061	.044	.016	-.042	-.045	.047	.039	.013
.1	.5	-.074	-.054	.095	.055	.035	-.063	-.058	.066	.050	.028
.2	.0	-.009	-.024	.031	.021	-.010	-.009	-.023	.028	.018	-.007
.2	.1	-.015	-.029	.032	.027	-.008	-.013	-.029	.029	.024	-.005
.2	.2	-.025	-.038	.036	.034	-.003	-.018	-.033	.031	.031	-.002
.2	.3	-.035	-.044	.045	.042	.006	-.027	-.041	.036	.039	.004
.2	.4	-.055	-.057	.063	.053	.019	-.044	-.057	.047	.049	.015
.2	.5	-.088	-.076	.101	.071	.046	-.072	-.078	.069	.066	.034

Biases, Variances and Covariances in Monte Carlo Study

q_1	q_2	$\kappa = 1.5$ (1)	(2)	(3)	(4)	(5)	$\kappa = 2.0$ (1)	(2)	(3)	(4)	(5)
n= 20											
.0	.0	-.014	-.040	.055	.028	-.010	-.010	-.042	.054	.028	-.007
.0	.1	-.022	-.049	.057	.035	-.006	-.016	-.048	.056	.034	-.004
.0	.2	-.030	-.054	.061	.041	-.001	-.024	-.055	.060	.040	.001
.0	.3	-.041	-.062	.068	.049	.006	-.034	-.062	.067	.049	.008
.0	.4	-.059	-.073	.082	.058	.017	-.049	-.071	.082	.060	.021
.0	.5	-.093	-.092	.110	.073	.037	-.080	-.090	.108	.074	.039
.1	.0	-.012	-.046	.055	.031	-.011	-.008	-.048	.055	.030	-.009
.1	.1	-.020	-.056	.057	.039	-.007	-.014	-.056	.056	.038	-.006
.1	.2	-.029	-.064	.061	.047	-.002	-.023	-.065	.060	.045	.000
.1	.3	-.042	-.075	.068	.057	.006	-.035	-.075	.067	.056	.008
.1	.4	-.064	-.091	.082	.068	.019	-.052	-.089	.082	.070	.023
.1	.5	-.105	-.120	.112	.087	.042	-.091	-.118	.110	.089	.045
.2	.0	-.008	-.053	.056	.035	-.012	-.003	-.057	.056	.033	-.011
.2	.1	-.016	-.066	.058	.044	-.009	-.010	-.068	.057	.043	-.008
.2	.2	-.026	-.077	.061	.054	-.003	-.020	-.081	.060	.053	-.002
.2	.3	-.042	-.093	.068	.066	.006	-.034	-.096	.067	.067	.008
.2	.4	-.068	-.118	.083	.083	.021	-.056	-.119	.082	.086	.024
.2	.5	-.123	-.167	.115	.111	.051	-.108	-.168	.113	.114	.054
n= 40											
.0	.0	-.010	-.017	.027	.014	-.005	-.006	-.018	.026	.014	-.005
.0	.1	-.012	-.018	.028	.018	-.003	-.009	-.019	.027	.017	-.003
.0	.2	-.018	-.022	.030	.022	.000	-.015	-.025	.029	.020	-.001
.0	.3	-.024	-.027	.034	.025	.003	-.019	-.028	.033	.023	.003
.0	.4	-.034	-.033	.042	.030	.009	-.030	-.036	.039	.029	.009
.0	.5	-.046	-.039	.056	.038	.020	-.046	-.046	.050	.036	.017
.1	.0	-.010	-.018	.027	.015	-.006	-.005	-.020	.026	.015	-.005
.1	.1	-.012	-.019	.028	.019	-.004	-.008	-.022	.027	.019	-.004
.1	.2	-.017	-.025	.030	.024	-.001	-.014	-.030	.029	.023	-.001
.1	.3	-.024	-.030	.034	.029	.004	-.020	-.034	.033	.028	.003
.1	.4	-.035	-.038	.042	.035	.011	-.032	-.045	.040	.035	.010
.1	.5	-.050	-.048	.058	.047	.024	-.051	-.059	.051	.045	.021
.2	.0	-.008	-.022	.027	.017	-.007	-.003	-.023	.027	.017	-.006
.2	.1	-.010	-.024	.028	.022	-.005	-.006	-.027	.027	.022	-.004
.2	.2	-.016	-.031	.030	.028	-.001	-.013	-.036	.029	.027	-.001
.2	.3	-.024	-.039	.034	.035	.004	-.019	-.043	.033	.034	.004
.2	.4	-.037	-.052	.042	.045	.013	-.034	-.059	.040	.045	.013
.2	.5	-.058	-.069	.061	.063	.031	-.059	-.083	.053	.061	.027

Biases, Variances and Covariances in Monte Carlo Study

q_1	q_2	κ = 3.0 (1)	κ = 3.0 (2)	κ = 3.0 (3)	κ = 3.0 (4)	κ = 3.0 (5)	κ = 4.0 (1)	κ = 4.0 (2)	κ = 4.0 (3)	κ = 4.0 (4)	κ = 4.0 (5)
n= 20											
.0	.0	-.009	-.035	.053	.028	-.007	-.007	-.037	.050	.027	-.005
.0	.1	-.016	-.042	.055	.034	-.004	-.012	-.042	.052	.032	-.002
.0	.2	-.024	-.049	.058	.040	.000	-.019	-.048	.055	.037	.001
.0	.3	-.037	-.060	.064	.047	.007	-.030	-.057	.061	.045	.008
.0	.4	-.055	-.073	.076	.056	.017	-.046	-.069	.071	.055	.018
.0	.5	-.083	-.092	.097	.069	.033	-.068	-.083	.090	.068	.034
.1	.0	-.007	-.039	.054	.030	-.008	-.005	-.042	.051	.030	-.007
.1	.1	-.014	-.048	.055	.037	-.005	-.010	-.049	.052	.037	-.004
.1	.2	-.022	-.057	.058	.045	-.001	-.017	-.057	.055	.042	.000
.1	.3	-.037	-.072	.064	.054	.007	-.029	-.069	.061	.052	.007
.1	.4	-.058	-.090	.076	.066	.018	-.048	-.086	.071	.066	.018
.1	.5	-.092	-.118	.099	.084	.038	-.076	-.109	.091	.086	.039
.2	.0	-.004	-.044	.054	.034	-.010	-.001	-.049	.053	.034	-.010
.2	.1	-.011	-.056	.056	.043	-.007	-.006	-.057	.054	.042	-.007
.2	.2	-.019	-.068	.058	.052	-.002	-.014	-.068	.056	.050	-.003
.2	.3	-.035	-.088	.064	.065	.006	-.027	-.085	.061	.063	.005
.2	.4	-.059	-.115	.076	.082	.020	-.049	-.111	.071	.085	.019
.2	.5	-.104	-.160	.102	.111	.047	-.086	-.149	.093	.118	.046
n= 40											
.0	.0	-.002	-.020	.026	.013	-.003	-.006	-.020	.026	.013	-.003
.0	.1	-.004	-.023	.027	.015	-.002	-.009	-.023	.027	.017	-.001
.0	.2	-.008	-.026	.029	.019	.001	-.012	-.026	.028	.020	.001
.0	.3	-.013	-.030	.032	.023	.004	-.016	-.029	.031	.024	.004
.0	.4	-.021	-.037	.038	.028	.010	-.024	-.035	.037	.029	.010
.0	.5	-.032	-.043	.049	.035	.019	-.036	-.044	.047	.036	.018
.1	.0	.000	-.024	.026	.014	-.004	-.004	-.023	.026	.015	-.004
.1	.1	-.003	-.027	.027	.017	-.002	-.007	-.028	.027	.019	-.002
.1	.2	-.007	-.033	.029	.022	.000	-.011	-.032	.028	.023	.000
.1	.3	-.013	-.039	.032	.027	.004	-.016	-.037	.031	.028	.004
.1	.4	-.024	-.048	.038	.034	.011	-.026	-.047	.037	.035	.010
.1	.5	-.038	-.060	.050	.045	.022	-.042	-.060	.048	.046	.021
.2	.0	.002	-.028	.027	.016	-.005	-.002	-.027	.027	.017	-.005
.2	.1	-.001	-.032	.028	.020	-.004	-.005	-.033	.027	.022	-.003
.2	.2	-.005	-.039	.029	.026	-.001	-.009	-.040	.029	.028	-.001
.2	.3	-.012	-.048	.032	.034	.004	-.014	-.047	.031	.035	.003
.2	.4	-.025	-.062	.038	.044	.012	-.027	-.061	.037	.045	.011
.2	.5	-.045	-.082	.051	.062	.027	-.048	-.084	.050	.063	.026

Biases, Variances and Covariances in Monte Carlo Study

q_1	q_2	κ = 6.0 (1)	κ = 6.0 (2)	κ = 6.0 (3)	κ = 6.0 (4)	κ = 6.0 (5)	κ = 8.0 (1)	κ = 8.0 (2)	κ = 8.0 (3)	κ = 8.0 (4)	κ = 8.0 (5)
n= 20											
.0	.0	-.002	-.040	.051	.025	-.006	-.008	-.035	.048	.026	-.004
.0	.1	-.007	-.046	.053	.030	-.003	-.013	-.042	.050	.031	-.001
.0	.2	-.014	-.053	.055	.037	.001	-.019	-.048	.053	.037	.003
.0	.3	-.025	-.064	.060	.044	.007	-.030	-.059	.059	.043	.009
.0	.4	-.040	-.075	.070	.052	.015	-.045	-.071	.068	.053	.019
.0	.5	-.059	-.089	.089	.066	.032	-.067	-.087	.086	.065	.033
.1	.0	.001	-.046	.052	.029	-.007	-.004	-.041	.049	.029	-.005
.1	.1	-.004	-.054	.053	.034	-.004	-.010	-.050	.050	.035	-.002
.1	.2	-.012	-.064	.055	.043	.000	-.016	-.058	.053	.043	.003
.1	.3	-.024	-.078	.060	.052	.007	-.029	-.073	.059	.051	.010
.1	.4	-.042	-.095	.070	.063	.017	-.047	-.091	.068	.066	.022
.1	.5	-.067	-.117	.091	.085	.038	-.075	-.117	.089	.084	.041
.2	.0	.006	-.053	.053	.033	-.009	.000	-.048	.050	.033	-.007
.2	.1	.001	-.063	.054	.040	-.007	-.005	-.059	.051	.041	-.005
.2	.2	-.008	-.077	.056	.051	-.003	-.012	-.071	.053	.052	.000
.2	.3	-.022	-.097	.060	.063	.005	-.026	-.091	.059	.064	.008
.2	.4	-.043	-.123	.070	.080	.017	-.047	-.118	.069	.086	.023
.2	.5	-.077	-.162	.092	.115	.045	-.084	-.161	.090	.115	.048
n= 40											
.0	.0	-.004	-.020	.025	.013	-.002	-.007	-.018	.026	.013	-.002
.0	.1	-.006	-.022	.026	.015	-.001	-.009	-.022	.026	.015	-.001
.0	.2	-.009	-.026	.027	.018	.001	-.012	-.025	.028	.018	.001
.0	.3	-.015	-.031	.031	.022	.004	-.019	-.031	.030	.022	.004
.0	.4	-.024	-.038	.036	.027	.010	-.028	-.039	.035	.027	.009
.0	.5	-.036	-.047	.046	.034	.018	-.036	-.044	.043	.034	.017
.1	.0	-.003	-.022	.026	.014	-.003	-.006	-.021	.026	.014	-.003
.1	.1	-.005	-.025	.026	.018	-.002	-.008	-.025	.027	.017	-.002
.1	.2	-.009	-.029	.027	.021	.000	-.012	-.029	.028	.021	.001
.1	.3	-.015	-.036	.031	.026	.004	-.018	-.036	.030	.026	.004
.1	.4	-.024	-.046	.036	.033	.011	-.028	-.047	.035	.032	.010
.1	.5	-.040	-.059	.047	.044	.021	-.039	-.056	.044	.043	.019
.2	.0	-.001	-.024	.026	.016	-.005	-.004	-.023	.027	.017	-.004
.2	.1	-.003	-.028	.027	.021	-.003	-.007	-.028	.027	.021	-.003
.2	.2	-.007	-.033	.028	.025	-.001	-.010	-.033	.028	.026	-.001
.2	.3	-.014	-.043	.031	.033	.003	-.017	-.043	.031	.033	.003
.2	.4	-.024	-.056	.037	.043	.011	-.028	-.059	.035	.043	.010
.2	.5	-.043	-.078	.048	.062	.026	-.043	-.074	.045	.061	.023

Biases, Variances and Covariances in Monte Carlo Study

q_1	q_2	$\kappa = 10.0$ (1)	(2)	(3)	(4)	(5)	$\kappa = 12.0$ (1)	(2)	(3)	(4)	(5)
n= 20											
.0	.0	-.005	-.034	.052	.026	-.004	-.006	-.037	.050	.026	-.004
.0	.1	-.009	-.038	.053	.031	-.002	-.011	-.044	.051	.030	-.001
.0	.2	-.015	-.045	.056	.037	.003	-.017	-.050	.054	.035	.003
.0	.3	-.023	-.051	.062	.044	.009	-.027	-.059	.058	.043	.008
.0	.4	-.037	-.063	.071	.054	.018	-.043	-.073	.066	.050	.016
.0	.5	-.056	-.077	.087	.066	.032	-.066	-.091	.082	.061	.029
.1	.0	-.001	-.041	.053	.029	-.005	-.004	-.042	.050	.029	-.005
.1	.1	-.005	-.047	.053	.035	-.004	-.009	-.050	.052	.034	-.002
.1	.2	-.013	-.056	.057	.044	.001	-.015	-.059	.054	.042	.002
.1	.3	-.022	-.066	.062	.053	.008	-.026	-.072	.058	.052	.008
.1	.4	-.038	-.083	.072	.067	.020	-.044	-.091	.066	.063	.017
.1	.5	-.064	-.107	.089	.085	.037	-.072	-.118	.083	.080	.034
.2	.0	.003	-.048	.055	.033	-.008	.001	-.049	.052	.033	-.007
.2	.1	-.001	-.056	.055	.041	-.006	-.003	-.061	.053	.040	-.005
.2	.2	-.008	-.069	.057	.052	-.001	-.011	-.073	.055	.051	-.001
.2	.3	-.019	-.083	.062	.065	.006	-.022	-.091	.058	.064	.006
.2	.4	-.038	-.109	.072	.085	.020	-.044	-.120	.066	.081	.017
.2	.5	-.072	-.148	.090	.112	.042	-.081	-.165	.084	.106	.039
n= 40											
.0	.0	-.007	-.023	.025	.013	-.002	-.001	-.016	.025	.013	-.001
.0	.1	-.009	-.025	.026	.015	-.001	-.003	-.019	.026	.015	.000
.0	.2	-.011	-.028	.027	.018	.001	-.006	-.022	.028	.018	.002
.0	.3	-.017	-.033	.030	.022	.004	-.011	-.027	.031	.021	.005
.0	.4	-.023	-.038	.034	.027	.009	-.019	-.034	.035	.026	.010
.0	.5	-.035	-.047	.043	.034	.016	-.029	-.042	.043	.032	.017
.1	.0	-.006	-.025	.025	.015	-.003	.001	-.019	.026	.015	-.002
.1	.1	-.008	-.028	.026	.017	-.002	-.002	-.022	.027	.018	-.001
.1	.2	-.011	-.032	.027	.022	.001	-.005	-.027	.028	.022	.001
.1	.3	-.016	-.039	.030	.027	.004	-.011	-.034	.031	.026	.005
.1	.4	-.024	-.046	.034	.034	.010	-.020	-.044	.035	.034	.011
.1	.5	-.038	-.059	.044	.045	.020	-.032	-.056	.044	.044	.020
.2	.0	-.004	-.028	.026	.017	-.005	.003	-.022	.027	.017	-.004
.2	.1	-.005	-.032	.027	.021	-.004	.001	-.027	.027	.021	-.002
.2	.2	-.008	-.037	.028	.027	-.001	-.003	-.033	.028	.026	.000
.2	.3	-.015	-.047	.030	.035	.003	-.009	-.042	.031	.032	.004
.2	.4	-.023	-.058	.034	.045	.009	-.020	-.057	.035	.042	.011
.2	.5	-.042	-.080	.044	.064	.023	-.036	-.076	.045	.059	.023

Subject Index

Author Index